PROTOZOOLOGY

PROTOZOOLOGY

PROTOZOOLOGY

By

RICHARD R. KUDO, D.Sc.

Visiting Professor of Zoology
Southern Illinois University
Professor Emeritus of Zoology
The University of Illinois
Past President and Honorary Member
The Society of Protozoologists

With three hundred and eighty-eight figures

Fifth Edition

CHARLES C THOMAS · PUBLISHER
Springfield · Illinois · U.S.A.

Published and Distributed Throughout the World by
CHARLES C THOMAS · PUBLISHER
BANNERSTONE HOUSE
301-327 East Lawrence Avenue, Springfield, Illinois, U.S.A.
NATCHEZ PLANTATION HOUSE
735 North Atlantic Boulevard, Fort Lauderdale, Florida, U.S.A.

First Edition, January, 1931
Second Edition, September, 1939
Third Edition, January, 1946
Third Edition, Second Printing, November, 1947
Third Edition, Third Printing, August, 1950
Fourth Edition, September, 1954
Fourth Edition, Second Printing, October, 1960
Fourth Edition, Third Printing, March, 1963
Fifth Edition, February, 1966

*With THOMAS BOOKS careful attention is given to all details of
manufacturing and design. It is the Publisher's desire to present books
that are satisfactory as to their physical qualities and artistic possibilities
and appropriate for their particular use. THOMAS BOOKS will be true
to those laws of quality that assure a good name and good will.*

Printed in the United States of America
B-7

"*The revelations of the Microscope are perhaps not
excelled in importance by those of the telescope.
While exciting our curiosity, our wonder
and admiration, they have proved of
infinite service in advancing our
knowledge of things
around us.*"
LEIDY

Preface

THE fifth edition of *Protozoology* maintains its original aim in setting forth introductory information on the common and representative genera and species of both free-living and parasitic protozoa, for advanced undergraduate and graduate students in zoology in colleges and universities.

In Part I, each chapter has been revised in the light of recently published information. However, since an attempt was made to limit the expansion of text to a minimum, the selection of material from so great a number of publications of the last ten years has been an extremely difficult task.

In Part II, the classification of all major groups and chapter arrangement were revised. New genera and species have been added to all groups. As in former editions, one species is given for each genus, but for common genera, several to many species are mentioned. Good illustrations are indispensable in this kind of work, since they are far more easily comprehended than lengthy descriptions. Therefore, some of the old figures were rearranged and new figures were added. The text-figures now number three hundred eighty-eight. The new figures were redrawn, as before, from the illustrations found in published papers and the indebtedness of the author is indicated by mentioning the names of investigators from whose works the illustrations were taken. In order to increase the reference value, all figures, except diagrams, are accompanied by scales of magnification.

The list of references appended to the end of each chapter has been enlarged by addition of recent publications and is intended to aid those who wish to obtain fuller informations than those which are given in this volume.

The author expresses his indebtedness to numerous authors of published papers for the materials which have been incorporated in the present work.

Carbondale, Illinois   R. R. KUDO

Contents

PROTOZOOLOGY

PROTOZOOLOGY

PART I: GENERAL BIOLOGY

Chapter 1
Introduction

PROTOZOA are unicellular animals. The body of a protozoan is morphologically a single cell and manifests all characteristics common to the living thing. The various activities which make up the phenomena of life are carried on by parts within the body or cell. These parts are comparable in function with the organs of a metazoan which are composed of a large number of cells grouped into tissues, and are therefore called **organellae.** Thus the protozoan is a complete organism somewhat unlike the cell of a metazoan, each of which is dependent upon other cells and cannot live independently. From this viewpoint, certain students of protozoology maintain that the protozoa are non-cellular, and not unicellular, organisms. Dobell (1911), for example, pointed out that the term "cell" is employed to designate (1) the whole protozoan body, (2) a part of a metazoan organism, and (3) a potential whole organism (a fertilized egg) which consequently resulted in a confused state of knowledge regarding living things, and, therefore, proposed to define a cell as a mass of protoplasm composing part of an organism, and further considered that the protozoan is a non-cellular but complete organism, differently organized as compared with cellular organisms, the metazoa and metaphyta. Although some writers (Hyman, 1940; Lwoff, 1951) follow this view, the majority of protozoologists continue to consider the protozoa as unicellular animals. Through the processes of organic evolution, they have undergone cytological differentiation and the metazoa histological differentiation.

In being unicellular, protozoa and protophyta are alike. The majority of protozoa may be distinguished from the majority of protophyta on the basis of dimensions, methods of nutrition, direction of division-plane, etc. While many protophyta possess nuclear material, it is not easy to detect it in many forms; on the other hand, all protozoa contain at least one easily observable nucleus. The binary fission of protozoa and protophyta is longitudinal and transverse respectively. Ciliata, however, multiply by transverse division. In general, the nutrition of protozoa is holozoic and of protophyta, holophytic or saprophytic; but there are large numbers of protozoa which nourish themselves by the latter methods. Thus an absolute and clean-cut separation of the two groups of unicellular organisms is not possible. Haeckel (1866) coined the name **Protista** to include these organisms in a single group, but this is not generally adopted, since it includes undoubted animals and plants, thus creating an equal amount of confusion between it and the animal or the plant. Calkins (1933) excluded chromatophore-bearing Mastigophora from his treatment of protozoa, thus placing organisms similar in every respect except the presence or absence of chromatophores, in two different (animal and plant) groups. This intermingling of characteristics between the two groups of unicellular microorganisms shows clearly their close interrelationship and suggests strongly their common ancestry.

Although the majority of protozoa are solitary and the body is composed of a single cell, there are several forms in which the organism is made up of more than one cell. These forms, which are called colonial protozoa (p. 208), are well represented by the members of Phytomastigia, in which the individuals are either joined by cytoplasmic threads or embedded in a common matrix. These cells are alike both in structure and in function, although in a few forms there may be a differentiation of the individuals into reproductive and vegetative cells. Unlike the cells in a metazoan which form tissues, these vegatative cells of colonial protozoa are not so dependent upon other cells as are the cells in metazoa; therefore, they do not form any true tissue. The reproductive cells produce zygotes through sexual fusion, which subsequently undergo repeated division and may produce a stage comparable with the blastula stage of a metazoan, but never reaching the gastrula stage. Thus, colonial protozoa are only cell-aggregates without histological differentiation and may thus be distinguished from the metazoa.

An enormous number of species of protozoa are known to man. From comparatively simple forms such as Amoeba, up to highly complicated organisms as represented by numerous ciliates, the protozoa vary exceedingly in their body organization, morphological characteristics, behavior, habitat, etc., which necessitates a taxonomic arrangement for proper consideration as set forth in detail in Chapters 8 to 42.

Relationship of Protozoology to Other Fields of Biological Science

A brief consideration of the relationship of protozoology to other fields of biology and its possible applications may not be out of place here. Since the protozoa are single-celled animals manifesting the characteristics common to all living things, they have been studied by numerous investigators with a view to discovering the nature and mechanism of various phenomena, the sumtotal of which is known collectively as life. Though the investigators generally have been disappointed in the results, inasmuch as the assumed simplicity of unicellular organisms has proved to be offset by the complexity of their cell-structure, nevertheless, discussion of any biological principles today must take into account the information obtained from studies of protozoa. Use of increasing numbers of axenic cultures of protozoa is bringing to light hitherto unknown biochemical information regarding the metabolic activities of these organisms. It is recognized that adequate knowledge of protozoa is essential for a thorough comprehension of biology.

Practically all students agree in assuming that the higher types of animals have been derived from organisms which existed in the remote past and which probably were somewhat similar to the primitive protozoa of the present day. Since there is no sharp distinction between protozoa and protophyta or between protozoa and metazoa, and since there are intermediate forms between the major classes of protozoa themselves, progress in protozoology contributes toward the advancement of our knowledge on the probable steps by which living things in general evolved.

Geneticists have undertaken studies on heredity and variation among protozoa. "Unicellular animals," wrote Jennings (1909), "present all the problems of heredity and variation in miniature. The struggle for existence in a fauna of untold thousands showing as much variety of form and function as any higher group, works itself out, with ultimate survival of the fittest, in a few days under our eyes, in a finger bowl. For studying heredity and variation we get a generation a day, and we may keep unlimited numbers of pedigreed stock in a watch glass that can be placed under the microscope." Morphological and physiological variations are encountered commonly in all forms. Whether variation is due to germinal or environmental conditions, is often difficult to determine. Studies on conjugation in Paramecium and several other ciliates by utilizing the mating types which was first noted by Sonneborn (1937, 1938) not only brought to light a wealth of important information regarding the genetics of protozoa, but also are revealing a close insight concerning the relationship between the nuclear and cytoplasmic factors of heredity in living organisms.

Parasitic protozoa are confined to one or more specific hosts. Through studies of the forms belonging to one and the same genus or species, the phylogenetic relation among the host animals may be established or verified. The mosquitoes belonging to the genera Culex and Anopheles, for instance, are known to transmit avian and human malaria parasites respectively. By observing certain intestinal protozoa in some monkeys, Hegner (1928) obtained evidence on the probable phylogenetic relationship between them and other higher mammals. The relation of various protozoa of the wood-roach to those of the termite, as revealed by Cleveland and his associates (1934), gives further proof that the Blattidae and the Isoptera are closely related.

Study of a particular group of parasitic protozoa and their hosts may throw light on the geographic condition of the earth which existed in the remote past. The members of the genus Zelleriella (p. 1032) are usually found in the colon of the frogs belonging to the family Leptodactylidae. Through an extensive study of these amphibians from South America and Australia, Metcalf (1920, 1929) found that the species of Zelleriella occurring in the frogs of the two continents are almost identical. He finds it more difficult to conceive of convergent or parallel evolution of both the hosts and the parasites, than to assume that there once existed between Patagonia and Australia a land connection over which frogs, containing Zelleriella, migrated.

Experimental studies of large protozoa have thrown light on the relation between the nucleus and the cytoplasm, and have furnished a basis for an understanding of regeneration in animals. In protozoa we find various types of nuclear divisions ranging from a simple amitotic division to a complex process comparable in every detail with the typical metazoan mitosis. A part of our knowledge in cytology is based upon studies of protozoa.

Through the efforts of various investigators in the past fifty years, it has now become known that some 27 species of protozoa occur in man. *Entamoeba histolytica, Balantidium coli,* and four species of Plasmodium, all of which are pathogenic to man, are widely distributed throughout the world. In certain restricted areas are found other pathogenic forms, such as Trypanosoma and Leishmania. Since all parasitic protozoa presumably have originated in free-living forms and since our knowledge of the morphology, physiology, and reproduction of the parasitic forms has largely been obtained in conjunction with the studies of the free-living organisms, a general knowledge of the entire phylum is necessary to understand these parasitic forms.

Recent studies have further revealed that almost all domestic animals are hosts to numerous parasitic protozoa, many of which are responsible for serious infectious diseases. Some of the forms found in domestic animals are morphologically indistinguishable from those occurring in man. *Balantidium coli* is considered as a parasite of swine, and man is its secondary host. Knowledge of protozoan parasites is useful to medical practitioners, just as it is essential to veterinarians inasmuch as certain diseases of animals, such as southern cattle fever, dourine, nagana, blackhead, coccidiosis, etc., are caused by protozoa.

Sanitary betterment and improvement are fundamental requirements in the modern civilized world. One of man's necessities is safe drinking water. The majority of protozoa live freely in various bodies of water and some of them are responsible, if present in sufficiently large numbers, for giving certain odors to the waters of reservoirs or ponds (p. 139). But these protozoa which are occasionally harmful are relatively small in number compared with those which are beneficial to man. It is generally understood that bacteria live on various waste materials present in the polluted water, but that upon reaching a certain population, they would cease to multiply and would allow the excess organic substances to undergo decomposition. Numerous holozoic protozoa, however, feed on the bacteria and prevent them from reaching the saturation population. Protozoa thus seem to help indirectly in the purification of the water. Protozoology therefore must be considered as part of modern sanitary science.

Young fish feed extensively on small aquatic organisms, such as larvae of insects, small crustaceans, annelids, etc., all of which depend largely upon protozoa and protophyta as sources of food supply. Thus the fish are indirectly dependent upon protozoa as food material. Indeed, protozoa are a part of the foundation of food chains in water. On the other hand, there are numbers of protozoa which live at the expense of fish. The Myxosporida are almost exclusively parasites of fish and sometimes cause death to large numbers of commerically important fishes (Kudo, 1920) (p. 780). Success in fish-culture, therefore, requires among other things a thorough knowledge of protozoa.

Since Russell and Hutchinson (1909, 1913) suggested some fifty years ago that protozoa are probably a cause of limitation of the numbers, and therefore the activities of bacteria in the soil and thus tend to decrease the amount of nitrogen which is given to the soil by the nitrifying bacteria, several investigators have brought out the fact that in the soils of temperate climate various sarcodinans, flagellates and less frequently ciliates, are present and active throughout the year. The exact relation between specific protozoa and bacteria in the soil is not yet clear in spite of the numerous experiments and observations, since various antibiotic-producing fungi are common inhabitants of the soil. All soil investigators should be acquainted with the biology and taxonomy of free-living protozoa (Micropredators and bacteria in soil, Singh, 1960).

It is a matter of common knowledge that the silkworm and the honey bee suffer from microsporidan infections (p. 810). Sericulture in south-western Europe suffered great damages in the middle of the nineteenth century because of the "pébrine" disease, caused by the microsporidan, *Nosema bombycis*. During the first decade of the present century, another microsporidan, *Nosema apis,* was found to cause Nosema-disease in honey bees. Methods of control have been developed and put into practice so that these microsporidan infections are at present not serious, even though they still occur. On the other hand, other Microsporida are now known to infect certain insects, such as dipterous and lepidopterous pests, which, when heavily infected, die sooner or later. Methods of destruction of these insects by means of chemicals are more and more used, but attention should also be given to biological control of them by means of protozoa and protophyta.

While the majority of protozoa lack permanent skeletal structures and their fossil forms are little known, there are at least two large groups in the Sarcodina which possess conspicuous shells or skeletal structures and which are found as fossils. They are Foraminiferida and Radiolarida. From early palaeozoic era down to the present day, the carbonate of lime which makes up the skeletons of numerous foraminifers has been left embedded in various rock strata as limestone and chalk. Although there is no distinctive foraminiferan fauna characteristic of a given geologic period, there are certain peculiarities of fossil Foraminiferida which distinguish one formation from the other. From this fact one can understand that knowledge of foraminiferous rocks is highly useful in checking up logs in oil well drilling. The skeletons of the Radiolarida are the main constituent of the ooze of littoral and deep-sea regions. They have been found abundantly in siliceous rocks of the palaeozoic and the mesozoic eras, and are also identified with the clays and other formations of the miocene period. Thus, knowledge of these two orders of Sarcodina, at least, is essential for the student of geology and micropaleontology.

History of Protozoology

Aside from a comparatively small number of large forms, protozoa are unobservable with the naked eye, so that one can easily understand why they were unknown prior to the invention of the microscope. Antony van Leeuwenhoek (1632-1723) is commonly recognized as the father of protozoology. Grinding lenses himself, Leeuwenhoek made more than 400 simple lenses, including one which, it is said, had a magnification of 270 times (Harting). Among the many things he discovered were various protozoa. According to Dobell (1932), Leeuwenhoek saw in 1674 for the first time free-living fresh-water protozoa. Between 1674 and 1716, he observed many protozoa which he reported to the Royal Society of London and which, as Dobell interpreted, were Euglena ("green in the middle, and before and behind white"), Vorticella, Stylonychia, Carchesium, Volvox, Coleps, Kerona, Anthophysis, Elphidium, etc. Huygens gave in 1678 "unmistakable descriptions of Chilodon(-ella), Paramecium, Astasia and Vorticella, all found in infusions" (Dobell).

Colpoda was seen by Buonanni (1691) and Harris (1696) rediscovered Euglena. In 1718 there appeared the first treatise on microscopic organisms, particularly of protozoa, by Joblot who emphasized the non-existence of abiogenesis by using boiled hay-infusions in which no Infusoria developed without exposure to the atmosphere. This experiment confirmed that of Redi who, some forty years before, had made his well-known experiments by excluding flies from a jar containing meat. According to Woodruff (1937), Joblot gave the first identifiable figure of Paramecium. Trembley (1744) studied division in some ciliates, including probably Paramecium, which generic name was coined by Hill in 1752. Noctiluca was first described by Baker (1753).

Rösel von Rosenhof (1755) observed an organism, which he called "der kleine Proteus," and also Vorticella, Stentor, and Volvox. The "Proteus" which Linnaeus named *Volvox chaos* (1758) and later renamed *Chaos protheus* (1767), cannot be identified with any of the known amoeboid organisms (Kudo, 1946, 1959). Wrisberg (1764) coined the term "Infusoria" (Dujardin; Woodruff). By using the juice of geranium, Ellis (1769) caused the extrusion of the "fins" (trichocysts) in Paramecium. Eichhorn (1783) observed the heliozoan Actinosphaerium. O. F. Müller described Ceratium a little later and published two works on the Infusoria (1773, 1786), although he included unavoidably some metazoa and protophyta in his monographs; some of his descriptions and figures of Ciliata were so well done that they are of value even at the present time. Lamarck (1816) named Folliculina.

At the beginning of the nineteenth century, the cyclosis in Paramecium was brought to light by Gruithuisen. Goldfuss (1817) coined the term **Protozoa,** including in it the coelenterates. Nine years later there appeared d'Orbigny's systematic study of the Foraminiferida, which he considered "microscopical cephalopods." In 1828, Ehrenberg began publishing his observations on protozoa, and in 1838, he summarized his contributions in *Die Infusionsthierchen als vollkommene Organismen,* in which he diagnosed genera and species so well that many of them still hold good. Ehrenberg excluded Rotatoria and Cercaria from Infusoria. Through the studies of Ehrenberg the number of known protozoa increased greatly; he, however, proposed the term "Polygastricha," under which he placed Mastigophora, Rhizopoda, Ciliata, Suctoria, desmids, etc., since he believed that the food vacuoles present in them were stomachs. This hypothesis became immediately the center of controversy, which incidentally, together with the then-propounded cell theory and improvements in microscopy, stimulated researches on protozoa.

Dujardin (1835) took pains in studying the protoplasm of various protozoa and found it alike in all. He named it **sarcode.** In 1841, he published an extensive monograph of various protozoa which came under his observations. The term Rhizopoda was coined by this investigator. The commonly used term **protoplasm** was employed by Purkinje (1840) in the same sense as it is used today. The protozoa was given a distinct definition by Siebold in 1845, as follows: "Die Thiere, in welchen die verschiedenen Systeme der Organe nicht scharf ausgeschieden sind, und deren unregelmässige Form und einfache Organization sich auf eine Zelle reduzieren lassen." Siebold subdivided protozoa into Infusoria and Rhizopoda. The sharp differentiation of protozoa as a group certainly inspired numerous microscopists. As a result, several students brought forward various group names, such as Radiolaria (J. Müller, 1858), Ciliata (Perty, 1852), Flagellata (Cohn, 1853), Suctoria (Claparède and Lachmann, 1858), Heliozoa, Protista (Haeckel, 1862, 1866), Mastigophora (Diesing, 1865), etc. Of Suctoria, Stein failed to see the real nature (1849), but his two monographs on Ciliata and Mastigophora (1854, 1859-1883) contain concise descriptions and excellent illustrations of numerous species. Haeckel who went a step further than Siebold by distinguishing between protozoa and metazoa, devoted ten years to his study of Radiolarida, especially those of the Challenger collection, and described in his celebrated monographs more than 4000 species.

In 1879, the first comprehensive monograph on the protozoa of North America was put forward by Leidy under the title of *Freshwater Rhizopods of North America,* which showed the wide distribution of many known forms of Europe and revealed a number of new and interesting forms. This work was followed by Stokes' *The Freshwater Infusoria of the United States,* which appeared in 1888. Bütschli (1880-1889) established Sarcodina and made an excellent contribution to the taxonomy of the then-known species of protozoa, which is still considered as one of the most important works in general protozoology. The painstaking researches by Maupas, on the conjugation of ciliates, corrected erroneous interpretation of the phenomenon observed by Balbiani some thirty years before and gave impetus to a renewed cytological study of protozoa. The variety in form and structure of the protozoan nuclei became the subject of intensive studies by several cytologists. Weismann put into words the immortality of the protozoa. Schaudinn contributed much toward the cytological and developmental studies of protozoa. Penard contributed a great deal to our knowledge of Sarcodina.

In the first year of the present century, Calkins in the United States and Doflein in Germany wrote modern textbooks of protozoology dealing with the biology as well as the taxonomy. Jennings devoted his time for nearly forty years to the study of genetics of protozoa. Recent development of bacteria-free culture technique in certain flagellates and ciliates, has brought to light important information regarding the nutritional requirements and metabolism of these organisms.

Today the protozoa are more and more intensively and extensively studied from both the biological and the parasitological sides, and important contributions appear continuously. Since all parasitic protozoa appear to have originated in free-living forms, the comprehension of the morphology, physiology, and development of the latter group is obviously fundamentally important for a thorough understanding of the former group.

Compared with the advancement of our knowledge on free-living protozoa, that on parasitic forms has been very slow. This is to be expected, of course, since the vast majority of them are so minute that the discovery of their presence has been made possible only through improvements in the microscope and in technique.

Here again Leeuwenhoek seems to have been the first to observe a parasitic protozoan, for he observed, according to Dobell (1932), in the fall of 1674, the oocysts of the coccidian *Eimeria stiedae*, in the contents of the gall bladder of an old rabbit; in 1681, *Giardia intestinalis* in his own diarrhœic stools; and in 1683, Opalina and Nyctotherus in the gut contents of frogs. The oral Trichomonas of man was observed by O. F. Müller (1773) who named it *Cercaria tenax* (Dobell, 1939). There is no record of anyone having seen protozoa living in other organisms, until 1828, when Dufour's account of the gregarine from the intestine of coleopterous insects appeared. Some ten years later, Hake rediscovered the oocysts of *Eimeria stiedae*. A flagellate was observed in the blood of trout by Valentin in 1841, and the frog trypanosome was discovered by Gluge (1842) and Gruby (1843), the latter author creating the genus Trypanosoma for it.

The gregarines were a little later given attention by Kölliker (1848) and Stein (1848). The year 1849 marks the first record of an amoeba being found in man, for Gros then observed *Entamoeba gingivalis* in the human mouth. Five years later, Davaine found in the stools of cholera patients two flagellates (Trichomonas and Chilomastix). Kloss, in 1855, observed the coccidian, *Klossia helicina*, in the excretory organ of Helix; and Eimer (1870) made an extensive study of Coccidia occurring in various animals. *Balantidium coli* was discovered by Malmsten in 1857. Lewis, in 1870, observed *Entamoeba coli* in India, and Lösch, in 1875, found *Entamoeba histolytica* in Russia. During the early part of the last century, an epidemic disease, pébrine, of the silkworm appeared in Italy and France, and a number of biologists became engaged in its investigation. Foremost of all, Pasteur (1870) made an extensive report on the nature of the causative organism, now known as *Nosema bombycis*, and also on the method of control and prevention. Perhaps this is the first scientific study of a parasitic protozoan which resulted in an effective practical method of control of its infection.

Lewis observed, in 1878, an organism which is since known as *Trypanosoma lewisi* in the blood of rats. In 1879, Leuckart created the group Sporozoa, including in it the gregarines and coccidians. Other groups under Sporozoa were soon definitely designated. They are Myxosporidia (Bütschli, 1881), Microsporidia and Sarcosporidia (Balbiani, 1882).

Parasitic protozoology received a far-reaching stimulus when Laveran (November, 1880) discovered the microgamete formation ("flagellation") of a malaria parasite in the human blood. Smith and Kilborne (1893) demonstrated that Babesia of the Texas fever of cattle in the southern United States was transmitted by the cattle tick from host to host, and thus revealed for the first time the close relationship which exists between an arthropod and a parasitic protozoan. Two years later, Bruce discovered *Trypanosoma brucei* in the blood of domestic animals suffering from "nagana" disease in Africa and later (1897) demonstrated by experiments that the tsetse fly transmits the trypanosome. Studies of malaria organisms continued and several important contributions appeared. Golgi (1886, 1889) studied the schizogony and its relation to the occurrence of fever, and was able to distinguish the types of fever. MacCallum (1897) observed the microgamete formation in Haemoproteus of birds and suggested that the "flagella" observed by Laveran were microgametes of Plasmodium. In fact, he later observed the formation of the zygote through fusion of a microgamete and a macrogamete of *Plasmodium falciparum*. Almost at the same time, Schaudinn and Siedlecki (1897) showed that anisogamy results in the production of zygotes in Coccidia. The latter author published later further observations on the life-cycle of Coccidia (1898, 1899).

Ross (1898, 1898a) revealed the development of *Plasmodium relictum* (*P. praecox*) in *Culex fatigans* and established the fact that the host birds become infected by this protozoan through the bites of the infected mosquitoes. Since that time, investigators too numerous to mention here (p. 720), studied the biology and development of the malarial organisms. Among the more recent findings is the exo-erythrocytic development, fuller information on which is now being sought. In 1902, Dutton found that the sleeping sickness in equatorial Africa was caused by *Trypanosoma gambiense.* In 1903, Leishman and Donovan discovered simultaneously *Leishmania donovani,* the causative organism of "kala-azar" in India.

Artificial cultivation of bacteria had contributed toward a very rapid advancement in bacteriology, and it was natural, as the number of known parasitic protozoa rapidly increased, that attempts to cultivate them in vitro should be made. Musgrave and Clegg (1904) cultivated, on bouillon-agar, small free-living amoebae from old faecal matter. In 1905, Novy and MacNeal cultivated successfully the trypanosome of birds in blood-agar medium, which remained free from bacterial contamination and in which the organisms underwent multiplication. Almost all species of Trypanosoma and Leishmania have since been cultivated in a similar manner. This serves for detection of a mild infection and also identification of the species involved. It was found, further, that the changes which these organisms underwent in the culture media were imitative of those that took place in the invertebrate host, thus contributng toward the life-cycle studies of them.

During and since World War I, it became known that numerous intestinal protozoa of man are widely present throughout the tropical, subtropical and temperate zones. Taxonomic, morphological and developmental studies on these forms have therefore appeared in an enormous number. Cutler (1918) seems to have succeeded in cultivating *Entamoeba histolytica,* though his experiment was not repeated by others. Barret and Yarborough (1921) cultivated *Balantidium coli* and Boeck (1921) cultivated *Chilomastix mesnili.* Boeck and Drbohlav (1925) succeeded in cultivating *Entamoeba histolytica,* and their work was repeated and improved upon by many investigators.

In the past twenty years, studies of protozoa have become intensive, especially, along biochemical line, as pure cultures of protozoa are becoming available in increasing numbers. In 1947, a group of protozoologists met in Chicago, Illinois, and organized the Society of Protozoologists "for the presentation and discussion of new or important facts and problems in protozoology and for the adoption of such measures as will tend to advance protozoological science." In 1954, the Society began publication of the *Journal of Protozoology.* (Early American protozoologists, Wenrich, 1956.)

References

ALLMAN, G. J., 1855. On the occurrence among the Infusoria of peculiar organs resembling thread-cells. Quart J. Micr. Sci., 3:177.

BAKER, H., 1753. Employment for the microscope. London.

BALBIANI, G., 1882. Sur les microsporidies ou psorospermies des articules. C. R. Acad. Sci., 95:1168.

BARRET, H. P., and YARBROUGH, N., 1921. A method for the cultivation of *Balantidium coli.* Am. J. Trop. Med., 1:161.

BOECK, W. C., 1921. *Chilomastix mesnili* and a method for its culture. J. Exper. Med., 33:147.

———— and DRBOHLAV, J., 1925. The cultivation of *Endamoeba histolytica.* Amer. J. Hyg., 5:371.

BRUCE, D., 1895. Preliminary report on the tsetse fly disease or nagana in Zululand. Umbobo.

———— 1897. Further report, etc. Umbobo.

BÜTSCHLI, O., 1880-1889. Protozoa. Bronn's Klassen und Ordnungen des Thierreichs. Vols. 1–3.

———— 1881. Myxosporidia. Zool. Jahrb. 1880, 1:162.

BUONANNI, F., 1691. Observationes circa Viventia etc. Rome.

CALKINS, G. N., 1901. The protozoa. Philadelphia.

———— 1933. The biology of the protozoa. 2ed. Philadelphia.

CLAPARÈDE, J. L. R. A. E., and LACHMANN, J., 1858-59. Études sur les Infusoires et les Rhizopodes. Vol. 1. Geneva.

CLEVELAND, L. R., HALL, S. R., and SANDERS, E. P., 1934. The wood-feeding roach, Cryptocercus, its Protozoa, and the symbiosis between protozoa and roach. Mem. Am. Acad. Arts & Sci., 17:185.

COHN, F. J., 1853. Beiträge zur Entwickelungsgeschichte der Infusorien. Zeitschr. wiss. Zool., 6:253.

COLE, F. J., 1926. The history of protozoology. London.

CUTLER, D. W., 1918. A method for the cultivation of *Entamoeba histolytica.* J. Path. Bact., 22:22.

DAVAINE, C., 1854. Sur des animalcules infusoires, etc. C. R. Soc. Biol., (Par.) 1:129.

DOBELL, C. ,1911. The principles of protistology. Arch. Protist., 23:269.

———— 1932. Antony van Leeuwenhoek and his "little animals." New York.

———— 1939. The common flagellate of the human mouth, *Trichomonas tenax* (O.F.M.): *its discovery and its nomenclature. Parasit.,* 31:138.

DOFLEIN, F., 1901. Die Protozoen als Parasiten und Krankheitserreger. Jena.

———— and REICHENOW, E., 1929. Lehrbuch der Protozoenkunde. 5 ed. Jena.

DONOVAN, C., 1903. The etiology of one of the heterogeneous fevers in India. Brit. M. J., 2:1401.

D'ORBIGNY, A., 1826. Tableau méthodique de la Classe des Céphalopodes. Ann. Sci. Nat., 7:245.

DUFOUR, L., 1828. Note sur la grégarine, etc. *Ibid.*, 13:366.

DUJARDIN, F., 1835. Sur les prétendus estomacs des animalcules infusoires et sur une substance appelée sarcode. Ann. Sci. Nat. Zool., 4:343.

———— 1841. Histoire naturelle des zoophytes. Infusoires. Paris.

DUTTON, J. E., 1902. Preliminary note upon a trypanosome occurring in the blood of man. Rep. Thomson Yates Lab., 4:455.

EHRENBERG, C. G., 1838. Die Infusionsthierchen als vollkommene Organismen. Leipzig.

EICHHORN, J. C., 1783. Zugabe zu meinen Beyträgen, etc. Danzig.

EIMER, T., 1870. Ueber die ei- und kugelförmigen sogenannten Psorospermien der Wirbelthiere. Würzburg.

ELLIS, J., 1769. Observations on a particular manner of increase in the animalcula, etc. Phil. Trans., 59:138.

GLUGE, G., 1842. Ueber ein eigenthümliches Entozoon im Blute des Frosches. Arch. Anat. Phys. wiss. Med., 148.

GOLDFUSS, G. A., 1817. Ueber die Entwicklungsstufen des Thieres. Nürnberg.

GOLGI, C., 1886. Sulla infezione malarica. Arch. Sci. Méd., 10:109.

———— 1889. Sul ciclo evolutio dei parassiti malarici nella febbre terzana, etc. *Ibid.*, 13:173.

GROS, G., 1849. Fragments d'helminthologie et de physiologie microscopique. Bull. Soc. Imp. Nat. Moscou, 22:549.

GRUBY, D., 1843. Recherches et observations sur une nouvelle espece d'hematozoaire, *Trypanosoma sanguinis.* C. R. Acad. Sci., 17:1134.

HAECKEL, E. H., 1862. Betrachtungen ueber die Grenzen und Verwandschaft der Radiolarien und ueber die Systematik der Rhizopoden im Allgemeinen. Berlin.

———— 1866. Generelle Morphologie der Organismen. Berlin.

HAKE, T. G., 1839. A treatise on varicose capillaries, as constituting the structure of carcinoma of the hepatic ducts, etc. London.

HARRIS, J., 1696. Some microscopical observations of vast numbers of animalcula seen in water. Phil. Trans., 19:254.

HEGNER, R., 1928. The evolutionary significance of the protozoan parasites of monkeys and man. Quart. Rev. Biol., 3:225.

HILL, J., 1752. An history of animals, etc. London.

HYMAN, L. H., 1940. The invertebrates: Protozoa through Ctenophora. New York.

JENNINGS, H. S., 1909. Heredity and variation in the simplest organisms. Am. Nat., 43:322.

JOBLOT, L., 1718. Descriptions et usages de plusieurs nouveaux microscopes, etc. Paris.

KLOSS, H., 1855. Ueber Parasiten in der Niere von Helix. Abh. Senckenb. Naturf. Ges., 1:189.

KÖLLIKER, A., 1848. Beiträge zur Kenntnis niederer Thiere. Zeitschr. wiss. Zool., 1:34.

Kudo, R. R., 1920. Studies on Myxosporidia. Illinois Biol. Monogr. 5:nos. 3, 4.

——— 1946. *Pelomyxa carolinensis* Wilson. I. Jour. Morphol., 78:317.

——— 1959. Pelomyxa and related organisms. Ann. N. Y. Acad. Sci., 78:474.

Laveran, A., 1880. Note sur un nouveau parasite trouvé dans le sang de plusieurs malades atteints de fièvre palustre. Bull Acad. Méd., 9:1235, 1268, 1346.

——— 1880a. Un nouveau parasite trouvé dans le sang des malades atteints de fièvre palustre. Bull. Mém. Soc. Méd. Hôpit. Paris, 17:158.

Leidy, J., 1879. Freshwater Rhizopods of North America. Rep. U. S. Geol. Survey, 12.

Leishman, W. B., 1903. On the possibility of the occurrence of trypanosomiasis in India. Brit. M. J., 1:1252.

Leuckart, R., 1879. Die Parasiten des Menschen. 2 ed. Leipzig.

Lewis, T., R., 1870. A report on the microscopic objects found in cholera evacuations, etc. Ann. Rep. San. Comm. Gov. India (1869) 6:126.

——— 1878. The microscopic organisms found in the blood of man and animals, etc. *Ibid.* (1877) 14:157.

Linnaeus, C., 1758. Systema Naturae. 10 ed. 1:820.

——— 1767. Systema Naturae. 12 ed. 1:1324.

Lösch, F., 1875. Massenhafte Entwickelung von Amöben im Dickdarm. Arch. path. Anat., 65:196.

Lwoff, A., 1951. Biochemistry and physiology of protozoa. New York.

MacCallum, W. G., 1897. On the flagellated form of the malarial parasite. Lancet, 2:1240.

Malmsten, P. H., 1857. Infusorien als Intestinal-Thiere beim Menschen. Arch. path. Anat., 12:302.

Metcalf, M. M., 1920. Upon an important method of studying problems of relationship and of geographical distribution. Proc. Nat. Acad. Sci., 6:432.

——— 1929. Parasites and the aid they give in problems of taxonomy, geographical distribution, and paleogeography. Smith. Misc. Coll., 81: no. 8.

Musgrave, W. E., and Clegg, M. T., 1904. Amebas: their cultivation and aetiologic significance. Dept. Inter., Biol. Lab. Bull., Manila, no. 18:1.

Novy, F. G., and MacNeal, W. J., 1905. On the trypanosomes of birds. J. Infect. Dis., 2:256.

Pasteur, L., 1870. Études sur la maladie des vers à soie. Paris.

Perty, M., 1852. Zur Kenntnis kleinster Lebensformen, etc. Bern.

Rösel von Rosenhof, A. J., 1755. Der kleine Proteus. Der Monat.-herausgeg. Insect.-Belust., 3:622.

Ross, R., 1898. Report on the cultivation of Proteosoma Labbé in grey mosquitoes. Gov. Print. Calcutta.

——— 1898a. Preliminary report on the infection of birds with Proteosoma by the bites of mosquitoes. *Ibid.*

Russell, E. J., and Hutchinson, H. B., 1909. The effect of

partial sterilization of soil on the production of plant food. J. Agr. Sci., 3:111.

———— ———— 1913. Part 2. *Ibid.*, 5:152.

SCHAUDINN, F., and SIEDLECKI, M., 1897. Beiträge zur Kenntnis der Coccidien. Verhandl. deut. zool. Ges., p. 192.

SIEBOLD, C. T. v., 1845. Bericht ueber die Leistungen in der Naturgeschichte der Würmer, etc. Arch. Naturg., 11:256.

SIEDLECKI, M., 1898. Étude cytologique et cycle évolutif de la coccidie de la seiche. Ann. Inst. Pasteur, 12:799.

———— 1899. Étude cytologique et cycle évolutif de *Adelea ovata* Schneider. *Ibid.*, 13:169.

SINGH, B. N., 1960. Inter-relationship between micropredators and bacteria in soil. Proc. IV Indian Sci. Congress, Part 2, 14 p.

SMITH, T., and KILBORNE, F. L., 1893. Investigations into the nature, causation, and prevention of Texas or southern cattle fever. Bull. Bur. Animal Ind., U. S. Dep. Agr., No. 1.

SONNEBORN, T. M., 1937. Sex, sex inheriance and sex determination in *Paramecium aurelia*. Proc. Nat. Acad. Sc., 23:378.

———— 1938. Mating types in *Paramecium aurelia,* etc. Proc. Am. Phil. Soc., 79:411.

STEIN, S. F. N. v., 1854. Die Infusionsthiere auf ihre Entwickelungsgeschichte untersucht. Leipzig.

———— 1859-83. Der Organismus der Infusionsthiere. Leipzig.

STOKES, A. C., 1888. A preliminary contribution toward a history of the fresh-water Infusoria of the United States. J. Trenton Nat. Hist. Soc., 1:71.

TREMBLEY, A., 1744. Observations upon several newly discovered species of freshwater polypi. Phil. Trans., 43:169.

VALENTIN, 1841. Ueber ein Entozoon im Blute von *Salmo fario.* Arch. Anat. Phys. wiss. Med., p. 435.

WENRICH, D. H., 1956. Some American pioneers in protozoology. J. Protozool., 3:1.

WOODRUFF, L. L., 1937. Louis Joblot and the protozoa. Sc. Monthly, 44:41.

———— 1939. Some pioneers in microscopy, with special reference to protozoology. Tr. N. Y. Acad. Sc., Ser. 2, 1:74.

WRISBERG, H. A., 1765. Observationum de Animalculis infusoriis Satura. Göttingen.

Chapter 2
Ecology

W ITH regard to their habitats, protozoa may be divided into
free-living forms and those living on or in other organisms.
The latter group includes those that hold relationship of various
kinds with the host animal and is referred to in the present work
as parasitic protozoa in broad sense.

Free-living Protozoa

The vegetative or trophic stages of free-living protozoa occur in
every type of fresh and salt water, soil or decaying organic matter. Even in the circumpolar regions or at extremely high altitudes, certain protozoa are found at times in fairly large numbers.
The factors which influence their distribution and population in a
given body of water are temperature, light, chemical composition,
acidity, kind and amount of food present in the water and degree
of adaptability of the individual protozoa to various environmental changes. Their early appearance as living organisms, their
adaptability to various habitats, and their capacity to remain viable in the encysted condition, probably account for the wide distribution of the protozoa throughout the world. The common
free-living amoebae, numerous testaceans and others of fresh waters, to mention a few, have been observed in innumerable places
of the world.

Temperature. The majority of protozoa are able to live only
within a certain range of temperature variation, although in the
encysted state they can withstand a far greater temperature fluc-
tuation. The lower limit of the temperature is marked by the
freezing of the protoplasm, and the upper limit by the destructive
chemical change within the body protoplasm. The temperature
toleration seems to vary among different species of protozoa; and
even in the same species under different conditions. For example,
Chalkley (1930) placed *Paramecium caudatum* in four culture
media (balanced saline, saline with potassium excess, saline with

20

calcium excess, and saline with sodium excess), all with pH from
5.8 or 6 to 8.4 or 8.6, at 40°C. for 2-16 minutes and found
that (1) the resistance varies with the hydrogen-ion concentration, maxima appearing in the alkaline and acid ranges, and a
minimum at or near about 7.0; (2) in a balanced saline, and in
saline with an excess of sodium or potassium, the alkaline maximum is the higher, while in saline with an excess of calcium, the
acid maximum is the higher; (3) in general, acidity decreases and
alkalinity increases resistance; and (4) between pH 6.6 and 7.6,
excess of potassium decreases resistance and excess of calcium increases resistance. Glaser and Coria (1933) cultivated *Paramecium caudatum* on dead yeast free from living organisms at
20-28°C. (optimum 25°C.) and noted that at 30°C. the organisms were killed. Doudoroff (1936), on the other hand, found
that in *P. multimicronucleatum* its resistance to raised temperature was low in the presence of food, but rose to a maximum
when the food was exhausted, and there was no appreciable
difference in the resistance between single and conjugating individuals.

The thermal waters of hot springs have been known to contain living organisms including protozoa. Glaser and Coria (1935)
obtained from the thermal springs of Virginia, several species
of Mastigophora, Ciliata, and an amoeba which were living in
the water, the temperature of which was 34-36°C., but did not
notice any protozoa in the water at 39-41°C. Uyemura (1936,
1937), made a series of studies on protozoa living in various
thermal waters of Japan, and reported that many species lived
at unexpectedly high temperatures. Some of the protozoa observed and the temperatures of the water in which they were found
are as follows: *Amoeba* sp., *Vahlkampfia limax, A. radiosa,* 30-
51°C.; *Amoeba verrucosa, Chilodonella* sp., *Lionotus fasciola,
Paramecium caudatum,* 36-40°C.; *Oxytricha fallax,* 30-56°C.

Under experimental conditions, it has been shown repeatedly
that many protozoa become accustomed to a very high temperature if the change be made gradually. Dallinger (1887) showed
a long time ago that *Tetramitus rostratus* and two other species
of flagellates became gradually acclimatized up to 70°C. in
several years. In nature, however, the thermal death point of most

Protozoology

of the free-living protozoa appears to lie between 36° and 40°C. and the optimum temperature, between 16° and 25°C.

On the other hand, the low temperature seems to be less detrimental to protozoa than the higher one. Many protozoa have been found to live in water under ice, and several haematochrome-bearing Phytomastigia undergo vigorous multiplication on snow in high altitudes, producing the so-called "red snow." Klebs (1893) subjected the trophozoites of Euglena to repeated freezing without apparent injury and Jahn (1933) found no harmful effect when Euglena cultures were kept without freezing at —0.2°C. for one hour, but when kept at —4°C. for one hour the majority were killed. Gaylord (1908) exposed *Trypanosoma gambiense* to liquid air for twenty minutes without apparent injury, but the organisms were killed after forty minutes' immersion.

Kühne (1864) observed that Amoeba and Actinophrys suffered no ill effects when kept at 0°C. for several hours as long as the culture medium did not freeze, but were killed when the latter froze. Molisch (1897) likewise noticed that Amoeba dies as soon as the ice forms in its interior or immediate vicinity. Chambers and Hale (1932) demonstrated that internal freezing could be induced in an amoeba by inserting an ice-tipped pipette at —0.6°C., the ice spreading in the form of fine featherly crystals from the point touched by the pipette. They found that the internal freezing kills the amoebae, although if the ice is prevented from forming, a temperature as low as —5°C. brings about no visible damage to the organism. At 0°C., Deschiens (1934) found the trophozoites of *Entamoeba histolytica* remained alive, though immobile, for fifty-six hours, but were destroyed in a short time when the medium froze at —5°C.

According to Greeley (1902), when *Stentor coeruleus* was slowly subjected to low temperatures, the cilia kept on beating at 0°C. for one to three hours, then cilia and gullet were absorbed, the ectoplasm was thrown off, and the body became spherical. When the temperature was raised, this spherical body is said to have undergone a reverse process and resumed its normal activitiy. If the lowering of temperature is rapid and the medium becomes solidly frozen, Stentor perishes. Efimoff

(1924) observed that Paramecium multiplied once in about thirteen days at 0°C., withstood freezing at —1°C. for thirty minutes but died when kept for fifty to sixty minutes at the same temperature. He further stated that *Paramecium caudatum, Colpidium colpoda,* and *Spirostomum ambiguum,* perished in less than 30 minutes, when exposed below —4°C., and that quick and short cooling (not lower than —9°C.) produced no injury, but if it is prolonged, Paramecium became spherical and swollen to four to five times normal size, while Colpidium and Spirostomum shrunk. Wolfson (1935) studied *Paramecium* sp. in gradually descending subzero-temperature, and observed that as the temperature decreases the organism often swims backward, its bodily movements cease at —14.2°C., but the cilia continue to beat for some time. While Paramecium recover completely from a momentary exposure to —16°C., long cooling at this temperature brings about degeneration. When the water in which the organisms are kept freezes, no survival was noted. *Plasmodium knowlesi* and *P. inui* in the blood of *Macacus rhesus* remain viable, according to Coggeshall (1939), for as long as seventy days at —76°C., if frozen and thawed rapidly.

The sporozoites of *Plasmodium vivax, P. falciparum,* and *P. ovale* are said to have survived at —70°C. for 375, 183 and seventy days respectively (Jeffery and Rendtorff, 1955). Jeffery (1957) later reported that the blood parasites and sporozoites of all four species of human Plasmodium survived low temperature preservation for at least two years with little if any loss in vitality. Fulton and Smith (1953) found *Entamoeba histolytica* in cultures that contained a single species of bacteria and 5 per cent glycerol, survived —79°C. for sixty-five days. Seemingly, during the complete cessation of metabolic activity in these cases, life continues. (Low temperature on protozoa, Luyet and Gehenio, 1940.)

Light. In the Phytomastigia which include chromatophore-bearing flagellates, the sun light is essential to photosynthesis. Therefore, sun light further plays an important rôle in those protozoa which are dependent upon chromatophore-possessing organisms as chief source of food supply. Hence the light is another factor concerned with the distribution of free-living protozoa.

Chemical Composition of Water. The chemical nature of the water is another important factor that influences the very existence of protozoa in a given body of water. Protozoa differ from one another in morphological as well as physiological characteristics. Individual protozoan species requires a certain chemical composition of the water in which it can be cultivated under experimental conditions, although this may be more or less variable among different forms (Needham *et al.*, 1937).

In their "biological analysis of water" Kolkwitz and Marsson (1908, 1909) distinguished four types of habitats for many aquatic plant, and a few animal, organisms, which were based upon the kind and amount of inorganic and organic matter and amount of oxygen present in the water: namely, katharobic, oligosaprobic, mesosaprobic, and ploysaprobic. **Katharobic** protozoa are those which live in mountain springs, brooks, or ponds, the water of which is rich in oxygen, but comparatively free from organic matter. **Oligosaprobic** forms are those that inhabit waters which are rich in mineral matter, but in which no purification processes are taking place. Many Phytomastigia, various testaceans and ciliates, such as Frontonia, Lacrymaria, Oxytricha, Stylonychia, Vorticella, etc. inhabit such waters. **Mesosaprobic** protozoa live in waters in which active oxidation and decomposition of organic matter are taking place. The majority of freshwater protozoa belong to this group: namely, numerous Phytomastigia, Zoomastigia, Heliozoida, and all orders of Ciliata. Finally **polysaprobic** forms are capable of living in waters which, because of dominance of reduction and cleavage processes of organic matter, contain at most a very small amount of oxygen and are rich in carbonic acid gas and nitrogenous decomposition products. The black bottom slime contains usually an abundance of ferrous sulphide and other sulphurous substances. Lauterborn (1901) called this *sapropelic.* Examples of polysaprobic protozoa are *Pelomyxa palustris, Euglypha alveolata, Pamphagus armatus*, Mastigamoeba, *Treponomas agilis, Hexamita inflata, Rhynchomonas nasuta, Heteronema acus,* Bodo, Cercomonas, Dactylochlamys, Odontostomatida, etc. The so-called "sewage organisms" abound in such habitat (Lackey, 1925).

Certain free-living protozoa which inhabit waters rich in decomposing organic matter are frequently found in the faecal matter of various animals. Their cysts either pass through the alimentary canal of the animal unharmed or are introduced after the faeces are voided, and undergo development and multiplication in the faecal infusion. Such forms are collectively called **coprozoic** protozoa. The coprozoic protozoa grow easily in suspension of old faecal matter which is rich in decomposed organic matter and thus show a strikingly strong capacity of adapting themselves to conditions different from those of the water in which they normally live. Some of the coprozoic protozoa which are mentioned in the present work are: *Scytomonas pusilla, Rhynchomonas nasuta, Cercomonas longicauda, C. crassicauda, Trepomonas agilis, Naegleria gruberi, Acanthamoeba hyalina, Sappinia diploidea, Chlamydophrys stercorea, Tillina magna,* etc. (Coprozoic protozoa in mammals, Noble, 1958).

As a rule, the presence of sodium chloride in the sea water prevents the occurrence of numerous species of fresh-water inhabitants. Certain species, however, have been known to live in both fresh and brackish water or salt water. Among the species mentioned in the present work, the following species have been reported to occur in both fresh and salt waters: Mastigophora: *Amphidinium lacustre, Ceratium hirundinella;* Sarcodina: *Lieberkühnia wagneri;* Ciliata: *Mesodinium pulex, Prorodon discolor, Lacrymaria olor, Amphileptus claparedei, Lionotus fasciola, Nassula aurea, Trochilioides recta, Chilodonella cucullulus, Trimyema compressum, Paramecium calkinsi, Colpidium campylum, Platynematum sociale, Cinetochilum margaritaceum, Pleuronema coronatum, Caenomorpha medusula, Spirostomum minus, S. teres, Climacostomum virens,* and *Thuricola folliculata;* Suctoria: *Metacineta mystacina, Endosphaera engelmanni.*

It seems probable that many other protozoa are able to live in both fresh and salt water, judging from the observations such as that made by Finley (1930) who subjected some fifty species of freshwater protozoa of Wisconsin to various concentrations of sea water, either by direct transfer or by gradual addition of the sea water. He found that *Bodo uncinatus, Uronema marinum, Pleuronema jaculans* and *Colpoda aspera* are able to live and reproduce even when directly transferred to sea water, that *Amoeba verru-*

cosa, Euglena, Phacus, Monas, Cyclidium, Euplotes, Lionotus, Paramecium, Stylonychia, etc., tolerate only a low salinity when directly transferred, but, if the salinity is gradually increased, they live in 100 per cent sea water, and that Arcella, Cyphoderia, Aspidisca, Blepharisma, *Colpoda cucullus,* Halteria, etc. could not tolerate 10 per cent sea water even when the change was gradual. Finley noted no morphological changes in the experimental protozoa which might be attributed to the presence of the salt in the water, except *Amoeba verrucosa,* in which certain structural and physiological changes were observed as follows: as the salinity increased, the pulsation of the contractile vacuole became slower. The body activity continued up to 44 per cent sea water and the vacuole pulsated only once in forty minutes, and after systole, it did not reappear for ten to fifteen minutes. The organism became less active above this concentration and in 84 per cent sea water the vacuole disappeared, but there was still a tendency to form the characteristic ridges, even in 91 per cent sea water, in which the organism was less fan-shaped and the cytoplasm seemed to be more viscous. Yocom (1934) found *Euplotes patella* able to live normally and multiply in up to 66 per cent of sea water; above that concentration no division was noticed, though the organism lived for a few days in up to 100 percent salt water, and *Paramecium caudatum* and *Spirostomum ambiguum* were less adaptive to salt water, rarely living in 60 per cent sea water. Frisch (1939) found that no freshwater Protozoa lived above 40 per cent sea water and that *Paramecium caudatum* and *P. multimicronucleatum* died in 33-52 per cent sea water. Hardin (1942) reports that *Oikomonas termo* will grow when transferred directly to a glycerol-peptone culture medium, in up to 45 per cent sea water; and cultures contaminated with bacteria and growing in a dilute glycerol-peptone medium will grow in 100 per cent sea water.

Hydrogen-ion Concentration. Closely related to the chemical composition is the hydrogen-ion concentration (pH) of the water. Some protozoa appear to tolerate a wide range of pH. The interesting proteomyxan, *Leptomyxa reticulata,* occurs in soil ranging in pH 4.3 to 7.8, and grows very well in non-nutrient agar between pH 4.2 and 8.7, provided a suitable bacterial strain is supplied as food (Singh, 1948); and according to Loefer and Guido (1950), a strain of *Euglena gracilis* var. *bacillaris* grows between pH 3.2 and 8.3. However, the majority of protozoa seem to prefer a certain range of pH for the maximum metabolic activity.

The hydrogen-ion concentration of freshwater bodies varies a great deal between highly acid bog waters in which various testaceans may frequently be present, to highly alkaline water in which such forms as Acanthocystis, Hyalobryon, etc., occur. In standing deep fresh water, the bottom region is often acid because of the decomposing organic matter, while the surface water is less acid or slightly alkaline due to the photosynthesis of green plants which utilize carbon dioxide. In some cases, different pH may bring about morphological differences. For example, in bacteria-free cultures of *Paramecium bursaria* in a tryptone medium, Loefer (1938) found that at pH 7.6-8.0 the length averaged 86 or 87μ, but at 6.0-6.3 the length was about 129μ. The greatest variation took place at pH 4.6 in which no growth occurred. The shortest animals at the acid and alkaline extremes of growth were the widest, while the narrowest forms (about 44 μ wide) were found in culture at pH 5.7-7.4. Lee and McCall (1959) found differences in size of body and food vacuole in *P. multimicronucleatum* at different pH. Many workers observed the pH ranges in which certain protozoa live, grow, and multiply, some of which are collected in Table 1.

Food. The kind and amount of food available in a given body of water also controls the distribution of protozoa. The food is ordinarily one of the deciding factors of the number of protozoa in a natural habitat. Species of Paramecium and many other holozoic protozoa cannot live in waters in which bacteria or minute protozoa do not occur. If other conditions are favorable, then the greater the number of food bacteria, the greater the number of protozoa. Noland (1925) studied more than sixty-five species of freshwater ciliates with respect to various factors and came to the conclusion that the nature and amount of available food has more to do with the distribution of these organisms than any other one

TABLE 1

PROTOZOA AND HYDROGEN-ION CONCENTRATIONS

Protozoa	pH Range in Which Growth Occurs	Optimum pH	Observers
A. In bacteria-free cultures			
Euglena gracilis	3.5–9.0	—	Dusi
	3.0–7.7	6.7	Alexander
	3.9–9.9	6.6	Jahn
	—	5.0–6.5	Schoenborn
E. deses	6.5–8.0	7.0	Dusi
	5.3–8.0	7.0	Hall
E. pisciformis	6.0–8.0	6.5–7.5	Dusi
	5.4–7.5	6.8	Hall
E. viridis	—	5.0	Schoenborn
Chilomonas paramecium	4.8–8.0	6.8	Mast and Pace
	4.1–8.4	4.9; 7.0	Loefer
Chlorogonium euchlorum	4.8–8.7	7.1–7.5	"
C. elongatum	4.8–8.7	7.1–7.5	"
C. teragamum	4.2–8.6	6.7–8.3	"
Colpidium campylum	—	5.4	Kidder
Glaucoma scintillans	—	5.6–6.8	"
G. ficara	4.0–9.5	5.1; 6.7	Johnson
Tetrahymena pyriformis	—	5.6–8.0	Kidder
Strain HS	4.0–9.0	7.25–7.3	Prescott
T. vorax	—	6.2–7.6	Kidder
Paramecium bursaria	4.9–8.0	6.7–6.8	Loefer
B. In cultures containing bacteria			
Carteria obtusa	—	3.5–4.5	Wermel
Trichomonas vaginalis	6.4–8.4	—	Bland *et al.*
Actinosphaerium eichhorni	—	7.2–7.6	Howland
Acanthocystis aculeata	7.4 or above	8.1	Stern
Paramecium caudatum	5.3–8.2	7.0	Darby
	6.0–9.5	7.0	Morea
	—	6.9–7.1	Wichterman
P. aurelia	5.7–7.8	6.7	Morea
	5.9–8.2	5.9–7.7	Phelps
	—	7.0–7.2	Wichterman
P. multimicronucleatum	4.8–8.3	7.0	Jones
	—	6.5–7.0	Wichterman
P. trichium	—	6.7–7.1	"
P. bursaria	—	7.1–7.3	"
P. polycaryum	—	6.9–7.3	"
P. calkinsi	—	6.5–7.8	"
P. woodruffi	—	7.0–7.5	"
Colpidium sp.	6.0–8.5	—	Pruthi
Colpoda cucullus	5.5–9.5	6.5; 7.5	Morea
Holophyra sp.	6.5–7.4	—	Pruthi
Plagiopyla sp.	6.9–7.5	—	"
Amphileptus sp.	6.8–7.5	7.1–7.3	"
Spirostomum ambiguum	6.8–7.5	7.4	Saunders
S. sp.	6.5–8.0	7.5	Morea
Stentor coeruleus	7.8–8.0	—	Hetherington
Blepharisma undulans	—	6.5	Moore
Gastrostyla sp.	6.0–8.5	—	Pruthi
Stylonychia pustulata	6.0–8.0	6.7; 8.0	Darby

factor. *Didinium nasutum* feeds almost exclusively on paramecia; therefore, it cannot live in the absence of the latter ciliate. As a rule, euryphagous protozoa which feed on a variety of food organisms are widely distributed, while stenophagous forms that feed on a few species of food organisms are limited in their distribution.

In nature, protozoa live in association with diverse organisms. The interrelationships which exist among them are not understood in most cases. For example, the relationship between *Entamoeba histolytica* and certain bacteria in successful *in-vitro* cultivation has not yet been comprehended. Certain strains of bacteria were found by Hardin (1944) to be toxic for *Paramecium multimicronucleatum*, but if *Oikomonas termo* was present in the culture, the ciliate was maintained indefinitely. This worker suggested that the flagellate may be able to "detoxify" the metabolic products produced by the bacteria (Food relation in ciliates, Fauré-Fremiet, 1950, 1951a).

The adaptability of protozoa to varied environmental conditions influences their distribution. The degree of adaptability varies a great deal, not only among different species, but also among the individuals of the same species. *Stentor coreuleus* which grows ordinarily under nearly anaerobic conditions, is obviously not influenced by alkalinity, pH, temperature or free carbon dioxide in the water (Sprugel, 1951).

Some protozoa inhabit soil of various types and localities. Under ordinary circumstances, they occur near the surface, their maximum abundance being found at a depth of about 10-12 cm. (Sandon, 1927). It is said that a very few protozoa occur in the subsoil. Here also one notices a very wide geographical distribution of apparently one and the same species. For example, Sandon found *Amoeba proteus* in samples of soil collected from Greenland, Tristan da Cunha, Gough Island, England, Mauritius, Africa, India, and Argentina. This amoeba is known to occur commonly in various parts of North America, Europe, Japan, and Australia. The majority of Testacea inhabit moist soil in abundance. Sandon observed *Trinema enchelys* in the soils of Spitzbergen, Greenland, England, Japan, Australia, St. Helena, Barbados, Mauritius, Africa, and Argentina.

Parasitic Protozoa

Some protozoa belonging to all groups live on or in other organisms. The Sporozoa and Cnidosporidia are exclusively parasites. The relationships between the hosts and the protozoa differ in various ways, which make the basis for distinguishing the associations into three types as follows: **commensalism, symbiosis,** and **parasitism.**

Commensalism is an association in which an organism, the commensal, is benefited, while the host is neither injured nor benefited. Depending upon the location of the commensal in the host body, the term ectocommensalism or endocommensalism is used. Ectocommensalism is often represented by protozoa which may attach themselves to any aquatic animals that inhabit the same body of water, as shown by various species of chonotrichs, peritrichs, and Suctoria. In other cases, there is a definite relationship between the commensal and the host. For example, *Kerona polyporum* is found on various species of Hydra, and many thigmotrichs are inseparably associated with certain species of mussels.

Endocommensalism is often difficult to distinguish from endoparasitism, since the effect of the presence of a commensal upon the host cannot be easily understood. On the whole, the protozoa which live in the lumen of the alimentary canal may be looked upon as endocommensals. These protozoa undoubtedly use part of the food material which could be used by the host, but they do not invade the host tissue. As examples of endocommensals may be mentioned: *Endamoeba blattae, Lophomonas blattarum, L. striata, Nyctotherus ovalis,* etc., of the cockroach; *Entamoeba coli, Iodamoeba bütschlii, Endolimax nana, Chilomastix mesnili,* etc., of the human intestine; numerous opalinids in the colon of Anura, etc. The term **parasitic Protozoa,** in its broad sense, includes the commensals also.

Symbiosis, on the other hand, is an association of two species of organisms, which is of mutual benefit. The cryptomonads belonging to Chrysidella ("Zooxanthellae") containing yellow or brown chromatophores, which live in Foraminiferida and Radiolarida, and certain algae belonging to Chlorella ("Zoochlorellae") containing green chromatophores, which occur in some freshwater protozoa, such as *Paramecium bursaria, Stentor amethystinus,* etc., are looked upon as holding symbiotic relationship with the respective protozoan host. Many species of the highly interesting Hypermastigida, which are present commonly and abundantly in various species of termites and the woodroach Cryptocercus, have been demonstrated by Cleveland to digest the cellulose material which makes up the bulk of woodchips the host insects take in and to transform it into glycogenous substances that are used partly by the host insects. If deprived of these flagellates by being subjected to oxygen under pressure or to a high temperature, the termites die, even though the intestine is filled with wood-chips. If removed from the gut of the termite, the flagellates perish (Cleveland, 1924, 1925). Recently, Cleveland (1949-1957) found that the molting hormone produced by Cryptocercus induces sexual reproduction in several flagellates inhabiting its hind-gut. Thus, the association here may be said to be an absolute symbiosis.

Parasitism is an association in which one organism (the parasite) lives at the expense of the other (the host). Here also ectoparasitism and endoparasitism occur, although the former is not commonly found. *Hydramoeba hydroxena* (p. 553) feeds on the body cells of Hydra which, according to Reynolds and Looper (1928), die on an average in 6.8 days as a result of the infection and the amoebae disappear in from four to ten days, if removed from a host Hydra. *Costia necatrix* often occurs in an enormous number, attached to various freshwater fishes especially in aquaria, by piercing through the epidermal cells and appears to disturb the normal functions of the host tissue. Ichthyophthirius, another ectoparasite of freshwater and marine fishes, go further by completely burying themselves in the epidermis and feed on the host's tissue cells and, not infrequently, contributes toward the death of the host fish.

The parasites absorb by osmosis the vital body fluid, feed on the host cells or cell-fragments by pseudopodia or cytostome, or enter the host tissues or cells themselves, living on the cytoplasm or in some cases on the nucleus of host cells. Consequently they bring about abnormal or pathological conditions upon the host which often succumbs to the infection. Endoparasitic protozoa of

Protozoology

man are *Entamoeba histolytica, Balantidium coli,* species of Plasmodium, Leishmania, and Trypanosoma, etc. The Sporozoa and Cnidosporidia as was stated before, are without exception coelozoic, histozoic, or cytozoic parasites.

Because of their modes of living, the endoparasitic protozoa cause certain morphological changes in the cells, tissues, or organs of the host. The active growth of *Entamoeba histolytica* in the glands of the colon of the victim, produces first slightly raised nodules which develop into abscesses and the ulcers formed by the rupture of abscesses, may reach 2 cm. or more in diameter, completely destroying the tissues of the colon wall. Similar pathological changes may also occur in the case of infection by *Balantidium coli.* In *Leishmania donovani,* the victim shows an increase in number of the large macrophages and mononucleates and also enlargement of the spleen. *Trypanosoma cruzi* brings about the degeneration of the infected host cells and an abundance of leucocytes in the infected tissues, followed by an increase of fibrous tissue. *T. gambiense,* the causative organism of African sleeping sickness, causes enlargement of lymphatic glands and spleen, followed by changes in meninges and an increase of cerebro-spinal fluid. Its most charastistic changes are the thickening of the arterial coat and the round-celled infiltration around the blood vessels of the central nervous system.

Malarial infection is invariably accompanied by an enlargement of the spleen ("Spleen index"); the blood becomes watery; the erythrocytes decrease in number; the leucocytes, subnormal; but mononuclear cells increase in number; pigment granules which are set free in the blood plasma at the time of merozoite-liberation are engulfed by leucocytes; and enlarged spleen contains large amount of pigments. In *Plasmodium falciparum,* the blood capillaries of brain, spleen and other viscera may completely be blocked by infected erythrocytes.

In histozoic Myxosporida of fishes, the tissue cells that are in direct contact with highly enlarging parasites, undergo various morphological changes. For example, the circular muscle fibers of the small intestine of *Pomoxis sparoides,* which surround a cyst of *Myxobolus intestinalis,* become modified a great deal and turn about 90° from the original direction, due undoubtedly to the stimulation exercised by the parasite (Fig. 1, *a*). In the case of another myxosporidan, *Thelohanellus notatus,* the connective tissue cells of the host fish surrounding the protozoan body, transform themselves into "epithelial cells" (Fig. 1, *b*), a state comparable to the formation of the ciliated epithelium from a layer of fibroblasts lining a cyst formed around a piece of ovary inplanted into the adductor muscle of Pecten as observed by Drew (1911).

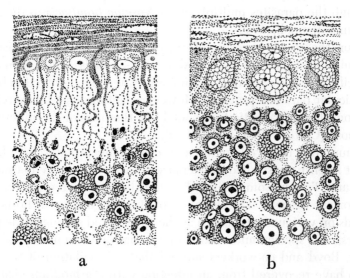

a   b

Fig. 1. Histological changes in host fish caused by myxosporidan infection, × 1920 (Kudo). *a,* portion of a cyst of *Myxobolus intestinalis,* surrounded by peri-intestinal muscle of the black crappie; *b,* part of a cyst of *Thelohanellus notatus,* enveloped by the connective tissue of the blunt-nosed minnow.

Practically all Microsporida are cytozoic, and the infected cells become hypertrophied enormously, producing in one genus the so-called Glugea cysts (Figs. 297, 299). In many cases, the hypertrophy of the nucleus of the infected cell is far more conspicuous than that of the cytoplasm (Figs. 297, 300) (Kudo, 1924).

When the gonads are parasitized heavily, the germ cells of the host animal often do not develop, thus resulting in parasitic castration. For example, the ciliate, *Orchitophrya stellarum,* a parasite in the male reproductive organ of *Asterias rubens,* was found by Vevers (1951) to break down completely all germinal tissues of the testes in the majority of the host starfish. In other cases, the protozoan does not invade the gonads, but there is no development of the germ cells. The microsporidan, *Nosema apis*, attacks solely the gut epithelium of the honey bee, but the ovary of an infected queen bee degenerates to varying degrees (Hassanein, 1951). Still in other instances, the protozoa invade developing ova of the host, but do not hinder the development of the embryo, though the parasites multiply, as in *Nosema bombycis* in the silkworm (Stempell, 1909).

For the great majority of parasitic protozoa, there exists a definite host-parasite relationship and animals other than the specific hosts possess a natural immunity against an infection by a particular parasitic protozoa. However, caution must be exercised in considering parasitic forms solely on the basis of the difference in host (Cairus, 1953; Saxe, 1954).

Immunity involved in diseases caused by protozoa has been most intensively studied in haemozoic forms, expecially Plasmodium and Trypanosoma, since they are the causative organisms of important diseases. Development of these organisms in hosts depends on various factors such as the species and strains of the parasites, the species and strains of vectors, and immunity of the host. Boyd and co-workers showed that reinoculation of persons who have recovered from an infection with *Plasmodium vivax* or *P. falciparum* with the same strain of the parasites, will not result in a second clinical attack, because of the development of homologous immunity, but with a different strain of the same species or different species, a definite clinical attack occurs, thus there being no heterologous tolerance. The homologous immunity was found to continue for at least three years and in one case for about seven years in *P. vivax*, and for at least four months in *P. faciparum* after apparent eradication of the infection. In the case of leishmaniasis, recovery from a natural or induced infection apparently develops a lasting immunity against reinfection with the same species of Leishmania.

It has been shown that in infections with avian, monkey and human Plasmodium or *Trypanosoma lewisi*, a considerable num-

ber of the parasites are destroyed during the developmental phase
of the infection and that after a variable length of time, resistance
to the parasites often develops in the host, as the parasites disappear from the peripheral blood and symptoms subside, though the
host still harbors the organisms. In malarious countries, the adults
and children show usually a low and a high rate of malaria infection respectively, but the latter frequently do not show symptoms
of infection, even though the parasites are detectable in the
blood. Apparently repeated infection produces tolerance which
can keep, as long as the host remains healthy, the parasites under
control. There seems to be also racial difference in the degree of
immunity against Plasmodium and Trypanosoma (Young *et al.*,
1955).

As to the mechanism of immunity, the destruction of the parasites by phagocytosis of the endothelial cells of the spleen, bone
marrow and liver and continued regenerative process to replace
the destroyed blood cells, are the two important phases in the cellular defense mechanism. Besides, there are indications that hu-
moral defense mechanism through the production of antibodies is
in active operation in infections by *Plasmodium knowlesi* and trypanosomes (Taliaferro, 1926; Maegraith, 1948; Culbertson,
1951). (Immunity, Taliaferro, 1941; Chandler, 1958.)

With regard to the origin of parasitic protozoa, it is generally
agreed among biologists that the parasite in general evolved
from the free-living form. The protozoan association with other
organisms was begun when various protozoa which lived attached
to, or by crawling on, submerged objects happened to transfer
themselves to various invertebrates which live in the same water.
These protozoa benefit by change in location as the host animal
moves about, and thus enlarging the opportunity to obtain a continued supply of food material. Such ectocommensals are found
abundantly; for example, the peritrichous ciliates attached to the
body and appendages of various aquatic animals such as larval
insects and microcrustaceans. Ectocommensalism may next lead
to ectoparasitism as in the case of Costia or Hydramoeba, and
then again instead of confining themselves to the body surface,
the protozoa may bore into the body wall from outside and ac-

tually acquire the habit of feeding on tissue cells of the attached animals as in the case of Ichthyophthirius.

The next step in the evolution of parasitism must have been reached when protozoa, accidentally or passively, were taken into the digestive system of the metazoa. Such a sudden change in environment appears to be fatal to most protozoa. But certain others possess extraordinary capacity to adapt themselves to an entirely different environment. For example, Dobell (1918) observed in the tadpole gut, a typical free-living limax amoeba, with characteristic nucleus, contractile vacuoles, etc., which was found in numbers in the water containing the faecal matter of the tadpole.

Tetrahymena pyriformis, a free-living ciliate, was found to occur in the body cavity of the larvae of *Theobaldia annulata* (after MacArthur) and in the larvae of *Chironomus plumosus* (after Treillard and Lwoff). Lwoff successfully inoculated this ciliate into the larvae of *Galleria mellonella* which died later from the infection. Janda and Jírovec (1937) injected bacteria-free culture of this ciliate into annelids, molluscs, crustaceans, insects, fishes, and amphibians, and found that only insects—all of fourteen species (both larvae and adults)—became infected by this ciliate. In a few days after injection, the haemocoele became filled with the ciliates. Of various organs, the ciliates were most abundantly found in the adipose tissue. The organisms were much larger than those present in the original culture. The insects, into which the ciliates were injected, died from the infection in a few days. The course of development of the ciliate within an experimental insect depended not only on the amount of the culture injected, but also on the temperature. At 1-4°C. the development was much slower than at 26°C.; but if an infected insect was kept at 32-36°C. for 0.5-3 hours, the ciliates were apparently killed and the insect continued to live. When glaucoma taken from *Dixippus morosus* were placed in ordinary water, they continued to live and underwent multiplication. The ciliate showed a remarkable power of withstanding the artificial digestion; namely, at 18°C. they lived four days in artificial gastric juice with pH 4.2; 2-3 days in a juice with pH 3.6; and a few hours in a juice with pH 1.0 Thompson (1958) inoculated axenic cultures of five species of

Tetrahymena into a variety of animals and found heavy growth of these ciliates in several insects. They were also able to live in chick embryos, guppies and tadpoles.

Cleveland (1928) observed *Tritrichomonas fecalis* in the faeces of a human subject for three years which grew well in faeces diluted with tap water, in hay infusions with or without free-living protozoa or in tap water with tissues at —3° to 37°C., and which, when fed *per os*, was able to live indefinitely in the gut of frogs and tadpoles. Reynolds (1936) found that *Colpoda steini*, a free-living ciliate of fresh water, occurs naturally in the intestine and other viscera of the land slug, *Agriolimax agrestis*, the slug forms being much larger than the free-living individuals. A pure culture of *Crithidia* sp., from the gut of the hemipteron *Euryophthalmus davisi*, when deposited on the chorio-allantoic membrane of chick and duck embryos, heavy infection resulted, the flagellates being found in the membrane, in phagocytic cells and in the cavity of the chorio-allantois (McGhee, 1959).

It may be further speculated that Vahlkampfia, Hydramoeba, and Endamoeba, are the different stages of the course the intestinal amoebae might have taken during their evolution. Obviously, endocommensalism in the alimentary canal was the initial phase of endoparasitism. When these endocommensals began to consume an excessive amount of food or to feed on the tissue cells of the host gut, they became the true endoparasites. Destroying or penetrating through the intestinal wall, they became first established in the body or organ cavities and then invaded tissues, cells or even nuclei, thus developing into pathogenic protozoa. The endoparasites developing in invertebrates which feed upon the blood of vertebrates as source of food supply, will have opportunities to establish themselves in the higher animals.

Hyperparasitism. Certain parasitic protozoa have been found to parasitize other protozoan or metazoan parasites. This association is names hyperparasitism. The microsporidan *Nosema notabilis* (p. 813) is an exclusive parasite of the myxosporidan *Sphaerospora polymorpha*, which is a very common inhabitant of the urinary bladder of the toad fish along the Atlantic and Gulf coasts. A heavy infection of the microsporidan results in the degeneration

Protozoology

and death of the host myxosporidan trophozoite (Kudo, 1944). Thus *Nosema notabilis* is a hyperparasite. (Organisms living on and in protozoa, Duboscq and Grassé, 1927, 1929; Georgévitch, 1936; Grassé, 1936; Kirby, 1932, 1938, 1941, 1941a, 1942, 1942a, 1942b, 1944, 1946.)

References

BLAND, P. B., *et al.*, 1932. Studies on the biology of *Trichomonas vaginalis*. Amer. J. Hyg., 56:492.

CAIRUS, J., 1953. Transfaunation studies on the host specificity of the enteric protozoa of Amphibia and various other vertebrates. Proc. Acad. Nat. Sci., Philadelphia, 105:45.

CHALKLEY, H. W., 1930. Resistance of Paramecium to heat as affected by changes in hydrogen-ion concentration and in inorganic salt balance in surrounding medium. U. S. Pub. Health, Rep., 45:481.

CHAMBERS, R., and HALE, H. P., 1932. The formation of ice in protoplasm. Proc. Roy. Soc. London, Series B, 110:336.

CHANDLER, A. C., 1958. Some considerations relative to the nature of immunity in *Trypanosoma lewisi* infections. J. Parasit., 44:129.

CLEVELAND, L. R., 1924. The physiological and symbiotic relationships between the intestinal protozoa of termites and their host, with special reference to *Reticulitermes flavipes* Kollar. Biol. Bull., 46:177.

———— 1925. The effects of oxygenation and starvation on the symbiosis between the termite, Termopsis, and its intestinal flagellates. *Ibid.*, 48:309.

———— 1926. Symbiosis among animals with special reference to termites and their intestinal flagellates. Gen. Rev. Biol., 1:51.

———— 1928. *Tritrichomonas fecalis* nov. sp. of man, etc. Amer. J. Hyg., 8:232.

———— 1949. Hormone-induced sexual cycles of flagellates. I. J. Morphol., 85:197.

———— 1950. II. *Ibid.*, 86:185.

———— 1950a. III. *Ibid.*, 86:215.

———— 1950b. IV. *Ibid.*, 87:317.

———— 1950c. V. *Ibid.*, 87:349.

———— 1957. Correlation between the moulting period of Cryptocercus and sexuality in its protozoa. J. Protozool., 4:168.

COGGESHALL, L. T., 1939. Preservation of viable malaria parasites in the frozen state. Proc. Soc., Exp. Biol., 42:499.

CULBERTSON, J. T., 1951. Immunological mechanisms in parasitic infections. In Most: Parasitic infections in man. New York.

DALLINGER, W. H., 1887. The president's address. J. Roy. Micro. Soc., London, 7:185.

DARBY, H. H., 1929. The effect of the hydrogen-ion concentration on the sequence of protozoan forms. Arch. Protist., 65:1.

DENNIS, E. W., 1932. The life-cycle of *Babesia bigemina*, etc., Univ. Cal. Publ. Zoology, 36:263.

DESCHIENS, R., 1934. Influence du froid sur les formes végétatives de l'amibe dysenterique. C. R. Soc. Biol., 115:793.

DOBELL, C., 1918. Are *Entamoeba histolytica* and *E. ranarum* the same species? Parasit., 10:294.

DOUDOROFF, M., 1936. Studies in thermal death in Paramecium. J. Exper. Zool., 72:369.

DREW, G. H., 1911. Experimental metaphasia. I. J. Exper. Zool., 10:349.

DUBOSCQ, O., and GRASSÉ, P. P., 1927. Flagellés et Schizophytes de *Calotermes (Glyptotermes) iridipennis*. Arch. zool. exp. gén., 66:451.

———— ———— 1929. Sur quelques protistes d'un Calotermes, etc. *Ibid.*, 68:8.

EFIMOFF, W. W., 1924. Ueber Ausfrieren und Ueberkaeltung der Protozoen. Arch. Protist., 49:431.

FAURÉ-FREMIET, E., 1950. Ecology of ciliate infusoria. Endeavour 9, 3 pp.

———— 1951. The marine sand-dwelling ciliates of Cape Cod. Biol. Bull., 100:59.

———— 1951a. Ecologie des Protistes littoraux. Ann. Biol., 27:205.

FINLEY, H. E., 1930. Toleration of freshwater Protozoa to increased salinity. Ecology, 11:337.

FRISCH, J. A., 1939. The experimental adaptation of Paramecium to sea water. Arch. Protist., 93:38.

FULTON, J. D., and SMITH, A. U., 1950. Preservation of *Entamoeba histolytica* at −79°C. in the presence of glycerol. Ann. Trop. Med. & Parasit., 47:240.

GAYLORD, H. R., 1908. The resistance of embryonic epithelium, etc. J. Infect. Dis., 5:443.

GEORGÉVITCH, J., 1936. Ein neuer Hyperparasit, *Leishmania esocis* n. sp. Arch. Protist., 88:90.

GLASER, R. W., and CORIA, N. A., 1933. The culture of *Paramecium caudatum* free from living microorganisms. J. Parasit., 20:33.

———— ———— 1935 . The culture and reactions of purified protozoa. Amer. J. Hyg., 21:111.

GRASSÉ, P. P., 1938. La vêture schizophytique des flagellés termiticoles, etc. Bull. Soc. zool. France, 63:110.

GREELEY, A. W., 1902. On the analogy between the effects of loss of water and lowering of temperature. Amer. J. Physiol., 6:122.

HARDIN, G., 1944. Symbiosis of Paramecium and Oikomonas Ecology, 25:304.

HASSANEIN, M. H., 1951. Studies on the effect of infection with *Nosema apis* on the physiology of the queen honey-bee. Quart. J. Micr. Sc., 92:225.

HOWLAND, RUTH, 1930. Micrurgical studies on the contractile vacuole. III. J. Exper. Zool., 55:53.

JAHN, T. L., 1933. Studies on the physiology of the euglenoid flagellates. IV. Arch. Protist., 79:249.

JANDA, V., and JÍROVEC, O., 1937. Ueber künstlich hervorgerufenen Parasitismus eines freilebenden Ciliaten *Glaucoma piriformis,* etc. Mém. Soc. Zool. Tchéc. Prague, 5:34.

JEFFREY, G. M., 1957. Extended low-temperature preservation of human malaria parasites. J. Parasit., 43:488.

—— and RENDTORFF, R. C., 1955. Preservation of viable human malaria sporozoites by low-temperature freezing. Exp. Parasit., 4:445.

KIDDER, G. W., 1941. Growth studies on ciliates. VII. Biol. Bull., 80:50.

KIRBY, H. JR., 1932, Flagellates of the genus Trichonympha in termites. Univ. Cal. Publ. Zool., 37:349.

—— 1938. The devescovinid flagellates, etc. *Ibid.,* 43:1.

—— 1941. Devescovinid flagellates of termites. I. *Ibid.,* 45:1.

—— 1941a. Organisms living on and in protozoa. Calkins and Summers: Protozoa in biological research.

—— 1942. Devescovinid flagellates of termites. II. Univ. Cal. Publ. Zool., 45:93.

—— 1942a. III. *Ibid.,* 45:167.

—— 1942b. A parasite of the macronucleus of Vorticella. J. Parasit., 28:311.

—— 1944. The structural characteristics and nuclear parasites of some species of Trichonympha in termites. Univ. Cal. Publ. Zool., 49:185.

—— 1946. *Gigantomonas herculea,* etc. *Ibid.,* 53:163.

KLEBS, G., 1893. Flagellatenstudien. Zeitschr. wiss. Zool., 55:265.

KOLKWITZ, R., and MARSSON, M., 1909. Oekologie der tierischen Sabrobien. Intern. Rev. Ges. Hydrobiol. u. Hydrogr., 2:126.

KUDO, R. R., 1924. A biologic and taxonomic study of the Microsporidia. Illinois Biol. Monogr., 9: nos. 3 and 4.

—— 1929. Histozoic Myxosporidia found in freshwater fishes of Illinois, U. S. A. Arch. Protist., 65:364.

—— 1944. Morphology and development of *Nosema notabilis* Kudo, parasitic in *Sphaerospora polymorpha* Davis, a parasite of *Opsanus tau* and *O. beta.* Illinois Biol. Monogr., 20:1.

KÜHNE, W., 1864. Untersuchungen ueber das Protoplasma und die Contractilität. Leipzig.

LACKEY, J. B., 1925. The fauna of Imhof tanks. Bull. N. J. Agr. Ex. St., No. 417.

LAUTERBORN, R., 1901. Die "sapropelische" Lebewelt. Zool. Anz., 24:50.

LEE, J. W., and McCALL, W., 1959. Effects of pH and viscosity on surface membranes in *Paramecium multimicronucleatum.* J. Protozool., 6:146.

LOEFER, J. B., 1935. Relation of hydrogen-ion concentration to growth of Chilomonas and Chlorogonium. Arch. Partist., 85:209.

—— 1938. Effect of hydrogen-ion concentration on the growth and morphology of *Paramecium bursaria. Ibid.,* 90:185.

—— 1939. Acclimatization of fresh-water ciliates and flagellates to media of higher osmotic pressure. Physiol. Zool., 12:161.

———— and GUIDO, VIRGINIA M., 1950. Growth and survival of *Euglena gracilis,* etc. Texas J. Sci., 2:225.

LUYET, B. J., and GEHENIO, P. M., 1940. The mechanism of injury and death by low temperature. A review. Biodynamica, 3: no. 60.

McGHEE, R. B., 1959. The infection of avian embryos with *Crithidia* species and *Leishmania tarentola.* J. Infect. Dis., 105:18.

MAEGRAITH, B., 1948. Pathological processes in malaria and blackwater fever. Springfield, Illinois.

MOLISCH, H., 1897. Untersuchungen ueber das Erfrieren der Pflanzen. Jena.

NEEDHUM, J. G., GALTSOFF, P. S., LUTZ, F. E., and WELCH, P. S., 1937. Culture methods for invertebrate animals. Ithaca, N.Y.

NOBLE, G. A., 1958. Coprozoic protozoa from Wyoming mammals. J. Protozool., 5:69.

NOLAND, L. E., 1925. Factors influencing the distribution of freshwater ciliates. Ecology, 6:437.

PHELPS, A., 1934. Studies on the nutrition of Paramecium. Arch. Protist., 82:134.

PRESCOTT, D. M., 1958. The growth rate of *Tetrahymena geleii* H S under optimal conditions. Physiol. Zool., 31:111.

REYNOLDS, B. D., 1936. *Colpoda steini,* a facultative parasite of the land slug, *Agriolimax agrestis.* J. Parasit., 22:48.

———— and LOOPER, J. B. 1928. Infection experiments with *Hydramoeba hydroxena* nov. gen. *Ibid.,* 15:23.

ROSENBERG, L. E., 1936. On the viability of *Trichomonas augusta.* Tr. Am. Micr. Soc., 55:313.

SANDON, H., 1927. The composition and distribution of the protozoan fauna of soil. Edinburgh.

SAXE, L. H., 1954. Transfaunation studies on the host specifiicity of the enteric protozoa of rodents. J. Protozool., 1:220.

SCHOENBORN, H. W., 1950. Nutritional requirements and the effect of pH on growth of *Euglena viridis* in pure culture. Tr. Am. Micr. Soc., 69:217.

SINGH, B. N., 1948. Studies on giant amoeboid organisms. I. J. Gen. Microbiol., 2:7.

SPRUGEL, G., JR., 1951. Vertical distribution of *Stentor coeruleus* in relation to dissolved oxygen levels in an Iowa pond. Ecology, 32:147.

STEMPELL, W., 1909. Ueber *Nosema bombycis.* Arch. Protist., 16: 281.

TALIAFERRO, W. H., 1926. Host resistance and types of infections in trypanosomiasis and malaria. Quart. Rev. Biol., 1:246.

———— 1941. The immunology of the parasitic Protozoa. In Calkins and Summers: Protozoa in biological research.

THOMPSON, J. C., 1958. Experimental infection of various animals with strains of the genus Tetrahymena. J. Protozool., 5:203.

UYEMURA, M., 1936. Biological studies of thermal waters in Japan. IV. Ecolog. St., 2:171.

—— 1937. V. Rep. Japan. Sci. A., 12:264.

VEVERS, H. G., 1951. The biology of *Asterias rubens*. II.   J. Mar. Biol. A. United Kingdom, 29:619.

WICHTERMAN, R., 1948. The hydrogen-ion concentration in the cultivation and growth of 8 species of Paramecium. Biol. Bull., 95:271.

WOLFSON, C., 1935. Observations on Paramecium during exposure to subzero temperatures. Ecology, 16:630.

YOCOM, H. B., 1934. Observations on the experimental adaptation of certain freshwater ciliates to sea water. Biol. Bull., 67:273.

YOUNG, M. D., *et al.*, 1955. Experimental testing of the immunity of negroes to *Plasmodium vivax*. J. Parasit., 41:315.

Chapter 3
Morphology

PROTOZOA range in size from submicroscopic to macroscopic, though they are on the whole minute microscopic animals. The parasitic forms, especially cytozoic parasites, are often extremely small, while free-living protozoa are usually of much larger dimensions. Noctiluca, Foraminiferida, Radiolarida, many ciliates such as Stentor, Bursaria, etc., represent larger forms. Colonial protozoa such as Carchesium, Zoothamnium, Ophrydium, etc., are even greater than the solitary forms. On the other hand, Plasmodium, Leishmania, and microsporidan spores may be mentioned as examples of the smallest forms. The unit of measurement employed in protozoology is, as in general microscopy, 1 micron (μ) which is equal to 0.001 mm.

Knowledge on morphology of protozoa is rapidly increasing because of improvement in the microscope and of introduction of new techniques and devices. While the ordinary microscopes that are now commonly called light microscopes remain a basic tool for observation, the phase contrast microscopy is widely used, as it reveals various intracellular structures in living organisms. Furthermore, since the introduction of thin sectioning, electron microscopy is revealing hitherto unknown structural detail of various protozoa.

The body form of protozoa is even more varied, and because of its extreme plasticity it frequently does not remain constant. Furthermore the form and size of a given species may vary according to the kind and amount of food as is discussed elsewhere (p. 131). From a small simple spheroidal mass up to large highly complex forms, all possible body forms occur. Although the great majority are without symmetry, there are some which possess a definite symmetry. Thus, bilateral symmetry is noted in Giardia, Hexamita, etc.; radial symmetry in Gonium, Cyclonexis, etc.; and universal symmetry, in certain Heliozoida; Volvox, etc.

The fundamental component of the protozoan body is the protoplasm which is without exception differentiated into the nucleus and the cytoplasm. Haeckel's (1868, 1870) monera are now considered as nonexistent, since improved microscopic technique has failed in recent years to reveal any anucleated protozoa. The nucleus and the cytoplasm are inseparably important to the well-being of a protozoan, as has been shown by numerous investigators since Verworn's pioneer work. In all cases, successful regeneration of the body is accomplished only by the nucleus-bearing portions and enucleate parts degenerate sooner or later, and, when the nucleus is taken out of a protozoan, both the nucleus and cytoplasm degenerate, which indicates their intimate association in carrying on the activities of the body. It appears certain that the nucleus controls the assimilative phase of metabolism which takes place in the cytoplasm in normal animals, while the cytoplasm is capable of carrying on the catabolic phase of the metabolism. Aside from the importance as the controlling center of metabolism, evidences point to the conclusion that the nucleus contains the genes or hereditary factors which characterize each species of protozoa from generation to generation, as in the cells of multicellular animals and plants.

The Nucleus

Because of a great variety of the body form and organization, the protozoan nuclei are of various forms, sizes and structures. At one extreme there is a small nucleus and, at the other, a large voluminous one; between these extremes is found almost every conceivable variety of form and structure. The majority of protozoa contain a single nucleus, though many may possess two or more throughout the greater part of their life-cycle. In several species, each individual possesses two similar nuclei, as in Hexamitidae, Protoopalina and Zelleriella. In Ciliata and Suctoria, two dissimilar nuclei, a macronucleus and a micronucleus, are present. The macronucleus is always larger than the micronucleus, and controls the trophic activities and regeneration processes of the organism, while the micronucleus is concerned with the reproductive activity. Certain protozoa possess numerous nuclei of similar structure, as, for example, in Pelomyxa, Mycetozoida, Actinosphaerium, Opalina, Cepedea, Myxosporida, Microsporida, etc.

The essential morphological components of the protozoan nucleus are the nuclear membrane, chromatin, plastin and nucleoplasm or nuclear sap. Their interrelationship varies sometimes from one developmental stage to another, and vastly among different species. Structually, they fall, in general, into one of the two types: vesicular and compact.

The **vesicular** nucleus (Fig. 2, *a, c, e*) consists of a nuclear membrane which is sometimes very delicate but distinct, nucleoplasm, achromatin and chromatin. Besides there is an intranuclear body which is, as a rule, more or less spherical and which appears to be of different make-ups as judged by its staining reactions among different nuclei. It may be composed of chromatin, of plastin, or of a mixture of both. The first type is sometimes called karyosome and the second, nucleolus or plasmosome. Absolute distinction between these two terms cannot be made as they are based solely upon the difference in affinity to nuclear stains which cannot be standardized and hence do not give uniformly the same result. Following Minchin (1912), the term **endosome** is advocated here to designate one or more conspicuous bodies other than the chromatin granules, present within the nuclear membrane (Fig. 2, *b, d*).

The vesicular nucleus of many small amoebae contains a conspicuous endosome that appears spherical in stained preparations. However, viewed under phase microscope, the endosome is found to undergo active form change when the organism is in motion and feeding.

When viewed in life the nucleoplasm is ordinarily homogeneous and structureless. But, upon fixation, there appear invariably achromatic strands or networks which seem to connect the endosome and the nuclear membrane (Fig. 2 *b, d*). Some investigators hold that these strands or networks exist naturally in life, but due to the similarity of refractive indices of the strands and of the nucleoplasm, they are not visible and that, when fixed, they become readily recognizable because of a change in these indices. In some nuclei, however, certain strands have been observed in life, as for example in the nucleus of the species of Barbulanympha (Fig. 176, *c*), according to Cleveland and his associates (1934). Others maintain that the achromatic structures prominent

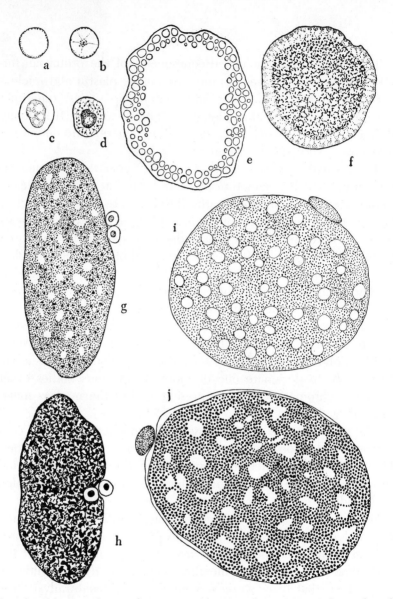

FIG. 2. *a–f*, vesicular nuclei; *g–j*, compact nuclei ×980. *a, b*, nuclei of *Entamoeba invadens* (*a*, in life; *b*, in stained organism); *c, d*, nuclei of *Amoeba spumosa* (*c*, in life, showing a large endosome; *d*, stained); *e, f*, nuclei of *A. proteus* (*e*, in life; *f*, a nucleus subjected to Feulgen's nucleal reaction); *g, h*, nuclei of *Paramecium aurelia* (*g*, in life under phase microscope, showing two vesicular micronuclei and compact macronucleus; *h*, Feulgen-stained nuclei); *i, j*, nuclei of *Frontonia leucas*, showing a micronucleus and macronucleus, both of which are compact (*i*, in life, showing many endosomes imbedded among the granules, *j*, nuclei stained with acidified methyl green.)

in fixed vesicular nuclei are mere artifacts brought about by fixation and do not exist in life and that the nucleoplasm is a homogeneous liquid matrix of the nucleus in which the chromatin is usually distributed as small granules. Frequently, larger granules of various sizes and forms may occur along the inner surface of the nuclear membrane. These so-called peripheral granules that occur in Amoeba, Entamoeba, Pelomyxa, etc., are apparently not chromatinic (Fig. 2, *a, e*). The vesicular nucleus is most commonly present in various orders of Sarcodina and Mastigophora.

Under electron microscope, the intranuclear space in *Amoeba proteus* is often occupied by a loosely aggregated filamentous material among which are found clusters of small well-formed helices that radiate in a stellate fashion from diffuse centers (Mercer, 1959; Roth, Obetz and Daniels, 1960; Pappas, 1956).

The **compact** nucleus (Fig. 2, *g–j*), on the other hand, contains a large amount of chromatin substance and a comparatively small amount of nucleoplasm, and is thus massive. The macronucleus of the Ciliophora is almost always of this kind. The variety of forms of the compact nuclei is indeed remarkable. It may be spherical, ovate, cylindrical, club-shaped, band-form, moniliform, horseshoe-form, filamentous, or dendritic. The nuclear membrane is always distinct, and the chromatin substance is usually of spheroidal form, varying in size among different species and often even in the same species. In the majority of species, the chromatin granules are small and compact (Fig. 2, *h, i*), though in some forms, such as *Nyctotherus ovalis* (Fig. 3), they may reach 20 μ or more in diameter in some individuals and while the smaller chromatin granules seem to be homogeneous, larger forms contain alveoli of different sizes in which smaller chromatin granules are suspended (Kudo, 1936).

Precise knowledge of chromatin or DNA (deoxyribonucleic acid) is still lacking. The determination of the chromatin depends upon the following tests: (1) artificial digestion which does not destroy this substance, while non-chromatinic parts of the nucleus are completely dissolved; (2) acidified methyl green which stains the chromatin bright green; (3) 10 per cent sodium chloride solution which dissolves, or causes swelling of, chromatin granules, while nuclear membrane and achromatic substances remain unattacked; and (4) in the fixed condition Feulgen's nucleal reaction (p. 1079). While the macronucleus of the majority of ciliates is uniform in structure throughout the nucleus, that of the ciliates belonging to Chlamydodontidae, Dysteriidae, Chonotrichida, is, ac-

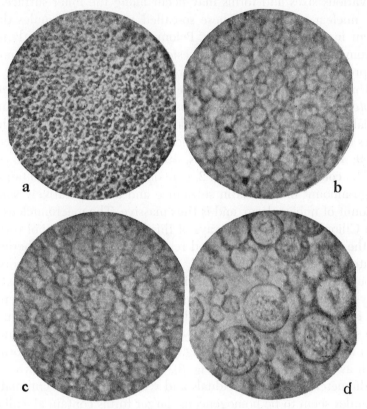

FIG. 3. Parts of four macronuclei of *Nyctotherus ovalis*, showing chromatin spherules of different size, ×650 (Kudo).

cording to Fauré-Fremiet (1957) composed of two parts: one with deoxyribonucleoprotein granules and the other with diffused deoxyribonucleic acid (DNA) and a large deoxyribonucleoprotein endosome. These two parts are in close juxtaposition. Fauré-Fremiet holds that these nuclei which he called "heteromerous" were derived from the "homeomerous" type found in many other ciliates (Action of methyl green, Pollister and Leuchtenberger, 1949).

There is no sharp demarcation between the vesicular and compact nuclei, since there are numerous nuclei the structures of which are intermediate between the two. Moreover, what appears to be a vesicular nucleus in life, may approach a compact nucleus when fixed and stained as in the case of Euglenoidida. Several experimental observations show that the number, size, and structure of the endosome in the vesicular nucleus, and the amount and arrangement of the chromatin in the compact nucleus, vary according to the physiological state of the whole organism. The macronucleus may be divided into two or more parts with or without connections among them and in *Dileptus anser* into more than 200 small nuclei, each of which is "composed of a plastin core and a chromatin cortex" (Calkins; Hayes).

In a compact nucleus, the chromatin granules or spherules fill, as a rule, the intranuclear space conpactly, in which one or more endosomes (Fig. 2, *i*) may occur. In many nuclei, these chromatin granules appear to be suspended freely, while, in others, a reticulum appears to make the background. The chromatin of compact nuclei gives a strong positive Feulgen's nucleal reaction. The macronuclear and micronuclear chromatin substances respond differently to Feulgen's nucleal reaction or to the so-called nuclear stains, as judged by the difference in the intensity or tone of color. In *Paramecium caudatum*, *P. aurelia*, Chilodonella, *Nyctotherus ovalis*, etc., the macronuclear chromatin is colored more deeply than the micronuclear chromatin, while in Colpoda, Urostyla, Euplotes, Stylonychia, and others, the reverse seems to be the case, which may support the validity of the assumption by Heidenhain that the two types of the nuclei of ciliata and Suctoria are made up of different chromatin substances—idiochromatin in the micronucleus and trophochromatin in the macronucleus—and in other classes of Protozoa, the two kinds of chromatin are present together in a single nucleus. The macronucleus and the micronucleus of vegetative *Paramecium caudatum* were found by Moses (1950) to possess a similar nucleic acid-protein composition; namely, similar concentrations of total protein, nonhistone protein, deoxyribonucleic acid and ribonucleic acid. Of the two latter nucleic acids, ribonucleic acid is said to be present in a larger amount than deoxyribonucleic acid in both nuclei. In post-conjugation stages of *Epistylis articulata,* Seshachar and Dass (1953) report evidence to indicate that within each macronuclear fragment conversion of DNA into RNA is taking place.

Chromidia. Since the detection of chromatin had solely depended on its affinity to certain nuclear stains, several investigators found extranuclear chromatin granules in many protozoa. Finding such granules in the cytoplasm of *Actinosphaerium eichhorni, Arcella vulgaris,* and others, Hertwig (1902) called them chromidia, and maintained that under certain circumstances, such as lack of food material, the nuclei disappear and the chromatin granules become scattered throughout the cytoplasm. In the case of *Arcella vulgaris,* the two nuclei break down completely to produce a chromidial-net which later reforms into smaller secondary nuclei. It has, however, been found by Bělař that the lack of food caused the encystment rather than chromidia-formation in Actinosphaerium and, according to Reichenow, Jollos observed that in Arcella the nuclei persisted, but were thickly covered by chromidial-net which could be cleared away by artificial digestion to reveal the two nuclei. In Difflugia, the chromidial-net is vacuolated or alveolated in the fall and in each alveolus appear glycogen granules which seem to serve as reserve food material for the reproduction that takes place during that season (Zuelzer), and the chromidia occurring in Actinosphaerium appear to be of a combination of a carbohydrate and a protein (Rumjantzew and Wermel, 1925).

Apparently, the widely distributed volutin (p. 140) and many inclusions or cytozoic parasites, such as Sphaerita (p. 1074), which occur occasionally in different Sarcodina, have, in some cases, been called chromidia. By using Feulgen's nucleal reaction, Reichenow (1928) obtained a diffused violet-stained zone in Chlamydofonas and held them to be dissolved volutin. Calkins (1933) found the chromidia of *Arcella vulgaris* negative to the nucleal reaction, but by omitting acid-hydrolysis and treating with fuchsin-sulphurous acid for eight to fourteen hours, the chromidia and the secondary nuclei were found to show a typical positive reaction and believed that the chromidia were chromatin. Thus the real nature of chromidia is still not clearly known, although many protozoologists are inclined to think that the substance is not chromatinic, but, in some way, is connected with the metabolism of the protozoan.

The Cytoplasm

The extranuclear part of the protozoan body is the cytoplasm. It is composed of a colloidal system, which may be homogeneous, granulated, vacuolated, reticulated, or fibrillar in optical texture, and is almost always colorless The chromatophore-bearing protozoa are variously colored, and those with symbiotic algae or cryptomonads are also greenish or brownish in color. Furthermore, pigment or crystals which are produced in the body may give protozoa various colorations. In several forms pigments are diffused throughout the cytoplasm. For example, many dinoflagellates are beautifully colored, which, according to Kofoid and Swezy, is due to a thorough diffusion of pigment in the cytoplasm.

Stentor coeruleus is beautifully blue-colored. This coloration is due to the presence of pigment *stentorin* (Lankester, 1873) which occurs as granules in the ectoplasm (Fig. 14). The pigment is highly resistant to various solvents such as acids and alkalis, and the sunlight does not affect its nature. It is destroyed by bleaching with chlorine gas or with potassium permanganate, followed by immersion in 5 per cent oxalic acid (Weisz, 1948). Møller (1962) suggests that stentorin and zoopurpurin, mentioned below, belong to the meso-naphthodianthrone group of compounds. In *Stentor niger*, there occur two kinds of pigments, one of which is reddish-violet (stentorol) and the other brown (Barbier *et al.*, 1956).

Several species of Blepharisma are rose- or purple-colored. This coloration is due to the presence of zoopurpurin (Arcichovskij, 1905) which is lodged in the ectoplasm in granular form. This pigment is soluble in alcohol, ether or acetone, and is destroyed by strong light (Giese, 1938). Inaba *et al.* (1958) found, in electron micrographs, these pigment granules occur just below the pellicle and mitochondria are present below this layer. Rudzinska and Granick (1953) found the ordinarily colorless *Tetrahymena pyriformis* (strain W) produce a reddish pigment when grown rapidly on solid media in darkness. They identified the pigment protoporphyrin.

The extent and nature of the cytoplasmic differention differ greatly among various groups. In the majority of protozoa, the cytoplasm is differentiated into the ectoplasm and the endoplasm. The **ectoplasm** is the cortical zone which is hyaline and homogeneous in Sarcodina and Sporozoa. In the Ciliophora, it is a permanent and distinct part of the body and contains several organelles. The **endoplasm** is more voluminous and fluid. It is granulated or alveolated and contains various organellae. While the alveolated cytoplasm is normal in forms such as the members of Heliozoida and Radiolarida; in other cases the alveolation of normally granulated or vacuolated cytoplasm indicates invariably the beginning of degeneration of the protozoan body. In Amoeba and other Sarcodina, the "hyaline cap" and "layer" (Mast) make up the ectoplasm, and the "plasmasol" and "plamagel" (Mast) compose the endoplasm (Fig. 48).

In numerous Sarcodina and certain Mastigophora, the body surface is naked and not protected by any form-giving organella. However, the surface layer is not only elastic, but solid, and therefore the name **plasma-membrane** or plasmalemma (Mast) may be applied to it. According to Bairati and Lehmann (1953), the plasmalemma of *Ameoba proteus* appears to be a sintered globular film (about 500 Å thick) formed by densely packed minute globular bodies (probably mucoproteids) in electron micrographs, and Pappas (1954) considered it composed in part of a neutral mucopolysaccharide or mucin. In *Amoeba striata, A. verrucosa* (Howland, 1924), Pelomyxa (Kudo, 1946, 1951, 1957), etc., the plasmalemma could be compared with pellicle.

The **pellicle** of a ciliate is thick and often variously ridged or sculptured. In many, linear furrows and ridges run longitudinally, obliquely, or spirally; and, in others, the ridges are combined with hexagonal or rectangular depressed areas. Still in others, such as Coleps, elevated platelets are arranged parallel to the longitudinal axis of the body. In certain peritrichous ciliates, such as *Vorticella monilata, Carchesium granulatum*, etc., the pellicle may possess nodular thickenings arranged in more or less parallel rows at right angles to the body axis.

While the pellicle always covers the protozoan body closely, there are other kinds of protective envelopes produced by protozoa which may cover the body rather loosely. These are the shell, test, lorica or envelope. The **shell** of various Phytomastigia is usually made up of cellulose which is widely distributed in the plant kingdom. It may be composed of a single or several layers, and may possess ridges or markings of various patterns on it. In addition to the shell, gelatinous substance may in many forms be produced to surround the shelled body or in the members of Volvocidae to form the matrix of the entire colony in which the individuals are embedded. In the dinoflagellates, the shell is highly developed and often composed of numerous plates which are variously sculptured.

In Arcella and allied forms, the shell is made up of chitinous material constructed in particular ways which characterize the different genera. Newly formed shell is colorless, but older ones become brownish, because of the presence of iron oxide. Difflugia and related genera form shells by gluing together small sandgrains, diatom-shells, debris, etc., with chitinous or pseudochitinous substances. Foraminifers seem to possess a remarkable selective power in the use of foreign materials, for the construction of their shells. According to Cushman (1933) *Psammosphaera fusca* uses sand-grains of uniform color but of different sizes, while *P. parva* uses grains of more or less uniform size but adds, as a rule, a single large acerose sponge spicule which is built into the test and which extends out both ways considerably. Cushman thinks that this is not accidental, since the specimens without the spicules are few and those with a short or broken spicules are not found. *P. bowmanni*, on the other hand, uses only mica flakes which are found in a comparatively small amount, and *P. rustica* uses acerose sponge spicules for the framework of the shell, skilfully fitting smaller broken pieces into polygonal areas. Other foraminiferans combine chitinous secretion with calcium carbonate and produce beautifully constructed shells (Fig. 4) with one or numerous pores. In the Coccolithidae, variously shaped platelets of calcium carbonate ornament the shell.

The silica is present in the shells of various protozoa. In Euglypha and related testaceans, siliceous scales or platelets are produced in the endoplasm and compose a new shell at the time of fission or of encystment together with the chitinous secretion. In many heliozoans, siliceous substance forms spicules, platelets, or combination of both which are embedded in the mucilaginous envelope that surrounds the body and, in some cases, a special clathrate shell composed of silica, is to be found. In some Radiolarida, isolated siliceous spicules occur as in Heliozoida, while in others the lateral development of the spines results in production of highly complex and the most beautiful shells with various ornamentations or incorporation of foreign materials. Many pelagic

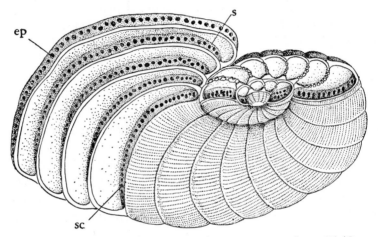

FIG. 4. Diagram of the shell of *Peneroplis pertusus*, × about 35 (Carpenter). *ep*, external pore; *s*, septum; *sc*, stolon canal.

radiolaridans possess numberous conspicuous radiating spines in connection with the skeleton, which apparently aid the organisms in maintaining their existence in the open sea.

Certain Protomonadida possess a funnel-like collar in the flagellated end and in some in addition a chitinous lorica surrounds the body. The lorica found in the Ciliophora is mostly composed of chitinous substance alone, especially in Peritrichida, although others produce test made up of gelatinous secretion containing foreign materials as in Stentor (p. 969). In the Tintinnida, the loricae are either solely chitinous in numerous forms not mentioned in the present work or composed of sand-grains or coccoliths cemented together by chitinous secretion in fresh-water forms.

Locomotor Organelles

Closely associated with the body surface are the organelles of locomotion: *pseudopodia, flagella,* and *cilia.* These organelles are not confined to protozoa alone and occur in various cells of metazoa. All protoplasmic masses are capable of movement which may result in change of their forms.

Pseudopodia. A pseudopodium is a temporary projection of part of the cytoplasm of those protozoa which do not possess a rigid pellicle. Pseudopodia are therefore a characteristic organella of Sarcodina, though many Mastigophora and other groups which lack a pellicle, are also able to produce them. According to their form and structure, four kinds of pseudopodia are usually distinguished.

1) **Lobopodium** is formed by an extension of the ectoplasm, accompanied by a flow of endoplasm as is commonly found in *Amoeba proteus* (Figs. 48; 186). It is finger- or tongue-like, sometimes branched, and its distal end is typically rounded. It is quickly formed and equally quickly retracted. In many cases, there are many pseudopodia formed from the entire body surface, in which the largest one will counteract the smaller ones and the organism will move in one direction; while in others, there may be a single pseudopodium formed, as in *Amoeba striata, A. guttual, Pelomyxa carolinensis* (Fig. 188, *b*), etc., in which case it is a broadly tonguelike extension of the body in one direction and the progressive movement of the organisms is comparatively rapid. The lobopodia may occasionally be conical in general shape, as in *Amoeba spumosa* (Fig. 187, *a*). Although ordinarily the formation of lobopodia is by a general flow of the cytoplasm, in some it is sudden and "eruptive," as in *Endamoeba blattae* or *Entamoeba histolytica* in which the flow of the plasmasol presses against the inner zone of the plasmagal and the accumulated pressure finally causes a break through the zone, resulting in a sudden extension of the protoplasmic flow at that point.

2) **Filopodium** is a more or less filamentous projection composed almost exclusively of the ectoplasm. It may sometimes be branched, but the branches do not anastomose. Many testaceans, such as Lecythium, Boderia, Plagiophrys, Pamphagus, Euglypha,

etc., form this type of pseudopodia. The pseudopodia of *Amoeba radiosa* may be considered as approaching this type rather than the lobopodia.

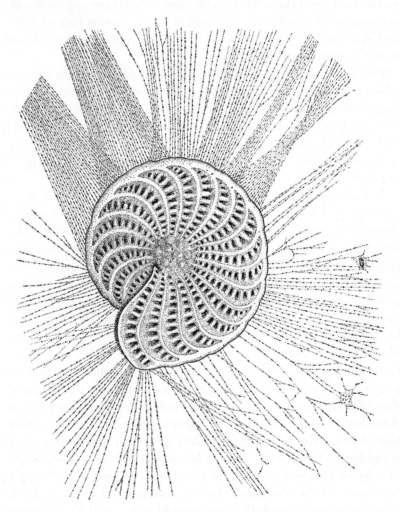

Fig. 5. Pseudopodia of *Elphidium strigilata,* × about 50
(Schulze from Kühn).

3) **Rhizopodium** is also filamentous, but branching and anastomosing. It is found in numerous Foraminiferida, such as Elphidium (Fig. 5), Peneroplis, etc., and in certain testaceans, such as

Lieberkühnia, Myxotheca, etc. The abundantly branching and anastomosing rhizopodia often produce a large network which serves almost exclusively for capturing prey.

4) **Axopodium,** unlike the other three types, is a more or less semi-permanent structure and composed of axial rod and cytoplasmic envelope. Axopodia are found in many Heliozoida, such as Actinophrys, Actinosphaerium, Camptonema, Sphaerastrum, and Acanthocystis. The axial rod, which is composed of a number of

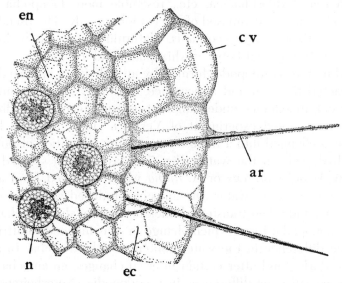

Fig. 6. Portion of *Actinosphaerium eichhorni*, ×800 (Kühn). *ar*, axial rod; *cv*, contractile vacuole; *ec*, ectoplasm; *en*, endoplasm; *n*, nucleus.

fibrils (Doflein; Roskin, 1925; Rumjantzew and Wermel, 1925), arises from the central body or the nucleus located in the approximate center of the body, from each of the nuclei in multinucleate forms, or from the zone between the ectoplasm and endoplasm (Fig. 6). Although semipermanent in structure, the axial rod is easily absorbed and reformed. In the genera of Heliozoida not mentioned above and in numerous radiolarians, the radiating filamentous pseudopodia are so extremely delicate that it is difficult to determine whether an axial rod exists in each or not, although they resemble axopodia in general appearance.

Electron microscopy reveals the axopodia of *Actinosphaerium nucleofilum* as ectoplasmic extensions composed of vacuoles, some of which contain mitochondria and granules. The axial rod is bire-fringent and consists of many fine filaments that are arranged lengthwise, enter deep into the endoplasm and end near a nucleus (Anderson and Beams, 1960).

There is no sharp demarcation between the four types of pseu-dopodia, as there are transitional pseudopodia between any two of them. For example, the pseudopodia formed by Arcella, Les-quereusia, Hyalosphaenia, etc., resemble more lobopodia than filopodia, though composed of the ectoplasm only. The pseudopo-dia of Actinomonas, Elaeorhanis, Clathrulina, etc., may be looked upon as transitional between rhizopodia and axopodia.

While the pseudopodia formed by an individual are usually of characteristic form and appearance, they may show an entirely different appearance under different circumstances. According to the often-quoted experiment of Verworn (1897, 1913), a limax amoeba changed into a radiosa amoeba upon addition of potassi-um hydroxide to the water (Fig. 7). Mast (1928a) showed that when *Amoeba proteus* or *A. dubia* was transferred from a salt medium into pure water, the amoeba produced radiating pseu-dopodia, and when transferred back to a salt medium, it changed into monopodal form, which change he was inclined to attribute to the difference in the water contents of the amoeba. In some cases, during and after certain internal changes, an amoeba may show conspicuous differences in pseudopodia (Neresheimer). As was stated before, pseudopodia occur widely in forms which are placed under classes other than Sarcodina during a part of their life-cycle. Care, therefore, should be exercised in using them for taxonomic consideration.

Flagella. The flagellum is a filamentous extension of the cyto-plasm and is ordinarily extremely fine and highly vibratile, so that it is difficult to recognize it distinctly in life under the microscope. It is more clearly observed under a darkfield or phase microscope. Lugol's solution (p. 1073) usually makes it more easily visible, though the organism is killed. In a small number of species, the flagellum can be seen in life under an ordinary microscope as a long filament, as for example in Peranema. As a rule, the number of flagella present in an individual is small, varying from one to eight and most commonly one or two; but in Hypermastigida there occur numerous flagella.

A flagellum appears to be composed of two parts: an elastic axial filament or **axoneme** and the contractile cytoplasmic sheath surrounding the axoneme (Fig. 8, *a*, *b*). In some flagella, both components extend the entire length and terminate in a bluntly rounded point, while in others the distal portion of the axoneme is apparently very thinly sheathed (Fig. 8, *c*).

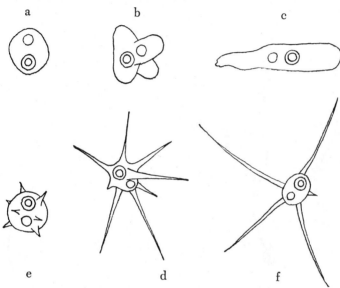

FIG. 7. Form-change in a limax-amoeba (Verworn). *a, b,* contracted forms; *c,* individual showing typical form; *d–f,* radiosa-forms, after addition of KOH solution to the water.

In some flagellates, stained flagella show numerous lateral fibrils (Fig. 8, *d*) (Fischer, 1894; Dellinger, 1909; Mainx, 1929; Petersen, 1929; etc.). These flagella or *ciliary flagella* have also been noticed by several observers in unstained organisms under darkfield microscope (Vlk, 1938; Pitelka, 1949). In recent years, the electron microscope has been used by some to observe the flagellar structure (Schmitt, Hall and Jakus, 1943; Brown, 1945; Pitelka, 1949; Chen, 1950), but in all cases, the organisms were air-dried on collodion films for examination so that the flagella disintegrated more or less completely at the time of observation.

Pitelka (1949) studied flagella of euglenoid organisms under light and electron microscopes. She found that the flagellum of *Euglena gracilis, Astasia longa* and *Rhabdomonas incurva,* consists of an axoneme, composed of nine fibrils, 350-600 Å in diameter, arranged in two compact, parallel bundles, and a sheath which is made up of fibrillar elements, a probably semi-fluid matrix and a limiting membrane. Under conditions always associated with death of the organism, the fibrils of the sheath fray out on

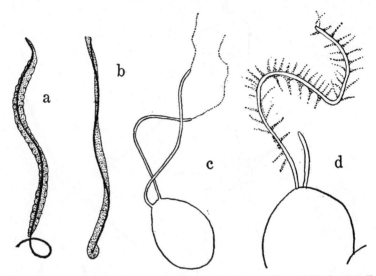

FIG. 8. Diagrams of flagella. *a,* flagellum of Euglena (Bütschli); *b,* flagellum of Trachelomonas (Plenge); *c,* flagella of *Polytoma uvella; d,* flagella of *Monas socialis* (Vlk).

one or more sides of the flagellum into fine lateral filaments or *mastigonemes.* The electron micrographs obtained by various investigators on supposedly one and the same flagellate present a varied appearance of the structure. Compare, for example, the micrographs of the frayed flagellum of *Euglena gracilis* by Brown (1945), Pitelka (1949) and Houwink (1951).

The anterior flagellum of *Peranema trichophorum* frays out into three strands during the course of disintegration as first observed by Dellinger (1909) and by several recent observers. It can be easily demonstrated by treating the organism with reagents such as acidified methyl green. Under electron microscope, Pitelka noted no frayed mastigonemes in the flagellum of Peranema, while Chen (1950) observed numerous mastigonemes extending out from all sides like a brush, except the basal portion of the flagellum.

The electron micrographs of the flagellum of trypanosomes reveal that it also consists of an axoneme and a sheath of cytoplasm. The axoneme is composed of a number of long parallel fibrils, 8 in *Trypanosoma lewisi,* each with estimated diameters of

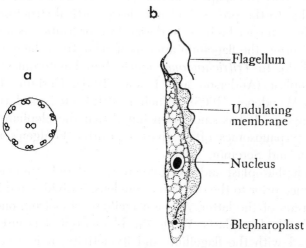

FIG. 9. *a,* cross-section of a flagellum or a cilium as viewed under electron microscope; *b,* a diagram showing the structure of a trypanosome (Kühn).

0.055-0.06μ (Kleinschmidt and Kinder, 1950), and up to 9 in *T. evansi,* with estimated diameters of 0.04-0.05μ (Kraneveld, Houwink and Keidel, 1951). The cytoplasmic sheath of the latter species was said to be cross-striated at about 0.05μ interval. Each flagellum of *Tritrichomonas muris* is composed of ten pairs (one central and nine peripheral) of fibrils (Anderson, 1955). In Trichonympha, the extracellular portion of the flagellum is made up in a similar way (Fig. 9, *a*), but the root portion is a tube of nine fibers and does not contain two central fibers (Pitelka and Schooley, 1958).

The frayed condition of a flagellum which had become detached from the organism or which is still attached to a moribund individual, as revealed by the darkfield microscope, may indicate a phase in disintegration of the flagellum. It is reasonable to assume that different flagella may have structural differences as revealed by the electron microscope, but evidence for the occurrence of mastigonemes on an active flagellum of a normally living organism appears not to be on hand.

A flagellum takes its origin in a **blepharoplast** imbedded in the cytoplasm. The blepharoplast is a small compact granule, but in certain parasitic flagellates, it may be comparatively large and ovoid or short rod-shaped, surrounded often by a halo. Whether this is due to the presence of a delicate cortical structure enveloping the compact body or to desiccation or fixation is unknown. In such forms, the flagellum appears to arise from the outer edge of the halo. In *Tritrichomonas muris,* four kinetosomes make a blepharoplast (Anderson and Beams, 1961), Certain observers such as Woodcock (1906), Minchin (1912), etc., used the term kinetonucleus. It has since been found that the blepharoplast of certain tyrpanosomes often gives a positive Feulgen's reaction (Bresslau and Scremin, 1924).

The blepharoplast and centriole are considered synonymous by some, since prior to the division of nucleus, it divides and initiates the division of the latter. A new flagellum arises from one of the daughter blepharoplasts. While the blepharoplast is inseparably connected with the flagellum and its activity, it is exceedingly small or absent in *Trypanosoma equinum* and in some strains of *T. evansi* (Hoare, 1939, 1954). Furthermore, this condition may be produced by exposure of normal individuals to certain chemical substances (Jírovec, 1929; Piekarski, 1949) or spontaneously (p. 273) without decrease in flagellar activity.

The flagellum is most frequently inserted near the anterior end of the body and directed forward, its movement pulling the organism forward. Combined with this, there may be a trailing flagellum which is directed posteriorly and serves to steer the course of movement or to push the body forward to a certain extent. In a comparatively small number of flagellates, the flagellum is inserted near the posterior end of the body and would push the body forward by its vibration. Under favorable conditions, flagellates regenerate lost flagella. For example, *Peranema trichophorum* from which its anterior flagellum was cut off, regenerated a new one in two hours (Chen, 1950).

In certain parasitic Mastigophora, such as Trypanosoma (Fig. 9), Trichomonas, etc., there is a very delicate membrane that extends out from the side of the body, a flagellum bordering its outer margin. When this membrane vibrates, it shows a characteristic undulating movement, as seen in *Trypanosoma rotatorium* of the frog, and is called the **undulating membrane.** In dinoflagellates, the transverse flagellum seems to be similarly constructed (Kofoid and Swezy). (Flagella, Sleigh, 1962.)

Cilia. The cilia are the organelle of locomotion found in the Ciliophora and opalinids. They aid in the ingestion of food and serve often as a tactile organelle. The cilia are fine and short processes of ectoplasm. They may be uniformly long or may be of different lengths, being longer at the extremities, on certain areas, in peristome or in circumoral areas. Ordinarily, the cilia are arranged in longitudinal, oblique, or spiral rows, being inserted either on the ridges or in the furrows. A cilium originates in a *kinetosome* which is embedded in the ectoplasm. A short distance to the right of kinetosomes, there occur a fibril, *kinetodesma* (Fig. 24). The ciliary row or *kinety* consists of the kinetosomes, kinetodesma and associated fibrils which complex is intimately connected with external ciliature. It is present even in aciliated stages of Ciliophora. Chatton and Lwoff named it *infraciliature.*

As to its structure, a cilium appears to be made up of axoneme and contractile sheath (Fig. 10, *a*). Gelei observed in flagella and cilia, lipoid substance in granular or rod-like forms which differed even among different individuals of the same species; and Klein (1929) found in many cilia of *Colpidium colpoda,* an argentophilous substance in granular form much resembling the lipoid structure of Gelei and called them "cross straition" of the contractile component (Fig. 10, *b, c*). In electron micrographs of a dried cilium of Paramecium, Jakus and Hall (1946) found that it consisted of a bundle of about 11 fibrils extending the full length (Fig. 10, *d*). These fibrils were about 300-500 Å in diameter. As there was no visible sheath, the two observers remarked that if a sheath exists, it must be very fragile and easily ruptured. Electron micrographs of sectioned cilia of various ciliates show clearly that a cilium is similar in structure to a flagellum; namely, two central and nine double peripheral fibers are enveloped by a membrane which is continuous with the pellicle (Fig. 9, *a*). The

central fibers end in a bulb just above a septum located at about
the level of pellicle, but nine peripheral fibrils extend inward to

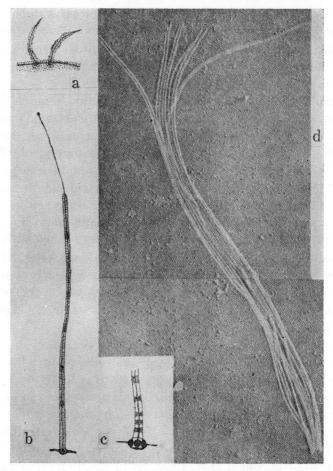

FIG. 10. *a,* cilia of Coleps; *b,* cilium of *Cyclidium glaucoma; c,* basal portion of a cilium of *Colpidium colpoda,* all in silver preparations (Klein); *d,*
electromicrograph of a dried cilium of Paramecium, shadow-cast with
chromium, ×11,000 (Jakus and Hall).

form a cylindrical body, the *kinestome* (Structure of cilia, Sleigh,
1962).

The cilia are often present more densely in a certain area than
in other parts of body and, consequently, such an area stands out

conspicuously, and is sometimes referred to as a *ciliary field*. If this area is in the form of a zone, it may be called a *ciliary zone*. Some authors use *pectinellae* for short longitudinal rows or transverse bands of close-set cilia. In a number of forms, such as Coleps, Stentor, etc., there occur, mingled among the vibratile cilia, immobile stiff cilia which are apparently solely tactile in function.

In the Hypotrichida, the cilia are largely replaced by cirri, although in some species both may occur. A **cirrus** is composed of a number of cilia arranged in two or three rows that fused into one structure completely (Figs. 11, *a;* 12, *a*), as was first demonstrated by Taylor. Klein also showed by desiccation that each marginal cirrus of Stylonychia was composed of seven to eight cilia. In some instances, the distal portion of a cirrus may show two or more branches. The cirri are confined to the ventral surface in Hypotrichida, and called frontal, ventral, anal, caudal, and marginal cirri, according to their location (Fig. 11, *b*). Unlike cilia, the cirri may move in any direction so that the organisms bearing them show various types of locomotion. Oxytricha, Stylonychia, etc., "walk" on frontals, ventrals, and anals, while swimming movement by other species is of different types.

In all Ciliata except Holotrichida, there are adoral membranellae. A **membranella** is a double ciliary lamella, fused completely into a plate (Fig. 12, *b*). A number of these membranellae occur along the peristome, forming the **adoral zone** of membranellae,which serves for bringing the food particles to the cytostome as well as for locomotion. The frontal portion of the zone, the so-called **frontal membrane** appears to serve for locomotion and Kahl considers that it is probably made up of three lamellae. The oral membranes which are often found in Holotrichida and Heterotrichida, are transparent thin membranous structures composed of one or two rows of cilia, that are more or less strongly fused. The membranes, located in the lower end of the peristome are sometimes called perioral membranes,and those in the cytopharynx, undulating membranes (Structure of undulating membrane, Fauré-Fremiet and Breton-Gorius, 1955).

In Suctoria, cilia are present only during the swimming stages, and, as the organisms become attached to substrate, tentacles de-

velop in their stead. The **tentacles** are concerned with food-capturing, and are either prehensile or suctorial. The prehensile tentacle appears to be essentially similar in structure to the axopo-

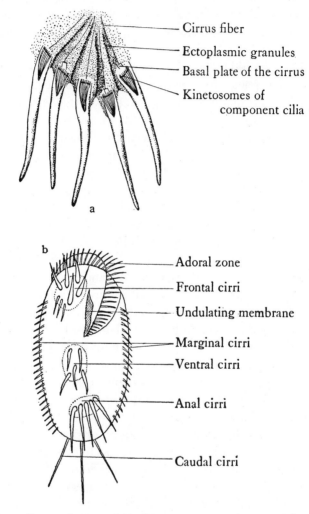

FIG. 11. *a,* five anal cirri of *Euplotes eurystomus* (Taylor); *b,* schematic
ventral view of Stylonychia to show the distribution of cirri.

dium (Roskin, 1925). The suctorial tentacles are tubular and this
type is interpreted by Collin as possibly derived from cytostome
and cytopharnyx of the ciliate (Fig. 13). Electron microscopic

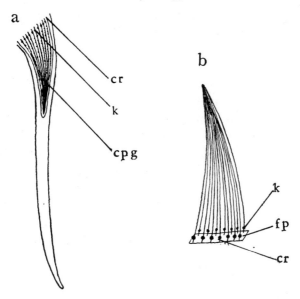

Fig. 12. Diagrams of cirrus and membranella of *Euplotes eurystomus*, ×1450 (Taylor). *a*, anal cirrus in side view; *b*, a membranella (*cpg*, coagulated protoplasmic granules; *cr*, ciliary root; *fp*, fiber plate, *k*, kinetosome).

studies of the tentacles of Ephelota show that the central canal of the suctorial tentacle is limited by a fine pellicle composed of many longitudinal fibrils and bearing sixteen to eighteen membrano-fibrillar ridges arranged radially in the lumen of the canal, which resemble the myonemes of Stentor. The prehensile tentacle contains four to six axial protein fibers, each consisting of a lamel-

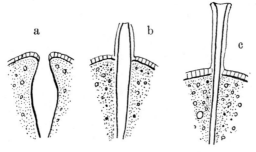

Fig. 13. Diagrams showing the possible development of a suctorian tentacle from a cytostome and cytopharynx of a ciliate (Collin).

lo-fibrillar bundle and separated from one another by membranes (Rouiller *et al.* 1956a) (Tentacles of Tokophrya, Rudzinska and Porter, 1954).

Although the vast majority of protozoa possess only one of the three organelles of locomotion mentioned above, a few may possess pseudopodia in one stage and flagella in another during their development. Among several examples may be mentioned Naegleriidae, *Tetramitus rostratus,* etc. Furthermore, there are some protozoa which possess two types of organellae at the same time. Flagellum or flagella and pseudopodia occur in many Phytomastigia and Rhizomastigida, and a flagellum and cilia are present in Ileonema (Fig. 303, *h, i*).

In the cytoplasm of protozoa there occur various structures, each of which will be considered here briefly.

Fibrillar Structures

One of the fundamental characteristics of the protoplasm is its contractility. If a fully expanded *Amoeba proteus* is subjected to a mechanical pressure, it retracts its pseudopodia and contracts into a more or less spherical form. In this response there is no special organelle, the whole body reacts. But in certain other protozoa, there are special organelles of contraction. Many Ciliophora are able to contract instantaneously when subjected to mechanical pressure, as will easily be noticed by following the movement of Stentor, Spirostomum, Trachelocerca, Vorticella, etc., under a dissecting microscope. The earliest observer of the contractile elements of protozoa appears to be Lieberkühn (1857) who noted the "muscle fibers" in the ectoplasm of Stentor which were later named myonemes (Haeckel) or neurophanes (Neresheimer).

The **myonemes** of Stentor have been studied by several investigators. According to Schröder (1906), there is a canal between each two longitudinal striae and in it occurs a long banded myoneme which measures in cross-section 3-7μ high by about 1μ wide and which appears cross-striated (Fig. 14). Roskin (1923) considers that the myoneme is a homogeneous cytoplasm (kinoplasm) and the wall of the canal is highly elastic and counteracts the contraction of the myonemes. All observers agree that the myoneme is a highly contractile organella.

Many stalked peritrichous ciliates have well-developed myonemes not only in the body proper, but also in the stalk. Kolt-zoff's (1911) studies show that the stalk is a pseudochitinous tube, enclosing an inner tube filled with granulated thecoplasm, which surrounds a central rod, composed of kinoplasm, on the surface of which are arranged skeletal fibrils (Fig. 15). The contraction of the stalk is brought about by the action of kinoplasm

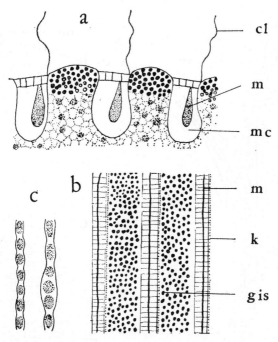

Fig. 14. Myonemes in *Stentor coeruleus* (Schröder). *a,* cross-section of the ectoplasm; *b,* surface view of three myonemes; *c,* two isolated myonemes (*cl,* cilium; *gis,* granules between striae; *k,* kinetosome; *m,* myoneme; *mc,* myoneme canal).

and walls, while elastic rods will lead to extension of the stalk. Myonemes present in the ciliates aid in the contraction of body, but those which occur in many Gregarinida aid apparently in locomotion, being arranged longitudinally, transversely and probably spirally (Roskin and Levinsohn, 1929) (Fig. 15, *c*). In certain Radiolarida, such as *Acanthometron elasticum* (Fig. 222, *c*), etc., each axial spine is connected with 10-30 myonemes (myophrisks) originating in the body surface. When these myonemes

contract, the body volume is increased, thus in this case function-
ing as a hydrostatic organella.

In *Isotricha prostoma* and *I. intestinalis,* Schuberg (1888) observed the nucleus suspended by ectoplasmic fibrils and called the
apparatus **karyophore.** In some forms these fibrils are replaced by
ectoplasmic membranes as in *Nyctotherus ovalis* (Zulueta; Kudo).

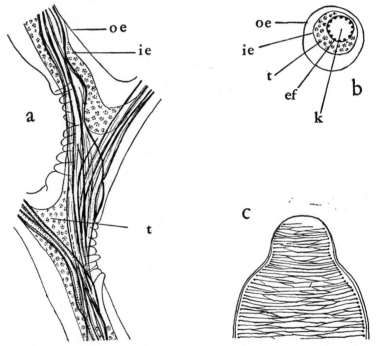

Fig. 15. *a, b,* fibrillar structures of the stalk of Zoothamnium (Koltzoff);
c, myonemes in Gregarina (Schneider). *ef,* elastic fiber; *ie,* inner envelope;
k, kinoplasm; *oe,* outer envelope; *t,* thecoplasm.

ten Kate (1927, 1928) studied fibrillar systems in Opalina,
Nyctotherus, Ichthyophthirius, Didinium, and Balantidium, and
found that there are numerous fibrils, each of which originates in
the kinetosome and takes a transverse or oblique course through
the endoplasm, ending in a kinetosome located on the other side
of the body. He further noted that the cytopharynx and nucleus
are also connected with these fibrils. ten Kate suggested **morpho-**

nemes for them, since he believed that the majority were form-retaining fibrils.

The well-coordinated movement of cilia in the ciliate has long been recognized, but it was Sharp (1914) who definitely showed that this ciliary coordination is made possible by a certain fibrillar system which he discovered in *Epidinium (Diplodinium) ecaudatum* (Fig. 16). Sharp recognized in this ciliate a complicated fibrillar system connecting all the motor organelles of the cytostomal region, and thinking that it was "probably nervous in function," as its size, arrangement and location did not suggest supporting or contractile function, he gave the name **neuromotor apparatus** to the whole system. This apparatus consists of a central motor mass, the *motorium* (which is stained red with Zenker fixation and modified Mallory's connective tissue staining), located in the ectoplasm just above the base of the left skeletal area, from which definite strands radiate: namely, one to the roots of the dorsal membranellae (a dorsal motor strand); one to the roots of the adoral membranellae (a ventral motor strand); one to the cytopharynx (a circum-oesophageal ring and oesophageal fibers); and several strands into the ectoplasm of the operculum (opercular fibers). Neuromotor system has since been observed in many other ciliates: Euplotes (Yocom; Taylor), Balantiduum (McDonald), Paramecium (Rees; Brown; Lund), Tintinnopsis (Campbell), Boveria (Pickard), Dileptus (Visscher), Chlamydodon (MacDougall), Entorhipidium and Lechriopyla (Lynch), Eupoterion (MacLennan and Connell), Metopus (Lucas), Troglodytella (Swezey), Oxytricha (Lund), Ancistruma and Conchophthirus (Kidder), Eudiplodinium (Fernandez-Galiano), etc. (Ciliata fibrillar system, Taylor, 1941.)

Euplotes, a common free-living hypotrichous ciliate, has been known for nearly sixty years to possess definite fibrils connecting the anal cirri with the anterior part of the body. Engelmann suggested that their function was more or less nervelike, while others maintained that they were supporting or contracting in function. Yocom (1918) traced the fibrils to the motorium, a very small bilobed body (about 8μ by 2μ) located close to the right anterior corner of the triangular cystostome (Fig. 17, *m*). Joining with its left end are five long fibers (*acf*) from the anal cirri which con-

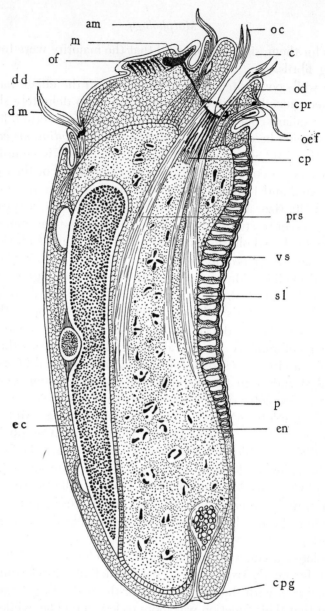

F<small>IG</small>. 16. A composite drawing from three median sagittal sections of *Epidinium ecaudatum*, fixed in Zenker and stained with Mallory's connective tissue stain, ×1200 (Sharp). *am*, adoral membranellae; *c*, cytostome; *cp*, cytopharynx; *cpg*, cytopyge; *cpr*, circumpharyngeal ring; *dd*, dorsal disk; *dm*, dorsal membrane; *ec*, ectoplasm; *en*, endoplasm; *m*, motorium; *oc*, oral cilia; *od*, oral disk; *oef*, oesophageal fibers; *of*, opercular fibers; *p*, pellicle; *prs*, pharyngeal retractor strands; *sl*, skeletal laminae; *vs*, ventral skeletal area.

verge and appear to unite with the motorium as a single strand. From the right end of the motorium extends the membranella-fiber anteriorly and then to left along the proximal border of the oral lip and the bases of all membranellae. Yocom further noticed

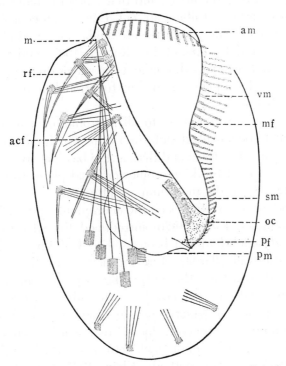

Fig. 17. Ventral view of *Euplotes eurystomus* (*E. patella*) showing neuromotor system, ×670 (Hammond). *acf*, fibril of anal cirrus; *am*, anterior adoral zone membranelle; *m*, motorium; *mf*, membranelle fibrils; *oc*, endoral cilia; *pf*, post-pharyngeal fibril; *pm*, post-pharyngeal membrane; *rf*, radiating fibrils; *sm*, suboral membranelles; *vm*, ventral adoral zone membranelles.

that within the lip there is a latticework structure whose bases very closely approximate the cytostomal fiber. Taylor (1920) recognized two additional groups of fibrils in the same organism: (1) membranella fiber plates, each of which is contiguous with a membranella basal plate, and is attached at one end to the membranella fiber; (2) dissociated fiber plates contiguous with the basal plates of the frontal, ventral and marginal cirri, to each of which are attached the dissociated fibers (*rf*). By means of micro-dissection needles, Taylor demonstrated that these fibers have nothing to do with the maintenance of the body form, since there results no deformity when Euplotes is cut fully two-thirds its width, thus cutting the fibers, and that when the motorium is destroyed or its attached fibers are cut, there is no coordination in the movements of the adoral membranellae and anal cirri. Hammond (1937) and Hammond and Kofoid (1937) find the neuromotor system continuous throughout the stages during asexual reproduction and conjugation so that functional activity is maintained at all times.

A striking feature common to all neuromotor systems, is that there seems to be a central motorium from which radiate fibers to different ciliary structures and that, at the bases of such motor organellae, are found the kinetosomes or basal plates to which the "nerve" fibers from the motorium are attached.

Independent of the studies on the neuromotor system of American investigators, Klein (1926) introduced the silver-impregnation method which had first been used by Golgi in 1873 to demonstrate various fibrillar structures of metazoan cells, to protozoa in order to demonstrate the cortical fibers present in ciliates, by dry-fixation and impregnating with silver nitrate. Klein (1926-1942) subjected ciliates of numerous genera and species to this method, and observed that there was a fibrillar system in the ectoplasm at the level of the kinetosomes which could not be demonstrated by other methods. Klein (1927) named the fibers **silver lines** and the whole complex, the **silverline system,** which vary among different species (Figs. 18-20). Gelei, Chatton and Lwoff, Jírovec, Lynch, Jacobson, Kidder, Lund, Burt, Raabe, and others, applied the silver-impregnation method to many other ciliates and confirmed Klein's observations.

The question whether the neuromotor apparatus and the silverline system are independent structures or different aspects of the same structure has been raised frequently. Turner (1933) found that in *Euplotes patella* (*E. eurystomus*) the silverline system is a regular latticework on the dorsal surface and a more irregular network on the ventral surface. These lines are associated with rows of rosettes from which bristles extend. These bristles are held to be sensory in function and the network, a sensory conductor system, which is connected with the neuromotor system. Turner maintained that the neuromotor apparatus in Euplotes is augmented by a distinct but connected external network of sensory fibrils. He, however, finds no motorium in this protozoan.

Lund (1933) also made a comparative study of the two systems

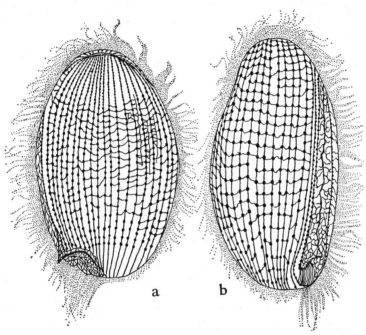

Fig. 18. The silverline system of *Ancistruma mytili,* ×1000 (Kidder). *a,* vental view; *b,* dorsal view.

in *Paramecium multimicronucleatum,* and observed that the silverline system of this ciliate consists of two parts. One portion is made up of a series of closely-set polygons, usually hexagons, but flattened into rhomboids or other quadrilaterals in the regions of the cytostome, cytopyge, and suture. This system of lines stains if the organisms are well dried. Usually the lines appear solid, but frequently they are interrupted to appear double at the vertices of the polygons which Klein called "indirectly connected" (pellicular) conductile system. In the middle of the anterior and posterior sides of the hexagons is found one granule or a cluster of two to four granules, which marks the outer end of the trichocyst. The second part which Klein called "directly connected" (subpellicular) conductile system consists essentially of the longitudinal lines connecting all kinetosomes in a longitudinal row of hexagons and of delicate transverse fibrils connecting granules of adjacent rows especially in the cytostomal region (Fig. 19).

By using Sharp's technique, Lund found the neuromotor system of *P. multimicronucleatum* constructed as follows: The subpellic-

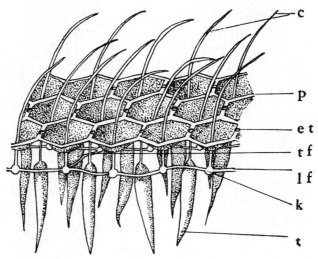

FIG. 19. Diagram of the cortical region of *Paramecium multimicronucleatum,* showing various organellae (Lund), *c,* cilia; *et,* tip of trichocyst; *k,* kinetosome; *lf,* longitudinal fibril; *p,* pellicle; *t,* trichocyst; *tf,* transverse fibril.

ular portion of the system is the longitudinal fibrils which connect the kinetosomes. In the cytostomal region, the fibrils of right and left sides curve inward forming complete circuits (the circular cytostomal fibrils) (Fig. 20). The postoral suture is separated at the point where the cytopyge is situated. Usually 40-50 fibrils radiate outward from the cytostome (the radial cytostomal fibrils). The pharyngeal portion is more complex and consists of (1) the oesophageal network; (2) the motorium and associated fibrils; (3) peniculi which are composed of two sets of four rows of kinetosomes, thus forming a heavy band of cilia in the cytopharynx; (4) oesophageal process; (5) paraoesophageal fibrils; (6) posterior

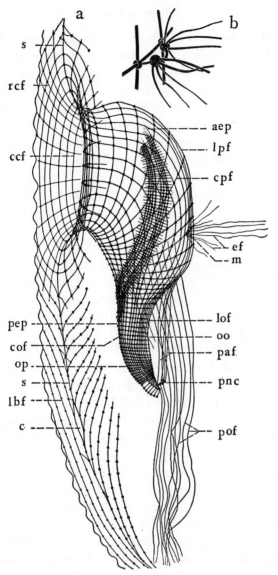

FIG. 20. The neuromotor system of *Paramecium multimicronucleatum* (Lund). *a*, oral network; *b*, motorium, ×1670. *aep*, anterior end of penniculus; *c*, cytopyge; *ccf*, circular cytostomal fibril; *cof*, circular oesophageal fibril; *cpf*, circular pharyngeal fibril; *ef*, endoplasmic fibrils; *lbf*, longitudinal body fibril; *lof*, longitudinal oesophageal fibrils; *lpf*, longitudinal pharyngeal fibril; *m*, motorium; *oo*, opening of oesophagus; *op*, oesophageal process; *paf*, paraoesophageal fibrils; *pep*, posterior end of penniculus; *pnc*, posterior neuromotor chain; *pof*, postoesophageal fibrils; *rcf*, radial cytostomal fibril; *s*, suture.

neuromotor chain, and (7) postoesophageal fibrils. Lund concludes that the so-called silverline system includes three structures: namely, the peculiarly ridged pellicle; trichocysts which
have no fibrillar connections among them or with fibrils, hence
not conductile; and the subpellicular system, the last of which is
that part of the neuromotor system that concerns with the body
cilia. ten Kate (1927) suggested that **sensomotor apparatus** is a
better term than the neuromotor apparatus. In *P. aurelia,* Metz *et
al.* (1953) found that the kinetosome was about 0.27µ in diameter and the longitudinal fibril or kinetodesma which is a bundle of
fibrils, each of which arises independently from a kinetosome,
passes to one side, and joins the main bundle (Fig. 21). (Silverline system, Klein, 1926–1942, 1958, Gelei, 1932; cytoplasmic fibils,
Jacobson, 1932, Taylor, 1941, Párducz, 1958; argyrome in Astomatida, Puytorac, 1951; peniculus and other gullet organelles,
Ehret and Powers, 1957; oral organelles, Yusa, 1957; EM study of
silverline and fibrillar systems in Tetrahymena, Pitelka, 1961.)

Protective or Supportive Organelles

The external structures as found among various protozoa which
serve for body protection, have already been considered (p. 53).
Here certain internal structures will be discussed. In Radiolarida,
there is a membranous structure, the **central capsule,** which divides the body into a central region and a peripheral zone. The
intracapsular portion contains the nucleus or nuclei, and is the
seat of reproductive processes, and thus the capsule is to be considered as a protective organella. The skeletal structures of Radiolarida vary in chemical composition and forms, and are arranged
with a remarkable regularity (p. 617).

In some of the astomatous ciliates, there are certain structures
which appear to serve for attaching the body to the host's organ,
but which seem to be supportive to a certain extent also. The peculiar organelle *furcula,* observed by Lynch in Lechriopyla (p.
876) is said to be concerned with either the neuromotor system or
protection. The members of the family Ophryoscolecidae (p. 982),
which are common commensals in the stomach of ruminants, have
conspicuous **endoskeletal plates** which arise in the oral region and
extend posteriorly. Dogiel (1923) believed that the skeletal

FIG. 21. View from inside of a fragment of *Paramecium aurelia* in electron micrograph (Metz, Pitelka and Westfall). Cilia are seen to pass through the ciliary rings in the pellicle and end at the kinetosomes. Kinetodesmal fibrils pass from the kinetosomes and join the kinetodesmal bundles. ×21,000.

plates of Cycloposthium and Ophryoscolecidae are made up of hemicellulose, "ophryoscolecin," which was also observed by Strelkow (1929). MacLennan found that the skeletal plates of *Polyplastron multivesiculatum* were composed of small, roughly primatic blocks of paraglycogen, each possessing a central granule.

In certain Zoomastigia, there occurs a flexible structure known as **axostyle,** which varies from a filamentous structure as in several Trichomonas, to a very conspicuous rod-like structure occurring in Parajoenia, Gigantomonas, etc. The axostyle of *Tritrichomonas muris* appears to be tubular (Anderson and Beams, 1959). The anterior end of the axostyle is very close to the anterior tip of the body, and it extends lengthwise through the cytoplasm, ending near the posterior end or extending beyond the body surface. In other cases, the axostyle is replaced by a bundle of **axostylar filaments** that are connected with the flagella (Lophomonas). The axostyle appears to be supportive in function, but in forms such as Saccinobaculus, it undulates and presumably aids in locomotion (p. 446).

In trichomonad flagellates there is often present along the line of attachment of the undulating membrane, a rod-like structure which has been known as **costa** (Kunstler) and which, according to Kirby's extensive study, appears to be most highly developed in Pseudotrypanosoma and Trichomonas. The staining reaction indicates that its chemical composition is different from that of flagella, blepharoplast, parabasal body, or chromatin. Anderson and Beams (1961) found that the costa of *Tritrichomonas muris* is a striated fiber running the entire length of the body and its slender proximal end appears to be attached to the kinetosome at the base of the recurrent flagellum. Its function seems to be to support the undulating membrane.

In the gymnostomatous ciliates, the cytopharynx is often surrounded by rod-like bodies, and the entire apparatus is often called **oral** or **pharyngeal basket,** which is considered as supportive in function. These rods are arranged to form the wall of the cytopharynx in a characteristic way. For example, the oral basket of *Chilodonella cucullulus* (Fig. 313, *c, d*) is made up of twelve long rods which are so completely fused in part that it appears to be a smooth tube; in other forms, the rods are evidently similar to the tubular trichocysts or trichites mentioned below. Electron microscopy reveals that these rods are rigid, elastic and birefringent; they are composed of fibrils which are comparatively uniform in diameter (150-200 Å) among different species (Rouiller *et al.*, 1956).

In numerous holotrichs, there occur unique organelles, **trichocysts**, which are imbedded in the ectoplasm, and usually arranged at right angles to the body surface, though in forms such as Cyclogramma, they are seemingly arranged obliquely. Under certain stimulations, the trichocysts "explode" and form long filaments which extend out into the surrounding medium. The shape of the trichocyst varies somewhat among different ciliates, being pyriform, fusiform or cyclindrical (Penard, 1922; Krüger, 1936). They appear as homogeneous refractile bodies. The extrusion of the trichocyst is easily brought about by means of mechanical pressure or of chemical (acid or alkaline) stimulation.

In forms such as Paramecium, Frontonia, etc., the trichocyst is elongate pyriform or fusiform. It is supposed that within an expansible membrane, there is a layer of swelling body which is responsible for the remarkable longitudinal extension of the membrane (Krüger) (Fig. 22, *a*). In other forms such as Prorodon, Didinium, etc., the tubular trichocyst or **trichites** are cylindrical in shape and the membrane is a thick capsule with a coiled thread, and when stimulated, the extrusion of the thread takes place. The trichites of *Prorodon teres* measure about 10-11μ long (Fig. 22, *d*) and when extruded, the whole measures about 20μ; those of *Didinium nasutum* are 15-20μ long and after extrusion, measure about 40μ in length (Fig. 22, *e*, *f*). In *Spathidium spathula* (Fig. 22, *c*), trichites are imbedded like a paling in the thickened rim of the anterior end. They are also distributed throughout the endoplasm and, according to Woodruff and Spencer, "some of these are apparently newly formed and being transported to the oral region, while others may well be trichites which have been torn away during the process of prey ingestion." Weinreb (1955) made a similar observation in *Homalozoon vermiculare*. Whether the numerous 12-20μ long needle-like structures which Kahl observed in Remanella (p. 857) are modified trichites or not, is not known.

Dileptus anser feeds on various ciliates through the cytostome,

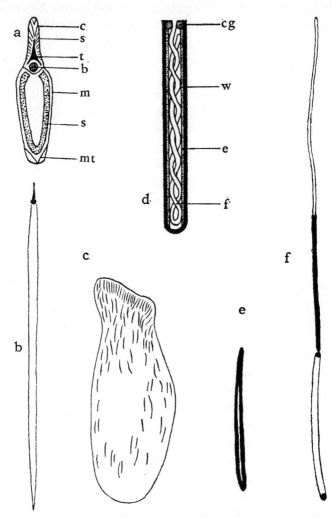

F<small>IG</small>. 22. a, a schematic drawing of the trichocyst of *Paramecium cau-datum* (Krüger) (b, base of the tip; c, cap; m, membrane; mt, membrane of extruded trichocyst; s, swelling body; t, tip); b, an extruded trichocyst, viewed under phase dark contrast, ×1800; c, trichites in *Spathidium spathula*, ×300 (Woodruff and Spencer); d, a diagram of the trichocyst of *Prorodon teres* (Krüger) (cg, capsule-granule; e, end-piece of filament; f, filament w, capsule wall); e, f, normal and extruded trichocysts of *Didi-nium nasutum* (Krüger).

located at the base of the proboscis, which possesses a band of long trichocysts on its ventral side. When food organisms come in contact with the ventral side of the proboscis, they give a violent jerk, and remain motionless. Visscher saw no formed elements discharged from the trichocysts, and, therefore, considered that these trichocysts contained a toxic fluid and named them toxicysts. But Krüger and Hayes (1938) found that the extruded trichocysts can be recognized.

Perhaps the most frequently studied trichocysts are those of Paramecium. They are elongate pyriform, with a fine tip at the broad end facing the body surface. The tip is connected with the pellicle (Fig. 19, *t*). Krüger found this tip is covered by a cap (Fig. 22, *a*) which can be seen under darkfield or phase microscope and which was demonstrated by Jakus (1945) in an electron micrograph (Fig. 23, *a*). When extruded violently, the entire structure is to be found outside the body of Paramecium. The extruded trichocyst is composed of two parts: the tip and the main body (Fig. 22, *b*). The tip is a small inverted tack, and may be straight, curved or bent. The main body or shaft is a straight rod, tapering gradually into a sharp point at the end opposite the tip. Extruded trichocysts measure 20-40µ or more in length, and do not show any visible structures, except a highly refractile granule present at the base of the tuck-shaped tip (Fig. 22, *b*). The electron microscope studies of the extruded trichocysts by Jakus (1945), Jakus and Hall (1946) and Wohlfarth-Bottermann (1950, 1953), show the shaft to be cross-striated (Fig. 23). Jakus considers that the main component of the trichocyst is a thin cylindrical membrane formed by close packing of longitudinal fibrils characterized by a periodic pattern (somewhat resembling that of collagen), and as the fibrils are in phase with respect to this pattern, the membrane appears cross-striated (Recent EM studies, Dragesco, 1952, Beyersdorfer and Dragesco, 1952, 1952a; Trichocysts of *Frontonia atra*, Rouiller and Fauré-Fremiet, 1957).

As to the mechanism of the extrusion, no precise information is available, though all observers agree that the contents of the triocyst suddenly increase in volume. Krüger maintains that the trichocyst cap is first lifted and the swelling body increases enormously in volume by absorbing water and lengthwise extension

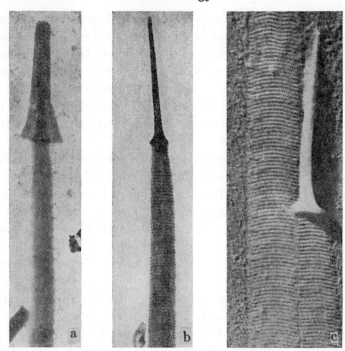

Fig. 23. Electromicrographs of extruded trichocysts of Paramecium. a, dried and stained with phosphotungstic acid, ×11,000 (Jakus); b, a similarly treated one, ×15,000 (Jakus); c, shadow-cast with chromium, ×16,000 (Jakus and Hall).

takes place, while Jakus is inclined to think that the membrane itself extends by the sudden uptake of water.

How are these organelles formed? Tönniges (1914) believes that the trichocysts of *Frontonia leucas* originate in the endosomes of the macronucleus and development takes place during their migration to the ectoplasm. Brodsky (1924) holds that the trichocyst is composed of colloidal excretory substances and is first formed in the vicinity of the macronucleus. Chatton and Lwoff (1935) find however in Gymnodinioides the trichocysts are formed only in tomite stage and each trichocyst arises from a *trichocystosome*, a granule formed by division of a kinetosome (Fig. 24, *a-c*). In Polyspira, the trichocyst formation is not confined to one phase, each kinetosome is said to give rise to two granules, one of which may detach itself, migrate into other part of the

body and develops into a trichocyst (*d*). In Foettingeria, the kinetosomes divide in young trophont stage into *trichitosomes* which
develop into trichites (*e*). The two authors note that normally
cilia-producing kinetosomes may give rise to trichocysts or trichites, depending upon their position (or environment) and the
phase of development of the organism.

Although the trichocyst was first discovered by Ellis (1769)
and so named by Allman (1855), nothing concrete is yet known

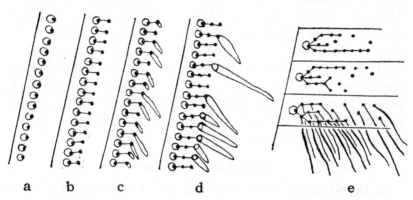

a   b   c   d   e

Fɪɢ. 24. Diagrams showing the formation of trichocysts in Gymnodinioides (a–c) and in Polyspira (d) and of trichites in Foettingeria (e) (Chat-
tion and Lwoff). a, a ciliary row, composed of kinetosomes, large satellite
corpuscles and kinetodesma (a solid line); b, each kinetosome divides into
two, producing trichocystosome; c, transformation of trichocystosomes
into trichocysts; d, formation of trichocyst from one of the two division
products of kinetosome; e, formation of trichites from the division products of kinetosomes.

as to their function. Ordinarily, the trichocysts are considered as a
defensive organelle as in the case of the oft-quoted example Paramecium, but, as Mast demonstrated, the extruded trichocysts of
this ciliate do not have any effect upon Didinium other than forming a viscid mass about the former to hamper the latter. On the
other hand, the trichocysts and trichites are clearly an offensive
organelle in capturing food organisms in organisms such as Dileptus, Didinium, Spathidium, etc.

Saunders (1925) considered that the extruded trichocysts of
Paramecium serve for attachment of the body to other objects. But

Wohlfarth-Bottermann (1950, 1953) saw *Paramecium caudatum* extruding up to 300 trichocysts without any apparent external stimulation and trichocyst-less individuals were able to adhere to foreign objects. This worker suggested that the trichocyst secretes calcium salt and probably also sodium and potassium, and thus may serve an osmoregulatory function. Some years ago, Penard (1922) considered that some trichocysts may be secretory organellae to produce material for loricae or envelope, with which view Kahl concurs, as granular to rod-shaped trichocysts occur in Metopus, Amphileptus, etc. Klein has called these ectoplasmic granules **protrichocysts,** and in Prorodon, Krüger observed, besides typical tubular trichocysts, torpedo-like forms to which he applied the same name. To this group may belong the trichocysts recognized by Kidder in *Conchophthirus mytili.* The trichocysts present in certain Cryptomonadina (Chilomonas and Cyathomonas) are probably homologous with the protrichocysts (Krüger, 1934; Hollande, 1942; Dragesco, 1951).

Hold-fast Organelle

In the Mastigophora, Ciliophora, and a few Sarcodina, there are forms which possess a **stalk** supporting the body or the lorica. In Peritrichida and Suctoria, a special area at the anterior end of the migratory individual, called *scopula,* secretes the stalk. The attached end is usually enlarged into a disc. The disc by which the stalk of the suctorian *Tokophrya infusionum* is attached to the substrate is said to be made up of numerous fibrils (Rudzinska and Porter, 1954). In some cases, as in Anthophysis, Maryna, etc., the dendritic stalks are made up of gelatinous substances rich in iron, which gives to them a reddish brown color. In parasitic protozoa, there are special organellae developed for attachment. Many genera of cephaline gregarines are provided with an **epimerite** by which the organisms are able to attach themselves to the gut epithelium of the host. In Astomatida, such as Intoshellina, Maupasella, Lachmannella, etc., simple or compex protrusible chitinous structures are often present in the anterior region; or a certain area of the body may be concave and serves for adhesion to the host, as in Rhizocaryum, Perezella, etc.; or, again, there may be a distinctive sucker-like organella in the anterior half of the body, as in Haptophyra, Steinella, etc. The antero-ventral part of *Giardia intestinalis* appears to serve as a sucker.

In the Myxosporida and Actinomyxida, there appear, during the development of the spore, one to six special cells which develop into **polar capsules** enclosing a more or less long spirally coiled delicate threat, the **polar filament** (Figs. 286, 293). The polar filament is considered as a temporary anchoring organella of the spore at the time of its germination after it gained entrance into the alimentary canal of a suitable host. In the Microsporida, the filament is coiled spirally close to the spore membrane. The **nematocysts** (Fig. 135, *b*) of certain dinoflagellates belonging to Nematoidium and Polykrikos, are almost identical in structure with those found in the coelenterates. They are distributed through the cytoplasm, and various developmental stages were noticed by Chatton, and Kofoid and Swezy, which indicates that they are characteristic structures of these dinoflagellates and not foreign in origin as had been held by some. The function of the nematocysts in these protozoans is not understood.

Parabasal Apparatus

In the cytoplasm of many parasitic flagellates, there is frequently present a conspicuous structure known as the **parabasal apparatus** (Janicki, 1911), consisting of the parabasal body and thread (Cleveland), which latter may be absent in some species. This structure varies greatly among different genera and species in appearance, structure and position within the body. It is usually connected with the blepharoplast and located very close to the nucleus, though not directly connected with it. It may be single, double, or multiple, and may be pyriform, straight or curved rod-like, bandform, spirally coiled or collar-like (Fig. 25). Kofoid and Swezy considered that the parabasal body is derived from the nuclear chromatin, varies in size according to the metabolic demands of the organism, and is a "kinetic reservoir." On the other hand, Duboscq and Grassé (1933) maintain that this body is the Golgi apparatus, since (1) acetic acid destroys both the parabasal body and the Golgi apparatus; (2) both are demonstrable with the same technique; (3) the parabasal body is made up of chromophile and chromophobe parts as is the Golgi apparatus; and (4)

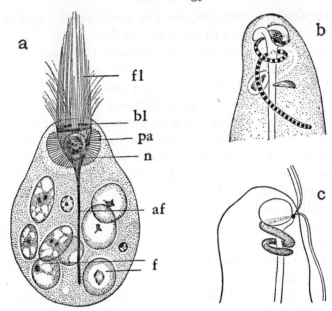

Fig. 25. Parabasal apparatus in: a, *Lophomonas blattarum* (Kudo); b, *Metadevescovina debilis;* c, *Devescovina* sp. (Kirby). af, axostylar filaments; bl, blepharoplasts; f, food particles; fl, flagella; n, nucleus; pa, parabasal apparatus.

there is a strong evidence that the parabasal body is secretory in function. According to Kirby (1931), the parabasal body could be stained with Delafield's haematoxylin or Mallory's triple stain after fixation with acetic acid-containing fixatives and the body does not show any evidence to indicate that it is a secretory organella. Moreover the parabasal body is discarded or absorbed at the time of division of the body and two new ones are formed.

The parabasal body of *Lophomonas blattarum* is discarded when the organism divides and two new ones are reformed from the blepharoplasts (Fig. 68), and its function appears to be supportive. Possibly not all so-called parabasal bodies are homologous or analogous. A fuller comprehension of the structure and function of the organella rests on further investigations.

Golgi Apparatus

With the discovery of wide distribution of the so-called Golgi apparatus in metazoan cells, a number of protozoologists also reported a homologous structure from many protozoa. It seems impossible to indicate just exactly what the Golgi apparatus is, since the so-called Golgi techniques, the important ones of which are based upon the assumption that the Golgi material is osmiophile and argentophile, and possesses a strong affinity to neutral red, are not specific and the results obtained by using the same method often vary a great deal. Some of the examples of the Golgi apparatus reported from protozoa are summarized in Table 2.

TABLE 2

GOLGI APPARATUS IN PROTOZOA

Protozoa	*Golgi Apparatus*	*Observers*
Chromulina, Astasia	Rings, spherules with a dark rim	Hall
Chilomonas	Granules, vacuoles	Hall
Euglenoidida	Stigma	Grassé
Euglena gracilis	Spherical, discoidal with dark rim; tend to group around or near nucleus	Brown
Peranema	Rings, globules, granules	Hall
Pyrsonymphia, Dinenympha	Rings, crescents, spherules; granules break down to form network near posterior end	Brown
Holomastigotes, Pyrsonympha, etc.	Parabasal bodies	Dubocsq and Grassé
Amoeba proteus (Fig. 26)	Rings, crescents, globules, granules	Brown
Endamoeba blattae	Spheres, rings, crescents	Hirschler
Monocystis, Gregarina	Spheres, rings, crescents	Hirschler
Aggregata, gregarines	Crescents, rings	Joyet-Lavergne
Adelea	Crescents, beaded grains	King and Gatenby
Blepharisma undulans	Rings in the cytoplasm	Moore
Vorticella, Lionotus, Paramecium, Dogielella, Nassula, Chilomonas, Chilodonella	The membrane of contractile vacuole and collecting canals	Nassonov

It appears that the Golgi bodies occurring in protozoa are small osmiophilic granules or larger spherules which are composed of osmiophile cortical and osmiophobe central substances. Frequently the cortical layer is of unequal thickness, and, therefore, crescentic forms appear. Ringform apparatus was noted in Chilodonella and Dogielella by Nassonov (1925) and network-like forms were observed by Brown in Pyrsonympha and Dinenympha. The Golgi apparatus of protozoa as well as of metazoa appears to be composed of a lipoidal material in combination with protein substance. The Golgi bodies of *Peranema trichophorum* as seen in electron micrographs are composed of some thirty compressed and compact vesicles that are closely grouped together and surrounded by smaller vesicles (Roth, 1959).

In line with the suggestion made for the metazoan cell, the Golgi apparatus of protozoa is considered as having something to do with secretion or excretion. Nassonov (1924) considers that

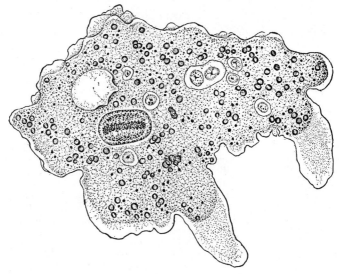

FIG. 26. The Golgi bodies in *Amoeba proteus* (Brown).

osmiophilic lipoidal substance, which he observed in the vicinity of the walls of the contractile vacuole and its collecting canals in many ciliates and flagellates, is homologous with the metazoan Golgi apparatus and secretes the fluid waste material into the vacuole from which it is excreted to the exterior. According to Brown, there is no blackening by osmic impregnation of the contractile vacuole in *Amoeba proteus* (Fig. 26), but fusion of minute vacuoles associated with crescentic Golgi bodies produces the vacuole. Das and Tewari (1955) saw in *A. verrucosa* three types of Golgi bodies: dark solid granules, small crescent or ringed vacuole, and a few larger vacuoles, and considered them as successive active phases of Golgi bodies.

Duboscq and Grassé (1933) maintain Golgi body as a source of energy which is utilized by motor organelles. Joyet-Lavergne points out that in certain Sporozoa, the Golgi body is composed of granules and may be the center of enzyme production. Similar to Golgi material, the so-called *vacuome*, which consists of neutral red-staining and osmiophile globules, has been reported to occur in many protozoa (Hall, 1931; Hall and Nigrelli, 1937). The exact morphological and physiological significance of these organelles and the relation between them must be looked for in future investigations (Golgi apparatus in protozoa, Alexeieff, 1928; MacLennan, 1941, Smythe, 1941; Grassé, 1952).

Mitochondria

Mitochondria possess a low refractive index, and are composed of substances easily soluble in alcohol, acetic acid, etc. Osmium tetroxide blackens them, but the color bleaches faster than Golgi bodies. Janus green B stains them even in 1:500,000 dilution, but stains also other inclusions such as Golgi bodies (in some cases) and certain bacteria. According to Horning (1926), Janus red is a more exclusive mitochondria stain, as it does not stain bacteria. The chemical composition of the mitochondria seems to be somewhat similar to that of Golgi body; namely, it is a protein compounded with a lipoidal substance. If the protein is small in amount, it is said to be unstable and easily attacked by reagents; on the other hand, if protein is relatively abundant, it is more stable and resistant to reagents.

Mitochondria occur as small spherical to oval granules, rod-like or filamentous bodies, and show a tendency to adhere to or remain near protoplasmic surfaces. In many cases, they are distributed without any definite order; in others, as in Paramecium or Opalina, they are regularly arranged between the kinetosomes (Horning). In *Tillina canalifera*, Turner (1940) noticed that the endoplasmic mitochondria are evenly distributed throughout the cytoplasm (Fig. 27, *b*), while the ectoplasmic mitochondria are arranged in regular cross rows, one in the center of each square formed by four cilia (Fig. 27). In *Homalozoon vermiculare*, granular mitochondria are distributed throughout the body, but occur in mass in pharyngeal region (Weinreb, 1955). In *Peranema tri-*

chophorum, Hall (1929) noted the peripheral mitochondria located along the spiral striae, which Chadefaud (1938) considered as mucus bodies. Roth (1959) remarked that the mitochondria in this organism were filled with oval structures similar to the cristae seen in metazoan mitochondria. Weisz (1949) maintains that stentorin and zoopurpurin already mentioned (p. 51) are mitochondria.

In certain protozoa, the mitochondria are not always demonstrable. For example, Horning finds in Monocystis the mitochon-

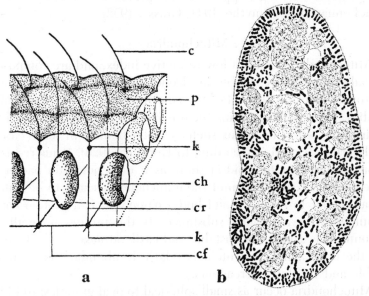

a   b

FIG. 27. Mitochondria in *Tillina canalifera* (Turner). a, diagram showing the ectoplasmic mitochondria (c, cilium; cf, coordinating fibril; ch, mitochondrion; cr, ciliary rootlet; k, kinetosome I and II; p, pellicle); b, a section showing mitochondria and food vacuoles.

dria are present during the asexual life-cycle as rod-shaped bodies, but at the beginning of spore formation, they diminish in size and number, and in the spore none exists. But they appear as soon as the sporozoites are set free. Thus it would appear that they are reformed *de novo.* On the other hand, Fauré-Fremiet, the first observer of mitochondria in protozoa, maintained they reproduce by division, which has since been confirmed by many. Horning found in Opalina the mitochondria are twisted filamentous structures which undergo multiple longitudinal fission in asexual division phase. Before encystment, the mitochondria divide repeatedly transversely and become spherical bodies. In zygotes, these spherical bodies fuse to produce longer forms which break up into elongated structures. Richardson and Horning are reported to have succeeded in causing the division of mitochondria in Opalina by changing pH of the medium. However, Torch (1955) reports that the elongated or dumbbell-shaped mitochondria in *Pelomyxa carolinensis* did not actually divide *in vivo* during prolonged observation.

As to the function of mitochondria, opinions vary. A number of observers hold that they are concerned with the digestive process. After studying the relationship between the mitochondria and food vacuoles of Amoeba and Paramecium, Horning suggested that the mitochondria are the seat of enzyme activity and it is even probable that they actually give up their own substance for this purpose. Mast (1926) described "beta granules" in *Amoeba proteus* which are more abundantly found around the contractile vacuole. Mast and Doyle (1935, 1935a) noted that these spherical to rod-like beta granules are plastic and stain like mitochondria and that there is a direct relation between the number of beta granules in the cytoplasm and the frequency of contraction of the contractile vacuole. They maintained that these granules "probably function in transferring substances from place to place in the cytoplasm." The beta granules are said to be composed of lipid and protein (Pappas, 1956). Similar granules occur in the species of Pelomyxa (Andresen, 1942; Wilbur, 1942; Kudo, 1951). Torch (1955) considers them to be mitochondria.

Electron microscope studies of various protozoa show the presence of mitochondria in all. In *Amoeba proteus* mitochondria occur throughout the cytoplasm, but more abundantly around the contractile vacuole. Mercer (1959) finds the mitochondria of this organism spherical or ellipsoidal with double walls. The inner wall invaginates internally into finger-like microvilli which seem to end blindly (Fig. 28). Chapman-Andresen and Nilsson (1960) suggest that mitochondria may be concerned with dehydration or digestion of the fluid in the vacuoles involved in pinocytosis.

The view that mitochondria may have something to do with the cell-respiration expressed by Kingsbury was further elaborated by Joyet-Lavergne through his studies on certain Sporozoa. That mitochondria are actively concerned with the development of the gametes of metazoa is well known. Vickerman (1962) noted under electron microscope morphological changes in the mitochondria at the time of encystment of limax amoebae and considered that this was probably associated with depressed respiratory activity during encystment. (Mitochondria in protozoa, MacLen-

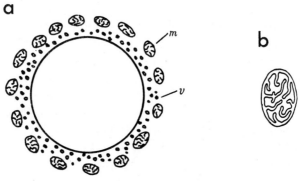

Fig. 28. *Amoeba proteus* (Mercer). a, diagram showing a contractile vacuole, surrounded by mitochondria (m) and small vacuoles (v); b, a mitochondrion.

nan, 1941, Grassé, 1952; Janus green B staining of mitochondria, Lazarow and Cooperstein, 1953.)

Numerous minute granules, less than 1μ in diameter, occur usually abundantly suspended in the cytoplasm. They can most clearly be noted under phase microscope. Mast named those found in Amoeba "alpha granules."

Contractile and Other Vacuoles

The majority of protozoa possess one or more vacuoles known as **contractile vacuoles.** They occur regularly in all freshwater-inhabiting Sarcodina, Mastigophora and Ciliophora. Marine or parasitic Sarcodina and Mastigophora do not ordinarily have a contractile vacuole. This organelle is present with a few exceptions in all marine and parasitic Ciliata, while it is wholly absent in Sporo-

zoa and Cnidosporidia. Although the contractile vacuole was reported to occur in a few sponges (Jepps, 1947), it seems to be a
characteristic structure of protozoa.

In various species of free-living amoebae, the contractile vacuole is formed by accumulation of water in one or more droplets
which finally fuse into one. It enlarges itself continuously until it
reaches a maximum size (*diastole*) and suddenly bursts through
the thin cytoplasmic layer above it (*systole*), discharging its content to outside. The location of the vacuole is not definite in such
forms and, therefore, it moves about with the cytoplasmic movements; and, as a rule, it is confined to the temporary posterior re-

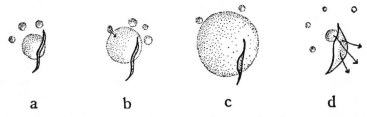

a　　　　　　b　　　　　　c　　　　　　d

FIG. 29. Diagrams showing the contractile vacuole, the accessory vacuoles and the aperture, during diastole and systole in Conchophthirus
(Kidder).

gion of the body. Although almost spherical in form, it may occasionally be irregular in shape, as in *Amoeba striata* (Fig. 186, *f*).
In many testaceans and heliozoans, the contractile vacuoles,
which are variable in number, are formed in the ectoplasm and
the body surface bulges out above the vacuoles at diastole. In
Mastigophora, the contractile vacuole appears to be located in the
anterior region.

In the Ciliophora there occur one to many contractile vacuoles,
which seem to be located in the deepest part of the ectoplasm and
therefore constant in position. Directly above each vacuole is
found a pore in the pellicle, through which the content of the
vacuole is discharged to outside. In the species of Conchophthirus, Kidder (1934) observed a narrow slit in the pellicle just posterior to the vacuole on the dorsal surface (Fig. 29). The margin
of the slit is thickened and highly refractile. During diastole, the
slit is nearly closed and, at systole, the wall of the contractile

vacuole appears to break and the slit opens suddenly, the vacuolar content pouring out slowly. When there is only one contractile vacuole, it is usually located either near the cytopharynx or,
more often, in the posterior part of the body. When several to
many vacuoles are present, they may be distributed without apparent order, in linear series, or along the body outline. When the
contractile vacuoles are deeply seated, there is a delicate duct
which connects the vacuole with the pore on the pellicle as in
Paramecium woodruffi or in Ophryoscolecidae. In Balantidium,
Nyctotherus, etc., the contractile vacuole is formed very close to
the permanent cytopyge located at the posterior extremity,
through which it empties its content.

In a number of ciliates there occur **collecting canals** besides the
main contractile vacuole. These canals radiate from the central
vacuole in Paramecium, Frontonia, Disematostoma, etc. But
when the vacuole is terminal, the collecting canals of course do
not radiate, in which case the number of the canals varies among
different species: one in Spirostomum, Stentor, etc.; two in Climacostomum, Eschaneustyla, etc., and several in Tillina. In Peritrichida, the contractile vacuole occurs near the posterior region of
the cytopharynx and its content is discharged through a canal into
the vestibule and in *Ophrydium ectatum,* the contractile vacuole empties its content into the cytopharynx through a long duct
(Mast).

Of numerous observations concerning the operation of the contractile vacuole, that of King (1935) on *Paramecium multimicronucleatum* (Figs. 30, 31) may be quoted here. In this ciliate,
there are two to seven contractile vacuoles which are located
below the ectoplasm on the aboral side. There is a permanent
pore above each vacuole. Leading to the pore is a short tube-like
invagination of the pellicle, with inner end of which the temporary membrane of the vacuole is in contact (Fig. 30, *a*). Each
vacuole has five to ten long collecting canals with strongly osmiophilic walls (Fig. 31), in which Gelei (1939) demonstrated
longitudinal fibrils, and each canal is made up of terminal portion,
a proximal injection canal, and an ampulla between them. Surrounding the distal portion, there is osmiophilic cytoplasm which
may be granulated or finely reticulated, and which Nassonov

(1924) interpreted as homologous with the Golgi apparatus of the metazoan cell. The injection canal extends up to the pore. The ampulla becomes distended first with fluid transported discontinuously down the canal and the fluid next moves into the injection canal. The fluid now is expelled into the cytoplasm just beneath the pore as a vesicle, the membrane of which is derived from that which closed the end of the injection canal. These fluid vesicles coalesce presently to form the contractile vacuole in full diastole and the fluid is discharged to exterior through the pore, which becomes closed by the remains of the membrane of the dis-

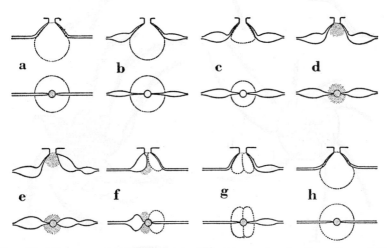

FIG. 30. Diagrams showing the successive stages in the formation of the contractile vacuole in *Paramecium multimicronucleatum* (King); upper figures are side views; lower figures front views; solid lines indicate permanent structures; dotted lines temporary structures. a, full diastole; b–d, stages of systole; e, content of ampulla passing into injection canal; f, formation of vesicles from injection canals; g, fusion of vesicles to form contractile vacuole; h, full diastole.

charged vacuole (Origin and movements of the pore, King, 1954).

Schneider (1960) made electron microscope studies of the contractile vacuoles and associated organelles of *Paramecium caudatum* and *P. aurelia*. Surrounding the collecting canal, there is a network of fine branching "nephridial" tubules with average diameter 200 Å. These tubules are connected peripherally directly with endoplasmic reticulum which extends throughout the body.

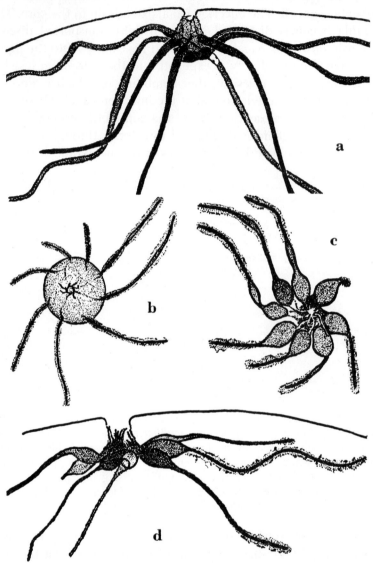

FIG. 31. Contractile vacuoles of *Paramecium multimicronucleatum,* ×1200 (King). a, early systole, side view; b, diastole, front view; c, complete systole, front view; d, systole, side view.

During diastole of the canal, the nephridial tubules open into it, but this connection is broken during systole. The osmiophilic wall of the terminal collecting canal is continuous with the wall of the ampulla, the injection canal and the contractile vacuole. Contractile fibrillar elements, arranged in flat, band-like bundles and of tubular structure (150-250 Å in diameter) begin at the top of the ampulla and extend over the injection canal and contractile vacuole to the excretory tube. The contraction of the vacuole is effected by the fibrils.

In *Tokophrya infusionum*, Rudzinska (1958) finds the contractile vacuole and excretory pore are connected by a tubule. The tubule opens into the vacuole through a papilla and its diameter varies; at diastole it is narrow, 25-30 mμ in diameter, and at systole it is widely open. It is supposed that the change in the diameter of the tubule is due to the contraction of many fine (180 Å thick) fibrils which are radially arranged around the canal, and which connect the channel and vacuolar membrane.

In *Haptophrya michiganensis*, MacLennan (1944) reported that accessory vacuoles appear in the wall of the contractile canal which extends along the dorsal side from the sucker to the posterior end, as the canal contracts (Fig. 32). The canal wall expands and enlarging accessory vacuoles fuse with one another, followed by a full expansion of the canal. Through several excretory pores with short ducts the content of the contractile canal is excreted to the exterior. The function of the contractile vacuole is considered in the following chapter (p. 142) (Comparative study of contractile vacuoles, Haye, 1930, Weatherby, 1941; structure and function, Kitching, 1956).

Various other vacuoles or vesicles occur in different protozoa. In the ciliates belonging to Loxodidae, there are variable numbers of **Müller's vesicles** or bodies, arranged in one to two rows along the aboral surface. These vesicles (Fig. 33, *a-c*) vary in diameter from 5 to 8.5μ and contain a clear fluid in which one large spherule or several small highly refractile spherules are suspended. In some, there is a filamentous connection between the spherules and the wall of the vesicle. Penard maintains that these bodies are balancing cell-organs and called the vesicle, the statocyst, and the spherules, the statoliths.

Another vacuole, known as concrement vacuole, is a character-istic organelle in certain ciliates (Bütschliidae, Paraisotrichidae, etc.). As a rule, there is a single vacuole present in an individual in the anterior third of body. It is spherical to oval and its structure appears to be highly complex. According to Dogiel (1929), the vacuole is composed of a pellicular cap, a permanent vacuolar wall, concrement grains and two fibrillar systems (Fig. 33, *d*).

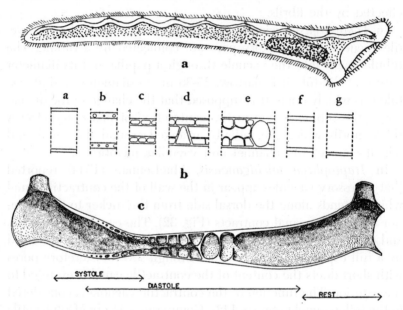

FIG. 32. Excretory canal of *Haptophrya michiganensis* (MacLennan). a, an individual in side view, showing a contraction wave passing down the canal; b, successive views of the same region of the contractile canal dur-ing a full pulsatory cycle (a-c, systole; d-g, diastole); c, diagram showing a contractile wave passing from left to right between two adjacent pores.

When the organism divides, the anterior daughter individual re-tains it, and the posterior individual develops a new one from the pellicle into which concrement grains enter after first appear-ing in the endoplasm. This vacuole shows no external pore. Dogiel believes that its function is sensory and has named the vacuole, the statocyst, and the enclosed grains, the statoliths.

Concretions occur also in the cytoplasm of many ciliates. They are refringent and appear to be composed of calcium carbonate, though their nature is unknown in some.

Food Vacuoles are conspicuously present in the holozoic protozoa which take in whole or parts of other organisms as food. The food vacuole is a space in the cytoplasm, containing the fluid medium which surrounds the protozoans and in which are sus-

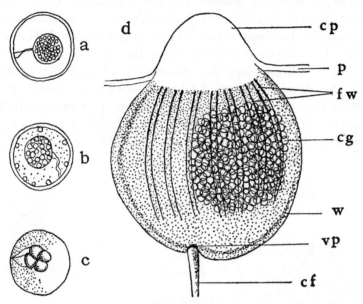

FIG. 33. a-c, Müller's vesicles in Loxodes (a, b) and in Remanella (c) (a, Penard; b, c, Kahl); d, concrement vacuole of Blepharoprosthium (Dogiel). cf, centripetal fibril; cg, concrement grains; cp, cap; fw, fibrils of wall; p, pellicle; vp, vacuolar pore; w, wall.

pended the food matter, such as various protophyta, other protozoa or small metazoa. In the Sarcodina and the Mastigophora, which do not possess a cytostome, the food vacuoles assume the shape of the food materials and, when these particles are large, it is difficult to make out the thin film of water which surrounds them. When minute food particles are taken through a cytostome, as is the case with the majority of ciliates, the food vacuoles are usually spherical and of approximately the same size within a single protozoan. In the saprozoic protozoa, which absorb fluid substances through the body surface, food vacuoles containing solid food, of course, do not occur.

Chromatophore and Associated Organelles

In the Phytomastigia and certain other forms which are green-colored, one to many **chromatophores** (Fig. 34) containing chlorophyll occur in the cytoplasm. The chromatophores vary in form among different species; namely, discoidal, ovoid, band-form, rod-like, cup-like, fusiform, network or irregularly diffused. The color of the chromatophore depends upon the amount and kinds of pigment which envelops the underlying chlorophyll substance. Thus, the chromatophores of Chrysomonadida are brown or orange, as they contain one or more accessory pigments, including phycochrysin, and those of Cryptomonadida are of various types of brown with very diverse pigmentation (Allen *et al.*, 1959). In

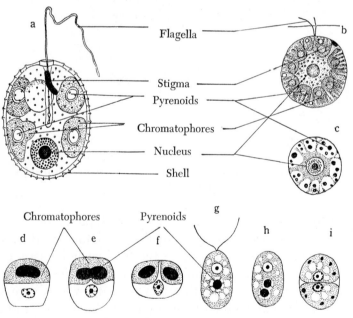

FIG. 34. a, *Trachelomonas hispida*, ×530 (Doflein); b, c, living and stained reproductive cells of *Pleodorina illinoisensis*, ×1000 (Merton); d-f, terminal cells of *Hydrurus foetidus*, showing division of chromatophore and pyrenoid (Geitler); g-i, *Chlamydomonas* sp., showing the division of pyrenoid (Geitler).

Chloromonadida, the chromatophores are bright green, containing an excess of xanthophyll. In dinoflagellates, they are dark yellow or brown, because of the presence of pigments: carotin, phylloxanthin, and peridinin (Kylin, 1927), the last of which is said to give the brown coloration. A few species of Gymnodinium contain blue-green chromatophores for which phycocyanin is held to be responsible. The chromatophores of Phytomonadida and Euglenoidida are free from any pigmentation, and therefore green. Aside from various pigments associated with the chromatophores, there are carotinoid pigments which occur often outside the chromatophores, and are collectively known as **haematochrome.** The haematochrome occurs in *Haematococcus pluvialis, Euglena sanguinea, E. rubra,* Chlamydomonas, etc. In Haematococcus, it increases in volume and in intensity when there is a deficiency in phosphorus and especially in nitrogen; and when nitrogen and phosphorus are present sufficiently in the culture medium, the haematochrome loses its color completely (Reichenow, 1909; Pringsheim, 1914). Steinecke also noticed that the frequent yellow coloration of phytomonads in moorland pools is due to development of carotin in the chromatophores as a result of deficiency in nitrogen. Johnson (1939) noted that the haematochrome granules of *Euglena rubra* become collected in the central portion instead of being scattered throughout the body when sunlight becomes weaker. Thus, this Euglena appears green in a weak light and red in a strong light. The chromatophores divide when the organism divides (Johnson, 1956), and, therefore, the number of chromatophores appear to remain about the same through generations (Fig. 34).

When chlorophyll-bearing protozoa are deprived of light for a certain length of time, the green coloration is lost, but when light becomes available, the color returns. Wolken and Palade (1953) find the chromatophores of *Euglena gracilis* a laminated structure as seen in electron micrographs and remain as remnants during darkness.

In association with the chromatophores are found the **pyrenoids** (Fig. 34) which are usually embedded in them. The pyrenoid is a viscous structureless mass of protein (Czurda), and may or may not be covered by tightly fitting starch-envelope, composed of several pieces or grains which appear to grow by apposition of new material on the external surface. A pyrenoid divides when it reaches a certain size, and also at the time of the division of the organism in which it occurs. As to its function, it is generally agreed that the pyrenoid is concerned with the formation of the starch and allied anabolic products of photosynthesis (Pyrenoid, Geitler, 1926).

Chromatophore-bearing protozoa usually possess also a **stigma** (Fig. 34) or eye-spot. The stigma may occur in exceptional cases in colorless forms, as in Khawkinea, Polytomella, etc. It is ordinarily situated in the anterior region and appears as a reddish or brownish red dot or short rod, embedded in the cortical layer of the cytoplasm. The color of stigma seems to be due to the presence of carotenoid pigment, astaxanthin (Lwoff, 1930), in a cytoplasmic network, but the exact nature is unknown (Goodwin and Jamikorn, 1954). The stigma is incapable of division and a new one is formed at the time of cell division.

In many species, the stigma possesses no accessory parts, but, according to Mast (1928), the pigment mass in Chlamydomonas, Pandorina, Eudorina, Euglena, Trachelomonas, etc., is in cupform, the concavity being deeper in the colonial than in solitary forms. There is a colorless mass in the concavity, which appears to function as a lens. In certain dinoflagellates, there is an **ocellus** (Fig. 130, *c, d, g, h*) which is composed of amyloid lens and a dark pigment mass (melanosome) that is sometimes capable of amoeboid change of form. The stigma is, in general, regarded as an organella for the perception of light intensity. Mast (1926) considers that the stigma in the Volvocidae is an organella which determines the direction of the movement.

Viewed in electron micrographs, the stigma of *Euglena gracilis* is composed of about forty to fifty packed granules arranged in a flat group, averaging 100-300 mμ in diameter. These pigment granules may contain β-carotine or more than one light-absorbing pigment (Wolken and Shin, 1958).

Other Cytoplasmic Inclusions

Aside from the cytoplasmic structures mentioned on the preceding pages, granules, particles, rods, filaments, etc., occur in protozoa. Chytrid fungus, *Sphaerita* (Dangeard, 1886) (Fig. 387), has long been known to be present in the cytoplasm of various free-living and parasitic protozoa. Sphaerita are small (up to 3μ) refractive spheroid bodies and occur singly or many in rounded masses. A similar chytrid *Nucleophaga* (Dangeard, 1895) (Fig. 387) occurs in the nuclei of many protozoa. Kirby (1932) described various particles from the termite flagellates, especially Trichonympha, which were considered to be parasites of unknown nature. In *Pelomyxa palustris,* two types of baciliform bodies are present.

Chatton and Lwoff (1929) noted in *Ellobiophrya donacis* cytoplasmic rods up to 20μ in length which they supposed to be symbiotic bacteria. Miyashita (1933) found numerous rods in the endoplasm of *Ptychostomum bacteriophilum* and interpreted them as symbiotic bacteria. Fauré-Fremiet (1950) found many minute (1μ in diameter) spherules in the endoplasm of *Remanella multinucleata,* and considered them as bacteria. The same author observed Feulgen-positive filaments (5-12μ by 0.5μ) in *Euplotes eurystomus* and *E. patella.* These filaments disappeared when the ciliates were exposed to a weak dose of penicillin, followed by death of the ciliates in about ten days. Fauré-Fremiet considered the filaments bacteria holding symbiotic relationship with Euplotes.

References

ALEXEIEFF, A., 1928. Sur la question des mitochondries et de l'appareil de Golgi chez les protistes. Arch. Protist., 60:269.

——— 1929. Nouvelles observations sur les chondriosomes chez les protozoaires. *Ibid.,* 65:45.

ALLEN, M. B., *et al.,* 1959. Photoreactive pigments in flagellates. Nature, 184:1047.

ANDERSON, E., 1955. The electron microscopy of *Trichomonas muris.* J. Protozool., 2:114.

——— and BEAMS, H. W., 1959. The cytology of Tritrichomonas as revealed by the electron microscope. J. Morphol., 104:205.

——— ——— 1960. The fine structure of the heliozoan *Actinosphaerium nucleofilum.* J. Protozool., 7:190.

——— ——— 1961. The ultrastructure of Trichomonas with special reference to the blepharoplast complex. J. Protozool., 8:71.

ARCICHOVSKIJ, V., 1905. Ueber das Zoopurpurin, ein neues Pigment der Protozoa (*Blepharisma lateritium*). Arch. Protist., 6:227.

BAIRATI, A., and LEHMANN, F. E., 1953. Structural and chemical properties of the plasmalemma of *Amoeba proteus*. Exper. Cell. Res., 5:220.

BARBIER, M., *et al.*, 1956. Sur les pigments du cilié *Stentor niger*. C. R. Acad. Sci., 242:2182.

BEERS, C. D., 1946. *Tillina magna:* micronuclear number, etc. Biol. Bull., 91:256.

BĚLAŘ, K., 1926. Der Formwechsel der Protistenkerne. Ergebn. u. Fortschr. Zool., 6:235.

BEYERSDORFER, K., and DRAGESCO, J., 1952. Microscopie électronique des trichocystes de Frontonia. Congr. microsc. électr., p. 657.

———— ———— 1952a. Étude comparative des trichocystes de sept espèces de Paramécies. *Ibid.*, p. 661.

BRESSLAU, E., and SCREMIN, L., 1924. Die Kerne der Trypanosomen und ihre Verhalten zur Nuclealreaktion. Arch. Protist., 48:509.

BRODSKY, A., 1924. Die Trichocysten der Infusorien. Arch. russ. Protist., 3:23.

BROWN, H. P., 1945. On the structure and mechanics of the protozoan flagellum. Ohio J. Sci., 45:247.

BROWN, V. E., 1930. The Golgi apparatus of *Amoeba proteus*. Biol. Bull., 59:240.

———— 1930a. The Golgi apparatus of Pyrsonympha and Dinenympha. Arch. Protist., 71:453.

———— 1930b. The neuromotor apparatus of Paramecium. Arch. zool. exper. gén., 70:469.

BURT, R. L., 1940. Specific analysis of the genus Colpoda with special reference to the standardization of experimental material. Tr. Am. Micr. Soc., 59:414.

CHADEFAUD, M., 1938. Nouvelles recherches sur l'anatomie comparé des Eugléniens: les Peranémines. Rev. Algol., 11:189.

CHAPMAN-ANDRESEN, C., and NILSSON, J. R., 1960. Electron micrographs of pinocytosis channels in *Amoeba proteus*. Exper. Cell. Res., 19:631.

CHATTON, E., and LWOFF, A., 1929. Contribution a l'étude de l'adaptation, etc. Bull. biol. France-Belg., 63:321.

———— ———— 1935. Les ciliés apostomes, etc. Arch. zool. exp. gén., 77:1.

———— ———— and LWOFF, M., 1929. Les infraciliature et la continuité génétique des systèmes ciliaries récessifs. C. R. Acad. Sci., 188:1190.

CHEN, Y. T., 1950. Investigations of the biology of *Peranema trichophorum*. Quart. J. Micr. Sci., 91:279.

CLEVELAND, L. R., HALL, S. R., SANDERS, E. P., and COLLIER, J., 1934. The woodfeeding roach Cryptocercus, its protozoa, etc. Mem. Am. Acad. Arts. Sci., 17:185.

CUSHMAN, J. A., 1933. Foraminifera: their classification and economic use. 2 ed. Sharon, Mass.

DANGEARD, P. A., 1886. Sur un nouveau genre de Chitridinés parasites des Rhizopodes et des flagellates. Bull. Soc. Bot. France, 33:240.

DAS, S. M. and TEWARI, H. B., 1955. Golgi apparatus in *Amoeba verrucosa.* Curr. Sci., 24:58.

DELLINGER, O. P., 1909. The cilium as a key to the structure of contractile protoplasm. J. Morphol., 20:171.

DIERKS, K. 1926. Untersuchungen ueber die Morphologie und Physiologie des *Stentor coeruleus.* Arch. Protist., 54:1.

DOFLEIN, F., 1916. Studien zur Naturgeschichte der Protozoen. VII. Zool. Jahrb., Anat., 39:335.

DOGIEL, V., 1923. Cellulose als Bestandteil des Skellettes bei einigen Infusorien. Biol. Zentralbl., 43:289.

—— 1929. Die sog. "Konkrementenvakuole" der Infusorien als eine Statocyste betrachtet. Arch. Protist., 68:319.

DRAGESCO, J., 1951. Sur la structure des trichocystes du flagellé cryptomonadine, *Chilomonas paramecium.* Bull. Micr. appl., 2 ser., 1:172.

—— 1952. Sur la structure des trichocystes toxiques des infusoires holotriches gymnostomes (Note préliminaire). Bull. microsc. appl., 2:92.

DUBOSCQ, O., and GRASSÉ, P.-P., 1933. L'appareil parabasal des flagellés. Arch. zool. exp. gén., 63:381.

EHRET, C. F., and POWERS, E. L. 1957. The organization of gullet organelles in *Paramecium bursaria.* J. Protozool., 4:55.

FAURÉ-FREMIET, E., 1957. Le macronucleus héteromere de quelques ciliés. J. Protozool., 4:7.

—— and BRETON-GORIUS, A., 1955. Microscopie électronique des membranelles vibratiles de quelques ciliés. C. R. soc. biol., 149:872.

—— and GAUCHERY, M., 1957. Concrétions minérales intracytoplasmiques chez les ciliés. J. Protozool., 4:96.

—— and ROUILLER, C., 1959. Le cortex de la vacuole contractile et son ultrastructure chez les ciliés. *Ibid.,* 6:29.

—— —— and GAUCHERY, M. 1956. Les structures myoïdes chez les ciliés. Étude au microscope électronique. Arch. d'anat. microsc. morph. exper., 45:139.

FERNANDEZ-GALIANO, D., 1955. El aparato neuromotor de *Eudiplodinium maggii* Flor. Bol. Real Soc. Esp. Hist. Nat., 53:53.

FISCHER, A., 1894. Ueber die Geisseln einiger Flagellaten. Jahresb. wiss. Bot., 26:187.

GEITLER, L., 1926. Zur Morphologie und Entwicklungsgeschichte der Pyrenoide. Arch. Protist., 56:128.

GELEI, J. v., 1926. Zur Kenntnis des Wimperapparates. Zeitschr. ges. Anat., Abt. 1, 81:530.

—— 1932. Die reizleitenden Elemente der Ciliaten, etc. Arch. Protist., 77:152.

GIESE, A. C., 1938. Reversible bleaching of Blepharisma. Tr. Am. Micr. Soc., 57:77.

GOODWIN, T. W., and JAMIKORN, M., 1954. Studies in carotenogenesis. J. Protozool., 1:216.

GRASSÉ, P.-P., 1952. Traité de Zoologie. I. fasc. 1.

GROSS, J. A., JAHN, T. L., and BERNSTEIN, E., 1955. The effect of antihistamines on the pigments of green protista. J. Protozool., 2:71.

HAECKEL, E., 1868. Monographie der Moneren. Jen. Zeit. Naturwiss., 4.

———— 1870. Studien ueber Moneren und andere Protisten. Leipzig.

HALL, R. P., 1929. Reaction of certain cytoplasmic inclusions to vital dyes and their relation to mictochondria, etc. J. Morphol. Physiol., 48:105.

———— and NIGRELLI, R. F., 1937. A note on the vacuome of *Paramecium bursaria* and the contractile vacuole of certain ciliates. Tr. Am. Micr. Soc., 56:185.

HAMMOND, D. M., 1937. The neuromotor system of *Euplotes patella* during binary fission and conjugation. Quart. J. Micr. Sci., 79:507.

———— and KOFOID, C. A. 1937. The continuity of structure and function in the neuromotor system of *Euplotes patella* during its life cycle. Proc. Am. Phil. Soc., 77:207.

HAYE, A., 1930. Ueber den Exkretionsapparat bei den Protisten, etc. Arch. Protist., 70:1.

HAYES, M. L., 1938. Cytological studies on *Dileptus anser*. Tr. Am. Micr. Soc., 57:11.

HERFS, A., 1922. Die pulsierende Vakuole der Protozoen, etc. Arch. Protist., 44:227.

HERTWIG, R., 1902. Die Protozen und die Zelltheorie. *Ibid.*, 1:1.

HOARE, C. A., 1939. Morphological and taxonomic studies on mammalian trypanosomes. VI. Parasitology, 30:529.

———— 1954. The loss of the kinetoplast in trypanosomes with special reference to *Trypanosoma evansi*. J. Protozool., 1:28.

HOLLANDE, A., 1942. Étude cytologique et biologique de quelques flagellés libres. Arch. zool. exp. gén., 83:1.

HORNING, E. S., 1926. Observations on mitochondria. Australian J. Exper. Biol., 3:149.

———— 1927. On the orientation of mitochondria on the surface cytoplasm of infusorians. *Ibid.*, 4:187.

———— 1929. Mitochondrial behavior during the life cycle of a sporozoan (Monocystis). Quart. J. Micr. Sci., 73:135.

HOUWINK, A. L., 1951. An E. M. study of the flagellum of *Euglena gracilis*. Proc. Kon. Nederl. Ak. Weten., Ser. C, 54:132.

———— 1952. Die Pellikularschuppen und die Geissel der *Physomonas vestita* Stokes. Z. wiss. mikrosk. Tech., 60:402.

HOWLAND, R. B., 1924. Dissection of the pellicle of *Amoeba verrucosa*. J. Exper. Zool., 40:263.

INABA, F., *et al.*, 1958. An electron-microscopic study on the pigment granules of Blepharisma. Cytologia, 25:72.

JACOBSON, I., 1932. Fibrilläre Differenzierungen bei Ciliaten. Arch. Protist., 75:31.

JAKUS, M. A., 1945. The structure and properties of the trichocysts of Paramecium. J. Exper. Zool., 100:457.

—— and HALL, C. E., 1946. Electron microscope observations of the trichocysts and cilia of Paramecium. Biol. Bull., 91:141.

JANICKI, C., 1911. Zur Kenntnis des Parabasalapparates bei parasitischen Flagellaten. Biol. Zentralbl., 31:321.

JEPPS, M. W., 1947. Contribution to the study of the sponges. Proc. Roy. Soc. London, Series B, 134:408.

JÍROVEC, O., 1929. Studien ueber blepharoplastlose Trypanosomen. Arch. Protist., 68:187.

JOHNSON, L. P., 1956. Observations on *Euglena fracta* sp. nov., etc. Tr. Am. Micr. Soc., 75:271.

KIDDER, G. W., 1933. On the genus Ancistruma Strand (Ancistrum Maupas). Biol. Bull., 64:1.

—— 1933a. *Conchophthirus caryoclada* sp. nov. *Ibid.*, 65:175.

—— 1934. Studies on the ciliates from freshwater mussels. I, II. *Ibid.*, 66:69,286.

KING, R. L., 1935. The contractile vacuole of *Paramecium multimicronucleatum*. J. Morphol., 58:555.

—— 1954. Origin and morphogenetic movements of the pores of the contractile vacuoles in *Paramecium aurelia*. J. Protozool., 1:121.

—— BEANS, H. W., et al., 1961. The ciliature and infraciliature of *Nyctotherus ovalis* Leidy. *Ibid.*, 8:98.

KIRBY, H. JR., 1931. The parabasal body in trichomonad flagellates. Tr. Am. Micr. Soc., 50:189.

—— 1932. Flagellates of the genus Trichonympha in termites. Univ. Cal. Publ. Zool., 37:349.

—— 1941. Organisms living on and in protozoa. In Calkins, G. N. and Summers, F. M.: Protozoa in biological research. Columbia University Press.

KITCHING, J. A., 1956. Contractile vacuoles of protozoa. Protoplasmatologia, 3:D3a.

—— 1956a. Food vacuoles. *Ibid.*, 3:D3b.

KLEIN, B. M., 1926. Ueber eine neue Eigentümlichkeit per Pellicula von *Chilodon uncinatus*. Zool. Anz., 67:160.

—— 1926a. Ergebnisse mit einer Silbermethode bei Ciliaten. Arch. Protist., 56:243.

—— 1927. Die Silberliniensysteme der Ciliaten. *Ibid.*, 58:55.

—— 1928. Die Silberliniensysteme der Ciliaten. *Ibid.*, 60:55 and 62:177.

—— 1929. Weitere Breiträge zur Kenntnis des Silberliniensystems der Ciliaten. *Ibid.*, 65:183.

—— 1930. Das Silberliniensystem der Ciliaten. IV. *Ibid.*,69:235.

—— 1942. Differenzierungsstufen des Silberlinien- oder neuroformativen Systems. *Ibid.*, 96:1.

——— 1958. The "dry" silver method and its proper use. J. Protozool., 5:99.

KLEINSCHMIDT, A., and KINDER, E., 1950. Elektronenoptische Befunde an Rattentrypanosomen. Zentralbl. Bakt. I Abt. Orig., 156:219.

KOFOID, C. A., and SWEZY, O., 1921. The free-swimming unarmored Dinoflagellata. Mem. Univ. Cal., 5:1.

KOLTZOFF, N. K., 1911. Untersuchungen ueber die Kontraktilität des Stieles von *Zoothamnium alternans.* Biol. Zeitschr. Moskau, 2:55.

KRANEVELD, F. C., HOUWINK, A. L., and KEIDLE, H. J. W., 1951. Electron microscopical investigations on trypanosomes. I. Proc. Kon. Nederl. Akad. Wetensch., C, 54:393.

KRÜGER, F., 1934. Bemerkungen über Flagellatentrichocysten. Arch. Protist., 83:321.

——— 1936. Die Trichocysten der Ciliaten im Dunkelfeld. Zoologica, 34 (H. 91):1.

KUDO, R. R., 1924. A biologic and taxonomic study of the Microsporidia. Illinois Biol. Monogr., 9:80.

——— 1936. Studies on *Nyctotherus ovalis* Leidy, etc. Arch. Protist., 87:10.

——— 1946. *Pelomyxa carolinensis* Wilson. I. J. Morphol., 78:317.

——— 1951. Observations on *Pelomyxa illinoisensis. Ibid.,*88:145.

——— 1957. *Pelomyxa palustrs* Greeff. I. J. Protozool., 4:154.

KYLIN, H., 1927. Ueber die karotinoiden Farbstoffe der Algen. Zeitschr. physiol. Chem., 166:39.

LAZAROW, A., and Cooperstein, S. J., 1953. Studies on the mechanism of Janus green B staining of mitochondria. Exper. Cell Res., 5:56.

LUND, E. E., 1933. A correlation of the silverline and neuromotor systems of Paramecium. Univ. Cal. Publ. Zool., 39:35.

LWOFF, M., and LWOFF, A., 1930. Détermination experimentale de la synthèse massive de pigment carotinoide par le flagellé *Haematococcus pluvialis.* C. R. Soc. Biol., 105:454.

LYNCH, J. E., 1930. Studies on the ciliates from the intestine of Strongylocentrotus. II. Uni. Cal. Publ. Zool., 33:307.

MACLENNAN, R. F., 1941. Cytoplasmic inclusions. In Calkins and Summers: Protozoa in biological research.

——— 1944. The pulsatory cycle of the contractile canal in the ciliate Haptophrya. Tr. Am. Micr. Soc., 63:187.

MAINX, F., 1928. Beiträge zur Morphologie und Physiologie der Eugleninen. Arch. Protist., 60:305.

MAST, S. O., 1926. Structure, movement, locomotion and stimulation in Amoeba. J. Morphol., 41:347.

——— 1928. Structure and function of the eye-spot in unicellular and colonial organisms. Arch. Protist., 60:197.

——— 1928a. Factors involved in changes in form in Amoeba. J. Exper. Zool., 51:97.

——— 1944. A new peritrich belonging to the genus Ophrydium. Tr. Am. Micr. Soc., 63:181.

——— and DOYLE, W. L., 1935. Structure, origin and function of cytoplasmic constituents in *Amoeba proteus*. Arch. Protist., 86:155.

——— ——— 1935a. II. *Ibid.*, 86:278.

MERCER, E. H., 1959. An electron microscopic study of *Amoeba proteus*. Proc. Roy. Soc. London, Series B, 150:216.

METZ, C. B., *et al.*, 1953. The fibrillar systems of ciliates as revealed by the electron microscope. I. Paramecium. Biol. Bull., 104:408.

MØLLER, K. M., 1962. On the nature of stentorin. C. R. Trav. Lab. Carlsberg, 32:471.

MOSES, M. J., 1950. Nucleic acids and proteins of the nuclei of Paramecium. J. Morphol., 87:493.

NASSONOV, D., 1924. Der Exkretionsapparat (kontractile Vacuole) der Protozoen als Homologen des Golgischen Apparatus der Metazoenzelle. Arch. mikr. Anat., 103:437.

——— 1925. Zur Frage ueber den Bau und die Bedeutung des Lipoiden Exkretionsapparates bei Protozoen. Ztschr. Zellforsch., 2:87.

OWEN, H. M., 1947. Flagellar structure. I. Tr. Am. Micr. Soc., 66:50.

——— 1949. II. *Ibid.*, 68:261.

PAPPAS, G. D., 1954. Structural and cytochemical studies of the cytoplasm in the family Amoebidae. Ohio J. Sci., 54:195.

PENARD, E., 1922. Études sur les infusoires d'eau douce. Geneva.

PETERSEN, J. B., 1929. Beiträge zur Kenntnis der Flagellatengeisseln. Bot. Tidsskr., 40:373.

PICKARD, EDITH A., 1927. The neuromotor apparatus of *Boveria teredinidi* Nelson, etc. Univ. Cal. Publ. Zool., 29:405.

PIEKARSKI, G., 1949. Blepharoplast und Trypaflavinwirkung bei *Trypanosoma brucei*. Zentralbl. Bakt., Orig., 153:109.

PINEY, A., 1931. Recent advances in microscopy. London.

PITELKA, DOROTHY R., 1949. Observations on flagellum structure in Flagellata. Univ. Cal. Publ. Zool., 53:377.

——— 1961. Fine structure of silverline and fibrillar systems of three tetrahymenid ciliates. J. Protozool., 8:75.

——— and SCHOOLEY, C. N., 1955. Comparative morphology of some protistan flagella. Univ. Cal. Publ. Zool., 61:79.

——— ——— 1958. The fine structure of the flagellar apparatus in Trichonympha. J. Morphol., 102:199.

POLLISTER, A. W., and LEUCHTENBERGER, C., 1949. The nature of the specificity of methyl green for chromatin. Proc. Nat. Acad. Sci., 35:111.

PRINGSHEIM, E., 1914. Die Ernährung von *Haematococcus pluvialis*. Beitr. Biol. Pflanz., 12:413.

PROVASOLI, L., HUNTER, S. H., and SCHATZ, A., 1948. Streptomycin-induced chlorophyll-less races of Euglena. Proc. Soc. Exper. Biol. Med., 69:279.

PUYTORAC, P. DE 1951. Sur la présence d'un argyrome chez quelques ciliés astomes. Arch. zool. exper. gén., 88 (N.-R.):49.

RAABE, Z., 1949. Studies on the family Hysterocinetidae Diesing. Ann. Mus. Zool. Polonici, 14:21.

REICHENOW, E., 1909. Untersuchungen an *Haematococcus pluvialis*, etc. Arb. kaiserl. Gesundh., 33:1.

—— 1928. Ergebnisse mit der Nuklealfärbung bei Protozoen. Arch. Protist. 61:144.

RICHARDSON, K. C., and HORNING, E. S., 1931. Cytoplasmic structures in binucleate opalinids with special reference to the Golgi apparatus. J. Morphol. Physiol., 52:27.

ROBBINS, W. J., *et al.*, 1953. Euglena and vitamin B$_{12}$. Ann. N.Y. Acad. Sci., 56:818.

ROSKIN, G., 1923. La structure des myonèmes des infusoires. Bull. biol. France et Belg., 57:143.

—— 1925. Ueber die Axopodien der Heliozoa und die Greiftentakel der Ephelotidae. Arch. Protist., 52:207.

—— and LEVINSOHN, L. B., 1929. Die Kontractilen und die Skelettelemente der Protozoen. I. *Ibid.*, 66:355.

ROTH, L. E., 1959. An electron-microscope study of the cytology of the protozoan *Peranema trichophorum*. J. Protozool., 6:107.

—— *et al.*, 1960. Electron microscope studies of mitosis in amoebae. J. Biophys. Biochem. Cytol., 8:207.

ROUILLER, C., and FAURÉ-FREMIET, E., 1957. L'ultrastructure des trichocystes fusiformes de *Frontonia atra*. Bull. microsc. appl., 7:135.

—— —— and GAUCHERY, M., 1956. The pharyngeal protein fibers of the ciliates. Proc. Stockholm Conf. Électr. Microscopy., p. 216.

—— —— 1956a. Les tentacules d'Ephelota; étude au microscope électronique. J. Protozool., 3:194.

RUDZINSKA, M. A., 1958. An electron microscope study of the contractile vacuole in *Tokophrya infusionum*. J. Biophys. Biochem. Cytol., 4:195.

—— and GRANICK, S., 1953. Protoporphyrin production of *Tetrahymena geleii*. Proc. Soc. Exper. Biol. & Med., 83:525.

—— and PORTER, K. R., 1954. Electron microscopic study of intact tentacles and disc in *Tokophrya infusionum*. Experientia, 10:460.

RUMJANTZEW, A., and WERMEL, E., 1925. Untersuchungen ueber den Protoplasmabau von *Actinosphaerium eichhorni*. Arch. Protist., 52:217.

SAUNDERS, J. T., 1925. The trichocysts of Paramecium. Proc. Cambridge Philos. Soc., Biol. Sci., 1:249.

SCHNEIDER, L., 1960. Elektronenmikroskopische Untersuchungen ueber das Nephridialsystem von Paramecium. J. Protozool., 7:75.

SCHRÖDER, O., 1906. Beiträge zur Kenntnis von *Stentor coeruleus* und *St. roeselii*. Arch. Protist., 8:1.

SCHUBERG, A., 1888. Die Protozoen des Wiederkäuermagens. I. Zool. Jahrb,. Abt. Syst., 3:365.

SESHACHAR, B. R., and DASS, C. M. S., 1953. Evidence for conversion of desoxyribonucleic acid to ribonucleic acid in *Epistylis articulata*. Exper. Cell Res., 5:248.

SHARP, R., 1914. *Diplodinium ecaudatum* with an account of its neuromotor apparatus. Univ. Cal. Publ. Zool., 13:43.

SLEIGH, M. A., 1962. The biology of cilia and flagella. New York.

SMYTH, J. D., 1941. The morphology of the osmiophile material in some ciliates. Proc. Roy. Irish Acad., 46:189.

STRELKOW, A., 1929. Morphologische Studien ueber oligotriche Infusorien aus dem Darme des Pferdes. I. Arch. Protist., 68:503.

TAYLOR, C. V., 1920. Demonstration of the function of the neuromotor apparatus in Euplotes by the method of micro-dissection. Univ. Cal. Publ. Zool., 19:403.

—— 1941. Ciliate fibrillar systems. In Calkins and Summers: Protozoa in biological research.

TEN KATE, C. G. B., 1927. Ueber das Fibrillensystem der Ciliaten. Arch. Protist., 57:362.

—— 1928. II. *Ibid.*, 62:328.

THON, K., 1905. Ueber den feineren Bau von *Didinium nasutum*. *Ibid.*, 5:282.

TIMOTHÉE, C., 1952. Le système fibrillaire *d'Epidinium ecaudatum* Crawley. Ann. Sci. Nat. Zool., 11 ser., 14:375.

TOBIE, ELEANOR J., 1951. Loss of the kinetoplast in a strain of *Trypanosoma equiperdum*. Tr. Am. Micr. Soc., 70:251.

TÖNNIGES, C., 1914. Die Trichocysten von *Frontonia leucas* und ihr chromidialer Ursprung. Arch. Protist., 32:298.

TORCH, R., 1955. Cytological studies on *Pelomyxa carolinensis*. etc. J. Protozool., 2:167.

TURNER, J. P., 1933. The external fibrillar system of Euplotes with notes on the neuromotor apparatus. Biol. Bull., 64:53.

—— 1937. Studies on the cilitate *Tillina canalifera* n. sp. Tr. Am. Micr. Soc., 56:447.

—— 1940. Cytoplasmic inclusions in the cilitate, *Tillina canalifera*. Arch. Protist., 93:255.

VERWORN, M., 1897. Die polar Eregung der lebenden Substanz durch den constanten Strom. Arch. ges. Physiol., 65:47.

—— 1913. Irritability. 264 pp. New Haven.

VICKERMAN, K., 1962. Patterns of cellular organization in Limax amoebae. An electron microscope study. Exper. Cell. Res., 26:497.

VISSCHER, J. P., 1926. Feeding reactions in the ciliate *Dileptus gigas*, etc. Biol. Bull., 45:113.

VLK, W., 1938. Ueber den Bau der Geissel. Arch. Protist., 90:448.

WEATHERBY, J. H., 1941. The contractile vacuole. In Calkins and Summers: Protozoa in biological research.

WEINREB, S., 1955. *Homalozoon vermiculare*. II. J. Protozool., 2:67.

WEISZ, P. B., 1948. The rôle of carbohydrate reserves in the regeneration of Stentor fragments. J. Exper. Zool., 108:263.

—— 1949. A cytochemical and cytological study of differentiation in normal and reorganizational stages of *Stentor coeruleus. J. Morphol.,* 84: 335.

—— 1950. On the mitochondrial nature of the pigmented granules in Stentor and Blepharisma. *Ibid.,* 86:177.

WETZEL, A., 1925. Vergleichend cytologische Untersuchungen an Ciliaten. Arch. Protist., 51:209.

WILBER, C. G., 1942. The cytology of *Pelomyxa carolinensis.* Tr. Am. Micr. Soc., 61:227.

—— 1945. Origin and function of the protoplasmic constituents in *Pelomyxa carolinensis.* Biol. Bull., 88:207.

WOHLFARTH-BOTTERMANN, K.-E., 1950. Funktion und Struktur der Parameciumtrichocysten. Wissenschaften, 37:562.

—— 1953. Experimentelle und elektronenoptische Untersuchungen zur Funktion der Trichocysten von *Paramecium caudatum.* Arch. Protist., 98:169.

WOLKEN, J. J., 1956. A molecular morphology of *Euglena gracilis* var. *bacillaris.* J. Protozool., 3:211.

—— and PALADE, G. E., 1953. An electron microscope study of two flagellates, etc. Ann. N.Y. Acad. Sci., 56:873.

—— and SHIN, E., 1958. Photomotion in *Euglena gracilis.* J. Protozool., 5:39.

WOODCOCK, H. M., 1906. The haemoflagellates: a review of present knowledge relating to the trypanosomes and allied forms. Quart. J. Micr. Sci., 50:151.

WOODRUFF, L. L., and SPENCER, H., 1922. Studies on *Spathidium spatula.* I. J. Exper. Zool., 35:189.

YOCOM, H. B., 1918. The neuromotor apparatus of *Euplotes patella.* Univ. Calif. Publ. Zool., 18:337.

YUSA, A., 1957. The morphology and morphogenesis of the buccal organelles in Paramecium, etc. J. Protozool., 4:128.

Chapter 4
Physiology

THE morphological consideration given in the last chapter indicates the presence of various organelles in protozoa. The life processes of the protozoan, as in all living organisms, are the sum-total of all physiological activities which are carried on by these organelles confined within a cell, unlike that in metazoa in which tissues, organs and systems handle various vital functions such as food-getting, digestion, assimilation, metabolism, growth, locomotion, reproduction, etc. Thus, morphologically as well as physiologically, protozoa are far more complex than the cell of metazoa.

Nutrition

Protozoa obtain nourishment in manifold ways which may be placed in two categories: autotrophic and heterotrophic. Autotrophic protozoa utilize inorganic substances, while heterotrophs require organic food materials. Autotrophs are also called holophytic protozoa, and heterotrophs may be divided into holozoic and saprozoic protozoa. Holophytic and saprozoic protozoa utilize dissolved foods; they are thus osmotrophs. Holzoic forms ingest solid foods in addition to food in fluid form, hence they are phagotrophs.

Information on the nutrition of protozoa is undergoing an accelerated progress in recent years, as axenic cultures of various protozoa in synthetic media are becoming available in increasing numbers (Lwoff, 1951).

Holophytic Nutrition. The holophytic protozoa are able to synthesize simple carbohydrates from carbon dioxide and water by chlorophyll contained in chromatophores in the presence of sun light (photosynthesis). The first products are sugars, from which starch and related products are formed. In the synthesis of fats and protein, mineral salts, nitrates, ammonium salts and vitamins are ultilized.

Aside from Phytomastigia, chromatophores occur in the ciliate *Cyclotrichium meunieri* (p. 844) (Powers, 1932; Bary and Stuckey, 1950). In a number of other cases, the organism itself is without chromatophores but is apparently not holozoic, because of the presence of chlorophyll-bearing organisms within it. For example, in Paulinella (p. 584) in which occur no food vacuoles, green bodies of peculiar shape are always present. The latter appear to be a species of alga which holds a symbiotic relationship with Paulinella. A similar relationship seems to exist between *Paramecium bursaria, Stentor polymorphus*, etc. and Zoochlorellae; *Paraeuplotes tortugensis* and zooxanthella and others. Pringsheim (1928) showed that organic matters are passed on from the zoochlorellae to their host, Paramecium and serve as food. Perhaps, in these instances, photosynthesis of the symbionts supplements holozoic nutrition of the hosts.

Some Phytomastigia combine photosynthesis with absorption of organic food in dissolved (Euglena) or solid (Chrysomonadida) form. In addition, vitamins are required and the organic food is the source of nitrogen. Thus, in these organisms, the product of photosynthesis is apparently inadequate and is supplemented by oxidation of organic substances as energy sources.

The photosynthesis and utilization of organic food may differ even in the organisms which are morphologically similar. *Euglena pisciformis* is able to grow in light, but is unable to grow in the dark, even if kept in organic media with vitamins and acetate. Therefore, they will perish when kept in the dark. On the other hand, *E. gracilis* is able to grow in the light or in the dark, by utilization of acetate in the medium as a carbon source. In the dark, the chlorophyll is lost, but when exposed to light it reappears. Experimentally, the bleaching of chlorophyll or *apochlorosis* may be brought about by treating the organisms with streptomycin in light or in dark, as first reported by Provasoli *et al.* (1948), or with antihistamines (Gross, Jahn and Berstein, 1955). Streptomycin is said to shatter chromatophores into fragments, which is irreversible, and the bleached organisms grow in acetate-containing organic media. This irreversible loss of chromatophores may also be produced by aureomycin, pyribenzamine and cultivation at higher (35-36°C.) temperature. Rosen and Gawlik (1961) consider that the bleaching by streptomycin is not caused by breakdown of chlorophyll or interference with replication of mature chlorophyll, but it inhibits the formation of a plastid or pigment precursor.

As will be mentioned later, there are many colorless Phytomastigia that are similar to chlorophyll-bearing forms except the absence of chromatophores. One can speculate that they have arisen as mutants from colored forms due to unequal distribution of chromatophores at the time of division or, as Provasoli and Hutner (1951) suggested, due to action of streptomycin, since Actinomyces producing the antibiotic are common in soil. As to various nutritional requirements of Phytomastigia, the reader is referred to Hutner and Provasoli (1951).

Holozoic Nutrition. This is the method by which all animals obtain their nourishment; namely, the protozoa require organic materials as source of food. It involves food-capture and ingestion, digestion and assimilation, and getting rid of indigestible portions, etc.

The methods of food-capturing vary among different forms. In the Sarcodina, the food organisms are captured by pseudopods and taken into the body at any point. The methods however vary. According to Rhumbler's (1910) oft-quoted observations, four methods of food-ingestion occur in amoebae (Fig. 35); namely, (1) by "import," in which the food is taken into the body upon contact, with very little movement on the part of the amoeba (*a*); (2) by "circumfluence," in which the cytoplasm flows around the food organism as soon as it comes in contact with it on all sides and engulfs it (*b*); (3) by "circumvallation," in which the amoeba without contact with the food, forms pseudopodia which surround the food on all sides and ingest it (*c*); (4) by "invagination," in which the amoeba touches and adheres to the food, and the ectoplasm in contact with it is invaginated into the endoplasm as a tube, the cytoplasmic membrane later disappears (*d-h*). The method of obtaining food of the intracorpuscular malaria organisms was assumed to be osmotrophic, but electron microscope studies by Rudzinska and Trager (1957, 1958, 1959, 1962) reveal that the schizonts of *Plasmodium lophurae, P. berghi* and *Babesia rodhaini* engulf portions of the cytoplasm of the host erythrocytes by invagination of their plasma membrane with subsequent formation of food vacuoles. Thus it is phagotrophic as in amoebae.

In a species of Hartmannella, Ray (1951) reports an agglutination of large numbers of motile bacteria over the body surface, which later form a large mass and are taken into a food cup.

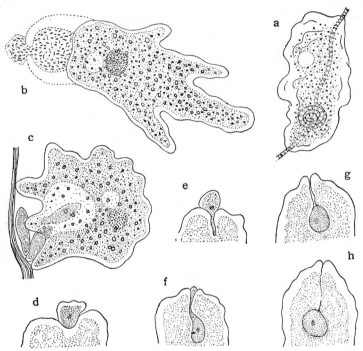

FIG. 35. Various ways by which amoebae capture food organisms. a, *Amoeba verrucosa* feeding on Oscillatoria by 'import' (Rhumbler); b, *A. proteus* feeding on bacterial glea by 'circumfluence'; c, on Paramecius by 'circumvallation' (Kepner and Whitlock); d-h, *A. verrucosa* ingesting a food particle by 'invagination' (Gross-Allermann).

In certain testaceans, such as Gromia, several rhizopodia cooperate in engulfing the prey and, in Lieberkühnia (Fig. 36), Verworn noted ciliates are captured by and digested in rhizopodia. Similar observation was made by Schaudinn in the heliozoan Camptonema in which several axopodia anastomose to capture a prey (Fig. 214, *d*). In the holozoic Mastigophora, such as Hypermastigida, which do not possess cytostome, the food-ingestion is by import or invagination as noted in *Trichonympha campanula* (Cleveland, 1925a; Emik, 1941) (Fig. 37, *a*) and *Lophomonas blattarum* (Kudo, 1926).

The food particles become attached to the pseudopod and are held there on account of its viscid nature. The sudden immobility of active organisms upon coming in contact with pseudopodia of certain forms, such as Actinophrys, Actinosphaerium, Gromia, Elphidium, etc., suggests, however, probable discharge of poisonous

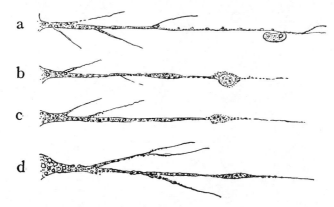

FIG. 36. Rhizopodium of Lieberkühnia, capturing and digesting *Colpidium colpoda* (Verworn).

substances. In the Suctoria which lack a cytostome, the tentacles serve as food-capturing organellae. The suctorial tentacle bears on its distal end a rounded knob which, when it comes in contact with an actively swimming ciliate, stops the latter immediately (*Parapodophrya typha,* Fig. 383, *a*). The prehensile tentacles of Ephelotidae are said to be similar in structure to the axopodia, in that each possesses a bundle of axial filaments around a cytoplasmic core (Roskin, 1925). These tentacles are capable of piercing through the body of a prey. In some suctorians, such as Choanophrya (Fig. 386, *a*), the tubular tentacles are clearly observable, and both solid and liquid food materials are sucked in through the cavity. The rapidity with which tentacles of a suctorian stop a very actively swimming ciliate is attributed to a certain substance secreted by the tentacles, which paralyses the prey.

In the cytostome-bearing Mastigophora, the lashing of flagella will aid in bringing about the food particles to the cytostome, where it is taken into the endoplasm. Chen (1950) observed Peranema feeding on immobile organisms. When the tip of the anterior flagellum comes in contact with an immobile Euglena, the whole flagellum beats actively and the body contracts, followed

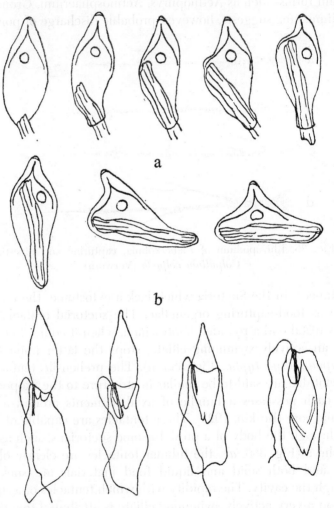

FIG. 37. a, eight outline sketches of a *Trichonympha campanula*, ingesting a large particle of food, ×150 (Emik); b, four outline sketches of a *Peranema trichophorum* feeding on an immobile Euglena (Chen).

by elongation. The process is repeated several times until the body touches Euglena. Then the cytostome stretches open, the oral rods move up, protrude from the body and become attached to Euglena. Peranema advances toward the prey and the whole Euglena is engulfed in two to fifteen minutes (Fig. 37, *b*).

In the ciliates, there are many types of cytostome and associated organelles, but the food-capturing seems to be in general of two kinds. When the cytostome is permanently open, the organism ingests continuously food particles that are small enough to pass the cytostome and cytopharynx, as in the case of Paramecium. The other type is carried on by organisms bearing cytostome which is ordinarily closed such as seen in Coleps, Didinium, Perispira (Dewey and Kidder, 1940), but which expands to often an extraordinary size when the ingestion of prey takes place.

Many protozoa appear to possess capacity to select what they consume. That Didinium depends on ciliates, particularly Paramecium as a chief food material, has already been mentioned. *Perispira ovum* has a closed mouth, but it opens it upon coming in contact with the flagellum of a euglenoid organism (Dewey and Kidder, 1940), and three species of Bresslaua ingested *Colpoda steinii;* but *B. sicaria* refused to ingest *Glaucoma scintillans,* while *B. vorax* and *B. insidiatrix* grew well on Glaucoma (Claff, Dewey and Kidder, 1941).

Singh (1945, 1960) found that certain soil amoebae fed on bacteria with orange pigment, but rejected those which produced red, green or blue pigment. A species of Acanthamoeba isolated from monkey kidney cultures (in which it existed as a natural contaminant) is said to be able to engulf and denucleate chicken erythrocytes, while leaving guinea pig erythrocytes untouched (Chi *et al.,* 1959). As to the mechanism of selectivity, nothing is known. (Cannibalism, Dawson, 1919, Lapage, 1922, Gelei, 1925a, Tanabe and Komada, 1932, Giese and Alden, 1938, Chen, 1950, Samuels, 1957, Steinberg, 1959; food of protozoa, Sandon, 1932.)

The ingested food particles are usually surrounded by a film of fluid in which the organisms live and the whole is known as **food vacuole.** The quantity of fluid taken in with the food varies greatly, and, generally speaking, it seems to be inversely proportional to the size, but proportional to the activity, of the food organisms. Food vacuoles composed entirely of surrounding liquid medium have occasionally been observed. Edwards (1925) noticed ingestion of fluid medium by an amoeba by forming food-cups under changed chemical composition. Brug (1928) reports seeing *Entamoeba histolytica* engulf liquid culture medium by formation of lip-like elevation of the ectoplasm.

Lewis (1931) observed ingestion of droplets of liquid by the macrophage and fibroplasts as well as sarcoma and carcinoma cells in the culture of rat omentum, and named it **pinocytosis.** He considered that by this process the cells are able to take in substances which cannot diffuse into them. Kirby (1932) figures ingestion of the brine containing no visible organisms by the cytostome of *Rhopalophrya salina* (Fig. 38). Mast and Doyle (1934)

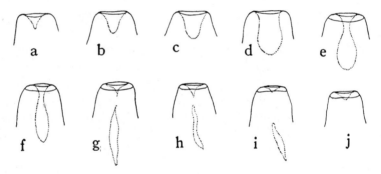

Fig. 38. Ingestion of brine by *Rhopalophrya salina* (Kirby).

state that if *Amoeba proteus, A. dubia, A. dofleini,* or *A. radiosa* is placed in an albumin solution, a hypertonic balanced salt solution, or a hypertonic solution of calcium gluconate it rapidly decreases in volume, and forms numerous tubes filled with fluid, which disintegrate sooner or later and release their fluid content in the cytoplasm. At times, fifty or more such tubes may be present, which indicates that the organism ingests considerable quantities of fluid in this way. The two authors consider that it is "a biological adaptation which serves to compensate for the rapid loss of water." Observing that the solutions of proteins and inorganic ions induced pinocytosis in *Amoeba proteus* and *Pelomyxa carolinensis,* Chapman-Andresen and Prescott (1956) considered that the pinocytosis may serve a nutritional process by ingestion of proteins in addition to the functions mentioned by Mast and Doyle (Pinocytosis, Chapman-Andresen, 1958; Holter, 1959).

The food vacuoles finally reach the endoplasm and in forms such as Amoebida the vacuoles are carried about by the moving endoplasm. In the ciliates, the fluid endoplasm shows often a definite rotation movement. In Paramecium, the general direction is along the aboral side to the anterior region and down the other side, with a short cyclosis in the posterior half of the body.

Some observers maintain that in ciliates there is a definite "digestive tubule" beginning with the cytostome and ending in the cytopyge, and the food vacuoles travel through it. Cosmovici (1931, 1932) saw such a canal in soluble starch-fed *Colpidium colpoda* upon staining with iodine, but Hall and Alvey (1933) could not detect such a structure in the same organism. Kitching (1938b) observed no such tubule in the peritrichous ciliates he studied, and concluded that the food vacuoles are propelled over the determined part of the course by the contraction of surrounding cytoplasm. In *Vorticella* sp., food vacuoles are formed one by one at the end of cytopharynx, migrate through different parts of the cytoplasm without order and food material is digested (Fig. 39, *a*). Old food vacuoles are defecated through a small papilla on the lower wall of the cytopharynx and thence to the outside (Hall and Dunihue, 1931) (Fig. 39, *b-d*).

As stated above, in a number of species the food organisms are paralyzed or killed upon contact with pseudopodia, tentacles or exploded trichocysts. In numerous other cases, the captured organism is taken into the food vacuole alive, as will easily be noted by observing Chilomonas taken in by *Amoeba proteus* or actively moving bacteria ingested by Paramecium. But the prey ceases to move in a very short time. It is generally believed that some substances are secreted into the food vacuole by the protoplasm of the organisms to stop the activity of the prey within the food vacuole. Engelmann (1878) demonstrated that the granules of blue litmus, when ingested by Paramecium or Amoeba, became red in a few minutes. Brandt (1881) examined the staining reactions of amoebae by means of haematoxylin, and found that the watery vacuoles contained an acid. Metschnikoff (1889) also showed that there appears an acid secretion around the ingested litmus grains in Mycetozoida.

Greenwood and Saunders (1894) found in Carchesium that ingestion of food particles stimulated the cytoplasm to secrete a mineral acid. According to Nirenstein (1925), the food vacuole

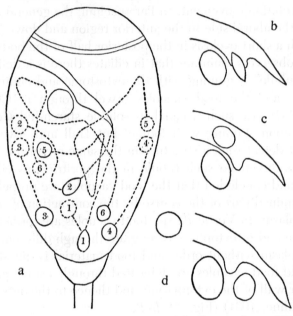

Fig. 39. Diagrams showing movements of food vacuoles in *Vorticella* sp. (Hall and Dunihue). a, diagram of the migration paths of six food vacuoles (vacuoles 1, 2, most recently formed; 3, 4, recently formed; 5, 6, formed some time before); b-d, stages in extrusion of a food vacuole (b, food vacuole entering gullet; c, a later stage; d, the food vacuole leaving cytostome, while another one is moving up toward the cytopyge).

in Paramecium undergoes change in reaction which can be grouped in two periods. The first is acid reaction and the second alkaline reaction, in which albumin digestion takes place. On the other hand, Khainsky (1910) observed that the food vacuole of ciliates, such as Paramecium, is acid during the entire period of protein digestion, and becomes neutral to finally alkaline when the solution of the food substance is ended. Metalnikoff (1912) found that in Paramecium, besides acid-alkaline reaction change, some vacuoles never show acid reaction and others occasionally show sustained acid reaction.

Shapiro (1927) studied the reaction change of the food vacuoles in *Paramecium caudatum* by using phenol red, neutral red, Congo red, and litmus, and found that when the organism is kept in a medium with pH 7, its food vacuoles are first alkaline (pH 7.6), soon reach a maximum acidity (pH 4.0), while still in the posterior half of the body. Later, the vacuoles show a decreased acidity, finally reaching pH 7.0. In *Vorticella* sp. and *Stylonychia pustulata,* the range of pH observed in the food vacuoles was said to be 4.5-7.0 and 4.8-7.0 respectively. The food vacuoles of Actinosphaerium, according to Howland (1928), possess at the beginning pH 6.0-7.0 for 5 to 10 minutes, but this soon changes to more acid (pH 4.3) in which digestion appears to be carried on. In older food vacuoles which are of less acid (pH 5.4-5.6), the digestion appears to be at an end. In the species of Bresslaua, Claff, Dewey and Kidder (1941) noted that a Colpoda taken into the food vacuole is instantly killed with a sudden release of an acid which shows pH 3.0-4.2. During digestion, the protoplasm of the prey becomes alkaline and the undigested residue becomes acid before extrusion.

Mast's observations (1942) on the food vacuoles in *Amoeba proteus* and *A. dubia* containing Chilomonas or Colpidium, indicate: (1) the fluid in the vacuoles becomes first acid and then alkaline; (2) the increase in the acidity of the fluid in the vacuole is not due to cytoplasmic secretion, but is probably due to respiration of the ingested organisms and chemical changes associated with their death, etc.; and (3) the death of the organisms taken in the food vacuoles is probably caused by the decrease in oxygen in the vacuoles, owing to the respiration of the organisms in them.

Arena (1941, 1942) found the maximum acidity of the fluid of food vacuoles in *Pelomyxa carolinensis* containing *Colpidium striatum* was pH 5.8 and was not fatal for the ciliate, but considered the possibility of the existence in the food vacuole of "some lethal agent" which kills the prey. The digestion of ingested ciliates by *Pelomyxa carolinensis* and *P. illinoisensis* as viewed in electron micrographs indicates that the rupture of the pellicle of the ciliates occurs, followed by disintegration of mitochondria, cytoplasmic matrix, nucleus, cilia and trichocysts. The thin (100 mμ thick) membrane of the food vacuole is involved in removal of water from the vacuole, in transfer of cytoplasmic fluid into the vacuole and in removal of digested products from the vacuole by pinocytosis.

Just exactly what processes take place in the food vacuole is not known. Nirenstein (1925) noticed the appearance of numerous neutral red-stainable granules around the food vacuole which passed into its interior and regarded them as carriers of a tryptic ferment, while Roskin and Levinsohn (1926) demonstrated the oxidase reaction in these granules. Hopkins and Warner (1946) believe that the digestion of food in *Entamoeba histolytica* is brought about by enzymes carried to the food vacuoles by "digestive spherules" which arise at the periphery of the nucleus, apparently due to the action of the substances diffusing from the nucleus into the cytoplasm. Harinasuta and Maegrath (1958) demonstrated the production of proteolytic enzyme in the cultures of two strains of *Entamoeba histolytica*.

As to the localization or distribution of enzymes within protozoan body, definite information is not yet available. In centrifuged *Amoeba proteus*, Holter and Kopac (1937) found the peptidase activity independent of all cytoplasmic inclusions that were stratified by centrifugal forces. Holter and Løvtrup (1949) found peptidase in centrifuged *Pelomyxa carolinensis* comparatively evenly distributed after centrifugation, possibly with a tendency to be concentrated in the lighter half, while proteinase was largely localized in the heavier half in which cytoplasmic granules were accumulated, and concluded that these two enzymes are bound, at least in part, to different cytoplasmic components. Holter and Lowy (1959) found acid phosphatase localized mainly in granular constituents. In *Stylonychia pustulata*, Hunter (1959) reported that aconitase was confined to mitochondria, zymohexase in cytoplasm and probably also in mitochondria, and phosphatase, lipase, urease, were localized in mitochondria. Predatory protozoa may utilize digestive enzymes of the victims. Doyle and Patterson (1942) noted *Didinium nasutum* utilizing dipepsidase in the paramecia which it ingested. A number of enzymes have been reported to occur in protozoa, some of which are listed in Table 3.

TABLE 3

ENZYMES IN PROTOZOA

Protozoa	*Enzymes*	*Observers*
Amoeba proteus	Peptidase	Holter and Kopac (1937); Holter and Doyle (1938); Andresen and Holter (1949); Holter and Løvtrup (1949)
	Proteinase	Andresen and Holter (1949) Holter and Løvtrup (1949)
	Amylase	Holter and Doyle (1938a)
A. dubia	Lipolytic substance	Dawson and Belkin (1928)
Pelomyxa palustris	Diastatic enzyme	Hartog and Dixon (1893); Stolc (1900)
	Pepsin-like enzyme	Hartog and Dixon (1893)
	Peptidase	Andresen and Holter (1949)
	Proteinase	"
P. carolinensis	Peptidase	"
	Proteinase	"
	Succinic dehydrogenase	Andresen, Engel and Holter (1951)
	Lipase	Wilber (1946)
Soil amoeba	"Amoebo-diastase," a trypsin-like enzyme	Mouton (1902)
Euglena gracilis	Proteolytic enzyme	Jahn (1931)
Trichomonas vaginalis	Lactic dehydrogenase	Kupferberg *et al.* (1953)
	Hexokinase	Wirtschafter and Jahn (1954)
	Aldolase	Wirtschafter and Jahn (1954)
Xylophagous Poly- and Hypermastigida	Cellulase	Trager (1932)
	Cellobiase	Cleveland *et al.* (1934)
Didinium nasutum	Dipeptidase	Doyle and Patterson (1942)
Tetrahymena pyriformis	Proteolytic enzyme	Lwoff (1932); Lawrie (1937)
	Peptidases	Kidder and Dewey (1951)
	Acetylcholinesterase	Seaman and Houlihan (1951)
	Urease	Seaman (1954)
Colpidium striatum	Proteolytic enzyme	Elliott (1933)
Paramecium caudatum	Peptidase	Holter and Doyle (1938)
	Amylase	"
P. multimicronucleatum	Dipeptidase	Doyle and Patterson (1942)
Frontonia sp.	Peptidase	Holter and Doyle (1938)
	Amylase	"
Balantidium coli	Diastase	Glaessner (1908)

These findings suffice to indicate that the digestion in protozoa is carried on by enzymes and its course appears to vary among different protozoa. The albuminous substances are digested and decomposed into simpler compounds by enzymes and absorbed by the surrounding cytoplasm. The power to digest starch into soluble sugars is widely found among various protozoa. It has been reported in Mycetozoida, Foraminiferida, Pelomyxa, Amoeba, Entamoeba, Ophryoscolecidae and other ciliates by several investigators.

The members of Vampyrella (p. 500) are known to dissolve the cellulose wall of algae, especially Spirogyra in order to feed on their contents. Pelomyxa (Stolc), Foraminiferida (Schaudinn), Amoeba (Rhumbler), Hypermastigida, Polymastigida (Cleveland), etc., have also been known for possessing the power of cellulose digestion. Many of the Hypermastigida and Polymastigida which lead symbiotic life in the intestine of the termite and of the wood roach, as demonstrated by Cleveland and his co-workers, digest by enzymes the cellulose which the host insect ingests. The assimilation products produced by an enormous number of these flagellates are seemingly sufficient to support the protozoa as well as the host. The ciliate commensals inhabiting the stomach of ruminants also apparently digest the cellulose, since the faecal matter as a rule does not contain this substance (Becker *et al.*, 1930; Weineck, 1934).

Dawson and Belkin (1928) injected oil into *Amoeba dubia* and found 1.4 to 8.3 percent digested. Mast (1938) reported that the neutral fat globules of Colpidium are digested by *Amoeba proteus* and transformed into fatty acid and glycerine which unite and form neutral fat. Chen (1950) found that when *Peranema trichophorum* was fed on almond oil (stained dark blue with Sudan black), Sudan III-stainable droplets gradually increased in number in five to ten hours, while ingested oil-droplets decreased in size, and considered that the droplets were "fat-substances" resynthesized from products of digestion of almond oil by this flagellate. The digestion of rice starch is followed by the appearance of increasing number of ovoid paramylum granules, and the digestion of casein results in the formation of oil droplets and paramylum bodies. (Food vacuoles, Kitching, 1956a).

As to how the digested substances in the food vacuoles are incorporated into the cytoplasm, Mast and Doyle (1935) believed that in *Amoeba proteus* the digested foods become transformed into fat globules, refringent bodies and crystals within the food vacuoles which later become distributed throughout the body. On the other hand, Andresen *et al.* (1950) maintain that the products of digestion are taken up by the cytoplasm in which various inclusion bodies are found.

In certain Sarcodina such as Amoeba and Pelomyxa, **refringent bodies** occur conspicuously in the cytoplasm. They were first noticed in *Pelomyxa palustris*, by Greeff (1874), who called them "Glanzkörper." Stolc (1900) and Leiner (1924) considered them as glycogen enclosed within a membrane and associated intimately with the carbohydrate metabolism of the organism, since their number was proportionate to the amount of food obtained by the organism. Veley (1905) on the other hand found them albuminoid in nature. Studies of the refringent bodies in *Amoeba proteus* led Mast and Doyle (1935, 1935a) to conclude that the outer layer is composed of a protein stroma impregnated with lipid containing fatty acid, which gives positive reaction for Golgi substance; the envelope is made up of a carbohydrate which is neither starch nor glycogen; and the refringent bodies function as reserve food, since they disintegrate during starvation. The same function was assigned to those occurring in *Pelomyxa carolinensis* by Wilbur (1945, 1945a), but Andresen and Holter (1945) do not agree with this view, as they observed the number of the refringent bodies ("heavy spherical bodies") remains the same in starvation.

The indigestible residue of the food is extruded from the body. The extrusion may take place at any point on the body surface in many Sarcodina by a reverse process of the ingestion of food. But in pellicle-bearing forms, the defecation takes place either through the cytopyge located in the posterior region of the body or through an aperture to the vestibule (Fig. 39, *b-d*). Permanent cytopyge is lacking in some forms. In *Fabrea salina*, Kirby (1934) noticed that a large opening is formed at the posterior end, the contents of food vacuoles are discharged, and the opening closes over. At first the margin of the body is left uneven, but soon the evenly rounded outline is restored. The same seems to be the case with Spriostomum (Fig. 40), Blepharisma, etc. (Cytopyge, Klein, 1939).

Saprozoic Nutrition. In this nutrition, the organisms utilize dis-

solved organic and inorganic substances. These substances are
simple compounds which originate in animal or vegetable matter
due to the decomposing activities of microorganisms. The substances are seemingly absorbed through the body surface and involves no special organelles. Perhaps the only instance this is said
to be carried on by a special organelle is the pusule in marine
dinoflagellates which, according to Kofoid and Swezy (1921),
appears to contain decomposed organic matter and to aid the or-

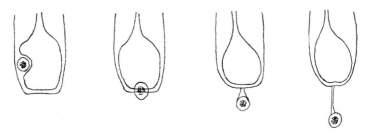

FIG. 40. Outline sketches showing the defecation process in
Spirostomum ambiguum (Blättner).

ganism in saprozoic nutrition. Numerous free-living flagellates
nourish themselves by this method.

The protozoa which live within the body of another organism
are able to nourish themselves by absorbing the digested, cell or
tissue, substances of the host and thus live saprozoically. Coelozoic protozoa belong for the most part to this group, as, for example,
astomatous ciliates, trypanosomes, etc. In Cnidosporidia, the host
cytoplasm is presumably hydrolyzed by enzymes before being
consumed by them. The parasitic protozoa which actually feed on
host's tissue cells such as *Entamoeba histolytica, Balantidium coli*,
etc., employ, of course, the holozoic nutrition.

Tentrahymena pyriformis and other species of the genus are
holozoic and saprozoic in nature, but they have become saprozoic
in liquid synthetic culture media free from any other organism.
Their food requirements are inorganic salts, amino acids, nucleic
acids, carbon source and growth factors (vitamins) (p. 116) and
are similar to those of higher metazoa in several respects. It appears that some protozoa employ, or are capable of carrying on,

more than one type of nutrition at the same or different times. This is called **mixotrophic nutrition** (Pfeiffer). One of the most striking examples is found in the species of the chrysomonad Ochromonas which are able to utilize all three types of nutrition (Pringsheim, 1952; Aaronson and Baker, 1959). (Nutritional requirements, Lwoff, 1951; Hutner and Lwoff, 1955.)

Differences in the kind and amount of food material may bring about conspicuous differences in form, size and structure in protozoa. For example, Kidder, Lilly and Claff (1940) noted in *Tetrahymena vorax* (Fig. 41), bacteria-feeders are tailed (50-75μ long), saprozoic forms are fusiform to ovoid (30-70μ long), forms feeding on sterile dead ciliates are fusiform (60-80μ long), and carnivores and cannibals are irregularly ovoid (100-250μ long), in the latter form of which a large "preparatory vacuole" becomes developed. In *Chilomonas paramecium,* Mast (1939) observed the individuals grown in sterile glucose-peptone solution were much smaller than those cultured in acetate-ammonium solution and moreover the former contained many small starch grains, but no fat, while the latter showed many larger starch grains and a little fat. *Amoeba proteus* when fed exclusively on Colpidium, became very large and extremely "fat" and sluggish, growing and multiplying slowly, but indefinitely; when fed on Chilomonas only, they grew and multiplied for several days, then decreased in number and soon died, but lived longer on Chilomonas cultured in the glucose-peptone. It is well known that protozoa as any other organism, show atypical or abnormal morphological and physiological peculiarities. In the case of carnivorous forms, the condition of food organisms may produce abnormalities in them, as was shown by Beers (1933) in Didinium fed on starved paramecia (Fig. 42) and Burbanck and Eisen (1960) noted that Didinium showed a high division rate when fed on *Paramecium aurelia* raised on a mixture of bacteria, but when fed on starved paramecia or on those raised on one of the five species of bacteria, the division rate fell, abnormalities appeared and died after three or four days. It was thought that this was possibly due to the lack of some substances related to the enzyme system.

Rudzinska (1953) finds that when *Tokophrya infusionum* are

fed on *Tetrahymena pyriformis* continuously for some forty-eight hours, giant individuals that are 100 times or more in body size as compared with individuals before heavy feeding began, appeared. She found that such individuals with short or no tenta-

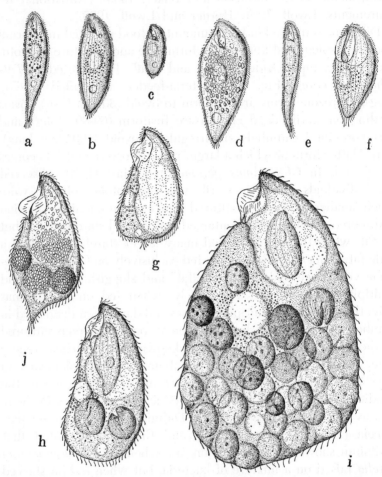

Fig. 41. Form and size variation in *Tetrahymena vorax*, due to differences in kind and amount of food material, as seen in life, ×400 (Kidder, Lilly and Claff). a, bacteria-feeder; b, c, saprozoic forms; d, individual which has fed on killed *Colpidium campylum;* e, starved individual from a killed-Colpidium culture; f-i, progressive form and size changes of saprozoic form in the presence of living Colpidium; j, a young carnivore which has been removed to a culture with living yeast.

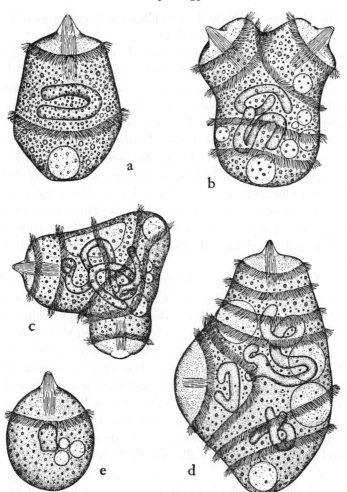

Fig. 42. *Didinium nasutum,* ×265 (Beers). a, normal fully grown animal; b-e, abnormal organisms which were fed on starved Paramecium.

cles, lose reproductive capacity rapidly; thus their life span became greatly shortened. But when the suctorian is kept for some time on starvation diet or at low temperature (4°C.), the animals continued to reproduce. She considered that a continuous and heavy feeding results in accumulation of catabolic products and thus in auto-intoxication; on the other hand, a population kept on a meager diet with intermittent starvation, would result in survival of the vigorous individuals which have a longer span of life in spite of inadequate food material.

Krascheninnikow (1959) found several "monsters" of *Balantidium coli* that had been cultured for four months in Dobell-Laidlaw's medium, and considered it probable that addition of acriflavine which is mutagenic for certain bacteria or gentian violet which inhibits division of sea urchin eggs, to the culture medium may have produced the abnormalities. In contrast to these observations, Wenzel (1955) obtained giant forms from a new amicronucleate *Spathidium ascendens* which were fed on *Colpidium campylum* and maintained that the gigantism develops spontaneously and is fixed genetically. These giant forms to which he gave a new species name *S. polymorphum*, show morphological and physiological alterations as compared with *S. ascendens.*

Some forty years ago, Robertson (1921-1927) reported that when two ciliates, Enchelys and Colpoda, are placed in a small amount of fresh culture medium, the rate of reproduction following a "lag period" is more than twice (up to ten times) that of a single animal in the same amount of the medium. He assumed that this acceleration was due to a certain agent or substance produced within the animal, which diffused into the culture medium. When more than one animal is confined in a limited amount of culture fluid, this substance is present in a higher concentration than with one animal, and an increased rate of division is the result. Robertson called this "allelocatalytic result," and the phenomenon, "allelocatalysis."

Soon a large number of observers came forward with varying results—some confirmatory, others contradictory. The vast majority of these observations including Robertson's own, were carried on ciliates which were grown in association with various bacteria, and naturally, the results lacked agreement. For a review of these observations, too numerous to mention here, the reader is referred to Allee (1931, 1934), Mast and Pace (1938) and Richards (1941). When bacteria-free cultivation became possible for some protozoa, it was hoped that this problem might be solved under controlled conditions. However, the results still lack agreement. For example, Phelps (1935) reported that in Tetrahymena (Glaucoma), the growth rate and the maximum yield were the same between two cultures: one started with 0.014 organism and the other, with 1600 organisms per ml. Thus there was no allelocatalysis. On the other hand, Mast and Pace (1938) noted a significant acceleration of the growth rate in Chilomonas when up to 50 organisms were inoculated into 0.4 cc. of culture fluid as compared to the growth rate in cultures with one or more Chilomonas inocula, and furthermore, a single Chilomonas showed an increased rate of reproduction as the volume of the culture fluid was reduced.

Respiration

In order to carry on manifold vital activities, the protozoa, as all other organisms, must transform the potential energy stored in highly complex compounds present in the cytoplasm into various forms of active energy by oxidation. They utilize carbohydrates, fats and proteins as energy sources through varied processes involving many enzymes. The oxygen involved in these processes appears to be brought into contact with the substance in one of the two ways. The great majority of free-living and certain parasitic forms absorb free molecular oxygen from the surrounding media. The absorption is carried on by the permeable body surface, since there seems no special organelle for it.

The polysaprobic protozoa are known to live in water containing no free oxygen. For example, Noland (1927) observed *Metopus es* in a pool, six feet in diameter and eighteen inches deep, filled with fallen leaves, which gave a strong odor of hydrogen sulphide. The water in it showed pH 7.2 at 14°C., and contained no dissolved oxygen, but 14.9 cc per liter of free carbon dioxide and 78.7 cc per liter of fixed carbon dioxide. Parasitic protozoa of metazoan digestive system live also in a medium containing no molecular oxygen. All these forms appear to possess the capacity to split complex oxygen-bearing substances to produce necessary oxygen.

The oxygen consumption of protozoa appears to vary not only among different species, but also among individuals of the same species. The respiratory quotient (R.Q.) or the ratio of volume of carbon dioxide produced to volume of oxygen consumed, has been used to find the type of material being oxidized, as definite R.Q. values are recognized for the main types of food substances: 1.0 for carbohydrates, 0.7 for fats, and 0.8 for proteins. If R.Q. value lies between 0.7 and 1.0, both fats and carbohydrates are being burned and R.Q. value above 1.0 may indicate synthesis and storage of fat derived from carbohydrates as was found by Pace and Lyman (1947) in young cultures of Tetrahymena. Green flagellates show generally low quotients, which had been interpreted as due to synthesis of carbohydrates from carbon dioxide (Mast *et al.*, 1936). The R.Q. is subject to variation further according to physiological state, stages in developmental cycle, culture media, temperature, etc.

A single *Paramecium caudatum* is said to consume in one hour at 21°C. from 0.0052 cc (Kalmus) to 0.00049 cc (Howland and Bernstein) of oxygen. The oxygen consumption of this ciliate in heavy suspensions (3×10^3 to 301×10^3 in 3 cc) and associated bacteria, ranged, according to Gremsbergen and Reynaerts-De Pont (1952), from 1000 to 4000 nM^3 per hour per million individuals at 23.5°C. The two observers considered that *P. caudatum* possesses a typical cytochrome-oxidase system. *Amoeba proteus*, according to Hulpieu (1930), succumbs slowly when the amount of oxygen in water is less than 0.005 per cent and also in excess, which latter confirms Pütter's observation on Spirostomum. According to Clark (1942), a normal *Amoeba proteus* consumes 1.4×10^{-3} mm^3 of oxygen per hour, while an enucleated amoeba only 0.2×10^{-3} mm^3. He suggests that "the oxygen-carriers concerned with 70 per cent of the normal respiration of an amoeba are related in some way to the presence of the nucleus."

In *Pelomyxa carolinensis*, the rate of oxygen consumption at 25°C. was found by Pace and Belda (1944) to be 0.244 ± 0.028 mm^3 per hour per mm^3 cell substance and does not differ greatly from that of *Amoeba proteus* and *Actinosphaerium eichhorni*. The temperature coefficient for the rate of respiration is nearly the same as that in Paramecium, varying from 1.7 at 15-25°C. to 2.1 at 25-35°C. Pace and Kimura (1946) further note in *Pelomyxa carolinensis* that carbohydrate metabolism is greater at higher than at lower temperature and that a cytochrome-

cytochrome oxidase system is the mechanism chiefly involved in oxidation of carbohydrate.

Several investigators studied the influence of abundance or lack of oxygen upon different protozoa. For example, Pütter (1905) demonstrated that several ciliates reacted differently when subjected to anaerobic condition, some perishing rapidly, others living for a considerable length of time. Death is said by Löhner to be brought about by a volume-increase due to accumulation of the waste products. When first starved for a few days and then placed in anaerobic environment, Paramecium and Colpidium died much more rapidly than unstarved individuals. Pütter, therefore, supposed that the difference in longevity of aerobic protozoa in anaerobic conditions was correlated with that of the amount of reserve food material such as protein, glycogen and paraglycogen present in the body. Pütter further noticed that Paramecium is less affected by anaerobic condition than Spirostomum in a small amount of water, and maintained that the smaller the size of body and the more elaborate the contractile vacuole system, the organisms suffer the less the lack of oxygen in the water, since the removal of catabolic products depends upon these factors

Wittner (1957, 1957a) subjected *Paramecium caudatum* to varying oxygen pressure at 1-27°C. and found that above 5°C. oxygen toxicity varied directly with pressure, but below this temperature, oxygen toxicity varied inversely as the temperature. He suggested that oxygen may be affecting two cellular processes, one of which is temperature limited below 5°C. and would decrease the death time as the temperature is lowered. Resistance to exposure to ozone differs among different species, but all become sluggish, followed by surface vesiculation and cytolysis (Giese and Christensen, 1954).

The Hypermastigida of termites are killed, according to Cleveland (1925), when the host animals are kept in an excess of oxygen. Jahn found that *Chilomonas paramecium* in bacteria-free cultures in heavily buffered peptone-phosphate media at pH 6.0, required for rapid growth carbon dioxide which apparently brings about a favorable intracellular hydrogen-ion concentration,

(Respiratory metabolism, Meldrum, 1934; Jahn, 1941; Seaman, 1955; Hutner and Provasoli, 1955).

Reserve Food Matter

The anabolic activities of protozoa result in the growth and increase in the volume of the organism, and also in the formation and storage of reserve food-substances which are deposited in the cytoplasm to be utilized later for growth or reproduction. The reserve food stuff varies according to types of nutrition. In holozoic and saprozoic protozoa, it is ordinarily glycogen or glycogenous substances. Thus, in saprozoic Gregarinida, there occur in the cytoplasm numerous refractile bodies which stain brown to brownish-violet in Lugol's solution, are insoluble in cold water, alcohol, and ether, become swollen and later dissolved in boiling water, and are reduced to a sugar by boiling in dilute sulphuric acid. It appears to be closely related to glycogen, but differs from it as it does not stain with Best's carmine and is not digested by amylase. Bütschli called them *paraglycogen bodies*. Göhre considered them as stabilized polymerization product of glycogen.

The glycogen is found very commonly in Sarcodina. It abounds usually in *Pelomyxa palustris* (Leiner, 1924; Waldner, 1956) and in Actinosphaerium (Rumjantzew and Wermel, 1925), etc. In the cysts of *Iodamoeba bütschlii*, glycogen body is always conspicuously present and is looked upon as a characteristic feature of the organism. The iodinophile vacuole of the spores of Myxobolidae is a well-defined vacuole containing glycogenous substance and is also considered as possessing a taxonomic value. In many ciliates, both free-living (Paramecium, Glaucoma, Vorticella, Stentor, etc.) and parasitic (Ophryoscolecidae, Nyctotherus, Balantidium (Fauré-Fremiet and Thaureaux, 1944)), glycogenous bodies are always present. According to MacLennan (1936), the development of the paraglycogen in Ichthyophthirius is associated with the mitochondria. In *Eimeria tenella*, glycogenous substance does apparently not occur in the schizonts, merozoites, or microgametocytes; but becomes apparent first in the macrogametocyte, and increases in amount with its development, a small amount being demonstrable in the sporozoites (Edgar *et al.*, 1944).

The anabolic products of the holophytic nutrition are starch, paramylum, oil and fats. The **paramylum** bodies are of various forms among different species, but appear to maintain a certain characteristic form within a species and can be used to a certain extent in taxonomic consideration. According to Heidt (1937), the paramylum of *Euglena sanguinea* (Fig. 43) is spirally coiled

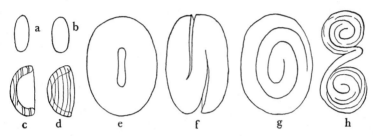

FIG. 43. a-d, two types of paramylum present in *Euglena gracilis* (Bütschli); *e-h*, paramylum of *E. sanguinea*, × 1100 (Heidt). (e, natural appearance; f, g, dried forms; h, strongly pressed body.)

which confirms Bütschli's observation. The paramylum appears to be a polysaccharide which is insoluble in boiling water, but dissolves in concentrated sulphuric acid, potassium hydroxide, and slowly in formaldehyde. It does not stain with either iodine or chlor-zinc-iodide and when treated with a dilute potassium hydroxide, the paramylum bodies become enlarged and frequently exhibit a concentric stratification.

In the Chrysomonadida, the reserve food material is in the form of refractile spheroid bodies which are known as **leucosin,** probably a carbohydrate which when boiled in water stains with iodine. **Oil** droplets occur in various protozoa and when there is a large number of oil-producing forms in a body of water, the water may develop various odors as indicated in Table 4.

Fats occur widely in protozoa. They appear usually as small refractile globules. Zingher (1934) found that in the Sarcodina and Ciliata he studied, each species showed morphological characteristics of the fatty substance it contained. Fat globules occur abundantly in Amoeba and Pelomyxa which are easily seen by staining with Sudan III. In *Tillina canalifera*, fat droplets, 1-2μ in diameter, are present especially in the region to the right of the cyto-

TABLE 4

PROTOZOA AND ODORS OF WATER

Protozoa	*Odor Produced*
Cryptomonas	candied violets
Mallomonas	aromatic, violets, fishy
Synura	ripe cucumber, muskmelon, bitter and spicy taste
Uroglenopsis	fishy, cod-liver oil-like
Dinobryon	fishy, like rockweed
Chlamydomonas	fishy, unpleasant or aromatic
Eudorina	faintly fishy
Pandorina	faintly fishy
Volvox	fishy
Ceratium	vile stench
Glenodinium	fishy
Peridinium	fishy, like clam-shells
Bursaria	Irish moss, salt marsh, fishy (Whipple, 1927)
Pelomyxa	ripe cucumber (Schaeffer, 1937)

pharynx (Turner, 1940). According to Panzer (1913), the fat content of *Eimeria gadi* was 3.55 per cent and Pratje (1921) reports that 12 per cent of the dry matter of *Noctiluca scintillans* appeared to be the fatty substance present in the form of granules and is said to give luminescence upon mechanical or chemical stimulation. But the chemical nature of these "photogenic" granules is still unknown at present (Harvey, 1952).

Hastings and Sweeney (1957) noted that cell-free extracts of *Gonyaulax polyhedra* emit light only in the presence of high concentrations of salts such as sodium chloride and of oxygen, and show an optimum activity at 24°C. and pH 6.6. An enzyme, purified by ammonium sulphate fractionation, requires for luminescence a factor present in boiled crude extracts. The two authors suggested that this factor be designated as Gonyaulax **luciferin,** and held that the luminescence involves an oxidation of the luciferin. A number of other dinoflagellates, such as Peridinium, Ceratium, Gymnodinium, etc., also emit luminescence. In other forms the fat may be hydrostatic in function, as is the case with a number of pelagic Radiolarida, many of which are also luminous. (Luminescence in protozoa, Harvey, 1952).

Another reserve food-stuff which occurs widely in protozoa, excepting Ciliophora, is the so-called **volutin** or metachromatic granule. It is apparently equally widely present in protophyta. In fact, it was first discovered in the protophytan *Spirillum volutans*. Meyer coined the name and held it to be made up of a nucleic acid. In *Trypanosoma cruzi*, ribonucleic acid occurs in volutin granules (von Brand *et al.*, 1959). It stains deeply with nuclear dyes. Reichenow (1909) demonstrated that if *Haematococcus pluvialis* (Fig. 44) is cultivated in a phosphorus-free medium, the

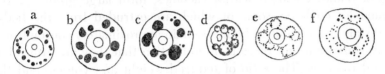

Fig. 44. *Haematococcus pluvialis*, showing the development of volutin in the medium rich in phosphorus and its disintegration in an exhausted medium, ×570 (Reichenow). a, second day; b, third day; c, fourth day; d, e, sixth day; f, eighth day.

volutin is quickly used up and does not reappear. If however, the organisms are cultivated in a medium rich in phosphorus, the volutin increases greatly in volume and, as the culture becomes old, it gradually breaks down. In *Polytomella agilis* (Fig. 117, c, d), Doflein (1918) showed that an addition of sodium phosphate resulted in an increase of volutin. Reichenow, Schumacher, and others, hold that the volutin appears to be a free nucleic acid, and is a special reserve food material for the nuclear substance. Sassuchin (1935) studied the volutin in *Spirillum volutans* and *Sarcina flava* and found that the volutin appears during the period of strong growth, nourishment and multiplication, disappears in unfavorable condition of nourishment and gives a series of characteristic carbohydrate reactions. Sassuchin considers that the volutin is not related to the nucleus, but is a reserve food material of the cell, and is composed of glycoprotein. (Volutin, Jírovec, 1926).

Starvation. As in all living things, when deprived of food, protozoa perish sooner or later. The changes noticeable under the microscope are: gradual loss of cytoplasmic movement, increasing number of vacuoles and their coalescence, and finally the disintegration of the body. In starved *Pelomyxa carolinensis*, Andresen and Holter (1945) noticed the following changes: the animals disintegrate in ten to twenty-five days at 22°C.; body volume decreases particularly during the early days of starvation and is about twenty to thirty per cent of the initial volume at the time of death; food vacuoles are extruded from the body in twenty-four to forty-eight hours; the cytoplasm becomes less viscous and many fluid vacuoles make their appearance; crystals and refringent bodies enclosed within vacuoles, form large groups as the vacuoles coalesce, some of which are extruded from the body; crystals and refringent bodies remain approximately constant during starvation and there is no indication that they are utilized as food reserves. The ratio of reduced weight and volume and the specific gravity remain reasonably constant during starvation (Zeuthen, 1948). Andresen (1945) found starved *Amoeba proteus* to show a similar change on the whole, except that the number of mitochondria decreased and in some cases dissolution of crystals occurred just before disintegration.

Excretion and Secretion

The catabolic waste material composed of water, carbon dioxide, and nitrogenous compounds such as urea, uric acid or ammonia, pass out of the body by diffusion through the surface or by way of the contractile vacuole. The protoplasm of the protozoa is generally considered to possess a molecular make-up which appears to be similar among those living in various habitats. In the freshwater protozoa the body of which is hypertonic to surrounding water, the water diffuses through the body surface and so increases the water content of the body protoplasm and interfere with its normal function. The contractile vacuole, which is invariably present in all freshwater forms, is the means of getting rid of this excess water from the body. On the other hand, marine or parasitic protozoa live in nearly isotonic media and there is no excess of water entering the body, hence the contractile vacuoles are not found in them. Just exactly why all ciliates and suctorians possess the contractile vacuole regardless of habitat, has not fully been explained. It is assumed that the pellicle of the ciliate is impermeable to salts and slowly permeable to water (Kitching,

1936). Tartar (1954) showed that in fragments of Paramecium lacking mouth and gullet, pulsation of the contractile vacuole continued, though it was slower than normal rate, which indicates that some water passes through pellicle

That the elimination of excess amount of water from the body is one of the functions of the contractile vacuole appears to be beyond doubt judging from the observations of Zuelzer (1907), Finley (1930) and others, on *Amoeba verrucosa* which lost gradually its contractile vacuole as sodium chloride was added to the water, losing the organella completely in the seawater concentration. Furthermore, marine amoebae develop contractile vacuoles *do novo* when they are transplanted to fresh water as in the case of *Vahlkampfia calkinsi* (Hogue, 1923) and *Amoeba biddulphiae* (Zuelzer, 1927). After studying an x-ray induced mutant of *Chlamydomonas moewusii* without contractile vacuoles, Guillard (1960) concluded that water elimination is the sole essential function of the contractile vacuole in this organism. Herfs (1922) studied the pulsation of the contractile vacuoles of *Paramecium caudatum* in fresh water as well as in salt water and obtained the following measurements:

Per cent NaCl in water	0	0.25	0.5	0.75	1.00
Contraction period in second	6.2	9.3	18.4	24.8	163.0
Excretion per hour in body volumes	4.8	2.82	1.38	1.08	0.16

The number of the contractile vacuoles present in a species is constant under normal conditions. The contraction period varies from a few seconds to several minutes in freshwater inhabitants, and is, as a rule, considerably longer in marine protozoa. Kitching (1938a) estimated that a quantity of water equivalent to the body volume is eliminated by freshwater protozoa in four to 45 minutes and by marine forms in about three to four hours. The size of contractile vacuole in diastole may vary. Botsford (1926) reported that the contractile vacuole in *Amoeba proteus* varied considerably within a short period of time in size and rate of contraction under seemingly identical conditions. The rate of contraction is subject to change with the temperature, physiological state of the organism, amount of food substances, etc. For example, Rossbach noted in the three ciliates listed below, the contraction was accelerated first rapidly and then more slowly with rise of the temperature:

	Time in seconds between two systoles at different temperature (C.)					
	5°	10°	15°	20°	25°	30°
Euplotes charon	61	48	31	28	22	23
Stylonychia pustulata	18	14	10–11	6–8	5–6	4
Chilodonella cucullulus	9	7	5	4	4	—

How much water enters through the body surface of protozoa is difficult to determine. In *Pelomyxa carolinensis,* 2 to 4 per cent of the total volume per hour of water enters through the body surface (Løvtrup and Pigón, 1951). Water also enters the protozoan body in food vacuoles. In *Vampyrella lateritia* which feeds on the cell contents of Spirogyra in a single feeding, many contractile vacuoles appear within the cytoplasm and evacuate the water that has come in with the food (Lloyd, 1926) and the members of Ophryoscolecidae show an increased number and activity of contractile vacuoles while feeding (MacLennan, 1933). The amount of water contained in food vacuoles seems, however, to be far smaller than the amount evacuated by contractile vacuoles (Gelei, 1925; Eisenberg, 1925). Other evidences such as the contractile vacuole continues to pulsate when cytostome-bearing protozoa are not feeding and its occurrence in astomatous ciliates, would indicate also that the water entering through this avenue is not of a large quantity. In Suctoria, the contractile vacuole pulsates faster at the time of ingesting the protoplasm of the prey and thus apparently aid in feeding by eliminating water from the body (Kitching, 1956; Hull, 1954). How much water is produced during the metabolic activity of the organisms is unknown, but it is considered to be a very small amount (Kitching, 1938). The mechanism by which the difference in osmotic pressure can be maintained at the body surface is unknown. It may be, as suggested by Kitching (1934), that the contractile vacuole extrudes water but retains the solutes or some osmotically active substances must be continuously produced within the body.

Detection of catabolic products present in the contractile vacuole, in the body protoplasm or in the culture medium is still a difficult task. Weatherby (1927) found in the spring water in which he kept a number of thoroughly washed Paramecium, urea and ammonia after thirty to thirty-six hours and supposed that the urea excreted by the organisms gave rise to ammonia. He found also urea in similar experiments with Spirostomum and Didinium (Weatherby, 1929). Doyle and Harding (1937) found Glaucoma excreting ammonia, and not urea. In Tetrahymena, Seaman (1955, 1959) reported that urea is produced in growing

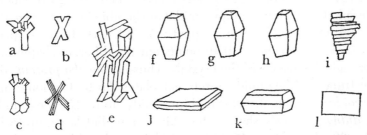

FIG. 45. Examples of crystals present in protozoa. a–e, in *Paramecium caudatum* (Schewiakoff), (a–d, ×1000, e, ×2600); f, in *Amoeba proteus;* g, in *A. discoides;* h–1, in *A. dubia* (Schaeffer).

cultures, but in older cultures only ammonia is recovered, and urea is no longer present, while Dewey *et al.* (1959) found neither urea nor urease activity in this ciliate. Carbon dioxide is obviously excreted by the body surface as well as by the contractile vacuole. (General reference, Weatherby, 1941; permeability of water in protozoa, Belda, 1942, Løvtrup and Pigón, 1951; physiology of contractile vacuole, Stempell, 1924, Fortner, 1926, Gaw, 1936, Kitching, 1938a, 1956).

Aside from the soluble substances, there often occur in the protozoan body insoluble substances in the forms of **crystals** and **granules** of various kinds. Schewiakoff (1894) first noticed that Paramecium often contained crystals (Fig. 45) composed of calcium phosphate, which disappeared completely in one or two days when the organisms were starved, and reappeared when food was given. Schewiakoff did not see the extrusion of these crystals, but considered that they were first dissolved and excreted by the contractile vacuoles, as they were seen collected around the vacuoles. When exposed to x-irradiation, the symbiotic Chlorella of *Paramecium bursaria* disappear gradually and crystals appear and persist in the cytoplasm of the ciliate (Wichterman, 1948a). These crystals varying in size from a few to 12μ, are found mainly in the posterior region of the body. Wichterman notes that the appearance or disappearance of crystals seems to be correlated with the absence or presence of symbiotic Chlorella and with the holozoic or holophytic (by the alga) nutrition of the organism.

In *Amoeba proteus,* Schubotz (1905) noted crystals of calcium phosphate which were bipyramidal or rhombic in form, were doubly refractile and measured about 2-5μ in length. In three species of Amoeba, Schaeffer (1920) points out the different shape, number and dimensions of the crystals. Thus in *Amoeba proteus,* they are truncate bipyramids, rarely flat plates, up to 4.5μ long; in A. *discoides,* abundant, truncate bipyramids, up to 2.5μ long; and in A. *dubia,* variously shaped (four kinds) few, but large, up to 10μ, 12μ, 30μ long (Fig. 45). Bipyramidal or plate-like crystals are especially abundant in *Pelomyxa illinoisensis* at all times (Kudo, 1951); the crystals of *P carolinesis* remain the same during the starvation of the organism (Andresen and Holter, 1945; Holter, 1950).

The crystals present in protozoa appear to be of varied chemical nature. Luce and Pohl (1935) noticed that at certain times amoebae in culture are clear and contain relatively a few crystals but, as the culture grows older and the water becomes more neutral, the crystals become abundant and the organisms become opaque in transmitted light. These crystals are tubular and six-sided, and vary in length from 0.5 to 3.5μ. They considered the crystals were composed of calcium chlorophosphate. Mast and Doyle (1935), on the other hand, noted in *Amoeba proteus* two kinds of crystals, plate-like and bipyramidal, which vary in size up to 7μ in length and which are suspended in alkaline fluid in viscous vacuoles. These two authors believed that the plate-like crystals are probably leucine, while the bipyramidal crystals consist of a magnesium salt of a substituted glycine. Other crystals are said to be composed of urate, carbonate, oxalate, etc. Griffin (1960) and Carlstrom and Møller 1961) report that these crystals probably consist of a tetragonal form of triuret (carbonyldiurea).

Another catabolic product is the **haemozoin** granules which occur in many haemosporidans and which appear to be composed of a derivative of the haemoglobin of the infected erythrocyte (p. 724). In certain Radiolarida, there occurs a brownish amorphous mass which is considered as catabolic waste material and, in Foraminiferida, the cytoplasm is frequently loaded with masses of brown granules which appear also to be catabolic waste and are extruded from the body periodically.

While intracellular secretions are usually difficult to recognize, because the majority remain in fluid form except those which produce endoskeletal structures occurring in Foraminiferida, Heliozoida, Radiolarida, certain parasitic ciliates, etc., the extracellular secretions are easily recognizable as loricae, shells, envelopes, stalks, collars, mucous substance, etc. Furthermore, many protozoa secrete, as was stated before, certain substances through the pseudopodia, tentacles or trichocysts which possess paralyzing or killing effect upon the preys.

Locomotion

Protozoa move about by means of pseudopodia, flagella or cilia, which may be combined with internal contractile elements.

Pseudopodial Movement. The movement by formation of pseudopodia, often called *amoeboid movement* has long been studied by numerous observers. The first attempt to explain the movement was made by Berthold (1886), who held that the difference in the surface tension was the cause of amoeboid movements, which view was supported by the observations and experiments of Bütschli (1894) and Rhumbler (1898). According to this view, when an amoeba forms a pseudopod, there probably occurs a diminution of the surface tension of the cytoplasm at that point, due to certain internal changes which are continuously going on within the body and possibly due also to external causes, and the internal pressure of the cytoplasm will then cause the streaming of the cytoplasm. This results in the formation of a pseudopod which becomes attached to the substratum and an increase in tension of the plasma-membrane draws up the posterior end of the amoeba, thus bringing about the movement of the whole body.

Jennings (1904) found that the movement of *Amoeba verruco-sa* (Fig 46, *a*) could not be explained by the surface tension theory, since he observed "in an advancing amoeba substance flows forward on the upper surface, rolls over at the anterior edge, coming in contact with the substratum, then remains quiet until the body of the amoeba has passed over it. It then moves upward at the posterior end, and forward again on the upper surface, continuing in rotation as long as the amoeba continues to progress."

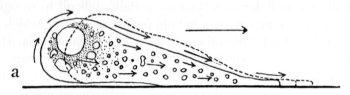

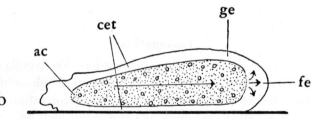

Fig. 46. a, diagram showing the movement of *Amoeba verrucosa* in side view (Jennings); b, a marine limax-amoeba in locomotion (Pantin from Reichenow). ac, area of conversion; cet, contracting ectoplasmic tube; fe, fluid ectoplasm; ge, gelated ectoplasm.

Thus *Amoeba verrucosa* may be compared with an elastic sac filled with fluid.

Dellinger (1906) studied the movement of *Amoeba proteus, A. verrucosa* and *Difflugia spiralis.* Studying in side view, he found that the amoeba (Fig. 47) extends a pseudopod, "swings it about, brings it into the line of advance, and attaches it" to the substratum and that there is then a concentration of the substance back of this point and a flow of the substance toward the anterior end. Dellinger held thus that "the movements of amoebae are due to the presence of a contractile substance," which was said to be located in the endoplasm as a coarse reticulum. Wilber (1946) pointed out that *Pelomyxa carolinensis* carries on a similar movement at times.

In the face of advancement of our knowledge on the nature of protoplasm, Rhumbler (1910) realized the difficulties of the surface tension theory and later suggested that the conversion of the ectoplasm to endoplasm and vice versa were the cause of the cytoplasmic movements, which was much extended by Hyman

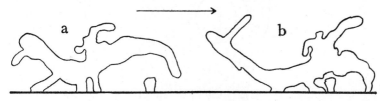

FIG. 47. Outline sketches of photomicrographs of *Amoeba proteus* during locomotion, as viewed from side (Dellinger).

(1917). Hyman considered that: (1) a gradient in susceptibility to potassium cyanide exists in each pseudopod, being the greatest at the distal end, and the most recent pseudopod, the most susceptible; (2) the susceptibility gradient (or metabolic gradient) arises in the amoebae before the pseudopod appears and hence the metabolic change which produces increased susceptibility, is the primary cause of pseudopodium formation; and (3) since the surface is in a state of gelation, amoeboid movement must be due to alterations of the colloidal state. Solation, which is brought about by the metabolic change, is regarded as the cause of the extension of a pseudopod, and gelation, of the withdrawal of a pseudopod and of active contraction. Schaeffer (1920) mentioned the importance of the surface layer which is a true surface tension film, the ectoplasm, and the streaming of endoplasm in the amoeboid movement.

Pantin (1923) studied a marine limax amoeba (Fig. 46, *b*) and came to recognize acid secretion and absorption of water at the place where the pseudopod was formed. This results in swelling of the cytoplasm and the pseudopod is formed. Because of the acidity, the surface tension increases and to lower or reduce this, concentration of substances in the "wall" of the pseudopod follows. This leads to the formation of a gelatinous ectoplasmic tube

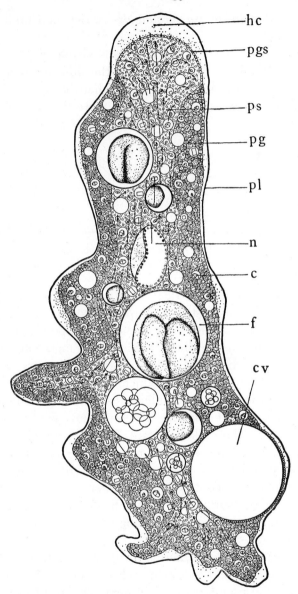

FIG. 48. Diagram of *Amoeba proteus*, showing the solation and gelation of the cytoplasm during amoeboid movement (Mast). c, crystal; cv, contractile vacuole f, food vacuole; hc, hyaline cap; n, nucleus pg, plasmagel; pgs, plasmagel sheet; pl, plasmalemma; ps, plasmasol.

which, as the pseudopod extends, moves toward the posterior region where the acid condition is lost, gives up water and contracts finally becoming transformed into endoplasm near the posterior end. The contraction of the ectoplasmic tube forces the endoplasmic streaming to the front.

This observation is in agreement with that of Mast (1923, 1926, 1931) who after a series of observations on *Amoeba proteus* came to hold that the amoeboid movement is brought about by "four primary processes; namely, attachment to the substratum, gelation of plasmasol at the anterior end, solation of plasmagel at the posterior end and the contraction of the plasmagel at the posterior end" (Fig. 48). As to how these processes work, Mast states: "The gelation of the plasmasol at the anterior end extends ordinarily the plasmagel tube forward as rapidly as it is

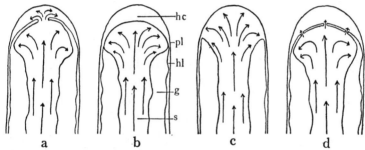

Fig. 49. Diagrams of varied cytoplasmic movements at the tip of a pseudopodium in *Amoeba proteus* (Mast). g, plasmagel; hc, hyaline cap; hl, hyaline layer; pl, plasmalemma; s, plasmasol.

broken down at the posterior end by solation and the contraction of the plasmagel tube at the posterior end drives the plasmasol forward. The plasmagel tube is sometimes open at the anterior end and the plamsasol extends forward and comes in contact with the plasmalemma at this end (Fig. 49, *a*), but at other times it is closed by a thin sheet of gel which prevents the plasmasol from reaching the anterior end (*b*). This gel sheet at times persists intact for considerable periods, being built up by gelation as rapidly as it is broken down by stretching, owing to the pressure of the plasmagel against it. Usually it breaks periodically at various places. Sometimes the breaks are small and only a few granules of plasmasol pass through and these gelate immediately and close the openings (*d*). At other times, the breaks are large and plasmasol streams through, filling the hyaline cap (*c*), after which the sol adjoining the plasmalemma gelates forming a new gel sheet. An amoeba is a turgid system, and the plasmagel is under continuous tension. The plasmagel is elastic and, consequently, is pushed out at the region where its elasticity is weakest and this results in pseudopodial formation. When an amoeba is elongated and undergoing movement, the elastic strength of the plasmagel is the highest at its sides, lowest at the anterior end and intermediate at the posterior end, which results in continuity of the elongated form and in extension of the anterior end. If pressure is brought against the anterior end, the direction of streaming of plasmasol is immediately reversed, and a new hyaline cap is formed at the posterior end which is thus changed into a new anterior end." The rate of amoeboid locomotion appears to be influenced by environmental factors such as pH, osmotic pressure, salt concentration, substratum, temperature, etc. (Mast and Prosser, 1932).

The sol and gel changes were, Mast held, associated with the changes in surface permeability and in water content or tonicity. Goldacre and Lorch (1950) consider that these changes are caused by folding (solation) and unfolding (gelation) of the protein molecules, which is brought about by energy that is supplied from adenosin triphosphate (ATP). The two authors injected ATP into the posterior region of amoebae and found locomotion accelerated. As Noland (1957) pointed out, there appears to be a similarity between contraction of vertebrate muscle and pseudopod formation. (Structure and chemistry of plasmalemma, Bairati and Lehmann, 1953; amoeboid movements, Noland, 1957.)

Flagellar Movement. Flagellum in motion is in a few instances observable as in Peranema, but in most cases it is so rapid that it is very difficult to see in life under an ordinary microscope. It can easily be seen under a phase or darkfield microscope. The first explanation of the flagellar movement was advanced by Bütschli who observed that the flagellum undergoes a series of lateral movements and, in so doing, a pressure is exerted on the water at right angles to its surface. This pressure can be resolved into two forces: one directed parallel, and the other at right angles, to the main body axis. The former will drive the organism forward, while the latter will tend to rotate the animal on its own axis.

Gray (1928), who gave an excellent account of the movement of flagella, points out that "in order to produce propulsion there must be a force which is always applied to the water in the same direction and which is independent of the phase of lateral movement. There can be little doubt that this condition is satisfied in flagellated organisms not because each particle of the flagellum is moving laterally to and fro, but by the transmission of the waves from one end of the flagellum to the other, and because the direction of the transmission is always the same. A stationary wave, as apparently contemplated by Bütschli, could not effect propulsion since the forces acting on the water are equal and opposite during the two phases of the movement. If however the waves are being transmitted in one direction only, definite propulsive forces are present which always act in a direction opposite to that of the waves."

Because of the nature of the flagellar movement, the actual process has often not been observed. Verworn observed long ago that in *Peranema trichophorum* the undulation of the distal portion of flagellum is accompanied by a slow forward movement, while undulation along the entire length is followed by a rapid forward movement. Krijgsman (1925) studied the movements of the long flagellum of *Monas* sp. (Fig. 50) which he found in soil cultures, under the darkfield microscope and stated: (1) when the organism moves forward with the maximum speed, the flagellum starting from *c 1*, with the wave beginning at the base, stretches back (*c 1–6*), and then waves back (*d, e*), which brings about the forward movement. Another type is one in which the flagellum bends back beginning at its base (*f*) until it coincides with the body axis, and in its effective stroke waves back as a more or less rigid structure (*g*); (2) when the organism moves forward with moderate speed, the tip of the flagellum passes through 45° or less (*h–j*); (3) when the animal moves backward, the flagellum undergoes undulation which begins at its base (*k–o*); (4) when the animal moves to one side, the flagellum becomes bent at right angles to the body and undulation passes along it from its base to tip (*p*); and (5) when the organism undergoes a slight lateral movement, only the distal end of the flagellum undulates (*q*).

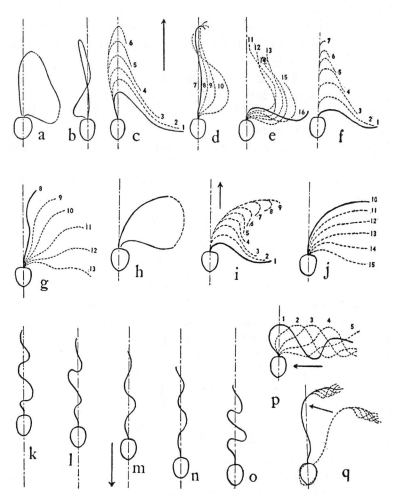

Fig. 50. Diagrams illustrating flagellar movements of *Monas* sp. (Krijgs-man). a–g, rapid forward movement (a, b, optical image of the movement in front and side view; c, preparatory and d, e, effective stroke; f, preparatory and g, effective stroke); h–j, moderate forward movement (h, optical image; i, preparatory and j, effective stroke); k–o, undulatory movement of the flagellum in backward movement p, lateral movement; q, turning movement.

Lowndes (1945) reports that the mechanical principle in flagellar movement of *Monas stigmatica* is the same as that of the screw or propeller. In the forward movement of this flagellate, waves pass along the long flagellum from base to tip in a spiral manner with an increase in velocity and amplitude. The force generated is transmitted to the surface of the cell at *a* (Fig. 51). This causes the organism to rotate and gyrate about the axis *d*. The edge *b* going below the surface of the paper and *c* being

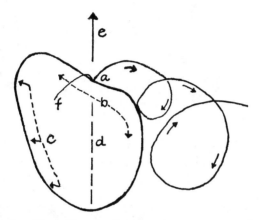

Fig. 51. A diagram showing the movements of
Monas stigmatica (Lowndes).

raised above it. Thus the organism is converted into rotating inclined plane and moves forward more or less in the direction indicated by the arrow *e*. The short flagellum *f* appears to act as a guiding or sensory organ during normal swimming. Though the chief function of the flagellum is for locomotion, it serves also for obtaining food, sensory function and attachment to substrate.

Ciliary Movement. Well-coordinated movement of numerous cilia of ciliates produce rapid locomotion. Individual cilium beats throughout its length and strikes the water so that the organism tends to move in a direction opposite to that of the effective beat, while the water moves in the direction of the beat (Fig. 52). In many ciliates, the cilia are arranged in longitudinal, or oblique rows and it is clearly noticeable that the cilia are not beating in the same phase, although they are moving at the same rate. A cilium (Fig. 52, *e*) in a single row is slightly in advance of the cilium behind it and slightly behind the one just in front of it, thus the cilia on the same longitudinal row beat metachronously. On the other hand, the cilia on the same transverse row beat synchronously.

Jennings (1906) suggested that the character and direction of the movement of a ciliate is determined by the body form and the activity of oral cilia, but Bullington (1925, 1930), after examin-

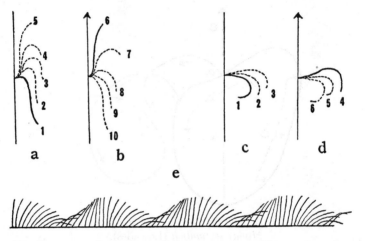

Fig. 52. Diagrams illustrating ciliary movements (Verworn). a–d, movement of a marginal cilium of *Urostyla grandis* (a, preparatory and b, effective stroke, resulting in rapid movement; c, preparatory, and d, effective stroke, bringing about moderate speed); e, metachronous movements of cilia in a longitudinal row.

ing some 160 species of ciliates, maintained that the spiral movement in ciliates depends upon the oblique stroke of all body cilia working together and beating in the same direction. In a ciliate with left spiral swimming, the cilia beat obliquely backward to the right for forward movement and beat obliquely forward to the left for backward movement.

The organized movements of cilia, cirri, membranellae and undulating membranes are probably controlled by the neuromotor or silverline system (p. 71) which appears to be conductile as judged by the results of micro-dissection experiments of Taylor

(p. 74). (Ciliary movement, Gray, 1928; spiral movement of ciliates, Bullington, 1925, 1930; movement of Paramecium, Dembowski, 1923, 1929a; Párducz, 1953, 1954; movement of Spirostomum, Blättner, 1926; mechanism of ciliary movement Jahn,
1961.)

In addition to the three types of locomotion, the protozoa
which possess myonemes are able to move by contraction of the
body or of the stalk, and others are said to combine this with secretion of mucous substances as found in Haemogregarina and
Gregarinida, although Beams *et al.* (1959) find no evidence for
mucous secretion.

Responses to Various Stimuli

Under natural conditions, the Protozoa do not behave always
in the same manner, because several stimuli act upon them usually in combination and predominating stimulus or stimuli vary
under different circumstances. Of various responses expressed by
a protozoan against a stimulus such as changes in body form, movement, structure, behavior, reproduction, etc., the movement is
the most clearly recognizable one and, therefore, free-swimming
forms, particularly ciliates, have been the favorite objects of study.
We consider the reaction to a stimulus in protozoa as the movement response, and this appears in one of the two directions:
namely, toward, or away from, the source of the stimulus. Here
we speak of positive or negative reaction. In forms such as
Amoeba, the external stimulation is first received by the body
surface and then by the whole protoplasmic body. In flagellated
or ciliated protozoa, the flagella or cilia act in part sensory; in
fact in a number of ciliates are found non-vibratile cilia which
appear to be sensory in function. In a comparatively small number of forms, there are specialized sensory organellae for reception of stimuli such as stigma, ocellus, statocysts, concretion vacuoles, etc.

In general, the reaction of a protozoan to any external stimulus
depends upon its intensity so that a certain chemical substance
may bring about entirely opposite reactions on the part of the protozoa in different concentrations and, even under identical conditions, different individuals of a given species may react differently.

(Irritability, Jennings, 1906; Mast, 1941; in Spirostomum, Blättner, 1926.)

Reaction to Mechanical Stimuli. One of the most common stimuli a protozoan would encounter in the natural habitat is that which comes from contact with a solid object. When an amoeba which Jennings observed, came in contact with the end of a dead algal filament at the middle of its anterior surface (Fig. 53, *a*), the

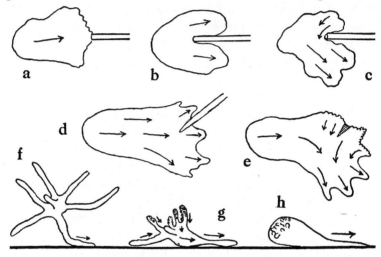

Fig. 53. Reactions of amoebae to mechanical stimuli (Jennings). a–c, an amoeba avoiding an obstacle; d, e, negative reaction to mechanical stimulation f–h, positive reaction of a floating amoeba.

amoeboid movements proceeded on both sides of the filament (*b*), but soon motion ceased on one side, while it continued on the other, and the organism avoided the obstacle by reversing a part of the current and flowing in another direction (*c*). When an amoeba is stimulated mechanically by the tip of a glass rod (*d*), it turns away from the side touched, by changing endoplasmic streaming and forming new pseudopods (*e*). Positive reactions are also often noted, when a suspended amoeba (*f*) comes in contact with a solid surface with the tip of a pseudopod, the latter adheres to it by spreading out (*g*). Streaming of the cytoplasm follows and it becomes a creeping form (*h*). Positive reactions toward solid bodies account of course for the ingestion of food particles.

In Paramecium, according to Jennings, the anterior end is more sensitive than any other parts, and while swimming, if it comes in contact with a solid object, the response may be either negative or positive. In the former case, avoiding movement (Fig. 54, *c*) follows and in the latter case, the organism rests with its anterior end or the whole side in direct contact with the object, in which position it ingests food particles through the cytostome.

Reaction to Gravity. The reaction to gravity varies among different protozoa, according to body organization, locomotor organellae, etc. Amoebae, Testacea and others which are usually found attached to the bottom of the container, react as a rule positively toward gravity, while others manifest negative reaction as in the case of Paramecium (Jensen; Jennings), which explains in part why Paramecium in a culture jar are found just below the surface film in mass, although the vertical movement of *P. caudatum* is undoubtedly influenced by various factors (Koehler, 1922, 1930; Dembowski, 1923, 1929, 1929a; Merton, 1935).

Reaction to Current. Free-swimming protozoa appear to move or orientate themselves against the current of water. In the case of Paramecium, Jennings observed the majority place themselves in line with the current, with anterior end upstream. Mycetozoida are said to exhibit also a well-marked positive reaction.

Reaction to Chemical Stimuli. When methylgreen, methylene blue, or sodium chloride is brought in contact with an advancing amoeba, the latter organism reacts negatively (Jennings). Jennings further observed various reactions of Paramecium against chemical stimulation. This ciliate shows positive reaction to weak solutions of many acids and negative reactions above certain concentrations. For example, Paramecium enters and stays within the area of a drop of 0.02 per cent acetic acid introduced to the preparation (Fig. 54, *a*); and if stronger acid is used, the organisms collect about its periphery where the acid is diluted by the surrounding water (*b*). The reaction to chemical stimuli is probably of the greatest importance for the existence of protozoa, since it leads them to proper food substances, the ingestion of which is the foundation of metabolic activities. In the case of parasitic protozoa, possibly the reaction to chemical stimuli results in their finding specific host animals and their distribution in different or-

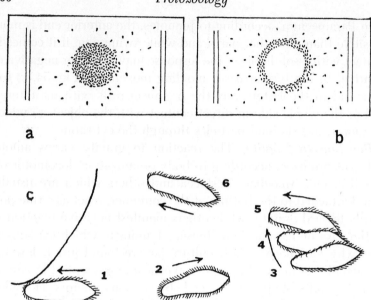

Fig. 54. Reactions of Paramecium (Jennings). a, collecting in a drop of
0.02% acetic acid; b, ring-formation around a drop of a stronger solution
of the acid c, avoiding reaction.

gans and tissues within the host body. Recent investigations tend
to indicate that chemotaxis plays an important rôle in the sexual
reproduction in protozoa. (Chemotaxis in Peranema, Chen,
1950.)

Reaction to Light Stimuli. Most protozoa seem to be indifferent
to the ordinary light, but when the light intensity is suddenly increased, there is usually a negative reaction. Verworn saw the direction of movements of an amoeba reversed when its anterior
end was subjected to a sudden illumination; Rhumbler observed
that an amoeba, which was in the act of feeding, stopped feeding
when it was subjected to strong light. According to Mast, *Amoeba
proteus* ceases to move when suddenly strongly illuminated, but
continues to move if the increase in intensity is gradual and if the
illumination remains constant, the amoeba begins to move. *Pelomyxa carolinensis* and *P. palustris* react negatively to light (Kudo,
1946, 1957).

The positive reaction to light is most clearly shown in stigma-
bearing Mastigophora, as is well observable in a jar containing Eu-

glena, Phacus, etc., in which the organisms collect at the place where the light is strongest. If the light is excluded completely, the organisms become scattered throughout the container, and inactive and sometimes encyst, although the mixotrophic forms would continue activities by saprozoic method. The positive reaction to light by chromatophore-bearing forms enables them to find places in the water where photosynthesis can be carried on to the maximum degree. (Photomotion in Euglena, Wolken and Shin, 1958.)

Tetrahymena pyriformis does not grow in a proteose-peptone medium exposed to strong light in the visible spectrum, but addition of riboflavin restores normal growth. The organisms in natural waters exposed to strong light apparently utilize riboflavin produced by bacteria (Phelps, 1959). Thus the process which leads to a definite response must be a complex one.

All protozoa seem to be more sensitive to ultraviolet rays. Inman found that amoeba shows a greater reaction to the rays than others and Hertel observed that Paramecium which was indifferent to an ordinary light, showed an immediate response (negative reaction) to the rays. MacDougall brought about mutations in Chilodonella by means of these rays (p. 275). Horváth (1950) exposed *Kahlia simplex* to ultraviolet rays and destroyed the micronucleus. The emicronucleate individuals lived and showed a greater vitality than normal individuals, as judged by the division rate at 34°C. Mazia and Hirshfield (1951) subjected *Amoeba proteus* to ultraviolet radiation and noticed that irradiation of the whole and nucleated half of amoebae delays division immediately following exposure; later progeny of the irradiated amoebae have a normal division rate; amputation of half of the cytoplasm greatly increases the radiation sensitivity as measured by delayed division or by the dose required for permanent inhibition of division (sterilization dose); individuals that have received this dose may survive for twenty to thirty days; and the survival time of an enucleate fragment is very much reduced by small (200-500 ergs/sq. mm) doses. The two workers consider that the overall radiation effect may have both nuclear and cytoplasmic components. By exposing *Pelomyxa carolinensis* to 2537 Å ultraviolet irradiation, Wilber and Slane (1951) found the effects variable; however, all recovered from a two minutes' exposure, none survived a 10-minute exposure, and 70 per cent of fat were released after two minutes' exposure. Giese *et al.* (1954) reported that starved *Colpidium colpoda* are much more sensitive to ultra-violet rays than are well fed ones. (Review, Giese, 1953; ultra-violet radiation on Astasia, Schoenborn, 1956.)

Zuelzer (1905) found the effect of radium rays upon various protozoa vary; in all cases, a long exposure was fatal to protozoa, the first effect of exposure being shown by accelerated movement. Halberstaedter and Luntz (1929, 1930) studied injuries and death of *Eudorina elegans* by exposure to radium rays. *Entamoeba histolytica* in culture when subjected to radium rays, Nasset and Kofoid (1928) noticed the following changes: the division rate rose two to four times by the exposure, which effect continued for not more than 24 hours after the removal of the radium and was followed by a retardation of the rate; radium exposure produced changes in nuclear structure, increase in size, enucleation or autotomy, which were more striking when a larger amount of radium was used for a short time than a smaller amount acting on for a long time; and the effects persisted for four to six days after the removal of the radium and then the culture gradually returned to normalcy. Halberstaedter (1914) reported that when exposed to beta rays, *Trypanosoma brucei* lost its infectivity, though remained alive.

Halberstaedter (1938) exposed *Trypanosoma gambiense* to x-rays and found that 12,000r rendered the organisms not infectious for mice, while 600,000r was needed to kill the flagellates. Emmett (1950) exposed *T. cruzi* to x-rays and noticed that dosages between 51,000r and 100,000r were necessary to destroy the infectivity of this trypanosome; the cultures, after exposure to 100,000r, appeared to be thriving up to three months.

When *Paramecium bursaria* were exposed to x-rays, Wichterman (1948) noted: dosages higher than 100,000r retard the locomotion of the ciliate; none survives 700,000r; the symbiotic Chlorella is destroyed by exposure to 300,000-600,000r; irradiation inhibits division temporarily, but the animals recover normal division rate after certain length of time; and mating types are not destroyed, though minor changes occur. By exposing *P. multimicronucleatum* to large dosages of x-radiation, Wichterman

(1959) noted the following changes which persisted in cultures: loss of all micronuclei, decrease in vitality and reproductive rate, and increase in x-radiation sensitivity, monstrosity and reduced body size.

X-radiation (150-300kr) inhibited growth and fission in *Stentor coeruleus*, but regeneration of removed parts proceeded even at the highest doses at normal rate, which led Kimball (1958) to conclude that delay in fission by radiation is a secondary consequence of inhibition of growth and not a consequence of blocks in the specific morphogenetic processes required for division. (X-radiation on Paramecium, Wichterman, 1961.)

In *Pelomyxa carolinensis*, Daniels (1951) observed: the median lethal dose of x-rays is 96,000r; with dosages 15,000-140,000r, the first plasmotomy is greatly delayed and the second plasmotomy is also somewhat delayed, but later plasmotomies show complete recovery; x-irradiation does not change the type of plasmotomy; and in individuals formed by plasmogamy of x-irradiated halves to non-irradiated halves, the nuclei divide simultaneously as in a normal individual. Elliott and Clark (1956) exposed *Tetrahymena pyriformis* (2,II) to 400,000r and mated them with non-irradiated individuals of various mating types in the same variety 2. The irradiated animals failed to produce a migratory nucleus, which resulted in the production of haploid exconjugants. (X-irradiation on Pelomyxa, Daniels, 1952, 1954, 1955, 1958, 1959; biochemical mutation in Astasia, Schoenborn, 1954; x-irradiation of Spathidium, Williams, 1962.)

Reaction to Temperature Stimuli. As was stated before, there seems to be an optimum temperature range for each protozoan, although it can withstand temperatures which are lower or higher than that range. As a general rule, the higher the temperature, the greater the metabolic activities, and the latter condition results in turn in a more rapid growth and more frequent reproduction. It has been suggested that change to different phases in the life-cycle of a protozoan in association with the seasonal change may be largely due to changes in temperature of the environment. In parasitic forms such as malaria parasites, the difference in the body temperature of hosts together with other factors may bring about specific stages in their life cycle.

Reaction to Electrical Stimuli. Since Verworn's experiments, several investigators studied the effects of electric current which is passed through protozoa in water. Amoeba shows negative reaction to the anode and moves toward the cathode either by reversing the cytoplasmic streaming (Verworn) or by turning around the body (Jennings). The free-swimming ciliates move mostly toward the cathode, but a few may take a transverse position (Spirostomum) or swim to the anode (Paramecium, Stentor, etc.). Of flagellates, Verworn noticed that Trachelomonas and Peridinium moved to the cathode, while Chilomonas, Cryptomonas, and Polytomella, swam to the anode. When *Paramecium caudatum* was exposed to a high-frequency electrostatic or electromagnetic field, Kahler *et al.* (1929) found the effect was primarily caused by a temperature increase in the organism. By subjecting *Pelomyxa carolinensis* to a direct current electric field, Daniel and May (1950) noted that the time required for the rupture of the body in a given current density is directly correlated with the size of the organism and that calcium increases the time required for rupture at a fixed body size and current density, but does not alter the size effect. (Galvanotaxis of Oxytricha, Luntz, 1935; of Arcella, Miller, 1932.)

Reaction to Antibiotics. The antibiotics apparently interfere with, and inactivate, enzymatic activities of microorganisms, resulting in inhibition of growth and reproduction. Bacteria are highly affected by them, while protozoa are not injured at antibacterial strength. The action of streptomycin on chlorophyll body has already been mentioned. But the use of antibiotics is limited to elimination of bacteria and maintaining axenic cultures of various protozoa.

Attempts are, however, being made to utilize various antibiotics for therapeutic purposes for infections by *Entamoeba histolytica*, *Trichomonas vaginalis*, etc. Fumagillin appears to be effective for control of *Nosema apis* infection in bees (Katznelson and Jamieson, 1952; Farrar, 1954). (Literature on streptomycin, Waksman, 1962; antibiotics on *Entamoeba histolytica*, Anderson *et al.*, 1953.)

References

AARONSON, S., and BAKER, H., 1959. A comparative biochemical study of two species of Ochromonas. J. Protozool., 6:282.

ALLEE, W. C., 1931. Animal aggregations. Chicago.

—— 1934. Recent studies in mass physiology. Biol. Rev., 9:1.

ANDERSON, H. H., et al., 1953. Amebiasis. Springfield, Thomas.

ANDRESEN, N., 1945. Cytoplasmic changes during starvation and during neutral red staining of the amoeba, etc. C. R. Lab. Carlsberg, Sér. Chim., 25:169.

—— CHAPMAN-ANRESEN, C., et al., 1950. The distribution of food in amoeba cytoplasm studied by means of autoradiography. Exper. Cell. Res., 1:139.

—— ENGEL, F., and HOLTER, H., 1951. Succinic dehydrogenase and cytochrome oxidase in *Chaos chaos*. C. R. Lab. Carlsberg, Sér. Chem., 27:408.

—— and HOLTER, H., 1945. Cytoplasmic changes during starvation of the amoeba *Chaos chaos* L. *Ibid.*, 25:107.

—— —— 1949. The genera of amoebae. Science, 110:114.

BAIRATI, A., and LEHMANN, F. E., 1953. Structural and chemical properties of the plasmalemma of *Amoeba proteus*. Exper. Cell Res., 5:220.

BARY, B. M., and STUCKEY, R. G., 1950. An occurrence in Wellington Harbour of *Cyclotrichium meunieri* Powers, etc. Tr. Roy. Soc. New Zealand, 78:86.

BEAMS, H. W., et al., 1959. Studies on the fine structure of a gregarine parasitic in the gut of the grasshopper, *Melonoplus differentialis*. J. Protozool., 6:136.

BECKER, E. R., et al., 1930. Experiments on the physiological relationship between the stomach Infusoria of ruminants and their hosts, etc. Iowa State Coll. J. Sci., 4:215.

BEERS, C. D., 1933. Diet in relation to depression and recovery in the ciliate *Didinium nasutum*. Arch. Protist., 79:101.

BELDA, W. H., 1942. Permeability to water in *Pelomyxa carolinensis*. II. Salesianum. 37:125.

BERNHEIMER, A. W., 1938. A comparative study of the crystalline inclusions of protozoa. Tr. Am. Micr. Soc., 57:336.

BERTHOLD, C., 1886. Studien ueber Protoplasmamechanik. Leipzig.

BLÄTTNER, H., 1926. Beiträge zur Reizphysiologie von *Spirostomum ambiguum*. Arch. Protist., 53:253.

BOTSFORD, E. F., 1926. Studies on the contractile vacuole of *Amoeba proteus*. J. Exper. Zool., 45:95.

BOZLER, E., 1924. Ueber die Morphologie der Ernährungsorganelle und die Physiologie der Nahrungsaufnahme bei *Paramecium caudatum*. Arch. Protist., 49:163.

Brug, S. L., 1928. Observations on a culture of *Entamoeba histolytica*. Med. Dienst Volksges. Ned. Indie, p. 1.

Burbanck, W. D., and Eisen, J. D., 1960. The inadequacy of mono-bacterially-fed *Paramecium aurelia* as food for *Didinium nasutum*. J. Protozool., 7:201.

Bütschli, O., 1885. Bemerkungen ueber einen dem Glykogen verwandten Körper in den Gregarinen. Ztschr. Biol., 21:603.

Bullington, W. E., 1925. A study of spiral movement in the ciliate Infusoria. Arch. Protist., 50:219.

——— 1930. A further study of spiraling in the ciliate Paramecium, etc. J. Exper. Zool., 56:423.

Calkins, G. N., 1933. The biology of the protozoa. 2 ed. Philadelphia.

——— and Summers, F. M., 1941. Protozoa in biological research. New York.

Carlstrom, D., and Møller, K. M., 1961. Further observations on the native and recrystallized crystals of *Amoeba proteus*. Exper. Cell Res., 24:393.

Chapman-Andresen, C., 1958. Pinocytosis of inorganic salts by *Amoeba proteus*. C. R. Lab. Carlsberg, 31:77.

——— and Prescott, D. M., 1956. Studies on pinocytosis in the amoebae *Chaos chaos* and *Amoeba proteus*. Ibid., 30:57.

Chen, Y. T., 1950. Investigations of the biology of *Peranema trichophorum*. Quart. J. Micr. Sci., 91: 279.

Chi, L., et al., 1959. Selective phagocytosis of nucleated erythrocytes by cytotoxic amoebae in cell culture. Science, 130:1763.

Claff, C. L., et al., 1941. Feeding mechanisms and nutrition in 3 species of Bresslaua. Biol. Bull., 81:221.

Clark, A. M., 1942. Some effects of removing the nucleus from Amoeba. Australian J. Exper. Biol., 20:241.

Cleveland, L. R. 1925. Toxicity of oxygen for protozoa *in vivo* and *in vitro*, etc. Biol. Bull., 48:455.

——— 1925a. The method by which *Trichonympha campanula*, a protozoon in the intestine of termites, ingests solid particles of wood for food. *Ibid.*, 48:282.

——— Hall, S. R. et. al. 1934. The wood-feeding roach Cryptocercus, its protozoa, and the symbiosis between protozoa and roach. Mem. Am. Acad. Arts Sci., 17:185.

Cosmovici, N. L., 1932. La nutrition et le rôle physiologique du vacuome chez les infusoires. Ann. Sci. Univ. Jassy, 17:294.

Daniel, G. E., and May, G. H., 1950. Observations on the reaction of *Pelomyxa carolinensis* subjected to a direct current electric field. Physiol. Zool., 23:231.

Daniels, E. W., 1951. Studies on the effect of x-irradiation upon *Pelomyxa carolinensis* with special reference to nuclear division and plasmotomy.

J. Exper. Zool., 117:189.

———— 1952. Cell division in the giant amoeba, *Pelomyxa carolinensis* following x-irradiation. I. J. Exper. Zool., 120:525.

———— 1954. II. *Ibid.*, 127:427.

———— 1955. X-irradiation of the giant amoeba, *Pelomyxa illinoisensis*. I. *Ibid.*, 130:183.

———— 1958. II. *Ibid.*, 137:425.

———— 1959. Micrurgical studies on irradiated Pelomyxa. Ann. N.Y. Acad. Sci., 78:662.

DAWSON, J. A., and BELKIN, M., 1928. The digestion of oil by *Amoeba dubia*. Proc. Soc. Exper. Biol., 25:790.

DE LA ARENA, J. F. 1941. El pH de las vacuolas digestivas. Mem. Soc. Cubana Hist. Nat., 15:345.

———— 1942. Liberacion experimental de ciliados in el cytoplasma de Amiba. *Ibid.*, 16:73.

DELLINGER, O. P., 1906. Locomotion of amoebae and allied forms. J. Exper. Zool., 3:337.

DEMBROWSKI, J., 1923. Ueber die Bewegungen von *Paramecium caudatum*. Arch. Protist., 47:25.

———— 1929. Die Vertikalbewegungen von *Paramecium caudatum*. I. *Ibid.*, 66:104.

———— 1929a. II. *Ibid.*, 68:215.

DEWEY, V., and KIDDER, G. W., 1940. Growth studies on ciliates. VI. Biol. Bull., 79:255.

———— *et al.*, 1957. Evidence for the absence of the urea cycle in Tetrahymena. J. Protozool., 4:211.

DOFLEIN, F., 1918. Studien zur Naturgeschichte der Protozoen. X. Zool. Jahrb. Anat., 41:1.

DOYLE, W. L., 1943. The nutrition of the protozoa. Biol. Rev., 18:119.

———— and HARDING, J. P., 1937. Quantitative studies on the ciliate Glaucoma. J. Exper. Biol., 14:462.

———— and PATTERSON, E. K., 1942. Origin of dipeptidase in a protozoan. Science, 95:206.

EDGAR, S. A., HERRICK, C. A., and FRASER, L. A., 1944. Glycogen in the life cycle of the coccidium *Eimeria tenella*. Tr. Am. Micr. Soc., 63:199.

EISENBERG, E., 1925. Recherches sur le fonctionnement de la vesicule pulsatile des infusoires, etc. Arch. Biol. Paris, 35:441.

ELLIOTT, A. M., and CLARK, G. M., 1956. The induction of haploidy in *Tetrahymena pyriformis* following x-irradiation. J. Protozool., 3:181.

EMIK, L. O., 1941. Ingestion of food by Trichonympha. Tr. Am. Micr. Soc., 60:1.

EMMETT, J., 1950. Effect of x-radiation on *Trypanosoma cruzi*. J. Parasitol., 36:45.

ENGELMANN, T. W., 1878. Flimmer und Protoplasmabewegung. Hermann: Handb. d. Physiologie, 1:349.

FARRAR, C. L., 1954. Fumagillin for Nosema control in package bees. Amer. Bee J. 94:52.

FAURÉ-FREMIET, E., and THAUREAUX, J., 1944. Les globules de "paraglycogène" chez *Balantidium elongatum* et *Vorticella monilata*. Bull. Soc. Zool. France, 69:3.

FORTNER, H., 1926. Zur Frage der diskontinuierlichen Exkretion bei Protisten. Arch. Protist., 56:295.

FRISCH, J. A., 1937. The rate of pulsation and the function of the contractile vacuole in *Paramecium multimicronucleatum*. *Ibid.*, 90:123.

GAW, H. Z., 1936. Physiology of the contractile vacuole in ciliates. I-IV. *Ibid.*, 87:185.

GELEI, G., 1939. Neuere Beiträge zum Bau und zu der Funktion des Exkretionssystems von Paramecium. *Ibid.*, 92:384.

GELEI, J., 1925. Nephridialapparat bei den Protozoen. Biol. Zentralb., 45:676.

———— 1925a. Ueber der Kannibalismus bei Stentoren. Arch. Protist., 52:405.

GIESE, A. C., 1953. Protozoa in photobiological research. Physiol. Zool., 26:1.

———— and ALDEN, R. H., 1938. Cannibalism and giant formation in Stylonychia. J. Exper. Zool., 78:117.

———— *et al.*, 1954. The effect of starvation on photoreactivation in *Colpidium colpoda*. Physiol. Zool., 27:71.

———— and CHRISTENSEN, E., 1954. Effects of ozone on organisms. Physiol. Zool., 27:101.

GÖHRE, E., 1943. Untersuchungen ueber den plasmatischen Feinbau der Gregarinen, etc. Arch. Protist., 96:295.

GRASSÉ, P.-P., 1952. Traité de Zoologie. I. Fasc. 1. Paris.

GRAY, J., 1928. Ciliary movement. Cambridge.

GREEFF, R., 1874. *Pelomyxa palustris* (Pelobius), ein amoebenartiger Organismus des suessen Wassers. Arch. mikr. Anat., 10:53.

GREENWOOD, M., and SAUNDERS, E. R., 1894. On the rôle of acid in protozoan digestion. J. Physiol., 16:441.

GROSS, J. A., 1955. A comparison of different criteria for determining the effects of antibiotics on *Tetrahymena pyriformis* E. J. Protozool., 2:42.

———— and JAHN, T. L., 1958. Some biological characteristics of chlorotic substrains of *Euglena gracilis*. *Ibid.*, 5:126.

———— *et al.*, 1955. The effect of antihistamines on the pigments of green protista. *Ibid.*, 2:71.

GUILLARD, R. R. L., 1960. A mutant of *Chlamydomonas moewusii* lacking contractile vacuoles. J. Protozool., 7:262.

HALBERSTAEDTLER, L., 1914. Experimentelle Untersuchungen an Trypanosomen, etc. Berl. klin. Woch., p. 252.

—— and LUNTZ, A., 1929. Die Wirkung der Radiumstrahlen auf *Eudorina elegans*. Arch Protist., 68:177.

—— —— 1930. Weitere Untersuchungen ueber die Wirkung von Radiumstrahlen, etc. *Ibid.*, 71:295.

HALL, R. P., 1939. The trophic nature of the plant-like flagellates. Quart. Rev. Biol., 14:1.

—— 1941. Food requirements and other factors. In Calkins and Summers (1941).

—— and ALVEY, C. H., 1933. The vacuome and so-called canalicular system of Colpidium. Tr. Am. Micr. Soc., 52:26.

—— and DUNIHUE, F. W., 1931. On the vacuome and food vacuoles in Vorticella, *Ibid.*, 50: 196.

HARINASUTA, C., and MAEGRAITH, B. G., 1958. The demonstration of proteolytic enzyme activity of *Entamoeba histolytica* by use of photographic gelatin film. Ann. Trop. Med. & Parasit., 52:508.

HARVEY, E. N., 1952. Bioluminescence. New York.

HASTINGS, J. W., and SWEENEY, B. M., 1957. The luminescent reaction in extracts of the marine dinoflagellate, *Gonyaulax polyedra*. J. Cell Comp. Physiol., 49:209.

HEIDT, K., 1937. Form und Struktur der Paramylonkörper von *Euglena sanguinea*. Arch. Protist., 88:127.

HERFS, A., 1922. Die pulsierende Vakuole der Protozoen, ein Schutzorgan gegen Aussüssung. *Ibid.*, 44:227.

HOGUE, MARY J., 1923. Contractile vacuoles in amoebae, etc. J. E. Mitchell Sci. Soc., 39:49.

HOLTER, H., 1950. The function of cell inclusions in the metabolism of *Chaos chaos*. Ann. N. Y. Acad. Sci., 50:1000.

—— 1959. Problems of pinocytosis, with special regard to amoebae. Ann N.Y. Acad. Sci., 78:524.

—— and DOYLE, W. L. 1938. Studies on enzymatic histochemistry. J. Cell. Comp. Physiol., 12:295.

—— —— 1938a. Ueber die Lokalisation der Amylase in Amoeben. C. R. Lab. Carlsberg., Sér. Chim., 22:219.

—— and KOPAC, M. J. 1937. Localization of peptidase in the ameba. J. Cell. Comp. Physiol., 10:423.

—— and LØVTRUP, S. 1949. Proteolytic enzymes in *Chaos chaos*. C. R. Lab. Carlsberg., Sér. Chim., 27:27.

—— and LOWY, B. A. 1959. A study of the properties and localization of acid phosphatase in the amoeba *Chaos chaos*. *Ibid.*, 31:105.

HOPKINS, D. L., 1938. The vacuoles and vacuolar activity in the marine amoeba, etc. Biodynamica, 34, 22 pp.

—— and WARNER, KAY L., 1946. Functional cytology of *Entamoeba histolytica*. J. Parasitol., 32:175.

HORVÁTH, J., 1950. Vitalitätsäusserung einer mikronukleuslosen Bodenziliate in der vegetativen Fortpflanzung. Oesterr. zool. Ztschr., 2:336.

HOWLAND, RUTH B., 1928. The pH of gastric vacuoles. Protoplasma, 5:127.
—— and BERNSTEIN, A., 1931. A method for determining the oxygen consumption of a single cell. J. Gen. Physiol., 14:339.
HULL, R. W., 1954. Feeding processes in *Solenophrya micraster* Penard. J. Protozool., 1:178.
HULPIEU, H. R., 1930. The effect of oxygen on *Amoeba proteus*. J. Exper. Zool., 56:321.
HUNTER, N. W., 1959. Enzyme systems of *Stylonychia pustulata*. II. J. Protozool., 6:100.
HUTNER, S. H., and PROVASOLI, L., 1951. The phytoflagellates. In: Lwoff (1951).
—— —— 1955. Comparative biochemistry of flagellates. In Hutner and Lwoff: Protozoa. II. New York, Academic Press.
HYMAN, LIBBY H., 1917. Metabolic gradients in Amoeba and their relation to the mechanism of amoeboid movement. J. Exper. Zool., 24:55.
JAHN, T. L., 1941. Respiratory metabolism. In Calkins and Summers (1941).
—— 1961. The mechanism of ciliary movement. I. J. Protozool., 8:369.
JENNINGS, H. S., 1904. Contributions to the study of the behavior of the lower organisms. Publ. Carnegie Inst. Washington, No. 16.
—— 1906. Behavior of the lower organisms. New York.
JÍROVEC, O., 1926. Protozoenstudien. I. Arch. Protist., 56:280.
JOHNSON, W. H., 1956. Nutrition of protozoa. Ann. Rev. Microbiol., 10: 193.
—— and MILLER, C. A., 1956. A further analysis of the nutrition of Paramecium. J. Protozool., 3:221.
JURAND, A., 1961. An electron microscope study of food vacuoles in *Paramecium aurelia*. J. Protozool., 8:125.
KAHLER, H. *et. al.,* 1929. The nature of the effect of a high-frequency electric field upon Paramecium. Publ. Health Report, 44:339.
KATZNELSON, H., and JAMIESON, C. A., 1952. Control of Nosema disease of honey bees with fumagillin. Science, 115:70.
KEPNER, W. A., and WHITLOCK, W. C., 1921. Food reactions of *Amoeba proteus*. J. Exper. Zool., 32:397.
KHAINSKY, A., 1910. Zur Morphologie und Physiologie einiger Infusorien, etc. Arch. Protist., 21:1.
KIDDER, G. W., 1951. Nutrition and metabolism of protozoa. Ann. Rev. Microbiol., 5:139.
—— and DEWEY, V. C., 1951. The biochemistry of ciliates in pure culture. In Lwoff (1951).
——, LILLY, D. M., and CLAFF, C. L., 1940. Growth studies on ciliates. IV. Biol. Bull., 78:9.
KIMBALL, R. F., 1958. Experiments with *Stentor coeruleus* on the nature of the radiation-induced delay in fission in the ciliates. J. Protozool., 5:151.
KIRBY, H. JR., 1932. Two protozoa from brine. Tr. Am. Micr. Soc., 51:8.
—— 1934. Some ciliates from salt marshes in California. Arch. Protist., 82:114.

KITCHING, J. A., 1934. The physiology of contractile vacuoles. I. J. Exper. Biol., 11:364.

—— 1936. II. *Ibid.*, 13:11.

—— 1938. III. *Ibid.*, 15:143.

—— 1938a. Contractile vacuoles. Biol. Rev., 13:403.

—— 1938b. On the mechanism of movement of food vacuoles in peritrich ciliates. Arch. Protist., 91:78.

—— 1956. Contractile vacuoles of protozoa. Protoplasmatologia, 3:D3a.

—— 1956a. Food vacuoles. *Ibid.*, 3:D3b.

KOEHLER, O., 1922. Ueber die Geotaxis von Paramecium. Arch. Protist., 45:1.

—— 1930. II. *Ibid.*, 70:279.

KOFOID, C. A., and SWEZY, O., 1921. The free-living unarmored Dinoflagellata. Mem. Univ. California, 5:1.

KRASCHENINNIKOW, S., 1959. Abnormal infraciliature of *Balantidium coli* and *B. caviae* (?) and some morphological observations on these species. J. Protozool., 6:61.

KRIJGSMAN, B. J., 1925. Beiträge zum Problem der Geisselbewegung. Arch. Protist., 52:478.

KUDO, R. R., 1926. Observations on *Lophomonas blattarum,* etc. *Ibid.*, 53:191.

—— 1946. *Pelomyxa carolinensis* Wilson. I. J. Morphol., 78:317.

—— 1951. Observations on *Pelomyxa illinoisensis. Ibid.*, 88:145.

—— 1957. *Pelomyxa palustris* Greeff. I. J. Protozool., 4:154.

KUPFERBERG, A. B., et al., 1953. Studies on the metabolism of *Trichomonas vaginalis.* Ann. N.Y. Acad. Sci., 56:1006.

LAPAGE, G., 1922. Cannibalism in *Amoeba vespertilio.* Quart. J. Micr. Sci., 66:669.

LEINER, M., 1924. Die Glycogen in *Pelomyxa palustris,* etc. Arch. Protist., 47:253.

LEWIS, W. H., 1931. Pinocytosis. Bull. Johns Hopkins Hosp., 49:17.

LLOYD, F. E., 1926. Some behaviours of *Vampyrella lateritia,* etc. Michigan Acad. Sci., 7:395.

LØVTRUP, S., and PIGÓN, A., 1951. Diffusion and active transport of water in the amoeba. C. R. Lab. Carlsberg, Sér. Chim., 28:1.

LOWNDES, A. G., 1945. Swimming of *Monas stigmatica.* Nature, 155:579.

LUCE, R. H., and POHL, A. W., 1935. Nature of crystals found in amoeba. Science 82:595.

LUNTZ, A., 1935. Untersuchungen ueber die Galvanotaxis der Einzelligen. I. Arch. Protist., 84:495.

LWOFF, A., 1932. Recherches biochemique sur la nutrition des Protozoaires. Monogr. Inst. Pasteur, 160 pp.

—— (edited by) 1951. Biochemistry and physiology of protozoa. New York.

MACLENNAN, R. F., 1933. The pulsation cycle of the contractile vacuoles, etc. Univ. Cal. Publ. Zool., 39:205.

———— 1936. Dedifferentiation and redifferentiation in Ichthyophthirius. II. Arch. Protist., 86:404.

MAST, S. O., 1923. Mechanics of locomotion in amoeba. Proc. Nat. Acad. Sci., 9:258.

———— 1926. Structure, movement, locomotion, and stimulation in amoeba. J. Morphol. Physiol., 41:347.

———— 1931. Locomotion in *Amoeba proteus*. Protoplasma 14:321.

———— 1938. Digestion of fat in *Amoeba proteus*. Biol. Bull., 75:389.

———— 1939. The relation between kind of food, growth and structure in Amoeba. *Ibid.*, 77:391.

———— 1941. Motor response in unicellular animals. In Calkins and Summers (1941).

———— 1942. The hydrogen ion concentration of the content of the food vacuoles and the cytoplasm in Amoeba, etc. Biol. Bull., 83:173.

———— and DOYLE, W. L., 1934. Ingestion of fluid by amoeba. Protoplasma, 20:555.

———— ———— 1935. Structure, origin and function of cytoplasmic constituents in *Amoeba proteus*. I. Arch. Protist., 86:155.

———— ———— 1935a. II. *Ibid.*, 86:278.

———— and PACE, D. M., 1938. The effect of substances produced by *Chilomonas paramecium* on the rate of reproduction. Physiol. Zool., 11:359.

———— and PROSSER, C. L., 1932. Effect of temperature, salts, and hydrogen ion concentration on rupture of the plasmagel sheet, etc. J. Cell. Comp. Physiol., 1:333.

MAZIA, D., and HIRSHFIELD, H. I., 1951. Nucleus-cytoplasm relationships in the action of ultraviolet radiation on *Amoeba proteus*. Exper. Cell. Res., 2:58.

MCCALLA, D. R., 1962. Chloroplasts of *Euglena gracilis* affected by furadantin. Science, 137:225.

MELDRUM, N. U., 1934. Cellular respiration. London.

MERTON. H., 1935. Versuche zur Geotaxis von Paramecium. Arch. Protist., 85:33.

METALNIKOFF, S., 1912. Contributions à l'étude de la digestion intracellulaire chez les protozoaires. Arch. zool. exper. gén., 9:373.

METCHNIKOFF, E., 1889. Recherches sur la digestion intracellulaire. Ann. Inst. Pasteur, 3:25.

MILLER, E. D. W., 1932. Reappropriation of cytoplasmic fragments. Arch. Protist., 78:635.

MOST, H. (edited by) 1951. Parasitic infections in man. New York.

MOUTON, H., 1912. Recherches sur la digestion chez les amibes et sur leur diastase intracellulaire. Ann. Inst. Pasteur., 16:457.

NASSET, ELIZABETH C., and KOFOID, C. A., 1928. The effects of radium and radium in combination with metallic sensitizers on *Entamoeba dysenteriae in vitro*. Univ. Cal. Publ. Zool., 31:387.

NIRENSTEIN, E., 1925. Ueber die Natur und Stärke der Säurebildung in den Nahrungsvakuolen von *Paramecium caudatum*. Ztschr. wiss. Zool., 125: 513.

NOLAND, L. E., 1927. Conjugation in the ciliate *Metopus sigmoides*. J. Morphol. Physiol., 44:341.

—— 1957. Protoplasmic streaming: a perennial puzzle. J. Protozool., 4:1.

ORMEROD, W. E., 1958. A comparative study of cytoplasmic inclusions (volutin granules) in different species of trypanosomes. J. Gen. Microbiol., 19:271.

PACE, D. M., and BELDA, W. H., 1944. The effect of food content and temperature on respiration in *Pelomyxa carolinensis* Wilson. Biol. Bull., 86:146.

—— and KIMURA, T. E., 1946. Relation between metabolic activity and cyanide inhibition in *Pelomyxa carolineneis*. Proc. Soc. Exper Biol., 62: 223.

PANTIN, C. F. A., 1923. On the physiology of amoeboid movement. I. J. Marine Biol. Ass., Plymouth, 13:24.

PANZER, T., 1913. Beitrag zur Biochemie der Protozoen. Hoppe Seylers Ztschr. phys. Chem., 86:33.

PAPPAS, G. D., 1954, Structural and cytochemical studies of the cytoplasm in the family Amoebidae. Ohio J. Sci., 54:195.

PÁRDUCZ, B., 1953. Zur Mechanik der Zilienbewegung. Acta Biol. Acad. Sci. Hungaricae, 4:177.

—— 1954. Reizphysiologische Untersuchungen an Ziliaten. II. *Ibid.*, 5: 169.

—— 1958. Das Interziliäre Fasersystem in seiner Beziehungen zu gewissen Fibrillenkomplexen der Infusorien. *Ibid.*, 8:191.

PHELPS, A., 1935. Growth of protozoa in pure culture. I. J. Exper. Zool., 70:109.

—— 1959. Effect of visible light on the growth of *Tetrahymena pyriformis*. Ecology, 40:512.

PIGON, P., 1959. Respiration of *Colpoda cucullus* during active life and encystment. J. Protozool., 6:303.

POWERS, P. B. A., 1932. *Cyclotrichium meunieri*, etc. Biol. Bull., 63:74.

PRATJE, A., 1921. Makrochemische, quantitative Bestimmung des Fettes und Cholesterins, sowie ihrer Kennzahlen bei *Noctiluca miliaris*. Biol. Zentralbl., 41:433.

PRINGSHEIM, E. G., 1923. Zur Physiologie der saprophytischer Flagellaten. Beitr. allg. Bot., 2:88.

—— 1928. Physiologische Untersuchungen an *Paramecium bursaria*, Arch. Protist., 64:289.

—— 1937. Beiträge zur Physiologie der saprophytischer Algen und Flagellaten. I. Planta, 26:631.

—— 1937a. Algenreinkulturen. Arch. Protist., 88:143.

—— 1952. On the nutrition of Ochromonas. Quart. J. Micro. Sci., 93:71.

—— and Hovasse, R., 1948. The loss of chromatophores in *Euglena gracilis*. The New Phytologist., 47:52.

Pütter, A., 1905. Die Atmung der Protozoen. Ztschr. allg. Physiol., 5:566.

—— 1908. Methoden zur Forschung des Lebens der Protisten. Tigerstedt's Handb. physiol. Methodik., 1:1.

Ray, D. L., 1951. Agglutination of bacteria: a feeding method in the soil ameba *Hartmannella* sp. J. Exper. Zool., 118:443.

Reeves, R. E., and Frye, W. W., 1960. Cultivation of *Entamoeba histolytica* with penicillin-inhibited *Bacteroides symbiosus* cells. J. Parasit., 46:187.

Reichenow, E., 1909. Untersuchungen an *Haematococcus pluvialis,* etc. Berlin. Sitz.-Ber. Ges. naturf. Freunde, p. 85.

Rhumbler, L., 1910. Die verschiedenartigen Nahrungsaufnahmen bei Amoeben als Folge verschiedener Colloidalzustände ihrer Oberflächen. Arch. Entw. Organ., 30:194.

Richards, O. W., 1941. The growth of the protozoa. In Calkins and Summers (1941).

Robertson, T. B., 1921. Experimental studies on cellular multiplication. I. Biochem. J., 15:595.

—— 1921a. II. *Ibid.,* 15:612.

—— 1924. The nature of the factors which determine the duration of the period of lag in cultures in Infusoria. Australian J. Exper. Biol., 1:105.

—— 1924a. The influence of washing upon the multiplication of isolated Infusoria and upon the allelocatalytic effect in cultures initially containing two Infusoria. *Ibid.,* 1:151.

—— 1924b. Allelocatalytic effect in cultures of Colpidium in hay infusion and in synthetic media. Biochem. J., 18:1240.

—— 1927. On some conditions affecting the viability of Infusoria and the occurrence of allelocatalysis therein. Australian J. Exper. Biol., 4:1.

Rosen, W. G., and Gawlik, S. R., 1961. Effect of streptomycin on chlorophyll accumulation in *Euglena gracilis*. J. Protozool., 8:90.

Roskin, G., 1925. Ueber die Axopodien der Heliozoa und die Greiftentakel der Ephelotidae. Arch. Protist., 52:207.

—— and Levinsohn, L., 1926. Die Oxydasen und Peroxydasen bei Protozoen. *Ibid.,* 56:145.

Rudzinska, M. A., 1953. Giant individuals and vigor of populations in *Tokophrya infusionum*. Ann. N. Y. acad. Sci., 56:1087.

—— and Trager, W., 1957. Intracellular phagotrophy by malaria parasites: an electron microscope study of *Plasmodium lophurae*. J. Protozool., 4:190.

—— —— 1959. Phagotrophy and two new structures in the malaria parasite *Plasmodium berghei*. J. Biophys. Biochem. Cytol., 6:103.

—— —— 1962. Intracellular phagotrophy in *Babesia rodhaini* as revealed by electron miscroscopy. J. Protozool., 9:279.

Rumjantzew, A., and Wermel, E., 1925. Untersuchungen ueber den Protoplasmabau, etc. Arch. Protist., 52:217.

SAMUELS, R., 1957. Studies of *Tritrichomonas batrachorum*. I. J. Protozool., 4-110.

SANDON, H., 1932. The food of protozoa. Publ. Fac. Sci. Egyptian University, 1:1.

SASSUCHIN, D. N., 1935. Zum Studium der Protisten- und Bakterienkerne. I. Arch. Protist., 84:186.

SCHAEFFER, A. A., 1920. Amoeboid movement. Princeton.

SCHEWIAKOFF, W., 1894. Ueber die Natur der sogennannten Exkretkörner der Infusorien. Ztschr. wiss. Zool., 57:32.

SCHOENBORN, H. W., 1954. Mutations in *Astasia longa* induced by radiation. J. Protozool., 1:170.

———— 1956. Protection against lethal damage induced by ultraviolet radiation. J. Protozool., 3:97.

SCHULZE, K. L., 1951. Experimentelle Untersuchungen ueber die Chlorellen-symbiose bei Ciliaten. Biol. Gen., Vienna, 19:281.

SEAMAN, G. R., 1955. Metabolism of free-living ciliates. In Hutner and Lwoff: Protozoa. II.

———— 1959. Cytochemical evidence for urease activity in Tetrahymena. J. Protozool., 6:331.

———— and HOULIHAN, R. K., 1951. Enzyme systems in *Tetrahymena geleii* S. II. J. Cell. Comp. Physiol., 37:309.

SHAPIRO, N. N., 1927. The cycle of hydrogen-ion concentration in the food vacuoles of Paramecium, Vorticella, and Stylonychia. Tr. Am. Micr. Soc., 46:45.

SINGH, B. N., 1945. The selection of bacterial food by soil amoebae, and the toxic effects of bacterial pigments and other products on soil protozoa. Brit. J. Exper. Path., 26:316.

———— 1960. Interrelationship between micropredators and bacteria in soil. Proc. 47th Indian Sci. Congr., Pt. 2, 14 pp.

STEINBERG, P., 1959. The cause of giantism and cannibalism in *Blepharisma undulans*. Biol. Rev. City Coll., N.Y., 21:4.

STEMPELL, W., 1924. Weitere Beiträge zur Physiologie der pulsierenden Vakuole von Paramecium. I. Arch. Protist., 48: 342.

STOLC, A., 1900. Beobachtungen und Versuche ueber die Verdauung und Bildung der Kohlenhydrate bei einen amoebenartigen Organismen, *Pelomyxa palustris*. Ztschr. wiss. Zool., 68:625.

TANABE, M., and KOMADA, K., 1932. On the cultivation of *Balantidium coli*. Keijo J. Med., 3:385.

TARTAR, V., 1954. Anomalies in regeneration of Paramecium. J. Protozool., 1:11.

TAYLOR, C. V., 1923. The contractile vacuole in Euplotes, etc. J. Exper. Zool., 37:259.

TRAGER, W., 1932. A cellulase from the symbiotic intestinal flagellates of termites, etc. Biochem. J., 26:1762.

TURNER, J. P., 1940. Cytoplasmic inclusions in the ciliate *Tillina canalifera*. Arch. Protist., 93:255.

VELEY, L. J., 1905. A further contribution to the study of *Pelomyxa palustris.* J. Linn. Soc. Zool., 29:374.

VERWORN, M., 1889. Psycho-physiologische Protisten-Studien. Jena.

—— 1903. Allgemeine Physiologie. 4 ed. Jena.

VON BRAND, T., et al., 1959. Chemical composition of the culture form of *Trypanosoma cruzi.* Exper. Parasit., 8:171.

WALDNER, H., 1956. Das Glykogen in *Pelomyxa palustris* Greeff. II. Zeitschr. vergl. Physiol., 38:334.

WEATHERBY, J. H., 1927. The function of the contractile vacuole in *Paramecium caudatum.* Biol. Bull., 52:208.

—— 1929. Excretion of nitrogenous substances in protozoa. Physiol. Zool., 2:375.

—— 1941. The contractile vacuole. In Calkins and Summers (1941).

WEINECK, E., 1934. Die Celluloseverdauung bei den Ciliaten des Wiederkäuermagens. Arch. Protist., 82:169.

WENZEL, F., 1955. Ueber eine Artenstehung innerhalb der Gattung Spathidium. Arch. Protist., 100:515.

WHIPPLE, G. C., 1927. The microscopy of drinking water. 4th ed. New York.

WICHTERMAN, R., 1948. The biological effects of x-rays on mating types and conjugation of *Paramecium bursaria.* Biol. Bull., 94:113.

—— 1948a. The presence of optically active crystals in *Paramecium bursaria* and their relationship to symbiosis. Anat. Rec., 101:97.

—— 1959. Mutation in the protozoan *Paramecium multimicronucleatum* as a result of x-irradiation. Science, 129:207.

—— 1961. Survival and reproduction of Paramecium after x-irradiation. J. Protozool., 8:158.

WILBER, C. G., 1945. Origin and function of the protoplasmic constituents in *Pelomyxa carolinensis.* Biol. Bull., 88:207.

—— 1945a. The composition of the refractive bodies in rhizopod, etc. Tr. Am. Micr. Soc., 64:289.

—— 1946. Notes on locomotion in *Pelomyxa carolinensis. Ibid.,* 65:318.

—— and SLANE, G. M., 1951. The effect of ultraviolet light on the protoplasm in *Pelomyxa carolinensis. Ibid.,* 70:265.

WILLIAMS, D. B., 1962. Sensitivity of *Spathidium spathula* to low doses of x-radiation. J. Protozool., 9:119.

WITTNER, M., 1957. Effects of temperature and pressure on oxygen poisoning of Paramecium. J. Protozool., 4:20.

—— 1957a. Inhibition and reversal of oxygen poisoning in Paramecium. *Ibid.,* 4:24.

WOLKEN, J. J., and SHIN, E., 1958. Photomotion in *Euglena gracilis.* J. Protozool., 5:39.

YOCOM, H. B., 1934. Observations on the experimental adaptation of certain freshwater ciliates to sea water. Biol. Bull., 67:273.

ZEUTHEN, E., 1948. Reduced weight and volume during starvation of the amoeba, etc. C. R. Lab. Carlsberg, Sér. Chim., 26:267.

ZINGHER, J. A., 1934. Beobachtungen an Fetteinschlüssen bei einigen Protozoen. Arch. Protist., 82:57.

ZUELZER, M., 1905. Ueber die Einwirkung der Radiumstrahlen auf Protozoen. *Ibid.*, 5:358.

—— 1907. Ueber den Einfluss des Meerwassers auf die pulsierende Vacuole. Berlin. Sitz.-Ber. Ges. naturf. Freunde, p. 90.

—— 1927. Ueber *Amoeba biddulphiae*, etc. Arch. Protist., 57:247.

ZUMSTEIN, H., 1900. Zur Morphologie und Physiologie der *Euglena gracilis*. Pringsheims Jahrb. wiss. Botanik., 34:149.

Chapter 5

Reproduction

THE mode of reproduction in protozoa is highly variable among different groups, although it is primarily a cell division. The reproduction is initiated by the nuclear division in nearly all cases, which will therefore be considered first.

Nuclear Division

Between a simple direct division on the one hand and a complicated indirect division which is comparable with the typical metazoan mitosis on the other hand, all types of nuclear division occur.

Amitosis. Although not so widely found as it was thought to be in former years, amitosis occurs normally and regularly in many forms. While the micronuclear division of the Ciliophora is mitotic (p. 201), the macronuclear division in invariably amitosis. The sole exception to this general statement appears to be the so-called promitosis reported by Ivanić (1938) in the macronucleus in the "Vermehrungsruhe" stage of *Chilodonella uncinata* in which chromosomes and spindle-fibers were observed. In *Paramecium caudatum* (Fig. 55), the micronucleus initiates the division by mitosis and the macronucleus elongates itself without any visible changes in its internal structure. The elongated nucleus becomes constricted through the middle and two daughter nuclei are produced.

It is assumed that the nuclear components undergo solation during division, since the formed particles of nucleus which are stationary in the resting stage manifest a very active Brownian movement. Furthermore, in some cases the nuclear components may undergo phase reversal, that is to say, the chromatin granules which are dispersed phase in the non-staining fluid dispersion medium in the resting nucleus, become dispersion medium in which the latter is suspended as dispersed phase. By using Feulgen's nucleal reaction, Reichenow (1928) demonstrated this reversal phenomenon in the division of the macronucleus of *Chilodonella cucullulus* (Fig. 56).

178

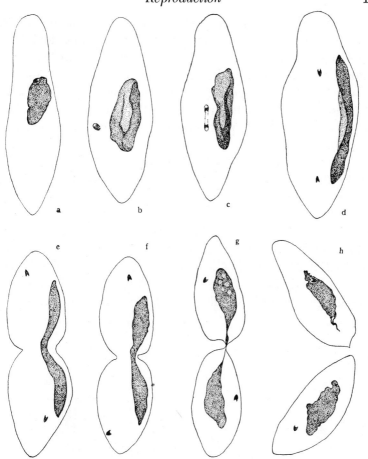

F<small>IG</small>. 55. Nuclear and cytoplasmic division of *Paramecium caudatum* as seen in stained smears, ×260 (Kudo).

The macronucleus becomes at the time of its division somewhat enlarged and its chromatin granules are more deeply stained than before. But chromosomes which characterize the mitotic division are entirely absent, although in a few forms in which mating types occur, the type difference and certain other characters, according to Sonneborn and Kimball, appear to be under control of genic constituents of the macronucleus. Since the number of chromatin granules appear approximately the same in the macronuclei of different generations of a given species, the reduced number of chromatin granules must be restored sometime before the next division takes place. Calkins (1926) is of the opinion that "each granule elongates and divides into two parts, thus doubling the number of chromomeres." Reichenow (1928) found that in *Chilodonella cucullulus* the lightly Feulgen positive endosome appeared to form chromatin granules and Kudo (1936) maintained that the large chromatin spherules of the macronucleus of *Nyctotherus ovalis* probably produce smaller spherules in their alveoli (Fig. 3).

When the macronucleus is elongated as in Spirostomum, Stentor, Euplotes, etc., the nucleus becomes condensed into a rounded form prior to its division. During the "shortening period" of the elongated macronuclei prior to division, there appear one to three characteristic zones which have been called by various names, such as nuclear clefts, reconstruction bands, reorganization bands, etc. In *Euplotes patella* (*E. eurystomus*), Turner (1930) observed prior to the division of the macronucleus a reorganization band consisting of a faintly staining zone ("reconstruction plane") and a

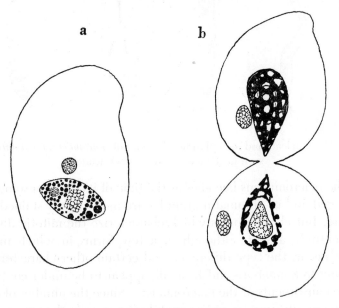

a   b

Fig. 56. The solation of chromatin during the macronuclear division of *Chilodonella cucullulus*, as demonstrated by Feulgen's nucleal reaction, ×1800 (Reichenow).

deeply staining zone ("solution plane"), appears at each end of the nucleus (Fig. 57, *a*) and as each moves toward the center, a more chromatinic area is left behind (*b-d*). The two bands finally meet in the center and the nucleus assumes an ovoid form. This is followed by a simple division into two. In the T-shaped macronucleus of *E. woodruffi*, according to Pierson (1943), a reorganization band appears first in the right arm and the posterior tip of the stem of the nucleus. When the anterior band reaches the junction of the arm and stem, it splits into two, one part moving along the

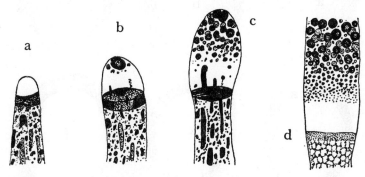

Fig. 57. Macronuclear reorganization before division in *Euplotes eurystomus*, ×240 (Turner). a, reorganization band appearing at a tip of the macronucleus; b–d, later stages.

left arm to its tip, and the other entering and passing down the stem to join the posterior band.

According to Summers (1935), a process similar to that of *E. eurystomus* occurs in *Diophrys appendiculata* and *Stylonychia pustulata;* but in *Aspidisca lynceus* (Fig. 58) a reorganization band appears first near the middle region of the macronucleus (*b*), divides into two and each moves toward an end, leaving between them a greater chromatinic content of the reticulum (*c-i*). Summers suggested that "the reorganization bands are local regions of karyolysis and resynthesis of macronuclear materials with the possibility of an elimination of physically or possibly chemically modified nonstaining substances into the cytoplasm." Weisz (1950a) finds that the nodes of the moniliform macronucleus of *Stentor coeruleus* contain different concentration of thymonucleic acid which is correlated with morphogenetic activity of individual

Protozoology

nodes, and that fusion of ill-staining nodes results in a return of strong affinity to methyl green. It appears, therefore, concentration of bandform or moniliform macronucleus prior to division may serve to recover morphogenetic potential prior to division.

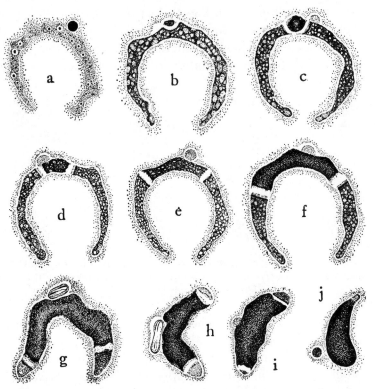

Fig. 58. Macronuclear reorganization prior to division in *Aspidisca lynceus*, ×1400 (Summers). a, resting nucleus; b–i, successive stages in reorganization process; j, a daughter macronucleus shortly after division.

(Reorganization under electron microscope, Fauré-Fremiet *et al.*, 1959.)

In a small number of ciliates, the macronucleus is distributed as many small bodies throughout the cytoplasm. In *Urostyla grandis*, the macronuclear material is lodged in 100 or more small bodies scattered in the cytoplasm. Prior to fission, all macronuclear bodies fuse with one another and form one macronucleus which then divides three times into eight and the latter are evenly distributed between the two daughter individuals, followed by divisions until the number reaches 100 or more (Raabe, 1947). On the other hand, in *Dileptus anser* (Fig. 310, *c*), "each granule divides where it happens to be and with the majority of granules both halves remain in one daughter cell after division" (Calkins). Hayes noticed a similar division, but at the time of simultaneous division prior to cell division, each macronucleus becomes elongated and breaks into several small nuclei.

The extrusion of a certain portion of the macronuclear material during division has been observed in a number of species. In *Uroleptus halseyi*, Calkins actually noticed each of the eight macronuclei is "purified" by discarding a reorganization band and an "x-body" into the cytoplasm before fusing into a single macronucleus which then divides into two nuclei. In the more or less rounded macronucleus that is commonly found in many ciliates, no reorganization band has been recognized. A number of observers have, however, noted that during the nuclear division there appears and persists a small body within the nuclear figure, located at the division plane as in the case of Loxocephalus (Behrend), Eupoterion (MacLennan and Connell) and even in the widely different protozoan, *Endamoeba blattae* (Kudo, 1926).

Kidder (1933) observed that during the division of the macronucleus of *Conchophthirus mytili* (Fig. 59), the nucleus "casts out a part of its chromatin at every vegetative division," which "is broken down and disappears in the cytoplasm of either daughter organism." A similar phenomenon has since been found further in *C. anodontae*, *C. curtus*, *C. magna* (Kidder), *Urocentrum turbo*, *Colpidium colpoda*, *C. campylum*, *Glaucoma scintillans* (Kidder and Diller), *Allosphaerium convexa* (Kidder and Summers), *Colpoda inflata*, *C. maupasi*, *Tillina canalifera*, *Bresslaua vorax* (Burt et al., 1941), *Disematostoma colpidioides* (Tuffrau and Savoie, 1961), *Ophryoglena singularis* (Canella and Trincas, 1961), etc. Beers (1946, 1963) noted chromatin extrusion from the macronucleus during division and in permanent cysts in *Tillina magna*, and in *Conchophthirus curtus*, a chromatin mass was extruded from the macronuclei while the two daughters were still united.

What is the significance of this phenomenon? Kidder and his associates believe that the process is probably elimination of waste substances of the prolonged cell-division, since chromatin extrusion does not take place during a few divisions subsequent to reorganization after conjugation in *Conchophthirus mytili* and since in Colpidium and Glaucoma, the chromatin elimination appears to be followed by a high division rate and infrequency of conjugation. Dass (1950) noticed a dark body between two daughter macronuclei of a ciliate designated by him as *Glaucoma*

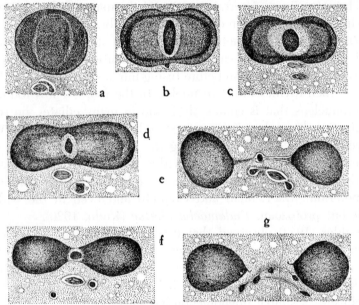

Fig. 59. Macronuclear division in *Conchophthirus mytili*, ×440 (Kidder).

pyriformis and considered it as surplus deoxyribonucleic acid about to be converted by the cytoplasm to ribonucleic acid necessary for active growth. Kaneda (1960) noticed also the extrusion of macronuclear substances into the cytoplasm during division of *Chlamydodon pedarius* and considered that the process may aid in restoring the organism to more normal condition which was lost due to high division rate and infrequent conjugation.

In *Paramecium aurelia*, Woodruff and Erdmann (1914) reported the occurrence of "endomixis." At regular intervals of about thirty days, the old macronucleus breaks down and disappears, while each of the two micronuclei divides twice, forming eight nuclei. Of these, six disintegrate. The animal then divides into two, each daughter individual receiving one micronucleus. This nucleus soon divides twice into four, two of which develop into two macronuclei, while the other two divide once more. Here the organism divides again into two individuals, each bearing one macronucleus and two micronuclei. This process, they maintained, is "a complete periodic nuclear reorganization without cell fusion in a pedigreed race of Paramecium." The so-called endomixis has since been reported to occur in many ciliates. However, as pointed out by Wilson (1928), Diller (1936), Sonneborn (1947) and others, there are several difficulties in holding that endomixis is a valid process. Diller considers that endomixis may have been based upon partial observations on hemixis (p. 244) or autogamy (p. 240). Sonneborn could not find any indication that this process occurs in numerous stocks and varieties of *Paramecium aurelia,* including the progeny of the strains studied by Woodruff, and maintained that endomixis does not occur in this species of Paramecium.

As has been stated already, two types of nuclei: macronucleus and micronucleus, occur in Ciliata and Suctoria. The macronucleus is the center of the whole metabolic activity of the organism and in the absence of this nucleus, the animal perishes. The waste substances which become accumulated in the macronucleus through its manifold activities, are apparently eliminated at the time of division, as has been cited above in many species. On the other hand, it is also probable that under certain circumstances, the macronucleus becomes impregnated with waste materials which cannot be eliminated through this process. Prior to and during conjugation (p. 223) and autogamy (p. 240), the macronucleus becomes transformed, in many species, into irregularly coiled thread-like structure (Fig. 87) which undergoes segmentation into pieces and finally is absorbed by the cytoplasm. New macronuclei are produced from some of the division-products of micronuclei by probably incorporating the old macronuclear material. In most cases, this supposition is not demonstrable. However, Kidder (1938) has shown, in the encysted *Paraclevelandia simplex,* an endocommensal of the colon of certain wood-feeding roaches, this is actually the case; namely, one of the divided micronuclei fuses directly with a part of macronucleus to form a macronuclear anlage which then develops into a macronucleus after passing through "ball-of-yarn" stage similar to that which appears in an exconjugant of Nyctotherus (Fig. 87).

Since the macronucleus originates in a micronucleus, it must contain all structures which characterize the micronucleus. Why then does it not divide mitotically as does the micronucleus? During conjugation or autogamy in a ciliate, the macronucleus degenerates, disintegrates and finally becomes absorbed in the cytoplasm. In *Paramecium aurelia,* Sonneborn (1940, 1942, 1947) (Fig. 60) observed that when the animal in conjugation is exposed to 38°C. from the time of the synkaryon-formation until before the second postzygotic nuclear division (*a-c*), the development of the two newly formed macronuclei is retarded and do not divide as usual with the result that one of the individuals formed by the second postzygotic division receives the newly formed macronucleus, while the other lacks this (*c*). In the latter, however, division continues, during which some of the original twenty to forty pieces of the old macronucleus that have been present in the cytoplasm segregate in approximately equal number at each division (*d, e*) until there is only one in the animal (*f*). Thereafter the macronucleus divides at each division (*g*). Sonneborn found this "macronuclear regeneration" in the varieties 1 and 4, but considered that it occurs in all stocks. Thus the macronucleus in this ciliate appears to be a compound structure with its 20-40 component parts, each containing all that is needed for development into a complete macronucleus. From these observations, Sonneborn concludes that the macronucleus in *P. aurelia* appears to undergo amitosis, since it is a compound nucleus composed of many "subnuclei" and since at fission all that is necessary to bring about genetically equivalent functional macronuclei is to segregate these multiple subnuclei into two random groups. Jurand *et al.* (1962) find a large number of structures, 0.5µ in diameter, in the macronucleus of *P. aurelia,* consisting of an outer RNA-containing coat and central DNA elements and consider them to be "subnuclei."

While the macronuclear division usually follows the micronuclear division, it takes place in the absence of the latter as seen in amicronucleate individuals of ciliates which possess normally a micronucleus. Amicronucleate ciliates have been found to occur naturally or produced experimentally in the following species: *Didinium nasutum* (Thon, 1905; Patten, 1921), *Oxytricha hymenostoma* (Dawson, 1919), *O. fallax, Urostyla grandis* (Woodruff 1921), *Paramecium caudatum* (Landis, 1920; Woodruff, 1921), *Tetrahymena pyriformis,* etc.

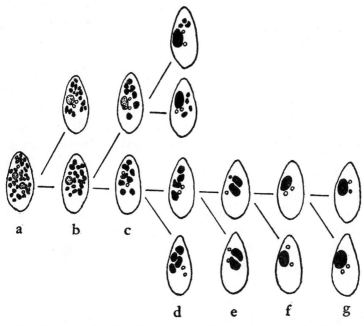

Fɪɢ. 60. Diagram showing the macronuclear regeneration in *Paramecium aurelia* (Sonneborn). a, an individual before the first division after conjugation or autogamy, containing two macronuclear (stippled) anlagen, two micronuclei (rings) and about 30 disintegrating (solid black) masses of the old macronucleus; b, two individuals formed by the first division, each containing one macronuclear anlage, two micronuclei and macronuclear masses; c, two individuals produced by the second division: one (above) with the new macronucleus, two micronuclei and macronuclear masses, and the other without new macronucleus; d–f, binary fissions in which the two micronuclei divide, but old macronuclear masses are distributed equally between the two daughters until there are one large regenerated macronucleus and two micronuclei; g, division following f, goes on in an ordinary manner.

Amicronucleate *Oxytricha fallax* which were kept under observation by Reynolds (1932) for twenty-nine months, showed the same course of regeneration as the normal individuals. Beers (1946b) saw no difference in vegetative activity between amicronucleates and normal individuals of *Tillina magna*. In *Euplotes patella,* amicronucleates arise from "double" forms (p. 281) with a single micronucleus, and Kimball (1941a) found that the micronucleus is not essential for continued life in at least some clones, though its absence results in a marked decrease in vigor. The bimicronucleate *Paramecium bursaria* which Woodruff (1931) isolated, developed in the course of seven years of cultivation, unimicronucleate and finally amicronucleate forms, in which no marked variation in the vitality of the race was observed. These data indicate that amicronucleates are capable of carrying on vegetative activity and multiplication, but are unable to conjugate or if cell-pairing occurs, the result is abortive, though Chen (1940c) reported conjugation between normal and amicronucleate individuals of *P. bursaria* (p. 226). Elliott and Nanny (1952) found that *Tetrahymena pyriformis* strain E was amicronucleate for at least twenty years.

More than half of the collection of *T. pyriformis* from South and Central America was found to be amicronucleates (Elliott and Hayes). Thus amicronucleates are not rare in nature. They of course cannot mate and therefore must be considered senile (Sonneborn, 1957). On the other hand, Horváth (1950) succeeded in destroying the micronucleus in *Kahlia simplex* (p. 161) and found the emicronucleates as vigorous as the normal forms, judged by the division rate, but were killed within 15 days by proactinomycin, while normal individuals resisted by encystment. This worker reasons that the emicronucleates are easily destroyed by unfavorable conditions and, therefore, ciliates without a micronucleus occur rarely in nature. Miyake (1956, 1957) cultured *Paramecium caudatum* in media containing urea, and found both amicronucleates and multimicronucleates. The former showed diminution of body size and decrease in vitality. He believes that the micronucleus is essential for both reproduction and trophic activity of the animal.

Other examples of amitosis are found in the vegetative nuclei in the trophozoite of Myxosporida, as for example *Myxosoma cato-stomi* (Fig. 61), *Thelohanellus notatus* (Debaisieux), etc., in which the endosome divides first, followed by the nuclear constriction. In *Streblomastix strix*, the compact elongated nucleus was found to undergo a simple division by Kofoid and Swezy.

Mitosis. The mitotic division of the protozoan nuclei is of manifold types as compared with the mitosis in the metazoan cell, in

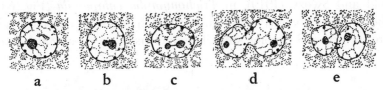

a   b   c   d   e

FIG. 61. Amitosis of the vegetative nucleus in the trophozoite of *Myxosoma catostomi*, ×2250 (Kudo).

which, aside from minor variations, the division is of a uniform pattern. Chatton, Alexeieff, and others, have proposed several terms to designate the various types of indirect nuclear division, but no one of these types is sharply defined. Grassé (1952) considered that the protozoan mitosis could be grouped into two types; namely, one in which equatorial plate and metaphase appear (orthomitosis) and the other in which these phases do not occur (pleuromitosis).

A veritable mitosis was noted by Dobell in the heliozoan *Oxnerella maritima* (Fig. 62), which possesses an eccentrically situated nucleus and a central centriole, from which radiate many axopodia (*a*). The first sign of the nuclear division is the slight enlargement, and migration toward the centriole, of the nucleus (*b*). The centriole first divides into two (*c*, *d*) and the nucleus becomes located between the two centrioles (*e*). Presently spindle fibers are formed and the nuclear membrane disappears (*f*, *g*). After passing through an equatorial-plate stage, the two groups of 24 chromosomes move toward the opposite poles (*g-i*). As the spindle fibers become indistinct, radiation around the centrioles becomes conspicuous and the two daughter nuclei are completely reconstructed to assume the resting phase (*j-l*). The mitosis of another heliozoan *Acanthocystis aculeata* is said to be very similar to the above. Aside from these two species, the centriole has been reported in many others, such as Hartmannella (Arndt), Euglypha, Monocystis (Bělař), Aggregata (Dobell; Bělař; Naville), and various Hypermastigida (Kofoid; Duboscq and Grassé; Kirby; Cleveland and his associates).

In numerous species the division of the centriole and a connecting strand between them, which has been called **desmose** (centrodesmose or paradesmose), have been observed. According to Kofoid and Swezy (1919), in *Trichonympha campanula* (Fig. 63), the prophase begins early, during which fifty-two chromosomes are formed and become split. The nucleus moves nearer the ante-

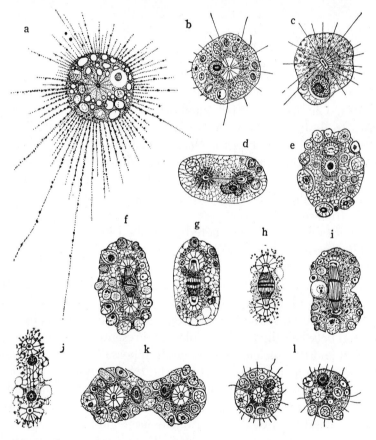

Fig. 62. Nuclear and cytoplasmic division in *Oxnerella maritima*, × about 1000 (Dobell). a, a living individual; b, stained specimen; c–g, prophase; h, metaphase i, anaphase; j, k, telophase; l, division completed.

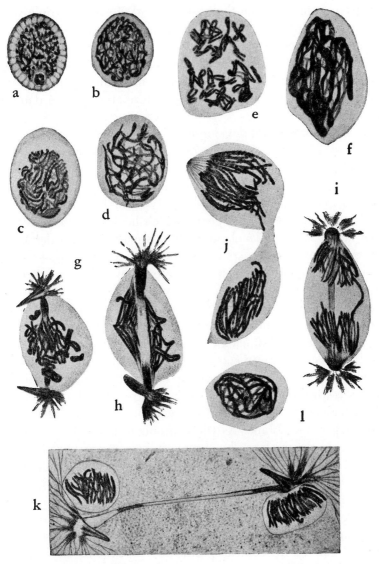

FIG. 63. Mitosis in *Trichonympha campanula,* × 800 (Kofoid and Swezy). a, resting nucleus; b–g, prophase; h, metaphase, i, j, anaphase; k, telophase; l, a daughter nucleus being reconstructed.

rior end where the centriole divides into two, between which develops a desmose. From the posterior end of each centriole, astral
rays extend out and the split chromosomes form loops and pass
through "tangled skein" stage. In the metaphase, the equatorial
plate is made up of V-shaped chromosomes as each of the split
chromosomes is still connected at one end, which finally becomes
separated in anaphase, followed by reformation of two daughter
nuclei.

As to the origin and development of the achromatic figure, various observations and interpretations have been advanced. Certain Hypermastigida possess very large filiform centrioles and a
large rounded nucleus. In Barbulanympha (Fig. 64), Cleveland
(1938a) found that the centrioles vary from 15 to 30μ in length
in the four species of the genus which he studied. They can be
seen, according to Cleveland, in life as made up of a dense hyaline protoplasm. When stained, it becomes apparent that the two
centrioles are joined at their anterior ends by a desmose and their
distal ends 20 to 30μ apart, each of which is surrounded by a special centrosome (*a*). In the resting stage no fibers extend from either centriole, but in the prophase, astral rays begin to grow out
from the distal end of each centriole (*b*). As the rays grow longer
(*c*), the two sets soon meet and the individual rays or fibers join,
grow along one another and overlap to form the central spindle (*d*).
In the resting nucleus, there are large irregular chromatin granules which are connected by fibrils with one another and also with
the nuclear membrane. As the achromatic figure is formed and
approaches the nucleus, the chromatin becomes arranged in a single spireme imbedded in matrix. The spireme soon divides longitudinally and the double spireme presently breaks up transversely
into paired chromosomes. The central spindle begins to compress
the nuclear membrane and the chromosomes become shorter and
move apart. The intra- and extra-nuclear fibrils unite as the process goes on (*e*), the central spindle now assumes an axial position,
and two groups of V-shaped chromosomes are drawn to opposite
poles. In the telophase, the chromosomes elongate and become
branched, thus assuming conditions seen in the resting nucleus.

Recently, Cleveland (1956) discovered that oxygen concentration of 70-80 per cent of an atmosphere destroys all chromo-

somes of Trichonympha of the woodroach, provided the treatment is carried out during the early stages of gametogenesis at which time the chromosomes are in the process of duplicating themselves. This treatment does not damage the cytoplasm or various organelles. The centrioles function in the production of

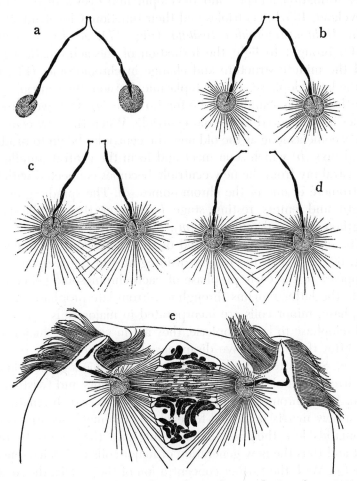

Fig. 64. Development of spindle and astral rays during the mitosis in Barbulanympha, ×930 (Cleveland). a, interphase centrioles and centrosomes; b, prophase centrioles with astral rays developing from their distal ends through the centrosomes; c, meeting of astral rays from two centrioles; d, astral rays developing into the early central spindle; e, a later stage showing the entire mitotic figure.

the achromatic figure, the flagella and parabasal bodies. Then follow the cytoplasmic division, thus producing two anucleate gametes which make some progress in the cytoplasmic differentiations characteristic of normal male and female gametes. It would thus appear that extranuclear centrioles are the center of spindle fiber formation. (Types and developmental cycles of centrioles, Cleveland, 1957b; centrioles and their function, Cleveland, 1960.)

In *Holomastigotoides tusitala* (Fig. 174, *a*, *b*), Cleveland (1949) brought to light the formation of the achromatic figure, and the minute structure and change in chromosomes (Fig. 65). In the late telophase, after cytoplasmic division, the centrioles follow the flagellar bands 4 and 5 for 1.5 turns (*a*). The two chromosomes are anchored to the old centriole. When the new centriole has become as long as the old one, the centrioles begin to produce astral rays (*b*) which soon meet and form the central spindle (*c*). An astral ray from the new centriole becomes connected with the centromere of one of the chromosomes (*d*). The spindle grows in length and enters resting stage (*e-j*), later the spindle fibers lengthen (*k*, *l*) and pull apart (*m*).

The chromosome is composed of the matrix and chromonema (Fig. 66), of which the former disintegrates in the telophase and reappears in the early prophase of each chromosome generation, while the latter remains throughout. From late prophase to mid-telophase, minor coils are incorporated in major coils (*a-c*); from mid-telophase to late telophase, they are in very loose majors (*d*); and after the majors have disappeared completely, they become free (*e*). Soon after cytoplasmic division, the majors become looser and irregular and finally disappear, while minors and twisting remain. Each chromosome presently divides into two chromatids (*f*) and a new matrix is formed for each. As the matrix contracts the chromatids lose their relational coiling and the minors become bent and thus the new generation of major coils makes its appearance (*g*). With the further concentration of the matrix, the majors become more conspicuous (*h*), the minors being incorporated into them. When most of the relational coiling has been lost and majors are close together, the chromosomal changes cease for days or weeks. This is the late prophase. After the resting stage, the achromatic figure commences to grow again (*i*, *j*) and the two

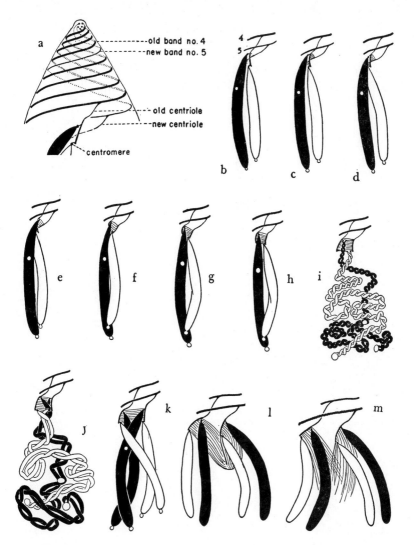

FIG. 65. Mitosis in *Holomastigotoides tusitala* (Cleveland). a, anterior region showing flagellar bands, centrioles, centromeres and chromosomes. b–h, telophase; i, j, prophase; k, metaphase; l, anaphase; m, telophase. b, c, new and old centrioles forming achromatic figure; d, one chromosome has shifted its connection from old to new centriole; e, f, flattening out of centrioles and achromatic figure; g, h, beginning of chromosomal twisting; i, chromosomes duplicated, producing many gyres of close-together relational coiling of chromatics, and centromeres duplicated; j, chromatids losing their relational coiling by unwinding; k, relational coiling disappeared achromatic figure elongating and separating sister chromatids; l, central spindle bent, chromatids in two groups; m, central spindle pulled apart.

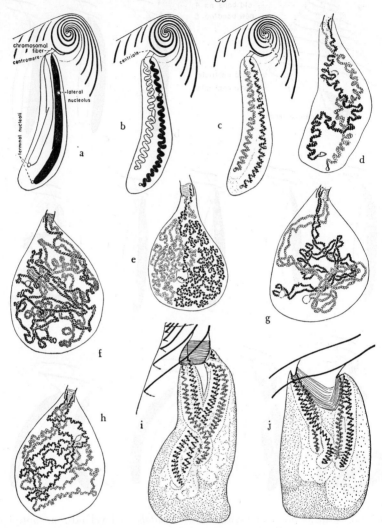

Fig. 66. Chromosomal changes in *Holomastigotoides tusitala*, ×1050 (Cleveland). a, telophase shortly after cytoplasmic division, new fifth band and new centriole are growing out and chromosomes are twisted; b, c, the same chromosome showing major and minor coils respectively; d, later telophase, showing minor coils; e, matrix completely disintegrated, showing minor coils; f, a prophase nucleus, showing division of chromosomes into two chromatids; g, later prophase, in which majors are developing with minors; h, later prophase; i, metaphase in which distal halves of the chromatids have not yet separated, showing minor coils; j, anaphase, showing major and minor coils of chromonemata.

groups of chromatids are carried to the poles, followed by transverse cytoplasmic division (Fig. 67). The coils remain nearly the same during metaphase to early telophase. Thus Cleveland showed the continuity of chromosomes from generation to generation. He finds that the resting stage of chromosomes varies in different types of cells: some chromosomes rest in interphase, some in early prophase and others in telophase, and that the centromere is an important structure associated with the movement of chromatids and in the reduction of chromosomes in meiosis. (For fuller information, the reader is referred to the profusely illustrated original papers, Cleveland, 1949, 1958, 1958a.)

In *Lophomonas blattarum,* the nuclear division (Fig. 68) is initiated by the migration of the nucleus out of the calyx. On the nuclear membrane is attached the centriole which probably originates in the blepharoplast ring; the centriole divides and the desmose which grows, now stains very deeply, the centrioles becoming more conspicuous in the anaphase when new flagella develop from them. Chromatin granules become larger and form a spireme, from which six to eight chromosomes are produced. Two groups of chromosomes move toward the opposite poles, and when the division is completed, each centriole becomes the center of formation of all motor organellae.

In some forms, such as Noctiluca (Calkins) Actinophrys (Bělař), etc., there may appear at each pole, a structureless mass of cytoplasm (centrosphere), but in a very large number of species there appear no special structures at poles and the spindle fibers become stretched seemingly between the two extremities of the elongating nuclear membrane. Such seems to be the case in Pelomyxa (Kudo) (Fig. 69), Cryptomonas (Bělař), Rhizochrysis (Doflein), Aulacantha (Borgert), and in micronuclear division of the majority of the Ciliata and Suctoria.

The behavior of the endosome during the mitosis differs among different species as are probably their functions. In *Eimeria schubergi* (Schaudinn), *Euglena viridis* (Tschenzoff), *Oxyrrhis marina* (Hall), *Colacium vesiculosum* (Johnson), *Haplosporidium limnodrili* (Granata), etc., the conspicuously staining endosome divides by elongation and constriction along with other chromatic elements, but in many other cases, it disappears during the early

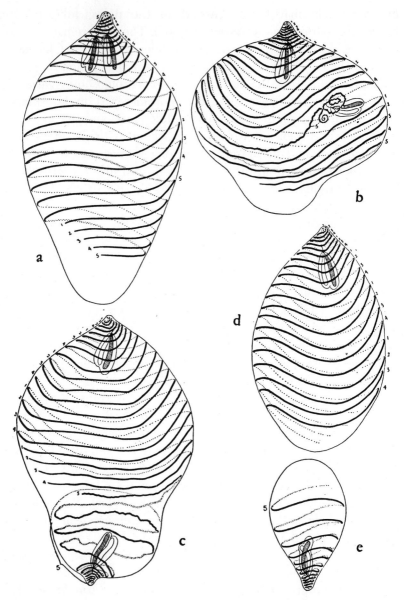

FIG. 67. Cytoplasmic division in *Holomastigotoides tusitala*, × about 430
(Cleveland). a, fifth flagellar band has separated from others; b, one
nucleus and fifth band moving toward posterior end; c, the movement of
the band and nucleus has been completed; d, e, anterior and posterior
daughter individuals, produced by transverse division.

part of division and reappears when the daughter nuclei are re-constructed as observed in Monocystis, Dimorpha, Euglypha, Pamphagus (Bělař), Acanthocystis (Stern), Chilomonas (Doflein), Dinenympha (Kirby), etc.

In the vegetative division of the micronucleus of *Conchopthirus anodontae,* Kidder (1934) found that prior to division the micro-

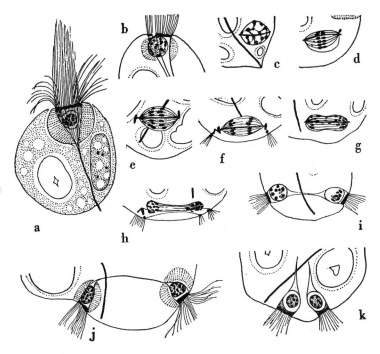

Fig. 68. Nuclear division in *Lophomonas blattarum,* ×1530 (Kudo). a, rest-ing nucleus; b, c, prophase d, metaphase; e–h, anaphase; i–k, telophase.

nucleus moves out of the pocket in the macronucleus and the chromatin becomes irregularly disposed in a reticulum; swelling continues and the chromatin condenses into a twisted band, a spi-reme, which breaks into many small segments, each composed of large chromatin granules. With the rapid development of the spindle fibers, the twelve bands become arranged in the equatori-al plane and condense. Each chromosome now splits longitudinal-ly and two groups of twelve daughter chromosomes move to op-posite poles and transform themselves into two compact daughter nuclei. A detailed study of micronuclear division (Fig. 70) of

Urostyla grandis was made by Raabe (1946). The micronucleus is a compact body in the interphase (*a*), but increases in size and the chromatin becomes grouped into small masses (*b, c*), which become associated into a spiral ribbon (*d-g*). The latter then breaks up into twelve segments that are arranged parallel to the

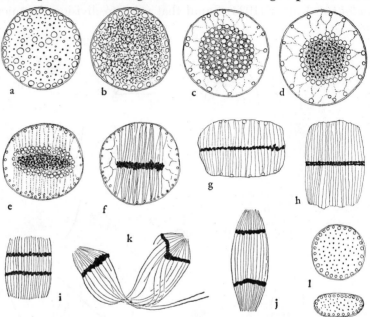

Fig. 69. Mitosis in *Pelomyxa carolinensis*, ×1150 (Kudo). a, c, l, in life; b, d–k, in acidified methyl green. a, b, resting nuclei; c–g, prophase; h, metaphase; i–k, anaphase; l, front and sided view of a young daughter nucleus.

axis of the elongating nucleus (*h-i*). Each segment condenses into a chromosome which splits longitudinally into two (*k*) and the two groups of chromosomes move to opposite poles (*l-p*). In *Zelleriella elliptica* (Fig. 376) and four other species of the genus inhabiting the colon of *Bufo valliceps,* Chen (1936, 1948) observed the formation of twenty-four chromosomes, each of which is connected with a fiber of the intranuclear spindle and splits lengthwise in the metaphase.

While in the majority of protozoan mitosis, the chromosomes split longitudinally, there are observations which suggest a

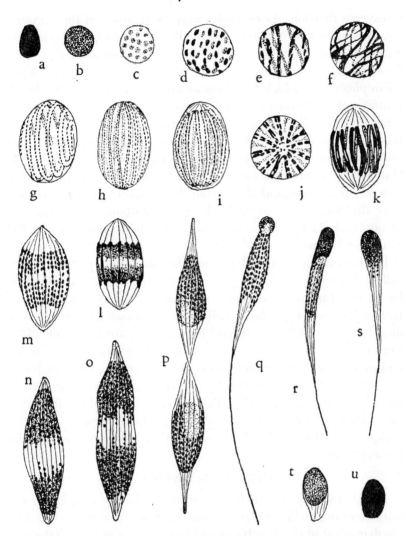

Fɪɢ. 70. Micronuclear division of *Urostyla grandis,* ×2100 (H. Raabe).
a, resting stage; b–j, prophase (b–e, stages in the formation of spireme;
f, g, spireme ribbon; h, i, twelve segments of ribbon arranged in the direction of the elongating nuclear axis; j, a polar view of the same); k, l, metaphase, condensation of the segments; m–o, anaphase; p, late anaphase;
q, a daughter nucleus in telophase r–t, reconstruction stages; u, a resting
daughter nucleus.

transverse division. As examples may be mentioned the chromosomal divisions in *Astasia laevis* (Bĕlař), *Entosiphon sulcatum* (Lackey), and a number of ciliates. In a small number of species, observations vary within a species, as, for example, in *Peranema trichophorum* in which the chromosomes were observed to divide transversely (Hartmann and Chagas) as well as longitudinally (Hall and Powell; Brown). It is inconceivable that the division of the chromosome in a single species of organism is haphazard. The apparent transverse division might be explained by assuming, as Hall (1937) showed in *Euglena gracilis*, that the splitting is not completed at once and the pulling force acting upon them soon after division, brings forth the long chromosomes still connected at one end. Thus, the chromosomes remain together before the anaphase begins.

In the instances considered on the preceding pages, the so-called chromosomes found in them, appear to be essentially similar in structure and behavior to typical metazoan chromosomes. In many other cases, the so-called chromosomes or "pseudochromosomes" are slightly enlarged chromatin granules which differ from the ordinary chromatin granules in their time of appearance and movement only. In these cases, it is, of course, not possible at present to determine how and when their division occurs before separating to the respective division pole. In Table 5 are listed the number of the "chromosomes" which have been reported by various investigators in the protozoa that are mentioned in the present work.

Cytoplasmic Division

The division of the nucleus is accompanied by division of extranuclear organelles such as chromatophores, pyrenoids, etc. The blepharoplast of the flagellates and kinetosomes of the ciliates undergo division, giving rise to daughter blepharoplasts and kinetosomes that become organized into characteristic locomotor organelles. (Morphogenesis in the apostomes, Chatton and Lwoff, 1935, Lwoff, 1950; mechanism of morphogenesis in ciliates, Fauré-Fremiet, 1948, Guilcher, 1950, Weisz, 1951, 1951a.)

Binary Fission. As in metazoan cells, the binary fission occurs very widely among the Protozoa. It is a division of the body through middle of the extended long axis into two nearly equal

TABLE 5

CHROMOSOMES IN PROTOZOA

Protozoa	Number of Chromosomes	Observers
Rhizochrysis scherffeli	22	Doflein
Haematococcus pluvialis	20–30	Elliott
Polytomella agilis	5	Doflein
Chlamydomonas spp.	10 (haploid)	Pascher
Pandorina morum	12 (haploid)	Dangeard; Coleman
Euglena pisciformis	12–15 (?)	Dangeard
E. viridis	30 or more	Dangeard
Phacus pyrum	30–40	Dangeard
Rhabdomonas incurva	About 12	Hall
Vacuolaria virescens	About 30	Fott
Syndinium turbo	5	Chatton
Anthophysis vegetans	8–10	Dangeard
Cercomonas longicauda	4–5	Dangeard
Collodictyon triciliatum	About 20	Bělař
Chilomastix gallinarum	About 12	Boeck and Tanabe
Eutrichomastix serpentis	5	Kofoid and Swezy
Dinenympha fimbriata	25–30	Kirby
Metadevescovina debilis	About 4	Light
Trichomonas tenax	3	Hinshaw
T. gallinae	6	Stabler
T. hominis	5 or 6	Bishop
T. vaginalis	5	Hawes
Tritrichomonas augusta	5	Kofoid and Swezy
	4 or 8	Kuczynski
	6	Samuels
T. batrachorum	4 or 8	Kuczynski
	6	Bishop
T. muris	6	Wenrich
Hexamita salmonis	5 or 6	Davis
Giardia intestinalis	4	Kofoid and Swezy
G. muris	4	Kofoid and Christiansen
Calonympha grassii	4 or 5	Janicki
Spirotrichonympha polygyra	2 doubles	Cup
	2	Cleveland
S. bispira	2	Cleveland
Lophomonas blattarum	16 or 8 doubles	Janicki
	8 or 6	Kudo
	12 or 6 doubles	Bělař
L. striata	12 or 6 doubles	Bělař
Barbulanympha laurabuda	40	Cleveland
B. ufalula	50	Cleveland
Rhynchonympha tarda	19	Cleveland
Urinympha talea	14	Cleveland
Staurojoenia assimilis	24	Kirby
Trichonympha campanula	52 or 26 doubles	Kofoid and Swezy
T. grandis	22	Cleveland
Plasmodiophora brassicae	8 (diploid)	Terby
Naegleria gruberi	14–16	Rafalko
N. bistadialis	16–18	Kühn

TABLE 5 (*Continued*)

Protozoa	Number of Chromosomes	Observers
Amoeba proteus	500–600	Liesche
Endamoeba disparata	About 12	Kirby
Entamoeba histolytica	6	Kofoid and Swezy; Uribe
E. coli	6	Swezy; Stabler
	4	Liebmann
E. gingivalis	5	Stabler; Noble
Endolimax nana	10	Dobell
Dientamoeba fragilis	4	Wenrich
	6	Dobell
Hydramoeba hydroxena	8	Reynolds and Threlkeld
Lesquereusia spiralis	175–200	Stump
Spirillina vivipara	12 (diploid)	Myers
Patellina corrugata	24 (diploid)	Myers
Pontigulasia vas	8–12	Stump
Actinophrys sol	44 (diploid)	Bělař
Oxnerella maritima	About 24	Dobell
Thalassicolla nucleata	4	Bělař
Aulacantha scolymantha	More than 1600	Borgert
	4 in gamogony	Bělař
Zygosoma globosum	12 (diploid)	Noble
Diplocystis schneideri	6 (diploid)	Jameson
Gregarina blattarum	6 (diploid)	Sprague
Nina gracilis	5 (haploid)	Léger and Duboscq
Actinocephalus parvus	8 (diploid)	Weschenfelder
Aggregata eberthi	12 (diploid)	Dobell; Bělař; Naville
Merocystis kathae	6 (haploid)	Patten
Adelea ovata	8–10 (diploid)	Greiner
Adelina deronis	20 (diploid)	Hauschka
Orcheobius herpobdellae	10–12	Kunze
Chloromyxum leydigi	4 (diploid)	Naville
Sphaerospora polymorpha	4 (diploid)	Kudo
Myxidium lieberkühni	4	Bremer
M. serotinum	4 (diploid)	Kudo
Sphaeromyxa sabrazesi	6	Debaisieux; Bělař
	4	Naville
S. balbianii	4	Naville
Myxobolus pfeifferi	4	Keysselitz; Mercier; Georgevitch
Protoopalina intestinalis	8 (diploid)	Metcalf
Zelleriella antilliensis	2 (?)	Metcalf
Z. intermedia	24	Chen
Didinium nasutum	16 (diploid)	Prandtl
Cyclotrichium meunieri	6	Powers
Chilodonella uncinata	4 (diploid)	Enrique; MacDougall
C. uncinata (tetraploid)	8; 4	MacDougall
Conchophthirus anodontae	12 (diploid)	Kidder
C. mytili	16 (diploid)	Kidder
Ancistruma isseli	About 5 (haploid)	Kidder
Paramecium aurelia	30–40	Diller
	About 35	Sonneborn

TABLE 5 (*Continued*)

Protozoa	Number of Chromosomes	Observers
P. caudatum	About 36	Penn
Tetrahymena pyriformis	10 (diploid)	Elliott and Hays; Ray
Blepharisma undulans japonicus	8 (diploid)	Suzuki
Stentor coeruleus	28 (diploid)	Mulsow
Tetratoxum unifasciculatum	About 14	Davis
Oxytricha bifaria	24 (diploid)	Kay
O. fallax	24 (diploid)	Gregory
Uroleptus halseyi	24 (diploid)	Calkins
Pleurotricha lanceolata	About 40 (dipl.)	Manwell
Stylonychia pustulata	6	Prowazek
Euplotes patella	6 (diploid)	Yocom; Ivanic
E. eurystomus	8 (diploid)	Turner
Vorticella microstoma	4	Finley
Carchesium polypinum	16 (diploid)	Popoff
Trichodina sp.	4–6	Diller

daughter individuals. In *Amoeba proteus,* Chalkley and Daniel found that there is a definite correlation between the stages of nuclear division and external morphological changes (Fig. 71).

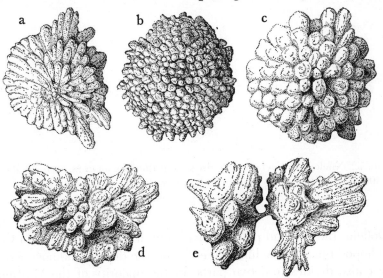

Fig. 71. External morphological changes during division of *Amoeba proteus,* as viewed in life in reflected light, × about 20 (Chalkley and Daniel). a, shortly before the formation of the division sphere; b, a later stage; c, prior to elongation; d, further elongation; e, division almost completed.

During the prophase, the organism is rounded, studded with fine pseudopodia and exhibits under reflected light a clearly defined hyaline area near its center (*a*), which disappears in the metaphase (*b*, *c*). During the anaphase, the pseudopodia rapidly become coarser; in the telophase, the elongation of body, cleft formation, and return to normal pseudopodia, take place.

In Testacea, one of the daughter individuals remains, as a rule, within the old test, while the other moves into a newly formed one, as in Arcella, Pyxidicula, Euglypha, etc. According to

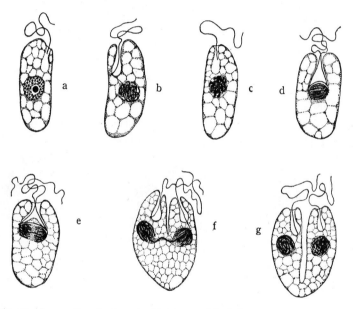

FIG. 72. Nuclear and cytoplasmic division in *Rhabdomonas incurva*, × about 1400 (Hall). a, resting stage; b, c, prophase; d, equatorial plate; e, f, anaphase; g, telophase.

Doflein, the division plane coincides with the axis of body in Cochliopodium, Pseudodifflugia, etc., and the delicate homogeneous test also divides into two parts. In the majority of the Mastigophora, the division is longitudinal, as is shown by that of *Rhabdomonas incurva* (Fig. 72). In certain dinoflagellates, such as Ceratium, Cochlodinium, etc., the division plane is oblique, while in forms such as Oxyrrhis (Dunkerly; Hall), the fission is transverse. In *Streblomastix strix* (Kofoid and Swezy, 1919), *Lophomonas striata* (Kudo, 1926b), *Spirotrichonympha bispira* (Cleveland, 1938), *Holomastigotoides tusitala* (Fig. 67) and others (Cleveland, 1947), and *Strombidium clavellinae* (Buddenbrock, 1922), the division takes place transversely but the polarity of the posterior individual is reversed so that the posterior end of the parent organism becomes the anterior end of the posterior daughter individual. In the ciliate Bursaria, Lund (1917), observed reversal of polarity in one of the daughter organisms at the time of division of normal individuals and also in those which regenerated after being cut into one-half the normal size.

In the Ciliophora, the division is as a rule transverse (Fig. 55), in which the body without any enlargement or elongation divides by constriction through the middle so that the two daughter individuals are about half as large at the end of division. Both individuals usually retain their polarity.

Multiple Division. In multiple division, the body divides into a number of daughter individuals, with or without residual cytoplasmic masses of the parent body. In this process, the nucleus may undergo either simultaneous multiple division, as in Aggregata, or more commonly, repeated binary fission, as in Plasmodium (Fig. 262) to produce large numbers of nuclei, each of which becomes the center of a new individual. The number of daughter individuals often varies, not only among the different species, but also within one and the same species. Multiple division occurs commonly in the Foraminiferida (Fig. 210); the Radiolarida (Fig. 221), and various groups of Sporozoa and Cnidosporidia in which the trophozoite multiplies abundantly by this method.

Budding. Multiplication by budding which occurs in the Protozoa is the formation of one or more smaller individuals from the parent organism. It is either exogenous or endogenous, depending upon the location of the developing buds or gemmules. Exogenous budding has been reported in Acanthocystis, Noctiluca (Fig. 130), Myxosporida, astomatous ciliates (Fig. 347), Chonotrichida, Suctoria (Fig. 379, *i*), etc. Endogenous budding has been found in Testacida, Gregarinida, Myxosporida as well as Suctoria. Collin observed a unique budding in *Tokophrya cyclopum* in which the entire body, except the stalk, transforms itself into a young bud. Hull (1954) made a detailed study of the bud formation in another suctorian *Solenophrya micraster* (Fig. 73).

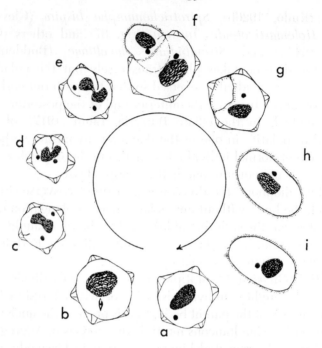

FIG. 73. Bud formation in the suctorian, *Solenophrya micraster* (Hull). a–e,
bud formation; f, emergence of a bud; h, i, free swimmers.

Plasmotomy. Occasionally the multinucleate body of a protozoan divides into two or more small, mutinucleate individuals, the
cytoplasmic division taking place independently of nuclear division. This has been called plasmotomy by Doflein. It has been observed in the trophozoites of several coelozoic myxosporidans,
such as *Chloromyxum leydigi, Sphaeromyxa balbianii,* etc. It occurs further in certain Sarcodina such as Mycetozoida (Fig. 181)
and Pelomyxa (Fig. 74). (Division types, Kormos and Kormos,
1958.)

Colony Formation

When the division is repeated without a complete separation of
the daughter individuals, a colonial form is produced. The component individuals of a colony may either have protoplasmic connections among them or be grouped within a gelatinous envelope
if completely separated. Or, in the case of loricate or stalked

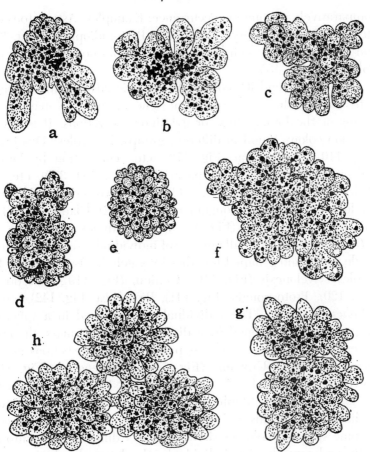

Fig. 74. Eight individuals of *Pelomyxa carolinensis,* seen undisturbed in culture dishes, in which mitotic stages occurred as follows, ×40 (Kudo): a, early prophase; b, c, later prophase; d, metaphase; e, f, early and late anaphase; g, h, late telophase to resting nuclei (g, plasmotomy into two individuals; h, plasmotomy into three daughters).

forms, these exoskeletal structures may become attached to one another. Although varied in appearance, the arrangement and relationship of the component individuals are constant, and this makes the basis for distinguishing the types of protozoan colonies, as follows:

Catenoid or linear colony. The daughter individuals are attached endwise, forming a chain of several individuals. It is of comparatively uncommon occurrence. Examples: Astomatous ciliates such as Radiophrya (Fig. 347), Protoradiophrya (Fig. 347) and dinoflagellates such as Ceratium, Haplozoon (Fig. 133) and Polykrikos (Fig. 135).

Arboroid or dendritic colony. The individuals remain connected with one another in a tree-form. The attachment may be by means of the lorica, stalk, or gelatinous secretions. It is a very common colony found in different groups. Examples: Dinobryon (Fig. 110), Hyalobryon (Fig. 110), etc. (connection by lorica); Colacium (Fig. 124), many peritrichs (Figs. 369, 370), etc. (by stalk); Poteriodendron (Fig. 142), Stylobryon (Fig. 111), etc. (by lorica and stalk); Spongomonas (Fig. 152), Cladomonas (Fig. 152) and Anthophysis (Fig. 111) (by gelatinous secretions).

Discoid colony. A small number of individuals are arranged in a single plane and grouped together by a gelatinous substance. Examples: Cyclonexis (Fig. 110), Gonium (Fig. 119), Platydorina (Fig. 120), Protospongia (Fig. 141), Bicosoeca (Fig. 142), etc.

Spheroid colony. The individuals are grouped in a spherical form. Usually enveloped by a distinct gelatinous mass, the component individuals may possess protoplasmic connections among them. Examples: Uroglena (Fig. 110), Uroglenopsis (Fig. 110), Volvox (Fig. 118), Pandorina (Fig. 120), Eudorina (Fig. 120), etc. Such forms as Stephanoon (Fig. 120) appear to be intermediate between this and the discoid type. The component cells of some spheroid colonies show a distinct differentiation into somatic and reproductive individuals, the latter developing from certain somatic cells during the course of development.

The *gregaloid* colony, which is sometimes spoken of, is a loose group of individuals of one species, usually of Sarcodina, which become attached to one another by means of pseudopodia in an irregular form.

Asexual Reproduction

The protozoa nourish themselves by certain methods, grow and multiply, by the methods described in the preceding pages. This phase of the life-cycle of a protozoan is the vegetative stage or the **trophozoite**. The trophozoite repeats its asexual reproduction process under favorable circumstances. Generally speaking, the

Sporozoa and Cnidosporidia increase to a much greater number by multiple division or schizogony and the trophozoites are called **schizonts.**

Under certain conditions, the trophozoite undergoes **encystment** (Fig. 75). Prior to encystment, the trophozoites cease to in-

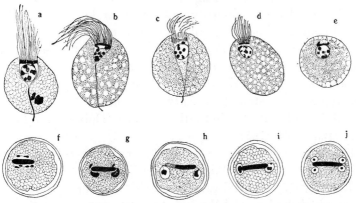

Fig. 75. Encystment of *Lophomonas blattarum,* ×1150 (Kudo).

gest, and extrude remains of, food particles, resulting in somewhat smaller forms which are usually rounded and less active. This phase is sometimes called the *precystic stage*. The whole organism becomes dedifferentiated; namely, various cell organs such as cilia, cirri, flagella, axostyle, etc., become usually absorbed. Finally the organism secretes substances which become solidified into a resistant wall, and thus the **cyst** is formed. In this condition, the protozoan is apparently able to maintain its vitality for a certain length of time under unfavorable conditions.

Protozoa appear to encyst under various conditions. Low temperature (Schmähl, 1926), evaporation (Bělař, 1921; Bodine, 1923; Garnjobst, 1928), change in pH (Koffman, 1924; Darby, 1929), low or high oxygen content (Brand, 1923; Rosenberg, 1938), accumulation of metabolic products (Bělař, 1921; Mast and Ibara, 1923; Beers, 1926) or of associated bacteria (Mouton, 1902; Bělař, 1921) and over-population (Barker and Taylor, 1931; Jones, 1951; Jeffries, 1956) in the water in which protozoa live, have been reported to bring about encystment. While lack of food in the culture has been noted by many observers (Oehler, 1916;

Penn, 1935; Kidder and Stuart, 1939; Claff, Dewey and Kidder, 1941; Singh, 1941; Beers, 1948; Strickland and Haagen-Smit, 1948; etc.) as a cause of encystment in a number of protozoa such as Blepharisma (Stolte, 1922), Polytomella (Kater and Burroughs, 1926), Didinium (Mast and Ibara, 1931) Uroleptus (Calkins, 1933), Colpoda (Garnjobst, 1947), etc., an abundance of food and adequate nourishment seem to be prerequisite for encystment in others. Particular food was found in some instances to induce encystment. For example, Singh (1948) employed for culture of *Leptomyxa reticulata*, forty strains of bacteria, of which fifteen led to the production of a large number of cysts in this sarcodinan. Encystment of *Entamoeba histolytica* is easily obtained by adding starch to the culture (Dobell and Liadlow, 1926; Balamuth, 1951).

The age of culture, if kept under favorable conditions, does not influence encystment. Didinium after 750 generations, according to Beers (1927), showed practically the same encystment rate as those which had passed through ten or twenty generations since the last encystment. When Leptomyxa, mentioned above, is cultured for more than a year, no encystment occurred, but young cultures, when supplied with certain bacteria, encysted (Singh, 1948).

In some cases, the organisms encyst temporarily in order to undergo nuclear reorganization and multiplication as in Colpoda (Fig. 76) (Kidder and Claff, 1938; Stuart, Kidder and Griffin, 1939; Pigon and Edstrom, 1961), Tillina (Beers, 1946), etc. In Ichthyophthirius, the organism encysts after leaving the host fish and upon coming in contact with a solid object, and multiplies into numerous "ciliospores" (MacLennan, 1937). *Pelomyxa carolinensis* (Illinois stock) has not encysted since its discovery in 1944, although the cultures were subjected to various environmental changes. However, Wenstrup (1945) and Musacchia (1947) reported encystment in their stocks. *P. illinoisensis* has been found to encyst and excyst frequently in thriving cultures, the shortest time between encystment and excystment at room temperature being fifty-three days (Kudo, 1951). *Bursalia truncatella* produces two types of cysts: permanent (stable) and temporary (unstable) (Beers, 1948; Miyake, 1957). The former re-

quire two to nine months dormancy to mature, while the latter
may excyst whenever transferred to a suitable cultural condition.
Miyake found that the stable cyst excysts in three days, one week
and two weeks after being held at 40°, 35° and 30°C respective-

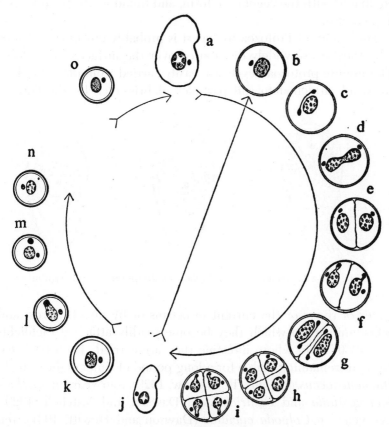

Fig. 76. Diagram showing the life cycle of *Colpoda cucullus* (Kidder and
Claff). a–j, normal reproductive activity repeated (j–b) under favorable
cultural conditions; k–o, resistant cyst (k–n, nuclear reorganization and
chromatin elimination).

ly. Thus, it appears that protozoa encyst under diverse environmental conditions and that unknown internal factors play as great
a part as do the external factors.

The cyst is covered by one to three membranes. Though generally homogenous, the wall of cyst may contain siliceous scales as

in Euglypha (Fig. 77). While chitinous substance is the common material of which the cyst wall is composed, cellulose makes up the cyst membrane of many Phytomastigia. Entz (1925) found the cysts of various species of Ceratium less variable in size as compared with the vegetative form, and found in all, glycogen, oil and volutin.

The ability of Protozoa to encyst is probably one of the reasons why they are so widely distributed over the surface of the globe. The minute protozoan cysts are easily carried from place to place by wind, attached to soil particles, debris, etc., by the flowing

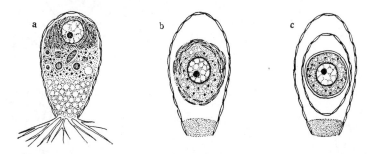

FIG. 77. Encystment of *Euglypha acanthophora,* ×320 (Kühn).

water of rivers or the current in oceans or by insects, birds, and other animals to which they become readily attached (Schlichting, 1960) or in whose intestine they are carried. The cyst is capable of remaining viable for a long period of time: eight years in *Haematococcus pluvialis* (Reichenow, 1929), four years in *Spathidium spathula* and *Oxytricha* sp. (Dawson and Mitchell, 1929), five years in *Colpoda cucullus* (Dawson and Hewitt, 1931), ten years in *Didinium nasutum* (Beers, 1937), etc.

When a cyst encounters a proper environment, redifferentiation takes place within the cyst. Various organellae which characterize the organism, are regenerated and reformed, and the young trophozoite excysts. The reserve food material which the organism stored within its body prior to encystment, will furnish needed energy for excystment.

The emerged organism returns once more to its trophic phase of existence. Experimental data indicate that excystment takes place under conditions such as addition of fresh culture medium (Kühn, 1915; Rosenberg, 1938; Jones, 1951; Bridgman, 1956), hypertonic solution (Ilowaisky, 1926), distilled water (Johnson and Evans, 1941), organic infusion (Mast, 1917; Beers, 1926; Barker and Taylor, 1933), and bacterial infusion (Singh, 1941; Beers, 1946a) to the culture medium. Change in pH (Koffman, 1924), lowering the temperature (Johnson and Evans, 1941) and increase in oxygen content (Brand, 1923; Finley, 1936) of the medium have also been reported as bringing about excystment. Excystment in *Colpoda cucullus* is said to be due to specific inducing substances present in plant infusion (Thimann and Barker, 1934; Haagen-Smit and Thimann, 1938). Excystment in *Colpoda duodenaria* took place in one hour and forty minutes, when placed in a medium composed of 0.3 Methyl alcohol and 0.0001 M potassium phosphate in distilled water at 24°C. (Strickland and Haagen-Smit, 1948). Experimenting with two soil amoebae, "species 4 and Z," Crump (1950) found that the excystment in species Z took place without the presence of bacteria and regardless of the age of the cysts, but species 4 excysted only in the presence of certain bacteria (*Aerobacter* sp. or "4036") and the excystment diminished with the age of cysts. Crump suggested that the two strains of bacteria appeared to produce some material which induced excystment in Amoeba species 4, which was confirmed by Singh *et al.* (1956). The latter investigators found further that certain amino acids and nucleotides cause excystment in soil amoebae (Singh *et al.*, 1958; Singh, 1960). In *Tillina magna*, Beers (1945) found, however, the primary excystment-inducing factor to be of an osmotic nature and inducing substances, a secondary one.

As to how an aperture or apertures are formed in the cyst wall prior to the emergence of the content, precise information is not yet on hand, though there are many observations. In the excystment in Didinium and Tillina, Beers (1935, 1945, 1945a) notes that an increased internal pressure due to the imbibition of water, results in the rupture of the cyst wall which had lost its rigidity and resistance (Fig. 78). Apertures in the cyst wall of *Pelomyxa illinoisensis* are apparently produced by pseudopodial pressure (Kudo, 1951). Seeing a similar aperture formation in the cyst of

Entamoeba histolytica, Dobell (1928) "imagined that the amoeba secretes a ferment which dissolves the cyst wall." According to Goodey (1913), the inner cyst membrane of Colpoda is composed of a carbohydrate ("cystose") which is digested by an enzyme ("cystase") at the time of excystment.

Although encystment seems to be an essential phase in the life

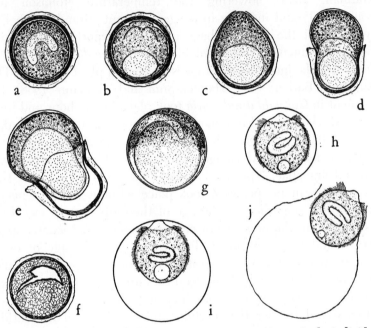

Fig. 78. Excystment in *Didinium nasutum,* as seen in a single individual, ×250 (Beers). a, resting cyst; b, appearance of "excystment" vacuole; c, rupture of the cyst membrane, the vacuole is becoming enlarged; d, e, emergence of the cyst content, the vacuole increasing in size; f, the empty outer cyst membrane; g, the free organism with the inner membrane; h, organism after discharge of vacuole; i, j, later stages of emergence.

cycle of protozoa in general, there are certain protozoa, including such common and widely distributed forms as the species of Paramecium, in which this phenomenon has not been definitely observed (p. 905). In some Sporozoa, encystment is followed by production of large numbers of spores, while in others there is no encystment. Here, at the end of active multiplication of trophozoite, sexual processes produce oocysts or sporonts in which the spores develop. These spores are protected by resistant membranes and are capable of remaining viable for a long period of time outside the host body.

Sexual Reproduction and Life-cycle

Besides reproducing by the asexual method, numerous protozoa reproduce themselves in a manner comparable with the sexual reproduction which occurs universally in the metazoa. Various types of sexual reproduction have been reported in literature, of which a few will be considered here. The sexual fusion or **syngamy** which is a complete union of two gametes, has been reported from various groups, while the conjugation which is a

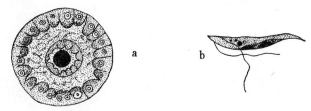

Fig. 79. a, macrogamete, and b, microgamete of *Volvox aureus,* ×1000 (Klein).

temporary union of two individuals for the purpose of exchanging the nuclear material, is found exclusively in the Ciliophora.

Sexual Fusion. The **gametes** which develop from trophozoites, may be morphologically alike (**isogametes**) or unlike (**anisogametes**), both of which are, in well-studied forms, physiologically different as judged by their behavior toward each other. If a gamete does not meet with another one, it perishes. Anisogametes are called **microgametes** and **macrogametes.** Difference between them is comparable in many instances (Figs. 79, 262) with that which exists between the spermatozoa and the ova of metazoa. The microgametes are motile, relatively small and usually numerous, while the macrogametes are usually not motile, much more voluminous and fewer in number. In the much-studied Coccidida and Haemosporida, for example, the two gametes are morphologically and physiologically differentiated, and sexual fusion always takes place only between two anisogametes. Therefore, they have sometimes been referred to as the male and female gametes (Fig. 79).

The isogamy is typically represented by the flagellate *Copromonas subtilis* (Fig. 80), in which there occurs, according to Dobell, a complete nuclear and cytoplasmic fusion between two isogametes. Each nucleus, after casting off a portion of its nuclear material, fuses with the other, thus forming a zygote containing a **synkaryon**. In *Trinema lineare* (Fig. 81), Dunkerly (1923) saw isogamy in which two individuals underwent a complete fusion within one test and encysted. In *Stephanosphaera pluvialis* (Fig. 82), both asexual and sexual reproductions occur (Hieronymus,

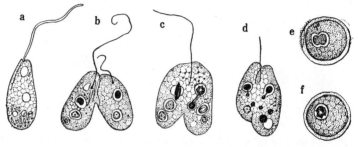

Fig. 80. Sexual fusion in *Copromonas subtilis*, ×1300 (Dobell).

1884). Each individual multiplies and develops into numerous biflagellate gametes, all of which are alike. Isogamy between two gametes results in the formation of numerous zygotes which later develop into trophozoites.

Anisogamy has been observed in certain Foraminiferida. It perhaps occurs in the Radiolarida also, although positive evidence has yet to be presented. Anisogamy seems to be more widely distributed. In *Pandorina morum*, Pringsheim observed that each cell develops asexually into a young colony or into anisogametes which undergo sexual fusion and encyst. The organism emerges from the cyst and develops into a young trophozoite. A similar life-cycle was found by Goebel in *Eudorina elegans*.

The wood-roach inhabiting flagellates belonging to Trichonympha, Oxymonas, Saccinobaculus, Notila and Eucomonympha, were found by Cleveland (1949a-1960) to undergo sexual reproduction when the host insect molts. It has been observed that the gamete-formation is induced by *ecdysone* (the molting hormone) produced by the prothoracic glands of the host insect. This sexual reproduction of Trichonympha, possessing twenty-four chromosomes, as observed and described by Cleveland, is briefly as follows (Figs. 83, 84): About three days before its host molts,

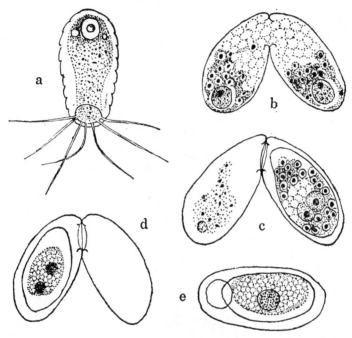

FIG. 81. Sexual fusion in *Trinema lineare*, ×960 (Dunkerly). a, an organism in life, with the resting nucleus and two contractile vacuoles; b, union of two individuals; c, fusion of the organisms in one test, surrounded by cyst membrane; d, older cyst; e, still older cyst with a single nucleus.

the haploid nucleus in the flagellate divides, in which two types of daughter chromosomes (or chromatids) become separated from each other: the dark-staining male gamete nucleus and light-staining female gamete nucleus (Fig. 83, *b-d*); in the meantime, a membrane is formed to envelop the organism (*b, d*). When the cytoplasmic division is completed (*e-g*), the two gametes "excyst" and become free in the host gut (*h*; Fig. 84, *b*). In the female gamete, there appear "fertilization granules" (Fig. 83, *h*), which gather at the posterior extremity (*i*), through which a fluid-

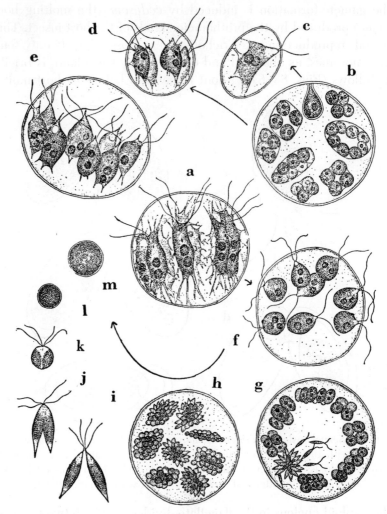

FIG. 82. The life-cycle of *Stephanosphaera pluvialis* (Hieronymus). a–e,
asexual reproduction; f–m, sexual reproduction.

filled vesicle ("fertilization cone") protrudes (Fig. 84, *a*). A male
gamete (*b*) comes in touch with a female gamete only at this point
(*c*), and enters the latter (*d-f*). The two gamete nuclei fuse into a
diploid synkaryon (*g, h*). The zygote and its nucleus begin immediately to increase in size, and undergo two meiotic divisions
(*i-k*), finally giving rise to vegetative individuals (Fig. 83, *a*).

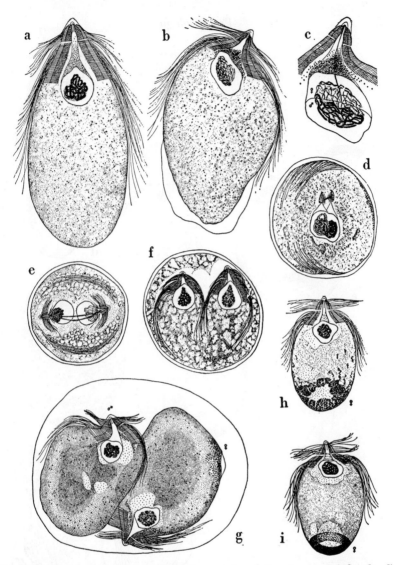

Fig. 83. Sexual reproduction in Trichonympha of Cryptocercus (Cleveland). a, vegetative individual; b, gametocyte in early stage of encystment; c, anterior end of the same organism (chromosomes have been duplicated, nuclear sleeve is opening at seams and granules are flowing into the cytoplasm); d, further separation of the male and female chromosomes; e, the nuclear division has been completed, few old flagella remain and new post rostral flagella are growing; f, the cytoplasmic division has begun at the anterior end; g, the gametes just before excystment, the female showing the developing ring of fertilization granules; h, a female gamete; i, a female gamete with a fertilization ring. a, ×350; b, ×320; c, ×600; d-i, ×280.

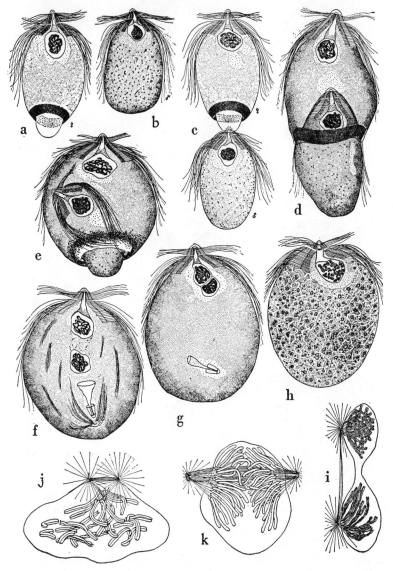

FIG. 84. Sexual reproduction in Trichonympha of Cryptocercus (Cleveland). a, a female gamete with a fertilization ring and cone; b, a male gamete; c–g, stages in fusion and fertilization; h, a zygote; i, telophase of the first meiotic division of the zygote nucleus; j, k, prophase and anaphase of the second meiotic division. a-g, ×280; h, ×215; i-k, ×600.

Cleveland *et al.* (1960) found further that when exogenous ecdysone is injected into a host woodroach that does not have the hormone of its own, gametogenesis is induced in its protozoa, although the host itself never undergoes ecdysis. (Microphotographs of fertilization processes in Trichonympha, Cleveland, 1958b,c; sexual cycles of the flagellates of the woodroach, Cleveland, 1956, 1957, 1957a, 1960.)

Among the Sporozoa, anisogamy is of common occurrence. In Coccidida, the process was well studied in *Eimeria schubergi* (Fig. 249), *Aggregata eberthi* (Fig. 252), *Adelea ovata* (Fig. 259), etc., and the resulting products are the **oocysts** (zygotes) in which the spores or sporozoites develop. Similarly in Haemosporida such as *Plasmodium vivax* (Fig. 262), anisogamy results in the formation of the **ookinetes** or motile zygotes which give rise to a large number of sporozoites. Among Myxosporida, a complete information as to how the initiation of sporogony is associated with sexual reproduction, is still lacking. Naville, however, states that in the trophozoite of *Sphaeromyxa sabrazesi* (Fig. 282), micro- and macro-gametes develop, each with a haploid nucleus. Anisogamy, however, is peculiar in that the two nuclei remain independent. The microgametic nucleus divides once and the two nuclei remain as the vegetative nuclei of the pansporoblast, while the macrogamete nucleus multiplies repeatedly and develops into two spores. Anisogamy has been suggested to occur in some members of Amoebida, particularly in *Endamoeba blattae* (Mercier, 1909). Cultural studies of various parasitic amoebae in recent years show, however, no evidence of sexual reproduction in them.

Conjugation. The conjugation is a temporary union of two individuals of one and the same species for the purpose of exchanging part of the nuclear material and occurs exclusively in the Ciliata and Suctoria. The two individuals which participate in this process may be either isogamous or anisogamous. In *Paramecium caudatum* (Fig. 85), the process of conjugation has been studied by many workers, including Bütschli (1876), Maupas (1889), Calkins and Cull (1907), and others. Briefly the process is as follows: Two similar individuals come in contact on their oral surface (*a*). The micronucleus in each conjugant divides twice (*b-e*), forming four micronuclei, three of which degenerate and do not

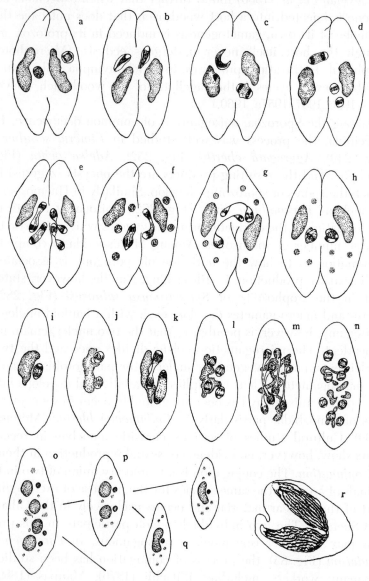

FIG. 85. Diagram illustrating the conjugation of *Paramecium caudatum.* a-q, × about 130 (Calkins); r, a synkaryon formation as in h, ×1200 (Dehorne).

take active part during further changes (*f-h*). The remaining micronucleus divides once more, producing a wandering pronucleus and a stationary pronucleus (*f, g*). The wandering pronucleus in each of the conjugants enters the other individual and fuses with its stationary pronucleus (*h, r*). The two conjugants now separate from each other and become exconjugants. In each exconjugant, the synkaryon divides three times in succession (*i-m*) and produces eight nuclei (*n*), four of which remain as micronuclei, while the other four develop into new macronuclei (*o*). Cytoplasmic fission follows then, producing first, two individuals with four nuclei (*p*) and then, four small individuals, each containing a micronucleus and a macronucleus (*a*). Jennings maintained that of the four smaller nuclei formed in the exconjugant (*o*), only one remains active and the other three degenerate. This active nucleus divides prior to the cytoplasmic division so that in the next stage (*p*), there are two developing macronuclei and one micronucleus which divides once more before the second and last cytoplasmic division (*q*). During these changes, the original macronucleus disintegrates, degenerates, and finally becomes absorbed in the cytoplasm. Vivier and Andre (1961) find that there occurs cytoplasmic communication at the area of contact of the conjugants, and that cilia and trichocysts disappear, but the kinetosomes remain intact. (Interchange of pronuclei, Diller, 1950.)

Although this is the general course of events in the conjugation of this ciliate, recent observations revealed a number of different nuclear behavior. For example, there may not be pronuclear exchange between the conjugants (cytogamy, p. 243), thus resulting in self fertilization (Diller, 1950a). In a number of races, Diller 1950) found that one of the two nuclei produced by the first division of the synkaryon degenerates, while the other nucleus divides three times, forming eight nuclei, and furthermore, an exconjugant may conjugate occasionally with another individual before the reorganization has been completed.

The conjugation of *P. bursaria* has also received attention of many workers. According to Chen (1946a), the first micronuclear division is a long process. One daughter nucleus degenerates and the other undergoes a second division. Here again one nucleus degenerates, while the other divides once more, giving rise to a wandering and a stationary pronucleus. Exchange of the wandering pronuclei is followed by the fusion of the two pronuclei in each conjugant. The synkaryon then divides. One of the two nuclei formed by this division degenerates, while the other gives rise to four nuclei by two divisions. The latter presently become differentiated into two micronuclei and two macronuclei, followed by a cytoplasmic division. The time two conjugants remain paired is said to be twenty to thirty-eight or more hours (Chen, 1946c). In this Paramecium also, various nuclear activities have been reported. Chen (1940a, c) found that conjugation between a micronucleate and an amicronucleate can sometimes occur. In such a case, the micronucleus in the normal individual divides three times, and one of the pronuclei migrates into the amicronucleate in which there is naturally no nuclear division. The single haploid nucleus ("hemicaryon") in each individual divides three times as mentioned above and four nuclei are produced. Thus amicronucleate becomes micronucleated. Conjugating pairs sometimes separate from each other in a few hours. Chen (1946c) found that when such pairs are kept in a depression slide, temporary pairing recurs daily for many days, though there is seemingly no nuclear change. Chen (1940) further observed that the micronucleus in this species is subject to variation in size and in the quantity of chromatin it contains, which gives rise to different (about eighty to several hundred) chromosome numbers during conjugation in different races, and that polyploidy is not uncommon in this ciliate. This investigator considers that polyploidy is a result of fusion of more than two pronuclei which he observed on several occasions. The increased number of pronuclei in a conjugant may be due to: (1) the failure of one of the two nuclei produced by the first or second division to degenerate; (2) the conjugation between a unimicronucleate and a bimicronucleate, or (3) the failure of the wandering pronucleus to enter the other conjugant; with this latter view Wichterman (1946) agrees. Apparently polyploidy occurs in other species also; for example, in *P. caudatum* (Calkins and Cull, 1907; Penn, 1937). (Development of macronucleus in exconjugants, Egelhaaf, 1955.)

In *P. trichium*, Diller (1948) reported that the usual process of conjugation is the sequence of three micronuclear divisions, producing the pronuclei (during which degeneration of nuclei may occur at the end of both the first and second division), cross- or self-fertilization and three divisions of the synkarya. Ordinarily, four of the eight nuclei become macronuclei, one remains as the micronucleus and the other three degenerate. The micronucleus divides at each of the two cytoplasmic divisions. Exchange of strands of the macronuclear skein may take place between the conjugants. Diller found a number of variations such as omission of the third prefertilization division, autogamous development, etc., and remarked that heteroploidy is pronounced and common.

In *P. aurelia* possessing typically two micronuclei, the process of conjugation was studied by Maupas (1889), Hertwig (1889), Diller (1936), Sonneborn (1947), etc., and is as follows: Soon after biassociation begins, the two micronuclei in each conjugant divide twice and produce eight nuclei, seven of which degenerate, while the remaining one divides into two gametic nuclei (Maupas, Woodruff, Sonneborn). Diller notes that two or more of the eight nuclei divide for the third time, but all but two degenerate; the two gametic nuclei may or may not be sister nuclei. All agree that there are two functional pronuclei in each conjugant. As in other species of Paramecium already noted, there is a nuclear exchange which results in the formation of a synkaryon in each conjugant. The synkaryon divides twice and the conjugants separate from each other at about this time. Two nuclei develop into macronuclei and the other two into micronuclei. Prior to the first cytoplasmic division of the exconjugant, the micronuclei divide once, but the macronucleus does not divide, so that each of the two daughters receives one macronucleus and two micronuclei. The original macronucleus in the conjugant becomes transformed into a skein which breaks up into twenty to forty small masses. These are resorbed in the cytoplasm as in other species. As to when these nuclear fragments are absorbed, depends upon the nutritive condition of the organism (Sonneborn); namely, under a poor nutritional condition the resorption begins and is completed early, but under a better condition this resorption takes place after several divisions.

During conjugation, reciprocal migration of a pronucleus thus occurs in all cases. During biassociation and even in autogamy (p.

240), there develops a conical elevation ("paroral cone") and the nuclear migration takes place through this region. Although there is ordinarily no cytoplasmic exchange between the conjugants, this may occur in some cases as observed by Sonneborn, (1943a, 1944). *P. aurelia* of variety 4, according to Sonneborn, do occasionally not separate after fertilization, but remain united by a thin strand in the region of the paroral cones. In some pairs, the strand enlarges into a broad band through which cytoplasm flows from one individual to the other. The first division gives off a normal single animal from each of the "parabiotic twins" and the two clones derived from the two individuals belong to the same mating type (p. 231). Elliott and Tremor (1958) demonstrated the existence of tubules through pellicle of the two conjugating Tetrahymena and believed that these tubules would allow the exchange of cytoplasmic particles up to 0.2μ in diameter, between the conjugants. As mentioned already, Vivier and Andre (1961) observed cytoplasmic communication in *Paramecium caudatum* in conjugation.

Conjugation between different species of Paramecium has been attempted by several workers. Müller (1932) succeeded in producing a few pairings between normal *P. caudatum* and exconjugant *P. multimicronucleatum*. The nuclear process ran normally in caudatum, which led Müller to believe that crossing might be possible, but without success. De Garis (1935) mixed "double animals" (p. 274) of *P. caudatum* and conjugating population of *P. aurelia*. Pairing between them occurred readily, in which the aurelia mates remained attached to caudatum for five to twelve hours. Four pairs remained together, but aurelia underwent cytolysis on the second day. The separated aurelia from other pairs died after showing "cloudy swelling" on the second or third day after biassociation. The caudatum double-animals on the other hand lived for two to twelve (average six) days during which there was neither growth nor division and finally perished after "hyaline degeneration." No information on nuclear behavior in these animals is available. Apparently, the different species of Paramecium are incompatible with one another.

In 1937, Sonneborn discovered that in certain races of *Paramecium aurelia*, there are two classes of individuals with respect to "sexual" differentiation and that the members of different classes conjugate with each other, while the members of each class do not. The members of a class are progeny of one of the two individuals formed by the first division of an exconjugant and thus possess the same macronuclear constitution. These classes or *caryonides* (Sonneborn, 1939) were designated by Sonneborn (1938) as **mating types.** Soon, a similar phenomenon was observed in other species of Paramecium; namely, *P. bursaria* (Jennings, 1938), *P. caudatum* (Gilman, 1939, 1950; Hiwatashi, 1949-1959), *P. trichium, P. calkinsi* (Sonneborn, 1938; Wichterman, 1951) and *P. multimicronucleatum* (Giese, 1939, 1957).

In *P. bursaria,* when organisms which belong to different mating types are brought together, they adhere to one another in large clumps ("agglutination") of numerous individuals (Fig. 86, *b*). After a few to several hours, the large masses break down into smaller masses (*c*) and still later, conjugants appear in pairs (*d*).

How widely mating types occur is not known at present. But as was pointed out by Jennings, the mating types may be of general occurrence among ciliates; for example, Maupas (1889) observed that in *Lionotus (Loxophyllum) fasciola, Leucophrys patula, Stylonychia pustulata,* and *Onychodromus grandis,* conjugation took place between the members of two clones of different origin, and not among the members of a single clone. In recent years, mating types were found to occur in ciliates other than Paramecium; namely, *Tetrahymena pyriformis* (Elliott and Hayes, 1953), *Oxytricha bifaria* (Siegel, 1956), *Stylonychia putrina* (Downs, 1959), *Euplotes patella* (Kimball, 1939) and *E. eurystomus* (Katashima, 1959).

In *Paramecium aurelia,* Sonneborn now distinguishes 16 varieties. What he called variety 16 is usually known as *P. multimicronucleatum.* Each variety consists of two interbreeding mating types. But variety 16 has four mating types (Giese, 1941, 1957). Under optimum breeding conditions two mating types of the same variety give 95 per cent immediate agglutination and conjugation. Mating reactions between the members of different varieties do not usually occur. However, exceptions were noted in certain combinations of mating types of the eight varieties 1, 3, 4, 5, 7, 8, 10 and 14 as indicated in Table 6. The intensity of mat-

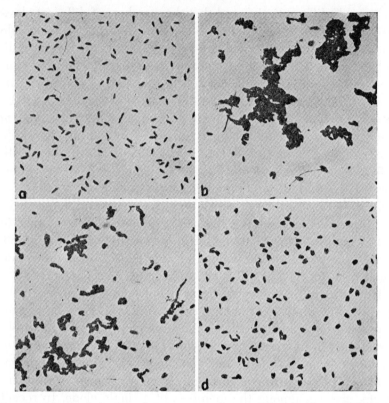

FIG. 86. Mating behavior of *Paramecium bursaria* (Jennings). a, individuals of a single mating type; b, 6 minutes after individuals of two mating types have been mixed; c, after about 5 hours, the large masses have been broken down into small masses; d, after 24 hours, paired conjugants.

ing reactions vary from forming conjugating pairs to weaker reaction. But the reactions appear to show that mating types of different varieties are related to one another. In the two pairs of varieties 1 and 5, and 4 and 8, the odd-numbered mating type of each variety mates with the even-numbered mating type of the other. From the data shown in the table, the odd-numbered mating types in the eight varieties form one similar (—) group and even-numbered mating types form another similar (+) group (Sonneborn and Dippell, 1946). The two mating types of each of the eight varieties not listed in Table 6, do not undergo intervarietal conjugation.

TABLE 6

THE SYSTEM OF SEXUAL REACTIONS AMONG MATING TYPES OF THOSE VARIETIES OF *P. aurelia* IN WHICH INTERVARIETAL REACTIONS OCCUR[a] (SONNEBORN, 1957)

Variety	1		3		4		5		7		8		10		14	
Mating Type	I	II	V	VI	VII	VIII	IX	X	XIII	XIV	XV	XVI	XIX	XX	XXVII	XXVIII
I	−	+++	−	−	−	−	−	++	−	−	±	±	−	−	−	−
II	C	−	+	−	−	−	++	−	++	−	±	−	−	−	−	−
V	0	C	−	+++	−	−	−	−	++	−	−	++	−	−	−	−
VI	0	0	C	−	−	−	−	−	±	−	−	−	−	−	−	−
VII	0	0	0	0	−	+++	−	−	±	−	−	+++	−	−	+	−
VIII	0	0	0	0	C	−	−	−	−	−	++	+++	+	−	−	−
IX	0	C	0	0	0	0	−	+++	−	−	++	−	+	−	−	−
X	C	0	0	0	0	0	C	−	±	−	−	−	−	−	−	−
XIII	0	C	0	0	0	0	0	0	−	+++	−	±	−	−	−	−
XIV	0	0	0	0	0	0	0	0	C	−	−	−	−	−	−	−
XV	0	0	0	0	0	C	0	0	0	0	−	+++	+	−	−	−
XVI	0	0	C	0	C	0	0	0	0	0	C	−	−	+++	+	?
XIX	0	0	0	0	0	0	0	0	0	0	0	0	−	+++	+	−
XX	0	0	0	0	0	0	0	0	0	0	0	0	C	−	−	+++
XXVII	0	0	0	0	0	0	0	0	0	0	0	0	0	0	−	+++
XXVIII	0	0	0	0	0	0	0	0	0	0	0	0	0	0	C	−

[a] Symbols above the diagonal refer only to occurrence of mating reactions (adhesion, agglutination); symbols below the diagonal refer to the occurrence of complete conjugation. Varieties 10 and 14 are newly discovered and still incompletely studied; their reactions may be more extensive than now known. +++ maximal reaction. ++ reduced mating reaction. + weak mating reaction. ± barely detectible reaction. − no mating reaction. C conjugants formed. 0 no conjugants formed.

The intervarietal mating reactions are (1) always less intense than intravarietal reaction; (2) dependent upon the degree of reactivity of the culture, and (3) different from the intravarietal reaction with respect to the conditions for optimum reaction. Furthermore in most cases, the progeny of intervarietal matings are not viable. Since the two mating types of a variety do not interbreed, as a rule, with those of other varieties, they constitute a distinct species. Yet it is practically impossible to differentiate morphologically sixteen varieties of *P. aurelia*. Sonneborn (1957) designated them **syngens** rather than species.

In *P. bursaria*, Jennings (1938, 1939) found three varieties. Varieties 1 and 3 contain four mating types each, while variety 2, eight mating types. Jennings and Opitz (1944) further found variety 4 (Russian), composed of two mating types and doubtful variety 5 under which several Russian clones were placed. Chen (1946a) added variety 6 (originating in Europe) containing four mating types.

In *P. caudatum*, there are sixteen varieties or syngens (Gilman, 1941, 1950; Hiwatashi, 1949; Sonneborn, 1957), and like *aurelia*, each variety is made up of two mating types. In *Tetrahymena pyriformis*, nine varieties with varying number of mating types are known: variety 1 with five mating types; variety 2 with eleven mating types; variety 3 with eight mating types; variety 4 with three mating types; variety 5 with two mating types; variety 6 with three mating types; varieties 7 and 8 with two mating types each, and variety 9 with five mating types (Elliott and Hayes, 1953, 1955; Elliott, 1957; Gruchy, 1955).

In *Euplotes patella*, Kimball (1939) found six mating types, and in *E. eurystomus*, Katashima (1959) found the mating types fall into five varieties, each having two complementary mating types except variety 5 which contained only one. Intervarietal mating does not occur, but exceptionally it takes place between types V and VII, and VI and VII. Downs (1959) reported finding two varieties: one with fifteen mating types and the other with eleven mating types in *Stylonychia putrina* and Siegel (1956) found nine mating types in *Oxytricha bifaria*.

Though the members of a clone are of the same mating type and therefore do not conjugate, a clone may undergo at very long intervals (some 2000 culture days), "self-differentiation" into two mating types which then conjugate (Jennings, 1941). Furthermore, Jennings and Opitz (1944) found that in *Paramecium bursaria* mating type R (variety 4) conjugated with E, K, L or M (variety 2), but all conjugants or exconjugants perished without multiplication. Chen (1946a) made a cytological study of them and observed that the nuclear changes which are seemingly normal during the first sixteen hours, become abnormal suddenly after that time, and the micronuclei divide only once and there is no nuclear exchange. The death of conjugants or exconjugants is possibly due to physiological incompatibility between the varieties upon coming in contact or probably due to "something that diffuses from one conjugant to the other."

Studies of mating types have revealed much information regarding conjugation. Conjugation usually does not occur in well-fed or extremely starved animals, and appears to take place shortly after the depletion of food. Temperature also plays a rôle in conjugation, as it takes place within a certain range of temperature which varies even in a single species among different varieties (Sonneborn). Light seems to have different effects on conjugation in different varieties of *P. aurelia*. The time between two conjugations also varies in different species and varieties. In *P. bursaria,* Jennings found that in some races the second conjugation would not take place for many months after the first, while in others such an "immature" period may be only a few weeks. In *P. aurelia,* in some varieties there is no "immature" period, while in others there is six to ten days' "immaturity."

The individuals that participate in conjugation show much viscous body surface. Boell and Woodruff (1941) found that the mating individuals of *Paramecium calkinsi* show a lower respiratory rate than not-mating individuals. Neither is the mechanism of conjugation understood at present. Kimball (1942) discovered in *Euplotes patella,* the fluid taken from cultures of animals of one type induces conjugation among the animals of other types Presumably certain substances are secreted by the organisms and become diffused in the culture fluid. Holz *et al.* (1959) noted differences in temperature-tolerance, generation time, nutritional requirements, osmoresistance, pigment production, sensitivity to high hydrogen or hydroxyl ion concentration occur among varieties and among mating types within varieties of *Tetrahymena pyriformis.*

In *Paramecium aurelia,* Sonneborn (1943) found that of the four races of variety 4, race 51 was a "killer," while the other three races, "sensitive." Fluid in which the killer race grew, kills the individuals of the sensitive races. *P. bursaria* designated as type T (variety 5) conjugates with none. But Chen (1945) found that its culture fluid induces conjugation among a small number of the individuals of one mating type of varieties 2, 3, 4 and 6, in which nuclear changes proceed as in normal conjugation. Furthermore, this fluid is capable of inducing autogamy in single animals. Other visible influences of the fluid on organisms are sluggishness of movement and darker coloration and distortion of the body. Chen (1955) further recognizes two killer substances (paramecin 22 and 34) in this Paramecium. Paramecin 22 produces various effects on sensitives: sluggishness, dark coloration, distortion, surface stickiness, and death, while paramecin 34 produces dark coloration, active movements, blister formation, bursting and death.

Boell and Woodruff (1941) noticed that in *P. calkinsi,* living individuals of one mating type will agglutinate with dead ones of the complementary mating type. A similar phenomenon was also observed by Metz (1946, 1947, 1948) who employed various methods of killing the animals. The pairs composed of living and formaldehyde-killed animals, behave much like normal conjugating pairs; there is of course no cross-fertilization, but the living member of the pair undergoes autogamy. While the "mating type substances" can be destroyed by exposure to 52°C., for five minutes; by x-irradiation; by exposure of formaldehyde-killed reactive animals to specific antisera or to 100°C., etc., Metz demonstrated that animals may be killed by many reagents which do not destroy these substances. Furthermore, all mating activities disappear when the animals are thoroughly broken up, which suggests that Paramecium might release some mating substance inhibitory agent. This agent was later found in this Paramecium (Metz and Butterfield, 1950). Metz (1948, 1954) points out that the mating reaction involves substances present on the surfaces of the cilia, and supposes that the interaction between two mating-type substances initiates a chain of reactions leading up to the process of conjugation and autogamy. Hiwatashi (1949a, 1950) using four groups (each composed of two mating types) of *P. caudatum*, confirmed Metz's observation and showed further (1960) that the mating reactivity is confined to anteroventral surface. Metz and Butterfield (1951) reported that non-proteolytic enzymes (lecithinase, hyaluronidase, lysozyme, ptyalin, ribonuclease) have no detectable effect on the mating reactivity of *P. calkinsi*; but proteolytic enzymes such as trypsin and chymotrypsin destroy the mating reactivity, and mating substance activity was not found in the digest of enzyme-treated organisms. The two observers believe that the mating reactivity is dependent upon protein integrity.

Recently Miyake (1958) induced conjugation in *P. caudatum* by addition of chemical agents to culture media. He finds K, Mg and heparin are highly effective under Ca poor condition. Some organic agents such as acetamide are said to increase the rate of conjugation induced by K or Mg, if used with these cations. Evidently these chemical agents induce conjugation in animals belonging to the same mating type without changing type. The process of the conjugation is reported to be the same as that occurring through mating reaction. The chemical agents can further induce intervarietal conjugation. Miyake considers that K, Mg and heparin liquify the cortical protoplasm only under Ca poor condition, as Ca increases the rigidity of the protoplasm. Partial liquifaction of the cortical protoplasm on the ventral side causes the "sticky" state on the body surface and brings two animals together.

When the ciliate possesses more than one micronucleus, the first division ordinarily occurs in all and the second may or may not take place in all, varying apparently even among individuals of the same species. This seems to be the case with the majority, although more than one micronucleus may divide for the third time to produce several pronuclei, for example, two in *Euplotes patella, Stylonychia pustulata;* two to three in *Oxytricha fallax* and two to four in *Uroleptus mobilis*. This third division is often characterized by long extended nuclear membrane stretched between the division products.

Ordinarily, the individuals which undergo conjugation appear to be morphologically similar to those that are engaged in the trophic activity, but in some species, the organism divides just prior to conjugation. According to Wichterman (1936), conjugation in *Nyctotherus cordiformis* (Fig. 87) takes place only among those which live in the tadpoles undergoing metamorphosis (*f-j*). The conjugants are said to be much smaller than the ordinary trophozoites, because of the preconjugation fission (*d-e*). The micronuclear divisions are similar to those that have been described for *Paramecium caudatum* and finally two pronuclei are formed in each conjugant. Exchange and fusion of pronuclei follow. In each exconjugant, the synkaryon divides once to form the micronucleus and the macronuclear anlage (*k-l*) which develops into the "spireme ball" and finally into the macronucleus (*m-o*).

A sexual process which is somewhat intermediate between the sexual fusion and conjugation, is noted in several instances. According to Maupas' (1888) classical work on *Vorticella nebulifera*, the ordinary vegetative form divides twice, forming four small individuals, which become detached from one another and swim about independently. Presently, each becomes attached to one side of a stalked individual. In it, the micronucleus divides three times and produces eight nuclei, of which seven degenerate; and the remaining nucleus divides once more. In the stalked form, the micronucleus divides twice, forming four nuclei, of which three degenerate, and the other dividing into two. During these changes, the two conjugants fuse completely. The wandering nucleus of the smaller conjugant unites with the stationary nucleus of the larger conjugant, the other two pronuclei degenerating. The synkaryon divides several times to form a number of nuclei, from some of which macronuclei are differentiated and exconjugant undergoes multiplication. In *Vorticella microstoma* (Fig. 88), Finley (1943) notes that a vegetative individual undergoes unequal division except the micronucleus which divides equally (*a*), and forms a large stalked macroconjugant and a small free microconjugant (*b*). The conjugation which requires eighteen

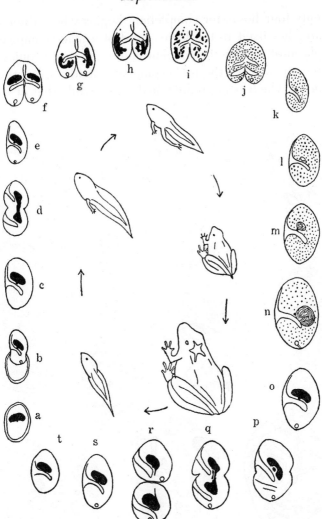

FIG. 87. The life-cycle of *Nyctotherus cordiformis* in *Hyla versicolor* (Wichterman). a, a cyst; b, excystment in tadpole; c, d, division is repeated until host metamorphoses; e, smaller preconjugant; f–j, conjugation; k, exconjugant; l, amphinucleus divides into 2 nuclei, one micronucleus and the other passes through the "spireme ball" stage before developing into a macronucleus; k–n, exconjugants found nearly exclusively in recently transformed host; o, mature trophozoite; p–s, binary fission stages; t, precystic stage.

to twenty-four hours for completion, begins when a microconjugant attaches itself to the lower third of a macroconjugant. The protoplasm of the microconjugant enters the macroconjugant (*c*). The micronucleus of the microconjugant divides three times, the last one of which being reductional (*d, e*), while that of the mac-

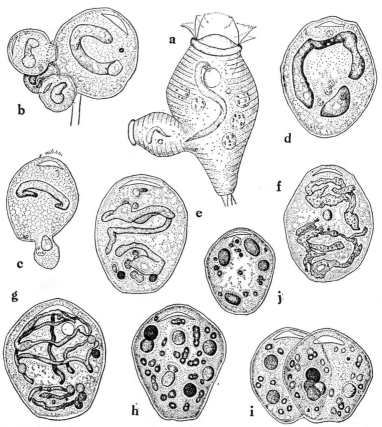

FIG. 88. Sexual reproduction in *Vorticella microstoma,* ×800 (Finley). a, preconjugation division which forms a macroconjugant and a microconjugant; b, a macroconjugant with three microconjugants; c, a microconjugant fusing with a macroconjugant; d, the micronucleus of the microconjugant divided into four nuclei; e, with 12 nuclei formed by divisions of the two micronuclei of conjugants; f, synkaryon; g, eight nuclei after three divisions of synkaryon; h, seven enlarging macronuclear anlagen and a micronucleus in division; i, first division; j, a daughter individual with a micronucleus, four macronuclear anlagen, and old macronuclear fragments.

roconjugant divides twice (one mitotic and one meiotic). Fusion of one of each produces a synkaryon (*f*) which divides three times. One of the division products becomes a micronucleus and the other seven macronuclear anlagen (*g, h*) which are distributed among the progeny (*i, j*). In *Epistylis articulata*, Seshachar and Dass (1953) found that DNA of the disintegrating old macronuclear fragments become converted into RNA.

Another example of this type has been observed in *Metopus es* (Fig. 89). According to Noland (1927), the conjugants fuse along the anterior end (*a*), and the micronucleus in each individual divides in the same ways as was observed in *Paramecium caudatum* (*b-e*). But the cytoplasm and both pronuclei of one conjugant pass into the other (*f*), leaving the degenerating macronucleus and a small amount of cytoplasm behind in the shrunken pellicle of the smaller conjugant which then separates from the other (*j*). In the larger exconjugant, two pronuclei fuse, and the other two degenerate and disappear (*g, h*). The synkaryon divides into two nuclei, one of which condenses into the micronucleus and the other grows into the macronucleus (*i, k-m*). This is followed by the loss of cilia and encystment.

While ordinarily two individuals participate in conjugation, three or four individuals are occasionally involved. For example, conjugation of three animals was observed in *P. caudatum* by Stein (1867), Jickeli (1884), Maupas (1889) and in *Blepharisma undulans* by Giese (1938) and Weisz (1950). Chen (1940b, 1948) made a careful study of such a conjugation which he found in *Paramecium bursaria* (Fig. 90). He found that the usual manner of association is conjugation between a pair with the third conjugant attached to the posterior part of one of them (*a*). Nuclear changes occur in all three individuals, and in each, two pronuclei are formed by three divisions (*c*). But the exchange of the pronuclei takes place only between two anterior conjugants (*c-e*) and autogamy (see below) occurs in the third individual.

Automixis. In certain Protozoa, the fusion occurs between two nuclei which originate in a single nucleus of an individual. This process has been called automixis by Hartmann, in contrast to the amphimixis (Weismann) which is the complete fusion of two nuclei originating in two individuals, as was discussed in the preceding pages. If the two nuclei which undergo a complete fusion are present in a single cell, the process is called **autogamy** but, if they are in two different cells, then **paedogamy.** The autogamy is of common occurrence in the myxosporidan spores. The young sporoplasm contains two nuclei which fuse together prior to or during the process of germination in the alimentary canal of a specific

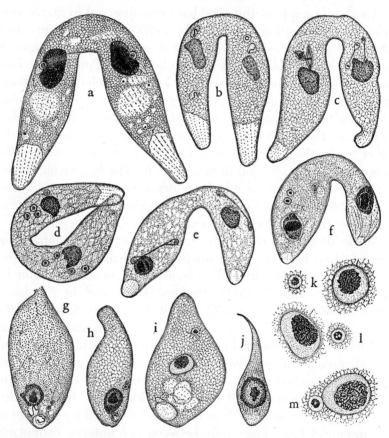

FIG. 89. Conjugation of *Metopus es* (Noland). a, early stage; b, first micronuclear division; c, d, second micronuclear division; e, third micronuclear division; f, migration of pronuclei from one conjugant into the other; g, large conjugant with two pronuclei ready to fuse; h, large conjugant with the synkaryon, degenerating pronuclei and macronucleus; i, large exconjugant with newly formed micronucleus and macronucleus; j, small exconjugant with degenerating macronucleus; k–m, development of two nuclei. a, ×290; b–j, ×250, k–m, ×590.

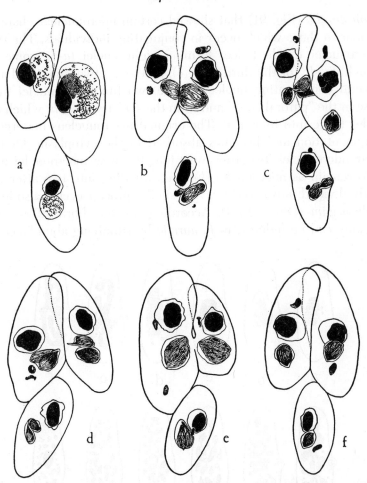

FIG. 90. Conjugation of three individuals in *Paramecium bursaria,* ×365 (Chen). a, late prophase of the first nuclear division (the individual on right is a member of a race with "several hundred chromosomes," while the other two belong to another race with "about 80 chromosomes"); b, anaphase of the third division (each individual contains 2 degenerating nuclei); c, beginning of pronuclear exchange between two anterior animals; d, e, synkaryon formation; f, after the first division of synkaryon, one daughter nucleus undergoing degeneration in all animals.

host fish, as for example in *Sphaeromyxa sabrazesi* (Figs. 281, 282) and *Myxosoma catostomi* (Fig. 280). It occurs also in Foraminiferida (Grell, 1954).

Diller (1934, 1936) observed autogamy in solitary *Parame-*

cium aurelia (Fig. 91) that showed certain micronuclear changes similar to those which occur in conjugating individuals. The two micronuclei divide twice, forming eight nuclei (*a-d*), some of which divide for the third time (*e*), producing two functional and several degenerating nuclei (*f*). The two functional nuclei then fuse in the "paroral cone" and form the synkaryon (*g, h*) which divides twice into four (*i, j*). The original macronucleus undergoes fragmentation and becomes absorbed in the cytoplasm. Of the four micronuclei, two transform into the new macronuclei and two remain as micronuclei (*k*), each dividing into two after the body divided into two (*l*). A very similar process occurs also in *P. polycaryum* (Diller, 1954). According to Sonneborn (1950), autogamy can be induced in *P. aurelia* by supplying abundance of

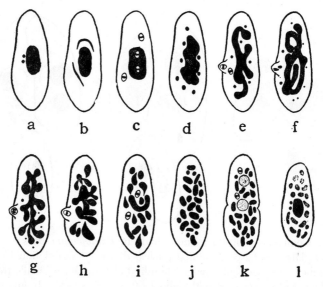

Fig. 91. Diagram illustrating autogamy in *Paramecium aurelia* (Diller). a, normal animal; b, first micronuclear division; c, second micronuclear division; d, individual with 8 micronuclei and macronucleus preparing for skein formation; e, two micronuclei dividing for the third time; f, two gamete-nuclei formed by the third division in the paroral cone; g, fusion of the nuclei, producing synkaryon; h, i, first and second division of synkaryon; j, with 4 nuclei, 2 becoming macronuclei and the other 2 remaining as micronuclei; k, macronuclei developing, micronuclei dividing; l, one of the daughter individuals produced by fission.

food and then by depriving of food. (Autogamy in Frontonia, Devi, 1961)

Another sexual process appears to have been observed by Diller (1934) in conjugating *Paramecium trichium* in which there was no nuclear exchange between the two conjugants. Wichterman (1940) observed a similar process in *P. caudatum* and named it **cytogamy.** Two small (about 200μ long) individuals of *P. caudatum* fuse on their oral surfaces. There occur three micronuclear divisions as in the case of conjugation, but there is no nuclear exchange between the members of the pair. The two gametic nuclei in each individual are said to fuse and form a synkaryon as in autogamy. Sonneborn (1941) finds the frequency of cytogamy in *P. aurelia* to be correlated with temperature. At 17°C., conjugation occurs in about 95 per cent of the pairs and cytogamy in about 5 per cent; but at 10° and 27°C., cytogamy takes place in 47 and 60 per cent respectively. In addition, there is some indication that sodium decreases and calcium increases the frequency of occurrence of cytogamy. A similar process was found to occur in *P. polycaryum* (Diller, 1958).

The paedogamy occurs in at least two species of Myxosporida, namely, *Leptotheca ohlmacheri* (Fig. 286) and *Unicapsula muscularis* (Fig. 285). The spores of these myxosporidans contain two uninucleate sporoplasms which are independent at first, but prior to emergence from the spore, they undergo a complete fusion to metamorphose into a uninucleate amoebula. Perhaps the classical example of the paedogamy is that which was found by Hertwig (1898) in *Actinosphaerium eichhorni*. The organism encysts and the body divides into numerous uninucleate secondary cysts. Each secondary cyst divides into two and remains together within a common cystwall. In each, the nucleus divides twice, and forms four nuclei, one of which remains functional, the remaining three degenerating. The paedogamy results in formation of a zygote in place of a secondary cyst. Bělař (1923) observed a similar process in *Actinophrys sol* (Fig. 92). This heliozoan withdraws its axopodia and divides into two uninucleate bodies which become surrounded by a common gelatinous envelope. Both nuclei divide twice and produce four nuclei, three of which degenerate. The two daughter cells, each with one haploid nucleus, undergo paedogamy and the resulting individual now contains a diploid nucleus.

In *Paramecium aurelia,* Diller (1936) found simple fragmentation of the macronucleus which was not correlated with any special micronuclear activity and which could not be stages in conjugation or autogamy. Diller suggests that if conjugation or autogamy is to create a new nuclear complex, as is generally held, it is conceivable that somewhat the same result might be achieved by "purification act" (through fragmentation on the part of the macronucleus itself, without involving micronuclei. He coined the term **hemixis** for this reorganization. Miyake (1955) reported the production of somewhat similar nuclear fragmentation in *P. caudatum* by addition of glucose and sucrose to the culture fluid.

Meiosis. In the foregoing sections, references have been made to the divisions which the nuclei undergo prior to sexual fusion or conjugation. In all metazoa, during the development of the ga-

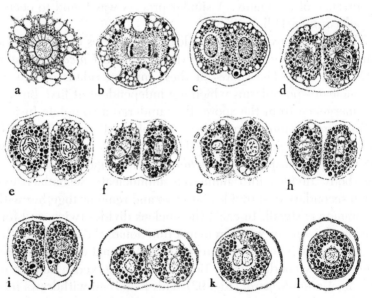

FIG. 92. Paedogamy in *Actinophrys sol,* ×460 (Bělař). a, withdrawal of axopodia; b, c, division into two uninucleate bodies, surrounded by a common gelatinous envelope; d–f, the first reduction division; g–i, the second reduction division; j–l, synkaryon formation.

metes, the gametocytes undergo reduction division or meiosis, by which the number of chromosomes is halved; that is to say, each fully mature gamete possesses half (haploid) number of chromosomes typical of the species (diploid). In the zygote, the diploid number is reestablished. In the protozoa in which sexual reproduction occurs during their life-cycle, meiosis presumably takes place to maintain the constancy of chromosome-number, but the process is understood only in a small number of species.

In conjugation, the meiosis seems to take place in the second

Fig. 93. Mitotic and meiotic micronuclear divisions in conjugating *Didinium nasutum*. (Prandtl, modified). a, normal micronucleus; b, equatorial plate in the first (mitotic) division; c, anaphase in the first division; d, equatorial plate in the second division; e, anaphase in the second (meiotic) division.

micronuclear division, although in some, for example, *Oxytricha fallax*, according to Gregory, the actual reduction occurs during the first division. Prandtl (1906) was the first to note a reduction in number of chromosomes in the protozoa. In conjugating *Didinium nasutum* (Fig. 93), he observed sixteen chromosomes in each of the daughter micronuclei during the first division, but only eight in the second division. Since that time, the fact that meiosis occurs during the second micronuclear division has been observed in *Chilodonella uncinata* (Enriques; MacDougall), *Carchesium polypinum* (Popoff), *Uroleptus halseyi* (Calkins), etc. (note the ciliates in Table 5 on p. 203). In various species of Paramecium and many other forms, the number of chromosomes appears to be too great to allow a precise counting, but the observations of Sonneborn, as quoted elsewhere (p. 278) and of Jennings (1942) on *P. aurelia* and *P. bursaria* respectively, indicate clearly the occurrence of meiosis prior to nuclear exchange during conjugation.

Information on the meiosis involved in the complete fusion of gametes is even more scanty and fragmentary. In *Monocystis rostrata* (Fig. 94), a parasite of the earthworm, Mulsow (1911) noticed that the nuclei of two gametocytes which encyst together, multiply by mitosis in which eight chromosomes are constantly present (*a-g*), but in the last division in gamete formation, each

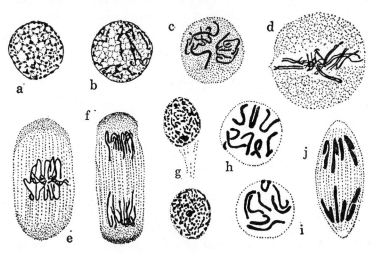

Fig. 94. Motisis and meiosis in *Monocystis rostrata* (Mulsow). a–g, mitosis; h–j, meiosis. a, a resting nucleus in the gametocyte; b, development of chromosomes; c, polar view of equatorial plate; d, longitudinal splitting of eight chromosomes; e, separation of chromosomes in two groups; f, late anaphase; g, two daughter nuclei; h, i, polar view of the equatorial plate in the last division; j, anaphase, the gamete nucleus is now haploid (4). a–c, ×1840; d–g, ×1400; h–j, ×3000.

daughter nucleus receives only four chromosomes *h-j*). In another species of Monocystis, Calkins and Bowling (1926) observed that the diploid number of chromosomes was ten and that haploid condition is established in the last gametic division thus confirming Mulsow's finding.

In the paedogamy of *Actinophrys sol* (Fig. 92), Bělař (1923) finds forty-four chromosomes in the first nuclear division, but after two meiotic divisions, the remaining functional nucleus contains only twenty-two chromosomes so that when paedogamy is completed the diploid number is restored.

In the coccidian, *Aggregata eberthi* (Fig. 252), according to

Dobell (1925), Naville (1925) and Bělař (1926) and in the gregarine, *Diplocystis schneideri,* according to Jameson (1920), there is no reduction in the number of chromosomes during the gamete-formation, but the first zygotic division is meiotic, twelve to six and six to three, respectively. A similar reduction takes place also in *Actinocephalus parvus* (eight to four, after Weschenfelder, 1938), *Gregarina blattarum* (six to three, after Sprague, 1941), *Adelina deronis* (twenty to ten, after Hauschka, 1943), *Stylocephalus longicollis* (eight to four, after Grell, 1940), etc. Trichonympha and other flagellates of the woodroach and Chlamydomonas also undergo postzygotic meiosis. Thus in these organisms, the zygote is the only stage in which the nucleus is diploid.

Some eighty years ago Weismann pointed out that a protozoan grows and multiplies by binary fission or budding into two equal or unequal individuals without loss of any protoplasmic part and these in turn grow and divide, and that thus in protozoa there is neither senescence nor natural death which occur invariably in metazoa in which germ and soma cells are differentiated. Since that time, the problem of potential immortality of protozoa has been a matter which attracted the attention of numerous investigators. Because of large dimensions, rapid growth and reproduction, and ease with which they can be cultivated in the laboratory, the majority of protozoa used in the study of the problem have been free-living freshwater ciliates that feed on bacteria and other microorganisms.

The very first extended study was made by Maupas (1888) who isolated *Stylonychia pustulata* on February 27, 1886, and observed 316 binary fissions until July 10. During this period, there was noted a gradual decrease in size and increasing abnormality in form and structure, until the animals could no longer divide and died (Fig. 95). A large number of isolation culture experiments have since been carried on numerous species of ciliates by many investigators. The results obtained are not in agreement. However, the bulk of obtained data indicates that the vitality of animals decreases with the passing of generations until finally the organisms suffer inevitable death, and that in the species in which conjugation or other sexual reproduction occurs, the declining vitality often becomes restored.

Perhaps the most thorough experiment was carried on by Calk-

ins (1919, 1933) with *Uroleptus mobilis.* Starting with an exconjugant on November 1, 1917, a series of pure-line cultures
was established by the daily isolation method. It was found that
no series lived longer than a year, but when two of the progeny of
a series were allowed to conjugate after the first seventy-five generations, the exconjugants repeated the history of the parent series, and did not die when the parent series died. In this way,

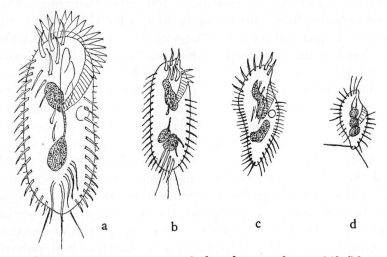

FIG. 95. Degeneration or aging in *Stylonychia pustulata.* ×340 (Maupas,
modified). a, Beginning stage with reduction in size and completely atrophied
micronucleus; b, c, advanced stages in which disappearance of the frontal
zone, reduction in size, and fragmentation of the macronucleus occurred; d,
final stage before disintegration.

lines of the same organism have lived for more than twelve years,
passing through numerous series. In a series, the average division
for the first sixty days was 15.4 divisions per ten days, but the
rate gradually declined until death. Woodruff and Spencer
(1924) also found the isolation cultures of *Spathidium spathula*
(fed on *Colpidium colpoda*) died after a gradual decline in the division rate, but were inclined to think that improper environmental conditions rather than internal factors were responsible
for the decline.

On the other hand, Woodruff (1932) found that 5071 generations produced by binary fission from a single individual of *Par-*

amecium aurelia between May 1, 1907 and May 1, 1915, did not manifest any decrease in vitality after eight years of continued asexual reproduction. Other examples of longevity of ciliates without conjugation are: Glaucoma for 2701 generations (Enriques, 1916), *Paramecium caudatum* for 3967 generations (Metalnikov, 1922), *Spathidium spathula* for 1080 generations (Woodruff and Moore, 1924), *Didinium nasutum* for 1384 generations (Beers, 1929), etc. With *Actinophrys sol*, Bělař (1924) carried on isolation cultures for 1244 generations for a period of thirty-two months and noticed no decline in the division rate. Hartmann (1921) made a similar observation on *Eudorina elegans*. It would appear that in these forms, the life continues indefinitely without apparent decrease in vital activity.

As has been noted in the beginning part of the chapter, the macronucleus in the ciliates undergoes, at the time of binary fission a reorganization process before dividing into two parts and undoubtedly, there occurs at the same time extensive cytoplasmic reorganization as judged by the degeneration and absorption of the old, and formation of the new, organellae. It is reasonable to suppose that this reorganization of the whole body structure at the time of division is an elimination process of waste material accumulated by the organism during the various phases of vital activities as was considered by Kidder and others (p. 184) and that this elimination, though not complete, enables the protoplasm of the products of division to carry on their metabolic functions more actively.

As the generations are multiplied, the general decline in vitality is manifest not only in the decreased division-rate, slow growth, abnormal form and function of certain organellae, etc., but also in inability to complete the process involved in conjugation. Jennings (1944) distinguished four successive periods in various clone cultures of *Paramecium bursaria;* namely, (1) a period of sexual immaturity during which neither sexual reaction nor conjugation occurs; (2) a period of transition during which weak sexual reactions appear in a few individuals; (3) a period of maturity in which conjugation takes place readily when proper mating types are brought together; and (4) a period of decline, ending in death. The length of the first two periods depends on the cultural conditions. Exconjugant clones that are kept in condition under which the animals multiply rapidly, reach maturity in three to five months, while those subjected to depressing condition require ten to fourteen months to reach maturity. The third period lasts for several years and is followed by the fourth period during which fission becomes slower, abnormalities appear, many individuals die and the clones die out completely. (Changes with age in *P. aurelia,* Sonneborn and Dippell, 1960).

Does conjugation affect the longevity of clones in *Paramecium bursaria?* A comparative study of the fate of exconjugants and non-conjugants, led Jennings (1944a) to conclude that (1) conjugation results in production of one of the following four types: (a) exconjugants perish without division, (b) exconjugants divide one to four times and then die, (c) exconjugants produce weak abnormal clones which may become numerous, and (d) exconjugants multiply vigorously and later undergo conjugation again; at times the latter are more vigorous than the parent clones, thus showing rejuvenescence through conjugation; (2) conjugation of young clones results in little or no mortality, while that of old clones results in high (often 100 per cent) mortality; (3) conjugation between a young and an old clone, results in the death of most or all of the exconjugants; (4) the two members of a conjugating pair have the same fate; and (5) what other causes besides age bring about the death, weakness or abnormality of the exconjugants, are not known.

It is probable that the process of replacing old macronuclei by micronuclear material which are derived from the products of fusion of two micronuclei of either the same (autogamy) or two different animals (conjugation), would perhaps result in a complete elimination of waste substances from the newly formed macronuclei, and divisions which follow this fusion may result in shifting the waste substances unequally among different daughter individuals. Thus in some individuals there may be a complete elimination of waste material and consequently a restored high vitality, while in others the influence of waste substances present in the cytoplasm may offset or handicap the activity of new macronuclei, giving rise to stocks of low vitality which will perish sooner or later. In addition in conjugation, the union of two haploid micronuclei produces diverse genetic constitutions which would be manifest in progeny in manifold ways. Experimental evidences indicate clearly such is actually the case.

In many ciliates, the elimination of waste substances at the time of binary fission and sexual reproduction (conjugation and autogamy), seemingly allow the organisms continued existence through a long chain of generations indefinitely. Jennings (1929, 1942) who reviewed the whole problem states: "Some protozoa are so constituted that they are predestined to decline and death after a number of generations. Some are so constituted that decline occurs, but this is checked or reversed by substitution of reserve parts for those that are exhausted; they can live indefinitely, but are dependent on this substitution. In some the constitution is such that life and multiplication can continue indefinitely without visible substitution of a reserve nucleus for an exhausted one; but whether this is due to the continued substitution, on a minute scale, of reserve parts for those that are outworn, cannot now be positively stated. This perfected condition, in which living itself includes continuously the necessary processes of repair and elimination, is found in some free cells, but not in all."

"In the absence of sexual processes," writes Sonneborn (1957), "paramecia grow old and either die or become genetically extinct by losing their germplasm (micronuclei). Survival depends upon the occurrence of fertilization before senility has gone too far. The fertilized animals are reinvigorated and start new young clones."

Regeneration

The capacity of regenerating the lost parts, though variable among different species, is characteristic of all protozoa from simple forms to those with highly complex organizations, as shown by observations of numerous investigators. It is now a well-established fact that when a protozoan is cut into two parts and the parts are kept under proper environmental conditions, the enucleated portion is able to carry on catabolic activities, but unable to undertake anabolic activities, and consequently degenerates sooner or later, while the nucleated portion regenerates. Brandt (1877) studied regeneration in *Actinosphaerium eichhorni* and found that only nucleate portions containing at least one nucleus regenerated and enucleate portions or isolated nuclei degenerated. Similarly Gruber (1886) found in *Amoeba proteus* the nucleate portion regenerated completely, while enucleate part became rounded and perished in a few days.

The parts which do not contain nuclear material may continue to show certain metabolic activities such as locomotion, contraction of contractile vacuoles, etc., for some time; for example, Grosse-Allermann (1909) saw enucleate portions of *Amoeba verrucosa* alive for twenty to twenty-five days, while Stolc (1910) found enucleate *Amoeba proteus* living for thirty days. Clark (1942, 1943) showed that *Amoeba proteus* lives for about seven days after it has been deprived of its nucleus. Enucleated individuals show a 70 per cent depression of respiration and are unable to digest food due to the failure of zymogens to be activated in the dedifferentiating cytoplasm. According to Brachet (1950), the enucleated half of an amoeba shows a steady decrease in ribonucleic acid content, while the nucleated half retains a much larger amount of this substance. Thus, it appears that the synthesis of the cytoplasmic particles containing ribonucleic acid is under the control of the nucleus.

In Arcella (Martini; Hegner) and Difflugia (Verworn; Penard), when the tests are partially destroyed, the broken tests remain unchanged. Verworn considered that in these testaceans test-forming activity of the nucleus is limited to the time of asexual reproduction of the organisms. On the other hand, several observers report in Foraminiferida the broken shell is completely regenerated at all times. Verworn pointed out that this indicates that here the nucleus controls the formation of shell at all times. In a radiolarian, *Thalassicolla nucleata,* the central capsule, if dissected out from the rest of body, will regenerate into a complete organism (Schneider).

An enormous number of regeneration experiments have been conducted on more than fifty ciliates by numerous investigators. Here also the general conclusion is that the nucleus is necessary for regeneration. In many cases, the macronucleus seems to be the only essential nucleus for regeneration, as judged by the continued division on record of several amicronucleate ciliates and by experiments such as Schwartz's (1935) in which there was no regeneration in *Stentor coeruleus* from which the whole macronucleus had been removed.

A remarkably small part of a protozoan is able to regenerate completely if nuclear material is included. For example, in *Stentor coeruleus*, pieces as small as 1/30th (Lillie, 1896) or 1/64th (Morgan, 1901) of the original specimen or about 70μ in diameter (Weisz, 1948) can regenerate. Burnside (1929) cut twenty-seven specimens of this ciliate belonging to a single clone, into two or more parts in such a way that some of the pieces contained a large portion of the nucleus, while others a small portion. These fragments regenerated and multiplied, giving rise to 268 individuals. No dimensional differences resulted from the different amounts of nuclear material present in cut specimens. Apparently, regulatory processes took place, and in all cases, normal size was restored. Thus biotypes of diverse sizes are not produced by causing inequalities in the proportions of nuclear material in different individuals.

When substantial portions of the posterior poles of starvation dwarfs of *S. coeruleus* are cut off, Tartar (1961) found that pieces as small as only 1/123rd the volume of large normal animals could regenerate completely and survive for over six days. Apparently, a part of the macronucleus or any node in moniliform macronucleus is equivalent to any other part in capacity to support regeneration of the nucleus and redifferentiation of the cell (Tartar, 1957). In addition to these restorative regenerations, there occur regeneration of various organelles at the time of asexual or sexual reproduction of all protozoa. (Morphogenesis in regeneration, Chatton and Lwoff, 1935; Balamuth, 1940; Summers, 1941; Fauré-Fremiet, 1948; Weisz, 1948, 1951; Tartar, 1954, 1961; Suzuki, 1957; Yow, 1958; Hashimoto, 1961.)

References

BALAMUTH, W., 1940. Regeneration in protozoa: a problem of morphogenesis. Quart. Rev. Biol., 15:290.

———— 1951. Biological studies on *Entamoeba histolytica*. III. J. Infect. Dis., 88:230.

BARKER, H. A., and TAYLOR, C. V., 1931. A study of the conditions of encystment of *Colpoda cucullus*. Physiol. Zool., 4:620.

———— ———— 1933. Studies on the excystment of *Colpoda cucullus. Ibid.*, 6:127.

BEERS, C. D., 1926. The life-cycle in the ciliate *Didinium nasutum* with reference to encystment. J. Morphol., 42:1.

———— 1927. Factors involved in encystment in the ciliate *Didinium nasutum.* J. Morphol. Physiol., 43:499.

———— 1928. Rhythms in Infusoria with special reference to *Didinium nasutum.* J. Exper. Zool., 51:485.

———— 1930. On the possibility of indefinite reproduction in the ciliate, etc. Am. Nat., 63:125.

———— 1931. Some effects of conjugation in the ciliate *Didinium nasutum.* J. Exper. Zool., 58:455.

———— 1935. Structural changes during encystment and excystment in the ciliate *Didinium nasutum.* Arch. Protist., 84:133.

———— 1937. The viability of 10-year old Didinium cysts. Am. Nat., 71:521.

———— 1945. Some factors affecting excystment in the ciliate *Tillina magna.* Physiol. Zool., 18:82.

———— 1945a. The excystment process in the ciliate *Didinium nasutum.* J. El. Mitchell Sci. Soc., 61:264.

———— 1946. History of the nuclei of *Tillina magna* during division and encystment. J. Morphol., 78:181.

———— 1946a. The excystment in *Didinium nasutum* with special reference to the rôle of bacteria. J. Exper. Zool., 103:201.

———— 1946b. *Tillina magna:* micronuclear number, etc. Biol. Bull., 91:256.

———— 1947. The relation of density of population to encystment in *Didinium nasutum.* J. El. Mitchell Sci. Soc., 63:141.

———— 1963. A comparison of two methods of chromatin extrusion from the macronucleus of *Conchophthirus curtus.* Tr. Am. Micr. Soc., 82:131.

BĚLAŘ, K., 1921. Untersuchungen ueber Thecamoeben der Chlamydophrys-Gruppe. Arch. Protist., 43:287.

———— 1923. Untersuchungen an *Actinophrys sol.* I. *Ibid.,* 46:1.

———— 1924. II. *Ibid.,* 48:371.

———— 1926. Der Formwechsel der Protistenkerne. Ergebn. u. Fortsch. Zool., 6:235.

BODINE, J. H., 1923. Excystation of *Colpoda cucullus.* J. Exper. Zool., 37:115.

BOELL, E. J., and WOODRUFF, L. L., 1941. Respiratory metabolism of mating types in *Paramecium calkinsi.* J. Exper. Zool., 87:385.

BRACHET, J., 1950. Un étude cytochimique des fragments nucléés et enucléés d'amibes. Experientia, 6:294.

BRAND, T. v., 1923. Die Encystierung bei *Vorticella microstoma* und hypotrichen Infusorien. Arch. Protist., 47:59.

BRIDGMAN, A. J., 1956. Studies on dried cysts of *Tillina magna.* J. Protozool., 4:17.

BUDDENBROCK, W. v. 1922. Ueber eine neue Strombidium-Art aus Heligoland. Arch. Protist., 45:129.

BÜTSCHLI, O., 1876. Studien über die ersten Entwicklungsvorgänge der Eizelle, die Zelltheilung und die Conjugation der Infusorien. Abh. Senk. Nat. Ges. Frankf., 10:1.

BURNSIDE, L. H., 1929. Relation of body size to nuclear size in *Stentor coeruleus*. J. Exper. Zool., 54:473.

BURT, R. L., KIDDER, G. W., and CLAFF, C. L., 1941. Nuclear reorganization in the family Colpodidae. J. Morphol., 69:537.

CALKINS, G. N., 1919. *Uroleptus mobilis*. II. J. Exper. Zool., 29:121.

—— 1933. The biology of the protozoa. 2nd ed. Philadelphia.

—— and BOWLING, R. C., 1926. Gametic meiosis in Monocystis. Biol. Bull., 51:385.

—— and CULL, S. W., 1907. The conjugation of *Paramecium aurelia* (*caudatum*). Arch. Protist., 10:375.

—— and SUMMERS, F. M., editors, 1941. Protozoa in biological research. New York.

CANELLA, I. R., and TRINCAS, L., 1961. Cicle vitale e nucleare di *Ophryoglena singularis* sp.n. Pub. Civico Muses Stor. Nat., Ferrara, 5:1.

CHALKLEY, H. W., 1936. The behavior of the karyosome and the "peripheral chromatin" during mitosis and interkinesis in *Amoeba proteus*, etc. J. Morphol., 60:13.

—— and DANIEL, G. E., 1933. The relation between the form of the living cell and the nuclear phases of division in *Amoeba proteus*. Physiol. Zool., 6:592.

CHATTON, E. and LWOFF, A., 1935. Les Ciliés Apostomes. I. Arch. zool. exper. gén., 77:1.

CHEN, T. T., 1936. Observations on mitosis in opalinids. Proc. Nat. Acad. Sci., 22:594.

—— 1940. Polyploidy and its origin in Paramecium. J. Hered., 31:175.

—— 1940a. Conjugation in *Paramecium bursaria* between animals with diverse nuclear constitutions. *Ibid.*, 31:185.

—— 1940b. Conjugation of three animals in *Paramecium bursaria*. Proc. Nat. Acad. Sci., 26:231.

—— 1940c. Conjugation in *Paramecium bursaria* between animals with very different chromosome numbers, etc. *Ibid.*, 26:243.

—— 1945. Induction of conjugation in *Paramecium bursaria*, etc. *Ibid.*, 31:404.

—— 1946. Conjugation in *Paramecium bursaria*. I. J. Morphol., 78:353.

—— 1946a. II. *Ibid.*, 79:125.

—— 1946b. Varieties and mating types in *Paramecium bursaria*. Proc. Nat. Acad. Sci., 32:173.

—— 1946c. Temporary pair formation in *Paramecium bursaria*. Biol. Bull., 91:112.

—— 1948. Chromosomes in Opalinidae, etc. J. Morphol., 83:281.

—— 1955. Paramecin 34, a killer substance produced by *Paramecium bursaria*. Proc. Soc. Exper. Biol. Med., 88:541.

CHEN, Y. T., 1944. Mating types in *Paramecium caudatum*. Am. Nat., 78:334.

CLARK, A. M., 1942. Some effects of removing the nucleus from Amoeba. Australian J. Exper. Biol., 20:241.

———— 1943. Some physiological functions of the nucleus in Amoeba, etc. *Ibid.*, 21:215.

CLEVELAND, L. R., 1938. Longitudinal and transverse division in two closely related flagellates. Biol. Bull., 74:1.

———— 1938a. Origin and development of the achromatic figure. *Ibid.*, 74:41.

———— 1949. The whole life cycle of chromosomes and their coiling systems. Tr. Am. Philos. Soc., 39:1.

———— 1949a. Hormone-induced sexual cycles of flagellates. I. J. Morphol., 85:197.

———— 1950. II. *Ibid.*, 86:185.

———— 1950a. III. *Ibid.*, 86:215.

———— 1950b. IV. *Ibid.*, 87:317.

———— 1950c. V. *Ibid.*, 87:349.

———— 1951. VI. *Ibid.*, 88:199.

———— 1951a. VII. *Ibid.*, 88:385.

———— 1956. Cell division without chromatin in Trichonympha and Barbulanympha. J. Protozool., 3:78.

———— 1956a. Brief accounts of the sexual cycles of the flagellates of Cryptocercus. *Ibid.*, 3:161.

———— 1957. Additional observations on gametogenesis and fertilization in Trichonympha. *Ibid.*, 4:164.

———— 1957a. Correlation between the molting period of Crypstocercus and sexuality in its protozoa. *Ibid.*, 4:168.

———— 1957b. Types and life cycles of centrioles of flagellates. *Ibid.*, 4:230.

———— 1958. A factual analysis of chromosomal movement in Barbulanympha. *Ibid.*, 5:47.

———— 1958a. Movement of chromosomes in Spirotrichonympha, etc. *Ibid.*, 5:63.

———— 1960. The centrioles of Trichonympha from termites and their functions in reproduction. *Ibid.*, 7:326.

————, BURKE, A. W., Jr., and KARLSON, P., 1960. Ecdysone induced modifications in the sexual cycles of the protozoa of Cryptocercus. *Ibid.*, 7:229.

————, HALL, S. R., *et al.*, 1934. The wood-feeding roach, Cryptocercus, etc. Mem. Amer. Acad. Arts & Sci., 17:185.

———— and NUTTING, W. L., 1955. Suppression of sexual cycles and death of the protozoa of Cryptocercus resulting from change of hosts during molting period. J. Exper. Zool., 130:485.

COLEMAN, A. W., 1959. Sexual isolation in *Pandorina morum*. J. Protozool., 6:249.

CRUMP, LETTICE M., 1950. The influence of bacterial environment on the excystment of amoebae from soil. J. Gen. Microbiol., 4:16.

CULTER, D. W., and CRUMP, L. M., 1935. The effect of bacterial products on amoebic growth. Brit. J. Exper. Biol., 12:52.

DANGEARD, P. A., 1889. Mémoire sur les algues. Le Botaniste, 1:127.

———— 1900. Observations sur le développment du *Pandorina morum*. *Ibid.*, 7:192.

DANIEL, G. E., and CHALKLEY, H. W., 1932. The influence of temperature upon the process of division in *Amoeba proteus*. J. Cell. Comp. Physiol., 2:311.

DARBY, H. H., 1929. The effect of the hydrogen-ion concentration on the sequence of protozoan forms. Arch. Protist., 65:1.

DASS, C. M. S., 1950. Chromatin elimination in *Glaucoma pyriformis*. Nature, 165:693.

DAVIS, T. G., 1941. Morphology and division in *Tetratoxum unifasciculatum*. Tr. Am. Micr. Soc., 60:441.

DAWSON, J. A., 1919. An experimental study of an amicronucleate Oxytricha. I. J. Exper. Zool., 29:473.

———— and HEWITT, D. C., 1931. The longevity of encysted Colpoda. Am. Nat., 65:181.

———— and MITCHELL, W. H., 1929. The vitality of certain infusorian cysts. *Ibid.*, 63:476.

DE GARIS, C. F., 1935. Lethal effects of conjugation between *Paramecium aurelia* and double-monsters of *P. caudatum*. Am. Nat., 69:87 .

DEVI, R. V., 1961. Autogamy in *Frontonia leucas*. J. Protozool., 8:277.

DILLER, W. F., 1934. Autogamy in *Paramecium aurelia*. Science, 79:57.

———— 1936. Nuclear reorganization processes in *Paramecium aurelia*, etc. J. Morphol., 59:11.

———— 1948. Nuclear behavior of *Paramecium trichium* during conjugation. *Ibid.*, 82:1.

———— 1950. An extra postzygotic nuclear division in *Paramecium caudatum*. Tr. Am. Micr. Soc., 69:309.

———— 1950a. Cytological evidence for pronuclear interchange in *Paramecium caudatum*. *Ibid.*, 69:317.

———— 1954. Autogamy in *Paramecium polycaryum*. J. Protozool., 1:60.

———— 1958. Studies on conjugation in *Paramecium polycaryum*. *Ibid.*, 5:282.

DOBELL, C., 1908. The structure and life history of *Copromonas subtilis*, etc. Quart. J. Micr. Sci., 52:75.

———— 1917. On *Oxnerella maritima*, etc. *Ibid.*, 62:515.

———— 1925. The life history and chromosome cycle of *Aggregata eberthi*. Parasitology, 17:1.

—— 1928. Researches on the intestinal Protozoa of monkeys and man. I, II. *Ibid.*, 20:357.

—— 1943. XI. *Ibid.*, 35:134.

—— and LAIDLAW, P. P., 1926. On the cultivation of *Entamoeba histolytica*, etc. *Ibid.*, 18:283.

DOWNS, L. E., 1959. Mating types and their determination in *Stylonychia putrina*. J. Protozool., 6:285.

EGELHAAF, A., 1955. Cytologisch-entwicklungsphysiologische Untersuchungen zur Konjugation von *Paramecium bursaria*. Arch. Protist., 100: 447.

ELLIOTT, A. M., and HAYES, R. E., 1953. Mating types in Tetrahymena. Biol. Bull., 105:269.

—— —— 1955. Tetrahymena from Mexico, Panama and Columbia, with special reference to sexuality. J. Protozool., 2:75.

—— and NANNY, D. L., 1952. Conjugation in Tetrahymena. Science, 116:33.

—— and TREMOR, J. W., 1958. The fine structure of the pellicle in the contact area of conjugating *Tetrahymena pyriformis*. J. Biophys. Biochem. Cytol., 4:839.

ENRIQUES, P., 1916. Duemila cinquecento generazioni in un infusorio, senza conjugazione ne partenogenesi, ne depressioni. Rev. Acad. Sci. Bologna, 20:67.

ENTZ, G., 1925. Ueber Cysten und Encystierung der Süsswasser-Ceratien. Arch. Protist., 51:131.

EVERRITT, M. G., 1950. The relationship of population growth, etc. J. Parasit., 36:586.

FAURÉ-FREMIET, E., 1948. Les mécanismes de la morphogenése chez les ciliés. Folia Bioth., 3:25.

—— *et al.*, 1957. La réorganization macronucléare chez les Euplotes. Exper. Cell Res., 12:135.

FINLEY, H. E., 1936. A method for inducing conjugation within Vorticella cultures. Tr. Am. Micr. Soc., 55:323.

—— 1943. The conjugation of *Vorticella microstoma*. *Ibid.*, 62:97.

—— and LEWIS, A. C., 1960. Observations on excystment and encystment of *Vorticella microstoma*. J. Protozool., 7:347.

FROSCH, P., 1897. Zur Frage der Reinzuchtung der Amoeben. Zentralbl. Bakt. I. Abt., 21:926.

GARNJOBST, L., 1928. Induced encystment and excystment in *Euplotes taylori*, etc. Physiol. Zool., 1:561.

—— 1947. The effect of certain deficient media on resting cyst formation in *Colpoda duodenaria*. *Ibid.*, 20:5.

GIESE, A. C., 1938. Size and conjugation in Blepharisma. Arch. Protist., 91:125.

—— 1939. Studies on conjugation in *Paramecium multimicronucleatum*. Am. Nat., 73:432.

——— 1939a. Mating types in *Paramecium caudatum*. Am. Nat., 73:445.

——— 1957. Mating types in *Paramecium multimicronucleatum*. J. Protozool., 4:120.

GILMAN, L. C., 1941. Mating types in diverse races of *Paramecium caudatum*. Biol. Bull. 80:384.

——— 1950. The position of Japanese varieties of *Paramecium caudatum* with respect to American varieties. Biol. Bull., 99:348.

GOODEY, T., 1913. The excystation of *Colpoda cucullus* from its resting cysts, etc. Proc. Roy. Soc., Series B, 86:427.

GRASSÉ, P.-P., 1952. Traité de Zoologie. I:Fasc.1. Paris.

GRELL, K. G., 1940. Der Kernphasenwechsel von *Stylocephalus (Stylorhynchus) longicollis*. Vor. Mitt., Zool. Anz., 130:41.

——— 1954. Die Generationswechsel der polythalamen Foraminifere *Rotaliella heterocaryotica*. Arch. Protist., 100:268.

GRUCHY, D. F., 1955. The breeding system and distribution of *Tetrahymena pyriformis*. J. Protozool., 2:178.

GUILCHER, Y., 1950. Contribution a l'étude des ciliés gemmipares, etc. Univ. de Paris thesis, Sér. A. no. 2369.

HAAGEN-SMIT, A. J., and THIMANN, K. V., 1938. The excystment of *Colpoda cucullus*. I. J. Cell. Comp. Physiol., 11:389.

HALL, R. P., 1923. Morphology and binary fission of *Menoidium incurvum*. Univ. Cal. Publ. Zool., 20:447.

——— 1937. A note on behavior of the chromosomes in Euglena. Tr. Am. Micr. Soc., 56:288.

HARTMANN, M., 1917. Ueber die dauernde rein agame Züchtung von *Eudorina elegans*, etc. Ber. preuss. Akad. Wiss., Phys.-Math. Kl., p. 760.

HASHIMOTO, K., 1961. Stomatogenesis and formation of cirri in fragments of *Oxytricha fallax*. J. Protozool., 8:433.

HAUSCHKA, T. S., 1943. Life history and chromosome cycle of the coccidian, *Adelina deronis*. J. Morphol., 73:529.

HERTWIG, R., 1889. Ueber die Conjugation der Infusorien. Abh. bayerl. Akad. Wiss., 17:151.

HINSHAW, H. C., 1926. On the morphology and mitosis of *Trichomonas buccalis*. Univ. Cal. Publ. Zool., 29:159.

HIWATASHI, K., 1949. Studies on the conjugation of *Paramecium caudatum*. I. Sc. Rep. Tohoku Univ. Ser. IV, 18:137.

——— 1949a. II. *Ibid.*, 18:141.

——— 1950. III. *Ibid.*, 18:270.

——— 1951. IV. *Ibid.*, 19:95.

——— 1955. V. *Ibid.*, 21:199.

——— 1955a. VI. *Ibid.*, 21:207.

——— 1958. Inheritance of mating types in variety 12 of *Paramecium caudatum*. *Ibid.*, 24:119.

——— 1959. Induction of conjugation by ethylenediamine tetracetic acid. *Ibid.*, 25:81.

—— 1960. Locality of mating reactivity on the surface of *Paramecium caudatum. Ibid.*, 27:93.

HOLZ, G. G. JR., *et al.*, 1959. Some physiological characteristics of the mating types and varieties of *Tetrahymena pyriformis*. J. Protozool., 6:149.

HORVÁTH, J., 1950. Vitalitätsausserung einer mikronucleuslose Bodenziliate in der vegetativen Fortpflanzung. Oesterr. zool. Ztschr., 2:333.

HULL, R. W., 1954. The morphology and life cycle of *Solenophrya micraster*. J. Protozool., 1:93.

ILOWAISKY, S. A., 1926. Material zum Studium der Cysten der Hypotrichen. Arch. Protist., 54:92.

IVANIČ, M., 1934. Ueber die Ruhestadienbildung und die damit am Kernapparate verbundenen Veränderungen bei *Lionotus cygnus*. Zool. Anz., 108:17.

—— 1938. Ueber die mit der Chromosomenbildung verbundene promitotische Grosskernteilung bei den Vermehrungsruhe Stadien von *Chilodon uncinatus*. Arch. Protist., 91:61.

JAMESON, A. P., 1920. The chromosome cycle of gregarines with special reference to *Diplocystis schneideri*. Quart. J. Micr. Sci., 64:207.

JEFFRIES, W. B., 1956. Studies on excystment in the hypotrichous ciliate *Pleurotricha lanceolata*. J. Protozool., 3:136.

JENNINGS, H. S., 1929. Genetics of the Protozoa. Bibliogr. Gen., 5:105.

—— 1938. Sex relation types and their inheritance in *Paramecium bursaria*. I. Proc. Nat. Acad. Sci., 24:112.

—— 1939. Genetics of *Paramecium bursaria*. I. Genetics, 24:202.

—— 1941. II. Proc. Am. Phil. Soc., 85:25.

—— 1942. III. Genetics, 27:193.

—— 1942a. Senescence and death in Protozoa and invertebrates. E. V. Cowdry's Problems of aging. 2 ed. Baltimore.

—— 1944. *Paramecium bursaria:* Life history. I. Biol. Bull., 86:131.

—— 1944a. II. J. Exper. Zool., 96:17.

—— and OPITZ, P., 1944. Genetics of *Paramecium bursaria*. IV. Genetics, 29:576.

——, RAFFEL, D., LYNCH, R. S., and SONNEBORN, T. M., 1932. The diverse biotypes produced by conjugation within a clone of *Paramecium aurelia*. J. Exper. Zool., 62:363.

JICKELI, C. F. 1884. Ueber die Kernverhältnisse der Infusorien. Zool. Anz., 7:491.

JOHNSON, W. H., and EVANS, F. R., 1940. Environmental factors affecting encystment in *Woodruffia metabolica*. Physiol. Zool., 13:102.

—— —— 1941. A further study of environmental factors affecting cystment in *Wooodruffia metabolica. Ibid.*, 14:227.

JONES, E. E., JR., 1951. Encystment, excystment and the nuclear cycle in the ciliate *Dileptus anser*. J. El. Mitchell Sci. Soc., 67:205.

JURAND, A., *et al.*, 1962. Studies on the macronucleus of *Paramecium aurelia*. J. Protozool., 9:122.

KANEDA, M., 1960. The structure and reorganization of the macronucleus during the binary fission of *Chlamydodon pedarius*. Japan. J. Zool., 12:477.

KATASHIMA, R., 1959. Mating types in *Euplotes eurystomus*. J. Protozool., 6:75.

KATER, J. M., and BURROUGHS, R. D., 1926. The cause and nature of encystment in *Polytomella citri*. Biol. Bull., 50:38.

KAY, M. M., 1946. Studies on *Oxytricha bifaria*. III. Tr. Am. Micr. Soc., 65:132.

KIDDER, G. W., 1933. Studies on *Conchophthirus mytili* de Morgan. I. Arch. Protist., 79:1.

———— 1938. Nuclear reorganization without cell division in *Paraclevelandia simplex*, etc. *Ibid.*, 91:69.

———— and CLAFF, C. L., 1938. Cytological investigations of *Colpoda cucullus*. Biol. Bull., 74:178.

———— and DILLER, W. F., 1934. Observations on the binary fission of four species of common free-living ciliates, etc. *Ibid.*, 67:201.

———— and STUART, C. A., 1939. Growth studies on ciliates. II. Physiol. Zool., 12:341.

———— and SUMMERS, F. M., 1935. Taxonomic and cytological studies on the ciliates associated with the amphipod family, etc. Biol. Bull., 68:51.

KIMBALL, R. F., 1939. Change of mating type during vegetative reproduction in *Paramecium aurelia*. J. Exper. Zool., 81:165.

———— 1939a. Mating types in Euplotes. Amer. Nat., 73:451.

———— 1941. The inheritance of mating type in the ciliate protozoan *Euplotes patella*. Genetics, 26:158.

———— 1941a. Double animals and amicronucleate animals, etc. J. Exper. Zool., 86:1.

———— 1942. The nature and inheritance of mating types in *Euplotes patella*. Genetics, 27:269.

———— 1943. Mating types in the ciliate Protozoa. Quart. Rev. Biol., 18:30.

KOFFMAN, M., 1924. Ueber die Bedeutung der Wasserstoffionenkonzentration für die Encystierung bei einigen Ciliatenarten. Arch. mikr. Anat., 103:168.

KOFOID, C. A., and SWEZY, O., 1919. Studies on the parasites of the termites. I. Univ. Cal. Publ. Zool., 20:1.

———— ———— 1919a. III. *Ibid.*, 20:41.

KORMOS, J., and KORMOS, K., 1958. Die Zellteilungstypen der Protozoen. Act. Biol. Acad. Sci. Hungaricae, 8:127.

KUDO, R. R. 1926. Observation on *Endamoeba blattae*. Amer. J. Hyg., 6:139.

———— 1926a. Observations on *Lophomonas blattarum*, etc. Arch. Protist., 53:191.

—— 1926b. A cytological study of *Lophomonas striata*. *Ibid.*, 55:504.

—— 1936. Studies on *Nyctotherus ovalis*, etc. *Ibid.*, 87:10.

—— 1947. *Pelomyxa carolinensis* Wilson. II. J. Morphol., 80:93.

—— 1951. Observations on *Pelomyxa illinoisensis*. *Ibid.*, 88:145.

KÜHN, A., 1915. Ueber Bau, Teilung und Encystierung von *Bodo edax*. Arch. Protist., 36:212.

LANDIS, E. M., 1920. An amicronucleate race of *Paramecium caudatum*. Anat. Rec., 54:453.

LIEBMANN, H., 1944. Beitrag zur Kenntnis der Kernteilung bei vegetativen Stadien von *Entamoeba coli*. Arch. Protist., 97:1.

LIESCHE, W., 1938. Die Kern- und Fortpflanzungsverhältnisse von *Amoeba proteus*. *Ibid.*, 91:135.

LILLIE, F. R., 1896. On the smallest parts of Stentor capable of regeneration. J. Morphol., 12:239.

LUND, E. J., 1917. Reversibility of morphogenetic processes in Bursaria. J. Exper. Zool., 24:1.

LWOFF, A., 1950. Problems of morphogenesis in ciliates. New York.

MacLENNAN, R. F., 1937. Growth in the ciliate Ichthyophthirius. I. J. Exper. Zool., 76:243.

MANWELL, R. D., 1928. Conjugation, division and encystment in *Pleurotricha lanceolata*. Biol. Bull., 54:417.

MAST, S. O., and IBARA, Y., 1923. The effect of temperature, food and the age of the culture on the encystment of *Didinium nasutum*. *Ibid.*, 45:105.

MAUPAS, E., 1888. Recherches expérimentales sur la multiplication des infusoires ciliés. Arch. zool. exper. (2), 6:165.

—— 1889. Le rejeunissement karyogamique chez les ciliés. *Ibid.*, 7:149.

METALNIKOV, S., 1922. Dix aus de culture des infusoires sans conjugasion. C. R. Acad. Sci., 175:776.

METZ, C. B., 1946. Effects of various agents on the mating type substance of *Paramecium aurelia* variety 4. Anat. Rec., 93:347.

—— 1947. Induction of "pseudo selfing" and meiosis in *Paramecium aurelia* by formalin killed animals of opposite mating type. J. Exper. Zool., 105:115.

—— 1948. The nature and mode of action of the mating type substances. Am. Nat., 82:85.

—— 1954. Mating substances and the physiology of fertilization in ciliates. In D. H. Wenrich: Sex in Microorganisms. p. 284.

—— and BUTTERFIELD, W., 1950. Extraction of a mating reaction inhibiting agent from *Paramecium calkinsi*. Proc. Nat. Acad. Sci., 36:268.

MIYAKE, A., 1955. Induction of macronuclear division by osmotic change in *Paramecium caudatum*. Physiol. & Ecol., Tokio, 6:87.

—— 1956. Artificially induced micronuclear variation in *Paramecium caudatum*. J. Inst. Polytechnics, Osaka City University, D, 7:147.

—— 1957. On the stability of bimicronucleate condition in *Paramecium caudatum*. *Ibid.*, D, 8:11.

—— 1957a. Artificial induction of excystment in *Bursaria truncatella*. Physiol. & Ecology, Tokio, 7:123.

—— 1958. Induction of conjugation by chemical agents in *Paramecium caudatum*. J. Inst. Polytech., Osaka City University, D, 9:251.

MORGAN, T. H., 1901. Regeneration of proportionate structure in Stentor. Biol. Bull., 2:311.

MOUTON, H., 1902. Recherches sur la digestion chez les amibes, etc. Ann. Inst. Pasteur, 16:457.

MÜLLER, W., 1932. Cytologische und vergleichend-physiologische Untersuchungen ueber Paramecium, etc. Arch. Protist., 78:361.

MULSOW, K., 1911. Ueber Fortpflanzungserscheinungen bei *Monocystis rostrata*, n. sp. *Ibid.*, 22:20.

MUSSACCHIA, X. J., 1947. Factors affecting encystment in *Pelomyxa carolinensis*. Anat. Rec., 99:116.

NAVILLE, A., 1925. Recherches sur le cycle sporogonique des Aggregata. Rev. Suiss. Zool., 32:125.

NOBLE, E. R., 1947. Cell division in *Entamoeba gingivalis*. Univ. Cal. Publ. Zool., 53:263.

NOLAND, L. E., 1927. Conjugation in the ciliate *Metopus sygmoides*. J. Morphol. Physiol., 44:341.

NUTTING, W. L., and CLEVELAND, L. R., 1958. Effects of glandular extirpations on Cryptocercus and the sexual cycles of its protozoa. J. Exper. Zool., 137:13.

OEHLER, R., 1916. Amoebenzucht auf reinem Boden. Arch. Protist., 37:175.

PATTEN, M. W., 1921. The life history of an amicronucleate race of *Didinium nasutum*. Proc. Soc. Exper. Biol., 18:188.

PENN, A. B. K., 1927. Reinvestigation into the cytology of conjugation in *Paramecium caudatum*. Arch. Protist., 89:46.

—— 1935. Factors which control encystment in *Pleurotricha lanceolata*. Arch. Protist., 84:101.

POWERS, E. L., 1943. The mating types of double animals in *Euplotes patella*. Am. Midland Nat., 30:175.

PRANDTL, H., 1906. Die Konjugation von *Didinium nasutum*. Arch. Protist., 7:251.

RAABE, H., 1946. L'appareil nucléaire d'*Urostyla grandis*. I. Ann. Univ. Marie Curie-Skl., Lublin, Sec. C, 1:18.

—— 1947. II. *Ibid.*, 1:151.

RAFALKO, J. S., 1947. Cytological observations on the amoeboflagellate, *Naegleria gruberi*. J. Morphol., 81:1.

RAY, C., Jr., 1956. Meiosis and nuclear behavior in *Tetrahymena pyriformis*. J. Protozool., 3:88.

Reichenow, E., 1928. Ergebnisse mit der Nuclealfärbung bei Protozoen. Arch. Protist., 61:144.

――― 1929. In: Doflein-Reichenow's Lehrbuch der Protozoenkunde. Jena.

Reynolds, Mary E. C., 1932. Regeneration in an amicronucleate infusorian. J. Exper. Zool., 62:327.

Rhumbler, L., 1888. Die verschiedenen Cystenbildungen und die Entwicklungsgeschichte der holotrichen Infusoriengattung Colpoda. Zeitschr. wiss. Zool., 46:449.

Rosenberg, L. E., 1938. Cyst stages of *Opisthonecta henneguyi*. Tr. Am. Micr. Soc., 57:147.

Schlichting, H. E., Jr., 1960. The role of waterfowl in the dispersal of algae. Tr. Am. Micr. Soc., 79:160.

Schmähl, O., 1926. Die Neubildung des Peristoms bei der Teilung von *Bursaria truncatella*. Arch. Protist., 54:359.

Seshachar, B. R., and Dass, C. M. S., 1953. Evidence for the conversion of deoxyribonucleic acid (DNA) to ribonucleic acid (RNA) in *Epistylis articulata*. Exper. Cell Res., 4:248.

Siegel, R. W., 1956. Mating types in Oxytricha and the significance of mating type system in ciliates. Biol. Bull., 110:352.

Singh, B. N., 1941. The influence of different bacterial food supplies on the rate of reproduction in *Colpoda steini*, etc. Ann. Appl. Biol., 27:65.

――― 1948. Studies on giant amoeboid organisms. I. J. Gen. Microb., 2:8.

――― 1960. Interrelationship between micropredators and bacteria in soil. Proc. 47th Indian Sci. Congress, Pt. 2, 14 pp.

――― et al., 1956. Occurrence and nature of an Amoeba excystment factor produced by *Aerobacter* sp. Nature, 177:621.

――― et al., 1958. The role of *Aerobacter* sp., *Escherichia coli* and certain amino acids in the excystment of *Schizopyrenus russelli*. J. Gen. Microbiol., 19:104.

Sokoloff, B., 1924. Das Regenerationsproblem bei Protozoen. Arch. Protist., 47:143.

Sonneborn, T. M., 1937. Sex, sex inheritance and sex determination in *Paramecium aurelia*. Proc. Nat. Acad. Sci., 23:378.

――― 1938. Mating types in *Paramecium aurelia*, etc. Proc. Am. Phil. Soc., 79:411.

――― 1939. *Paramecium aurelia*: mating types and groups, etc. Am. Nat., 73:390.

――― 1940. The relation of macronuclear regeneration in *Paramecium aurelia* to macronuclear structure, etc. Anat. Rec., 78:53.

――― 1941. The occurrence, frequency and causes of failure to undergo reciprocal cross-fertilization, etc. Ibid., 81, Suppl.:66.

――― 1942. Sex hormones in unicellular organisms. Cold Spr. Harb. Symp. Quant. Biol., 10:111.

——— 1942a. Inheritance in ciliate Protozoa. Am. Nat., 76:46.

——— 1943. Gene and cytoplasm. I. Proc. Nat. Acad. Sci., 29:329.

——— 1943a. II. *Ibid.*, 29:338.

——— 1944. Exchange of cytoplasm at conjugation in *Paramecium aurelia*, variety 4. Anat. Rec., 89:49.

——— 1947. Recent advances in the genetics of Paramecium and Euplotes. Adv. Genetics, 1:263.

——— 1950. The cytoplasm in heredity. Heredity, 4:11.

——— 1954. The relation of autogamy to senescence and rejuvenescence in *Paramecium aurelia*. J. Protozool., 1:38.

——— 1957. Breeding systems, reproductive methods and species problems in protozoa. In The Species Problems. AAAS Publ. 50, p. 155.

——— and DIPPELL, R. V., 1943. Sexual isolation, mating types, and sexual responses to diverse conditions in variety 4, *Paramecium aurelia*. Biol. Bull., 85:36.

——— ——— 1946. Mating reactions and conjugation between varieties of *Paramecium aurelia*, etc. Physiol. Zool., 19:1.

——— ——— 1960. Cellular changes with age in Paramecium. AIBS Symposium 6:285.

SPRAGUE, V., 1941. Studies on *Gregarina blattarum*, etc., Ill. Biol. Monogr., 18, no. 2.

STEIN, F., 1867. Der Organismus der Infusionsthiere. Pt. 2:1.

STOLTE, H. A., 1922. Verlauf, Ursachen und Bedeutung der Encystierung bei Blepharisma. Verh. deutsch. zool. Gesell., 27:79.

STRICKLAND, A. G. R., and HAAGEN-SMIT, A. J., 1948. The excystment of *Colpoda duodenaria*. Science, 107:204.

STUART, C. A., KIDDER, G. W., and GRIFFIN, A. M., 1939. Growth studies on ciliates. III. Physiol. Zool., 12:348.

SUMMERS, F. M., 1935. The division and reorganization of the macronuclei of *Aspidisca lynceus*, etc. Arch. Protist., 85:173.

——— 1941. The Protozoa in connection with morphogenetic problems. In Calkins and Summers: Protozoa in biological research.

SUZUKI, S., 1957. Morphogenesis in the regeneration of *Blepharisma undulans japonicus*. Bull. Yamagata University, Nat. Sci., 4:86.

SWEZY, O., 1922. Mitosis in the encysted stages of *Entamoeba coli*. Univ. Cal. Publ. Zool., 20:313.

TARTAR, V., 1953. Chimeras and nuclear transplantations in ciliates, *Stentor coeruleus* × *S. polymorphus*. J. Exper. Zool., 124:63.

——— 1954. Anomalies in regeneration of Paramecium. J. Protozool., 1:11.

——— 1957. Equivalence of macronuclear nodes. J. Exper. Zool., 135:387.

——— 1961. The biology of Stentor. Pergamon Press. N.Y.

——— and CHEN, T. T., 1941. Mating reactions of enucleate fragments in *Paramecium bursaria*. Biol. Bull., 80:130.

TAYLOR, C. V., and STRICKLAND, A. G. R., 1938. Reactions of *Colpoda duodenaria* to environmental factors. I. Arch. Protist., 90:398.

THIMANN, K. V., and BARKER, H. A., 1934. Studies on the excystment of *Colpoda cucullus*. II. J. Exper. Zool., 69:37.

—— and HAAGEN-SMIT, A. J., 1937. Effects of salts on emergence from the cyst in protozoa. Nature, 140:645.

THON, K., 1905. Ueber den feineren Bau von *Didinium nasutum*. Arch. Protist., 5:282.

TUFFRAU, M., and SAVOIE, A., 1961. Étude morphologique du cilié hyménostome *Disematostoma colpidioides*. J. Protozool., 8:64.

TURNER, J. P., 1930. Division and conjugation in *Euplotes patella*, etc. Univ. Cal. Publ. Zool., 33:193.

VIVIER, E., and ANDRE, J., 1961. Données structurales et ultra-structurales nouvelles sur la conjugaison de *Paramecium caudatum*. J. Protozool., 8:416.

WEISZ, P. B., 1948. Time, polarity, size and nuclear content in the regeneration of Stentor fragments. J. Exper. Zool., 107:269.

—— 1950. Multiconjugation in Blepharisma. Biol. Bull., 98:242.

—— 1950a. A correlation between macronuclear thymonucleic acid concentration and the capacity of morphogenesis in Stentor. J. Morphol., 87:275.

—— 1951. An experimental analysis of morphogenesis in *Stentor coeruleus*. J. Exper. Zool., 116:231.

—— 1951a. A general mechanism of differentiation based on morphogenetic studies in ciliates. Am. Nat., 85:293.

WENRICH, D. H., 1939. Studies on *Dientamoeba fragilis*. III. J. Parasit., 25:43.

WENSTRUP, E. J., 1945. Encystment and excystment in Chaos. Science, 101:407.

WESCHENFELDER, R., 1938. Die Entwicklung von *Actinocephalus parvus*. Arch. Protist., 91:1.

WICHTERMAN, R., 1936. Division and conjugation in *Nyctotherus cordiformis*, etc. J. Morphol., 60:563.

—— 1940. Cytogamy: a sexual process occurring in living joined pairs of *Paramecium caudatum*, etc. *Ibid.*, 66:423.

—— 1946. Further evidence of polyploidy in the conjugation of green and colorless *Paramecium bursaria*. Biol. Bull., 91:234.

—— 1946a. Direct observation of the transfer of pronuclei in living conjugants of *Paramecium bursaria*. Science, 104:505.

—— 1951. The ecology, cultivation, structural characteristics and mating types of *Paramecium calkinsi*. Proc. Penn. Acad. Sci., 25:51.

—— 1953. Biology of Paramecium. Blakiston, Philadelphia.

WILSON, E. B., 1928. The cell in development and heredity. New York.

WOLFF, E., 1927. Un facteur de l'enkystment des amibes d'eau douce. C. R. Soc. Biol., 96:636.

WOODRUFF, L. L., 1921. Micronucleate and amicronucleate races of Infusoria. J. Exper. Zool., 34:329.

—— 1931. Micronuclear variation in *Paramecium bursaria*. Quart. J. Micr. Sci., 74:537.

—— 1932. *Paramecium aurelia* in pedigree culture for 25 years. Tr. Am. Micr. Soc., 51: 196.

—— and Erdmann, R., 1914. A normal periodic reorganization process without cell fusion in Paramecium. J. Exper. Zool., 17:425.

—— and Spencer, H., 1921. The survival value of conjugation in the life history of *Spathidium spathula*. Proc. Soc. Exper. Biol., 18:303.

Yow, F. W., 1958. A study of the regeneration pattern of *Euplotes eurystomus*. J. Protozool., 5:84.

Chapter 6
Variation and Heredity

IT IS generally recognized that individuals of all species of organism vary in morphological and physiological characteristics. Protozoa are no exception, and manifest a wide variation in size, form, structure, and physiological characters among the members of a single species. The different groups within a species are spoken of as the races, varieties, strains, etc. It is well known that dinoflagellates show a great morphological variation in different localities. Wesenberg-Lund (1908) noticed a definite seasonal morphological variation in *Ceratium hirundinella* in Danish lakes, while Schröder (1914) found at least nine varieties of this organism (Fig. 96) occurring in various bodies of water in Europe, and List (1913) reported that the organisms living in shallow ponds possess a marked morphological difference from those living in deep ponds. *Cyphoderia ampulla* is said to vary in size among those inhabiting the same deep lakes; namely, individuals from the deep water may reach 200μ in length, while those from the surface layer measure only about 100μ long.

In many species of foraminifers, the shell varies in thickness according to the part of ocean in which the organisms live. Thus the strains which live floating in surface water have a much thinner shell than those that dwell on the bottom. For example, according to Rhumbler, *Orbulina universa* inhabiting surface water has a comparatively thin shell, 1.28-18μ thick, while individuals living on the bottom have a thick shell, up to 24μ in thickness. According to Uyemura, a species of Amoeba living in thermal waters, showed a distinct dimensional difference in different springs. It measured 10-40μ in diameter in sulphurous water and 45-80μ in ferrous water; in both types of water the amoebae were larger at 36-40°C. than at 51°C.

Such differences or variations appear to be due to the influence of diverse environmental conditions, and will continue to exist under these conditions; but when the organisms of different varie-

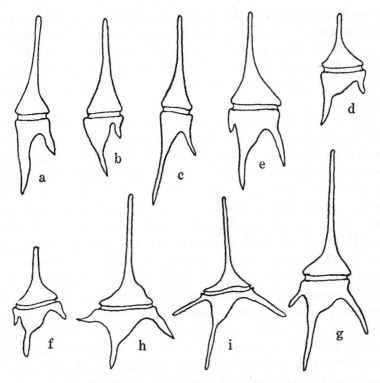

F<small>IG</small>. 96. Varieties of *Ceratium hirundinella* from various European waters (Schröder). a, *furcoides*-type (130–300μ. by 30–45μ.); b, *brachyceroides*-type (130–145μ. by 30–45μ.); c, *silesiacum*-type (148–280μ. by 28–34μ.); d, *carinthiacum*-type (120–145μ. by 45–60μ.); e, *gracile*-type (140–200μ. by 60–75μ.); f, *austriacum*-type (120–160μ. by 45–60μ.); g, *robustum*-type (270–310μ. by 45–55μ.); h, *scotticum*-type (160–210μ. by 50–60μ.); i, *pibur-gense*-type (180–260μ. by 50–60μ.).

ties are subjected to a similar environment, the strain differences usually disappear sooner or later. That the differences in kind and amount of foods bring about extremely diverse individuals in *Te-trahymena vorax* and *Chilomonas paramecium* in bacteria-free cultures has already been mentioned (p. 131).

While in many species, the races or varieties have apparently been brought about into being under the influence of environmental conditions, in others the inherited characters persist for a long period, and still in others the biotype may show different inherited characters. To the last-mentioned category belongs perhaps a strain of *Tetrahymena pyriformis* in which, according to Furgason (1940), a pure-line bacteria-free culture derived from a single individual was found to be composed of individuals differing in shape and size which became more marked in older cultures.

The first comprehensive study dealing with the variation in size and its inheritance in asexual reproduction of protozoa was conducted by Jennings (1909). From a "wild" lot of *Paramecium caudatum*, eight races or biotypes with the relative mean lengths of 206, 200, 194, 176, 142, 125, 100, and 45μ were isolated. It was found that within each clone derived from a single parent, the size of individuals varies greatly (which is attributable to growth, amount of food, and other environmental conditions), any one of which may give rise to progeny of the same mean size. Thus selection within the pure race has no effect on the size, and the differences brought about merely by environment are not inherited. Jennings (1916) examined the inheritance of the size and number of spines, size of shell, diameter of mouth, and size and number of teeth of the testacean *Difflugia corona*, and showed that "a population consists of many hereditarily diverse stocks, and a single stock, derived from a single progenitor, gradually differentiates into such hereditarily diverse stocks, so that by selection marked results are produced." Root (1918) with *Centropyxis aculeata*, Hegner (1919) with *Arcella dentata*, and Reynolds (1924) with *A. polypora*, obtained similar results. Jennings (1937) studied the inheritance of teeth in *Difflugia corona* in normal fission and by altering through operation, and found that operated mouth or teeth were restored to normal form in three or four generations and that three factors appeared to determine the character and number of teeth: namely, the size of the mouth, the number and arrangement of teeth in the parent, and "something in the constitution of the clone (its genotype) which tends toward the production of a mouth of a certain size, with teeth of a certain form, arrangement, and number."

Races or strains have been recognized in almost all well studied protozoa. For example, Ujihara (1914) and Dobell and Jepps (1918) noticed five races in *Entamoeba histolytica* on the basis of differences in the size of cysts. Spector (1936) distinguished two races in the trophozoites of this amoeba. The large strain was found to be pathogenic to kittens, but the small strain was not. Meleney and Frye (1933, 1935) and Frye and Meleny (1939) also hold that there is a small race in *Entamoeba histolytica* which has a weak capacity for invading the intestinal wall and not pathogenic to man. Sapiro, Hakansson and Louttit (1942) similarly notice two races which can be distinguished by the diameters of cysts, the division line being 10μ and 9μ in living and balsam-mounted specimens respectively. The race with large cysts gives rise to trophozoites which are more actively motile, ingest erythrocytes, and culture easily, is pathogenic to man and kitten, while the race with small cysts develops into less actively motile amoebae which do not ingest erythrocytes and are difficult to culture, is not pathogenic to hosts, thus not being histozoic.

It is interesting to note, however, that Cleveland and Sanders (1930) found the diameter of the cysts produced in a pure-line culture of this sarcodinan, which had originated in a single cyst, varied from 7 to 23μ. Furthermore, the small race of Frye and Meleney mentioned above was later found by Meleney and Zuckerman (1948) to give rise to larger forms in culture, which led the last two observers to consider that the size range of the strains of this amoeba is a characteristic which may change from small to large or *vice versa* under different environmental conditions. Recently, Burrows (1957) came to the conclusion that the small race constitutes a distinct species, *E. hartmanni* Prowazek (1912), which view was supported by Goldman (1959) who found antigenic difference by microfluorimetry between the large forms (*E. histolytica*) and the small forms (*E. hartmanni*).

Investigations by Boyd and his co-workers and others show that the species of Plasmodium appear to be composed of many strains which vary in diverse physiological characters. In an extended study on *Trypanosoma lewisi*, Taliaferro (1921-1926) found that this flagellate multiplies only during the first ten days in the blood of a rat after inoculation, after which the organisms do not reproduce. In the adult trypanosomes, the variability for total length in a population is about 3 per cent. Inoculation of the same pure line into different rats sometimes brings about small but significant differences in the mean size and passage through a rat-flea generally results in a significant variability of the pure line. It is considered that some differences in dimensions among strains are apparently due to environment (host), but others cannot be considered as due to this cause, since they persist when several strains showing such differences are inoculated into the same host. The two strains of *T. cruzi* isolated from human hosts and maintained for twenty-eight and forty-one months by Hauschka (1949), showed well-defined and constant strain-specific levels of virulence, different degrees of affinity for certain host tissues, unequal susceptibility to the quinoline-derivative Bayer 7602, and a difference in response to environmental temperature. The five strains of *Trichomonas gallinae* studied by Stabler (1948) were found to possess a marked variation in virulence to its hosts.

According to Kidder and his associates, the six strains (H, E, T, T-P, W, GHH) of *Tetrahymena pyriformis* and the two strains (V, PP) of *T. vorax* differ in biochemical reactions. They found the appearance of a biochemical variation between a parent strain (T) and a daughter strain (T-P) during a few years of separation and a greater difference in the reactions between the two species than that between the strains of each species. These strains show further differences in antigenic relationships. Five strains of *pyriformis* contain qualitatively identical antigens, but differ quantitatively with respect to amount, concentration or distribution of antigenic materials. The sixth strain (T) contains all the antigens of the other five strains and additional antigens. The two strains of *vorax* are said to be nearly identical antigenically. The antigenic differences between the two species were marked, since there is no cross-reaction within the standard testing time. In these cases, thus, some aspects of the physiological difference among different strains are understood.

Jollos (1921) subjected *Paramecium caudatum* to various environmental influences such as temperature and chemicals, and found that the animals develop tolerance which is inherited through many generations even after removal to the original environment. For example, one of the clones which tolerated only 1.1 per cent of standard solution of arsenic acid, was cultivated in gradually increasing concentrations for four months, at the end of which the tolerance for this chemical was raised to 5 per cent. After being removed to water without arsenic acid, the tolerance changed as follows: twenty-two days, 5 per cent; forty-six days, 4.5 per cent; 151 days, 4 per cent; 166 days, 3 per cent; 183 days, 2.5 per cent; 198 days, 1.25 per cent and 255 days, 1 per cent. As the organisms reproduced about once a day, the acquired increased tolerance to arsenic was inherited for about 250 generations.

There are also known inherited changes in form and structure which are produced under the influence of certain environmental conditions. Jollos designated these changes long-lasting modifications (*Dauermodifikationen*) and maintained that a change in environmental conditions, if applied gradually, brings about a change, not in the nucleus, but in the cytoplasm, of the organism which when transferred to the original environment, is inherited for a number of generations. These modifications are lost usually during sexual processes at which time the whole organism is reorganized.

The long-lasting morphological and physiological modifications induced by chemical substances have long been known in parasitic protozoa. Werbitzki (1910) discovered that *Trypanosoma brucei* loses its blepharoplast when inoculated into mice which have been treated with pyronin, acridin, oxazin and allied dyes, and Piekarski (1949) showed that trypaflavin and organic metal compounds which act as nuclear poisons and interfere with nuclear division, also bring about the loss of blepharoplast in this trypanosome. Laveran and Roudsky (1911) found that the dyes mentioned above have a special affinity for, and bring about the destruction by auto-oxidation of, the blepharoplast. Such trypanosomes lacking a blepharoplast behave normally and remain in that condition during many passages through mice. When subjected to small doses of certain drugs repeatedly, species of Trypanosoma often develop into drug-fast or drug-resistant strains which resist doses of the drug greater than those used for the treatment of the disease for which they are responsible. These modifications may also persist for several hundred passages through host animals and invertebrate vectors, but are eventually lost.

Long-lasting modifications have also been produced by several investigators by subjecting protozoa to various environmental influences during the nuclear reorganization at the time of fission, conjugation, or autogamy. In Stentor (Popoff) and Glaucoma (Chatton), long-lasting modifications appeared during asexual divisions. Calkins (1924) observed a double-type *Uroleptus mobilis*

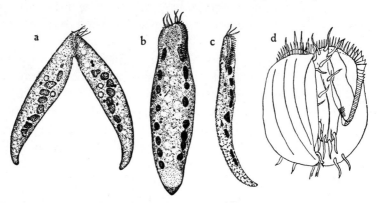

F_IG. 97. a–c, *Uroleptus mobilis* (Calkins) (a, a pair in conjugation; b, an individual from the third generation by division of a double organism which had been formed by the coalescence of a conjugating pair; c, a product of reversion); d, a double animal of *Euplotes patella* (Kimball).

(Fig. 97, *b*) which was formed by a complete fusion of two conjugants. This abnormal animal underwent fission 367 times for 405 days, but finally reverted back to normal forms, without reversion to double form. The double animal of *Euplotes patella* (*d*) is, according to Kimball (1941) and Powers (1943), said to be formed by incomplete division and rarely through conjugation. De Garis (1930) produced double animals in *Paramecium caudatum* through inhibition of division by exposing the animals to cyanide vapor or to low temperatures.

Jennings (1941) outlined five types of long-lasting inherited changes during vegetative reproduction, as follows: (1) changes that occur in the course of normal life history, immaturity to sexual maturity which involves many generations; (2) degenerative changes resulting from existence under unfavorable conditions; (3) adaptive changes or inherited acclimatization or immunity; (4) changes which are neither adaptive nor degenerative, occurring under specific environmental conditions; and (5) changes in form, size, and other characters, which are apparently not due to environment.

Whatever exact mechanism by which the long-lasting modifications are brought about may be, they are difficult to distinguish from permanent modification or mutation, since they persist for hundreds of generations, and cases of mutation have in most instances not been followed by sufficiently long enough pure-

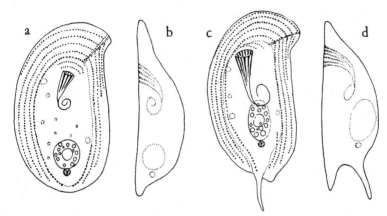

FIG. 98. *Chilodonella uncinata* (MacDougall). a, b, ventral and side view of normal individual; c, d, ventral and side view of the tailed mutant.

line cultures, to definitely establish them as such (Jollos, 1934; Sonneborn, 1947).

Jollos observed that if Paramecium were subjected to environmental change during late stages of conjugation, certain individuals, if not all, become permanently changed. Possibly the recombining and reorganizing nuclear materials are affected in such a way that the hereditary constitution or genotype becomes altered.

MacDougall subjected *Chilodonella uncinata* to ultraviolet rays and produced many changes which were placed in three groups: (1) abnormalities which caused the death of the organism; (2) temporary variations which disappeared by the third generation; and (3) variations which were inherited through successive generations and hence considered as mutations. The mutants were triploid, tetraploid, and tailed diploid forms (Fig. 98), which bred true for a variable length of time in pure-line cultures, either being lost or dying off finally. The tailed form differed from the normal form in the body shape, in the number of ciliary rows and contractile vacuoles, and in movement, but during conjugation it showed the diploid number of chromosomes as in the typical form. The tailed mutant remained true and underwent twenty conjugations during ten months.

By exposing *Paramecium multimicronucleatum* to high dosages of x-irradiation, Wichterman (1959) observed a number of changes which he considered mutations as they persisted in cultures; namely, loss of micronuclei, decrease in vitality and reproductive rate, increase in x-radiation sensitivity, monstrosity, and reduced body size, and was inclined to think that x-irradiation altered sets of genes in the macronucleus.

Kimball (1950) exposed *Paramecium aurelia* to beta particles from plaques containing P^{32} and obtained many clones which multiplied more slowly than normal animals or died, which conditions were interpreted by him to be due to mutational changes induced in the micronuclei by the radiation. Kimball found that the radiation was less effective if given just before the cytoplasmic division than if given at other times during the division interval and that exposure of the organisms to ultraviolet ray of wave length 2537 Å inactivates the Kappa (p. 286).

The loss of the blepharoplast in trypanosomes mentioned above occurs also spontaneously in nature. A strain of *Trypanosoma evansi* which had been maintained in laboratory animals for five years, suddenly lost the blepharoplast (Wenyon, 1928) which condition remained for some 17 years (Hoare, 1954). Hoare and Bennett (1937) found five camels out of 100 they examined infected by the same species of trypanosome that was without a blepharoplast. One strain inoculated into laboratory animals has retained this peculiarity for nearly three years. Hoare (1954) considers that failure to divide at the time of division results in the formation of ablepharoplast forms which breed true.

In sexual reproduction, the nuclei of two individuals participate in producing new combinations which would naturally bring about diverse genetic constitutions. The new combination is accomplished either by sexual fusion in Sarcodina, Mastigophora, and Sporozoa, or by conjugation in Ciliophora.

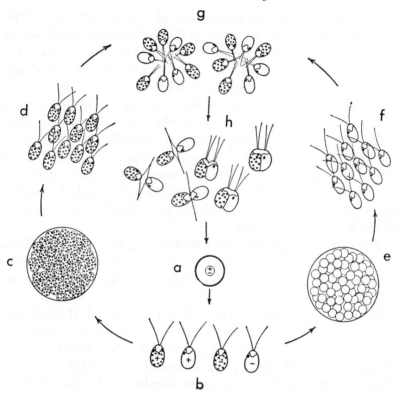

Fig. 99. The life cycle of *Chlamydomonas reinhardi*, showing the segregation of mating type, denoted by plus and minus signs, and of the marker y_1 (the dotted cells are y_1^+. the undotted are y_1^- (Sager). a, mature zygote; b, 4 offspring from a single zygote; c, plus colony on agar; d, motile plus cells in liquid; e, minus colony on agar; f, motile minus cells in liquid; g, clumping reaction; h, pairing and fusion to form zygote.

The genetics of sexual fusion is known only in a few forms. The species of Chlamydomonas (Fig. 99) are easily cultivated. They are haploid in all stages of development except the newly formed zygote which is diploid. The zygotic division is meiotic and gives rise to usually four biflagellated haploid vegetative forms. Of the four, two are of one mating type (+) and the other two are of another type (−), and they develop into colonies on agar medium. When the cultures are flooded, the two mating type gametes appear. After clumping reaction, + and − pairing and fusion follow, resulting in the production of zygotes. Pascher (1916) found hybrid zygotes gave rise to the lines identical with one or the other parent type, though some lines showed combination of parental characters. In recent years, various mutants were produced by exposure to ultraviolet radiation in especially *C. reinhardi*. These mutant characteristics segregate in Mendelian ratio, producing F_1 2:2 ratios. However, Streptomycin-resistant character seems to depend upon a determiner which is passed along in the gametes of one mating type only and appears not to be controlled by chromosomal elements (Sager, 1955, 1960). (Genetics of Chlamydomonas, Lewin, 1953; nutritional control of sexuality, Sager and Granick, 1954; linkage of heritable characteristics, Ebersold and Levine, 1959.)

The genetics of protozoa in which conjugation occurs has been most thoroughly studied in *Paramecium aurelia* due mainly to the painstaking and long-continued studies by Sonneborn and his associates.

In variety 1 of this ciliate, when types I and II conjugate, among a set of exconjugants some produce all of one mating type, others all of the other mating type and still others both types (one of one type and the other of the other type). In the last mentioned exconjugants, the types segregate usually at the first division, since of the two individuals produced by the first division, one and all its progeny, are of one mating type, and the other and all its progeny are of the other mating type. A similar change was also found to take place at autogamy. Sonneborn therefore considers that the mating types are determined by macronuclei, as judged by segregation at first or sometimes second division in exconjugants and by the influence of temperature during conjugation and the first division.

When the members of the stock containing both types I and II (two-type condition) conjugate with those of the stock containing one type (one-type condition), all the descendants of the hybrid exconjugants show two-type condition, which shows the dominancy of two-type condition over one-type condition. The factor for the two-type condition may be designated A and that for the one-type condition a. The parent stocks are AA and aa, and all F_1 hybrids Aa. When the hybrids (Aa) are backcrossed to recessive parent (aa) (158 conjugating pairs in one experiment), approximately one-half (81) of the pairs gave rise to two-type condition (Aa) and the remaining one-half (77) of the pairs to one-type condition (aa), thus showing a typical Mendelian result. When F_1 hybrids (Aa) were interbred by 120 conjugating pairs, each exconjugant in eighty-eight of the pairs gave rise to two-type condition and each exconjugant in thirty-two pairs produced one-type condition, thus approximating an expected Mendelian ratio of three dominants to one recessive. That the F_2 dominants are composed of two-thirds heterozygotes (Aa) and one-third homozygotes (AA) was confirmed by the results obtained by allowing F_2 dominants to conjugate with the recessive parent stock (aa). Of nineteen pairs of conjugants, six pairs gave rise to only dominant progeny, which shows that they were homozygous (AA) and their progeny heterozygous (Aa), while thirteen pairs produced one-half dominants and one-half recessives, which indicates that they were heterozygous (Aa) and their progeny half homozygous (aa) and half heterozygous (Aa). Thus the genic agreement between two conjugants of a pair and the relative frequency of various gene combinations as shown in these experiments confirm definitely the occurrence of meiosis and chromosomal exchange during conjugation which have hitherto been considered only on cytological ground. (Sonneborn, 1939).

Further studies revealed that in *P. aurelia* there are two different systems of mating type determination and inheritance (Sonneborn, 1947, 1957). In one group (A) of varieties, individuals of the same caryonides are usually of the same mating type, while those belonging to different caryonides may be of the same or different mating types. For example, in variety 1 conjugation, all four caryonides may be of mating type I or type II, or there may be any of the possible assortments of type I or II. It appears clear that in group A varieties, mating types are determined by the macronuclei as was mentioned above. Sonneborn (1957) placed in group A varieties 1, 3, 5, 9 and 11 (and probably 10, 12 and 14 also).

On the other hand, in group B of varieties 2, 4, 6 and 8, the mating type is determined by the cytoplasm and mating types change rarely after fertilization. For example, in a conjugating pair of variety 4 (mating types VII and VIII), the exconjugant clones which derived their cytoplasm from the VII parent are usually of type VII and those which derived their cytoplasm from the VIII parent are usually of type VIII. Thus in group A mating type is a caryonidal trait and in group B it is a clonal trait.

In *Euplotes patella*, Kimball (1942) made various matings with respect to the inheritance of the mating type. The results obtained can be explained if it is assumed that mating types I, II and V, are determined by different heterozygous combinations of three allelic genes which if homozygous determine mating types III, IV and VI. Upon this supposition, type I has one allele in common with type II, and this allele is homozygous in type IV. It has one allele in common with type V, and this allele is homozygous in type VI. Type II has one allele in common with type V and this is homozygous in type III. These alleles were designated by Kimball, mt^1, mt^2, and mt^3. The genotypes of the six mating types may be indicated as follows: mt^1mt^2 (I), mt^1mt^3 (II), mt^3mt^3 (III), mt^1mt^1 (IV), mt^2mt^3 (V), and mt^2mt^2 (VI).

There is no dominance among these alleles, the three heterozygous combinations determining three mating types being different from one another and from the three determined by homozygous combination. Kimball (1939, 1941) had shown that the fluid obtained free of Euplotes from a culture of one mating type will induce conjugation among animals of certain other mating types. When all possible combinations of fluids and animals are made, it was found that the fluid from any of the heterozygous types induces conjugation among animals of any types other then its own and the fluid from any of the homozygous types induces conjugation only among animals of the types which do not have the same allele as the type from which the fluid came. These reactions may be explained by an assumption that each of the mating type alleles is responsible for the production by the animal of a specific conjugation-inducing substance. Thus the two alleles in a heterozygote act independently of each other; each brings about the production by the animal of a substance of its own. Thus heterozygous animals are induced to conjugate only by the fluids from individuals which possess an allele not present in the heterozygotes.

The double animals of *Euplotes patella* (p. 281) conjugate with double animals or with single animals in appropriate mixtures and at times a double animal gives rise by binary fission to a double and two single animals instead of two animals (Fig. 100). Powers (1943) obtained doubles of various genotypes for mating types which were determined by observing the mating type of each of the two singles that arose from the doubles. Doubles of type IV

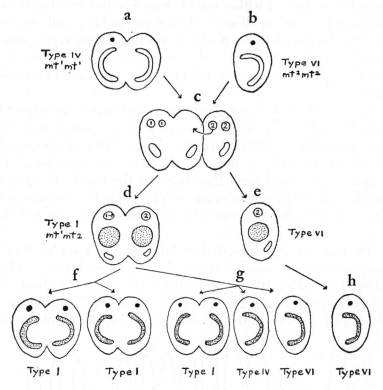

Fɪɢ. 100. Diagram showing conjugation between a double (type IV) and a single (type VI) of *Euplotes patella* (Powers). a, a double organism with one micronucleus (genotype mt¹mt¹); b, a normal single with a micronucleus (genotype mt²mt²); c, conjugation of the single with the amicronucleate half of the double (one of the pronuclei produced in the single, migrates into the double, while the two pronuclei of the double undergo autogamy); d, the exconjugant double is shown to be type I (mt¹mt²); e, exconjugant single remains type VI; f, the double divides into two type I doubles; g, occasionally the anterior half of the double is widely "split," and division produces a double and two singles, the latter testing as type IV and type VI; h, line of exconjugant single. Newly formed macronuclei are stippled.

(mt^1mt^1) with a single micronucleus (Fig. 100, *a*) were mated with singles of type VI (mt^2mt^2) (*b*). The double exconjugants (*d*) were "split" into their component singles belonging to mating types IV and VI (*g*), while the doubles were type I (*f*). Thus it was found that the phenotype of a double animal with separate nuclei was the same as though the alleles present in the nuclei were located within one nucleus. The fact that loss of one micronucleus had no effect on the type of doubles, tends to show that the micronucleus has no direct effect on mating types. Sonneborn's view that the macronucleus is the determininer of the mating types in *Paramecium aurelia* appears to hold true in Euplotes also.

The relation between the cytoplasm and nucleus in respect to inheritance has become better known in recent years in some ciliates. Sonneborn (1934) crossed two clones of *Paramecium aurelia* differing markedly in size and division rate, and found the difference persisted for a time between the two F_1 clones produced from the two members of each hybrid pair of exconjugants, but later both clones became practically identical in size and division rate (Sonneborn, 1947).

De Garis (1935) succeeded in bringing about conjugation in *Paramecium caudatum*, between the members of a large clone (198µ long) (Fig. 101, *a*) and of a small clone (73µ long) (*b*). The exconjugants of a pair are different only in the cytoplasm as the nuclei are alike through exchange of a haploid set of chromo-

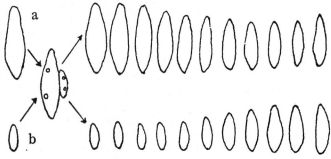

Fig. 101. Diagram showing the size changes in two clones derived from a pair of conjugants of *Paramecium caudatum*, differing in size (a, b). Gradual change in dimensions in each clone during 22 days resulted in intermediate size (De Garis).

somes. The two exconjugants divide and give rise to progeny which grow to size characteristic of each parent clone, division continuing at the rate of once or twice a day. However, as division is repeated, the descendants of the large clone become gradually smaller after successive fissions, while the descendants of the small clone become gradually larger, until at the end of twenty-two days (in one experiment) both clones produced individuals of intermediate size (about 135µ long) which remained in the generations that followed. Since the exconjugants differed in the cytoplasm only, it must be considered probable that at first the cytoplasmic character was inherited through several vegetative divisions, but ultimately the influence of the new nucleus gradually changed the cytoplasmic character. The ultimate size between the two clones is however not always midway between the mean sizes of the two parent clones, and is apparently dependent upon the nuclear combinations brought about by conjugation. It has also become known that different pairs of conjugants between the same two clones give rise to diverse progeny, similar to those of sexual reproduction in metazoa, which indicates that clones of *Paramecium caudatum* are in many cases heterozygous for size factors and recombination of factors occurs at the time of conjugation.

In *P. aurelia,* Kimball (1939) observed that there occasionally occurs a change of one mating type into another following autogamy. When the change is from type II to type I, not all animals change type immediately. Following the first few divisions of the product of the first division after autogamy there are present still some type II animals, although ultimately all become transformed into type I. Here also the cytoplasmic influence persists and is inherited through vegetative divisions.

Jennings (1941) in his excellent review writes: "The primary source of diversities in inherited characters lies in the nucleus. But the nucleus by known material interchanges impresses its constitution on the cytoplasm. The cytoplasm retains the constitution so impressed for a considerable length of time, during which it assimilates and reproduces true to its impressed character. It may do this after removal from contact with the nucleus to which its present constitution is due, and even for a time in the presence of another nucleus of different constitution. During this period, cytoplasmic inheritance may occur in vegetative reproduction. The new cells produced show the characteristics due to this cytoplasmic constitution impressed earlier by a nucleus that is no longer present. But in time the new nucleus asserts itself, impressing its own constitution on the cytoplasm. Such cycles are repeated as often as the nucleus is changed by conjugation."

Since the first demonstration some forty years ago of "cytoplasmic inheritance" in higher plants, many cytoplasmic factors have been observed in various plants (Michaelis and Michaelis,

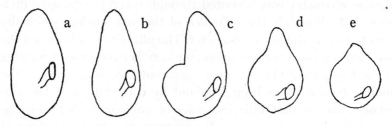

Fig. 102. *Paramecium aurelia*. The changes leading up to death when the sensitives are exposed to the killer stock 51 (variety 4) (Sonneborn).

1948). Information on similar phenomena in metazoa and protozoa is of recent origin.

As was already mentioned (p. 234), Sonneborn found in four races of variety 4 of *Paramecium aurelia* a pair of characters which he designated as "killer" and "sensitive." The killers contain certain particles in their cytoplasm which are capable of liberating a toxic agent. When the sensitive races are exposed to the fluid in which the killer race 51 lived, they show after hours a hump first on the oral and later on the aboral surface toward the posterior end which becomes enlarged, while the anterior part of the body gradually wastes away. The whole body becomes smaller and rounded; and finally the organisms perish (Fig. 102). Further works revealed that varieties 2 and 8 also contain killers, and that different stocks of killers show different killing actions on sensitives such as spinning motion, paralysis, vacuolization and rapid lysis in addition to hump formation described above, all terminating in death. The toxic agent is discrete particles and a sensitive animal may be killed by one particle. Sonneborn (1946) named it "paramecin" or "p particle" which presumably originates in endoplasmic kappa or allied particles.

Sensitives can be mated to the killers, however, without injury if proper precaution is taken, since the toxic agent does not affect them during conjugation. The two exconjugants obtain identical genotype, but their progeny are different; that is one is a killer and the other is a sensitive. F_2 progeny obtained by selfing show no segregation. Therefore, the difference between the killer and the sensitive is due to a cytoplasmic difference and not to a genic difference.

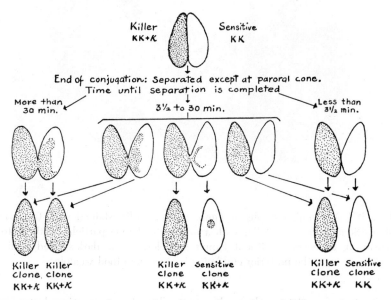

Fig. 103. Diagram showing the effects of transfers of different amounts of the cytoplasm between mates in conjugation of KK+kappa killers and KK sensitives in *Paramecium aurelia* (Sonneborn).

Sonneborn further noted that the thin cytoplasmic paroral strand which appears between conjugating pair that ordinarily breaks off within a minute, occasionally may remain for a long time, and if the strand persists as long as thirty minutes, there occurs an interchange of cytoplasm between the pair (Fig. 103). When this happens, both exconjugants produce killer clones. In F_2 no segregation takes place. Thus killers can introduce the kill-

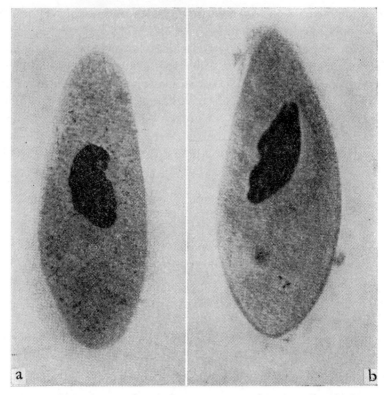

Fig. 104. Photomicrographs of *Paramecium aurelia*, stained with Giemsa's stain (Sonneborn). a, a killer with a number of kappa particles in the cytoplasm; b, a sensitive without kappa particles, a few dark-stained bodies near the posterior end being bacteria in a food vacuole.

er trait to sensitives through a cytoplasmic connection between them. It was supposed that the killers contain a cytoplasmic genic factor or a *plasmagene* which determines the killer trait and Sonneborn called it **kappa.** Preer (1948) demonstrated that this kappa is a particle which can be recognized in Giemsa-stained specimens (Fig. 104). It was further found that killers can be irreversibly transformed into hereditary sensitives by eliminating kappa particles by exposure to high temperature (Sonneborn, 1946), x-irradiation (Preer, 1948b) or nitrogen mustard (Geckler, 1949) and that sensitives can be transformed to hereditary killers by placing them in concentrated suspensions of broken bodies of killers (Sonneborn, 1948a). Therefore, it became clear that kappa is a self-multiplying cytoplasmic body which is produced when some are already present.

Killer races of variety 2 differ from each other and from that of variety 4 mentioned above, in the effects produced on sensitives before the latter are killed. These sensitives possess a gene different from that of the killers and cannot be changed into killers by immersing it to kappa suspensions of broken bodies of killers. When this sensitive is mated with a killer, F_2 generation produced by selfing among the killer F_1 clones, shows segregation of sensitives and killers in the ratio of a single gene difference. In the presence of dominant gene K, kappa is maintained, but in recessive k homozygotes, kappa cannot be maintained and any kappa carried over from killers is rapidly lost. Thus it is evident, Sonneborn points out, that the plasmagene kappa is dependent on gene K.

Dippell (1948, 1950) found a number of killer mutants in variety 4. She showed through breeding analysis that these mutations have brought about no change in any gene affecting kappa or the killer trait, but have been in every case due to changes in kappa. In a mutant which was capable of producing two types of killing, there were two kinds of kappa which she succeeded in separating in different animals and their progeny. Thus it became apparent that kappa can undergo mutation, that various mutant kappas can multiply in animals with the original genome, and that the kappas are determined by themselves and not by nuclear genes.

According to Preer (1948), the kappa particles (Fig. 104) in the killer race G are about 0.4μ long, and those in a mutant Gml only about 0.2-0.3μ long, while in other strains they measure as much as 0.8μ in length. Preer (1948a, 1950) further observed that the kappa particles contain desoxyribonucleic acid and vary in form (rod-like or spherical), size and number in different races of killers, and that an increase, reduction or destruction of the kappas, as determined by indirect methods, was correlated with the observed number of the stained particles. Preer and Stark (1953) reported later that kappa particles are rod-shaped,

0.6-10μ long, differ among different strains of host animals, and may occur as singles, doubles or enlarged forms with a refractile body, etc.

As to the nature of kappa or p particle opinions vary among different authors. Various views have been advanced to date: a derivative of alga (Altenburg), a rickettsia (Preer), bacteria (Preer, Gores, Dippell), virus (Luria), etc. Sonneborn (1961) concludes his discussion of kappa and other cytoplasmic particles in *P. aurelia* by stating that they are "infectious, intracellular parasites" of unknown biological affinity. (Kappa and related particles in Paramecium, Sonneborn, 1959, 1961.)

The application of antigen-antibody reactions to free-living protozoa began some forty years ago. Bernheimer and Harrison (1940, 1941) pointed out the antigenic dissimilarity of three species of Paramecium in which the members of a clone differ widely in their susceptibility to the immobilizing action of a given serum. Strains of *Tetrahymena pyriformis* differ in antigenic reactions, as has already been mentioned (p. 272). Sonneborn and his co-workers have studied serological reactions in *Paramecium aurelia* (Sonneborn, 1943, 1950).

When a rabbit is inoculated intraperitoneally with a large number of a strain of *P. aurelia,* its serum immobilizes in a high dilution, the organisms of the same strain, but not of other strains. Such a serologically distinct strain is called a *serotype* or antigenic type. It was found that a clone originating in a homozygous individual gives rise to a series of various serotypes. Race 51 gave rise to eight serotypes: A, B, C, D, E, G, H and J, and race 29, to seven serotypes: A, B, C, D, F, H and J. When a serotype is exposed to its antiserum, it changes into other types, which course Sonneborn was able to control by temperature and other conditions. For example, serotype D (stock 29) may be changed by its antiserum to type B at 32°C. and to type H at 20°C., types B, F and H are convertible one into the other and all other types can be transformed to any of the three; and serotypes A and B (stock 51) are convertible one into the other, and other types can be changed to A or B. The antigenic types are inherited, if the cultures are kept at 26°-27°C. with food enough to allow one division a day. When induced or spontaneous changes of serotype occur, crosses made among different serotypes of the same strain reveal no effective gene differences among them; thus all serotypes of a strain possess apparently an identical genic constitution. Sonneborn finds serotype A of stock 29 is not exactly the same as the type A of stock 51. When these are crossed, it is found that the difference between two antigens is controlled by a pair of allelic genes of which the 51A-gene is dominant over the 29A-gene. On th basis of these observations, it has been concluded that nuclear genes control the specificity of the physical basis of cytoplasmic inheritance in these antigenic traits, and hereditary transformations of serotype are cytoplasmic "mutations" of hitherto unknown type.

Preer and Preer (1959) studied four antigens extractable from one strain of *P. aurelia* by gel-diffusion method and found: antigen 1, heatstable, derived from trichocysts; antigen 2, not associated with any cellular particulate in homogenates; antigen 3, obtained from mitochondria; antigen 4, correlated with serotype is derived mainly from the cilia and probably from the body wall and has the properties of a protein. Hiwatashi (1951, 1952) used antiserum in separating one mating type from its partner and also in producing double monsters in *P. caudatum*. Thus, we are just beginning to realize how complex the serological reactions are.

In the inheritance of the killer trait and of serotype, both traits are cytoplasmically determined and inherited; hereditary changes are brought about by environmental conditions; and the traits are dependent for their maintenance upon nuclear genes. However, the specific type of killer trait is controlled by the kind of kappa present, not by the genes, while the specific type of A antigen is determined by the nuclear genes. The transformation of the killer to the sensitive is made irreversible, but that of serotypes is not. The various types of killer character are not mutually exclusive, as different kinds of kappa can coexist in the same organism and its progeny, each kind of kappa controlling production of its corresponding kind of paramecin, while in serotype, two kinds of antigen substances cannot coexist, thus being mutually exclusive. The physical basis of the killer trait lies in the visible Feulgen-positive kappa particles, while no such particles have so far been found in association with the serotype. (Serological studies, Beale, 1954; Padnos, 1962.)

References

ALTENBURG, E., 1948. The rôle of symbionts and autocatalysts in the genetics of the ciliates. Am. Nat., 82:252.

BEALE, G. H., 1954. The genetics of Paramecium. Cambridge Univ. Press.

BERNHEIMER, A. W., and HARRISON, J. A., 1940. Antigen-antibody reactions in Paramecium: the aurelia group. J. Immunol., 39: 73.

―――― ―――― 1941. Antigenic differentiation among strains of *Paramecium aurelia. Ibid.*, 41:201.

BURROW, R. B., 1957. *Endamoeba hartmanni.* Amer. J. Hyg., 65:172.

CALKINS, G. N., 1925. *Uroleptus mobilis.* J. Exper. Zool., 41:191.

CHEN, T. T., 1955. Paramecin 34, a killer substance produced by *Paramecium bursaria.* Proc. Soc. Exper. Biol. & Med., 88:541.

CLEVELAND, L. R., and SANDERS, E. P., 1930. Encystation, multiple fission without encystment, etc. Arch. Protist., 70: 223.

DE GARIS, C. F., 1930. Genetic results from conjugation of double monsters and free individuals of *Paramecium caudatum.* Anat. Rec., 47:393.

―――― 1930a. Nucleus versus cytoplasm in the heredity of *Paramecium caudatum* as shown by conjugation of double monsters. *Ibid.*, 47:393.

―――― 1935. Heritable effects of conjugation between free individuals and double monsters in diverse races of *Paramecium.* J. Exper. Zool., 71:209.

DIPPELL, R. V., 1948. Mutation of the killer plasmagene, Kappa, in variety 4 of *Paramecium aurelia.* Amer. Nat., 82:43.

―――― 1950. Mutation of the killer cytoplasmic factor in *Paramecium aurelia.* Heredity, 4:165.

DOBELL, C., and JEPPS, M. W., 1918. A study of the diverse races of *Entamoeba histolytica* distinguishable from one another by the dimensions of their cysts. Parasitology, 10:320.

EBERSOLD, W. T., and LEVINE, R. P., 1959. A genetic analysis of linkage group 1 of *Chlamydomonas reinhardi.* Zeitschr. Vererbungsl., 90:74.

FRYE, W. W., and MELENEY, H. E., 1938. The pathogenicity of a strain of small race *Entamoeba histolytica.* Amer. J. Hyg., 27:580.

FURGASON, W. H., 1940. The significant cytostomal pattern of the "Glaucoma-Colpidium group," and a proposed new genus and species, *Tetrahymena geleii.* Arch. Protist., 94:224.

GECKLER, R. P., 1949. Nitrogen mustard inactivation of the cytoplasmic factor, kappa, in Paramecium. Science, 110:89.

GOLDMAN, M., 1959. Microfluorimetric evidence of antigenic difference between *Entamoeba histolytica* and *E. hartmanni.* Proc. Soc. Exper. Biol. & Med., 102:189.

HAUSCHKA, T. S., 1949. Persistence of strain-specific behavior in two strains of *Trypanosoma cruzi* after prolonged transfer through inbred mice. J. Parasit., 35:593.

HEGNER, R. W., 1919. Heredity, variation, and the appearance of diversities during the vegetative reproduction of *Arcella dentata*. Genetics, 4:95.

HIWATASHI, K., 1951. A new method for separating animals of one mating type from their partners in a conjugating mixture of Paramecium. Ann. Zool. Japon., 24:83.

———— 1952. Double monsters of *Paramecium caudatum*. Sci. Rep., Tohoku Univ. IV. Biology, 19: 275.

HOARE, C. A., 1940. Recent studies on the kinetoplast in relation to heritable variation in trypanosomes. J. Roy. Micr. Soc., 60: 26.

———— 1943. Biological races in parasitic Protozoa. Biol. Rev., 18:137.

———— 1954. The loss of the kinetoplast in trypanosomes with special reference to *Trypanosoma evansi*. J. Protozool., 1:28.

———— and BENNETT, S. C. J., 1937. Morphological and taxonomic studies on mammalian trypanosomes. III. Parasitology, 29,43.

———— ———— 1939. IV. *Ibid.*, 30:529.

JENNINGS, H. S., 1909. Heredity and variation in the simplest organisms. Amer. Nat., 43:322.

———— 1916. Heredity, variation and the results of selection in the uniparental reproduction of *Difflugia corona*. Genetics, 1:407.

———— 1929. Genetics of the Protozoa. Bibliogr. Genetica, 5:105.

———— 1937. Formation, inheritance and variation of the teeth in *Difflugia corona*. J. Exper. Zool., 77:287.

———— 1938. Sex reaction types and their interrelations in *Paramecium bursaria*. Proc. Nat. Acad. Sci., 24:112.

———— 1939. Genetics of *Paramecium bursaria*. I. Genetics, 24:202.

———— 1941. Inheritance in Protozoa. In Calkins and Summers (1941): Protozoa in biological research. New York.

———— *et al.*, 1932. The diverse biotypes produced by conjugation within a clone of Paramecium. J. Exper. Zool., 63: 363.

JOLLOS, V., 1913. Experimentelle Untersuchungen an Infusorien. Biol. Zentralbl., 33:222.

———— 1921. Experimentelle Protistenstudien. I. Arch. Protist., 43:1.

———— 1934. Dauermodifikationen und Mutationen bei Protozoen. *Ibid.*, 83: 197.

KIDDER, G. W., *et al.*, 1945. Antigenic relationship in the genus Tetrahymena. Physiol. Zool., 18:415.

KIMBALL, R. F., 1939. A delayed change of phenotype following a change of genotype in *Paramecium aurelia*. Genetics, 24:49.

———— 1939a. Mating types in Euplotes. Am. Nat., 73:451.

———— 1941. Double animals and amicronucleate animals in *Euplotes patella* with particular reference to their conjugation. J. Exper. Zool., 86:1.

———— 1942. The nature and inheritance of mating types in *Euplotes patella*. Genetics, 27:269.

———— 1947. The induction of inheritable modification in reaction to antiserum in *Paramecium aurelia*. Genetics, 32:486.

——— 1950. The effect of radiations on genetic mechanism of *Paramecium aurelia.* J. Cell. Comp. Physiol., 35 (sup. 1): 157.

Lewin, R. A., 1953. The genetics of *Chlamydomonas moewusii.* J. Genet., 51:543.

List, T., 1913. Ueber die Temperal- und Lokalvariation von *Ceratium hirundinella,* etc. Arch. Hydrobiol., 9:81.

MacDougall, M. S., 1929. Modifications in *Chilodon uncinatus* produced by ultraviolet radiation. J. Exper. Zool., 54:95.

Meleney, H. E., and Zuckerman, L. K., 1948. Note on a strain of small race *Entamoeba histolytica* which became large in culture. Amer. J. Hyg., 47:187.

Michaelis, P., and Michaelis, G., 1948. Ueber die Konstanz des zytoplasmons bei Epilobium. Planta, 35:467.

Padnos, M., 1962. Serological studies on Protozoa. I. J. Protozool., 9:7.

Pascher, A., 1916. Ueber die Kreuzung einzelliger, haploider Organismen: Chlamydomonas. Ber. deut. bot. Gasell., 34:228. Also 1918. *Ibid.* 36: 163.

Piekarski, G., 1949. Blepharoplast und Trypaflavinwirkung bei *Trypanosoma brucei.* Zentralbl. Bakt., I. Orig., 153:109.

Powers, E. L. 1943. The mating types of double animals in *Euplotes patella.* Am. Midland Nat., 30:175.

Preer, J. R., Jr., 1948. The killer cytoplasmic factor kappa: its rate of reproduction, the number of particles per cell, and its size. Am. Nat., 82:35.

——— 1948a. Microscopic bodies in the cytoplasm of "killers" of *Paramecium aurelia* and evidence for the identification of these bodies with cytoplasmic factor, kappa. Genetics, 33:625.

——— 1950. Microscopically visible bodies in the cytoplasm of the "killer" strain of *Paramecium aurelia, Ibid.,* 35:344.

——— and Preer, L. B., 1959. Gel diffusion studies on the antigens of isolated cellular components of Paramecium. J. Protozool., 6:88.

——— and Stark, P., 1953. Cytological observations on the cytoplasmic factor "Kappa" in *Paramecium aurelia.* Exper. Cell. Res., 5:478.

Prowazek, S. v., 1912. Weitere Beitrag zur Kenntnis der Entamoeben. Arch. Protist., 26:241.

Reynolds, B. D., 1924. Interactions of protoplasmic masses in relation to the study of heredity and environment in *Arcella polypora.* Biol. Bull., 46:106.

Root, F. M., 1918. Inheritance in the asexual reproduction in *Centropyxis aculeata.* Genetics, 3:173.

Sager, R. A., 1955. Inheritance in the green alga *Chlamydomonas reinhardi.* Genetics, 40:476.

——— 1960. Genetic systems in Chlamydomonas. Science, 132:1459.

——— and Granick, S., 1954. Nutritional control of sexuality in *Chlamydomonas reinhardi.* J. Gen. Physiol., 37:729.

SAPIRO, J. J., *et. al.*, 1942. The occurrence of two significantly distinct races of *Entamoeba histolytica*. Amer. J. Trop. Med., 22:191.

SCHOENBORN, H. W., 1954. Mutations in *Astasia longa* induced by radiation. J. Protozool., 1:170.

SCHRÖDER, B., 1914. Ueber Planktonepibionten. Biol. Zentralbl., 34:328.

SONNEBORN, T. M., 1937. Sex, sex inheritance and sex determination in *Paramecium aurelia*. Proc. Nat. Acad. Sci., 23:378.

———— 1939. *Paramecium aurelia:* mating types and groups; etc. Am. Nat., 73:390.

———— 1942. Inheritance in ciliate protozoa. *Ibid.*, 76:46.

———— 1943. Gene and cytoplasm. I, II. Proc. Nat. Acad. Sci., 29:329.

———— 1943a. Development and inheritance of serological characters in variety 1 of *Plasmecium aurelia*. Genetics, 28:80.

———— 1946. Experimental control of the concentration of cytoplasmic genetic factors in Paramecium. Cold Springs Harbor Symp. Quant. Biol., 11:236.

———— 1947. Recent advances in the genetics of Paramecium and Euplotes. Adv. Genetics, 1:263.

———— 1948. Introduction to symposium on plasmagenes, genes and characters in *Paramecium aurelia*. Am. Nat., 82:26.

———— 1950. The cytoplasm in heredity. Heredity, 4:11.

———— 1957. Breeding systems, reproductive methods, and species problems in protozoa. In The species problem. A.A.A.S., p. 155.

———— 1959. Kappa and related particles in Paramecium. In Advances in virus research. p. 229.

———— 1961. Kappa particles and their bearings on host-parasite relations. In Perspectives in virology. M. Pollard, editor, 2:5.

———— and LYNCH, R. S., 1934. Hybridization and segregation in *Paramecium aurelia*. J. Exper. Zool., 67:1.

STABLER, R. M., 1948. Variations in virulence of strains of *Trichomonas gallinae* in pigeons. J. Parasit., 34:147.

TALIAFERRO, W. H., 1926. Variability and inheritance of size in *Trypanosoma lewisi*. J. Exper. Zool., 43:429.

———— 1929. The immunology of parasitic infections. New York.

———— and HUFF, C. G., 1940. The genetics of the parasitic Protozoa. AAAS Publ., 12:57.

UJIHARA, K., 1914. Studien ueber die Amoebendysenterie. Ztschr. Hyg., 77:329.

WENYON, C. M., 1928. The loss of the parabasal body in trypanosomes. Tr. Roy. Soc. Trop. Med. Hyg., 22:85.

WESENBERG-LUND, C., 1908. Plankton investigations of the Danish lakes. Copenhagen.

WICHTERMAN, R., 1959. Mutation in the protozoan *Paramecium multimicronucleatum* as a result of x-irradiation. Science, 129:207.

PART II: TAXONOMY AND
SPECIAL BIOLOGY

Chapter 7

Major Groups and Phylogeny of Protozoa

THE protozoa are grouped into two subphyla: Plasmodroma and Ciliophora. In subphylum Plasmodroma are placed those protozoa which possess one to many nuclei of one kind and flagella or pseudopodia or no such organelles of locomotion. The group is subdivided into four classes: Mastigophora (p. 303), Sarcodina (p. 496), Sporozoa (p. 627) and Cnidosporidia (p. 774). Subphylum Ciliophora is characterized by possession of two kinds of nuclei (macronucleus and micronucleus) and of cilia or similar organelles at least at one stage of development. It is subdivided into Ciliata (p. 827) and Suctoria (p. 1036).

In classifying protozoa, the natural system would be one which is based upon the phylogenetic relationships among them in conformity with the doctrine that the present day organisms have descended from primitive ancestral forms through organic evolution. Unlike metazoa, the great majority of protozoa now existing do not possess skeletal structures, which condition also seemingly prevailed among their ancestors, and when they die, they disintegrate and leave nothing behind. The exceptions are Foraminiferida (p. 589) and Radiolarida (p. 616) which produce multiform varieties of skeletal structures composed of inorganic substances and which are found abundantly preserved as fossils in the earliest fossiliferous strata. These fossils show clearly that the two classes of Sarcodina were already well-differentiated groups at the time of fossilization. The sole information the palaeontological record reveals for our reference is that the differentiation of the major groups of protozoa must have occurred in an extremely remote period of the earth history. Therefore, consideration of phylogeny of protozoa had to depend for the most part upon the data obtained through morphological, physiological, and developmental observations of the present-day forms.

The older concept which found its advocates until the beginning of the present century, holds that the Sarcodina are the most

297

primitive of protozoa. It was supposed that at the very beginning of the living world, there came into being undifferentiated masses of protoplasm which later became differentiated into the nucleus and the cytoplasm. The Sarcodina represented by amoebae and allied forms do not have any further differentiation and lack a definite body wall, they are, therefore, able to change body form by forming pseudopodia. These pseudopods are temporary cytoplasmic processes and formed or withdrawn freely, even in the more or less permanent axopodia. On the other hand, flagella and cilia are permanent cell-organs possessing definite structural plans. Thus from the morphological viewpoint, the advocates of this concept maintained that the Sarcodina are the protozoa which were most closely related to ancestral forms.

This concept is however difficult to follow, since it does not agree with the general belief that the plants came into existence before the animals; namely, holophytic organisms living on inorganic substances anteceded holozoic organisms living on organic substances. Therefore, from the physiological standpoint the Mastigophora which include a vast number of chlorophyll-bearing forms, must be considered as more primitive than the holozoic Sarcodina.

The class Mastigophora is composed of Phytomastigia (chromatophore-bearing flagellates and closely related colorless forms) and Zoomastigia (colorless flagellates). Of the former, chrysomonads (p. 305) are mostly naked, and are characterized by possession of 1-2 flagella, 1-2 yellow chromatophores and leucosin. Though holophytic nutrition is general, many are also able to carry on holozoic nutrition. Numerous chrysomonads produce pseudopodia of different types; some possess both flagellum and pseudopodia; others such as Chrysamoeba (Fig. 107) may show flagellate and amoeboid forms (Klebs; Scherffel); still others, for example, members of Rhizochrysidina (p. 319), may lack flagella completely, though retaining the characteristics of Chrysomonadida. When individuals of Rhizochrysis (p. 319) divide, Scherffel (1901) noticed unequal distribution of the chromatophore resulted in the formation of colorless and colored individuals (Fig. 113, *a, b*). Pascher (1917) also observed that in the colonial chrysomonad, Chrysarachnion (p. 321), the division of component individuals produces many in which the chromatophore is entirely lacking (Fig. 113, *c, d*). Thus these chrysomonads which lack chromatophores, resemble Sarcodina rather than the parent Chrysomonadida.

Throughout all groups of Phytomastigia there occur forms which are morphologically alike except for the presence or absence of chromatophores. For example, Cryptomonas (p. 326) and Chilomonas (p. 326), the two genera of Cryptomonadida, are so morphologically alike that had it not been for the chromatophore, the former can hardly be distinguished from the latter. Other examples are Euglena, Astasia, and Khawkinea; Chlorogonium and Hyalogonium; Chlamydomonas and Polytoma; etc.

The chromatophores of various Phytomastigia degenerate readily under experimental conditions. For instance, Zumstein (1900) and recently Pringsheim and Hovasse (1948) showed that *Euglena gracilis* loses its green coloration even in light if cultured in fluids rich in organic substances; in a culture fluid with a small amount of organic substances, the organisms retain green color in light, lose it in darkness; and when cultured in a pure inorganic culture fluid, the flagellates remain green even in darkness. Therefore, it would appear reasonable to consider that the morphologically similar forms with or without chromatophores such as are cited above, are closely related to each other phylogenetically, that they should be grouped together in any scheme of classification, and that the apparent heterogeneity among Phytomastigia is due to the natural course of events. The newer concept which is at present followed widely is that the Mastigophora are the most primitive unicellular animal organisms.

Of Mastigophora. Phytomastigia are to be considered on the same ground more primitive than Zoomastigia. According to the studies of Pascher, Scherffel and others, Chrysomonadida appear to be the nearest to ancestral forms from which other groups of Phytomastigia arose. Among Zoomastigia, Rhizomastigida possibly gave rise to Protomonadida, from which Polymastigida and Hypermastigida later arose. The last-mentioned group is the most highly advanced one of Mastigophora in which an increased number of flagella is an outstanding characteristic.

As to the origin of Sarcodina, many arose undoubtedly from various Zoomastigia, but there are indications that they may have evolved directly from Phytomastigia. As was stated already, Rhizochrysidina possess no flagella and the chromatophore often degenerates or is lost through unequal distribution during division, apparently being able to nourish themselves by methods other than holophytic nutrition. Such forms may have given rise to Amoebida. Some chrysomonads such as Cyrtophora (p. 310) and Palatinella, have axopodia, and it may be considered that they are closer to the ancestral forms from which Heliozoida arose through stages such as shown by Actinomonas (p. 399), Dimorpha (p. 399), and Pteridomonas (p. 400) than any other forms. Another chrysomonad, Porochrysis (p. 311), possesses a striking resemblance to Testacida. The interesting marine chrysomonad, Chrysothylakion (p. 321) that produces a brownish calcareous test from which extrudes anastomosing rhizopodial network, resembling a monothalamous foraminiferan, and forms such as Distephanus (p. 318) with siliceous skeletons, may depict the ancestral forms of Foraminiferida and Radiolarida, respectively. The flagellate origin of these two groups of Sarcodina is also seen in the appearance of flagellated swarmers in many of them during their development. The Mycetozoida show also flagellated phase during their life cycle, which perhaps suggests their origin in flagellated organisms. In fact, in the chrysomonad Myxochrysis (p. 311), Pascher (1917) finds a multinucleate and chromatophore-bearing organism (Fig. 107, *e-j*) that stands intermediate between Chrysomonadida and Mycetozoida. Thus there are a number of morphological, developmental, and physiological observations which suggest the flagellate origin of various members of Sarcodina.

Sporozoa appear to be also polyphyletic. The occurrence of flagellated microgametes in many forms suggest their derivation from flagellates. Léger and Duboscq even considered them to have arisen from Bodonidae (p. 428) on the basis of flagellar arrangement. Obviously, Gregarinida are the most primitive of the three groups. The occurrence of such a form as Selenococcidium (p. 680), would indicate the gregarine-origin of the Coccidida and the members of Haemogregarinidae (p. 705) suggest the probable origin of the Haemosporida in the Coccidida.

The Cnidosporidia are characterized by multinucleate tropho-zoites and by the spore in which at least one coiled filament oc-curs. Some consider them as having evolved from Mycetozoida-like organisms, because of the similarity in multinucleate Tropho-zoites. The polar filament is entirely different from a flagellum in structure and does not appear to have been derived from the latter.

The Ciliata and Suctoria are distinctly separated from the other groups. They possess probably the most complex body organiza-tion seen among protozoa. All ciliates possess cilia or cirri which differ from flagella essentially only in size. In Coleps, Urotricha, Trimyema, Anophrys, etc., which have, in addition to numerous cilia, a long flagellum-like process at the posterior end, and Ileo-nema that possesses an anterior vibratile flagellum and numerous cilia, which also indicates flagellated organisms as their ancestors. It is reasonable to assume that Holotricha are the ciliates from which Spirotricha and Peritrichida evolved. The Suctoria are obviously very closely related to Ciliata and most probably arose from holotrichous ancestors by loss of cilia during adult stage and by developing tentacles in some forms from cystotomes as was suggested by Collin (Fig. 13). (General reference, Franz, 1919; Lwoff, 1951.)

References

BÜTSCHLI, O., 1883-1887. Bronn's Klassen und Ordnungen des Thierreichs. 1.

DOFLEIN, F., and REICHENOW, E., 1949-1953. Lehrbuch der Protozoen-kunde. 6th ed.

FRANZ, V., 1919. Zur Frage der phylogenetischen Stellung der Protisten, besonders der Protozoen. Arch. Protist., 39:263.

GRASSÉ, P.P., 1952-1953. Traité de Zoologie. 1, fasc. 1, 2.

LWOFF, A., 1951. Biochemistry and physiology of protozoa. New York.

MINCHIN, E. A., 1912. Introduction to the study of the protozoa. London.

PASCHER, A., 1912. Ueber Rhizopoden- und Palmellastadien bei Flagellaten, etc. Arch. Protist., 25:153.

———— 1916. Rhizopodialnetz als Fangvorrichtung bei einer Plasmodialen Chrysomonade. *Ibid.*, 37:15.

———— 1916a. Fusionsplasmodien bei Flagellaten und ihre Bedeutung für die Ableitung der Rhizopoden von den Flagellaten. *Ibid.*, 37:31.

———— 1917. Flagellaten und Rhizopoden in ihren gegenseitigen Bezie-hungen. *Ibid.*, 38:1.

—— 1942. Zur Klärung einiger gefärbter und farbloser Flagellaten und ihrer Einrichtungen zur Aufnahme animalischer Nahrung. *Ibid.*, 96:75.

PRINGSHEIM, E. G., and HOVASSE, R. 1948. The loss of chromatophores in *Euglena gracilis.* New Phytologist, 47:52.

SCHERFFEL, A. 1901. Kleiner Beitrag zur Phylogenie einiger Gruppen niederer Organismen. Bot. Zeit., 59:143.

ZUMSTEIN, H., 1900. Zur Morphologie und Physiologie der *Euglena gracilis.* Jahrb. wiss. Botanik, 34:149.

Chapter 8
Phylum **Protozoa** Goldfuss
Subphylum 1. **Plasmodroma** Doflein

THE Plasmodroma possess pseudopodia which are used for locomotion and food-getting or flagella that serve for locomotion. The body structure is less complicated than that of Ciliophora. In some groups are found various endo- and exo-skeletons. The nucleus is of one kind, but vary in number, one to many. All types of nutrition occur. Sexual reproduction is by syngamy or automixis; asexual reproduction is by binary or multiple fission, budding or plasmotomy. The majority of Mastigophora and Sarcodina are free-living, but parasitic forms occur. The members of Sporozoa and Cnidosporidia are without exception parasites.

The subphylum is subdivided into four classes: Mastigophora, Sarcodina (p. 496), Sporozoa (p. 627) and Cnidosporidia (p. 774).

Class 1. **Mastigophora** Diesing

The Mastigophora includes those protozoa which possess one to several flagella. Aside from this common characteristic, this class makes a very heterogeneous assemblage and seems to prevent a sharp distinction between protozoa and protophyta, as it includes Phytomastigia which are often dealt with by botanists.

In the majority of Mastigophora, each individual possesses one to four flagella during the vegetative stage, although species of Polymastigida may possess up to eight or more flagella and of Hypermastigida a greater number of flagella. The palmella stage (Fig. 105) is common among the Phytomastigia and the organism is capable in this stage not only of metabolic activity and growth, but also of reproduction. In this respect, this group shows also a close relationship to algae.

All three types of nutrition, carried on separately or in combination, are to be found among the members of Mastigophora. In holophytic forms, the chlorophyll is contained in the chromatophores which are of various forms among different species and which differ in colors, from green to red. The difference in color appears to be due to the pigments which envelope the chlorophyll body (p. 102). Many forms adapt their mode of nutrition to changed environmental conditions; for instance, from holophytic to saprozoic in the absence of the sunlight. Holozoic, saprozoic and holophytic nutrition are said to be combined in such a form as Ochromonas. In association with chromatophores, there occurs refractile granules or bodies, the pyrenoids, which are connected with starch-formation.

In less complicated forms, the body is naked except for a slight cortical differentiation of the ectoplasm to delimit the body surface and is capable of forming pseudopodia. In others, there occurs a thin plastic pellicle produced by the cytoplasm, which covers the body surface closely. In still others, the body form is constant, being encased in a shell, test, or lorica, which is composed of chitin, pseudochitin, or cellulose. Not infrequently, a gelatinous secretion envelops the body. In three families of Protomonadida there is a collar-like structure located at the anterior end, through which the flagellum protrudes.

The great majority of Mastigophora possess a single nucleus, and only a few are multinucleated. The nucleus is vesicular and contains a conspicuous endosome. Contractile vacuoles are always present in the forms inhabiting fresh water. In simple forms, the contents of the vacuoles are discharged directly through the body surface to the exterior; in others there occurs a single contractile vacuole near a reservoir which opens to the exterior through the so-called cytopharynx. In the Dinoflagellida, there are apparently no contractile vacuoles, but non-contractile pusules (p. 130) occur in some forms. In chromatophore-bearing forms, there occurs usually a stigma which is located near the base of the flagellum and seems to be the center of phototactic activity of the organism which possesses it.

Asexual reproduction is, as a rule, by longitudinal fission, but in some forms multiple fission also takes place under certain circumstances, and in others budding may take place. Colony-formation (p. 208), due to incomplete separation of daughter individuals, is widely found among this group. Sexual reproduction has been reported in a number of species.

The Mastigophora are free-living or parasitic. The free-living forms are found in fresh and salt waters of every description; many are free-swimming, others creep over the surface of submerged objects, and still others are sessile. Together with algae, the Mastigophora compose a major portion of plankton life which makes the nutritional foundation for the existence of all higher aquatic organisms. The parasitic forms are ecto- or endo-parasitic, and the latter inhabit either the digestive tract or the circulatory system of the host animal. Trypanosoma, a representative genus of the latter group, includes important disease-causing parasites of man and of domestic animals.

Mastigophora are divided into two subclasses: Phytomastigia and Zoomastigia (p. 397).

Subclass 1. **Phytomastigia** Doflein

The Phytomastigia possess the chromatophores and their usual method of nutrition is holophytic, though some are holozoic, saprozoic or mixotrophic; the majority are conspicuously colored; some that lack chromatophores are included in this group, since their structure and development resemble closely those of typical Phytomastigia.

Some observers consider the types of flagella as one of the characters in taxonomic consideration (Petersen, 1929; Vlk, 1938; Owen, 1949; etc.). Owen found, for example, "lash flagella" (with a terminal filament) in some species of Phytomonadida, Rhizomastigida, Protomonadida and Polymastigida and simple flagella in the forms included in Chrysomonadida, Cryptomonadida, Euglenoidida and Dinoflagellida.

Phytomastigia are subdivided into six orders: Chrysomonadida, Cryptomonadida (p. 325), Phytomonadida (p. 330), Euglenoidida (p. 350), Chloromonadida (p. 365) and Dinoflagellida (p. 370).

Order 1. **Chrysomonadida** Stein

The chrysomonads are minute organisms and are plastic, since the majority lack a definite cell-wall. Chromatophores are yellow to brown and usually discoid, though sometimes reticulated, in form. Metabolic products are leucosin and fats. One to two flagella are inserted at or near the anterior end of body where a stigma is present.

Many chrysomonads are able to form pseudopodia for obtaining food materials which vary among different species. Nutrition, though chiefly holophytic, is also holozoic or saprozoic. Contractile vacuoles are invariably found in freshwater forms, and are ordinarily of simple structure.

Under conditions not fully understood, the chrysomonads lose their flagella and undergo division with development of mucilag-

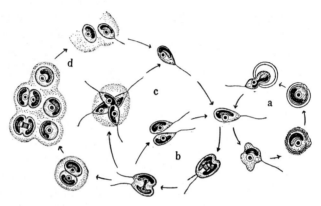

Fig. 105. The life-cycle of Chromulina, × about 200 (Kühn). a, encystment; b, fission; c, colony-formation d, palmella-formation.

inous envelope and thus transform themselves often into large bodies known as the **palmella phase** and undertake metabolic activities as well as multiplication (Fig. 105). Asexual reproduction is, as a rule, by longitudinal division during either the motile or the palmella stage. Incomplete separation of the daughter individuals followed by repeated fission, results in numerous colonial forms mentioned elsewhere (p. 208). Some very closely resemble higher algae. Sexual reproduction is unknown in this group. Encystment occurs commonly; the cyst is often enveloped by a silicious wall possessing an opening with a plug. The chrysomonads inhabit both fresh and salt waters, often occurring abundantly in plankton. (Taxonomy, Doflein, 1923; Schiller, 1925a; Pascher, 1926; Conrad, 1936; Scherffel, 1926; Hollande, 1952; nutrition, Hunter *et al.* 1953).

The order is subdivided into three suborders: Euchrysomonadina, Rhizochrysidina (p. 319) and Chrysocapsina (p. 321).

Suborder 1. **Euchrysomonadina** Pascher

With one to two flagella; motile stage dominant; with calcareous or silicious shell or skeleton. (Five families.)

Family 1. **Chromulinidae** Engler

Minute forms, naked or with sculptured shell; with a single flagellum; often with rhizopods; a few colonial; free-swimming or attached.

Genus **Chromulina** Cienkowski. Oval; round in cross-section; amoeboid; one to two chromatophores; palmella stage often large; in fresh water. Numerous species. The presence of a large number of these organisms gives a golden-brown color to the surface of the water. Electron microscope study of *C. psammobia* shows the presence of a short second internal flagellum in close association with the stigma (Rouiller and Fauré-Fremiet, 1958). (Development, Doflein, 1923; species, Schiller, 1929; Pascher, 1929; Conrad, 1930.)

C. pascheri Hoefeneder (Fig. 106, *a, b*). 15-20μ in diameter.

Genus **Oikomonas** Kent. Similar to Chromulina, but without chromatophores; uninucleate; cyst common; in stagnant water, moist soil; coprozoic.

O. termo (Ehrenberg) (Fig. 106, *m*). Spherical or oval; anterior end lip-like; flagellum about twice the body length; a contractile vacuole; body 5-20μ in diameter; stagnant water. (Axenic culture, Hardin, 1942; bacterial food, Hardin, 1944, 1944a.)

Genus **Pseudochromulina** Doflein. Spheroid body amoeboid; cytoplasm granulated; two contractile vacuoles anterior; a single flagellum about the body length; a yellow tray-like chromatophore with upturned edge; stigma and pyrenoid absent; nucleus central; cyst ovoid, with asymmetrical siliceous wall with an aperture tube (Doflein, 1921).

P. asymmetrica D. Body 3-4μ in diameter; cytoplasm with fat and probably leucosin; cyst 4μ by 3μ; aperture tube about 1μ; fresh water (Doflein, 1921).

Genus **Chrysamoeba** Klebs. Body naked; flagellate stage ovoid with two chromatophores, sometimes slender pseudopodia at the same time; flagellum may be lost and the organism becomes amoeboid, resembling *Rhizochrysis* (p. 319); standing fresh water.

C. radians K. (Fig. 107, *a, b*). Flagellated form measures 8 by
3.5μ; amoebid stage about 8-10μ by 3-4μ, with 10-20μ long radiating pseudopodia; cyst 7μ in diameter (Doeflein, 1922).

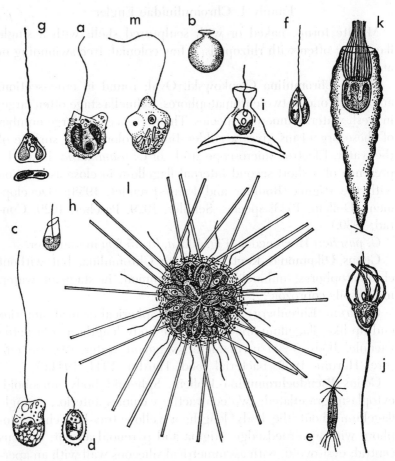

FIG. 106. a, b, *Chromulina pascheri*, ×670 (Hofeneder); c, *Chrysapsis
sagene*, ×1000 (Pascher); d, *Chrysococcus ornatus*, ×600 (Pascher); e,
Mallomonas litomosa, ×400 (Stokes); f, *Pyramidochrysis modesta*, ×670
(Pascher); g, *Sphaleromantis ochracea*, ×600 (Pascher); h, *Kephyrion ovum*,
×1600 (Pascher); i, *Chrysopyxis cyathus*, ×600 (Pascher); j, *Cyrtophora
pedicellata*, ×400 (Pascher); k, *Palatinella cyrtophora*, ×400 (Lauterborn);
l, *Chrysosphaerella longispina*, ×600 (Lauterborn); m, *Oikomonas termo*,
×1330 (Lemmermann).

Genus **Chrysapsis** Pascher. Solitary; plastic or rigid; chromato-phore diffused or branching; with stigma; amoeboid movement; holophytic, holozoic; fresh water. (Several species.)

C. sagene P. (Fig. 106, *c*). Anterior region actively plastic; stigma small; 8-14μ long; flagellum about 30μ long.

Genus **Chrysococcus** Klebs. Shell spheroidal or ovoidal, smooth or sculptured and often brown-colored; through an opeinng a flagellum protrudes; one to two chromatophores; one of the daughter individuals formed by binary fission leaves the parent shell and forms a new one; fresh water. Lackey (1938) found several species in Scioto River, Ohio.

C. ornatus Pascher (Fig. 106, *d*). 14-16μ by 7-10μ.

Genus **Mallomonas** Perty (*Pseudomallomonas* Chodat). Body elongated; with silicious scales and often spines; two chromato-phores rod-shaped; fresh water. Numerous species (Pascher, 1921; Conrad, 1927, 1930)

M. litomosa Stokes (Fig. 106, *e*). Scales very delicate, needle-like projections at both ends; flagellum as long as body; 24-32μ by 8μ.

Genus **Microglena** Ehrenberg. Body ovoid to cylindrical; with a firm envelope in the surface of which are embedded many lenti-cular masses of silica (Conrad, 1928); a single flagellum at ante-rior end; a reservoir around which four to eight contractile vacu-oles occur; a sheet-like chromatophore; stigma; leucosin; fresh water.

M. ovum Conrad (Fig. 108, *a*). 31-38μ by 18-25μ (Conrad, 1928).

Genus **Pyramidochrysis** Pascher. Body form constant; pyriform with three longitudinal ridges; flagellate end drawn out; a single chromatophore; two contractile vacuoles; fresh water.

P. modesta P. (Fig. 106, *f*). 11-13μ long.

Genus **Sphaleromantis** Pascher. Triangular or heart-shaped; highly flattened; slightly plastic; two chromatophores; two con-tractile vacuoles; stigma large; long flagellum; fresh water.

S. ochracea P. (Fig. 106, *g*). 6-13μ long.

Genus **Kephyrion** Pascher. With oval or fusiform lorica; body fills posterior half of lorica; one chromatophore; a single short flagellum; small; fresh water. Species (Conrad, 1930)

K. ovum P. (Fig. 106, *h*). Lorica up to 7μ by 4μ.

Genus **Chrysopyxis** Stein. With lorica of various forms, more or less flattened; one to two chromatophores; a flagellum; attached to algae in fresh water.

C. cyathus Pascher (Fig. 106, *i*). One chromatophore; flagellum twice body length; lorica 20-25μ by 12-15μ.

Genus **Cyrtophora** Pascher. Body inverted pyramid with six to eight axopodia and a single flagellum; with a contractile stalk; a single chromatophore; a contractile vacuole; fresh water.

C. pedicellata P. (Fig. 106, *j*). Body 18-22μ long; axopodia 40-60μ long; stalfl 50-80μ long.

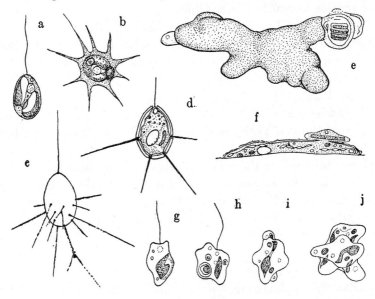

Fig. 107. a, flagellate and b, amoeboid phase of *Chrysamoeba radians*, ×670 (Klebs); c, surface view and d, optical section of *Porochrysis aspergillus* ×400 (Pascher); e–j, *Myxochrysis paradoxa* (Pascher). e, a medium large plasmodium with characteristic envelop; the large food vacuole contains protophytan, Scenedesmus, ×830; f, diagrammatic side view of a plasmodium, engulfing a diatom; moniliform bodies are yellowish chromatophores, ×1000; g–i, development of swarmer into plasmodium (stippled bodies are chromatophores), ×1200.

Genus **Palatinella** Lauterborn. Lorica tubular; body heart-shaped; anterior border with sixteen to twenty axopodia; a single flagellum; a chromatophore; several contractile vacuoles; fresh water.

P. cyrtophora L. (Fig. 106, *k*). Lorica 80-150µ long; body 20-25µ by 18-25µ; axopodia 50µ long.

Genus **Chrysosphaerella** Lauterborn. In spherical colony, individual cell, oval or pyriform, with two chromatophores; imbedded in gelatinous mass; fresh water.

C. longispina L. (Fig. 106, *l*). Individuals up to 15µ by 9µ; colony up to 250µ in diameter; in standing water rich in vegetation.

Genus **Porochrysis** Pascher. Shell with several pores through which rhizopodia are extended; a flagellum passes through an apical pore; a single small chromatophore; leucosin; a contractile vacuole; fresh water.

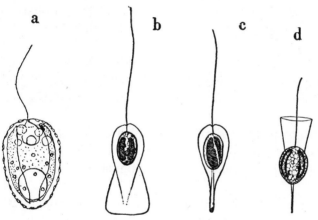

FIG. 108. a, *Microglena ovum*, ×680 (Conrad); b, c, two views of *Angulochrysis erratica*, ×900 (Lackey); d, *Stylochromonas minuta*, ×1200 (Lackey).

P. aspergillus P. (Fig. 107, *c, d*). Shell about 35µ long by 25µ wide; chromatophore very small; a large leucosin grain; fresh water.

Genus **Myxochrysis** Pascher. Body multinucleate, amoeboid; with yellowish moniliform chromatophores, many leucosin granules and contractile vacuoles; holozoic; surrounded by a brownish envelope which conforms with body form; flagellated swarmers develop into multinucleate plasmodium; plasmotomy and plasmogamy; fresh water (Pascher, 1916a).

M. paradoxa P. (Fig. 107, *e-j*). Plasmodium 15-18µ or more in diameter; in standing water.

Genus **Angulochrysis** Lackey. Body ovoid; colorless, thin lorica rounded anteriorly and flattened posteriorly into "wings"; a single flagellum long; no cytostome; two bright yellow-brown chromatophores; no stigma; swims with a slow rotation; marine (Lackey, 1940).

A. erratica L. (Fig. 108, *b, c*). Body up to 12µ long; lorica up-to 30µ high; flagellum about four times the body length; Woods Hole.

Genus **Stylochromonas** L. Body ovoid, sessile with a stiff stalk; with a large collar at anterior end; a single flagellum; two golden brown chromatophores; no stigma; marine (Lackey, 1940).

S. minuta L. (Fig. 108, *d*). Body 5-8µ long; collar about 6µ high; flagellum about twice the body length.

Family 2. **Syncryptidae** Poche

Solitary or colonial chrysomonads with two equal flagella; with or without pellicle (when present, often sculptured); some possess stalk.

Genus **Syncrypta** Ehrenberg. Spherical colonies; individuals with two lateral chromatophores, embedded in a gelatinous mass; two contractile vacuoles; without stigma; cysts unknown; fresh water.

S. volvox E. (Fig. 109, *a*). 8-14µ by 7-12µ; colony 20-70µ in diameter; in standing water.

Genus **Synura** Ehrenberg (*Synuropsis*, Schiller). Spherical or ellipsoidal colony composed of two to fifty ovoid individuals arranged radially; body usually covered by short bristles; two chromatophores lateral; no stigma; asexual reproduction of individuals is by longitudinal division; that of colony by bipartition; cysts spherical; fresh water. (Taxonomy, Korshikov, 1929; Kufferath, 1946)

S. uvella E. (Fig. 109, *b*). Cells oval; bristles conspicuous; 20-40µ by 8-17µ; colony 100-400µ in diameter; if present in large numbers, the organism is said to be responsible for an odor of the water resembling that of ripe cucumber.

S. adamsi Smith (Fig. 109, *c*). Spherical colony with individuals radiating; individuals long spindle, 42-47μ by 6.5-7μ; two flagella up to 17μ long; in fresh water pond.

Genus **Hymenomonas** Stein. Solitary; ellipsoid to cylindrical; membrane brownish, often sculptured; two chromatophores; without stigma; a contractile vacuole anterior; fresh water.

H. roseola S. (Fig. 109, *d*). 17-50μ by 10-20μ.

Genus **Derepyxis** Stokes. With cellulose lorica, with or without a short stalk; body ellipsoid to spherical with one to two chromatophores; two equal flagella; fresh water.

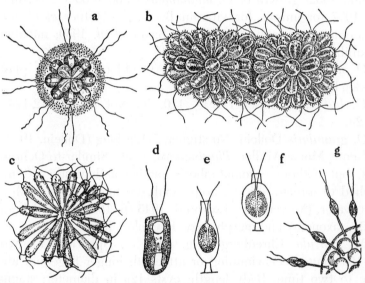

Fig. 109. a, *Syncrypta volvox*, ×430 (Stein); b, *Synura uvella*, ×500 (Stein); c, *S. adamsi*, ×280 (Smith); d, *Hymenomonas roseola*, ×400 (Klebs); e, *Derepyxis amphora*, ×540 (Stokes); f, *D. ollula*, ×600 (Stokes); g, *Stylochrysallis parasitica*, ×430 (Stein).

D. amphora S. (Fig. 109, *e*). Lorica 25-30μ by 9-18μ; on algae in standing water.

D. ollula S. (Fig. 109, *f*). Lorica 20-25μ by 15μ.

Genus **Stylochrysalis** Stein. Body fusiform; with a gelatinous stalk attached to Volvocidae; two equal flagella; two chromatophores; without stigma; fresh water.

S. parasitica S. (Fig. 109, *g*). Body 9-11μ long; stalk about 15μ long; on phytomonads.

Family 3. **Ochromonadidae** Pascher

With two unequal flagella; no pellicle and plastic; contractile vacuoles simple; with or without a delicate test; solitary or colonial; free-swimming or attached.

Genus **Ochromonas** Wyssotzki. Solitary or colonial; body surface delicate; posterior end often drawn out for attachment; one to two chromatophores; usually with a stigma; encystment; fresh water. (Many species, Doflein, 1921, 1923; nutrition, Pringsheim, 1952; growth of *O. malhamensis* above 35°C., Hutner *et al.* 1957; organisms for vitamin B_{12} assay, Hutner *et al.* 1958; thiamine assay in biologic fluids, Baker *et al.* 1959; nutritional flexibility, Aaronson and Baker, 1959)

O. mutabilis Klebs (Fig. 110, *a*). Ovoid to spherical; plastic, 15-30μ by 8-22μ.

O. ludibunda Pascher (Fig. 110, *b*). Not plastic; 12-17μ by 6-12μ.

O. granularis Doflein. No stigma; 5-12μ long (Doflein, 1922).

Genus **Monas** Müller (*Physomonas,* Kent). Similar to Ochromonas, but without chromatophores; active and plastic; often attached to foreign objects; body small, up to 20μ long; fresh and salt water. (Movements, Lowndes, 1945 (p. 155); cysts, Scherffel, 1924; taxonomy and morphology, Reynolds, 1934)

M. guttula Ehrenberg (Fig. 110, *c*). Spherical to ovoid; 14-16μ long; free-swimming or attached; longer flagellum about one to two times body length; cysts 12μ in diameter; stagnant water.

M. elongata Stokes (Fig. 110, *d*). Elongate; about 11μ long; free-swimming or attached; anterior end obliquely truncate; fresh water.

M. socialis Kent (Fig. 110, *e*). Spherical; 5–10μ long; among decaying vegetation in fresh water.

M. vestita (Stokes). Spherical; about 13.5μ in diameter; stalk about 40μ long; pond water. Reynolds (1934) made a careful study of the organism.

M. sociabilis Meyer. Body 8-10μ long by 5μ; two unequal flagella; the longer one is as long as the body and the shorter one about one-fourth; twenty to fifty individuals form a spheroid colony, resembling a detached colony of *Anthophysis;* polysaprobic.

Genus **Uroglena** Ehrenberg. Spherical or ovoidal colony, composed of ovoid or ellipsoidal individuals arranged on periphery of a gelatinous mass; all individuals connected with one another by gelatinous processes running inward and meeting at a point; with a stigma and a plate-like chromatophore; asexual reproduction of

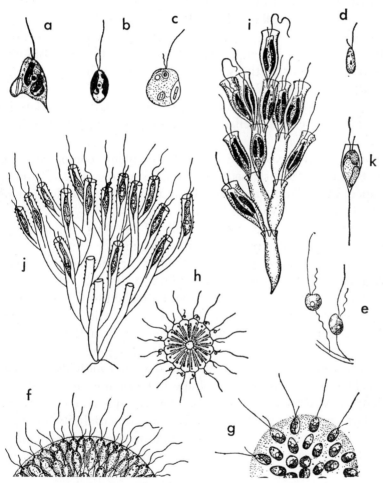

FIG. 110. a, *Ochromonas mutabilis,* ×670 (Senn); b, *O. ludibunda,* ×540 (Pascher); c. *Monas guttula,* ×620 (Fritsch); d, *M. elongata,* ×670 (Stokes); e, *M. socialis,* ×670 (Kent); f, *Uroglena volvox,* ×430 (Stein); g, *Uroglenopsis americana,* ×470 (Lemmermann); h, *Cyclonexis annularis,* ×540 (Stokes); i, *Dinobryon sertularia,* ×670 (Scherffel); j, *Hyalobryon ramosum,* ×540 (Lauterborn); k, *Stylopyxis mucicola,* ×470 (Bolochonzew).

individuals by longitudinal fission, that of colony by bipartition; cysts spherical with spinous projections, and a long tubular process; fresh water. One species.

U. volvox E. (Fig. 110, *f*). Cells 12-20μ by 8-13μ; colony 40-400μ in diameter, in standing water.

Genus **Uroglenopsis** Lemmermann. Similar to *Uroglena,* but individuals without inner connecting processes.

U. americana Calkins (Fig. 110, *g*). Each cell with one chromatophore; 5-8μ long; flagellum up to 32μ long; colony up to 300μ in diameter; when present in abundance, the organism gives an offensive odor to the water (Calkins). (Morphology, development, Troitzkaja, 1924.)

U. europaea Pascher. Similar to the last-named species; but chromatophores two, cells up to 7μ long; colony, 150-300μ in diameter.

Genus **Cyclonexis** Stokes. Wheel-like colony, composed of ten to twenty wedge-shaped individuals; young colony funnel-shaped; chromatophores two, lateral; no stigma; reproduction and encystment unknown; fresh water.

C. annularis S. (Fig. 110, *h*). Cells 11-14μ long; colony 25-30μ in diameter; in marshy water with sphagnum.

Genus **Dinobryon** Ehrenberg. Solitary or colonial; individuals with vase-like, hyaline, but sometimes, yellowish cellulose test, drawn out at its base; elongated and attached to the base of test with its attenuated posterior tip; one to two lateral chromatophores; usually with a stigma; asexual reproduction by binary fission; one of the daughter individuals leaving test as a swarmer, to form a new one; in colonial forms daughter individuals remain attached to the inner margin of aperture of parent tests and there secrete new tests; encystment common; the spherical cysts possess a short process; Ahlstrom (1937) studied variability of North American species and found the organisms occur more commonly in alkaline water than elsewhere; fresh water. Numerous species.

D. sertularia E. (Fig. 110 *i*). 23-43μ by 10-14μ.

D. divergens Imhof. 26–65μ long; great variation in different localities.

Genus **Hyalobryon** Lauterborn. Solitary or colonial; individual body structure similar to that of *Dinobryon*; lorica in some cases tubular, and those of young individuals are attached to the exterior of parent tests; fresh water.

H. ramosum L. (Fig. 110, *j*). Lorica 50-70µ long by 5-9µ in diameter; body up to 30µ by 5µ; on vegetation in standing fresh water.

Genus **Stylopyxis** Bolochonzew. Solitary; body located at bottom of a delicate stalked lorica with a wide aperture; two lateral chromatophores; fresh water.

S. mucicola B. (Fig. 110, *k*). Lorica 17-18µ long; stalk about 33µ long; body 9-11µ long; fresh water.

Genus **Stokesiella** Lemmermann. Body attached by a fine cytoplasmic thread to a delicate and stalked vase-like lorica; two contractile vacuoles; fresh water.

S. dissimilis (Stokes) (Fig. 111, *a*). Solitary; lorica about 28µ long.

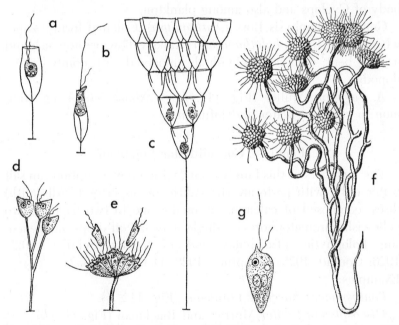

Fig. 111. a, *Stokesiella dissimilis*, ×500 (Stokes); b, *S. leptostoma*, ×840 (Stokes); c, *Stylobryon abbotti*, ×480 (Stokes); d, *Dendromonas virgaria*, a young colony, ×670 (Stein); e, *Cephalothamnium cyclopum*, ×440 (Stein); f, g, *Anthophysis vegetans* (f, part of a colony, ×230; g, an individual, ×770) (Stein).

S. leptostoma (S.) (Fig. 111, *b*). Lorica about 17µ long; often in groups; on vegetation.

Genus **Stylobryon** Fromentel. Similar to *Stokesiella;* but colonial; on algae in fresh water.

S. abbotti Stokes (Fig. 111, *c*). Lorica campanulate; about 17µ long; main stalk about 100µ high; body oval or spheroidal; flagella short.

Genus **Dendromonas** Stein. Colonial; individuals without lorica, located at end of branched stalks; fresh water among vegetation.

D. virgaria Weisse (Fig. 111, *d*). About 8µ long; colony 200µ high; pond water.

Genus **Cephalothamnium** Stein. Colonial; without lorica, but individuals clustered at the end of a stalk which is colorless and rigid; fresh water.

C. cyclopum S. (Fig. 111, *e*). Ovoid; 5-10µ long; attached to body of Cyclops and also among plankton.

Genus **Anthophysis** Bory (*Anthophysa*). Colonial forms, somewhat similar to *Cephalothamnium;* stalks yellow or brownish and usually bent; detached individuals amoeboid with pointed pseudopodia.

A. vegetans Müller (Fig. 111, *f, g*). About 5-6µ long; common in stagnant water and infusion.

Family 4. **Coccolithidae** Lohmann

The members of this family occur, with a few exceptions, in salt water only; with perforate (tremalith) or imperforate (discolith) discs, composed of calcium carbonate; one to two flagella; two yellowish chromatophores; a single nucleus; oil drops and leucosin; holophytic. (Taxonomy and phylogeny, Schiller, 1925, 1926; Conrad, 1928; Kamptner, 1928; Deflandre, 1952a)
Examples:
Pontosphaera haeckeli Lohmann (Fig. 112, *a*).
Discosphaera tubifer Murray and Blackman (Fig. 112, *b*).

Family 5. **Silicoflagellidae** Borgert

Exclusively marine planktons; with siliceous skeleton whicn envelops the body. Example: *Distephanus speculum* (Müller) (Fig. 112, *c*) (Deflandre, 1952).

Suborder 2. **Rhizochrysidina** Pascher

No flagellate stage is known to occur; the organism possesses pseudopodia; highly provisional group, based wholly upon the absence of flagella; naked or with test; various forms; in some species chromatophores are entirely lacking, so that the organisms resemble some members of the Sarcodina. Several genera.

Genus **Rhizochrysis** Pascher. Body naked and amoeboid; with one to two chromatophores; fresh water.

R. scherffeli P. (Figs. 112, *d;* 113, *a, b*). 10-14µ in diameter; one to two chromatophores: branching rhizopods; fresh water.

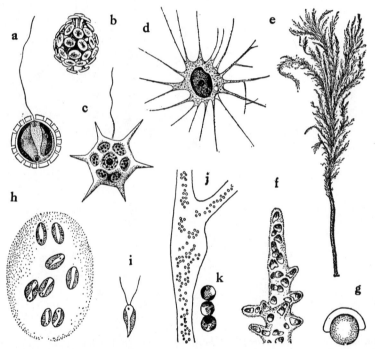

FIG. 112. a, *Pontosphaera haeckeli,* ×1070 (Kühn); b, *Discosphaera tubifer,* ×670 (Kühn); c, *Distephanus speculum,* ×530 (Kühn); d, *Rhizochrysis scherffeli,* ×670 (Doflein); e–g, *Hydrurus foetidus* (e, entire colony; f, portion; g, cyst), e (Berthold), f, ×330, g, ×800 (Klebs); h, i, *Chrysocapsa paludosa,* ×530 (West); j, k, *Phaeosphaera gelatinosa* (j, part of a mass, ×70; k, three cells, ×330) (West).

Genus **Chrysidiastrum** Lauterborn. Naked; spherical; often several in linear association by pseudopodia; one yellow-brown chromatophore; fresh water.

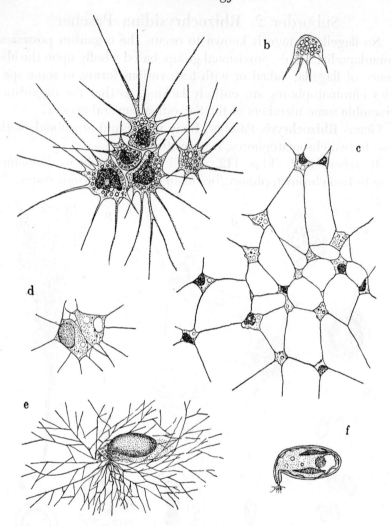

Fig. 113. a, b, *Rhizochrysis scherffeli*, ×500 (Scherffel). a, 4 chromato-phore-bearing individuals and an individual without chromatophore; b, the last-mentioned individual after 7 hours. c, d, *Chrysarachnion insidians* (Pascher). c, part of a colony composed of individuals with and without chromatophore, ×1270; d, products of division, one individual lacks chromatophore, but with a leucosin body, ×2530. e, f, *Chrysothylakion vorax* (Pascher). e, an individual with anastomosing rhizopodia and "excretion granules," ×870; f, optical section of an individual; the cytoplasm contains two fusiform brownish chromatophores, a spheroid nucleus, a large leucosin body and contractile vacuole, × about 1200.

C. catenatum L. Cells 12-14μ in diameter (Pascher, 1916a).

Genus **Chrysarachnion** Pascher. Amoeboid organism; with a chromatophore, leucosin grain and contractile vacuole; many individuals arranged in a plane and connected by extremely fine rhizopods, the whole forming a cobweb network. Small animals are trapped by the net; chromatophores are small; nutrition both holophytic and holozoic; during division the chromatophore is often unevenly distributed so that many individuals without any chromatophore are produced; fresh water (Pascher, 1916a).

C. insidians P. (Fig. 113, *c, d*). Highly amoeboid individuals 3-4μ in diameter; chromatophore pale yellowish brown, but becomes bluish green upon death of organisms; a leucosin grain and a contractile vacuole; colony made up of 200 or more individuals.

Genus **Chrysothylakion** Pascher. With retort-shaped calcareous shell with a bent neck and an opening; shell reddish brown (with iron) in old individuals; through the aperture are extruded extremely fine anastomosing rhizopods; protoplasm which fills the shell is colorless; a single nucleus, two spindle-form brown chromatophores, several contractile vacuoles and leucosin body; marine water.

C. vorax. P. (Fig. 113, *e, f*). The shell measures 14-18μ long, 7-10μ broad, and 5-6μ high; on marine algae.

Suborder 3. **Chrysocapsina** Pascher

Palmella stage prominent; flagellate forms transient; colonial; individuals enclosed in a gelatinous mass; one to two flagella, one chromatophore, and a contractile vacuole; one group of relatively minute forms and the other of large organisms.

Genus **Hydrurus** Agardh. In a large (1-30 cm. long) branching gelatinous cylindrical mass; cells yellowish brown; spherical to ellipsoidal; with a chromatophore; individuals arranged loosely in gelatinous matrix; apical growth resembles much higher algae; multiplication of individuals results in formation of pyrimidal forms with a flagellum, a chromatophore, and a leucosin mass; cyst may show a wing-like rim; cold freshwater streams.

H. foetidus Kirschner (Figs. 34, *d-f*; 112, *e-g*). Olive-green, feathery tufts, 1-30 cm. long, develops an offensive odor; sticky to touch; occasionally encrusted with calcium carbonate; in running fresh water.

Genus **Chrysocapsa** Pascher. In a spherical to ellipsoidal gelatinous mass; cells spherical to ellipsoid; one to two chromatophores; with or without stigma; fresh water.

C. paludosa P. (Fig. 112, *h, i*). Spherical or ellipsoidal with cells distributed without order; with a stigma; two chromatophores; swarmer pyriform with two flagella; cells 11μ long; colony up to 100μ in diameter.

Genus **Phaeosphaera** West and West. In a simple or branching cylindrical gelatinous mass; cells spherical with a single chromatophore; fresh water.

P. gelatinosa W. and W. (Fig. 112, *j, k*). Cells 14-17.5μ in diameter.

References

AARONSON, S., and BAKER, H., 1959. A comparative biochemical study of two species of Ochromonas. J. Protozool., 6:282.

AHLSTROM, E. H., 1936. The deep-water plankton of Lake Michigan, exclusive of the Crustacea. Tr. Am. Micr. Soc., 55:286.

—— 1937. Studies on variability in the genus Dinobryon. *Ibid.*, 56:139.

BAKER, H., *et al.*, 1959. Assay of thiamine in biologic fluids. Clin. Chem., 5:13.

BÜTSCHLI, O., 1883-1887. Mastigophora. Bronn's Klassen und Ordnungen des Thierreichs. 1, pt. 2.

CONRAD, W., 1926. Recherches sur les flagellates de nos eaux saumâtres. II. Arch. Protist., 56:167.

—— 1927. Essai d'une monographie des genres Mallomonas Perty (1852) et Pseudomallomonas Chodat (1920). *Ibid.*, 59:423.

—— 1928. Le genre Microglena. *Ibid.*, 60:415.

—— 1928a. Sur les Coccolithophoracées d'eau douce. *Ibid.*, 63:58.

—— 1930. Flagellates nouveaux ou peu connus. I. *Ibid.*, 70:657.

DEFLANDRE, G. 1952. Classe des Silicoflagellidés. In Grassé (1952), p. 425.

—— 1952a. Classe des Coccolithophoridés. *Ibid.*, p. 440.

DOFLEIN, F. 1921. Mitteilungen über Chrysomonadien aus dem Schwarzwald. Zool. Anz., 53:153.

—— 1922. Untersuchungen über Chrysomonadinen. I, II. Arch. Protist., 44:149.

—— 1923. III. *Ibid.*, 45:267.

—— and REICHENOW, E., 1949-1953. Lehrbuch der Protozoenkunde. 2 vols. 6th ed. Jena.

FRITSCH, F. E., 1935. The structure and reproduction of the algae. Cambridge.

GRASSÉ, P.-P., 1952. Traité de Zoologie. I:1.

HARDIN, G., 1942. An investigation of the physiological requirements of a pure culture of the heterotrophic flagellate, *Oikomonas termo* Kent. Physiol. Zool., 15:466.

——— 1944. Physiological observations and their ecological significance: etc. Ecology, 25:192.

——— 1944a. Symbiosis of Paramecium and Oikomonas. *Ibid.*, 25:304.

HOLLANDE, A., 1952. Classe des Chrysomonadines. In Grassé (1952), p. 471.

HUTNER, S. H., *et al.*, 1953. Nutrition of some phagotrophic fresh-water chrysomonads. Ann. N.Y. Acad. Sci., 56:852.

——— *et al.*, 1957. Growing *Ochromonas malhamensis* above 35°C. J. Protozool., 4:259.

——— *et al.*, 1958. Microbial assays. Anal. Chem., 30:849.

KAMPTNER, E., 1928. Ueber das System und die Phylogenie der Kalkflagellaten. Arch. Protist., 64:19.

KENT, S., 1880-1882. A manual of Infusoria. London.

KORSHIKOV, A. A., 1929. Studies on the chrysomonads. I. Arch. Protistenk., 67:253.

KUFFERATH, H., 1946. Position systematique du genre Synura Ehrenberg. Bull. Soc. Roy. Bot. Belgique, 78:46.

LOWNDES, A. G., 1945. Swimming of *Monas stigmatica*. Nature, 155:579.

OWEN, H. M. 1947. Flagellar structure. I. Tr. Am. Micr. Soc., 66:50.

——— 1949. II. *Ibid.*, 68:261.

PASCHER, A., 1914. Flagellatae: Allgemeiner Teil. In Die Süsswasserflora Deutschlands. Part 1.

——— 1916. Studien über die rhizopodiale Entwicklung der Flagellaten. Arch. Protist., 36:81.

——— 1916a. Rhizopodialnetz als Fangvorrichtung bei einer plasmodialen Chrysomonade. *Ibid.*, 37:15.

——— 1916b. Fusionplasmodien bei Flagellaten und ihre Bedeutung für die Ableitung der Rhizopoden von den Flagellaten. *Ibid.*, 37:31.

——— 1917. Flagellaten und Rhizopoden in ihren gegenseitigen Beziehungen. *Ibid.*, 38:584.

——— 1921. Neue oder wenig bekannte Protisten. *Ibid.*, 44:119.

——— 1929. XXI. *Ibid.*, 65:426.

PRINGSHEIM, E. G., 1952. On the nutrition of Ochromonas. Quart. J. Micro. Sci., 93:71.

REYNOLDS, B. D., 1934. Studies on monad flagellates. I and II. Arch. Protist., 81:399.

ROUILLER, C., and FAURÉ-FREMIET, E., 1958. Structure fine d'un flagellé chrysomonadien: *Chromulina psammobia*. Exper. Cell. Res., 14:47

SCHERFFEL, A., 1924. Ueber die Cyste von Monas. Arch. Protist., 48:187.

324 *Protozoology*

———— 1901. Kleiner Beitrag zur Phylogenie einiger Gruppen niederer Organismen. Bot. Zeit., 59:143.

———— 1927. Beitrag zur Kenntnis der Chrysomonadineen. II. Arch. Protist., 57:331.

SCHILLER, J., 1925. Die planktonischen Vegetationen des adriatischen Meeres, A. *Ibid.*, 51:1.

———— 1925a. B. *Ibid.*, 53:59.

———— 1926. Ueber Fortpflanzung, geissellose Gattungen und die Nomenklatur der Coccolithophoraceen, etc. *Ibid.*, 53:326.

———— 1926a. Der thermische Einfluss und die Wirkung des Eises auf die planktonischen Herbstvegetationen, etc. *Ibid.*, 56:1.

———— 1929. Neue Chryso- und Cryptomonaden aus Altwässern der Donau bei Wien. *Ibid.*, 66:436.

SMITH, G. M., 1950. The freshwater algae of the United States. 2 ed. New York.

STEIN, F., 1878. Der Organismus der Infusionsthiere. 3 Abt. Leipzig.

———— 1883. Der Organismus der Flagellate oder Geisselinfusorien. Parts 1, 2. Leipzig.

STOKES, A. C., 1888. A preliminary contribution toward a history of the freshwater Infusoria of the United States. J. Trenton Nat. Hist. Soc., 1:71.

TROITZKAJA, O. V., 1924. Zur Morphologie und Entwicklungsgeschichte von *Uroglenopsis americana.* Arch. Protist., 49:260.

WEST, G. S., and FRITSCH, F. E., 1927: A treatise on the British freshwater algae. Cambridge.

Chapter 9
Order 2. **Cryptomonadida** Ehrenberg

THE cryptomonads differ from the chrysomonads in having a constant body form. Pseudopodia are very rarely formed, as the body is covered by a pellicle. The majority show dorso-ventral differentiation, with an oblique longitudinal furrow. One to two unequal flagella arise from the furrow or from the cytopharynx. If two flagella are present, both may be directed anteriorly or one posteriorly. These organisms are free-swimming or creeping.

One or two chromatophores are usually present. They are discoid or band-form. The color of chromatophores varies: yellow, brown, red, olive-green; the nature of the pigment is not well understood, but it is said to be similar to that which is found in the Dinoflagellida (Pascher). One or more spherical pyrenoids which are enclosed within a starch envelope appear to occur outside the chromatophores. Nutrition is mostly holophytic; a few are saprozoic or holozoic. Assimilation products are solid discoid carbohydrates which stain blue with iodine in Cryptomonas or which stain reddish violet by iodine in Cryptochrysis; fat and starch are produced in holozoic forms which feed upon bacteria and small protozoa. The stigma is usually located near the insertion point of the flagella. Contractile vacuoles, one to several, are simple and are situated near the cytopharynx. A single vesicular nucleus is ordinarily located near the middle of the body.

Asexual reproduction takes place in either the active or the non-motile stage. Sexual reproduction is unknown. Some cryptomonads form palmella stage and others gelatinous aggregates. In the suborder Phaeocapsina, the palmella stage is permanent. Cysts are spherical, and the cyst wall is composed of cellulose. The cryptomonads occur in fresh or sea water, living also often as symbionts in marine organisms.

Cryptomonadida are divided into two suborders: Eucryptomonadina and Phaeocapsina (p. 328).

Suborder 1. **Eucryptomonadina** Pascher

Flagellate forms are predominant. Two families.

Family 1. **Cryptomonadidae** Stein

Body truncate anteriorly; two anterior flagella; an oblique furrow near anterior region.

Genus **Cryptomonas** Ehrenberg. Elliptical body with a firm pellicle; anterior end truncate, with two flagella; dorsal side convex, ventral side slightly so or flat; nucleus posterior; "cytopharynx" with granules, considered trichocysts by some observers (Hollande, 1942, 1952); two lateral chromatophores vary in color from green to blue-green, brown or rarely red; holophytic; with small starch-like bodies which stain blue in iodine; one to three contractile vacuoles anterior; fresh water. Several species. (Morphology and taxonomy, Hollande, 1942, 1952.)

C. ovata E. (Fig. 114, *a*). 30-60μ by 20-25μ; among vegetation.

Genus **Chilomonas** Ehrenberg. Similar to *Cryptomonas* in general body form and structure, but colorless because of the absence of chromatophores; without pyrenoid; "cytopharynx" deep, lower half surrounded by granules, considered by Hollande (1942) and Dragesco (1951) as trichocysts; one contractile vacuole anterior; nucleus in posterior half; endoplasm usually filled with polygonal starch grains; saprozoic; fresh water.

C. paramecium E. (Fig. 114, *b*). Posteriorly narrowed, slightly bent "dorsally"; 30-40μ by 10-15μ; saprozoic; widely distributed in stagnant water. (Cytology, Mast and Doyle, 1935; Hollande, 1942; Dragesco, 1951; bacteria-free culture, Mast and Pace, 1933; metabolism, Mast and Pace, 1929; Pace, 1941; growth substances, Pace, 1944, 1947; Mast and Pace, 1946; effects of vitamins, Pace, 1947; Oxidative metabolism, Holz, 1954; induced resistance to sulfonamides, Hall, 1957.)

C. oblonga Pascher. Oblong; posterior end broadly rounded; 20-50μ long.

Genus **Chrysidella** Pascher. Somewhat similar to *Cryptomonas*—but much smaller; yellow chromatophores much shorter; often found in Foraminiferida and Radiolarida as symbionts ("zoox-

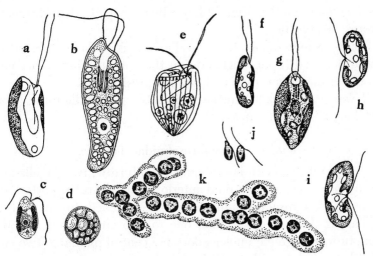

Fig. 114. a, *Cryptomonas ovata,* ×800 (Pascher); b, *Chilomonas para-mecium,* ×1330 (Bütschli); c, d, *Chrysidella schaudinni,* ×1330 (Winter); e, *Cyathomonas truncata,* ×670 (Ulehla); f, *Cryptochrysis commutata,* ×670 (Pascher); g, *Rhodomonas lens,* ×1330 (Ruttner); h, *Nephroselmis olvacea,* ×670 (Pascher); i, *Protochrysis phaeophycearum,* ×800 (Pascher); j, k, *Phaeothamnion confervicolum,* ×600 (Kühn).

anthellae"). Several species. (Culture, McLaughlin and Zahl, 1957.)

C. schaudinni Winter (Fig. 114, *c,d*). Body less than 10µ long; in the foraminiferan *Peneroplis pertusus.*

Genus **Cyathomonas** Fromentel. Body small, somewhat oval; without chromatophores; much compressed; anterior end oblique-ly truncate; with two equal or subequal anterior flagella; colorless; nucleus central; anabolic products, stained red or reddish violet by iodine; contractile vacuole usually anterior; a row of refractile granules, protrichocysts, close and parallel to anterior margin of body; asexual reproduction by longitudinal fission; holozoic; in stagnant water and infusion. One species.

C. truncata Ehrenberg (Fig. 114, *e*). 15-25µ by 10-15µ.

Genus **Cryptochrysis** Pascher. Furrow indistinctly granulated; two or more chromatophores brownish, olive-green, or dark green, rarely red; pyrenoid central; two equal flagella; some lose flagella and may assume amoeboid form; fresh water.

C. commutata P. (Fig. 114, *f*). Bean-shaped; two chromatophores; 19μ by 10μ.

Genus **Rhodomonas** Karsten. Furrow granulated; chromatophore one, red (upon degeneration the coloring matter becomes dissolved in water); pyrenoid central; fresh water.

R. lens Pascher and Ruttner (Fig. 114, *g*). Spindle-form; about 16μ long; in fresh water.

Family 2. **Nephroselmidae** Pascher

Body reniform; with lateral equatorial furrow; two flagella arising from furrow, one directed anteriorly and the other posteriorly.

Genus **Nephroselmis** Stein. Reniform; flattened; furrow and cytopharynx distinct; no stigma; one to two chromatophores, discoid, brownish green; nucleus dorsal; a central pyrenoid; two contractile vacuoles; with reddish globules; fresh water.

N. olvacea S. (Fig. 114, *h*). 20-25μ by 15μ.

Genus **Protochrysis** Pascher. Reniform; not flattened; with a distinct furrow, but without cytopharynx; a stigma at base of flagella; one to two chromatophores, brownish yellow; pyrenoid central; two contractile vacuoles; fission seems to take place during the resting stage; fresh water.

P. phaeophycearum P. (Fig. 114, *i*). 15-17μ by 7-9μ.

Suborder 2. **Phaeocapsina** Pascher

Palmella stage predominant; perhaps border-line forms between brown algae and cryptomonads. Example: *Phaeothamnion confervicolum* Lagerheim (Fig. 114, *j*, *k*) which is less than 10μ long.

References

DRAGESCO, J., 1951. Sur la structure des trichocystes du flagellé cryptomonadine, *Chilomonas paramecium*. Bull. micr. appl., 2 sér. 1:172.

FRITSCH, F. E., 1935. The structure and reproduction of the algae. Cambridge.

HALL, R. P., 1957. Induced resistance to sulfonamides in *Chilomonas paramecium*. J. Protozool., 4:42.

HOLLANDE, A., 1942. Étude cytologique et biologique de quelques flagellés libres. Arch. zool. exper. gén., 83:1.

—— 1952. Classe des Cryptomonadines. In Grassé (1952), p. 286.

Holz, G. G., Jr., 1954. The oxidative metabolism of a cryptomonad flagellate, *Chilomonas paramecium*. J. Protozool., 1:114.

Mast, S. O., and Doyle, W. L., 1935. A new type of cytoplasmic structure in the flagellate *Chilomonas paramecium*. Arch. Protist., 85:145.

────── and Pace, D. M., 1933. Synthesis from inorganic compound of starch, fats, proteins and protoplasm in the colorless animals, *Chilomonas paramecium*. Protoplasma, 20:326.

────── ────── 1939. The effect of calcium and magnesium on metabolic processes in Chilomonas. J. Cell. Comp. Physiol., 14: 261.

────── ────── 1946. The nature of the growth-substance produced by *Chilomonas paramecium*. Physiol. Zool., 19:223.

McLaughlin, J. J. A., and Zahl, P. A., 1957. Studies in marine biology. II. In vitro culture of Zooxanthellae. Proc. Soc. Exper. Biol. & Med., 95:115.

Pace, D. M., 1941. The effects of sodium and potassium on metabolic processes in *Chilomonas paramecium*. J. Cell. Comp. Physiol., 18:243.

────── 1944. The relation between concentration of growth promoting substance and its effect on growth in *Chilomonas paramecium*. Physiol. Zool., 17:278.

────── 1947. The effects of vitamins and growth-promoting substance on growth in *Chilomonas paramecium*. Exper. Med. Surg. 5:140.

Pascher, A., 1913. Cryptomonadinae. Süsswasserflora Deutschlands. 2.

West, G. S., and Fritsch, F. E., 1927. A treatise on the British freshwater algae. Cambridge.

Chapter 10

Order 3. **Phytomonadida** Blochmann

THE phytomonads are small, more or less rounded, green flagellates, with a close resemblance to the algae. They show a definite body form, and most of them possess a cellulose membrane, which is thick in some and thin in others. There is a distinct opening in the membrane at the anterior end, through which one to two (or four or more) flagella protrude. The majority possess numerous grass-green chromatophores, each of which contains one or more pyrenoids. The method of nutrition is mostly holophytic or mixotrophic; some colorless forms are, however, saprozoic. The metabolic products are usually starch and oils. Some phytomonads are red in color, owing to the presence of haematochrome. The contractile vacuoles may be located in the anterior part or scattered throughout the body. The nucleus is ordinarily centrally located, and its division seems to be mitotic, chromosomes having been definitely noted in several species.

Asexual reproduction is by longitudinal fission, and the daughter individuals remain within the parent membrane for some time. Sexual reproduction seems to occur widely. Colony formation also occurs, especially in the family Volvocidae. Encystment and formation of the palmella stage are common among many forms. The phytomonads have a much wider distribution in fresh water than in salt water.

The phytomonads are subdivided into seven families: Chlamydomonadidae, Trichlorididae (p. 336), Carteriidae (p. 336), Chlorasteridae (p. 339), Polyblepharididae (p. 339), Phacotidae (p. 339), and Volvocidae (p. 340).

Family 1. **Chlamydomonadidae** Bütschli

Solitary; spheroid, oval, or ellipsoid; with a cellulose membrane; two flagella; chromatophores, stigma, and pyrenoids usually present. Cytology (Hollande, 1942).

Genus **Chlamydomonas** Ehrenberg. Spherical, ovoid or elongated; sometimes flattened; two flagella; membrane often thickened at anterior end; a large chromatophore, containing one or more pyrenoids; stigma; a single nucleus; two contractile vacuoles anterior; asexual reproduction and palmella formation; sexual reproduction isogamy or anisogamy; fresh water. Numerous species (Pascher, 1921, 1925, 1929, 1930, 1932; Skortzow, 1929; Pringsheim, 1930; Pascher and Jahoda, 1928; Gerloff, 1940; behavior and chemotaxis, Tsubo, 1961)

C. monadina Stein (Fig. 115, *a-c*). 15-30μ long; fresh water; Landacre noted that the organisms obstructed the sand filters used in connection with a septic tank, together with the diatom Navicula.

C. angulosa Dill. About 20μ by 12-15μ; fresh water.

C. epiphytica Smith (Fig. 115, *d*). 8-9μ by 7-8μ; in freshwater lakes.

C. globosa Snow (Fig. 115, *e*). Spheroid or ellipsoid; 5-7μ in diameter; in freshwater lakes.

C. gracilis S. (Fig. 115, *f*). 10-13μ by 5-7μ; fresh water.

Genus **Haematococcus** Agardh (*Sphaerella* Sommerfeldt). Spheroidal or ovoid with a gelatinous envelope; chromatophore peripheral and reticulate, with two to eight scattered pyrenoids; several contractile vacuoles; haematochrome frequently abundant in both motile and encysted stages; asexual reproduction in motile form; sexual reproduction isogamy; fresh water.

H. pluvialis (Flotow) (Figs. 44; 115, *g*). Spherical; with numerous radial cytoplasmic processes; chromatophore U-shape in optical section; body 8–50μ, stigma fusiform, lateral; fresh water. Reichenow (1909) noticed the disappearance of haematochrome if the culture medium was rich in nitrogen and phosphorus. In bacteria-free cultures, Elliott (1934) observed four types of cells: large and small flagellates, palmella stage and haematocysts. Large flagellates predominate in liquid cultures, but when conditions become unfavorable, palmella stage and then haematocysts develop. When the cysts are placed in a favorable environment after exposure to freezing, desiccation, etc., they give rise to small flagellates which grow into palmella stage or large flagellates. No syngamy of small flagellates was noticed. Haematochrome appears during certain phases in sunlight and its appearance is accelerated by sodium acetate under sunlight. (Sexuality, Schulze, 1927.)

Genus **Sphaerellopsis** Korschikoff (*Chlamydococcus* Stein). With

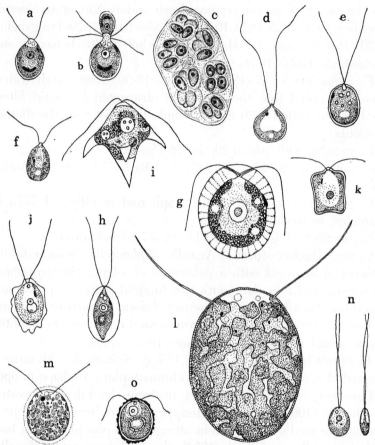

Fig. 115. a–c, *Chlamydomonas monadina,* ×470 (Goroschankin) (a, typical organism; b, anisogamy; c, palmella stage); d, *C. epiphytica,* ×1030 (Smith); e, *C. globosa,* ×2000 (Snow); f, *C. gracilis,* ×770 (Snow); g, *Haematococcus pluvialis,* ×500 (Reichenow); h, *Sphaerellopsis fluviatilis,* ×490 (Korschikoff); i, *Brachiomonas westiana,* ×960 (West); j, *Lobomonas rostrata,* ×1335 (Hazen); k, *Diplostauron pentagonium,* ×1110 (Hazen); l, *Gigantochloris permaxima,* ×370 (Pascher); m, *Gloeomonas ovalis,* ×330 (Pascher); n, *Scourfieldia complanata,* ×1540 (West); o, *Thorakomonas sabulosa,* ×670 (Korschikoff).

gelatinous envelope which is usually ellipsoid with rounded ends; body elongate fusiform or pyriform, no protoplasmic processes to envelope; two equally long flagella; chromatophore large; a pyrenoid; with or without stigma; nucleus in anterior half; two contractile vacuoles; fresh water.

S. *fluviatilis* Stein (Fig. 115, *h*). 14-30μ by 10-20μ; fresh water.

Genus **Brachiomonas** Bohlin. Lobate; with horn-like processes, all directed posteriorly; contractile vacuoles; ill-defined chromatophore; pyrenoids; with or without stigma; sexual and asexual reproduction; fresh, brackish or salt water.

B. *westiana* Pascher (Fig. 115, *i*). 15-24μ by 13-23μ; brackish water.

Genus **Lobomonas** Dangeard. Ovoid or irregularly angular; chromatophore cup-shaped; pyrenoid; stigma; a contractile vacuole; fresh water.

L. *rostrata* Hazen (Fig. 115, *j*). 5-12μ by 4-8μ

Genus **Diplostauron** Korschikoff. Rectangular with raised corners; two equally long flagella; chromatophore; one pyrenoid; stigma; two contractile vacuoles anterior; fresh water.

D. *pentagonium* Hazen (Fig. 115, *k*). 10-13μ by 9-10μ.

Genus **Gigantochloris** Pascher. Unusually large form, equalling in size a colony of *Eudorina;* flattened; oval in front view; elongate ellipsoid in profile; membrane radially striated; two flagella widely apart, less than body length; chromatophore in network; numerous pyrenoids; often without stigma; in woodland pools.

G. *permaxima* P. (Fig. 115, *l*). 70-150μ by 40-80μ by 25-50μ.

Genus **Gloeomonas** Klebs. Broadly ovoid, nearly subspherical; with a delicate membrane and a thin gelatinous envelope; two flagella widely apart; chromatophores numerous, circular or oval discs; pyrenoids (?); stigma; two contractile vacuoles anterior; fresh water.

G. *ovalis* K. (Fig. 115, *m*). 38-42μ by 23-33μ; gelatinous envelope over 2μ thick.

Genus **Scourfieldia** West. Whole body flattened; ovoid in front view; membrane delicate; two flagella two to five times body length; a chromatophore; without pyrenoid or stigma; contractile vacuole anterior; nucleus central; fresh water.

S. *complanata* W. (Fig. 115, *n*). 5.2-5.7μ by 4.4-4.6μ; fresh water.

Genus **Thorakomonas** Korschikoff. Flattened; somewhat irregularly shaped or ellipsoid in front view; membrane thick, enclustered with iron-bearing material, deep brown to black in color; protoplasmic body similar to that of *Chlamydomonas;* a chromatophore with a pyrenoid; two contractile vacuoles; standing fresh water.

T. sabulosa K. (Fig. 115, *o*). Up to 16µ by 14µ.

Genus **Coccomonas** Stein. Shell smooth; globular; body not filling intracapsular space; stigma; contractile vacuole; asexual reproduction into four individuals; fresh water. Species (Conrad 1930).

C. orbicularis S. (Fig. 116, *a*). 18-25µ in diameter; fresh water.

Genus **Chlorogonium** Ehrenberg. Fusiform; membrane thin and adheres closely to protoplasmic body; plate-like chromatophores usually present, sometimes ill-contoured; one or more pyrenoids; numerous scattered contractile vacuoles; usually a stigma; a central nucleus; asexual reproduction by two successive transverse fissions during the motile phase; isogamy reported; fresh water. (Sexuality, Schulze, 1927; nutrition, Loefer, 1935; electron microscope study, Haller and Rouiller, 1961.)

C. euchlorum E. (Fig. 116, *b*). 25-70µ by 4-15µ; in stagnant water.

Genus **Phyllomonas** Korschikoff. Extremely flattened; membrane delicate; two flagella; chromatophore often faded or indistinct; numerous pyrenoids; with or without stigma; many contractile vacuoles; fresh water.

P. phacoides K. (Fig. 116, *c*). Leaf-like; rotation movement; up to 100µ long; in standing fresh water.

Genus **Sphaenochloris** Pascher. Body truncate or concave at flagellate end in front view; sharply pointed in profile; two flagella widely apart; chromatophore large; pyrenoid; stigma; contractile vacuole anterior; fresh water.

S. printzi P. (Fig. 116, *d*). Up to 18µ by 9µ.

Genus **Korschikoffia** Pascher. Elongate pyriform with an undulating outline; anterior end narrow, posterior end more bluntly rounded; plastic; chromatophores in posterior half; stigma absent; contractile vacuole anterior; two equally long flagella; nucleus nearly central; salt water.

K. guttula P. (Fig. 116, *e*). 6-10µ by 5µ; brackish water.

Genus **Furcilla** Stokes. U-shape, with two posterior processes; in side view somewhat flattened; anterior end with a papilla; two flagella equally long; one to two contractile vacuoles anterior; oil droplets; fresh water.

F. lobosa S. (Fig. 116, *f*). 11-14µ long; fresh water.

Genus **Hyalogonium** Pascher. Elongate spindle-form; anterior end bluntly rounded; posterior end more pointed; two flagella; protoplasm colorless; with starch granules; a stigma; asexual reproduction results in up to eight daughter cells; fresh water.

H. klebsi P. (Fig. 116, *g*). 30-80µ by up to 10µ; stagnant water.

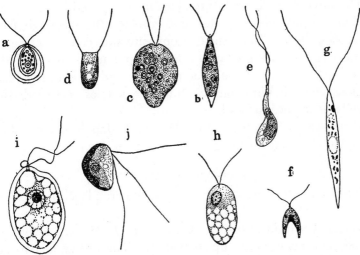

FIG. 116. a, *Coccomonas orbicularis*, ×500 (Stein); b, *Chlorogonium euchlorum*, ×430 (Jacobsen); c, *Phyllomonas phacoides*, ×200 (Korschikoff); d, *Sphaenochloris printzi*, ×600 (Printz); e, *Korschikoffia guttula*, ×1670 (Pascher); f, *Furcilla lobosa*, ×670 (Stokes); g, *Hyalogonium klebsi*, ×470 (Klebs); h, *Polytoma uvella*, ×670 (Dangeard); i, *Parapolytoma satura*, ×1600 (Jameson); j, *Trichloris paradoxa*, ×990 (Pascher).

Genus **Polytoma** Ehrenberg (*Chlamydoblepharis* Francé; *Tussetia* Pascher). Ovoid; no chromatophores; membrane yellowish to brown; pyrenoid unobserved; two contractile vacuoles; two flagella about body length; stigma if present, red or pale-colored; many starch bodies and oil droplets in posterior half of body; asexual reproduction in motile stage; isogamy (Dogiel, 1935); saprozoic; in stagnant fresh water.

P. uvella E. (Figs. 8, *c*; 116, *h*). Oval to pyriform; stigma may be absent; 15-30µ by 9-20µ (Cytology, Entz, 1918; Hollande, 1942; enzyme synthesis, Cirillo, 1956; adaptation, Cirillo, 1957).

Genus **Parapolytoma** Jameson. Anterior margin obliquely truncate, resembling a cryptomonad, but without chromatophores; without stigma and starch; division into four individuals within envelope; fresh water.

P. satura J. (Fig. 116, *i*). About 15µ by 10µ; fresh water.

Family 2. **Trichlorididae**

The phytomonads with three flagella are placed here.

Genus **Trichloris** Scherffel and Pascher. Bean-shaped; flagellated side flattened or concave; opposite side convex; chromatophore large, covering convex side; two pyrenoids surrounded by starch granules; a stigma near posterior end of chromatophore; nucleus central; numerous contractile vacuoles scattered; three flagella near anterior end.

T. paradoxa S. and P. (Fig. 116, *j*). 12-15µ broad by 10-12µ high; flagella up to 30µ long.

Family 3. **Carteriidae**

Solitary phytomonads with four flagella are placed in this family.

Genus **Carteria** Diesing (*Corbierea, Pithiscus,* Dangeard). Ovoid, chromatophore cup-shaped; pyrenoid; stigma; two contractile vacuoles; fresh water. Numerous species (Pascher, 1925, 1932; Schiller, 1925.)

C. cordiformis Carter (Fig. 117, *a*). Heart-shaped in front view; ovoid in profile; chromatophore large; 18-23µ by 16-20µ.

C. ellipsoidalis Bold. Ellipsoid; chromatophore; a small stigma; division into two, four, or eight individuals in encysted stage; 6-24µ long; fresh water, Maryland (Bold, 1938).

Genus **Pyramimonas** Schmarda (*Pyramidomonas,* Stein). Small pyramidal or heart-shaped body; with bluntly drawn-out posterior end; usually four ridges in anterior region; four flagella; green chromatophore cup-shaped; with or without stigma; a large pyre-

noid in the posterior part; two contractile vacuoles in the anterior
portion; encystment; fresh water. Several species (Geitler, 1925)

P. *tetrarhynchus* S. (Fig. 117, *b*). 20-28μ by 12-18μ; fresh water;
Wisconsin (Smith, 1933).

P. *montana* Geitler. Bluntly conical; anterior end fourlobed or
truncate; posterior end narrowly rounded; plastic; pyriform nucleus anterior, closely associated with four flagella; stigma; two
contractile vacuoles anterior; chromatophore cup-shaped, granular, with scattered starch grains and oil droplets; a pyrenoid with
a ring of small starch grains; 17-22.5μ long (Geitler, 1925); 12-20μ

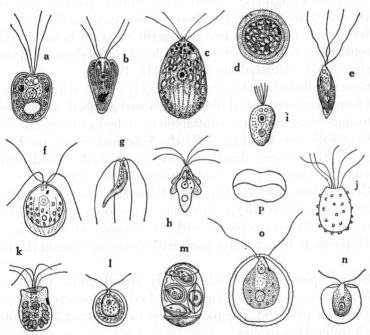

Fig. 117. a, *Carteria cordiformis*, ×600 (Dill); b, *Pyramimonas tetrarhynchus*, ×400 (Dill); c, d, *Polytomella agilis*, ×1000 (Doflein) (d, a
cyst); e, *Spirogonium chlorogonioides*, ×670 (Pascher); f, *Tetrablepharis
multifilis*, ×670 (Pascher); g, *Spermatozopsis exultans*, ×1630 (Pascher);
h, *Chloraster gyrans*, ×670 (Stein); i, *Polyblepharides singularis*, ×870
(Dangeard); j, k, *Pocillomonas flos aquae*, ×920 (Steinecke); l, m, *Phacotus lenticularis*, ×430 (Stein); n. *Pteromonas angulosa*, ×670 (West); o, p,
Dysmorphococcus variabilis, ×1000 (Bold).

by 8-16μ (Bold); flagella about body length; fresh water, Maryland (Bold, 1938).

Genus **Polytomella** Aragão. Ovoid to ellipsoid, with a delicate membrane; a small anterior papilla, where four equally long flagella arise; often "vacuole" in the posterior half of body; starch grains; nucleus small; two contractile vacuoles anterior; with or without a stigma; cysts contain starch, but when dried, is replaced by oil droplets; fresh water. (Axenic culture, Pringsheim, 1955.)

P. agilis A. (Fig. 117, *c*, *d*). Numerous starch grains; 8-18μ by 5-9μ; flagella 12-17μ long; fresh water; hay infusion (Aragão, 1910; Doflein, 1916).

P. caeca Pringsheim. Ovoid with bluntly pointed posterior end; 17-20μ by 10-12μ; membrane delicate; a small papilla at anterior end; no stigma; two contractile vacuoles below papilla; cytoplasm ordinarily filled with starch grains; fresh water (Pringsheim, 1937). (Metabolism, Wise, 1955, 1959.)

Genus **Collodictyon** Carter. Body highly plastic; with longitudinal furrows; posterior end bluntly narrowed or lobed; no apparent cytostome; four flagella; a contractile vacuole; fresh water.

C. triciliatum C. (Fig. 157, *d*). Spherical ovoid or heart-shaped; 27-60μ long; flagella as long as the body; pond water. (Cytology, Rhodes, 1919; food ingestion, Bělař, 1921.)

Genus **Medusochloris** Pascher. Hollowed hemisphere with 4 processes, each bearing a flagellum at its lower edge; a lobed plate-shaped chromatophore; without pyrenoid. One species.

M. phiale P. In salt water pools with decaying algae in the Baltic.

Genus **Spirogonium** Pascher. Body spindle-form; membrane delicate; flagella a little longer than body; chromatophore conspicuous; a pyrenoid; stigma anterior; two contractile vacuoles; fresh water. One species.

S. chlorogonioides P. (Fig. 117, *e*). Body up to 25μ by 15μ.

Genus **Tetrablepharis** Senn. Ellipsoid to ovoid; pyrenoid present; fresh water.

T. multifilis Klebs (Fig. 117, *f*). 12-20μ by 8-15μ; stagnant water.

Genus **Spermatozopsis** Korschikoff. Sickle-form; bent easily, occasionally plastic; chromotophore mostly on convex side; a distinct stigma at more rounded anterior end; flagella equally long; two contractile vacuoles anterior; fresh water infusion.

S. exultans K. (Fig. 117, *g*). 7-9μ long; also biflagellate; in fresh water with algae, leaves, etc.

Family 4. Chlorasteridae

Genus **Chloraster** Ehrenberg. Similar to *Pyramimonas,* but anterior half with a conical envelope drawn out at four corners; with five flagella; fresh or salt water.

C. gyrans E. (Fig. 117, *h*). Up to 18μ long; standing water; also reported from salt water.

Family 5. Polyblepharididae Dangeard

Genus **Polyblepharides** Dangeard. Ellipsoid or ovoid; flagella six to eight, shorter than body length; chromatophore; a pyrenoid; a central nucleus; two contractile vacuoles anterior; cysts; a questionable genus; fresh water.

P. singularis D. (Fig. 117, *i*). 10-14μ by 8-9μ.

Genus **Pocillomonas** Steinecke. Ovoid with broadly concave anterior end; covered with gelatinous substance with numerous small projections; six flagella; chromatophores disc-shaped; two contractile vacuoles anterior; nucleus central; starch bodies; without pyrenoid.

P. flos aquae S. (Fig. 117, *j*, *k*). 13μ by 10μ; fresh water pools.

Family 6. Phacotidae Poche

The shell typically composed of two valves; two flagella protrude from anterior end; with stigma and chromatophores; asexual reproduction within the shell; valves may become separated from each other owing to an increase in gelatinous contents.

Genus **Phacotus** Perty. Oval to circular in front view; lenticular in profile; protoplasmic body does not fill dark-colored shell completely; flagella protrude through a foramen; asexual reproduction into two to eight individuals; fresh water.

P. lenticularis (Ehrenberg) (Fig. 117, *l*, *m*). 13-20μ in diameter; in stagnant water.

Genus **Pteromonas** Seligo. Body broadly winged in plane of suture of two valves; protoplasmic body fills shell; chromatophore cup-shaped; one or more pyrenoids; stigma; two contractile vacuoles; asexual reproduction into two to four individuals; sexual reproduction by isogamy; zygotes usually brown; fresh water. Several species.

P. angulosa (Lemmermann) (Fig. 117, *n*). With a rounded wing and four protoplasmic projections in profile; 13-17µ by 9-20µ; fresh water.

Genus **Dysmorphococcus** Takeda. Circular in front view; anterior region narrowed; posterior end broad; shell distinctly flattened posteriorly, ornamented by numerous pores; sutural ridge without pores; two flagella; two contractile vacuoles; stigma, pyrenoid, cup-shaped chromatophore; nucleus; multiplication by binary fission; fresh water.

D. variabilis T. (Fig. 117, *o*, *p*). Shell 14-19µ by 13-17µ; older shells dark brown; fresh water; Maryland (Bold, 1938).

Family 7. **Volvocidae** Ehrenberg

An interesting group of colonial flagellates; individual similar to Chlamydomonadidae, with two equally long flagella (one in *Mastigosphaera;* four in *Spondylomorum*), green chromatophores, pyrenoids, stigma, and contractile vacuoles; body covered by a cellulose membrane and not plastic; colony or coenobium is discoid or spherical; exclusively fresh water inhabitants.

Genus **Volvox** Linnaeus. Often large spherical or subspherical colonies, consisting of a large number of cells which are differentiated into somatic and reproductive cells; somatic cells numerous, embedded in gelatinous matrix, and contains a chromatophore, one or more pyrenoids, a stigma, two flagella and several contractile vacuoles; in some species cytoplasmic connection occurs between adjacent cells; generative cells few and large. Reproduction is by parthenogenesis or true sexual fusion. In parthenogenetic colonies, the gametes are larger in size and fewer in number as compared with the macrogametes of the female colonies. Sexual fusion is anisogamy (Fig. 79) and sexual colonies may be monoecious or dioecious. Zygotes are usually yellowish to brownish red in color and covered by a smooth, ridged or spinous wall. Fresh water. Many species. Smith (1944) made a comprehensive study of eighteen species on which the following species descriptions are based.

V. globator L. (Fig. 118, *a*, *b*). Monoecious. Sexual colonies 350-500μ in diameter; 5000-15,000 cells, with cytoplasmic connections; three to seven microgametocytes, each of which develops into over 250 microgametes; ten to forty macrogametes; zygotes 35-45μ in diameter, covered with many spines with rounded tip. Parthenogenetic colonies 400-600μ in diameter; four to ten gametes, 10-13μ in diameter; young colonies up to 250μ. Europe and North America.

V. aureus Ehrenberg (Figs. 79; 118 *c-e*). Dioecious. Male colonies 300-350μ in diameter; 1000-1500 cells, with cytoplasmic connections; numerous microgametocytes; clusters of some thirty-two microgametes, 15-18μ in diameter. Female colonies 300-400μ; 2000-3000 cells; ten to fourteen macrogametes; zygotes 40-60μ with smooth surface. Parthenogenetic colonies up to 500μ; four to twelve gametes; young colonies 150μ in diameter. Europe and North America. (Sexual differentiation, Mainx, 1929.)

V. terius Meyer. Dioecious. Male colonies up to 170μ in diameter; 180-500 cells, without cytoplasmic connections; about fifty microgametocytes. Female colonies up to 500μ; 500-2000 cells; two to twelve macrogametes; zygotes 60-65μ with smooth wall. Parthenogenetic colonies up to 600μ in diameter; 500-2000 cells; two to twelve gametes. Europe and North America.

V. spermatosphaera Powers (Fig. 118, *f-h*). Dioecious. Male colonies up to 100μ in diameter; cells, without connection, up to 128 microgametocytes. Female colonies up to 500μ in diameter; six to sixteen macrogametes; zygotes 35-45μ, with smooth membrane. Parthenogenetic colonies up to 650μ in diameter; eight to ten gametes; young colonies ellipsoid, up to 100μ in diameter. North America (Powers, 1908).

V. weismannia P. (Fig. 118, *i*). Male colonies 100-150μ in diameter; 250-500 cells; 6-50 microgametocytes; clusters of microgametes (up to 128) discoid, 12-15μ in diameter. Female colonies up to 400μ; 2000-3000 cells; eight to twenty-four macrogametes; zygotes 30-50μ in diameter, with reticulate ridges on shell. Parthenogenetic colonies up to 400μ; 1500-3000 cells;

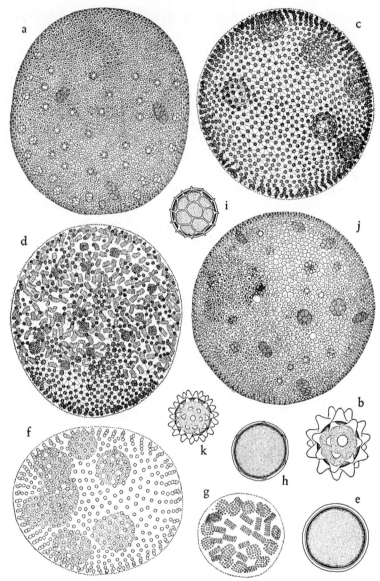

FIG. 118. Species of Volvox (Smith). a, b, *Volvox globator* (a, a female colony, ×150; b, a zygote, ×370); c–e, *V. aureus* (c, a young partheno-genetic colony; d, a mature male colony, ×125; e, a zygote, ×370); f–h, *V. spermatosphaera*: f, a parthenogenetic colony, ×185; g, a mature male colony, ×370; h, a zygote, ×370); i, a zygote of *V. weismannia*, ×370; j, k, *V. perglobator* (j, a male colony, ×150; k, a zygote, ×370).

eight or ten gametes; 40-60μ in diameter; young colonies 100-200μ in diameter. North America (Powers, 1908).

V. *perglobator* P. (Fig. 118, *j*, *k*). Dioecious. Male colonies 300-450μ in diameter 5000-10,000 cells, with delicate cytoplasmic connections; sixty to eighty microgametocytes. Female colonies 300-550μ in diameter; 9000-13,000 cells; 50-120 macrogametes; zygotes 30–34μ, covered with bluntly pointed spines. Parthenogenetic colonies as large as 1.1 mm; three to nine gametes; young colonies 250-275μ in diameter. North America.

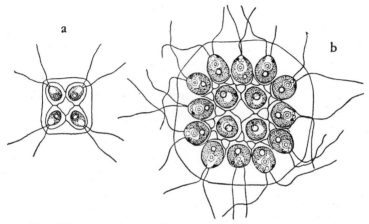

FIG. 119. a, *Gonium sociale*, ×270 (Chodat); b, *G. pectorale*, ×670 (Hartmann).

Genus **Gonium** Müller. Four or sixteen individuals arranged in one plane; cell ovoid or slightly polygonal; with two flagella arranged in the plane of coenobium; with or without a gelatinous envelope; protoplasmic connections among individuals occur occasionally; asexual reproduction through simultaneous divisions of component cells; sexual reproduction isogamy; zygotes reddish; fresh water. (Colony formation, Hartmann, 1924.)

G. *sociale* Dujardin (Fig. 119, *a*). Four individuals from a discoid colony; cells 10-22μ by 6-16μ wide; in open waters of ponds and lakes.

G. *pectorale* M. (Fig. 119, *b*). Sixteen (rarely four or eight) individuals form a colony; four cells in center; twelve peripheral, closely arranged; cells 5-14μ by 10μ; colony up to 90μ in diameter; fresh water. (Morphology, genetics, Stein, 1958.)

G. formosum Pascher. Sixteen cells in a colony further apart; peripheral gelatinous envelope reduced; cells similar in size to those of *G. sociale* but colony somewhat larger; fresh water lakes.

Genus **Stephanoon** Schewiakoff. Spherical or ellipsoidal colony, surrounded by gelatinous envelope, and composed of eight or sixteen biflagellate cells, arranged in two alternating rows on equatorial plane; fresh water.

S. askenasii S. (Fig. 120, *a*). Sixteen individuals in ellipsoidal colony; cells 9μ in diameter; flagella up to 30μ long; colony 78μ by 60μ.

Genus **Platydorina** Kofoid. Thirty-two cells arranged in a slightly twisted plane; flagella directed alternately to both sides; dioecious; fresh water.

P. caudata K. (Fig. 120, *b*). Individual cells 10-15μ long; colony up to 165μ long by 145μ wide, and 25μ thick; dioecious; anisogamy; macrogametes escape from female colonies and remain attached to them or swim about until fertilized by microgametes; zygotes become thickly walled (Taft, 1940).

Genus **Spondylomorum** Ehrenberg. Sixteen cells in a compact group in four transverse rings; each with four flagella; asexual reproduction by simultaneous division of component cells; fresh water. One species.

S. quaternarium E. (Fig. 120, *c*). Cells 12-26μ by 8-15μ; colony up to 60μ long.

Genus **Chlamydobotrys** Korschikoff. Colony composed of eight or sixteen individuals; cells with two flagella; chromatophore; stigma; no pyrenoid; fresh water. (Species, Pascher, 1925; culture, Schulze, 1927.)

C. stellata K. (Fig. 120, *d*). Colony composed of eight individuals arranged in two rings; individuals 14-15μ long; colony 30-40μ in diameter; Maryland (Bold, 1933).

Genus **Stephanosphaera** Cohn. Spherical or subspherical colony, with eight (rarely four or sixteen) cells arranged in a ring; cells pyriform, but with several processes; two flagella on one face; asexual reproduction and isogamy (p. 220); fresh water.

S. pluvialis C. (Figs. 82; 120, *e*). Cells 7-13μ long; colony 30-60μ in diameter. (Culture and sexuality, Schulze, 1927).

Genus **Pandorina** Bory. Spherical or subspherical colony of usually sixteen (sometimes eight or thirty-two) biflagellate individuals, closely packed within a gelatinous, but firm and thick matrix; individuals often angular; with stigma and chromatophores; asexual reproduction through simultaneous division of component individuals; anisogamy; zygotes colored and covered by a smooth wall; fresh water. One species.

P. morum Müller (Fig. 120, *f*). Cells 8-17μ long; colony 20-40μ, up to 250μ, in diameter; cell with a single cup-shaped chromatophore; one or more pyrenoids; two contractile vacuoles

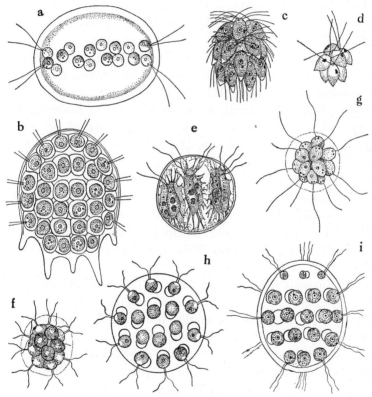

FIG. 120. a, *Stephanoon askenasii,* ×440 (Schewiakoff); b, *Platydorina caudata,* ×280 (Kofoid); c, *Spondylomorum quaternarium,* ×330 (Stein); d, *Chlamydobotrys stellata,* ×430 (Korschikoff); e, *Stephanosphaera pluvialis,* ×250 (Hieronymus); f, *Pandorina morum,* ×300 (Smith); g, *Mastigosphaera gobii,* ×520 (Schewiakoff); h, *Eudorina elegans,* ×310 (Goebel); i, *Pleodorina illinoisensis,* ×200 (Kofoid).

and a stigma; fresh water. Coleman (1959) reports fifteen separate pairs of mating types.

Genus **Mastigosphaera** Schewiakoff. Similar to *Pandorina;* but individuals with a single flagellum which is 3.5 times the body length; fresh water.

M. gobii S. (Fig. 120, *g*). Individual 9μ long; colony 30-33μ.

Genus **Eudorina** Ehrenberg. Spherical or ellipsoidal colony of usually thirty-two or sometimes sixteen spherical cells; asexual reproduction similar to that of *Pandorina;* sexual reproduction with thirty-two to sixty-four spherical green macrogametes and numerous clustered microgametes which when mature, unite with the macrogametes within the colony; reddish zygotes with a smooth wall; fresh water. (Colony formation, Hartmann, 1924.)

E. elegans E. (Fig. 120, *h*). Cells 10-24μ in diameter; colony 40-150μ in diameter; in ponds, ditches and lakes. (Culture and morphology, Hartmann, 1921; response to light, Luntz, 1935).

Genus **Pleodorina** Shaw. Somewhat similar to *Eudorina*, being composed of thirty-two, sixty-four, or 128 ovoid or spherical cells of two types: small somatic and large generative, located within a gelatinous matrix; sexual reproduction similar to that of *Eudorina;* fresh water.

P. californica S. Spherical colony with sixty-four or 128 cells, of which one-half to two-thirds are reproductive cells; vegetative cells 13-15μ; reproductive cells up to 27μ; colony up to 450μ, both in diameter. (Variation, Tiffany, 1935; in Ukraine, Swirenko, 1926.)

P. illinoisensis Kofoid (Figs. 34, *b, c;* 120, *i*). Thirty-two cells in ellipsoid colony, four vegetative and twenty-eight reproductive individuals; arranged in five circles, four in each polar circle, eight at equator and eight on either side of equator; four small vegetative cells at anterior pole; vegetative cells 10-16μ in diameter; reproductive cells 19-25μ in diameter; colony up to 160μ by 130μ.

References

Aragão, H. B., 1910. Untersuchungen über *Polytomella agilis* n.g., n. sp. Mem. Inst. Oswaldo Cruz, 2:42.

Bělař, K., 1921. Protozoenstudien. III. Arch. Protist., 43:431.

Bold, H. C., 1938. Notes on Maryland algae. Bull. Torrey Bot. Club., 65: 293.

CIRILLO, V. P., 1956. Induced enzyme synthesis in the phytoflagellate, Polytoma. J. Protozool., 3:69.

—— 1957. Long-term adaptation to fatty acids by phytoflagellate, *Polytoma uvella. Ibid.*, 4:60.

CONRAD, W., 1930. Flagellates nouveaux ou peu connus. I. Arch. Protist., 70:657.

CROW, W. B., 1918. The classification of some colonial Chlamydomonads. New Phytol., 17:151.

DANGEARD, P., 1900. Observations sur la structure et le développement du *Pandorina morum.* Le Botaniste, 7:192.

DOFLEIN, F., 1916. Polytomella agilis. Zool. Anz., 47:273.

DOGIEL, V., 1935. Le mode de conjugaison de *Polytoma uvella.* Arch. zool. exper. gén., 77 (N. et R.):1:1.

ELLIOTT, A. M., 1934. Morphology and life history of *Haematococcus pluvialis.* Arch. Protist., 82:250.

ENTZ, G., JR., 1913. Cytologische Beobachtungen an *Polytoma uvella.* Verh. deutsch. zool. Ges. Ver. Berlin, 23:249.

—— 1918. Ueber die mitotische Teilung von *Polytoma uvella.* Arch. Protist., 38:324.

FRITSCH, F. E., 1935. The structure and reproduction of the algae.

GEITLER, L., 1925. Zur Kenntnis der Gattung Pyramidomonas. Arch. Protist., 52:356.

GERLOFF, J., 1940. Beiträge zur Kenntnis der Variabilität und Systematik der Gattung Chlamydomonas. *Ibid.*, 94:311.

HARPER, R. A., 1912. The structure and development of the colony in Gonium. Tr. Am. Micr. Sic., 31:65.

HARTMANN, M., 1921. Untersuchungen über die Morphologie und Physiologie des Formwechsels der Phytomonadien. III. Arch. Protist., 43:223.

—— 1924. Ueber die Veränderung der Koloniebildung von *Eudorina elegans* und *Gonium pectorale* unter dem Einfluss äusserer Bedingungen. IV. *Ibid.*, 49:375.

HOLLANDE, A., 1942. Étude cytologique et biologique de queleques flagellés libres. Arch. zool. exper. gén., 83:1.

JANET, C., 1912, 1922, 1923. Le Volvox. I, II and III Memoires. Paris.

KOFOID, C. A., 1900. Plankton studies. II, III. Ann. Mag. Nat. Hist., Ser. 7, 6:139.

LOEFER, J. B., 1935. Effect of certain carbohydrates and organic acids on growth of Chlorogonium and Chilomonas. Arch. Protist., 84:456.

—— 1935a. Effect of certain nitrogen compounds on growth of Chlorogonium and Chilomonas. *Ibild.*, 85:74.

LUNTZ, A., 1935. Ueber die Regulation der Reizbeantwortung bei koloniebildenden grünen Einzelligen. *Ibid.*, 86:90.

MAINX, F., 1929. Ueber die Geschlechterverteilung bei *Volvox aureus. Ibid.*, 67:205.

MAST, S. O., 1928. Structure and function of the eye-spot in unicellular and colonial organisms. *Ibid.*, 60:197.

PASCHER, A., 1921. Neue oder wenig bekannte Protisten. Arch. Protist., 44:119.

―――― 1925. Neue oder wenig bekannte Protisten. XVII. *Ibid.*, 51:549.

―――― 1925a. XVIII. *Ibid.*, 52:566.

―――― 1927. Volvocales—Phytomonadinae. Die Süsswasserflora. Pt. 4.

―――― 1929. Neue oder wenig bekannte Protisten. Arch. Protist., 65:426.

―――― 1930. Neue Volvocalen. *Ibid.*, 69:103.

―――― 1932. Zur Kenntnis der einzelligen Volvocalen. *Ibid.*, 76:1.

―――― and JAHODA, R., 1928. Neue Polyblepharidinen und Chlamydomonadinen aus den Almtümpeln um Lunz. *Ibid.*, 61:239.

PAVILLARD, J., 1952. Classe de Phytomonadines ou Volvocales. In Grassé (1952), p. 154.

POWERS, J. H., 1907. New forms of Volvox. Tr. Am. Micr. Soc., 27:123.

―――― 1908. Further studies in Volvox with descriptions of three new species. *Ibid.*, 28:141.

PRINGSHEIM, E. G., 1930. Neue Chlamydomonadaceen, etc. Arch. Protist., 69:95.

―――― 1937. Zur Kenntnis saprotropher Algen und Flagellaten. II. *Ibid.*, 88:151.

―――― 1955. The genus Polytomella. J. Protozool., 2:137.

REICHENOW, E., 1909. Untersuchungen an *Haematococcus pulvialis* nebst Bemerkungen über andere Flagellaten. Arb. kaiserl. Gesundh., 33:1.

RHODES, R. C., 1919. Binary fission in *Collodictyon triciliatum* Carter. Univ. Cal. Publ. Zool., 19:201.

SCHILLER, J., 1925. Die planktonischen Vegetationen des adriatischen Meeres. B. Arch. Protist., 53:59.

SCHULZE, B., 1927. Zur Kenntnis einiger Volvocales. *Ibid.*, 58:508.

SHAW, W. R., 1894. Pleodorina, a new genus of the Volvocideae. Bot. Gaz., 19:279.

SKVORTZOW, B. W., 1929. Einige neue und wenig begannte Chlamydomonadaceae aus Manchuria. Arch. Protist., 66:160.

SMITH, G. M., 1944. A comparative study of the species of Volvox. Tr. Am. Micr. Soc., 63:265.

―――― 1950. The freshwater algae of the United States. New York.

STEIN, J. R., 1958. A morphologic and genetic study of *Gonium pectorale*. Amer. J. Botany, 45:665.

SWIRENKO, 1926. Ueber einige neue und interessante Volvocineae, etc. Arch. Protist., 55:191.

TAFT, C. E., 1940. Asexual and sexual reproduction in *Platydorina caudata*. Tr. Am. Micr. Soc., 59:1.

TIFFANY, L. H., 1935. Homothallism and other variations in *Pleodorina californica*. Arch. Protist., 85:140.

Tsubo, Y., 1961. Chemotaxis and sexual behavior in Chlamydomonas. J. Protozool., 8:114.

West, G. S., and Fritsch, F. E., 1927. A treatise on the British freshwater algae. Cambridge.

Wise, D. L., 1955. Carbon sources for *Polytomella caeca*, J. Protozool., 2: 156.

———— 1959. Carbon nutrition and metabolism of *Polytomella caeca*. *Ibid.*, 6:19.

Chapter 11
Order 4. **Euglenoidida** Blochmann

THE Body is, as a rule, elongated; some are plastic, others have a definite body form with a well-developed, striated or variously sculptured pellicle. At the anterior end, there is an opening through which a flagellum protrudes. In holophytic forms the so-called cytostome and cytopharynx, if present, are apparently not concerned with the food-taking, but seem to give a passage-way for the flagellum and also to excrete the waste fluid matters which become collected in one or more contractile vacuoles located near the reservoir. In holozoic forms, a well-developed cytostome and cytopharynx are present. Ordinarily there is only one flagellum, but some possess two or three. Chromatophores are present in the majority of the Euglenidae, but absent in two families. They are green, vary in shape, such as spheroidal, band-form, cup-form, discoidal, or fusiform, and usually possess pyrenoids. Some forms may contain haematochrome. A small but conspicuous stigma is invariably present near the anterior end of the body in chromatophore-bearing forms.

Reserve food material is the paramylum body, fat, and oil, the presence of which depends naturally on the metabolic condition of the organism. The paramylum body assumes diverse forms in different species, but is, as a rule, constant in each species, and this facilitates specific identification to a certain extent. Nutrition is holophytic in chromatophore-possessing forms, which, however, may be saprozoic, depending on the amount of light and organic substances present in the water. The holozoic forms feed upon bacteria, algae, and smaller Protozoa.

The nucleus is, as a rule, large and distinct and contains almost always a large endosome. Asexual reproduction is by longitudinal fission; sexual reproduction has been reported in a few species. Encystment is common. The majority inhabit fresh water, but some live in brackish or salt water, and a few are parasitic. (Taxonomy, Mainx, 1928; Hollande, 1942, 1952a; Jahn, 1946; Pringsheim, 1950.)

Family 1. **Euglenidae** Stein

Body plastic ("euglenoid"), but, as a rule, more or less spindle-form during locomotion. The flagellum arises from a blepharo-plast located in the cytoplasm at the posterior margin of the reservoir. Between the blepharoplast and the "cytostome," the flagellum shows a swelling which appears to be photosensitive (Mast, 1938). Many observers consider that the basal portion of the flagellum is bifurcated and ends in two blepharoplasts, but Hollande (1942), Pringsheim (1948) and others, hold that in addition to a long flagellum arising from a blepharoplast, there is present a short flagellum which does not extend beyond the neck of the reservoir and often adheres to the long flagellum, producing the appearance of bifurcation. Reproduction asexual. (Culture and physiology, Mainx, 1928; cytology, Günther, 1928; Hollande, 1942.)

Genus **Euglena** Ehrenberg. Short or elongated spindle, cylindrical, or band-form; pellicle usually marked by longitudinal or spiral striae; some with a thin pellicle highly plastic; others regularly spirally twisted; stigma usually anterior; chromatophores numerous and discoid, band-form, or fusiform; pyrenoids may or may not be surrounded by starch envelope; paramylum bodies which may be two in number, one being located on either side of nucleus, and rod-like to ovoid in shape or numerous and scattered throughout; contractile vacuole small, near reservoir; asexual reproduction by longitudinal fission; sexual reproduction reported in *Euglena sanguinea;* common in stagnant water, especially where algae occur; when present in large numbers, the active organisms may form a green film on the surface of water and resting or encysted stages may produce conspicuous green spots on the bottom of pond or pool; in fresh water. Numerous species. (Pascher, 1925; Johnson, 1944; Conrad and Van Meel, 1952; Gojdics, 1953; Perez-Reyes and Gómez, 1958.)

E. pisciformis Klebs (Fig. 121, *a*). 20-35μ by 5-10μ; spindle-form with bluntly pointed anterior and sharply attenuated posterior end; slightly plastic; a body-length flagellum, active; a few chromatophores; division into two or four individuals in encysted stage (Johnson, 1944).

E. viridis Ehrenberg (Fig. 121, *b*). 40-65μ by 14-20μ; anterior end rounded, posterior end pointed; fusiform during locomotion; highly plastic when stationary; flagellum as long as the body; pellicle obliquely striated; chromatophores more or less bandform, radially arranged; nucleus posterior; nutrition holophytic, but also saprozoic. Multiplication in thin-walled cysts (Johnson). Chromosome cycle (Saito, 1961).

E. acus E. (Fig. 121, *c*). 50-175μ by 8-18μ; body long spindle or cylinder, with a sharply pointed posterior end; flagellum short, about one-fourth the body length; spiral striation of pellicle very delicate; numerous discoid chromatophores; several paramylum bodies, rod-form and 12-20μ long; nucleus central; stigma distinct; movement sluggish.

E. spirogyra E. (Fig. 121, *d*). 80-125μ by 10-35μ; cylindrical; anterior end a little narrowed and rounded, posterior end drawn out; spiral striae, made up of small knobs, conspicuous; many discoid chromatophores; two ovoidal paramylum bodies, 18-45μ by 10-18μ, one on either side of centrally located nucleus; flagellum about one-fourth the body length; stigma prominent; sluggish.

E. oxyuris Schmarda (Fig. 121, *e*). 150-500μ by 20-40μ; cylindrical; almost always twisted, somewhat flattened; anterior end rounded, posterior end pointed; pellicle with spiral striae; numerous discoid chromatophores; two ovoid paramylum bodies, 20-40μ long, one on either side of nucleus, and also small bodies; stigma large; flagellum short; sluggish.

E. sanguinea E. (Fig. 121, *f*). 80-170μ by 25-45μ; posterior end bluntly rounded; flagellum about the body length; pellicle striated; elongate chromatophores lie parallel to the striae; haematochrome granules scattered in sun light and collected in the central area in darkness.

E. deses E. (Fig. 121, *g*). 85-170μ by 10-20μ; elongate; highly plastic; faint striae; stigma distinct; nucleus central; chromatophores discoid with pyrenoid; several small rod-shaped paramylum scattered; flagellum less than one-fourth the body length.

E. gracilis Klebs (Fig. 121, *h*). 35-55μ by 6-25μ; cylindrical to elongate oval; highly plastic; flagellum about the body length; fusiform chromatophores variable in number (Gross and Villaire, 1960); nucleus central; pyrenoids; fresh water. (Electron microscope studies, Wolken, 1956; Pitelka and Schooley, 1958; mass culture, Bach, 1960; pigments, Goodwin *et al.*, 1954; Goodwin and Gross, 1958; Greenblatt and Schiff, 1959; vitamin B12 assay, Hutner *et al.*, 1956.)

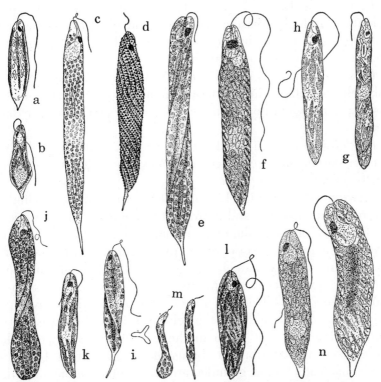

FIG. 121. Species of Euglena (Johnson). a, *Euglena pisciformis*, ×855; b, *E. viridis*, ×400; c, *E. acus*, ×555; d, *E. spirogyra*, ×460; e, *E. oxyuris*, ×200; f, *E. sanguinea*, ×400; g, *E. deses*, ×315; h, *E. gracilis*, ×865; i, *E. tripteris*, with optical section of body, ×345; j, *E. ehrenbergi*, ×145; k, *E. terricola*, ×345; l, *E. sociabilis*, ×320; m, two individuals of *E. klebsi*, ×335; n, two individuals of *E. rubra*, ×355.

E. tripteris Dujardin (Fig. 121, *i*). 70-120µ by 12-16µ; elongate; three-ridged, rounded anteriorly and drawn out posteriorly; pellicle longitudinally striated; only slightly plastic; stigma prominent; discoid chromatophores numerous; two paramylum bodies, rod-shaped and one on either side of the nucleus; flagellum about three-fourths the body length.

E. *ehrenbergi* Klebs (Fig. 121, *j*). 170-400µ by 15-40µ; cylindrical and flattened, posterior end rounded; plastic, often twisted; spiral striation; numerous small discoid chromatophores; stigma conspicuous; two paramylum bodies elongate, up to over 100µ long; flagellum about one-half the body length or less.

E. *terricola* Dangeard (Fig. 121, *k*). 65-95µ by 8-18µ; pellicle thin and highly plastic; nucleus central; chromatophores long (20-30µ) rods; paramylum bodies small and annular; flagellum about one-third the body length.

E. *sociabilis* D. (Fig. 121, *l*). 65-112µ by 15-30µ; cylindrical; delicate pellicle; highly plastic; numerous elongate chromatophores; paramylum bodies discoid; flagellum slightly longer than body.

E. *klebsi* Mainx (Fig. 121, *m*). 45-85µ by 5-10µ; form highly plastic; chromatophores discoid; paramylum bodies rod-shaped, up to several; flagellum short.

E. *rubra* Hardy (Fig. 121, *n*). 70-170µ by 25-36µ; cylindrical; rounded anteriorly and drawn out posteriorly; spiral striation; nucleus posterior; flagellum longer than body; stigma about 7µ in diameter; many fusiform chromatophores aligned with the body striae; numerous haematochrome granules, 0.3-0.5µ in diameter: ovoid paramylum bodies; reproductive and temporary cysts and protective cysts, 34-47µ in diameter, with a gelatinous envelope.

Johnson (1939) found that the color of this Euglena was red in the morning and dull green in the late afternoon, due to the difference in the distribution of haematochrome within the body. When haematochrome granules are distributed throughout the body, the organism is bright-red, but when they are condensed in the center of the body, the organism is dull green. When part of the area of the pond was shaded with a board early in the morning, shortly after sunrise all the scum became red except the shaded area. When the board was removed, the red color appeared in eleven minutes while the temperature of the water remained 21°C. In the evening, the change was reversed. Johnson and Jahn (1942) later found that green-red color change could be induced by raising the temperature of the water to 30–40°C. and by irradiation with infrared rays or visible light. The two workers hold that the function of haematochrome may be protective, since it migrates to a position which shields the chromatophores from very bright light. If this is true, it is easy to find the species thriving in hot weather in shallow ponds where temperature of the water rises to 35-45°C. In colder weather, it is supposed that this *Euglena* is less abundant and it exists in a green phase, containing a few haematochrome granules.

Genus **Khawkinea** Jahn and McKibben. Similar to Genus *Euglena,* but without chromatophores and thus permanently colorless; fresh water.

K. halli J. and M. 30-65μ by 12-14μ; fusiform; pellicle spirally striated; plastic; flagellum slightly longer than body; stigma 2–3μ in diameter, yellow-orange to reddish-orange, composed of many granules; numerous (25-100) paramylum bodies elliptical or polyhedral; cysts 20-30μ in diameter; putrid leaf infusion; saprozoic (Jahn and McKibben, 1937).

K. ocellata (Khawkine). Similar to above; flagellum one and a half to two times body length; fresh water.

Genus **Phacus** Dujardin. Highly flattened; asymmetrical; pellicle firm; body form constant; prominent longitudinal or oblique striation; flagellum and a stigma; chromatophores without pyrenoid (Pringsheim) are discoid and green; holophytic; fresh water. Numerous species (Skvortzov, 1937; Pochmann, 1942; Conrad, 1943; Allegre and Jahn, 1943; morphology and cytology, Krichenbauer, 1937; Conrad, 1943).

P. pleuronectes Müller (Fig. 122, *a*). 45-100μ by 30-70μ; short posterior prolongation slightly curved; a prominent ridge on the convex side, extending to posterior end; longitudinally striated; usually one circular paramylum body near center; flagellum as long as body.

P. longicauda (Ehrenberg) (Fig. 122, *b*). 120-170μ by 45-70μ; usually slightly twisted; a long caudal prolongation; flagellum about one-half the body length; stigma prominent; many chromatophores; a discoidal paramylum body central; pellicle longitudinally striated.

P. pyrum (E.) (Fig. 122, *c*). About 30-50μ by 10-20μ; circular in cross-section; with a medium long caudal prolongation; pellicle obliquely ridged; stigma inconspicuous; two discoid paramylum bodies; flagellum as long as the body.

P. acuminata Stokes (Fig. 122, *d*). About 30-40μ by 20-30μ; nearly circular in outline; longitudinally striated; usually one small paramylum body; flagellum as long as the body.

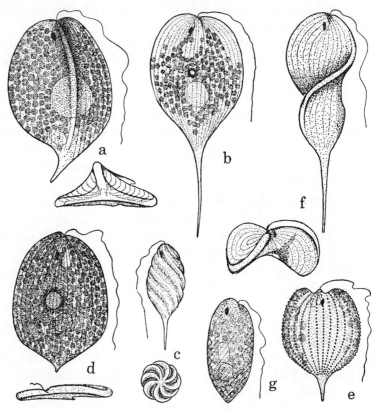

FIG. 122. Species of Phacus (Allegre and Jahn). a, *Phacus pleuronectes* and an end view, ×800; b, *P. longicauda*, ×500; c, *P. pyrum* and an end view, ×880; d, *P. acuminata* and an end view, ×1300; e, *P. monilata*, ×800; f, *P. torta*, and an end view, ×800; g, *P. oscillans*, ×1400.

P. monilata S. (Fig. 122, *e*). 40-55μ by 32-40μ; a short caudal projection; pellicle with minute knobs arranged in longitudinal rows; discoid chromatophores; flagellum about the body length.

P. torta Lemmermann (Fig. 122, *f*). 80-100μ by 40-45μ; body twisted, with a long caudal prolongation; longitudinal striae on pellicle; chromatophores discoid; one large circular paramylum body; flagellum about one-half the body length.

P. oscillans Klebs (Figs. 122, *g*). 15-35μ by 7-10μ; rounded anteriorly and bluntly pointed posteriorly; striation oblique; one or two paramylum bodies; flagellum about as long as the body.

Genus **Lepocinclis** Perty (*Crumenula* Dujardin). Body more or less ovo-cylindrical; rigid with spirally striated pellicle; often with a short posterior spinous projection; stigma sometimes present; discoidal chromatophores numerous and marginal; paramylum bodies usually large and ring-shaped, laterally disposed; without pyrenoids; fresh water. Many species (Pascher, 1925, 1929; Conrad, 1934; Skvortzov, 1937).

L. ovum Ehrenberg (Fig. 123, *a*). Body 20-40μ long.

Genus **Trachelomonas** Ehrenberg. With a lorica which often possesses numerous spines; sometimes yellowish to dark brown, composed of ferric hydroxide impregnated with a brown manganic compound (Pringsheim, 1948); a single long flagellum protrudes from the anterior aperture, the rim of which is frequently thickened to form a collar; chromatophores either two curved plates or numerous discs; paramylum bodies small grains; a stigma and pyrenoid; multiplication by fission, one daughter individial retains the lorica and flagellum, while the other escapes and forms a new one; cysts common; fresh water. Numerous species (Palmer, 1902, 1905, 1925, 1925a; Pascher, 1924, 1925, 1925a, 1926, 1929; Gordienko, 1929; Conrad, 1932; Skvortzov, 1937; Balech, 1944; Conrad and Van Meel, 1952).

T. hispida (Perty) (Figs. 34, *a;* 123, *b*). Lorica oval, with numerous minute spines; brownish; eight to ten chromatophores; 20-42μ by 15-26μ; many varieties.

T. urceolata Stokes (Fig. 123, *c*). Lorica vasiform, smooth with a short neck; about 45μ long.

T. piscatoris Fisher (Fig. 123, *d*). Lorica cylindrical with a short neck and with numerous short, conical spines; 25-40μ long; flagellum one to two times body length.

T. verrucosa Stokes (Fig. 123, *e*). Lorica spherical, with numerous knob-like attachments; no neck; 24-25μ in diameter.

T. vermiculosa Palmer (Fig. 123, *f*). Lorica spherical; with many sausage-form markings; 23μ in diameter.

Genus **Cryptoglena** Ehrenberg. Body rigid, flattened, two band-form chromatophores lateral; a single flagellum; nucleus posterior; among freshwater algae. One species.

C. pigra E. (Fig. 123, *g*). Ovoid, pointed posteriorly; flagellum short; stigma prominent; 10-15μ by 6-10μ; standing water.

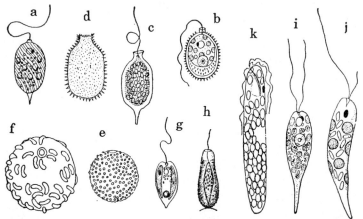

Fig. 123. a, *Lepocinclis ovum*, ×430 (Stein); b, *Trachelomonas hispida*, ×430 (Stein); c, *T. urceolata*, ×430 (Stokes); d, *T. piscatoris*, ×520 (Fisher); e, *T. verrucosa*, ×550 (Stokes); f, *T. vermiculosa*, ×800 (Palmer); g, *Cryptoglena pigra*, ×430 (Stein); h, *Ascoglena vaginicola*, ×390 (Stein); i, *Eutreptia viridis*, ×270 (Klebs); j, *E. marina*, ×670 (da Cunha); k, *Euglenamorpha hegneri*, ×730 (Wenrich).

Genus **Ascoglena** Stein. Encased in a flexible, colorless to brown lorica, attached with its base to foreign object; solitary; without stalk; body ovoidal, plastic; attached to test with its posterior end; a single flagellum; a stigma; numerous chromatophores discoid; with or without pyrenoids; reproduction as in *Trachelomonas;* fresh water.

A. vaginicola S. (Fig. 123, *h*). Lorica about 43μ by 15μ.

Genus **Colacium** Ehrenberg. Stalked individuals form colony; frequently attached to animals such as copepods, rotifers, etc.; stalk mucilaginous; individual cells pyriform, ellipsoidal or cylindrical; without flagellum; a single flagellum only in free-swimming stage; discoidal chromatophores numerous; with pyrenoids;

multiplication by longitudinal fission; also by swarmers, possessing a flagellum and a stigma; fresh water. Several species.

C. vesiculosum E. (Fig. 124). Solitary or colonial, made up of two to eight individuals; flagellate form ovoid to spindle; 22μ by 12μ; seven to ten elongate chromatophores along the periphery; flagellum up to twice the body length; a stigma; many paramylum bodies; palmella stage conspicuous; stalked form (Johnson, 1934).

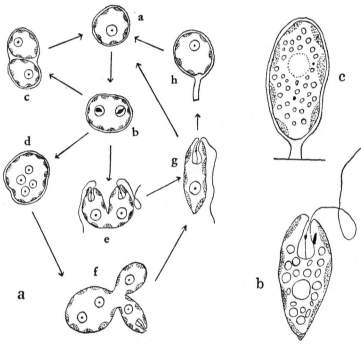

Fig. 124. *Colacium vesiculosum* (Johnson). a, diagram showing the life cycle (a–d, palmella stage; e. formation of flagellate stage; f, formation of flagellate stage by budding of Palmella stage; g, flagellate stage; h, attached stage); b, flagellate and c, stalked form on a crustacean, ×1840.

Genus **Eutreptia** Perty (*Eutreptiella* da Cunha). With two flagella at anterior end; pellicle distinctly striated; plastic; spindle-shaped during movement; stigma; numerous discoid chromatophores; pyrenoids absent; paramylum bodies spherical or sub-cylindrical; multiplication as in Euglena; cyst with a thick stratified wall; fresh or salt water.

E. viridis P. (Fig. 123, *i*). 50-70µ by 5-13µ; in fresh water; a variety was reported from brackish water ponds.

E. marina da Cunha (Fig. 123, *j*). Flagella unequal in length; longer one is as long as body, shorter one is one-third as long; body 40-50µ by 8-10µ; salt water.

Genus **Euglenamorpha** Wenrich. Body form and structure similar to those of *Euglena*, but with three flagella; in gut of frog tadpoles. One species.

E. hegneri W. (Fig. 123, *k*). 40-50µ long (Wenrich, 1924).

Family 2. **Astasiidae** Bütschli

Similar to Euglenidae in body form and general structure, but without chromatophores; body highly plastic, although usually elongate spindle.

Genus **Astasia** Dujardin. Body plastic, although ordinarily elongate; fresh water or parasitic (?) in microcrustaceans. Many species (Pringsheim, 1942; Christen, 1959; axenic culture, Schoenborn, 1946; lethal effect by radiation and protection against lethal damage, Schoenborn, 1953, 1956; metabolism, Hutner and Lee, 1962.)

A. klebsi Lemmermann (Fig. 125, *a*). Spindle-form; posterior portion drawn out; flagellum as long as the body; plastic; saprozoic; paramylum bodies oval; 40-50µ by 13-20µ; stagnant water.

A. acus Christen. Body form resembles *Euglena acus* (p. 352), but without chromatophores and stigma; paramylum bodies; 61-68µ by 6.5-8µ; fresh water pond. Other species (Skuja, 1956; Christen, 1958, 1959.)

Genus **Urceolus** Mereschkowsky (*Phialonema*, Stein). Body colorless; plastic; flask-shaped; striated; a funnel-like neck; posterior region stout; a single flagellum protrudes from funnel and reaches inward the posterior third of body; fresh or salt water.

U. cyclostomus (Stein) (Fig. 125, *b*). 25-50µ long; fresh water.

U. sabulosus (Stokes) (Fig. 125, *c*). Spindle-form; covered with minute sand-grains; about 58µ long; fresh water.

Genus **Petalomonas** Stein. Oval or pyriform; not plastic; pellicle often with straight or spiral furrows; a single flagellum; paramylum bodies; a nucleus; holozoic or saprozoic. Many species in fresh water and a few in salt water. Species (Shawhan and Jahn, 1947; Christen, 1959).

P. mediocanellata S. (Fig. 125, *d*). Ovoid with longitudinal furrows on two sides; flagellum about as long as the body; 21–26µ long.

Genus **Rhabdomonas** Fresenius. Rigid body, cylindrical and not flattened, more or less arched; pellicle longitudinally ridged; a flagellum through aperture at the anterior tip; fresh water (Pringsheim, 1942). (Species, Pascher, 1925; relation to *Menoidium*, Pringsheim, 1942.)

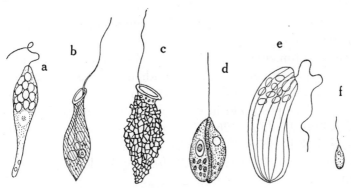

Fig. 125. a, *Astasia klebsi*, ×500 (Klebs); b, *Urceolus cyclostomus*, ×430 (Stein); c, *U. sabulosus*, ×430 (Stokes); d, *Petalomonas mediocanellata*, ×1000 (Klebs); e, *Rhabdomonas incurva*, ×1400 (Hall); f, *Scytomonas pusilla*, ×430 (Stein).

R. incurva F. (Figs. 72; 125, *e*). Banana-shaped; longitudinal ridges conspicuous; flagellum as long as the body; 15-25µ by 7-8µ (Hall, 1923); 13-15µ by 5-7µ (Hollande, 1952a); common in standing water.

Genus **Scytomonas** Stein. Oval or pyriform, with a delicate pellicle; a single flagellum; a contractile vacuole with a reservoir; holozoic on bacteria; longitudinal fission in motile stage; stagnant water and coprozoic.

S. pusilla S. (Fig. 125, *f*). About 150 long. (Cytology, Schüssler, 1917.)

Genus **Copromonas** Dobell. Elongate ovoid; with a single flagellum; a small cytostome at anterior end; holozoic on bacteria;

permanent fusion followed by encystment (p. 218); coprozoic in faecal matters of frog, toad, and man; several authors hold that this genus is probably identical with *Scytomonas* which was incompletely described by Stein.

C. subtilis D. (Fig. 80). 7-20μ long. Golgi body (Gatenby and Singh, 1938).

Family 3. **Anisonemidae** Schewiakoff

Colorless body plastic or rigid with a variously marked pellicle; two flagella, one directed anteriorly and the other usually posteriorly; contractile vacuoles and reservoir; stigma absent; paramylum bodies usually present; free-swimming or creeping. Genera (Christen, 1959).

Genus **Anisonema** Dujardin. Generally ovoid; more or less flattened; asymmetrical; plastic or rigid; a slit-like ventral furrow; flagella at anterior end; cytopharynx long; contractile vacuole anterior; nucleus posterior; in fresh water. Several species.

A. acinus D. (Fig. 126, *a*). Rigid; oval; somewhat flattened; pellicle slightly striated; 25-40μ by 16-22μ.

A. truncatum Stein (Fig. 126, *b*). Rigid; elongate ovoid; 60μ by 20μ.

A. emarginatum Stokes (Fig. 126, *c*). Rigid; 14μ long; flagella long.

Genus **Peranema** Dujardin. Elongate, with a broad rounded or truncate posterior end during locomotion; highly plastic when stationary; delicate pellicle shows a fine striation; expansible cytostome with a thickened ridge and two oral rods at anterior end; aperture through which the flagella protrude is also at anterior end; a free flagellum, long and conspicuous, tapers toward free end; a second flagellum adheres to the pellicle; nucleus central; a contractile vacuole, anterior, close to the reservoir; holozoic; fresh water.

P. trichophorum (Ehrenberg) (Fig. 126, *d*). 40–70μ long; body ordinarily filled with paramylum or starch grains derived from Astasia, Rhabdomonas, Euglena, etc., which coinhabit the culture; holozoic; very common in stagnant water. (Cell inclusion, Hall, 1929; structure and behavior, Chen, 1950; development, Lackey, 1929; flagellar apparatus, Lackey, 1933; Pitelka, 1945; food intake, Hall, 1933; Hollande, 1942; Hyman, 1936; Chen, 1950; culture and nutrition, Storm and Hutner, 1953; electron microscope study, Roth, 1959.)

P. granulifera Penard. Much smaller in size. 8–15µ long; elongate, but plastic; pellicle granulated; standing water.

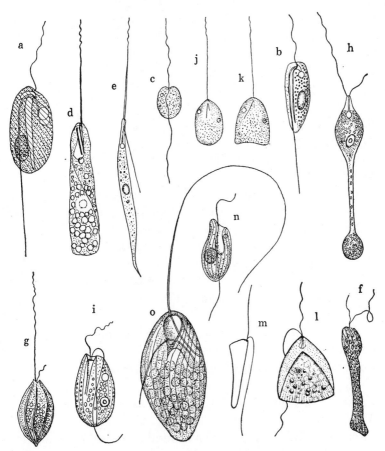

FIG. 126. a, *Anisonema acinus*, ×400 (Klebs); b, *A. truncatum*, ×430 (Stein); c, *A. emerginatum*, ×530 (Stokes); d, *Peranema trichophorum*, ×670; e, *Heteronema acus*, ×430 (Stein); f, *H. mutabile*, ×120 (Stokes); g, *Tropidoscyphus octocostatus*, ×290 (Lemmermann); h, *Distigma proteus*, ×430 (Stein); i, *Entosiphon sulcatum*, ×430 (Stein); j, *Notosolenus apocamptus*, ×120 (Stokes); k, *N. sinuatus*, ×600 (Stokes); l, m, front and side views of *Triangulomonas rigida*, ×935 (Lackey); n, *Marsupiogaster striata*, ×590 (Schewiakoff); o, *M. picta* (Faria, da Cunha and Pinto).

Genus **Heteronema** Dujardin. Plastic; rounded or elongate; flagella arise from anterior end, one directed forward and the other trailing; cytostome near base of flagella; holozoic; fresh water. Several species.

H. acus (Ehrenberg) (Fig. 126, *e*). Extended body tapers towards both ends; anterior flagellum as long as body, trailing one about one-half; contractile vacuole anterior; nucleus central; 40–50µ long; fresh water. (Morphology, reproduction, Loefer, 1931.)

H. mutabile (Stokes) (Fig. 126, *f*). Elongate; highly plastic; longitudinally striated; about 250µ long; in cypress swamp.

Genus **Tropidoscyphus** Stein. Slightly plastic; pellicle with eight longitudinal ridges; two unequal flagella at anterior pole; holozoic or saprozoic; fresh or salt water.

T. octocostatus S. (Fig. 126, *g*). 35-63µ long; fresh water, rich in vegetation.

Genus **Distigma** Ehrenberg. Plastic; elongate when extended; body surface without any marking; two flagella unequal in length, directed forward; cytostome and cytopharynx located at anterior end; endoplasm usually transparent; holozoic. Several species (Pringsheim, 1942).

D. proteus E. (Fig. 126, *h*). 50-110µ long when extended; nucleus central; stagnant water; infusion. (Cytology, Hollande, 1937.)

Genus **Entosiphon** Stein Oval, flattened; more or less rigid; flagella arise from a cytostome, one flagellum trailing; protrusible cytopharynx a long conical tubule almost reaching posterior end; nucleus centro-lateral; fresh water.

E. sulcatum (Dujardin) (Fig. 126, *i*). About 20µ long (Lackey, 1929, 1929a).

E. ovatum Stokes. Anterior end rounded; ten to twelve longitudinal striae; about 25-28µ long.

Genus **Notosolenus** Stokes. Free-swimming; rigid oval; ventral surface convex, dorsal surface with a broad longitudinal groove; flagella anterior; one long, directed anteriorly and vibratile; the other shorter and trailing; fresh water with vegetation.

N. apocamptus S. (Fig. 126, *j*). Oval with broad posterior end; 6-11µ long.

N. sinuatus S. (Fig. 126, *k*). Posterior end truncate or concave; about 22µ long.

Genus **Triangulomonas** Lackey. Rigid body, triangular, with convex sides; one surface flat, the other elevated near the anterior end; pellicle brownish; a mouth at anterior end with cytopharynx and reservoir; two flagella, one trailing; salt water.

T. rigida L. (Fig. 126, *l, m*). Body 18µ by 15µ; anterior flagellum as long as the body; posterior flagellum one and a half times the body length; Woods Hole (Lackey, 1940).

Genus **Marsupiogaster** Schewiakoff. Oval; flattened; asymmetrical; cytostome occupies entire anterior end; cytopharynx conspicuous, one-half body length; body longitudinally striated; two flagella, one directed anteriorly, the other posteriorly; spherical nucleus; contractile vacuole anterior; fresh or salt water.

M. striata Schewiakoff (Fig. 126, *n*). About 27µ by 15µ; fresh water; Hawaii.

M. picta Faria, da Cunha and Pinto (Fig. 126, *o*). In salt water; Rio de Janeiro.

Order 5. **Chloromonadida** Klebs

The chloromonads are of rare occurrence and consequently not well known. The majority possess small discoidal grass-green chromatophores with a large amount of xanthophyll which on addition of an acid become blue-green. No pyrenoids occur. The metabolic products are fatty oil. Starch or allied carbohydrates are absent. Stigma is also not present. Genera (Poisson and Hollande, 1943; Hollande, 1952).

Genus **Gonyostomum** Diesing (*Rhaphidomonas*, Stein). With a single flagellum; chromatophores grass-green; highly refractile trichocyst-like bodies in cytoplasm; fresh water. A few species.

G. semen D. (Fig. 127, *a*). Sluggish animals; about 45-60µ long; among decaying vegetation.

Genus **Vacuolaria** Cienkowski (*Coelomonas* Stein). Highly plastic; without trichocyst-like structures; anterior end narrow; two flagella; cyst with a gelatinous envelope. One species.

V. virescens C. (Fig. 127, *b*). 50-70µ by 18-25µ; fresh water. (Cytology, Fott, 1935; Poisson and Hollande, 1943.)

Genus **Trentonia** Stokes. Bi-flagellate as in the last genus; but flattened; anterior margin slightly biolobed. One species.

T. flagellata S. (Fig. 127, *c*). Slow-moving organism; encystment followed by binary fission; about 60μ long; fresh water.

Genus **Thaumatomastix** Lauterborn. Colorless; pseudopodia formed; two flagella, one extended anteriorly, the other trailing;

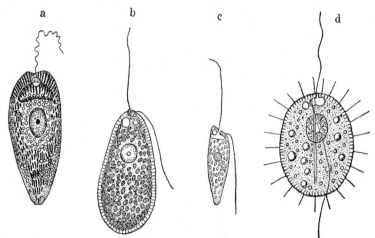

Fig. 127. a, *Gonyostomum semen*, ×540 (Stein); b, *Vacuolaria virescens*, ×460 (Senn); c, *Trentonia flagellata*, ×330 (Stokes); d, *Thaumatomastix setifera*, ×830 (Lauterborn).

holozoic; perhaps a transitional form between the Mastigophora and the Sarcodina. One species.

T. setifera L. (Fig. 127, *d*). About 20-35μ by 15-28μ; fresh water.

References

ALLEGRE, C. F., and JAHN, T. L., 1943. A survey of the genus Phacus Dujardin. Tr. Am. Micr. Soc., 62:233.

BACH, M. K., 1960. Mass culture of *Euglena gracilis*. J. Protozool., 7:50.

BALECH, E., 1944. Trachelomonas de la Argentina. An. Mus. Argent. Cien. Nat., 41:221.

CHEN, Y. T., 1950. Investigations of the biology of *Peranema trichophorum*. Quart. J. Micr., Sci., 91:279.

CHRISTEN, H. R., 1958. Farblose Euglenalen aus dem Hypolimnion des Hausersees. Z. Hydrobiologie, 20:141.

——— 1959. New colorless Eugleninae. J. Protozool., 6:292.

CONRAD, W., 1932. Flagellates nouveaux ou peu connus. III. Arch. Protist., 78:463.

——— 1934. Matériaux pour une monographie du genre Lepocinclis. *Ibid.*, 82:203.

——— 1943. Notes protistologiques. XXVIII. Bull. Mus. Roy. d'Hist. Natur. Belgique, 19, no. 6.

——— and VAN MEEL, L., 1952. Matériaux pour une monographie de Trachelomonas Ehrenberg, etc. Mém. Inst. Roy. Sci. Nat. Belgique, 124:1.

DA CUNHA, A. M., 1913. Sobre um novo genero de "Euglenoidea." Brazil Medico, 27:213.

DANGEARD, P., 1901. Recherches sur les Eugléniens. La Bot., 8:97.

EPSTEIN, H. T., and SCHIFF, J. A., 1961. Studies of chloroplast development in Euglena. 4. J. Protozool., 8:427.

FOTT, B., 1935. Ueber den inneren Bau von *Vacuolaria viridis*. Arch. Protist., 84:242.

FRITSCH, F. E., 1935. The structure and reproduction of the algae.

GATENBY, J. B., and SINGH, B. N., 1938. The Golgi apparatus of *Copromonas subtilis* and *Eugena* sp. Quart. J. Micr. Sci., 80:567.

GOJDICS, M., 1953. The genus Euglena. Madison, Wisconsin.

GOODWIN, T. W., and GROSS, J. A., 1958. Carotenoid distribution in bleached substrains of *Euglena gracilis*. J. Protozool., 5:292.

GORDIENKO, M., 1929. Zur Frage der Systematik der Gattung Trachelomonas. Arch. Protist., 65:258.

GREENBLATT, C. L., and SCHIFF, J. A., 1959. A phenophytin-like pigment in dark-adapted *Euglena gracilis*. J. Protozool., 6:23.

GROSS, J. A., and VILLAIRE, M., 1960. Chloroplast development and numbers in relation to culture age in Euglena. Tr. Am. Mic. Soc., 79:144.

GÜNTHER, F., 1928. Ueber den Bau und die Lebensweise der Euglenen, etc. Arch. Protist., 60:511.

HALL, R. P., 1923. Morphology and binary fission of *Menoidium incurvum*. Univ. Calif. Publ. Zool., 20:447.

——— 1929. Reaction of certain cytoplasmic inclusions to vital dyes and their relation to mitochondria and Golgi apparatus in the flagellate *Peranema trichophorum*. J. Morphol. Physiol., 48:105.

——— 1933. The method of ingestion in Peranema, etc. Arch. Protist., 81:308.

——— 1934. A note on the flagellar apparatus of Peranema, etc. Tr. Am. Micr. Soc., 53:237.

——— 1937. A note on behavior of chromosomes. *Ibid.*, 56:288.

HOLLANDE, A., 1937. Quelques données nouvelles sur la cytologie d'une Astasiacée peu connu: *Distigma proteus*. Bull. Soc. zool. Fr., 62:236.

——— 1942. Études cytologique et biologique de quelques flagellés libres. Arch. zool. exp. gén., 83:1.

——— 1952. Classe de Chloromonadines In Grassé (1952), p. 227.

——— 1952a. Classe des Eugleniens. *Ibid.*, p. 239.

HUTNER, S. H., *et al.*, 1956. A sugar-containing basal medium for vitamin B₁₂—assay with Euglena, etc. J. Protozool., 3:101.

—— and LEE, J. W., 1962. On the metabolism of *Astasia longa*. *Ibid.*, 9:74.

HYMAN, L. H., 1936. Observations on Protozoa. II. Quart. J. Micr. Sci., 79:50.

JAHN, T. L., 1946. The euglenoid flagellates. Quart. Rev. Biol., 21:246.

—— and McKIBBEN, W. R., 1937. A colorless euglenoid flagellate, *Khawkinea halli* n.g., n.sp. Tr. Am. Micr. Soc., 56:48.

JOHNSON, D. F., 1934. Morphology and life history of *Colacium vesiculosum*. Arch. Protist., 83:241.

JOHNSON, L. P., 1939. A study of *Euglena rubra*. Tr. Am. Micr. Soc., 58:42.

—— 1944. Euglena of Iowa. *Ibid.*, 63:97.

—— 1956. Observations on *Euglena fracta* sp. nov., etc. Tr. Am. Micr. Soc., 75:271.

—— and JAHN, T. L., 1942. Cause of the green-red color change in *Euglena rubra*. Physiol. Zool., 15:89.

KRICHENBAUER, H., 1937. Beitrag zur Kenntnis der Morphologie und Entwicklungsgeschichte der Gattungen Euglena und Phacus. Arch. Protist., 90:88.

LACKEY, J. B., 1929. Studies on the life history of Euglenida. I. *Ibid.*, 66:175.

—— 1929a. II. *Ibid.*, 67:128.

—— 1933. III. Biol. Bull., 65:238.

—— 1940. Some new flagellates from the Woods Hole area. Am. Midland Nat., 23:463.

LEMMERMANN, E., 1913. Eugleninae. Süsswasserflora Deutschlands. Pt. 2.

LOEFER, J. B., 1931. Morphology and binary fission of *Heteronema acus*. Arch. Protist., 74:449.

MAINX, F., 1928. Beiträge zur Morphologie und Physiologie der Eugleninen. I, II. *Ibid.*, 60:305.

PALMER, T. C., 1902. Five new species of Trachelomonas. Proc. Acad. Nat. Sci., Philadelphia, 54:791.

—— 1905. Delaware valley forms of Trachelomonas. *Ibid.*, 57:665.

—— 1925. Trachelomonas: etc. *Ibid.*, 77:15.

—— 1925a. Nomenclature of Trachelomonas. *Ibid.*, 77:185.

PASCHER, A., 1913. Chloromonadinae. Süsswasserflora Deutsch. Pt. 2.

—— 1924. Neue oder wenig bekannte Protisten. XIII. Arch. Protist., 48:492.

—— 1925. XV. *Ibid.*, 50:486.

—— 1925a. XVII. *Ibid.*, 51:549.

—— 1926. XIX. *Ibid.*, 53:459.

—— 1929. XXI. *Ibid.*, 65:426.

PEREZ-REYES, R., and GÓMEZ, E. S., 1958. Euglenae de Valle de Mexico. I. Rev. Latinoamer. Microbiol., 1:303.

PITELKA, DOROTHY R., 1945. Morphology and taxonomy of flagellates of the genus Peranema Dujardin. J. Morphol., 76:179.

——— and SCHOOLEY, C. N., 1958. The pellicular fine structure of *Euglena gracilis*. J. Protozool., 4, Suppl: 10.

POCHMANN, A., 1942. Synopsis der Gattung Phacus. Arch. Protist., 95:81.

POISSON, R., and HOLLANDE, A., 1943. Considérations sur la cytologie, la mitose et les affinités des Chloromonadies. Ann. Sc. Nat. Ser. Bot. Zool., 5:147.

PRINGSHEIM, E. G., 1942. Contribution to our knowledge of saprophytic Algae and Flagellata. III. New Phytologist, 41:171.

——— 1948. Taxonomic problems in the Euglenineae. Biol. Rev., 23:46.

——— and HOVASSE, R., 1948. The loss of chromatophores in *Euglena gracilis*. New Phytologist, 47:52.

——— ——— 1950. Les relations de parenté entre Astasiacées et Euglènacées. Arch. zool. exper. gén., 86:499.

ROTH, L. E., 1959. An electron-microscope study of the cytology of the protozoan *Peranema trichophorum*. J. Protozool., 6:107.

SAITO, M., 1961. Studies on the mitosis of Euglena. I. J. Protozool., 8:300.

SCHOENBORN, H. W., 1946. Studies on the nutrition of colorless euglenoid flagellates. II. Physiol. Zool., 19:430.

——— 1953. Lethal effect of ultraviolet and x-radiation on the protozoan flagellate *Astasia longa*. Physiol. Zool., 26:312.

——— 1956. Protection against lethal damage induced by ultraviolet radiation. J. Protozool., 3:97.

SCHÜSSLER, H., 1917. Cytologische und entwicklungsgeschichtliche Protozoenstudien. I. Arch. Protist., 38:117.

SHAWHAN, F. M., and JAHN, T. L., 1947. A survey of the genus Petalomonas. Tr. Am. Micr. Soc., 66:182.

SKUJA, H., 1956. Taxonomische und biologische Studien ueber das Phytoplankton schwedischer Binnengewaesser. Nov. Acta Reg. Soc. Sci. Upsaliensis, 4 (16):412.

SKVORTZOV, B. V., 1937. Contributions to our knowledge of the freshwater algae of Rangoon, Burma, India. I. Arch. Protist., 90:69.

STERN, A. I., et al., 1960. Isolation of ergosterol from *Euglena gracilis;* distribution among mutant strains. J. Protozool., 7:52.

STOKES, A. C., 1888. A preliminary contribution toward a history of the freshwater Infusoria of the United States. J. Trenton Nat. Hist. Soc., 1:71.

STORM, J., and HUTNER, H. S., 1953. Nutrition of Peranema. Ann. N. Y. Acad. Sci., 56:901.

WOLKEN, J. J., 1956. A molecular morphology of *Euglena gracilis* var. *bacillaris*. J. Protozool., 3:211.

Chapter 12

Order 6. **Dinoflagellida** Bütschli

THE dinoflagellates make one of the most distinct groups of the Mastigophora, inhabiting mostly marine water, and to a lesser extent fresh water. In the general appearance, the arrangement of the two flagella, the characteristic furrows, and the possession of brown chromatophores, they are closely related to the Cryptomonadida.

The body is covered by an envelope composed of cellulose which may be a simple smooth piece, or may be composed of two valves or of numerous plates, that are variously sculptured and possess manifold projections. Differences in the position and course of the furrows and in the projections of the envelope produce numerous asymmetrical forms. The furrows, or grooves, are a transverse annulus and a longitudinal sulcus. The **annulus** is a girdle around the middle or toward one end of the body. It may be a complete, incomplete or sometimes spiral ring. While the majority show a single transverse furrow, a few may possess several. The part of the shell anterior to the annulus is called the **epitheca** and that posterior to the annulus the **hypotheca.** If the envelope is not developed, the terms **epicone** and **hypocone** are used (Fig. 128). The **sulcus** may run from end to end or from one end to the annulus. The two flagella arise typically from the furrows, one being transverse and the other longitudinal.

The **transverse flagellum** which is often band-form, encircles the body and undergoes undulating movements, which in former years were looked upon as ciliary movements (hence the name Clioflagellata). In the suborder Prorocentrinea, this flagellum vibrates freely in a circle near the anterior end. The **longitudinal flagellum** often projects beyond the body and vibrates. Combination of the movements of these flagella produces whirling movements characteristic of the organisms.

The majority of dinoflagellates possess a single somewhat massive nucleus with evenly scattered chromatin, and usually several

370

endosomes. There are two kinds of vacuoles. One is often surrounded by a ring of smaller vacuoles, while the other is large,
contains pink-colored fluid and connected with the exterior by a
canal opening into a flagellar pore. The latter is known as the **pusule** which may function as a digestive organella (Kofoid and
Swezy). In many freshwater forms, a stigma is present, and in
Pouchetiidae, there is an ocellus composed of an amyloid lens and
a dark pigment-ball. Trichocysts occur in some and nematocysts
characterize Polykrikos. The majority of planktonic forms possess

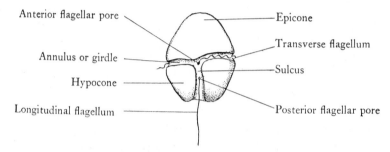

Anterior flagellar pore

Annulus or girdle

Hypocone

Longitudinal flagellum

Epicone

Transverse flagellum

Sulcus

Posterior flagellar pore

Fig. 128. Diagram of a typical naked dinoflagelate (Lebour).

a large number of small chromatophores which are usually dark
yellow, brown or sometimes slightly greenish and are located in
the periphery of the body, while bottom-dwelling and parasitic
forms are, as a rule, colorless, because of the absence of chromatophores. A few forms contain haematochrome. The method of nutrition is holophytic, holozoic, saprozoic, or mixotrophic. In holophytic forms, anabolic products are starch, oil, or fats.

Asexual reproduction is by binary or multiple fission or budding
in either the active or the resting stage and differs among different
groups. Encystment is of common occurrence. In some forms, the
cyst wall is formed within the test. The cysts remain alive for
many years; for example, Ceratium cysts were found to retain
their vitality in one instance for six and one-half years. Conjugation and sexual fusion have been reported in certain forms, but
definite knowledge on sexual reproduction awaits further investigation.

The dinoflagellates are abundant in the plankton of the sea and play an important part in the economy of marine life as a whole. A number of parasitic forms are also known. Their hosts include various diatoms, protozoa, various invertebrates and fish.

Some dinoflagellates inhabiting various seas multiply suddenly in enormous numbers within certain areas, and bring about distinct discolorations of water, often referred to as "red tide" or "red water." Occasionally, the red water causes the death of a large number of fishes and of various invertebrates. According to Galtsoff (1948, 1949), the red water which appeared on the west coast of Florida in 1946 and 1947, was due to the presence of an enormous number of *Gymnodinium brevis* (Davis, 1947) and this dinoflagellate seemed in some manner to have been closely correlated with the fatal effect on animals entering the discolored water. Ketchum and Keen (1948) found the total phosphorus content of the water containing dense Gymnodinium populations to be 2.5 to ten times the maximum expected in the sea, and the substance associated with Gymnodinium and other dinoflagellates causes nose and throat irritations in man. Woodcock (1948) observed that similar irritations can be produced by breathing air artificially laden with small drops of the red water containing 56×10^6 dinoflagellates per liter. The irritant substance passed through a fine bacterial filter, and was found to be very stable, remaining active in stored red water for several weeks. (Red tide in the Gulf of Mexico, Galtsoff, 1948; Lasker and Smith, 1954; Lackey and Hynes, 1955; distribution and taxonomy, Kofoid, 1906, 1907, 1909, 1931; Kofoid and Swezy, 1921; Prescott, 1928; Playfair, 1919; Wailes, 1934; Thompson, 1947, 1950; Balech, 1949, 1951, 1956, 1958, 1959; Rampi, 1950; Chatton, 1952; Graham, 1954; Hulburt, 1957; Conrad and Kufferath, 1954; Ballantine, 1961; locomotion, Peter, 1929.)

The Dinoflagellida is subdivided into three suborders: Prorocentrina, Peridiniina (p. 374) and Cystoflagellina (p. 392).

Suborder 1. **Prorocentrina** Poche

Test bivalve; without any groove; with yellow chromatophores; two flagella anterior, one directed anteriorly, the other vibrates in a circle; fresh or salt water.

Family **Prorocentridae** Kofoid

Genus **Prorocentrum** Ehrenberg. Elongate oval; anterior end bluntly pointed, with a spinous projection at pole; chromatophores small, yellowish brown; salt water. Species (Schiller, 1918, 1928).

P. micans E. (Fig. 129, *a*). 36-52μ long; a cause of "red water."

P. triangulatum Martin. Triangular with rounded posterior end; shell-valves flattened; one valve with a delicate tooth; surface covered with minute pores; margin striated; chromatophores yellow-brown, irregular, broken up in small masses; 17-22μ. Martin (1929) found it extremely abundant in brackish water in New Jersey.

Genus **Exuviaella** Cienkowski. Subspherical or oval; no anterior projection, except two flagella; two lateral chromatophores, large,

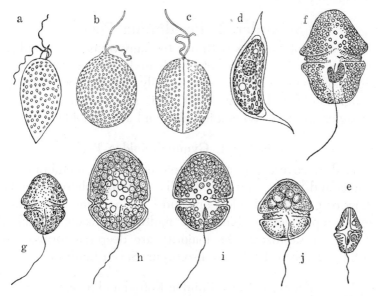

Fig. 129. a, *Prorocentrum micans*, ×420 (Schütt); b, c, *Exuviaella marina*, ×420 (Schütt); d, e, *Cystodinium steini*, ×370 (Klebs); f, *Glenodinium cinctum*, ×590 (Schilling); g, *G. pulvisculum*, ×420 (Schilling); h, *G. uliginosum*, ×590 (Schilling); i, *G. edax*, ×490 (Schilling); j, *G. neglectum*, ×650 (Schilling).

brown, each with a pyrenoid and a starch body; nucleus posterior; salt and fresh water. In *E. baltica*, Braarud *et al.* (1958) found under electron microscope the theca is spinous and the cell wall is made up of cellulose fibrils. The spines may play an important role in obtaining nourishment. Several species (Schiller, 1918, 1928).

E. marina C. (Fig. 129, *b*, *c*). 36–50μ long.

E. apora Schiller. Compressed, oval; striae on margin of valves; chromatophores numerous yellow-brown, irregular in form; 30-32u by 21-26μ (Schiller); 17-22μ by 14-19u (Lebour; Martin); common in brackish water, New Jersey.

E. compressa (Stein). Flattened ellipsoid test; anterior end with a depression through which two flagella emerge; two chromatophores pale or deep green, each with a pyrenoid; nucleus posterior; no stigma; 22-26μ by 15-18μ by 11-12μ; fresh and salt water (Thompson, 1950).

Suborder 2. **Peridiniina** Poche

Typical dinoflagellates with one to many transverse annuli and a sulcus; two flagella, one of which undergoes a typical undulating movement, while the other usually directed posteriorly. Following Kofoid and Swezy (1921), this suborder is divided into two superfamilies: Gymnodinioidea and Peridinioidea (p. 386).

Superfamily 1. **Gymnodinioidea** Poche

Naked or covered by a single piece cellulose membrane with annulus and sulcus, and two flagella; chromatophores abundant, yellow or greenish platelets or bands; stigma sometimes present; asexual reproduction, binary or multiple division; holophytic, holozoic, or saprozoic; the majority are deep-sea forms; a few coastal or fresh water forms also occur. Seven families.

Family 1. **Cystodiniidae** Kofoid and Swezy

Genus **Cystodinium** Klebs. In swimming phase, oval, with extremely delicate envelope; annulus somewhat acyclic; cyst-membrane drawn out into two horns. Species (Pascher, 1928; Thompson, 1949).

C. steini K. (Fig. 129, *d, e*). Stigma beneath sulcus; chromatophores brown; swarmer about 45μ long; freshwater ponds.

Genus **Glenodinium** Ehrenberg (*Glenodiniopsis, Stasziecella,* Woloszynska). Spherical; ellipsoidal or reniform in end-view; annulus a circle; several discoidal, yellow to brown chromatophores; horseshoe- or rod-shaped stigma in some; often with gelatinous envelope; fresh water. Many species (Thompson, 1950).

G. cinctum E. (Fig. 129, *f*). Spherical to ovoid; annulus equatorial; stigma horseshoe-shaped; 43μ by 40μ. (Morphology and reproduction, Lindemann, 1929.)

G. pulvisculum Stein (Fig. 129, *g*). No stigma; 38μ by 30μ.

G. uliginosum Schilling (Fig. 129, *h*). 36-48μ by 30μ.

G. edax S. (Fig. 129, *i*). 34μ by 33μ.

G. neglectum S. (Fig. 129, *j*). 30-32μ by 29μ.

Family 2. **Pronoctilucidae** Lebour

Genus **Pronoctiluca** Fabre-Domergue. Body with an anteroventral tentacle and sulcus; annulus poorly marked; salt water.

P. tentaculatum (Kofoid and Swezy) (Fig. 130, *a*). About 54μ long; off California coast.

Genus **Oxyrrhis** Dujardin. Subovoidal, asymmetrical posteriorly; annulus incomplete; salt water.

O. marina D. (Fig. 130, *b*). 10-37μ long. (Division, Dunkerly, 1921; Hall, 1925.)

Family 3. **Pouchetiidae** Kofoid and Swezy

Ocellus consists of lens and melanosome (pigment mass); sulcus and annulus somewhat twisted; pusules usually present; cytoplasm colored; salt water (pelagic).

Genus **Pouchetia** Schütt. Nucleus anterior to ocellus; ocellus with red or black pigment mass with a red, brown, yellow, or colorless central core; lens hyaline; body surface usually smooth; holozoic, saprozoic; encystment common; salt water. Many species (Schiller, 1928a).

P. fusus S. (Fig. 130, *c*). About 94μ by 41μ; ocellus 27μ long.

P. maxima Kofoid and Swezy (Fig. 130, *d*). 145μ by 92μ; ocellus 20μ; off California coast.

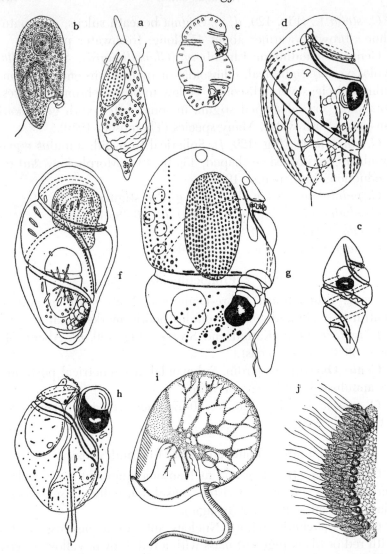

FIG. 130. a, *Pronoctiluca tentaculatum,* ×730 (Kofoid and Swezy) b, *Oxyrrhis marina,* ×840 (Senn); c. *Pouchetia fusus,* ×340 (Schütt); d, *P. maxima,* ×330 (Kofoid and Swezy); e, *Protopsis ochrea,* ×340 (Wright); f, *Nematodinium partitum,* ×560 (Kofoid and Swezy); g. *Proterythropsis crassicaudata,* ×740 (Kofoid and Swezq); h, *Erythropsis cornuta,* ×340 (Kofoid and Swezy; i, j, *Noctiluca scintillans* (i, side view; j, budding process), ×140 (Robin).

Genus **Protopsis** Kofoid and Swezy. Annulus and sulcus similar to those of *Gymnodinium* or *Gyrodinium;* with a simple or compound ocellus; no tentacles; body not twisted; salt water. A few species.

P. ochrea (Wright) (Fig. 130, *e*). 55μ by 45μ; ocellus 22μ long; Nova Scotia.

Genus **Nematodinium** Kofoid and Swezy. With nematocysts; girdle more than one turn; ocellus distributed or concentrated, posterior; holozoic; salt water.

N. partitum K. and S. (Fig. 130, *f*). 91μ long; off California coast.

Genus **Proterythropsis** Kofoid and Swezy. Annulus median; ocellus posterior; a stout rudimentary tentacle; salt water. One species.

P. crassicaudata K. and S. (Fig. 130, *g*), 70μ long; off California.

Genus **Erythropsis** Hertwig. Epicone flattened, less than one-fourth hypocone; ocellus very large, composed of one or several hyaline lenses attached to or imbedded in a red, brownish or black pigment body with a red, brown or yellow core, located at left of sulcus; sulcus expands posteriorly into ventro-posterior tentacle; saprozoic; salt water. Several species.

E. cornuta (Schütt) (Fig. 130, *h*). 104μ long; off California coast (Kofoid and Swezy).

Family 4. **Noctilucidae** Kent

Contractile tentacle arises from sulcal area and extends posteriorly; a flagellum; this group has formerly been included in the Cystoflagellata; studies by recent investigators, particularly by Kofoid, show its affinity with the present suborder; holozoic; salt water.

Genus **Noctiluca** Suriray. Spherical, bilaterally symmetrical; peristome marks the median line of body; cytostome at the botton of peristome; with a conspicuous tentacle and a short flagellum; cytoplasm greatly vacuolated, and cytoplasmic strands connect the central mass with periphery; specific gravity is less than that of sea water, due to the presence of an osmotically active substance with a lower specific gravity than sodium chloride, which appears to be ammonium chloride (Goethard and Heinsius); certain granules are luminescent (Fig. 131); cytoplasm colorless or blue-green; sometimes tinged with yellow coloration in center; swarmers formed by budding, and each possesses one flagellum, annulus, and tentacle; widely distributed in salt water; holozoic. One species.

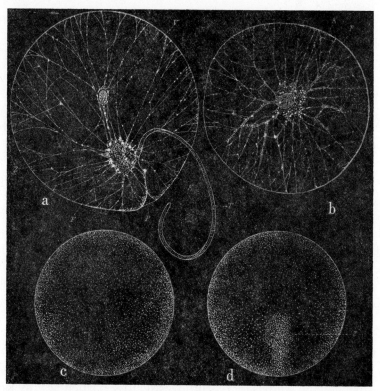

FIG. 131. *Noctiluca scintillans,* as seen under darkfield microscope (Pratje). a, an active individual; b, a so-called "resting stage," with fat droplets in the central cytoplasm, prior to either division or swarmer formation; c, d, appearance of luminescent individuals (F, fat-droplets; K, nucleus; P, peristome; T, tentacle; V, food body; Z, central protoplasm).

N. scintillans (Macartney) (*N. miliaris* S.) (Fig. 130, *i, j;* 131). Usually 500–1000μ in diameter, with extremes of 200μ and 3 mm. Gross (1934) observed that complete fusion of two swarmers (isogametes) results in cyst formation from which trophozoites develop. Acid content of the body fluid is said to be about pH 3.

(Nuclear division, Calkins, 1898; morphology and physiology, Goor, 1918; Kofoid, 1920; Pratje, 1921; feeding, Hofker, 1930; luminescence, Harvey, 1952; Nicol, 1958.)

Genus **Pavillardia** Kofoid and Swezy. Annulus and sulcus similar to those of *Gymnodinium*; longitudinal flagellum absent; stout finger-like mobile tentacle directed posteriorly; salt water. One species.

P. tentaculifera K. and S. 58μ by 27μ; pale yellow; off California.

Family 5. Gymnodiniidae Kofoid

Naked forms with simple but distinct 1/2-4 turns of annulus; with or without chromatophores; fresh or salt water.

Genus **Gymnodinium** Stein. Pellicle delicate; subcircular; bilaterally symmetrical; numerous discoid chromatophores varicolored (yellow to deep brown, green, or blue) or sometimes absent; stigma present in few; many with mucilaginous envelope; salt, brackish, or fresh water. Numerous species (Schiller, 1928a; Hulburt, 1957; cultivation and development, Lindemann, 1929).

G. aeruginosum S. (Fig. 132, *a*). Green chromatophores; 20-32μ by 13-25μ (Thompson, 1950); ponds and lakes.

G. rotundatum Klebs (Fig. 132, *b*). 32-35μ by 22-25μ; fresh water.

G. palustre Schilling (Fig. 132, *c*). 45μ by 38μ; fresh water.

G. agile Kofoid and Swezy (Fig. 132, *d*). About 28μ long; along sandy beaches. McLaughlin and Zahl (1959) found a similar organism in axenic cultures of zooxanthellae in coelenterates (Cassiopeia, Condylactis), though there appeared to be a slight difference in size

Genus **Hemidinium** Stein. Asymmetrical; oval; annulus about half a turn, only on left half. One species.

H. nasutum S. (Fig. 132, *e*). Sulcus posterior; chromatophores yellow to brown; with a reddish brown oil drop; nucleus posterior; transverse fission; 24-28μ by 16-17μ; fresh water.

Genus **Amphidinium** Claparède and Lachmann. Form variable; epicone small; annulus anterior; sulcus straight on hypocone or also on part of epicone; with or without chromatophores; mainly holophytic, some holozoic; coastal or fresh water. Numerous species (Schiller, 1928a; Hulburt, 1957).

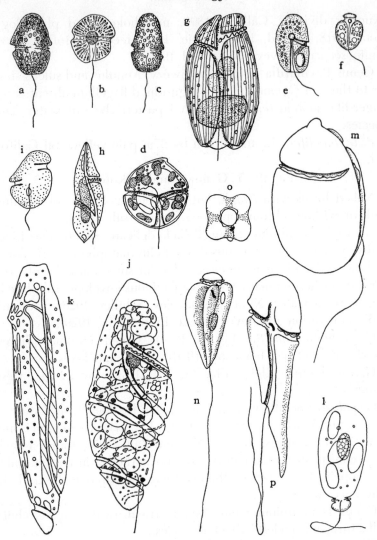

FIG. 132. a, *Gymnodinium aeruginosum*, ×500 (Schilling); b, *G. rotundatum*, ×360 (Klebs); c, *G. palustre*, ×360 (Schilling); d, *G. agile*, ×740 (Kofoid and Swezy); e, *Hemidinium nasutum*, ×670 (Stein); f, *Amphidinium lacustre*, ×440 (Stein); g, *A. scissum*, ×880 (Kofoid and Swezy); h, *Gyrodinium biconicum*, ×340 (Kofoid and Swezy); i, *G. hyalinum*, ×670 (Kofoid and Swezy); j, *Cochlodinium atromaculatum*, ×340 (Kofoid and Swezy); k, *Torodinium robustum*, ×670 (Kofoid and Swezy); l, *Massartia nieuportensis*, ×670 (Conrad); m, *Chilodinium cruciatum*, ×900 (Conrad); n, o, *Trochodinium prismaticum*, ×1270 (Conrad); p, *Ceratodinium asymmetricum*, ×670 (Conrad).

A. lacustre Stein (Fig. 132, *f*). 30μ by 18μ; in fresh and salt (?) water.

A. scissum Kofoid and Swezy (Fig. 132, *g*). 50-60μ long; along sandy beaches.

A. fusiforme Martin. Fusiform, twice as long as broad: circular in cross-section; epicone rounded conical; annulus anterior; hypocone two to two and a half times as long as epicone; sulcus obscure; body filled with yellowish green chromatophores except at posterior end; stigma dull orange, below girdle; nucleus ellipsoid, posterior to annulus; pellicle delicate; 17-22μ by 8-11μ in diameter. Martin (1929) found that it was extremely abundant in parts of Delaware Bay and gave rise to red coloration of the water ("Red water").

Genus **Gyrodinium** Kofoid and Swezy. Annulus descending left spiral; sulcus extending from end to end; nucleus central; pusules; surface smooth or striated; chromatophores rarely present; cytoplasm colored; holozoic; salt or fresh water. Many species (Schiller, 1928a; Hulburt, 1957).

G. biconicum K. and S. (Fig. 132, *h*). 68μ long; salt water; off California.

G. hyalinum Schilling (Fig. 132, *i*). About 24μ long; fresh water.

Genus **Cochlodinium** Schütt. Twisted at least 1.5 turns, annulus descending left spiral; pusules; cytoplasm colorless to highly colored; chromatophores rarely present; holozoic; surface smooth or striated; salt water. Numerous species (Schiller, 1928a).

C. atromaculatum Kofoid and Swezy (Fig. 132, *j*). 183-185μ by 72μ; longitudinal flagellum 45μ long; off California.

Genus **Torodinium** Kofoid and Swezy. Elongate; epicone several times longer than hypocone; annulus and hypocone form augur-shaped cone; sulcus long; nucleus greatly elongate; salt water. Two species (Schiller, 1928).

T. robustum K. and S. (Fig. 132, *k*). 67-75μ long; off California.

Genus **Massartia** Conrad. Cylindrical; epicone larger (nine to ten times longer and three times wider) than hypocone; no sulcus; with or without yellowish discoid chromatophore (Thompson, 1950; Conrad and Kufferath, 1954; Hulburt, 1957). (Axenic culture, Hulburt *et al.*, 1960.)

M. nieuportensis C. (Fig. 132, *l*). 28-37μ long; brackish water.

Genus **Chilodinium** Conrad. Ellipsoid; posterior end broadly rounded, anterior end narrowed and drawn out into a digitform process closely adhering to body; sulcus, apex to one-fifth from posterior end; annulus oblique, in anterior one-third (Conrad, 1926.)

C. cruciatum C. (Fig. 132, *m*). 40-50µ by 30-40µ; with trichocysts; brackish water.

Genus **Trochodinium** Conrad. Somewhat similar to *Amphidinium*; epicone small, button-like; hypocone with four longitudinal rounded ridges; stigma; without chromatophores.

T. prismaticum C. (Fig. 132, n, *o*). 18-22µ by 9-12µ epicone 5-7µ in diameter; brackish water (Conrad, 1926).

Genus **Ceratodinium** Conrad. Cuneiform; asymmetrical, colorless, more or less flattened; annulus complete, oblique; sulcus on half of epicone and full length of hypocone; stigma.

C. asymmetricum C. (Fig. 132, *p*). 68-80µ by about 10µ; brackish water (Conrad, 1926).

Family 6. **Blastodiniidae** Kofoid and Swezy

All parasitic in or on plants and animals; in colony forming genera, there occur **trophocyte** (Chatton) by which organism is attached to host and more or less numerous **gonocytes** (Chatton). (Taxonomy, Chatton, 1920; Reichenow, 1930; Hollande and Enjumet, 1955.)

Genus **Blastodinium** Chatton. In the gut of copepods; spindleshaped, arched, ends attenuated; envelope (not cellulose) often with two spiral rows of bristles; young forms binucleate; when present, chromatophores in yellowish brown network; swarmers similar to those of *Gymnodinium;* in salt water. Many species.

B. spinulosum C. (Fig. 133, *a*). About 235µ by 33-39µ; swarmers 5-10µ; in *Palacalanus parvus, Clausocalanus arcuicornis* and *C. furcatus.*

Genus **Oodinium** Chatton. Spherical or pyriform; with a short stalk; nucleus large; often with yellowish pigment; on Salpa, Annelida, Siphonophora, marine fishes, etc.

O. poucheti Lemmermann (Fig. 133, *b, c*). Fully grown individuals up to 170µ long; bright yellow ochre; mature forms become detached and free, dividing into numerous gymnodinium-like swarmers; on the tunicate, *Oikopleura dioica.*

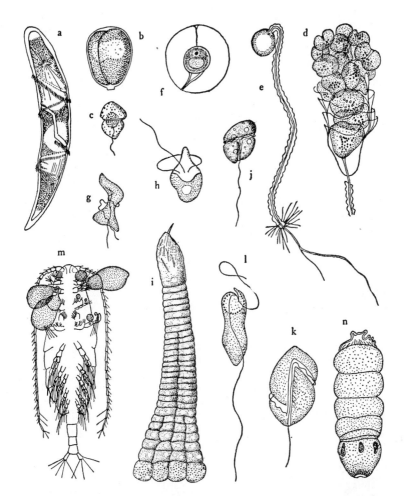

Fig. 133. a, *Blastodinium spinulosum,* ×240 (Chatton); b, c, *Oodinium poucheti* (c, a swarmer) (Chatton); d, e, *Apodinium mycetoides* (d, swarmer-formation, ×450; e, a younger stage, ×640) (Chatton); f, *Chytriodinium parasiticum* in a copepod egg (Dogiel); g, *Trypanodinium ovicola,* ×1070 (Chatton); h, *Duboscqella tintinnicola* (Duboscq and Collin); i, j, *Haplozoon clymenellae* (i, mature colony, ×300; j. a swarmer, ×1340) (Shumway); k, *Syndinium turbo,* ×1340; (Chatton); l, *Paradinium poucheti,* ×800 (Chatton); m, *Ellobiopsis chattoni* on *Calanus finmarchicus* (Caullery); n, *Paraellobiopsis coutieri* (Collin).

O. ocellatum Brown (Fig. 134, *a, b*). Attached to the gill filaments of marine fish by means of cytoplasmic processes; oval in
form; 12μ by 10μ to 104μ by 80μ, average 60μ by 50μ; nucleus
spherical; many chromatophores and starch grains; a stigma.
When grown, the organism drops off the gill and becomes enlarged
to as much as 150μ in diameter. Soon the cytoplasmic processes

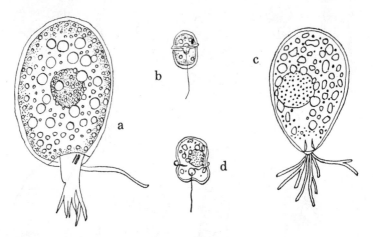

Fɪɢ. 134. a, *Oodinium ocellatum* recently detached from host gill; b, a
free living flagellate form, ×760 (Nigrelli); c, d, *O. limneticum*, ×800
(Jacobs).

and the broad flagellum are retracted and the aperture of shell
closes by secretion of cellulose substance. The body divides up to
128 cells, which become flagellated and each divides once more.
These flagellates, 12μ by 8μ, reach the gills of fish and become
attached (Brown, 1931; Nigrelli, 1936).

O. limneticum Jacobs. (Fig. 134, *c, d*). Pyriform; 12μ by 7.5μ
to 20μ by 13μ; light green chromatophores variable in size and
shape; no stigma; without flagella; filopodia straight or branched;
the organism grows into about 60μ long in three days at 25°C.;
observed maximum, 96μ by 80μ; starch becomes abundant;
fission takes place in cyst; flagellate forms measure about 15μ
long; ectoparasitic on the integument of freshwater fishes in
aquaria (Jacobs, 1946).

Genus **Apodinium** Chatton. Young individuals elongate, spherical or pyriform; binucleate; adult colorless; formation of numerous swarmers in adult stage is peculiar in that lower of the two individuals formed at each division secretes a new envelope, and delays its further division until the upper one has divided for the second time, leaving several open cups; on tunicates.

A. *mycetoides* C. (Fig. 133, *d*, *e*). On gill-slits of *Fritillaria pellucida*.

Genus **Chytriodinium** Chatton. In eggs of planktonic copepods; young individuals grow at the expense of host egg and when fully formed, body divides into many parts, each producing four swarmers. Several species.

C. *parasiticum* (Dogiel) (Fig. 133, *f*). In copepod eggs; Naples.

Genus **Trypanodinium** Chatton. In copepod eggs; swarmer-stage only known.

T. *ovicola* C. (Fig. 133, *g*). Swarmers biflagellate; about 15µ long.

Genus **Duboscqella** Chatton. Rounded cell with a large nucleus; parasitic in Tintinnidae. One species.

D. *tintinnicola* Lohmann (Fig. 133, *h*). Intracellular stage oval, about 100µ in diameter with a large nucleus; swarmers biflagellate.

Genus **Haplozoon** Dogiel in gut of polychaetes; mature forms composed of variable number of cells arranged in line or in pyramid; salt water. Many species.

H. *clymenellae* (Calkins). (*Microtaeniella clymenellae* C.) (Fig. 133 *i*, *j*). In the intestine of *Clymenella torquata;* colonial forms consist of 250 or more cells; Woods Hole (Shumway, 1924).

Genus **Syndinium** Chatton. In gut and body cavity of marine copepods; multinucleate round cysts in gut considered as young forms; multinucleate body in host body cavity with numerous needle-like inclusions.

S. *turbo* C. (Fig. 133, *k*). In *Paracalanus parvus*, *Corycaeus venustus*, *Calanus finmarchicus;* swarmers about 15µ long.

Genus **Paradinium** Chatton. In body-cavity of copepods; multinucleate body without inclusions; swarmers formed outside the host body.

P. *poucheti* C. (Fig. 133, *l*). In the copepod, *Acartia clausi;* swarmers about 25µ long, amoeboid.

Genus **Ellobiopsis** Caullery. Pyriform; with stalk; often a septum near stalked end; attached to anterior appendages of marine
copepods.

E. chattoni C. (Fig. 133, *m*). Up to 700µ long; on antennae and
oral appendages of *Calanus finmarchicus, Pseudocalanus elongatus* and *Acartia clausi.* (Development, Steuer, 1928.)

Genus **Paraellobiopsis** Collin. Young forms stalkless; spherical;
mature individuals in chain-form; on Malacostraca.

P. coutieri C. (Fig. 133, *n*). On appendages of *Nebalia bipes.*

Family 7. **Polykrikidae** Kofoid and Swezy

Two, four, eight, or sixteen individuals permanently joined; individuals similar to *Gymnodinium;* sulcus however extending entire body length; with nematocysts (Fig. 135, *b*); greenish to
pink; nuclei about one-half the number of individuals; holozoic;
salt water. (Nematocysts, Hovasse, 1951.)

Genus **Polykrikos** Bütschli. With the above-mentioned characters; salt or brackish water. Species (Schiller, 1928; Hulburt,
1957).

P. kofoidi Chatton (Fig. 135, *a, b*). Greenish grey to rose;
composed of two, four, eight, or sixteen individuals; with nematocysts; each nematocyst possesses presumably a hollow thread, and
discharges under suitable stimulation its content; a binucleate colony composed of four individuals about 110µ long; off California.

P. barnegatensis Martin. Ovate, nearly circular in cross-section,
slightly concave ventrally; composed of two individuals; constriction slight; beaded nucleus in center; annuli descending left spiral, displaced twice their width; sulcus ends near anterior end; cytoplasm colorless, with numerous oval, yellow-brown chromatophores; nematocysts absent; 46µ by 31.5µ; in brackish water of
Barnegat Bay.

Superfamily 2. **Peridinioidea** Poche

The shell composed of epitheca, annulus and hypotheca, which
may be divided into numerous plates; body form variable. Three
families.

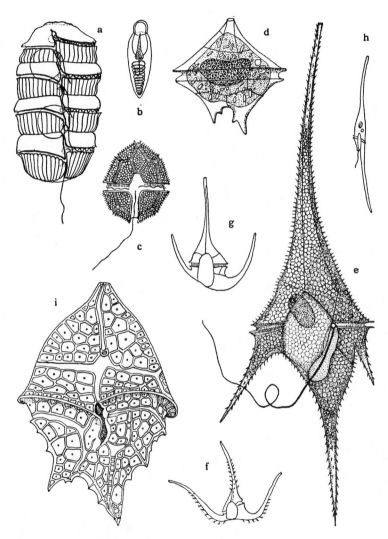

FIG. 135. a, b, *Polykrikos kofoidi* (a, colony of four individuals, ×340; b, a nematocyst, ×1040) (Kofoid and Swezy); c, *Peridinium tabulatum*, ×460 (Schilling); d, *P. divergens*, ×340 (Calkins); e, *Ceratium hirundinella*, ×540 (Stein); f. *C. longipes*, ×100 (Wailes); g, *C. tripos*, ×140 (Wailes); h, *C. fusus*, ×100 (Wailes); i, *Heterodinium scrippsi*, ×570 (Kofoid and Adamson).

Family 1. **Peridiniidae** Kent

Shell composed of numerous plates; annulus usually at equator, covered by a plate known as **cingulum**; variously sculptured and finely perforated plates vary in shape and number among different species; in many species certain plates drawn out into various processes, varying greatly in different seasons and localities even among one and the same species; these processes seem to retard descending movement of organisms from upper to lower level in water when flagellar activity ceases; chromatophores numerous small platelets, yellow or green; some deep-sea forms without chromatophores; chain formation in some forms; mostly surface and pelagic inhabitants in fresh or salt water.

Genus **Peridinium** Ehrenberg. Subspherical to ovoid; reniform in cross-section; annulus slightly spiral with projecting rims; hypotheca often with short horns and epitheca drawn out; colorless, green, or brown; stigma usually present; cysts spherical. In *P. trochoideum*, prior to division, the organism emerges from the theca and after division, a new theca is formed by each daughter; the cysts that often appear in culture, are red and smooth-walled (Braarud, 1958). Salt or fresh water. Numerous species. (Species and variation, Böhm, 1933; Diwald, 1939; Balech, 1959; chromatophore and pyrenoids, Geitler, 1926.)

P. tabulatum Claparède and Lachmann (Fig. 135, *c*). 48µ by 44µ; fresh water.

P. divergens Ehrenberg (Fig. 135, *d*). About 45µ in diameter; yellowish, salt water.

Genus **Ceratium** Schrank. Body flattened; with one anterior and one to four posterior horn-like processes; often large; chromatophores yellow, brown, or greenish; color variation conspicuous; fission is said to take place at night and in the early morning; fresh or salt water. Numerous species; specific identification is difficult due to a great variation (p. 268). (Biology and morphology, Entz, 1927; encystment, Entz, 1925.)

C. hirundinella Müller (Figs. 96; 135, *e*). One apical and two to three antapical horns; seasonal and geographical variations (p. 268); chain-formation frequent; 95-700µ long; fresh and salt water. Numerous varieties. (Reproduction, Entz, 1921, 1931;

Hall, 1925a; Borgert, 1935; holozoic nutrition, Hofeneder, 1930.)

C. longipes Bailey (Fig. 135, *f*). About 210μ by 51-57μ; salt water.

C. tripos Müller (Fig. 135, *g*). About 225μ by 75μ; salt water. Wailes (1928) observed var. *atlantica* in British Columbia; Martin (1929) saw it in Barnegat Inlet, New Jersey. Nuclear division (Schneider, 1924).

C. fusus (Ehrenberg) (Fig. 135, *h*). 300-600μ by 15-30μ, salt water; widely distributed; British Columbia (Wailes), New Jersey (Martin), etc.

Genus **Heterodinium** Kofoid. Flattened or spheroidal; two large antapical horns; annulus submedian; with post-cingular ridge; sulcus short, narrow; shell hyaline, reticulate, porulate; salt water. Numerous species.

H. scrippsi K. (Fig. 135, *i*). 130-155μ long; Pacific and Atlantic (tropical).

Genus **Dolichodinium** Kofoid and Adamson. Subconical, elongate; without apical or antapical horns; sulcus only one-half the length of hypotheca; plate porulate; salt water.

D. lineatum Kofoid and Michener (Fig. 136, *a*). 58μ long; eastern tropical Pacific.

Genus **Goniodoma** Stein. Polyhedral with a deep annulus; epitheca and hypotheca slightly unequal in size, composed of regularly arranged armored plates; chromatophores small brown platelets; fresh or salt water.

G. acuminata (Ehrenberg) (Fig. 136, *b*). About 50μ long; salt water.

Genus **Gonyaulax** Diesing. Spherical, polyhedral, fusiform, elongated with stout apical and antapical prolongations, or dorsoventrally flattened; apex never sharply attenuated; annulus equatorial; sulcus from apex to antapex, broadened posteriorly; plates 1-6 apical, 0-3 anterior intercalaries, six precingulars, six annular plates, six postincingulars, one posterior intercalary and one antapical; porulate; chromatophores yellow to dark brown, often dense; without stigma; fresh, brackish or salt water. Numerous species (Kofoid, 1911; Whedon and Kofoid, 1936).

G. polyedra Stein (Fig. 136, *c*). Angular, polyhedral; ridges along sutures, annulus displaced one to two annulus widths, regu-

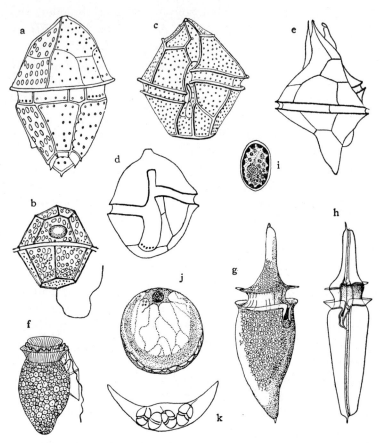

FIG. 136. a, *Dolichodinium lineatum,* ×670 (Kofoid and Adamson); b, *Goniodoma acuminata,* ×340 (Stein); c, *Gonyaulax polyedra,* ×670 (Kofoid); d, *G. apiculata,* ×670 (Lindemann); e, *Spiraulax jolliffei,* right side of theca, ×340 (Kofoid); f, *Dinophysis acuta,* ×580 (Schütt); g, h, *Oxyphysis oxytoxoides,* ×780 (Kofoid); i, *Phytodinium simplex,* ×340 (Klebs); j. k, *Dissodinium lunula,* j, primary cyst (Dogiel); k, secondary cyst with 4 swarmers (Wailes), ×220.

larly pitted; salt water. "Very abundant in the San Diego region in the summer plankton, July-September, when it causes local outbreaks of 'red water,' which extend along the coast of southern and lower California" (Kofoid, 1911; Allen, 1946). The organisms occurred also in abundance (85 per cent of plankton) in pools of sea water off the beach of Areia Branca, Portugal, and caused "red water" during the day and an extreme luminescence when agitated at night (Santos-Pinto, 1949).

When cultured under alternating light and dark periods of twelve hours each, Sweeney and Hastings (1958) found that at least 85 per cent of all divisions which occur in a day took place during a five-hour period toward the end of the dark period; and a maximum number of newly divided organisms are found at about the time the light period begins. (Luminescent reaction, Hastings and Sweeney, 1957.)

G. apiculata (Penard) (Fig. 136, *d*). Ovate, chromatophores yellowish brown; 30-60µ long; fresh water.

G. monilata Howell. About 37µ long; 57µ in diameter; chain-formation. Howell (1953) considered it as the cause of a red tide which spread on the east coast of Florida in late summer, 1951.

Genus **Spiraulax** Kofoid. Biconical; apices pointed; sulcus not reaching apex; no ventral pore; surface heavily pitted; salt water.

S. jolliffei (Murray and Whitting) (Fig. 136, *e*). 132µ by 92µ; California (Kofoid, 1911a).

Genus **Woloszynskia** Thompson (1950). An apparently intermediate form between Gymnodinioidea and Peridinioidea.

W. limnetica Bursa. Subspherical; 28-63µ by 30-60µ; epicone dome-shaped; hypocone broadly rounded; annulus straight or displaced; flagella long, up to twice the body length; five to nine chromatophores spherical or subspherical, but in some many and clavate; pyriform nucleus apical; stigma brown red, V- or U-shaped; numerous starch grains; cysts spherical or subspherical; body wall reticulated; fresh water (Bursa, 1958).

Family 2. **Dinophysidae** Kofoid

Genus **Dinophysis** Ehrenberg. Highly compressed; annulus widened, funnel-like, surrounding small epitheca; chromatophores yellow; salt water. Several species (Schiller, 1928). (Morphology and taxonomy, Tai and Skogsberg, 1934.)

D. acuta E. (Fig. 136, *f*). Oval; attenuated posteriorly; 54-94µ long; widely distributed; British Columbia (Wailes).

Genus **Oxyphysis** Kofoid. Epitheca developed; sulcus short; sulcal lists feebly developed; sagittal suture conspicuous; annulus impressed; salt water (Kofoid, 1926).

O. oxytoxoides K. (Fig. 136, g, *h*). 63-68µ by 15µ; off Alaska.

Family 3. **Phytodiniidae** Klebs

Genus **Phytodinium** Klebs. Spherical or ellipsoidal; without furrows; chromatophores discoidal, yellowish brown.

P. simplex K. (Fig. 136, *i*). Spherical or oval; 42-50μ by 30-45μ fresh water.

Genus **Dissodinium** Klebs (*Pyrocystis* Paulsen). Primary cyst, spherical, uninucleate; contents divide into eight to sixteen crescentic secondary cysts which become set free; in them are formed two, four, six, or eight Gymnodinium-like swarmers; salt water.

D. lunula (Schütt) (Fig. 136, *j*, *k*). Primary cysts 80-155μ in diameter; secondary cysts 104-130μ long; swarmers 22μ long; widely distributed; British Columbia (Wailes).

Suborder 3. **Cystoflagellina** Haeckel

Since Noctiluca which had for many years been placed in this suborder, has been removed, according to Kofoid, to the second suborder, the Cystoflagellina becomes a highly ill-defined group and includes two peculiar marine forms: ***Leptodiscus medusoides***

Fig. 137. a, *Leptodiscus medusoides,* ×50 (Hertwig); b, *Craspedotella pileolus,* ×110 (Kofoid).

Hertwig (Fig. 137, *a*). and ***Craspedotella pileolus*** Kofoid (Fig. 137, *b*), both of which are medusoid in general body form.

References

ALLEN, W. E., 1946. Significance of "red water" in the sea. Turtox news, 24:49.

BALECH, E., 1949. Etude de quelques espèces de Peridinium, souvent confondues. Hydrobiologia, 1:390.

―――― 1951. Deuxième contribution à la connaissance des Peridinium. *Ibid.*, 3:305.

―――― 1956. Étude des Dinoflagellés du sable de Roscoff. Rev. Algol., 2:29.

—— 1958. Plancton de la Campaña Antártica Argentina. 1954-1955. Physis, 21:75.

BALLANTINE, D., 1961. *Gymnodinium chukwanii* n. sp. and other marine dinoflagellates, etc. J. Protozool., 8:217.

Böhm, A., 1933. Beobachtungen an adriatischen Peridinium-Arten. Arch. Protist., 80:303.

BORGERT, A., 1935. Fortpflanzungsvorgänge und Heteromorphismus bei marinen Ceratien, etc. *Ibid.*, 86:318.

BRAARUD, T., 1958. Observations on *Peridinium trochoideum* in culture. Nytt Mag. Bot., 6:39.

—— *et al.*, 1958. A note on the thecal structure of *Exuviaella baltica. Ibid.*, 6:43.

BROWN, E. M., 1931. Note on a new species of dinoflagellate from the gills and epidermis of marine fishes. Proc. Zool. Soc. London, 1:345.

BURSA, A., 1958. The freshwater dinoflagellate *Woloszynskia limnetica* n.sp. Membrane and protoplasmic structure. J. Protozool., 5:299.

CALKINS, G. N., 1898. Mitosis in *Noctiluca milliaris.* J. Morph., 15. 58 pp.

CHATTON, E., 1920. Les Péridiniens parasites: Morphologie, reproduction, ethologie. Arch. zool. exper. gén., 59:1.

—— 1952. Classe des Dinoflagellés ou Péridiniens. In Grassé (1952), p. 310.

CONRAD, W., 1926. Recherches sur les flagellates de nos eaux saumâtres. I. Arch. Protist, 55:63.

—— and KUFFERATH, H., 1954. Recherches sur les eaux saumâtres des environs de Lilloo. II. Mém. Inst. Roy. Sci. Nat. Belgique, No. 127.

DAVIS, C. C., 1947. *Gymnodinium brevis* n. sp., a cause of discolored water and animal mortality in the Gulf of Mexico. Bot. Gaz., 109:358.

DIWALD, K., 1939. Ein Beitrag zur Variabilität und Systematik der Gattung Peridinium. Arch. Protist., 93:121.

DUNKERLY, J. S., 1921. Nuclear division in the dinoflagellate, *Oxyrrhis marina.* Proc. Roy. Phys. Soc., Edinburgh, 20:217.

EDDY, S., 1930. The freshwater armored or thecate dinoflagellates. Tr. Am. Micr. Soc., 49:1.

ENTZ, G., 1921. Ueber die mitotische Teilung von *Ceratium hirundinella.* Arch. Protist., 43:415.

—— 1925.. Ueber Cysten und Encystierung der Süsswasser-Ceratien. *Ibid.*, 51:131.

—— 1927. Beiträge zur Kenntnis der Peridineen. *Ibid.*, 58:344.

—— 1931. Analyse des Wachstums und Teilung einer Population sowie eines Individuums des Protisten Ceratium, etc. *Ibid.*, 74:310.

FRITSCH, F. E., 1935. The structure and reproduction of the algae.

GALTSOFF, P. S., 1948. Red tide: etc. Spec. Sc. Rep., U. S. Fish Wildl. Service, no. 46.

—— 1949. The mystery of the red tide. Sc. Monthly, 68:109.

GEITLER, L., 1926. Ueber Chromatophoren und Pyrenoide bei Peridineen, Arch. Protist., 53:343.

GOOR, A. C. J. VAN, 1918. Die Cytologie von *Noctiluca miliaris*. *Ibid.*, 39: 147.

GRAHAM, H. W., 1943. *Gymnodinium catenatum*, etc. Tr. Am. Micr. Soc., 62:259.

———— 1954. Dinoflagellates of the Gulf of Mexico. U. S. Department Inter., Fish. Bull., 89:223.

GROSS, F., 1934. Zur Biologie und Entwicklungsgeschichte von *Noctiluca miliaris*. Arch. Protist., 83:178.

HALL, R. P., 1925. Binary fission in *Oxyrrhis marina*. Univ. Cal. Publ. Zool., 26:281.

———— 1925a. Mitosis in *Ceratium hirundinella*, etc. *Ibid.*, 28:29.

HARVEY, E. N., 1952. Bioluminescence. New York.

HASTINGS, J. W., and SWEENEY, B. M., 1957. The luminescent reaction in extracts of the marine dinoflagellate *Gonyaulax polyedra*. J. Cell Comp. Physiol., 49:209.

HOFENEDER, H., 1930. Ueber die animalische Ernährung von Ceratium, etc. Arch. Protist., 71:1.

HOFKER, J., 1930. Ueber *Noctiluca scintillans*. *Ibid.*, 71:57.

HOLLANDE, A., and EUJUMET, M., 1955. Parasites et cycles évolutif des Radiolaires et des Acanthaires. Bull. Stat. d'Aquic. Pêche, Castiglione. Nouv. Ser., 7:152.

HOVASSE, R., 1951. Contribution a l'étude de la cnidogénèse chez les Péridiniens. I. Arch. zool. exper. gén., 87:299.

HOWELL, J. F., 1953. *Gonyaulax monilata* n. sp., etc. Tr. Am. Micr. Soc., 72:153.

HULBURT, E. M., 1957. The taxonomy of unarmored Dinophyceae of shallow embayments on Cape Cod, Massachusetts. Biol. Bull., 112:196.

———— et al., 1960. *Katodinium dorsalisulcum*, a new species of unarmored Dinophyceae. J. Protozool., 7:323.

JACOBS, D. L., 1946. A new parasitic dinoflagellate from fresh water fish. Tr. Am. Micr. Soc., 65:1.

KETCHUM, B. H., and KEEN, J., 1948. Unusual phosphorus concentrations in the Florida "red tide" sea water. J. Mar. Res., 7:17.

KOFOID, C. A., 1907. The plate of Ceratium, etc. Zool. Anz., 32:177.

———— 1909. On *Peridinium steinii*, etc. Arch. Protist., 14:25.

———— 1911. Dinoflagellata of the San Diego Region. IV. Univ. Cal. Publ. Zool., 8:187.

———— 1911a. V. *Ibid.*, 8:295.

———— 1920. A new morphological interpretation of Noctiluca, etc. *Ibid.*, 19:317.

———— 1926. On *Oxyphysis oxytoxoides*, etc. *Ibid.*, 28:203.

———— 1931. Report of the biological survey of Mutsu Bay. XVIII. Sc. Rep. Tohoku Imp. Uni., Biol., 6:1.

——— and ADAMSON, A. M., 1933. The Dinoflagellata: the family Hetero-diniidae, etc. Mem. Mus. Comp. Zool. Harvard, 54:1.

——— and SWEZY, O., 1921. The free-living unarmored Dinoflagellata. Mem. Univ. California, 5:1.

LACKEY, J. B., and HYNES, J. A., 1955. The Florida Gulf coast red tide. Engineer. Progr. Univ. Florida, 9:no. 2 (23 p).

LASKER, R., and SMITH, F. G. W., 1954. Red tide. In: Gulf of Mexico—its origin, waters and marine life. U.S. Dept. Interior, Fish. Bull., 89:173.

LEBOUR, MARIE V., 1925. The dinoflagellates of northern seas. London.

LINDEMANN, E., 1929. Experimentelle Studien über die Fortpflanzungser-scheinungen der Süsswasserperidineen auf Grund von Reinkulturen. Arch. Protist., 68:1.

MARTIN, G. W., 1929. Dinoflagellates from marine and brackish waters of New Jersey. Univ. Iowa Stud. Nat. Hist., 12, no. 9.

McLAUGHLIN, J. J. A., and ZAHL, P. A., 1959. Axenic zooxanthellae from various invertebrate hosts. Ann. N.Y. Acad. Sci., 77:55.

NICOL, J. A. C., 1958. Observations on luminescence in Noctiluca. J. Mar. Biol. Assoc., United Kingdom, 37:535.

NIGRELLI, R. F., 1936. The morphology, cytology and life-history of *Oodinium ocellatum* Brown, etc. Zoologica, 21:129.

PASCHER, A., 1928. Von einer neue Dinococcale, etc. Arch. Protist., 63:241.

PETERS, N., 1929. Ueber Orts- und Geisselbewegung bei marinen Dino-flagellaten. *Ibid.*, 67:291.

PLAYFAIR, G. I., 1919. Peridineae of New South Wales. Proc. Linn. Soc. N. S. Wales, 44:793.

PRATJE, A., 1921. *Noctiluca miliaris* Suriray. Beiträge zur Morphologie, Physiologie und Cytologie. I. Arch. Protist., 42:1.

PRESCOTT, G. W., 1928. The motile algae of Iowa. Univ. Iowa Stud. Nat. Hist., 12:5.

RAMPI, L., 1950. Péridiniens rares ou nouveaux pour la Pacifique Sud-Equatorial. Bull. l'Inst. Océanogr., no. 974.

REICHENOW, E., 1930. Parasitische Peridinea. In: Grimpe's Die Tierwelt der Nord- und Ost-See. Pt. 19, II, d3.

SANTOS-PINTO, J. d., 1949. Um caso de "red water" motivado por abundan-cia anormal de *Gonyaulax poliedra*. Bol. Soc. Port. Ci. Nat., 17:94.

SCHILLER, J., 1918. Ueber neue Prorocentrum- und Exuviella-Arten, etc. Arch. Protist., 38:250.

——— 1928. Die planktischen Vegetationen des adriatischen Meers. I. *Ibid.*, 61:45.

——— 1928a. II. *Ibid.*, 62:119.

SCHILLING, A., 1913. Dinoflagellatae (Peridineae). Die Süsswasserflora Deutschlands. Pt. 3.

SCHNEIDER, H., 1924. Kern und Kernteilung bei *Ceratium tripos*. Arch. Protist., 48:302.

SHUMWAY, W., 1924. The genus Haplozoon, etc. J. Parasit., 11: 59.

STEUER, A., 1928. Ueber *Ellobiopsis chattoni* Caullery, etc. Arch. Protist., 60:501.

SWEENEY, B. M., and HASTINGS, J. W., 1958. Rhythmic cell division in populations of *Gonyaulax polyedra*. J. Protozool., 5:217.

TAI, L.-S., and SKOGSBERG, T., 1934. Studies on the Dinophysoidae, etc. Arch. Protistenk, 82:380.

THOMPSON, R. H., 1947. Freshwater dinoflagellates of Maryland. Chesapeake Biol. Lab. Publ., no. 67.

—— 1949. Immobile Dinophyceae. I. Am. J. Bot., 36:301.

—— 1950. A new genus and new records of freshwater Pyrrophyta, etc. Lloydia, 13:277.

WAILES, G. H., 1928. Dinoflagellates and Protozoa from British Columbia. Vancouver Mus. Notes, 3.

—— 1934. Freshwater dinoflagellates of North America. *Ibid.*, 7, Suppl., 11.

WHEDON, W. F., and KOFOID, C. A., 1936. Dinoflagellates of the San Francisco region. Univ. Cal. Publ. Zool., 41:25.

WOODCOCK, A. H., 1948. Note concerning human respiration irritation with high concentration of plankton and mass mortality of marine organisms. J. Mar. Res., 7:56.

Chapter 13
Subclass 2. Zoomastigia Doflein

THE Zoomastigia lack chromatophores and their body organizations vary greatly from a simple to a very complex type. The majority possess a single nucleus which is, as a rule, vesicular in structure. Characteristic organellae such as parabasal body, axostyle, etc., are present in numerous forms and myonemes are found in some species. Nutrition is holozoic or saprozoic (parasitic). Asexual reproduction is by longitudinal fission; sexual reproduction is known in a few forms. Encystment occurs commonly. The Zoomastigia are free-living or parasitic in various animals.

The Zoomastigia are subdivided into five orders: Rhizomastigida, Protomonadida (p. 405), Polymastigida (p. 438), Trichomonadida (p. 457) and Hypermastigida (p. 480).

Order 1. Rhizomastigida Bütschli

A number of borderline forms between the Sarcodina and the Mastigophora are placed here. Flagella vary in number from one to several and pseudopods also vary greatly in number and in appearance. Two families.

Family 1. Multiciliidae Poche

Genus **Multicilia** Cienkowski. Generally spheroidal, but amoeboid; with forty to fifty flagella, long and evenly distributed; one or more nuclei; holozoic; food obtained by means of pseudopodia; multiplication by fission; fresh or salt water.

M. marina C. (Fig. 138, *a*). 20-30μ in diameter; uninucleate; salt water.

M. lacustris Lauterborn. (Fig. 138, *b*). Multinucleate; 30-40μ in diameter; fresh water.

Family 2. Mastigamoebidae

With one to three or rarely four flagella and axopodia or lobopodia; uninucleate; flagellum arises from a basal granule which is connected with the nucleus by a rhizoplast; binary fission in both

397

trophic and encysted stages; sexual reproduction has been reported in one species; holozoic or saprozoic; the majority are free-living, though a few parasitic.

Genus **Mastigamoeba** Schulze (*Mastigina,* Frenzel). Monomastigote, uninucleate, with finger-like pseudopodia; flagellum long and connected with nucleus; fresh water, soil or endocommensal. (Species, Klug, 1936.)

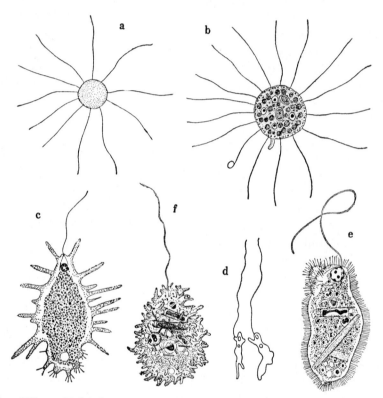

Fig. 138. a, *Multicilia marina,* ×400 (Cienkowski); b, *M. lacustris,* ×400 (Lauterborn); c, *Mastigamoeba aspera,* ×200 (Schulze); d, *M, longifilum,* ×340 (Stokes); e, *M. setosa,* ×370 (Goldschmidt); f, *Mastigella vitrea,* ×370 (Goldschmidt).

M. aspera S. (Fig. 138, *c*). Subspherical or oval; during locomotion elongate and narrowed anteriorly, while posterior end rounded or lobed; numerous pseudopods slender, straight; nucleus near flagellate end; two contractile vacuoles; 150-200μ by about 50μ; in ooze of pond.

M. longfilum Stokes (Fig. 138, *d*). Elongate, transparent; flagellum twice body length; pseudopods few, short; contractile vacuole anterior; body 28μ long when extended, contracted about 10μ; stagnant water.

M. setosa (Goldschmidt) (Fig. 138, *e*). Up to 140μ long.

M. hylae (Frenzel) (Fig. 139, *a*). In the hind-gut of the tadpoles of frogs and toads; 80-135μ by 21-31μ; flagellum about 10μ long (Becker, 1925). Development (Ivanić, 1936).

Genus **Mastigella** Frenzel. Flagellum apparently not connected with nucleus; pseudopods numerous, digitate; body form changes actively and continuously; contractile vacuole.

M. vitrea Goldschmidt (Fig. 138, *f*). 150μ long; sexual reproduction (Goldschmidt).

Genus **Actinomonas** Kent. Generally spheroidal, with a single flagellum and radiating pseudopods; ordinarily attached to foreign object with a cytoplasmic process, but swims freely by withdrawing it; nucleus central; several contractile vacuoles; holozoic.

A. mirabilis K. (Fig. 139, *b*). Numerous simple filopodia; about 10μ in diameter; flagellum 20μ long; fresh water.

Genus **Dimorpha** Gruber. Ovoid or subspherical; with two flagella and radiating axopodia, all arising from an eccentric centriole; nucleus eccentric; pseudopods sometimes withdrawn; fresh water. (Species, Pascher, 1925; taxonomic status, Bovee, 1960.)

D. mutans G. (Fig. 139, *c*). 15-20μ in diameter; flagella about 20-30μ long.

D. floridanis Bovee. Flagellate phase 18-20μ long by 6-7μ wide; two flagella 10-12μ long, one directed anteriorly and the other bent laterally; heliozoan forms 8-10μ with thirty-two axopods; fresh water lake (Bovee, 1960).

Genus **Tetradimorpha** Hsiung. Spherical with radiating axopodia; four flagella originate in a slightly depressed area; nucleus central. When disturbed, all axopodia turn away from the flagellated pole and are withdrawn into body, and the organism undergoes swimming movement; fresh water ponds.

T. radiata H. (Fig. 139, *d, e*). Body 27-38μ in diameter; axopodia 27-65μ long; flagella 38-57μ long (Hsiung, 1927).

Genus **Pteridomonas** Penard. Small, heart-shaped; usually attached with a long cytoplasmic process; from opposite pole there arises a single flagellum, around which occurs a ring of extremely fine filopods; nucleus central; a contractile vacuole; holozoic; fresh water.

P. pulex P. (Fig. 139, *f*). 6-12μ broad.

Genus **Histomonas** Tyzzer. Actively amoeboid; mostly rounded,

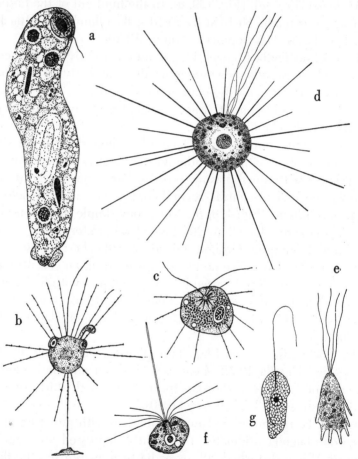

FIG. 139. a, *Mastigamoeba hylae*, ×690 (Becker); b, *Actinomonas mirabilis*, ×1140 (Griessmann); c, *Dimorpha mutans*, ×940 (Blochmann); d, e, *Tetradimorpha radiata* (Hsiung) (d, a typical specimen, ×430; e, swimming individual, ×300); f, *Pteridomonas pulex*, ×540 (Penard); g, *Rhizomastix gracilis*, ×1340 (Mackinnon).

sometimes elongate; a single nucleus; an extremely fine flagellum arises from a blepharoplast, located close to nucleus; axostyle (?) sometimes present; in domestic fowls. (One species.)

H. meleagridis (Smith) (*Amoeba meleagridis* S.) (Fig. 140). Actively amoeboid organism; usually rounded; 8-12μ (average 10-14μ) in the largest diameter; nucleus circular or pyriform with usually a large endosome; a fine flagellum; food vacuoles contain bacteria, starch grains and erythrocytes; binary fission; during division flagellum is discarded; cysts unobserved; in young turkeys, chicks, grouse, and quail. Bayon and Bishop (1937) successfully cultured the organism from hen's liver. (Morphology of the cultured forms, Bishop, 1938.)

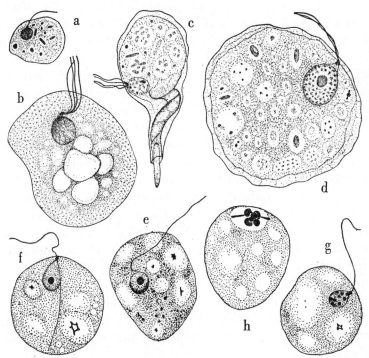

FIG. 140. *Histomonas meleagridis.* a–d, from host animals (Wenrich); e–h, from cultures (Bishop). a, b, organisms in caecum of chicken (in a Tyzzer slide); c, an individual from pheasant showing "ingestion tube" with a bacterial rod; d, a large individual from the same source, all ×1765; e, an amoeboid form; f, a rounded form with axostyle (?); g, h, stages in nuclear division, ×2200.

This organism is the cause of enterohepatitis known as "blackhead," an infectious disease, in young turkeys and also in other fowls, in which it is often fatal. Smith (1895) discovered the organism and considered it an amoeba (1910). It invades and destroys the mucosa of the intestine and caeca as well as the liver tissues. Trophozoites voided in feces by infected birds may become the source of new infection when taken in by young birds with drink or food. Tyzzer (1920) found the organism to possess a flagellate stage and established the genus *Histomonas* for it. Tyzzer and Fabyan (1922) and Tyzzer (1934) demonstrated that the organism is transmissible from bird to bird in the eggs of the nematode *Heterakis gallinae,* which method appears to be a convenient and reliable one for producing *Histomonas* infection in turkeys (McKay and Morehouse, 1948). Desowitz (1950) noticed in a Heterakis two enlarged gut cells filled with amoebulae which he suggested might be a stage of this protozoan. Niimi (1937) reported that the organism enters through the mouth of the nematode and invades its eggs. Dobell (1940) points out the similarity between this flagellate and *Dientamoeba fragilis* (p. 551). Wenrich (1943) made a comparative study of forms found in the caecal smears of wild ring-neck pheasants and of chicks. The organisms measured 5-30μ in diameter and possessed one to four flagella, though often there were no flagella.

Lund (1959) described a non-pathogenic strain which does not invade the host tissue, but lives in the lumen of the caecum and which resembled closely what Wenrich had seen, and reported that the rectal inoculation of "several thousand" organisms into young turkeys on two or three consecutive days brought about protection against moderate rectal challenges with pathogenic strain. (Axenic culture, Lesser, 1961; blood-induced blackhead, McGuire and Morehouse, 1958; viability in nematode eggs, Farr, 1961; occurrence in the adults and eggs of Heterakis, Gibbs, 1962.)

Genus **Rhizomastix** Alexeieff. Body amoeboid; nucleus central; blepharoplast located between nucleus and posterior end; a long fiber runs from it to anterior end and continues into the flagellum; without contractile vacuole; division in spherical cyst.

R. gracilis A. (Fig. 139, *g*). 8-14μ long; flagellum 20μ long; in intestine of axolotles and tipulid larvae.

References

BAYON, H. P., and BISHOP, A., 1937. Cultivation of *Histomonas meleagridis* from the liver lesions of a hen. Nature, 139:370.

BECKER, E. R., 1925. The morphology of *Mastigina hylae* (Frenzel) from the intestine of the tadpole. J. Parasit., 11:213.

BISHOP, A., 1938. *Histomonas meleagridis*, etc. Parasit., 30:181.

BOVEE, E. C., 1960. Studies on the helioflagellates. I. Arch. Protist., 104: 503.

DESOWITZ, R. S., 1950. Protozoan hyperparasitism of *Heterakis gallinae*. Nature, 165: 1023.

DOBELL, C., 1940. Research on the intestinal Protozoa of monkeys and man. X. Parasit., 32:417.

FARR, M. M., 1961. Further observations on survival of the protozoan parasite, *Histomonas meleagridis*, and eggs of poultry nematodes in feces of infected birds. Cornell Vet., 51:3.

GIBBS, B. J., 1962. The occurrence of the protozoan parasite *Histomonas meleagridis* in the adults and eggs of the cecal worm *Heterakis gallinae*. J. Protozool., 9:288.

HSIUNG, T.-S., 1927. *Tetradimorpha radiata*, etc. Tr. Am. Micr. Soc., 46:208.

KLUG, G., 1936. Neue oder wenig bekannte Arten der Gattungen Masti-gamoeba, etc. Arch. Protist., 87:97.

LEMMERMANN, E., 1914. Pantostomatinae. Süsswasserflora Deutschlands. Pt. 1.

LESSER, E., 1961. *In vitro* cultivation of *Histomonas meleagridis* free of demonstrable bacteria. J. Protozool., 8:228.

LUND, E. E., 1959. Immunizing action of a non-pathogenic strain of Histomonas against blackhead in turkeys. J. Protozool., 6:182.

McGUIRE, W. C., and MOREHOUSE, N. F., 1958. Blood-induced blackhead. J. Parasit., 44:292.

McKAY, F., and MOREHOUSE, N. F., 1948. Studies on experimental black-head infection in turkeys. J. Parasit., 34:137.

NIIMI, D., 1937. Studies on blackhead. II. J. Japan. Soc. Vet. Sc., 16:183.

PASCHER, A., 1925. Neue oder wenig bekannte Protisten. XV. Arch. Protist., 50:486.

SMITH, T., 1895. An infectious disease among turkeys caused by protozoa. Bull. Bur. Animal Ind., U. S. Dep. Agr., no. 8.

——— 1910. *Amoeba meleagridis*. Science, 32:509.

——— 1915. Further investigations into the etiology of the protozoan dis-ease of turkeys known as blackheads, etc. J. Med. Res., 33:243.

Tyzzer, E. E., 1919. Developmental phases of the protozoan of "blackhead" in turkeys. *Ibid.*, 40:1.

—— 1920. The flagellate character and reclassification of the parasite producing "blackhead" in turkey, etc. J. Parasit., 6: 124.

—— 1934. Studies on histomoniasis, etc. Proc. Am. Acad. Arts Sc., 69:189.

—— and Fabyan, M., 1920. Further studies on "blackhead" in turkeys, etc. J. Infect. Dis., 27:207.

—— —— 1922. A further inquiry into the source of the virus in black-head of turkeys, etc. J. Exper. Med., 35:791.

Wenrich, D. H., 1943. Observations on the morphology of Histomonas from pheasants and chickens. J. Morphol., 72:279.

Chapter 14
Order 2. **Protomonadida** Blochmann

THE protomonads possess one or two flagella and are composed of a heterogeneous lot of protozoa, mostly parasitic, whose affinities to one another are very incompletely known. The body is in many cases plastic, having no definite pellicle, and in some forms amoeboid. The method of nutrition is holozoic, or saprozoic. Reproduction is, as a rule, by longitudinal fission, although budding or multiple fission has also been known to occur, while sexual reproduction, though reported in some forms, has not been confirmed. Seven families.

Family 1. **Phalansteriidae** Kent

With one flagellum; collared body in jelly.

Genus **Phalansterium** Cienkowski. Small, ovoid; one flagellum and a small collar; numerous individuals are embedded in gelatinous substance, with protruding flagella; fresh water.

P. digitatum Stein (Fig. 141, *a*). Cells about 17μ long; oval; colony dendritic; fresh water among vegetation.

Family 2. **Codosigidae** Kent

Small flagellates; delicate collar surrounds a flagellum; ordinarily sedentary forms; if temporarily free, organisms swim with flagellum directed backward; holozoic on bacteria or saprozoic; often colonial; free-living in fresh water. (Feeding process, Lapage, 1925.)

Genus **Codosiga** Kent (*Codonocladium* Stein; *Astrosiga* Kent). Individuals clustered at end of a simple or branching stalk; fresh water.

C. utriculus Stokes (Fig. 141, *b*). About 11μ long; attached to freshwater plants.

C. disjuncta (Fromentel) (Fig. 141, *c*). In stellate clusters; cells about 15μ long; fresh water.

Genus **Monosiga** Kent. Solitary; with or without stalk; occasionally with short pseudopodia; attached to freshwater plants. Several species.

M. ovata K. (Fig. 141, *d*). 5-15µ long; with a short stalk.

M. robusta Stokes (Fig. 141, *e*). 13µ long; stalk very long.

Genus **Desmarella** Kent. Cells united laterally to one another; fresh water.

D. moniliformis K. (Fig. 141, *f*). Cells about 6µ long; cluster composed of two to twelve individuals; standing fresh water.

D. irregularis Stokes. Cluster of individuals irregularly branching, composed of more than fifty cells; cells 7-11µ long; pond water.

Genus **Protospongia** Kent. Stalkness individuals embedded irregularly in a jelly mass, collars protruding; fresh water.

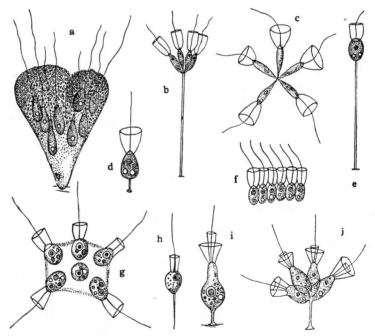

FIG. 141. a, *Phalansterium digitatum*, ×540 (Stein); b, *Codosiga utriculus*, ×1340 (Stokes); c, *C. disjuncta*, ×400 (Kent); d, *Monosiga ovata*, ×800 (Kent); e, *M. robusta*, ×770 (Stokes); f. *Desmarella moniliformis*, ×800 (Kent); g, *Protospongia haeckeli*, ×400 (Lemmermann); h, an individual of *Sphaeroeca volvox*, ×890 (Lemmermann); i, *Diplosiga francei*, ×400 (Lemmermann); j, *D. socialis*, ×670 (Francé).

P. haeckeli K. (Fig. 141, *g*). Body oval; 8μ long; flagellum 24-32μ long; six to sixty cells in a colony.

Genus **Sphaeroeca** Lauterborn. Somewhat similar to the last genus; but individuals with stalks and radiating; gelatinous mass spheroidal; fresh water.

S. volvox L. (Fig. 141, *h*). Cells ovoid, 8-12μ long; stalk about twice as long; flagellum long; contractile vacuole posterior; colony 82-200μ in diameter; fresh water.

Genus **Diplosiga** Frenzel (*Codonosigopsis* Senn). With two collars; without lorica; a contractile vacuole; solitary or clustered (up to four); fresh water.

D. socialis F. (Fig. 141, *j*). Body about 15μ long; usually four clustered at one end of stalk (15μ long).

D. francei Lemmermann (Fig. 141, *i*). With a short pedicel; 12μ long; flagellum as long as body.

Family 3. **Bicosoecidae** Poche

Small monomastigote; with lorica; solitary or colonial; collar may be rudimentary; holozoic; fresh water. (Taxonomy and morphology, Grassé and Deflandre, 1952.)

Genus **Bicosoeca** James-Clark. With vase-like lorica; body small, ovoid with rudimentary collar, a flagellum extending through it; protoplasmic body anchored to base by a contractile filament (flagellum?); a nucleus and a contractile vacuole; attached or free-swimming. (Lorica, Rovinow, 1958.)

B. socialis Lauterborn (Fig. 142, *a*). Lorica cylindrical, 23μ by 12μ; body about 10μ long; often in groups; free-swimming in fresh water.

B. kepneri Reynolds. Body pyriform: 10μ by 6μ; lorica about 1.5 times the body length; flagellum about 30μ long (Reynolds, 1927).

Genus **Salpingoeca** James-Clark. With a vase-like chitinous lorica to which stalked or stalkless organism is attached; fresh or salt water. Numerous species (Pascher, 1925, 1929). (Morphology, Hofeneder, 1925.)

S. fusiformis Kent (Fig. 142, *b*). Lorica short vase-like, about 15-16μ long; body filling lorica; flagellum as long as body; fresh water.

Genus **Diplosigopsis** Francé. Similar to *Diplosiga* but with lorica; solitary; fresh water on algae.

D. affinis Lemmermann (Fig. 142, *c*). Chitinous lorica, spindle-form, about 15μ long; body not filling lorica; fresh water.

Genus **Histiona** Voigt. With lorica; but body without attaching filament; anterior end with lips and sail-like projection; fresh water. (Morphology, Pascher, 1943.)

H. zachariasi V. (Fig. 142, *d*). Lorica cup-like; without stalk; about 13μ long; oval body 13μ long; flagellum long; standing fresh water.

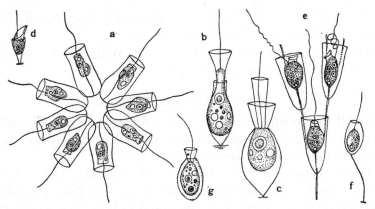

FIG. 142. a, *Bicosoeca socialis*, ×560 (Lauterborn); b, *Salpingoeca fusiformis*, ×400 (Lemmermann); c, *Diplosigopsis affinis*, ×590 (Francé); d, *Histiona zachariasi*, ×440 (Lemmermann); e, *Poteriodendron petiolatum*, ×440 (Stein); f, *Codonoeca inclinata*, ×540 (Kent); g, *Lagenoeca ovata*, ×400 (Lemmermann).

Genus **Poteriodendron** Stein. Similar to *Bicosoeca;* but colonial; lorica vase-shaped: with a prolonged stalk; fresh water. (Flagellar movement, Geitler, 1942.)

P. petiolatum (S.) (Fig. 142, *e*). Lorica 17-50μ high; body 21-35μ long; flagellum twice as long as body; contractile vacuole terminal; standing fresh water.

Genus **Codonoeca** James-Clark. With a stalked lorica; a single flagellum; one to two contractile vacuoles; fresh or salt water.

C. inclinata Kent (Fig. 142, f). Lorica oval; aperture truncate; about 23μ long; stalk twice as long; body oval, about 17μ long; flagellum 1.5 times as long as body; contractile vacuole posterior; standing fresh water.

Genus **Lagenoeca** Kent. Resembles somewhat *Salpingoeca;* with lorica; but without any pedicel between body and lorica; solitary; free-swimming; fresh water.

L. ovata Lemmermann (Fig. 142, *g*). Lorica oval, 15μ long; body loosely filling lorica; flagellum 1.5 times body length; fresh water.

Genus **Stelexomonas** Lackey. A single collar longer than body; vesicular nucleus median; a contractile vacuole terminal; individuals are enclosed in arboroid, dichotomously branching tubes; fresh water.

S. dichotoma L. Body ovoid, 10μ by 8μ; flagellum up to 25μ long; collar 12μ long; the dichotomous tube infolded and wrinkled where branched; organisms are not attached to the tube (Lackey, 1942).

Family 4. **Trypanosomatidae** Doflein

Body characteristically leaf-like, though changeable to a certain extent; a single nucleus and a blepharoplast from which a flagellum arises (Figs. 9; 143); basal portion of the flagellum forms the outer margin of undulating membrane which extends along one side of body; exclusively parasitic; a number of important

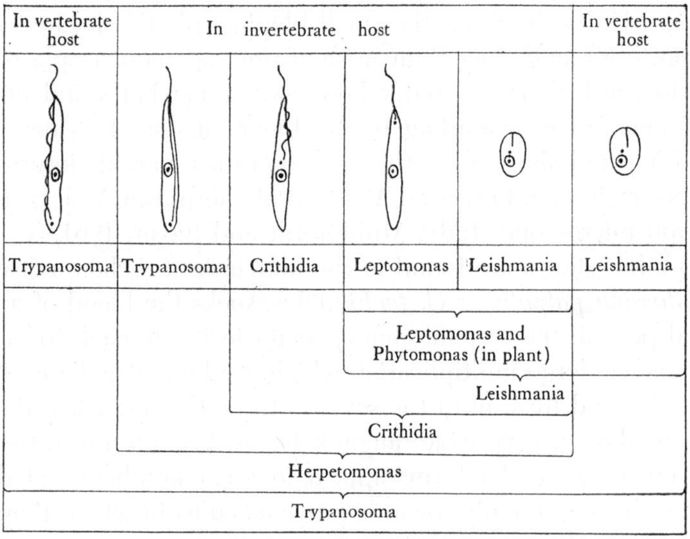

FIG. 143. Diagram illustrating the morphological differences among the genera of Trypanosomatidae (Wenyon)

parasitic protozoa which are responsible for serious diseases of man and domestic animals in various parts of the world are included in it. (Morphology and taxonomy, Grassé, 1952, Clark, 1959; metabolism, von Brand, 1960.)

Genus **Trypanosoma** Gruby. Parasitic in the circulatory system of vertebrates; highly flattened, pointed at flagellate end, and bluntly rounded, or pointed, at other; polymorphism due to differences in development common; nucleus central; near aflagellate end, there is a blepharoplast from which the flagellum arises and nuns toward opposite end, making the outer boundary of the undulating membrane; in most cases flagellum extends freely beyond body; many with myonemes; multiplication by binary or multiple fission. The organism is carried from host to host by blood-sucking invertebrates and undergoes a series of changes in the digestive system of the latter (Fig. 144). A number of forms are pathogenic to their hosts and the diseased condition is termed *trypanosomiasis* in general. (Classification of mammalian trypanosomes, Hoare, 1936; Krampitz, 1961; morphology, Clark, 1959; cultivation and metabolism, Bowman *et al.*, 1960.)

T. gambiense Dutton (Fig. 145, *a-d*). The trypanosome, as it occurs in the blood, lymph or cerebro-spinal fluid of man, is extremely active; body elongate, tapering towards both ends and sinuous; 15-30μ by 1-3μ; the small blepharoplast is located near the posterior end; flagellum arises from the blepharoplast and runs forward along the outer border of somewhat spiral undulating membrane, extending freely; binary fission; between long (dividing) and short (recently divided) forms, various intermediates occur; in man in central Africa. (Polymorphism, Wijers, 1959; electron microscope study, Mühlpfordt and Bayer, 1961.)

No other stages are found in the human host. When a "tsetse" fly, *Glossina palpalis* or *G. tachinoides*, sucks the blood of an infected person, the trypanosomes remain in its stomach for a few days and undergo multiplication which produces flagellates of diverse size and form until the seventh to tenth days when the organisms show a very wide range of forms. From tenth to twelfth days on, long slender forms appear in great numbers and these migrate back gradually towards proventriculus in which they be-

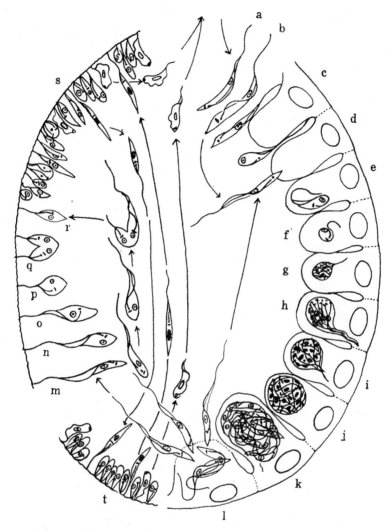

FIG. 144. The life-cycle of *Trypanosoma lewisi* in the flea, *Ceratophyllus fasciatus* (Minchin and Thomson, modified). a, trypanosome from rat's blood; b, individual after being in flea's stomach for a few hours; c–l, stages in intracellular schizogony in stomach epithelium; m–r, two ways in which rectal phase may arise from stomach forms in rectum; s, rectal phase, showing various types; t, secondary infection of pylorus of hind-gut, showing forms similar to those of rectum.

come predominant forms. They further migrate to the salivary
glands and attach themselves to the duct-wall in crithidia form.
Here the development continues for two to five days and the
flagellates finally transform themselves into small trypanosomes

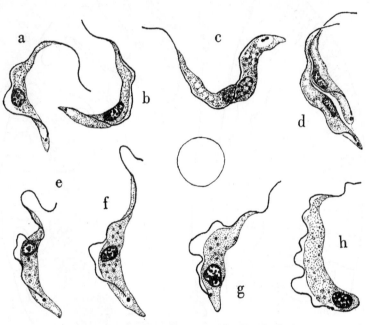

FIG. 145. a–d, *Trypanosoma gambiense;* e–h, *T. rhodesiense,* in stained
blood smears of experimental rats, ×2300. An erythrocyte of rat is shown
for comparison. a, b, typical forms; c, d, division stages; e, f, typical forms;
g, h, post-nuclear forms.

which are now infective. These metacyclic trypanosomes pass
down through the ducts and hypopharynx. When the fly bites a
person, the trypanosomes enter the victim. In addition to this so-
called cyclic transmission, mechanical transmission may take
place. (Culture, Weinman, 1953; distribution of Glossina in
Congo, Evans, 1953.)

Trypanosoma gambiense is a pathogenic protozoan which
causes Gambian or Central African sleeping sickness. The disease
occurs in, and confined to, central Africa within a zone on both
sides of the equator where the vectors, *Glossina palpalis* and *G.
tachinoides* (on the west coastal region) live. Many wild animals

have been found naturally infected by the organisms and are considered to be reservoir hosts. Among the domestic animals, the pigs appear to be one of the most significant, as they themselves are said not to suffer from infection. Though under control now, the sleeping sickness remains to be endemic in many areas (Morris, 1960).

The chief lesions of infection are in the lymphatic glands and in the central nervous system. In all cases, there is an extensive small-cell infiltration of the perivascular lymphatic tissue throughout the central nervous system.

T. rhodesiense Stephens and Fantham (Fig. 145, *e-h*). Morphologically similar to *T. gambiense,* but when inoculated into rats, the position of the nucleus shifts in certain proportion (usually less that 5 per cent) of individuals toward the posterior end, near or behind the blepharoplast, together with the shortening of body. Some consider this trypanosome as a virulent race of *T. gambiense* or one transmitted by a different vector, others consider it a human strain of *T. brucei.* Ashcroft (1958) failed to isolate this trypanosome from twenty-two species of wild animals in Tanganyika and concluded that proof of wild animals acting as reservoir hosts of this organism is still lacking.

The disease caused by this trypanosome appears to be more virulent and runs a course of only a few months. It is known as Rhodesian or East African sleeping sickness. The organism is confined to south-eastern coastal areas of Africa and transmitted by *Glossina morsitans* and *G. pallidipes.* (Prophylactic action of diamidines, De Andrade Silva and Caseiro, 1957.)

T. cruzi Chagas (*Schizotrypanum cruzi* C.) (Fig. 146). A small curved (C or U) form about 20µ long; nucleus central; blepharoplast conspicuously large, located close to sharply pointed nonflagellate end; multiplication takes place in the cells of nearly every organ of the host body; upon entering a host cell, the trypanosome loses its flagellum and undulating membrane, and assumes a leishmania form which measures 2 to 5µ in diameter; this form undergoes repeated binary fission, and a large number of daughter individuals are produced; they develop sooner or later into trypanosomes which, through rupture of host cells, become liberated into blood stream. (Life cycle, Elkeles, 1951, Pereira Da

Silva, 1959; enzymes, Baernstein, 1953; strains, Herrer and Diaz, 1955.)

This trypanosome is the causative organism of Chagas' disease or American trypanosomiasis which is mainly a children's disease, and is widely distributed in South and Central America and as far north as Mexico in North America. In the infected person, the heart and skeletal muscles show minute cyst-like bodies.

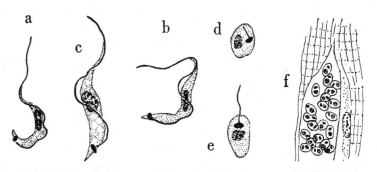

FIG. 146. *Trypanosoma cruzi* in experimental *rats*. a–c, flagellate forms in blood; d, e, cytozoic forms, all ×2300; f, a portion of infected cardiac muscle, ×900.

The transmission of the organism is carried on apparently by numerous species of reduviid bugs, bed bugs and certain ticks, though the first named bugs belonging to genus Triatoma (cone-nosed or kissing bugs) especially *T. megista (Panstrongylus megistus)*, are the chief vectors. When *P. megistus* (nymph or adult) ingests the infected blood, the organisms undergo division in the stomach and intestine, and become transformed into crithidia forms which continue to multiply. In eight to ten days the meta-cyclic or infective trypanosomes make their appearance in the rectal region and pass out in the faeces of the bug at the time of feeding on host. The parasites gain entrance to the circulatory system when the victim scratches the bite-site or through the mu-cous membrane of the eye (Brumpt, 1912; Denecke and von Hall-er, 1939; Weinstein and Pratt, 1948).

Cats, dogs, opossums, monkeys, armadillos, bats, foxes, squir-rels, mice, skunks, wood rats, racoons, guinea pigs, etc., have been found to be naturally infected by *T. cruzi*. Diamond and Rubin (1958) inoculated subcutaneously or intracardially pigs, lambs, kids and calves with *T. cruzi* isolated from a racoon in Maryland and found that all became infected without showing any clinical symptoms. Thus, these animals seem to serve as reservoir hosts.

No cases of Chagas' disease have been reported from the United States until 1955, but Wood (1934) found a San Diego wood rat (*Neotoma fuscipes macrotis*) in the vicinity of San Diego, California, infected by *Trypanosoma cruzi* and Packchanian (1942) observed in Texas, one nine-banded armadillo (*Dasypis novemcinctus*), eight opossums (*Didelphys virginiana*), two house mice (*Mus musculus*), and thirty-two wood rats (*Neotoma micropus micropus*), naturally infected by *Trypanosoma cruzi*. It has now become known through the studies of Kofoid, Wood, and others that *Triatoma protracta* (California, New Mexico), *T. rubida* (Arizona, Texas), *T. gerstaeckeri* (Texas), *T. heidemanni* (Texas), *T. longipes* (Arizona), etc., are naturally infected by *T. cruzi*. Wood and Wood (1941) consider it probable that human cases of Chagas' disease may exist in southwestern United States. In fact, the organisms from a naturally infected *Triatoma heidemanni* were shown by Packchanian (1943) to give rise to a typical Chagas' disease in a volunteer. Finally, Woody and Woody (1955) reported an indigenous infection by *T. cruzi* in an infant in the United States. (Reduviid bugs, Usinger, 1944; Chagas's disease in the United States, Packchanian, 1950; in Brazil, Dias, 1949; division, Noble *et al.*, 1953; trypanostatic agents, Grossowicz and Rasooly, 1958; serodiagnosis, Fife and Muschel, 1959.)

T. brucei Plimmer and Bradford (Fig. 147, *a*). Polymorphic; 15-30µ long (average 20µ); transmitted by various species of tsetse flies, Glossina; the most virulent of all trypanosomes; the cause of the fatal disease known as "nagana" among mules, donkeys, horses, camels, cattle, swine, dogs, etc., which terminates in the death of the host animal in from two weeks to a few months; wild animals are equally susceptible; the disease occurs, of course, only in the region in Africa where the tsetse flies live. (Culture, Packchanian, 1959.)

T. theileri Laveran (Fig. 147, *b*). Large trypanosome which occurs in blood of cattle; sharply pointed at both ends; 60-70µ long; myonemes are well developed. (Cytology, Hartmann and Nöller, 1918; culture, Ristic and Trager, 1958; Simpson and Green, 1959.)

T. *americanum* Crawley. In American cattle; 17-25μ or longer; only crithidia forms develop in culture. Crawley (1909, 1912) found it in 74 per cent and Glaser (1922a) in 25 per cent of cattle they examined. The latter worker considered that this organism was an intermediate form between Trypanosoma and Crithidia.

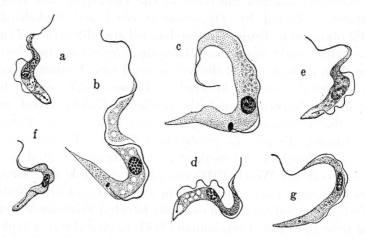

FIG. 147. a, *Trypanosoma brucei;* b. T. *theileri;* c, T. *melophagium;* d, T. *evansi;* e, T. *equinum* f, T. *equiperdum;* g. T. *lewisi;* all ×1330 (several authors).

T. *melophagium* (Flu) (Fig. 147, c). A trypanosome of the sheep; 50-60μ long with attenuated ends; transmitted by *Melophagus ovinus.*

T. *evansi* (Steel) (Fig. 147, d). In horses, mules, donkeys, cattle, dogs, camels, elephants, etc.; infection in horses seems to be usually fatal and known as "surra"; about 25μ long; monomorphic; transmitted by tabanid flies; widely distributed. The bilirubin content of the blood of infected horses is said to rise with the progress of the disease (Ibragimov, 1957); ablepharoplast forms are capable of multiplication in mice (Inoki *et al.,* 1960); transmission (Nieschulz, 1928).

T. *equinum* Vages (Fig. 147, e). In horses in South America, causing an acute disease known as "mal de Caderas"; other domestic animals do not suffer as much as do horses; 20-25µ long; without blepharoplast. Hoare (1954) suggested that this is a mutant of *T. evansi* which lost its blepharoplast.

T. equiperdum Doflein (Fig. 147, *f*). In horses and donkeys; causes "dourine," a chronic disease; widely distributed; 25-30µ long; no intermediate host; transmission takes place directly from host to host during sexual act. (Nuclear division, Roskin and Schisch., 1928; in California, Carr, 1943; electron microscope study, Anderson *et al.*, 1956.)

T. hippicum Darling. In horses and mules in Panama; the cause of "murrina" or 'derrengadera"; 16-18µ long; posterior end obtuse; mechanically transmitted by flies; experimentally various domestic and wild animals are susceptible, but calf is refractory (Darling, 1910, 1911). (Serological tests, Taliaferro and Taliaferro, 1934.)

T. lewisi Kent (Figs. 144; 147, *g*). In the blood of rats; widely distributed; about 30µ long; body slender with a long flagellum; transmitted by the flea *Ceratophyllus fasciatus*, in which the organism undergoes multiplication and form change (Fig. 144); when a rat swallows freshly voided faecal matter of infected fleas containing the metacyclic organisms, it becomes infected. Many laboratory animals are refractory to this trypanosome, but guinea pigs are susceptible (Laveran and Mesnil, 1901; Coventry, 1929). The intermediate host in Hawaii is said to be the rat flea, *Xenopsylla vexabilis hawaiiensis* (Kartman, 1954). (Variation and inheritance of size, Taliaferro, 1921, 1921a, 1923; reproduction-inhibiting reaction product, Taliaferro, 1924, 1932; nuclear division, Wolcott, 1952.)

T. neotomae Wood (? *T. triatomae*, Kofoid and McCulloch). In wood rats, *Neotoma fuscipes annectens* and *N. f. macrotis;* resembles *T. lewisi;* about 29µ long; blepharoplast large, rod-form; free flagellum relatively short; the development in the vector flea *Orchopeas w. wickhami*, similar to that of *T. lewisi;* experimentally Norway rats are refractory (and wood rats are refractory to *T. lewisi* (Fae D. Wood, 1936); comparative morphology of trypanosomes which occur in California rodents and shrews (Davis, 1952).

T. duttoni Thiroux. In the mouse; similar to *T. lewisi,* but rats are said not to be susceptible, hence considered as a distinct species; transmission by fleas. (Antibodies, Taliaferro, 1938.)

T. peromysci Watson. Similar to *T. lewisi;* in Canadian deer mice, *Peromyscus maniculatus* and others.

T. nabiasi Railliet. Similar to *T. lewisi;* in rabbits, *Lepus domesticus* and *L cuniculus.*

T. paddae Laveran and Mesnil. In Java sparrow, *Munia oryzivora.*

T. noctuae (Schaudinn). In the owl *Athene noctura.*

Numerous other species occur in birds (Novy and MacNeal, 1905; Laveran and Mesnil, 1912; Wenyon, 1926; Bennett and Fallis, 1960). Bennett (1961) reported that there exists no vertebrate host specificity. A number of reptiles and amphibians are also hosts for trypanosomes (Roudabush and Coatney, 1937). Transmission is by blood-sucking arthropods or leeches.

T. rotatorium Mayer (Fig. 148, *a*). In tadpoles and adults of various species of frog; between a slender form with a long projecting flagellum measuring about 35µ long and a very broad one without free portion of flagellum, various intermediate forms are to be noted in a single host; blood vessels of internal organs, such as kidneys, contain more individuals than the peripheral vessels; nucleus central, hard to stain; blepharoplast small; undulating membrane highly developed; myonemes prominent; multiplication by longitudinal fission; the leech, *Placobdella marginata,* has been found to be the transmitter in some localities. *In vitro* culture (Wallace, 1956).

T. inopinatum Sergent and Sergent (Fig. 148, *b*). In blood of various frogs; slender; 12-20µ long; larger forms 30-35µ long; blepharoplast comparatively large; transmitted by leeches.

Numerous species of Trypanosoma have been reported from the frog, but specific identification is difficult; it is better and safer to hold that they belong to one of the two species mentioned above until their development and transmission become better known.

T. diemyctyli Tobey (Fig. 148, *c*). In blood of the newt, *Triturus viridescens;* a comparatively large form; slender; about 50µ by 2-5µ; flagellum 20-25µ long; with well developed undulating membrane. (Morphology and development in the leech *Batra-*

chobdella picta, Barrow, 1953, 1954; life cycle influenced by environmental factors, Barrow, 1958.)

T. ambystomae Lehmann. In the salamander *Ambystoma gracile* Lehmann, 1954, 1955, 1958).

T. granulosae L. Polymorphic; 41-59 by 7μ; blepharoplast rod-shaped; nucleus with an endosome; in the newt, *Taricha granulosa* (Lehmann, 1959a). (Cultivation of urodele trypanosomes, Lehmann, 1959.)

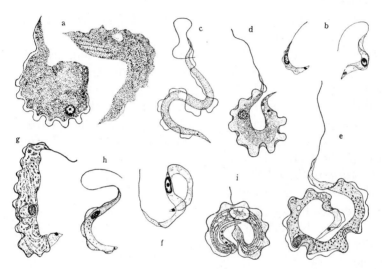

Fig. 148. a, *Trypanosoma rotatorium* ×750 (Kudo); b, *T. inopinatum,* ×1180 (Kudo); c, *T. diemyctyli,* ×800 (Hegner); d, *T. giganteum,* ×500 (Neumann); e, *T. granulosum,* ×1000 (Minchin); f, *T. remaki,* ×1650 (Kudo); g, *T. percae,* ×1000 (Minchin); h, *T. danilewskyi,* ×1000 (Laveran and Mesnil); i, *T, rajae,* ×1600 (Kudo).

Both fresh and salt water fishes are hosts to different species of trypanosomes; what effect these parasites exercise upon the host fish is not understood; as a rule, only a few individuals are observed in the peripheral blood of the host. (Transmission, Robertson, 1911; species, Neumann, 1909; Laveran and Mesnil, 1912; Wenyon, 1926; Laird, 1951; Baker, 1960, 1961.)

T. granulosum Laveran and Mesnil (Fig. 148, *e*). In the eel *Anguilla vulgaris;* 70-80μ long.

T. giganteum Neumann (Fig. 148, *d*). In *Raja oxyrhynchus;* 125-130μ long.

T. remaki Laveran and Mesnil (Fig. 148, *f*). In *Esox lucius, E. reticulatus* and probably other species; 24-33μ long (Kudo, 1921).

T. percae Brumpt (Fig. 148, *g*). In *Perca fluviatilis;* 45-50μ long.

T. danilewskyi Laveran and Mesnil (Fig. 148, *h*). In carp and goldfish; widely distributed; 40μ long. Development in the leech, *Hemiclepsis marginata* (Qadri, 1962).

T. rajae L. and M. (Fig. 148, *i*). In various species of Raja; 30-35μ long (Kudo, 1923).

T. balistes Saunders. In the common Triggerfish, *Balistes capriscus;* 43.5μ long by 3μ broad (Saunders, 1959).

Genus **Crithidia** Léger. Parasitic in arthropods and other invertebrates; blepharoplast located between central nucleus and flagellum bearing end (Fig. 143); undulating membrane not so well developed as in *Trypanosoma;* it may lose the flagellum and form a leptomonas or rounded leishmania stage which leaves host intestine with faecal matter and becomes the source of infection in other host animals.

C. euryophthalmi McCulloch (Fig. 149, *a-c*). In gut of *Euryophthalmus convivus;* California coast.

C. gerridis Patton (Fig. 149, *d*). In intestine of water bugs, Gerris and Microvelia; 22-45μ long. Becker (1923) saw this in *Gerris remigis.* Laird (1959) established genus Blastocrithidia for this species.

C. hyalommae O'Farrell (Fig. 149, *e, f*). In body cavity of the cattle tick, *Hyalomma aegyptium* in Egypt; the flagellate through its invasion of ova is said to be capable of infecting the offspring while it is still in the body of the parent tick.

C. fasciculata Léger. In Anopheles and Culex mosquitoes. (Growth and nutrition, Kidder and Dutta, 1958; use in assay of pteridines, Nathan *et al.*, 1958; contractile vacuole, Cosgrove and Kessel, 1958; anaerobic metabolism, Schwartz, 1961.)

C. flexonema (Wallace *et al.* 1960) was found in the gut of the water strider, *Gerris remigis.*

Genus **Leptomonas** Kent Exclusively parasitic in invertebrates; blepharoplast very close to flagellate end; without undulating membrane (Fig. 143); non-flagellate phase resembles *Leishmania.*

L. ctenocephali Fantham (Fig. 149, *g, h*). In hindgut of the dog flea, *Ctenocephalus canis;* widely distributed. (Morphology, Yamasaki, 1924.)

Genus **Phytomonas** Donovan. Morphologically similar to *Leptomonas* (Fig. 143); in the latex of plants belonging to the families Euphorbiaceae, Asclepiadaceae, Apocynaceae, Sapotaceae and

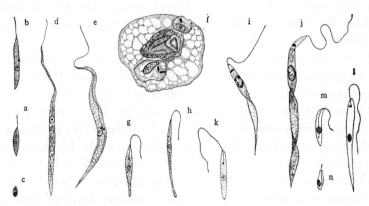

FIG. 149. a–c, *Crithidia euryophthalmi* (a, b, in mid-gut; c, in rectum), ×880 (McCulloch); d, *C. gerridis,* ×1070 (Becker); e, f, *C. hyalommae,* ×1000 (O'Farrell); g, h, *Leptomonas ctenocephali,* ×1000 (Wenyon); i, j, *Phytomonas elmassiani* (i, in milkweed, *Asclepias* sp.; j, in gut of a suspected transmitter, *Oncopeltus fasciatus*), ×1500 (Holmes); k, *Herpetomonas muscarum,* ×1070 (Becker); l–n, *H. drosophilae,* ×1000 (Chatton and Léger).

Utricaceae; transmitted by hemipterous insects; often found in enormous numbers in localized areas in host plant; infection spreads from part to part; infected latex is clear fluid, owing to the absence of starch grains and other particles, and this results in degeneration of the infected part of the plant. Several species.

P. davidi (Lafront). 15-20μ by about 1.5μ; posterior portion of body often twisted two or three times; multiplication by longitudinal fission; widely distributed; in various species of Euphorbia.

P. elmassiani (Migone) (Fig. 149, *i, j*). In various species of milkweeds; 9-12μ long; transmitted by *Oncopeltus fasciatus* (Holmes, 1924) and *O. famelicus* (Ruiz, 1958); in South and North America and Africa. (Developmental cycle, Vickerman, 1962.)

Genus **Herpetomonas** Kent. Ill-defined genus (Fig. 143); exclusively invertebrate parasites; Trypanosoma-, Crithidia-, Leptomonas-, and Leishmania-forms occur during development. Several species. (Species in insects, Drbohlav, 1925.)

H. muscarum (Leidy) (*H. muscae-domesticae*, Burnett) Fig. 149, *k*). In the digestive tube of flies belonging to the genera: Musca, Calliphora, Cochliomyia, Sarcophaga, Lucilia, Phormia, Phaenicia, etc.; up to 30μ by 2-3μ. (Effect on experimental animals, Claser, 1922; comparative study, Becker, 1923a; axenic culture and morphology, Wallace and Clark, 1959).

H. drosophilae (Chatton and Alilaire) (Fig. 149, *l-n*). In intestine of *Drosophila confusa;* large leptomonad forms 21-25μ long, flagellum body-length; forms attached to rectum 4-5μ long.

Genus **Leishmania** Ross. In man or dog, the organism is an ovoid body with a nucleus and a blepharoplast; 2-5μ in diameter; with vacuoles and sometimes a rhizoplast near the blepharoplast; intracellular parasite in the cells of reticulo-endothelial system; multiplication by binary fission. In the intestine of blood-sucking insects or in blood-agar cultures, the organism develops into leptomonad form (Fig. 150, *d-f*) which multiplies by longitudinal fission. Nuclear division (Roskin and Romanowa, 1928).

There are known at present three "species" of Leishmania which are morphologically alike. They do not show any distinct differential characteristics either by animal inoculation experiments or by culture method or agglutination test.

Species of Phlebotomus (sand-flies) have long been suspected as vectors of Leishmania. When a Phlebotomus feeds on kala-azar patient, the leishmania bodies become flagellated and undergo multiplication so that by the third day after the feeding, there are large numbers of Leptomonas flagellates in the mid-gut. These flagellates migrate forward to the pharynx and mouth cavity on the fourth or fifth day. On the seventh to ninth days (after the fly is fed a second time), the organisms may be found in the proboscis. But the great majority of the attempts to infect animals and man by the bite of infected Phlebotomus have failed, although in a number of cases small numbers of positive infection have been reported. Adler and Ber (1941) have finally succeeded in producing cutaneous leishmaniasis in five out of nine human volunteers on the site of bites by laboratory-bred *P. papatasii* which were fed on the flagellates of *Leishmania tropica* suspended in three parts 2.7 per cent saline and one part defibrinated blood and kept at a temperature of 30°C. Swaminath, Shortt and Anderson (1942) also succeeded in producing kala-azar infections in three out of five volunteers through the bites of infected *P. argentipes.*

L. *donovani* Laveran and Mesnil (*L. infantum* Nicolle) (Fig. 150). As seen in stained spleen puncture smears, the organism is

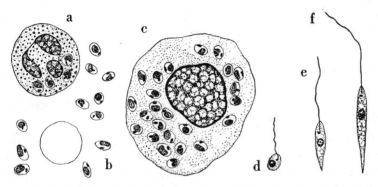

Fig. 150. *Leishmania donovani,* ×1535. a, an infected polymorpho-nuclear leucocyte; b, organisms scattered in the blood plasm; c, an infected mono-cyte; d–f, flagellate forms which develop in blood-agar cultures.

rounded (1-3µ) or ovoid (2-4µ by 1.5-2.5µ); cytoplasm homo-geneous, but often with minute vacuoles; nucleus comparatively large, often spread out and of varied shapes; blepharoplast stains more deeply and small; number of parasites in a host cell varies from a few to over 100. (Electron microscope study, Chang, 1956; nuclear division in leptomonad form, Chakraborty *et al.,* 1962; nutrition, Actor, 1960; leishmaniasis in golden hamster, Ritterson, 1955.)

This is the cause of *kala-azar* or visceral leishmaniasis which is widely distributed in Europe (Portugal, Spain, Italy, Malta, Greece, and southern Russia), in Africa (Morocco, Algeria, Tuni-sia, Libya, Abyssinia, Sudan, northern Kenya and Nigeria), in Asia (India, China, Turkestan, etc.), and in South America. The para-site is most abundantly found in the macrophages, mononuclear leucocytes, and polymorphonuclears of the reticuolo-endothelial system of various organs such as spleen, liver, bone marrow, in-testinal mucosa, lymphatic glands, etc. The most characteristic histological change appears to be an increase in number of large macrophages and mononuclears. The spleen and liver become enlarged due in part to increased fibrous tissue and macrophages.

The organism is easily cultivated in blood-agar media (p. 1065). After two days, it becomes larger and elongate until it measures 14-20μ by 2μ. A flagellum as long as the body develops from the blepharoplast and it thus assumes leptomonad form (Fig. 150 *f*) which repeats longitudinal division. Dogs are naturally infected with *L. donovani* and may be looked upon as a reservoir host. Vectors are *Phlebotomus argentipes* and other species of Phlebotomus. (Bionomics of *P. argentipes*, Smith, 1959; electron microscope study of the basal appartus of the flagellum, Pyne and Chakraborty, 1958; pathology and treatment, Manson-Bahr, 1959.)

L. tropica (Wright). This is the causative organism of the Oriental sore or cutaneous leishmaniasis. It has been reported from Africa (mainly regions bordering the Mediterranean Sea), Europe (Spain, Italy, France, and Greece), Asia (Syria, Palestine, Armenia, Southern Russia, Iraq, Iran, Arabia, Turkestan, India, Indo-China, and China), and Australia (northern Queensland). The organisms are present in the endothelial cells in and around the cutaneous lesions, located on hands, feet, legs, face, etc.

L. tropica is morphologically indistinguishable from *L. donovani*, but some believe that it shows a wider range of form and size than the latter. In addition to rounded or ovoid forms, elongate forms are often found, and even leptomonad forms have been reported from the scrapings of lesions. The insect vectors are *Phlebotomus papatasii, P. sergenti* and others. Direct transmission through wounds in the skin also takes place. The lesion appears first as a small papula on skin; it increases in size and later becomes ulcerated. Microscopically an infiltration of corium and its papillae by lymphocytes and macrophages is noticed; in ulcerated lesions leishmania bodies are found in the peripheral zone and below the floor of the ulcers.

L. brasiliensis Vianna. This organism causes Espundia, Bubos, or American or naso-oral leishmaniasis, which appears to be confined to South and Central America. It has been reported from Brazil, Peru, Paraguay, Argentina, Uruguay, Boliva, Venezuela, Ecuador, Colombia, Panama, Costa Rica, and Mexico.

Its morphological characteristics are identical with those of *L. tropica,* and a number of investigators combine the two species into one. However, *L. brasiliensis* produces lesions in the mucous membrane of the nose and mouth. Vectors appear to be *Phlebotomus intermedius, P. panamensis, P. longipalpis* and other species

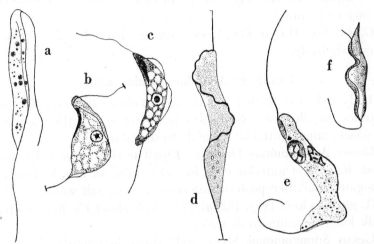

FIG. 151. a, a neutral red stained and b, a fixed and stained *Cryptobia helicis,* ×2200 (Kozloff); c, stained specimen of the same organism, ×1690 (Bělař); d, a living and e, stained *C. borreli,* ×1730 (Mavor); f, *C. cyprini,* ×600 (Plehn).

of the genus. Direct transmission through wounds is also possible. Fuller and Geiman (1942) find *Citellus tridecemlineatus* a suitable experimental animal.

Family 5. **Cryptobiidae** Poche

Biflagellate trypanosome-like protomonads; one flagellum free, the other marks outer margin of undulating membrane; blepharoplast an elongated rod-like structure, often referred to as the parabasal body; all parasitic.

Genus **Cryptobia** Leidy (*Trypanoplasma* Laveran and Mesnil). Parasitic in the reproductive organ of molluscs (Leidy, 1846) and other invertebrates; also in the blood of fishes.

C. helicis L. (Fig. 151, *a-c*). In the reproductive organ of various species of pulmonate snails: *Triodopsis albolabris, T. tridentata, Anguispira alternata* (Leidy, 1846), *Helix aspersa,* and

Monadenia fidelis (Kozloff, 1948); 16-26.5μ by 1.5-3.3μ. (Morphology and culture, Schindera, 1922.)

C. *borreli* (Laveran and Mesnil) (Fig. 151, *d,e*). In blood of various freshwater fishes such as Catostomus, Cyprinus, etc.; 20-25μ long (Mavor, 1915).

C. *cyprini (Plehn)* (Fig. 151, *f*). In blood of carp and goldfish; 10-30μ long; rare.

C. *grobbeni* (Keysselitz). In coelenteric cavity of Siphonophora; about 65μ by 4μ.

Family 6. **Amphimonadidae** Kent

Body naked or with a gelatinous envelope; two equally long anterior flagella; often colonial; one to two contractile vacuoles; free-swimming or attached; mainly fresh water.

Genus **Amphimonas** Dujardin. Small oval or rounded amoeboid; flagella at anterior end; free-swimming or attached by an elongated stalk-like posterior process; fresh or salt water.

A. *globosa* Kent (Fig. 152, *a*). Spherical; about 13μ in diameter; stalk long, delicate; fresh water.

Genus **Spongomonas** Stein. Individuals in granulated gelatinous masses; two flagella; one contractile vacuole; colonial; with pointed pseudopodia in motile stage; fresh water.

S. *uvella* S. (Fig. 152, *b*). Oval; 8-12μ long; flagella two to three times as long; colony about 50μ high; fresh water.

Genus **Cladomonas** Stein. Individuals are embedded in dichotomous dendritic gelatinous tubes which are united laterally; fresh water.

C. *fruticulosa* S. (Fig. 152, *c*). Oval; about 8μ long; colony up to 85μ high.

Genus **Rhipidodendron** Stein. Similar to *Cladomonas*, but tubes are fused lengthwise; fresh water.

R. *splendidum* S. (Fig. 152, *d, e*). Oval; about 13μ long; flagella about two to three times body length; fully grown colony 350μ high.

Genus **Spiromonas** Perty. Elongate; without gelatinous covering; spirally twisted; two flagella anterior; solitary; fresh water.

S. *augusta* (Dujardin) (Fig. 152, *f*). Spindle-form; about 10μ long; stagnant water.

Genus **Diplomita** Kent. With transparent lorica; body attached to bottom of lorica by a retractile filamentous process; a rudimentary stigma(?); fresh water.

D. socialis K. (Fig. 152, *g*). Oval flagellum about two to three times the body length; lorica yellowish or pale brown; broadly spindle in form; about 15µ long; pond water.

Genus **Streptomonas** Klebs. Free-swimming; naked; distinctly keeled; fresh water.

S. cordata (Perty) (Fig. 152, *h*). Heart-shaped; 15µ by 13µ; rotation movement.

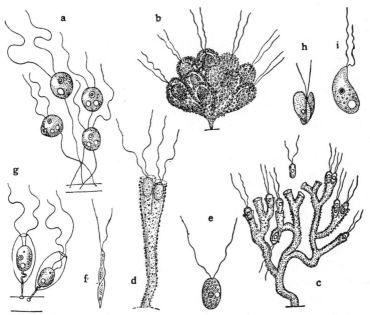

FIG. 152. a, *Amphimonas globosa*, ×540 (Kent); b, *Spongomonas uvella*, ×440 (Stein); c, *Cladomonas fruticulosa*, ×440 (Stein); d, e, *Rhipidodendron splendidum* (d, a young colony, ×440; e, a free-swimming individual, ×770) (Stein); f, *Spiromonas augusta*, ×1000 (Kent); g, *Diplomita socialis*, ×1000 (Kent); h, *Streptomonas cordata*, ×890 (Lemmermann); i, *Dinomonas vorax*, ×800 (Kent).

Genus **Dinomonas** Kent. Ovate or pyriform, plastic, free-swimming; two flagella, equal or sub-equal, inserted at anterior extremity, where large oral aperture, visible only at time of food ingestion, is also located, feeding on other flagellates; in infusions.

D. vorax K. (Fig. 152, *i*). Ovoid, anterior end pointed; 15-16µ long; flagella longer than body;hay infusion and stagnant water.

Family 7 **Bodonidae** Bütschli

With two flagella; one directed anteriorly and the other posteriorly and trailing; flagella originate in anterior end which is drawn out to a varying degree; one to several contractile vacuoles; asexual reproduction by binary fission; holozoic or saprozoic (parasitic). (Morphology and taxonomy, Hollande, 1942, 1952.)

Genus **Bodo** Ehrenberg (*Prowazekia* Hartmann and Chagas). Small, ovoid, but plastic; cytostome anterior; nucleus central or anterior; flagella connected with two blepharoplasts in some species; encystment common; in stagnant water and coprozoic. Numerous species. (Cytology, Bělař, 1920; Hollande, 1936; E.M. study, Pitelka, 1961.)

B. caudatus Dujardin (Fig. 153, *a, b*). Highly flattened, usually tapering posteriorly; 11-22µ by 5-10µ; anterior flagellum about body length, trailing flagellum longer; blepharoplast; cysts spherical; stagnant water.

B. edax Klebs (Fig. 153, *c*). Pyriform with bluntly pointed ends; 11-15µ by 5-7µ; stagnant water.

Genus **Pleuromonas** Perty. Naked, somewhat amoeboid; usually attached with trailing flagellum; active cytoplasmic movement; fresh water.

P. jaculans P. (Fig. 153, *d*). Body 6-10µ by about 5µ; flagellum two to three times body length; four to eight young individuals are said to emerge from a spherical cyst; stagnant water.

Genus **Rhynchomonas** Klebs (*Cruzella* Faria, da Cunha and Pinto). Similar to *Bodo*, but there is an anterior extension of body, in which one of the flagella is embedded, while the other flagellum trails; a single nucleus; minute forms; fresh or salt water; also sometimes coprozoic.

R. nasuta (Stokes) (Fig. 153, *e*). Oval, flattened; 5-6µ by 2-3µ; fresh water and coprozoic.

R. marina (F., C. and P.). In salt water.

Genus **Proteromonas** Kunstler (*Prowazekella* Alexeieff). Elongated pyriform; two flagella from anterior end, one directed anteriorly and the other, posteriorly; nucleus anterior; encysted stage is remarkable in that it is capable of increasing in size to a marked degree; exclusively parasitic; in gut of various species of lizards. (Species, Grassé, 1926, 1952.)

P. lacertae (Grassi) (Fig. 153, *f*). Elongate, pyriform; 10-30μ long; gut of lizards belonging to the genera Lacerta, Tarentola, etc.

Genus **Retortamonas** Grassi (*Embadomonas* Mackinnon). Body plastic, usually pyriform or fusiform, drawn out posteriorly; a large cytostome toward anterior end; nucleus anterior; two flagel-

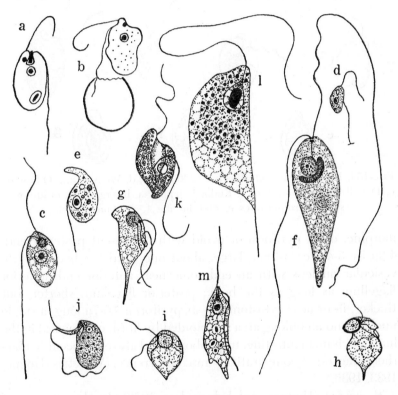

Fig. 153. a, b, *Bodo caudatus*, ×1500 (Sinton); c, *B. edax*, ×1400 (Kühn); d, *Pleuromonas jaculans*, ×650 (Lemmermann); e, *Rhinchomonas nasuta*, ×1800 (Parisi); f, *Proteromonas lacertae*, ×2500 (Kühn); g, *Retortamonas gryllotalpae*, ×2000 (Wenrich); h, *R. blattae*, ×2000 (Wenrich); i, *R. intestinalis*, ×2000 (Wenrich); j. *Phyllomitus undulans*, ×1000 (Stein); k, *Colponema loxodes*, ×650 (Stein); l, *Cercomonas longicauda*, ×2000 (Wenyon); m, *C. crassicauda*, ×2000 (Dobell).

la; cysts pyriform or ovoid; parasitic in the intestines of various animals. (Taxonomy, Wenrich, 1932; Kirby and Honigberg, 1950.)

R. *gryllotalpae* G. (Fig. 153, g). About 7-14µ (average 10µ) long; in intestine of the mole cricket, *Gryllotalpa gryllotalpa.*

R. *blattae* (Bishop) (Fig. 153, h). About 6-9µ long; in colon of cockroaches.

R. *intestinalis* (Wenyon and O'Connor) (Figs. 153, i; 154). Poly-

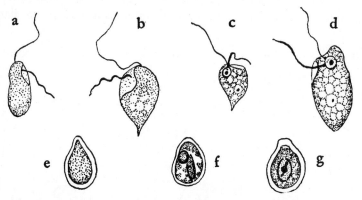

Fig. 154. *Retortamonas intestinalis,* ×2300 (a, b, d, Wenyon and O'Connor; c, Dobell and O'Connor; e, g, Kudo; f, Jepps). a, b, organisms in life; c, d, stained trophozoites e, cyst in life; f, g, stained cysts.

morphic, often pyriform or ovoid with drawn-out posterior end; 4-9µ by 3-4µ; cytostome large, about one third the body length; vesicular nucleus with an endosome near anterior end; anterior flagellum as long as the body; posterior flagellum shorter, but thicker, in or near cytostome; cysts pyriform; 4.5-7µ long; a single nucleus and an oblong area surrounded by fibril; commensal in the lumen of human intestine; trophoboites and also cysts occur in diarrhoeic faeces; of comparatively rare occurrence. (Varieties, Hogue, 1933, 1936.)

R. *caviae* (Hegner and Schumaker, 1928). In the caecum of guinea pigs; stained trophozoites 4-7µ by 2.4-3.2µ (H. and S.), 4.4-7.7µ by 4-4.3µ (Nie, 1950); stained cysts 3.4-5.2µ by 3.3-3.6µ (H. and S.) 4.5-5.7µ by 3.4-3.7µ (Nie).

Genus **Phyllomitus** Stein. Oval; highly plastic; cytostome large and conspicuous; two unequal flagella, each originates in a blepharoplast; fresh water or coprozoic.

P. undulans S. (Fig. 153, *j*), Ovoid; 21-27μ long; trailing flagellum much longer than anterior one; stagnant water.

Genus **Colponema** Stein. Body small; rigid; ventral furrow conspicuous, wide at anterior end; one flagellum arises from anterior end and the other from middle of body; fresh water.

C. loxodes S. (Fig. 153, *k*). 18-30μ by 14μ; cytoplasm with refractile globules.

Genus **Cercomonas** Dujardin. Biflagellate, both flagella arising from anterior end of body; one directed anteriorly and the other runs backward over body surface, becoming a trailing flagellum; plastic; pyriform nucleus connected with the blepharoplast of flagella; spherical cysts uninucleate; fresh water or coprozoic.

C. longicauda D. (Fig. 153, *l*). Pyriform or ovoid; posterior end drawn out; 18-36μ by 9-14μ; flagella as long as body; pseudopodia; fresh water and coprozoic.

C. crassicauda D. (Fig. 153, *m*). 10-16μ by 7-10μ; fresh water and coprozoic.

References

ACTOR, P., 1960. Protein and vitamin intake and visceral leishmaniasis in the mouse. Exper. Parasit., 10:1.

ADLER, S., and BER, M., 1941. Transmission of *Leishmania tropica* by the bite of *Phlebotomus papatasii*. Indian J. Med. Res. 29:803.

ANDERSON, E., *et al.*, 1956. Electron microscope observations of *Trypanosoma equiperdum*. J. Parasit., 42:11.

ASHCROFT, M. T., 1958. An attempt to isolate *Trypanosoma rhodesiense* from wild animals. Trans. Roy. Soc. Trop. Med. & Hyg., 52:276.

BAERNSTEIN, H. D., 1953. The enzyme systems of the culture form of *Trypanosoma cruzi*. Ann. N. Y. Acad. Sci., 56:982.

BAKER, J. R., 1960. Trypanosomes and dactylosomes from the blood of fresh-water fish in east Africa. Parasitology, 50:515.

——— 1961. Trypanosomes of African fresh-water fish. *Ibid.*, 51:263.

BARROW, J. H. JR., 1953. The biology of *Trypanosoma diemyctyli*. I. Tr. Am. Micr. Soc., 72:197.

——— 1954. II. *Ibid.*, 73:242.

——— 1958. III. Factors influencing the cycle in the vertebrate host. J. Protozool., 5:161.

BECKER, E. R., 1923. Observations on the morphology and life cycle of *Crithidia gerridis*, etc. J. Parasit., 9:141.

——— 1923a. Observations on the morphology and life history of *Herpetomonas muscae-domesticae* in North American muscoid flies. *Ibid.*, 9:199.

BĚLAŘ, K., 1920. Die Kernteilung von Prowazekia. Arch. Protist., 41:308.

BENNETT, G. F., 1961. On the specificity and transmission of some avian trypanosomes. Canad. J. Zool., 39:17.

—— and FALLIS, A. M., 1960. Blood parasites of birds in Algonquin Park, Canada, etc. *Ibid.*, 38: 261.

BOWMAN, I. B. R., *et al.*, 1960. The cultivation and metabolism of trypanosomes in the presence of trehalose with observations on trehalase in blood serum. Exper. Parasit., 10:274.

BRAND, T. v., 1960. Der Stoffwechsel der Trypanosomen. Ergebnisse Biol., 22:30.

BRUMPT, E., 1912. Pénétration du *Schizotrypanum cruzi* a travers la muqueuse oculaire saine. Bull. Soc. Path. Exot., 5:723.

CARR, A. K., 1943. Dourine appears for the first time in California. Bull. Cal. Dept. Agr., 32:46.

CHAKRABORTY, J., *et al.*, 1962. Cytology of the flagellate form of *Leishmania donovani.* J. Parasit., 48:131.

CHANG, P. C. H., 1956. The ultrastructure of *Leishmania donovani.* J. Parasit., 42:126.

CLARK, T. B., 1959. Comparative morphology of four genera of Trypanosomatidae. J. Protozool., 6:227.

COSGROVE, W. B., and KESSEL, R. G., 1958. The activity of the contractile vacuole of *Crithidia fasciculata.* J. Protozool., 5:296.

COVENTRY, F. A., 1929. Experimental infections with *Trypanosoma lewisi* in the guinea pigs. Amer. J. Hyg., 9:247.

CRAWLEY, H., 1909. Studies on blood and blood parasites. I-III. Bull. Bur. Animal Ind., U. S. Dept. Agr., No. 119.

—— 1912. *Trypanosoma americanum*, a common blood parasite of American cattle. *Ibid.*, No. 145.

DARLING, S. T., 1910. Equine trypanosomiasis in the canal zone. Bull. Soc. Path. Exot., 3:381.

—— 1911. Murrina: etc. J. Infect. Dis., 8:467.

DAVIS, B. S., 1952. Studies on the trypanosomes of some California mammals. Univ. Cal. Publ. Zool., 57:145.

DE ANDRARE SILVA, M. A., and CASEIRO, A., 1957. Prophylactic action of diamidines against *Trypanosoma rhodesiense.* Ann. Inst. Med. Trop., 14:171.

DENECKE, K., and HALLER, E. v., 1939. Recherches expérimentale sur le mode de transmission et le course de l'infection par *Trypanosoma cruzi* chez les souris. Ann. Parasit., 17:313.

DIAMOND, L. S., and RUBIN, R., 1958. Experimental infection of certain farm mammals with a North American strain of *Trypanosoma cruzi* from the racoon. Exper. Parasit., 7:383.

DIAS, E., 1949. Consideracões sôbre a doenca de Chagas. Mem. Inst. Oswaldo Cruz, 47:679.

DRBOHLAV, J., 1925. Studies on the relation of insect flagellates to Leishmaniasis. I-III. Amer. J. Hyg., 5:580.

ELKELES, G., 1951. On the life cycle of *Trypanosoma cruzi.* J. Parasit., 37:379.

EVANS, M. J. C., 1953. Dispersion géographique des glossines au Congo Belge. Mém. Inst. Roy. Sci. Nat. Belgique, Ser. II, 48:1.

FIFE, E. H. JR., and MUSCHEL, L. H., 1959. Fluorescent-antibody technique for serodiagnosis of Trypanosoma infection. Proc. Soc. Exper. Biol. & Med., 101:540.

FULLER, H. S., and GEIMAN, Q. M., 1942. South American cutaneous leishmaniasis in experimental animals. J. Parasit., 28:429.

GEITLER, L., 1942. Beobachtungen über die Geisselbewegung der Bicoecacee Poteriodendron. Arch. Protist., 96:119.

GLASER, R. W., 1922. *Herpetomonas muscae-domesticae,* its behavior and effect in laboratory animals. J. Parasit., 8:99.

———— 1922a. A study of *Trypanosoma americanum. Ibid.,* 8:136.

GRASSÉ, P.-P., 1926. Contribution à l'étude des flagellés parasites. Arch. zool. exper. gén., 65:345.

———— 1952. Traité de zoologie. I. Fasc. 1. Paris.

———— and DEFLANDRE, G., 1952. Ordre des Bicoecidea. In Grassé (1952), p. 599.

GROSSOWICZ, N., and RASOOLY, G., 1958. Evaluation of some trypanostatic agents by means of *Herpetomonas culicidarum.* J. Protozool., 5:249.

HARTMANN, M., and NÖLLER, W., 1918. Untersuchungen ueber die Cytologie von *Trypanosoma theileri.* Arch. Protist., 38:355.

HEGNER, R. W., and SCHUMAKER, E., 1928. Some intestinal amoebae and flagellates from the chimpanzee, etc. J. Parasit., 15:31.

HERRER, A., and DIAZ, J., 1955. Trypanosomiasis americana en el Peru. Rev. Med. Exper., Lima, 9:92.

HOARE, C., 1936. Morphological and taxonomic studies on mammalian trypanosomes. I. Parasitol., 28:98.

———— 1954. The loss of the kinetoplast in trypanosomes, etc. J. Protozool., 1:28.

HOFENEDER, H., 1925. Ueber eine neue Caspedomonadine. Arch. Protist., 51:192.

HOGUE, MARY J., 1933. A new variety of *Retortamonas intestinalis* from man. Amer. J. Hyg., 18:433.

———— 1936. Four races of *Retortamonas intestinalis. Ibid.,* 23:80.

HOLLANDE, A., 1936. Sur la cytologie d'un flagellé du genre Bodo. C. R. Soc. Biol., 123:651.

———— 1942. Études cytologique et biologique de quelques flagellés libres. Arch. zool. exper. gén., 83:1.

———— 1952. Ordre des Bodonides. In: Grassé (1952), p. 669.

HOLMES, F. O., 1924. Herpetomonad flagellates in the latex of milkweed in Maryland. Phytopathology, 14:146.

———— 1925. The relation of *Herpetomonas elmassiani* to its plant and insect hosts. Biol. Bull., 49:323.

INOKI, S., *et al.*, 1960. Multiplication ability of the akinetoplastic form of *Trypanosoma evansi*. Biken's J., 3:123.

IVANIĆ, M., 1936. Zur Kenntnis der Entwicklungsgeschichte bei *Mastigina hylae*. Arch. Protist., 87:225.

KARTMAN, L., 1954. Observations on *Trypanosoma lewisi*, etc. J. Parasit., 40:571.

KIDDER, G. W., and DUTTA, B. N., 1958. The growth and nutrition of *Crithidia fasciculata*. J. Gen. Microbiol., 18:621.

KIRBY, H., and HONIGBERG, B., 1950. Intestinal flagellates from a wallaroo, *Macropus robustus*. Univ. Cal. Publ. Zool., 55:35.

KOZLOFF, E. N., 1948. The morphology of *Cryptobia helicis*. J. Morphol., 83:253.

KRAMPITZ, H. E., 1961. Kritisches zur Taxonomie und Systematik parasitischer Säugetier-Trypanosomen, etc. Z. Tropenmed. Parasit., 12:117.

KUDO, R. R., 1921. On some protozoa parasitic in fresh water fishes of New York. J. Parasit., 7:166.

——— 1923. Skate trypanosome from Woods Hole. *Ibid.*, 9:179.

LACKEY, J. B., 1942. Two new flagellate Protozoa from the Tennessee River. Tr. Am. Micr. Soc., 61:36.

——— 1959. Morphology and biology of a species of Protospongia. *Ibid.*, 78:202.

LAIRD, M., 1951. Studies on the trypanosomes of New Zealand fish. Proc. Zool. Soc., London, 121:285.

——— 1959. Blastocrithidia n.g., etc. Canad. J. Zool., 37: 749.

LAPAGE, G., 1925. Notes on the choanoflagellate, *Codosiga botrytis*. Quart. J. Micr. Sci., 69:471.

LAVERAN, A., and MESNIL, F., 1901. Sur les flagellés à membrane ondulante des poissons, etc. C. R. Acad. Sci., 133:670.

——— ——— 1912. Trypanosomes et trypanosomiases. 2 ed. Paris.

LEHMANN, D. L., 1954. A new species of trypanosome from the salamander, etc. J. Parasit., 40:656.

——— 1955. Notes on the biology of *Trypanosoma ambystomae*. I. The cultivation. J. Protozool., 2:28.

——— 1958. II. *Ibid.*, 5:96.

——— 1959. The cultivation of some trypanosomes from urodeles. *Ibid.*, 6:340.

——— 1959a. *Trypanosoma granulosae* n. sp. from the newt, *Taricha granulosa*. *Ibid.*, 6:167.

LEIDY, J., 1846. Description of a new genus and species of Entozoa. Proc. Acad. Nat. Sci. Philadelphia, 3:100.

LEMMERMANN, E., 1914. Protomastiginae. Süsswasserflora Deutschlands. H. 1.

MANSON-BAHR, P. E. C., 1959. East African kala-azar with special reference to the pathology, prophylaxis and treatment. Trans. Roy. Soc. Trop. Med. & Hyg., 53:123.

Mavor, J. W., 1915. On the occurrence of a trypanosome, probably *Trypanoplasma borreli*, etc. J. Parasit., 2:1.

Minchin, E. A., and Thomson, J. D., 1915. The rat trypanosome, *Trypanosoma lewisi*, in its relation to the rat flea, etc. Quart. J. Micr. Sci., 60: 463.

Morris, K. R. S., 1960. Studies on the epidemiology of sleeping sickness in East Africa. II. Trans. Roy. Soc. Trop. Med. & Hyg., 54:71

Mühlpfordt, H.; and Bayer, M., 1961. Elektronenmikroskopische Untersuchungen an Protozoen (*Trypanosoma gambiense*). Z. Tropenmed. Parasit., 12:333.

Nathan, H. A., *et al.*, 1958. Assay of pteridines with *Crithidia fasciculata*. J. Protozool., 5:134.

Nelson, R., 1922. The occurrence of protozoa in plants affected with mosaic and related diseases. Tech. Bull. Bot. Stat. Michigan Agr. Coll., no. 58.

Neumann, R. O., 1909. Studien ueber protozoische Parasiten im Blut von Meeresfischen. Zeitschr. Hyg. Infekt. 64:1.

Nie, D., 1950. Morphology and taxonomy of the intestinal protozoa of the guinea-pig, *Cavia porcella*. J. Morphol., 86:381.

Nieschulz, O., 1928. Zoologische Beiträge zum Surraproblem. XVII. Arch. Protist., 61:92.

Noble, E. R., *et al.*, 1953. Cell division in trypanosomes. Tr. Am. Micr. Soc., 72:236.

Novy, F. G., and MacNeal, W. J., 1905. On the trypanosomes of birds. J. Infect. Dis., 2:256.

Packchanian, A., 1942. Reservoir hosts of Chagas' disease in the State of Texas. Am. J. Trop. Med., 22:623.

———— 1943. Infectivity of the Texas strain of *Trypanosoma cruzi* to man. *Ibid.*, 23:309.

———— 1950. The present status of Chagas' disease in the United States. Rev. Soc. Mexicana Hist. Nat., 10:91.

———— 1959. On the cultivation of *Trypanosoma brucei in vitro*. Amer. J. Trop. Med. & Hyg., 8:168.

Pascher, A., 1925. Neue oder wenig bekannte Protisten. XVII. Arch. Protist., 51:549.

———— 1929. XXI. *Ibid.*, 65:426.

———— 1943. Eine neue Art der farblosen Flagellatengattung Histiona aus den Uralpen. *Ibid.*, 96:288.

Pereira da Silva, L. H., 1959. Observacoes sobre o cyclo do *Trypanosoma cruzi*. Rev. Inst. Med. Trop. Sao Paulo, 1:99.

Pitelka, D. A., 1961. Observations on the kinetoplast-mitochondrion and the cytostome of Bodo. Exper. Cell Res., 25:87.

Pyne, C. K., and Chakraborty, J., 1958. Electron microscope studies on the basal apparatus of the flagellum in the protozoan, *Leishmania donovani*. J. Protozool., 5:264.

QADRI, S. S., 1962. An experimental study of the life cycle of *Trypanosoma danilewskyi* in the leech, *Hemiclepsis marginata*. J. Protozool., 9:254.

REYNOLDS, B. D., 1927. *Bicosoeca kepneri*. Tr. Am. Micr. Soc., 46:54.

RISTICS, M., and TRAGER, W., 1958. Cultivation at 37°C. of a trypanosome (*Trypanosoma theileri*) from cows with depressed milk production. J. Protozool., 5:146.

RITTERSON, A. L., 1955. Studies on leishmaniasis in the golden hamster. J. Parasit., 41:603.

ROBERTSON, M., 1911. Transmission of flagellates living in the blood of certain freshwater fishes. Philos. Trans., B, 202:29.

ROBINOW, C. F., 1956. Observations on vase-shaped, iron-containing houses of two colorless flagellates of the family Bicoecidae. J. Biophys. Biochem. Cytol., Suppl., 2:233.

ROSKIN, G., and ROMANOWA, K., 1928. Die Kernteilung bei *Leishmania tropica*. Arch. Protist., 60:482.

―――― and SCHISCHLIAIEWA, S., 1928. Die Kernteilung bei Trypanosomen. *Ibid.*, 60:460.

ROUDABUSH, R. L., and COATNEY, G. R., 1937. On some blood Protozoa of reptiles and amphibians. Tr. Am. Micr. Soc., 56:291.

RUIZ, A., 1958. Contribución al estudio del género Phytomonas Donovan en Costa Rica. II. Rev. Biol. Trop., 6:272.

SAUNDERS, D. C., 1959. *Trypanosoma balistes* n. sp., etc. J. Parasit., 45:623.

SCHINDERA, M., 1922. Beiträge zur Biologie, Agglomeration und Züchtung von *Trypanoplasma helicis*. Arch. Protist., 45:200.

SCHWARTZ, J. B., 1961. Anaerobic metabolism of *Crithidia fasciculata*. J. Protozool., 8:9.

SIMPSON, C. F., and GREEN, J. H., 1959. Cultivation of *Trypanosoma theileri* in liquid medium at 37°C. Cornell Vet., 49:192.

SMITH, R. O. A., 1959. Bionomics of *Phlebotomus argentipes*. Bull. Calcutta Sch. Trop. Med., 7:17.

SWAMINATH, C. S., SHORTT, E., and ANDERSON, A. P., 1942. Transmission of Indian kala-azar to man by the bites of *Phlebotomus argentipes*. Indian J. Med. Res., 30:473.

TALIAFERRO, W. H., 1921. Variation and inheritance in size in *Trypanosoma lewisi*. I. Proc. Nat. Acad. Sci., 7:138.

―――― 1921a. II. *Ibid.*, 7:163.

―――― 1923. A study of size and variability, throughout the course of "pure line" infections, with *Trypanosoma lewisi*. J. Exper. Zool., 37:127.

―――― 1924. A reaction product in infections with *Trypanosoma lewisi* which inhibits the reproduction of the trypanosomes. J. Exper. Med., 39:171.

―――― 1926. Variability and inheritance of size in *Trypanosoma lewisi*. J. Exper. Zool., 43:429.

―――― 1932. Trypanocidal and reproduction-inhibiting antibodies to *Trypanosoma lewisi* in rats and rabbits. Amer. J. Hyg., 16:32.

―――― 1938. Ablastic and trypanocidal antibodies against *Trypanosoma duttoni*. J. Immunol., 35:303.

―――― and TALIAFERRO, L. G., 1934. Complement fixation, precipitin, adhesion, mercuric chloride and Wassermann tests in equine trypanosomiasis of Panama. *Ibid.*, 26:193.

TOBEY, E. N., 1906. Trypanosomiasis of the newt. J. Med. Res., 15:147.

USINGER, R. L., 1944. The Triatominae of North and Central America and the West Indies and their public health significance. U. S. Publ. Health Bull., no. 288.

VIANNA, G., 1911. Sobre uma nova especie de Leishmania. Brazil Medico, no. 41.

VICKERMAN, K., 1962. Observations on the life cycle of *Phytomonas elmassiani* (Migone) in East Africa. J. Protozool., 9:26.

WALLACE, F. G., 1956. Cultivation of *Trypanosoma ranarum* on a liquid medium. J. Protozool., 3:47.

―――― and CLARK, T. B., 1959. Flagellate parasites of the fly, *Phaenicia sericata*. *Ibid.*, 7:390.

―――― *et al.*, 1960. Two new species of flagellates cultivated from insects of the genus Gerris. *Ibid.*, 7:390.

WEINMAN, D., 1953. African sleeping sickness trypanosomes: cultivation and properties of the culture forms. Ann. N. Y. Acad. Sci., 56:995.

WEINSTEIN, P. P., and PRATT, H. D., 1948. The laboratory infection of *Triatoma neotomae* Neiva with *Trypanosoma cruzi*, etc. J. Parasit., 34:231.

WENRICH, D. H., 1932. The relation of the protozoan flagellate, *Retortamonas gryllotalpae*, etc. Tr. Am. Micr. Soc., 51:225.

WENYON, C. M., 1911. Oriental sore in Bagdad, etc. Parasitology, 4:273.

―――― 1926. Protozoology. London and Baltimore.

WIJERS, D. J. B., 1959. Polymorphism in *Trypanosoma gambiense* and *T. rhodesiense*, etc. Ann. Trop. Med. & Parasit., 53:59.

WOLCOTT, G. B., 1952. Mitosis in *Trypanosoma lewisi*. J. Morphol., 90:189.

WOOD, FAE D., 1934. Natural and experimental infection of *Triatoma protracta* Uhler and mammals in California with American human trypanosomiasis. Am. J. Trop. Med., 14:497.

―――― 1936. *Trypanosoma neotomae* sp. nov., etc. Univ. Cal. Publ. Zool., 41:133.

―――― and WOOD, S. F., 1941. Present knowledge of the distribution of *Trypanosoma cruzi* in reservoir animals and vectors. Am. J. Trop. Med., 21:335.

WOODY, N. C., and WOODY, H. B., 1955. American trypanosomiasis (Chagas' disease); first indigenous case in the United States. J. A. M. A., 159:676.

YAMASAKI, S., 1924. Ueber *Leptomonas ctenocephali*, etc. Arch. Protist., 48:137.

Chapter 15
Order 3. **Polymastigida** Blochmann

THE flagellates placed in this order possess three to eight flagella, although in one family, there may be a dozen or more flagella on each individual. Undulating membrane does not occur; axial filament or axostyle may occur in some; uninucleate or binucleate and in one genus multinucleate. The order is a heterogeneous group.

These flagellates are small organisms inhabiting the digestive tract of various animals. Xylophagous forms living in the termite intestine appear to hold symbiotic relationship with the host. Eight families.

Family 1. **Trimastigidae** Kent

Three flagella; one anterior flagellum, two trailing flagella.

Genus **Trimastix** Kent. Ovate or pyriform; naked; free-swim-

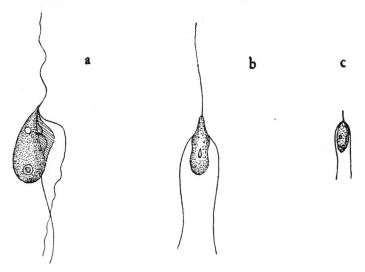

Fig. 155. a, *Trimastix marina*, ×1250 (Kent); b, *Dallingeria drysdali*, ×2000 (Kent); c, *Macromastix lapsa,* ×1500 (Stokes).

ming; with a laterally produced membranous border; three flagella (one anterior flagellum vibrating, two trailing); salt water. (Species, Grassé, 1952a.)

T. marina K. (Fig. 155. *a*). About 18μ long; salt water.

Genus **Dallingeria** Kent. Free-swimming or attached; with trailing flagella; body small; with drawn-out anterior end; fresh water with decomposed organic matter.

D. drysdali K. (Fig. 155, *b*). Small; elongate oval; less than 6μ long; stagnant water.

Genus **Macromastix** Stokes. Free-swimming, somewhat like *Dallingeria,* but anterior region not constricted; three flagella from anterior end; one contractile vacuole; fresh water.

M. lapsa S. (Fig. 155, *c*). Ovoid 5;5μ long; anterior flagellum one-half and trailing flagella two to three times body length; pond water.

Family 2 **Tetramitidae** Bütschli

With four flagella; one or two flagella may trail.

Genus **Tetramitus** Perty. Ellipsoidal or pyriform; free-swimming; cytostome at anterior end; four flagella unequal in length; a contractile vacuole; holozoic; fresh or salt water or parasitic. (Species, Klug, 1936.)

T. rostratus P. (Fig. 157, *a*). Body form variable, usually ovoid and narrowed posteriorly 18-30μ by 8-11μ; stagnant water. Bunting (1922, 1926) observed an interesting life cycle of what appeared to be this organism which she had found in cultures of the caecal content of rats (Fig. 156). (Nuclear division, Bunting and Wenrich, 1929.)

T. pyriformis Klebs (Fig. 157, *b*). Pyriform, with pointed posterior end; 11-13μ by 10-12μ; stagnant water.

T. salinus Entz (Fig. 157,*c*). Two anterior flagella, two long trailing flagella; nucleus anterior; cytostome anterior to nucleus; a groove to posterior end; cytopharynx temporary and length variable; 20-30μ long (Entz); 15-19μ long (Kirby). Kirby observed it in a pool with a high salinity at Marina, California.

Genus **Costia** Leclerque. Ovoid in front view, pyriform in profile; toward the right side, there is a shallow depression which leads into cytostome (?) and from which extend two long and two short flagella (only two flagella (Andai, 1933)); contractile vacuole posterior; encystment; ectoparasitic in freshwater fishes.

C. necatrix Henneguy (Fig. 157, *e-j*). 10-20μ by 5-10μ (Henneguy), 5-18μ by 2.5-7μ (Tavolga and Nigrelli, 1947); nucleus central; uninucleate cyst, spherical, 7-10μ in diameter; when present

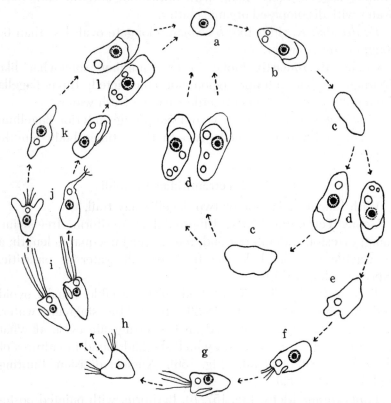

FIG. 156. Diagram illustrating the life-cycle of *Tetramitus rostratus* (Bunting). a, cyst; b, vegetative amoeba; c, division; d, after division; e, f, stages in transformation to flagellate form g, fully formed flagellate; h, flagellate prior to division; i, flagellate after division; j–l, transformation stages to amoeba.

in large numbers, the epidermis of the fish appears to be covered by a whitish coat. Davis (1943) found a similar organism which measured 9-14μ by 5-8μ, on trout, *Salmo irideus* and *Salvelinus fontinalis,* and named it *Costia pyriformis.*

Genus **Enteromonas** da Fonseca (*Tricercomonas* Wenyon and O'Connor). Spherical or pyriform, though plastic; three anterior flagella; the fourth flagellum runs along the flattened body surface and extends a little freely at the posterior tip of body; nucleus anterior; no cytostome; cyst ovoid and with four nuclei when mature; parasitic in mammals. da Fonseca (1915) originally observed only three flagella and no cysts; four flagella and encysted forms were noticed in Tricercomonas by Wenyon and O'Conner (1917); in da Fonseca's original-preparations, Dobell (1935) observed four flagella as well as cysts and concluded that Enteromonas and Tricercomonas are one and the same flagellate.

E. hominis da F. (*T. intestinalis* W. and O.) (Figs. 157, *k*; 158, *a-d*). Trophozoites 4-10µ by 3-6µ; nucleus circular or pyriform, with a large endosome, near anterior end; four flagella take their

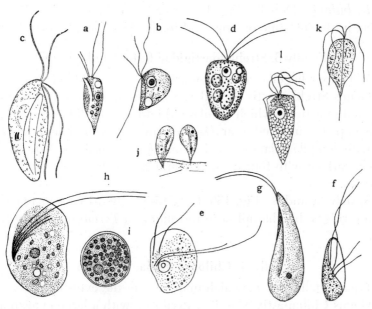

FIG. 157. a, *Tetramitus rostratus*, ×620 (Lemmermann); b, *T. pyriformis*, ×670 (Klebs); c. *T. salinus*, ×1630 (Kirby); d, *Collodictyon triciliatum*, ×400 (Carter); e–j, *Costia necatrix* (e, f, ×800 (Weltner); g–i, ×1400 (Moroff); j, two individuals attached to host integument ×500 (Kudo); k, *Enteromonas hominis*, ×1730 (Wenyon and O'Connor); 1, *Copromastix prowazeki*, ×1070 (Aragão).

origin in blepharoplasts located close to nucleus; cytoplasm vacuolated or reticulated, contains bacteria; cysts ovoid, 6-8μ by
4-6μ; with one, two or four nuclei; commensal in the lumen of
human intestine; found in diarrhoeic stools. Widely distributed.

E. caviae Lynch. Similar to the species mentioned above, but
slightly smaller; in the caecum of guinea-pigs (Lynch, 1922) (Cytology, Nie, 1950.)

Genus **Copromastix** Argão. Four anterior flagella equally long;
body triangular or pyramidal; coprozoic.

C. *prowazeki* A. (Fig. 157, *l*). About 16-18μ long ; in human
and rat faeces.

Genus **Karotomorpha** Travis (*Tetramastic* Alexeieff). Elongate
pyriform; body more or less rigid; four unequal flagella at the anterior end, in two groups; nucleus anterior; without cytostome;
parasitic in the intestine of Amphibia. (Species, Travis, 1934).

K. *bufonis* (Dobell) (Fig. 158, *e*). Spindle in shape; 12-16μ by
2-6μ; in the intestine of frogs and toads. (Cytology, Grassé, 1926.)

Family 3. **Streblomastigidae** Kofoid and Swezy

With four flagella; a rostellum; body surface ridged.

Genus **Streblomastix** K. and S. Spindle-form; with a **rostellum,**
the anterior tip of which is enlarged into a sucker-like cup; below
the cup are inserted four (Kidder) or six (Kofoid and Swezy)
equally long flagella; extremely elongate nucleus below rostellum;
body surface with four or more spiral ridges; in termite gut. One
species.

S. *strix* K. and S. (Fig. 158, *f*, *g*). 15-52μ by 2-15μ; four to eight
spiral ridges; blepharoplast in rostellum; in *Termopsis angusticollis*.

Family 4. **Chilomastigidae** Wenyon

Four flagella, one of which undulates in the cytostome.

Genus **Chilomastix** Alexeieff. Pyriform; with a large cytostomal
cleft at anterior end; nucleus anterior; three anteriorly directed
flagella; short fourth flagellum undulates within the cleft; cysts
common; in intestine of vertebrates. (Several species.)

C. *mesnili* Wenyon (Fig. 158, *h-k*). The trophozoite is oval or
pyriform; 5-20 (10-15)μ long; jerky movements; a large cytostomal cleft near anterior end; nucleus, vesicular, often without en-

dosome; three anterior flagella about 7-10μ long; the fourth flagellum short, undulates in the cleft which ridge is marked by two fibrils. The cyst pyriform; 7-10μ long; a single nucleus; two cytostomal fibrils and a short flagellum; commensal in the caecum and colon (some consider also in small intestine) of man. Both trophozoites and cysts occur in diarrhoeic faeces. It is widely distributed and very common. (Cytology, Kofoid and Swezy, 1920; cultivation, Boeck, 1921; vitality *in vitro,* Wysocka and Wegner, 1955.)

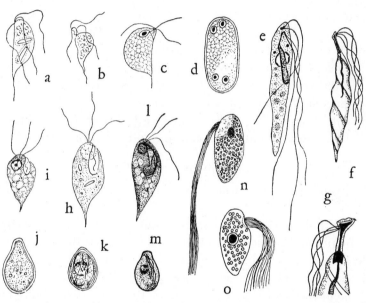

FIG. 158. a–d, *Enteromonas hominis,* ×1730 (Wenyon and O'Connor) (a, b, living and c, stained trophozoites; d, a stained cyst); e, *Karotomorpha bufonis,* ×2000 (Grassé); f, *Streblomastix strix,* ×1030; g, anterior end of the organism, showing the rostellum, blepharoplast, sucking cup and flagella (Kidder); h–k, *Chilomastix mesnili,* ×1530 (h, living and i, stained trophozoites; j, a fresh cyst; k, a stained cyst) l, a stained trophozoite, and m, a stained cyst of *C. gallinarum,* ×1330 (Boeck and Tanabe); n, *Callimastix frontalis,* ×1500 (Braune); o, *C. equi,* ×1100 (Hsiung).

C. intestinalis Kuczynski. In guinea-pigs; 13-27μ by 5-11μ (Geiman, 1935); 8.8-28μ by 6.6-11 (Nie, 1950).

C. bettencourti da Fonseca. In rats and mice.

C. cuniculi da F. In rabbits.

C. caprae da F. In goat.

C. gallinarum Martin and Robertson (Fig. 158, *l, m*). 11-20μ by 5-6μ; in the caeca of turkeys and chicks. (Morphology, Boeck and Tanabe, 1926.)

Family 5. **Callimastigidae** da Fonseca

Flagella twelve or more; in stomach of ruminants or in caecum and colon of horse.

Genus **Callimastix** Weissenberg. Ovoid; compact nucleus central or anterior; twelve to fifteen long flagella near anterior end, vibrate in unison. Weissenberg (1912) considered this genus to be related to *Lophomonas* (p. 485), but organism lacks axial organellae; in Cyclops and alimentary canal of ruminants and horse.

C. cyclopis W. In body-cavity of *Cyclops* sp.

C. frontalis Braune (Fig. 158, *n*). twelve flagella; about 12μ long; flagella 30μ long; in cattle, sheep and goats.

C. equi Hsiung (Fig. 158, *o*). Twelve to fifteen flagella; 12-18μ by 7-10μ; nucleus central; in caecum and colon of horse.

Family 6. **Polymastigidae** Bütschli

Typically four flagella arise as two pairs; axostyle or axial filament.

Genus **Polymastix** Bütschli. Pyriform; four flagella arise from two blepharoplasts located at anterior end; cytostome and axostyle inconspicuous; body often covered by a protophytan; commensals in insects. (Species, Grassé, 1926, 1952.)

P. melolonthae Grassi (Fig. 159, *a*). 10-15μ by 4-8μ; body covered by *Fusiformis melolonthae* (Grassé, 1926); in the intestine of Melolontha, Oryctes, Cetonia, Rhizotrogus, Tipula, etc.

Genus **Monocercomonoides** Travis (*Monocercomonas* Grassi). Small; four flagella inserted in pairs in two places; two directed anteriorly and the other two posteriorly; axostyle filamentous; parasitic. (Taxonomy, Travis, 1932.)

M. melolonthae Grassi (Fig. 159, *b*). Ovoid; 4-15μ long; in the larvae of *Melolontha melolontha*, etc.

Genus **Chilomitus** da Fonseca. Elongate oval; pellicle well developed; aboral surface convex; cytostome near anterior end, through which four flagella originating in a bi-lobed blepharoplast, protrude; rudimentary axostyle; nucleus and parabasal body below the cytostome (da Fonseca, 1915).

C. caviae da F. (Fig. 159, *c*). In the caecum of guinea-pigs; stained trophozoites 6-14μ by 3.1-4.6μ; cytoplasm contains siderophilic bodies of unknown nature (Nie, 1950).

Genus **Cochlosoma** Kotlán. Body small, oval; sucker in the anterior half; six flagella; axostyle filamentous; parasitic (Kotlán, 1923).

C. rostratum Kimura (Fig. 159, *d*). In the colon of domestic

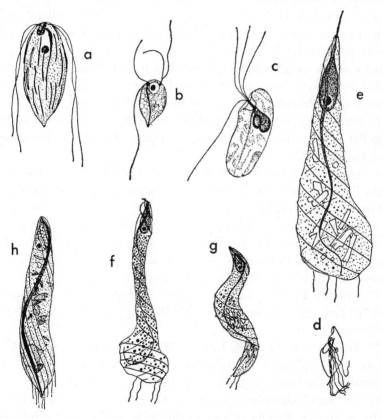

FIG. 159. a, *Polymastix melolonthae*, ×2000 (Grassé); b, *Monocercomonoides melolonthae*, ×2000 (Grassé); c, *Chilomitus caviae* (Nie); d, *Cochlosoma rostratum*, ×1465 (Kimura); e, *Pyrsonympha vertens*, ×730; f, *P. granulata*, ×500 (Powell); g, *Dinenympha gracilis*, ×730; h, *D. fimbriata*, ×625 (Kirby).

ducks, *Anas platyrhynchus* and *Carina moschata;* 6-10μ by 4-6.5μ (Kimura, 1934). McNeil and Hinshaw (1942) observed this organism in the intestine of young poults and in the region of caecal tonsil in adults.

Family 7. **Pyrsonymphidae** Grassi

Four to twelve flagella; well-developed axostyle; with or without proboscis; uninucleate, except Proboscidiella which is multinucleate.

Genus **Pyrsonympha** Leidy. Large; club-shaped, the posterior end is rounded; body surface with four to eight flagellar cords which are arranged lengthwise or slightly spirally; flagella extend freely posteriorly; blepharoplast at the anterior tip; often with a short process for attachment; axostyle a narrow band, may be divided into parts; large pyriform nucleus anterior; in termite gut. (Species, Koidzumi, 1921; nuclear division, Cleveland, 1938.)

P. vertens L. (Fig. 159, *e*). About 100-150μ long; four to eight flagellar cords; in *Reticulitermes flavipes*. (Cytology, Duboscq and Grassé, 1925.)

P. granulata Powell (Fig. 159, *f*). 40-120μ by 5-35μ; four to eight flagellar cords; in *Reticulitermes hesperus* (Powell, 1928).

Genus **Dinenympha** Leidy. Medium large; spindle form; four to eight flagellar cords adhering to body which are spirally twisted about one turn; the flagella free at the posterior end; axostyle varies from cord to band; pyriform nucleus, anterior, with a large endosome; in termite gut. (Species, Koidzumi, 1921.)

D. gracilis L. (Fig. 159, *g*). 24-50μ by 6-12μ; body flattened and twisted; ends attenuated; with adhering protophytes; in *Reticulitermes flavipes*.

D. fimbriata Kirby (Fig. 159, *h*). 52-64μ by 8-18μ; 4-8 flagellar cords; with adherent protophytes; axostyle varies in width; in *Reticulitermes hesperus* (Kirby, 1924).

Genus **Saccinobaculus** Cleveland. Elongate to spherical; four, eight, or twelve flagella adhere to the body, and project out freely; axostyle is an extremely large paddle-like body and undulates, serving as cell-organ of locomotion; posterior end of axostyle enclosed in a sheath; in woodroach gut. (Sexual cycle, Cleveland, 1956.)

S. ambloaxostylus C. (Fig. 160, *a-c*). 65-110μ by 18-26μ; in

Cryptocercus punctulatus. (Sexual reproduction, Cleveland, 1950a.)

Genus **Notila** Cleveland. Body elongate, plastic; four flagella, the attached portion of which shows attached granules (Fig. 160, *i*); axostyle large, paddle-like, much broader than that of *Pyrsonympha;* no axostyler sheath at posterior end, but with large granules or spherules embedded in it; in *Cryptocercus punctulatus.* (Sexual cycle, Cleveland, 1956.)

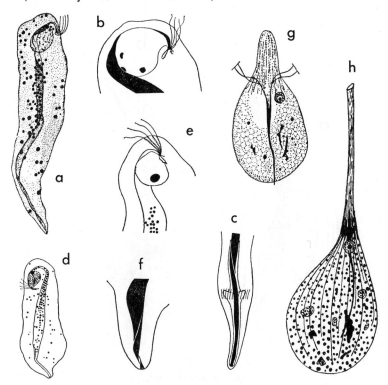

FIG. 160. a–c, *Saccinobaculus ambloaxostylus* (Cleveland) (a, whole organism, ×600; b, anterior and c, posterior portion of trophozoites); d–f, *Notila proteus* (Cleveland) (d, diploid individual, ×360; e, anterior and f, posterior end of the organism); g, h, *Oxymonas dimorpha* (Cornell) (g, a motile form, ×900; h, an attached aflagellated form, ×460).

N. proteus C. (Fig. 160, *d-f*). Size not given; gametogenesis and sexual fusion, induced by the molting hormone of the host; diploid number of chromosomes twenty-eight (Cleveland, 1950b).

Genus **Oxymonas** Janicki. Attached phase with a conspicuous rostellum, the anterior end of which forms a sucking-cup for attachment; pyriform. In motile phase, rostellum is less conspicuous; two blepharoplasts located near the anterior extremity of axostyle, give rise to two flagella each; axostyle conspicuous; xylophagous; in termite and woodroach; sexual reproduction in some (Cleveland, 1950).

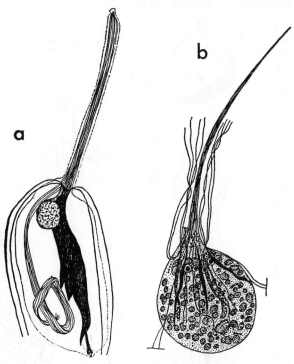

FIG. 161. a, *Oxymonas grandis*, ×265 (Cleveland); b, *Proboscidiella kofoidi*, ×600 (Kirby).

O. dimorpha Connell (Fig. 160, *g*, *h*). Subovoid; delicate pellicle; axostyle slightly protruding; a pair of long anterior flagella from two blepharoplasts, connected by rhizoplast; nucleus anterior. When attached to intestine, rostellum elongate, flagella disappear; 17μ by 14μ to 195μ by 165μ; in *Neotermes simplicicornis*; California and Arizona (Connell, 1930).

O. grandis Cleveland (Fig. 161, *a*). Body 76μ by 31μ to 183μ by 79μ; rostellum varies 30-200μ in length; nucleus without an endosome, anterior, about 20-23µ in diameter; axostyle consists of a staining part and a non-staining part; in the intestine of *Neotermes dalbergiae* and *N. tectonae* (Cleveland, 1935).

Genus **Proboscidiella** Kofoid and Swezy (*Microrhopalodina*, Grassi and Foà; *Kirbyella* Zeliff). Attached and motile forms similar to *Oxymonas*; but multinucleate; four flagella from each karyomastigont (p. 315); rostellum with filaments which extend posteriorly as axostyles; in termite gut (Kofoid and Swezy, 1926; Zeliff, 1930a).

P. kofoidi Kirby (Fig. 161, *b*). Average size 66µ by 46µ; rostellum as long as, or longer than, the body; karyomastigonts two to nineteen or more (average eight); each mastigont with two blepharoplasts from which extend four flagella; in *Cryptotermes dudleyi* (Kirby, 1928).

Family 8. **Hexamitidae** Kent

Flagella six or eight; with axial filament; bilaterally symmetrical; binucleate.

Genus **Hexamita** Dujardin (*Octomitus* Prowazek). Pyriform; two nuclei near anterior end; six anterior and two posterior flagella; two axostyles; one to two contractile vacuoles in free-living forms; cytostome obscure; endoplasm with refractile granules; encystment; in stagnant water or parasitic. (In oyster, Scheltema, 1962.)

H. inflata D. (Fig. 162, *a*). Broadly oval; posterior end truncate; 13-25µ by 9-15µ; in stagnant water.

H. intestinalis D. (Fig. 162, *b*, *c*). 10-16µ long; in intestine of frogs, also in midgut of *Trutta fario* and in rectum of *Motella tricirrata* and *M. mustela* in European waters. (Morphology, Schmidt, 1920.)

H. salmonis (Moore) (Fig. 162, *d*). 10-12µ by 6-8µ; in intestine of various species of trout and salmon; schizogony in epithelium of pyloric caeca and intestine; cysts; pathogenic to young host fish (Moore, 1922, 1923; Davis, 1925). (Chemotherapy, Yasutake *et al.*, 1961.)

H. periplanetae (Bělař). 5-8µ long; in intestine of cockroaches.

H. cryptocerci Cleveland (Fig. 162, *e*). 8-13µ by 4-5.5µ; in *Cryptocercus punctulatus*.

H. meleagridis McNeil, Hinshaw and Kofoid (Fig. 163, *a*).

Body 6-12μ by 2-5μ. It causes a severe catarrhal enteritis in young turkeys. Experimentally it is transmitted to young quail, chicks, and duckling (McNeil, Hinshaw and Kofoid, 1941). (Life cycle, Slavin and Wilson, 1960.)

H. sp. Hunninen and Wichterman (1938) (Fig. 163, *b*). Average

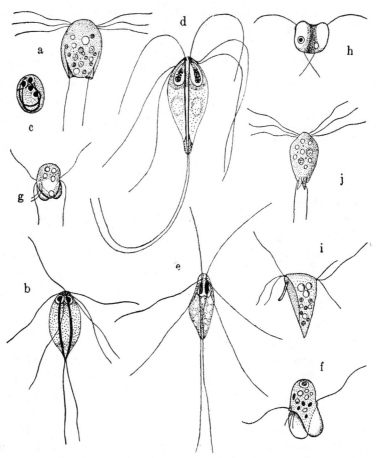

Fig. 162. a, *Hexamita inflata,* ×600 (Klebs); b, c, trophozoite and cyst of *H. intestinalis,* ×1600 (Alexeieff); d, *H. salmonis,* ×2100 (Davis); e, *H. cryptocerci,* ×1600 (Cleveland); f, *Trepomonas agilis,* ×1070 (Klebs); g, *T. rotans,* ×710 (Lemmermann); h, *Gyromonas ambulans,* ×530 (Seligo); i, *Trigonomonas compressa,* ×490 (Klebs); j, *Urophagus rostratus,* ×800 (Klebs).

dimensions 10μ by 5.5μ; found in the reproductive organs of the trematode, *Deropristis inflata,* parasitic in the eel; heavily infected eggs are said not to develop.

Genus **Giardia** Kunstler (*Lamblia* Blanchard). Pyriform to ellipsoid; anterior end broadly rounded, posterior end drawn out; bi-

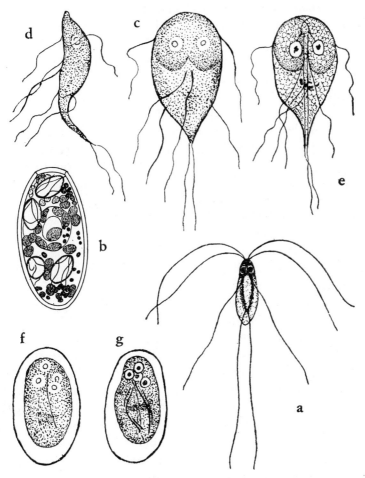

Fig. 163. a, *Hexamita meleagridis,* ×1875 (McNeil *et al.*); b, an egg of *Deropristis inflata* containing *Hexamita,* ×770 (Hunninen and Wichterman); c–g, *Giardia intestinalis,* ×2300 (c, front and d, side view of living organisms; e, stained trophozoite; f, fresh and g, stained mature cysts).

laterally symmetrical; dorsal side convex, ventral side concave or flat, with a sucking disc in anterior half; two nuclei; two axostyles; eight flagella in four pairs; cysts oval to ellipsoid; with two or four nuclei and fibrils; in the intestine of various vertebrates. Many species. (Criteria for species differentiation, Simon, 1921; Hegner, 1922; cytology and taxonomy, Filice, 1952.)

G. intestinalis Lambl (*G. enterica*, Grassi; *G. lamblia*, Stiles) (Fig. 163, *c-g*). When the flagella lash actively, the organism shows a slight forward movement with a sidewise rocking motion. The trophozoite is broadly pyriform, not plastic; 9-20µ by 5-10µ; sucking disc acts as attachment organella; cytoplasm hyaline; two needle-like axostyles; two vesicular nuclei near anterior margin; eight flagella in four pairs; two flagella originate near the anterior end of axostyles, cross each other and follow the anterolateral margin of the disc, becoming free; two originating in anterior part of axostyles, leave the body about one-third from the posterior tip; two (ventral) which are thicker than others, originate in axostyles at nuclear level and remain free; two (caudal) flagella arise from the posterior tips of axostyles; a deeply staining body may be found in cytoplasm.

The cysts are ovoid and refractile; 8-14µ by 6-10µ; cyst wall thin; contents do not fill the wall; two or four nuclei, axostyles, fibrils and flagella are visible in stained specimens.

This flagellate inhabits the lumen of the duodenum and other parts of small intestine and colon of man. Both trophozoites and cysts are ordinarily found in diarrhoeic faeces. In severe cases of infection, an enormous number of the organisms attach themselves to the mucous membrane of the intestine which may result in abnormal functions of the host tissues. In some cases, the flagellate has been reported from the gall bladder. The stools often contain unusual amount of mucus. Although there is no evidence that *G. intestinalis* attacks the intestinal epithelium, experimental observations point to its pathogenicity (Tsuchiya and Andrews, 1930; Payne *et al.*, 1960). Rendtorff (1954) found that when 1-1,000,000 cysts in saline solution were fed to forty volunteers, twenty-one became infected. Even ten to twenty-five cysts caused infection. The prepatent period varied from six to fifteen days. The infection did not bring about any clinical illness, and disappeared spontaneously in five to forty-one days. (Cytology, Kofoid and Swezy, 1922.)

G. *duodenalis* (Davaine). In the intestine of rabbits; 13-19μ by 8-11μ (Hegner, 1922).

G. *canis* Hegner. In dogs; 12-17μ by 7.6-10μ; cysts oval, 9-13μ by 7-9μ (Hegner, 1922).

G. *muris* (Grassi). In rats and mice; 7-13μ by 5-10μ (Simon, 1922). (Freezing and survival, Bemrick, 1961a.)

G. *simoni* Lavier. In the small intestine of rats; 14-19μ by 7-10.5μ (Lavier, 1924); 11-16μ by 5-8μ (Nieschulz and Krijgsman, 1925).

G. *ondatrae* Travis. In the intestine of the muskrat, *Ondatra zibethica;* 13μ by 7μ (Travis, 1939); 10μ by 5.5μ (Waters *et al.*).

G. *caviae* Hegner. In the intestine of guinea-pigs; 8-14μ by 5.5-10μ (Hegner, 1923).

Genus **Trepomonas** Dujardin. Free-swimming; flattened; more or less rounded; cytostomal grooves on posterior half, one on each side; eight flagella (one long and three short flagella on each side) arise from anterior margin of groove; near anterior margin there is a horseshoe-form structure, in which two nuclei are located; fresh water, parasitic, or coprozoic.

T. *agilis* D. (Fig. 162, *f*). More or less ovoid; 7-30μ long; one long and three short flagella on each side; rotation movement; stagnant water; also reported from intestine of amphibians.

T. *rotans* Klebs (Fig. 162, *g*). Broadly oval; posterior half highly flattened; two long and two short flagella on each of two cytostomes; stagnant water.

Genus **Gyromonas** Seligo. Free-swimming; small; form constant, flattened; slightly spirally coiled; four flagella at anterior end; cytostome not observed; fresh water.

G. *ambulans* S. (Fig. 162, *h*). Rounded; 8-15μ long; standing water.

Genus **Trigonomonas** Klebs. Free-swimming; pyriform, plastic; cytostome on either side, from anterior margin of which arise three flagella; flagella six in all; two nuclei situated near anterior end; movement rotation; holozoic; fresh water.

T. *compressa* K. (Fig. 162, *i*). 24-33μ by 10-16μ; flagella of different lengths; standing water (Klug, 1936).

Genus **Urophagus** Klebs. Somewhat similar to *Hexamita;* but a single cytostome; two moveable posterior processes; holozoic; stagnant water.

U. rostratus (Stein) (Fig. 162, *j*). Spindle-form; 16-25μ by 6-12μ.

References

ALEXEIEFF, A., 1912. Sus quelques noms de genre des flagellés, etc. Zool. Anz., 39:674.

ANDAI, G., 1933. Ueber *Costia necatrix.* Arch. Protist., 79:284.

BEMRICK, W. J., 1961. A note on the incidence of three species of Giardia in Minnesota. J. Parasitol., 47:87.

——— 1961a. Effect of low temperatures on trophozoites of *Giardia muris. Ibid.,* 47:573.

BOECK, W. C., 1921. *Chilomastix mesnili* and a method for its culture. J. Exper. Med., 33:147.

——— and TANABE, M., 1926. *Chilomastix gallinarum,* morphology, division and cultivation. Amer. J. Hyg., 6:319.

BUNTING, M., 1926. Studies of the life-cycle of *Tetramitus rostratus.* J. Morphol. Physiol., 42:23.

——— and WENRICH, D. H., 1929. Binary fission in the amoeboid and flagellate phases of *Tetramitus rostratus. Ibid.,* 47:37.

CLEVELAND, L. R., 1935. The intranuclear achromatic figure of *Oxymonas grandis* sp. nov. Biol. Bull., 69:54.

——— 1938. Mitosis in Pyrsonympha. Arch. Protist., 91: 452.

——— 1950. Hormone-induced sexual cycle of flagellates. II. J. Morphol., 86:185.

——— 1950a. III. *Ibid.,* 86:215.

——— 1950b. IV. *Ibid.,* 87:317.

———, HALL, S. R., SANDERS, E. P., and COLLIER, J., 1934. The wood-feeding roach, Cryptocercus, its Protozoa, etc. Mem. Am. Acad. Arts and Sci., 17:185.

CONNELL, F. H., 1930. The morphology and life-cycle of *Oxymonas dimorpha,* etc. Univ. Cal. Publ. Zool., 36:51.

DA FONSECA, O. O. R., 1915. Sobre os flagellados dos mammiferos do Brazil. Brazil Medico, 29:281.

DAVIS, H. S., 1925. *Octomitus salmonis,* a parasitic flagellate of trout. Bull. Bur. Fisher., 42:9.

——— 1943. A new polymastigine flagellate, *Costia pyriformis,* parasitic on trout. J. Parasit., 29:385.

FILICE, F. P., 1952. Studies on the cytology and life history of a Giardia from the laboratory rat. Univ. Cal. Publ. Zool., 57:53.

GALLI-VALERIO, B., 1903. Notes de parasitologie. Centralbl. Bakt. I. Orig., 35:81.

GEIMAN, Q. M., 1935. Cytological studies of the Chilomastix of man and other mammals. J. Morphol., 57: 429.

GRASSÉ, P.-P., 1926. Contribution a l'étude des flagellés parasites. Arch. zool. exper. gén., 65:345.

HEGNER, R. W., 1922. A comparative study of the Giardias living in man, rabbit, and dog. Amer. J. Hyg., 3:442.

———— 1923. Giardias from wild rats and mice and *Giardia caviae* sp. n. from the guinea-pig. *Ibid.*, 3:345.

HSIUNG, T. S., 1930. A monograph on the Protozoa of the large intestine of the horse. Iowa State Coll. J. Sci., 4:356.

HUNNINEN, A. V., and WICHTERMAN, R., 1938. Hyperparasitism: a species of Hexamita found in the reproductive systems, etc. J. Parasit., 24:95.

KIMURA, G. G., 1934. *Cochlosoma rostratum* sp. nov., etc. Tr. Am. Micr. Soc., 53:102.

KIRBY, H. JR., 1924. Morphology and mitosis of *Dinenympha fimbriata.* Univ. Cal. Publ. Zool., 26:199.

———— 1926. On *Staurojoenina assimilis* sp.n. *Ibid.*, 29:25.

———— 1928. A species of Proboscidiella from Kalotermes, etc. Quart. J. Micr. Sci., 72:355.

———— 1932. Two Protozoa from brine. Tr. Am. Micr. Soc., 51:8.

KOFOID, C. A., and CHRISTIANSEN, E. B., 1915. On binary and multiple fission in *Giardia muris.* Univ. Cal. Publ. Zool., 16:30.

———— and SWEZY, O., 1920. On the morphology and mitosis of *Chilomastix mesnili*, etc. Univ. Cal. Publ. Zool., 20:117.

———— ———— 1922. Mitosis and fission in the active and encysted phases of *Giardia enterica*, etc. *Ibid.*, 20:199.

———— ————.1926. On *Proboscidiella multinucleata*, etc. *Ibid.*, 20:301.

KOIDZUMI, M., 1921: Studies on the intestinal Protozoa found in the termites of Japan. Parasitology, 13:235.

KOTLÁN, A., 1923. Zur Kenntnis der Darmflagellaten aus der Hausente und anderen Wasservögeln. Centralbl. Bakt. I. Orig., 90:24.

LAMBL, W., 1859. Mikroskopische Untersuchungen der Darm-Excrete. Vierteljahrschr. prakt. Heilk., 61:1.

LAVIER, G., 1924. Deux espèces de Giardia, etc. Ann. Parasit., 2:161.

LYNCH, K. M., 1922. *Tricercomonas intestinalis* and *Enteromonas caviae* n. sp., etc. J. Parasit., 9:29.

MARTIN, C. H., and ROBERTSON, M., 1911. Further observations on the caecal parasites of fowls. Quart. J. Micr. Sci., 57:53.

MCNEIL, E., and HINSHAW, W. R., 1942. *Cochlosoma rostratum* from the turkey. J. Parasit., 28: 349.

———— ———— and KOFOID, C. A., 1941. *Hexamita meleagridis* sp. nov. from the turkey. Amer. J. Hyg., 34:71.

MOORE, E., 1922. *Octomitus salmonis*, a new species of intestinal parasite in trout. Tr. Am. Fish. Soc., 52:74.

———— 1923. Diseases of fish in State hatcheries. Rep. Bur. Prev. Stream Poll., New York, 12:18.

NIE, D., 1950. Morphology and taxonomy of the intestinal Protozoa of the guinea-pig, *Cavia porcella.* J. Morphol., 86:381.

NIESCHULZ, O., and KRIJGSMAN, B. J., 1925. Ueber *Giardia simoni* Lavier. Arch. Protist., 52:166.

PAYNE, F. J., *et al.*, 1960. Association of *Giardia lamblia* with disease. J. Parasit., 46:742.

POWELL, W. N., 1928. On the morphology of Pyrsonympha with a description of three new species, etc. Univ. Cal. Publ. Zool., 31:179.

RENDTORFF, R. C., 1954. The experimental transmission of human intestinal protozoan parasites. II. *Giardia lamblia* cysts given in capsules. Amer. J. Hyg., 59:209.

SCHELTEMA, R. S., 1962. The relationship between the flagellate protozoan Hexamita and the oyster *Crassostrea virginica.* J. Parasit., 48:137.

SCHMIDT, W., 1920. Untersuchungen über *Octomitus intestinalis.* Arch. Protist., 40:253.

SIMON, C. E., 1921. *Giardia enterica:* etc. Amer. J. Hyg., 1:440.

———— 1922. A critique of the supposed rodent origin of human giardiasis. *Ibid.*, 2:406.

SLAVIN, D., and WILSON, J. E., 1960. A fuller conception of the life cycle of *Hexamita meleagridis.* Poultry Sci., 39:1559.

TAVOLGA, W. N., and NIGRELLI, R. F., 1947. Studies on *Costia necatrix.* Tr. Am. Micr. Soc., 66: 366.

TRAVIS, B. V., 1932. A discussion of synonymy in the nomenclature of certain insect flagellates, etc. Iowa State Coll. J. Sci., 6:317.

———— 1934. Karotomorpha, a new name for Tetramastix, etc. Tr. Am. Micr. Soc., 53:277.

———— 1939. Descriptions of five new species of flagellate protozoa of the genus Giardia. J. Parasit., 25:11.

TSUCHIYA, H., and ANDREWS, J., 1930. A report on a case of giardiasis. Amer. J. Hyg., 12:297.

WATERS, P. C., FIENE, A. R., and BECKER, E. R., 1940. Strains in *Giardia ondatrae* Travis. Tr. Am. Micr. Soc., 59:160.

WEISSENBERG, R., 1912. *Callimastix cyclopis* n.g., n.sp., etc. Berlin. Sitz.-Ber. Ges. naturf. Freunde, p. 299.

WENRICH, D. H., 1932. The relation of the protozoan flagellate *Retortamonas gryllotalpae*, etc. Tr. Am. Micr. Soc., 51:225.

WYSOCKA, F., and WEGNER, Z., 1955. Observations on the vitality of **Chilomastix mesnili in vitro** (Polish). **Biul. Inst. Med. Morskiej w** Gdansku, 6:255.

YASUTAKE, W. T., *et al.*, 1961. Chemotherapy of hexamitiasis in fish. J. Parasit., 47:81.

ZELIFF, C. C., 1930. A cytological study of Oxymonas, etc. Amer. J. Hyg., 11:714.

———— 1930a. *Kirbyella zeteki*, etc., *Ibid.*, 11:740.

Chapter 16
Order 4. **Trichomonadida** Kirby

FOLLOWING Kirby (1947), these flagellates which had been formerly placed in Polymastigida are separated from it and grouped here. They possess a mastigont of three to six flagella. One flagellum is typically differentiated as a trailing flagellum which may border an undulating membrane. The *mastigont* is associated with a nucleus, an axostyle and a parabasal body, although in some multinucleated forms the nucleus may be dissociated from the complex.

The majority of Trichomonadida inhabit the digestive tract of various animal hosts and nutrition is holozoic or saprozoic. Asexual reproduction by fission; sexual reproduction and encystment are unknown. Four families.

Family 1. **Monocercomonadidae** Kirby

The trailing flagellum is either free or adherent, but without cresta or undulating membrane; parasitic.

Genus **Monocercomonas** Grassi (*Eutrichomastix*, Kofoid and Swezy). Three anterior flagella and one trailing flagellum which may be partially attached to body or free; no undulating membrane; parabasal body bacilliform; in the intestine of invertebrates and vertebrates. (Many species.)

M. axostylis (Kirby) (Fig. 164, *a*). Elongate, ellipsoid or pyriform; axostyle projects beyond the body; 5-10.5μ by 2-3.5μ; 3 anterior flagella 5-10μ long; in the gut of *Masutitermes kirbyi* (Kirby, 1931).

M. verrens Honigberg. Average dimensions 7μ by 5μ; in the faeces of Tapir (Honigberg, 1947).

Genus **Tricercomitus** Kirby. Small; three anterior flagella; a long trailing flagellum, adhering to body; nucleus anterior, without endosome; blepharoplast large, with a parabasal body and an axial filament; parasitic.

T. termopsidis K. (Fig. 164, *b*). 4-12μ by 2-3μ; anterior flagella

457

6-20μ long; trailing flagellum 19-65μ long; in gut of *Zootermopsis angusticollis, Z. nevadensis* and *Z. laticeps;* California and Arizona. (Culture and encystment, Trager, 1934.)

Genus **Hexamastix** Alexeieff. Body similar to *Monocercomonas* but with six flagella, of which one trails; axostyle often conspicuous; parabasal body prominent.

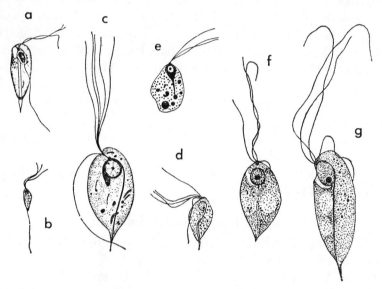

FIG. 164. a, *Monocercomonas axostylis,* ×2000; b, *Tricercomitus termopsidis,* ×665; c, *Hexamastix termopsidis,* ×2670 (all Kirby); d, *H. batrachorum;* e, *Protrichomonas legeri,* ×1000 (Alexeieff); f,g, *Pseudotrichomonas keilini,* ×2200 (Bishop).

H. termopsidis Kirby (Fig. 164, *c*). Ovoidal or pyriform; 5-11μ long; flagella 15-25μ long; in gut of *Zootermopsis angusticollis* and *Z. nevadensis;* California (Kirby, 1930).

H. caviae and *H. robustus* were described by Nie (1950) from the caecum of guinea-pigs.

H. batrachorum Alexeieff (Fig. 164, *d*). Oval or spindle form; 8-14μ by 4-8μ; flagella about body length; about 11μ by 7μ; five anterior flagella and one trailing flagellum (Honigberg and Christian, 1954); in the gut of *Triturus vulgaris, T. helveticus* and *Ambystoma maculatum.* (Species in lizards, Honigberg, 1955.)

Genus **Protrichomonas** Alexeieff. Three anterior flagella of equal length, arising from a blepharoplast located at anterior end; parasitic.

P. legeri (A.) (Fig. 164, *e*). In oesophagus of the marine fish, *Box boops*.

Genus **Pseudotrichomonas** Bishop. Body form, structure and movement, are exactly like those of *Tritrichomonas* (p. 470), but free-living in freshwater pond (Bishop, 1939).

P. keilini B. (Fig. 164, *f*, *g*). When alive 7-11μ by 3-6μ; highly plastic; young cultures contain more globular forms, while old cultures more elongated organisms; three unequally long anterior flagella; undulating membrane short, does not extend more than one-half the body and without a free flagellum; cytostome; holozoic, feeding on bacteria; nucleus anterior; axostyle filamentous, invisible in life; no cysts; in a pond in Lincolnshire, England. Bishop (1935) cultivated this flagellate in serum-saline medium, in hay infusion and in pond or rain water with boiled wheat grains at 4-31°C. (Bishop, 1936, 1939).

Genus **Coelotrichomastix** Hollande. Dorsal side convex, ventral side concave; three anterior flagella, one borders the undulating membrane; free-living.

C. convexa H. (Fig. 166, *a*). In life, 10-22μ by 8-10μ; dorsal side strongly convex, ventral side concave; three free flagella nearly equally long, fourth flagellum borders the undulating membrane; spherical nucleus voluminous, with a large endosome; free-living and coprozoic (Hollande, 1939; Grassé, 1952a).

Genus **Hypotrichomonas** Lee. Three unequal anterior flagella; one recurrent flagellum borders a weak undulating membrane; no costa; capitulum of stout axostyle, spatulate or scoop-shaped; parabasal apparatus rod-or V-shaped body or filaments; in reptiles (Lee, 1960).

H. acosta (Moskowitz). 7.3μ by 4.8μ; in various reptiles. Lee considers this an intermediate stage between Monocercomonadidae and Trichomonadidae.

Family 2. **Devescovinidae** Doflein

Usually three anterior flagella and a trailing stout flagellum; near base of trailing flagellum an elongated **cresta** (becoming **a** large internal membrane in some species) (Fig. 165); trailing flagellum lightly adheres to body surface along edge of cresta; axostyle; parabasal body of various forms; single nucleus anterior; without undulating membrane; generally xylophagous. (Cytology and morphogenesis, Kirby, 1944.)

Genus **Devescovina** Foà. Elongate body, usually pointed poste-

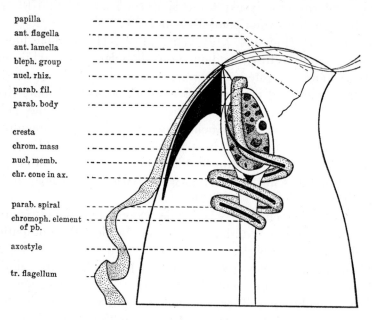

papilla
ant. flagella
ant. lamella
bleph. group
nucl. rhiz.
parab. fil.
parab. body

cresta
chrom. mass
nucl. memb.
chr. cone in ax.

parab. spiral
chromoph. element of pb.

axostyle

tr. flagellum

Fig. 165. A diagrammatic view of the anterior part of *Devescovina lemniscata*, showing the cresta and other organellae (Kirby).

riorly; three anterior flagella about the body length; trailing flagellum, slender to band-form, about 1-1.5 times the body length; cresta; parabasal body spiraled around axostyle or nucleus; in termite intestine. Many species (Kirby, 1941, 1949).

D. lemniscata Kirby (Figs. 165, 166, *b*). 21-51µ by 9-17µ; trailing flagellum a band; cresta long, 7-9µ; in *Cryptotermes hermsi* and many species of the genus; species of Neotermes, Glyptotermes and Kalotermes (Kirby, 1926a).

Genus **Parajoenia** Janicki. Medium large; with rounded extremities; three anterior flagella and trailing flagellum long; cresta of moderate size; parabasal body well developed with its anterior end close to blepharoplast; stout axostyle expanded anteriorly into leaf-like capitulum, bearing a longitudinal keel; in intestine of termites.

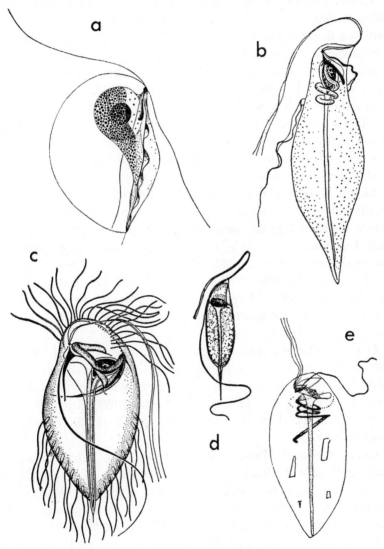

Fig. 166. a, *Coelotrichomastix convexa*, ×1310 (Hollande); b, *Devescovina lemniscata*, ×1600 (Kirby); c, *Parajoenia grassii*, with attached "spiro-chaetes," ×1150 (Kirby); d, *Foaina nana*, ×1150 (Kirby); e, *Metadeves-covina debilis*, ×1130 (Light).

P. grassii J. (Fig. 166, *c*). 29-59μ by 12-33μ; trailing flagellum stout, cordlike; cresta about 9μ long; in *Neotermes connexus* (Kirby, 1937, 1942a).

Genus **Foaina** Janicki (*Janickiella*, Duboscq and Grassé; *Paradevescovina, Crucinympha*, Kirby). Small to medium large; three anterior flagella; trailing flagellum about twice the body length; cresta slender, 2.5-17μ long; parabasal body single, in some with rami; in intestine of termites. Many species (Kirby, 1942a, 1949).

F. nana Kirby (Fig. 166, *d*). 6-18μ by 4.5-8.5μ; trailing flagellum a moderately stout cord, two to three times the body length; cresta slender, 8.5μ long; filament part of the parabasal body reaching the middle of body; in *Cryptotermes hermsi* and many species of the genus; also species of Glyptotermes, Rugitermes, and Procryptotermes (Kirby, 1942a).

Genus **Macrotrichomonas** Grassi. Large; three anterior flagella; trailing flagellum well developed, 1-1.5 times the body length; cresta a broad internal membrane, 21-86μ long; parabasal body coiled around the axostyle, one to thirteen times; in termite gut. (Several species, Kirby, 1942, 1949.)

M. pulchra G. (Fig. 167, a). 44-91μ by 21-41μ; trailing flagellum band-form, cresta large; parabasal body coiled closely four to ten times; in *Glyptotermes parvulus*, and many other species of the genus (Kirby, 1942).

Genus **Metadevescovina** Light. Moderately large; three anterior flagella; a short trailing flagellum; cresta small; parabasal body loosely coiled around axostyle; anterior end of axostyle in a loop; in termite gut. (Many species, Light, 1926; Kirby, 1945.)

M. debilis L. (Fig. 166, *e*). 30-70μ by 15-30μ; in *Kalotermes hubbardi*.

Genus **Caduceia** Franca. Large; three long anterior flagella; trailing flagellum slender, shorter than body; cresta relatively small, 1-12μ long; parabasal body coiled around axostyle two to twenty times; nucleus relatively large; axostyle terminates in filament; in termites. (Several species, Kirby, 1942, 1949.)

C. bugnioni Kirby (Fig. 167, *b*). 48-80μ by 18-40μ; in *Neotermes greeni* (Kirby, 1942).

Genus **Hyperdevescovina** Kirby. Similar to *Caduceia;* but cresta very small; stout axostyle projects from the body; in Proglyp-

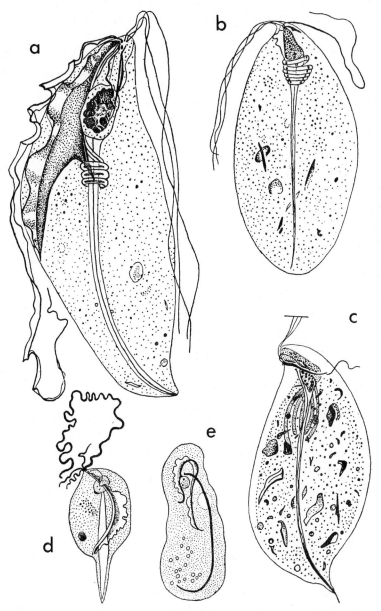

FIG. 167. a, *Macrotrichomonas pulchra,* ×1600 (Kirby); b, *Caduceia bugnioni,* ×930 (Kirby); c, *Pseudodevescovina uniflagellata,* ×1190 (Kirby); d,e, *Gigantomonas herculea* (Dogiel) (d, ×530; e, "amoeboid" form (Myxomonas), ×400).

totermes, Neotermes; New Zealand and South Africa. (Many species, Kirby, 1949.)

H. calotermitis (Nurse). 52-114μ by 30-65μ; projecting portion
of the axostyle 45-63μ; in *Proglyptotermes browni;* New Zealand.

Genus **Pseudodevescovina** Sutherland. Large; three short anterior flagella; one short trailing flagellum; axostyle stout; cresta of
moderate size; parabasal body large, divided into a number of attached cords; in termite gut. (Several species, Kirby, 1945.)

P. uniflagellata S. (Fig. 167, c). 52-95μ by 26-60μ; three delicate flagella, 30μ long; trailing flagellum a little stouter; cresta
11-20μ long; main parabasal body C-shaped, with seven to nineteen attached cords; in *Kalotermes insularis* (Kirby, 1936, 1945).

Genus **Bullanympha** Kirby. Flagella and cresta similar to those
in *Pseudodevescovina;* axostyle similar to that in *Caduceia;* proximal part of parabasal body bent in U-form around the nucleus
and attached voluminous distal portion coiled around the axostyle; in termite gut (Kirby, 1938, 1949).

B. silvestrii K. (Fig. 171, i). 50-138μ by 35-100μ; cresta about
5.8μ long; distal portion of parabasal body coils around axostyle
about twice; in *Neotermes erythraeus.*

Genus **Gigantomonas** Dogiel (*Myxomonas* D.). Medium large;
3 anterior flagella; a long and stout trailing flagellum; cresta conspicuously large; large axostyle; in termite gut. According to
Kirby (1946), the so-called undulating membrane is a large cresta;
in aflagellate phase (Myxomonas) the nuclear division takes place.

G. herculea D. (*M. polymorpha* D.) (Fig. 167, d, e). 60-75μ by
30-35μ; in the intestine of *Hodotermes mossambicus* (Kirby,
1946).

Family 3. **Calonymphidae** Grassi

Trichomonadida with a permanent polymonad or multinuclear
organization. Each nucleus is associated with a blepharoplast
(from which a flagellum extends), a parabasal body and an axial
filament. Janicki called this complex *karyomastigont* (Fig. 168, a)
and the complex which does not contain a nucleus, *akaryomasti-
gont* (Fig. 168, e).

Genus **Calonympha** Foà. Body rounded; large; numerous long
flagella arise from anterior region; numerous nuclei; karyomas-

tigonts and akaryomastigonts; axial filaments form a bundle; in termite gut (Foà, 1905).

C. grassii F. (Fig. 168, *a*). 69-90μ long; in *Cryptotermes grassii*.

Genus **Stephanonympha** Janicki. Oval, but plastic; numerous nuclei spirally arranged in the anterior half; karyomastigonts; axial filaments form a bundle; in termite gut (Janicki, 1911).

S. nelumbium Kirby (Fig. 168, *b*). 45μ by 27μ; in *Cryptotermes hermsi.*

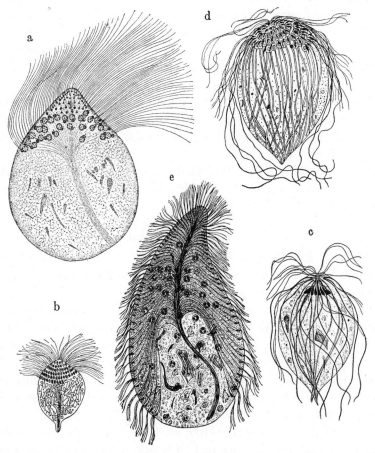

Fig. 168. a, *Calonympha grassii*, ×900 (Janicki); b, *Stephanonympha nelumbium*, ×400 (Kirby); c, *Coronympha clevelandi*, ×1000 (Kirby); d, *Metacoronympha senta*, ×485 (Kirby); e, *Snyderella tabogae*, ×350 (Kirby).

Genus **Coronympha** Kirby. Pyriform with eight or sixteen nuclei, arranged in a single circle in anterior region; eight or sixteen karyomastigonts; axostyles distributed; in termite gut (Kirby, 1929, 1939).

C. clevelandi K. (Fig. 168, *c*). 25-53μ by 18-46μ; in *Kalotermes clevelandi*.

Genus **Metacoronympha** Kirby. Pyriform; one hundred or more karyomastigonts arranged in spiral rows meeting at the anterior end; each karyomastigont is composed of nucleus, blepharoplast, cresta, three anterior flagella, a trailing flagellum, and an axostyle; axostyle as in the last genus; in termite gut (Kirby, 1939).

M. senta K. (Fig. 168, *d*). 22-92μ by 15-67μ; karyomastigonts about 66-345 (average 150) in usually six spiral rows; in *Kalotermes emersoni* and four other species of the genus.

Genus **Snyderella** Kirby Numerous nuclei scattered through the cytoplasm; akaryomastigonts close together and extend through the greater part of peripheral region; axial filaments in a bundle; in termite gut (Kirby, 1929).

T. tobogae K. (Fig. 168, *e*). Pyriform; rounded posteriorly; bluntly conical anteriorly; 77-172μ by 53-97μ; in *Cryptotermes longicollis*.

Family 4. **Trichomonadidae** Wenyon

Anterior flagella variable in number; undulating membrane bordered by a flagellum; axostyle often conspicuous and protrudes from the posterior end; costa.

Genus **Trichomonas** Donné. Pyriform; typically with four free anterior flagella; fifth flagellum along the outer margin of the undulating membrane; costa at the base of the membrane; axostyle developed, often protruding beyond the posterior end of the body; encystment has not been definitely observed; all parasitic. (Numerous species, Wenrich, 1944; Cytology and morphogenesis, Kirby, 1944; division process, Kuczynski, 1918.)

T. hominis (Davaine) (Fig. 169, *a*). Active flagellate, undergoing a jerky or spinning movement; highly plastic, but usually ovoid or pyriform; 5-20μ long; cytostome near anterior end; four anterior flagella equally long; fifth flagellum borders undulating membrane which is seen in life; in degenerating individuals the membrane may undulate, even after loss of flagella, simulating amoeboid movement; axostyle straight along the median line; vacuolated cytoplasm with bacteria; commensal in the colon and ileum of man; found in diarrhoeic stools. Wenrich (1944) states that in all twenty cases which he studied, some or most of the in-

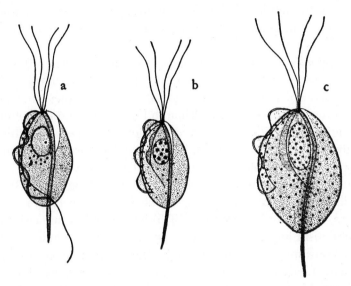

Fig. 169. Diagrams showing the species of Trichomonas which live in man, ×2500 (modified after Wenrich). a, *Trichomonas hominis;* b, *T. tenax;* c, *T. vaginalis.*

dividuals showed five anterior flagella and two unequal blepharoplasts.

Since encysted forms have not yet been found, transmission is assumed to be carried on by trophozoites. According to Dobell (1934), he became infected by an intestinal Trichomonas of a monkey (*Macacus nemestrinus*) by swallowing "a rich two-day culture" plus bacteria which were mixed with 10 cc. of sterilized milk on an empty stomach. The presence of Trichomonas in his stools was established on the sixth day by culture and on the thirteenth day by microscopical examination after taking in the cultures. The infection which lasted for about four and a half years, did not cause any ill effects upon him. The organism is killed after five minutes' exposure to N/20 HCl at 37°C., but at 15-22°C., is able to survive, though in small numbers, up to fifteen minutes after exposure to the acid (Bishop, 1930). This flagellate is widely distributed and of common occurrence, especially in tropical and subtropical regions.

T. tenax Müller (*T. elongata* Steinberg; *T. buccalis* Goodey) (Fig. 169, *b*). Similar to the last mentioned species; commensal in the tartar and gum of human mouth. (Nomenclature, Dobell, 1939.)

T. vaginalis Donné (Fig. 169, *c*). Broadly pyriform; 10-30μ by 10-20μ; cytoplasm contains many granules and bacteria; cytostome inconspicuous; nutrition parasitic and holozoic; parasitic in human reproductive organ. Although the organism does not enter the vaginal tissues, many observers believe it to be responsible for certain diseases of the vagina. Trussell and Johnson (1945) maintain that it is capable of inciting an inflammatory reaction in the vaginal mucous membrane and according to Hogue (1943), this flagellate produces a substance which injures the cells in tissue culture. Asami and Nakamura (1955) inoculated axenic culture of the organism into nine volunteers and came to conclude that vigorous consumption of glycogen in the vaginal mucosa by the flagellate facilitates the bacterial invasion and brings about the vaginitis. More recent observers such as Bertrand and Deulier, Catterall and Nicol (1960), Chappaz (1960), etc., consider that *T. vaginalis* is pathogenic. The organism occurs also in the male urethra (Feo, 1944). (Morphology, Reuling, 1921; Wenrich, 1939, 1944, 1944a, 1947; comprehensive monograph Trussell, 1947; transmission, Burch *et al.*, 1959a; metabolism, Wirtschafter and Jahn, 1956; diagnosis, Burch *et al.*, 1959, 1959a; strains, Reardon *et al.*, 1961; pathogenicity, Honigberg, 1961.)

Because of the morphological similarity of these three species of human Trichomonas, a number of workers maintain that they may be one and the same species. Dobell (1934) inoculated a rich culture of Trichomonas obtained from his stools into the vagina of a monkey (*Macacus rhesus*) and obtained a positive infection which was easily proven by culture, but unsatisfactorily by microscopical examination of smears. The infection thus produced lasted over three years and did not bring about any ill effect on the monkey. He considers that *T. vaginalis* and *T. hominis* are synonyms and that there occur diverse strains different in minor morphological characters and physiological properties. Andrews (1929) noted the organism obtained from vaginal secretion was larger than *T. hominis* and its undulating membrane extended for one-half or two-thirds the body length, but when cultured *in vitro*, the organisms became smaller in size and the undulating membrane protruded beyond the body as a free flagellum. On the other hand, Stabler and his co-workers (1941, 1942) failed to obtain infections in volunteers by inoculating intravaginally with cultures of *T. hominis*. Shinohara (1958) noted a high specificity in agglomeration test between *T. tenax* and *T. hominis* and found later (1958a) that in 10 per cent bile solution, *T. tenax* became immobile within thirty minutes, while *T. hominis* remained mobile for sixty minutes. Wenrich (1944) who made comparative studies of these flagellates, considers that there exist distinctly recognizable morphological differences among them as shown in Fig. 169. Jírovec *et al.* (1959) are of the same opinion.

T. macacovaginae Hegner and Ratcliffe. In the vagina of *Macacus rhesus*. Dobell (1934) held that this is identical with *T. vaginalis* and *T. hominis*.

T. microti Wenrich and Saxe (Fig. 170, *a*). In the caecum of rodents, *Microtus pennsylvanicus, Peromyscus leucopus, Rattus norvegicus, Mesocricetus auratus;* 4-9μ long; four free flagella; a blepharoplast; undulating membrane medium long; axostyle conspicuous. Ring (1959) found this flagellate in the caecum of an *Apodemus sylvaticus;* dimensions were 5.4-7.4μ by 2.7-4.1μ.

T. gallinae (Rivolta) (*T. columbae* Rivolta and Delprato) (Fig. 170, *b-d*). Pyriform; 6-19μ by 2-9μ; ovoid nucleus anterior together with a blepharoplast and parabasal body; axostyle protrudes a little; cytoplasmic granules; four anterior flagella 8-13μ long; autotomy; in the upper digestive tract of pigeon and also turkey, chicken, and dove. Experimentally it is transferable to quail, bob-white, hawk, canary, etc., and often fatal to hosts. (Species, Travis, 1932; morphology, Stabler, 1941; pathology, Levine and Brandly, 1940; transmission, Levine *et al.*, 1941; distribution, Barnes, 1951; Stabler, 1951; Locke and Herman, 1961; growth factors, Shorb and Lund, 1959; pathogenicity, Honigberg, 1961.)

T. linearis Kirby (Fig. 170, *e*). Elongate spindle in form; 9-24μ by 3-8μ; in the intestine of *Orthognathotermes wheeleri;* Panama. (Other species in termites, Kirby, 1931.)

T. limacis Dujardin. In the intestine and liver-tubules of slugs, *Deroceras agreste* (Dujardin, 1841) and *Limax flavus* (Kozloff, 1945); subspherical to ellipsoidal; 11-17μ by 8-13μ; four anterior

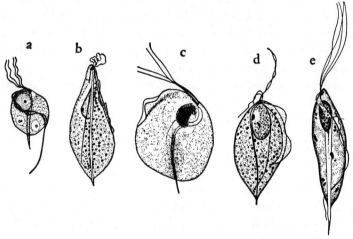

FIG. 170. a, *Trichomonas microti,* ×2000 (Wenrich and Saxe); b–d, *T. gallinae,* ×1765 (Stabler) (b, from domestic pigeon; c, from turkey; d, from red-tailed hawk); e, *T. linearis,* ×2000 (Kirby).

flagella; undulating membrane extends to posterior end, with free flagellum (Kozloff).

Genus **Tritrichomonas** Kofoid. Similar to *Trichomonas* in appearance, behavior and structure, but with only three anterior flagella; parasitic. (Many species.)

T. foetus (Riedmüller) (Fig. 171, *a, b*). In the genitalia of cattle; pathogenic; 10-15μ long; transmission by sexual act, from cow to bull or bull to cow and also by "natural contamination" (Andrews and Miller, 1936) from cow to cow. Infection brings about permanent or temporary suspension of the conception or the death of foetus. Sheep are susceptible (Andrews and Rees, 1936). (Morphology, Wenrich and Emmerson, 1933; Morgan and Noland, 1943; Kirby, 1951; effect on tissue culture, Hogue, 1938; effect on reproductibility of cow, Bartlett, 1947, 1948; McTackett, 1961; under freezing temperature, Levine and Marquardt, 1955; Fitzgeralds and Levine, 1961; relation to the nasal Tritrichomonas of swine, Buttrey, 1956; Doran, 1957; culture, Sanders, 1957; metabolism, Doran, 1958, 1959; Lindblom, 1961).

T. fecalis Cleveland. 5μ by 4μ to 12μ by 6μ; average dimensions 8.5μ by 5.7μ; axostyle long, protruding one-third to one-half the body length from the posterior end; of three flagella, one is longer and less active than the other two; in the faeces of man. Its remarkable adaptability observed by Cleveland was noted elsewhere (p. 37).

T. suis (Gruby and Delafond). Spindle to Pyriform in form; undulating membrane and costa reach near the posterior end of body; axostyle stout, ending in a sharp process; three anterior flagella about as long as the body; nucleus oval and near the anterior end; parabasal body a single elongate structure, not straight; length about 11μ and breadth 3.5μ. In nasal cavity, stomach, small intestine and caecum of pigs. The optimum pH of axenic culture media was between 6.5 and 8.0 (Hibler *et al.*, 1960). Hibler and his co-workers reported the existence of two smaller species in pigs: *T. rotunda* and *Trichomonas buttreyi*.

T. muris (Grassi) (Fig. 171, *c*, *d*). Fusiform; 10-16μ by 5-10μ; three anterior flagella short, posterior flagellum extends beyond body; axostyle large, its tip protruding; in the caecum and colon of mice (Mus, Peromyscus) (Wenrich, 1921) and ground squirrel (*Citellus lateralis chrysodeirus*) (Kirby and Honigberg, 1949). The organism has been found within nematodes which coinhabit the host intestine. For example, Theiler and Farber (1932) found the flagellate in the chyle-stomach of the nematodes, *Aspicularis tetraptera* and *Syphacia obvelata* inhabiting the colon of the mouse, and Becker (1933) noted two active individuals of this flagellate within the egg shell of the last-named nematode. (Morphology and division, Kofoid and Swezy, 1915, Wenrich, 1921, Ring, 1959; electron microscope study, Anderson, 1955, Anderson and Beams, 1959, 1961.)

T. caviae (Davaine). In the caecum and large intestine of guinea-pigs. Ovoid or pyriform; 5-22μ long; undulating membrane long; axostyle protrudes; spherical cysts about 7μ in diameter (Galli-Valerio, 1903; Wenyon, 1926). (Cytology and reproduction, Grassé and Faure, 1939.)

T. batrachorum (Perty) (Fig. 171, *e*). Ovoid; 14-18μ by 6-10μ (Alexeieff); in culture, 7-22μ by 4-7μ (Bishop, 1931); axostyle without granules; parabasal body V-shaped (Honigberg, 1950); in the colon of frogs and toads. Bishop (1934) succeeded in infecting the tadpoles of *Rana temporaria* and *Bufo vulgaris* by feeding them cultures which were free from cysts. Ingestion of food is by pseudopods and cannibalism is not uncommon (Samuels, 1957).

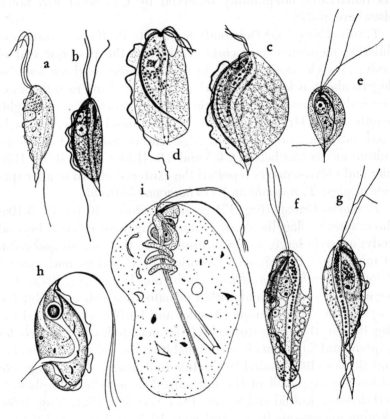

Fig. 171. a, *Tritrichomonas foetus* in life, ×1330 (Morgan and Noland); b, a stained organism, ×1765 (Wenrich and Emmerson); c,d, *T. muris*, ×2000 (Wenrich); e, *T. batrachorum*, ×1465 (Bishop); f,g, *T. augusta*, ×1455 (Samuels); h, *T. brevicollis*, ×2000 (Kirby); i, *Bullanympha silvestrii*, ×780 (Kirby).

(Morphology and mitosis, Samuels, 1957, 1957a; anomalies, Samuels, 1959; structure and hosts, Honigberg, 1953.)

T. augusta (Alexeieff) (Fig. 171, *f*, *g*). Elongate spindle; 15-27μ by 5-13μ; parabasal body rod- or sausage-shaped (Honigberg, 1950); the axostyle protrudes, and contains dark-staining granules; in the colon of frogs, toads and lizards (Kulda, 1957). (Morphology and division, Kofoid and Swezy, 1915; Samuels, 1941; viability, Rosenberg, 1936; in frog liver lesions, Stabler and Pennypacker, 1939; morphological variation, Buttrey, 1954; transfaunation study, Cairn, 1953.)

T. brevicollis Kirby (Fig. 171, *h*). Ovoid, undulating membrane curved around end; 10-17μ by 4-8μ; in the intestine of *Kalotermes brevicollis;* Panama.

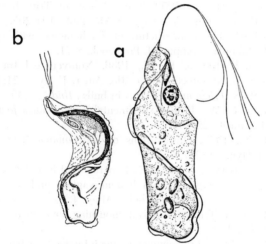

FIG. 172. a, *Pentatrichomonoides scroa*, ×1500; b, *Pseudotrypanosoma giganteum*, ×435 (Kirby).

Genus **Pentatrichomonas** Mesnil. Similar to *Trichomonas*, but with five free anterior flagella.

P. bengalensis Chatterjee. 9-20μ by 7-14μ; in human intestine. Kirby (1943, 1945a) observed that of the five flagella, four arise from the end of a columnar (1-2μ long) extension, while the fifth flagellum is a little shorter and takes its origin about 1μ behind the extension.

Genus **Pentatrichomonoides** Kirby. Five anterior flagella and the undulating membrane; axostyle very slightly developed; fusiform parabasal body; nucleus separated from the anterior blepharoplast; in termite gut.

P. scroa K. (Fig. 172, *a*). 14-45μ by 6-15μ; in *Cryptotermes dudleyi* and *Lobitermes longicollis.*

474 *Protozoology*

Genus **Pseudotrypanosoma** Grassi. Large, elongate; three anterior flagella; undulating membrane; slender axostyle; band-like structure between nucleus and blepharoplast; parabasal body long, narrow; in termite gut.

P. giganteum G. (Fig. 172, *b*). 55-111μ long (Grassi); 145-205μ by 20-40μ; anterior flagella about 30μ long (Kirby); in gut of *Porotermes adamsoni* and *P. grandis*.

References

ANDERSON, E., 1955. The electron microscopy of *Trichomonas muris*. J. Protozool., 2:114.

—— and BEAMS, H. W., 1959. The cytology of Tritrichomonas as revealed by the electron microscope. J. Morphol., 104:205.

—— —— 1961. The ultrastructure of Trichomonas with special reference to blepharoplast complex. J. Protozool., 8:71.

ANDREWS, J., and MILLER, F. W., 1936. Non-venereal transmission of *Trichomonas foetus* infection in cattle. Amer. J. Hyg., 24:433.

—— —— 1938. *Trichomonas foetus* in bulls., *Ibid.*, 28:40

—— and REES, C. W., 1936. Experimental *Trichomonas foetus* infection in sheep. J. Parasit., 22:108.

ANDREWS, M. N., 1929. Observations on *Trichomonas vaginalis*, etc. J. Trop. Med. Hyg., 32:237.

ASAMI, K., and NAKAMURA, M., 1955. Experimental inoculation of bacteria-free *Trichomonas vaginalis* into human vaginas and its effect on the glycogen content of vaginal epithelia. Amer. J. Trop. Med. Hyg., 4:254.

BARNES, W. B., 1951. Trichomoniasis in mourning doves in Indiana. Indiana Audub. Quart., 29:8.

BARTLETT, D. E., 1947. *Trichomonas foetus* infection and bovine reproduction. Am. J. Vet. Res., 8:343.

—— 1948. Bovine venereal trichomoniasis: etc. Proc. U. S. Livestock Sanit., A. 1947, p. 170.

BECKER, E. R., 1933. Two observations on helminths. Tr. Am. Micr. Soc., 52:361.

BĚLAŘ, K., 1921. Protozoenstudien. III. Arch. Protist., 43:431.

BERTRAND, P., and LEULIER, J., 1960. Essais cliniques sur la trichomonase des partenaires des femmes infestées. Gynaecologia, 149:93.

BISHOP, A., 1930. The action of HCl upon cultures of Trichomonas. Parasitology, 22:230.

—— 1931. The morphology and method of division of Trichomonas. *Ibid.*, 23:129.

—— 1934. The experimental infection of Amphibia with cultures of Trichomonas. *Ibid.*, 26:26.

—— 1935. Observations upon a "trichomonas" from pond water. *Ibid.*, 27:246.

—— 1936. Further observations upon a "Trichomonas" from pond water. *Ibid.*, 28:443.

—— 1939. A note upon the systematic position of *"Trichomonas" keilini*. *Ibid.*, 31:469.

BURCH, T. A., *et al.*, 1959. Diagnosis of *Trichomonas vaginalis* vaginitis. Am. J. Obst. & Gynec., 77:309.

—— *et al.*, 1959a. Epidemiological studies on human trichomoniasis. Amer. J. Trop. Mes. Hyg., 8:312.

BUTTREY, B. W., 1954. Morphological variations in *Tritrichomonas augusta* from Amphibia. J. Morphol., 94:125.

—— 1956. A morphological description of a Tritrichomonas from the nasal cavity of swine. J. Protozool., 3:8.

CAIRNS, J., 1953. Transfaunation studies on the host-specificity of the enteric protozoa of Amphibia and various other vertebrates. Proc. Acad. Nat. Sci. Philadelphia, 105:45.

CATTERALL, R. D., and NICOL, C. S., 1960. Is trichomonas infestation a venereal disease? Gynaecologia, 149:87.

CHAPPEZ, G., 1960. Etiologie de la trichomonase féminine. *Ibid.*, 149:1.

DOBELL, C., 1934. Researches on the intestinal protozoa of monkeys and man. VI. Parasitology, 26:531.

—— 1935. VII. *Ibid.*, 27:564.

—— 1939. The common flagellate of the human mouth, *Trichomonas tenax* (O.F.M.): etc. *Ibid.*, 31:138.

—— and O'CONNOR, F. W., 1921. The intestinal protozoa of man. London.

DOGIEL, V., 1916. Researches on the parasitic protozoa from the intestine of termites. I. J. russ. zool., 1:1.

DONNÉ, A., 1836. Animalcules observés dans les matières purulentes et le produit des sécrétions des organes génitaux de l'homme et de la famme. C. R. Acad. Sci., 3:385.

DORAN, D. J., 1957. Studies on trichomonads. I. J. Protozool., 4:182.

—— 1958. II. *Ibid.*, 5:89.

—— 1959. III. *Ibid.*, 6:177.

DUBOSCQ, O., and GRASSÉ, P., 1925. Appareil de Golgi, mitochondries, etc. C. R. Soc. Biol., 93:345.

FEO, L. G., 1944. The incidence and significance of *Trichomonas vaginalis* in the male. Am. J. Trop. Med., 24:195.

FITZGERALD, P. R., and LEVINE, N. D., 1961. Effect of storage temperature, equilibration time, and buffers in survival of *Tritrichomonas foetus* in the presence of glycerol at freezing temperature. J. Protozool., 8:21.

FOÀ, A., 1905. Due nuovi flagellati parassiti (Nota prelim.). Rend. Acc. Lincei, 14:542.

GRASSÉ, P. P., 1926. Contribution à l'étude des flagellés parasites. Arch. zool. exper. gén., 65:345.

—— 1952. Traité de zoologie. I. Fasc. 1. Paris.

—— 1952a. Ordre des Trichomonadines. In Grassé (1952), p. 704.

―――― and FAURE, A., 1939. Quelques données nouvelles sur la cytologie et la reproduction de *Trichomonas caviae*. Bull. biol. France et Belg., 73:1.

GRASSI, B., 1917. Flagellati nei Termitidi. Mem. R. Acc. Lincei, Ser. 5, 12:331.

―――― and FOÀ, A., 1911. Intorno ai protozoi dei termitidi (n.p.). Rend. R. Acc. Lincei, S. V. Cl. Sc. fils, 20:725.

HAWES, R. S., 1947. On the structure, division, and systematic position of *Trichomonas vaginalis* Donné, with a note on its methods of feeding. Quart. J. Micr. Sci., 88:79.

HIBLER, C. P., *et al.*, 1960. The morphology and incidence of the trichomonads of swine, etc. J. Protozool., 7:159.

HOGUE, M. J., 1938. The effect of *Trichomonas foetus* on tissue culture cells. Amer. J. Hyg., 28:288.

―――― 1943. The effect of *Trichomonas vaginalis* on tissue culture cells. *Ibid.*, 37:142.

HOLLANDE, A. V., 1939. Sur un genre nouveau de Trichomonadide libre: etc. Bull. soc. zool. France, 64:114.

HONIGBERG, B. M., 1947. The characteristics of the flagellate *Monocercomonas verrens* sp.n., from *Tapirus malayanus*. Univ. Cal. Publ. Zool., 53:227.

―――― 1950. On the structure of the parabasal body in *Tritrichomonas batrachorum*, etc. J. Parasit., 36:89.

―――― 1953. Structure, taxonomic status and host list of *Tritrichomonas batrachorum*. *Ibid.*, 39:191.

―――― 1955. Structure and morphogenesis of two new species of Hexamastix from lizards. *Ibid.*, 41:1.

―――― 1961. Comparative pathogenicity of *Trichomonas vaginalis* and *T. gallinae* to mice. I. J. Parasitol., 47:545.

―――― and CHRISTIAN, H. H., 1954. Characteristics of *Hexamastix batrachorum*. *Ibid.*, 40:508.

―――― and LEE, J. J., 1959. Structure and division of *Trichomonas tenax*. Amer. J. Hyg., 69:177.

JANICKI, C., 1911. Zur Kenntnis des Parabasalapparats bei parasitischen Flagellaten. Biol. Centralbl., 31:321.

―――― 1915. Untersuchungen an parasitischen Flagellaten. II. Ztschr. wiss. Zool., 112:573.

JÍROVEC, O., *et al.*, 1959. Ueber einige Probleme der vaginalen Mikrobiologie und der geschlichtlichen trichomoniasis. J. Hyg. Epidemiol. Micr. & Immunol., 3:195.

KEAN, B. N., 1960. Conjugal trichomoniasis. Gynaecol., 149:97.

KIRBY, H. JR., 1926. The intestinal flagellates of the termite, *Cryptotermes hermsi*. Univ. Cal. Publ. Zool., 29:103.

―――― 1929. Snyderella and Coronympha, etc. *Ibid.*, 31:417.

―――― 1930. Trichomonad flagellates from termites. I. *Ibid.*, 33:393.

―――― 1931. II. *Ibid.*, 36:171.

———— 1936. Two polymastigote flagellates of the genera Pseudodevescovina and Caduceia. Quart. J. Micr. Sci., 79:309.

———— 1937. The devescovinid flagellate *Parajoenia grassii* from a Hawaiian termite. Univ. Cal. Publ. Zool., 41:213.

———— 1938. Polymastigote flagellates of the genus Foaia Janicki, etc. Quart. J. Micr. Sci., 81:1.

———— 1939. Two new flagellates from termites in the genera Coronympha Kirby, etc. Proc. Cal. Acad. Sci., 22:207.

———— 1941. Devescovinid flagellates of termites. I. Univ. Cal. Publ. Zool., 45:1.

———— 1942. II. *Ibid.*, 45:93.

———— 1942a. III. *Ibid.*, 45:167.

———— 1943. Observations on a trichomonad from the intestine of man. J. Parasit., 29:422.

———— 1944. Some observations on cytology and morphogenesis in flagellate protozoa. J. Morphol., 75:361.

———— 1945. The structure of the common intestinal trichomonad of man. J. Parasit., 31:163.

———— 1946. *Gigantomonas herculea.* Univ. Cal. Publ. Zool., 53:163.

———— 1947. Flagellate and host relationships of trichomonad flagellates. J. Parasit., 33:214.

———— 1949. Devescovinid flagellates of termites. V. Univ. Cal. Publ. Zool., 45:319.

———— 1951. Observations on the trichomonad flagellate of the reproductive organs of cattle. J. Parasit., 37:445.

———— and HONIGBERG, B., 1949. Flagellates of the caecum of ground squirrels. Univ. Cal. Publ. Zool., 53:315.

KOFOID, C. A., and SWEZY, O., 1915. Mitosis and multiple fission in trichomonad flagellates. Proc. Amer. Acad. Arts & Sci., 51:289.

KOZLOFF, E. N., 1945. The morphology of *Trichomonas limacis.* J. Morphol., 77:53.

KUCZYNSKI, M. H., 1918. Ueber die Teilungsvorgänge verschiedener Trichomonaden, etc. Arch. Protist., 39:107.

KULDA, J., 1957. *Tritrichomonas augusta* as a parasite of some lizards in Czechoslovakia. Vest. Cesk. Spol. Zool., 21:209.

LEE, J. J., 1960. *Hypotrichomonas acosta* gen. nov. from reptiles. I. J. Protozool., 7:393.

———— and PIERCE, S., 1960. II. *Ibid.*, 7:402.

LEIDY, J., 1877. On intestinal parasites of *Termes flavipes.* Proc. Acad. Nat. Sci., Philadelphia, p. 146.

LEVINE, N. D., BOLEY, L. E. and HESTER, H. R., 1941. Experimental transmission of *Trichomonas gallinae* from the chicken to other birds. Amer. J. Hyg., 33:23.

———— and BRANDLY, C. A., 1940. Further studies on the pathogenicity of *Trichomonas gallinae* for baby chicks. Poultry Sci., 19:205.

—— and MARQUARDT, W. C., 1955. The effect of glycerol and related compounds on survival of *Tritrichomonas foetus* at freezing temperatures. J. Protozool., 2:100.

LIGHT, S. F., 1926. On *Metadevescovina debilis* g.n., sp.n. Univ. Cal. Publ. Zool., 29:141.

LINDBLOM, G. P., 1961. Carbohydrate metabolism of Trichomonas, etc. J. Protozool., 8:139.

LOCKE, L. N., and HERMAN, C. M., 1961. Trichomonad infection in mourning doves, *Zenaidura macroura*, in Maryland. Chesapeake Sci., 2:45.

McTACKETT, A. R., 1961. Trichomoniasis, a cause of infertility in cattle. Queensland Agr. J., 87:381.

MORGAN, B. B., and NOLAND, L. E., 1943. Laboratory methods for differentiating *Trichomonas foetus* from other Protozoa in the diagnosis of trichomoniasis in cattle. J. Am. Vet. Med. Assoc., 102:11.

REARDON, L. V., *et al.*, 1961. Differences in strains of *Trichomonas vaginalis*, etc. J. Parasit., 47:527.

REES, C. W., 1938. Observations on bovine venereal trichomoniasis. Vet. Med., 33:321.

REULING, F., 1921. Zur Morphologie von *Trichomonas vaginalis* Arch. Protist., 42:347.

RING, M., 1959. Studies on the parasitic protozoa of wild mice from Berkshire with a description of a new species of Trichomonas. Proc. Zool. Soc. London, 132:381.

ROSENBERG, L. E., 1936. On the viability of *Tritrichomonas augusta*. Tr. Am. Micr. Soc., 55:313.

SAMUELS, R., 1941. The morphology and division of *Trichomonas augusta*. *Ibid.*, 60:421.

—— 1957. Studies of *Tritrichomonas batrachorum*. I. J. Protozool., 4:110.

—— 1957a. II Tr. Am. Micr. Soc., 76:295.

—— 1959. III. *Ibid.*, 78:49.

SANDERS, M., 1957. Replacement of serum for in vitro cultivation of *Tritrichomonas foetus*. J. Protozool., 4:118.

SHINOHARA, T., 1958. Agglomeration test on *Tritrichomonas tenax* and *T. hominis*. Keio Igaku, 35:151.

—— 1958a. Effect of several digestive fluids upon the survival of *Trichomonas tenax* and *T. hominis. Ibid.*, 35:383.

SHORB, M. S., and LUND, P. G., 1959. Requirement of trichomonads for unidentified growth factors, saturated and unsaturated fatty acids. J. Protozool., 6:122.

STABLER, R. M., 1941. The morphology of *Trichomonas gallinae* (=*columbae*). J. Morphol., 69:501.

—— 1951. Effect of *Trichomonas gallinae* from diseased mourning doves on clean domestic pigeons. J. Parasit., 37:473.

—— and ENGLEY, F. B., 1946. Studies on *Trichomonas gallinae* infections in pigeon squabs. J. Parasit., 32:225.

———, FEO, L. G., and RAKOFF, A. E., 1941. Implantation of intestinal trichomonads (*T. hominis*) into the human vagina. Amer. J. Hyg., 34:114.

——— and PENNYPACKER, M. I. 1939. A brief account of *Trichomonas augusta,* etc. Tr. Am. Micr. Soc., 58:391.

SUTHERLAND, J. L., 1933. Protozoa from Australian termites. Quart. J. Micr. Sci., 76:145.

THEILER, H., and FARBER, S. M., 1932. *Trichomonas muris,* parasitic in Oxyurids of the white mouse. J. Parasit., 19:169.

TRAGER, W., 1934. A note on the cultivation of *Tricercomitus termopsidis,* etc. Arch. Protist., 83:264.

TRAVIS, B. V., 1932. *Trichomonas phasiani,* a new flagellate from the ring-necked pheasant, etc. J. Parasit., 18:285.

TRUSSELL, R. E., 1947. *Trichomonas vaginalis* and trichomoniasis. Springfield, Illinois.

——— and JOHNSON, G., 1945. *Trichomonas vaginalis* Donné. Recent experimental advances. Puerto Rico J. P. H. Trop. Med., 20:289.

WENRICH, D. H., 1944. Comparative morphology of the trichomonad flagellates of man. Am. J. Trop. Med., 24:39.

——— 1944a. Morphology of the intestinal trichomonad flagellates in man and of similar forms in monkeys, cats, dogs and rats. J. Morphol., 74:189.

——— 1947. The species of Trichomonas in man. J. Parasit., 33:177.

——— and EMMERSON, M. A., 1933. Studies on the morphology of *Tritrichomonas foetus* (Riedmüller) from American cows. J. Morphol., 55:193.

——— and SAXE, L. H., 1950. *Trichomonas microti* n.sp. J. Parasit., 36:261.

WENYON, C. M., 1926. Protozoology. 1. London and Baltimore.

WIRTSCHAFTER, S. K., 1954. Evidence for the existence of the enzymes hexokinase and aldolase in the protozoan parasite *Trichomonas vaginalis.* J. Parasit., 40:360.

——— and JAHN, T. L., 1956. The metabolism of *Trichomonas vaginalis.* J. Protozool., 3:83 and 86.

Chapter 17

Order 5. Hypermastigida Grassi and Foà

ALL members of this order are inhabitants of the alimentary canal of termites, cockroaches, and woodroaches. The cytoplasmic organization is of high complexity, although there is only a single nucleus. Flagella are numerous and have their origin in the blepharoplasts located in the anterior region of body. In many species which are xylophagous, there exists a true symbiotic relationship between the most termite and the protozoans (p. 31). Method of nutrition is either holozoic or saprozoic (parasitic). Bits of wood, starch grains, and other food material are taken in by means of pseudopodia (p. 120).

Asexual reproduction is by binary fission; multiple division has also been noted in some species under certain conditions, while sexual reproduction has been observed in many species. Encystment occurs in some genera of Lophomonadidae and certain species inhabiting woodroaches in which moulting of the host insect leads to encystment and sexual reproduction. The protozoan fauna of the colon is lost at the time of molting of the host insect, but newly molted individuals regain the fauna by proctodeal feeding (Andrews, 1930).

The number of protozoa present in the colon of the termite is usually very enormous. The total weight of all protozoa present in a termite worker has been estimated to be from about one-seventh to one-fourth (Hungate, 1939) or one-third (Katzin and Kirby, 1939) to as much as one-half (Cleveland, 1925) of the body weight of the host. The relationship between the termite and its intestinal flagellate fauna, has been studied by several observers. Kirby (1937) notes that certain groups of flagellates occur only in certain groups of termites, while others are widely distributed. Flagellates of one host termite introduced into individuals of another species survive for a limited time only (Light and Sanford, 1928; Cleveland, Hall et al., 1934; Dropkin, 1941, 1946).

(Taxonomy, Koidzumi, 1921; Kirby, 1926; Bernstein, 1928.) Eight families.

Family 1. **Holomastigotidae** Janicki

Flagella arranged in spiral rows.

Genus **Holomastigotes** Grassi. Body small; spindle-shaped; few spiral rows reach from anterior to posterior end; nucleus anterior, surrounded by a mass of dense cytoplasm; saprozoic; in the termite gut.

H. elongatum G. (Fig. 173, *a*). In gut of *Reticulitermes lucifugus, R. speratus, R. flaviceps,* and *Macrohodotermes massambicus;* up to 70µ by 24µ (Grassi, 1892).

Genus **Holomastigotoides** Grassi and Foà. Large; pyriform; spiral rows of flagella as in the last genus, but more numerous (12-40 rows); a mass of dense cytoplasm surrounds ovoid nucleus near the anterior end; in termite gut (Grassi and Foà, 1911). (Cytology, Cleveland, 1949.)

H. hartmanni Koidzumi (Fig. 173, *b*). 50-140µ long; in *Coptotermes formosanus.*

H. tusitala Cleveland (Figs. 65; 66; 67; 174, *a, b*). In the hindgut of *Prorhinotermes simplex;* largest species in this host; elongate pyriform; five flagellar bands, arise at the anterior end and spiral the body five and one-half times; dimorphic with respect to chromosome numbers, 2 and 3; 130-200µ long. Cleveland's observation on its chromosome cycle has been mentioned elsewhere (p. 194).

Genus **Spirotrichonympha** Grassi and Foà (1911). Moderately large; elongate pyriform; flagella deeply embedded in cytoplasm in anterior region, arising from one to several spiral bands; mass of dense cytoplasm conical and its base indistinct; nucleus spherical; in termite gut (Development, Duboscq and Grassé, 1928.)

S. leidyi Koidzumi (Fig. 173, *c*). In *Coptotermes formosanus;* 15-50µ by 8-30µ.

S. pulchella Brown (Fig. 173, *d*). 36-42µ by 14-16µ; in *Reticulitermes hageni.*

S. bispira Cleveland. In *Kalotermes simplicicornis;* 59-102µ by 32-48µ; two flagellar bands in thirty-four spiral turns; resting nucleus with two chromosomes; the cytoplasmic division is unique in that portion of the anterior end shifts its position to the posterior end, where a new flagellar band develops; thus the division is longitudinal (Cleveland, 1938).

Genus **Spirotrichonymphella** Grassi. Small; without spiral ridges; flagella long; saprozoic, not wood-feeding; in termite gut.

S. pudibunda G. In *Porotermes adamsoni;* Australia. Multiple fusion (Sutherland).

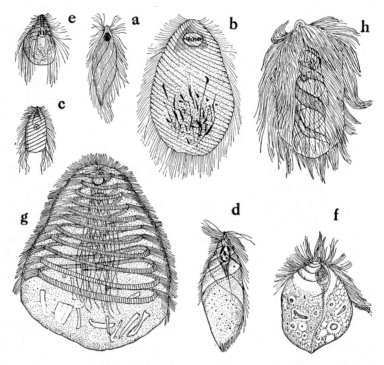

Fig. 173. a, *Holomastigotes elongatum,* ×700 (Koidzumi); b, *Holomastigotoides hartmanni,* ×250 (Koidzumi); c, *Spirotrichonympha leidyi,* ×400 (Koidzumi); d, *S. pulchella,* ×900 (Brown); e, *Microspirotrichonympha porteri,* ×250 (Koidzumi); f, *M. ovalis,* ×600 (Brown); g, *Macrospironympha xylopletha,* ×300 (Cleveland et al.); h, *Leptospironympha eupora,* ×1050 (Cleveland et al.).

Genus **Microspirotrichonympha** Koidzumi (*Spironympha* Koidzumi). Small, surface not ridged; spiral rows of flagella only on anterior half; a tubular structure between nucleus and anterior extremity; a mass of dense cytoplasm surrounds nucleus; with or without axial rod; in termite gut (Koidzumi, 1917, 1921).

M. porteri K. (Fig. 173, *e*). In *Reticulitermes flaviceps;* 20-55μ by 20-40μ.

M. ovalis (Brown) (Fig. 173, *f*). 36-48μ by about 40μ; in *Reticulitermes hesperus* (Brown, 1931).

Genus **Spirotrichosoma** Sutherland. Pyriform or elongate; below operculum, two deeply staining rods from which flagella arise and which extend posteriorly into two spiral flagellar bands; without axostyle; nucleus anterior, median; wood chips always present; in *Stolotermes victoriensis;* Australia.

S. capitatum S. 87μ by 38μ; flagellar bands closely spiral, reaching the posterior end.

Twelve other species were observed in Stolotermes by Cleveland and Day (1958).

Genus **Macrospironympha** Cleveland *et al.* Broadly conical; flagella on two broad flagellar bands which make ten to twelve spiral turns, two inner bands; axostyles thirty-six to fifty or more; during mitosis nucleus migrates posteriorly; encystment, in which only nucleus and centrioles are retained, takes place at each ecdysis of host; in *Cryptocercus punctulatus.* (Sexual cycle, Cleveland, 1956a.)

M. xylopletha C. *et al.* (Fig. 173, *g*). 112-154μ by 72-127μ.

Genus **Leptospironympha** Cleveland *et al.* Cylindrical; small; flagella on two bands winding spirally along body axis; axostyle single, hyaline; in *Cryptocercus punctulatus.* (Several species.) (Sexual reproduction, Cleveland, 1951, 1956.) Cleveland and Day (1958) observed six species from Stolotermes.

L. eupora C. *et al.* (Fig. 173, *h*). 30-38μ by 18-21μ.

Genus **Apospironympha** Cleveland and Day. Pyriform body, with two primary flagellar bands in rostrum and anterior part of body; many secondary flagellar bands elsewhere; anterior part of the primary bands give off flagella and fine secondary bands which extend to or near the posterior end of body. (Four species.)

A. lata C. and D. In *Stolotermes africanus;* body 120-180μ by 75-140μ (Cleveland and Day, 1958).

Genus **Rostronympha** Duboscq, Grassé and Rose. Form variable, ovoid to medusoid; with or without a long contractile attaching organelle like a trunk, constricted in three places and of annulated surface; spiral ridges from which flagella arise, do not reach the posterior half; posterior half with attached spirochaetes; xylophagous; in the intestine of Anacanthotermes in Algier.

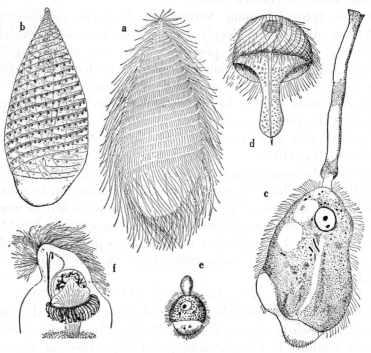

FIG. 174. a, b, *Holomastigotoides tusitala* (Cleveland) (a, surface view; b, flagellar bands, parabasal bodies, thin axostyles); c–e, *Rostronympha magna* (Duboscq and Grassé) (c, a large individual with the completely extended trunk, with axostyle, ×500; d, a small medusoid form, ×1000; e, a young individual with posteriorly attached spirochaetes, ×500); f, anterior end of *Joenia annectens* (Duboscq and Grassé).

R. magna D., G. and R. (Fig. 174, *c-e*). Large individuals, 135-180μ by 110-135μ, with the trunk-like extension reaching a length of 180μ; the body proper is divided into two parts; the posterior portion may be drawn out like the manubrium of a medusa; axostyle conspicuous; in the gut of *Anacanthotermes ochraceus* of Algier (Duboscq and Grassé, 1943).

Family 2. **Lophomonadidae** Kent

With one anterior flagellar tuft.

Genus **Lophomonas** Stein. Ovoid or elongate; small: a vesicular nucleus anterior; axostyle composed of many filaments; cysts common; in colon of cockroaches.

L. blattarum S. (Figs. 25, *a;* 68; 75; 175, *a-e*). Small pyriform, plastic; bundle of axostylar filaments may project beyond the posterior end; active movements; binary or multiple fission; 25-30μ long; encystment; holozoic; in the colon of cockroaches, *Blatta orientalis* in particular; widely distributed (Kudo, 1926). (Cytology, Janicki, 1910; Bělař, 1926; Kudo, 1926.)

L. striata Bütschli (Fig. 175 *f-h*). Elongate spindle; body with obliquely arranged needle-like structures which some investigators believe to be a protophytan (to which Grassé gave the name, *Fusiformis lophomonadis*); bundle of axial filaments short, never protruding; movement sluggish; cyst spherical with needle-like structures; in same habitat as the last species. Beams and his coworkers (1960) found under electron microscope the needle-like structures or rods are located on the external surface of the body of the organism, held possibly by an agglutination reaction. The rods appear to divide by transverse fission. They suggested that these are bacteria holding probably symbiotic relationship with the flagellate. (Cytology, Grassé, 1926; Kudo, 1926a, 1954.)

Genus **Eulophomonas** Grassi and Foà. Similar to *Lophomonas*, but flagella vary from five to fifteen or a little more in number; in termite gut.

E. kalotermitis Grassi. In *Kalotermes flavicollis;* this flagellate has not been observed by other workers.

Genus **Prolophomonas** Cleveland *et al.* Similar to *Eulophomas;* established since Eulophomonas had not been seen by recent observers; it would become a synonym "if Eulophomonas can be found in *K. flavicollis*" (Cleveland *et al.*).

P. tocopola C. *et al.* (Fig. 175, *i*). 14-19μ by 12-15μ; in *Cryptocercus punctulatus.*

Genus **Joenia** Grassi. Ellipsoidal; anterior portion capable of forming pseudopodia; flagellar tufts in part directed posteriorly; surface covered by numerous immobile short filamentous processes, nucleus spherical, anterior; posterior to it a conspicuous axostyle composed of numerous axial filaments, a parabasal apparatus surrounding it; xylophagous; in termite gut (Grassi, 1885).

J. *annectens* G. (Figs. 174, *f*; 175, *j*). In *Kalotermes flavicollis.* (Parabasal apparatus, Duboscq and Grassé, 1928a.)

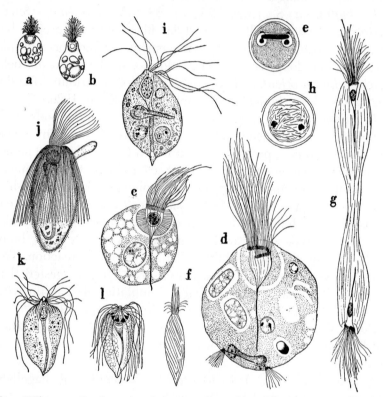

Fig. 175. a–e, *Lophomonas blattarum* (a, b, in life, ×320; c, a stained specimen; d, a trophozoite in which the nucleus is dividing; e, a stained cyst, ×1150) (Kudo); f–h, *L. striata* (f, in life, ×320; g, a stained dividing individual; h, a stained cyst, ×1150) (Kudo); i, *Prolophomonas tocopola,* ×1200 (Cleveland *et al.*); j, *Joenia annectens* (Grassi and Foà); k, *Microjoenia pyriformis,* ×920 (Brown); l, *Torquenympha octoplus,* ×920 (Brown).

Genus **Joenina** Grassi. More complex in structure than that of *Joenia;* flagella inserted at anterior end in a semi-circle; parabasal bodies two elongated curved rods; xylophagous (Grassi, 1917).

J. *pulchella* G. In *Porotermes adamsoni.*

Genus **Joenopsis** Cutler. Oval; large; a horseshoe-shaped pillar at anterior end, flagella arising from it; some directed anteriorly, others posteriorly; parabasal bodies long rods; a strong axostyle; xylophagous; in the termite gut (Cutler, 1920).

J. polytricha C. In *Archotermopsis wroughtoni;* 95-129μ long.

Genus **Microjoenia** Grassi. Small, pyriform; anterior end flattened; flagella arranged in longitudinal rows; axostyle; parabasal body simple; in termite gut (Grassi, 1892).

M. pyriformis Brown (Fig. 175, *k*). 44-52μ by 24-30μ; in *Reticulitermes hageni* (Brown, 1930).

Genus **Mesojoenia** Grassi and Foà. Large; flagellar tuft spreads over a wide area; distinct axostyle, bent at posterior end; two parabasal bodies; in termite gut (Grassi and Foà, 1911).

M. decipiens G. In *Kalotermes flavicollis.*

Genus **Torquenympha** Brown. Small; pyriform or top-form; axostyle; radially symmetrical; eight radially arranged parabasal bodies; nucleus anterior; in termite gut (Brown, 1930).

T. octoplus B. (Fig. 175, *l*). 15-26μ by 9-13μ; in *Reticulitermes hesperus.*

Family 3 **Hoplonymphidae** Light

With two anterior flagellar tufts.

Genus **Hoplonympha** Light. Slender fusiform, covered with thick, rigid pellicular armor; each of two flagellar tufts arises from a plate connected with blepharoplast at anterior end; nucleus near anterior extremity, more or less triangular in form; in termite gut (Light, 1926).

H. natator L. (Fig. 176, *a, b*). 60-120μ by 5-12μ; in *Kalotermes simplicicornis.*

Genus **Barbulanympha** Cleveland *et al.* Acorn-shaped: small, narrow, nuclear sleeve between centrioles; number of rows of flagella greater at base; large chromatin granules; numerous (80-350) parabasals; axostylar filaments 80-350; flagella 1500-13,000; different species show different number of chromosomes during mitosis; in gut of *Cryptocercus punctulatus.* (Four species) (Gametogenesis, Cleveland, 1953, 1954; reorganization in zygote, Cleveland, 1954a; meiosis, Cleveland, 1954b; unusual **be-**

havior of gametes and centrioles, Cleveland, 1955, 1957, 1957a; sexual cycle, Cleveland, 1956.)

B. ufalula C. *et al.* (Figs. 64; 176, *c*). 250-340μ by 175-275μ; fifty chromosomes; flagellated area 36-41μ long; centriole 28-35μ long.

B. laurabuda C. *et al.* 180-240μ by 135-170μ; forty chromo-

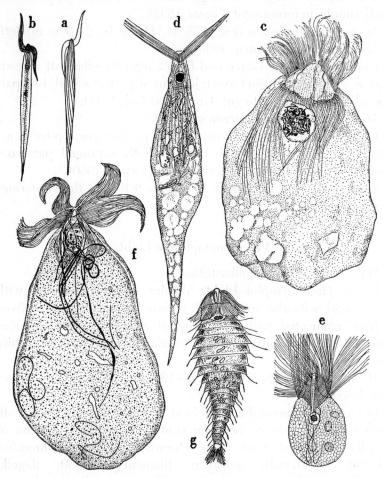

Fig. 176. a, b, *Hoplonympha natator*, ×450 (Light); c, *Barbulanympha ufalula*, ×210 (Cleveland *et al.*); d, *Urinympha talea*, ×350 (Cleveland *et al.*); e, *Staurojoenina assimilis*, ×200 (Kirby); f, *Idionympha perissa*, ×250 (Cleveland *et al.*); g. *Teratonympha mirabilis*, ×200 (Dogiel).

somes; flagellated area 29-33μ long; centrole 24-28μ long.

Genus **Rhynchonympha** Cleveland *et al.* Elongate; number of flagellar rows same throughout; axial filaments somewhat larger and longer, about thirty; thirty parabasals: 2400 flagella: in *Cryptocercus punctulatus*. (Sexual cycle, Cleveland, 1952, 1956.)

R. tarda C. *et al.* (Fig. 177, *f*). 130-215μ by 30-70μ.

Genus **Urinympha** Cleveland *et al.* Narrow, slender; flagellated area, smaller than that of the two genera mentioned above; flagella move as a unit; about twenty-four axial filaments; twenty-four parabasals; 600 flagella; in gut of *Cryptocercus punctulatus* Cleveland, 1951a). (Sexual cycle, Cleveland, 1956.)

U. talea C. *et al.* (Fig. 176,*d*). 75-300μ by 15-50μ; sexual reproduction (Cleveland, 1951a).

Family 4 **Staurojoeninidae** Grassi

With four anterior flagellar tufts.

Genus **Staurojoenina** Grassi. Pyriform to cylindrical; anterior region conical; nucleus spherical, central; four flagellar tufts from anterior end; ingest wood fragments; in termite gut (Grassi, 1917).

S. assimilis Kirby (Fig. 176, *e*). 105-190μ long; in *Kalotermes minor* (Kirby, 1926).

Genus **Idionympha** Cleveland *et al.* Acorn-shaped; axostyles eight to eighteen fine parabasals grouped in four areas; pellicle non-striated; nucleus nearer anterior end than that of Staurojoenina; flagellated areas smaller; in gut of *Cryptocercus punctulatus*.

I. perissa C. *et al.* (Fig. 176, *f*). 169-275μ by 98-155μ.

Family 5 **Kofoidiidae** Light

With several anterior flagellar tufts (loriculae).

Genus **Kofoidia** Light. Spherical; flagellar tufts composed of eight to sixteen *loriculae* (permanently fused bundles of flagella); without either axostyle or parabasal body; between oval nucleus and bases of flagellar tufts, there occurs a chromatin collar; in termite gut (Light, 1927).

K. loriculata L. (Fig. 177, *a, b*). 60-140μ in diameter; in *Kalotermes simplicicornis*.

Family 6. **Trichonymphidae** Kent

Flagella not in tufts; posterior portion of the body not flagellated.

Genus **Trichonympha** Leidy (*Leidyonella* Frenzel; *Gymnonympha* Dobell; *?Leidyopsis*, Kofoid and Swezy). Anterior portion consists of nipple and bell, both of which are composed of two layers; a distinct axial core; nucleus central; flagella located in longitudinal rows on bell; xylophagous; in the intestine of termites and woodroach. Many species. The species inhabiting the woodroach undergo sexual reproduction at the time of molting of the host (Cleveland, 1949a, 1956, 1957) (p. 218). (Species, Leidy. 1877, Kirby, 1932, 1944; nomenclature, Cleveland, 1938; Dobell, 1939; mineral ash, MacLennan and Murer, 1934; flagellar apparatus under electron microscope, Pitelka and Schooley, 1958; metabolism Gutierrez, 1956.)

T. campanula Kofoid and Swezy (Figs. 63; 177, *c*). 144-313μ by 57-144μ; wood particles are taken in by posterior region of the body (Fig. 37, *a*); in *Zootermopsis angusticollis, Z. nevadensis* and *Z. laticeps* (Kofoid and Swezy, 1919).

T. agilis Leidy (Fig. 177, *d*). 55-115μ by 22-45μ; in *Reticulitermes flavipes, R. lucifugus, R. speratus, R. flaviceps, R. hesperus, R. tibialis* (Leidy, 1877).

T. grandis Cleveland *et al.* 190-205μ by 79-88μ; in *Cryptocercus punctulatus.* (Gametogenesis and fertilization, Cleveland, 1957; photographs of fertilization, Cleveland, 1958.)

Genus **Pseudotrichonympha** Grassi and Foà. Two parts in anterior end as in *Trichonympha;* head organ with a spherical body at its tip and surrounded by a single layer of ectoplasm; bell covered by two layers of ectoplasm; nucleus lies freely; body covered by slightly oblique rows of short flagella; in termite gut (Grassi and Foà, 1911).

P. grassii Koidzumi. In *Coptotermes formosanus;* spindle-form; 200-300μ by 50-120μ (Koidzumi, 1921).

Genus **Deltotrichonympha** Sutherland. Triangular; with a small dome-shaped "head"; composed of two layers; head and neck with long active flagella; body flagella short, arranged in five longitudinal rows; flagella absent along posterior margin; nucleus large oval, located in anterior third; cytoplasm with wood chips; in termite gut. (One species.)

D. operculata S. Up to 230μ long, 164μ wide, and about 50μ thick; in the gut of *Mastotermes darwiniensis;* Australia.

Genus **Mixotricha** Sutherland. Large; elongate; anterior tip spirally twisted and motile; body surface with a coat of flagella in closely packed transverse bands (insertion and movement are entirely different from those of Trichonympha) except the posterior end; three short flagella at anterior end; nucleus, 20μ by 2μ, connected with blepharoplasts by prolonged tube which encloses nucleus itself; cytoplasm with scattered wood chips; in termites. (One species. Taxonomic position undetermined.)

M. paradoxa S. About 340μ long, 200μ broad and 25μ thick; in the intestine of *Mastotermes darwiniensis;* Australia.

Family 7 **Eucomonymphidae** Cleveland *et al.*

Flagella distributed over the entire body surface.

Genus **Eucomonympha** Cleveland *et al.* Body covered with flagella arranged in two longer rostral and shorter post-rostral) zones; rostral tube very broad, filled with hyaline material; nucleus at base of rostrum; in gut of *Cryptocercus punctulatus.* (Sexual cycle, Cleveland, 1956.)

E. imla C. *et al.* (Fig. 177, *e*). 100-165μ by 48-160μ; attached forms more elongate than free individuals. (Sexual reproduction, Cleveland, 1950.)

Family 8 **Teranymphidae** Koidzumi

The body with many transverse ridges to give a segmented appearance.

Genus **Teranympha** K. (*Teratonympha* K.; *Calonympha* Dogiel). Large and elongate; transversely ridged, and presents a metameric appearance; each ridge with a single row of flagella; anterior end complex, containing a nucleus; reproduction by longitudinal fission; in termite gut (Koidzumi, 1917, 1921; Dogiel, 1917).

T. mirabilis K. (Fig. 176, *g*). 200-300μ or longer by 40-50μ; in *Reticulitermes speratus.* (Mitosis, Cleveland, 1938a.)

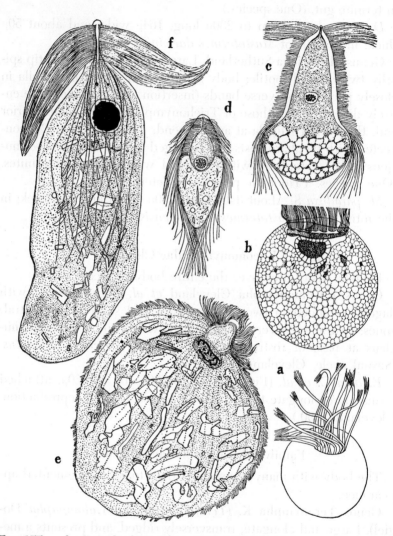

FIG. 177. a, b, *Kofoidia loriculata*, ×175, ×300 (Light); c, *Trichonympha campanula*, ×150 (Kofoid and Swezy); d, *T. agilis*, ×410 (Kirby); e, *Eucomonympha imla*, ×350 (Cleveland *et al.*); f, *Rhynchonympha tarda*, ×350 (Cleveland *et al.*).

References

ANDREWS, B. J., 1930. Method and rate of protozoan refaunation in the termite, etc. Univ. Cal. Publ. Zool., 33:449.

BEAMS, H. W., *et al.*, 1960. Electron microscope studies on *Lophomonas striata* with special reference to the nature and position of the striations. J. Protozool., 7:91.

BĚLAŘ, K., 1926. Der Formwechsel der Protistenkerne. Ergebn. u. Fortschr. Zool., 6:235.

BERNSTEIN, T., 1928. Untersuchungen an Flagellaten aus dem Darmkanal der Termiten aus Turkestan. Arch. Protist., 61:9.

BROWN, V. E., 1930. Hypermastigote flagellates from the termites Reticulitermes: etc. Univ. Cal. Publ. Zool., 36:67.

———— 1930a. On the morphology of Spirotrichonympha with a description of two new species, etc. Arch. Protist., 70:517.

———— 1931. The morphology of Spironympha, etc. J. Morphol. Physiol., 51:291.

CLEVELAND, L. R., 1925. The effects of oxygenation and starvation on the symbiosis between the termite, Termopsis, and its intestinal flagellates. Biol. Bull., 48:455.

———— 1938. Longitudinal and transverse division in two closely related flagellates. *Ibid.*, 74:1.

———— 1938a. Morphology and mitosis of Teranympha. Arch. Protist., 91:442.

———— 1949. The whole life cycle of chromosomes and their coiling systems. Tr. Am. Philos. Soc., 39:1.

———— 1949a. Hormone-induced sexual cycles of flagellates. I. J. Morphol., 85:197.

———— 1950. V. *Ibid.*, 87:349.

———— 1951. VI. *Ibid.*, 88:199.

———— 1951a. VII. *Ibid.*, 88:385.

———— 1952. VIII. *Ibid.*, 91:269.

———— 1953. IX. *Ibid.*, 93:371.

———— 1954. X. *Ibid.*, 95:189.

———— 1954a. XI. *Ibid.*, 95:213.

———— 1954b. XII. *Ibid.*, 95: 557.

———— 1955. XIII. *Ibid.*, 97:511.

———— 1956. XIV. Arch. Protist., 101:99.

———— 1956a. Brief accounts of the sexual cycles of the flagellates of Cryptocercus. J. Protozool., 3:161.

———— 1957. Additional observations on gametogenesis and fertilization in Trichonympha. J. Protozool., 4:164.

———— 1957a. Achromatic figure formation by multiple centrioles of Barbulanympha. *Ibid.*, 4:241.

—— 1958. Photographs of fertilization in *Trichonympha grandis*. *Ibid.*, 5:115.

—— and Day, M. 1958. Spirotrichonymphidae of Stolotermes. Arch. Protist., 103:1.

—— *et al.* 1934. The wood-feeding roach, Cryptocercus, its protozoa, etc. Mem. Amer. Acad. Arts & Sci., 17:185.

CUTLER, D. W., 1920. Protozoa parasitic in termites. II. Quart. J. Micr. Sci., 64:383.

DOBELL, C., 1939. On "Teranympha" and other monstrous latin parasites. Parasitology, 31:255.

DOGIEL, V. A. 1917. *Cyclonympha strobila* n. g., n. sp. J. Microbiol., 4:47.

—— 1922. Untersuchungen an parasitischen Protozoen aus dem Darmkanal der Termiten. II, III. Arch. Soc. Russ. Protist., 1:226.

DROPKIN, V. H., 1937. Host-parasite relations in the distribution of Protozoa in termites. Univ. Cal. Publ. Zool., 41:189.

—— 1941. Host specificity relations of termite Protozoa. Ecology, 22:200.

—— 1946. The use of mixed colonies of termites in the study of host-symbiont relations. J. Parasit., 32:247.

DUBOSCQ, O., and GRASSÉ, P., 1928. Notes sur les protistes parasites des termites de France. V. Arch. zool. exper. gén., 67 (N.-R.):159.

—— —— 1928a. L'appareil parabasal de *Joenia annectens*. C. R. Soc. biol., 99: 1118.

—— —— 1943. Les flagellés de l'*Anacanthotermes ochraceus*. Arch. zool. exper. gén., 82:401.

—— —— and ROSE, M., 1937. La flagellé de l'*Anacanthotermes ochraceus* du Sud-Algerien. C. R. Acad. Sci., 205:574.

GRASSÉ, P. P. 1926. Sur la nature des cotes cuticulaire des Polymastix et du *Lophomonas striata*. C. R. Soc. Biol., 94:1014.

—— 1952. Traité de Zoologie. I. Fasc.1. Paris.

—— and HOLLANDE, A. 1945. La structure d'une hypermastigine complexe *Staurojoenina caulleryi*. Ann Sci. Nat. Bot. Zool., 7:147.

GRASSI, B., 1885. Intorno ad alcuni protozoi parassiti delle termiti. Atti Accad. Gioenia Sci. Nat. Catania, Ser. 3, 18:235.

—— 1892. Conclusioni d'una memoria sulla societa dei termiti. Atti R. Accad. Lincei, Ser. 5, 1:33.

—— 1917. Flagellati viventi nei termiti. Mem. R. Accad. Lincei, 12:331.

—— and FOÀ, A., 1911. Intorno di protozoi dei termitidi. Atti R. Accad. Lincei, Ser. 5, 20:725.

GUTIERREZ, J., 1956. The metabolism of cellulose-digesting symbiotic flagellates of the genus Trichonympha from the termite Zootermopsis. J. Protozool., 3:39.

HUNGATE, R. E., 1939. Experiments on the nutrition of Zootermopsis. III. Ecology, 20:230.

JANICKI, C., 1910. Untersuchungen an parasitischen Flagellaten. I. Ztschr. wiss. Zool., 95:245.

———— (1915). II. *Ibid.*, 112:573.

KATZIN, L. I., and KIRBY, H. JR., 1939. The relative weights of termites and their protozoa. J. Parasit., 25:444.

KIRBY, H. JR., 1926. On *Staurojoenina assimilis*, etc. Univ. Cal. Publ. Zool., 29:25.

———— 1932. Flagellates of the genus Trichonympha. *Ibid.*, 37:349.

———— 1937. Host-parasite relations in the distribution of protozoa in termites. *Ibid.*, 41:189.

———— 1944. The structural characteristics and nuclear parasites of some species of Trichonympha in termites. *Ibid.*, 49:185.

KOFOID, C. A., and SWEZY, O., 1919. Studies on the parasites of termites. III. *Ibid.*, 20:41.

———— ———— 1919a. IV. *Ibid.*, 20:99.

KOIDZUMI, M., 1917. Studies on the Protozoa harboured by the termites of Japan. Rep. Invest. on termites, 6:1.

———— 1921. Studies, on the intestinal Protozoa found in the termites of Japan. Parasitology, 13:235.

KUDO, R. R., 1926. Observations on *Lophomonas blattarum*, etc. Arch. Protist., 53:191.

———— 1926a. A cytological study of *Lophomonas striata*. *Ibid.*, 55:504.

———— 1954. On the cytoplasmic fibrils of *Lophomonas striata*. J. Protozool., 1:80.

LEIDY, J., 1877. On intestinal parasites of *Termes flavipes*. Proc. Acad. Nat. Sci., Philadelphia, p. 146.

LIGHT, S. F., 1926. *Hoplonympha natator*. Univ. California Publ. Zool., 29:123.

———— 1927. Kofoidia, a new flagellate, from a California termite. *Ibid.*, 29:467.

———— and SANFORD, M. F., 1928. Experimental transfaunation of termites. *Ibid.*, 31:269.

MACLENNAN, R. F., and MURER, H. K., 1934. Localization of mineral ash in the organelles of Trichonympha, etc. J. Morphol., 56:231.

PITELKA, D. R., and SCHOOLEY, C. N., 1958. The fine structure of the flagellar apparatus in Trichonympha. J. Morphol., 102:199.

SUTHERLAND, J. L., 1933. Protozoa from Australian termites. Quart. J. Micr. Sci., 76:145.

SWEZY, O., 1923. The pseudopodial method of feeding by trichomonad flagellates parasitic in wood-eating termites. Univ. Cal. Publ. Zool., 20:391.

Chapter 18

Class 2. **Sarcodina** Hertwig and Lesser

T HE members of this class possess a comparatively thin pellicle and, therefore, are capable of forming pseudopods. The term "amoeboid" is often used to describe their body form change and movement. Pseudopods serve for both locomotion and food-capturing. Skeletal structures are variously developed in some orders. Thus, in Testacida and Foraminiferida, there are well-developed tests or shells that usually have apertures or pores, through which the pseudopods are extruded, while Heliozoida and Radiolarida, skeletons of various forms and materials are found.

The cytoplasm is usually differentiated into the ectoplasm and the endosplasm, but this differentiation is not always constant. In the endoplasm are found the nucleus, food vacuoles and various granules. The majority of Sarcodina are uninculeate, but Amoebida, Foraminiferida Mycetozoida include multinucleate forms. Sarcodina are typically holozoic. They feed on other protozoa, small metazoa and various protophyta which are conspicuously observable in the food vacuoles. The methods of ingestion have been considered before (p. 117). Contractile vacuoles are invariably present in species inhabiting fresh water, but absent in parasitic forms or in those which live in marine water.

Asexual reproduction is usually by binary (or rarely multiple) fission, budding, or plasmotomy. Definite proof of sexual reproduction has been noted in a comparatively small number of species. Encystment is common in the majority of Sarcodina, but is unknown in some species. (Taxonomy, Loeblich and Tappan, 1961, 1961a.)

The class Sarcodina is subdivided into two subclasses; Rhizopoda (with lobopodia, rhizopodia or filopodia) and Actinopoda (with axopodia) (p. 604).

Subclass 1. **Rhizopoda** Siebold

The name Rhizopoda has often been used to designate the entire class, but it is used here as one of the two subclasses which includes the following five orders: Proteomyxida (with radiating pseudopods); Mycetozoida (with rhizopods and forming plasmodium) (p. 507); Amoebida (with lobopods) (p. 518); Testacida (with single chambered test) (p. 565), and Foraminiferida (with one to many chambered test) (p. 589).

Order 1. **Proteomyxida** Lankester

A number of incompletely known organisms are placed in this group, and therefore, the grouping of genera is provisional. Pseudopods are filopods which often branch or anastomose with one another. In this respect the Proteomyxida show affinity to Mycetozoida. The majority lead parasitic life in algae or higher plants in fresh or salt water. (Taxonomy, Valkanov, 1940.) (Three families.)

Family 1. **Labyrinthulidae** Haeckel

Small fusiform protoplasmic masses are grouped in network of sparingly branched and anastomosing filopodia; individuals encyst independently; with or without flagellate stages.

Genus **Labyrinthula** Cienkowski. Minute forms feeding on various species of algae in fresh or salt water; often brightly colored due to carotin. Jepps (1931) found these organisms common in marine aquaria. Young (1943) considers the six known species as actually three species and two varieties, while Watson (1951) holds that only one species, *L. macrocytis*, should be reognized.

L. cienkowskii Zopf (Fig. 178, *a*). Attacks Vaucheria in fresh water.

L. macrocystis Cienkowski. Renn (1934, 1936) found a species in the diseased leaf-tissue of the 'spotting and darkening' eelgrass, *Zostera marina*, along the Atlantic coast of the United States. Young (1943) identified the organism which he studied as *L. macrocystis*, and noted that its hosts included various algae and three genera of Naiadaceae: Zostera, Ruppia and Zannichellia.

The 'net-plasmodium' contains fusiform cells which average in size 18μ by 4μ and which multiply by binary fission; many cells encyst together within a tough, opaque membrane. The growth is best at 14-24°C. and at 12-22 per cent chlorinity (Young). Watson and Ordal (1951) cultivated the organism on agar and sea water with various bacteria, and found that the organism is fusiform in young cultures; highly motile; filamentous projections are formed from the flat mucoid lamellae, secreted by the organism, and ex-

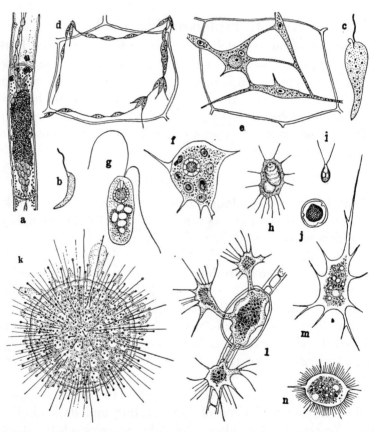

Fig. 178. a, *Labyrinthula cienkowskii*, ×200 (Doflein); b–e, *Labyrintho-myxa sauvageaui* (b, c, flagellate forms, ×100; d, e, amoeboid forms, ×500) (Duboscq); f, g, *Pseudospora volvocis*, ×670 (Robertson); h–j, *Protomonas amyli* (Zopf); k, l, *Vampyrella lateritia*, ×530 (k (Leidy), l (Doflein)); m, n, *Nuclearia delicatula*, ×300 (Cash).

pand to form passways over which the organism travels; holozoic, saprozoic.

Genus **Labyrinthomyxa** Duboscq. Body fusiform; amoeboid and flagellate phases, variable in size; flagellate stage penetrates the host cell membrane; in plants.

L. sauvageaui D. (Fig. 178, *b-e*). Fusiform body 7-11μ long; pseudoplasmodium-formation; amoeboid stage 2.5-14μ long; flagellate stage 7-18μ long; parasitic in *Laminaria lejolisii* at Roscoff, France.

Family 2. **Pseudosporidae** Berlese

Solitary and heliozoa-like in appearance; with flagellate swarmers.

Genus **Pseudospora** Cienkowski. Body minute; parasitic in algae and Mastigophora (including Volvocidae); organism nourishes itself on host protoplasm, grows and multiplies into a number of smaller individuals, by repeated division; the latter biflagellate, seek a new host and transform themselves into amoeboid stage; encystment common. (Morphology and development, Schussnig, 1929.)

P. volvocis C. (Fig. 178, *f*, *g*). Heliozoan form about 12-30μ in diameter; pseudopodia radiating; cysts about 25μ in diameter; in species of Volvox. (Morphology, Roskin, 1927.)

P. parasitica C. Attacks Spirogyra and allied algae.

P. eudorini Roskin. Heliozoan forms 10-12μ in diameter; radiating pseudopodia two to three times longer; amoeboid within host colony; cysts 15μ in diameter; in *Eudorina elegans*.

Genus **Protomonas** Cienkowski. Body irregularly rounded with radiating filopodia; food consists of starch grains; division into biflagellate organisms which become and unite to form pseudoplasmodium; fresh or salt water.

P. amyli C. (Fig. 178, *h-j*). In fresh water.

Family 3. **Vampyrellidae** Doflein

Filopodia radiate from all sides or formed from a limited area; flagellate forms do not occur; the organism is able to bore through the cellulose membrane of various algae and feeds on protoplasmic contents; body often reddish because of the presence of carotin; multinucleate; multiplication in encysted stage into uni- or multi-nucleate bodies; cysts often also reddish.

Genus **Vampyrella** Cienkowski. Heliozoa-like; endoplasm vacuolated or granulated, with carotin granules; numerous vesicular nuclei and contractile vacuoles; multinucleate cysts, sometimes with stalk; 50-700µ in diameter. Several species.

V. lateritia (Fresenius) (Fig. 178, *k*, *l*). Spherical; orange-red except the hyaline ectoplasm; feeds on Spirogyra and other algae in fresh water. On coming in contact with an alga, it often travels along it and sometimes breaks it at joints, or pierces individual cell and extracts chlorophyll bodies by means of pseudopodia; multiplication in encysted condition; 30-40µ in diameter. (Behavior, Lloyd, 1926, 1929.)

Genus **Nuclearia** Cienkowski. Subspherical, with sharply pointed fine radiating pseudopodia; actively moving forms vary in shape; with or without a mucous envelope; with one or many nuclei; fresh water.

N. delicatula C. (Fig. 178, *m*, *n*). Form changes, variable; four to ten nuclei; bacteria often adhering to the gelatinous envelope; up to 60µ in diameter. (Cultivation in Petri dishes with Oscillatoria, Blanc-Brude, *et al.*, 1955.)

N. simplex C. Uninucleate; 30µ in diameter.

Genus **Arachnula** Cienkowski. Body irregularly chain-form with filopodia extending from ends of branches; numerous nuclei and contractile vacuoles; feeds on diatoms and other microorganisms.

A. impatiens C. (Fig. 179, *a*). 40-350µ in diameter.

Genus **Chlamydomyxa** Archer. Body spheroidal; ectoplasm and endoplasm well differentiated; endoplasm often green-colored due to the presence of green spherules; numerous vesicular nuclei; one to two contractile vacuoles; secretion of an envelope around the body is followed by multiplication into numerous secondary cysts; cyst wall cellulose; in sphagnum swamp.

C. montana Lankester (Fig. 179, *b*, *c*). Rounded or ovoid; cytoplasm colored; about 50µ in diameter; when moving, elongate with extremely fine pseudopodia which are straight or slightly curved and which are capable of movement from side to side; non-contractile vacuoles at bases of grouped pseudopods; in active individual there is a constant movement of minute fusiform

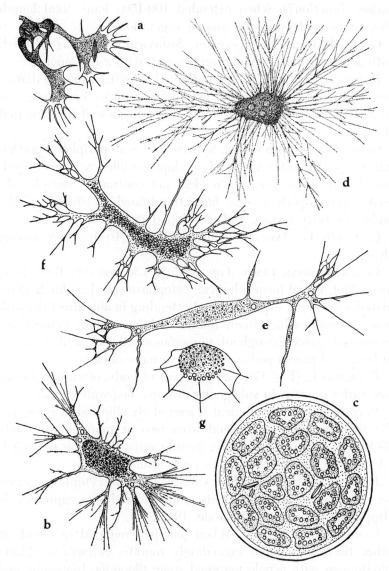

Fig. 179. a, *Arachnula impatiens,* ×670 (Dobell); b, c, *Chalmydomyxa montana;* b, ×270 (Cash); c, ×530 (Penard); d, *Rhizoplasma kaiseri* (Verworn); e, *Biomyxa vagans,* ×200 (Cash); f, *Penardia mutabilis,* ×200 (Cash); g, *Hyalodiscus rubicundus,* ×370 (Penard).

bodies (function?); when extended 100-150μ long; total length 300μ or more; fresh water among vegetation.

Genus **Rhizoplasma** Verworn. Spherical or sausage-shaped; with anastomosing filopodia; orange-red; with a few nuclei.

R. kaiseri V. (Fig. 179, *d*). Contracted form 0.5-1 mm. in diameter; with one to three nuclei; pseudopodia up to 3 cm. long; extended body up to 10 mm. long; originally described from Red Sea.

Genus **Chondropus** Greeff. Spherical to oval; peripheral portion transparent but often yellowish; endoplasm filled with green, yellow, brown bodies; neither nucleus nor contractile vacuoles observed; pseudopods straight, fine, often branched; small pearl-like bodies on body surface and pseudopodia.

C. viridis G. Average diameter 35-45μ; fresh water among algae.

Genus **Biomyxa** Leidy (*Gymnophrys* Cienkowski). Body form inconstant; initial form spherical; cytoplasm colorless, finely granulated, capable of expanding and extending in any direction, with many filopodia which freely branch and anastomose; cytoplasmic movement active throughout; numerous small contractile vacuoles in body and pseudopodia; with one or more nuclei.

B. vagans L. (Fig. 179, *e*). Main part of body, of various forms; size varies greatly; in sphagnous swamps, bog-water, etc.

B. cometa (C.) Subspherical or irregularly ellipsoidal; pseudopodia small in number, formed from two or more points; body 35-40μ, or up to 80μ or more; pseudopodia 400μ long or longer. Cienkowski maintained that this was a *moneran*.

B. merdaria Hollande. Size variable, 10-95μ; cytoplasm granular, undifferentiated; filaments, some branching; coprozoic in Hippopotamus faeces (Hollande, 1942).

Genus **Penardia** Cash. When inactive, rounded or ovoid; at other times expanded; exceedingly mobile; endoplasm chlorophyll-green with a pale marginal zone; filopodia, branching and anastomosing, colorless; nucleus inconspicuous; one or more contractile vacuoles, small; fresh water.

P. mutabilis C. (Fig. 179, *f*). Resting form 90-100μ in diameter; extended forms (including pseudopodia) 300-400μ long.

Genus **Hyalodiscus** Hertwig and Lesser. Discoid, though outline varies; endoplasm reddish, often vacuolated and sometimes shows filamentous projections reaching body surface; a single nucleus; ectoplasmic band of varying width surrounds the body completely; closely allied to Vampyrella; fresh water.

H. rubicundus H. and L. (Fig. 179, *g*). 50-80µ by about 30µ; polymorphic; when its progress during movement is interrupted by an object, the body doubles back upon itself, and moves on in some other direction; fresh water ponds among surface vegetation.

H. simplex Wohlfarth-Bottermann. 50-70µ in diameter; movement by broad hyaline pseudopod; in fresh water among algae (Wohlfarth-Bottermann, 1960).

Genus **Leptomyxa** Goodey. Multinucleate, thin amoeboid organisms; multinucleate cysts formed by condensation of protoplasm; free-living in soil (Goodey, 1915).

L. reticulata G. (Fig. 180, *a-c*). Body composed of a thin transparent protoplasm; when fully extended, 3 mm. or more in length; superficially resembles an endosporous mycetozoan, but no reversible cytoplasimc movement; multinucleate with eight to twenty to several hundred nuclei; nuclei, 5-6µ in diameter, with a large endosome; nuclear division simultaneous, but not synchronous; plasmotomy; plasmogamy; cysts multinucleate, by local condensation of protoplasm; widely distributed in British soil (Singh, 1948, 1948a). McLennan (1930) found a similar organism in and on the root of diseased hops in Tasmania.

Genus **Megamoebomyxa** Nyholm. Extremely large amoeboid organism; when contracted, lobulate, with adhering detritus; when cultured at 8-10°C. on debris, filopodia are formed and form-change occurs; lobate during locomotion; "nutrient chiefly detritus"; Marine. (One species, Nyholm, 1950.)

M. argillobia N. (Fig. 180, *d*). An opaque white organism; up to 25 mm. long; polymorphic; in marine sediment, rich in debris at the depth of 45-70 m.; Gullmar Fjord, Sweden.

Genus **Reticulomyxa** Nauss. Highly polymorphic, multinucleate amoeboid organism; rhizopodia radiating from a central mass of undifferentiated granular protoplasm with many non-contractile vacuoles; plasmotomy usually into three, after discarding extraneous particles and migrating to new site; when transferred to

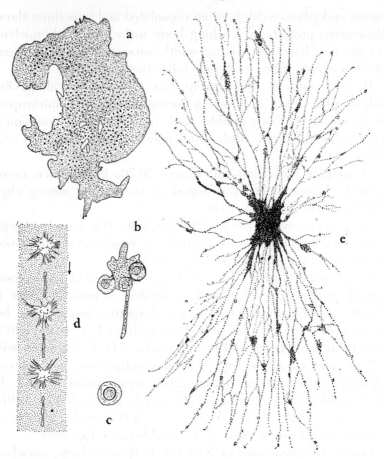

Fig. 180. a–c, *Leptomyxa reticulata*, ×73 (Singh) (a, a trophozoite; b, cyst-formation; c, a cyst); d, an individual of *Megamoebomyxa argillobia*, showing the changes of body form, ×2/3 (Nyholm); e, a young trophozoite of *Reticulomyxa filosa*, ×3 (Nauss).

fresh dish of water, "spore-like" bodies are dispersed; fresh water among decaying leaves. Nauss (1949) points out its affinity to Proteomyxida, Mycetozoida and Foraminiferida.

R. filosa N. (Fig. 180, *e*). On moist blotting paper the central mass is an elevated body, but in water it spreads into a broad sheet, four to six mm. in diameter; pseudopodia may be up to ten times the diameter of the central white mass; encystment occurs when subjected to lower temperature or when cultured with algae; food consists of "worms," rotifers and organic debris.

References

BLANC-BRUDE, R., *et al.*, 1955. Sur la biologie de *Nuclearia delicatula*. Bull. microsc. appl., (2) 5:113.

CASH, J., 1905, 1909. The British freshwater Rhizopoda and Heliozoa. 1, 2. London.

―― and WAILES, G. H., 1915–1918. 3, 4. London.

CIENKOWSKI, L., 1863. Das Plasmodium. Pringsheim's Jahrb. Bot., 3:400.

―― 1867. Ueber den Bau und die Entwicklung der Labyrinthuleen. Arch. mikr. Anat., 3:274.

DOBELL, C., 1913. Observations on the life-history of Cienkowski's Arachnula. Arch. Protist., 31:317.

DOFLEIN, F., and REICHENOW, E., 1949–1953. Lehrbuch der Protozoenkunde. 6 ed. Jena.

DUBOSCQ, O., 1921. *Labyrinthomyxa sauvageaui*, etc. C. R. Soc. biol., 84:27.

GOODEY, T., 1915. A preliminary communication of three new proteomyxan rhizopods from soil. Arch. Protist., 35:80.

HOLLANDE, A., 1942. Contribution a l'étude morphologique et cytologique des genre Biomyxa, etc. Arch. zool. expér. gén., 82, N. et R. 119.

JEPPS, M. W., 1931. Note on a marine Labyrinthula. J. Marine Biol. Assn. United Kingdom, 17:833.

―― 1956. The protozoa, Sarcodina. Edinburgh.

KÜHN, A., 1926. Morphologie der Tiere in Bildern. H.2, T.2. Rhizopoden. Jena.

LEIDY, J., 1879. Freshwater Rhizopods of North America. Rep. U. S. Geol. Survey, 12.

LLOYD, F. E., 1926. Some behaviours of *Vampyrella lateritia*, etc. Papers Mich. Acad. Sci., 6:275.

―― 1929. The behavior of *Vampyrella lateritia*, etc. Arch. Protist., 67:219.

LOEBLICH, A. R., JR., and TAPPAN, H., 1961. Suprageneric classification of the Rhizopodea. J. Paleontol., 35:245.

―― ―― 1961a. Remarks on the systematics of the Sarkodina, etc. Proc. Biol. Soc., Washington, 74:213.

McLENNAN, E. I., 1930. A disease of hops in Tasmania and an account of a proteomyxan organism, etc. Australian J. Exper. Biol., 7:9.

NAUSS, R. N., 1949. *Reticulomyxa filosa*, etc. Bull. Torrey Bot. Club, 76:161.

NYHOLM, K.-G., 1950. A marine nude rhizopod type *Megamoebomyxa argillobia*. Zool. Bidrag. Uppsala, 29:93.

PENARD, E., 1902. Faune rhizopodique du bassin du Léman. Geneva.

RENN, C. E., 1935. A mycetozoan parasite of *Zostera marina*. Nature, 135:544.

——— 1936. The wasting disease of *Zostera marina*. Biol. Bull., 70:148.

ROSKIN, G., 1927. Zur Kenntnis der Gattung Pseudospora. Arch. Protist., 59:350.

SCHUSSNIG, B., 1929. Beiträge zur Entwicklungsgeschichte der Protophyten. IV. *Ibid.*, 68:555.

SINGH, B. N., 1948. Studies on giant amoeboid organisms. I. J. Gen. Microbiol., 2:7.

——— 1948a. II. *Ibid.*, 2:89.

VALKANOV, A., 1929. Protistenstudien. IV. Arch. Protist., 67:110.

——— 1940. Die Heliozoen und Proteomyxien. *Ibid.*, 93:225.

WATSON, S. W., and ORDAL, E. J., 1951. Studies on Labyrinthula. Univ. Washington Oceanogr. Lab., Tech. Rep., 3, 37 pp.

WOHLFARTH-BOTTERMANN, K. E., 1960. Protistenstudien. X. Protoplasma, 52:58.

YOUNG, E. L., 1943. Studies on Labyrinthula, etc. Am. J. Bot., 30: 586.

ZOPF, W., 1887. Handbuch der Botanik (A. Schenk), 3:24.

Chapter 19
Order 2. **Mycetozoida** de Bary

THE Mycetozoida had been considered to be closely related to the fungi, being known as Myxomycetes, or Myxogasteres, the 'slime molds.' Through extended studies of their development, de Bary showed that they are more closely related to the protozoa than to the protophyta, although they stand undoubtedly on the border-line between these two groups of microorganisms. The Mycetozoida occur on dead wood or decaying vegetable matter of various kinds.

The most conspicuous part of a mycetozoan is its **plasmodium** which is formed by fusion of many **myxamoebae,** thus producing a large multinucleate body (Fig. 181, *a*). The greater part of the cytoplasm is granulated, although there is a thin layer of hyaline and homogeneous cytoplasm surrounding the whole body. The numerous vesicular nuclei are distributed throughout the granular cytoplasm. Many small contractile vacuoles are present in the peripheral portion of the plasmodium. The nuclei increase in number by division as the body grows; the division seems to be amitotic during the growth period of the plasmodium, but is mitotic prior to the spore-formation. The granulation of the cytoplasm is due to the presence of enormous numbers of granules which in some forms are made up of carbonate of lime. The plasmodium is usually colorless, but sometimes yellow, green, or reddish, because of the numerous droplets of fluid pigment present in the cytoplasm.

The food of Mycetozoida varies among different species. The great majority feed on decaying vegetable matter, but some, such as Badhamia, devour living fungi. Thus the Mycetozoida are holozoic or saprozoic in their mode of nutrition. Pepsin has been found in the plasmodium of Fuligo and is perhaps secreted into the food vacuoles, into which protein materials are taken. The

507

plasmodium of Badhamia is said to possess the power of cellulose digestion.

When exposed to unfavorable conditions, such as desiccation,

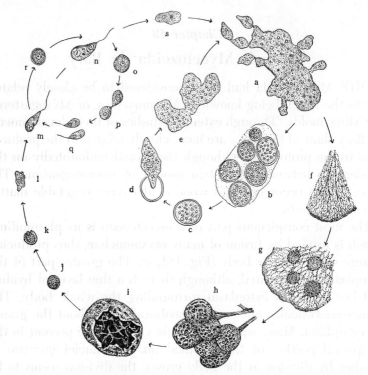

FIG. 181. The life-cycle of the endosporous mycetozoan (de Bary, Lister, and others). a, plasmodium-formation by fusion of numerous myxamoebae; b, c, formation of sclerotium; d, e, germination of sclerotium and formation of plasmodium; f, portion of a plasmodium showing streaming protoplasmic thickenings; g, h, formation of sporangia; i, a sporangium opened, showing capillitum; j, a spore; k, germination of spore; l, myxamoeba; m, n, myxoflagellates; o–q, multiplication of myxoflagellate; r, microcyst; s, myxamoeba. Variously magnified.

the protoplasmic movement ceases gradually, foreign bodies are extruded, and the whole plasmodium becomes divided into numerous **sclerotia** or cysts, each containing ten to twenty nuclei and being surrounded by a resistant wall (b). These cysts may live as long as three years. Upon return of favorable conditions, the contents of the sclerotia germinate, fuse together, and thus again produce plasmodia (*c-e*).

When lack of food material occurs, the plasmodium undergoes changes and develops **sporangia**. The first indication of this process is the appearance of lobular masses of protoplasm in various parts of the body (*f*, *g*). These masses are at first connected with the streaming protoplasmic thickenings, but later become completely segregated into young sporangia. During the course of sporangium-formation, foreign bodies are thrown out of the body, and around each sporangium there is secreted a wall which, when mature, possesses a wrinkled appearance (*h*). The wall continues down to the substratum as a slender stalk of varying length, and in many genera the end of a stalk spreads into a network over the substratum, which forms the base, **hypothallus,** for the stalk. With these changes the interior of the sporangium becomes penetrated by an anastomosing network, **capillitium,** of flat bands which are continuous with the outer covering (*i*). Soon after the differentiation of these protective and supporting structures, the nuclei divide simultaneously by mitosis and the cytoplasm breaks up into many small bodies. These uninucleate bodies are the **spores** which measure 3-20μ in diameter and which soon become covered by a more or less thick cellulose membrane (*j*), variously colored in different species.

The mature sporangium breaks open sooner or later and the spores are carried, and scattered, by the wind. When a spore falls in water, its membrane ruptures, and the protoplasmic contents emerge as an amoebula (*k*, *l*). The amoebula possesses a single vesicular nucleus and contractile vacuoles, and undergoes a typical amoeboid movement. It presently assumes an elongate form and one flagellum or two unequally long flagella (Elliott, 1948; Kerr, 1960) develop from the nucleated end, thus forming a **myxoflagellate** (*m*, *n*) which undergoes a peculiar dancing movement and is able to form short, pointed pseudopodia from the posterior end. It feeds on bacteria, grows and multiplies by binary fission (*o-q*). After a series of division, the myxoflagellate may encyst and becomes a **microcyst** (*r*). When the microcyst germinates, the content develops into a myxamoeba (*s*) which, through fusion with many others, produces the plasmodium mentioned above. This is the life-cycle of a typical endosporous myce-tozoan.

In the genus Ceratiomyxa in which spores are formed on the surface of **sporophores,** the development is briefly as follows: the plasmodium lives on or in decayed wood and presents a horn-like appearance. The body is covered by a gelatinous hyaline sub-stance, within which the protoplasmic movements may be noted. The protoplasm soon leaves the interior and accumulates at the surface of the mass; at first as a close-set reticulum and then into a mosaic of polygonal cells, each containing a single nucleus. Each of these cells moves outward at right angles to the surface, still enveloped by the thin hyaline layer, which forms a stalk below. These cells are spores which become ellipsoid and covered by a membrane when fully formed. The spore is uninucleate at first, but soon becomes tetranucleate. When a spore reaches the water, its content emerges as an amoebula which divides three times, forming 8 small bodies, each of which develops a flagellum and becomes a myxoflagellate. The remaining part of the de-velopment is presumably similar to that of the endosporous form. (Morphology, de Bary, 1864, 1864; MacBride, 1922; Jahn, 1928; MacBride and Martin, 1934; culture, Cohn, 1953; Sobels and Cohn, 1953.)

The Mycetozoida is subdivided into three suborders: Eumyce-tozoina (spores develop into myxoflagellates; myxoamoebae fuse completely and form plasmodia; sporangia are formed), Phyto-myxina (no sporangia are formed; mostly parasitic in plants) (p. 514) and Acrasina (no flagellated stage; prior to spore formation, pseudoplasmodia may occur) (p. 515).

Numerous mycetozoan genera and species are reported. Here a few examples are given. (American species, Hagelstein, 1944.)

Suborder 1. **Eumycetozoina** Zopt

Spores develop within sporangia
  Spores violet or violet-brown
    Sporangia with lime
      Lime in small granular form ....Family 1 Physaridae

Genus **Badhamia** Berkeley (Fig. 182 *a, b*).

Capillitium, a course network with lime throughout.

Genus **Fuligo** Haller (Fig. 182, *c, d*)

Axenic culture (Lazo, 1961).

Capillitium, a delicate network of threads with vesicular expansions filled with granules of lime.

Lime in crystalline form ....Family 2 Didymiidae

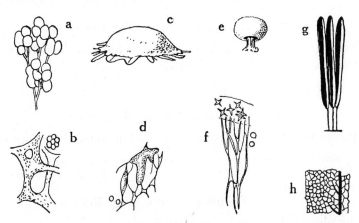

Fig. 182. a, b, *Badhamia utricularis* Berkeley (a, cluster of sporangia, ×4; b, part of capillitium and spore-cluster, ×140) (Lister); c, d, *Fuligo septica* Gmelin (c, a group of sporangia, ×⅓; d, part of capillitium and two spores, ×120) (Lister); e, f, *Didymium effusum* Link (e, sporangium, ×12; f, portion of capillitium and wall of sporangium showing the crystals of calcium carbonate and two spores, ×200) (Lister); g, h, *Stemonitis splendens* Rostafinski (g, three sporangia, ×2; h, columella and capillitium, ×42) (Lister).

Genus **Didymium** Schrader (Fig. 182, *e, f*)

Flagellum formation by myxamoebae of *D. nigripes* (Kerr, 1960).

Lime crystals stellate, distributed over the wall of sporangium.

Sporangia without lime
  Sporangia stalked ....Family 3. Stemonitidae

Genus **Stemonitis** Gleditsch (Fig. 182, *g, h*).

Sporangium-wall evanescent; capillitium arising from all parts of columella to form a network.

Sporangium combined into aethalium ..Family 4. Amaurochaetidae

Genus **Amaurochaete** Rostafinski (Fig. 183, *a, b*).

With irregularly branching thread-like capillitium.

Spores variously colored, except violet
  Capillitium absent or not forming a system of uniform threads.
    Sporangium-wall membranous; with minute round granules ......
    ....Family 5. Cribrariidae

Genus **Cribraria** Persoon (Fig. 183, *c*).

Sporangia stalked; wall thickened and forms a delicate persistent network expanded at the nodes.

Sporangia solitary; stalked ....Family 6. Liceidae

Genus **Orcadella** Wingate (Fig. 183 *d*)

Sporangia stalked, furnished with a lid of thinner substance.

Sporangium-wall membranous without granular deposits ....
    ....Family 7. Tubulinidae

Genus **Tubulina** Persoon (Fig. 183 *e*)

Sporangia without tubular extensions.

Many sporangia more or less closely fused to form large bodies
    (aethalia); sporangium-wall incomplete and perforated ......
    ....Family 8. Reticulariidae

Genus **Reticularia** Bulliard (Fig. 183, *f*)

Walls of convoluted sporangia incomplete, forming tubes and folds with numerous anastomosing threads.

Sporangia forming aethalium ....Family 9. Lycogalidae

Genus **Lycogala** Micheli (Fig. 183, *g*)

Capillitium a system of uniform threads
  Capillitium threads with spiral or annular thickenings ....
    ....Family 10. Trichiidae

Genus **Trichia** Haller (Fig. 183, *h-j*)

Capillitium abundant, consisting of free elasters with spiral thickenings.

Capillitium combined into an elastic network with thickenings in
    forms of cogs, half-rings, spines, or warts. Family 11. Arcyriidae

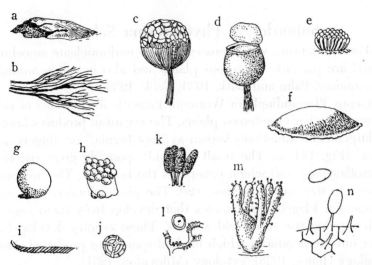

FIG. 183. a, b, *Amaurochaete fuliginosa* MacBride (a, group of sporangia, ×½; b, capillitium, ×10) (Lister); c, empty sporangium of *Cribraria aurantiaca* Schrader, ×20 (Lister); d, sporangium of *Orcadella operculata* Wingate, ×80 (Lister); e, cluster of sporangia of *Tubulina fragiformis* Persoon, ×3 (Lister); f, aethalium of *Reticularia lycoperdon* Bull., ×1 (Lister); g, aethalium of *Lycogala miniatum* Persoon ×1 (Lister); h–j, *Trichia affinis* de Bary (h, group of sporangia, ×2; i, elater, ×250; j, spore ×400) (Lister);k, l, *Arcyria punicea* Persoon (k, four sporangia, ×2; l, part of capillitium, ×250 and a spore, ×560) (Lister); m, n, *Ceratiomyxa fruticulosa* MacBride (m, sporophore, ×40; n, part of mature sporophore, showing two spores, ×480) (Lister).

Genus **Arcyria** Wiggers (Fig. 183, *k*, *l*)

Sporangia stalked; sporangium-wall evanescent above, persistent and membranous in the lower third.

Capillitium abundant; sporangia normally sessile . . . .
. . . . .Family 12. Margaritidae

Genus **Margarita** Lister

Capillitium profuse, long, coiled hair-like.

Spores develop on the surface of sporophores
Spores white; borne singly or filiform stalk . . . .
. . . . .Family 13. Ceratiomyxidae

Genus **Ceratiomyxa** Schröter (Fig. 183, *m*, *n*)

Suborder 2. **Phytomyxina** Schröter

These organisms which possess a large multinucleate amoeboid body, are parasitic in various plants and also in a few animals. (Taxonomy, Palm and Burk, 1933; Cook, 1933.)

Genus **Plasmodiophora** Woronin. Parasitic in the roots of cabbage and other cruciferous plants. The organism produces knotty enlargements, sometimes known as "root-hernia," or "fingers and toes" (Fig. 184, *a*). The small (haploid) spore (*b*) gives rise to a myxoflagellate (*c-f*) which penetrates the host cell. The organism grows in size and multiples (*g-h*). The plasmodium divides into sporangia. Flagellated gametes that develop from them fuse in pairs, giving rise to diploid zygotes. These zygotes develop further into plasmodia in which haploid spores are produced. Morphology (Jones, 1928); cytology (Milovidov, 1931).

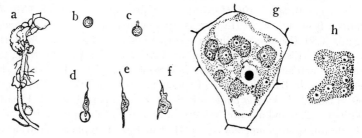

FIG. 184. *Plasmodiophora brassicae*. a, root-hernia of cabbage; b, a spore, ×620; c–e, stages in germination of spore, ×620; f, myxamoeba, ×620 (Woronin); g, a host cell with several young parasites, ×400; h, an older parasite, ×400 (Nawaschin).

P. brassicae W. (Fig. 184). In *Brassica* spp.

Genus **Sorosphaera** Schröter. Parasitic in *Veronica* spp.

Genus **Tetramyxa** Goebel. In Ruppia, Zannichellia, etc.

Genus **Octomyxa** Couch, Leitner and Whiffen. In *Achlya glomerata*.

Genus **Sorodiscus** Lagerheim and Winge. In Chara, Callitriche, etc.

Genus **Polymyxa** Ledingham. In Triticum, etc.

Genus **Membranosorus** Ostenfeld and Petersen. In *Heteranthera dubia*.

Genus **Spongospora** Brunchorst. Parasitic in Solanum; the diseased condition of potatoes is known as powdery or corky scab.

Genus **Ligniera** Maire and Tison. In Alisma, Juncus, etc.

Suborder 3. **Acrasina** van Tieghem

The spores give rise to uninucleate amoebae; no flagellated cells; pseudoplasmodia are produced; free-living in soil, dung, vegetative matter, etc.

Genus **Guttulina** Cienkowski. Irregular fruiting bodies which may be stalked; spores with smooth membrane.

Genus **Guttulinopsis** Olive. Pseudospores with crinkled membrane.

Genus **Dictyostelium** Brefeld. Spores give rise to amoebae which feed on bacteria and multiply; aggregation of the amoebae takes place under certain conditions which results in pseudoplasmodia; after migration, pseudoplasmodia develop spores. (Six species, Bonner, 1959.)

D. discoideum Raper. The elliptical spores give rise to amoebae with a nucleus and one or more contractile vacuoles. As amoebae grow by feeding on bacteria, they multiply in number. Cannibalism is said to be common. The size varies 5 to 16μ in diameter. Aggregation of the amoebae occurs in dense population, when food is depleted, culture media dry up or the temperature rises. The aggregative movement is induced by acrasin, a chemical substance, secreted by some of the amoebae which seem to lead the movement (Bonner, 1947, 1949). Completion of aggregation is a long process and takes four to eight hours. At the end of migration through contraction of component amoebae, pseudoplasmodium rounds up, followed by growth of the principal vertical axis on which numerous spores become differentiated on the upper edge of pre-spore mass which differentiates into two groups: stalk-forming and basal disc cells (Raper, 1940; Bonner, 1959). (Variation, Sussman and Sussman, 1953; distribution of Dictyostelium, Singh, 1947, 1947a; environmental factors on aggregation, Bradley *et al.* 1956.)

Other genera are **Polysphondylium** (Brefeld), **Acytostelium** (Brefeld), **Acrasis** (van Tieghem), etc.

References

BONNER, J. T., 1944. A descriptive study of the development of the slime mold, *Dictyostelium discoideum*. Amer. J. Bot., 31:175.

———— 1947. Evidence for formation of cell aggregates by chemotaxis in the development of the slime mold *Dictyostelium discoideum*. J. Exper. Zool., 106:1.

———— 1949. The demonstration of acrasin in the later stages of the development of the slime mold *Dictyostelium discoideum*. *Ibid.*, 110:259.

———— 1959. The cellular slime molds. Princeton Univ. Press.

BRADLEY, S. G., *et al.*, 1956. Environmental factors affecting the aggregation of the cellular slime mold, etc. J. Protozool., 3:33.

COHN, A. L., 1953. The isolation and culture of opsimorphic organisms. I. Ann. N. Y. Acad. Sci., 56:938.

COOK, W. R. I., 1933. A monograph of the Plasmodiophorales. Arch. Protist., 80:179.

DE BARY, A., 1864. Die Mycetozoa. Leipzig.

———— 1884. Vergleichende Morphologie und Biologie der Pilze, Mycetozoen, und Bacterien. Leipzig.

ELLIOTT, E. W., 1948. The sperm-cells of Myxomycetes. J. Washington Acad. Sci., 38:133.

HAGELSTEIN, R., 1944. The Mycetozoa of North America. New York.

JAHN, E., 1901–1920. Myxomycetenstudien. I–X. Ber. deutsch. bot. Ges., 19, 20, 22–26, 29, 36 and 37.

———— 1928. Myxomycetenstudien. XII. *Ibid.*, p. 80.

JONES, P. M., 1928. Morphology and cultural study of *Plasmodiophora brassicae*. Arch. Protist., 62:313.

KARLING, J. S., 1942. The Plasmodiophorales. New York.

KERR, N. S., 1960. Flagella formation by myxamoebae of the true slime mold, *Didymium nigripes*. J. Protozool., 7:103.

LAZO, W. R., 1961. Obtaining the slime mold *Fuligo septica* in pure culture. J. Protozool., 8:97.

LISTER, A., 1925. A monograph on the Mycetozoa. 3 ed. London.

MACBRIDE, T. H., 1922. North American slime molds. 2 ed. New York.

———— and MARTIN, G. H., 1934. The Myxomycetes. New York.

MILOVIDOV, P. F., 1931. Cytologische Untersuchungen an *Plasmodiophora brassicae*. Arch. Protist., 73:1.

OLIVE, L. S., *et al.*, 1961. Variation in the cellular slime mold *Acrasis rosea*. J. Protozool., 8:467.

PALM, B. T., and BURK, M., 1933. The taxonomy of the Plasmodiophoraceae. *Ibid.*, 79:262.

RAPER, K. B., 1940. Pseudoplasmodium formation and organization in *Dictyostelium discoideum*. J. E. Mitchell Sci. Soc., 56:241.

SINGH, B. N., 1947. Studies on soil Acrasieae. I. J. Gen. Microbiol., 1:11.
—— 1947a. II. *Ibid.*, 1:361.
SOBELS, J. C., and COHN, A. L., 1953. The isolation and culture of opsimorphic organisms. II. Ann. N. Y. Acad. Sci., 56:944.
SUSSMAN, R. R., and M., 1953. Cellular differentiation in Dictyosteliaceae, etc. *Ibid.*, 56:949.

Chapter 20

Order 3. **Amoebida** Ehrenberg

THE Amoebida show a very little cortical differentiation. There is no thick pellicle or test, surrounding the body, although in some a delicate pellicle occurs. The cytoplasm is more or less distinctly differentiated into the ectoplasm and the endoplasm. The ectoplasm is hyaline and homogeneous, and appears tougher than the endoplasm. In the endoplasm, which is granulated or vacuolated, are found one or more nuclei, various food vacuoles, crystals, and other inclusions. In the freshwater forms, there is at least one distinctly visible contractile vacuole. The pseudopodia are lobopodia, and ordinarily both the ectoplasm and endoplasm are found in them. They are formed by streaming or fountain movement of the cytoplasm. In some members of this order, the formation of pseudopodia is eruptive or explosive, since the granules present in the endoplasm break through the border line between the two cytoplasmic layers and suddenly flow into the pseudopodia. Asexual reproduction is ordinarily by binary fission, although multiple fission may occasionally take place. Encystment is of common occurrence. Sexual reproduction, which has been reported in a few species, has not been confirmed.

The Amoebida inhabit all sorts of fresh, brackish, and salt waters. They are also found in moist soil and on moist ground covered with decaying leaves. Many are inhabitants of the digestive tract of various animals and some are pathogenic to the hosts.

The taxonomic status of the group is highly uncertain and confusing, since their life-histories are mostly unknown and since numerous protozoa other than the members of this group often possess amoeboid stages. Four families.

Family 1. **Naegleriidae**

The members of the two genera placed in this family possess both amoeboid and flagellate phases (*diphasic*). In the former, the organism undergoes amoeboid movement by means of lobopodia and in the latter the body is more or less elongated. Binary fission seems to take place during the amoeboid phase. Thus these are diphasic amoebae, in which the amoeboid stage predominates over the flagellate.

Genus **Naegleria** Alexeieff. Minute flagellate stage with two flagella; amoeboid stage resembles Vahlkampfia (p. 527), with lobopodia; cytoplasm differentiated; vesicular nucleus with a

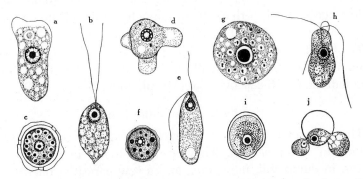

Fig. 185. a–c, trophozoite, flagellate phase and cyst (all stained) of *Naegleria gruberi*, ×750 (Alexeieff); d–f, similar stages of *N. bistadialis*, ×750 (Kühn); g–j, trophozoite, flagellate phase, cyst, and excystment of *Trimastigamoeba philippinensis*, ×950 (Whitmore).

large endosome; contractile vacuole conspicuous; food vacuoles contain bacteria; cysts uninucleate; free-living in stagnant water and often coprozoic. (Taxonomy and cytology, Rafalko, 1947; Singh, 1952.)

N. gruberi (Schardinger) (Fig. 185, *a-c*). Amoeboid stage 10-36μ by 8-18μ; cyst uninucleate; cyst wall with several apertures; flagellate stage 18μ by 8μ; stagnant water and often coprozoic. Chang (1958, 1958a) cultivated this amoeba in a buffered sucrose agar with its original bacterial associate, *Proteus mirabilis*. Flagella were formed by filamentous protrusion of endoplasm, one to three pairs of flagella being produced from a single protrusion. A body, held to be "parabasal body" was always present at the base of the protrusion or flagella which showed alternating light and dark bands. The transformation from flagellate to amoeboid stage took place by absorption of the flagella, the shedding of one or more flagella and the absorption of the rest, or by casting-off a small part of the body to which the flagella were attached.

N. bistadialis (Puschkarew) (Fig. 185, *d-f*). Similar in size; but cyst with a smooth wall.

Genus **Trimastigamoeba** Whitmore. Flagellate stage bears four flagella, two pairs each arising from a single blepharoplast (Bovee, 1959); vesicular nucleus with a large endosome; amoeboid stage small, less than 20μ in diameter; uninucleate cyst with a smooth wall; stagnant water.

T. philippinensis W. (Fig. 185, *g-j*). Amoeboid stage 16-18μ in diameter; clavate forms 30-40μ long; oval cysts 13-14μ by 8-12μ; flagellate stage 16-22μ by 6-8μ. (Morphology, Bovee, 1959.)

Family 2. **Amoebidae** Bronn

These amoebae do not have flagellate stage and are exclusively amoeboid (*monophasic*). They are free-living in fresh or salt water, in damp soil, moss, etc., and a few parasitic; one, two or many nuclei; contractile vacuoles in freshwater forms; multiplication by binary or multiple fission or plasmotomy: encystment common. (Genera, Leidy, 1879; Penard, 1902; taxonomy, Singh, 1952; nutrition and life-span, Danielli, 1959; nutrition, locomotion and pseudopods, Bovee, 1960; pinocytosis. Chapman-Andresen, 1963.)

Genus **Amoeba** Ehrenberg (*Proteus* Müller; *Amiba* Bory). Amoeboid; a vesicular nucleus, either with many spherical granules or with a conspicuous endosome; usually one contractile vacuole; pseudopodia are lobopodia, never anastomosing with one another; holozoic; in fresh, brackish or salt water. Numerous species. (Nomenclature, Schaeffer, 1926; Mast and Johnson, 1931; Kudo, 1952, 1959.)

A. proteus (Pallas) (Figs. 2, *e, f;* 26; 35, *b, c;* 47-49; 71; 186, *a, b*). Up to 600μ or longer in largest diameter; creeping with a few large lobopodia, showing longitudinal ridges; ectoplasm and endoplasm usually distinctly differentiated; typically uninucleate; nucleus discoidal but polymorphic; endoplasmic crystals truncate bi-pyramid, up to 4.5µ long (Schaeffer, 1916); nuclear and cytoplasmic divisions show a distinct correlation (p. 205); fresh water. (Cytology, Mast, 1926; Mast and Doyle, 1935, 1935a; Pappas, 1954; nuclear division, Chalkley, 1936; Liesche, 1938; electron microscope studies, Pappas, 1956; Roth *et al.*, 1960.)

A. discoides Schaeffer (Figs. 45, *g*; 186, *c*). About 400µ long during locomotion; a few blunt, smooth pseudopodia; crystals

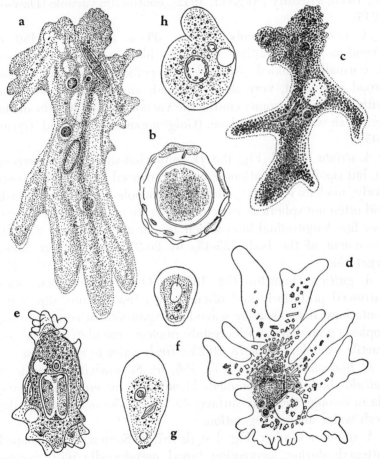

FIG. 186. a, b, *Amoeba proteus* (a, ×130 (Schaeffer), b, cyst (Doflein)); c, *A. discoides*, ×130 (Schaeffer); d, *A. dubia*, ×130 (Schaeffer); e, *A. verrucosa*, ×200 (Cash); f, *A. striata*, ×400 (Penard); g, *A. guttula*, ×800 (Penard); h, *A. limicola*, ×530 (Penard).

abundant, truncate bipyramidal, about 2.5μ long (Schaeffer); endoplasm with numerous coarse granules; fresh water.

A. *dubia* S. (Figs. 45, *h-l;* 186, *d*). About 400μ long; numerous pseudopodia flattened and with smooth surface; crystals, few large, up to 30μ long and of various forms among which at least four types are said to be distinct (Schaeffer); contractile vacuole one or more; fresh water. Nuclear division (Dawson *et al.*, 1935); viscosity (Angerer, 1942); contractile vacuole (Dawson, 1945).

A. *verruscosa* Ehrenberg (Figs. 35, *a, d-h;* 46, *a;* 186, *e*). Ovoid in general outline with wart-like expansions; body surface usually wrinkled, with a definite pellicle; pseudopodia short, broad and blunt, very slowly formed; nucleus ovoid, vesicular, with a large endosome; contractile vacuole; up to 200μ in diameter; fresh water among algae. (Golgi apparatus, Das and Tewari, 1955.)

A. *striata* Penard (Fig. 186, *f*). Somewhat similar to A. *verrucosa*, but small; body flattened; ovoid narrowed and rounded posteriorly; nucleus vesicular; contractile vacuole comparatively large and often not spherical; extremely delicate pellicle shows three or four fine longitudinal lines which appear and disappear with the movement of the body; 25-45μ by 20-35μ; fresh water among vegetation.

A. *guttula* Dujardin (Fig. 186, *g*). Ovoid during locomotion, narrowed posteriorly and often with a few minute, nipple-like dentations; movement by wave-like expansions of ectoplasm; endoplasm granulated, with crystals; nucleus vesicular; a single contractile vacuole; 30-35μ by 20-25μ; fresh water in vegetation.

A. *limicola* Rhumbler (Fig. 186, *h*). Somewhat similar to A. *guttula;* body more rounded; locomotion by eruption of cytoplasm through the body surface; 45-55μ by 35μ; nucleus vesicular; fresh water among vegetation.

A. *spumosa* Gruber (Fig. 2, *c, d;* 187, *a*). Somewhat fan-shaped; flattened; during locomotion broad pseudopodia with pointed end; temporary posterior region with nipple-like projections; a small number of striae become visible during movement, showing there is a very thin pellicle; endoplasm always vacuolated, the vacuoles varying in size (up to 30μ in diameter); vesicular nucleus with an endosome; 50-125μ long during locomotion; fresh water.

A. vespertilio Penard (Fig. 187, *b, c*.) Pseudopodia conical, comparatively short, connected at base by web-like expansions of

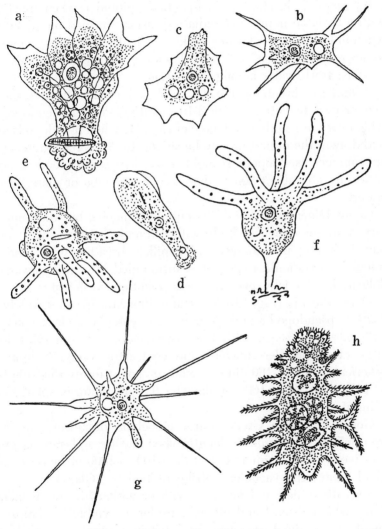

FIG. 187. a, *Amoeba spumosa,* ×400 (Penard); b, c, *A. vespertilio,* ×300 (Penard); d–f, *A. gorgonia,* ×400 (Penard); g, *A. radiosa,* ×500 (Penard); h, *Dinamoeba mirabilis,* ×250 (Leidy).

ectoplasm; endoplasm colorless, with numerous granules and food particles; a single vesicular nucleus with a large endosome; contractile vacuoles; 60-100μ long; fresh water. (Cannibalism, Lapage, 1922; contractile vacuole, Hyman, 1936; morphology and biology, Raabe, 1951.)

A. *gorgonia* P. (Fig. 187, *d-f*). Body globular when inactive with a variable number of radiating "arms," formed on all sides; when in locomotion, clavate; nucleus vesicular, with a large endosome; rounded forms 40-50μ in diameter; clavate individuals up to 100μ; fresh water among vegetation.

A. *radiosa* Ehrenberg (Fig. 187, *g*). Small, usually inactive, globular or oval in outline; with 3-10 radiating slender pseudopodia which vary in length and degree of rigidity; when pseudopods are withdrawn, the organism may be similar to A. *proteus* in general appearance; pseudopods straight, curved or spirally coiled; size varies, usually about 30μ in diameter, up to 120μ or more; fresh water.

Genus **Dinamoeba** Leidy. Essentially Amoeba, but the temporary posterior region of body with retractile papillae; body surface including pseudopods and papillae, bristling with minute spicules or motionless cils; often surrounded by a thick layer of delicate hyaline jelly, even during locomotion; fresh water.

D. *mirabilis* L. (Fig. 187, *h*). Oval to limaciform; spheroid when floating; pseudopodia numerous, conical; ectoplasm clear, usually with cils; endoplasm with food vacuoles, oil (?) spherules and large clear globules; nucleus and contractile vacuole obscure; spherical forms 64-160μ in diameter; creeping forms 152-340μ by 60-220μ; cyst about 160μ in diameter (Groot, 1936); in sphagnous swamp.

Genus **Pelomyxa** Greeff. Large amoeboid organisms, ranging from 0.5 to 4 or 5 mm. in length when clavate and moving progressively; nuclei numerous less than 100 to 1000 or more; many small contractile vacuoles; refringent bodies ("Glanzkörper") of various dimension and number; with or without bacterial inclusions (which Penard and others consider as symbiotic); holozoic on plant or animal organisms or detritus; plasmotomy simple or multiple; in fresh water. Several species (Kudo, 1946). (Nomenclature, Schaeffer, 1926; Mast and Johnson, 1931; Rice, 1945; Kudo,

1946, 1952, 1957, 1959; Wilber, 1947; Andresen, 1956; pinocytosis and food vacuoles under electron microscope, Roth, 1960).

P. palustris G. (*P. villosa* Leidy) (Fig. 188, *a*). Large; 2-3 mm. or larger in diameter; sluggish, with usually one broad pseudopodium; undifferentiated cytoplasm with many nuclei and various inclusions such as fragments of plant bodies, numerous small sand particles, etc., which brings about opacity and dark coloration of body; in addition bacteria (*Cladothrix pelomyxae* Veley, *Myxococcus pelomyxae* Keller and *Bacterium parapelomyxae* Keller) occur in the cytoplasm which some observers consider as symbionts; cyst with two to three envelopes (Stolc, see Kudo, 1951);

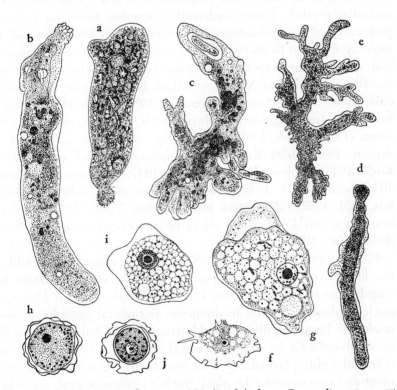

FIG. 188. a, *Pelomyxa palustris*, ×160 (Leidy); b, c, *P. carolinensis*, ×45 (Kudo) (b, an individual in locomotion; c, feeding form); d, e, *P. illinoisensis*, ×40 (Kudo) (d, an individual in locomotion; e, a more or less stationary animal); f, *Vahlkampfia patuxent*, ×660 (Hogue); g, h, *Acanthamoeba castellanii*, ×1270 (Hewitt); i, j, *A. hyalina*, ×840 (Dobell).

feeds on plant and inorganic debris; polysaprobic in still stagnant water, buried in mud. Central Europe, Great Britain and North America. (Distribution, Kudo, 1947, 1957; morphology, Greeff, 1874; Leidy, 1879; Hollande, 1945; Leiner, Wohlfeil and Schmidt, 1954, Kudo, 1957; bacteria in the cytoplasm, Leiner and Wohlfeil, 1953, 1954; glycogen, Waldner, 1956; locomotion, Okada, 1930a; Mast, 1934; plasmotomy, Okada, 1930; Kudo, 1957; laboratory cultures, Hollande, 1945; Kudo, 1957.)

P. carolinensis Wilson (Figs. 69; 74; 188, *b*, *c*). Monopodal forms 1-5 mm. long; polypodal forms 1-2 mm. in diameter; locomotion active; nuclei up to 1000 or more, circular in front view, about 20μ in diameter and ellipsoid in profile; fluid and food vacuoles, crystals, many contractile vacuoles; feeds on various protozoa and invertebrates; easily cultivated in laboratory; plasmotomy into two to six individuals; nuclear division simultaneous and synchronous; experimental plasmogamy; no encystment in the Illinois stock, but New Jersey stock is said to encyst (Musacchia, 1950); North America. Prescott (1956) obtained clone cultures from two fragments, each of which contained only one nucleus. (Distribution, Kudo, 1946; morphology, Wilson, 1900; Andresen, 1942, 1956; Kudo, 1946; plasmotomy, Schaeffer, 1938; Kudo 1949; nuclear division, Kudo, 1947; locomotion, Wilber, 1946; permeability, Belda, 1942-43; effect of x-irradiation, Daniels, 1951, 1952, 1952a, 1954; mitochondria, Torch, 1955; respiration, Pace and Frost, 1952; effect of surface-active agents, Nardone *et al.*, 1956.)

P. illinoisensis Kudo (Fig. 188, *d*, *e*). The organism resembles the last-named species, but much smaller in size; 500-1000μ in length; clavate forms seldom exceed 1.5 mm.; several hundred nuclei, spherical, 14-16μ in diameter; peripheral granules of the nuclei are large and often discoid, irregularly distributed; crystals occur abundantly in all physiological conditions; chalky white in reflected light; plasmotomy into two to five daughters; encystment and excystment take place freely in cultures; cysts measure 250-350μ in diameter with usually two membranes, a multinucleate amoeba emerges from a cyst after several weeks (Kudo, 1950, 1951). (Other species of Pelomyxa, Kudo, 1951; nuclear division, McClellan, 1959; x-irradiation studies, Daniels, 1955; Daniels and Roth, 1961.)

Genus **Vahlkampfia** Chatton and Lalung-Bonnaire. Small amoebae; vesicular nucleus with a large endosome and peripheral chromatin; with polar caps during nuclear division; snail-like movement with one broad pseudopodium; cysts with a perforated wall; fresh water or parasitic. (Nuclear division, Jollos, 1917.)

V. limax Dujardin. 30-40μ long; fresh water.

V. patuxent Hogue (Fig. 188, *f*). In the alimentary canal of the oyster; about 20μ long during the first few days of artificial cultivation, but later reaching as long as 140μ in diameter; ordinarily one large broad fan-shaped pseudopodium composed of the ectoplasm; in culture, pseudopodium composed of the ectoplasm; in cuture, pseudopodium-formation eruptive; holozoic on bacteria; multiplication by fission or budding; encystment rare; cysts uninucleate.

Genus **Hartmannella** Alexeieff. Small amoebae, with moderately or well-developed ectoplasm; vesicular nucleus with a large endosome; mitotic figure ellipsoidal or cylindrical, without polar caps. Cysts rounded; wall smooth or slightly wrinkled in one species. Several species. Volkonsky (1933) distinguishes four groups. (Species and morphology, Singh, 1952; nuclear division, Jollos, 1917.)

H. hyalina Dangeard. 20-25μ in diameter; ectoplasm well developed; endoplasm vacuolated; slender pseudopodia extend in different directions; Hartmann and Chagas observed a centriole in the endosome.

H. astromyxis Ray and Hayes. 18-42μ in diameter; pseudopods variable; one to several contractile vacuoles; cysts, 14-32μ in diameter, circular and biconvex; the inner wall stellate with 3-9 (5-7) drawn-out points; during encystment, endosomic materials are extruded into the cytoplasm (Ray and Hayes, 1954). Fresh water.

Genus **Acanthamoeba** Volkonsky. Small amoebae similar to *Hartmannella;* ectoplasm is not well developed; mitotic figure at the end of metaphase, a straight or concave spindle with sharply pointed poles. Cysts enveloped by two membranes, the outer envelope being highly wrinkled and mammillated. Several species. Neff (1957) obtained axenic cultures of a species from soil. The presence of Acanthamoeba in tissue cultures of trypsinized monkey-kidney cells has been reported by several observers since 1957 (Chi *et al.*, 1959).

A. *castellanii* (Douglas) Fig. 188, *g*, *h*). In association with fungi and certain bacteria; Hewitt obtained the organism from agar cultures of sample soil taken from among the roots of white clover; co-existing with yeast-like fungi, *Flavobacterium trifolium* and *Rhizobium* sp.; 12-30µ in diameter; some cysts are said to remain viable at 37°C. for 6 days.

A. *hyalina* (Dobell and O'Connor) (Fig. 188, *i*, *j*). According to Volkonsky, the organism described by Dobell and O'Connor as *Hartmannella hyalina*, is transferred to this genus. Small amoeba; 9-17µ in diameter when rounded; a single contractile vacuole; binary fission; mitotic figure a sharply pointed spindle. Cysts spherical; 10-15µ in diameter; with a smooth inner and a much wrinkled outer wall; easily cultivated from old faeces of man and animals; also in soil and fresh water.

Genus **Sappinia** Dangeard. With two closely associated nuclei.

S. *diploidea* (Hartmann and Nägler). Coprozoic in the faeces of different animals; pseudopodia short, broad, and few; highly vacuolated endoplasm with two nuclei, food vacuoles, and a contractile vacuole; surface wrinkled; the nuclei divide simultaneously; during encystment, two individuals come together and secrete a common cyst wall; two nuclei fuse so that each individual possesses a single nucleus; finally cytoplasmic masses unite into one; each nucleus gives off reduction bodies (?) which degenerate; two nuclei now come in contact without fusion, thus producing binucleate cyst supported by a stalk.

Noble (1958) found this coprozoic organism in the elk faeces; average diameter 45µ in rounded individuals; when extended it may measure up to 60µ long. Some authors consider this a member of Mycetozoida.

Family 3. **Endamoebidae** Calkins

Parasitic amoebae; the trophozoites are usually relatively small and live in the alimentary canal of various host animals; multiplication by binary fission; encystment common; generic differentiation is based upon the nuclear characteristics. (Nomenclature, Dobell, 1919, 1938; Hemming, 1951, 1954.)

Genus **Endamoeba** Leidy (1879). Nucleus spheroidal to ovoid; membrane thick; in life, filled with numerous granules of uniform dimensions along its peripheral region; upon fixation, a fine chromatic network becomes noticeable in their stead; central portion coarsely reticulated; with several endosomes between the

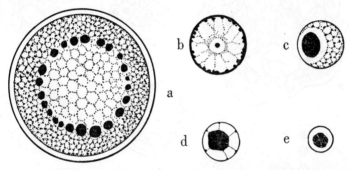

FIG. 189. Diagram showing the stained nuclei of the trophozoites of five genera of parasitic amoebae. a, Endamoeba; b, Entamoeba; c, Iodamoeba; d, Endolimax; e, Dientamoeba.

two zones (Fig. 189, *a*); in some, cytoplasm becomes prominently striated during locomotion; in the intestine of invertebrates.

Endamoeba was once considered a synonym of Entamoeba by the International Commission of Zoological Nomenclature in its opinion 99 (1928), but the Commission through opinion 312 (1954) now recognizes Endamoeba and Entamoeba as different and distinct genera.

E. blattae Bütschli (Fig. 190). In the colon of cockroaches; 10-150μ in diameter; rounded individuals with broad pseudopodia, show a distinct differentiation of cytoplasm; elongated forms with a few pseudopodia, show ectoplasm only at the extremities of the pseudopods; endoplasm of actively motile trophozoites shows a distinct striation, a condition not seen in other amoebae; fluid-filled vacuoles occur in large numbers; amoebae feed on starch grains, yeast cells, and bacteria, all of which coexist in the host organ; cysts, 20-50μ in diameter, commonly seen in the colon contents, with often more than sixty nuclei. The life-cycle of this amoeba is still unknown. Mercier (1909) held that when the multinucleate cysts gain entrance to the host intestine through its mouth, each of the cyst-nuclei becomes the center of a gamete; when the cyst-membrane ruptures, the gametes are set

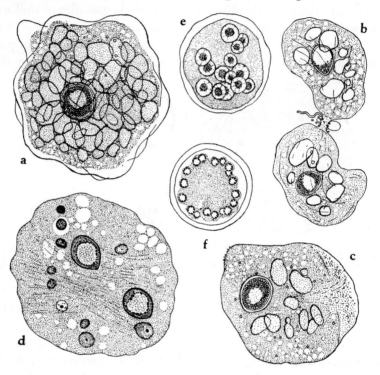

Fig. 190. *Endamoeba blattae.* a–c, trophozoites in life, ×530; d, a stained binucleate amoeba; e, f, stained and fresh cysts, ×700 (Kudo).

free and anisogamy takes place, resulting in formation of numerous zygotes which develop into the habitual trophozoites. (Morphology, Leidy, 1879; Kudo, 1926; Morris, 1936; Meglitsch, 1940.)

E. *disparata* Kirby. In colon of *Microtermes hispaniolae;* 20-40µ long; active; xylophagous (Kirby, 1927).

E. *majestas* K. (Fig. 191, *a*). In the same habitat; 65-165µ in diameter; many short pseudopodia; cytoplasm filled with food particles (Kirby, 1927).

E. *simulans* K. (Fig. 191, *b*). In the gut of *Microtermes panamaensis;* 50-150µ in diameter.

E. sabulosa K. In the same habitat; small 19-35µ in diameter.

E. pellucida, E. granosa, E. lutea and *E. suggrandis* were described from the colon of *Cubitermes* sp. of Africa (Henderson, 1941).

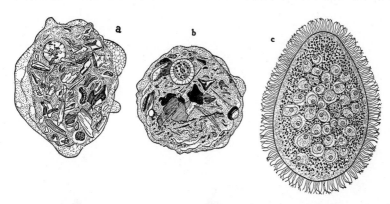

FIG. 191. a, *Endamoeba majestus,* ×420 (Kirby); b, *E. simulans,* ×420 (Kirby); c, *Entamoeba paulista* in Zelleriella, ×290 (Stabler and Chen).

Genus **Entamoeba** Casagrandi and Barbagallo (1895) (*Poneramoeba,* Lühe). Nucleus vesicular, with a comparatively small endosome, located in or near the center and with varying number of peripheral nonchromatinic granules attached to the nuclear membrane (Fig. 189, *b*); chromatin in the endosome and in peri-endosomal region. The genus was established by the two Italian authors who were unaware of the existence of the genus *Endamoeba* (p. 529). Numerous species in vertebrates and invertebrates; one species in protozoa; one species free-living.

E. histolytica Schaudinn (1903) (Figs. 192, 193). The trophozoite is an active amoeba and measures 7-35 (9-20) µ in diameter; cytoplasm usually well differentiated; eruptive formation of large lobopodia, composed largely of ectoplasm; when fresh, active monopodal progressive movement; the vesicular nucleus appears in life as a ring, difficult to recognize; food vacuoles contain erythrocytes, tissue cell fragments, leucocytes, etc.; stained nucleus shows a membrane, comparatively small peripheral granules, a centrally located small endosome and an indistinct network with a few scattered chromatin granules. The trophozoite multiplies by binary fission. The amoeba lives in the lumen and in the tissues of the wall of the colon, and brings about characteristic ulceration of the colon which is typically accompanied by symptoms of *amoebic dysentery.* Through the portal vein, the amoeba may in-

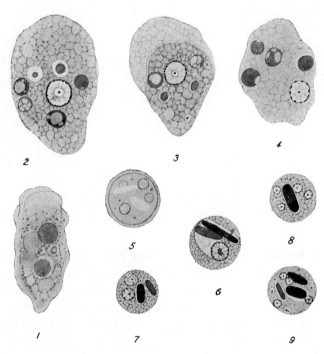

FIG. 192. *Entamoeba histolytica,* ×1150 (Kudo). 1, a living trophozoite; 2–4, stained trophozoites; 5, a fresh cyst; 6–9, stained cysts.

vade the liver in which it produces abscess, and other organs such as lung, brain, testis, etc. The infection in these organs is referred to as *amoebiasis.* (Ingestion of erythrocytes, Kradolfer and Gross, 1958; biology of amoebic hepatitis, Beheyt *et al.,* 1961.)

Under certain circumstances not well understood, the amoebae remain small after division. Such amoebae are sluggish and known as the precystic forms. The precystic amoeba secretes presently a resistant wall and becomes encysted. The highly re-

fractile cyst is spherical and measures 5-20µ in diameter. At first
it contains a single nucleus which divides twice. The mature **cyst**
contains four nuclei. In addition the cyst contains diffused glycogen and elongated refractile rod-like bodies with rounded extremities which stain deeply with haematoxylin (hence called *chromatoid bodies*). These inclusions are absorbed and disappear as the
cyst matures. No further changes take place in the cyst as long as
it remains outside the host's intestine. The trophozoites are found
in dysenteric or diarrhoeic faeces, but formed faeces usually contain cysts. The cyst is the stage by which the organism begins its
life in a new host. (Encystment *in vitro,* Fukushima, 1960.)

The life-cycle of *Entamoeba histolytica* in human host is unknown. The amoeba has, however, been cultivated in vitro by numerous investigators since the first successful cultivation by
Boeck and Drbohlav (1925) (p. 1067). The excystment of cysts and
metacystic development have also been observed and studied
especially by Dobell (1928) and Cleveland and Sanders (1930) in
cultures. Snyder and Meleney (1941) found that bacteria-free
cysts usually excyst when suspended in various media with living
bacteria and in the absence of bacteria, excystment was observed
only in the presence of the reducing agents, cysteine or neutral-
ized thioglycollic acid or under conditions of reduced oxygen tension. According to Dobell, in the process of excystation, a single
tetranucleate amoeba emerges from a cyst through a minute pore
in the cyst wall. The tetranucleate metacystic amoeba produces a
new generation of trophozoites by a diverse series of nuclear and
cytoplasmic divisions (Fig. 193) which result in production of
eight uninucleate amoebulae. These amoebulae are young trophozoites which grow into larger ones. No sexual phenomena
have been observed during these changes. It is supposed that
when viable cysts reach the lower portion of the small intestine or
the colon, the changes stated above take place in the lumen and
the young uninucleate amoebulae initiate an infection.

While the description of *Entamoeba histolytica* given above
applies in general, diversities in dimensions of trophozoites and
cysts, and in pathogenicity in human host as well as in experimental animals have been reported. A number of observers are

inclined to think that there are several varieties or races of this amoeba, as has already been mentioned (p. 270).

It is generally believed that the large races are histozoic and therefore, pathogenic, while the small races are coelozoic and nonpathogenic. Faust and Read (1959) cultured a large race of the amoeba from a symptomless carrier in Balamuth medium and

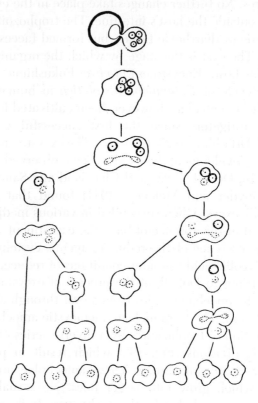

Fig. 193. Diagram showing excystment and a common way by which a metacystic amoeba of *Entamoeba histolytica* divides into 8 uninucleate amoebulae (Dobell).

found the organisms ingested starches from various foods. The two authors were inclined to think that the organisms which produce amylase and maltase obtain nourishment by ingesting starch in green plantain and yucca, and do not need to invade the intestinal mucosa to survive.

Burrows (1957) is of the opinion that the small (5-11μ) race is non-pathogenic and is different from the large (20-40μ) histozoic forms, and proposed to designate the small race as *E. hartmanni* Prowazek (1912). Goldman (1959) agrees with this view, as he found antigenic difference between the two organisms as detected by microfluorimetry.

Entamoeba histolytica, is commonly known as the "the dysentery amoeba," was first definitely recognized by Lösch in Russia in 1875. It is now known to be widely distributed in tropical, subtropical and temperate regions alike, although it is more prevalent in warmer regions. The incidence of infection depends mainly on the sanitary conditions of the community, since the cysts of the organism are voided from host in faeces. Faecal examinations which have been carried on by numerous investigators in different countries of the world, reveal that the incidence of infection is as high as over 50 per cent in some areas. According to Craig (1934), 49,336 examinations made by many observers in various parts of the United States show that the infection rate varied from 0.2 to 53 per cent, averaging 11.6 per cent, which justifies Craig's (1926) earlier estimate that about 10 per cent of the general population harbor this protozoan. An acute infection by *E. histolytica* is accompanied by dysentery, while in chronic cases or in convalescence, the host may void infectious cysts without suffering from the infection himself. Such a person is known as a **cyst-carrier** or **-passer.** Moreover, this amoeba parasitizes cats and dogs and therefore these animals may serve as carriers.

The trophozoite if voided in faeces perish in a comparatively short time. The dissemination of infection is thus exclusively carried on by the cyst. Viable cysts may be transmitted (1) by contamination of food through contact with contaminated water or through unsanitary habit of food handlers who are cyst-carriers; (2) by droppings of flies and cockroaches which, as noted below, contain viable cysts for a comparatively long time after feeding on faeces containing cysts and by soiled appendages of these insects which may directly transfer the cysts to food by walking on it; and (3) by contaminated water in which the cysts live considerably longer than in faeces (p. 537). Longevity of trophozoites outside the human host (Simitch and Petrovitch, 1953).

The seriousness of water-borne infection in crowded areas is easily realized when one recalls the outbreak (some 1400 cases) of amoebic dysentery and amoebiasis which originated in Chicago in 1933, where defective plumbing in certain establishments contaminated the water system with the cysts of *Entamoeba histolytica* (Bundesen *et al.*, 1936) and the development of some 100 cases of amoebic dysentery among firemen who drank contaminated water in connection with the 1934 fire of the Union Stockyards in Chicago (Hardy and Spector), although in the latter instance, some workers believe that severe amoebic infections may have resulted from already existing dormant infections aided by the newly formed association with bacteria. Another example is the outbreak of amoebiasis among the employees of a plant as the water was contaminated with cysts. More than half of some 1500 persons became infected, including thirty-one clinical cases and four fatalities (Brooke *et al.* 1955). (Bacteria in culture, Buonomini and Mignani, 1960.)

The cysts remain viable for a considerable length of time outside the human intestine, if environmental conditions are favorable. Since information regarding the viability and longevity of the cyst is highly important from the epidemiological standpoint, many papers have dealt with it. In testing the viability of the cyst, the following two tests have been used by the majority of investigators.

(a) Eosin-staining test. Kuenen and Swellengrebel (1913) first used a dilute solution of eosin (1:1000). It has since been used by Wenyon and O'Connor, Root, Boeck, and many others. Solutions used vary from 1:10,000 (Root) to 1:100 (Boeck). A small amount of fresh cyst-containing material and a drop of eosin solution are mixed on a slide, then dead cysts will appear stained reddish under the microscope, while living cysts remain unstained. Whether or not unstained cysts might be dead or uninfectious is unknown. But as Wenyon and O'Connor wrote, "if we accept the eosin test as a criterion and regard all unstained cysts as living, the error in judgment will be on the safe side." Root found neutral red in 1:10,000 dilution to give a slightly larger proportion of stained cysts than eosin. Frye and Meleney's (1936) comparative study leads one to look upon this method as a fairly dependable one.

(b) Cultivation test. Improved cultural technique now brings about easily excystment of viable cysts in a proper culture medium. For example, Yorke and Adams (1926) obtained in twenty-four hours "a plentiful growth of vegetative forms" from cysts in Locke-egg-serum medium (p. 1067). Snyder and Meleney (1941) note recently that the excystation does not take place in various culture media unless living bacteria were added or oxygen concentration of the media was decreased. Animal infection method has not been used much, as experimental animals (cats) show individual difference in susceptibility. Some of the published results are summarized below. The testing method used is indicated by: *a* for eosin test or *b* for cultivation test and is given after the name of the investigators.

1. Cyst in faeces kept in a covered container. All cysts disappeared in three days at 37°C.; at 27–30°C. half of the cysts found dead by the fourth and all dead by the ninth day (Kuenen and Swellengrebel; *a*). Alive for three weeks (Thomson and Thomson; *a*). Remain unchanged for several weeks if kept "cool and moist" (Dobell). All dead within ten days at 16–20° or 0°C. (Yorke and Adams; *b*).

2. Cysts kept in water emulsion. All alive on the ninth, but almost all dead on the thirteenth day (Kuenen and Swellengrebel; *a*). Viable for twenty-five days (Thomson and Thomson; *a*). Cysts in running water for fifteen days, excysted in pancreatic juice (Penfold, Woodcock and Drew). Viable for thirty days (Wenyon and O'Connor; *a*); for five week (Dobell); for 153 days (Boeck; *a*). Alive for ten and seventeen days at 16–20° and 0°C. respectively (Yorke and Adams; *b*); for three, ten, thirty, and ninety days at 30°, 20°, 10° and 0°C. respectively (Chang and Fair; *b*).

3. Cysts in relation to high temperatures. Cyst are killed at 68°C. in five minutes (Boeck; *a*); at 50°C. in five minutes (Yorke and Adams; *b*). Dipping in boiling water for five to ten seconds kills the cysts (Kessel; *a*).

4. Cysts in relation to desiccation. Desiccation kills cysts instantly (Kuenen and Swellengrebel; Wenyon and O'Connor, Do-

bell, etc.). Therefore, the cysts carried in dust are most probably not viable under ordinary circumstances.

5. Cysts in relation to chemicals.

HgCl₂. 0.1 per cent solution kills cysts in four hours (Kuenen and Swellengrebel; *a*); kills readily (Lin; *b*). 1:2500 solution kills cysts in thirty minutes at 20–25°C. (Yorke and Adams; *b*).

Creolin. 1:250 solution kills cysts in five to ten minutes (Kuenen and Swellengrabel; *a*).

Alcohol. 50 per cent alcohol kills cysts immediately (Kuenen and Swellengrebel; *a*); in one hour (Kessel; *a*).

Formaldehyde. Cysts treated in 1 per cent solution for four hours were apparently dead, though not stained with eosin (Wenyon and O'Connor). 0.5 per cent solution kills cysts in thirty minutes at 20–25° or 37°C. (Yorke and Adams; *b*).

Cresol. 1:20, 1:30, and 1:100, killed the cysts immediately, in one minute and in thirty minutes respectively (Wenyon and O'Connor; *a*).

Phenol. 1:40 and 1:100 killed cysts in fifteen minutes and seven hours respectively (Wenyon and O'Connor; *a*). 1 per cent solution of phenol or lysol kills cysts in thirty minutes at 20–25° or 37°C. (Yorke and Adams; *b*).

HCl. 7.5 per cent solution at 20–25°C. and 5 per cent at 37°C. kill the cysts in thirty minutes (Yorke and Adams; *b*).

NaOH. 2.5 per cent solution kills cysts in thirty minutes at 20–25° or 37°C. (Yorke and Adams; *b*).

Chlorine. 1:10,000 solution did not have any effect on cysts after several hours (Wenyon and O'Connor; *a*). 0.2 per cent and 0.5 per cent solutions kill the cysts in seven days and seventy-two hours respectively (Kessel; *a*). 0.5 per cent and 1 per cent solutions kill the cysts in thirty-six to forty-eight and twelve to twenty-four hours respectively (Lin; *b*). 1/64 of a saturated solution of chlorine (about 0.7 weight per cent) at 20–25°C. and 1/320 solution at 37°C. killed the cysts in thirty minutes (Yorke and Adams; *b*). Exposure to the residual chlorine five, eight and even ten parts per million for thirty minutes allowed cysts to remain viable (Becker *et al.*). Thus the cysts of *E. histolytica* are resistant to chlorinated water far above the concentration which is used ordinarily in water treatment.

Potassium permanganate. 2 per cent solution kills the cysts in three days (Kessel; *a*). 1:500 solution kills cysts in twenty-four to forty-eight hours (Lin; *b*). 1 per cent solution does not kill cysts at 20–25° or 37°C. in thirty minutes (Yorke and Adams; *b*).

Emetin hydrochloride and yatren. 5 per cent solutions of the two drugs did not have any effects upon cysts at 20–25° or 37°C. in thirty minutes (Yorke and Adams; *b*).

Antibiotics. The majority of antibiotics appear to inhibit the growth of bacteria, which results in the death of the amoeba in culture. Prodigiosin, however, according to Balamuth and Brent (1950), kills the amoebae when added in the dilution of 1:400,000, while bacterial flora, oxidation-reduction potentials and pH are not affected by it.

6. Cysts in relation to passage through the intestine of insects. Wenyon and O'Connor found that the cysts of *E. histolytica* survived as long as twenty-four hours in the intestines of flies, *Musca domestica*, Calliphora, and Lucilia, and living cysts were voided for sixteen hours after feeding on faecal material containing cysts. Roubaud using *Musca domestica*, found also unaltered cysts for over twenty-four hours (but rarely after forty hours) after taking the cysts in its gut, and if a fly drowned in water, the cysts remained viable for about a week. Root (1921) using *Musca domestica, Calliphora erythroecephala* (and *Fannia canicularis, Lucilia caesar,* and *Chrysomyia macellaria*) found that about half the cysts were dead after fifteen hours and last living cysts were found after forty-nine hours in the intestines of these flies after feeding on cyst-containing material, and that when the flies which ingested cysts were drowned in water, about half the cysts were found dead in three days and last living cysts were noticed on the seventh day. Frye and Meleney (1932) found cysts in the intestines of flies which were caught in four of twelve houses where infected subjects lived.

Macfie (1922) reported that the cysts of *Entamoeba histolytica* he observed in the intestine of *Periplaneta americana* appeared unharmed. Tejera (1926) reports successful experimental infection in two kittens that were fed on the droppings of cockroaches (sp. ?) caught in a kitchen, which contained cysts resembling those of *E. histolytica*. Frye and Meleney (1936) observed that the cysts passed through the intestine of *Periplaneta americana* in as early as ten to twelve hours and remained in the intestine for as long as seventy-two hours, after feeding on experimental material. Cysts which stayed in the cockroach intestine for forty-eight hours gave good cultures of trophozoites in egg-horse-serum-Ringer medium. (Monoxenic cultivation with *Trypanosoma cruzi*, Phillips and Rees, 1950; axenic culture, Diamond, 1961; electron microscope study, Miller *et al.*, 1961; diagnostic technique Brooke, 1958; treatment, Powell *et al.*, 1959; Cooper *et al.*, 1958; Brooke *et al.*, 1958; Young and Freed, 1956, 1957.)

In addition to *E. histolytica*, there are now known four other intestinal amoebae living in man. They are *E. coli, Endolimax nana, Iodamoeba bütschii* and *Dientamoeba fragilis*. In Table 7 are given the characteristics necessary for distinguishing *E. histolytica* from the other four intestinal amoebae.

E. coli (Grassi) (Fig. 194). The trophozoite measures 15-40μ in diameter; average individuals 20-35μ; cytoplasm not well differentiated; movement sluggish; endoplasm granulated, contains micro-organisms and faecal debris of various sizes in food vacuoles; erythrocytes are not ingested, though in a few cases (Tyzzer and Geiman) and in culture (Dobell, etc.), they may be taken in as food particles (see below); nucleus, 5-8μ in diameter, seen in vivo; compared with *E. histolytica*, the endosome is somewhat large (about 1μ in diameter) and located eccentrically; peripheral granules more conspicuous. The precystic form, 10-30μ in diameter, resembles that of *E. histolytica*. Separation of the two species of amoebae by this stage is ordinarily impossible.

The cyst is spherical or often ovoid, highly refractile; 10-30μ in diameter; immature cyst contains one, two or four nuclei, one or more large glycogen bodies with distinct outlines, but comparatively small number of acicular, filamentous or irregular chromatoid bodies with sharply pointed extremities; when mature the cyst contains eight nuclei and a few or no chromatoid bodies. The trophozoites and small number of cysts occur in diarrhoeic or semiformed faeces and the formed faeces contain mostly cysts.

The cysts remain viable up to sixteen days in aerated tap water kept at 8°C. (Rendtorff and Holt, 1954). Rendtorff (1954) fed twenty-six volunteers the cysts varying in number up to 10,000. Eight acquired infections. The prepatent periods of infections

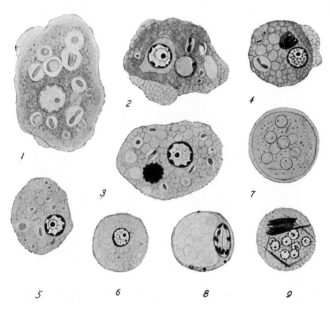

Fig. 194. *Entamoeba coli,* ×1150 (Kudo). 1, a living amoeba; 2–5, stained trophozoites; 3, an amoeba infected by *Sphaerita;* 6, a precystic amoeba; 7, a fresh cyst; 8, a stained young cyst with a large glycogen vacuole; 9, a stained mature cyst.

ranged from six to twenty-two days (average ten days). The eight men who became infected were kept under observation for 68-162 days, during which no spontaneous cure was seen.

This amoeba lives in the lumen of the colon and does not enter the tissues of the wall. As noted above, it has been observed in a few instances to ingest erythrocytes, but there is no evidence to show that it takes them in from living tissues. This amoeba is therefore considered a commensal. The abundant occurrence of the trophozoite in diarrhoeic faeces is to be looked upon as a re-

Table 7.—*Differential diagnosis of the intestinal amoebae of man*

	Entamoeba histolytica	*Entamoeba coli*	*Iodamoeba bütschlii*	*Endolimax nana*	*Dientamoeba fragilis*
Trophozoite					
1. Living specimens					
a. Diameter	7–35μ	10–40μ	6–25μ	6–18μ	4–18μ
b. Movement	Active progressive movement; eruptive formation of pseudopodia	Less active amoeboid movement	Less active amoeboid movement	Progressive movement	Progressive movement
c. Cytoplasm	Hyaline; erythrocytes tissue cells, taken in as food	Granulated; bacteria, yeasts, faecal debris in food vacuoles	Granulated; bacteria in food vacuoles	Hyaline; bacteria in food vacuoles	Hyaline; bacteria in food vacuoles
d. Nucleus	Faintly visible ring	Ring of coarse granules	Faintly seen	Rarely seen	Faintly seen
2. Stained specimens					
a. Nucleus	Fig. 192, 2–4	Fig. 194, 2–5	Fig. 197, 2–5	Fig. 198, b	Fig. 199, c, d
b. Inclusions	Erythrocytes, fragments of tissue cells	Bacteria, faecal debris, etc.	Bacteria	Bacteria	Bacteria
Cyst					
1. Living specimens					
a. Form	Spherical; circular in outline	Circular, often oval	Of various forms	Often oval to ellipsoid	Unseen
b. Diameter	5–20μ	10–30μ	6–15μ	5–12μ	—
2. Lugol-treated specimens	Cytoplasm greenish yellow; glycogen diffused; 1, 2, or 4 nuclei	Cytoplasm yellowish brown; glycogen body often big; 1, 2, 4, or 8 nuclei	Cytoplasm yellowish; large glycogen body sharply outlined; 1 nucleus	Cytoplasm greenish yellow; glycogen scanty, diffused; 1, 2, or 4 nuclei	—
3. Stained specimens	1, 2, or 4 nuclei; chromatoid bodies with rounded ends	1, 2, 4, or 8 nuclei; chromatoid bodies few, acicular or irregular with pointed ends	One nucleus; conspicuous glycogen vacuole; no chromatoid body	1, 2, or 4 nuclei; chromatoid bodies very small if present	—

sult and not the cause of the intestinal disturbance. This amoeba is of common occurrence and widely distributed throughout the world.

Nothing is known about its life-cycle in the human intestine. Cultivation of cysts in vitro indicates, according to Dobell (1938), the following changes: The cyst content usually emerges as a single multinucleate amoeba through a large opening in the cyst wall. Prior to or during the emergence, the amoeba may divide. Normal mature cysts "frequently lose" one to four of their original eight nuclei before germination, thus becoming "infranucleate" (with 4-7 nuclei). Unlike in *E. histolytica*, there is no nuclear division in the metacystic stages. By a series of binary divisions with random nuclear distribution, uninucleate amoebulae are finally produced. These are young amoebae which develop into large trophozoites. Here also, there is no sexual phenomenon in the life-cycle. A similar organism was reported from the intestine of Sudanese lizards (Neal, 1954). (Nomenclature and morphology, Dobell, 1919, 1938.)

E. gingivalis Gros (*E. buccalis* Prowazek) (Fig. 195). This amoeba lives in carious teeth, in tartar and debris accumulated around the roots of teeth, and in abscesses of gums, tonsils, etc. The tro-

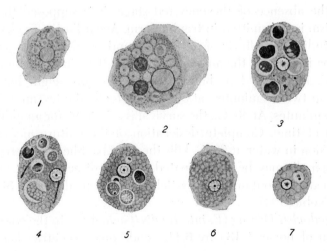

Fig. 195. *Entamoeba gingivalis*, ×1150 (Kudo), 1, 2, living amoebae; 3–7 stained amoebae.

phozoite is as active as that of *E. histolytica;* 8-30μ (average 10-20μ) in diameter; cytoplasm well differentiated; monopodal progressive movement in some individuals; endoplasm hyaline, but vacuolated, and contains ordinarily a large number of pale greenish bodies (which are probably nuclei of leucocytes, pus cells or other degenerating host cells) and bacteria in food vacuoles; nucleus, 2-4μ in diameter, appears as a ring; when stained it shows a small central endosome and small peripheral granules closely attached to the membrane. Stabler (1940) observed five chromosomes during binary fission. Encysted forms have not been observed in this amoeba. Kofoid and Johnstone (1930) reported having seen the same organism in the mouth of monkeys (Rhesus and Cynomolgus) from southeast Asia. (Effects of penicillin on culture, Clayton and Ball, 1954.)

E. gingivalis is the very first parasitic amoeba that has become known to man. Gros (1849) found it in Russia in the tartar on the surface of the teeth. Some observers maintain that this amoeba is the cause of pyorrhoea alveolaris, but evidence for such an assumption seems to be still lacking. It has been found in the healthy gums and even in false teeth (Lynch). Therefore, it is generally consdiered as a commensal. It is widely distributed and of common occurrence.

In the absence of the encysted stage, it is supposed that the organism is transmitted in trophic form. According to Kock (1927) who studied the effects of desiccation and temperatures upon the amoeba in culture, the amoeba is killed at 0°C. in eighteen hours, at 5°C. in twenty-four hours, at 10°C. in forty-eight hours, at 45°C. in twenty minutes, at 50°C. in fifteen minutes, an at 55°C. in two minutes. At 40°C., the survival is said to be for an indefinite length of time. Complete desiccation of the culture medium or immersion in water at 60°C. kills the amoeba. She considered that *E. gingivalis* may be disseminated both by direct contact and by intermediate contaminated articles. (Nuclear division, Stabler, 1940; Noble, 1947.)

E. gedoelsti Hsiung (*E. intestinalis* (Gedoelst)). In the colon and caecum of horse; 6-13μ by 6-11μ; endosome eccentric; bacteria-feeder.

E. equi Fantham. 40-50μ by 23-29μ; nucleus oval; cysts tetranucleate, 15-24μ in diameter; seen in the faeces of horse; Fantham reports that the endoplasm contained erythrocytes.

E. bovis Liebetanz. 5-20μ in diameter; uninucleate cysts, 4-15μ in diameter; in the stomach of cattle and gnu, *Connochaetes taurinus* (Mackinnon and Dibb, 1938). (Morphology, Noble, 1950.)

E. ovis Swellengrebel. Cyst uninucleate; in the intestine of sheep.

E. caprae Fantham. In goat intestine.

E. polecki Prowazek. In the colon of pigs; 10-12μ in diameter; cyst uninucleate, 5-11μ in diameter.

E. debliecki Nieschulz (Fig. 196, *a*). 5-10μ in diameter; cysts uninucleate; in the intestine of pigs and goats. Two races (Hoare, 1940); morphology (Nieschulz, 1924); Entamoebae of domestic animals (Noble and Noble, 1952).

E. venaticum Darling. In the colon of dog; similar to *E. histolytica;* since the dog is experimentally infected with the latter, this amoeba discovered from spontaneous amoebic dysentery cases of dogs, in one of which were noted abscesses of liver, is probably *E. histolytica.*

E. cuniculi Brug. Similar to *E. coli* in both trophic and encysted stages; in the intestine of rabbits.

E. cobayae Walker (*E. caviae* Chatton). Similar to *E. coli;* in the intestine of guinea-pigs (Nie, 1950).

E. muris Grassi (Fig. 196, *b, c*). In the caecum of rats and mice; trophozoite 8-30μ; cytoplasm with rod-shaped or fusiform bacteria and flagellates coinhabiting the host's organ; nucleus 3-9μ in diameter and resembles closely that of *E. coli;* cysts 9-20μ in diameter, with eight nuclei when mature. Nuclear division (Wenrich, 1940); food habits (Wenrich, 1941); inhibition by estrucomycin (Arcamone, 1957).

E. citelli Becker (Fig. 196, *d, e*). In the caecum and colon of the striped ground squirrel, *Citellus tridecemlineatus;* rounded trophozoites 10-25μ in diameter; nucleus 4-6μ in diameter, with a comparatively large endosome which varies in position from central to peripheral; cysts with eight nuclei, about 15μ in diameter.

E. gallinarum Tyzzer. In the caeca of chicken, turkeys and possibly other fowls; trophozoites 9-25 (16-18)μ; cysts octonucleate, 15μ by 12μ.

E. testudinis Hartmann. In intestine of turtles, *Testudo graeca,*
T. argentina, T. calcarata and *Terrapene carolina.*

E. barreti Taliaferro and Holmes (Fig. 196, *f*). In the colon of
snapping turtle, *Chelydra serpentina;* trophozoites 14-23 (18)μ
long. Cultivation (Barret and Smith, 1924).

E. terrapinae Sanders and Cleveland (Fig. 196, *g, h*). Tro-
phoizoites 10-15μ long; cysts 8-14μ in diameter, tetranucleate
when mature; upon excystment, the cyst content divides into four
uninucleate amoebulae; in the colon of *Chrysemys elegans* (Sanders and Cleveland, 1930).

E. invadens Rodhain (Figs. 2, *a, b;* 196, *i, j*). Resembles
E. histolytica. Trophozoites measure 15.9μ in average diameter

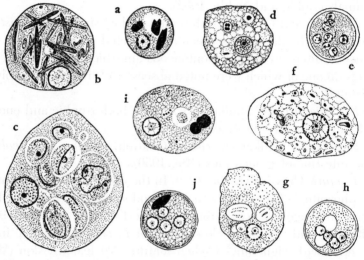

FIG. 196. a, a stained cyst of *Entamoeba debliecki,* ×1330 (Hoare); b, c,
E. muris, ×1330 (Wenrich) (b, with fusiform bacilli; c, with *Tritricho-*
monas muris); d, e, stained trophozoite and cyst of *E. citelli,* ×880 (Becker);
f, a stained trophozoite of *E. barreti,* ×1330 (Taliaferro and Holmes); g, h,
stained trophozoite and cyst of *E. terrapinae,* ×1665 (Sanders and Cleveland); i, j, stained trophozoite and cyst of *E. invadens,* ×1045 (Geiman
and Ratcliffe).

(9.2-38.6μ by 9-30μ); active locomotion; feed on leucocytes, liver
cells, epithelial cell debris, bacteria, etc.; nucleus simliar to that
of *E. histolytica.* Cysts 13.9μ (11-20μ) in diameter; one to four

nuclei; glycogen vacuole; chromatoid bodies acicular, rod-like or cylindrical.

Hosts include various reptiles: *Varanus salvator, V. varius, Tiliqua scincoides, Pseudoboa clelia, Lampropeltis getulus, Ancistrodon mokasen, Natrix rhombifer. N. sipedon, N. sipedon sipedon, N. cyclopion, Python sebae, Rachidelus brazili,* etc.

Meerovitch (1958) found a painted turtle, *Chrysemys picta,* naturally infected with this amoeba and through infection experiments showed its high pathogenicity to snakes. He suggested that this amoeba is a natural non-pathogenic parasite of turtles. He (1961) later reported that this amoeba does not infect snakes in nature. Barrow and Stockton (1960) report that eight species of snakes experimentally infected with *E. invadens* did not show any pathological effect when the snakes were kept at 13°C., but when kept at 25°C., showed mild to severe pathological effect. Zoological gardens in Antwerp (Rodhain), Philadelphia (Geiman and Ratcliffe) and London (Hill and Neal, 1954). (Relationship to other species of the genus, McConnachie, 1955.)

The amoeba produces lesions in the stomach, duodenum, ileum, colon and liver in host animals. Time for excystation in host's intestine (jejunum and ileum) five to fourteen hours; time for metacystic development in host's intestine seven to twenty-four hours; the excysted amoeba with four nuclei, each of which divides once, divides finally into eight amoebulae; optimum temperature for culture 20-30°C. (Geiman and Ratcliffe, 1936). Ratcliffe and Geiman (1938) observed spontaneous and experimental amoebiasis in thirty-two reptiles.

E. ranarum (Grassi). In colon of various species of frogs; resembles *E. histolytica;* 10-50μ in diameter; cysts are usually tetranucleate, but some contain as many as sixteen nuclei; amoebic abscess of the liver was reported in one frog. Comparison with *E. histolytica* (Dobell, 1918); life cycle (Sanders, 1931).

E. (?) phallusiae Mackinnon and Ray. In the intestine of the ascidian, *Phallusia mamillata;* 15-30μ by 10-15μ; nucleus about 5μ in diameter, structure not well defined; cysts uninucleate, about 20μ in diameter; parasitic nutrition.

E. minchini Mackinnon. In gut of tipulid larvae; 5-30μ in diameter; cyst nuclei up to ten in number.

E. apis Fantham and Porter. In *Apis mellifica;* similar to *E. coli.*

E. thomsoni Lucas. In the colon of cockroaches; when rounded 7-30 (15-25)μ in diameter; usually attached to debris by a knob-like process, highly adhesive; cytoplasm poorly differentiated; vesicular nucleus with peripheral granules; endosome variable, with loosely aggregated granules and a central dot; cysts 8-16μ in diameter, with one to four nuclei (Lucas, 1927).

E. aulastomi Nöller. In the gut of the horse-leech, *Haemopis sanguisuga;* cysts with four nuclei. Morphology and development (Bishop, 1932).

E. paulista (Carini) (*Brumptina paulista* C.) (Fig. 191, *c*). In the cytoplasm of many species of opalinids; trophozoites 5.3-14.3μ in diameter; cysts about 9.4μ in diameter, uninucleate; no effect upon host ciliates even in case of heavy infection (Stabler and Chen, 1936; Chen and Stabler, 1936). Carini and Reichenow (1935): trophozoites 8-14μ in diameter; cysts 8-12μ; either identical with *E. ranarum* or a race derived from it.

E. moshkovskii Tshalaia. This amoeba resembles *E. histolytica,* but appears to be free-living in the sewage. The organism is actively motile and assumes limacid form during locomotion. The nucleus resembles closely that of *E. histolytica.* No contractile vacuole occurs. It measures 9-29μ or more, but commonly 11-13μ in largest diameter. Cysts are spherical and 7-17μ in diameter. When mature it contains four nuclei, glycogen vacuoles and chromatoid bodies.

Since its discovery by Tschalaia (1941) in Russia, it has been found in the United States (Wright, Cram and Nolan, 1942), Brazil (Amaral and Azzi Leal, 1949), England (Neal, 1950), Canada (Lachance, 1959) and Costa Rica (Ruiz, 1960). All observers noted a striking morphological resemblance between this amoeba and *E. histolytica.* Neal (1953) could not infect guinea pigs, rats or frog tadpole with this amoeba and considered a distinct species. (Two strains, Hirschlerowa and Swiecicki, 1960.)

Genus **Iodamoeba** Dobell. Vesicular nucleus, with a large endosome rich in chromatin, a layer of globules surrounds the endosome and do not stain deeply, and achromatic strands between the endosome and membrane (Fig. 189, *c*); cysts ordinarily uninucleate, contain a large glycogenous vacuole which stains conspicuously with iodine; in intestine of man and mammals (Dobell, 1919).

I. bütschlii (Prowazek) (*I. williamsi* P.) (Fig. 197). The trophozoite is 6-25μ (average 8-15μ) in diameter; fairly active with progressive movement, when fresh; cytoplasm not well differentiated; endoplasm granulated, contains bacteria and yeasts in food vacu-

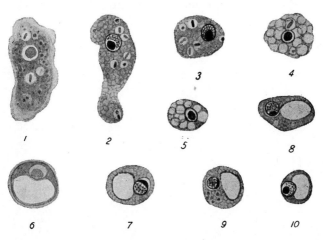

Fig. 197. *Iodamoeba bütschlii*, ×1150 (Kudo). 1, a living amoeba; 2–5, stained trophozoites; 4, 5, somewhat degenerating trophozoites; 6, a fresh cyst; 7–10, stained cysts.

oles; the nucleus (3-4μ in diameter) visible *in vivo;* the large endosome about one-half the diameter of nucleus, surrounded by small spherules.

The cysts are spherical, ovoid, ellipsoid, triangular, pyriform or square; rounded cysts measure about 6-15μ in the largest diameter; a large glycogen body which becomes conspicuously stained with Lugol's solution (hence formerly called "iodine cysts") persists; nucleus with a large, usually eccentric endosome.

The trophozoites and cysts are ordinarily present in diarrhoeic faeces, while the formed faeces contain cysts only. This amoeba apparently lives in the lumen of the colon and does not seem to attack host's tissues and is, therefore, considered to be a commensal. (Nomenclature, Dobell, 1919; nuclear structure, Wenrich, 1937a.)

I. suis O'Connor. In colon of pig; widely distributed; indistinguishable from *I. bütschlii;* it is considered by some that pigs are probably reservoir host of *I. bütschlii.*

Genus **Endolimax** Kuenen and Swellengrebel. Small; vesicular nucleus with a comparatively large irregularly shaped endosome, composed of chromatin granules embedded in an achromatic ground mass and several achromatic threads connecting the en-

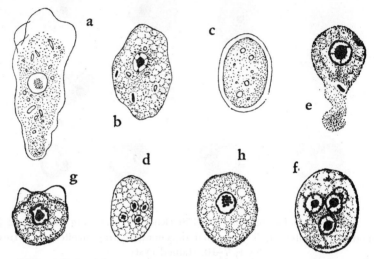

FIG. 198. a–d, *Endolimax nana,* ×2300 (Kudo) (a, b, living and stained trophozoites; c, d, fresh and stained cysts); e, f, stained trophozoite and cyst of *E. clevelandi,* ×3000 (Gutierrez-Ballesteros and Wenrich); g, h, stained trophozoites of *Martinezia baezi,* ×1700 (Hegner and Hewitt).

dosome with membrane (Fig. 189, *d*); commensal in hindgut in man and animals. Several species.

E. nana (Wenyon and O'Connor) (Fig. 198, *a-d*). The trophozoite measures 6-18μ in diameter; fairly active monopodal movement by forming a broad pseudopodium; when stationary pseudopodia are formed at different points; endoplasm is granulated and contains bacteria as food particles; the vesicular nucleus, 1.5-3μ in diameter, is composed of a delicate membrane with a few chromatin granules and a large irregularly shaped endosome.

The cyst is usually ovoid; young cyst contains one or two nuclei; mature cyst with four nuclei; indistinctly outlined glycogen body may be present while immature; dimensions 5-12μ (majority 7-10μ) in diameter.

The trophozoites are found in diarrhoeic or semifluid faeces together with the cysts, and formed faeces contain cysts. This amoeba is coelozoic in the lumen of the upper portion of colon and is considered to be a commensal. Cytology and life-history (Dobell, 1943).

E. caviae Hegner. In the caecum of guinea-pigs. (Morphology, Hegner, 1926; Nie, 1950.)

E. gregariniformis Tyzzer. In the caeca of fowls; 4-12μ in diameter; cysts uninucleate (Tyzzer, 1920).

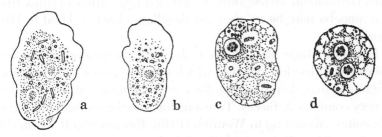

FIG. 199. *Dientamoeba fragilis,* ×2300 (Kudo). a, b, living bi- and uninucleate trophozoites; c, d, stained uni- and bi-nucleate trophozoites.

E. clevelandi Gutierrez-Ballesteros and Wenrich (Fig. 198, *e, f*). In the rectal contents of *Pseudemys floridana mobilensis;* trophozoites 5-14μ in diameter; cysts tetranucleate, 4.5-10μ large.

E. ranarum Epstein and Ilovaisky. In the colon of frogs; cysts octonucleate, up to 25μ in diameter.

E. blattae Lucas. In the colon of cockroaches; trophozoites 3-15μ long; cysts, 7-11μ in diameter and with one to three nuclei (Lucas, 1927).

Genus **Dientamoeba** Jepps and Dobell. Small amoeba; number of binucleate trophozoites often greater than that of uninucleate forms; nuclear membrane delicate; endosome consists of several chromatin granules embedded in plasmosomic substances and connected with the membrane by delicate strands (Fig. 189, *e*); in colon of man (Jepps and Dobell, 1918).

D. fragilis J. and D. (Fig. 199). The trophozoite is actively amoeboid; 4-18μ (average 5-12μ) in diameter; progressive movement; cytoplasm well differentiated; endoplasm granulated contains bacteria in food vacuoles; nucleus only faintly visible; one or two nuclei, the ratio is variable; in some material binucleate forms may be 80% or more (Jepps and Dobell), while in others uninucleate forms may predominate (Kudo, 1926a; Wenrich, 1937); nucleus is made up of a delicate membrane and a large endosome (more than one-half the dimeter of nucleus) in which are embedded four to eight chromatin granules along the periphery. According to Dobell (1940), the binucleate condition represents an arrested telophase stage of mitosis and the chromatin granules are in reality chromosomes, probably six in number. Comparison with *Histomonas meleagridis* (p. 401) led this author to think that this amoeba may be an aberrant flagellate closely related to Histomonas.

Encysted stage has not been observed. Degenerating trophozoites often develop vacuoles which coalesce into a large one and organisms may then resemble *Blastocystis hominis* (p. 1073) which is very common in faeces. Transmission may be carried on by trophozoites. According to Wenrich (1940), this amoeba if left in the faeces remains alive up to forty-eight hours at room temperature, but disappears probably by disintegration in two hours at 3.5°C. Since all attempts to bring about experimental infection by mouth or by rectum failed, Dobell considered that the amoeba may be transmitted from host to host in the eggs of nematodes such as Trichuris or Ascaris, as in the case of Histomonas. Burrow and Swerdlow (1956) also suggested that the amoeba may be carried in the eggs of the pinworm, *Enterobius vermicularis*.

The amoeba inhabits the lumen of the colon. There is no indication that it is histozoic or cytozoic. Some workers attribute certain intestinal disturbances to this amoeba, but no definite evidence for its pathogenicity is available at present. It seems to be widely distributed, but not as common as the other intestinal amoebae mentioned above, although in some areas it appears to be common. (Nuclear division, Wenrich, 1936, 1939, 1944a; Dobell, 1940.)

Genus **Martinezia** Hegner and Hewitt. The nucleus consists of a wrinkled membrane, a large compact or granular endosome and heavy peripheral beads; cysts unknown; parasitic.

M. baezi H. and H. (Fig. 198, *g, h*). In the intestine of iguanas, *Ctenosaura acanthura;* 8-21μ by 6.5-16μ; nucleus about 4μ in diameter; two nuclei in about 3 per cent of the organisms; cysts not seen.

Genus **Dobellina** Bishop and Tate. Trophozoite: small amoeba; ectoplasm and endoplasm differentiated; usually monopodal; nucleus one to many; nucleus with a large central endosome and an achromatic nuclear membrane; nuclear divisions mitotic and simultaneous; no solid food vacuoles; no contractile vacuole; with refringent granules. Cysts: spherical; thin-walled; devoid of glycogen and of chromatoid bodies; two or more nuclei; parasitic (Bishop and Tate, 1939).

D. mesnili (Keilin) (Fig. 200, *a-c*). Uninucleate amoebae as small as 3.6μ in diameter; multinucleate forms 20-25μ by 10-15μ; cysts 8-11μ in diameter; in the space between the peritrophic membrane and the epithelium of the gut in the larvae of *Trichocera hiemalis, T. annulata,* and *T. regelationis* (winter gnats).

Genus **Hydramoeba** Reynolds and Looper. Nucleus vesicular with a large central endosome composed of a centriole (?) and chromatin granules embedded in an achromatic mass, achromatic strands radiating from endosome to membrane; a ring made up of numerous rod-shaped chromatin bodies in the nuclear-sap zone; one or more contractile vacuoles; apparently the most primitive parasitic amoeba; parasitic on Hydra and freshwater medusa.

H. hydroxena (Entz) (Fig. 200, *d-j*). Parasitic in various species of Hydra; first observed by Entz; Wermel found 90 per cent of Hydra he studied in Russia were infected by the amoeba; Reynolds and Looper (1928) stated that infected Hydra die on an average in 6.8 days and that the amoebae disappear in four to ten days if removed from a host hydra. More or less spheroidal, with blunt pseudopods; 60-380μ in diameter; nucleus shows some twenty refractile peripheral granules in life; contractile vacuoles; food vacuoles contain host cells; multiplication by binary fission. (Nuclear division, Reynolds and Threlkeld, 1929.)

Ito (1949) found this organism in *Hydra japonica, H. magnipapillata, Palmathydra robusta,* etc. in Japan. The trophozoites measured 26-210μ long with a nucleus, 10-12μ in diameter. Early infection occurs on the tip of tentacles and spreads to the body proper (Fig. 200, *g-j*). Since the tentacles remain contracted, the host hydra cannot feed on food organisms and becomes "depressed." The amoebae finally enter the coelenteric cavity and feed on the endoderm cells. The host hydra becomes spherical. At 25°C. death of the hydra may occur in one week. Encystment takes place soon after the death of the host or occasionally when the organisms become detached from the host; cysts are spherical, measure 27.5-29μ, and contain one or more nuclei, nematocysts and a large vacuole (*f*).

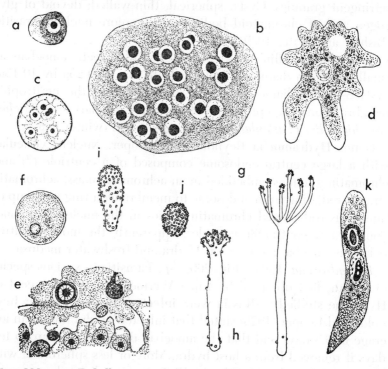

FIG. 200. a–c, *Dobellina mesnili* (Bishop and Tate) (a, b, stained trophozoites, ×2200; c, cyst, ×1760); d–j, *Hydramoeba hydroxena* (d, f-j, Ito; e, Reynolds and Looper) (d, a trophozoite in life, ×330; e, a trophozoite feeding on cells of a Hydra in section, ×470; f, living cyst, ×530; g-j, stages of infection in Hydra, ×6.5); k, *Paramoeba pigmentifera* with its nucleus in center, ×800 (Janicki).

The organism found by Rice (1960) on the fresh-water medusa, *Craspedacusta sowerbii* is said to be capable of destroying its adult hosts in six days and is also pathogenic to *Hydra cauliculata* which when starved is less resistant to the attack of the amoeba. Stiven (1962) reports differences in susceptibility to infection by species of Hydra: *H. pseudoligactis* and *H. oligactis* are most susceptible, *Chlorohydra viridissima* is resistant and *H. littoralis* is nearly completely immune to this amoeba.

Family 4. **Paramoebidae** Poche

Genus **Paramoeba** Schaudinn. The amoeba possesses a nucleus and "secondary" nucleus, both of which multiply by division simultaneously; both nuclei contain DNA; the secondary nucleus is considered by Grell (1961) "parasitic"; parasitic or free-living. (Cytology, Grell, 1961.)

P. pigmentifera (Grassi) (Fig. 200, *k*). About 30μ long; sluggish; cytoplasm distinctly differentiated; secondary body larger than the nucleus; flagellated swarmers are said to occur; parasitic in coelom of Chaetognatha such as *Sagitta claparedei*, *Spadella bipunctata*, *S. inflata*, and *S. serratodentata*. (Cytology, Janicki, 1928, 1932.)

P. eilhardi Schaudinn. Amoeboid form 50-60μ; rounded form 30-40μ; two nuclei are closely associated; among algae, Villefranche (Grell, 1961).

References

AMARAL, A. D. F., and AZZI LEAL, R., 1949. Sôbre uma Endamoeba semelhante à *Endamoeba histolytica* encontrada em material de esgôto. Rev. Paulist. Med., 34:173.

ANDRESEN, N., 1942. Cytoplasmic components in the amoeba, *Chaos chaos* L. C. R. Lab. Carlsberg, Sér. chim., 24:139.

——— 1956. Cytological investigations on the giant amoeba. *Ibid.*, 29:435.

ANGERER, C. A., 1942. The action of cupric chloride on the protoplasmic viscosity of *Amoeba dubia*. Physiol. Zool., 15:436.

BALAMUTH, W., and BRENT, M. M., 1950. Biological studies on *Entamoeba histolytica*. IV. Proc. Soc. Exper. Biol. Med., 75:374.

BARRET, H. P., and SMITH, N. M., 1924. The cultivation of an Endamoeba from the turtle, *Chelydra serpentina*. Amer. J. Hyg., 4:155.

BARROW, J. H., and STOCKTON, J. J., 1960. The influence of temperature on the host-parasite relationship, etc. J. Protozool., 7:377.

BECKER, E. R., 1926. *Endamoeba citelli* sp. nov., etc. Biol. Bull., 50:414.

———, BURKS, C., and KALEITA, E., 1946. Cultivation of *Endamoeba his-*

tolytica in artificial media from cysts in drinking water subjected to chlorination. Am. J. Trop. Med., 26:783.

BEHEYT, P., *et al.*, 1961. Le diagnostic biologique de l'amibiase hepatique aigue. Ann. Soc. Belge Med. Trop., 41:93.

BELDA, W. H., 1942. Permeability to water in *Pelomyxa carolinensis*. I. Salesianum, 37:68.

———— 1942a. II. *Ibid.*, 37:125.

———— 1943. III. *Ibid.*, 38:17.

BISHOP, A., 1932. *Entamoeba aulastomi*. Parasitology, 24:225.

———— 1937. Further observations upon *Entamoeba aulastomi*. *Ibid.*, 29:57.

———— and TATE, P., 1939. The morphology and systematic position of *Dobellina mesnili*, etc. *Ibid.*, 31:501.

BOVEE, E. C., 1959. Studies on Amoebo-flagellates. I. J. Protozool., 6:69.

———— 1960. Studies concerning the effects of nutrition on morphology of amebas. I. Am. Midland Nat., 63:257.

BROOKE, M. M., 1958. Amebiasis: methods in laboratory diagnosis. U. S. Public Health Service, 67 pp.

———— *et al.*, 1955. Studies of a water-borne outbreak of amebiasis, South Bend, Indiana. III. Amer. J. Hyg., 62:214.

———— *et al.*, 1958. Mass therapy in attempted control of amebiasis in a mental institution. Publ. Health Service Rep., 73:499.

BUNDESEN, H. N., *et al.*, 1936. Epidemic amebic dysentery: etc. Nat. Inst. Health Bull., no. 166.

BUONOMINI, G., and MIGNANI, E., 1960. Studi sulla biologia di *Entamoeba histolytica*. IV. Boll. Ist. Sieroterap. Milan, 39:121.

BURROWS, R. B., 1957. *Endamoeba hartmanni*. Amer. J. Hyg., 67:172.

———— and SWERDLOW, M. A., 1956. *Enterobius vermicularis* as a probable vector of *Dientamoeba fragilis*. Am. J. Trop. Med. Hyg., 5:258.

CARINI, A., 1933. Parasitisme des Zellerielles par des microorganismes nouveaux (Brumptina n.g.). Ann. Parasit., 11:297.

———— 1943. Novas observações em batráquios e ofidios, etc. Arqu. Biol., 27:1.

———— and REICHENOW, E., 1935. Ueber Amoebeninfektion in Zelleriellen. Arch. Protist., 84:175.

CASAGRANDI, O., and BARBAGALLO, P., 1895. Ricerche biologiche e clinique sull' *Amoeba coli*. I, II. Nota prelim. Bull. Accad. Gioenia Sc. Nat. Catania, 39:4 and 41:7.

CASH, J., 1905. The British freshwater Rhizopoda and Heliozoa. 1. London.

CHALKLEY, H. W., 1936. The behavior of the karyosome and the "peripheral chromatin" during mitosis, etc. J. Morphol., 60:13.

———— and DANIEL, G. E., 1933. The relation between the form of the living cell and the nuclear phases of division in *Amoeba proteus*. Physiol. Zool., 6:592.

CHANG, S. L., 1958. Cultural, cytological and ecological observations on the amoeba stage of *Naegleria gruberi*. J. Gen. Microbiol., 18:565.

—— 1958a. Cytological and ecological observations on the flagellate transformation of *Naegleria gruberi*. *Ibid.*, 18:579.

CHAPMAN-ANDRESEN, C., 1962. Studies on pinocytosis in amoebae. C. R. Trav. Lab. Carlsberg, 33:73.

CHEN, T. T., and STABLER, R. M., 1936. Further studies on the Endamoeba parasitizing opalinid ciliates. Biol. Bull., 70:72.

CHI, L., *et al.*, 1959. Selective phagocytosis of nucleated erythrocytes by cytotoxic amebae in cell culture. Science, 130:1763.

CLAYTON, J. P., JR., and BALL, G. H., 1954. Effects of penicillin on *Entamoeba gingivalis* in cultures with bacteria from the human mouth. J. Parasit., 40:347.

CLEVELAND, L. R., and SANDERS, E. P., 1930. Encystation, multiple fission without encystment, excystation, etc. Arch. Protist., 70:223.

COOPER, J., *et al.*, 1958. Therapy of amebiasis carriers with oleandomycin-tetracycline (Signemycin). Antibiot. Med. & Clin. Therapy, 5:302.

CRAIG, C. F., 1934. Amebiasis and amebic dysentery. Springfield, Ill.

DANGEARD, P. A., 1900. Étude de la karyokinèse chez l'*Amoeba hyalina*. Le Bot., Ser. 7:49.

DANIELLI, J. F., 1959. Some alternative states of Amoeba with special reference to life-span. Gerontologia, 3:76.

DANIELS, E. W., 1951. Studies on the effect of x-irradiation upon *Pelomyxa carolinensis* with special reference to nuclear division and plasmotomy. J. Exper. Zool., 117:189.

—— 1952. Some effects on cell division in *Pelomyxa carolinensis* following x-irradiation, etc. *Ibid.*, 120:509.

—— 1952a. Cell division in the giant amoeba, *Pelomyxa carolinensis*, following x-irradiation. I. *Ibid.*, 120:525.

—— 1954. II. *Ibid.*, 127:427.

—— 1955. X-irradiation of the giant amoeba, *Pelomyxa illinoisensis*. I. J. Exper. Zool., 130:183.

—— and Roth, L. E. 1961. III. Rad. Res., 14:66.

DAS, S. M., and TEWARI, H. B., 1955. Golgi apparatus in *Amoeba verrucosa*. Current Sci., 24:58.

DAWSON, J. A., 1945. Studies on the contractile vacuole of *Amoeba dubia*. J. Exp. Zool., 100:179.

—— *et al.*, 1935. Mitosis in *Amoeba dubia*. Biol. Bull., 69:447.

DIAMOND, L. S., 1961. Axenic culture of *Entamoeba histolytica*. Science, 134:336.

DOBELL, C., 1918. Are *Entamoeba histolytica* and *E. ranarum* the same species? Parasitology, 10:294.

—— 1919. Amoebae living in man. London.

—— 1928. Researches on the intestinal protozoa of monkeys and man. I, II. Parasitology, 20:359.

—— 1938. IX. *Ibid.*, 30:195.

—— 1940. X. *Ibid.*, 32:417.

———— 1943. XI. *Ibid.*, 35:134.

———— and O'CONNOR, F. W. 1921. The intestinal protozoa of man. London.

DOUGLAS, M., 1930. Notes on the classification of the amoeba, etc. Am. J. Trop. Med. Hyg., 33:258.

ENTZ, G. JR., 1912. Ueber eine neue Amoebe auf Süsswasser-Polypen (*Hydra oligactis*). Arch. Protist., 27:19.

EYLES, D. E., *et al.*, 1954. Amebic infections in dogs. J. Parasit., 40:163.

FAUST, E. C., and READ, T. R., 1959. Capacity of *Entamoeba histolytica* of human origin to utilize different types of starches in its metabolism. Am. J. Trop. Med. Hyg., 8:293.

FUKUSHIMA, J., 1960. Studies on encystment of *Entamoeba histolytica* in vitro. J. Osaka City Med. Center, 9:131.

GEIMAN, Q. M., and RATCLIFFE, H. L., 1936. Morphology and life-cycle of an amoeba producing amoebiasis in reptiles. Parasitology, 28:208.

GOLDMAN, M., 1959. Microfluorimetric evidence of antigenic difference between *Entamoeba histolytica* and *E. hartmanni*. Proc. Soc. Exper. Biol. & Med., 102:189.

GREEFF, R., 1874. *Pelomyxa palustris* (Pelobius), ein amoebenartiger Organismus des süssen Wassers. Arch. mikr. Anat., 10:53.

GREIDER, M. H., *et al.*, 1958. Electron microscopy of *Amoebe proteus*. J. Protozool., 5:139.

GRELL, K. G., 1961. Ueber den "Nebenkoerper" von *Paramoeba eilhardi* Schaudinn. Arch. Protist., 105:303.

GROOT, A. A. DE, 1936. Einige Beobachtungen an *Dinamoeba mirabilis*. Arch. Protist., 87:427.

GUTIERREZ-BALLESTEROS, E., and WENRICH, D. H., 1950. *Endolimax clevelandi*, n. sp. from turtle. J. Parasit., 36:489.

HARDY, A. V., and SPECTOR, B. K., 1935. The occurrence of infestations with *E. histolytica* associated with water-borne epidemic diseases. Publ. Health Rep. Washington, 50:323.

HARTMANN, M., and CHAGAS, C., 1910. Ueber die Kernteilung von *Amoeba hyalina*. Mem. Inst. Oswaldo Cruz, 2:159.

HEGNER, R. W., 1926. *Endolimax caviae*, etc. J. Parasit., 12:146.

HEMMING, F., 1951. Report on the investigation of the nomenclatorial problems associated with the generic names "*Endamoeba*," etc. Bull. Zool. Nomenclature, 2:277.

HENDERSON, J. C., 1941. Studies of some amoebae from a termite of the genus Cubitermes. Univ. Cal. Publ. Zool., 43:357.

HEWITT, R., 1937. The natural habitat and distribution of *Hartmannella castellanii*, etc. J. Parasit., 23:491.

HILL, W. C. O., and NEAL, R. A., 1954. An epizootic due to *Entamoeba invadens* at the Gardens of the Zoological Society of London. Proc. Zool. Soc. London, 123:731.

HOARE, C. A., 1940. On an Entamoeba occurring in English goats. Parasitology, 32: 226.

HOGUE, MARY J., 1921. Studies on the life history of *Vahlkampfia patuxent,* etc. Amer. J. Hyg., 1:321.

HOLLANDE, A., 1945. Biologie et reproduction des rhizopodes des genres Pelomyxa et Amoeba, etc. Bull. Biol. France et Belg., 79:31.

HYMAN, LIBBIE H., 1936. Observations on Protozoa. I. Quart. J. Micr. Sci., 79:43.

INTERNATIONAL COMM. ZOOL. Nomenclature 1954. Opinions and declarations., 7: Pt. 1.

ITO, T., 1949. On *Hydramoeba hydroxena* discovered in Japan. Sc. Rep. Tohoku Univ., Ser. 4, 18:205.

—— 1950. Further notes on *Hydramoeba hydroxena* from Japan. Mem. Ehime University, Sec. 2, 1:27.

JANICKI, C., 1928. Studien an Genus Paramoeba Schaud. Neue Folge. I. Zeitschr. wiss. Zool., 131:588.

—— 1932. II. *Ibid.,* 142:587.

JEPPS, MARGARET W., and DOBELL, C., 1918. *Dientamoeba fragilis,* etc. Parasitology, 10:352.

JOLLOS, V., 1917. Untersuchungen zur Morphologie der Amoebenteilung. Arch. Protist., 37:229.

KELLER, H., 1949. Untersuchungen über die intrazellulären Bakterien von *Pelomyxa palustris.* Ztschr. Naturforsch., 46:293.

KIRBY, H. JR., 1927. Studies on some amoebae from the termite Microtermes, etc. Quart. J. Micr. Sci., 71:189.

—— 1945. *Entamoeba coli* versus *Endamoeba coli.* J. Parasit., 31: 177.

KOCH, D. A., 1927. Relation of moisture and temperature to the viability of *Endamoeba gingivalis* in vitro. Univ. Cal. Publ. Zool., 31:17.

KOFOID, C. A., and JOHNSTONE, H. G., 1930. The oral amoeba of monkeys. *Ibid.,* 33:379.

KRADOLFER, F., and GROSS, F., 1958. Die Ingestion von Erythrozyten durch *Entamoeba histolytica.* Schweiz Zeitschr. allg. Path. & Bakteriol., 21:1014.

KUDO, R. R., 1926. Observations on *Endamoeba blattae.* Amer. J. Hyg., 6:139.

—— 1926a. Observations on *Dientamoeba fragilis.* Am. J. Trop. Med., 6:299.

—— 1946. *Pelomyxa carolinensis* Wilson. I. J. Morphol., 78:317.

—— 1947. II. *Ibid.,* 80:93.

—— 1949. III. *Ibid.,* 85: 163.

—— 1950. A species of Pelomyxa from Illinois. Tr. Am. Micr. Soc., 69:368.

—— 1951. Observations on *Pelomyxa illinoisensis.* J. Morphol., 88:145.

—— 1952. The genus Pelomyxa. Tr. Am. Micr. Soc., 71:108.

———— 1957. *Pelomyxa palustris* Greeff. I. Cultivation and general observation. J. Protozool., 4:154.

———— 1959. Pelomyxa and related organisms. Ann. N.Y. Acad. Sci., 78:474.

KUENEN, W. A., and SWELLENGREBEL, N. H., 1913. Die Entamoeben des Menschen und ihre praktische Bedeutung. Centralbl. Bakt. I. Orig., 71:378.

LACHANCE, P. J., 1959. A Canadian strain of *Entamoeba moshkovskii* Chalaia, 1941. Canad. J. Zool., 37:415.

LAPAGE, G., 1922. Cannibalism in *Amoeba vespertilio*. Quart. J. Micr. Sci., 66:669.

LEIDY, J., 1879. Freshwater rhizopods of North America. Rep. U. S. Geol. Survey Terr., 12.

LEINER, M., and WOHLFEIL, M., 1953. *Pelomyxa palustris* Greeff und ihre symbiontischen Bakterien. Arch. Protist., 98:227.

———— ———— 1954. Das symbiontische Bakterium in *Pelomyxa palustris* Greeff. III. Zeitschr. Morph. u. Oekol. Tiere, 42:529.

———— ———— and Schmidt, D. 1954. *Pelomyxa palustris* Greeff. Ann. Sci. Nat., Zool., 11 Ser., 16:537.

LIESCHE, W., 1938. Der Kern- und Fortpflanzungsverhältnisse von *Amoeba proteus*. Arch. Protist., 91: 135.

LUCAS, C. L. T. 1927. Two new species of amoeba found in cockroaches, etc. Parasitology, 19:223.

LÜHE, M., 1908. Generationswechsel bei Protozoen. Schr. phys.-ökon. Gesellsch. Königberg, 49:418.

MACKINNON, D. L., and RAY, H. N., 1931. An amoeba from the intestine of an ascidian at Plymouth. J. Mar. Biol. Assn. United Kingdom, 17:583.

———— and DIBB, M. J., 1938. Report on intestinal protozoa of some mammals, etc. Proc. Zool. Soc. London, B, 108:323.

MAST, S. O., 1926. Structure, movement, locomotion and stimulation in Amoeba. J. Morphol., 14:347.

———— 1934. Amoeboid movement in *Pelomyxa palustris*. Physiol. Zool., 7:470.

———— 1938. Amoeba and Pelomyxa vs. Chaos. Turt. News, 16:56.

———— and DOYLE, W. L., 1935. Structure, origin and function of cytoplasmic constituents in *Amoeba proteus*. I. Arch. Protist., 86:155.

———— ———— 1935a. II. *Ibid.*, 86:278.

———— and JOHNSON, P. L., 1931. Concerning the scientific name of the common large amoeba, usually designated *Amoeba proteus*. Ibid., 75:14.

McCLELLAN, J. F., 1959. Nuclear division in *Pelomyxa illinoisensis* Kudo. J. Protozool., 6:322.

McCOONNACHIE, E., 1955. Studies on *Entamoeba invadens* Rodhain, 1934, in vitro and its relationship to some other species of Entamoeba. Parasitology, 45:452.

MEEROVITCH, E., 1958. A new host of *Entamoeba invadens* Rodhain. Canad. J. Zool., 36:423.

——— 1961. Infectivity and pathogenicity of polyxenic and monoxenic *Entamoeba invadens* to snakes, etc. J. Parasit., 47:791.

MEGLITSCH, P. A., 1940. Cytological observations on *Endamoeba blattae*. Illinois Biol. Monogr., 14: no. 4.

MERCER, E. H., 1959. An electron microscope study of *Amoeba proteus*. Proc. Roy. Soc. London, Series B. Biol. Sci. 150:216.

MERCIER, L., 1909. Le cycle évolutif d'*Amoeba blattae*. Arch. Protist., 16:164.

MILLER, J. H., *et al.*, 1961. An electron microscope study of *Entamoeba histolytica*. J. Parasit., 47:577.

MORRIS, S., 1936. Studies of *Endamoeba blattae*. J. Morphol., 59:225.

MUSACCHIA, X. J., 1950. Encystment in *Pelomyxa carolinensis*. St. Louis Univ. Stud., Sec. C., 1, 6 pp.

NARDONE, R. M., *et al.* 1956. The effect of surface-active agents on the permeability, survival and pseudopod formation of *Pelomyxa carolinensis*. J. Protozool., 3:119.

NEAL, R. A., 1950. A species of Entamoeba from sewage. Tr. Roy. Soc. Trop. Med. & Hyg., 44:9.

——— 1953. Studies on the morphology and biology of *Entamoeba moshkoviskii* Tshalaia, 1941. Parasitology, 43:253.

——— 1954. Amoebae found in the intestine of lizards from Sudan. *Ibid.*, 44:422.

NEFF, R. J., 1957. Purification, axenic cultivation and description of a soil amoeba, *Acanthamoeba* sp. J. Protozool., 4:176.

NIE, D., 1950. Morphology and taxonomy of the intestinal Protozoa of the guinea-pig, *Cavia porcella*. J. Morphol., 86:381.

NIESCHULZ, O., 1924. Ueber *Entamoeba deblicki* mihi, eine Darmamoebe des Schweines. Arch. Protist., 48:365.

NOBLE, E. R., 1947. Cell division in *Entamoeba gingivalis*. Univ. Cal. Publ. Zool., 53:263.

——— 1950. On the morphology of *Entamoeba bovis*. *Ibid.*, 57:341.

NOBLE, G. A., and NOBLE, E. R., 1952. Entamoebae in farm mammals. J. Parasit., 38:571.

——— 1958. Coprozoic protozoa from Wyoming mammals. J. Protozool., 5:69.

OKADA, Y. K., 1930. Transplantationsversuche an Protozoen. Arch. Protist., 69:39.

——— 1930a. Ueber den Bau und die Bewegungsweise von Pelomyxa. *Ibid.*, 70:131.

PACE, D. M., and FROST, B. L., 1952. The effects of ethyl alcohol on growth and respiration in *Pelomyxa carolinensis*. Biol. Bull., 103:97.

PAPPAS, G. D., 1954. Structural and cytochemical studies of the cytoplasm in the family Amoebidae. Ohio J. Sci., 54:195.

——— 1956. The fine structure of the nuclear envelope of *Amoeba proteus*. J. Biophys. Biochem. Cytol., 2:431.

PENARD, E., 1902. Faune rhizopodique du bassin du Léman. Geneva.

PHILLIPS, B. P., and REES, C. W., 1950. Growth of *Entamoeba histolytica* with live and heat treated *Trypanosoma cruzi*. Am. J. Trop. Med., 30: 185.

POWELL, S. T., *et al.*, 1959. Hepatic amoebiasis. Tr. Roy. Soc. Trop. Med. & Hyg., 53:190.

PRESCOTT, D. M., 1956. Mass and clone culturing of *Amoeba proteus* and *Chaos chaos*. C. R. Lab. Carlsberg, Chim., 30:1.

PROWAZEK, S. v., 1912. Weitere Beitrag zur Kenntnis der Entamoeben. Arch. Protist., 26:241.

RAABE, H., 1951. *Amoeba vespertilio* Penard; etc. Bull. Int. Acad. Pol. Sci. et Lett., Sèr. B., p. 353.

RAFALKO, J. S., 1947. Cytological observations on the amoeboflagellate, *Naegleria gruberi*. J. Morphol., 81:1.

RATCLIFFE, H. L., and GEIMAN, Q. M., 1938. Spontaneous and experimental amebic infection in reptiles. Arch. Path., 25:160.

RAY, D. L., and HAYES, R. E., 1954. *Hartmannella astromyxis*: a new species of free-living amoeba. J. Morphol., 95:159.

RENDTORFF, R. C., 1954. The experimental transmission of human intestinal protozoan parasites. I. Amer. J. Hyg., 59:196.

——— and HOLT, C. J., 1954. III. Attempts to transmit *Entamoeba coli* and *Giardia lamblia* cysts by flies. *Ibid.*, 60:320.

REYNOLDS, B. D., and LOOPER, J. B., 1928. Infection experiment with *Hydramoeba hydroxena*. J. Parasit., 15:23.

——— and THRELKELD, W. L., 1929. Nuclear division in *Hydramoeba hydroxena*. Arch. Protist., 68:305.

RICE, N. E., 1945. *Pelomyxa carolinensis* (Wilson) or *Chaos chaos* (Linnaeus)? Biol. Bull., 88: 139.

——— 1960. *Hydramoeba hydroxena*, a parasite on the fresh-water medusa, *Craspedacusta sowerii* Lankester, and its pathogenicity for *Hydra cauliculata* Hyman. J. Protozool., 7:151.

RODHAIN, J., 1934. *Entamoeba invadens* n. sp., etc. C. R. Soc. Biol., 117: 1195.

ROOT, F. M., 1921. Experiments on the carriage of intestinal Protozoa of man by flies. Amer. J. Hyg., 1:131.

ROTH, L. E., 1960. Electron microscopy of pinocytosis and food vacuoles in Pelomyxa. J. Protozool., 7:176.

——— *et al.*, 1960. Electron microscopic studies of mitosis in amoebae. I. J. Biophys. Biochem. Cytol., 8:207.

RUIZ, A., 1960. *Entamoeba moshkovskii* Tshalaia, 1941, en Costa Rica. Rev. Biol. Trop., 8: 253.

SANDERS, E. P., 1931. The life-cycle of *Entamoeba ranarum*. Arch. Protist., 74:365.

——— and CLEVELAND, L. R., 1930. The morphology and life-cycle of *Entamoeba terrapinae*, etc. *Ibid.*, 70:267.

SCHAEFFER, A. A., 1916. Notes on the specific and other characteristics of *Amoeba proteus.* etc. *Ibid.,* 37:204.

———— 1926. Taxonomy of the amebas. Papers Dep. Mar. Biol., Carnegie Inst. Washington, 24.

———— 1937. Rediscovery of the giant ameba of Roesel, etc. Turt. News, 15:114.

———— 1938. Significance of 3-daughter division in the giant amoeba. *Ibid.,* 16:157.

SCHAUDINN, F., 1896. Ueber den Zeugungskreis von *Paramoeba eilhardi.* etc. Math. naturwiss. Mitt., 1:25.

———— 1903. Untersuchungen ueber die Fortpflanzung einiger Rhizopoden. Arb. kaiserl. Gesundh.-Amte, 19:547.

SIMITCH, T., and PETROVITCH, Z., 1953. Longévité de la forme végétative de *Entamoeba dysenteriae* dans divers milieux. Arch. Inst. Pasteur Algerie, 31:375.

SINGH, B. N., 1952. Nuclear divisions in nine species of small free-living amoebae, etc. Phil. Tr. Roy. Soc. London, Series B, 236:405.

SNYDER, T. L., and MELENEY, H. E., 1941. The excystation of *Endamoeba histolytica* in bacteriologically sterile media. Am. J. Trop. Med., 21:63.

STABLER, R. M., 1940. Binary fission in *Entamoeba gingivalis.* J. Morphol., 66:357.

———— and CHEN, T. T., 1936. Observations on an Endamoeba parasitizing opalinid ciliates. Biol. Bull., 70:56.

STIVEN, A. E., 1962. Experimental studies on the epidemiology of the host-parasite system Hydra and *Hydramoeba hydroxena.* I. Physiol. Zool., 35:166.

TALIAFERRO, W. H., and HOLMES, F. O., 1924. *Endamoeba barreti,* etc. Amer. J. Hyg., 4: 155.

TSHALAIA, L. E., 1941. On a species of Entamoeba detected in sewage effluents (in Russian). Med. Parasitol. (Moscow), 10:244.

———— 1947. Contribution to the study of *Entamoeba moshkovskii. Ibid.,* 16:66.

TORCH, R., 1955. Cytological studies on *Pelomyxa carolinensis* with special reference to the mitochondria. J. Protozool., 2:167.

TYZZER, E. E., 1920. Amoebae of the caeca of the common fowl and of the turkey. J. Med. Res., 41:199.

VOLKONSKY, M., 1931. *Hartmannella castellanii,* etc. Arch. zool. exper. gén., 72:317.

WALDNER, H., 1956. Das Glykogen in *Pelomyxa palutris* Greeff. II. Zeitschr. vergl. Physiol., 38:334.

WALKER, E. L., 1908. The parasitic amoebae of the intestinal tract of man and other animals. J. Med. Res., 17:379.

WENRICH, D. H., 1936. Studies on *Dientamoeba fragilis.* I. Jour. Parasit., 22:76.

———— 1937. II. *Ibid.,* 23:183.

―― 1937a. Studies on *Iodamoeba bütschlii* with special reference to nuclear structure. Proc. Am. Philos. Soc., 77:183.

―― 1939. Studies on *Dientamoeba fragilis*. III. J. Parasit., 25:43.

―― 1940. Nuclear structure and nuclear division in the trophic stages of *Entamoeba muris*. J. Morphol., 66:215.

―― 1941. Observations on the food habits of *Entamoeba muris* and *Entamoeba ranarum*. Biol. Bull., 81:324.

―― 1944. Studies on *Dientamoeba fragilis*. IV. J. Parasit., 30:322.

―― 1944a. Nuclear structure and nuclear division in *Dientamoeba fragilis*. J. Morphol., 74:467.

WENYON, C. M., 1926. Protozoology. 1. London and Baltimore.

WERMEL, E., 1925. Beiträge zur Cytologie der *Amoeba hydroxena* Entz. Arch. russ. Protist., 4:95.

WILBER, C. G., 1942. The cytology of *Pelomyxa carolinensis*. Tr. Am. Micr. Soc., 61:227.

―― 1945. Origin and function of the protoplasmic constituents in *Pelomyxa carolinensis*. Biol. Bull., 88:207.

―― 1946. Notes on locomotion in *Pelomyxa carolinensis*. Tr. Am. Micr. Soc., 65:318.

―― 1947. Concerning the correct name of the rhizopod, *Pelomyxa carolinensis*. *Ibid.*, 66:99.

WILSON, H. V., 1900. Notes on a species of Pelomyxa. Am. Nat., 34:535.

WRIGHT, W. H., *et al.*, 1942. Preliminary observations on the effect of sewage treatment processes on the ova and cysts of intestinal parasites. Sewage works J., 14:1274.

YORKE, W., and ADAMS, A. R. D., 1926. Observations on *Entamoeba histolytica*. I. Ann. Trop. Med. Parasit., 20:279

YOUNG, M.D., 1959. The treatment of symptomatic amebiasis with puromycin. Antibiotic Med. & Clin. Therapy, 6:222.

―― and FREED, J. E., 1956. The effect of puromycin against *Entamoeba histolytica* and other intestinal parasites. South. Med. J., 49:537.

―― ―― 1957. The supressive and prophylactic effects of puromycin against intestinal protozoa. Am. J. Trop. Med. & Hyg., 6:808.

Chapter 21

Order 4. **Testacida** Schultze

THE Testacida or Thecamoeba comprises those amoeboid organisms which are enveloped by a simple shell or test, within which the body can be completely withdrawn. The shell has usually a single aperture through which pseudopodia protrude, and varies in shape and structure, although a chitinous or pseudochitinous membrane forms the basis of all. It may be thickened, as in Arcella and others, or composed of foreign bodies cemented together as in Difflugia, while in Euglypha siliceous platelets or scales are formed in the endoplasm and deposited in the shell.

The cytoplasm is ordinarily differentiated into the ectoplasm and endoplasm. The ectoplasm is conspicuously observable at the aperture of the shell where filopodia or slender ectoplasmic lobopodia are produced. The endoplasm is granulated or vacuolated and contains food vacuoles, contractile vacuoles and nuclei. In some forms there are present regularly in the cytoplasm numerous basophilic granules which are known as 'chromidia' (p. 50).

Asexual reproduction is either by longitudinal fission in the forms with thin tests, or by transverse division or budding, while in others multiple division occurs. Encystment is common. Sexual reproduction by amoeboid or flagellate gametes has been reported in some species.

The testaceans are mostly inhabitants of fresh water, but some live in salt water and others are semi-terrestrial, being found in moss or moist soil, especially peaty soil. (Biology of soil-inhabiting forms, Volz, 1929; ecology, Hoogenraad, 1935; in New England, Heal, 1961; Africa, Gauthier-Lièvre and Thomas, 1958; in soil, Chardez, 1960, 1960a; Madagascar, Decloitre, 1956; Germany, Groepietsch, 1958; Prague, Stepanek, 1952.) Four families.

Family 1. **Gromiidae** Eimer and Fickert

These organisms are usually placed in Foraminiferida. They form rhizopods, but their shells appear to be chitinous and most

565

of them live in fresh water. Until more information becomes available, they are placed here provisionally.

Genus **Gromia** Dujardin (*Allogromia, Rhynchogromia, Diplogromia,* Rhumbler). Thin test rigid or flexible, smooth or slightly coated with foreign bodies; spherical to elongate ellipsoid; aperture terminal; one or more nuclei; contractile vacuoles; many filopodia, branching and anastomosing; cytoplasm with numerous motile granules; fresh or salt water. Many species. (Electron microscope study, Hedley and Bertand, 1962.)

G. fluvialis D. (Fig. 201, *a*). Test spherical to subspherical; smooth or sparsely covered with siliceous particles; yellowish cytoplasm fills the test; aperture not seen; a large nucleus and numerous contractile vacuoles; filopodia long, often enveloping test; 90-250μ long; on aquatic plants, in moss or soil.

G. ovoidea (Rhumbler) (Fig. 201, *b*). In salt water.

G. nigricans (Penard) (Fig. 201, *c*). Test large, circular in cross-section; a single nucleus; 220-400μ long; in pond water among vegetation.

Genus **Microgromia** Hertwig and Lesser. Test small, hyaline, spherical or pyriform, not compressed; aperture terminal, circular; filopodia long straight or anastomosing, arising from a peduncle; a single nucleus and contractile vacuole; solitary or grouped. (Morphology, Valkanov, 1930.)

M. socialis (Archer) (Fig. 201, *d*). Cytoplasm bluish; contractile vacuole near aperture; filopodia arise from a peduncle, attenuate, branching, anastomosing; often numerous individuals are grouped; multiplication by fission and also by swarmers; 25-35μ in diameter; among vegetation in fresh water.

Genus **Microcometes** Cienkowski. Body globular, enclosed within a transparent, delicate, light yellowish and pliable envelope with three to five apertures, through which long branching filopodia extend; body protoplasm occupies about one-half the space of envelope; one to two contractile vacuoles; fresh water.

M. paludosa C. (Fig. 201, *e*). About 16-17μ in diameter; fresh water among algae (Valkanov, 1931; Jepps, 1934).

Genus **Artodiscus** Penard. Body globular, plastic; covered by envelope containing small grains of various kinds; nucleus eccentric; a few pseudopodia extend through pores of the envelope; movement very rapid; fresh water.

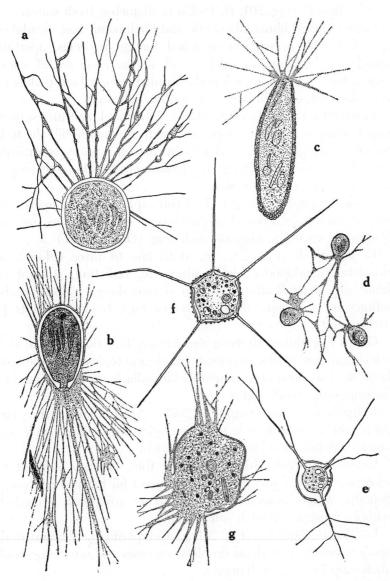

Fig. 201. a, *Gromia fluvialis*, ×120 (Dujardin); b, *G. ovoidea*, ×50 (Schultze); c, *G. nigricans*, ×200 (Cash and Wailes); d, *Microgromia socialis*, ×170 (Cash); e, *Microcometes paludosa*, ×670 (Penard); f, *Artodiscus saltans*, ×670 (Penard); g, *Schultzella diffluens*, ×120 (Rhumbler).

A. saltans P. (Fig. 201, *f*). 18-23μ in diameter; fresh water.

Genus **Lieberkühnia** Claparède and Lachmann. Test ovoidal or spherical, with or without attached foreign particles; aperture usually single, lateral or subterminal; one or more nuclei; many contractile vacuoles; pseudopodia formed from a long peduncle, reticulate, often enveloping test; fresh or salt water.

L. wagneri C. and L. (Fig. 202, *a*). Spheroidal; aperture subterminal, oblique, flexible; cytoplasm slightly yellowish, fills the test; 80-150 vesicular nuclei; nuclei 6μ in diameter; many contractile vacuoles; pseudopodia long, anastomosing; 60-160μ long; among algae in fresh and salt water.

Genus **Diplophrys** Barker. Test thin, spherical; two apertures, one at each pole; cytoplasm colorless; a single nucleus; several contractile vacuoles; filopodia radiating. (One species.)

D. archeri B. (Fig. 202, *b*). With one to three colored oil droplets; pseudopodia highly attenuate, radiating, straight or branched; multiplication into two or four daughter individuals; solitary or in groups; diameter 8-20μ; on submerged plants in fresh water.

Genus **Lecythium** Hertwig and Lesser. Test thin, flexible, colorless; aperture elastic, terminal; colorless cytoplasm fills the test; large nucleus posterior; numerous filopodia long, branching, not anastomosing; fresh water.

L. hyalinum (Ehrenberg) (Fig. 202, *c*). Spheroidal; aperture circular with a short flexible neck; a single contractile vacuole; diameter 20-45μ; in submerged vegetation.

Genus **Schultzella** Rhumbler. Test thin, delicate, difficult to recognize in life, easily broken at any point for formation of pseudopodia which branch and anastomose; irregularly rounded; without foreign material; salt water.

S. diffluens (Grubler) (Fig. 201, *g*). Cytoplasm finely granulated; opaque, colorless; with oil droplets, vacuoles and numerous small nuclei; up to 220μ in diameter.

Genus **Myxotheca** Schaudinn. Amoeboid; spherical or hemispherical, being flattened on the attached surface; a thin pseudochitinous test with foreign bodies, especially sand grains; pseudopodia anastomosing; salt water. (Nucleus, Föyn, 1936.)

M. arenilega S. (Fig. 202, *d*). Test yellow, with loosely attached foreign bodies; cytoplasm bright red due to the presence of highly refractile granules; 1-2 nuclei, 39-75µ in diameter; body diameter 160-560µ. (Life cycle, Grell, 1958.)

Genus **Dactylosaccus** Rhumbler. Test sausage-shape and variously twisted; pseudopodia filiform, anastomosing; salt water.

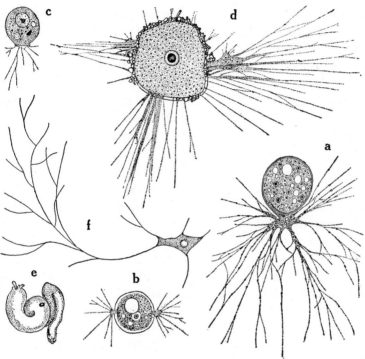

Fig. 202. a, *Lieberkühnia wagneri,* ×160 (Verworn); b, *Diplophrys archeri,* ×930 (Hertwig and Lesser); c, *Lecythium hyalinum,* ×330 (Cash and Wailes); d, *Myxotheca arenilega,* ×70 (Schaudinn); e, *Dactylosaccus vermiformis,* ×15 (Rhumbler); f, *Boderia turneri* (Wright).

D. vermiformis R. (Fig. 202, *e*). Test smooth; pseudopodia rise from small finger-like projections; one to two nuclei; body 4 mm. by 340µ; salt water.

Genus **Boderia** Wright. Body form changeable; often spherical, but usually flattened and angular; filopodia long; test extremely delicate, colorless; salt water.

B. turneri W. (Fig. 202, *f*). Body brown to orange; active cytoplasmic movement; one to ten nuclei; multiple division (?); 1.56-6.25 mm. in diameter; in shallow water.

Family 2. **Arcellidae** Schultze

Simple shells; form lobopodia or filopodia.

Genus **Arcella** Ehrenberg. Test transparent, chitinous, with hexagonal markings, colorless to brown (when old); in front view circular, angular, or stellate; in profile plano-convex or semicircular; variously ornamented; aperture circular, central, inverted like a funnel; protoplasmic body does not fill the test and connected with the latter by many ectoplasmic strands; slender lobopodia, few, digitate, simple or branched; two or more nuclei; several contractile vacuoles; fresh water. Numerous species. (Taxonomy and morphology, Deflandre, 1928; variation and heredity, Jollos, 1924.)

A. vulgaris E. (Fig. 203, *a, b*). Height of test about one-half the diameter; dome of hemispherical test evenly convex; aperture circular, central; colorless, yellow, or brown; protoplasmic body conforms with the shape of, but does not fill, the test; lobopodia hyaline; two vesicular nuclei; several contractile vacuoles; test 30-100µ in diameter; in the ooze and vegetation in stagnant water and also in soil. Of several varieties, two may be mentioned; var. *angulosa* (Perty), test smaller, 30-40µ in diameter, faceted, forming a five- to eight-sided figure, with obtuse angles; var. *gibbosa* (Penard), test gibbous, surface pitted with circular depressions of uniform dimensions; 45-50µ up to 100µ in diameter.

A. discoides E. (Fig. 203, *c*). Test circular in front view, planoconvex in profile; diameter about three to four times the height; test coloration and body structure similar to those of *A. vulgaris;* test 70-260µ in diameter; in fresh water.

A. mitrata Leidy (Fig. 203, *d*). Test balloon-shaped or polyhedral; height exceeds diameter of base; aperture circular, crenulated and usually evarted within inverted funnel; protoplasmic body spheroidal, with 'neck' to aperture and cytoplasmic strands to test; six or more slender lobopodia; test 100-145µ high, 100-152µ in diameter; in fresh water among vegetation.

A. catinus Penard (Fig. 203, *e, f*). Test oval or quadrate, not circular, in front view; aperture oval; dome compressed; lateral margin with six or eight facets; test 100-120μ in diameter and about 45μ high; fresh water among vegetation.

A. dentata Ehrenberg (Fig. 203, *g-i*). Test circular and dentate in front view, crown-like in profile; diameter more than twice the height; aperture circular, large; colorless to brown; about 95μ in diameter, aperture 30μ in diameter; fifteen to seventeen spines; in the ooze of freshwater ponds.

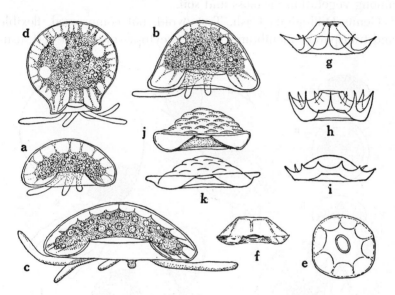

FIG. 203. a, b, *Arcella vulgaris*, ×170; ×230 (Leidy); c, *A. discoides*, ×170 (Leidy); d, *A. mitrata*, ×140 (Leidy); e, f, *A. catinus*. ×170 (Cash); g–i, *A. dentata*, ×170 (Leidy); j, k, *A. artocrea*, ×170 (Leidy).

A. artocrea Leidy (Fig. 203, *j, k*). Height of test one-fourth to one-half the diameter; dome convex; surface mammillated or pitted; border of test everted and rising one-fourth to one-half the height of test; about 175μ in diameter; fresh water.

Genus **Pyxidicula** Ehrenberg. Test patelliform; rigid, transparent, punctate; aperture circular, almost the entire diameter of test; cytoplasm similar to that of Arcella; a single nucleus; one or more contractile vacuoles; fresh water.

P. operculata (Agardh) (Fig. 204, *a, b*). Test smooth, colorless to brown; a single vesicular nucleus; pseudopodia short, lobose or digitate; 20μ in diameter; on vegetation.

Genus **Pseudochlamys** Claparède and Lachmann, Test discoid, flexible when young; body with a central nucleus and several contractile vacuoles.

P. patella C. and L. (Fig. 204, *c*). Young test hyaline, older one rigid and brown; often rolled up like a scroll; a short finger-like pseudopodium between folds; 40-45μ in diameter; in fresh water among vegetation, in moss and soil.

Genus **Difflugiella** Cash. Test ovoid, not compressed, flexible and transparent membrane; colorless cytoplasm fills the test, usu-

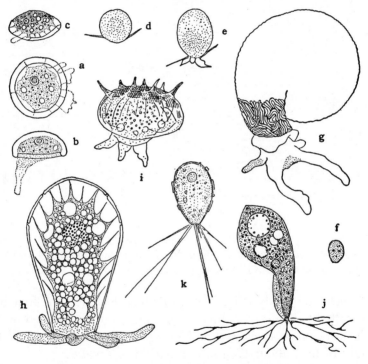

Fig. 204, a, b, *Pyxidicula operculata*, ×800 (Penard); c, *Pseudochlamys patella*, ×330 (Cash); d, e, *Difflugiella apiculata*, ×270 (Cash); f, *Cryptodifflugia oviformis*, ×320 (Cash); g, *Lesquereusia spiralis*, ×270 (West); h, *Hyalosphenia papilio*, ×330 (Leidy); i, *Corycia coronata*, ×170 (Penard); j, *Pamphagus mutabilis*, ×330 (Leidy); k, *Plagiophrys parvipunctata*, ×330 (Penard).

ally with chlorophyllous food material; median pseudopodia lobate or digitate with aciculate ends, while lateral pseudopods long, straight, and fine, tapering to a point; fresh water. (One species.)

D. apiculata C. (Fig. 204, *d, e*). About 40μ by 28μ; among vegetation.

Genus **Cryptodifflugia** Penard. Small test yellowish to brownish; Difflugia-like in general appearance, compressed; with or without foreign bodies; pseudopodia long, acutely pointed; fresh water.

C. oviformis P. (Fig. 204, *f*). Test ovoid; without foreign bodies; crown hemispherical; aperture truncate; cytoplasm with chlorophyllous food particles; 16-20μ by 12-15μ; in marshy soil.

Genus **Lesquereusia** Schlumberger. Test compressed, oval or globular in profile, narrowed at bent back; semispiral in appearance; with curved or comma-shaped rods or with sand-grains (in one species); body does not fill up the test; pseudopodia simple or branched; fresh water.

L. spiralis (Ehrenberg) (Fig. 204, *g*). Aperture circular; border distinct; cytoplasm appears pale yellow; a single nucleus; 96-188μ by 68-114μ; mitosis requires about forty-five to fifty minutes: number of chromosomes 175-200 (Stump, 1959); in marsh water.

Genus **Hyalosphenia** Stein. Test ovoid or pyriform; aperture end convex; homogeneous and hyaline, mostly compressed; crown uniformly arched; protoplasm partly filling the test; several blunt pseudopodia simple or digitate. (Several species.)

H. papilio Leidy (Fig. 204, *h*). Test yellowish; transparent; pyriform or oblong in front view; a minute pore on each side of crown and sometimes one also in center; aperture convex; in narrow lateral view, elongate pyriform, aperture a shallow notch; with chlorophyllous particles and oil globules; 110-140μ long; in fresh water among vegetation.

Genus **Corycia** Dujardin. Envelope extremely pliable, open at base, but when closed, sack-like; envelope changes its shape with movement and contraction of body; with or without spinous projections.

C. coronata Penard (Fig. 204, *i*). six to twelve spines; 140μ in diameter; in moss.

Genus **Pamphagus** Bailey. Test hyaline membranous, flexible; aperture small; body fills the envelope completely; spherical nucleus large; contractile vacuoles; filopodia long, delicate, branching, but not anastomosing; fresh water. (Species, Hoogenraad, 1936.)

P. mutabilis B. (Fig. 204, *j*). Envelope 40-100μ by 28-68μ.

Genus **Plagiophrys** Claparède and Lachmann. Envelope thin, hyaline, changeable with body form; usually elongate-oval with rounded posterior end; narrowed at other half; envelope finely punctated with a few small plates; aperture round; cytoplasm clear; nucleus large; pseudopods straight filopodia, sometimes branching; fresh water.

P. parvipunctata Penard (Fig. 204, *k*). Envelope 50μ long.

Genus **Leptochlamys** West. Test ovoid, thin transparent chitinous membrane, circular in optical section; aperture end slightly expanded with a short neck; aperture circular, often oblique; body fills test; without vacuoles; pseudopodium short, broadly expanded and sometimes cordate; fresh water.

L. ampullacea W. (Fig. 205, *a*). Nucleus large, posterior; with green or brown food particles; test 45-55μ by 36-40μ in diameter; aperture 15-17μ; among algae.

Genus **Chlamydophrys** Cienkowski. Test rigid, circular in cross-section; aperture often on drawn-out neck; body fills the test; zonal differentiation of cytoplasm distinct; nucleus vesicular; refractile waste granules; pseudopodia branching; fresh water or coprozoic. Species (Bělař, 1926); plasmogamy and division (Bělař, 1926).

C. stercorea C. (Fig. 205, *k*). Test 18-20μ by 12-15μ; mature cysts yellowish brown, 12-15μ in diameter; multiplication by budding; coprozoic and fresh water.

Genus **Cochliopodium** Hertwig and Lesser. Test thin, flexible, expansible and contractile; with or without extremely fine hairlike processes; pseudopodia blunt or pointed, but not acicular. (Several species.)

C. bilimbosum (*Auerbach*) (Fig. 205, *b*). Test hemispherical; pseudopodia conical with pointed ends; test 24-56μ in diameter; fresh water among algae.

Genus **Amphizonella** Greeff. Test membranous with a double marginal contour; inner membrane smooth, well-defined; outer serrulate; aperture inverted; a single nucleus; pseudopodia blunt, digitate, and divergent.

A. *violacea* G. (Fig. 205, *c*). Test patelliform; violet-tinted; with chlorophyllous corpuscles and grains; sluggish; average diameter 160μ; fresh water.

Genus **Zonomyxa** Nüsslin. Test rounded pyriform, flexible, chi-

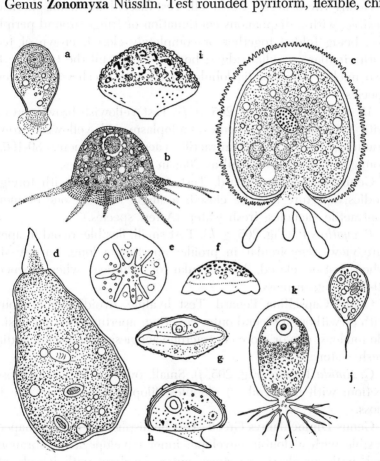

Fig. 205. a, *Leptochlamys ampullacea*, ×330 (West); b, *Cochliopodium bilimbosum*, ×670 (Leidy); c, *Amphizonella violacea*, ×270 (Greeff); d, *Zonomyxa violacea*, ×200 (Penard); e, f, *Microcorycia flava*, ×240 (Wailes); g, h, *Parmulina cyathus*, ×500 (Penard); i, *Diplochlamys leidyi* ×270 (Brown); j, *Capsellina timida*, ×270 (Wailes); k, *Chlamydophrys stercorea*, ×670 (Wenyon).

tinous, violet-colored; endoplasm vacuolated, with chlorophyllous particles; several nuclei; pseudopodia simple, not digitate; fresh water.

Z. violacea N. (Fig. 205, *d*). A single lobular pseudopodium with acuminate end; four nuclei; diameter 140-160μ; actively motile forms 250μ or longer; among sphagnum.

Genus **Microcorycia** Cockerell. Test discoidal or hemispherical, flexible, with a diaphanous continuation or fringe around periphery, being folded together or completely closed; crown of test with circular or radial ridges; body does not fill the test; one to two nuclei; pseudopodia lobular or digitate; fresh water. A few species.

M. flava (Greeff) (Fig. 205, *e*, *f*). Test yellowish brown; crown with few small foreign bodies; endoplasm with yellowish brown granules; two nuclei; contractile vacuoles; diameter 80-100μ; young individuals as small as 20μ; in moss.

Genus **Parmulina** Penard. Test ovoid, chitinoid with foreign bodies; aperture may be closed; a single nucleus; one or more contractile vacuoles; fresh water. (A few species.)

P. cyathus P. (Fig. 205, *g*, *h*). Test small, flexible; ovoid in aperture view, semicircular in profile; aperture a long, narrow slit when test is closed, but circular or elliptical when opened; 40-55μ long; in moss.

Genus **Capsellina** Penard. Test hyaline, ovoid, membranous; with or without a second outer covering; aperture long slit; a single nucleus; one or more contractile vacuoles; filose pseudopodia; fresh water.

C. timida Brown (Fig. 205, *j*). Small, ovoid; elliptical in cross-section; with many oil (?) globules; filopodium; 34μ by 25μ; in moss.

Genus **Diplochlamys** Greeff. Test hemispherical or cup-shaped, flexible with a double envelope; inner envelope a membranous sack with an elastic aperture; outer envelope with loosely attached foreign bodies; aperture large; nuclei up to 100; pseudopodia few, short, digitate or pointed; fresh water. (Several species.)

D. leidyi G. (Fig. 205, *i*). Test dark gray; inner envelope projecting beyond outer aperture; nuclei up to twenty in number; diameter 80-100μ.

Family 3. **Difflugiidae** Taránek

Shells with foreign bodies.

Genus **Difflugia** Leclerc. Test variable in shape, but generally circular in cross-section; composed of cemented quartz-sand, diatoms, and other foreign bodies; aperture terminal; often with zoochlorellae; cytoplasmic body almost fills the test; typically a single

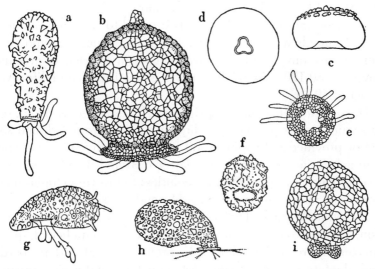

FIG. 206. a, *Difflugia oblonga*, ×130 (Cash); b, *D. urceolata*, ×130 (Leidy); c, d, *D. arcula*, ×170 (Leidy); e, *D. lobostoma*, ×130 (Leidy); f, *D. constricta*, ×200 (Cash); g, *Centropyxis aculeata*, ×200 (Cash); h, *Campuscus cornutus*, ×170 (Leidy); i, *Cucurbitella mespiliformis*, ×200 (Wailes).

nucleus; many contractile vacuoles; pseudopodia cylindrical, simple or branching; end rounded or pointed; fresh water, woodland soil, etc. (Species, Leidy, 1879; Stepanek, 1952; Decroitre, 1960b.)

D. oblonga Ehrenberg (*D. pyriformis*, Perty) (Fig. 206, *a*). Test pyriform, flask-shaped, or ovoid; neck variable in length; fundus rounded, with occasionally one to three conical processes; aperture terminal, typically circular; test composed of angular sand-grains, diatoms; bright green with chlorophyllous bodies; 60-580μ by 40-240μ; in the ooze of fresh water ponds, ditches and bogs; also in moist soil. (Several varieties, Chardez, 1960b.)

D. urceolata Carter (Fig. 206, *b*). A large ovoid, rotund test, with a short neck and a rim around aperture; multinucleate; 200-230μ by 150-200μ; in ditches, ponds, sphagnous swamps, etc. (Chardez, 1960b).

D. arcula Leidy (Fig. 206, *c*, *d*). Test hemispherical, base slightly concave, but not invaginated; aperture triangular, central, trilobed; test yellowish with scattered sand-grains or diatoms; diameter 100-140μ; in sphagnous swamp, moss, soil, etc.

D. lobostoma L. (Fig. 206, *e*). Test ovoid to subspherical; aperture terminal; with three to six lobes; test usually composed of sand-grains, rarely with diatoms; endoplasm colorless or greenish; diameter 80-120μ; in fresh water. Sexual fusion and life cycle (Goette, 1916).

D. constricta (Ehrenberg) (Fig. 206, *f*). Test laterally ovoid, fundus more or less prolonged obliquely upward, rounded, and simple or provided with spines; soil forms generally spineless; aperture antero-inferior, large, circular or oval and its edge inverted; test composed of quartz grains; colorless to brown; cytoplasm colorless; 80-340μ long; in the ooze of ponds and in soil.

D. corona Wallich. Test ovoid to spheroid, circular in cross-section; crown broadly rounded, with a variable number of spines, aperture more or less convex in profile, central and its border multidentate or multilobate; test with fine sand-grains, opaque; cytoplasm colorless; pseudopodia numerous, long, branching or bifurcating; 180-230μ by about 150μ; in fresh water. (Genetics, Jennings, 1916, 1937.)

Genus **Centropyxis** Stein. Test circular, ovoid, or discoid; aperture eccentric, circular or ovoidal, often with a lobate border; with or without spines; cytoplasm colorless; pseudopodia digitate; fresh water. (Species, Deflandre, 1929.)

C. aculeata S. (Fig. 206, *g*). Test variable in contour and size; with four to six spines; opaque or semitransparent; with fine sand-grains or diatom shells; pseudopodia sometimes knotted or branching; when encysted, the body assumes a spherical form in wider part of test; granulated, colorless or with green globules; diameter 100-150μ; aperture 50-60μ in diameter.

Genus **Campascus** Leidy. Test retort-shaped with curved neck, rounded triangular in cross-section; aperture circular, oblique, with a thin transparent discoid collar; nucleus large; one or more contractile vacuoles; body does not fill the test; fresh water.

C. cornutus L. (Fig. 206, *h*). Test pale-yellow, retort-form; with a covering of small sand particles; triangular in cross-section; a single nucleus and contractile vacuole; filopodia straight; 110-140μ long; aperture 24-28μ in diameter; in the ooze of mountain lakes.

Genus **Cucurbitella** Penard. Test ovoid with sand-grains, not compressed; aperture terminal, circular, surrounded by a four-lobed annular collar; cytoplasm grayish, with zoochlorellae; nucleus large; one to many contractile vacuoles; pseudopodia numerous, digitate; fresh water. (Species, Gauthier-Lievre and Thomas, 1960.)

C. mespiliformis P. (Fig. 206, *i*). 115-140μ long; diameter 80-105μ; in the ooze or on vegetation in ponds and ditches.

Genus **Plagiopyxis** Penard. Test subcircular in front view; ovoid in profile; aperture linear or lunate; cytoplasm gray, with a single nucleus and a contractile vacuole; fresh water. (Species, Thomas, 1958.)

P. callida P. (Fig. 207, *a*). Test gray, yellowish, or brown; large nucleus vesicular; pseudopodia numerous, radiating, short, pointed or palmate; diameter 55-135μ; in vegetation.

Genus **Pontigulasia** Rhumbler. Test similar to that of *Difflugia*, but with a constriction of neck and internally a diaphragm made of the same substances as those of the test.

P. vas (Leidy) (Fig. 207, *b*). Round or ovoid test; constriction deep and well-marked; with sand-grains and other particles; aperture terminal; 125-170μ long; fresh water ponds. Stump (1943) made a study of the nuclear division of the organism. During metaphase eight to twelve "chromosomes" form a well-defined equatorial plate; average time for completion of the division was found to be eighty minutes.

Genus **Phryganella** Penard. Test spheroidal or ovoid, with sand-grains and minute diatom shells; aperture terminal, round; pseudopodia drawn out to a point; fresh water.

P. acropodia (Hertwig and Lesser) (Fig. 207, *c*). Test circular in aperture view; hemispherical in profile; yellowish or brownish, semi-transparent, and covered with sand-grains and scales; in front view sharply pointed pseudopodia radiating; colorless endoplasm usually with chlorophyllous bodies; 30-50µ in diameter.

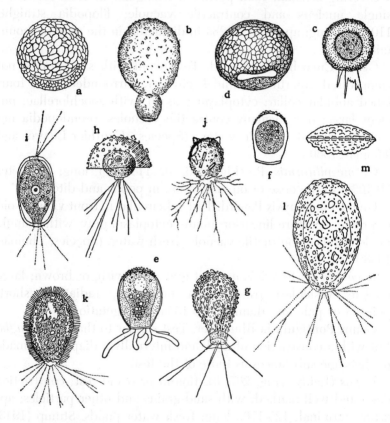

FIG. 207. a, *Plagiopyxis callida*, ×200 (Wailes); b, *Pontigulasia vas* ×200 (Cash); c, *Phryganella acropodia*, ×190 (Cash); d, *Bullinula indica*, ×130 (Wailes); e, f, *Heleopera petricola*, ×190 (Cash); g, *Nadinella tenella*, ×400 (Penard); h, *Frenzelina reniformis*, ×600 (Penard); i, *Amphitrema flavum*, ×360 (Cash and Wailes); j, *Pseudodifflugia gracilis*, ×330 (Cash); k, *Diaphoropodon mobile*, ×270 (Cash and Wailes); l, m, *Clypeolina marginata*, ×330 (Cash and Wailes).

Genus **Bullinula** Penard. Test ellipsoidal, flattened on one face, with silicious plates; on the flattened surface, ∞-shaped aperture; a single nucleus; pseudopodia digitate or spatulate, simple or branched; fresh water.

B. indica P. (Fig. 207, *d*). Test dark brown; 120-250μ in diameter. Distribution and morphology (Hoogenraad, 1933).

Genus **Heleopera** Leidy. Test variously colored; fundus hemispherical, with sand-grains; surface covered with amorphous scales, often overlapping; aperture truncate, narrow, elliptic notches in narrow lateral view; a single nucleus; pseudopodia variable in number, thin digitate or branching; fresh water. (Several species.)

H. petricola L. (Fig. 207, *e, f*). Test variable in size and color, strongly compressed; fundus rough with sand-grains of various sizes; aperture linear or elliptic, convex in front view; pseudopodia slender, branching; 80-100μ long; in boggy places.

Genus **Averintzia** Schouteden. Test similar to that of *Heleopera*, but small aperture elliptical; test thickened around aperture; fresh water.

A. cyclostoma (Penard). Test dark violet, with sand-grains of different sizes; elliptical in cross-section; pseudopodia unobserved; 135-180μ long; in sphagnum and aquatic plants.

Genus **Nadinella** Penard. Test chitinous, thin, hyaline, with foreign bodies and collar around aperture; filopodia; fresh water.

N. tenella P. (Fig. 207, *g*). 50-55μ long; fresh water lakes.

Genus **Frenzelina** Penard. Two envelopes, outer envelope hemispherical, thin, rigid, covered with siliceous particles; inner envelope round or ovoid, drawn out at aperture, thin, hyaline and covering the body closely; aperture round, through which a part of body with its often branching straight filopods extends; cytoplasm with diatoms, etc.; a nucleus and a contractile vacuole; fresh water.

F. reniformis P. (Fig. 207, *h*). Outer envelope 26-30μ in diameter; fresh water lakes.

Genus **Amphitrema** Archer. Test ovoid, symmetrical, compressed; composed of a transparent membrane, with or without adherent foreign bodies; two apertures at opposite poles; with zoochlorellae; nucleus central; one to several contractile vacuoles;

straight filopodia, sparsely branched, radiating; fresh water. (Several species.)

A. *flavum* A. (Fig. 207, *i*). Test brown, cylindrical with equally rounded ends in front view; elliptical in profile; ovoid with a small central oval aperture in end view; 45-77μ by 23-45μ; in sphagnum.

Genus **Pseudodifflugia** Schlumberger. Test ovoid, usually rigid, with foreign bodies; circular or elliptical in cross-section; aperture terminal; granulated cytoplasm colorless or greyish; nucleus posterior; a contractile vacuole; filopodia long, straight or branching; fresh water. (Several species.)

P. *gracilis* S. (Fig. 207, *j*). Test yellowish or brownish; subspherical, with sand-grains; aperture without neck; 20-65μ long.

Genus **Diaphoropodon** Archer. Test ovoid, flexible, with minute foreign bodies and a thick covering of hyaline hair-like projections; pseudopodia long, filose, branching; fresh water.

D. *mobile* A. (Fig. 207, *k*). Test brown; of various shapes; aperture terminal; body does not fill the test; nucleus large; one to two contractile vacuoles; 60-120μ long; projections 8-10μ long; in vegetation.

Genus **Clypeolina** Penard. Test ovoid, compressed, formed of a double envelope; outer envelope composed of two valves with scales and particles; inner envelope a membranous sack; long filopodia, often branching; fresh water.

C. *marginata* P. (Fig. 207, *l, m*). Outer test-valves yellow to dark brown; lenticular in cross-section; wide terminal aperture; endoplasm with many small globules; a single nucleus and contractile vacuole; 80-150μ long.

Family 4. **Euglyphidae** Wallich

Tests are made up of scales or platelets.

Genus **Euglypha** Dujardin (*Pareuglypha*, Penard). Test hyaline, ovoid, composed of circular, oval, or scutiofrm siliceous imbricated scales, arranged in longitudinal rows; aperature bordered with regularly arranged denticulate scales; usually with spines; one to two nuclei large, placed centrally; filopodia dichotomously branched; contractile vacuoles; fresh water. Numerous species.

(Division and encystment, Ivanić, 1934; African species, Thomas and Gauthier-Lievre, 1959.)

E. acanthophora Ehrenberg (*E. alveolata* D.) (Fig. 77). Test ovoid, or slightly elongate; three to seven scales protruding around the circular aperture; scales elliptical; body almost fills the test; 50-100µ long.

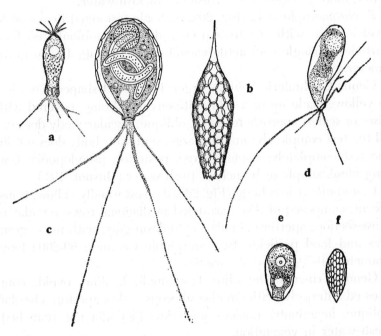

Fig. 208. a, *Euglypha cristata*, ×330 (Wailes); b, *E. mucronata*, ×330 (Wailes); c, *Paulinella chromatophora*, ×1000 (Wailes); d, *Cyphoderia ampulla*, ×200 (Cash); e, f, *Corythion pulchellum*, ×350 (Wailes).

E. cristata Leidy (Fig. 208, *a*). Test small, elongate with a long neck, fundus with three to eight spines; scales oval; aperture circular, bordered by a single row of five to six denticulate scales; cytoplasm colorless; nucleus posterior; reserve scales are said to be collected around the exterior of aperture, unlike other species in which they are kept within the cytoplasm; 30-70µ long; 12-23µ in diameter; aperture 6-12µ; scales 4.5-9.5µ by 2.5-6.5µ; spines 10-15µ long.

E. mucronata L. (Fig. 208, *b*). Test large; fundus conical, with one to two terminal spines (12-44µ long); aperture circular, bordered by a single row of six to eight denticulate scales; 100-150µ long, diameter 30-60µ; aperture 15-20µ in diameter.

Genus **Paulinella** Lauterborn. Test small ovoid, not compressed; with siliceous scales in alternating transverse rows; aperture terminal; body does not fill the test completely; nucleus posterior; among vegetation in fresh or brackish water.

P. chromatophora L. (Fig. 208, *c*). Scales arranged in eleven to twelve rows; with one to two curved algal symbionts; no food particles; a single contractile vacuole; 20-32µ long; 14-23µ in diameter.

Genus **Cyphoderia** Schlumberger Test retort-shaped; colorless to yellow; made up of a thin chitinous membrane, covered with discs or scales; aperture terminal, oblique, circular; body does not fill the test completely; nucleus large, posterior; body does not fill the test completely; nucleus large, posterior; pseudopodia, few, long filose, simple or branched; fresh water (Husnot, 1943).

C. ampulla (Ehrenberg) (Fig. 208, *d*). Test usually yellow, translucent, composed of discs, arranged in diagonal rows; circular in cross-section; aperture circular; cytoplasm gray, with many granules and food particles; two contractile vacuoles; 60-200µ long; diameter 30-70µ (Several varieties.)

Genus **Trinema** Dujardin. Test small, hyaline, ovoid, compressed anteriorly, with circular siliceous scales; aperture circular, oblique, invaginate; nucleus posterior; filopodia not branched; fresh water in vegetation.

T. enchelys (Ehrenberg) (Fig. 209, *a*). One to two contractile vacuoles; pseudopodia attenuate, radiating; 30-100µ long; 15-60µ wide; scales 4-12µ in diameter.

T. lineare Penard (Fig. 81). Test transparent; scales indistinct; about 35µ by 17µ; filopodia. (Sexual fusion, Dunkerly, 1923.)

Genus **Corythion** Taránek. Test small, hyaline, composed of small oval siliceous plates; compressed; elliptical in cross-section; aperture subterminal, ventral or oblique, and circular or oval; numerous filopodia; fresh water. (Tests, Decloitre, 1960.)

C. pulchellum Penard (Fig. 208, *e, f*). Aperture lenticular; cytoplasm colorless; two to three contractile vacuoles; 25-35µ by 15-20µ; aperture 7-10µ by 3-4µ.

Genus **Plascocista** Leidy. Test ovoid, hyaline, compressed; lenticular in cross-section; with oval or subcircular siliceous scales; apperture wide, linear, with flexible undulate borders; nucleus large, posterior; often with zoochlorellae; filopodia branching and many, generally arising from a protruded portion of cytoplasm; fresh water.

P. spinosa (Carter) (Fig. 209, *b*). Margin of test with spines, either singly or in pairs; 116-174μ by 70-100μ; in sphagnum.

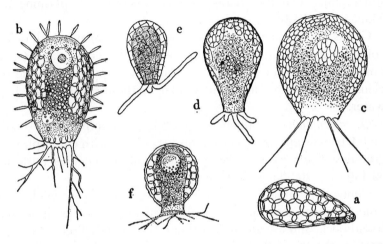

Fig. 209. a, *Trinema enchelys*, ×330 (Wailes); b, *Placocista spinosa*, ×200 (Wailes); c, *Assulina seminulum*, ×400 (Wailes); d, *Nebela collaris*, ×200 (Cash); e, *Quadrula symmetrica*, ×200 (Cash); f, *Sphenoderia lenta*, ×330 (Leidy).

Genus **Assulina** Ehrenberg. Test colorless or brown; ovoid; with elliptical scales, arranged in diagonal rows; aperture oval, terminal bordered by a thin chitinous dentate membrane; nucleus posterior; contractile vacuoles; filopodia divergent, sometimes branching; fresh water.

A. seminulum E. (Fig. 209, *c*). Body does not fill the test; with numerous food particles; pseudopodia few, straight, divergent, slender, seldom branched; 60-150μ by 50-75μ; in sphagnum.

Genus **Nebela** Leidy. Test thin, ovate or pyriform; with circular or oval platelets of uniform or various sizes; highly irregular; endoplasm with oil globules; nucleus posterior; body does not fill the test, and is connected with the latter by many ectoplasmic strands at fundus end; pseudopodia blunt, rarely branched; fresh water. Numerous species. (Taxonomy, Jung, 1942a; Gauthier-Lievre, 1958.)

N. collaris (Ehrenberg) (Fig. 209, *d*). Test pyriform, fundus obtuse in profile; aperture without any notch; endoplasm with chlorophyllous food particles; pseudopodia digitate, short, usually three to six in number; about 130μ by 85-90μ; in marshes among sphagnum. (Feeding habit, binary fission and plasmogamy, MacKinlay, 1936.)

Genus **Quadrula** Schulze. Test pyriform, hemispherical, or discoidal; with quadrangular siliceous or calcareous platelets, arranged generally in oblique series, not overlapping; a single nucleus; body and pseudopodia similar to those of *Difflugia;* fresh water.

Q. symmetrica Wallich (Fig. 209, *e*). Compressed, smaller platelets near aperture; cytoplasm very clear, with chlorophyllous granules; three to five pseudopodia digitate; nucleus posterior; 80-140μ by 40-96μ; in sphagnum.

Genus **Sphenoderia** Schlumberger. Test globular or oval, sometimes slightly compressed; hyaline, membranous, with a short broad neck, and a wide elliptical aperture; scales circular, oval, or hexagonal, arranged in alternating series; cytoplsm colorless; one to two contractile vacuoles; filopodia, fine branching; fresh water.

S. lenta S. (Fig. 209, *f*). Hyaline test ovoid or globular; scales circular or broadly oval; aperture terminal, surrounded by a thin chitinous collar, one side inclined inwards; nucleus large; cytoplasm colorless; two contractile vacuoles; 30-64μ by 20-46μ; aperture 10-22μ in diameter.

References

Bělař, K., 1921. Untersuchungen ueber Thecamoeben der Chlamydophrys-Gruppe. Arch. Protist. 43:287.

Breuer, R., 1916. Fortpflanzung und biologische Erscheinungen einer Chlamydophrys-Form auf Agarkulturen. *Ibid.*, 37:65.

Cash, J., 1905. The British freshwater Rhizopoda and Heliozoa. 1.

—— 1909. 2.

—— and Wailes, G. H., 1915. 3.

—— —— 1918. 4.

CHARDEZ, D., 1960. Introduction a l'étude des Thécamoebiens du Sol. Bull. Inst. Agron. Sta. Rech. Gembloux, 28:118.

————— 1960a. Étude comparée des Thécamoebien, etc. *Ibid.*, 28:132.

————— 1960b. Étude sur deux Difflugia. Hydrobiologia, 16:118.

Decloitre, L., 1956. Materiaux pour une faune Thécamoebienne de Madagascar. Mem. Inst. Sci. Madagascar. Ser. A. Biol. Animale, 11:2.

————— 1960. Structure de la theque du genre Corythion. Hydrobiol., 16: 215.

————— 1960a. Mise au point de systematique dans le genre Lesquereusia. *Ibid.*, 14:278.

————— 1960b. Mise au point de systematique dans le genre Difflugia. *Ibid.*, 14:386.

DEFLANDRE, G., 1928. Le genre Arcella. Arch. Protist., 64:152.

————— 1929. Le genre Centropyxis. *Ibid.*, 67:322.

DUNKERLY, J. S., 1923. Encystation and reserve food formation in *Trinema lineare*. Tr. Roy. Soc. Edinburgh, 53:297.

FÖYN, B., 1936. Ueber die Kernverhältnisse der Foraminifere *Myxotheca arelilega*. Arch. Protist., 87:272.

GAUTHIER-LIEVRE, L., 1958. Additions aux Nebela d'Afrique. Bull. Soc. Hist. Nat. Afrique du Nord, 48:494.

————— and THOMAS, R., 1958. Le genre Difflugia, etc. Arch. Protist., 103: 241.

————— ————— 1960. Le genre Cucurbitella Penard. *Ibid.*, 104:569.

GOETTE, A., 1916. Ueber die Lebenscyclus von *Difflugia lobostoma*. *Ibid.*, 37:93.

GROSPIETSCH, T., 1958. Beitraege zur Rhizopodenfauna Deutschlands. I. Thekamoeben des Rhoen. Hydrobiologia, 10:305.

HEAL, O. W., 1961. The distribution of testate amoebae in some fens and bogs in northern England. J. Linn. Soc., London, Zool., 44:369.

HEDLEY, R. H., and BERTAUD, W. S., 1962. Electron microscopic observations of *Gromia oviformis*. J. Protozool., 9:79.

HEGNER, R. W., 1920. The relation between nuclear number, chromatin mass, etc. J. Exper. Zool., 30:1.

HOOGENRAAD, H. R., 1933. Einige Beobachtungen an *Bullinula indica*. Arch. Protist., 79:119.

————— 1935. Studien ueber die sphagnicolen Rhizopoden der niederländischen Fauna. *Ibid.*, 84:1.

————— 1936. Was ist *Pamphagus mutabilis* Bailey? *Ibid.*, 87:417.

HUSNOT, P., 1943. Contribution à l'étude des Rhizopodes de Bretagne. Les Cyphoderia, etc. 143 pp. Paris.

IVANIĆ, M., 1934. Ueber die gewöhnliche Zweiteilung, multiple Teilung und Encystierung bei zwei Euglypha-Arten. Arch. Protist., 82:363.

JENNINGS, H. S., 1916. Heredity, variation and the results of selection in the uniparental reproduction of *Difflugia corona*. Genetics, 1:407.

—— 1937. Formation, inheritance and variation of the teeth in *Difflugia corona*. J. Exper. Zool., 77:287.

JEPPS, MARGARET W., 1934. On *Kibisidytes marinus*, etc. Quart. J. Micr. Sci., 77:121.

JOLLOS, V., 1924. Untersuchungen ueber Variabilität und Vererbung bei Arcellen. Arch. Protist., 49:307.

JUNG, W., 1942. Südchilenische Thekamoeben. *Ibid.*, 95:253.

—— 1942a. Illustrierte Thekamoeben-Bestimmungstabellen. I. *Ibid.*, 95: 357.

LEIDY, J., 1879. Freshwater Rhizopods of North America. Rep. U. S. Geol. Surv. Terr., 12.

MACKINLAY, ROSE B., 1936. Observations on *Nebela collaris*, etc. J. Roy. Micr. Soc., 56:307.

PENARD, E., 1890. Études sur les rhizopods d'eau douce. Mém. soc. phys. hist. nat., Geneva, 31:1.

—— 1902. Faune rhizopodique du bassin du Léman. Geneva.

—— 1905. Sarcodinés des Grands Lacs. Geneva.

STEPANEK, M., 1952. Testacea of the pond of Hradek at Kunratice (Prague). Acta Mus. Nat. Prag., 8, B:1.

STUMP, A. B., 1943. Mitosis and cell division in *Pontigulasia vas*. J. El. Mitch. Sci. Soc., 59:14.

—— 1959. Mitosis in the rhizopod *Lesquereusia spiralis*. J. Protozool., 6:185.

THOMAS, R., 1958. Le genre Plagiopyxis Penard. Hydrobiol., 10:198.

—— and GAUTHIER-LIEVRE, L. 1959. Note sur quelques Euglyphidae, etc. Bull. Soc. hist. nat. Afrique du Nord, 50:204.

VALKANOV, A., 1930. Morphologie und Karyologie der *Microgromia elegantula*. Arch. Protist., 71:241.

—— 1931. Beitrag zur Morphologie und Karyologie der *Microcometes paludosa*. *Ibid.*, 73:367.

VOLZ, P., 1929. Studien zur Biologie der bodenbewohnenden Thekamoeben. *Ibid.*, 69:348.

Chapter 22
Order 5. **Foraminiferida** d'Orbigny

THE Foraminiferida are comparatively large protozoa, living almost exclusively in the sea. They were very abundant in geologic times and the fossil forms are important in applied geology (p. 9). The majority live on ocean bottom, moving about sluggishly over the mud and ooze by means of their pseudopodia. Some are attached to various objects on the ocean floor, while others are pelagic. Bradshaw (1959) examined planktonic foraminifers from north and equatorial Pacific ocean and found four groups of fauna: a cold-water, a transition and two warm-water faunas, which were differentiated by characteristic temperature and salinity values. The distribution of inorganic phosphate may influence the abundance of these organisms.

The cytoplasm is ordinarily not differentiated into the two zones and streams out through the apertures, and in perforated forms through the numerous pores, of the shell, forming rhizopodia which are fine and often very long and which anastomose with one another to present a characteristic appearance (Fig. 5). The streaming movement of the cytoplasm in the pseudopodia are quite striking; the granules move toward the end of a pseudopodium and stream back along its periphery; in others the granules may move out on one side of a pseudopod and move back on the other side. The body cytoplasm is often loaded with brown granules which are apparently waste matter and in some forms such as *Peneroplis pertusus* these masses are extruded from the body from time to time, especially prior to the formation of a new chamber.

The test of the Foraminiferida varies greatly in form and structure. It may show various colorations—orange, red, brown, etc. The majority measure less than one millimeter, although larger forms may frequently reach several millimeters. The test may be siliceous or calcareous and in some forms, various foreign materials, such as sand-grains, sponge-spicules, etc. which are more or

589

less abundantly found where these organisms live, are loosely or compactly cemented together by pseudochitinous or gelatinous substances. Certain forms show a specific tendency in the selection of foreign materials for the test (p. 53). Siliceous tests are comparatively rare, being found in some species of Miliolidae inhabiting either the brackish water or deep sea. Calcareous tests are sometimes imperforated, but even in such cases those of the young are always perforated. By far the majority of the Foraminiferida possess perforated calcareous tests. The thickness of the shell varies considerably, as do also the size and number of apertures, among different species. Frequently, the perforations are very small in the young and later become large and coarse, while in others the reverse may be the case.

The form of the shell varies greatly. In some there is only one chamber composed of a central body and radiating arms which represent the material collected around the pseudopodia, as in Rhabdammina (Fig. 212, *a*), or of a tubular body alone, as in Hyperammina (Fig. 212, *d*). The polythalamous forms possess shells of various spirals. The first chamber is called the **proloculum** which may be formed either by the union of two gametes or by asexual reproduction. The former is ordinarily small and known as the **miscrospheric** proloculum, while the latter, which is usually large, is called the **megalospheric** proloculum. To the proloculum are added many chambers which may be closely or loosely coiled or not coiled at all. These chambers are ordinarily undivided, but in many higher forms they are divided into chamberlets. The chambers are delimited by the suture on the exterior of the shell. The septa which divide the chambers are perforated by one or more foramina known as stolon canals, through which the protoplasm extends throughout the chambers. The last chamber has one or more apertures of variable sizes, through which the cytoplasm extends to the exterior as pseudopodia. The food of Foraminiferida consists mostly of diatoms and algae, though pelagic forms are known to capture other protozoa and microcrustaceans.

All species of Foraminiferida manifest a more or less distinct tendency toward a dimorphism: the **megalospheric form** (*gamout*) has a large proloculum and is uninucleate, while the **microspheric form** (*agamont*) possesses a small proloculum, is multinucleate and is large. In some species, there is a difference in the direction of the spiral chambers of tests (Myers). For example, in *Discorbis opercularis,* the microspheric form has clockwise rotation of the chambers, and the megalospheric form shows counterclockwise rotation. The megalospheric forms are said to be much more numerous than the microspheric forms, especially in pelagic species. It is possible that, as Myers (1938) pointed out, the flagellate gametes are set free in open water and have a minimum of opportunity for syngamy.

Lister (1895) observed the development of the megalospheric form in Elphidium by asexual reproduction from the microspheric form. He noticed flagellated swarmers in megalospheric tests and considered them as gametes which through syngamy gave rise to microspheric individuals. Studies by Myers (1935–1940) confirm the correctness of this view, except that in some species the gametes are amoeboid. In *Spirillina vivipara* (Fig. 210, A, *1–5*), the mature microspheric form (*1*) which measures 125–152μ in diameter, becomes surrounded by an envelope composed of substrate debri and viscous substance. Within the "multiple fission cyst," nuclear and cytoplasmic fissions form numerous small uninucleate megalospheric individuals which produce tests and emerge from the cyst (*2*). They grow into mature megalospheric forms which measure 60–72μ in diameter. Two to four such individuals become associated and transform into "fertilization cyst" (*3*). The nucleus in each individual divides twice or occasionally three times and thus formed multinucleate bodies escape from the tests within the cyst envelope where many gametocytes are produced by multiple fissions. Each gametocyte which contain twelve chromosomes divides into two amoeboid haploid gametes by meiosis. Gametes developed from different parents presumably undergo fusion in pairs and zygotes are produced (*4*). Each zygote becomes proloculum in which the nucleus divides twice and when the coiled tubular chamber of test grows to about three-quarters of a whorl, young microspheric individuals escape from the cyst and lead independent existence (*5*).

In *Discorbis patelliformis* (Fig. 210, B, *1–5*), the same investigator noticed no fertilization cyst during the sexual reproduction, but two megalospheric individuals come in contact and flagellate

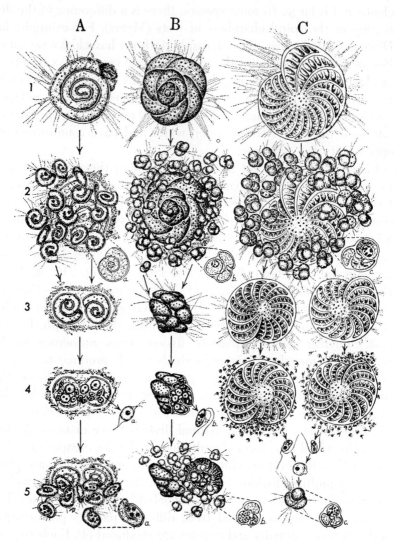

FIG. 210. Developmental cycles of Foraminiferida (Myers). A, *Spirillina vivipara;* B, *Discorbis patelliformis;* C. *Elphidium crispa.* 1, microspheric forms; 2, megalospheric forms, a–c, enlarged views of young megalospheric forms; 3, beginning of sexual reproduction; 4, gamete and zygote formation, a–c, gametes; 5, young microspheric forms, a–c, enlarged views of one in each species.

gametes are produced in them. The zygotes develop within the space formed by the dissolution of septa between chambers and tests; the zygote nucleus divides repeatedly within each zygote and forms about forty nuclei before a test is secreted. In *Elphidium crispa* (Fig. 210, C, *1–5*), there is no direct association of

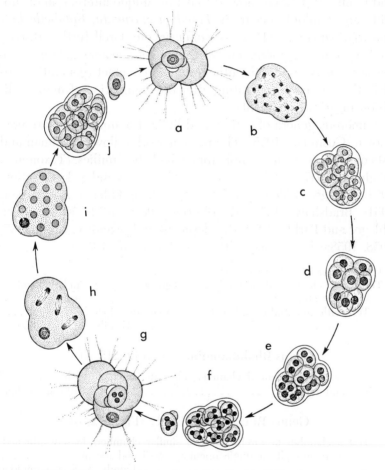

Fig. 211. The developmental cycle of *Rotaliella heterocaryotica* (Grell). a, grown gamont; b, mitosis; c, autogamy; d, zygotes; e, binucleate agamonts; f, 4-nucleated agamonts (each with 1 somatic and 3 generative nuclei); g, grown agamont (generative nuclei in proloculum); h, first meiosis; i, end of second meiosis with the degenerating somatic nucleus; j, formation of young agamonts.

megalospheric individuals during sexual reproduction. The flagellated gametes produced in each, are set free in the water and the fusion of the gametes depends entirely upon chance meeting.

In *Patellina corrugata* and *Discorbis vilardeboanus,* Le Calvez (1950) found the gametes are haploid and the agamonts diploid and Grell (1954-1958) showed that this unique alternation of diploidy and haploidy occurs in *Patellina corrugata, Rotaliella heterocaryotica* (Fig. 211), *R. roscoffensis,* etc. Grell further discovered in young agamonts one of the nuclei becomes enlarged (somatic nucleus), while the others remain compact (generative nuclei), the former degenerating after the completion of nuclear division in grown agamonts.

Numerous genera of extinct and living Foraminiferida are now known (Cushman, 1955). The present work follows Cushman and lists one genus as an example for each of the families. (Taxonomy, Höglund, 1947; Cushman, 1948, 1955; G. Vander Foundation, 1961; ecology, Phleger and Walton, 1950; Phleger and Parker, 1951; Bradshaw, 1959; distribution, Post, 1951; Illing, 1952; Phleger and Parker, 1954; development and sexuality, Grell, 1957, 1958, 1958a; laboratory cultivation, Lee *et al.*, 1961.)

Test entirely or in part arenaceous
  Test single-chambered or rarely an irregular group of similar chambers
    loosely attached
    Test with a central chamber, two or more arms; fossil and recent . . . .
    . . . . .Family 1: Astrorhizidae

Genus **Rhabdammina** Sars (Fig. 212, *a*)

Test without a central chamber, elongate, open at both ends; fossil
  and recent . . . . .Family 2. Rhizamminidae

Genus **Rhizammina** Brady (Fig. 212, *b*)

Test a chamber or rarely series of similar chambers loosely attached,
  with normally a single opening; fossil and recent . . . .
  . . . . .Family 3. Saccamminidae

Genus **Saccammina** Sars (Fig. 212, *c*)

Test two-chambered, a proloculum and long undivided tubular second
  chamber
    Test with the second chamber, simple or branching, not coiled;
      mostly recent and also fossil . . . . . .Family 4. Hyperamminidae

Genus **Hyperammina** Brady (Fig. 212, *d*)

Test with the second chamber usually coiled at least in young
 Test of arenaceous material with much cement, usually yellowish
 or reddish brown; fossil and recent. Family 5. Ammodiscidae

Genus **Ammodiscus** Reuss (Fig. 212, *e*)

Test of siliceous material, second chamber partially divided; fossils
 only ....Family 6. Silicinidae

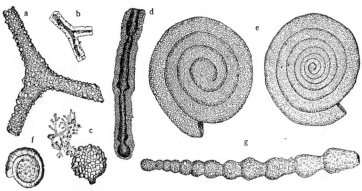

FIG. 212. a, *Rhabdammina abyssorum*, ×5 (Kühn); b, *Rhizammina algae-formis*, fragment of, ×14 (Cushman); c, *Saccammina sphaerica*, ×8 (Rhumbler); d, *Hyperammina subnodosa*, ×4 (Brady); e, *Ammodiscus incertus*, ×20 (Kühn); f, *Silicina limitata*, ×13 (Cushman); g, *Reophax nodulosus*, ×3 (Brady).

Genus **Silicina** Bornemann (Fig. 212, *f*)

Test typically many-chambered
 Test with all chambers in a rectilinear series; fossil and recent ......
 ....Family 7. Reophacidae

Genus **Reophax** Montfort (Fig. 212, *g*)

Test planispirally coiled at least in young
 Axis of coil, short; many uncoiled forms; fossil and recent ......
 ....Family 8. Lituolidae

Genus **Lituola** Lamarck (Fig. 213, *a*)

Axis of coil usually long, all close-coiled
 Interior not labyrinthic; fossil only ....Family 9. Fusulinidae

Genus **Fusulina** Fisher (Fig. 213, *b*)

Interior labydinthic; fossil only ....Family 10. Loftusiidae

Genus **Loftusia** Brady

Test typically biserial at least in young of microspheric form; fossil and recent ....Family 11. Textulariidae

Genus **Textularia** Defrance (Fig. 213, c)

Test typically triserial at least in young of microspheric form
Aperture usually without a tooth, test becoming simpler in higher forms; fossil and recent ....Family 12. Verneuilinidae

Genus **Verneuilina** d' Orbigny (Fig. 213, d)

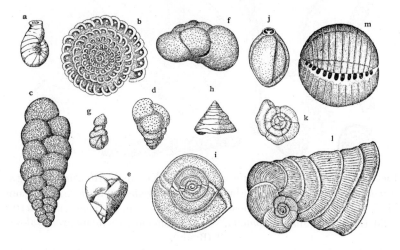

FIG. 213. a, *Lituola nautiloidea* (Cushman); b, section through a Fusulina (Carpenter); c. *Textularia agglutinans*, ×90 (Rhumbler); d. *Verneuilina propinqua*, ×8 (Brady); e, *Valvulina triangularis*, (d'Orbigny); f, *Trochammina inflata*, ×32 (Brady); g, *Placopsilina cenomana* (Reuss); h, *Tetrataxis palaeotrochus*, ×15 (Brady); i, *Spiroloculina limbata*, ×20 (Brady); j, *Triloculina trigonula*, ×15 (Brady); k, *Fischerina helix*, ×32 (Heron-Allen and Earland); 1, *Vertebralina striata*, ×40 (Kühn); m, *Alveolinella mello*, ×35 (Brady).

Aperture typically with a tooth, test becoming conical in higher forms; fossil and recent ....Family 13. Valvulinidae

Genus *Valvulina* d' Orbigny (Fig. 213, e)

Test with whole body labyrinthic, large, flattened, or cylindrical; recent ....Family 14. Neusinidae

Genus **Neusina** Goës

Test trochoid at least while young
    Mostly free, typically trochoid throughout; fossil and recent ....
    ....Family 15. Trochamminidae

Genus **Trochammina** Parker and Jones (Fig. 213, *f*)

Attached; young trochoid, later stages variously formed; fossil and
    recent ....Family 16. Placopsilinidae

Genus **Placopsilina** d'Orbigny (Fig. 213, *g*)

Free; conical, mostly of large size; fossil only ....
    ....Family 17. Orbitolinidae

Genus **Tetrataxis** Ehrenberg (Fig. 213, *h*)

Test coiled in varying planes, wall imperforate, with arenaceous
    portion only on the exterior; fossil and recent ....
    ....Family 18. Miliolidae (in part)

Genus **Spiroloculina** d'Orbigny (Fig. 213, *i*)

Test calcareous, imperforate, porcellaneous
    Test with chambers coiled in varying planes, at least in young; aperture
    large, toothed; fossil and recent ..Family 18. Miliolidae (in part)

Genus **Triloculina** d'Orbigny (Fig. 213, *j*)

Test trochoid; fossil and recent ....Family 19. Fischerinidae

Genus **Fischerina** Terquem (Fig. 213, *k*)

Test planispiral at least in young
    Axis very short, chambers usually simple; fossil and recent ....
    ....Family 20. Ophthalmidiidae

Genus **Vertebralina** d'Orbigny (Fig. 213, *l*)

Axis short, test typically compressed and often discoid, chambers
    mostly with many chamberlets; fossil and recent ....
    ....Family 21. Peneroplidae

Genus **Peneroplis** Montfort (Figs. 4; 214)

Axis typically elongate, chamberlets developed; mainly fossil ......
    ....Family 22. Alveolinellidae

Genus **Alveolinella** Doubillé (Fig. 213, *m*)

Test globular, aperture small, not toothed; recent only . . . .
. . . . .Family 23. Keramosphaeridae

Genus **Keramosphaera** Brady

Test calcareous, perforate
Test vitreous with a glassy lustre, aperture typically radiate, not trochoid

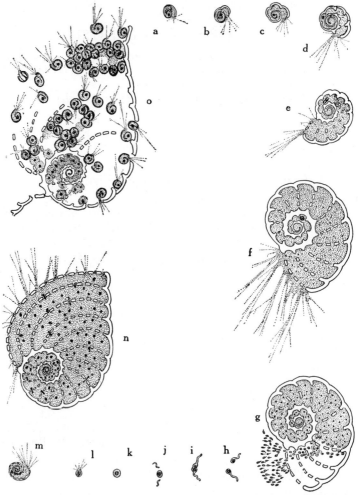

Fig. 214. Diagram illustrating the life-cycle of *Peneroplis pertusus* (Winter). a–f, megalospheric generation; g, gamete formation; h–k, isogamy; l–n, microspheric generation; o, multiple division.

Test planispirally coiled or becoming straight, or single-chambered; fossil and recent ....Family 24. Lagenidae

Genus **Lagena** Walker and Jacob (Fig. 215, *a*)

Test biserial or elongate spiral; fossil and recent ....
..............................Family 25. Polymorphinidae

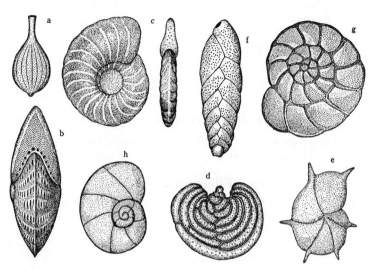

FIG. 215. a, *Lagena striata*, ×50 (Rhumbler); b, *Elphidium strigilata*, ×40 (Kühn); c, *Operculina ammonoides*, ×50 (Kühn); d, *Pavonina flabelliformis*, ×30 (Braly); e, *Hantkenina alabamensis*, ×40 (Cushman); f, *Bolivina punctata*, ×100 (Kühn); g, *Rotalia beccarii*, ×40 (Kühn); h, *Asterigerina carinata*, ×30 (d'Orbigny from Kühn).

Genus **Polymorphina** d'Orbigny

Test not vitreous; aperture not radiating
    Test planispiral, occasionally trochoid, then usually with processes along the suture lines, septa single, no canal system; fossil and recent ....Family 26. Nonionidae

Genus **Elphidium** Montfort (Figs. 5; 210, C; 215, *b*)
(*Polystomella*, Lamarck)

Test planispiral, at least in young, generally lenticular, septa double, canal system in higher forms; fossil and recent ....
..................................Family 27. Camerinidae

Genus **Operculina** d'Orbigny (Fig. 215, *c*)

Test generally biserial in at least microspheric form, aperture usually large, without teeth; fossil and recent ....
..............................Family 28. Heterohelicidae

Genus **Pavonina** d'Orbigny (Fig. 215, *d*)

Test planispiral, bi- or tri-serial with elongate spines and lobed aperture; fossil and recent ....Family 29, Hantkeninidae

Genus **Hantkenina** Cushman (Fig. 215, *e*)

Test typically with an internal tube, elongate
Aperture generally loop-shaped or cribrate; fossil and recent ...
..............................Family 30. Buliminidae

Genus **Bolivina** d'Orbigny (Fig. 215, *f*)

Aperture narrow, curved, with an overhanging portion; mostly fossil, also recent ....Family 31. Ellipsoidinidae

Genus **Ellipsoidina** Seguenza

Test trochoid, at least in young of microspheric form, usually coarsely perforate; when lenticular, with equatorial and lateral chambers
Test trochoid throughout, simple; aperture ventral
No alternating supplementary chambers on ventral side; fossil and recent ....Family 32. Rotaliidae

Genus **Rotalia** Lamarck (Fig. 215, *g*)

Genus **Spirillina** Ehrenberg (Fig. 210, A)

Genus **Patellina** Williamson

Genus **Discorbis** Lamarck (Fig. 210, B)

Genus **Rotaliella** Grell (1954)

Alternating supplementary chambers on ventral side; fossil and recent ....Family 33. Amphisteginidae

Genus **Asterigerina** d'Orbigny (Fig. 215, *h*)

Test trochoid and aperture ventral in young
With supplementary material and large spines, independent of chambers; fossil and recent ....Family 34. Calcarinidae

Genus **Calcarina** d'Orbigny (Fig. 216, *a*)

With later chambers in annular series or globose with multiple apertures, but not covering earlier ones; fossil and recent . . . .
. . . . .Family 35. Halkyardiidae

Genus **Halkyardia** Heron-Allen and Earland (Fig. 216, *b*)

With later chambers somewhat biserial; aperture elongate in the axis of coil; fossil and recent . . . .Family 36. Cassidulinidae

Genus **Cassidulina** d'Orbigny (Fig. 216, *c*)

With later chambers becoming involute, very few making up the exterior in adult; aperture typically elongate, semicircular; in a few species circular; fossil and recent . . . .
. . . . .Family 37. Chilostomellidae

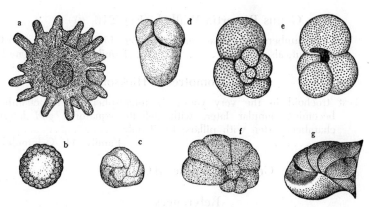

FIG. 216. a, *Calcarina defrancei*, ×25 (Brady); b, *Halkyardia radiata*, ×15 (Cushman); c, *Cassidulina laevigata*, ×25 (Brady); d, *Allomorphina trigona*, ×40 (Brady); e, *Globigerina bulloides*, ×30 (Kühn); f, *Anomalina punctulata* (d'Orbigny); g, *Rupertia stabilis*, ×50 (Brady).

Genus **Allomorphina** Reuss (Fig. 216, *d*)

With chambers mostly finely spinose and wall cancellated, adapted, for pelagic life, globular forms with the last chamber completely involute; aperture umbilicate or along the suture; fossil and recent . . . . .Family 38. Globigerinidae

Genus **Globigerina** d'Orbigny (Fig. 216, *e*)

Early chambers globigerine, later ones spreading and compressed; fossil and recent . . . . .Family 39. Globorotaliidae

Genus **Globorotalia** Cushman

Test trochoid at least in young, aperture peripheral or becoming dorsal
Mostly attached, dorsal side usually flattened; fossil and recent
.............................Family 40. Anomalinidae

Genus **Anomalina** d'Orbigny (Fig. 216, *f*)

Later chambers in annular series; fossil and recent ....
............................Family 41. Planorbulinidae

Genus **Planorbulina** d'Orbigny

Test trochoid in very young, later growing upward
Later chambers in loose spiral; fossil and recent ....
................................Family 42. Rupertiidae

Genus **Rupertia** Wallich (Fig. 216, *g*)

Later chambers in masses or branching, highly colored; mostly
recent, also fossil ....Family 43. Homotremidae

Genus **Homotrema** Hickson

Test trochoid in the very young of microspheric form, chambers
becoming annular later, with definite equatorial and laterial
chambers, often with pillars; fossil only ....
...................................Family 44. Orbitoididae

Genus **Orbitoides** d'Orbigny

References

BRADSHAW, J. S., 1959. Ecology of living planktonic Foraminifera in the north and equatorial Pacific Ocean. Contr. Cushman Found. Foram. Res., 10:25.

BRADY, B. H. 1884. Report on the Foraminifera dredged by H.M.S. *Challenger,* during the years 1873–1876. Rep. Voy. Chall., 9.

CUSHMAN, J. A., 1955. Foraminifera: their classification and economic use. 5th ed. Cambridge, Mass.

GEORGE VANDERBILT FOUNDATION. 1961. An index to the genera and species of the Foraminifera, 1890–1950. Stanford Univ. 393 pp.

GRELL, K. G., 1954. Die Generationswechsel der polythalamen Foraminifere *Rotaliella heterocaryotica.* Arch. Protist., 100:268.

——— 1957–1958. Untersunchungen ueber die Fortpflanzung und Sexualität der Foraminiferen. I, II, III. *Ibid.,* 102:147, 291, 449.

——— 1958. Studien zum Differenzierungsproblem an Foraminiferen. Die Naturwiss., 45:25.

HÖGLUND, H., 1947. Foraminifera in the Gullmar Fjord and the Akagerak. Zool. Bidrag. Fr. Uppsala, 26:1.

ILLING, M. A., 1952. Distribution of certain Foraminifera within the littoral zone on the Bahama Banks. Ann. Mag. Nat. Hist., 5:275.

LE CALVEZ, J., 1950. Recherches sur les Foraminifères. II. Arch. zool. expér. gén., 87:211.

LEE, J. J., *et al.*, 1961. Growth and physiology of Foraminifera in the laboratory: Part I. Collection and maintenance. Micropaleontology, 7:461.

MYERS, E. H., 1935. The life history of *Patellina corrugata*, etc. Bull. Scripps Inst. Oceanogr., Univ. Cal. Tech. Ser., 3:355.

—— 1936. The life-cycle of *Spirillina vivipara* Ehrenberg, with notes on morphogenesis, etc. J. Roy. Micr. Soc., 56:126.

—— 1938. The present state of our knowledge concerning the life cycle of the Foraminifera. Proc. Nat. Acad. Sci., 24:10.

—— 1940. Observations on the origin and fate of flagellated gametes in multiple tests of Discorbis. J. Mar. Biol. Assn. United Kingdom, 24:201.

PHLEGER, F. B., 1951. Ecology of Foraminifera, northwest Gulf of Mexico. I. Mem. Geol. Soc. America, 46:1.

—— and PARKER, F. L., 1951. II. *Ibid.*, 46:89.

—— —— 1954. Gulf of Mexico Foraminifera. U.S. Dept. Inter., Fish. Bull., 89:235.

—— and WALTON, W. R., 1950. Ecology of marsh and bay Foraminifera, Barnstable, Mass. Am. J. Sci. 248:274.

POST, RITA J., 1951. Foraminifera of the south Texas coast. Publ. Inst. Mar. Sci. 2:165.

RHUMBLER, L., 1904. Systematische Zusammenstellung der rezenten Reticulosa (Nuda u. Foraminifera). I. Arch. Protist., 3:181.

Chapter 23
Subclass 2. **Actinopoda** Calkins

THE Actinopoda are divided into two orders: Heliozoida and Radiolarida (p. 616).

Order 1. **Heliozoida** Haeckel

The Heliozoida are, as a rule, spherical in form with many radiating axopodia. The cytoplasm is differentiated, distinctly in Actinosphaerium, or indistinctly in other species, into the coarsely vacuolated ectoplasm and the less transparent and vacuolated endoplasm. The food of Heliozoida consists of living protozoa or protophyta; thus their mode of obtaining nourishment is holozoic. A large organism may sometimes be captured by a group of Heliozoida which gather around the prey. When an active ciliate or a small rotifer comes in contact with an axopodium, it seems to become suddenly paralyzed and, therefore, it has been suggested that the pseudopodia contain some poisonous substances. The axial filaments of the axopodia disappear and the pseudopodia become enlarged and surround the food completely. Then the food matter is carried into the main part of the body and is digested. The ectoplasm contains several contractile vacuoles and numerous refractile granules which are scattered throughout. The endoplasm is denser and usually devoid of granules. In the axopodium, the cytoplasm undergoes streaming movements. The hyaline and homogeneous axial filament runs straight through both the ectoplasm and the endoplasm, and terminates in a point just outside the nuclear membrane. When the pseudopodium is withdrawn, its axial filament disappears completely, through the latter sometimes disappears without the withdrawal of the pseudopodium itself. In Acanthocystis, the nucleus is eccentric (Fig. 219, *b*), but there is a central granule, or centroplast, in the center of the body from which radiate the axial filaments of the axopodia. In multinucleate Actinosphaerium, the axilia filaments terminate at the periphery of the endoplasm. In Camptonema, an axial filament arises from each of the nuclei (Fig. 217,*d*).

The skeletal structure of the Heliozoida varies among different species. The body may be naked, covered by a gelatinous mantle, or provided with a lattice-test with or without spicules. The spicules are variable in form and location and may be used for specific differentiation. In some forms there occur colored bodies bearing chromatophores, which are considered as holophytic Mastigophora (p. 30) living in the helioboans as symbionts.

The Heliozoida multiply by binary fission or budding. Incomplete division may result in the formation of colonies, as in Rhaphidiophrys. In Actinosphaerium, nuclear phenomena have been studied by several investigators (p. 243). In Acanthocystis and Oxnerella (Fig. 62), the central granule behaves somewhat like the centriole in a metazoan mitosis. Budding has been known in numerous species. In Acanthocystis, the nucleus undergoes amitosis several times, thus forming several nuclei, one of which remains in place while the other migrates toward the body surface. Each peripheral nucleus becomes surrounded by a protruding cytoplasmic body which becomes covered by spicules and which is set free in the water as a bud. These small individuals are supposed to grow into larger forms, the central granules being produced from the nucleus during the growth. Formation of swarmers is known in a few genera and sexual reproduction occurs in some forms. The Heliozoida live chiefly in fresh water, although some inhabit the sea. (Taxonomy and morphology, Penard, 1905, 1905a; Cash and Wailes, 1921; Roskin, 1929; Valkanov, 1940.) Nine families.

Family 1. **Actinocomidae** Poche

Genus **Actinocoma** Penard. Body spherical; one or more contractile vacuoles; nucleus with a thick membrane, central; filopodia, not axopodia, simple or in brush-like groups; fresh water.

A. ramosa P. (Fig. 217, *a*). Average diameter 14-26μ.

Family 2 **Actinophryidae** Claus

Axopods radiating; cytoplasm is highly vacuolated.

Genus **Actinophrys** Ehrenberg. Spheroidal; cytoplasm highly

vacuolated, especially ectoplasm; with often symbiotic zoochlorellae; nucleus central; one to many contractile vacuoles; axopodia straight, numerous, axial filaments terminate at surface of
the nucleus; "sun animalcules;" fresh water.

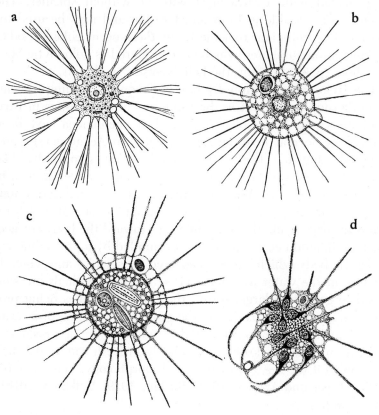

Fig. 217. a, *Actinocoma ramosa*, ×630 (Penard); b, *Actinophrys sol*, ×400
(Kudo); c, *Actinosphaerium eichhorni*, ×45 (Kudo); d, *Camptonema
nutans*, ×350 (Schaudinn).

A. *sol* E. (Figs. 92; 217, *b*). Spherical; ectoplasm vacuolated; endoplasm granulated with numerous small vacuoles; a large central nucleus; solitary but may be colonial when young; diameter
variable, average being 40-50μ; among plants in still fresh water.
(Reproduction, morphology and physiology, Bělař, 1923, 1924;
food habit, Looper, 1928.)

A. vesiculata Penard. Ectoplasm with saccate secondary vesicles, extending out of body surface between axopodia; nucleus central, with many endosomes; 25-30µ in average diameter; fresh water.

Genus **Actinosphaerium** Stein. Spherical; ectoplasm consists almost entirely of large vacuoles in one or several layers; endoplasm with numerous small vacuoles; numerous nuclei; axopodia end in the inner zone of ectoplasm (Fig. 6).

A. eichhorni Ehrenberg (Figs. 6;217, *c*). Numerous nuclei scattered in the periphery of endoplasm; two or more contractile vacuoles, large; axial filaments arise from a narrow zone of dense cytoplasm at the border line between endoplasm and ectoplasm; body large, diameter 200-300µ, sometimes up to 1 mm.; nuclei 12-20µ in diameter; among vegetation in freshwater bodies. (Nuclear change, Speeth, 1919; morphology, Rumjantzew and Wermel, 1925; transplantation, Okada, 1930; plasmogamy, Kuhl, 1953.)

A. nucleofilum Barrett. Similar to *eichhorni* in general appearance; nuclei are similar to that of *Actinophrys sol*, 4-8µ in diameter; filaments of pseudopods arise from the nuclei which are located in the peripheral zone of the endoplasm; body 230-400µ in diameter; fresh water (Barrett, 1958). (Structure under electron microscope, Anderson and Beams, 1960.)

A. arachnoideum Penard. Ectoplasm irregularly vacuolated; no distinct endoplasmic differentiation; nuclei smaller in number; pseudopodia of two kinds; one straight, very long and the other filiform, and anastomosing; 70-80µ in diameter; fresh water.

Genus **Camptonema** Schaudinn. Spheroidal; axial filaments of axopodia end in nuclei about fifty in number; vacuoles numerous and small in size; salt water.

C. nutans S. (Fig. 217, *d*). About 150µ in diameter.

Genus **Oxnerella** Dobell. Spherical; cytoplasm indistinctly differentiated; eccentric nucleus with a large endosome; axial filaments take their origin in the central granule; no contractile vacuole; nuclear division typical mitosis (Fig. 62).

O. maritima D. (Fig. 62). Small, 10-22µ in diameter; solitary, floating or creeping; salt water.

Family 3. **Ciliophryidae** Poche

Genus **Ciliophrys** Cienkowski. Spherical with extremely fine radiating filopodia, giving the appearance of a typical heliozoan, with a single flagellum which is difficult to distinguish from the numerous filopodia, but which becomes conspicious when the pseudopodia are withdrawn; fresh or salt water.

C. infusionum C. (Fig. 218, *a*). 25-30μ long; freshwater infusion.

C. marina Caullery. About 10μ in diameter; salt water.

Family 4. **Lithocollidae** Poche

Genus **Lithocolla** Schulze. Spherical body; outer envelope with usually one layer of sand-grains, diatoms, etc.; nucleus eccentric.

L. globosa S. (Fig. 218, *b*). Body reddish with numerous small colored granules; nucleus large; central granule unknown; envelope 35-50μ in diameter; in lakes, ponds, and rivers; also in brackish water.

Genus **Astrodisculus** Greeff. Spherical with gelatinous envelope, free from inclusions, sometimes absent; no demarcation between two regions of the cytoplasm; pseudopodia fine without granules; fresh water.

A. radians G. (Fig. 218, *c*). Outer surface usually with adherent foreign bodies and bacteria; cytoplasm often loaded with green, yellow, or brown granules; nucleus eccentric; a contractile vacuole; diameter 25-30μ including envelope; in pools and ditches.

Genus **Actinolophus** Schulze. Body pyriform, enveloped in a gelatinous mantle; stalked; apparently hollow; axopodia long, numerous; nucleus eccentric; salt water.

A. pedunculatus S. (Fig. 218, *d*). Diameter about 30μ; stalk about 100μ long.

Genus **Elaeorhanis** Greeff. Spherical; mucilaginous envelope with sand-grains and diatoms; cytoplasm with a large oil globule; nucleus eccentric; one or more contractile vacuoles; pseudopodia not granulated, sometimes forked; fresh water.

E. cincta G. (Fig. 218, *e*). Bluish with a large yellow oil globule; without any food particles; no central granule; pseudopodia rigid,

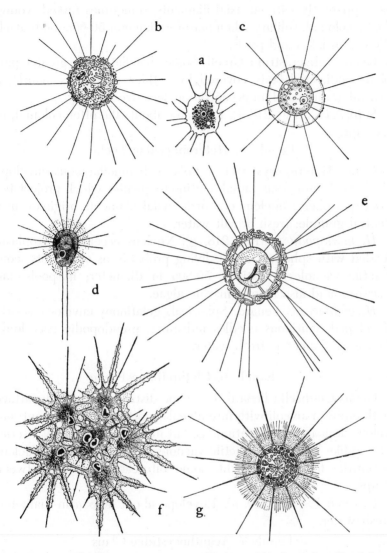

FIG. 218. a, *Ciliophrys infusionum,* ×400 (Bütschli); b, *Lithocolla globosa,* ×250 (Penard); c, *Astrodisculus radians,* ×600 (Penard); d, *Actinolophus pedunculatus,* ×400 (Schultze); e, *Elaeorhanis cincta,* ×300 (Penard); f, *Sphaerastrum fockei,* ×300 (Stubenrauch); g, *Heterophrys myriopoda,* ×270 (Penard).

but apparently without axial filaments, sometimes forked; young forms colonial; solitary when mature; diameter 50-60μ; body itself 25-30μ; in lakes and pools.

Genus **Sphaerastrum** Greeff. Somewhat flattened; greater part of axopodia and body covered by a thick gelatinous mantle; a central granule and an eccentric nucleus; fresh water.

S. fockei G. (Fig. 218, *f*). Diameter about 30μ; often colonial; in swamps.

Family 5. **Heterophryidae** Poche

Genus **Heterophrys** Archer. Spherical; mucilaginous envelope thick, with numerous radial, chitinous spicules which project beyond periphery; nucleus eccentric; axial filaments originate in a central granule; fresh or salt water.

H. myriopoda A. (Fig. 218, *g*). Nucleus eccentric; cytoplasm loaded with spherical algae, living probably as symbionts; contractile vacuoles indistinct; 50-80μ in diameter; in pools and marshes; and also among marine algae.

H. glabrescens Penard. Spherical; gelatinous envelope poorly developed; chitinous needles indistinct; pseudopodia very long; 11-15μ in diameter; fresh water.

Family 6. **Clathrellidae** Poche

Genus **Clathrella** Penard. Envelope distinct, polygonal; surface with uniform alveoli with interalveolar portion extending out; envelope appears to be continuous, but in reality formed by a series of cup-like bodies; contractile vacuole large; voluminous nucleus eccentric; filopodia straight, some bifurcated, arising between "cups."

C. foreli P. (Fig. 219, *a*). Envelope about 40-55μ in diameter; fresh water.

Family 7. **Acanthocystidae** Claus

Genus **Acanthocystis** Carter. Spherical; siliceous scales arranged tangentially and radiating siliceous spines with pointed or bifurcated ends; nucleus eccentric; a distinct central granule in which the axial filaments terminate. (Several species.) (Electron microscope study, Petersen and Hansen, 1960.)

A aculeata Hertwig and Lesser (Fig. 219, *b*). Tangential scales stout and pointed; spines curved and nail-headed; cytoplasm greyish; a single contractile vacuole; diameter 35-40µ; spines about one-third the body diameter; in fresh water. (Morphology and reproduction, Stern, 1924.)

Genus **Pompholyxophrys** Archer. Spherical; outer mucilaginous

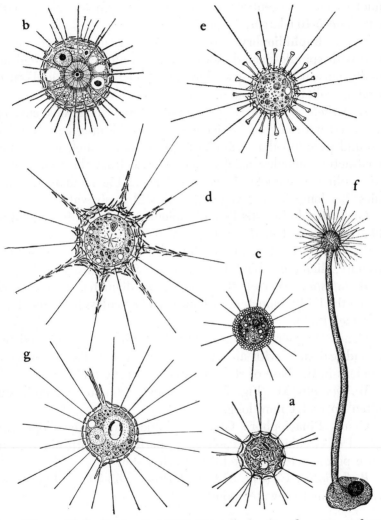

FIG. 219. a, *Clathrella foreli*, ×250 (Penard); b, *Acanthocystis aculeata*, ×300 (Stern); c, *Pompholyxophrys punicea*, ×260 (West); d, *Raphidiophrys pallida*, ×300 (Penard); e, *Raphidocystis tubifera*, ×500 (Penard); f, *Wagnerella borealis*, ×75 (Kühn); g, *Pinaciophora fluviatilis*, ×250 (Penard).

envelope with minute colorless spherical granules arranged in concentric layers; nucleus eccentric; contractile vacuoles; pseudopodia long, straight, acicular; fresh water.

P. punicea A. (Fig. 219, *c*). Body colorless or reddish, with usually many colored granules and green or brown food particles; nucleus large, eccentric; solitary, active; diameter 25-35μ; outer envelope 5-10μ larger; in pools.

Genus **Raphidiophrys** Archer. Spherical; mucilaginous envelope with spindle-shaped or discoidal spicules which extend normally outwards along pseudopodia; nucleus and endoplasm eccentric; solitary or colonial; fresh water. (Several species.)

R. pallida Schulze (Fig. 219, *d*). Outer gelatinous envelope crowded with curved lenticular spicules, forming accumulations around pseudopodia; ectoplasm granulated; nucleus eccentric; contractile vacuoles; axial filaments arise from the central granule; solitary; diameter 50-60μ nucleus 12-15μ in diameter; spicules 20μ long; among vegetation in still fresh water.

Genus **Raphidocystis** Penard. Spicules of various forms, but unlike those found in the last genus.

R. tubifera P. (Fig. 219, *e*). Spicules tubular with enlarged extremity; diameter about 18μ; envelope 25μ; fresh water.

R. infestans Wetzel. Body 20-40μ in diameter; thin axopodia twice the body diameter; without radial spicules; feeds on ciliates (Wetzel, 1925).

Genus **Wagnerella** Mereschkowsky. Spherical, supported by a cylindrical stalk with an enlarged base; small siliceous spicules; nucleus in the base of stalk; multiplication by budding.

W. borealis M. (Fig. 219, *f*). About 180μ in diameter; stalk often up to 1.1 mm. long; salt water.

Genus **Pinaciophora** Greeff. Spherical; outer envelope composed of circular discs, each being perforated with nineteen minute pores; cytoplasm reddish; fresh water.

P. fluviatilis G. (Fig. 219, *g*). Diameter 45-50μ, but somewhat variable; in freshwater ponds.

Family 8. **Clathrulinidae** Claus

Genus **Clathrulina** Cienkowski. Envelope spherical, homogeneous, with numerous regularly arranged orenings; with a

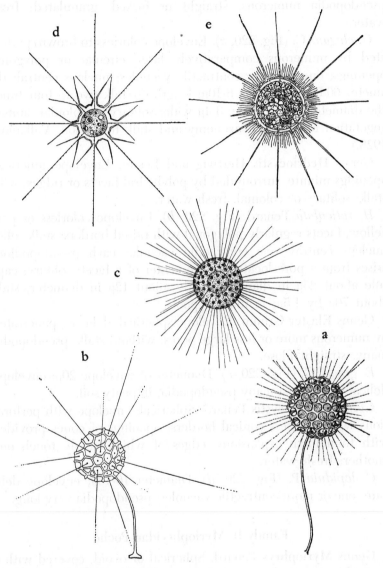

FIG. 220. a, *Clathrulina elegans*, ×250 (Leidy); b, *Hedriocystis reticulata*,
×500 (Brown); c, *Elaster greeffi*, ×680 (Penard); d, *Choanocystis lepidula*,
×690 (Penard); e, *Myriophrys paradoxa*, ×300 (Penard).

stalk; protoplasm central, not filling the capsule; nucleus central; pseudopodia numerous, straight or forked, granulated; fresh water.

C. elegans C. (Fig. 220, *a*). Envelope colorless to brown, perforated by numerous comparatively large circular or polygonal openings; one or more contractile vacuoles; nucleus central; diameter 60-90μ, openings 6-10μ; length of stalk two to four times the diameter of envelope, 3-4μ wide; solitary or colonial; among vegetation in ponds. (Taxonomy and stalk formation, Valkanov, 1928.)

Genus **Hedriocystis** Hertwig and Lesser. Envelope spherical, openings minute, surrounded by polyhedral facets or ridges; with stalk; solitary or colonial; fresh water.

H. reticulata Penard (Fig. 220, *b*). Envelope colorless or pale yellow, facets regularly polygonal with raised borders; stalk solid, nucleus central; one contractile vacuole; each pseudopodium arises from a pore located in the center of a facet; solitary; capsule about 25μ in diameter; body about 12μ in diameter; stalk about 70μ by 1.5μ; in marshy pools.

Genus **Elaster** Grimm. Envelope spherical, delicate, penetrated by numerous more or less large pores; without stalk; pseudopodia many, straight filose.

E. greeffi G. (Fig. 220, *c*). Diameter of envelope 20μ; envelope delicate, colorless; many pseudopodia; in peaty soil.

Genus **Choanocystis** Penard. Spherical; envelope with perforations which possess conical borders; openings of cones provided with funnel-like expansions, edges of which nearly touch one another; fresh water.

C. lepidula P. (Fig. 220, *d*). Diameter 10-13μ; envelope delicate; one or more contractile vacuoles; pseudopodia very long.

Family 9. **Myriophryidae** Poche

Genus **Myriophrys** Penard. Spherical or ovoid, covered with a protoplasmic envelope containing scales (?), surrounded by numerous fine processes; endoplasm vesicular; a large nucleus eccentric; a large contractile vacuole; long pseudopodia granulated and attenuated toward ends.

M. paradoxa P. (Fig. 220, *e*). Average diameter 40μ; in fresh-water swamps.

References

ANDERSON, E., and BEAMS, H. W., 1960. The fine structure of the heliozoan, *Actinosphaerium nucleofilum*. J. Protozool., 7:190.

BARRETT, J. M., 1958. Some observations on *Actinosphaerium nucleofilum* n.sp., a new freshwater actinophryid. *Ibid.*, 5:205.

BĚLAŘ, K., 1923. Untersuchungen an *Actinophrys sol.* I. Arch. Protist., 46:1
―――― 1924. II. *Ibid.*, 48:371.

CASH, J., and WAILES, G. H., 1921. The British freshwater Rhizopoda and Heliozoa. 5. London.

HARTMANN, M., 1952. Polyploide (polyenergide) Kerne bei Protozoen. Arch. Protist., 98:125.

KUHL, W., 1953. Untersuchungen ueber die Cytodynamik der Plasmogamie und temporaeren "Plasmabruecken" bei *Actinosphaerium eichhorni*, etc. Protoplasma, 42:133.

LEIDY, J., 1879. Freshwater Rhizopods of North America. Rep. U. S. Geol. Surv. Terr., 12.

OKADA, Y. K., 1930. Transplantationsversuche an Protozoen. Arch. Protist., 69:39.

PENARD, E., 1905. Les Héliozoaires d'eau douce. Geneva.
―――― 1905a. Les Sarcodinés des grands lacs. Geneva.

PETERSEN, J. B., and HANSEN, J. B., 1960. Elektonenmikroskopische Unter-suchungen von zwei Arten der Heliozoen-Gattung Acanthocystis. Arch. Protist., 104:547.

ROSKIN, G., 1929. Neue Heliozoa-Arten. I. Arch. Protist., 66:201.

RUMJANTZEW, A., and WERMEL, E., 1925. Untersuchungen ueber den Protoplasmabau von *Actinosphaerium eichhorni*. *Ibid.*, 52:217.

SPEETH, C., 1919. Ueber Kernveränderungen bei Actinosphaerium in Hun-ger- und Encystierungskulturen. *Ibid.*, 40:182.

STERN, C., 1924. Untersuchungen ueber Acanthocystideen. *Ibid.*, 48:437.

VALKANOV, A., 1928. Protistenstudien. III. *Ibid.*, 64:446.
―――― 1940. Die Heliozoen und Proteomyxien. *Ibid.*, 93:225.

WETZEL, A., 1925. Zur Morphologie und Biologie von *Raphidocystis in-festans* n. sp., etc. *Ibid.*, 53:135.

Chapter 24
Order 2. **Radiolarida** Müller

THE Radiolarida are pelagic in various oceans. A vast area of the ocean floor is known to be covered with the ooze made up chiefly of radiolaridan skeletons. They seem to have been equally abundant during former geologic ages, since rocks composed of their skeletons occur in various geological formations. Thus this group is the second group of protozoa important to geologists.

The body is generally spherical, although radially or bilaterally symmetrical forms are also encountered. The cytoplasm is divided distinctly into two regions which are sharply delimited by a membranous structure known as the **central capsule.** This is a single or double perforated membrane of pseudochitinous or mucinoid nature. Although its thickness varies a great deal, the capsule is ordinarily very thin and only made visible after addition of reagents. Its shape varies according to the form of the organism; thus in spherical forms it is spherical, in discoidal or lenticular forms it is more or less ellipsoidal, while in a few cases it shows a number of protruding processes. The capsule is capable of extension as the organism grows and of dissolution at the time of multiplication. The cytoplasm on either side of the capsule communicates with the other side through pores which may be large and few or small and numerous. The intracapsular portion of the body is the seat of repoduction, while the extracapsular region is nutritive and hydrostatic in function. The intracapsular cytoplasm is granulated, often greatly vacuolated, and is stratified either radially or concentrically. It contains one or more nuclei, pigments, oil droplets, fat globules, and crystals. The nucleus is usually of vesicular type, but its form, size, and structure, vary among different species and also at different stages of development even in one and the same species.

A thin assimilative layer, or matrix, surrounds the central capsule. In Tripylina, waste material forms a brownish mass known as phaeodium, around the chief aperture (astropyle) of the capsule.

Then there is a highly alveolated region, termed calymma, in which the alveoli are apparently filled with a mucilaginous secretion of the cytoplasm. Brandt showed that the vertical movement of some Radiolarida is due to the formation and expulsion of a fluid which consists of water saturated with carbon dioxide. Under ordinary weather and temperature conditions, the interchange between the alveoli and the exterior is gradual and there is a balance of loss and gain of the fluid, so that the organisms float on the surface of the sea. Under rough weather conditions or at extraordinary high temperatures, the pseudopodia are withdrawn, the alveoli burst, and the organisms descend into deeper water, where the alveoli are reformed.

The Radiolarida feed on microplankton such as copepods, diatoms, and various protozoa. The food is taken in through pseudopodia and passed down into the deeper region of calymma where it is digested in food vacuoles. The Radiolarida can, however, live under experimental conditions without solid food if kept under light. This is ordinarily attributed to the action of the yellow corpuscles which are present in various parts of the body, although they are, as a rule, located in the calymma. In Actipylina, they are found only in intracapsular cytoplasm, and in Tripylina they are absent altogether. They are spherical bodies, About 15μ in diameter, with a cellulose wall, two chromatophores, a pyrenoid, starch, and a single nucleus. They appear to multiply by fission. These bodies are considered as zooxanthellae (p. 30). In the absence of organic food material, the Radiolarida live probably by utilizing the products of holophytic nutrition of these symbiotic organisms.

The axopodia arise from either the extracapsular or the intracapsular portion and radiate in spherical forms in all directions, as in Heliozoida. In Actipylina, myonemes are present in certain pseudopodia and produce circular groups of short, rod-like bodies clustered around each of the radial spines (Fig. 222, c). They connect the peripheral portion of the body with the pseudopodial covering of the spicule and possess a great contractile power, supposedly with hydrostatic function (p. 69).

The skeletal structure of Radiolarida varies considerably from simple to complex and has a taxonomic value. The chemical nature of the skeleton is used in distinguishing the major subdivisions of the order. In the Actipylina it seems to be made up of strontium sulphate, while in the three other groups, Peripylina, Monopylina, and Tripylina, it consists fundamentally of siliceous substances. The skeleton of the Actipylina is sharply marked from others in form and structure. The majority of this group possess twenty rods radiating from center. The rod-shaped skeletons emerge from the body in most cases along five circles, which are

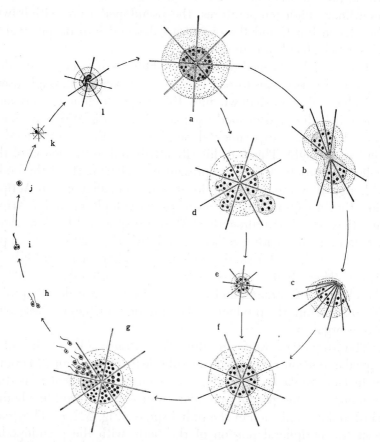

FIG. 221. Diagram illustrating the probable life-cycle of *Actipylina* (Kühn). a, mature individual; b, c, binary fission; d, e, multiplication by budding; f, mature individual similar to a; g, formation of swarmers; h–j, supposed, but not observed, union of two swarmers producing a zygote; k, l, young individuals.

comparable to the equatorial, two tropical and two circumpolar circles of the globe, which arrangement is known as **Müller's law,** since J. Müller first noticed it in 1858.

The life-cycle of the Radiolarida is very incompletely known (Fig. 221). Binary or multiple fission or budding has been seen in some Peripylina, Actipylina, and Tripylina. Multiple division is also known to occur in Thalassophysidae in which it is the sole known means of reproduction. The central capsule becomes very irregular in its outline and the nucleus breaks up into numerous chromatin globules. Finally the capsule and the intracapsular cytoplasm become transformed into numerous small bodies, each containing several nuclei. Further changes are unknown. Swarmer-formation is known in some forms. In Thalassicolla, the central capsule becomes separated from the remaining part of the body and the nuclei divide into a number of small nuclei, around each of which condenses a small ovoidal mass of cytoplasm. They soon develop flagellum. In the meantime the capsule descends to a depth of several hundred meters, where its wall bursts and the flagellates are liberated (g). Both isoswarmers and anisoswarmers occur. The former often contain a crystal and a few fat globules. Of the latter, the macroswarmers possess a nucleus and refringent spherules in the cytoplasm. Some forms possess two flagella, one of which is coiled around the groove of the body, which makes them resemble certain dinoflagellates. Further development is unknown. (Nuclear relationship, Hertwig, 1930; nuclear division in Aulacantha, Borgert, 1909; Grell, 1953; morphology and cytology, Hollande and Eujumet, 1953; syndinian parasites, Hovasse and Brown, 1953; form of skeletons, Thompson, 1942.)

Enormous numbers of species of Radiolaria are known. An outline of the classification is given below, together with a few examples, of the genera.

Skeleton composed of strontium sulphate . . . . Suborder 1 Actipylina
Skeleton composed of other substances
    Central capsule uniformly perforated, skeleton either tangential to the
        capsule or radiating without reaching the intracapsular region . .
        . . . . Suborder 2 Peripylina (p. 621).
    Central capsule not uniformly perforated
        Capsule monaxonic, bears at one pole a perforated plate forming
            the base of an inward-directed cone . . . .
        . . . . Suborder 3 Monopylina (p. 623).

Capsule with three openings: one astropyle and two parapyles ......
........................Suborder 4 Tripylina (p. 623).

Suborder 1. Actipylina Hertwig

Radial spines, 10–200, not arranged according to Müller's law.
Spines radiate from a common center, ancestral forms (Haeckel) ......
.....................................Family 1. Actineliidae

Genus **Actinelius** (Fig. 222, *a*)

10–16 spines irregularly set ....Family 2. Acanthociasmidae

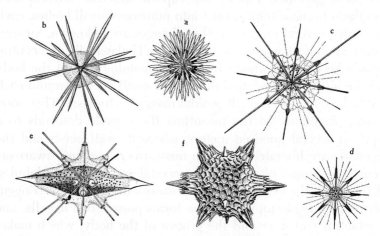

Fig. 222. a, *Actinelius primordialis,* ×25 (Haeckel); b, *Acanthociasma planum,* ×65 (Mielck); c, *Acanthometron elasticum* (Hertwig); d, *Acanthonia tetracopa,* ×40 (Schewiakoff); e, *Amphilonche hydrometrica,* ×130 (Haeckel); f, *Hexaconus serratus,* ×100 (Haeckel).

Genus **Acanthociasma** (Fig. 222, *b*)

Radial spines, few, arranged according to Müller's law
  Without tangential skeletons
    Spines more or less uniform in size
      Spicules circular in cross-section ......Family 3. Acanthometridae

Genus **Acanthometron** (Fig. 222, *c*)

Spicules cruciform in cross-section ......Family 4. Acanthoniidae

Genus **Acanthonia** (Fig. 222, *d*)

Two opposite spines much larger ......Family 5. Amphilonchidae

Genus **Amphilonche** (Fig. 222, *e*)

With tangential skeletons
  Twenty radial spines of equal size, shell composed of small plates, each
  with one pore ....Family 6. Sphaerocapsidae

Genus **Sphaerocapsa**

Two or six larger spines
  Two enormously large conical sheathed spines ....
  ....Family 7. Diploconidae

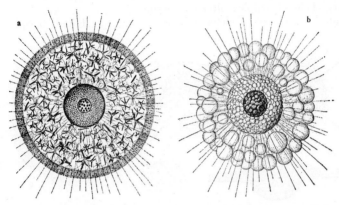

Fig. 223. a, *Lampoxanthium pandora*, ×20 (Haeckel); b, *Thalassicolla
nucleata*, ×15 (Huth).

Genus **Diploconus**

  Six large spines ....Family 8. Hexalaspidae

Genus **Hexaconus** (Fig. 222, *f*)

Suborder 2. **Peripylina** Hertwig

Solitary, skeleton wanting or simple spicules; mostly spherical
  Nucleus spherical with smooth membrane
    Vacuoles intracapsular ....Family 1. Physematiidae

Genus **Lampoxanthium** (Fig. 223, *a*)

    Vacuoles extracapsular ....Family 2. Thalassicollidae

Genus **Thalassicolla** (Fig. 223, *b*)

  Nuclear membrane not smoothly contoured
    Nuclear wall branching out into pouches, structure similar to the
    last ....Family 3. Thalassophysidae

Genus **Thalassophysa**

Nuclear wall crenate
  Huge double spicule .... Family 4. Thalassothamnidae

Genus **Thalassothamnus**

A latticed skeleton, with branching and thorny spines ....
.................................Family 5. Orosphaeridae

Genus **Orosphaera**

Solitary, skeleton complex, often concentric
  Central capsule and skeleton spherical .... Family 6. Sphaeroidae

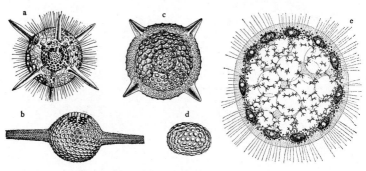

Fig. 224. a, *Hexacontium asteracanthion,* ×130; b, *Pipetta tuba,* ×100;
c, *Staurocyclia phacostaurus,* ×130; d, *Cenolarus primordialis,* ×100; e,
Sphaerozoum ovodimare, ×30 (Haeckel).

Genus **Hexacontium** (Fig. 224, *a*)

Central capsule and skeleton elliptical or cylindrical ....
.....................................Family 7. Prunoidae

Genus **Pipetta** (Fig. 224, *b*)

Central capsule and skeleton discoidal or lenticular ....
.....................................Family 8. Discoidae

Genus **Staurocyclia** (Fig. 224, *c*)

Similar to the above, but flattened .... Family 9. Larcoidae

Genus **Cenolarus** (Fig. 224, *d*)

Colonial, individuals with anastomosing extracapsular cytoplasm, embedded
    in a jelly mass
  Without latticed skeleton, but with siliceous spicules arranged tan-
    gentially to central capsule .... Family 10. Sphaerozoidae

Genus **Sphaerozoum** (Fig. 224, *e*)

Central capsule of each individual enclosed in a latticed skeleton ....
.....................................Family 11. Collosphaeridae

Genus **Collosphaera**

Suborder 3. **Monopylina** Hertwig

Without a skeleton ....Family 1. Nassoidae

Genus **Cystidium** (Fig. 225, *a*)

With skeleton
  Without a complete latticed skeleton
    Skeleton a basal tripod ....Family 2. Plectoidae

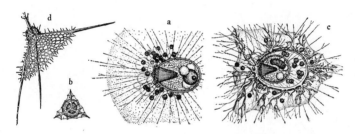

FIG. 225. a, *Cystidium princeps,* ×120; b, *Triplagia primordialis,* ×25; c,
Lithocircus magnificus, ×100; d, *Dictophimus hertwigi,* ×80 (Haeckel).

Genus **Triplagia** (Fig. 225, *b*)

Skeleton a simple or multiple sagittal ring ...Family 3. Stephoidae

Genus **Lithocircus** (Fig. 225, *c*)

With a complete latticed skeleton
  Lattice skeleton single, without constriction ..Family 4. Cyrtoidae

Genus **Dictyophimus** (Fig. 225, *d*)

  Lattice skeleton multiple ....Family 5. Botryoidae

Genus **Phormobothrys**

Suborder 4. **Tripylina** Hertwig

Without skeleton; with isolated spicules
  Skeleton consists of radial hollow rods and the tangential needles
  ....Family 1. Aulacanthidae

Genus **Aulacantha** (Fig. 226, *a*)

With foreign skeletons covering body surface ....
............................Family 2. Caementellidae

Genus **Caementella** (Fig. 226, *b*)

With skeleton
1–2 (concentric) usually spherical skeletons
  Outer lattice skeleton with triangular or areolar meshes ....
  ....Family 3. Sagosphaeridae

Genus **Sagenoscene**

One lattice skeleton with hollow radial bars ....
............................Family 4. Aulosphaeridae

Genus **Aulosphaera** (Fig. 226, *c*)

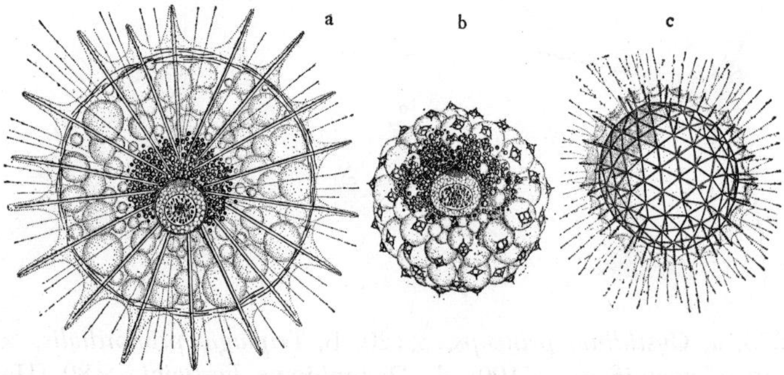

FIG. 226. a, *Aulacantha scolymantha*, ×30 (Kühn); b, *Caementella stapedia*,
×65 (Haeckel); c, *Aulosphaera labradoriensis*, ×10 (Haecker).

Two concentric lattice skeletons connected by radial bars ....
............................Family 5. Cannosphaeridae

Genus **Cannosphaera**

One skeleton, simple, but variable in shape; bilaterally symmetrical
  Skeleton with fine diatomaceous graining ..Family 6. Challengeridae

Genus **Challengeron** (Fig. 227, *a*)

Skeleton smooth or with small spines ......Family 7. Medusettidae

Genus **Medusetta** (Fig. 227, *b*)

One skeleton; spherical or polyhedral, with an opening and with radiating spines
  Skeleton spherical or polyhedral, with uniformly large round pores
  ....Family 8. Castanellidae

Genus **Castanidium** (Fig. 227, *c*)

Skeleton similar to the last, but the base of each radial spine surrounded by pores ....Family 9. Circoporidae

Genus **Circoporus** (Fig. 227, *d*)

Skeleton flask-shaped with one to two groups of spines ....
...................................Family 10. Tuscaroridae

Genus **Tuscarora** (Fig. 227, *e*)

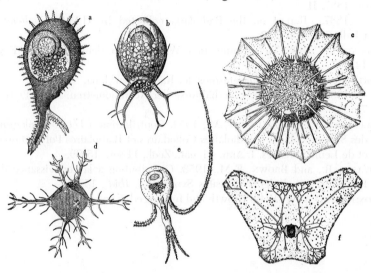

Fig. 227. a, *Challengeron wyvillei*, ×105 (Haeckel); b, *Medusetta ansata*, ×230 (Borgert); c, *Castanidium murrayi*, ×25 (Haecker); d, *Circoporus octahedrus*, ×65 (Haeckel); e, *Tuscarora murrayi*, ×7 (Haeckel); f, *Coelodendrum ramosissimum*, ×10 (Haecker).

Central portion of skeleton consists of two valves
   Valves thin, each with a conical process which divides into branched tubes ....Family 11. Coelodendridae

Genus **Coelodendrum** (Fig. 227, *f*)

References

BRANDT, K., 1905. Zur Systematik der koloniebildenden Radiolarien. Zool. Jahrb. Suppl., 8:311.
BORGERT, A., 1902. Mitteilungen ueber die Tripyleen-Ausbeute der Plankton-Expedition. I. Zool. Jahrb. Syst., 15:563.
——— 1904. II.*Ibid.*, 19:733.

—— 1905. Die Tripyleen Radiolarien der Plankton-Expedition. Ergebn. Plankton-Exp. Humboldt-Stiftung, 3:95.

—— 1909. Untersuchungen ueber dies Fortpflanzung der Tripyleen Radiolarien, speziell von *Aulacantha scolymantha*. II. Arch. Protist., 14:134.

—— 1913. Die Tripyleen Radiolarien der Plankton-Expedition. II. Ergebn. Plank.-Exped. Humboldt-Stiftung, 3:539.

GRELL, K. G., 1953. Die Chromosomen von *Aulacantha scolymantha*. Arch. Protist., 99:1.

HAECKEL, E., 1862. Die Radiolarien. Eine Monographie. I.

—— 1887. II

—— 1887a. Report on the Radiolaria collected by H.M.S. *Challenger*. Chall. Rep. Zool., 18.

HAECKER, V., 1908. Tiefseeradiolarien. Wiss. Ergebn. deutsch. Tiefsee-Exp., 14:337.

HERTWIG, R., 1879. Der Organismus der Radiolarien. Jena.

—— 1930. Ueber die Kernverhältnisse der Acanthometren. Arch. Protist., 71:33.

HOLLANDE, A., and EUJUMET, M., 1953. Contribution a l'étude biologique des Sphaerocollides (Radiolaires Collodaires et Radiolaires Polycyttaires) et de Leurs parasites. I. Ann. sci. nat., Zool., 11 ser., 15:99.

HOVASSE, R., and BROWN, E M., 1953. Contribution a la connaissance des Radiolaires et de leurs parasites Syndiniens. *Ibid.*, 15:405.

THOMPSON, D'A. W. 1942. Growth and form. 2 ed. Cambridge.

Chapter 25

Class 3. **Sporozoa** Leuckart

T HE protozoa placed in this class do not have any organelles of locomotion and produce spores toward the end of their life cycle. The spore consists of one or more sporozoites encased in a resistant spore membrane. In Haemosporida that have two hosts, the sporozoites do not have any protective membrane. Sporozoa are all parasites.

Both asexual and sexual reproductions are well known in many species. Asexual reproduction is by repeated binary or multiple fission or budding of intracellular trophozoites. The multiple division in a host cell produces far greater number of individuals than that of protozoans belonging to other classes and often is referred to as **schizogony.** The sexual reproduction is by isogamous or anisogamous fusion or autogamy and marks in many cases the beginning of **sporogony** or spore-formation.

As Cnidosporidia which was included in this class in former edition, is given a separate class-standing, class Sporozoa now includes four orders: Gregarinida, Coccidida (p. 678), Haemosporida (p. 717) and Haplosporida (p. 762).

Order 1. **Gregarinida** Lankester

The gregarines are chiefly coelozoic parasites in invertebrates, especially arthropods and annelids. They obtain their nourishment from the host organ-cavity through osmosis. The vast majority of gregarines do not undergo schizogony and an increase in number is carried on solely by sporogony. In a small group, however, schizogony takes place and this is used as the basis for grouping them into two suborders: Eugregarinina and Schizogregarinina (p. 666).

Suborder 1. **Eugregarinina** Doflein

This suborder includes the majority of the so-called **gregarines** which are common parasites of arthropods. When the spore gains entrance into a proper host, it germinates and the sporozoites emerge and enter the epithelial cells of the digestive tract. There they grow at the expense of the host cells which they leave soon and to which they become attached by various organellae of attachment (Fig. 240). These trophozoites become detached later from the host cells and move about in the lumen of the gut. This stage, **sporadin,** is ordinarily most frequently recognized. It is usually large and vermiform. The body is covered by a definite pellicle and its cytoplasm is clearly differentiated into the ectoplasm and endoplasm. The former contains myonemes (p. 70) which enable the organisms to undergo gliding movements (Watson, 1916).

In one group, Acephalinoidea, the body is of a single compartment, but in the other group, Cephalinoidea, the body is divided into two compartments by an ectoplasmic septum. The smaller anterior part is the **protomerite** and the larger posterior part, the **deutomerite,** contains a single nucleus. In Pileocephalus (Fig. 241, *s*) the nucleus is said to be located in the protomerite and according to Goodrich (1938) both the protomerite and deutomerite of *Nina gracilis* contain a nucleus. The endoplasm contains numerous spherical or ovoidal bodies which are called zooamylum or paraglycogen grains and which are apparently reserve food material (p. 138). The protomerite may possess an attaching process with hooks or other structures at its anterior border; this is called the **epimerite.** The epimerite is usually not found on detached sporadins. Goodrich observed recently that in Nina the protomerite is a knob-like part of the gregarine when contracted, but expands freely and used as a mobile sucker for attachment to the gut epithelium of the host Scolopendra. Presently multiple filiform epimerite grows at the free edge of the sucker and penetraits between the host cells. (Cytology, Göhre, 1943.)

Many gregarines are solitary, others are often found in an endwise association of two or more sporadins. This association is called **syzygy.** The anterior individual is known as the **primite** and the posterior, the **satellite.** What differences exist between the two individuals that become associated is not well known. But Mühl (1921) reported in *Gregarina cuneata*, the granules in the primite and the satellite stained differently with neutral red.

Sporadins usually encyst in pairs and become gametocytes. This process following bi-association was observed in a number of species; for example, in *Leidyana erratica* (Watson, 1916), *Gregarina blattarum* (Sprague, 1941) (Fig. 228), etc. Within the cyst-membrane, the nucleus in each individual undergoes repeated division, forming a large number of small nuclei which by a process of budding transform themselves into numerous gametes. The

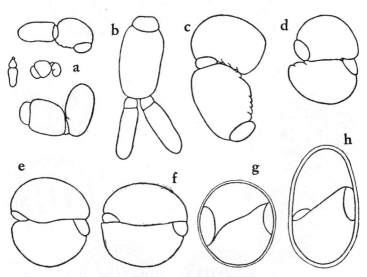

Fɪɢ. 228. Encystment in *Gregarina blattarum*, ×60 (Sprague). a, a trophozoite with epimerite and 3 pairs of syzygy; b, association of three individuals; c–h, encystment as seen in a single pair in about one hour.

gametes may be isogamous or anisogamous. Each of the gametes in one gametocyte appears to unite with one formed in the other, so that a large number of zygotes are produced. In some species such as *Nina gracilis* the microgametes enter the individual in which macrogametes develop, and the development of zygotes takes place, thus producing the so-called **pseudocyst.** The zygote becomes surrounded by a resistant membrane and its content develops into the sporozoites, thus developing into aspore. The spores germinate when taken into the alimentary canal of a host animal and the life-cycle is repeated.

According to Wenyon, in a typical Eugregarinina, *Lankesteria*

Protozoology

culicis (Fig. 229) of *Aedes aegypti,* the development in a new host begins when a larva of the latter ingests the spores which had been set free by infected adult mosquitoes in the water. From

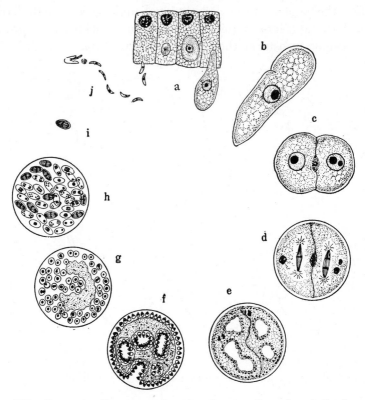

FIG. 229. Diagram illustrating the developmental cycle of *Lankesteria culicis* (Wenyon). a, entrance of sporozoite into the mid-gut epithelium and growth of trophozoites; b, mature trophozoite found in the lumen of gut; c, association of two gametocytes prior to encystment; d–f, gamete formation; g, zygote formation; h, development of spores from zygotes; i, a spore; j, emergence of eight sporozoites from a spore in a new host gut.

each spore are liberated eight sporozoites (*j*), which enter the epithelial cells of the stomach and grow (*a*). These vegetative forms leave the host cells later and become mingled with the food material present in the stomach lumen of the host (*b*). When the larva pupates, the sporadins enter the Malpighian tubules, where they encyst (*c*). The repeated nuclear division is followed by formation of large numbers of gametes (*d-f*) which unite in pairs (*g*). The zygotes thus formed develop into spores, each possessing eight sporozoites (*h*). Meanwhile the host pupa emerges as an adult mosquito, and the spores which become set free in the lumen of the tubules pass into the intestine, from which they are discharged into water. Larvae swallow the spores and acquire infection.

Eugregarinina are divided into two superfamilies: Acephalinoidea (trophozoite not septate) and Cephalinoidea (trophozoite septate) (p. 000).

Superfamily **Acephalinoidea** Kölliker

The acephalines are mainly found in the body cavity and organs associated with it. The infection begins by the ingestion of mature spores by a host, in the digestive tract of which the sporozoites are set free and undergo development or make their way through the gut wall and reach the coelom or various organs such as seminal vesicles. Young trophozoites are intracellular, while more mature forms are either intracellular or extracellular. Recent infection experiments on *Eisenia foetida* with two acephaline gregarines, *Apolocystis elongata* and *Nematocystis elmassiani* by Miles (1962) indicate that the host annelids acquire infection by ingesting spores and the dissemination of the spores in infected hosts occur only upon death and decay of the host body. Acephaline gregarines (Berlin, 1924; Bhatia and Chatterjee, 1925; Bhatia and Setna, 1926; Bhatia, 1929; Troisi, 1933). (Ten families.)

Family 1. **Monocystidae** Bütschli

Trophozoites spheroidal to cylindrical; anterior end not differentiated; solitary; spores biconical, without any spines, with 8 sporozoites. (Genera, Loubatières, 1955.)

Genus **Monocystis** Stein. Trophozoites variable in form; motile; incomplete sporulation in cyst; spore biconical, symmetrical; in coelom or seminal vesicles of oligochaetes. (Numerous species, Berlin, 1924.)

M. ventrosa Berlin (Fig. 230, *a-c*). Sporadins 109-183μ by

72-135μ; nucleus up to 43μ by 20μ; cysts 185-223μ by 154-182μ; spores 17-25 by 8-19μ; in *Lumbricus rubellus, L. castaneus* and *Helodrilus foetidus.*

M. lumbrici Henle (Fig. 230, *d, e*). Sporadins about 200μ by

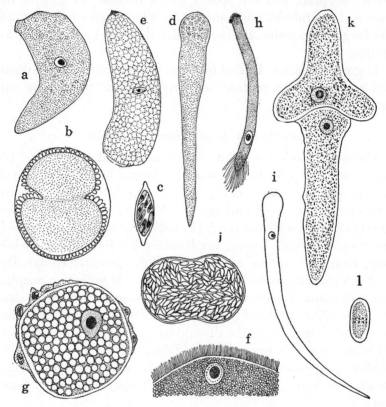

FIG. 230. a–c, *Monocystis ventrosa* (a, ×260; b, ×150; c, ×830) (Berlin); d, e, *M. lumbrici,* ×280 (Berlin); f. *Apolocystis gigantea,* ×90 (Troisi); g, *A. minuta,* with attached phagocytes, ×770 (Troisi); h, *Nematocystis vermicularis,* ×80 (Hesse); i, j, *Rhabdocystis claviformis* (i, ×220; j, ×270) (Boldt); k, l, *Enterocystis ensis* (k, ×140) (Zwetkow).

60-70μ; cysts about 162μ in diameter; in *Lumbricus terrestris, L. rubellus,* and *L. castaneus* (Berlin, 1924).

M. rostrata Mulsow (Figs. 94; 231). Elongate oval; average dimensions 450μ by 220μ; anterior end often drawn out into a process; pellicle thick, longitudinally striated; cysts about 750μ in diameter; spores 23μ by 9μ; in the seminal vesicles of *Lumbricus terrestris*. Mulsow (1911) found vegetative stages in autumn and winter and sporogony in spring. Meiosis in the last pre-gametic division (p. 246).

M. chagasi Adler and Mayrink. In *Phlebotomus longipalpis*, wild and laboratory bred; during oviposition, the surface of the fly eggs becomes contaminated with the spores discharged from infected accessory glands; the larvae emerging from these eggs are exposed to infection (Adler and Mayrink, 1961).

Genus **Apolocystis** Cognetti de Martiis. Trophozoites spherical; without principal axis marked by presence of any special periph-

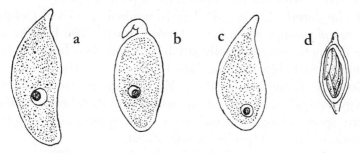

FIG. 231. *Monocystis rostrata* (Mulsow). a–c, trophozoites, ×90; d, spore, ×850.

eral organ; solitary; spore biconical; in seminal vesicles or coelom of various oligochaetes. (Many species.)

A. gigantea Troisi (Fig. 230, *f*). In seminal vesicles of *Helodrilus foetidus* and *Lumbricus rubellus;* late October to March only; fully grown trophozoites 250-800μ in diameter; whitish to naked eyes; pellicle thickly covered by 10-15μ long 'hairs'; endoplasm packed with spherical paraglycogen grains (3μ in diameter); nucleus 35-43μ in diameter; cysts 400-800μ in diameter; spores 19μ by 8.6μ (Troisi, 1933).

A. minuta Troisi (Fig. 230, *g*). In seminal vesicles of *Lumbricus terrestris*, *L. castaneus* and *L. rubellus;* mature trophozoites 40-46μ in diameter; endoplasm yellowish brown, packed with spherical paraglycogen grains (5.3-7μ in diameter); nucleus 10μ in diameter; cysts 68-74μ by 55-65μ; spores of three sizes, 11μ by 5.5μ. 18.8μ by 7μ and 21.6μ by 9.8μ.

Genus **Nematocystis** Hesse. Trophozoites elongate, cylindrical and shaped like a nematode; solitary. (Many species, Bhatia and Chatterjee, 1925.)

N. vermicularis H. (Fig. 230, *h*). In seminal vesicles of *Lumbricus terrestris, L. rubellus, Helodrilus longus, Pheretima barbadensis;* trophozoites 1 mm. by 100μ; cylindrical, both ends with projections; nucleus oval; endoplasm alveolated, with paraglycogen grains; sporadins become paired lengthwise; cysts and spores unknown.

Genus **Rhabdocystis** Boldt. Trophozoites elongate, gently curved; anterior end swollen, club-shaped; posterior end attenuated; spores with sharply pointed ends. (One species.)

R. claviformis B. (Fig. 230, *i, j*). In seminal vesicles of *Octolasium complanatum;* sporadins extended, up to 300μ by 30μ; pellicle distinctly longitudinally striated; zooamylum bodies 2-4μ in diameter; cysts biscuit-form, 110μ by 70μ; spores 16μ by 8μ.

Genus **Enterocystis** Zwetkow. Early stages of trophozoites in syzygy; sporadins in association ensiform; cysts spherical without ducts; spores elongate ovoid, with eight sporozoites; in gut of ephemerid larvae. (Species, Noble, 1938a.)

E. ensis Z. (Fig. 230, *k, l*). Sporadins in syzygy 200-510μ long; cysts 200-350μ in diameter; spores elongate ovoid; in gut of larvae of *Caenis* sp.

Genus **Echinocystis** Bhatia and Chatterjee. Body nearly spherical with two spine-like structures extending out from the body surface; solitary; spores biconical with equally truncated ends; in the seminal vesicles of earthworms (Bhatia and Chatterjee, 1925).

E. globosa B. and C. Body 740μ by 65μ; spines sometimes unequally long; observations on spores incomplete; in the sperm sacs of *Pheretima heterochaeta.*

Family 2. **Rhynchocystidae** Bhatia

Trophozoites ovoid, spherical or elongate, with a conical or cylindro-conical trunk at anterior end; solitary; spore biconical, with eight sporozoites.

Genus **Rhynchocystis** Hesse. Trophozoites ovoid or cylindrical; plastic epimerite, conical or cylindro-conical trunk; in seminal vesicles of oligochates. (Many species, Bhatia and Chatterjee, 1925; Troisi, 1933.)

R. pilosa Cuénot (Fig. 232, *a*). In seminal vesicles of *Lumbricus terrestris, L. castaneus* and *Helodrilus foetidus;* 217μ by 25.5μ; pellicle with close, longitudinal ridges from which arise 'hairs' up to 40μ in length; endoplasm viscous, packed with oval (3μ by 2μ) paraglycogen bodies; cysts ovoid, 95μ by 84μ; spores 13.3μ by 5μ (Troisi, 1933).

R. porrecta Schmidt (Fig. 232, *b, c*). In seminal vesicles of *Lumbricus rubellus* and *Helodrilus foetidus;* extremely long with

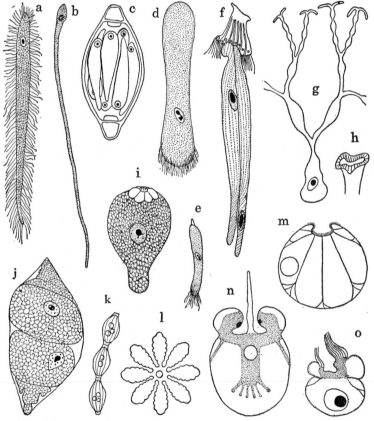

Fig. 232. a, *Rhynchocystis pilosa*, ×200 (Hesse); b, c, *R. porrecta*: b, ×170 (Hesse); c, spore, ×1330 (Troisi); d, e, *Zygocystis wenrichi* (d, ×45; e, ×450) (Troisi); f, *Pleurocystis cuenoti*, ×190 (Hesse); g, h, *Aikinetocystis singularis* (h, ×320) (Gates); i–k, *Stomatophora coronata* (i, j, ×430; k, ×870) (Hesse); l, *Astrocystella lobosa*, ×120; m, *Craterocystis papua*, ×65; n, *Choanocystella tentaculata*, ×570; o, *Choanocystoides costaricensis*, ×470 (Martiis).

an enlarged head; up to 2.5 mm. by 32-36µ; sluggish; endoplasm granulated, filled with oval (4µ by 2-3µ) paraglycogen grains; nucleus 17-25µ in diameter; spores 27.7-28µ by 12µ sporozoites 13-18µ by 3-5µ (Troisi, 1933).

Family 3. Zygocystidae Bhatia

Trophozoites in association; spores biconical, with peculiar thickenings at extremities; with eight sporozoites; in seminal vesicles or coelom of oligochaetes.

Genus **Zygocystis** Stein. Sporadins pyriform, two to three in syzygy; in seminal vesicles or coelom of oligochaetes. Several species.

Z. wenrichi Troisi (Fig. 232, *d, e*). In seminal vesicles of *Lumbricus rubellus* and *Helodrilus foetidus;* sporadins up to 1.5 mm. by 250µ in diameter; pellicle with longitudinal ridges which become free and form a 'tuft of hairs' at the posterior end; cysts 500-800µ by 300-500µ; spores 28µ by 13µ.

Genus **Pleurocystis** Hesse. Trophozoites in longitudinal or lateral association; spores biconical. (One species.)

P. cuenoti H. (Fig. 232, *f*). In the ciliated seminal horn of *Helodrilus longus* and *H. caliginosus;* 2 mm. by 300µ; pellicle striated longitudinally, obliquely near the posterior end; cysts 1.5-2 mm. in diameter; spores 28.5µ by 12µ (Hesse, 1909).

Family 4. Aikinetocystidae Bhatia

Trophozoites solitary or in syzygy; branching dichotomously, branches with sucker-like organellae of attachment; spores biconical.

Genus **Aikinetocystis** Gates. Trophozoites cylindrical or columnar, with a characteristic, regular dichotomous branching at attached end, with sucker-like bodies borne on ultimate branches; solitary or two (3-8) individuals in association; spores biconical.

A. singularis G. (Fig. 232, *g, h*). In coelom of *Eutyphoeus foveatus. E. rarus, E. peguanus* and *E. spinulosus* (of Burma); trophozoites up to 4 mm. long; number of branches eight or sixteen, each with an irregular sucker; ovoid nucleus near rounded end; spores of two sizes, 20-23µ long and 7-8µ long; a few cysts found, ovoid and about 600µ long.

Family 5. **Stomatophoridae** Bhatia

Trophozoites spherical to cylindrical or cup-shaped; with a sucker-like epimerite; solitary; spores navicular, ends truncate; eight sporozoites; in seminal vesicles of Pheretima (Oligochaeta).

Genus **Stomatophora** Drzewecki. Trophozoites spherical or ovoid; anterior end with a sucker-like epimeritic organella with a central spine; spores navicular. (Several species.)

S. *coronata* (Hesse) (Fig. 232, *i-k*). In seminal vesicles of *Pheretima rodericensis*, *P. hawayana* and *P. barbadensis;* trophozoites spherical, ovoid or elliptical, about 180μ by 130μ; endoplasm with ovoid paraglycogen grains; cysts ellipsoid or fusiform, 70-80μ by 50-60μ; spores in two sizes, 11μ by 6μ and 7μ by 3μ and in chain.

Genus **Astrocystella** Cognetti de Martiis. Trophozoites solitary; stellate with 5-9 lobes radiating from the central part containing nucleus; anterior surface with a depression.

A. *lobosa* C. (Fig. 232, *l*). In the seminal vesicles of *Pheretima beaufortii* (New Guinea); diameter about 200μ; spores fusiform.

Genus **Craterocystis** C. Trophozoites solitary; rounded; a sucker-like depression on anterior end; myonemes well developed, running from concave to convex side.

C. *papua* C. (Fig. 232, *m*). In the prostate and lymphatic glands of *Pheretima wendessiana* (New Guinea); trophozoites about 360-390μ in diameter.

Genus **Choanocystella** C. (*Choanocystis* C.). Trophozoites solitary; rounded or ovate; anterior end with a mobile sucker and a tentacle bearing cytoplasmic 'hairs'; myonemes.

C. *tentaculata* C. (Fig. 232, *n*). In the seminal vesicles of *Pheretima beaufortii* (New Guinea); trophozoites 50μ by 36μ.

Genus **Choanocystoides** C. Trophozoites solitary, rounded or cup-shaped; anterior end with a mobile sucker, bordered by cytoplasmic filaments.

C. *costaricensis* C. (Fig. 232, *o*). In seminal vesicles of *Pheretima heterochaeta* (Costa Rica); trophozoites 40-45μ in diameter; nucleus ovoid, large, 12μ in diameter.

Genus **Zeylanocystis** Dissanaike. Mature trophozoites bordered by papillae and filaments that may be absorbed to form a rim;

biassociating trophozoites are rounded; spores navicular with truncated ends (Dissanaike, 1953).

Z. burti D. (Fig. 233). In seminal vesicles of *Pheretima peguana;* trophozoites 70-95µ in diameter; cysts 140-190µ; spores 14-15µ by 7µ.

Genus **Beccaricystis** Cognetti de Martiis. Mature trophozoites elongate, cylindrical, with a sucker-like depression at anterior end; nucleus at its bottom. (One species.)

B. loriai C. (Fig. 234, *a*). In seminal vesicles of *Pheretima ser-*

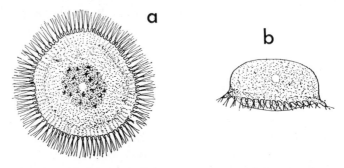

FIG. 233. *Zeylanocystis burti.* a, posterior and b, side views of trophozoites. ×375 (Dissanaike).

mowaiana; trophozoites cylindrical, with wart-like growths, myonemes run lengthwise with radially arranged transverse fibrils; about 100µ long.

Genus **Albertisella** C. Mature trophozoites cup-shaped, with anterior sucker with a smooth wall; nucleus at its bottom. (One species.)

A. crater C. In seminal vesicles of *Pheretima sermowaiana.*

Family 6. **Schaudinnellidae** Poche

Parasitic in the digestive system of oligochaetes; spores spherical; trophozoites do not encyst; male trophozoites producing microgametes and female, macrogametes; zygotes or amphionts (spores) rounded.

Genus **Schaudinnella** Nusbaum. Trophozoites elongate spindle, free in lumen or attached to gut wall; sporadins male or female; spherical macrogametes and fusiform microgametes; zygotes or amphionts encapsulated, pass out of host or enter gut epithelium, dividing to produce many sporozoites (autoinfection).

S. henleae N. (Fig. 234, *b, c*). In gut of *Henlea leptodera;* mature trophozoites about 70µ by 9µ; attached trophozoite with a

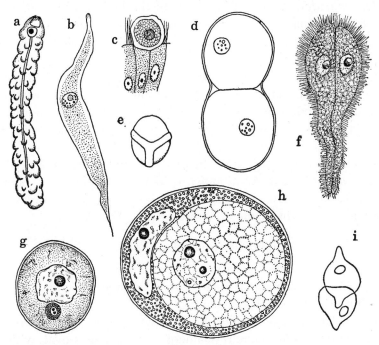

FIG. 234. a, *Beccaricystis loriai*, ×570 (Cognetti); b, c, *Schaudinnella henleae* (b, ×885; c, ×1000) (Nusbaum); d, e, *Diplocystis schneideri* (d, ×14; e, spore, ×2000) (Kunstler); f, *Urospora chiridotae,* ×200 (Pixell-Goodrich); g–i, *Gonospora minchini* (g, a young trophozoite in host egg; h, a mature trophozoite, ×330; i, sporadins in association, ×80) (Goodrich and Pixell-Goodrich).

clear wart-like epimerite; female and male sporadins; macrogametes, 5-7.5µ in diameter; microgametes, spindle-form, 1-1.25µ long; sporozoites rounded oval, 2.5-3µ in diameter.

Family 7. **Diplocystidae** Bhatia

Coelomic or gut parasites of insects; trophozoites solitary or associated early in pairs; spores round or oval, with eight sporozoites.

Genius **Diplocystis** Kunstler. Trophozoites spherical to oval; association of two individuals begin early in spherical form; spores round or oval, with eight sporozoites; in the intestine and coelom of insects.

D. schneideri K. (Fig. 234, *d, e*). In the body cavity of *Periplaneta americana;* young stages in gut epithelium; cysts up to two mm. in diameter; spores 7-8μ in diameter; sporozoites 8μ long. Meiosis (p. 247).

Genus **Lankesteria** Mingazzini. Trophozoites more or less spatulate; spherical cyst formed by two laterally associated sporadins in rotation; spores oval, with flattened ends, with eight sporozoites; in the gut of tunicates, flatworms and insects. (Several species.)

L. culicis Ross (Fig. 229). In gut and Malpighian tubules of *Aedes aegypti* and *A. albopictus;* mature trophozoites about 150-200μ by 31-41μ; cysts spherical, in Malpighian tubules of host, about 30μ in diameter; spores 10μ by 6μ.

Family 8. **Urosporidae** Woodcock

Coelomic parasites in various invertebrates; sporadins associative; spores with unequal ends; with or without epispore of various forms, with eight sporozoites.

Genus **Urospora** Schneider. Large; frequently in lengthwise association of two individuals of unequal sizes, spores oval, with a filamentous process at one end; in body cavity or blood vessel of Tubifex, Nemertinea, Sipunculus, Synapta, and Chiridota. (Several species.)

U. chiridotae (Dogiel) (Fig. 234, *f*). In blood vessel of *Chiridota laevis* (in Canada); paired trophozoites up to about 1 mm. long; with stiff 'hairs' (Goodrich, 1925).

U. hardyi Goodrich. In the coelom of *Sipunculus nudus;* spores about 16μ long, process 4-6μ long, with eight sporozoites; thinwalled cysts 0.5-2 mm. in diameter; active phagocytosis by host cells of cysts and some trophozoites, producing brownish masses, 5 by 2 mm. or more in diameter, which are crowded together in the posterior region of the host.

Genus **Gonospora** Schneider. Trophozoites polymorphic, oval, pyriform or vermiform; cysts spherical; spore with a funnel at one end, rounded at the other; in gut, coelom or ova of polychaetes.

G. minchini Goodrich and Pixell-Goodrich (Figs. 234, *g-i;* 235, *g*). In coelom of *Arenicola ecaudata;* young trophozoites live in host eggs which float in the coelomic fluid; fully grown trophozoites leave eggs in which they grew up to 200µ long, and encyst

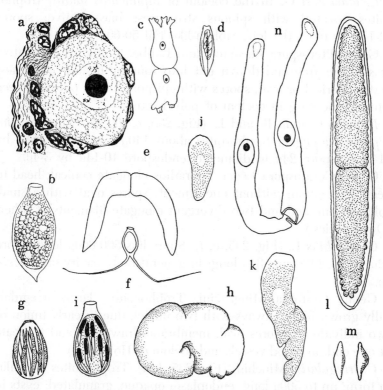

Fig. 235. a, b, *Lithocystis brachycercus*, ×1330 (Pixell-Goodrich); c, d, *Pterospora maldaneorum* (c, ×40; d, ×530) (Labbé); e, f, *Ceratospora mirabilis* (e, ×45; f, ×670) (Léger); g, *Gonospora minchini*, ×2000 (Goodrich); h, i, *Cystobia irregularis* (h, ×65; i, ×770) (Minchin); j–m, *Allantocystis dasyhelei* (j–l, ×500; m, ×560) (Keilin); n, *Ganymedes anaspides*, ×570 (Huxley).

together in pairs; spores without well-developed funnel, 8-10µ long (Goodrich and Goodrich, 1920).

Genus **Lithocystis** Giard. Trophozoites large, ovoid or cylindrical; attached for a long period to host tissue; pellicle with hairlike processes; endoplasm with calcium oxalate cystals; spores ovoid, with a long process at one end; in coelom of echinids.

L. brachycercus Goodrich (Fig. 235, *a*, *b*.) In the coelom of
Chiridota laevis; fully grown spherical trophozoites up to 200μ in
diameter; spores with a short flattened tail; in Canada (Goodrich,
1925).

L. lankesteri G. In the coelom of *Sipunculus nudus;* trophozoites covered with spinous structures; biassociative; spores
12-14μ by 6-8μ; the long ribbon-like tail 50-60μ long.

Genus **Pterospora** Racovitza and Labbé. Sporadins associative
or solitary; free end drawn out into four bifurcated processes;
cysts spherical or oval; spores with epispore drawn out into three
lateral processes; in coelom of polychaetes.

P. maldaneorum R. and L. (Fig. 235, *c*, *d*). In coelom of *Lio-
cephalus liopygue;* trophozoites about 140μ long; cysts 288μ by
214μ; epispore 24μ in diameter; endospore 10-14μ by 3-4μ.

Genus **Ceratospora** Léger. Sporadins elongate conical, head to
head association; without encystment; spores oval with a small
collar at one end and two divergent elongate filaments at other.
(One species.)

C. mirabilis L. (Fig. 235, *e*, *f*). Sporadins 500-600μ long; spore
12μ by 8μ, filaments 34μ long; in general body cavity of *Glycera*
sp.

Genus **Cystobia** Mingazzini. Trophozoites, large, irregular;
fully grown forms always with two nuclei, due to early union of
two individuals; spores oval, membrane drawn out and truncate
at one end; in blood vessels and coelom of Holothuria.

C. irregularis (Minchin) (Fig. 235, *h*, *i*). Trophozoites irregular
in form; up to 500μ long; endoplasm opaque, granulated; cysts in
connective tissue of vessels; spore ovoid, epispore bottle-like, 25μ
long; in blood vessel of *Holothuria nigra.*

Family 9. **Allantocystidae** Bhatia

Trophozoites elongate cylindrical; cysts elongate, sausage-like;
spores fusiform, sides slightly dissimilar.

Genus **Allantocystis** Keilin. Sporadins, head to head association; cysts sausage-like; in dipterous insect. (One species.)

A. dasyhelei K. (Fig. 235, *j-m*). In gut of larval *Dasyhelea obscura;* full-grown sporadins 65-75μ by 20-22μ; cysts 140-150μ by
20μ; spores 18μ by 6.5μ (Keilin, 1920).

Family 10. **Ganymedidae** Huxley

Trophozoites only known; mature individuals biassociative; posterior end of primite with a cup-like depression to which the epimeritic organella of satellite fits; cysts spherical; spores unknown.

Genus **Ganymedes** Huxley. Characters of the family; Huxley considers it as an intermediate form between Acephalinoidea and Cephalinoidea.

G. anaspides H. (Fig. 235, *n*). In gut and liver-tube of the crustacean, *Anaspides tasmaniae* (of Tasmania); trophozoites in association, 70-300μ by 60-130μ; cysts 85-115μ in diameter.

Superfamily 2. **Cephalinoidea** Delage

The body of a trophozoite is divided into the protomerite and deutomerite by an ectoplasmic septum; inhabitants of the alimentary canal of invertebrates, especially arthropods. (Taxonomy and distribution, Watson, 1916; Pinto, 1919; Kamm, 1922, 1922a; Chakravarty, 1959; terminology, Filipponi, 1949.) (Fourteen families.)

Family 1. **Lecudinidae** Kamm

Epimerite simple, symmetrical; non-septate; spores ovoidal, thickened at one pole; solitary; in gut of polychaetes and termites. Undoubtedly intermediate forms between Acephalinoidea and Cephalinoidea.

Genus **Lecudina** Mingazzini (*Doliocystis,* Léger, 1893). Epimerite simple, knob-like; in polychaetes. (Species, Kamm, 1922.)

L. pellucida (Kölliker) (Fig. 236, *a*). In *Nereis cultrifera* and *N. beaucordrayi;* trophozoites ellipsoid; spores 7μ by 5μ (Ellis, 1913).

Genus **Polyrhabdina** Mingazzini. Trophozoites flattened, ovoidal; epimerite with a corona of processes with split ends, deeply stainable; in polychaetes (Spionidae).

P. spionis (Kölliker) (Fig. 236, *b*). In *Scololepis fuligionosa;* 100μ by 35μ; epimerite with a corona of eight to ten processes; cysts unknown. Mackinnon and Ray (1931) report var. *bifurcata,* the epimerite of which there is a "knob-shaped structure with a circlet of fourteen to sixteen minute teeth at its base, and at its crown, two much larger diverging, clawlike processes."

Genus **Kofoidina** Henry. Epimerite rudimentary; development

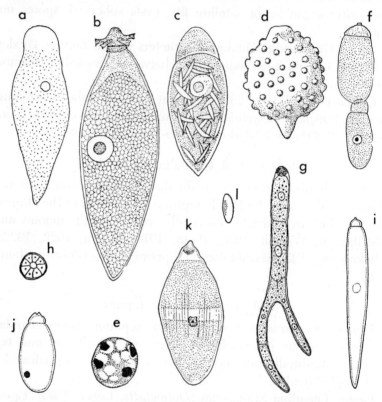

FIG. 236. a, *Lecudina pellucida* (Kölliker); b, *Polyrhabdina spionis*, ×800 (Reichenow); c, *Sycia inspinata* (Léger); d, e, *Zygosoma globosum* (d, ×60; e, ×1260) (Noble); f, *Cephaloidophora olivia*, ×190 (Kamm); g, h, *Carcinoecetes hesperus* (g, ×200; h, ×780) (Ball); i, *Stenophora larvata*, ×50 (Leidy); j, *S. robusta*, ×130 (Ellis); k, l, *Fonsecaia polymorpha* (k, ×220; l, ×430) (Pinto).

intracellular; two to fourteen sporadins in association; cysts and spores unknown (Henry, 1933).

K. ovata H. In mid-gut of *Zootermopsis angusticollis and Z. nevadensis;* syzygy 153-672μ long; sporadins 41-105μ long.

Genus **Sycia** Léger. Epimerite knobbed, bordered by a thick ring; protomerite subspherical; deutomerite conical, with navicular inclusions; in marine annelids (Léger, 1892).

S. inspinata L. (Fig. 236, *c*). In *Audouinia lamarcki.*

Genus **Zygosoma** Labbé. Trophozoites with wart-like projections; epimerite a simple knob; spores oval; in gut of marine annelids.

Z. globosum Noble (Fig. 236, *d, e*). Trophozoites 250-500μ by 200-380μ; epimerite a large globule; cysts 400μ by 360μ, without ducts; spores oval, with four sporozoites, 9μ by 7μ; reduction zygotic, twelve to six chromosomes; in gut of *Urechis caupo* in California.

Genus **Ulivina** Mingazzini. Elongate ellipsoid; epimerite simple; spores unknown; in gut of polychaetes.

U. rhynchoboli Crawley. Sporadins up to 700μ long; in the gut of *Rhynchobolus americanus* (Crawley, 1903).

Genus **Sphaerocystis** Léger. Sporadins solitary; without protomerite; spherical.

S. simplex L. Sporadins 100-140μ in diameter; protomerite in young trophozoites; spherical cysts in which individuals are not associative, 100μ in diameter; spores ovoid, 10.5μ by 7.5μ; in gut of *Cyphon pallidulus.*

Family 2. **Cephaloidophoridae** Kamm

Development intracellular; early association; cysts without sporoducts; spores ovoidal, with equatorial line; in gut of Crustacea.

Genus **Cephaloidophora** Mawrodiadi. Sporadins biassociative, early; epimerite rudimentary; cysts without sporoducts; spores in chain, ovoidal.

C. olivia (Watson) (Fig. 236, *f*). Biassociated sporadins up to 218μ long; individuals up to 118μ by 36μ; cysts spheroidal, 60μ in diameter; spores (?); in gut of *Libinia dubia;* Long Island.

C. nigrofusca (Watson). Sporadins, ovoid to rectangular, up to 125μ by 75μ; cysts and spores (?); in gut of *Uca pugnax* and *U. pugilator.*

Genus **Carcinoecetes** Ball. Sporadins in syzygy of two or more individuals; epimerite rudimentary; cysts without sporoducts; spores round to ovoidal, not in chain; in gut of Crustacea (Ball, 1938). Species (Ball, 1959).

C. hesperus B. (Fig. 236, *g, h*). Two to six sporadins in association; sporadins up to 320μ by 9μ; cysts about 140μ by 123μ, attached to the wall of hind-gut; spores 8.6μ by 7.7μ, with eight radially arranged sporozoites; in gut of the striped shore crab, *Pachygrapsus crassipes;* in California.

C. bermudensis B. In the mid- and hind-gut of *Pachygrapsus transversus;* in Bermuda (Ball, 1951).

C. mithraxi B. In the gut of *Mithrax forceps;* in Bermuda.

C. calappae B. In the gut of *Calappa flammea;* in Bermuda.

Genus **Caridohabitans** Ball. Epimerite, functional, transparent, crescent-shaped, concave anteriorly; early development intracellular; nucleus with irregularly distributed peripheral granules, without central endosome; early syzygy; cyst unobserved; in gut of crustaceans (Ball, 1959).

C. setnai B. In mid-gut of *Peneus semisulcatus.*

Family 3. **Stenophoridae** Léger and Duboscq

Development intracellular; sporadins solitary; with a simple epimerite or none; cysts open by rupture; spores ovoid, with or without equatorial line, not extruded in chain; in Diplopoda.

Genus **Stenophora** Labbé. With or without simple epimerite; spores ovoid with equatorial line, not in chain. (Species, Watson, 1916; Pinto, 1919; Ganapati and Narasimhamurti, 1956; Rodgi and Ball, 1961.)

S. larvata (Leidy) (Fig. 236, *i*). Sporadins up to 800μ by 23μ; protomerite small; in gut of *Spirobolus spinigerus.*

S. robusta Ellis (Fig. 236, *j*). Sporadins 140-180μ by 67μ; cysts and spores both unobserved; in gut of *Parajulus venustus, Orthomorpha gracilis* and *O.* sp.; Colorado.

Genus **Fonsecaia** Pinto. Spores elongate ovoid; without equatorial line.

F. polymorpha Pinto (Fig. 236, *k, l*). Sporadins 170μ long; spores 18μ by 8μ; in gut of *Orthomorpha gracilis;* Brazil.

Genus **Hyalosporina** Chakravarty. Sporont solitary; epimerite small, tongue-like; anisogametes; cyst without ducts; spores oval, with a hyaline membrane. (One species, Chakravarty, 1935, 1936.)

H. cambolopsisae C. (Fig. 237, *a-d*). Trophozoites 247-1111μ by

37-111μ; cysts oval, 292-390μ by 263-375μ; spores 8μ by 6μ; in the gut of the milliped, *Cambolopsis* sp.

Family 4. **Monoductidae** Ray and Chakravarty

As in the last family solitary; but cyst with a single sporoduct or none; spore with eight sporozoites.

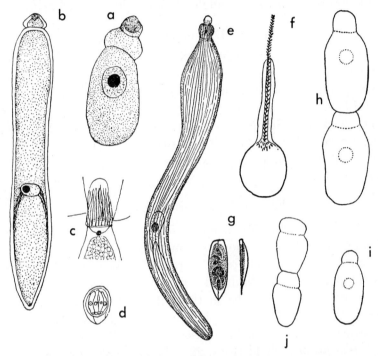

Fig. 237. a–d, *Hyalosporina cambolopsisae* (Chakravarty) (a, intracellular trophozoite, ×1110; b, a mature individual with fibrils tethering the nucleus, ×120; c, anterior part of attached form, ×2330; d, spore, ×1110); e–g, *Monoductus lunatus* (Ray and Chakravarty) (e, cephalin, ×240; f, cyst, ×120; g, two views of spore, ×2330); h, *Gregarina blattarum*, ×55 (Kudo); i, *G. locustae*, ×65 (Leidy); j, *G. oviceps*, ×30 (Crawley).

Genus **Monoductus** R. and C. Sporadins solitary epimerite a small elevation with prongs attached to its base; anisogamy; cyst with a single sporoduct; spores flattened fusiform, with dissimilar ends, each with eight sporozoites. (One species.)

M. lunatus R. and C. (Fig. 237, *e-g*). Cephalins 225-445μ by

33-47μ; epimerite with about 16 prongs; nucleus parachute-shaped, with myonemes attached at posterior margin; sporadins develop posterior pseudopodial processes before association; cysts spherical, 225-230μ in diameter, voided by host; development completed in three to four days outside the host body, with one duct; spores 10.25μ by 4μ, truncate at one end, attenuated at other and discharged in a single chain; in gut of *Diplopoda* sp.

Family 5. **Didymophyidae** Léger

Two to three sporadins in association; satellite without septum.

Genus **Didymophyes** Stein. Epimerite a small pointed papilla; cysts spherical, open by rupture; spores ellipsoidal. Species (Cordua, 1953).

D. gigantea S. Sporadins slender, 1 cm. by 80-100μ; 2 deutomerites; cysts spherical, 600-700μ in diameter; spores oval, 6.5μ by 6μ; in gut of larvae of *Oryctes nasicornis*, *O.* sp., and *Phyllognathus* sp. (Léger, 1892).

Family 6. **Gregarinidae** Labbé

Sporadins in association; epimerite simple, symmetrical; cysts with or without ducts; spores symmetrical.

Genus **Gregarina** Dufour. Sporadins biassociative; epimerite small, globular or cylindrical; spores dolioform to cylindrical; cysts open by sporoducts; in the gut of arthropods. Numerous species (Watson, 1916). (Morphology and physiology, Mühl, 1921; in African mountain cockroaches, Harrison, 1955.)

G. blattarum Siebold (Figs. 228; 237, *h*). Sporadins in syzygy, 500-1100μ by 160-400μ; cysts spherical or ovoidal; eight to ten sporoducts; spores cylindrical to dolioform, truncate at ends, 8-8.5μ by 3.5-4μ; in the midgut of cockroaches, especially *Blatta orientalis*. (Reproduction, Schiffmann, 1919; Sprague, 1941.)

G. locustae Lankester (Fig. 237, *i*). Sporadins 150-350μ long; in *Dissosteria carolina*.

G. oviceps Diesing (Fig. 237, *j*). Sporadins up to 500μ by 225μ; in syzygy; spherical cysts 250μ in diameter; two to five sporoducts up to 1 mm. long; spores dolioform, 4.5μ by 2.25μ; in *Gryllus abbreviatus* and *G. americanus* (Leidy, 1853).

G. polymorpha Hammerschmidt. Cylindrical sporadins up to

350μ by 100μ; in syzygy; protomerite dome-shaped; deutomerite cylindrical, rounded posteriorly; a small nucleus with an endosome; in the intestine of larvae and adults of *Tenebrio molitor* ("meal worm"). When the host insects are starved at 22°C., the gregarines disappear within forty days, but at 20°C. they remain at the same level for at least three months (von Brand, 1953).

G. *rigida* Hall. Sporadins 28μ by 20μ up to 424μ by 196μ; syzygy; spherical cysts 212-505μ in diameter; in the species of Melanoplus (grasshoppers) (Kamm, 1920; Allegre, 1948). A variety of this species was found by Tuzet and Rambier (1953) as the cause of death of a whole brood of *Ephippigerida nigromarginata* through intestinal occlusion. (Electron microscope study, Beams *et al.*, 1959.)

G. *garnhami* Canning. In the gut of the locust, *Schistocerca gregaria* and possibly of *Locusta migratoria* and *Anacrydium aegyptium* (Canning, 1956).

Genus **Protomagalhaesia** Pinto. Sporadins cylindrical; in syzygy, protomerite of satellite draws in the posterior end of primite; cysts without ducts; spores dolioform, with spines at ends.

P. *serpentula* (Magalhães) (Fig. 238, *a*). Sporadins up to 1.2 mm. by 180μ; in gut and coelom of *Blatta orientalis*.

P. *marottai* Filipponi. In *Scaurus striatus;* resembles the last species (Filipponi, 1952, 1953).

Genus **Gamocystis** Schneider. Septate only in trophozoites; sporadins non-septate; in syzygy; spore formation partial; with sporoducts; spores cylindrical. A few species.

G. *tenax* S. (Fig. 238, *b*). Association head to head; spherical cysts with fifteen or more ducts; spore cylindrical, with rounded ends; in gut of *Blattella lapponica* (Schneider, 1875).

Genus **Garnhamia** Crusz. Epimerite papillate to acicular; no protomerite; aseptate; biassociation occurs while primite is still attached to host's mid-gut epithelium; cyst with sporoducts; ovoid spores extruded in chain (Crusz, 1957).

G. *aciculata* (Bhatia) (Fig. 238, *c*). In the mid-gut of Ceylon silverfish, *Peliolepisma calva*.

Genus **Hyalospora** Schneider. Sporadins in syzygy; cytoplasm yellowish orange; epimerite a simple knob; cysts open by rupture; spores fusiform.

H. affinis S. Trophozoites 300μ long; cysts, yellow, 60μ in diameter; spores 8.7μ by 6μ; in gut of *Machilis cylindrica* (Labbé, 1899).

Genus **Tettigonospora** Smith. Similar to *Hyalospora*, but cytoplasm opaque white; spores spherical. (One species, Smith, 1930.)

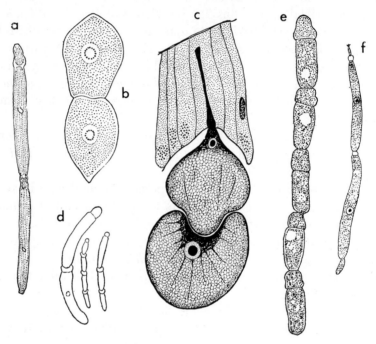

Fig. 238. a, *Protomagalhaesia serpentula*, ×35 (Pinto); b, *Gamocystis tenax* (Schneider); c, *Garnhamia aciculata*, ×520 (Crusz); d, *Hirmocystis harpali*, ×50 (Watson); e, *H. termitis*, ×85 (Henry); f, *Uradiophora cuenoti* in syzygy, ×65 (Mercier).

T. stenopelmati S. Sporadins 225-542μ by 118-225μ; spherical cysts 434-551μ in diameter, wall 17-66μ thick; spores 4.8-5μ in diameter; in the *mid-gut* of *Stenopelmatus fuscus* and *S. pictus* ("Jerusalem crickets").

Genus **Hirmocystis** Labbé. Sporadins associative, two to twelve or more; with a small cylindrical papilla-like epimerite; cysts without ducts; spores ovoidal.

H. harpali Watson (Fig. 238, *d*). Total length of association up to 1060μ; sporadins up to 560μ by 80μ; cysts unknown; in gut of *Harpalus pennsylvanicus erythropus* (Watson, 1916).

H. termitis (Leidy) (Fig. 238, *e*). Association 614-803μ long; epimerite simple sphere; cysts rare; spores (?); in *Zootermopsis angusticollis, Z. nevadensis*, etc. (Henry, 1933).

Genus **Uradiophora** Mercier. Sporadins in syzygy; deutomerite with small process; epimerite an elongate papilla; cysts oval without ducts; spores spherical, in chains (Mercier, 1911).

U. cuenoti M. (Fig. 238, *f*). Two to four sporadins in syzygy; individuals up to 700μ long; cysts ovoid, 44μ long; spores 4μ in diameter; in gut of *Atyaephrya desmaresti*.

Genus **Pyxinioides** Trégouboff. Sporadins biassociative; epimerite with 16 longitudinal furrows, small cone at end.

P. balani (Kölliker). Primite up to 130μ; satellite 60μ long; in gut of *Balanus amphitrite* and *B. eburneus*.

Genus **Anisolobus** Vincent. Sporadins in syzygy; epimerite lacking; protomerite of primite expanded to form sucker-like organella; cysts ellipsoid, with thick envelope; with six to eight sporoducts; spores barrel-shaped. (One species.)

A. dacnecola V. (Fig. 239, *a*). In the mid-gut of the coleopteran *Dacne rufifrons;* two sporadins in syzygy 100-300μ by 20-50μ; cysts without envelope, 130-150μ by 80-90μ; sporoducts 40-50μ long; spores in chain, dolioform, 6μ by 4μ (Vincent, 1924).

Genus **Heliospora** Goodrich. Elongated, septate; spores more or less spherical, with equatorial ray-like processes (Goodrich, 1949).

H. longissima (Siebold) (Fig. 239, *b-e*). Trophozoites elongate filiform, up to 228μ long; no intracellular stage; epimerite small, and is retained until the sporadins roll up for encystment; spherical cyst thinly walled and ruptures easily; microgametes flagellated; spores 7-8μ in diameter, with eight sporozoites and bear six long ray-like processes at the equator; in the gut of *Gammarus pulex*.

Genus **Rotundula** Goodrich. Rotund; button-like epimerite; precocious association; cyst without duct; spores, small, spherical or subspherical (Goodrich, 1949).

R. gammari (Diesing) (Fig. 239, *f*). Cysts 40-50μ; microgametes flagellate, 4μ in diameter; spores spherical, 5-6μ in diameter; in the gut of *Gammarus pulex*.

Family 6. Leidyanidae

Similar to the last two families, but sporadins are solitary and epimerite simple knob-like; cysts with several sporoducts.

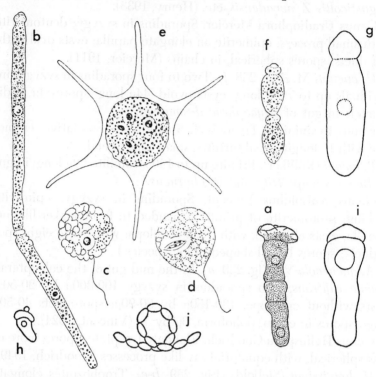

FIG. 239. a, *Anisolobus dacnecola*, ×270 (Vincent); b–e, *Heliospora longissima* (Goodrich) (b, a pair, ×330; c, microgamete; d, zygote; e, a spore with 4 nuclei, ×2665); f, *Rotundula gammari* in syzygy, ×330 (Goodrich); g, *Leidyana erratica*, ×170 (Watson); h-j, *Lepismatophila thermobiae* (h, i, ×85; j, spores, ×200) (Adams and Travis).

Genus **Leidyana** Watson. Solitary; epimerite a simple globular sessile knob; cysts with ducts; spores dolioform (Watson, 1915).

L. erratica (Crawley) (Fig. 239, g). Sporadins up to 500μ by 160μ; cysts about 350μ in diameter; membrane about 30μ thick; one to twelve sporoducts; spores extruded in chains, 6μ by 3μ; in gut of *Gryllus abbreviatus* and *G. pennsylvanicus*.

Family 8. **Lepismatophilidae**

Sporadins solitary; epimerite a simple knob; spores ovoid with or without processes.

Genus **Lepismatophila** Adams and Travis. Epimerite a simple knob; cysts without ducts; spores ellipsoidal, smooth, in chain. (One species, Adams and Travis, 1935.)

L. thermobiae A. and T. (Fig. 239, *h-j*). Sporadins 67-390µ by 30-174µ; cysts white to black ellipsoidal to subspherical, 244-378µ by 171-262µ; spores brown, 13.6 by 6.8µ; in ventriculus of the firebrat, *Thermobia domestica*.

Genus **Colepismatophila** Adams and Travis. Similar to the last genus; but larger; spores in wavy chains, hat-shaped, with two curved filamentous processes attached at opposite ends. (One species.)

C. watsonae A. and T. (Fig. 248, *h, i*). Sporadins 92-562µ by 55-189µ; cysts 226-464µ by 158-336µ; spores 16.5µ by 9.7µ; processes 21µ long; in ventriculus of *Thermobia domestica* (Adams and Travis, 1935).

Family 9. **Menosporidae** Léger

Sporadins solitary; epimerite a large cup, bordered with hooks, with a long neck; cysts without sporoducts; spores crescentic, smooth.

Genus **Menospora** Léger. With the characters of the family.

M. polyacantha L. (Fig. 240, *a, b*). Sporadins 600-700µ long; cysts 200µ in diameter; spores 15µ by 4µ; in gut of *Agrion puella*.

Family 10. **Dactylophoridae** Léger

Sporadins solitary; epimerite complex, digitate; cysts dehiscence by pseudocyst; spores cylindrical; in gut of chilopods.

Genus **Dactylophorus** Balbiani. Protomerite wide, bordered by digitiform processes; spores cylindrical.

D. robustus Léger (Fig. 240, *c, d*). Sporadins 700-800µ long; cysts spherical 200µ in diameter; spores 11µ by 4.3µ; in gut of *Cryptops hortensis*.

Genus **Echinomera** Labbé. Epimerite an eccentric cone with eight or more digitiform processes; cysts without sporoducts; spores cylindrical.

E. magalhaesi (Pinto) (Fig. 240, *e*). Sporadins up to 300μ by 70μ; in gut of *Scolopendra* sp.

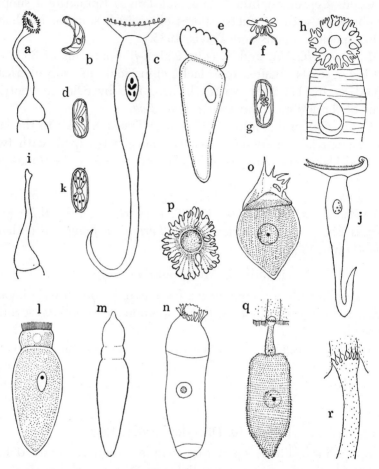

Fig. 240. a, b, *Menospora polyacantha* (Léger); c, d, *Dactylophorus robustus* (c, ×130; d, ×900) (Léger); e, *Echinomera magalhaesi,* ×130 (Pinto); f, g, *Rhopalonia hispida* (g, ×830) (Léger); h, *Dendrorhynchus systeni,* ×770 (Keilin); i, *Trichorhynchus pulcher* (Schneider); j, k, *Grebneckiella gracilis* (j, ×10) (Schneider); l, *Seticephalus elegans,* ×450 (Pinto); m, *Acutispora macrocephala,* ×65 (Crawley); n, *Metamera schubergi,* ×270 (Duke); o, p, *Hentschelia thalassemae* (o, ×230; p, ×620) (Mackinnon and Ray); q, r, *Lecythion thalassemae* (q, ×270; r, ×930) (Mackinnon and Ray).

Genus **Rhopalonia** Léger. Epimerite spherical, with ten or more digitiform processes; pseudocysts; spores cylindrical.

R. hispida (Schneider) (Fig. 240, *f*, *g*). Endoplasm yellowish orange; cysts 200-250µ in diameter; spores 16µ by 6.5µ; in gut of *Geophiles* sp. and *Stigmatogaster gracilis*.

Genus **Dendrorhynchus** Keilin. Elongate; epimerite a disc, surrounded by numerous ramified papillae; transverse fibrils conspicuous; cysts elliptical; spores fusiform.

D. systeni K. (Fig. 240, *h*). Sporadins 255µ by 18.5-20µ; spores 18-19µ by 7µ; in mid-gut of larvae of *Systenus* sp., a dolichopodid fly, found in decomposed sap of elm tree.

Genus **Trichorhynchus** Schneider. Protomerite prolonged anteriorly into a long neck, dilated at tip; pseudocyst; spores cylindrical to ellipsoidal.

T. pulcher S. (Fig. 240, *i*). Cysts 303-316µ in diameter; spores 9.7µ by 5.8µ; in gut of *Scutigera* sp. and *S. forceps* (Watson, 1916).

Genus **Grebneckiella** Bhatia (*Nina*, Grebnecki; *Pterocephalus*, Schneider). Protomerite made up of two long narrow horizontal lobes fused and upturned spirally at one end, peripheral portion with many teeth, from which project long filaments; spores in chain; in gut of myriapods. (Species, Watson, 1916.)

G. gracilis G. (Fig. 240, *j*, *k*). 1.5-5 mm. long; cyst spherical; spores ellipsoidal; in the gut of *Scolopendra cingulata* and *S. subspinipes* (Goodrich, 1938).

Genus **Seticephalus** Kamm. Protomerite with closely set brush-like bristles.

S. elegans (Pinto) (Fig. 240, *l*). Sporadins up to 75µ by 35µ; cysts and spores unknown; in gut of *Scolopendra* sp.

Genus **Acutispora** Crawley. Solitary; pseudocyst; spore biconical, with a thick blunt endosporal rod at each end. One species (Crawley, 1903).

A. macrocephala C. (Fig. 240, *m*). Sporadins up to 600µ long; cysts spherical, 410µ in diameter; spores navicular, slightly curved, 19µ by 4µ; in gut of *Lithobius forficatus*.

Genus **Metamera** Duke. Epimerite eccentric, bordered with many branched digitiform processes; cysts without ducts; spores biconical (Duke, 1910).

M. schubergi D. (Fig. 240, *n*). Sporadins 150µ by 45µ; spores

9μ by 7μ; in gut of the leeches, *Glossosiphonia complanata* and *Placobdella marginata*.

M. reynoldsi Jones. Sporadins with epimerite measure 280μ by 50μ; cysts spherical; dehiscence by rupture; spore biconical, 5μ by 3μ with eight sporozoites; in the stomach diverticula and intestine of *Glossosiphonia complanata*.

Genus **Hentschelia** Mackinnon and Ray. Epimerite with a short neck, umbrella-like with its margin divided into four to five lobes, each fluted on anterior surface; two sporadins encyst together; gametes anisogamous; flagellate and non-flagellate; zygote gives rise to a spherical spore with eight sporozoites. (One species.)

H. thalassemae M. and R. (Fig. 240, *o*, *p*). Cephalins 75-98μ by 30-45μ; in gut of *Thalassema neptuni* (Mackinnon and Ray, 1931).

Genus **Lecythion** Mackinnon and Ray. Epimerite a low cone, surrounded by fourteen to fifteen petal-shaped lobes, with a neck; cysts and spores unknown.

L. thalassemae M. and R. (Fig. 240, *q*, *r*). Cephalins 135μ by 52μ; epimerite about 27μ long; in gut of *Thalassema neptuni*.

Family 11. **Stylocephalidae** Ellis

Sporadins solitary; epimerite varied; pseudocysts; hat-shaped spores in chains.

Genus **Stylocephalus** Ellis. Epimerite nipple-like; cysts covered with papillae; in arthropods and molluscs.

S. giganteus E. (Fig. 241, *a*). Sporadins 1.2-1.8 mm. long; cysts spherical, 450μ in diameter; spores subspherical black, 11μ by 7μ; in *Eleodes* sp., *Asida opaca*, *A.* sp., and *Eusattus* sp. (Coleoptera) (Ellis, 1912).

Genus **Bulbocephalus** Watson. Epimerite a dilated papilla located in middle of a long neck (Watson, 1916a).

B. elongatus W. (Fig. 241, *b*). Sporadins up to 1.6 mm. by 50μ; nucleus diagonal; cysts and spores unknown; in gut of Cucujus larva (a coleopteran).

Genus **Sphaerorhynchus** Labbé. Epimerite a small sphere at end of a long neck.

S. ophioides Schneider. Cephalins 1.3 mm. long; epimerite 220μ long; terminal part 8.5μ; sporadins 3-4 mm. long; in gut of *Acis* sp.

Genus **Cystocephalus** Schneider. Epimerite a large lance-shaped papilla with a short neck; spore hat-shaped.

C. algerianus S. (Fig. 241, *c, d*). Sporadins 3-4 mm. long; spores 10-10.5μ long; in gut of *Pimelia* sp. (Labbé, 1899).

Genus **Lophocephalus** Labbé. Epimerite sessile crateriform disc with crenulate periphery, surrounded by digitiform processes.

L. insignis (Schneider) (Fig. 241, *e*). Sporadins 1 mm. long; cysts rounded; 430μ by 330μ; pseudocysts; spores 10μ long; in gut of *Helops striatus*.

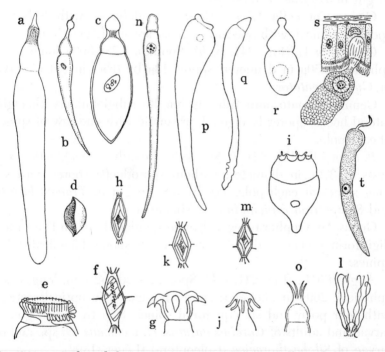

Fig. 241. a, *Stylocephalus giganteus*, ×65 (Ellis); b, *Bulbocephalus elongatus*, ×15 (Watson); c, d, *Cystocephalus algerianus* (c, ×6; d, ×930) (Schneider); e, *Lophocephalus insignis* (Schneider); f, *Acanthospora polymorpha*, ×1670 (Léger); g, h, *Corycella armata* (h, ×860) (Léger); i, *Prismatospora evansi*, ×50 (Ellis); j, k, *Ancyrophora gracilis* (k, ×1250) (Léger); l, m, *Cometoides capitatus* (m, ×1330) (Léger); n, o, *Actinocephalus acutispora* (Léger); p. *Amphoroides calverti*, ×130 (Watson); q, *Asterophora philica*, ×65 (Leidy); r, *Steinina rotunda*, ×130 (Watson); s, *Pileocephalus striatus*, ×180 (Léger and Duboscq); t, *Stylocystis praecox*, ×80 (Léger).

Family 12. **Acanthosporidae** Léger

Sporadins solitary; epimerite complex; cysts without sporoducts; spores with equatorial and polar spines.

Genus **Acanthospora** Léger. Epimerite simple conical knob; spores with spines.

A. polymorpha L. (Fig. 241, *f*). Sporadins polymorphic; up to 1 mm. long; protomerite cylindro-conical; deutomerite ovoidal; endoplasm yellowish brown; cyst 500-700µ in diameter; spore with four spines at each pole and six at equatorial plane, 8µ by 4.4µ; in gut of *Hydrous ceraboides*.

Genus **Corycella** Léger. Epimerite globular, with 8 hooks; spores biconical, with one row of polar spines (Léger, 1892).

C. armata L. (Fig. 241, *g, h*). Sporadins 280-300µ long; cysts spherical, 250µ in diameter; spores 13µ by 6.5µ; in gut of larva of *Gyrinus natator*.

Genus **Prismatospora** Ellis. Epimerite subglobular with eight lateral hooks; spores hexagonal, prismatic with one row of spines at each pole.

P. evansi E. (Fig. 241, *i*). Sporadins broadly conical, 400µ long; cysts of 370µ in diameter; without sporoducts; spores with six long spines at each pole, 11µ by 5.8µ in gut of *Tramea lacerta* and *Sympetrum rubicundulum;* Michigan.

Genus **Ancyrophora** Léger. Epimerite globular with five to ten digitiform processes directed posteriorly; spores biconical, with spines.

A. gracilis L. (Fig. 241, *j, k*). Sporadins 200µ-2 mm. long; cysts spherical, 200µ in diameter; spores hexagonal in optical section, with four polar and six equatorial spines, 8.5µ by 5µ; in gut of larvae and adults of *Carabus auratus, C. violaceus, C.* sp., and of larvae of *Silpha thoracica* (Coleoptera) (Léger, 1892).

Genus **Cometoides** Labbé. Epimerite globular with six to fifteen long filaments; spores with polar spines and two rows of equatorial spines.

C. capitatus (Léger) (Fig. 241, *l, m*). Sporadins up to 2 mm long, active; epimerite with twelve to fifteen filaments, 32-35µ long; cysts 300µ in diameter; spores 5.1µ by 2.5µ; in gut of larvae of *Hydrous* sp. (Coleoptera) (Watson, 1916).

Family 13. **Actinocephalidae** Léger

Sporadins solitary; epimerite variously formed; cysts without sporoducts; spores irregular, biconical or cylindro-biconical; in gut of insects.

Genus **Actinocephalus** Stein. Epimerite sessile or with a short neck, with eight to ten simple digitiform processes at its apex; spores biconical.

A. acutispora Léger (Fig. 241, *n, o*). Sporadins 1-1.5 mm. long; cysts ovoid, 550-600μ by 280μ; spores, acutely pointed, of two sizes, 4.5μ by 2.8μ and 6.4μ by 3.6μ; in gut of the coleopteran *Silpha laevigata*.

A. parvus Wellmer. Sporadins 180μ by 50μ; cysts rounded, 62-112μ in diameter; spores spindle-form, 6-7.5μ by 3-3.8μ; 8 diploid chromosomes; the first division in the zygote is meiotic; in the gut of larvae of dog-flea, *Ctenocephalus canis*. Development (Weschenfelder, 1938).

Genus **Amphoroides** Labbé. Epimerite a globular sessile papilla; protomerite cup-shaped; spores curved; in myriapods.

A. calverti (Crawley) (Fig. 241, *p*). Sporadins up to 1670μ by 120μ; cysts spherical, 380μ in diameter; spores unknown; in gut of *Callipus lactarius*.

Genus **Asterophora** Léger. Epimerite a thick horizontal disc with a milled border and a stout style projecting from center; spore cylindro-biconical; in Neuroptera and Coleoptera.

A. philica (Leidy) (Fig. 241, *q*). Sporadins 300μ-2 mm. long; cysts and spores unknown; in gut of *Nyctobates pennsylvanica* (Crawley, 1903).

Genus **Steinina** Léger and Duboscq. Solitary; epimerite a short motile digitiform process, changing into a flattened structure; spore biconical; in Coleoptera (Léger and Duboscq, 1904).

S. rotunda Watson (Fig. 241, *r*). Sporadins 180-250μ long; in gut of *Amara augustata* (Coleoptera) (Watson, 1915).

Genus **Pileocephalus** Schneider. Epimerite lance-shaped, with a short neck.

P. striatus Léger and Duboscq (Fig. 241, *s*). Sporadins 150μ long; nucleus in protomerite; cysts spherical; in gut of larvae of *Ptychoptera contaminata*.

Genus **Stylocystis** Léger. Epimerite a sharply pointed, curved process; spores biconical (Léger, 1899).

S. praecox L. (Fig. 241, *t*). Sporadins up to 500μ long; cysts ovoidal, 200μ long; spores 8μ by 5μ in gut of larval *Tanypus* sp.

Genus **Discorhynchus** Labbé. Epimerite a large spheroidal papilla with collar and short neck; spores biconical, slightly curved.

D. truncatus (Léger) (Fig. 242, *a*, *b*). Sporadins 300μ long; cysts spherical, 140μ in diameter; in gut of larvae of *Sericostoma* sp.

Genus **Anthorhynchus** Labbé. Epimerite a large flattened fluted disc; spores biconical, chained laterally.

A. sophiae (Schneider) (Fig. 242, *c*, *d*). Cephalins up to 2 mm. long, with 200μ long epimerite; protomerite 150μ long, endoplasm opaque; spores 7μ by 5μ; in gut of *Phalangium opilio*.

Genus **Sciadiophora** Labbé. Epimerite a large sessile disc with crenulate border; protomerite with numerous vertical laminations; spores biconical.

S. phalangii Léger (Fig. 242, *e-g*). Sporadins 2-2.5 mm long; protomerite with fifteen to sixteen plates; cysts 500μ in diameter; spores 9μ by 5μ; in gut of *Phalangium crassum* and *P. cornutum* (Arachnida).

Genus **Amphorocephalus** Ellis. Epimerite a sessile peripherally fluted disc set upon a short neck; protomerite constricted superficially; spores unknown (Ellis, 1913).

A. amphorellus E. (Fig. 242, *h*). Sporadins 500-970μ long; in gut of *Scolopendra heros*.

Genus **Pyxinia** Hammerschmidt. Epimerite a crenulate crateriform disc; with a style in center; spores biconical. (Species, Vincent, 1922.)

P. crystalligera Frenzel. In the gut of the beetles, *Dermestes vulpinus* and *D. peruvianus;* cephalins 345μ by 90μ to 620μ by 55μ; cysts less than 200μ in diameter; spores 12μ by 3μ. According to Kozloff (1958), the cyst matures in 15-17 hours at 25°C. after being voided in faeces. Prior to dehiscence, a clear area appears on the upper side of the cyst and the contents begin to shrink away, except in the region of the clear area. The spores emerge through the clear area and press against the cyst wall, resulting in the formation of a conical papilla. With the continued pressure of the spores, the papilla ruptures at its tip and the spores emerge in a continuous thread until dehiscence is completed. The chain of the spores may reach a length of 11 mm. (Development and cytology, Kozloff, 1953, 1958.)

P. bulbifera Watson (Fig. 242, *i*). Sporadins up to 850µ by 260µ; in gut of *Dermestes lardarius* (Watson, 1916a).

Genus **Schneideria** Léger. Epimerite sessile, a thick horizontal disc with milled border; a style arising from center; sporadins without protomerite; spores biconical (Léger, 1892).

S. mucronata L. (Fig. 242, *j*). Sporadins 700–800µ long; agile;

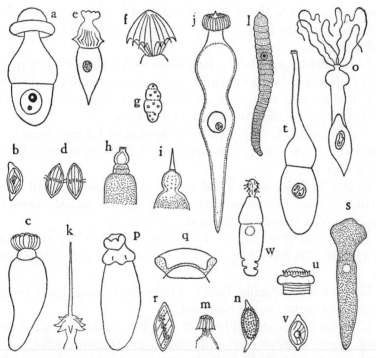

Fig. 242. a, b, *Discorhynchus truncatus* (a, ×130) (Léger); c, d, *Anthorhynchus sophiae* (c, ×15; d, ×1330) (Schneider); e–g, *Sciadiophora phalangii* (g, spore, ×1040) (Léger); h, *Amphorocephalus amphorellus* (Ellis); i, *Pyxinia bulbifera* (Watson); j, *Schneideria mucronata*, ×75 (Léger); k, *Beloides firmus* (Léger); l, *Taeniocystis mira*, ×85 (Léger); m, n, *Stictospora provincialis* (Léger); o, *Bothriopsis histrio* (Léger); p–r, *Coleorhynchus heros* (p, ×14) (Schneider); s, *Legeria agilis* (Schneider); t–v, *Phialoides ornata* (t, ×45; v, ×930) (Léger); w, *Geneiorhynchus aeschnae*, ×60 (Crawley).

polymorphic; cysts 270μ by 190μ; spores fusiform, 15μ by 9μ; in intestinal caeca of larvae of *Bibio marci*.

Genus **Beloides** Labbé. Epimerite bordered by pointed lateral processes and apical style; spores biconical (Labbé, 1899).

B. firmus (Léger) (Fig. 242, *k*). Style 80μ long; cysts 180-200μ in diameter; spores 14.5μ by 6μ; in gut of larvae of *Dermestes lardarius*.

Genus **Taeniocystis** Léger. Epimerite sessile or with a short neck; eight to ten digitiform processes at its apex; deutomerite divided by septa into many chambers; spores biconical.

T. mira L. (Fig. 242, *l*). Sporadins tapeworm-like; 400-500μ long; epimerite with six to eight curved hooks; cysts spherical, 130μ in diameter; spores 7μ by 3μ; in gut of larval *Ceratopogon solstitialis*.

Genus **Stictospora** Léger. Epimerite with a short neck, a spherical crateriform ball with twelve posteriorly-directed laminations set close to neck; cysts with a gelatinous envelope; without ducts; spores biconical, slightly curved (Léger, 1893).

S. provincialis L. (Fig. 242, *m, n*). Sporadins 1-2 mm. long; cysts 800μ in diameter; in gut of larvae of *Melolontha* sp. and *Rhizotrogus* sp.

Genus **Bothriopsis** Schneider. Epimerite sessile, small, oval, with 6 or more filamentous processes directed upward; spores biconical; cysts spherical (Schneider, 1875).

B. histrio S. (Fig. 242, *o*). Epimerite with six filaments, 80-90μ long; cysts 400-500μ long; spores 7.2μ by 5μ; in gut of *Hydaticus* sp.

Genus **Coleorhynchus** Labbé. Epimerite discoid, lower border over deutomerite; spores biconical.

C. heros (Schneider) (Figs. 242, *p-r*; 248, *j*). Sporadins 2-3 mm. long; in gut of *Nepa cinerea*. Development (Poisson, 1939).

Genus *Legeria* Labbé. Protomerite wider than deutomerite; epimerite unknown; cysts without duct; spores cylindro-biconical (Labbé, 1899).

L. agilis (Schneider) (Fig. 242, *s*). In gut of the larvae of *Colymbetes* sp.

Genus **Phialoides** Labbé. Epimerite a cushion set peripherally with stout teeth, surrounded by a wider collar; with a long neck; cysts spherical, without ducts; spores biconical.

P. ornata Léger (Fig. 242, *t-v*). Sporadins 500µ long; cysts 300-400µ in diameter; spores 10.5µ by 6.7µ; in gut of larvae of *Hydrophilus piceus.*

Genus **Geneiorhynchus** Schneider. Epimerite a tuft of short bristles at end of neck; spores cylindrical.

G. aeschnae Crawley (Fig. 242, *w*). Sporadins 420µ long; cysts and spores unknown; in *Aeschna constricta.*

Family 14. **Porosporidae** Labbé

When naked or well-protected sporozoites enter the stomach and mid-gut of a specific crustacean host, they develop into typical cephaline gregarines; one, two, or more sporadins become associated and encyst. Repeated nuclear and cytoplasmic division results in formation of an enormous number of **gymnospores** in hind-gut. Some observers consider this change as schizogony, and hence include the family in the suborder Schizogregarinina. When the gymnospores are voided in the faeces of crustaceans and come in contact with molluscan host, they enter, or are taken in by phagocytosis of, the epithelial cells of the gills, mantle or digestive system. These gymnospores are found especially in abundance in the lacunae of the gills. Presently they become paired and fuse (Hatt); the zygotes develop into naked or encapsulated sporozoites within the phagocytes of the molluscan host, which when taken in by a crustacean host, develop into cephaline gregarines.

Genus **Porospora** Schneider. Sporozoites formed in molluscan phagocytes without any protective envelope (Hatt, 1931).

P. gigantea (van Beneden) (Fig. 243, *a-f*). Sporadins in *Homarus gammarus,* up to 10 mm. long; cysts 3-4 mm. in diameter; gymnospores spherical, 8µ in diameter (Hatt), containing some 1500 merozoites; in molluscan hosts, *Mytilus minimus* and *Trochocochlea mutabilis,* they develop into naked sporozoites (17µ long) which are usually grouped within phagocytes.

Genus **Nematopsis** Schneider. Development similar to that of *Porospora* (Hatt); but each sporozoite in a double envelope. (Species in Gulf of Mexico, Sprague, 1954.)

N. legeri de Beauchamp (*Porospora galloprovincialis* Léger and Duboscq) (Fig. 243, *g-n*). Sporadins in a crustacean, *Eriphia spinifrons,* in linear or bifurcated syzygy 75-750µ long; cysts about

80μ in diameter; gymnospores 7μ in diameter, composed of fewer, but larger merozoites; permanent spores with a distinct one-piece shell (endospore) and a less conspicuous epispore, about 14-15μ long and circular in cross-section, develop in numerous species of molluscan hosts: *Mytilus galloprovincialis, M.*

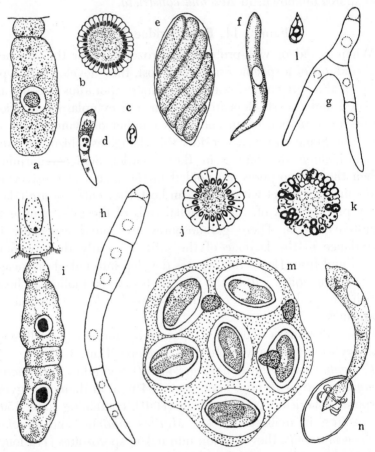

FIG. 243. a–f, *Porospora gigantea* (Hatt). a, a cephalin attached to Homarus gut, ×1250; b, gymnospores; c, d, developing sporozoites in mollusc; e, sporozoites enveloped by phagocyte; f, a sporozoite, ×2250. g–n, *Nematopsis legeri* (Hatt). g, h, trophozoites in Eriphia; i, associated trophozoites attached to gut-epithelium, ×1250; j, gymnospores; k, gymnospores after entering molluscan body; l, a young sporozoite, ×2250; m, cyst in mollusc with six spores; n, germination of a spore in Eriphia gut, ×1250.

minimus, Lasea rubra, Cardita calyculata, Chiton caprearum, Trochocochlea turbinata, T. articulata, T. mutabilis, Phorcus richardi, Gibbula divaricata, G. rarilineata, G. adamsoni, Pisania maculosa, Cerithium rupestre, Columbella rustica, and *Conus mediterraneus* in European waters (Hatt, 1931).

N. ostrearum Prytherch. Sporadins in syzygy in the mud crabs, *Panopeus herbsti* and *Eurypanopeus depressus,* 220-342µ; cysts 80-190µ in diameter; gymnospores 4µ in diameter; spores produced in the oyster, *Ostrea virginica,* 16µ by 11-12µ (Prytherch, 1940). Landau and Galtsoff (1951) showed that the organism is

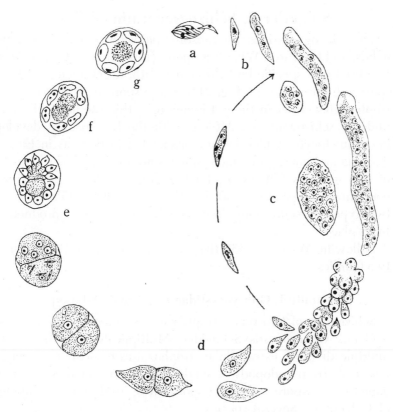

Fɪɢ. 244. The life-cycle of *Schizocystis gregarinoides,* ×1000 (Léger). a, germinating spore; b, growth of schizonts; c, schizogony; d, two gametocytes and their association; e, stages in gamete formation; f, zygote formation; g, cyst containing zygotes, each of which develops into a spore shown in a.

widely distributed among the oysters along the Atlantic and Gulf coasts, but found no evidence to suppose that the organism is destructive to the host mollusc.

N. panopei Ball. Sporadins up to 210μ by 14μ; protomerite about one-fifteenth the body length; epimerite on young individuals only; syzygy often multiple, as in other species; cysts 88μ by 74μ, free in the lumen or attached to the wall of the hind-gut; gymnospores about 6.5μ in diameter; in the gut of *Panopeus herbsti* and *P. occidentalis;* in Bermuda. Molluscan host unknown (Ball, 1951). (Species, Sprague and Orr, 1955).

Suborder 2. **Schizogregarinina** Léger

The schizogregarines are intestinal parasites of arthropods, annelids, and tunicates. When the spore gains entrance to the digestive tract of a specific host through mouth, it germinates and the sporozoites are set free (Fig. 244). These sporozoites develop into trophozoites either in the gut-lumen or within the host cells, and undergo schizogony (*c*). After growth the trophozoites develop into gametocytes which become associated in pairs as in Eugregarinina and encyst, in which gametes are produced (*d-e*). Fusion of two gametes follows (*f*). Each zygote develops into a spore containing sporozoites (*g, a*). In other forms (Fig. 245), there are two types of schizogony (*b-e, f-g*) before sexual reproduction takes place.

Following Weiser (1955), the schizogregarines are divided into two families.

Family 1. **Ophryocystidae** Léger and Duboscq

Schizogony of two types; trophozoites vermicular in form.

Genus **Ophryocystis** Schneider. Multiplication by binary or multiple division; extracellular; trophozoites conical, attached to host cells by pseudopods; a single spore in a pair of spheroidal gametocytes; spore with eight sporozoites; in Malpighian tubules of Coleoptera. Several species.

O. mesnili Léger (Fig. 246, *a-e*). In *Tenebrio molitor;* schizonts one to four nuclei; gametocytes 11μ in diameter; pairs 16-17μ by 11μ; spores biconical, 11μ by 7μ.

Genus **Mattesia** Naville. Schizogony in the fat body cell; trophozoites without pseudopodial processes; cyst usually with two spores, sometimes one (Naville, 1930; Weiser, 1954).

M. dispora N. (Fig. 245). In adipose tissue cells of larvae of flour moth, *Ephestia kuhniella* and *Plodia interpunctella* (pupa and adult also); schizonts 2.5-12μ long; cyst 8-12μ in diameter with

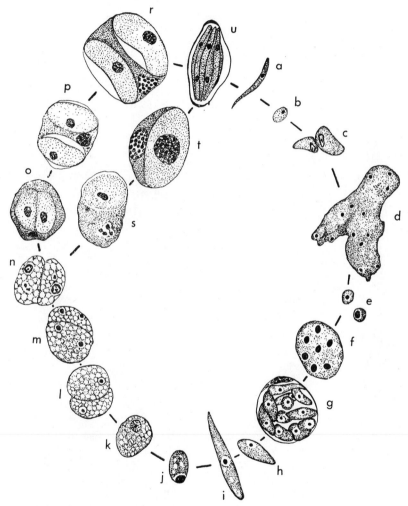

FIG. 245. The life-cycle of *Mattesia dispora* (Weiser). a, sporozoite; b-e, stages of the first schizogony; f, g, the second schizogony; h-j, trophozoites; k, gametocyte; l-n, syzygy of gametocytes; o-r, disporous sporogony; s, t, monosporous forms; u, a mature spore.

two spores, each with eight sporozoites; spores 14μ by 7.5μ (Naville, 1938); 11μ by 6μ (maximum 13.5μ by 8μ) (Musgrave and Mackinnon). Highly pathogenic according to Musgrave and Mackinnon.

Genus **Lipocystis** Grell. Schizogony and sporogony intracellu-

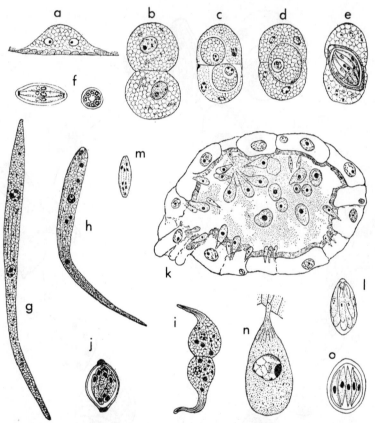

Fig. 246. a–e, *Ophryocystis mesnili* (a, trophozoite attached to Malpighian tubule; b–e, sporogony), ×1330 (Léger); f, two views of spore of *Lipocystis polyspora*, ×1485 (Grell); g–j, *Machadoella triatomae* (g, a trophozoite, ×1420; h, i, single and associated sporadins, ×710; j, spore, ×1920) (Reichenow); k, l, *Caulleryella pipientis* (k, gut of *Culex pipiens* with trophozoites, ×200; l, spore, ×1200) (Buschkiel); m, *Lipotropha macrospora*, ×800 (Keilin); n, o, *Merogregarina amaroncii*, ×1000 (Porter).

lar; gamete formation on the surface of cytomeres; isogamy; cyst produces 200-300 spores, each with eight sporozoites. (One species, Grell, 1938.)

L. polyspora G. (Fig. 246, *f*). Spores elongate ellipsoid, about 10μ by 4μ; in the fat body of *Panorpa communis*.

Genus **Machadoella** Reichenow. Nematode-like, rigid; simple rounded anterior end; thick pellicle, longitudinally striated; schizogony in vermiform stage; head to head association of gametocytes; cysts with three to six spores, each with eight sporozoites.

M. triatomae R. (Fig. 246, *g–j*). Schizonts about 55μ long; gametocytes 100–120μ long; schizogony into six to eight merozoites; cysts with three to six spores; spore 10-11μ by 7-7.5μ; in Malpighian tubules of *Triatoma dimidiata* (of Guatemala) (Reichenow, 1935).

Genus **Triboliocystis** Dissanaike. Trophozoites vermicular; gametocysts usually with eight spores (Dissanaike, 1955).

T. Garnhami D. Trophozoites fusiform, 13-20μ by 2.3-3.4μ; spores navicular with rounded membranous poles, 13-14μ by 6.6-7.7μ; in fat body of *Tribolium castaneum*.

Family 2. **Caulleryellidae** Weiser

Schizogony of one type; trophozoites broadly lanceolate.

Genus **Caulleryella** Keilin. Schizogony extracellular in host's intestinal cavity; one type of schizogony; each gametocyte gives rise to eight gametes and cyst with eight spores; spore with eight sporozoites; in the gut of dipterous larvae.

C. pipientis Buschkiel (Fig. 246, *k, l*). Average trophozoites 50-60μ by 23-26μ; with paraglycogen grains; schizogony produces 30-38 merozoites; in gut of larvae of *Culex pipiens*.

Genus **Lipotropha** Keilin. Schizogony and sporogony intracellular; cyst contains sixteen spores, each with eight sporozoites; in fat body of Systenus larvae. One species.

L. macrospora K. (Fig. 246, *m*). Spores about 13.5μ by 3μ.

Genus **Merogregarina** Porter. Schizogony intracellular; trophozoites attached to gut epithelum by a proboscidiform organella; two gametocytes giving rise to one spore containing eight sporozoites.

M. amaroucii P. (Fig. 246, *n. o*). In gut of the ascidian, *Amaroucium* sp.; extracellular; trophozoites with epimerite, 27-31μ long; spore about 14μ long.

Genus **Schizocystis** Léger. Mature trophozoite multinucleate; ovoid or cylindrical with differentiated anterior end; schizogony by multiple division; trophozoites become associated, encyst, and produce numerous (up to 32) spores, each with eight sporozoites; in Diptera, Annelida, and Sipunculoida (Léger, 1909).

S. gregarinoides L. (Fig. 244). In gut of larvae of *Ceratopogon solstitialis;* mature schizonts up to 400μ by 15μ; curved or spirally coiled; gametocytes 30-50μ long; cysts ovoid, 16-32μ long; spores biconical, 8μ by 4μ.

Genus **Siedleckia** Caullery and Mesnil. Trophozoites ribbonlike, multinucleate; uninucleate anisogametes; zygote forms a spore with ten to fourteen ovoid sporozoites; in *Scoloplos* spp. (Caullery and Mesnil, 1898).

S. caulleryi Chatton and Villeneuve (Fig. 247, *a, b*). In the gut of *Scoloplos armiger.*

Genus **Syncystis** Schneider. Schizogony and sporogony extracellular; young trophozoites elongate, amoeboid; mature schizonts more or less spheroidal, producing some 150 merozoites; cysts spherical, producing about 130 spores. (One species.)

S. mirabilis S. (Fig. 247, *c, d*). In coelomic fluid and fat bodies. of *Nepa cinerea;* merozoites, 7μ long; cysts spherical; spores navicular, three to four spines at ends, 10μ by 6μ with eight sporozoites.

Genus **Selenidium** Giard. Schizogony intracellular; many spores produced by a pair of extracellular gametocytes; spore with four or more sporozoites; in gut of annelids. (Generic status, Mackinnon and Ray, 1933.)

S. potamillae Mackinnon and Ray (Fig. 247, e-g). Trophozoites euglenoid, average size 40μ by 15μ; longitudinal striae; cysts oblong, producing many spores; spore, spherical with four (up to ten) sporozoites; in gut of the polychaete, *Potamilla reniformis* (Mackinnon and Ray, 1933).

Genus **Meroselenidium** Mackinnon and Ray. Schizogony intracellular, initiated by formation of small masses which give rise to merozoites; about twenty spores from a pair of gametocytes;

spores with 16 sporozoites. (One species, Mackinnon and Ray, 1933.)

M. keilini M. and R. (Fig. 247, *h-j*). Large schizonts about 150μ by 30μ; sporadins free in gut 200-300μ by 40-70μ; paired gametocytes 85μ by 40μ; spores 26-28μ by 14-16μ, bivalue (?), transverse ridges; in gut of *Potomilla reniformis*.

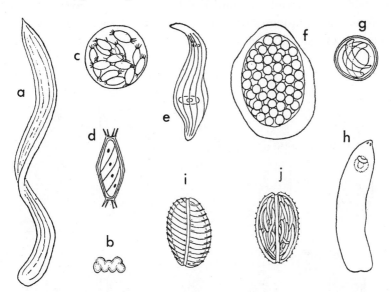

FIG. 247. a, b, *Siedleckia caulleryi* (a, trophozoite in life; b, cross-section view) (Chatton and Villeneuve); c, d, *Syncystis mirabilis* (c, cyst, ×470) (Steopoe); e–g, *Selenidium potamillae* (e, ×420; f, cyst with spores, ×330; g, spore) (Mackinnon and Ray); h–j, *Meroselenidium keilini* (h, sporadin, ×670; i, j, different views of spore, ×930) (Mackinnon and Ray).

Genus **Selenocystis** Dibb. Sporadins leaf-like with a median ridge; biassociation with posterior ends, forming an elongated cyst, attached to the host epithelium by a foot-like organelle; isogametes with a short flagellum; spores with four or eight sporozoites. (One species, Dibb, 1938.)

S. foliata (Ray) (Fig. 248, *a-e*). Trophozoites 30-250μ long; pellicle with sixteen to twenty-four striae; the broader end with which the organism is attached to the host epithelium depressed; surrounding this depression, a number of about 8μ long refrin-

gent filaments occur; while one organism is still attached, biassociation by posterior ends takes place; 26-226μ by 9-34μ isogametes; subspherical spores about 8.5μ in diameter, with four or
eight sporozoites; in the gut of the polychaete, *Scolelepis fuliginosa.*

Genus **Spirocystis** Léger and Duboscq. Schizogony intracellular; schizonts curved, one end highly narrowed; mature schizonts
snail-like, with numerous nuclei; repeated schizogony (?); no biassociation of gametocytes; gametes in chlorogogen cells, somatic

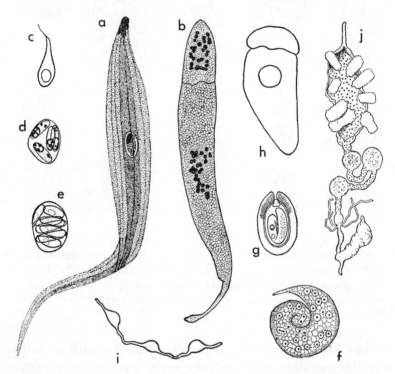

FIG. 248. a–e, *Selenocystis foliata* (a, a mature trophozoite, ×665
(Ray); b, migration of nuclei of gametocytes to the surface of cyst,
×565; c, gamete in life; d, e, spores with 4 and 8 sporozoites, ×1130
(Dibb)); f, g, *Spirocystis nidula* (f, ×770; g, ×500) (Léger and Duboscq); h, i, *Colepismatophila watsonnae* (h, ×85; i, spores, ×200)
(Adams and Travis); j, the digestive tube of *Nepa cinerea* with 8
trophozoites attached to the stomach (opened) epithelium and 3 cysts
of *Coleorhynchus heros* (Poisson).

and visceral peritonium; association of two gametes produces a spore. (One species.)

S. *nidula* L. and D. (Fig. 248, *f*, *g*). In coelom and gut epithelium of *Lumbricus variegatus;* multinucleate schizont about 35μ long; microgametes fusiform or ovoid, 7μ by 3μ; macrogametes ovoid or spherical, 11μ in diameter; fusion of two gametes produces one spore which is thick-walled, 35μ long and contains one sporozoite, up to 40μ long.

References

ADAMS, J. A., and TRAVIS, B. V., 1935. Two new species of gregarine Protozoa from the firebrat, etc. J. Parasit., 21:56.

ADLER, S, and MAYRINK, W., 1961. A gregarine, *Monocystis chagasi* n.sp., of *Phlebotomus longipalpis*. Rev. Inst. Med. Trop., Sao Paulo, 3:230.

ALLEGRE, C. F., 1948. Contributions to the life history of a gregarine parasitic in grasshoppers. Tr. Am. Micr. Soc., 67:211.

BALL, G H., 1938. The life history of *Carcinoecetes hesperus,* etc. Arch. Protist., 90:299.

——— 1951. Gregarines from Bermuda marine crustaceans. Univ. Cal. Publ. Zool., 47:351.

——— 1959. Some gregarines from crustaceans taken near Bombay, India. J. Protozool., 6:8.

BEAMS, H. W., *et al.*, 1959. Studies on the fine structure of a gregarine, etc. J. Protozool., 6:136.

BERLIN, H., 1924. Untersuchungen ueber Monocystideen in den Vesiculae seminales der schwedischen Oligochaeten. Arch. Protist., 48:1.

BHATIA, B. L., 1929. On the distribution of gregarines in oligochaetes. Parasitology, 21:120.

——— 1930. Synopsis of the genera and classification of haplocyte gregarines. *Ibid.*, 22:156.

——— 1938. Fauna of British India. Sporozoa. London.

——— and CHATTERJEE, G. B., 1925. On some gregarine parasites of Indian earthworms. Arch. Protist., 52:189.

——— and SETNA, S. B., 1926. On some more gregarine parasites of Indian earthworms. *Ibid.*, 53:361.

BUSCHKIEL, M. 1921. *Caulleryella pipientis,* etc. Zool. Jahrb. Anat., 43:97.

CALKINS, G. N., and BOWLING, R. C., 1926. Gametic meiosis in Monocystis. Biol. Bull., 51:385

CANNING, E. U., 1956. A new gregarine of locust, etc. J. Protozool., 3:50.

CAULLERY, M., and MESNIL, F., 1898. Sur un sporozoaire aberent (Siedleckia n. g.). C. R. Soc. Biol., 5:1093.

CHAKRAVARTY, M, 1935. Studies on Sporozoa from Indian millipeds. IV. Arch. Protist., 86:211.

—— 1936. V. *Ibid.*, 88:116.

—— 1959. Systematic position of some genera and classification of the suborder Cephalina. Proc. Zool. Soc., 12:71.

CHATTON, E., and VILLENEUVE, F., 1938. Le sexualité et le cycle évolutif des Siedleckia, etc. C. R. Acad. Sci., 203:505.

CODREANU, M., 1940. Sur quatre grégarines nouvelles du genre Enterocystis, etc. Arch. zool. expér. et gén., 81 (N-R):113.

COGNETTI DE MARTIIS, L. 1911, Contributo alla conoscenza delle Monocistidee e dei loro fenomeni riproduttivi. Arch. Protist., 23:205.

—— 1921. Resultats de l'expedition scientifique Neerlandaise a la Nouvelleguinée. XIII. Zoologie, 4:501.

—— 1923. Sul genera Monocysts. Monit. Zool. Ital. Firenze, 34:250.

—— 1925. Sulla classificazione e sui caratteri tassonomici delle Monocistidee degli oligocheti. *Ibid.*, 36:219.

—— 1926. Due nuove Gregarine Monocistidee a miocito profondo. Boll. Musei Zool. e Anat. Com. Sec. Ser. 6:17.

CORDUA, C-A., 1953. Untersuchungen ueber den Gregarineninfektion der Dungkaefer. Arch. Protist., 98:469.

CRAWLEY, H., 1903. List of polycystid gregarines of the United States. Proc. Acad. Nat. Sci., Philadelphia, 55:41.

—— 1903a. II. *Ibid.*, 55:632.

—— 1907. III. *Ibid.*, 59:220.

CRUSZ, H., 1957. Gregarine protozoa of silverfish, etc. J. Parasit., 43:90.

DIBB, M. J., 1838. *Selenocystis foliata* (Ray) from *Scolelepis fuliginosa*, and its identity with *Haplozoon* sp. Parasitology, 30:296.

DISSANAIKE, A. S., 1953. Acephaline gregarines of the Ceylon earthworm, *Pheretima peguana*. Ceylon J. Sci., Sec. B., 25:161.

—— 1955. A new schizogregarine *Triboliocystis garnhami* n.g., n.sp., etc. J. Protozool., 2:150.

DOFLEIN, F., and REICHENOW, E., 1949–1953. Lehrbuch der Protozoenkunde. 6 ed. Jena.

DUKE, H. L., 1910. Some observations on a new gregarine (*Metamera schubergi* n.g., n.sp.). Quart. J. Micr. Sci., 55:261.

ELLIS, M. M., 1912. A new species of polycystid gregarine from the United States. Zool. Anz., 39:25.

—— 1913. A descriptive list of the cephaline gregarines of the New World. Tr. Am. Micr. Soc., 32:259.

—— 1913a. New gregarines from the United States. Zool. Anz., 41:462.

—— 1914. An acanthosporid gregarine from North American dragonfly nymphs. Tr. Am. Micr. Soc., 33:215.

FILIPPONI, A., 1949. Gregarine policistidae parasite di *Laemostenus algerinus* con considerazioni sulla nomenclatura nelle gregarine. Riv. Parassitol., 10:245.

—— 1951. Studi sugli Stylocephalidae: III and IV. Estra. Rend., Ist. Super Sanita, 14:103 and 247.

———— 1952. *Protomagalhaeusia marottai* n.sp. (Gregarinidae) parassita di *Scanrus striatus. Ibid.*, 15:465.

———— 1953. Sul grado di stabilita di *Protomagalhaeusia marottai.* Riv. Parassitol., 14:137.

GANAPATI, P. N., and NARASIMHAMURTI, C. C., 1956. On a new cephaline gregarine *Stenophora xenoboli* n.sp. J. Zool. Soc. India, 8:165.

GATES, G. E., 1933. On a new gregarine from the coelom of a Burmese earthworm, *Pheretima compta.* Biol. Bull., 65:508.

GÖHRE, E., 1943. Untersuchungen ueber den plasmatischen Feinbau der Gregarinen, etc. Arch. Protist., 96:295.

GOODRICH, E. S., and GOODRICH, H. L. M. P., 1920. *Gonospora minchini,* n.sp., etc. Quart. J. Micr. Sci, 65:157.

GOODRICH, H. P., 1925. Observations on the gregarines of Chiridota. *Ibid.,* 69:619.

———— 1938. Nina: a remarkable gregarine. *Ibid.,* 81:107.

———— 1949. Heliospora n.g. and Rotundula n.g., etc. *Ibid.,* 90:27.

———— 1950. Sporozoa of Sipunculus. *Ibid.,* 91:469.

GRELL, K. G., 1938. Untersuchungen an Schizogregarinen. I. Arch. Protist., 91:526.

HARRISON, A. D., 1955. Four new species of gregarines from mountain cock-roaches of the Cape Peninsula. Ann. S. African Mus., 41:387.

HATT, P., 1931. L'évolution des porosporides chez les mollusques. Arch. zool. exper. gén., 72:341.

HESSE, E., 1909. Contribution à l'étude des monocystidees des Oligochaetes. *Ibid.,* 3:27.

JAMESON, A. P., 1920. The chromosome cycle of gregarines, with special reference to *Diplocystis schneideri.* Quart. J. Micr. Sci., 64:207.

JONES, A. W., 1943. *Metamera reynoldsi* n.sp., etc. Tr. Am. Micr. Soc. 62:254.

KAMM, M. W., 1920. The development of gregarines and their relation to the host tissues. III. J. Parasit., 7:23.

———— 1922. Studies on gregarines. II. Illinois Biol. Monogr., 7:1.

———— 1922a. A list of the new gregarines described from 1911 to 1920. Tr. Am. Micr. Soc., 41:122.

KEILIN, D., 1920. On two new gregarines, etc. Parasitology, 12:154.

KOZLOFF, E. N., 1953. The morphogenesis of *Pyxinia crystalligera* Frenzel, etc. J. Morphol., 92:39.

———— 1958. Development and dehiscence of gametocysts of the gregarine *Pixinia crystalligera.* J. Protozool., 5:171.

LABBÉ, A., 1899. Sporozoa. In Das Tierreich. Part 5.

LANDAU, HELEN and GALTSOFF, P. S., 1951. Distribution of Nematopsis infection on the oyster grounds of the Chesapeake Bay and in other waters of the Atlantic and Gulf states. Texas J. Sci., 3:115.

LÉGER, L., 1892. Recherches sur les grégarines. Tabl. zool., 3:1.

———— 1893. Sur une grégarine nouvelle des acridiens d'Algerie. C. R. Acad. Sci., 117:811.

—— 1906. Étude sur *Taeniocystis mira* Léger, etc. Arch. Protist., 7:307.

—— 1907. Les schizogrégarines des trachéates. I. *Ibid.*, 8:159.

—— 1909. II. *Ibid.*, 18:83.

—— and Duboscq, O., 1915. Études sur *Spirocystis nidula*, etc. *Ibid.*, 35: 199.

—— —— 1925. Les porosporidies et leur évolution. Trav. St. zool. Wimereux, 9:126.

Leidy, J., 1853. On the organization of the genus Gregarina of Dufour. Tr. Am. Philos. Soc., n.s., 10:233.

Loubatières, R., 1955. Contribution à l'étude des Grégarinomorphes Monocystidae parasites des oligochaètes, etc. Ann. Sci. Nat. Zool., 17:73.

Mackinnon, D. L., and Ray, H. N., 1931. Observations on dicystid gregarines from marine worms. Quart. J. Micr. Sci., 74:439.

—— —— 1933. The life cycle of two species of "Selenidium" from the polychaete worm *Potamilla reniformis*. Parasitology, 25:143.

Mercier, L., 1911. *Cephaloidophora cuenoti* n. sp., etc. C. R. Soc. Biol., 71:51.

Miles, H. B., 1962. The mode of transmission of the acephaline gregarine parasites of earthworms. J. Protozool., 9:305.

Mühl, D., 1921. Beitrag zur Kenntnis der Morphologie und Physiologie der Mehlwurmgregarinen. Arch. Protist., 43:361.

Mulsow, K., 1911. Ueber Fortpflanzungserscheinungen bei *Monocystis rostrata* n. sp., *Ibid.*, 22:20.

Musgrave, A. J., and Mackinnon, D. L., 1938. Infection of *Plodia interpunctella* with a schizogregarine, *Mattesia dispora*. Proc. Roy. Entom. Soc. London (A), 13:89.

Naville, A., 1930. Recherches cytologiques sur les schizogrégarines. Ztschr. Zellf. mikr. Anat., 11:375.

—— 1931. Les Sporozoaires. Mem. d'hist. nat. Geneva, 41:1.

Noble, E. R., 1938. The life-cycle of *Zygosoma globosum* sp. nov., a gregarine parasite of *Urechis caupo*. Univ. Cal. Publ. Zool., 43:41.

—— 1938a. A new gregarine from *Urechis caupo*. Tr. Am. Micr. Soc., 57:142.

Pinto, C., 1918. Contribuição as estudo das gregarines. Trav. Inst. Oswaldo Cruz, 113 pp.

Poisson, R., 1939. A propos de *Coleorhynchus heros*, grégarine parasite de la Nèpe cendrée. Bull. biol. France et Belg., 73:275.

Prytherch, H. F., 1940. The life cycle and morphology of *Nematopsis ostrearum*, etc. J. Morphol., 66:39.

Ray, H. N., 1930. Studies on some Sporozoa in polychaete worms. I. Parasitology, 22:370.

—— and Chakravarty, M. 1933. Studies on Sporozoa from Indian millipedes. II. Arch. Protist., 81:352.

Reichenow, E., 1932. Sporozoa. Grimpe's Die Tierwelt der Nord- und Ostsee. 21:pt. 2-g.

——— 1935. *Machadoella triatomae,* etc. Arch. Protist., 84:431.

RODGI, S. S. and BALL, G. H., 1961. New species of gregarines from millipeds of Mysore State, India. J. Protozool., 8:162.

SCHAUDINN, F., 1900. Untersuchungen ueber Generationswechsel bei Coccidien. Zool. Jahrb. Anat., 13:197.

SCHIFFMANN, O., 1919. Ueber die Fortpflanzung von *Gregarina blattarum* und *G. cuneata.* Arch. Protist., 40:76.

SCHNEIDER, A., 1875. Contributions à l'histoire des grégarines des invertebres de Paris et de Roscoff. Arch. zool. exper., 4:493.

SMITH, L. M., 1929. *Coccospora stenopelmati,* etc. Univ. Cal. Publ. Zool., 33:57.

——— 1930. Further observations on the protozoan Tettigonospora. *Ibid.,* 33:445.

SPRAGUE, V., 1941. Studies on *Gregarina blattarum* with particular reference to the chromosome cycle. Illinois Biol. Monogr., 18:no. 2.

——— 1954. Protozoa. U.S. Dep. Inter., Fishery Bull., 89:243.

——— and ORR, P. E. Jr., 1955. *Nematopsis ostrearum,* etc. J. Parasit., 41:89.

TROISI, R. L., 1933. Studies on the acephaline gregarines of some oligochaete annelids. Tr. Am. Micr. Soc., 52:326.

TUZET, O. and RAMBIER, J., 1953. Recherches sur les Grégarines des Orthopteroides. Ann. Sci. Nat. Zool. 11 sér., 15:247.

VINCENT, M., 1922. On the life history of a new gregarine: *Pyxinia anobii,* etc. Parasitology, 14:299.

——— 1924. On a new gregarine *Anisolobus dacnecola,* etc. *Ibid.,* 16:44.

WATSON, M., 1915. Some new gregarine parasites from Arthropoda. J. Parasit., 2:27.

——— 1916. Studies on gregarines. Illinois Biol. Monogr., 2:213.

——— 1916a. Observations on polycystid gregarines from Arthropoda. J. Parasit., 3:65.

WEISER, J., 1954. Zur systematischen Stellung der Schizogregarinen der Mehlmotte, *Ephestia kuhniella.* Arch. Protist., 100:127.

——— 1955. A new classification of the Schizogregarina. J. Protozool., 2:6 and 88.

——— 1955a. Zur Entwicklung der Schizogregarine *Syncystis mirabilis* (in Czek.). Czek. Parasit., 2:181.

WENYON, C. M., 1926. Protozoology. 1, 2. London and Baltimore.

WESCHENFELDER, R., 1938. Die Entwicklung von *Actinocephalus parvus.* Arch. Protist., 91:1.

ZWETKOW, W. N., 1926. Eine neue Gregarinengattung *Enterocystis ensis,* etc., Arch. Russ. Protist., 5:45.

Chapter 26

Order 2. **Coccidida** Leuckart

THE Coccidida show a wide zoological distribution, attacking all vertebrates and higher invertebrates alike. The majority are parasites of the eptihelium of the digestive tract and its associated glands. Asexual reproduction is by schizogony and sexual reproduction by anisogamy in the majority of species. Both kinds of reproduction take place in one and the same host body, with the exception of such forms as Aggregata in which alternation of generations and of hosts occurs. Order Coccidida is subdivided into two suborders; Eimeriina (gametocytes similar; many microgametes) and Adeleina (gametocytes dissimilar; a few microgametes) (p. 701).

Suborder 1. **Eimeriina** Léger

These coccidians are, as a rule, intracellular parasites of the gut epithelium. Both asexual (schizogonic) and sexual (sporogonic) generations occur in one host, although in some there is also alternation of hosts. The life-cycle of *Eimeria schubergi*, a gut parasite of the centipede, *Lithobius forficatus,* as observed by Schaudinn, is as follows (Fig. 249). The infection begins when the mature oocysts of the coccidian gain entrance into the host through the mouth. The sporozoites escape from the spores and make their way through the micropyle of the oocyst into the gut lumen (*p*). By active movement they reach and enter the epithelial cells (*a*). These schizonts grow into large rounded bodies and their nuclei multiply in number. The newly formed nuclei move to the body surface, and each becomes surrounded by a small mass of cytoplasm, forming a merozoite. When the host cells rupture, the merozoites are set free in the gut lumen, make their way into new host cells and repeat the development (*b*). Instead of going into schizonts, some merozoites transform themselves into macro- or micro-gametocytes (*c*). Each macrogametocyte contains refractile bodies, and becomes a mature macrogamete, after extruding a

678

part of its nuclear material (*d, e*). In the microgametocyte, the nucleus divides several times and each division-product assumes a compact appearance (*f-h*). The biflagellate comma-shaped microgametes thus produced, show activity when freed from the host

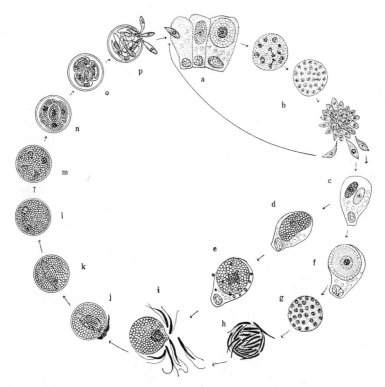

Fig. 249. The life-cycle of *Eimeria schubergi,* ×400 (Schaudinn) a, entrance of a sporozoite in the gut epithelial cell of host and growth of schizont; b, schizogony; c, macro- and micro-gametocyte; d, e, formation of macrogamete; f–h, formation of microgametes; i, mature gametes prior to fusion, j, k, fertilization; l–n, spore-formation; o, oocyst containing four mature spores, each with two sporozoites; p, germination of spores in host's gut.

cells (*i*). A microgamete and a macrogamete unite to form a zygote which secretes a membrane around itself (*j*). This stage is known as the **oocyst.** The nucleus divides twice and produces four nuclei (*k-m*). Each of the four nuclei becomes the center of a **sporoblast** which secretes a membrane and transforms itself into a spore (*n*). Its nucleus, in the meantime, undergoes a division, and two **sporozoites** develop in the spore (*o*). Oocysts leave the host in the faecal matter and become the source of infection. Four families.

Family 1. **Selenococcidiidae** Poche

Vermiform body and gametic differentiation place this family on the borderline between Coccidida and Gregarinida.

Genus **Selenococcidium** Léger and Duboscq. Nucleus of vermiform trophozoite divides three times, producing eight nuclei; trophozoite becomes rounded after entering gut-epithelium and divides into eight schizonts; this is apparently repeated; schizonts develop into gametocytes; microgametocyte produces numerous microgametes; gametic union and sporogony (?). (One species.)

S. intermedium L. and D. (Fig. 250). Octonucleate vermiform schizont 60-100μ long, and divides into vermicular merozoites in gut cells; parasitic in gut lumen of European lobster.

Genus **Ovivora** Mackinnon and Ray. Trophozoites large and

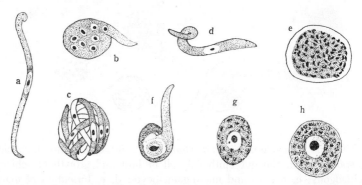

Fig. 250. *Selenococcidium intermedium*, ×550 (Léger and Duboscq). a, schizont in host gut; b, c, schizogony; d, microgametocyte; e, microgametes; f, macrogametocyte; g, macrogamete; h, zygote (oocyst).

vermiform (Fig. 251, *a*); gametocytes spherical (*c*); large macrogametocytes; small microgametocytes, giving rise to numerous biflagellate microgametes (*d*); oocyst membrane delicate or lacking; ovoid spores contain variable (averaging 12?) number of sporozoites; schizogony produces many merozoites; one host. (One species, Mackinnon and Ray, 1937.)

O. thalassemae (Lankester) (Fig. 251). In the egg of the echiurid worm, *Thalassema neptuni;* merozoites about 10μ long (*b*); macrogametocytes (*c*) 40-75μ in diameter; microgametocytes (*c*) 23-65μ; chromosome reduction, 14 to 7, in the zygote; spores (*f*) 15.5μ by 13.5μ (Mackinnon and Ray, 1937).

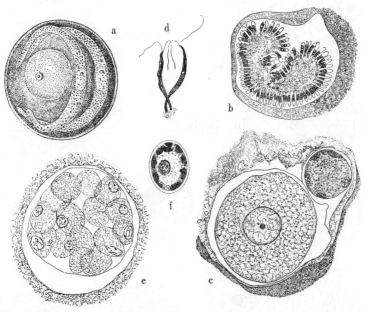

Fig. 251. *Ovivora thalassemae* (Mackinnon and Ray). a, two mature organisms in host egg, seen in reflected light, ×250; b, schizonts in sectioned egg; c, micro- and macro-gametocytes in an egg, ×500; d, two maturing microgametes still attached to cytoplasmic residuum, ×1075; e, cyst with zygotes in some of which nuclei are dividing, ×500; f, a spore with 10 nuclei, ×900.

Family 2. Aggregatidae Labbé

Anisogamy results in production of zygotes which become transformed into many spores, each with two to thirty sporozoites; in schizogony **cytomeres** first appear and then merozoites; alternation of generations and of hosts which are marine annelids, molluscs and crustaceans.

Genus **Aggregata** Frenzel. Schizogony in a crustacean and sporogony in a cephalopod; zygote produces many spores, each with three sporozoites. Many species. (Cytology, Moroff, 1908.)

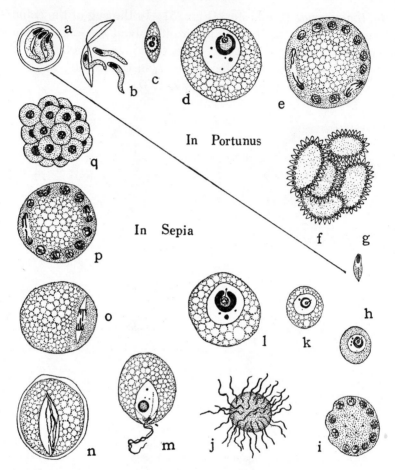

In Portunus

In Sepia

Fig. 252. The life-cycle of *Aggregata eberthi* (Dobell). a, a mature spore; b, germination of spore; c–f, schizogony; g, a merozoite, swallowed by Sepia; h–j, development of microgametes; k–l, development of macrogamete; m, fertilization; n, o, first zygotic division, chromosomes reduced in number from 12 to 6; p, q, development of sporoblasts, each of which develops into a spore with three sporozoites.

A. *eberthi* (Labbé) (Fig. 252). Schizogony in *Portunus depurator* and sporogony in *Sepia officinalis*. Spores (*a*) germinate in the crab gut, each liberating three sporozoites (*b*) which grow and produce merozoites (10μ by 2μ) by schizogony in peri-intestinal connective tissue cells (6 chromosomes) (*c–f*); when host crab is eaten by a

cuttlefish, merozoites penetrate gut wall and develop into micro-
and macro-gametocytes (*h, k*) and further into gametes (*j–l*); anisogamy (*m*) produces zygotes; zygote nucleus contains twelve chromosomes which become divided into two groups of six in the first
division (*n, o*); repeated nuclear division (*p*) forms many sporoblasts (*q*), each transforms itself into a spherical spore with three
sporozoites (Dobell, 1925; Naville, 1925; Bĕlař, 1926).

Genus **Merocystis** Dakin. Sporogony in the kidney of the
whelk, Buccinum; schizogony unknown, in another host (possibly
a crab); microgametocytes produce first cytomeres which in turn
form microgametes; anisogamy gives rise to zygotes, zygote forms
many sporoblasts, each developing into a spore; spore spherical,
with two sporozoites. One species.

M. kathae D. (Fig. 253, *a, b*). In the kidney of *Buccinum unda-
tum;* spores spherical, about 14µ in dimeter. Patten (1935)
studied its life cycle and found that during microgametogenesis
and sporogony, six chromosomes occur. She added that meiosis
occurs in the zygote which is the only diploid stage as in *Aggregata eberthi.*

Genus **Pseudoklossia** Léger and Duboscq. Anisogamy and sporogony in the kidney of marine mussels; oocyst or zygote produces numerous spores; spore with two sporozoites; no residual
body; schizogony unknown, in another host (Léger and Duboscq,
1915, 1917).

P. pectinis L. and D. (Fig. 253, *c*). In kidney of *Pecten maximus*
in France; association of two sporozoites which are 3.5µ in diameter.

Genus **Caryotropha** Siedlecki. Both schizogony and sporogony
take place in a host. (One species.)

C. mesnili S. In coelom (in floating bundles of spermatogonia)
of the polychaete, *Polymnia nebulosa;* schizogony in bundle of
spermatogonia, in which cytomeres with ten to sixteen nuclei and
then merozoites are formed; schizogony repeated; gametocytes
undergo development also in the same host cells; microgametes
become set free in coelom, where union with macrogametes takes
place; each oocyst forms about sixteen spores; spore with usually
twelve sporozoites; cysts are extruded with the reproductive cells
of the host worm.

684 *Protozoology*

Genus **Myriospora** Lermantoff. Anisogamy and sporogony in marine snails; schizogony unknown; oocyst forms numerous spores each with two sporozoites. (One species.)

M. trophoniae L. In the polychaete, *Trophonia plumosa;* macrogametes, vermiform, up to 800µ long, later ovoid; microgametocyte forms first about 100 cytomeres, each with some twenty nuclei; microgametes comma-shaped; anisogamy; oocyst with several hundred spores, each with about twenty-four sporozoites.

Genus **Hyaloklossia** Labbé. Schizogony unknown; sporogony in the kidney of marine mussels; oocyst in the organ-cavity; spherical spores of two kinds: smaller one with two spirally coiled sporozoites and the other with four to six sporozoites. (One species.)

H. pelseneeri Léger. Spherical oocysts 75-80µ in diameter; spores 8µ and 11-12µ in diameter; in kidney of *Tellina* sp. and *Donax* sp.

Genus **Angeiocystis** Brasil. Schizogony unknown; sporogony in polychaetes; oocyst forms four spores; spore oval, with about thirty sporozoites and residual body at a pole. One species.

A. audouiniae B. In the cardiac body of *Audouinia tentaculata;* macrogametes vermiform, up to 65µ long.

Family 3. **Dobelliidae** Ikeda

Numerous microgametes develop from each microgametocyte; the union of gametocytes begins early.

Genus **Dobellia** Ikeda. Schizonts sexually differentiated: microschizonts and macroschizonts; young schizonts binucleate; association of two gametocytes begins early as in Adeleina (p. 701), but many microgametes are formed in each microgametocyte (One species, Ikeda, 1914.)

D. binucleata I. In the gut of *Petalostoma minutum;* mature oocyst 20-25µ in diameter, with a thin wall, contains some 100 sporozoites without any spore membrane around them.

Family 4. **Eimeriidae** Léger

Macro- and micro-gametocytes develop independently; microgametocyte produces many gametes; an oocyst from a pair of anisogametes; oocyst with variable number of spores containing one to many sporozoites, which condition is used as basis of generic differentiation. Oocysts found in the faeces of hosts are usually immature; time needed for completion of spore formation depends upon the species, temperature, moisture, etc. Becker (1934) recommends the following bactericidal solutions in which oocysts develop to maturity: 1 per cent formaldehyde, 1 per cent chromic acid or 2–4 per cent potassium dichromate. The oocysts of *Eimeria tenella* and other chicken coccidia are said to survive one to two years in 2.5 per cent solution of potassium dichromate. Duncan (1959a) tested sporulation of the oocysts of *E. labbeana* in various chemical substances and found that the maturing oocysts proceeded rapidly in 15 per cent sodium chloride solution. Two per cent of potassium dichromate at room temperature (20–26°C.) gave consistently a high percentage of sporulation. At temperature below 4°C. or above 37°C. no sporulation occurred and the amount of light has no effect on sporulation. (Avian coccidia, Levine, 1953a; sporulation of avian Eimeria, Edgar, 1955; in domestic animals, Levine, 1957; coccidia in mink, Levine, 1948; McTaggart, 1960; coccidia in deer-mice, Levine and Ives, 1960.)

Genus **Eimeria** Schneider (*Coccidium* Leuckart). Zygote or oocyst produces four spores, each with two sporozoites. (Numerous species, Levine and Becker, 1933; Boughton and Volk, 1938; Hardcastle, 1943; Pellerfy, 1956; Levine, Ivens and Kruidenier, 1957; Levine *et al.*, 1959; Svanbaev, 1957, 1958; host specificity, Becker, 1933; structure of oocyst wall, Henry, 1932; drug resistance, Cuckler and Malanga, 1955, 1956.)

E. schubergi (Schaudinn) (Fig. 249). In the gut of *Lithobius forficatus;* oocysts spherical, 22-25μ in diameter.

E. stiedae (Lindemann) (*Coccidium oviforme* Leuckart) (Fig. 253, *d-k*). In the epithelium of the bile-duct and liver (with white nodules) of wild and domestic rabbits; schizonts ovoid or spherical, 15-18μ in diameter; merozoites 8-10μ long; oocysts ovoid to ellipsoid, often yellowish, micropylar end flattened; mature oocysts 28-40μ by 16-25μ; sporulation in sixty to seventy hours; heavy infection is believed to be fatal to young animals, which may occur in an epidemic form. Transmission and comparison with *E. perforans* (Uhlhorn, 1926).

E. perforans Leuckart (Fig. 253, *l, m*). In the small intestine of rabbits; oocysts with equally rounded ends, 24–30µ by 14–20µ; sporulation in forty-eight hours at 33°C.; the thermal death point of immature oocysts 51°C. (Becker and Crouch, 1931); pathogenic. Other species (Pérard, 1925; Becker, 1934). Lund (1950) found 17 per cent of coccidian infection among 1200 faecal specimens collected from twenty-three commercial rabbitries in southern California.

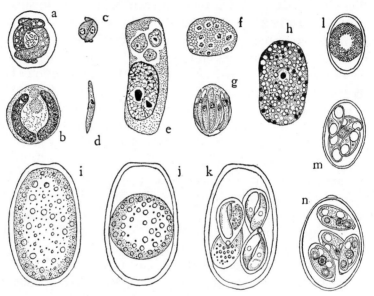

Fig. 253. a, b, Spores of *Merocystis kathae*, ×1000 (Foulon); c, *Pseudoklossia pectinis*, two sporozoites of a spore, ×1470 (Léger and Duboscq); d–k, *Eimeria stiedae* (d, a trophozoite; e, host cell with three trophozoites; f, g, schizogony; h, macrogametocyte, ×1270 (Hartmann); i–k, oocysts, ×830 (Wasilewski); l, m, *E. perforans*, ×750 (Pérard); n, *E. faurei*, ×800 (Wenyon).

E. zürnii (Rivolta). In the gut of cattle; oocysts spherical to ellipsoidal, 12-28µ by 10-20µ; sporulation in twenty-three (at 30°C.) to seventy-two (at 20°C.) hours. The organisms are common in range calves during their first fall and winter season (Marquardt *et al.*, 1960). Sporulation and subclinical infections (Marquardt, 1962).

E. bovis (Züblin) (*E. smithi* Yakimoff and Galouzo). In the gut of cattle; oocysts 23-34µ by 17-23µ; sporulation in three to five days in shallow dishes, and two weeks in deep dishes (Becker). (Development, Hammond *et al.*, 1946; resistance of calves to re-infection, Senger *et al.*, 1959.)

E. alabamensis Christensen. In cattle; schizogony and sporogony take place in the host cell nuclei (Christensen, 1941; Davis *et al.*, 1957).

E. ellipsoidalis (Becker and Frye) (Fig. 254, *a*). In the faeces of calf; oocysts ellipsoidal, 20-26μ by 13-17μ; sporulation in eighteen days (Becker and Frye, 1929).

E. cylindrica Wilson. In the faeces of cattle; oocysts cylindrical, 19-27μ by 12-15μ; sporulation in two to ten days.

E. brasiliensis Torres and Ildefonso. In the faeces of cattle; oocysts elliptical, 32-40μ by 23-27μ; spores 18.2μ by 8μ (Marquardt, 1959).

E. wyomingensis Huizinga and Winger. In the faeces of cattle; oocysts pyriform, 37-45μ by 26-31μ; spores 19μ by 3μ (Huizinga and Winger, 1942).

E. faurei Moussu and Marotel (Fig. 253, *n*). In the gut of sheep and goat; oocysts ovoid, 20-40μ by 17-26μ; sporulation in twenty-four to forty-eight hours.

E. arloingi Marotel. In the gut of sheep and goat; oocysts with a cap, ovoid, 25-35μ by 18-25μ; sporulation in three days.

E. intricata Spiegel. In the gut of sheep and goat; oocysts with thick wall, with or without cap, ellipsoidal, 42-60μ by 30-36μ; sporulation in about nine days. Species in North American sheep (Christensen, 1938); in Rocky Mountain Bighorn sheep (Honess, 1942).

E. debliecki Douwes (Fig. 254, *b*). In the gut of pigs; 30-82 per cent infection in California (Henry); oocysts 12-29μ by 12-20μ; sporulation in seven to nine days. (Development, Nöller and Frenz, 1922.)

E. scabra Henry. In the caecal contents of pigs; oocysts, brown, ellipsoidal, 22-36μ by 16-26μ. Henry (1931) recognized two other species in California swine.

E. caviae Sheather. In the gut of guinea pigs; oocysts subspherical to ellipsoid, 13-26μ by 13-22μ (Sheather, 1924). (Morphology and development, Lapage, 1940.)

E. canis Wenyon (Fig. 254, *c*). In the gut of dogs; oocysts, ellip-

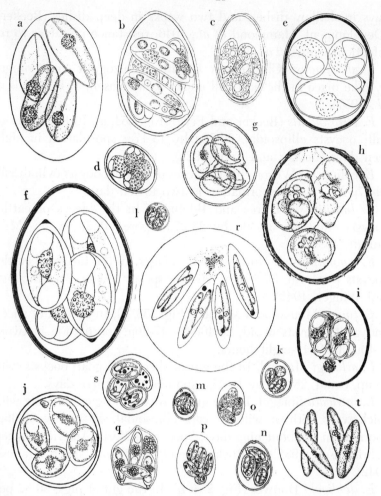

FIG. 254. Oocysts of Eimeria. a, *Eimeria ellipsoidalis*, ×1500 (Becker and Frye); b, *E. debliecki*, ×1070 (Wenyon); c, *E. canis*, ×650 (Wenyon); d, *E. falciformis*, ×730 (Wenyon); e, *E. separata*; f, *E. miyairii*, ×2000 (Becker, Hall and Hager); g, *E. mephitidis*, ×1000 (Andrews), h, *E. cynomysis*, ×1000 (Andrews); i, *E. citelli*, ×1360 (Kartchner and Becker), j, *E. monacis*, ×1630 (Fish); k, *E. tenella*, ×600 (Tyzzer); l, *E. mitis*, ×430 (Tyzzer); m, *E. acervulina*, ×430 (Tyzzer); n, *E. maxima*, ×470 (Tyzzer); o, *E. ranarum*, ×670; p, *E. prevoti*, ×670 (Laveran and Mesnil); q, *E. ranae*, ×670 (Dobell); r, *E. sardinae*, ×600, s, *E. clupearum*, ×600 (Thomson and Robertson); t, *E. brevoortiana* (Hardcastle).

soidal, 18-45μ by 11-28μ; spores 9.5μ by 2.5μ; sporulation in twenty-four hours.

E. *felina* Nieschulz. In the gut of cat; oocysts 21-26μ by 13-17μ.

E. *falciformis* (Eimer) (Fig. 254, *d*). In the gut of mice; oocysts spherical to ovoid, 16-21μ by 11-17μ; sporulation in three days.

E. *nieschulzi* Dieben. In the small intestine of rats; oocysts 16-26.4μ by 13-21μ; sporulation in sixty-five to seventy-two hours. (Growth-promoting potency of feeding stuffs, Becker, 1941; Becker, Manresa and Smith, 1943; Excystment, Landers, 1960.)

E. *separata* Becker and Hall (Fig. 254, *e*). In the caecum and colon of rats; oocysts 13-19.5μ by 11-17μ; sporulation in twenty-seven to thirty-six hours.

E. *miyairii* Ohira (Fig. 254, *f*). In the small intestine of rats; oocysts 16.5–29μ by 16–26μ; sporulation in 96–120 hours. Unsporulated oocysts perish in fifteen seconds at 53°C. and in twenty-four hours at 41°C.; sporulated oocysts are killed in two minutes at 52°C. (Reinhardt and Becker, 1933). Structure of oocyst wall (Henry, 1932); Eimeria in rodents (Fish, 1930; Henry, 1932a; Roudabush, 1937a; Levine *et al.*, 1957).

E. *mephitidis* Andrews (Fig. 254, *g*). In the faeces of the common skunk; oocysts oval to spherical, 17-25μ by 16-22μ; wall 1μ thick; a circular micropyle; spores with a rostrum, 10-12μ by 7-9μ; extended sporozoites 10-14μ by 4-5μ; other stages unknown (Andrews, 1928).

E. *cynomysis* A. (Fig. 254, *h*). In the faeces of the prairie dog; oocysts oval, 33-37μ by 28-32μ; a double fibrous wall, 1.5-2.5μ thick; the inner wall slightly orange-yellow; micropyle 5-6μ in diameter; spores, broad pyriform, 13-17μ by 8-12μ.

E. *citelli* Kartchner and Becker (Fig. 254, *i*). In the caecal contents of the striped ground squirrel, *Citellus tridecemlineatus*; subspherical to ellipsoidal oocysts 15-23μ by 14-19μ.

E. *monacis* Fish (Fig. 254, *j*). In the intestine of the woodchuck, *Marmota monax*; spherical to subspherical oocysts 20μ by 18μ (Fish, 1930), 14-20μ in diameter (Crouch and Becker, 1931); wall comparatively thick; sporulation completed in sixty to sixty-four hours in 2 per cent potassium bichromate at room temperature.

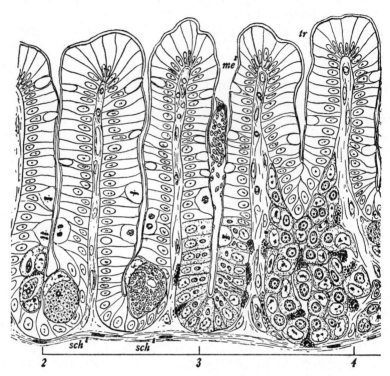

Fig. 255. Diagram illustrating the development of *Eimeria tenella* in the caecal glands of chick (Tyzzer). The numbers below indicate the days of infection. *ma,* macrogamete; *me,* merozoite (*me*[1], *me*[2], *me*[3], generation 1, 2, 3 merozoites respectively); *mi,* microgametocyte; *oo,* oocyst; *ret. oo* and *ret. sch,* oocysts and schizonts which failed to escape; *sch*[1], *sch*[2], schizonts of generation 1 and 2; *tr,* young growing trophozoites. (Continue to upper left of Fig. 256.)

Crouch and Becker found two other species: *E. perforoides* and *E. os,* in the woodchuck in Iowa. Emeria in lemming (Levine, 1952).

E. tenella (Railliet and Lucet) (Figs. 254, *k;* 255; 256). In the caeca, colon and lower small intestine of chicken; a cause of acute coccidiosis characterized by haemorrhage (Tyzzer); in the caecal contents of California quail (Henry); oocysts 19.5-26μ by 16.5-23μ; sporulation in forty-eight hours. (Sporulation and temperature, Edgar, 1954.)

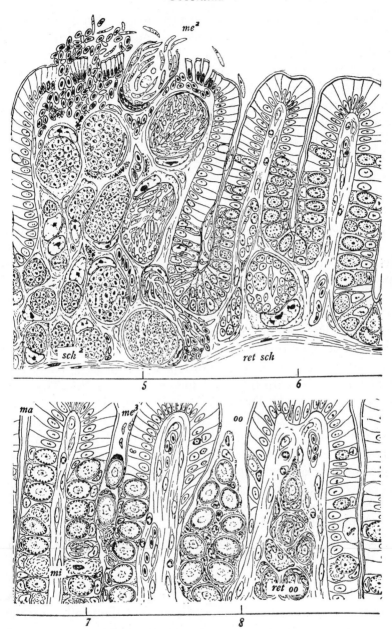

FIG. 256. Continuation of the diagram shown in Fig. 255 (Tyzzer). From the right end of the upper figure continue to the left of the lower figure; for explanation see Figure 255.

Tyzzer's observation on experimental infection in chicken is as follows (Figs. 255, 256): When a large number of oocysts are fed to chicken, the sporozoites emerge from the oocysts and spores, in as early as twenty hours and are found on the surface of the caecal mucosa. Toward the end of the second day, growing trophozoites are found in the gland epithelial cells; they undergo schizogony (Fig. 255, sch^1) by the middle of the third day. A single first generation schizont is estimated to produce about 900 pyriform merozoites which measure 2-4μ by 1-1.5μ and occur in the gland lumen (me^1). As these merozoites invade the epithelial cells of the fundi of the glands and become trophozoites, the infected host cells increase in size, become rounded and no longer form a continuous layer (*tr*). These trophozoites (Fig. 256, *sch*) grow to much greater dimensions (up to as much as 45μ in diameter) than those of the first generation and multiply into merozoites (me^2) by the fifth day. These merozoites are much larger and more elongated than those of the first generation and measure 16μ by 2μ. The haemorrhage in the affected mucosa which begins usually with the growth of the second generation trophozoites, increases in volume so that by the fifth day after infection, a great portion of the mucosa sloughs off, which coincides with the liberation of the merozoites. The merozoites formed in the host cells located in the deeper part of the mucosa are unable to become free and appear to grow into multinucleate forms (*ret sch*). When the liberated merozoites enter epithelial cells, most of them develop into macrogametocytes (*ma*) and microgametocytes (*mi*), while comparatively small numbers become trophozoites and form by budding a few, large third generation merozoites (me^3). Mature oocysts (*oo*) are found on seven to eight days after infection.

According to Pattillo (1959), the sporozoites enter the surface epithelium of the caeca at the tips of villi and proceed across the basement membrane into the tunica propria from which they advance to the epithelial cells lining the gland fundi. Challey and Burns (1959) note, on the other hand, that the sporozoites pass through the surface epithelium of the caecal mucosa into the lamina propria, where they are engulfed by macrophage and transported to the cells of the glands of Lieberkühn.

McLoughlin and Chester (1959) made a comparative study of six poultry coccidiostats, and found all efficient, but under the conditions of the experiments they carried, glycarbylamide and nicarbazin appeared to be more effective than four other compounds. (Eimeria species in chicken, Tyzzer, 1929, 1932; Henry, 1931a; Boles and Becker, 1954; development, Tyzzer, 1929, 1932; Scholtyseck, 1953; Nath, Dutta and Sagar, 1960; economic importance, Foster, 1949; Brockett and Bliznick, 1950; pathological changes, Tyzzer, 1929, 1932; Mayhew, 1937; Greven, 1953; statistical study of infections, Fish, 1931; mortality of hosts, Mayhew, 1933; killing oocysts, Fish, 1931a; control measures, Andrews, 1933; Rubin *et al.*, 1956; in wild fowls, Hasse, 1939; anticoccidial agents, Horton-Smith and Long, 1959.)

E. mitis Tyzzer (Fig. 254, *l*). In the anterior small intestine of chicken; oocysts subspherical; 16.2μ by 15.5μ; sporulation in forty-eight hours (Tyzzer, 1929).

E. acervulina T. (Fig. 254, *m*). In the anterior small intestine of chicken, and in California quail (Henry); oocysts oval, 17.7-20.2μ by 13.7-16.3μ; sporulation in twenty hours; associated with serious chronic coccidiosis (Tyzzer, 1929). When oocysts are fed to chicken, spores are released from the oocysts in the gizzard and sporozoites emerge in the duodenum and jejunum. The digestive fluid acts apparently on the spores (Doran and Farr, 1962). (Effect on host, Moynihan, 1950; Morehouse and McGuire, 1958; cytochemistry, Pattillo and Becker, 1955.)

E. maxima T. (Fig. 254, *n*). In the small intestine of chicken; oocysts oval, 21.5-42.5μ by 16.5-29.8μ (Tyzzer, 1929).

E. necatrix Johnson. In the small intestine (schizonts) and caeca (oocysts) of chicken; a cause of chronic coccidiosis; oocysts obovate, 13-23μ by 11-18μ; sporulation in forty-eight hours (Tyzzer, 1932). (Oocysts measurement, Becker *et al.*, 1956; sporozoite transportation, Van Doorninck and Becker, 1957.)

E. praecox J. In the upper third of the small intestine of chicken; oocysts ovoid, 20-25μ by 15.5-20μ; sporulation in forty-eight hours.

E. meleagridis Tyzzer. In the caeca of turkey; apparently nonpathogenic; oocysts, ellipsoidal, 19-30μ by 14.5-23μ (Tyzzer, 1927, 1932). (Coccidiosis in turkey, Hawkins, 1952.)

E. meleagrimitis T. In the lower small intestine of turkey; somewhat similar to *E. mitis;* oocysts, 16.5-20.5μ by 13.2-17.2μ (Tyzzer, 1929).

E. adenoeides Moore and Brown. In the ileum, caeca and rectum of turkeys; oocysts about 25.6μ by 16.5μ; highly pathogenic to young turkeys (Moore and Brown, 1950).

E. truncata Railliet and Lucet. In the kidney of geese; oocysts truncate at one pole, ovoid, 14-23μ by 13-18μ; some observers find this coccidian fatal to young geese.

E. anseris Kotlan. In the intestine of geese; oocysts spherical or pyriform, 11-16μ in diameter. (Coccidia in Canada goose, Levine, 1952a; Hanson *et al.*, 1957.)

E. labbeana Pinto. In the gut of domestic pigeon; oocysts sometimes light brown, 15-26μ by 14-24μ. (In birds, Henry, 1932b.)

E. dispersa Tyzzer. In the small intestine of bob-white quail and pheasant; oocysts ovate, 18.8-22.8μ (quail), smaller in pheasant, without polar inclusion; sporulation in about twenty-four hours. (Size of oocysts, Duncan, 1959.)

E. amydae Roudabush. In the intestine of *Amyda spinifera;* oocysts oval with a thin wall, 17-24μ by 12-17μ; elliptical spores about 11-16μ long (Roudabush, 1937).

E. chrysemydis Deeds and Jahn. In the intestine of *Chrysemys marginata;* oval oocysts 21-27μ by 13-18μ; fusiform spores 12-14μ by 5-8μ (Deeds and Jahn, 1939). (Other reptilian species, Roudabush, 1937.)

E. ranarum (Labbé) (Fig. 254, *o*). In the gut epithelium (nuclei) of frogs; oocysts about 17μ by 12μ.

E. prevoti (Laveran and Mesnil) (Fig. 254, *p*). In the gut epithelium of frogs; oocysts about 17μ by 12μ.

E. ranae Dobell (Fig. 254, *q*). In the gut of frogs; oocysts 22μ by 18μ.

Species of Eimeria are often parasitic in fishes used for human consumption, and thus may appear in faecal matter. A few examples will be mentioned here.

E. sardinae (Thélohan) (*E. oxyspora* Dobell) (Fig. 254, *r*). In the testis of sardine; spherical oocyst 30–50μ (Thélohan, 1890; Dobell, 1919).

E. clupearum (Thélohan) (*E. wenyoni* Dobell) (Fig. 254, *s*). In the liver of herring, mackerel, and sprat; spherical oocysts 18-33µ in diameter (Thélohan, 1894; Dobell, 1919). Taxonomy (Thomson and Robertson, 1926).

E. gadi Fiebiger. In the swim-bladder of *Gadus virens, G. morrhua,* and *G. aeglefinus;* schizogony and sporogony; germination of spores takes place in the bladder of the same host individual, bringing about a very heavy infection; oocysts 26-28µ; pathogenic (Fiebiger, 1913).

E. brevoortiana Hardcastle (Fig. 254, *t*). Schizogony in the epi-

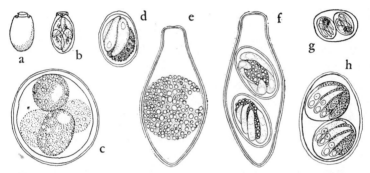

Fig. 257. Oocysts of Coccidia, a, b, *Jarrina paludosa,* ×800 (Léger and Hesse); c, d, oocyst and spore of *Wenyonella africana,* ×1330 (Hoare), e, f, a young and a mature oocyst of *Isospora hominis,* ×1400 (Dobell); g, *I. bigemina;* h, *I. rivolta,* ×930 (Wenyon).

thelium of the pyloric caeca and sporogony in the testis of the menhaden, *Brevoortiana tyrannus;* mature oocysts, spherical, 17.5-30µ in diameter or ovoid, 21-30µ by 15-27.5µ (Hardcastle, 1944).

Genus **Jarrina** Léger and Hesse. Oocysts ovoid, one end rounded and the other drawn out into a short neck; four spores, each with two sporozoites (Léger and Hesse, 1922).

J. paludosa L. and H. (Fig. 257, *a, b*). In the gut of *Fulica atra* and *Gallinula chloropus;* oocysts 15µ by 11µ; sporulation in fifteen days.

Genus **Wenyonella** Hoare. Oocysts with four spores, each with four sporozoites. Three species.

W. africana H. (Fig. 257, *c, d*). In the small intestine of *Boae-*

don lineatus ("brown snake") in Uganda; oocysts ovoid or sub-spherical, 18.5-19.2µ by 16-17.6µ; spores ovoid, 9.6µ by 8µ; sporulation in five to six days.

W. gallinae Ray. In the epithelium of the lower intestine of chicken; oval oocysts, 29.5-33.5µ by 20-23µ; spores 18.8µ by 8µ; sporozoites club-shaped; sporulation in four to six days at 28°C. (Ray, 1945).

Genus **Isospora** Schneider. Oocyst produces two spores, each containing four sporozoites. Avian Isospora (Boughton, Boughton and Volk, 1938); known species (Pellerdy, 1957); in Mustelidae (Prasad, 1961).

I hominis (Rivolta) (Fig. 257, *e*, *f*). In man. Its life cycle is unknown, but most probably the schizogony, gametogenesis and sexual fusion occur in the intestinal epithelium. Oocysts have only been seen in the stools of infected persons.

The oocyst is asymmetrically fusiform; 20-33µ by 10-16µ; wall is made up of two membranes which are highly resistant to chemicals; when voided in faeces, the contents either fill up the oocyst or appear as a spherical mass, composed of refractile granules of various sizes; nucleus appears as a clear circular area; when the faecal specimen is kept in a covered container at room temperature, the protoplasmic mass divides into two spherical sporoblasts in about twenty-four hours and each sporoblast develops in another twenty-four hours into a spore (10-16µ by 7-10µ) containing four sporozoites. Further changes take place when the oocyst finds its way into the human intestine in contaminated food or water.

I. hominis has been observed in widely separated regions, but appears not to be of common occurrence. As to its effect on the human host, very little is known. Connal (1922) described the course of an accidental oral infection by viable mature oocysts, as follows: The incubation period was about six days, the onset sudden, and the duration over a month. The cure was spontaneous. The symptoms were diarrhoea, abdominal discomfort, flatulence, lassitude, and loss of weight. During the first three weeks of the illness no oocysts were found, but then oocysts appeared in the stools for nine days. On the tenth day they were not seen, but reappeared on the eleventh and twelfth days, after which they were not found again. The acute signs of illness abated within one week of the finding of the oocysts. The faeces contained a large amount of undigested material, particularly fat which gave it a thick oily consistency, showing signs of slow gaseous formation.

Matsubayashi and Nozawa (1948) found six cases of infection in Japan. A volunteer ingested some 3000 oocysts. Eight days later diarrhoea developed, followed by a rise of temperature above 39°C., which lasted for ten days. On the following day, the diarrhoea subsided, but later returned and was especially pronounced on the seventeenth day, after which it disappeared completely. Oocysts were discharged regularly since the ninth day for thirty-two days. About a month after the cessation of oocyst-production, the person ingested again some 2500 cysts, but no infection resulted, which the two authors attributed to the immunity produced during the first infection. Another volunteer showed a similar course of infection. The symptoms disappeared without medication after the termination of oocyst discharge. Thus, the coccidiosis of man appears to be a self-limited one. Attempts to infect common laboratory animals with this coccidian have so far failed (Foner, 1939; Herrlich and Liebmann, 1944; Rita and Vida, 1949). (History, Dobell, 1919; human species, Dobell, 1926; incidence, Magath, 1935; Barksdale and Routh, 1948; Mukherjee, 1948; Jeffrey, 1956; Jarpa *et al.*, 1960.)

I. belli Wenyon. Wenyon (1923) recognized second species in man; oocyst with delicate envelope; size similar to the above-mentioned species; spores 12-14µ by 7-9µ. Elsdon-Dew and Freedman (1953) observed a number of cases in Africa.

I. bigemina Stiles (Fig. 257, *g*). In the gut of cat and dog; oocysts 10-14µ by 7-9µ. Host affinities (Becker, 1954).

I. rivolta Grassi (Fig. 257, *h*). In the gut of cat and dogs; oocysts 20-25µ by 15-20µ.

I. felis Wenyon (Fig. 258, *a*). In cat and dog; oocysts 39-48µ by 26-37µ. Hitchcock (1955) found that there are two schizogonic phases. The first cycle occurred in from second to fourth day, the second cycle lasted from fifth to sixth day and sexual phase took place on seventh and eighth days.

I. suis Biester. In swine faeces; oocysts subspherical, about 22.5µ by 19.4µ; sporulation in four days.

I. lacazei Labbé. In the small intestine of passerine birds (sparrows, blackbirds, finches, etc); oocysts subspherical or ovoidal, 18.5-30μ by 18-29.2μ; spores, 16.5-18.5μ by 10.3-12.4μ; heavily infected sparrows show definite symptoms of infection; sporulation in twenty-four hours (Henry, 1932b). Sparrows and other common small birds have been known to be free from Eimeria infection, while the barnyard fowls are seldom infected by Isospora (Boughton, 1929). (Significance of size variation in oocysts, Boughton, 1930; Henry, 1932b; development, Chakravarty and Kar, 1944; morphology, Levine and Mohan, 1960.)

I. *buteonis* Henry. In the duodenal contents of several species of hawks: *Buteo borealis, B. swainsoni, Accipiter cooperii,* and *Asio flammeus;* oocysts irregular in form with a thin wall, 16-19.2μ by 12.8-16μ; spores 9.6-13μ by 8-10.4μ (Henry, 1932b).

I. *lieberkühni* (Labbé) (Fig. 258, *b*). Oocyst about 40μ long; in the kidney of frogs. (Development, Nöller, 1923.)

I. *xantusiae* Amrein. Oocysts spherical to subspherical, without micropyle, 25-27μ in largest diameter; spores 15μ by 10μ, with a knob at one end; sporulation time twelve to twenty-four hours; in the small intestine of lizards, *Xantusia vigilis* and *X. henshawi* (Amrein, 1952).

Genus **Cyclospora** Schneider. Development similar to that of *Eimeria;* oocyst with two spores, each with two sporozoites and covered by a bi-valve shell.

C. *caryolytica* Schaudinn (Fig. 258, *c*). In the gut of the mole; sporozoites enter and develop in the nuclei of gut epithelial cells; oocyst oval, about 15μ by 11.5μ. (Development, Tanabe, 1938.)

Genus **Dorisiella** Ray. Zygote develops into two spores with a delicate envelope; each spore contains eight sporozoites when mature; microgametocytes migratory.

D. *scolelepidis* R. (Fig. 258, *d*). In the gut of the polychaete, *Scolelepis fuliginosa;* zygote contents divide into two oval spores, 12-16μ by 6-10μ; spore with eight sporozoites (Ray, 1930).

Genus **Caryospora** Léger. Oocyst develops into a single spore with eight sporozoites and a residual mass; membrane thick and yellow.

C. *simplex* L. (Fig. 258, *e, f*). In the gut-epithelium of *Vipera aspis;* oocyst thick-walled, 10-15μ in diameter.

C. *hermae* Bray. Oocyst wall colorless, 21-24μ by 20-22μ; in the gut of the sand snake, *Psammophis sibilans philipsi.* Bray (1960) described two other species from the same host.

Genus **Cryptosporidium** Tyzzer. Lumen-dwelling minute organisms; oocyst with four sporozoites.

C. muris T. (Fig. 258, *g-i*). In the peptic glands of the mouse; both schizogony and sporogony in the mucoid material on surface of the epithelium; oocysts 7μ by 5μ; four sporozoites, 12-14μ long (Tyzzer, 1910).

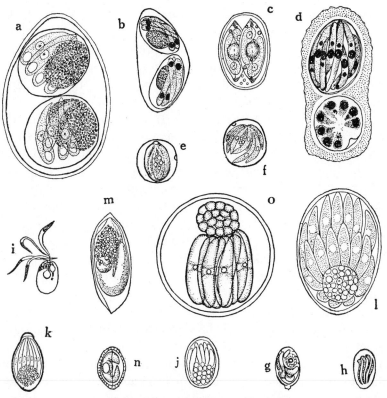

FIG. 258. a, *Isospora felis,* ×930 (Wenyon); b, *I. lieberkuhni,* ×660 (Laveran and Mesnil); c, *Cyclospora caryolytica,* ×1330 (Schaudinn); d, *Dorisiella scolelepidis,* oocyst with two spores, ×1400 (Ray); e, f, *Caryospora simplex,* ×800 (Léger); g-i, *Cryptosporidium muris* (g, h, oocysts; i, emergence of four sporozoites), ×1030 (Tyzzer); j, *Pfeifferinella ellipsoides,* ×1330 (Wasielewski); k, *P. impudica,* ×800 (Léger and Hollande); l, *Lankesterella minima,* a mature cyst in endothelial cell, ×1000 (Nöller); m, *Barrouxia ornata,* ×1330 (Schneider); n. *Echinospora labbei,* ×1000 (Léger); o, an oocyst of *Tyzzeria natrix,* ×2400 (Matsubayashi).

C. parvum T. In the glands of small intestine of the mouse; oocysts with four sporozoites, 4.5µ in diameter (Tyzzer, 1912).

Genus **Pfeifferinella** Wasielewski. Macrogamete with a "reception tubule" by which microgamete enters; oocyst produces directly eight sporozoites.

P. ellipsoides W. (Fig. 258, *j*). In the liver of *Planorbis corneus*; oocysts oval, 13-15µ long.

P. impudica Léger and Hollande. (Fig. 258, *k*). In the liver of *Limax marginatus*; oocysts ovoid, 20µ by 10µ.

Genus **Lankesterella** Labbé (*Atoxoplasma*, Garnham, 1950). Oocyst produces thirty-two or more sporozoites directly without spore-formation; in endothelial cells of cold-blooded vertebrates; mature sporozoites enter erythrocytes in which they are transmitted to a new host individual by bloodsucking invertebrates. (Species in birds, Lainson, 1959.)

L. minima (Chaussat) (Fig. 258, *l*). In frogs; transmitted by the leech (*Placobdella marginata*); frog acquires infection through introduction of sporozoites by a leech; sporozoites make their way into the blood capillaries of various organs; there they enter endothelial cells; schizogony produces numerous merozoites which bring about infection of many host cells; finally macro- and micro-gametocytes are formed; anisogamy produces zygotes which transform into oocysts, in which a number of sporozoites develop; these sporozoites are set free upon disintegration of cyst wall in the blood plasma and enter erythrocytes (Nöller); oocyst oval, about 33µ by 23µ.

L. garnhami Lainson. In English sparrows; young birds become infected as early as six days of age, possibly by ingestion of infected mites, *Dermanyssus gallinae*. Schizogony takes place in lymphoid-macrophage cells of the spleen, bone marrow and liver of host birds; gametocytes develop in similar cells of the liver, lungs, and kidneys; zygote develops into an oocyst containing a large number of sporozoites which invade the lymphocytes and monocytes of the peripheral blood (Lainson, 1959, 1960). (Electron microscope study, Garnham *et al.*, 1962.)

Genus **Schellackia** Reichenow. Schizogony and sporogony in the intestinal wall of host lizards; zygote produces eight sporozoites which enter the blood cells; as mites suck in infected blood, the sporozoites become collected in the gut cells. When the lizard takes in the infected mites, the infection begins.

S. *bolivari* R. In the mid-gut of the lizards, *Acanthodactylus vulgaris* and *Psammodromus hispanicus;* development somewhat similar to that of *Eimeria schubergi* (Fig. 249); oocysts spherical, 15-19μ in diameter, with eight sporozoites (Reichenow, 1919).

S. *occidentalis* Bonorris and Ball. In lizards, *Sceloporus occidentalis becki, S. o. biseriatus* and *Uta stansburiana hesperis;* the mite, *Geckobiella texana.* The mites feed on the blood of the lizard containing infected erythrocytes. When lizards ingest these mites, the sporozoites migrate into the intestinal epithelial cells of the lizards where schizogony and sporogony take place; sporozoites are said to appear in the lizard's peripheral blood in thirty to forty-five days after the ingestion of the mites (Bonorris and Ball, 1955).

Genus **Tyzzeria** Allen (? *Koidzumiella,* Matsubayashi). Oocyst contains eight naked sporozoites.

T. *perniciosa* (Allen). In the small intestine of *Anas domesticus;* oocysts 10-13.3μ by 9-10.8μ; highly pathogenic (Allen, 1936).

T. *natrix* Matsubayashi (Fig. 258, *o*). In the intestine of *Natrix tigrina;* spherical oocysts 12-18μ in diameter.

Genus **Barrouxia** Schneider. Oocyst with numerous spores, each with a single sporozoite; spore membrane uni- or bi-valve, with or without caudal prolongation. (Development, Schellack and Reichenow, 1913.)

B. *ornata* S. (Fig. 258, *m*). In gut of *Nepa cinerea;* oocysts spherical, 34-37μ in diameter, with many spores; spore with one sporozoite and bivalve shell, 17-20μ by 7-10μ.

Genus **Echinospora** Léger. Oocyst with four to eight spores, each with a sporozoite; endospore with many small spinous projections.

E. *labbei* L. (Fig. 258, *n*). In the gut of *Lithobius mutabilis;* oocyst spherical, 30-40μ in diameter; spores, 11μ by 9.4μ, with bi-valve shell; sporulation completed in about twenty days.

Suborder 2. **Adeleina** Léger

The members of this suborder are characterized by an early association of the micro- and macro-gametocytes (Fig. 259) and the small number of microgametes produced by each microgametocyte. The zygote becomes an oocyst which produces sporoblasts, each of which develops into a spore with two or four sporozoites. Two families.

Family 1. **Adeleidae** Léger

Typically in the gut epithelium of invertebrates.

Genus **Adelea** Schneider. Zygote develops into a thinly walled oocyst with numerous flattened spores, each with two sporozoites; in anthropods.

A. ovata S. (Fig. 259). In the gut of *Lithobius forficatus;* merozoites 17-22μ long; oocysts elongate oval, 40-50μ by 30-40μ; 17-33 or more spores; spores circular, flattened, 20μ by 4μ (Hesse, 1910a). (Life cycle, Schellack and Reichenow, 1913, 1915.)

Genus **Adelina** Hesse. Oocyst thick-walled; spores spherical, comparatively small in number; in the gut or coelom of arthropods and oligochaetes (Hesse, 1910, 1910a).

A. dimidiata (Schneider) (Fig. 260, *a*). In the gut of *Scolopendra cingulata* and other myriapods; oocysts with three to seventeen spores (Schellack, 1913).

A. octospora H. (Fig. 260, *b*). Spherical oocyst contains spores; in the coelom of *Slavina appendiculata* (Hesse, 1910a).

A. deronis Hauschka and Pennypacker. In peritoneum of *Dero limosa;* oocyst contains twelve (10-14) spores; meiosis at the first zygotic nuclear division; haploid chromosome number 10; the life cycle is completed in eighteen days at room temperature (Hauschka, 1943).

A. sericesthis Weiser and Beard. In the fat body of the larvae of *Sericesthis pruinosa* (Weiser and Beard, 1959).

Genus **Klossia** Schneider. Oocyst with numerous spherical spores, each with three to ten sporozoites. Several species. (Life cycle, Nabih, 1938.)

K. helicina S. In the kidneys of various land-snails, belonging to genera Helix, Succinea, and Vitrina; oocyst with a double envelope 120-180μ in diameter; spores 12μ in diameter, with five to six sporozoites (Debaisieux, 1911). (Cytology and development, Naville, 1927.)

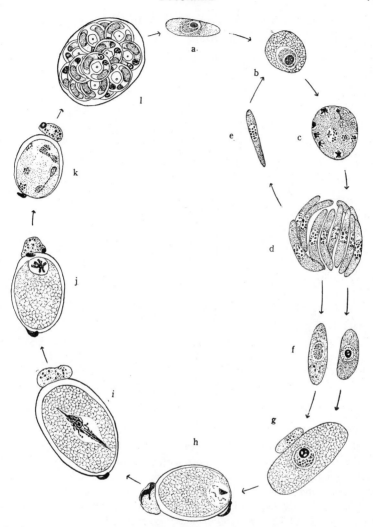

FIG. 259. The life-cycle of *Adelea ovata*, ×600 (Schellack and Reichenow).
a, schizont entering the gut epithelium of the host centipede; b–d, schizogony; e, larger form of merozoite; f, microgametocyte (left) and macrogametocyte (right); g, association of gametocytes; h, i, fertilization; j, zygote; k,
nuclear division in zygote; l, mature oocyst with many spores.

K. perplexeus Levine, Iven and Kruidenier (1955a). In the faeces of a deer mouse, *Peromyscus maniculatus;* oocysts 42-53μ by 36-44μ; twelve spores (each with four sporozoites) packed in an oocyst.

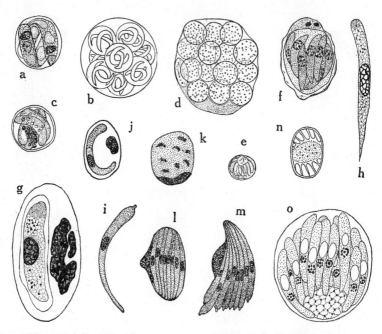

Fig. 260. a, *Adelina dimidiata,* a spore, ×1000 (Schellack); b, *A. octospora,* oocyst, ×1000 (Hesse); c, *Orcheobius herpobdellae,* ×550 (Kunze); d, e, *Klossiella muris* (d, renal cell of host with 14 sporoblasts; e, spore), ×280 (Smith and Johnson); f, *Legerella hydropori,* oocyst, ×1000 (Vincent); g, h, *Haemogregarina* of frog, ×1400 (Kudo); i–m, *H. simondi,* in the blood of the sole, *Solea vulgaris,* ×1300 (Laveran and Mesnil); n, *Hepatozoon muris,* spore, ×420 (Miller); o, *Karyolysus lacertae,* ×700 (Reichenow).

Genus **Orcheobius** Schuberg and Kunze. Macrogametes vermiform; oocyst with 25-30 spores, each with four (or six) sporozoites.

O. herpobdellae S. and K. (Fig. 260, c). In the testis of *Herpobdella atomaria;* mature macrogametes 180μ by 30μ; microgametes 50μ by 12μ; schizogony in April and May; sporogony in June and July.

Genus **Klossiella** Smith and Johnson. Microgametocyte produces two microgametes; oocyst with many spores, each with numerous sporozoites; in the kidney of mammals (Smith and Johnson, 1902).

K. muris S. and J. (Fig. 260, *d, e*). Oocyst with twelve to fourteen spherical spores; about thirty to thirty-four sporozoites in a spore, 16μ by 13μ; spores discharged in the host's urine; in the epithelium of the tubules and glomeruli in the kidney of the mouse, *Mus musculus.*

K. cobayae Seidelin. Oocyst with eight to twenty spores; spore with about thirty sporozoites; in the kidney of guinea pig. Bonciu *et al.*, (1957) found that infection occurs in guinea pigs as early as three weeks of age. In adults the infection rate varied thirteen to sixty-five per cent, being higher in summer than in winter.

K. equi Seibold and Thorson. Oocysts with as many as forty ovoid spores; spores 8-10μ by 4-5μ, containing up to twelve sporozoites. In renal epithelial cells of American jack, *Equus asinus.*

Genus **Legerella** Mesnil. Oocyst contains numerous sporozoites; spores entirely lacking; in arthropods (Mesnil, 1900).

L. hydropori Vincent (Fig. 260, *f*). In the epithelium of Malpighian tubules of *Hydroporus palustris;* oocysts ovoid, 20-25μ long, with sixteen sporozoites which measure 17μ by 3μ (Vincent, 1927).

Genus **Chagasella** Machado. Oocyst with three spores, each with four or six (or more) sporozoites; in hemipterous insects.

C. hartmanni (Chagas). In the gut of *Dysdercus ruficollis;* oocysts with three spores about 45μ in diameter; spore with four sporozoites, about 35μ by 15μ (Machado, 1911).

Genus **Ithania** Ludwig. Microgametocyte produces four microgametes; oocyst with one to four spores, each with nine to thirty-three sporozoites. (One species, Ludwig, 1947.)

I. wenrichi L. In the epithelial cells of the gastric caeca and mid-gut of the larvae of the crane-fly, *Tipula abdominalis;* oocysts 34-63μ by 22-50μ.

Family 2. **Haemogregarinidae** Léger

With two hosts: vertebrates (circulatory system) and invertebrates (digestive system).

Genus **Haemogregarina** Danilewsky. Schizogony takes place in blood cells of vertebrates; when gametocytes are taken into gut of leech or other blood-sucking invertebrates, sexual reproduction takes place; microgametocyte develops two or four microgametes; sporozoites formed without production of spores.

H. stepanowi D. (Fig. 261). Schizogony in *Emys orbicularis* and sexual reproduction in *Placobdella catenigera;* sporozoites introduced into blood of the chelonian host by leech (*a*), and enter erythrocytes in which they grow (*d-g*); schizogony in bone-marrow, each schizont producing twelve to twenty-four merozoites (*h*); schizogony repeated (*i*); some merozoites produce only six merozoites (*j, k*) which become gametocytes (*l-o*); gametogony occurs in leech; four microgametes formed from each microgametocyte and become associated with macrogametocytes in gut of leech (*p-r*); zygote (*s*) divides three times, and develops into eight sporozoites (*t-w*).

Haemogregarines are found commonly in various birds (Aragão, 1911), reptiles (Ball, 1958; MacKerras, 1961), amphibians (Fig. 260, *g, h*) (Roudabush and Coatney, 1937; Lehmann, 1959) and fishes (Fig. 260, *i-m*) (Saunders, 1955, 1958, 1959, 1960).

Genus **Hepatozoon** Miller. Schizogony in the cells of liver, spleen, and other organs of vertebrates; merozoites enter erythrocytes or leucocytes and develop into gametocytes; in blood-sucking arthropods (ticks, mites), micro- and macro-gametes develop and unite in pairs; zygotes become oocysts which increase in size and produce sporoblasts, spores, and sporozoites.

H. muris (Balfour) (Fig. 260, *n*). In various species of rat; several specific names were proposed on the basis of difference in host, locality, and effect on the host, but they are so indistinctly defined that specific separation appears to be impossible. Schizogony in the liver of rat; young gametocytes invade mononuclear leucocytes and appear as haemogregarines; when blood is taken in by the mite, *Laelaps echidninus,* union of two gametes produces vermicular body which penetrates gut-epithelium and reaches peri-intestinal tissues and grows; becoming surrounded by a cyst-membrane, cyst content breaks up into a number of sporoblasts and then into spores, each of which contains a number of sporozoites; when a rat devours infected mites, it becomes infected.

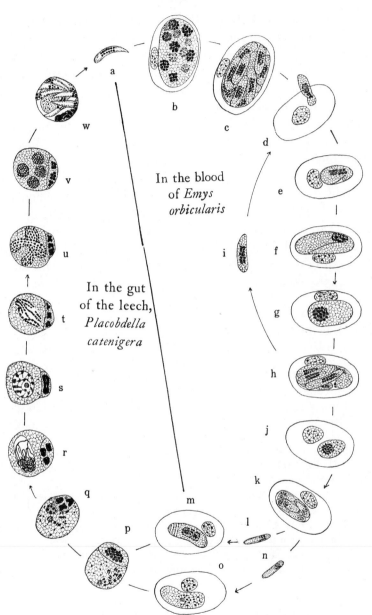

In the blood
of *Emys*
orbicularis

In the gut
of the leech,
Placobdella
catenigera

Fig. 261. The life-cycle of *Haemogregarina stepanowi*, ×1200 (Reichenow). a, sporozoite; b–i, schizogony; j–k, gametocyte-formation; l, m, micro-gametocytes; n, o, macrogametocytes; p, q, association of gametocytes; r, fertilization; s–w, division of the zygote nucleus to form eight sporozoites.

H. griseisciuri Clark. In the Eastern grey squirrel, *Sciurus carolinensis;* schizogony in the spleen, liver and bone marrow; sporogony in the mites, *Echinolaelaps echidninus* and *Euhaemogamasus ambulans* (Clark, 1958). Herman and Price (1955) report that the organism they observed in the same host was indistinguishable from *H. muris,* but named it *H. sciuri.*

Genus **Karyolysus** Labbé. Sporoblasts formed in the oocysts in gut-epithelium of a mite, vermiform sporokinetes, enter host ova and become mature; when young mites hatch, spores in gut-epithelium are cast off and discharged in faeces; a lizard swallows spores; liberated sporozoites enter endothelial cells in which schizogony takes place; merozoites enter erythrocytes as gametocytes which when taken in by a mite complete development in its gut.

K. lacertae (Danilewsky) (Fig. 260, *o*). In *Lacerta muralis;* sexual reproduction in *Liponyssus saurarum;* sporokinetes 40-50μ long; spores 20-25μ in diameter (Reichenow, 1913, 1921).

References

ALLEN, ENA A., 1936. *Tyzzeria pernisiona,* etc. Arch. Protist., 87:262.

AMREIN, Y. U., 1952. A new species of Isospora, *I. xantusiae,* from southern California lizards. J. Parasit., 38:147.

ANDREWS, J., 1928. New species of Coccidia from the skunk and prairie dog. J. Parasit., 14:193.

—— 1933. The control of poultry coccidiosis by the chemical treatment of litter. Amer. J. Hyg., 14:466.

—— and TSUCHIYA, H., 1931. The distribution of coccidial oocysts on a poultry farm in Maryland. Poultry Sci., 10:320.

ARAGÃO, H. B., 1911. Beobachtungen ueber Haemogregarinen von Vögeln Mem. Inst. Oswaldo Cruz. 3:54.

BANER, J. R., *et al.,* 1959. *Lankesterella corvi* n. sp., a blood parasite of the English Rook, *Corvus f. frugilegus.* J. Protozool., 6:233.

BALL, G. H., 1958. A haemogregarine from a water snake, *Natrix piscator,* taken in the vicinity of Bombay, India. *Ibid.,* 5:274.

BARKSDALE, W. L., and ROUTH, C. F., 1948. *Isospora hominis* infection among American personnel in southwest Pacific. Am. J. Trop. Med., 28:639.

BECKER, E. R., 1933. Cross-infection experiments with Coccidia of rodents and domesticated animals. J. Parasit., 19:230.

—— 1934. Coccidia and coccidiosis. Ames, Iowa.

—— 1941. Effect of parenteral administration of vitamin B$_1$, etc. Proc. Soc. Exper. Biol., 46:494.

—— 1954. The host affinities of *Isospora bigemina*-type Coccidia. Proc. Iowa Acad. Sci., 61:463.

—— and CROUCH, H. B., 1931. Some effects of temperature upon development of the oocysts of Coccidia. *Ibid.*, 28:529.

—— and FRYE, W. W., 1929. *Eimeria ellipsoidalis* n. sp., a new coccidium of cattle. J. Parasit., 15:175.

—— and HALL, P. R., 1931. *Eimeria separata*, a new species of coccidian from the Norway rat. Iowa State Coll. J. Sci., 6:131.

—— —— and HAGER, A., 1932. Quantitative, biometric and host-parasite studies on *Eimeria miyairii*, etc. *Ibid.*, 6:299.

—— JESSEN, R. J., *et al.*, 1956. A biometrical study of the oocyst of *Eimeria necatrix*, a parasite of the common fowl. J. Protozool., 3:126.

——, MANRESA, M. JR., and SMITH, L., 1943. Nature of *Eimeria nieschulzi*-growth-promoting potency of feeding stuffs. V. *Ibid.*, 17:257.

BĚLAŘ, K., 1926. Zur Cytologie von *Aggregata eberthi*. Arch. Protist., 53:312.

BOLES, J. I., and BECKER, E. R., 1954. The development of *Eimeria brunetti* Levine in the digestive tract of chicken. Iowa State Coll. J. Sci., 29:1.

BONORRIS, J. S., and BALL, G. H., 1955. *Schellackia occidentalis* n. sp., a blood-inhabiting coccidian found in lizards in southern California. J. Protozool., 2:31.

BONCIU, C., *et al.*, 1957. Contribution à l'étude de l'infection naturelle des cobayes avec des sporozoaires du genre Klossiella. Arch. Roum. Path. Exper. et Microbiol., 16:131.

BOUGHTON, D. C., 1929. A note on coccidiosis in sparrows and poultry. Poultry Sci., 8:184.

—— 1930. The value of measurements in the study of a protozoan parasite *Isospora lacazei*. Amer. J. Hyg., 11:212.

—— 1933. Diurnal gametic periodicity in avian Isospora. *Ibid.*, 18:161.

—— 1937. Notes on avian coccidiosis. Auk, 54:500.

——, BOUGHTON, RUTH B., and VOLK, J., 1938. Avian hosts of the genus Isospora. Ohio J. Sci., 38:149.

—— and VOLK, J. J., 1938. Avian hosts of Eimerian Coccidia. Bird Band., 9:139.

BRACKETT, S., and BLIZNICK, A., 1950. The occurrence and economic importance of coccidosis in chickens. 78 pp. Pearl River, N. Y.

—— —— 1952. The reproductive potential of 5 species of Coccidia of the chicken, etc. J. Parasit., 38:133.

BRAY, R. S., 1960. On the parasitic protozoa of Liberia. III. J. Protozool., 7:314.

CHAKRAVARTY, M. and KAR, A., 1944. Studies on Coccidia of Indian birds. I. J. Dep. Sci., Calcutta Univ., 1:78.

—— —— 1947. A study of the Coccidia of Indian birds. Proc. Roy. Soc. Edinburgh, 62:225.

CHALLEY, J. R., and BURNS, W. C., 1959. The invasion of the cecal mucosa by *Eimeria tenella* sporozoites and their transport by macrophages. J. Protozool., 6:238.

CHRISTENSEN, J. F., 1938. Species differentiation in the Coccidia from the domestic sheep. J. Parasit., 24:453.

—— 1941. The oocysts of coccidia from domestic cattle in Alabama, etc. J. Parasit., 27:203.

CLARK, G. M. 1958. *Hepatozoon griseisciuri* n. sp., etc. J. Parasit., 44:52.

CONNAL, A., 1922. Observations on the pathogenicity of *Isospora hominis,* etc. Tr. Roy. Soc. Trop. Med. Hyg., 16:223.

CROUCH, H. B., and BECKER, E. R., 1931. Three species of Coccidia from the woodchuck, etc. Iowa State Coll. J. Sci., 5:127.

CUCKLER, A. C., and MALANGA, C. M., 1955. Studies on drug resistance in Coccidia. J. Parasit., 41:302.

—— —— 1956. The effect of Nicarbazin on the development of immunity, etc. *Ibid.*, 42:593.

DAVIS, L. R., *et al.*, 1957. The endogenous development of *Eimeria alabamensis,* etc. J. Protozool., 4:219.

DEBAISIEUX, P., 1911. Recherches sur les coccidies. I. La Cellule, 27:89.

DEEDS, O. J., and JAHN, T. L., 1939. Coccidian infections of western painted turtles, etc. Tr. Am. Micr. Soc., 58:249.

DOBELL, C., 1919. A revision of the Coccidia parasitic in man. Parasitology, 11:147.

—— 1925. The life-history and chromosome cycle of *Aggregata eberthi.* *Ibid.*, 17:1.

DORAN, D. J., and FARR, M. M., 1962. Excystation of the poultry coccidium *Eimeria acevulina.* J. Protozool., 9:154.

DUNCAN, S., 1959. The size of the oocysts of *Eimeria labbeana.* J. Parasit., 45:191.

—— 1959a. The effects of some chemical and physical agents on the oocysts of *Eimeria labbeana. Ibid.*, 45:193.

EDGAR, S. A., 1954. Effect of temperature on the sporulation of oocysts of *Eimeria tenella.* Tr. Am. Micr. Soc., 73:237.

—— 1955. Sporulation of oocysts at specific temperatures and notes on the prepatent period of several species of avian coccidia. J. Parasit., 41:214.

ELSTON-DEW, R., and FREEDMAN, L. 1953. Coccidiosis in man: experiences in Natal. Tr. Roy. Soc. Trop. Med. Hyg., 47:209.

FIEBIGER, J., 1913. Studien ueber die Schwimmblasencoccidien der Gadus-Arten (*Eimeria gadi* n. sp.). Arch. Protist., 31:95.

FISH, F., 1930. Coccidia of rodents: etc. J. Parasit., 17:98.

—— 1931. Quantitative and statistical analysis of infections with *Eimeria tenella* in the chicken. Amer. J. Hyg., 14:560.

—— 1931a. The effect of physical and chemical agents on the oocysts. of *Eimeria tenella.* Science, 73:292.

Foner, A., 1939. An attempt to infect animals with *Isospora belli*. Tr. Roy. Soc. Trop. Med. Hyg., 33:357.

Foster, A. O., 1949. The economic losses due to coccidiosis. Ann. N.Y. Acad. Sci., 52:434.

Garnham, P. C. C., *et al.*, 1962. The fine structure of *Lankesterella garnhami*. J. Protozool., 9:107.

Gill, B. S., 1954. Speciation and viability of poultry coccidia in 120 faecal samples preserved in 2.5% potassium dichromate solution. Ind. J. Vet. Sci. & Animal Husb., 24:245.

Grell, K. G., 1953. Entwicklung und Geschlechtsbestimmung von *Eucoccidium dinophili*. Arch. Protist., 99:156.

Greven, U., 1953. Zur Pathologie der Gefluegelcoccidiose. *Ibid.*, 98:342.

Haase, A., 1939. Untersuchungen ueber die bei deutschen Wildhühnern vorkommenden Eimeria-Arten. Arch. Protist., 92:329.

Hammond, D. M., *et al.*, 1946. The endogenous phase of the life cycle of *Eimeria bovis*. J. Parasit., 32:409.

Hanson, H. C., *et al.*, 1957. Coccidia of North American wild geese and swans. Canad. J. Zool., 35:715.

Hardcastle, A. B., 1943. A check list and host index of the species of the genus Eimeria. Proc. Helm. Soc., 10:35.

——— 1944. *Eimeria brevoortiana*, etc. J. Parasit., 30:60.

Hawkins, P. A., 1952. Coccidiosis of the turkey. Tech. Bull. Michigan Agr. Exper. Stat., 226, 87 pp.

Henry, Dora P., 1931. A study of the species of Eimeria occurring in swine. Univ. Cal. Publ. Zool., 36:115.

——— 1931a. Species of Coccidia in chickens and quail in California. *Ibid.*, 36:157.

——— 1932. The oocyst wall in the genus Eimeria. *Ibid.*, 37:269.

——— 1932a. Observations on coccidia of small mammals in California, etc. *Ibid.*, 37:279.

——— 1932b. *Isospora buteonis* sp. nov. from the hawk and owl, and notes on *Isospora lacazei* in birds. *Ibid.*, 37:291.

Herman, C. M., and Price, D. L., 1955. The occurrence of Hepatozoon in the gray squirrel *(Sciurus carolinensis)*. J. Protozool., 2:48.

Herrlich, A., and Liebmann, H., 1943. Zur Kenntnis der menschlichen Coccidien. Ztschr. Hyg. Infektionskr., 125:331.

——— ——— 1944. Die menschliche Coccidiose, etc. *Ibid.*, 126:22.

Hesse, E., 1910. Protozoaires nouveaux parasites des animeaux d'eau douce. II. Ann. Univ. Grenoble, 23:396.

——— 1910a. Sur le genre Adelea, etc. Arch. zool. exper. gén., 7(N-R):15.

Hitchcock, D. L., 1955. The life cycle of *Isospora felis* in the kitten. J. Protozool., 41:383.

Honess, R. F., 1942. Coccidia infesting the Rocky Mountain Bighorn sheep in Wyoming, etc. Bull. Univ. Wyoming Agr. Exper. Stat., no. 249.

HORTON-SMITH, C., and LONG, P. L., 1959. The effects of different anti-coccidial agents on the intestinal coccidiosis of the fowl. J. Comp. Path & Therap., 69:192.

HUIZINGA, H., and WINGER, R. N., 1942. *Eimeria wyomingensis,* a new coccidium from cattle. Tr. Am. Micr. Soc., 61:131.

IKEDA, I., 1914. Studies on some sporozoan parasites of Sipunculoides. II. Arch. Protist., 33:205.

JARPA, A., *et al.,* 1960. Isosporosis humana. Bol. Chileno Parasit., 15:50.

JEFFERY, G. M., 1956. Human coccidiosis in South Carolina. J. Parasit., 42:491.

KARTCHNER, J. A., and BECKER, E. R., 1930. Observations on *Eimeria citelli,* etc. J. Parasit., 17:90.

LAINSON, R., 1959. Atoxoplasma Garnham, 1950, as a synonym for Lankesterella Labbé, 1899. J. Protozool., 6:360.

——— 1960. The transmission of Lankesterella in birds by the mite *Dermanyssus gallinae. Ibid.,* 7:321.

LANDERS, E. J. JR., 1960. Studies on excystation of coccidial oocysts. J. Parasit., 46:195.

LAPAGE, G., 1940. The study of coccidiosis *(Eimeria caviae)* in the guinea-pig. Vet. J., 96:144, 190, 242, 280.

LÉGER, L., 1911. *Caryospora simplex,* coccidie monosporée et la classification des coccidies. Arch. Protist., 22:71.

——— and DUBOSCQ, O., 1915. *Pseudoklossia glomerata* n.g., n. sp., coccidie de lamellibranche. Arch. zool. exper. gén., 55 (N-R):7.

——— ——— 1917. *Pseudoklossia pectinis* n. sp., etc. *Ibid.,* 56 (N-R):88.

——— and HESSE, E., 1922. Coccidies d'oiseaux palustres le genre Jarrina n.g. C. R. Acad. Sci., 174:74.

LEHMANN, D. L., 1959. The description of *Haemogregarina boyli* n. sp. from the yellow-legged frog, *Rana boyli boyli.* J. Parasit., 45:198.

LEVINE, N. D., 1948. Eimeria and Isospora of the mink. *Ibid.,* 34:386.

——— 1952. *Eimeria dicrostonicis* n. sp., a protozoan parasite of the lemming, etc. Tr. Illinois Acad. Sci., 44:205.

——— 1952a. *Eimeria magnalabia* and *Tyzzeria* sp. from the Canada goose. Cornell Vet., 42:247.

——— 1953. A review of the coccidia from the avian orders Galliformes, etc. Amer. Midland Natur., 49:696.

——— 1957. Protozoan diseases of laboratory animals. Proc. Animal Care Panel., 7:98.

——— and BECKER, E. R. 1933. A catalog and host-index of the species of the coccidian genus Eimeria. Iowa State Coll. J. Sci., 8:83.

——— BRAY, R. S., *et al.,* 1959. Coccidia of Liberian rodents. J. Protozool., 6:215.

——— and IVES, V. 1960. Eimeria and Tyzzeria from deer mice, etc. J. Parasit., 46:207.

————— ————— and KRUIDENIER, F. J., 1955. *Dorisiella arizonensis* n. sp., etc. J. Protozool., 2:52.

————— ————— ————— 1955a. Two new species of Klossia from a deer mouse and a bat. J. Parasit., 41:623.

————— ————— ————— 1957. New species of Eimeria from Arizona rodents. J. Protozool., 4:80.

————— and MOHAN, R. N., 1960. *Isospora* sp. from cattle, etc. J. Parasit., 46:733.

LICKFELD, K. G., 1959. Untersuchungen ueber das Katzencoccid *Isospora felis*. Arch. Protist., 103:427.

LUDWIG, F. W., 1947. Studies on the protozoan fauna of the larvae of the crane-fly. *Tipula abdominalis*. II. Tr. Am. Micr. Soc., 66:22.

LUND, E. E., 1950. A survey of intestinal parasites in domestic rabbits in six counties in southern California. J. Parasit., 36:13.

MACHADO, A., 1911. Sobro um novo coccidio do intestino de um hemiptero. Brazil Med., no. 39.

MACKERRAS, M. J., 1961. The haematozoa of Australian reptiles. Aust. J. Zool., 9:61.

MACKINNON, D. L., and RAY, H. N., 1937. A coccidian from the eggs of *Thalassema neptuni*. Parasitology, 29:457.

MAGATH, T. B., 1935. The coccidia of man. Am. J. Trop. Med., 15:91.

MARQUARDT, W. C., 1959. The morphology and sporulation of the oocyst of *Eimeria brasiliensis* Torres and Ildefonso Ramos 1939, of cattle. Amer. J. Vet. Res., 20:742.

————— 1962. Subclinical infections with coccidia in cattle, etc. J. Parasit., 48:270.

————— *et al.*, 1960. The effect of physical and chemical agents on the oocyst of *Eimeria zurnii*. J. Protozool., 7:186.

MARSHALL, E. K., 1950. Infection with *Isospora hominis*, etc. J. Parasit., 36:500.

MATSUBAYASHI, H., 1937. Studies on parasitic protozoa in Japan. II. Annot. Zool. Japon., 16:255.

————— and NOZAWA, T., 1948. Experimental infection of *Isospora hominis* in man. Am. J. Trop. Med., 28:683.

MAYHEW, R. L., 1933. Studies on coccidiosis. V. Poultry Sci., 12:206.

————— 1937. IX. Tr. Am. Micr. Soc., 56:431.

McLOUGHLIN, D. K., and CHESTER, D. K., 1959. The comparative efficacy of six anticoccidial compounds. Poultry Sci., 38:353.

McTAGGART, H. S., 1960. Coccidia from mink in Britain. J. Parasit., 46:201.

MILLER, W. W., 1908. *Hepatozoon perniciosum*, etc. U.S.P.H. Serv., Hyg. Lab. Bull., no. 46.

MOORE, E. N., and BROWN, J. A., 1951. A new coccidium pathogenic for turkey, *Eimeria adenoeides* n. sp. Cornell Veter., 41:124.

MOREHOUSE, N. F., and McGUIRE, W. C., 1958. The pathogenicity of *Eimeria acervulina*. Poultry Sci., 37:665.

MOROFF, T., 1908. Die bei den Cephalopoden volkommenden Aggregata-Arten, etc. Arch. Protist., 11:1.

MOYNIHAN, I. W., 1950. The rôle of the protozoan parasite, *Eimeria acervulina*, etc. Canada J. Comp. Med. Vet. Sci., 14:74.

MUKHERJEE, H. N., 1948. Incidence of coccidiosis in the Arakan. Indian Med. Gaz., 82:735.

NABIH, A., 1938. Studien über die Gattung Klossia, etc. Arch. Protist., 91:474.

NATH, V., *et al.*, 1960. The life-cycle of *Eimeria tenella*, etc. Res. Bull. Panjab Univ., 11:227.

NAVILLE, A., 1925. Recherches sur le cycle sporogonique des Aggregata. Rev. Suisse Zool., 32:125.

——— 1927. Recherches sur le cycle évolutif et chromosomique de *Klossia helicis*. Arch. Protist., 57:427.

NÖLLER, W., 1923. Zur Kenntnis eines Nierencoccids. *Ibid.*, 47:101.

——— and FRENZ, O., 1922. Zur Kenntnis des Ferkelkokzids und seiner Wirkung. Deutsch. tierärztl. Wochenschr., 30:1.

PATTEN, R., 1935. The life history of *Merocystis kathae* in the whelk, *Buccinum undulatum*. Parasitology, 27:399.

PATTILLO, W. H., 1959. Invasion of the cecal mucosa of the chicken by sporozoites of *Eimeria tenella*. J. Parasit., 45:253.

——— and BECKER, E. R., 1955. Cytochemistry of *Eimeria brunetti* and *E. acervulina* of the chicken. J. Morphol., 96:61.

PELLÉRDY, L., 1956. Catalogue of the genus Eimeria. Acta Veter., Acad. Sci. Hungaricae, 6:75.

——— 1957. Catalogue of the genus Isospora. *Ibid.*, 7:209.

PÉRARD, C., 1925. Recherches sur les coccidies et les coccidioses du lapin. Ann. Inst. Pasteur, 39:505.

PRASAD, H., 1961. The coccidia of the zorille *Ictonyx (Zorilla) capensis*. J. Protozool., 8:55.

PRATT, I., 1940. The effect of *Eimeria tenella* upon the blood sugar of the chicken. Tr. Am. Micr. Soc., 59:31.

RAY, H. N., 1930. Studies on some Sporozoa in polychaete worms. II. Parasitology, 22:471.

REICHENOW, E., 1913. *Karyolysus lacertae, etc.* Arb. kais. Gesundh. 45:317.

——— 1919. Der Entwicklungsgang der Haemococcidien Karyolysus, etc. Sitz-Ber. Gesell. naturf. Fr. Berlin, p. 440.

——— 1921. Die Haemococcidien der Eidechsen. Arch. Protist., 42:179.

REINHARDT, J. F., and BECKER, E. R., 1933. Time of exposure and temperature as lethal factors in the death of the oocysts, etc. Iowa State Coll. J. Sci., 7:505.

RITA, G., and VIDA, B. L. D., 1949. Coccidiosi umana da Isospora. Riv. Parassit., 10:117.

ROUDABUSH, R. L., 1937. Some coccidia of reptiles found in North America. J. Parasit., 23:354.

———— 1937a. The endogenous phases of the life cycle of *Eimeria nieschulzi*, etc. Iowa State Coll. J. Sci., 11:135.

———— 1937b. Two Eimeria from the flying squirrel, etc. J. Parasit., 23:107.

———— and COATNEY, G. R., 1937. On some blood Protozoa of reptiles and amphibians. Tr. Am. Micr. Soc., 56:291.

RUBIN, R., *et al.*, 1956. The efficacy of nicarbazin as a prophylactic drug in cecal coccidiosis of chickens. Poultry Sci., 35:856.

SAUNDERS, D. C., 1955. The occurrence of *Haemogregarina bigemina* and *H. achiri* n. sp., in marine fish from Florida. J. Parasit., 41:171.

———— 1958. The occurrence of *Haemogregarina bigemina* and *H. dasyatis* n. sp. in marine fish from Bimini, etc. Tr. Am. Micr. Soc., 77:404.

———— 1959. *Haemogregarina bigemina* from marine fish of Bermuda. *Ibid.*, 78:374.

SCHAUDINN, F., 1900. Untersuchungen ueber den Generationswechsel bei Coccidien. Zool. Jahrb. Abt. Morphol., 13:197.

SCHELLACK, C., 1913. Coccidien-Untersuchungen. II. Arb. kais. Gesundh., 45:269.

———— and REICHENOW, E., 1913. I. *Ibid.*, 44:30.

SCHNEIDER, A., 1885. Tablettes zoologiques. 1.

SCHOLTYSECK, E., 1953. Beitrag zur Kenntnis des Entwicklungsganges des Huehnercoccids *Eimeria tenella*. Arch. Protist., 98:415.

SEIBOLD, H. R., and THORSON, R. E., 1955. *Klossiella equi* n. sp. from the kidney of an American jack. J. Parasit., 41:285.

SENGER, C. M., *et al.*, 1959. Resistance of calves to reinfection with *Eimeria bovis*. J. Protozool., 6:51.

SHUMARD, R. F., 1957. Studies on ovine coccidiosis. I. J. Parasit., 43:548.

SMITH, T., and JOHNSON, H. P., 1902. On a coccidium *(Klossiella muris* g. et sp. nov.) parasitic in the renal epithelium of the mouse. J. Exper. Med., 6:303.

SVANBAEV, S. K., 1957. The fauna and morphology of sheep and goat coccidia of western Kasakhstan (Russian). Trudy Inst. Zool. Acad. Sci Kazakh S.S.R., 7:252.

———— 1958. Coccidia of rodents of central Kazakhstan (Russian). *Ibid.*, 9:183.

TANABE, M., 1938. On three species of Coccidia of the mole, *Mogera wogura coreana*, etc. Keijo J. Med., 9:21.

THÉLOHAN, P., 1890. Sur deux coccidies nouvelles, parasites de l'épinoche et de la sardine. C. R. Acad. Sci., 110:1214.

———— 1894. Nouvelles recherches sur les coccidies. Arch. zool. exp., 2:541.

THOMSON, J. G., and ROBERTSON, A., 1926. Fish as the source of certain Coccidia recently described as intestinal parasites of man. Brit. Med. J., p. 282.

TYZZER, E. E., 1910. An extracellular coccidian, *Cryptosporidium muris,* etc. J. Med. Res., 18:487.

—— 1912. *Cryptosporidium parvum,* etc. Arch. Protist., 26:394.

—— 1927. Species and strains of Coccidia in poultry. J. Parasit., 13:215.

—— 1929. Coccidiosis in gallinaceous birds. Amer. J. Hyg., 10:1.

—— 1932. Coccidiosis in gallinaceous birds. II. *Ibid.,* 15:319.

UHLHORN, E., 1926. Uebertragungsversuche von Kaninchenoccidien auf Hühnerkücken. Arch. Protist., 55:101.

VAN DOORNINCK, W. M., and BECKER, E. R., 1957. Transport of sporozoites of *Eimeria necatrix* in macrophage. J. Parasit., 43:40.

VINCENT, M., 1927. On *Legerella hydropori* n. sp., etc. Parasitology, 19:394.

WEISER, J., and BEARD, R. L., 1959. *Adelina sericesthis* n. sp., a new coccidian parasite of Scarabaeid larvae. J. Insect Path., 1:99.

WENYON, C. M., 1933. Coccidiosis of cats and dogs and the status of the Isopora of man. Ann. Trop. Med. Parasitol., 17:231.

—— 1926. Protozoology. 2. London and Baltimore.

ZÜBLIN, E., 1908. Beitrag zur Kenntnis der roten Ruhr der Rinde, etc. Schweiz. Arch. Tierheilk., 50:123.

Chapter 27

Order 3. **Haemosporida** Danilewsky

THE development of the Haemosporida is, on the whole, similar to that of the Coccidida in that they undergo asexual reproduction of schizogony, and also sexual reproduction resulting in sporozoite-formation; but the former takes place in the blood if vertebrates and the latter in the alimentary canal of some blood-sucking invertebrates. Thus one sees that the Haemosporida remain always within the body of one of the two hosts; hence, the sporozoites do not possess any protective envelope. However, the developmental cycles of Coccidida and Haemosporida resemble each other so closely that Doflein (1901) combined them into one order (Coccidiomorpha) which scheme had found many followers. In the present work, they are treated as orders.

The Haemosporida are minute intracorpuscular parasites of vertebrates. The malarial parasites of man are typical members of this order. The development of *Plasmodium vivax* is briefly as follows (Fig. 262). An infected female anopheline mosquito introduces **sporozoites** into human blood when it feeds on it through skin (*a*). The sporozoites are fusiform and 6-15μ long. They are capable of slight vibratory and gliding movement when seen under the microscope after removal from mosquitoes. After about seven to ten days of exo-erythrocytic development (p. 721), the organisms are found in erythrocytes (*c, d*) and are called **schizonts.** At the beginning the schizonts are small rings. They grow and finally divide into twelve to twenty-four or more **merozoites** (*e, f*) which are presently set free in the blood plasma (*g*). This **schizogony** requires forty-eight hours. The freed merozoites will, if not ingested by leucocytes, enter and repeat schizogony in the erythrocytes. After repeated and simultaneous schizogony in geometric progression, large numbers of infected erythrocytes will be destroyed at intervals of forty-eight hours, apparently setting free ever-increasing amounts of toxic substances into the blood. This is

717

the cause of the regular occurrence of a characteristic **paroxysm** on every third day.

In the meanwhile, some of the merozoites develop into **gameto-cytes** instead of undergoing schizogony (*h-k*). When fully formed they are differentiated into **macro-** and **micro-gametocytes,** but remain as such while in the human blood. When a female an-opheline mosquito takes in the blood containing gametocytes, the microgametocyte develops into four to eight **microgametes** (*k, l*), and the macrogametocyte into a **macrogamete** (*i, m*) in its stomach. An **ookinete** (zygote) is formed when a microgamete fuses with a macrogamete (*m, n*). The ookinetes are motile. Garn-ham *et al.,* (1962) noted under electron microscope fifty-five to sixty-five hollow peripheral fibrils which appear to be contractile, are present in the ookinetes of *P. cynomolgi* and *P. gallinaceum.* As the ookinetes come in contact with the stomach epithelium, they enter it and become rounded into **oocysts** which lie between the base of the epithelium and the outer membrane of the stom-ach (*o*). Within the oocysts, repeated nuclear division produces numerous sporozoites (*p*). When fully mature, the oocyst ruptures and the sporozoites are set free in the haemolymph through which they migrate to the salivary glands (*q, r*). The sporozoites make their way through the gland epithelium and finally to the duct of hypopharynx. They are ready to infect a human victim when the mosquito pierces the skin with its proboscis for another blood meal. Thus the sexual reproduction occurs in the mosquito (primary host) and the asexual reproduction, in man (secondary host). (Origin of Haemosporida, Manwell, 1955, 1961; location of the parasites in host erythrocytes, Wolcott *et al.,* 1958; chromo-somes, Wolcott, 1954, 1955, 1957; electron microscopy of oocysts, Duncan *et al.,* 1960.) Three families.

Family 1. **Plasmodiidae** Mesnil

With pigment granules; schizogony in peripheral blood of ver-tebrates.

Genus **Plasmodium** Marchiafava and Celli. Schizogony in erythrocytes and also probably in endothelial cells of man, mam-mals, birds, and reptiles; sexual reproduction in blood-sucking in-sects; widely distributed. (Numerous species.)

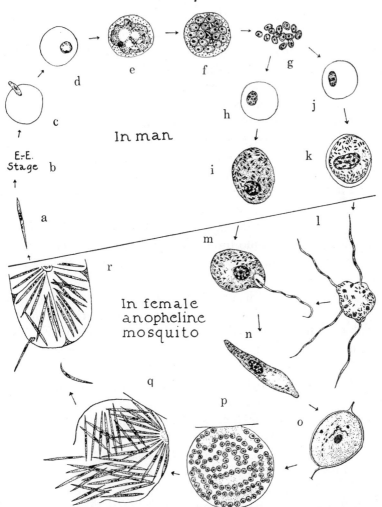

In man

E.-E.
Stage b

In female
anopheline
mosquito

Fɪɢ. 262. The life-cycle of *Plasmodium vivax* (Kudo). a, sporozoite entering human blood; b, exoerythrocytic stage; c, the initiation of the erythrocytic development; d, a young schizont ("ring form"); e–g, schizogony; h, i, macrogametocytes; j, k, microgametocytes; l, microgamete-formation in the stomach of a mosquito; m, union of the gametes; n, zygote or ookinete; o, rounding up of an ookinete in the stomach wall; p, oocyst in which sporozoites are developing; q, mature oocyst ruptured and sporozoites are set free in the haemolymph; r, sporozoites entering the salivary gland cells.

In all species, the infection in a vertebrate host begins under natural condition with the inoculation of the sporozoites by a vector mosquito. The form, size and structure of the sporozoites vary widely within a species so that identification of the species in this stage appears to be impossible (Boyd, 1935). Until some thirty years ago, it had been generally believed that the sporozoites upon entering the blood, penetrate and enter immediately the erythrocyte and begin intra-corpuscular development, which process Schaudinn (1902) reported to have seen in life. In this the eminent pioneer protozoologist was in error, since no one has up to the present time been able to confirm his observation. Et. and Edm. Sergent (1922) were the first to find that quinine given in large doses to the canaries on the day the birds were bitten by Culex mosquitoes infected with *Plasmodium relictum*, did not prevent infection in the birds. During the course of studies on *P. vivax* in cases of general paresis, Yorke and MacFie (1924) discovered that if quinine was given before the inoculation of infected blood, no infection resulted, but if the sporozoites were inoculated, quinine did not prevent infection. Similar observations were made on other species of malarial organisms. James (1931) suggested the possibility that the sporozoites are carried away from peripheral to visceral circulation and develop in the cells of the reticulo-endothelial system.

Boyd and Stratman-Thomas (1934) found that the peripheral blood of a person who had been subjected to the bites of fifteen anopheline mosquitoes infected by *Plasmodium vivax*, did not become infectious to other persons by subinoculation until the ninth day and that the parasites were not observed before the eleventh day in the stained films of the peripheral blood. Warren and Coggeshall (1937) observed that when suspensions of the sporozoites of *P. cathemerium* obtained from infected *Culex pipiens*, were inoculated into canaries, the blood was not infectious for seventy-two hours, but emulsions made from the spleen, liver and bone marrow contained infectious parasites which brought about infection by subinoculations in other birds. These and many similar observations cannot be satisfactorily explained if one follows Schaudinn's view. The fact that *P. elongatum* is capable of undergoing schizogony in the leucocytes and reticulo-endothelial cells in addition to erythrocytes of host birds had been observed by Raffaele (1934) and Huff and Bloom (1935).

As to the nature of development of Plasmodium during the pre-patent period, James and Tate (1938) showed that there occur schizonts and schizogonic stages in the endothelial cells of the spleen, heart, liver, lung, and brain of the birds infected by *P. gallinaceum* (Fig. 263). They suggested the term **exoerythrocytic** (E-E) to this schizogony in contrast to the well-known **erythrocytic schizogony.** Huff and his co-workers made a series of detailed studies of pre-erythrocytic stages of this avian species. According

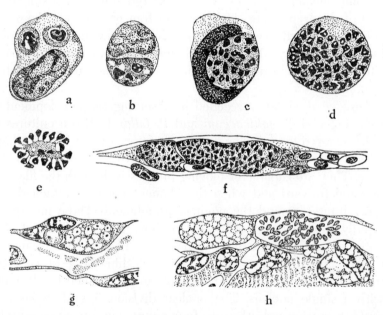

FIG. 263. Exoerythrocytic schizogony in avian Plasmodium. a–f, *P. gallinaceum* in smears from chicks (James and Tate). a, monocyte from lung, infected by 2 young schizonts; b, monocyte from liver, with a growing tri-nucleate schizont; c, monocyte from lung, with a large multinucleate schizont; d, large mature schizont containing many mature merozoites, free in lung; e, portion of broken schizont from lung, showing the attached developing merozoites. (×1660). f, a capillary of brain blocked by 3 large schizonts (×740). g, h, *P. cathemerium* in sections of organs of canaries (Porter; ×1900). g, capillary in the brain, showing an endothelial cell infected with a uninucleate and a multinucleate schizont; h, a multinucleate schizont and a group of merozoites found in a capillary of heart muscle.

to Huff and Coulston (1944), the sporozoites that are inoculated into the skin of chickens, are engulfed by phagocytes in 0.5-6 hours. In heterophile leucocytes, the sporozoites are apparently killed, but in the cells of lymphoid-macrophage system they develop into *cryptozoites* (Huff, Coulston and Cantrell, 1943) by assuming a spheroid shape and increasing in size for the first thirty-six hours, during which time there is a rapid repeated division of the nucleus. The schizogony is completed in thirty-six to forty-eight hours, each giving rise to 75-150 merozoites. These merozoites enter new lymphoid-macrophage and endothelial cells and become *metacryptozoites* which undergo schizogony similar to that of the cryptozoite. After three or four generations, the merozoites enter erythrocytes, and thus the erythrocytic stages appear in five to ten days. In avian and some mammalian malaria parasites, the E-E schizogony continues during the duration of infection.

Huff *et al.* (1960) succeeded in observing the development of E-E stages of *P. gallinaceum* and *P. fallax* in tissue cultures of chick embryo *in vivo*. The freed merozoites are pyriform in shape with a filament (1.2 times the body length) emerging from one pole. They are capable of moving progressively with the filament directed forward and when the filament becomes attached, the body rotates around the point of attachment. These merozoites rapidly enter a host cell by first probing with the filament. The entrance may occur actively or passively. Phagocytosis by macrophages and fibroblasts occurs. Once within the host cell, the merozoites become rounded and now may be called trophozoites with a single nucleus. The nuclear division is said to be very rapid, requiring only three to four minutes and is repeated with the growth of body.

E-E stages were further discovered in saurian Plasmodium (Thompson and Huff, 1944; Garnham, 1950) and in mammalian malaria organisms (Shortt and Garnham, 1948). In *P. cynomolgi*, Shortt and Garnham report that the E-E stages occur in the parenchymatous cells of the liver of host monkeys and are inclined to think that there is one generation only. The earliest forms were seen on the fifth day after the inoculation of the sporozoites. They are rounded bodies, about 10μ in diameter and contain about fifty chromatin granules of irregular shape. They grow in size to about 35μ in diameter, and divide in eight to nine days into some 1000 merozoites, each measuring about 1μ. These merozoites presumably invade the erythrocyte. In *P. vivax*, the E-E stages develop in the parenchymatous cells of the liver also and resemble those of *P. cynomolgi*. The forms found on the seventh day after sporozoite-inoculation were slightly larger (about 42μ in diameter) than those of *P. cynomologi*, and when mature, gave rise to 800-1000 merozoites. Garnham and co-workers (1955) studied further the pre-erythrocytic stages of *P. falciparum* and *P. ovale*. In the latter species, the pre-erythrocytic schizonts develop also apparently in parenchyma cells of the liver. The schizonts were oval to irregular in shape on the 5th day and measured 28-60μ in diameter. As the organisms grow, they become highly irregular in form and reach a length of 85μ on the ninth day which give rise to some 1500 merozoites. These merozoites are spherical uninucleate bodies which measure about 1.8μ in diameter. (E-E stages and development, Huff, 1947, 1948; Garnham, 1948.)

The incubation period of Plasmodium infections in man varies due to various factors such as the strain, vitality and number of the sporozoites injected by the mosquitoes, the varied susceptibility on the part of host, etc. Boyd and co-workers found that the incubation periods for the three species of human Plasmodium which they studied were, as follows: In *P. vivax*, eight to twenty-one days (the majority 11-14 days) after the bites of infected mosquitoes, but in one case as long as 304 days; in *P. malariae*, four to five weeks, with the onset of fever lagging three to twelve days behind; and in two strains of *P. falciparum*, one, six to twenty-five days and the other, nine to thirteen days; in another observation, *P. falciparum* was observable in the peripheral blood in five to nine days and the onset of fever in seven to twelve days, and Jeffery *et al.* (1959) noted the incubation period varied from eleven to 13.5 days.

The paroxysm of malaria is usually divisible into three stages: chill or rigor stage, high temperature or febrile stage (104°F. or over) and sweating or defervescent stage. The time of paroxysm corresponds, as was stated already, with the time of liberation of merozoites from erythrocytes, and is believed to be due to extrusion of certain substance into the blood plasma. The nature of this material is however unknown at present. In the grown schizonts as well as in gametocytes of Plasmodium, are found invariably yellowish brown to black pigment granules which vary in form, size and number among different species. They are usually called **haemozoin** granules and are apparently the catabolic products formed within the parasites. The pigment of *P. gallinaceum* and *P. cynomolgi* has been identified with haematin (ferri protoporphyrin) (Rimington and Fulton, 1947). The pigment possesses certain taxonomic significance, as will be described elow. The infected erythrocytes, if stained deeply, may show a punctate appearance. These dots are small and numerous in the erythrocytes infected by *P vivax* and *P. ovale,* and are known as Schüffner's (1899) dots, while those in the host cells infected by *P. falciparum* are few and coarse and are referred to as Maurer's (1902) dots. No dots occur in the erythrocytes infected by *P. malariae.* (Pathology, Maegraith, 1948; splenomegaly, Darling, 1924, 1926, Russell, 1935, 1952a, Hackett, 1944; histopathology, Taliaferro and Mulligan, 1937; character of paroxysm, Kitchen and Putnam, 1946; blood proteins during infection, Boyd and Proske, 1941; stippling of erythrocytes, Thomson, 1928.)

The condition which brings about the formation of gametocytes is not known at present. The gametocytes appear in the peripheral blood at various intervals after onset of fever, and remain inactive while in the human blood. The gametocytes of *P. falciparum* were found by Jeffery and co-workers (1959) in six to twenty-one days after the appearance of the first asexual parasites.

The assumption that the macrogametocytes undergo parthenogenesis under certain conditions and develop into schizonts as advocated by Grassi, Schaudinn and others, does not seem to be supported by factual evidence. The initiation of further development appears to be correlated with a lower temperature and also a change in pH of the medium (Manwell). If living mature microgametocytes of human Plasmodium taken from an infected person are examined microscopically under a sealed cover glass at room temperature (18-22°C.), development takes place in a short while and motile microgametes are produced ("exflagellation"). Similar changes take place when the gametocytes are taken into

the stomach of mosquitoes belonging to genera other than Anopheles, but no sexual fusion between gametes occurs in them and
all degenerate sooner or later. In the stomach of an anopheline
mosquito, however, the sexual reproduction of human Plasmodium continues, as has been stated before.

All species are transmitted by adult female mosquitoes. The
males are not concerned, since they do not take blood meal. The
species of Plasmodium which attack man are transmitted only by
the mosquitoes placed in genus Anopheles, while the majority of
the avian species of Plasmodium are transmitted by those which
belong to genera Culex, Aedes, and Theobaldia. The chief vectors
of the human malarial parasites in North America are *A. quadrimaculatus* (eastern, southern and middle-western States), *A.
punctipennis* (Widely distributed), *A. crucians* (southern and
south-eastern coastal area), *A. walkeri* (eastern area), and *A. maculipennis freeborni* (Pacific coast). Boyd and co-workers observed
that (1) *A. quadrimaculatus* and *A. punctipennis* were about
equally susceptible to *Plasmodium vivax;* (2) *A. quadrimaculatus*
was susceptible to several strains of *P. falciparum*, while *A. punctipennis* varied from highly susceptible to refractory to the same
strains; (3) *A. quadrimaculatus* was more susceptible to all three
species of Plasmodium than coastal or inland *A. crucians*. Thus *A.
quadrimaculatus* is the most dangerous malaria vector in the
United States as it shows high susceptibility to all human Plasmodium. *A. pseudopunctipennis* distributed from south-western
United States to Argentina and *A. albimanus* occurring in Central
America, are but a few out of many anopheline vectors of human
Plasmodium in the areas indicated. (Host-parasite relation, Boyd
and Coggeshall, 1938; malaria vectors of the world, Komp, 1948;
susceptibility of Anopheles to malaria, King, 1916, Boyd and
Kitchen, 1936; epidemiology in North America, Boyd, 1941; in
Brazil, Boyd, 1926; in Jamaica, Boyd and Aris, 1929; in Cuba,
Carr and Hill, 1942; in Trinidad and British West Indies, Downs,
Gillette and Shannon, 1943; in Puerto Rico, Earle, 1930, 1939; in
Haiti, Paul and Bellerive, 1947; in Philippine Islands, Russell,
1934, 1935; in India, Russell and Jacob, 1942; general picture,
Russell, 1952, 1952a; mosquito control, Russell, 1952a.)

The time required for completion of sexual reproduction of

Plasmodium in mosquitoes varies according to various conditions such as species and strain differences in both Plasmodium and Anopheles, temperature, etc. Boyd and co-workers showed that when the anophelines which fed on patients infected by *P. vivax* were allowed to feed on other persons, their infectivity was as follows: one to ten days after infective feeding, 87.2 per cent; eleven to twenty days, 93.8 per cent; twenty-one to thirty days, 78 per cent; thirty-one to forty days, 66 per cent; forty-one to fifty days, 20 per cent; and over fifty days, none. In a similar experiment with *P. falciparum,* during the first ten days the infection rate was 84 per cent, but thereafter the infectivity rapidly diminished until there was no infection after forty days. It is generally known that the development of the parasites in mosquitoes depends a great deal on temperature. Although the organisms may survive freezing temperature in mosquitoes (Coggeshall), sporozoite-formation is said not to take place at temperatures below 16°C. or above 35°C. (James). According to Stratman-Thomas (1940), the development of *Plasmodium vivax* in *Anopheles quadrimaculatus* is completed within the temperature range of 15-17° to 30°C. It varies from eight to thirty-eight days after infective feeding. The optimum temperature is said to be 28°C. at which the development is completed in the shortest time. A period of twenty-four hours at 37.5°C. will sterilize all but a very small per cent of *Anopheles quadrimaculatus* of their *Plasmodium vivax* infection. This has a bearing on the transmission of *Plasmodium vivax* in summer months. In certain localities oocysts may survive the winter and complete their development in the following spring. (Duration of infection in Anopheles, Boyd and St.-Thomas, 1943a, Boyd, St.-Thomas and Kitchen, 1936; blood meal volume, Jeffery, 1956.)

There are three long-recognized species of human Plasmodium. They are *P. vivax, P. falciparum* and *P. malariae.* To these *P. ovale* is here added. Each species appears to be represented by numerous strains or races as judged by the differences in virulence, immunological responses, incubation period, susceptibility to quinine, etc. (Boyd, 1934, 1940, 1940a; Boyd and Kitchen, 1948). Wolcott (1955) found the nuclei of all four species of human malaria parasites show two chromosomes in late schizogonic stages.

Malaria has been, and still is, perhaps the most serious protozoan disease of man in many countries. In India alone, malaria has been held to be the cause of over a million deaths annually among nearly 100 million persons who suffer from it (Sinton, 1936). In south-east Asia, China, parts of South America and equatorial Africa, malaria appears to be still prevelent. Russell (1952) estimated the total number of victims of malaria infection throughout the world at more than a quarter of a billion with about 1 per cent mortality rate. World Health Organization is succeeding in bringing about practically a complete eradication of this disease in several countries.

In the United States, the disease had been prevalent in places in southern States. In 1935, more than 150,000 malaria cases occurred, but ten years later the number declined to 60,000 cases. In 1955, the total number of malaria cases diminished to 522 and in 1958, only ninty-four cases occurred (Brody and Dunn, 1959).

In malarious countries, the disease is a serious economic and social problem, since it affects the majority of population and brings about a large number of persistent sickness, the loss of man power and retardation of both mental and physical development among children. (History of malaria, Ross, 1928, Boyd, 1941, Russell, 1943; general reference, Boyd, 1949, Russell, West and Manwell, 1946, Pampana and Russell, 1955; antimalarial drugs, Russell, 1952a, Coatney *et al.*, 1958; preservation of sporozoites by freezing, Jeffery and Rendtorff, 1955; chromosomes of human malaria, Wolcott, 1955; malaria therapy, Boyd and Stratman-Thomas, 1933, Mayne and Young, 1941; genetic factors in resistance to malaria, Allison, 1961.)

P. vivax (Grassi and Feletti) (Fig. 264). The benign tertian malaria parasite; schizogony completed in forty-eight hours and paroxysm every third day. *Ring forms:* About one-quarter to one-third the diameter of erythrocytes; unevenly narrow cytoplasmic ring is stained light blue (in Giemsa) and encloses a vacuole; nucleus stained dark-red, conspicuous. *Growth period:* Irregular amoeboid forms; host cell slightly enlarged; Schüffner's dots begin to appear. *Grown schizonts:* In about twenty-six hours after paroxysm; occupy about two-thirds of the enlarged erythrocytes, up to 12μ in diameter, which are distinctly paler than uninfected ones; Schüffner's dots more numerous; brownish haemozoin granules; a large nucleus. *Schizogonic stages:* Repeated nuclear division produces twelve to twenty-four or more merozoites; mul-

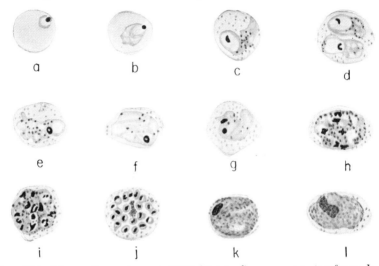

Fig. 264. *Plasmodium vivax*, ×1535 (Original). a, young ring-form; b, c, growing schizonts; d, two schizonts in an erythrocyte; e, f, large schizonts; g–i, schizogonic stages; j, fully developed merozoites; k, macrogametocyte; l, microgametocyte.

tinucleate schizons about 8-9μ in diameter; haemozoin granules in loose masses; merozoites about 1.5μ long. *Gametocytes:* Time required for development of ringform into a mature gametocyte is estimated to be about four days; smoothly rounded body, occupying almost whole of the enlarged erythocytes; brown haemozoin granules numerous. Macrogametocytes are about 9-10μ in diameter, stain more deeply and contain a small compact nucleus; microgametocytes are a little smaller (7-8μ in diameter), stain less deeply and contain a less deeply staining large nucleus. This species is said to invade reticulocytes rather than erythrocytes (Kitchen, 1938). Boyd (1935a) distinguished five series of erythrocytic organisms on the basis of nuclear and cytoplasmic characteristics. The organisms of series A give rise by schizogony to organisms of series B or D which in turn produce series C (micro-gametocytes) or series E (macrogametocytes). Onset of infection is said to occur usually when the parasite density is less than 100 per mm³ (Boyd, 1944). (Incubation period, Boyd and Stratman-Thomas, 1933c, 1934; concentration of organisms, Ferrebee and Geiman, 1946; immunity, Boyd and Stratman-Thomas, 1933 a, b; Boyd and Kitchen, 1936a, Boyd, 1947; susceptibility, Boyd and

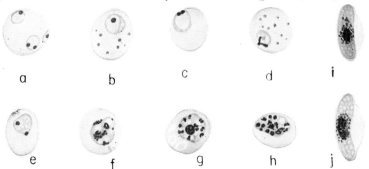

a   b   c   d   i

e   f   g   h   j

FIG. 265. *Plasmodium falciparum,* ×1535 (Original). a, three ring-forms in an erythrocyte; b, a somewhat grown schizont in an erythrocyte with Maurer's dots; c–f, growing and schizogonic stages; g, h, merozoite formation; i, macrogametocyte; j, microgametocyte.

Stratman-Thomas, 1933c, 1934; varieties, Wilcox *et al.,* 1954; nucleus, Wolcott, 1954.)

The benign malaria fever parasite is the commonest and the most widely distributed species in the tropical and subtropical regions as well as in the temperate zone. It has been reported as far north as the Great Lakes region in North America; England, southern Sweden and northern Russia in Europe; and as far south as Argentina, Australia, and Natal in the southern hemisphere. Generally speaking this species predominates in the spring and early summer over the other species.

P. falciparum Welch (*Laverania malariae* Grassi and Feletti; *P. tenue* Stephens) (Fig. 265). The subtertian, malignant tertian or aestivo-autumnal fever parasite; schizogonic cycle is somewhat irregular, though generally about forty-eight hours. *Ring forms:* Much smaller than those of *P. vivax;* about 1μ in diameter; marginal forms and multiple (2-6) infection common; nucleus often rod-form or divided into two granules; in about twelve hours after paroxysm, all schizonts disappear from the peripheral blood. *Growth and schizogonic stages:* These are almost exclusively found in the capillaries of internal organs; as schizonts mature, Maurer's dots appear in the infected erythrocytes; when about 5μ in diameter, nucleus divides repeatedly and eight to twenty-four or more small merozoites are produced; haemozoin granules dark brown or black and usually in a compact mass; infected erythrocytes are not enlarged. *Gametocytes:* Mature forms sausage-shaped ("crescent"), about 10-12μ by 2-3μ; appear in the peripheral blood. Macrogametocytes stain blue and contain a compact nucleus and coarser granules, grouped around nucleus; microgametocytes stain less deeply blue or reddish, and contain a large lightly staining nucleus and scattered smaller haemozoin granules. The organism invades both mature and immature erythrocytes (Kitchen, 1939). (Cytological study of microgametocytes and microgametes, MacDougall, 1947; different strains, Kitchen and Putnam, 1943; induced infection, Boyd and Kitchen, 1937; incubation period, Boyd and Kitchen, 1937b, Boyd and Matthews, 1939; immunity Boyd and Kitchen, 1945; E-E stages, Shortt *et al.*, 1951; Jeffery *et al.*, 1952; gametocyte density, Jeffery and Eyles, 1955.)

The subtertian fever parasite is widely distributed in the tropics. In the subtropical region, it is more prevalent in late summer or early autumn. It is relatively uncommon in the temperate zone. The malignancy of the fever brought about by this parasite is attributed in part to decreased elasticity of the infected erythrocytes which become clumped together into masses and which adhere to the walls of the capillaries of internal organs especially brain, thus preventing the circulation of blood through these capillaries.

P. malariae Laveran (Fig. 266). The quartan malaria parasite; schizogony in seventy-two hours and paroxysm every fourth day. *Ring forms:* Similar to those of *P. vivax. Growth period:* Less amoeboid, rounded; in about six to ten hours haemozoin granules begin to appear; granules are dark brown; in twenty-four hours, schizonts are about one-half the diameter of erythrocytes which remain normal in size; schizonts often stretched into "band-form" across the erythrocytes; no dots comparable with Schüffner's or Maurer's dots. *Mature and segmenting schizonts:* In about forty-eight hours, schizonts nearly fill the host cells; rounded; haemozoin granules begin to collect into a mass; nuclear divisions produce six to twelve merozoites which are the largest of the three

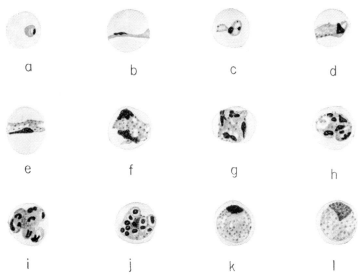

Fig. 266. *Plasmodium malariae,* ×1535 (Original). a, ring-form; b–e, band-form schizonts; f–i, schizogonic stages; j, merozoite formation; k, macrogametocyte; l, microgametocyte.

species and may often be arranged in a circle around a haemozoin mass. *Gametocytes:* Circular; with haemozoin granules. Macrogametocytes stain more deeply and contain a small, more deeply staining nucleus and coarser granules; microgametocytes stain less deeply and contain a larger lightly stained nucleus and finer and numerous granules. The organism invades most frequently mature red corpuscles (Kitchen, 1939).

The quartan fever parasite is distributed in the tropics and subtropics, though it is the least common of the three species. As a rule, in an area where the three species of Plasmodium occur, this species seems to appear later in the year than the other two.

P. ovale Stevens (Fig. 267). The Ovale or mild tertian fever parasite; schizogony in about forty-eight hours; its morphological characters resemble both *P. vivax* and *P. malariae*. *Ring forms:* Similar to those of the two species just mentioned; Schüffner's dots appear usually conspicuous. *Growth period:* Infected erythrocytes in stained smears are more or less oval with irregularly fimbriated margin (which are artefacts according to Garnham *et al.* (1955); slightly enlarged; not actively amoeboid, sometimes

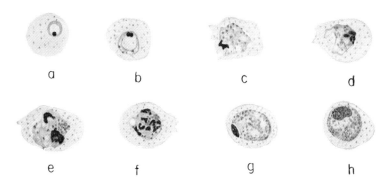

a   b   c   d

e   f   g   h

Fig. 267. *Plasmodium ovale*, ×1535 (Original). a, ring-form; b, c, growing schizonts; d–f, schizogonic stages; g, macrogametocyte; h, microgametocyte.

in band-form; with dark brown haemozoin granules; Schüffner's dots abundant. *Schizogonic stages:* six to twelve large merozoites. *Gametocytes:* Resemble closely those of *P. malariae;* host cells with Schüffner's dots and slightly enlarged.

This organism appears to be confined to Africa and Asia (Philippine Islands and India). Several malariologists doubt the validity of the species. After observing the pre-erythrocytic development, Garnham *et al.* (1955) came to conclude that this is a distinct species. (Strains, Jeffery, Wilcox and Young, 1955.)

The malarial parasites are ordinarily studied in stained blood films (p. 1082). Table 8 will serve for differential diagnosis of the three common species.

Several species of Plasmodium have been described from primates and monkeys, some of which resemble the human species. It appears certain that some of these monkey parasites are transmitted to man by mosquito vectors (Contacos *et al.*, 1962; Whar-

TABLE 8

DIFFERENTIAL DIAGNOSIS OF THREE SPECIES OF HUMAN PLASMODIUM

	P. vivax	*P. falciparum*	*P. malariae*
Ring forms	About $\frac{1}{4}$–$\frac{1}{3}$ the diameter of erythrocytes; a single granular nucleus.	About $\frac{1}{6}$–$\frac{1}{5}$ the diameter of erythrocytes; marginal forms and multiple (2–6) infection common.	Similar to those of *P. vivax;* cytoplasm slightly denser.
Infected erythrocytes	Much enlarged, up to 12μ in diameter, paler than normal (7.5μ in diameter) erythrocytes; Schüffner's dots.	Normal, some are distorted or contracted in later schizogonic period; Maurer's dots.	Not enlarged; sometimes slightly smaller than uninfected ones; no dots.
Growing schizonts	Irregularly amoeboid; vacuolated; paler; small yellowish brown haemozoin granules.	Partly grown ring forms often with rod-shaped or 2 granular nuclei; further development not seen in peripheral blood.	Not amoeboid; oval, rounded, band-form rarely irregular; less vacuolated cytoplasm deeper blue; dark brown granules.
Fully grown schizonts	Irregular in form; about $\frac{2}{3}$ the enlarged erythrocytes; vacuolated; brown haemozoin granules.	Only in internal organs; $\frac{1}{2}$–$\frac{2}{3}$ of erythrocytes; dark haemozoin in compact mass.	Nearly filling erythrocytes; rounded; cytoplasm deeper blue; dark brown pigment granules.
Schizogonic stages	12–24 or more merozoites; irregularly arranged in much enlarged host cells.	Only in internal organs; 8–24 or more small merozoites; irregularly arranged; dark pigment.	6–12 merozoites which are the largest of all, typically arranged in a circle.
Gametocytes	Almost filling enlarged erythrocytes; rounded or oval; with brown pigment granules.	Sausage-shaped; haemozoin dark brown; in the peripheral blood.	Filling normal-sized erythrocytes; round or ovoid, much smaller than those of *P. vivax;* dark brown pigment.

ton *et al.*, 1962). Here a few species will be mentioned. (Other species, Aberle, 1945.)

P. kochi (Laveran) (Fig. 268, *a-f*). In the monkeys belonging to the genera: Callicebus, Cercocebus, Cercopithecus, Erythrocebus, and Papio; schizogony in forty-eight hours; organism resembles *P. vivax;* infected erythrocytes become enlarged and sometimes stippling like Schüffner's dots occurs; eight to fourteen merozoites; gametocytes large and spheroid. *Culicoides adersi* and *C. fulvi-*

thorax are vectors (Garnham *et al.,* 1962). (Developmental cycle, Garnham, 1948a.)

P. *brasilianum* Gonder and Berenberg-Gossler (Fig. 268, *g-l*). In New World monkeys belonging to the genera: Alouatta, Ateles, Cacajao and Cebus; schizogony in seventy-two hours; it resembles P. *malariae;* no enlargement of infected erythrocytes; band-form schizonts; number of merozoites vary according to the difference in hosts, averaging eight to ten; gemetocytes rounded, comparatively small in number (Taliaferro and Taliaferro, 1934). (Haematology, Taliaferro and Klüver, 1940.)

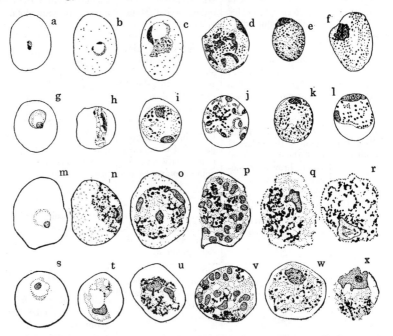

FIG. 268. Plasmodium of monkeys. Column 1, ring-forms; 2, 3, growing trophozoites; 4, segmenting schizonts; 5, macrogametocytes; 6, microgametocytes. a–f, *Plasmodium kochi,* ×1665 (Gonder and Berenberg-Gossler); g–l, P. *brasilianum,* ×1665; m–r, P. *cynomolgi,* ×2000; s–x, P. *knowlesi,* ×2000 (Taliaferro and Taliaferro).

P. *cynomolgi* Mayer (Fig. 268, *m-r*). In *Macaca irus (Macacus cynomolgus);* schizogony in forty-eight hours; eight to twenty-two merozoites; infected erythrocytes slightly enlarged and stippled; vectors are Anopheles. When infected salivary glands of

Anopheles were injected intravenously, rhesus monkeys showed E-E stages in liver after five days up to 120 days (Eyles, 1960). It is capable of producing low-level infections in man (Coatney *et al.*, 1961; Schmidt *et al.*, 1961; Beye *et al.*, 1961). (Schizogony, Wolfson and Winter, 1946, Taliaferro and Mulligan, 1937; morphology, Mulligan, 1935; cellular changes in host, Taliaferro and Mulligan, 1937.)

P. knowlesi Sinton and Mulligan (Fig. 268, *s-x*). In *Macaca irus;* experimentally man is susceptible; schizogonic cycle in twenty-four hours; chromosome number 2 (Wolcott, 1957); six to sixteen merozoites; infected erythrocytes are somewhat distorted. (Morphology and development, Brug, 1934, Mulligan, 1935, Taliaferro and Taliaferro, 1949; infections in man, Milam and Coggeshall, 1938.)

P. berghei Vincke and Lips. In the tree rat, *Thamnomys surdaster* of Congo (Vincke and Lips, 1948). White mice, white rats, cotton rats, the field vole (*Microtus guntheri*) and the golden hamster (*Mesocricetus auratus*) are susceptible; mosquito vector, *Anopheles dureni* (Mercado and Coatney, 1951). McGhee (1954) infected the erythrocytes of duck and goose embryos with this species, although the infection was transitory and the parasites could not be found after four days. (Parasitemia and characteristics in mice, Greenberg and Kendrick, 1957, 1957a, 1958; phagotrophy and structures under electron microscope, Rudzinska and Trager, 1959; liver lipids of infected rats, von Brand and Mercado, 1958.)

Many species of Plasmodium have been reported from numerous species of birds in which are observed clinical symptoms and pathlogical changes similar to those which exist in man with malaria infection. In recent years, the exoerythrocytic stages have been intensively studied in these forms. According to Hegner and co-workers the erythrocytes into which merozoites enter are often the most immature erythrocytes (polychromatophilic erythroblasts). The species of avian Plasmodium are transmitted by adult female mosquitoes belonging to Culex, Aedes or Theobaldia only. Transmission by Anopheles appears to be exceptional. For example, Hunninen and his co-workers (1950) found that *P. relictum* in wild-caught English sparrows developed fully in *Anopheles qua-*

drimaculatus and *A. crucians* as well as *Culex pipiens*. Some of the common species are briefly mentioned here. (Avian Plasmodium, Manwell, 1935a, Hewitt, 1940b; avain hosts, Wolfson, 1941; distribution, Manwell and Herman, 1935, Herman, 1938, Hewitt, 1940a, Wood and Herman, 1943, Hunninen and Young, 1950, Manwell, 1955a, Mohammed, 1958.)

P. relictum Grassi and Feletti (*P. praecox* G. and F.; *P. inconstans* Hartman) (Fig. 269, *a*). In English sparrow (*Passer domesticus*) and other passerine birds, also in mourning doves and pigeons (Coatney 1938); schizogony varies in different strains, in twelve, twenty-four, thirty or thirty-six hours; eight to fifteen or sixteen to thirty-two merozoites from a schizont; two chromosomes (Wolcott, 1957); gametocytes rounded, with small pigment granules; host-cell nucleus displaced; canaries (*Serinus canaria*) susceptible; many strains; transmitted by Culex, Aedes and Theobaldia; widely distributed. (Duration of infection, Manwell, 1934, Bishop, Tate and Thorpe, 1938; variety, Manwell, 1940; in *Culex pipiens*, Huff, 1934; development in birds, Mudrow and Reichenow, 1944; relationship of E-E and erythrocytic stages, Sergent, 1949, Becker, 1960; development of oocyst, Ball and Chao, 1957; susceptibility and resistance of pigeons and culicine mosquitoes, Huff *et al.*, 1959; electron microscope study of oocyst, Duncan *et al.*, 1960.)

P. vaughani Novy and McNeal (Fig. 269, *b*). In robin (*Turdus m. migratorius*) and starling (*Sturnus v. vulgaris*); four to eight (usually four) merozoites from a schizont, ordinarily with two pigment granules; schizogony in about twenty-four hours; gametocytes elongate; host-cell nucleus not displaced. (In Malaya, Laird, 1962.)

P. cathemerium Hartman (Fig. 269, *c*). In English sparrow, cowbird, red-winged blackbird, and other birds; schizogony in twenty-four hours, segmentation occurs at 6-10 p.m.; six to twenty-four merozoites from a schizont; mature schizonts and gametocytes about 7-8μ in diameter; gametocytes rounded; haemozoin granules in microgametocytes longer and more pointed than those present in macrogametocytes; canaries susceptible; numerous strains; common; transmitted by many species of Culex and Aedes (Harman, 1927). (Relapse, Manwell, 1929; acquired immunity, Cannon and Taliaferro, 1931; in ducks, Hegner and West, 1941; cultivation, Hewitt, 1939; effect of plasmochin, Wampler, 1930; effect of lowered temperature of duck embryo on the de-

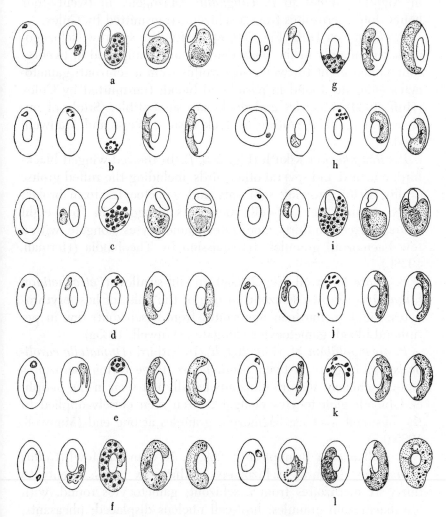

Fig. 269. a, *Plasmodium relictum;* b, *P. vaughani;* c, *P. cathemerium;* d, *P. rouxi;* e, *P. elongatum;* f, *P. circumflexum;* g, *P. polare;* h, *P. nucleophilum;* i, *P. gallinaceum;* j, *P. hexamerium;* k, *P. oti;* l, *P. lophurae.* Columns 1, ring-forms; 2, growing schizonts; 3, segmenting schizonts; 4, macrogametocytes; and 5, microgametocytes. × about 1400 (Several authors; from Hewitt, modified).

velopment of the parasites, McGhee, 1959; electron microscope study of oocysts formed in *Culex fatigans,* Duncan *et al.,* 1960.)

P. rouxi Sergent and Catanel (Fig. 269, *d*). In English sparrow in Algeria; similar to *P. vaughani;* schizogony in twenty-four hours; four merozoites from a schizont; transmitted by Culex.

P. elongatum Huff (Fig. 269, *e*). In English sparrow; schizogony occurs mainly in the bone marrow, and completed in twenty-four hours; eight to twelve merozoites from a schizont; gametocytes elongate, found in peripheral blood; transmitted by Culex (Huff, 1930). Canaries and ducks are susceptible. (Study of nucleus, Chen, 1944; culture in duck tissues, Weiss and Manwell, 1960.)

P. circumflexum Kikuth (Fig. 269, *f*). In the red-winged blackbird, cowbird and several other birds, including the ruffed grouse (Fallis, 1946); growing schizonts and gametocytes form broken rings around the host-cell nucleus; schizogony in forty-eight hours; thirteen to thirty merozoites; gametocytes elongate, with a few haemozoin granules; transmission by Theobaldia (Herman, 1938b).

P. polare Manwell (Fig. 269, *g*). In cliff swallow (*Petrochelidon l. lunifrons*); grown schizonts at one of the poles of host erythrocytes; eight to fourteen merozoites from a schizont; few in peripheral blood; gametocytes elongate (Manwell, 1935a).

P. nucleophilum M. (Fig. 269, *h*). In catbird (*Dumatella carolinesis*); schizogony in twenty-four hours; three to ten merozoites from a schizont; mature schizonts usually not seen in the peripheral blood; gametocytes elongate, often seen closely applied to the host-cell nucleus; haemozoin granules at one end (Manwell, 1935a).

P. gallinaceum Brumpt (Fig. 269, *i*). In domestic fowl (*Gallus domesticus*) in India; schizogony in thirty-six hours; twenty to thirty-six merozoites from a schizont; gametocytes round, with few haemozoin granules; host-cell nucleus displaced; pheasants, geese, partridges and peacocks are susceptible, but canaries, ducks, guinea fowls, etc., are refractory; transmitted by Aedes (Brumpt, 1935). E-E stages in chick embryo introduced into the haemocoele of *Aedes aegypti* brought about infection in chicks when exposed to the mosquitoes (Weathersby, 1960). (Vectors, Russell and Mohan, 1942; E-E stages under electron microscope, Meyer and Musacchio, 1960; E-E stages, Huff *et al.*, 1960.)

P. hexamerium Huff (Fig. 269, *j*). In bluebird (*Sialia s. sialis*) and Maryland yellow-throats; schizogony in forty-eight or seventy-two hours; grown schizonts often elongate; six merozoites from a schizont; gametocytes elongate (Huff, 1935). (E-E schizogony, Manwell, 1951; chemotherapy, Manwell and Khabir, 1954; glucose consumption, Khabir and Manwell, 1955.)

P. oti Wolfson (Fig. 269, *k*). In eastern screech owl (*Otus asio naevius*); eight merozoites from a schizont; body outlines irregular, rough; gametocytes elongate. Manwell (1949) considers this species identical with *P. hexamerium.*

P. lophurae Coggeshall (Fig. 269, *l*). In fire-back pheasant (*Lophura i. igniti*) from Borneo, examined at New York Zoological Park; eight to eighteen merozoites from a schizont; gametocytes large, elongate; host-cell nucleus not displaced; canaries are refractory, but chicks and especially ducks are highly susceptible (Coggeshall, 1938, 1941; Wolfson, 1940); young ducklings succumb less readily to its infection than older ducks (Becker, 1950). Experimentally *Aedes aegypti, A. albopictus* and *Anopheles quadrimaculatus* serve as vectors, but not *Culex pipiens* (Jeffery, 1944). In experimental infection in white pekin ducks, Trager (1948) found that young females with inactive ovaries and males developed heavy infections, while in the females with active ovaries, the parasites underwent little or no multiplication, and that the average free biotin and bound biotin-active lipid material of the blood plasma before inoculation, were highest in the egg-laying females, lowest in the males and intermediate in the females with inactive ovaries. Rudzinska and Trager (1957) discovered through electron microscope study of sectioned infected duckling erythrocytes that the trophozoites take in the cytoplasm of the host cell by phagotrophy as well as probably by diffusion. McGhee (1954) succeeded in infecting mice and young rats with this species and found (1956) that the mean numbers of merozoites temporarily decreased. (Characteristics, Terzian, 1941; cultivation, Trager, 1950; E-E stages in blood-inoculated turkey,

Becker and Manresa, 1950, 1953; host resistance, Jordan, 1957; schizogony under electron microscope, Rudzinska and Trager, 1961; serum alterations, Sherman and Hull, 1960, 1960a.)

A number of lizards have recently been found to be infected by Plasmodium. A few species are described here briefly. (Species, Thompson and Huff, 1944a; Laird, 1951.)

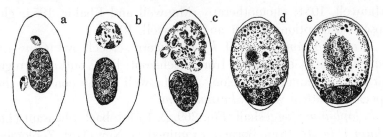

FIG. 270. *Plasmodium mexicanum,* ×1780 (Peláez *et al.*). a, b, young and growing trophozoites in host's erythrocyte; c, segmenting schizont; d, macrogametocyte; e, microgametocyte.

P. mexicanum Thompson and Huff (Fig. 270). In *Sceloporus ferrariperezi* of Mexico; experimentally S. *olivaceous, S. undulatus, Crotaphytus collaris, Phrynosoma cornutum* and *P. asio,* become infected; in erythrocytes and normoblasts, and in all types of circulating cells; host cells not hypertrophied; schizonts round to elongate; ten to forty merozoites; gametocytes 12-16μ by 6-7.7μ, only in haemoglobin-containing cells which becomes enlarged and distorted. (Thompson and Huff, 1944); a mite, *Hirstella* sp., was considered to be a possible vector (Peláez, Reyes and Barrera, 1948).

P. rhadinurum T. and H. In the erythrocytes of *Iguana iguana rhinolopha* in Mexico; schizonts extremely polymorphic with one or two long processes; four to five merozoites; gametocytes 6.5-7.7μ; vector unknown (Thompson and Huff, 1944a).

P. floridense T. and H. In the erythrocytes of *Sceloporus undulatus* in Florida; young trophozoites pyriform; 6-21 (12) merozoites; chromosome number 2 (Wolcott, 1957); gametocytes 7.5-8.0μ in diameter; vector unknown (Thompson and Huff, 1944a).

P. lygosomae Laird. In New Zealand skink, *Lygosoma moco* (Laird, 1951).

Family 2. **Haemoproteidae** Doflein

Schizogony occurs in the endothelial cells of vertebrates; merozoites enter circulating blood cells and develop into gametocytes; if blood is taken up by specific blood-sucking insects, gametocytes develop into gametes which unite to form zygotes that undergo changes similar to those stated above for the family Plasmodiidae.

Genus **Haemoproteus** Kruse. Gametocytes in erythrocytes, with pigment granules, halter-shaped when fully formed (hence *Hal-*

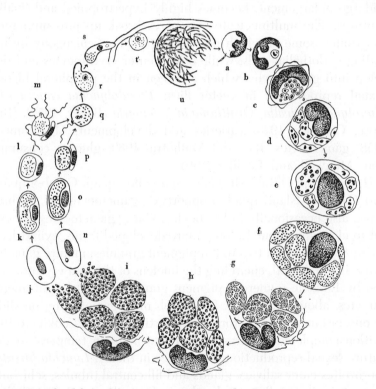

FIG. 271. The life-cycle of *Haemoproteus columbae*. (Several authors). a, a sporozoite entering an endothelial cell of the pigeon; b, growth of a schizont; c, segmentation of multinucleate schizont into uninucleate cytomeres; d–i, development of cytomeres to produce merozoites; j–m, development of microgametes; n–p, development of macrogamete; q, fertilization; r, s, ookinetes; t, a young oocyst in the stomach wall of a fly; u, a ruptured mature oocyst with sporozoites. a–k, n, o, in the pigeon; l, m, p–u, in *Pseudolynchia maura.*

teridium Labbé); schizogony in endothelial cells of viscera of birds and reptiles. (Species, Cerny, 1933, Coatney and Roudabush, 1937, Mohammed, 1958; transmission experiments, Nöller, 1920.)

H. columbae Kruse. (Fig. 271). In pigeons (*Columba livia*), etc.; widely distributed; young schizonts, minute and uninucleate, are in the endothelial cells of lungs and other organs, grow into large multinucleate bodies which divide into fifteen or more uninucleate cytomeres (Aragão). Each cytomere now grows and its nucleus divides repeatedly. The host cell in which many cytomeres undergo enlargement, becomes highly hypertrophied and finally ruptures. The multinucleate cytomeres break up into numerous merozoites, some of which possibly repeat the schizogony by invading endothelial cells, while others enter erythrocytes and develop into gametocytes which are seen in the peripheral blood; sexual reproduction in vector flies: *Pseudolynchia canariensis*, *Microlynchia pusilla*, *Ornithomyia avicularia* (Baker, 1957). (Relapse, Coatney, 1933; varieties and development, Mohammed, 1958; gametocytes, Ray and Malhotra, 1960; glucose consumption, Manwell and Loeffler, 1961.)

H. lophortyx O'Roke. In California valley quail, Gambel quail, and Catalina Island quail (Lophortyx); gametocytes in erythrocytes, also occasionally in leucocytes; young gametocytes, spherical to elongate, about 1μ long; more developed forms, cylindrical, about 8μ by 2μ with two to ten pigment granules; mature gametocytes, haltershaped, encircling the nucleus of the host erythrocyte, 18μ by 1.5-2.5μ; numerous pigment granules; four to eight microgametes, about 13.5μ long, from each microgametocyte; on slide in one instance, gamete-formation, fertilization and ookinete formation completed in fifty-two minutes at room temperature; in nature sexual reproduction takes place in the fly, *Lynchia hirsuta;* sporozoites enter salivary glands and fill central tubules; schizonts present in lungs, liver and spleen of quail after infected flies sucked blood from the bird; merozoites found in endothelial cells of capillaries of lungs, in epithelial cells of liver and rarely in peripheral blood cells; how merozoites enter blood cells is unknown; schizonts seldom seen in circulating blood; infected birds show pigment deposits in spleen and lungs (O'Roke, 1934). (Duration of infection, Herman and Bischoff, 1949.)

H. metchnikovi (Simond). In the Indian river tortoise, *Trionyx indicus* and the yellow-bellied terrapin, *Pseudemys elegans* (Hewitt, 1940).

H. sacharovi Novy and McNeal. In mourning dove and pigeons.

H. nettionis (Johnston and Cleland). In wood ducks, domestic ducks and geese (Herman, 1954; Fallis and Wood, 1957).

Genus **Leucocytozoon** Danilewsky. Schizogony in the endothelial cells as well as visceral cells of birds; sexual reproduction in

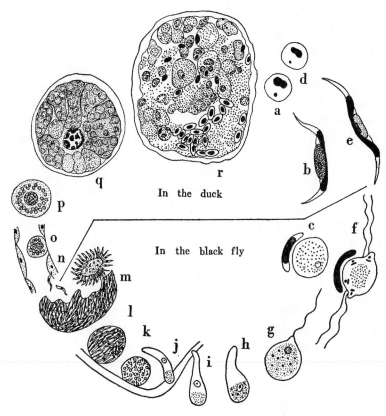

Fig. 272. The life-cycle of *Leucocytozoon simondi* (Brumpt, modified). a–c, development of macrogamete; d–f, development of microgametes; g, fertilization; h, ookinete; i, j, ookinete piercing through the stomach wall; k–m, development of sporozoites; n, sporozoites entering endothelial cells; o–r, schizogony.

blood-sucking insects; gametocytes in spindle-shaped host cells. (Several species, Cerny, 1933, Coatney and Roudabush, 1937; sporogony, Fallis and Bennett, 1962.)

L. *simondi* Mathis and Léger (*L. anatis* Wickware) (Fig. 272). Mathis and Léger (1910) described this species from the teal duck (*Querquedula crecca*) in Tonkin, China. Wickware (1915) saw *L. anatis* in ducks in Canada. O'Roke (1934) carried on experimental studies on the developmental cycle with the form which he found in wild and domestic ducks in Michigan. Herman (1938) observed the organism in common black ducks (*Anas rubripes tristis*), red-breasted merganser (*Mergus serrator*), and blue-winged teal (*Querquedula discors*) and considered *L. anatis* as identical with *L. simondi*. Huff (1942) studied the schizogony and gametocytes, and maintained the species he studied in mallard ducks (*Anas p. platyrhynchus*) and domestic ducks from Wisconsin, to be *L. simondi*. Levine and Hanson (1953) noted the organism in the Canada goose.

According to O'Roke, the vector is the black fly, *Simulium venustum*, in which the sexual reproduction takes place. Gametocytes develop into mature gametes in one to two minutes after blood is obtained from an infected duck; macrogametes about 8μ in diameter; four to eight microgametes, 15.7-24.1μ long, from a single microgametocyte; zygotes are found in stomach contents of fly in ten to twenty minutes after sucking in the infected blood of bird; motile ookinetes abundant after five hours, measure 33.3μ by 3-4.6μ twenty-two hours after sucking duck blood, oocysts are found on outer wall of stomach; sporozoites mature probably in twenty-four to forty-eight hours; five days after a duck has been bitten by infected black flies, schizogonic stages are noticed in endothelial cells of capillaries of lungs, liver, spleen; on about seventh day gametocytes appear in blood; liver and spleen become hypertrophied; the infection among ducklings is said to be highly fatal and appears often suddenly. In addition to the Simulium mentioned above, *Simulium parnassum*, *S. croxtoni*, *S. euryadminiculum* and *S. rugglesi* (Fallis *et al.*, 1951, 1956), appear to be the vectors.

Mathis and Léger: Macrogametocytes, oval; 14-15μ by 4.5-5.5μ; several vacuoles in darkly stained cytoplasm. Microgametocytes, oval; slightly smaller; cytoplasm stains less deeply. Infected host cells about 48μ long; nucleus elongate.

Huff found that (1) young schizonts are in macrophages of, and also extracellularly in, the spleen and liver; (2) two types of schizonts occur: one, "hepatic schizonts" in hepatic cells which cause no distortion or alteration of the host cell, and the other, "megaloschizonts" in the blood vessels of, or extravascularly in, the heart, spleen, liver and intestine; (3) megaloschizonts become divided into many cytomeres which give rise to numerous merozoites; (4) young gametocytes occur in lymphocytes, monocytes, myelocytes and late polychromatophile erythroblasts; (5) the cells in which fully grown gametocytes occur, appear to be macrophages. Cook (1954) found that the gametocytes develop exclusively in erythrocytic series and both the round and elongate forms appeared to be mature forms. (Life history and effect on host blood, Fallis *et al.*, 1951, 1956; development in ducklings, Chernin, 1952; megaloschizonts, Cowan, 1955.)

L. smithi Laveran and Lucet (1905). In turkey. Byrd (1959) found it to be common among free-ranging and pen-raised wild turkeys in Cumberland state forest, Virginia. Vector appears to be *Prosimulium hirtipes*. No symptom or pathogenicity was noted.

L. bonasae Clarke (1935). Found in ruffed grouse, *Bonasa umbellus,* and transmitted by black flies: *Simulium latipes* and *S. aureum.* Ordinarily symptomless (Fallis and Bennett, 1958).

Family 3. **Babesiidae** Poche

Minute non-pigmented parasites of the erythrocytes of mammals; transmission by ticks.

Genus **Babesia** Starcovici (*Piroplasma*, Patton). In erythrocytes of mammals; pear-shaped, arranged in pairs; vectors are ticks. As to the development of the organism, Dennis (1932) reported the occurrence of sexual reproduction as follows (Fig. 273): When a tick takes in infected blood into gut lumen, isogametes, 5.5-6μ long, are produced; isogamy results in motile club-shaped ookinetes, 7-12μ long, which pass through gut wall and invade larger ova (one to two, in one case about fifty, ookinetes per egg); each **ookinete** rounds itself up into a **sporont** 7.5-12μ in diameter, which grows in size and whose nucleus divides repeatedly; thus are produced multinucleated (four to thirty-two nuclei) amoeboid **sporokinetes,** up to 15μ long, which now migrate throughout embryonic tissue cells of tick, many of which cells develop into salivary gland cells; sporokinetes develop into sporozoites before or

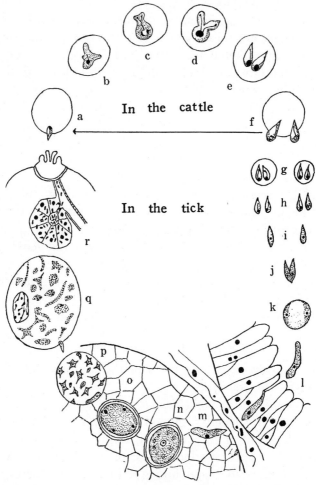

Fig. 273. The life-cycle of *Babesia bigemina* (Dennis). a–f, division in erythrocytes of cattle; g, h, gametocytes; i, isogametes; j, fertilization; k, zygote; l, ookinete penetrating through the gut wall; m, ookinete in host egg; n–p, sporoblast-formation; q, sporokinetes in a large embryonic cell; r, sporozoites in salivary gland.

after hatching of host tick; sporozoites bring about an infection to cattle when they are inoculated by tick at the time of feeding.

Other workers report that there is no sexual reproduction in Babesia. For example, Regendanz and Reichenow (1933), after a study of *B. canis* in female ticks, stated that when the ticks ingest the infected blood of dog, some of the organisms succeed in entering the gut epithelium of the tick. There they grow into large amoeboid bodies which divide by repeated division into a large number of minute organisms. They enter the developing eggs and remain dormant while the embryos develop. When the tick larvae molt, the organisms enter the salivary gland and there they multiply and develop. Infection of new host is produced by adult ticks. Regendanz (1936) found a similar development of *B. bigemina* in the tick, *Boophilus microplus.* (Taxonomy, Toit, 1918; species, Levine, 1961.)

B. bigemina (Smith and Kilborne) (Figs. 273; 274, *a-d*). The causative organism of the haemoglobinuric fever, Texas fever or redwater fever of cattle; the very first demonstration that an arthropod plays an important rôle in the transmission of a protozoan parasite; the infected cattle contain in their erythrocytes oval or pyriform bodies with a compact nucleus and vacuolated cytoplasm; the division is peculiar in that it appears as a budding process at the beginning. The hosts include in addition to cattle, water buffalo, deer, zebu, etc. The vectors are *Boophilus (Margaropus) annulatus, B. microplus* and *B. australis.* Texas fever of cattle once caused a considerable amount of damage to the cattle industry. Rees (1934) maintains that there is a somewhat smaller species, *B. argentina* Lignieres. (Initial development of infection, Hoyte, 1961.)

B. bovis Starcovici (Fig. 274, *e-h*). In European cattle; amoeboid form usually rounded, though sometimes stretched; 1-1.5μ in diameter; paired pyriform bodies make a larger angle, 1.5-2μ long; transmitted by *Ixodes ricinus.*

B. divergens (M'Fadyean and Stockman, 1911). In cattle in Great Britain. The vector is *Ixodes ricinus.*

B. canis (Piana and Galli-Valerio). Pyriform bodies 4.5-5μ long; the organism causes malignant jaundice in dogs; widely distributed; transmitted by the ticks: *Haemaphysalis leachi, Rhipice-*

phalus sanguineus, and *Dermacentor reticulatus* (Regendanz and Reichenow, 1933). In addition to dogs, wolf and jackal appear to be natural hosts. (Electron microscope study, Bayer and Dennig, 1961.)

B. tachyglossi Backhouse and Bolliger (1959). In the echidna, *Tachyglossus aculeatus.*

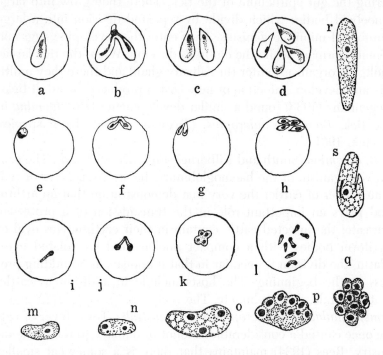

Fig. 274. a–d, *Babesia bigemina,* ×3000 (Nuttall); e–h, *B. bovis,* ×3000 (Nuttall); i–l, *Theileria-parva,* ×3000 (Nuttall); m–s, *Dactylosoma ranarum* (m–q, schizogony; r, s, gametocytes), ×2700 (Nöller).

B. equi Laveran and *B. caballi* Nuttall. They produce haemoglobinuric fever in horses; widely distributed.

Species of Babesia occur also in sheep, goats and pigs. (References, Levine, 1961.)

Genus **Theileria** Bettencourt, França and Borges. Schizogony takes place in endothelial cells of capillaries of viscera of mammals; certain forms thus produced enter erythrocytes and appear in the peripheral circulation.

T. parva Theiler (Fig. 274, *i-l*). In the cattle in Africa, cause of African coast fever; intracorpuscular forms 1-2µ in diameter; transmitted by the tick, *Rhipicephalus evertsi* and *R. appendiculatus* (Reichenow, 1937). (In tissue culture, Brocklesby and Hawking, 1958; macroschizonts, Barnett *et al.*, 1961.)

Genus **Dactylosoma** Labbé. In blood of reptiles and amphibians; schizogony and gametocytes in erythrocytes; invertebrate hosts unknown.

D. ranarum Kruse (Fig. 274, *m-s*). In frogs and toads; schizonts 4-9µ in diameter; four to six merozoites, 2-3µ by 1-1.5µ; gametocytes 5-8µ by 1.5-3µ. (Morphology and development, Jakowska and Nigrelli, 1956.)

Genus **Babesiosoma** Jakowska and Nigrelli. Somewhat similar to the last genus, but the cytoplasm in all stages is less granular and more vacuolated; no endosome in nucleus; reproduction by budding, binary fission or schizogony; not more than four merozoites in a rosette or cross-shaped; in erythrocytes of cold-blooded vertebrates (Jakowska and Nigrelli, 1956).

B. jahni (Nigrelli). In the erythrocyte of the common newt, *Triturus irridescens.*

References

ABERLE, S. D., 1945. Primate malaria. Office Inform. Nat. Res. Council, 171 pp.

ALLISON, A. C., 1961. Genetic factors in resistance to malaria. Ann. N.Y. Acad. Sci., 91:710.

ANDREWS, J. M., 1951. Nation-wide malaria eradication projects in the Americas. J. Nat. Mal. Soc., 10:99.

——— QUINBY, G. E., and LANGMUIR, A. D., 1950. Malaria eradication in the United States. Am. J. Pub. Health, 40:1405.

ARANTES, J. B., 1914. Contribuição para o estudo do Toxoplasma. Dissert., Coll. Med., Rio de Janeiro.

ATCHLEY, F. O., 1951. *Leucocytozoon andrewsi*, etc. J. Parasit., 37:483.

BACKHOUSE, T. C., and BOLLIGER, A., 1959. *Babesia tachyglossi* n. sp., from the echidna *Tachyglossus aculeatus.* J. Protozool., 6:320.

BAKER, J. R., 1957. A new vector of *Haemoproteus columbae* in England. J. Protozool., 4:204.

BALL, G. H., and CHAO, J., 1957. Development in vitro of isolated oocysts of *Plasmodium relictum.* J. Parasitol., 43:409.

BARNETT, S. F., *et al.*, 1961. Studies on macroschizonts of *Theileria parva.* Rev. Vet. Sci., 2:11.

BAYER, M. E., and DENNIG, K. H., 1961. Elektronenoptische Untersuchungen an *Babesia canis*. Zeitschr. Tropenmed. u. Parasit., 12:28.

BECKER, E. R., 1950. Mortality in relation to age in young white Pekin ducks with blood-induced *Plasmodium lophurae* infection. Proc. Iowa Acad. Sci., 57:435.

———— 1961. Some unfinished investigation of malaria in pigeons. J. Protozool., 8:1.

———— and MANRESA, M., JR., 1950. Phanerozoites in turkeys succumbing with blood-induced *Plasmodium lophurae* infection. Iowa State Coll. J. Sci., 24:353.

BEYE, H. K., *et al.*, 1961. Simian malaria in man. Am. J. Trop. Med. Hyg., 10:311.

BISHOP, ANN, *et al.*, 1938. The duration of *Plasmodium relictum* infection in canaries. Parasitology, 30:388.

BOYD, M. F., 1926. Studies of the epidemiology of malaria in the coastal lowlands of Brazil, etc. Amer. J. Hyg. Monogr. Ser. No. 5.

———— 1930. An introduction to malariology. Cambridge, Mass.

———— 1934. Observations on naturally induced malaria. South. Med. J., 27:155.

———— 1935. The comparative morphology of the sporozoites of the human species of Plasmodium. J. Parasit., 21:255.

———— 1935a. On the schizogonic cycle of *Plasmodium vivax*. Am. J. Trop. Med., 15:605.

———— 1940. On strains or races of the malaria parasites. *Ibid.*, 20:69.

———— 1940a. Observations on naturally and artifically induced quartan malaria. *Ibid.*, 20:749.

———— 1941. An historical sketch of the prevalence of malaria in North America. *Ibid.*, 21:223.

———— 1944. On the parasite density prevailing at certain periods in vivax malaria infection. J. Nat. Mal. Soc., 3:159.

———— 1947. A review of studies on immunity to vivax malaria. *Ibid.*, 6:12.

———— 1949. Malariology. A comprehensive survey of all aspects of this group of diseases from a global standpoint. 2 vols. Philadelphia.

———— and ARIS, F. W., 1929. A malaria survey of the island of Jamaica, B.W.I. Am. J. Trop. Med., 9:309.

———— and COGGESHALL, L. T., 1938. A résumé of studies on the host-parasite relation in malaria. Tr. 3rd Int. Congr. Trop. Med. Mal., 2:292.

———— and KITCHEN, S. F., 1936. The comparative susceptibility of *Anopheles quadrimaculatus*, etc. Am. J. Trop. Med., 16:67.

———— 1936a. On the efficiency of the homologous properties of acquired immunity to *Plasmodium vivax*. *Ibid.*, 16:447.

———— 1937. Observations on induced falciparum malaria. *Ibid.*, 17:213.

———— 1937a. A consideration of the duration of the intrinsic incubation period in vivax malaria, etc. *Ibid.*, 17:437.

———— ———— 1937b. The duration of the intrinsic incubation period in falciparum malaria, etc. *Ibid.*, 17:845.

———— ———— 1945. On the heterologous value of acquired immunity to *Plasmodium falciparum*. J. Nat. Mal. Soc., 4:301.

———— ———— 1948. On the homogeneity or heterogeneity of *Plasmodium vivax* infections acquired in highly endemic region. Am. J. Trop. Med., 28:29.

———— and MATTHEWS, C. B., 1939. An observation on the incubation period of *Plasmodium falciparum*. Am. J. Trop. Med., 19:69.

———— and PROSKE, H. O., 1941. Observations on the blood proteins during malaria infections. *Ibid.*, 21:245.

———— and STRATMAN-THOMAS, W. K., 1933. A controlled technique for the employment of naturally induced malaria in the therapy of paresis. Amer. J. Hyg., 17:37.

———— ———— 1933a. Studies on benign tertian malaria. I. *Ibid.*, 17:55.

———— ———— 1933b. III. *Ibid.*, 18:482.

———— ———— 1933c. IV. *Ibid.*, 18:485.

———— ———— 1934. V. *Ibid.*, 19:541.

———— ———— 1934a. On the duration of infectiousness in Anopheles harboring *Plasmodium vivax*. *Ibid.*, 19:539.

———— ———— and KITCHEN, S. F., 1936. On the duration of infectiousness in Anopheles harboring *P. falciparum*. Am. J. Trop. Med., 16:157.

BRAND, T. v., and MERCADO, T. I., 1958. Quantitative and histochemical studies on liver lipids of rats infected with *Plasmodium berghei*. Amer. J. Hyg., 67:311.

BRAY, R. S., 1959. On the parasitic protozoa of Liberia. II. J. Protozool., 6:13.

BROCKLESBY, D. W., and HAWKING, F., 1958. Growth of *Theileria annulata* and *T. parva* in tissue culture. Tr. Roy. Soc. Trop. Med. Hyg., 52:414.

BRODY, J. A., and DUNN, F. L., 1959. Malaria surveillance in the United States, 1958. Am. J. Trop. Med. Hyg., 8:635.

BRUG, S. L., 1934. Observations on monkey malaria. Revista Malario., 13:121.

BRUMPT, E., 1935. Paludisme aviaire: *Plasmodium gallinaceum* n. sp., de la poule domestique. C. R. Acad. Sci., 200:783.

BYRD, M. A., 1959. Observations on Leucocytozoon in pen-raised and free-ranging wild turkeys. J. Wildlife Manag., 23:145.

CANNON, P. R., and TALIAFERRO, W. H., 1931. Acquired immunity in avian malaria. III. J. Prev. Med., 5:37.

CARR, H. P., and HILL, R. B., 1942. A malaria survey of Cuba. Am. J. Trop. Med., 22:587.

CERNY, W., 1933. Studien an einigen Blutprotozoen aus Vögeln. Arch. Protist., 81:318.

CHEN, T. T., 1944. The nuclei in avian malaria parasites. I. Amer. J. Hyg., 40:26.

CHERNIN, E., 1952. Parasitemia in primary *Leucocytozoon simondi* infections. J. Parasit., 38:499.

CLARKE, C. H. D., 1938. Organisms of a malarial type in ruffed grouse, etc. J. Wildlife Manag., 2:146.

CLARKE, D. H., 1952. The use of phosphorus 32 in studies on *Plasmodium gallinaceum*. I, II. J. Exper. Med., 96:439.

COATNEY, C. R., 1933. Relapse and associated phenomena in the Haemoproteus infection of the pigeon. Amer. J. Hyg., 18:133.

——— 1938. A strain of *Plasmodium relictum* from doves and pigeons, etc. *Ibid.*, 27:380.

——— and ROUDABUSH, R. L., 1937. Some blood parasites from Nebraska birds. Am. Midland Natural., 18:1005.

——— and YOUNG, M. D., 1941. The taxonomy of the human malaria parasites, etc. A.A.A. Sci., Publ., No. 15:19.

——— *et al.*, 1958. Chloroquine or pyrimethamine in salt as a suppresive against sporozoite-induced vivax malaria. Bull. World Health Organiz., 19:53.

——— *et al.*, 1961. Transmission of the M strain of *Plasmodium cynomolgi* to man. Am. J. Trop. Med. Hyg., 10:673.

COGGESHALL, L. T., 1938. *Plasmodium lophurae*, etc. Amer. J. Hyg., 27:615.

——— 1941. Infection of *Anopheles quadrimaculatus* with *Plasmodium cynomolgi*, and with *P. lophurae*. Am. J. Trop. Med., 21:525.

CONTACOS, P. G., *et al.*, 1962. Man to man transfer of two strains of *Plasmodium cynomolgi* by mosquito bite. Am. J. Trop. Med. Hyg., 11:186.

COOK, ALICE R., 1954. The gametocyte development of *Leucocytozoon simondi*. Proc. Helminth. Soc. Washington, 21:1.

COWAN, A. B., 1955. The development of megaloschizonts of *Leucocytozoon simondi*. J. Protozool., 2:158.

DARLING, S. T., 1924. The spleen index in malaria. South. Med. J., 17:590.

——— 1926. Splenic enlargement as a measure of malaria. Ann. Clin. Med., 4:695.

DENNIS, E. W., 1932. The life cycle of *Babesia bigemina*, etc. Univ. Cal., Publ. Zool., 36:263.

DOFLEIN, F., 1901. Die Protozoen als Parasiten und Krankheits-erreger. Jena.

DOWNS, W. G. *et al.*, 1943. A malaria survey of Trinidad and Tobago British West Indies. J. Nat. Mal. Soc., 2:1.

DUNCAN, D., *et al.*, 1960. Electron microscope observations on malarial oocysts (*Plasmodium cathemerium*). J. Protozool., 7:18.

EARLE, W. C., 1930. Malaria in Puerto Rico. Am. J. Trop. Med., 10:207.

——— 1939. The epidemiology of malaria with special reference to Puerto Rico. P. R. J. Pub. Health Trop. Med., 15:3.

EYLES, D. E., 1960. The erythrocytic cycle of *Plasmodium cynomolgi*, etc. Am. J. Trop. Med. Hyg., 9:543.

FALLIS, A. M., 1946. *Plasmodium circumflexum* in ruffed grouse in Ontario. J. Parasit., 32:345.

———— *et al.*, 1951. Life history of *Leucocytozoon simondi*, etc. Canad. J. Zool., 29:305.

———— *et al.*, 1956. Further observations on the transmission and development of *Leucocytozoon simondi*. Canad. J. Zool., 34:389.

———— and BENNETT, G. F., 1959. Transmission of *Leucocytozoon bonasae* Clarke, etc. *Ibid.*, 36:533.

———— ———— 1962. Observations on the sporogony of *Leucocytozoon mirandae*, etc. *Ibid.*, 40:395.

————and WOOD, D. M., 1957. Biting midges as intermediate hosts for Haemoproteus of ducks. *Ibid.*, 34:425.

FERREBEE, J. W., and GEIMAN, Q. M., 1946. Studies on malaria parasites. III. J. Infect. Dis., 78:173.

GARNHAM, P. C. C., 1948. Exoerythrocytic schizogony in malaria. Trop. Dis. Bull., 45:831.

———— 1948a. The developmental cycle of *Hepatocystes (Plasmodium) kochi* in the monkey host. Tr. Roy. Soc. Trop. Med. Hyg., 41:601.

———— 1950. Blood parasites of East African vertebrates, etc. Parasitology, 40:328.

———— BRAY, R. S., *et al.*, 1955. The pre-erythrocytic stage of *Plasmodium ovale*. Tr. Roy. Soc. Trop. Med. Hyg., 49:158.

———— BIRD, R. G., and BAKER, J. R., 1962. Electron microscope studies of motile stages of malaria parasites. III. *Ibid.*, 56:116.

————— HEISCH, R. B., *et al.*, 1962. The midge as a host of monkey malaria parasites; unusual site of development between eyes and brain of insects. *Ibid.*, 56:1.

GONDER, R., and BERENBERG-GOSSLER, H., 1908. Untersuchungen über Malariaplasmodien der Affen. Malaria, 1:47.

GREENBERG, J., and KENDRICK, L. P., 1957. Parasitemia and survival in inbred strains of mice infected with *Plasmodium berghei*. J. Parasit., 43:413.

———— ———— 1957a. Some characteristics of *Plasmodium berghei*. *Ibid.*, 43:420.

———— ———— 1958. Parasitemia and survival in mice infected with *Plasmodium berghei*. *Ibid.*, 44:492.

HACKETT, L. W., 1944. Spleen measurement in malaria. J. Nat. Mal. Soc., 3: 121.

HARTMAN, E., 1927. Three species of bird malaria. Arch. Protist., 60:1.

———— and WEST, E., 1941. Modification of *Plasmodium cathemerium* when transferred from canaries into ducks. Amer. J. Hyg., 34:27.

HERMAN, C. M., 1938. The relative incidence of blood protozoa in some birds from Cape Cod. Tr. Am. Micr. Soc., 57:132.

———— 1938a. *Leucocytozoon anatis* Wickware, a synonym for *L. simondi* Mathis and Léger. J. Parasit., 24:472.

—— 1938b. Mosquito transmission of avian malaria parasites. Amer. J. Hyg., 27:345.

—— 1944. The blood parasites of North American birds. Bird Banding, 15:89.

—— 1954. Haemoproteus infections in water fowl. Proc. Helminth. Soc., Washington, 21:37.

—— and BISCHOFF, A. I., 1949. The duration of Haemoproteus infection in California quail. California Fish. Game, 35:293.

HEWITT, R., 1938. The cultivation of *Plasmodium cathemerium* for one asexual generation on inspissated egg and rabbit serum. Amer. J. Hyg., 27:341.

—— 1939. Splenic enlargement and infarction in canaries infected with a virulent strain of *Plasmodium cathemerium*. *Ibid.*, 30:49.

—— 1940. *Haemoproteus metchnikovi*, etc. Arch. Protist., 26:273.

—— 1940a. Studies on blood Protozoa obtained from Mexican wild birds. *Ibid.*, 26:287.

—— 1940b. Bird malaria. Amer. J. Hyg., Monogr. Ser. No. 15.

HOYTE, H. M. D., 1961. Initial development of infections with *Babesia bigemina*. J. Protozool., 8:462.

HUFF, C. G., 1930. *Plasmodium elongatum,* n. sp., etc. Amer. J. Hyg., 11:385.

—— 1934. Comparative studies on susceptible and insusceptible *Culex pipiens* in relation to infections with *Plasmodium cathemerium* and *P. relictum*. *Ibid.*, 19:123.

—— 1935. *Plasmodium hexamerium* n. sp., etc. *Ibid.*, 22:274.

—— 1942. Schizogony and gametocyte development in *Leucocytozoon simondi*, etc. J. Infect. Dis., 71:18.

—— 1947. Life cycle of malaria parasites. Ann. Rev. Microbiol., 1:43.

—— 1948. Exoerythrocytic stages of malarial parasites. Am. J. Trop. Med., 28:527.

—— and BLOOM, W., 1935. A malarial parasite infecting all blood and blood-forming cells of birds. J. Infect. Dis., 57:315.

—— and COULSTON, F., 1944. The development of *Plasmodium gallinaceum* from sporozoite to erythrocytic trophozoite. *Ibid.*, 75:231.

—— —— 1946. The relation of natural and acquired immunity of various avian hosts to the cryptozoites and metacryptozoites of *Plasmodium gallinaceum* and *P. relictum*. *Ibid.*, 78:99.

—— —— 1948. Symposium on exoerythrocytic forms of malarial parasites. II. J. Parasit., 34:264.

—— —— and CANTRELL, W., 1943. Malarial cryptozoites. Science, 97:286.

—— *et al.*, 1959. Susceptibility and resistance of avian and mosquito hosts to strains of *Plasmodium relictum* isolated from pigeons. J. Protozool., 6:46.

—— *et al.*, 1960. The morphology and behavior of living exoerythrocytic stages of *Plasmodium gallinaceum* and *P. fallax* and their host cells. J. Biophys. Biochem. Cytol., 7:93.

HUNNINEN, A. V., and YOUNG, M. D., 1950. Blood protozoa of birds at Columbia, South Carolina. J. Parasit., 36:258.

—— et al., 1950. The infection of anopheline mosquitoes by native avian malaria. J. Nat. Mal. Soc., 9:145.

JAMES, S. P., and TATE, P., 1938. Exoerythrocytic schizogony in *Plasmodium gallinaceum*. Parasitology, 30:128.

JEFFERY, G. M., 1944. Investigations on the mosquito transmission of *Plasmodium lophurae* Coggeshall. Amer. J. Hyg., 40:251.

—— and EYLES, D. E., 1955. Infectivity of mosquitoes to *Plasmodium falciparum*, etc. Am. J. Trop. Med. Hyg., 4:781.

—— and RENDTORFF, R. C., 1955. Preservation of viable human malaria sporozoites by low-temperature freezing. Exper. Parasitol., 4:445.

—— et al., 1952. Exo-erythrocytic stages of *Plasmodium falciparum*. Am. J. Trop. Med. Hyg., 1:917.

—— et al. 1955. A comparison of west African and west Pacific strains of *Plasmodium ovale*. Tr. Roy. Soc. Trop. Med. Hyg., 49:168.

—— et al., 1959. Early activity in sporozoite-induced *Plasmodium falciparum* infections. Ann. Trop. Med. & Parasit., 53:51.

JORDAN, H. B., 1957. Host resistance and regulation of the development of *Plasmodium lophurae* in pheasants, coots and domestic pigeons. J. Parasit., 43:395.

KHABIR, P. A., and MANWELL, R. D., 1955. Glucose consumption of *Plasmodium hexamerium. Ibid.*, 41:595.

KIKUTH, W., 1931. Immunobiologische und chemotherapeutische Studien an verschiedenen Stämmen von Vogelmalaria. Zentralbl. Bakt. Abt. I. Orig., 121:401.

KING, W. V., 1916. Experiments on the development of malaria parasites in three American species of Anopheles. J. Exper. Med., 23:703.

KITCHEN, S. F., 1938. The infection of reticulocytes by *Plasmodium vivax*. Am. J. Trop. Med., 18:347.

—— 1939. The infection of mature and immature erythrocytes by *P. falciparum* and *P. malariae. Ibid.*, 19:47.

—— and PUTNAM, P., 1943. Morphological studies of *Plasmodium falciparum* gametocytes of different strains in naturally induced infections. *Ibid.*, 23:163.

—— —— 1946. Observations on the character of the paroxysm in vivax malaria. J. Nat. Mal. Soc., 5:57.

KOMP, W. H. W., 1948. The anopheline vectors of malaria of the world. Proc. 4th Intern. Congr. Trop. Med. Malaria, p. 644.

KRUSE, W., 1890. Ueber Blutparasiten. Virchow's Arch., 121:359.

LAIRD, M., 1951. *Plasmodium lygosomae*, etc. J. Parasit., 37:183.

LAVERAN, A., 1899. Les hématozoaires endoglobulaires. Cinq. soc. biol. jub., p. 124.

—— and Lucet, 1905. Deux hématozoaires de la perdrix et du dindon. C. R. Acad. Sci., 191: 673.

Levine, N. D., 1961. Protozoan parasites of domestic animals and of man. Minneapolis.

—— and Hanson, H. C., 1953. Blood parasites of the Canada goose. J. Wildlife Manag., 17:185.

Lewert, R. M., 1952. Nucleic acids in plasmodia and the phosphorus partition of cells infected with *Plasmodium gallinaceum*. J. Infect. Dis., 91:125.

MacDougall, M. S., 1947. Cytological studies of Plasmodium: the male gamete. J. Nat. Mal. Soc., 6:91.

Maegraith, B., 1948. Pathological process in malaria and black water fever. Springfield, Illinois.

Manresa, M., Jr., 1953. The occurrence of phanerozoites of *Plasmodium lophurae* in blood-inoculated turkeys. J. Parasit., 39:1.

Manwell, R. D., 1929. Relapse in bird malaria. Am. J. Hyg., 9: 308.

—— 1934. The duration of malarial infection in birds. *Ibid.*, 19:532.

—— 1935. *Plasmodium vaughani* (Novy and MacNeal). *Ibid.*, 21:180.

—— 1935a. How many species of avian malaria parasites are there? Am. J. Trop. Med., 15:265.

—— 1940. Life-cycle of *Plasmodium relictum* var. *matutinum*. Am. J. Trop. Med., 20:859.

—— 1941. Avian toxoplasmosis with invasion of the erythrocytes. J. Parasit., 27:245.

—— 1949. *Plasmodium oti* and *P. hexamerium*. *Ibid.*, 35:561.

—— 1951. Exoerythrocytic schizogony in *Plasmodium hexamerium*. Amer. J. Hyg., 53:244.

—— 1955. Some evolutionary possibilities in the history of the malaria parasites. Indian J. Malariol., 9:247.

—— 1955a. The blood protozoa of 17 species of sparrows and other Fringillidae. J. Protozool., 2:21.

—— 1961. Protozoology. New York.

—— and Herman, C., 1935. The occurrence of the avian malarias in nature. Am. J. Trop. Med., 15:661.

—— and Khabir, P. A., 1954. Further studies in the chemotherapy of *Plasmodium hexamerium* infections in ducks. J. Protozool., 1:105.

—— and Loeffler, C. A. 1961. Glucose consumption by *Haemoproteus columbae*. J. Parasit., 47:285.

Maurer, G., 1922. Die Malaria perniciosa. Centralbl. Bakt. Orig., 32:695.

Mayer, M., 1907. Ueber Malaria beim Affen. Med. Klin. Berlin., 3:3579.

—— 1908. Ueber Malariaparasiten bei Affen. Arch. Protist., 12:314.

Mayne, B., and Young, M. D., 1941. The technic of induced malaria as used in the South Carolina State Hospital. Ven. Dis. Inform., 22:271.

McGhee, R. B., 1954. The infection of duck and goose erythrocytes by the mammalian malaria parasite, *Plasmodium berghei*. J. Protozool., 1:145.

—— 1956. A study of changes occurring in *Plasmodium lophurae* after three years of continuous existence in mice. *Ibid.*, 3:122.

—— 1959. The effect of lowered temperatures of the host on the reproductive activity of *Plasmodium cathemerium*. *Ibid.*, 6:84.

MERCADO, T. I., and COATNEY, G. R., 1951. The course of the blood-induced *Plasmodium berghei* infection in white mice. J. Parasit., 37:479.

MEYER, H., and MUSACCHIO, O., 1960. Electron microscope study of the exoerythrocytic form of *Plasmodium gallinaceum*, etc. J. Protozool., 7:222.

MILAM, D. F., and COGGESHALL, L. T., 1938. Duration of *Plasmodium knowlesi* infections in man. Am. J. Trop. Med., 18:331.

MOHAMMED, A. H. H., 1958. Systematic and experimental studies on protozoal blood parasites of Egyptian birds. 2 vols. Cairo University Press.

MUDROW, L., and REICHENOW, E., 1944. Endotheliale und erythrocytäre Entwicklung von *Plasmodium praecox*. Arch. Protist., 97:101.

MULLIGAN, H. W., 1935. Description of two species of monkey Plasmodium isolated from *Silenus irus*. *Ibid.*, 84:285.

NIGRELLI, R. F., 1930. *Dactylosoma jahni* sp. nov., etc. Ann. Protist., 3:1.

NÖLLER, W., 1920. Die neueren Ergebnisse der Haemoproteus-Zuchtung. *Ibid.*, 41:149.

NOVY, F. G., and MACNEAL, W. J., 1904. Trypanosomes and bird malaria. Amer. Med., Sec. I, 8:932.

O'ROKE, E. C., 1934. A malaria-like disease of ducks caused by *Leucocytozoon anatis* Wickware. Univ. Michigan Sch. Forest. Cons. Bull., No. 4.

PAMPANA, E. J., and RUSSELL, P. F., 1955. Malaria—a world problem. World Health Organiz., Geneva.

PATTON, W. H., 1895. The name of the Southern or splenic fever parasite. Am. Nat., 29:498.

PAUL, J. H., and BELLERIVE, A., 1947. A malaria reconnaissance of the Republic of Haiti. J. Nat. Mal. Soc., 6:41.

PELÁEZ, D., REYES, R. P., and BARRERA, A., 1948. Estudios sobre hematozoarios. I. An. Esc. Noc. Cienc. Biol., 5:197.

PORTER, R. J., 1942. The tissue distribution of exoerythrocytic schizonts in sporozoite-induced infections with *Plasmodium cathemerium*. J. Infect. Dis., 71:1.

—— and HUFF, C. G., 1940. Review of the literature on exoerythrocytic schizogony in certain malarial parasites and its relation to the schizogonic cycle in *Plasmodium elongatum*. Am. J. Trop. Med., 20:869.

RAFFAELE, G., 1934. Un ceppo italiano di *Plasmodium elongatum*. Riv. Malariol., 13:332.

—— 1934a. Sul comportamento degli sporozoiti nel sangue del-l'ospite. *Ibid.*, 13:395, 706.

RATCLIFFE, H. L., 1927. The relation of *Plasmodium vivax* and *P. praecox* to the red blood cells of their respective hosts as determined by sections of blood cells. Am. J. Trop. Med., 7:383.

—— 1928. The relation of *P. falciparum* to the human red blood cell as determined by sections. *Ibid.*, 8:559.

RAY, H. N., and MALHATRA, M. N., 1960. Further observations on the mode of development of gametocytes of *Haemoproteus columbae*, etc. Bull. Calcutta Sch. Trop. Med., 8:57.

REES, C. W., 1934. Characteristics of the piroplasms: *Babesia argentina* and *B. bigemina* in the United States. J. Agr. Res., 48:427.

REGENDANZ, P., 1936. Ueber den Entwicklungsgang von *Babesia bigemina* in der Zecke *Boophilus microplus*. Centralbl. Bakt. Parasitenk., Orig., 137:423.

—— and REICHENOW, E., 1933. Die Entwicklung von *Babesia canis* in *Dermacenter reticulatus*. Arch. Protist., 79:50.

REICHENOW, E., 1937. Ueber die Entwicklung von *Theileria parva*, etc. Centralbl. Bakt. Orig., 140:223.

RIMINGTON, C., and FULTON, J. D., 1947. The pigment of the malarial parasites, *P. knowlesi* and *P. gallinaceum*. Biochem. J., 41: 619.

ROSS, R., 1928. Studies on malaria. London.

RUDZINSKA, M. A., and TRAGER, W., 1957. Intracellular phagotrophy by malaria parasites: an electron microscope study of *Plasmodium lophurae*. J. Protozool., 4:190.

—— —— 1959. Phagotrophy and two new structures in the malaria parasite *Plasmodium berghei*. J. Biophys. Biochem. Cytol., 6:103.

—— —— 1961. The rôle of the cytoplasm during reproduction in a malarial parasite (*Plasmodium lophurae*), etc. J. Protozool., 8:307.

RUIZ, A. 1959. Sobre la presencia de un Dactylosoma en *Bufo marinus*. Rev. Biol. Trop., 7:113.

RUSSELL, P. F., 1934. Malaria and Culicidae in the Philippine Islands: History and critical bibliography, 1893–1933. Dept. Agr. Comm., P.I., Tech. Bull., No. 1.

—— 1935. The small spleen in malaria survey. Am. J. Trop. Med., 15:11.

—— 1935a. Epidemiology of malaria in the Philippines. Am. Pub. Health Assn., 26:1.

—— 1943. Malaria and its influence on world health. Bull. New York Acad. Med., 19:599.

—— 1952. The present status of malaria in the world. Am. J. Trop. Med. Hyg., 1:111.

—— 1952a. Malaria: basic principles briefly stated. Springfield, Illinois.

—— and JACOB, V. P., 1942. On the epidemiology of malaria in the Nilgiris district, Madras Presidency. J. Med. Inst. India, 4:349.

—— and MOHAN, B. N., 1942. Some mosquito hosts to avian plasmodia with special reference to *P. gallinaceum*. J. Parasit., 28:127.

——, WEST, L. S., and MANWELL, R. D., 1946. Practical malariology. Philadelphia.

SCHAUDINN, F., 1902. Studien über krankheitserregende Protozoen. II. *Plasmodium vivax* Grassi. Arb. kaiserl. Gesundh., 19:169.

SCHMIDT, L. H., *et al.*, 1961. The transmission of *Plasmodium cynomolgi* to man. Am. J. Trop. Med. Hyg., 10:677.

SCHÜFFNER, W., 1899. Beitrag zur Kenntnis der Malaria. Deutsche Arch. klin. Med., 64:428.

SERGENT, Ed., 1949. Sur deux cycles évolutifs insexués des Plasmodium chez les paludéens. C. R. Acad. Sci., Paris, 229:455.

SERGENT, ED. and ET., and CATANEL, A., 1928. Sur un parasite nouveau du paludisme des oiseaux. *Ibid.*, 186:809.

SERGENT, ET. and ED., 1922. Étude expérimentale du paludisme des oiseaux. Arch. Inst. Pasteur Afr. Nord, 2:320.

SHERMAN, I. V., and HULL, R. W., 1960. Serum alterations in avian malaria. J. Protozool., 7:171.

——————— 1960a. The pigment (hemozoin) and proteins of *Plasmodium lophurae*. *Ibid.*, 7:409.

SHORTT, H. E., and GARNHAM, P. C. C., 1948. The pre-erythrocytic development of *P. cynomolgi* and *P. vivax*. Tr. Roy. Soc. Trop. Med. Hyg., 41:785.

——— *et al.*, 1951. The pre-erythrocytic stage of *Plasmodium falciparum*. Tr. Roy. Soc. Trop. Med. Hyg., 44:405.

SIMMONS, J. S., CALLENDER, G. R., *et. al.*, 1939. Malaria in Panama. Amer. J. Hyg., Monogr. Ser. No. 13.

SINTON, J. A., and MULLIGAN, H. W., 1932. A critical review of the literature relating to the identification of the malarial parasites recorded from monkeys of the families Cercopithecidae and Colobidae. Rec. Mal. Surv. India, 3:357.

SMITH, T., and KILBORNE, F. L., 1893. Investigations into the nature, causation and prevention of Texas or Southern cattle fever. U.S.D. Agr., Bur. An. Ind. Bull., No. 1.

STARCOVICI, C., 1893. Bemerkungen über den durch Babes entdeckten Blutparasiten, etc. Centralbl. Bakt. I. Orig., 14:1.

STRATMAN-THOMAS, W. K., 1940. The influence of temperature on *Plasmodium vivax*. Am. J. Trop. Med., 20:703.

TALIAFERRO, W. H., and KLÜVER, C., 1940. The hematology of malaria (*P. brasilianum*) in Panamanian monkeys. 1, 2. J. Infect. Dis., 67:121.

——— and MULLIGAN, H. W., 1937. The histopathology of malaria with special reference to the function and origin of the macrophages in defence. Indian Med. Res., Memoires, 29:1.

——— and LUCY, G., 1934. Morphology, periodicity and course of infection of *P. brasilianum* in Panamanian monkeys. Amer. J. Hyg., 20:1.

——————— 1947. Asexual reproduction of *P. cynomolgi* in rhesus monkeys. J. Infect. Dis., 80:78.

——————— (1949). Asexual reproduction of *P. knowlesi* in rhesus monkeys. *Ibid.*, 85:107.

TERZIAN, L. A., 1941. Studies on *P. lophurae,* a malarial parasite in fowls. I. Amer. J. Hyg., 33:1.

THOMPSON, P. E., and HUFF, C. G., 1944. A saurian malarial parasite, *P. mexicanum,* n. sp., etc. J. Infect. Dis., 74:48.

———— ———— 1944a. Saurian malarial parasites of the United States and Mexico. *Ibid.,* 74:68.

THOMSON, J. G., 1928. Stippling of the red cells in malaria. Proc. Roy. Soc. Med., 21: 464.

TOIT, P. J. D., 1918. Zur Systematik der Piroplasmen. Arch. Protist., 39:84.

TRAGER, W., 1948. The resistance of egg-laying ducks to infection by the malaria parasite *Plasmodium lophurae.* J. Parasit., 34:389.

———— 1950. Studies on the extracellular cultivation of an intracellular parasite (avian malaria). I. J. Exper. Med., 92: 349.

———— 1954. Coenzyme A and the malaria parasite *Plasmodium lophurae.* J. Protozool., 1:231.

VINCKE, I. H., and LIPS, M., 1948. Un nouveau plasmodium d'un rougeur sauvage du Congo, *P. berghei,* n. sp. Ann. Soc. Belge méd. trop., 28:97.

WAMPLER, F. J., 1930. A preliminary report on the early effects of plasmochin on *P. cathemerium.* Arch. Protist., 69:1.

WARREN, A. J., and COGGESHALL, L. T., 1937. Infectivity of blood and organs in canaries after inoculation with sporozoites. Amer. J. Hyg., 26:1.

WEATHERSBY, A. B., 1960. Experimental infection of *Aedes aegypti* with exoerythrocytic stages of *Plasmodium gallinaceum.* Exper. Parasitol., 9:334.

WEISS, M. L., and MANWELL, R. D., 1960. In vitro cultivation of *Plasmodium elongatum* in duck tissues. J. Protozool., 7:342.

WENYON, C. M., 1926. Protozoology. Vol. 2. London and Baltimore.

WHARTON, R. H., et al., 1962. *Anopheles leucosphyrus* identified as a vector of monkey malaria in Malaya. Science, 137:758.

WICKWARE, A. B., 1915. Is *Leucocytozoon anatis* the cause of a new disease in ducks? Parasitology, 8:17.

WILCOX, A., et al., 1954. The Donaldson strain of malaria. II. Amer. J. Trop. Med. Hyg., 3:638.

WOLCOTT, G. B., 1954. Nuclear structure and division in the malaria parasite, *Plasmodium vivax.* J. Morphol., 94:353.

————1955. Chromosomes of the four species of human malaria. J. Heredity, 46:53.

————1957. Chromosome studies in the genus Plasmodium. J. Protozool., 4:48.

———— et al., 1958. On the position of malarial parasites with relation to erythrocytes. Tr. Roy. Soc. Trop. Med. Hyg., 52:87.

WOLFSON, F., 1936. *Plasmodium oti* n. sp., etc. Amer. J. Hyg., 24:94.

——— 1940. Virulence and exoerythrocytic schizogony in four species of Plasmodium in domestic ducks. J. Parasit., 26: Suppl.: 28.

——— 1941. Avian hosts for malaria research. Quart. Rev. Biol., 16:462.

——— and WINTER, M. W., 1946. Studies of *P. cynomolgi* in the rhesus monkey, *Macaca mulatta.* Amer. J. Hyg., 44:273.

WOOD, S. F., and HERMAN, C. M., 1943. The occurrence of blood parasites in birds from southwestern United States. J. Parasit., 29:187.

YORKE, W., and MACFIE, J. W. S., 1924. Observations on malaria made during treatment of general paresis. Tr. Roy. Soc. Trop. Med. Hyg., 18:13.

Chapter 28
Order 4. **Haplosporida** Caullery and Mesnil

THE members of this order are characterized by production of simple spores. The spore is usually subspherical to ellipsoidal in form and covered by a resistant membrane which may be ridged or drawn out into a more or less long process. In a few species the spore possesses a lid at one pole. The sporoplasm is uninucleate and fills the intrasporal cavity. These organisms are cytozoic, histozoic or coelozoic parasites of invertebrates and lower vertebrates. Very little is known about their life cycle.

The development of a haplosporidan, *Ichthyosporidium giganteum,* as worked out by Swarczewsky, is as follows (Fig. 275): The spores germinate in the alimentary canal of the host fish and the emerged amoebulae make their way to the connective tissue of various organs (*a*). These amoebulae grow and their nuclei multiply in number, thus forming plasmodia. The plasmodia divide into smaller bodies, while the nuclei continue to divide (*b-e*). Presently the nuclei become paired (*f, g*) and the nuclear membranes disappear (*h*). The plasmodia now break up into numerous small bodies, each of which contains one set of the paired nuclei (*i, j*). This is the sporont (*j*) which develops into two spores by further differentiation (*k-o*).

Thus the development of this group as known at present does not show any close relationship with three orders just considered. However, because of the formation of simple spores, the group is included in Sporozoa.

Genus **Haplosporidium** Caullery and Mesnil. After growing into a large form, plasmodium divides into uninucleate bodies, each of which develops into a spore; spore truncate with a lid at one end; envelope sometimes prolonged into processes; in aquatic annelids and molluscs. (Revision and species of the genus and restoration of Minchinia, Sprague, 1963, 1963a.)

H. chitonis (Lankester) (Fig. 276, *a, b*). In liver and connective tissue of *Craspidochilus cinereus;* spores oval, 10µ by 6µ; envelope with two prolonged projections.

H. limnodrili Granata (Fig. 276, *c*). In gut epithelium of *Limnodrilus udekemianus;* spores 10-12µ by 8-10µ.

H. nemertis Debaisieux (Fig. 276, *d*). In connective tissue of *Lineus bilineatus;* spores oval with a flat operculum, but without any projections of envelope, 7µ by 4µ.

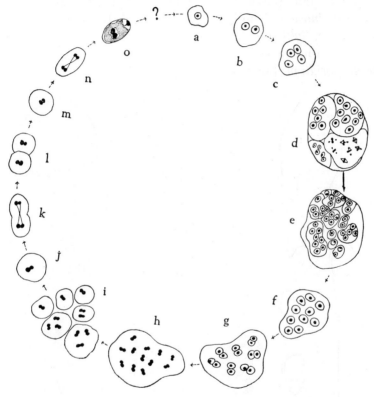

FIG. 275. The development of *Ichthyosporidium giganteum* (Swarczewsky). a–e, schizogony; f–n, sporogony; o, stained spore, × about 1280.

H. heterocirri C. and M. (Fig. 276, *e*). In gut epithelium of *Heterocirrus viridis;* mature organisms 50-60µ by 30-40µ; spores 6.5µ by 4µ.

H. scolopli C. and M. (Fig. 276, *f*). In *Scoloplos mülleri;* fully grown form 100-150µ by 20-30µ; spores 10µ by 6.5u.

H. vejdovskii C. and M. (Fig. 276, *g*). In a freshwater oligo-chaete, *Mesenchytraeus flavus;* spores 10-12µ long.

H. pickfordi Barrow. In the digestive gland of snails: *Lymnaea emerginata, Physa parkeri* and *Heliosoma campanulata;* sporocyst 25-60µ, contains 78-221 spores; spores 8.5-10µ by 3.8-4.6µ (Barrow, 1961).

Genus **Urosporidium** Caullery and Mesnil. Similar to *Haplo-sporidium*, but spherical spore with a long projection.

U. fuliginosum C. and M. (Fig. 276, *h, i*). In the coelom of the polychaete, *Syllis gracilis;* rare.

Genus **Anurosporidium** Caullery and Chappellier. Similar to *Haplosporidium*, but operculate spore spherical.

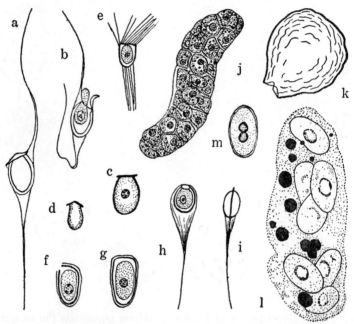

FIG. 276. a, b, *Haplosporidium chitonis,* ×1000 (Pixell-Goodrich); c, *H. limnodrili,* ×1000 (Granata); d, *H. nemertis,* ×1000 (Debaisieux); e, *H. heterocirii,* ×1000 (Caullery and Mesnil); f, *H. scolopli,* ×1000 (Caullery and Mesnil); g, *H. vejdovskii,* ×1000 (Caullery and Mesnil); h, i, *Urosporidium fuliginosum,* ×1000 (Caullery and Mesnil); j, k, *Bertramia asperospora* (j, cyst with spores; k, empty cyst), ×1040 (Minchin); l, m, *Coelosporidium periplanetae* (l, trophozoite with spores and chromatoid bodies), ×2540 (Sprague).

A. *pelseneeri* C. and C. In sporocyst of a trematode parasitic in *Donax trunculus;* schizogony intracellular; cysts extracellular, with up to 200 spores; spores about 5μ long.

Genus **Bertramia** Caullery and Mesnil. Parasitic in aquatic worms and rotifers; sausage-shaped bodies in coelom of host; spherical spores which develop in them, possess a uninucleate sporoplasm and a well-developed membrane.

B. *asperospora* (Fritsch) (Fig. 276, *j*, *k*). In body cavity of rotifers: Brachionus, Asplanchna, Synchaeta, Hydatina, etc.; fully grown vermicular body 70-90μ with 80-150 spores.

B. *capitellae* C. and M. In the annelid *Capitella capitata;* spores 2.5μ in diameter.

B. *euchlanis* Konsuloff. In coelom of rotifers belonging to the genus Euchlanis.

Genus **Ichthyosporidium** Caullery and Mesnil. In fish; often looked upon as Microsporida, as the organism develops into large bodies in body muscles, connective tissue, or gills, which appear as conspicuous "cysts," that are surrounded by a thick wall and contain numerous spores (Distribution, Porter and Vinall, 1956.)

I. *giganteum* (Thélohan) (Fig. 275). In various organs of *Crenilabrus melops* and *C. ocellatus;* cysts 30μ-2 mm. in diameter; spores 5-8μ long.

I. *hertwigi* Swarczewsky. In *Crenilabrus paro;* cysts 3-4 mm. in diameter in gills; spores 6μ long.

I. sp. Porter and Vinall. In the muscle and areolar tissue of *Hyphessobrycon innesi;* the cause of "neon" fish disease (Porter and Vinall, 1956).

Genus **Coelosporidium** Mesnil and Marchoux. In coelom of Cladocera or Malpighian tubules of cockroach; body small, forming cysts; spores resemble microsporidian spores; but without a polar filament.

C. *periplanetae* (Lutz and Splendore) (*C. blattellae* Crawley) (Fig. 276, *l*, *m*). In lumen of Malpighian tubules of cockroaches; common; spores 5.5-7.5μ by 3-4μ. (Cytology, Sprague, 1940.)

Genus **Coleospora** Gibbs. Trophozoites intra- and extra-cellular, multinucleate; five to ten zygotes formed in each; each zygote develops into a spore containing a sporoplasm; binucleate spore elongate, curved, with corrugated membrane (Gibbs, 1959).

C. binucleata G. In the Malpighian tubules of the beetle, *Gonocephalum arenarium;* spores 10.5μ by 2.5μ (Gibbs, 1959).

Genera of Unknown Taxonomic Position

Genus **Sarcocystis** Lankester. The organisms placed in this genus are all parasitic in the muscles in man, other mammals, birds and reptiles and had been known for a long time. The infected host muscles are characterized by the presence of opaque white masses (Miescher's tubes) (Fig. 277) which may measure up

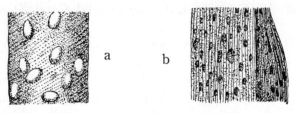

Fig. 277. a, *Sarcocystis tenella* in the oesophagus of sheep; b, S. *miescheriana* in the muscle of pig; ×1 (Schneidemühl from Doflein).

to several millimeters in length. The parasitic mass is surrounded by a wall and the body is divided into numerous compartments which contain crescentic or banana-shaped bodies (Rainey's corpuscles) (Fig. 278). These bodies have been considered as naked spores. Hence Sarcocystis, the only genus of order Sarcosporida (Bütschli, 1882), had been included in Sporozoa.

Recently, Ludvik (1956-1960) showed that in *Sarcocystis tenella* and S. *miescheriana* the so-called spores are not spores, but vegetative forms. They are capable of movement by twisting or flexing of the body. At the slightly pointed (anterior) end, there is a polar ring, and within it a hollow cone named conoid. From the ring, twnety-two to twenty-six fibrils run to the other end of the body. A nucleus, mitochondria and central granules are found in the cytoplasm. In addition, he found that these bodies multiply by longitudinal division. Thus according to Ludvik's observations, Sarcocystis does not belong to Sporozoa, and its affinity to other genera of protozoa remains unknown.

The life cycle of Sarcocystis is still not known. After seeing Sarcocystis in some mammals and birds in Panama Canal Zone, Darling (1915) suggested that they may be some protozoa, especially Microsporida of insects, which entered a vertebrate host by way of insect droppings on grass and cannot complete their development by being "permanently sidetracked." However, Scott (1920, 1930) showed that insects including honey bees infected by *Nosema apis* had no relationship to the infection of lamb by *S. tenella*.

Spindler and Zimmerman (1946) reported isolation of Aspergillus-like fungus from *S. miescheriana* which when fed or injected into young pigs, twenty-five of fifty experimental animals developed infection after four to six months, and come to conclude that Sarcocystis of pigs, sheep and ducks are fungi related to Aspergillus (Spindler, 1947). However, morphological characters of the organisms do not support Spindler's view, and furthermore, attempts by others to repeat and confirm these observations have failed. Therefore, the weight of evidence is that Sarcocystis are not fungi.

Many species have been reported from various host animals, mainly on the basis of the difference in host species. (Species, Wenyon, 1926, Babudieri, 1932; in lizards, Ball, 1944; morphology, Moroff, 1915, Noguer and Galiano, 1928, Ludvik, 1960; Sarcocystis in Panama, Darling, 1915; in cattle, Wang, 1950; distribution, Holz, 1957; Sebek, 1960.)

S. lindemanni (Rivolta). A few cases of Sarcocystis infection have been reported from man in muscle cells of larynx (Baraban and St. Remy), of biceps and tongue (Darling), of heart (Manifold), of breast (Vasudevan), etc. There seem to be dimensional discrepancies of organisms observed by different investigators. The dimensions of parasitic masses and of trophozoites are as follows: Parasites 1.6 mm. by 170μ and banana-shaped trophozoites 8-9μ long (Baraban and St. Remy); parasites 84μ by 27μ and trophozoites 4.25μ by 1.75μ (Darling); parasites spherical, 500μ in diameter and trophozoites over 10μ long (Manifold); parasites 5.3 cm. by 320μ and trophozoites 8.33μ by 1.6μ (Vasudevan). The parasitic masses are oval to spindle in form and imbedded in the muscle cells which are distended, and may appear white-streaked to naked eye. Seen in sections, the body is divided into compartments. Gilmore, Kean and Posey (1942) have recently found three bodies in sectioned heart muscles of an eleven year old child who died from an unknown cause, and considered them as sarcosporidian bodies. They measured 25μ by 19μ, 57μ by 30μ, and 41μ by 25μ in cross sections; there were no septa within the bodies; minute bodies present in the masses were mostly rounded and about 1μ in diameter, though a few were crescentic. The questions such as what species infect man, how man becomes infected, etc., are unanswered at present.

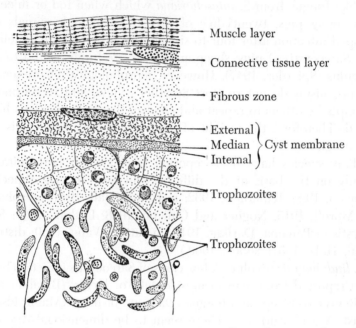

Fig. 278. Portion of a cyst of *Sarcocystis tenella* in sheep, × about 1000 (Alexeieff).

S. tenella Railliet (Figs. 277, *a*; 278). In the muscles of tongue, pharynx, oesophagus, larynx, neck, heart, etc., of sheep; parasitic masses 40μ-2 cm long; trophozoites sickleform (Alexeieff, 1913; Scott, 1943). (Electron microscope study, Ludvik, 1956.)

S. miescheriana (Kühn) (Fig. 277, *b*). In the muscles of pig; cysts up to 3-4 mm. by 3 mm.; envelope striated; trophozoites reniform. Musfeldt (1950) found fifteen of 264 pig diaphragms examined were infected by a Sarcocystis. The pigs were all garbage-fed animals. Sarcocystis infections were also noticed in the rats from one of the piggeries from which infected pigs were obtained. (Electron microscope study, Ludvik, 1961.)

S. *bertrami* Doflein. In the muscles of horse; similar to S. *miescheriana;* parasitic mass up to 9-10 mm.; envelope striated.

S. *muris* Blanchard. In body muscles of rats and mice; parasitic masses up to 3 cm. long; trophozoites 13-15µ by 2.5-3µ; transmissible to guinea pig (Negri) which shows experimental infection in muscles in 50-100 days after feeding on infected muscles.

S. *rileyi* Stiles. In muscles of various species of ducks and other birds; parasites in muscle, opaque white in color and measure up to 5 mm. by 2 mm.; trophozoites are sausage-shaped and 8-10µ by about 3µ. (Distribution, Erickson, 1940; Quortrup and Shillinger, 1941; Herman and Bolander, 1944; Salt, 1958.)

Genus **Toxoplasma** Nicolle and Manceaux. Typically minute intracellular parasites of various tissue cells of man, mammals and birds; small trophozoites are crescentic or banana-shaped with one end more attenuated than the other, measuring 5-10µ by 2-4µ; electron microscope reveals the presence of a conoid and of a number of fibrils (toxonemes) extending for some distance posteriorly. Thus the organism is somewhat similar in structure to that of the trophozoites of Sarcocystis just mentioned. Multiplication is by longitudinal division or by *endodyogeny* in which two small daughter cells are formed by internal budding and grow in size, finally becoming free by rupture of the parent cell (Goldman, Carver and Sulzer, 1958) or by schizogony by which four daughter cells are produced (Gavin *et al.,* 1962), and a variable number of the parasites form cysts or pseudocysts. Toxoplasma do not possess any organelles of locomotion, although they appear to be capable of gliding or body flexion (Manwell and Drobeck, 1953). Thus, this genus which appears to have a structural similarity to Sarcocystis cannot be assigned to any taxon.

T. *gondii* N. and M. (Fig. 279). This organism was first described by Nicolle and Manceaux (1909) from a rodent, *Ctenodactylus gundi* in North Africa. Since that time, the same organism has been found in many wild and domestic mammals, birds and man. Apparently, it is very widely distributed. Clinical form of toxoplasmosis is fortunately comparatively rare in man, but infections without manifesting symptoms are said to be quite common. The natural mode of transmission is not yet known, although congenital transmission occurs. Experimental infections can be established by inoculation or by feeding infected tissues to various laboratory animals.

The introduced organisms multiply in number and then by way of the circulatory system, invade the tissues of various organs in which they continue to multiply and produce local necrosis. As the infection continues, the parasites begin to disappear gradually from various organs, which is considered to be due to the formation of antibodies in the blood. Usually, the parasites disappear rapidly from certain organs such as liver, spleen, lung, etc., but more slowly from the heart, and the parasites remain in brain for a long time as cysts.

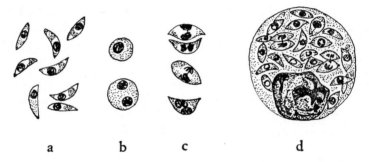

a   **b**   **c**   **d**

FIG. 279. *Toxoplasma gondii.* × about 1750. (Chatton and Blanc) a, isolated organisms; b, 2 trophozoites; c, organisms undergoing binary fission; d, a host cell with many organisms which developed by repeated binary fission.

Toxoplasma occurs in a variety of wild and domestic animals (Orio et al., 1959; Grünberg, 1959; Jaksch, 1959; Pridham and Belcher, 1958). The congenital toxoplasmosis in newborn infants is manifest by various symptoms such as encephalitis, hepatomegaly, jaundice, etc., and of high mortality (Feldman and Miller, 1956). (Infection in man, Sabin, 1942, Schwarz *et al.*, 1948, Remington *et al.*, 1960; nuclear division, Gangi and Manwell, 1961; cysts, Jacobs *et al.*, 1960; cysts in meat from swine, cattle and sheep, Jacob *et al.*, 1960a; slow freezing, Eyles *et al.*, 1956; diagnosis, Brooke, 1954, Eichenwald, 1962.)

References

ALEXEIEFF, A., 1913. Recherches sur Sarcosporidies. I. Arch. zool. expér. gén., 51:521.

BABUDIERI, B., 1932. I Sarcosporidi e le Sarcosporidiose. Arch. Protist., 78:421.

BALL, G. H., 1944. Sarcosporidia in southern California lizards. Tr. Am. Micr. Soc., 63:144.

BARABAN, L., and ST. REMY, G., 1894. Sur une cas de tubes psorospermiques observés chez l'homme. C. R. Soc. Biol., 10:201.

BARROW, J. H., 1961. Observations of a haplosporidian, *Haplosporidium pickfordi* sp. nov. in fresh water snails. Tr. Am. Micr. Soc., 80: 319.

BROOKE, M. M., 1954. The laboratory diagnosis of Toxoplasmosis. Bull. Publ. Health Lab., 12:109.

CAULLERY, M., and MESNIL, F., 1905. Recherches sur les Haplosporidies. Arch. zool. exper. gén., 4:101.

CRAWLEY, H., 1914. The evolution of *Sarcocystis muris* in the intestinal cells of the mouse. Proc. Acad. Nat. Sci., Philadelphia, 66:432.

―――― 1916. The zoological position of Sarcosporidia. *Ibid.,* 68:379.

CROSS, J. B., 1947. A cytological study of Toxoplasma, etc. J. Infect. Dis., 80:278.

DARLING, S. T., 1909. Sarcosporidiosis, with report of a case in man. Arch. Int. Med., 3:183.

―――― 1915. Sarcosporidia encountered in Panama. J. Parasit., 1:113.

―――― 1919. Sarcosporidiosis in an East Indian. *Ibid.,* 6:98.

DUBLIN, I. N., and WILCOX, A., 1947. Sarcocystis in *Macaca mulatta. Ibid.,* 33:151.

EICHENWALD, H. F., 1962. Advances in the laboratory diagnosis of toxoplasmosis. Ann. N.Y. Acad. Sci., 98:740.

ERICKSON, A. B., 1940. Sarcocystis in birds. Auk, 57:514.

EYLES, D. E., *et al.,* 1956. Preservation of *Toxoplasma gondii* by freezing. J. Parasit., 42:408.

GANGI, D. P., and MANWELL, R. D., 1961. Some aspects of the cytochemical anatomy of *Toxoplasma gondii.* J. Parasit., 47:291.

GAVIN, M. A., *et al.,* 1962. Electron microscope studies of reproducing and interkinetic Toxoplasma. J. Protozool., 9:222.

GIBBS, A. J., 1959. *Coleospora binucleata* gen. nov., sp. nov., etc. Parasitology, 49:522.

GILMORE, H. R., JR., *et al.,* 1942. A case of sarcosporidiosis with parasites found in heart. Am. J. Trop. Med., 22:121.

GOLDMAN, M., *et al.,* 1958. Reproduction of *Toxoplasma gondii* by internal budding. J. Parasit., 44:161.

GRÜNBERG, W., 1959. Toxoplasmose beim Bennett-Kaenguruh und einem Klippschliefer. Wien. Tieraerztl. Monatschr. 46:586.

HERMAN, C. M., and BOLANDER, G. L., 1943. A parasite in the muscles of ducks in California. California Fish and Game, 29:148.

HOGAN, M. J., 1951. Ocular toxoplasmosis. New York.

HOLZ, J., 1957. Die Verbreitung der Sarcosporidien in Wirtsorganisms. Hemera Zoa, 64:47.

JACOBS, L., 1953. The biology of Toxoplasma. Am. J. Trop. Med. Hyg., 2:365.

———— 1956. Increasing knowledge on human toxoplasmosis. Amer. J. Clin. Pathol., 26:168.

———— and JONES, F. E., 1950. The parasitemia in experimental toxoplasmosis. J. Infect. Dis., 87:78.

JAKSCH, W., 1959. Klinische and serologische Beobachtungen bei Toxoplasmose von Bennett-Kaenguruhs in einen Tiergarten. Wien. Tieraerztl. Monatsschr., 46:593.

KAUFMAN, H. E., *et al.*, 1959. Strain differences of *Toxoplasma gondii.* J. Parasit., 45:189.

LAMBERT, S. W., 1927. Sarcosporidial infection of the myocardium in man. Am. J. Path., 3:663.

LUDVIK, J., 1956. Vergleichende elektronenoptische Untersuchungen an *Toxoplasma gondii* und *Sarcocystis tenella.* Centralbl. Bakt. Parasitenk., Orig., 166:60.

———— 1958. Elektronenoptische Befunde zur Morphologie der Sarkosporidien. *Ibid.,* 172:330.

———— 1959. Die Silberfibrillenstruktur bei *Toxoplasma gondii* und *Sarcocystis tenella.* Zeitschr. Parasitenk., 19:311.

———— 1961. The electron microscopy of *Sarcocystis miescheriana* Kuhn. J. Protozool., 7:128.

MANTZ, F. A., JR., *et al.* 1949. Toxoplasmosis in Panama: report of two additional cases. Am. J. Trop. Med., 29:895.

MANWELL, R. D., 1939. Toxoplasma or exoerythrocytic schizogony in malaria? Riv. Malariol., 18:76.

———— and DROBECK, H. P., 1953. The behavior of Toxoplasma, with notes on its taxonomic status. J. Parasit., 39:577.

MOROFF, T., 1915. Zur Kenntnis der Sarkosporidien. Arch. Protist., 35:256.

MUSFELDT, I. W., 1950. A report on infection by *Sarcocystis* sp. in swine from Vancouver, Canada. Canad. J. Comp. Med. Vet. Sci., 14:126.

NICOLLE, C., and MANCEAUX, L., 1909. Sur un protozoaire nouveau du Gondi (Toxoplasma n. gen.). Arch. Inst. Pasteur, Tunis, 4:97.

NOGUER, D. S. A., and GALIANO, E. F., 1928. Contribucion al estudio de los Sarcosporidios. Mem. Real Acad. Cienc. y Artes, Barcelona, 21:257.

ORIO, J., *et al.*, 1959. Contributions à l'étude de la toxoplasmose en Afrique equatoriale. Bull. Soc. Pathol. Exot., 51:607.

PORTER, A., and VINALL, H. F., 1956. A protozoan parasite (*Ichthyosporidium* sp.) of the Neon fish *Hyphessobrycon innesi.* Proc. Zool. Soc. London, 126:397.

PRIDHAM, T. J., and BELCHER, J., 1958. Toxoplasmosis in mink. Canadian J. Comp. Med. & Vet. Sci., 22:99.

QUORTRUP, E. R., and SHILLINGER, J. E. 1941. 3000 wild bird autopsies in western lake areas. J. Amer. Vet. Med. Assoc., 99:382.

REMINGTON, J. S., *et al.*, 1960. Toxoplasmosis in the adult. New England J. Med., 262:180, 237.

SABIN, A. B., 1942. Toxoplasmosis. A recently recognized disease of human beings. In De Sanctis: Advances in pediatrics.

SALT, W. R., 1958. *Sarcocystis rileyi* in sage grouse. J. Parasit., 44:511.

SCHWARZ, G. A., *et al.*, 1948. Toxoplasmic encephalomyelitis, a clinical report of six cases. Pediatrics, 1:478.

SCOTT, J. W., 1920. Notes and experiments on *Sarcocystis tenella*. III. J. Parasit., 6:157.

———— 1930. The Sarcosporidia: a critical review. *Ibid.*, 16:111.

———— 1943. Life history of Sarcosporidia, with particular reference to *Sarcocystis tenella*. Bull. Univ. Wyoming Exper. Stat., no. 259.

———— and O'ROKE, E. C., 1920. *Sarcocystis tenella*. Bull. Agr. Exper. Station, Univ. Wyoming, No. 124.

SEBEK, Z., 1960. Sarcocystis in insectivora and rodents (Czech.). Zool. Listy, 23:1.

SPINDLER, L. A., 1947. A note on the fungoid nature of certain internal structures of Miescher's sacs, etc. Proc. Helm. Soc. Washington, 14:28.

———— and ZIMMERMAN, H. E. JR., 1945. The biological status of Sarcocystis. J. Parasit., 31: suppl.: 13.

———— ———— and JAQUETTE, D. S., 1946. Transmission of Sarcocystis to swine. Proc. Helm. Soc. Wash., 13:1.

SPRAGUE, V., 1940. Observations on *Coelosporidium periplanetae* with special reference to the development of the spore. Tr. Am. Micr. Soc., 59:460.

———— 1963. Revision of genus Haplosporidium and restoration of genus Minchinia. J. Protozool., 10:263.

———— 1963a. *Minchinia louisiana* n. sp., a parasite of *Panopeus herbstii*. *Ibid.*, 10:267.

SWARCZEWSKY, B., 1914. Ueber den Lebenscyklus einiger Haplosporidien. Arch. Protist., 33:49.

TAKOS, J. T., 1957. Notes on Sarcosporidia of birds in Panama. J. Parasit., 43:183.

TEICHMANN, E., 1912. Sarcosporidia. Prowazek's Handbuch der pathog. Protozoen. Part 3:345.

WANG, H., 1950. Notes on bovine sarcosporidiosis. J. Parasit., 36:416.

WEINMAN, D., 1952. Toxoplasma and toxoplasmosis. Ann. Rev. Microbiol., 6:281.

WEISSENBERG, R., 1921. Fischhaplosporidien. Prowazek's Handbuch der pathogenen Protozoen. Part 3:1391.

Chapter 29

Class 4. **Cnidosporidia** Doflein

THE members of this class possess resistant spores which are of unique structure. Each spore contains one to six polar filaments and one to many sporoplasms. The membrane which envelops these structures may be a single piece or made up of two or more valves.

In Myxosporida and Actinomyxida, there appear several cells which participate in the formation of the spore. These cells give rise to one, two or many sporoplasms or generative cells, capsulogenous cells which develop polar filaments and shell-valves. Such processes do not occur in Microsporida. Because of this, Lom and Vavra (1962) proposed to subdivide Cnidosporidia into two groups. The asexual reproduction is repeated binary or multiple fission, budding or plasmotomy. Isogamous, anisogamous and autogamous reproduction have been reported in a number of species. In many forms, the zygote is the sporont from which one to many spores develop.

The Cnidosporidia are exclusively parasites of the lower vertebrates and invertebrates. No secondary or intermediate host has been found. Cnidosporidian infections may occasionally occur in an epidemic form among such economically important animals as the silkworms, honey bees and commercial fishes. (History and economic importance, Auerbach, 1910; Kudo, 1920, 1024.) Four orders.

Order 1. **Myxosporida** Bütschli

The spores of Myxosporida are of various shape and dimensions. The spore membrane appears to be chitinous (Kudo, 1921) and is composed of two, three, four or six valves. In a typical bivalve spore, the two valves meet in a **sutural plane** which is either curved or twisted (in three genera) or more or less straight. The membrane or shell may possess various markings or processes. The **polar capsule,** containing a coiled **polar filament,** varies in number from one to six. Except in the family Myxidiidae, in which one polar capsule is situated near each of the poles of the spore, the polar capsules are grouped at one end which is ordinarily designated as the anterior end of the spore. Below or between (in Myxidiidae) the polar capsules, there is a **sporoplasm.** Ordinarily a young spore possesses two sporoplasm nuclei which fuse into one (autogamy) prior to germination. In Myxobolidae there is a glycogenous substance in a vacuole which stains mahogany red with iodine and is known as the **iodinophilous** (iodophile) **vacuole.**

The Myxosporida are exclusively parasites of lower vertebrates, especially fishes. Both fresh and salt water fishes have been found to harbor, or to be infected by, Myxosporida in various regions of the world. A few occur in Amphibia and Reptilia, but no species has been found to occur in either birds or mammals. When a spore gains entrance into the digestive tract of a specific host fish, the **sporoplasm** leaves the spore as an **amoebula** which penetrates through the gut-epithelium and, after a period of migration, enters the tissues of certain organs, where it grows into a trophozoite at the expense of the host tissue cells, and the nucleus divides repeatedly. Some nuclei become surrounded by masses of dense cytoplasm and becomes the **sporonts** (Fig. 280). The sporonts grow and their nuclei divide several times, forming six to eighteen daughter nuclei, each with a small mass of cytoplasm. The number of the nuclei thus produced depends upon the structure of the mature spore, and also upon whether one or two spores develop in a sporont. When the sporont develops into a single spore, it is called a monosporoblastic sporont, and if two spores are formed within a sporont, which is usually the case, the sporont is called disporoblastic, or **pansporoblast.** The spore-formation begins usually in the central area of the large trophozoite, which continues to grow. The surrounding host tissue becomes degenerated or modified and forms an envelope that is often large enough to be visible to the naked eye (Figs. 283, 285). This is ordinarily referred to as a **myxosporidan cyst.** If the site of infection is near the body surface, the large cyst breaks and the mature spores become set free in the water. In case the infection is confined to internal organs, the spores will not be set free while the host fish lives. Upon its death and disintegration of the body, however, the liberated spores become the source of new infection.

The more primitive Myxosporida are coelozoic in the host's organs, such as the gall bladder, uriniferous tubules of the kidney,

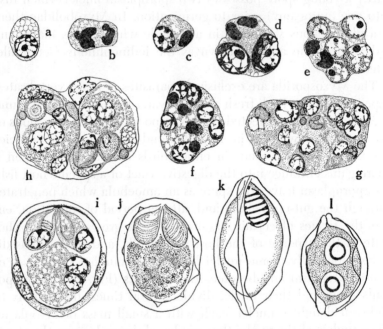

FIG. 280. Sporogony in *Myxosoma catostomi,* ×2130 (Kudo). a, sporont or pansporoblast; b–h, development of two sporoblasts within the sporont; i, a nearly mature spore; j–l, views of spore.

urinary bladder, etc. In these forms, the liberated amoebulae make their way into the specific organ and there grow into multinucleate amoeboid trophozoites which are capable of forming pseudopodia of various types. They multiply by exogenous or endogenous budding or plasmotomy. One to many spores develop in a trophozoite.

Almost all observers agree in maintaining the view that the two nuclei of the sporoplasm or two uninucleate sporoplasms fuse into one (autogamy or paedogamy) before germination, but as to the nuclear as well as cytoplasmic changes prior to, and during, spore-formation, there is a diversity of opinions. For illustration, the development of *Sphaeromyxa sabrazesi* (p. 795) as studied by two investigators may be taken as an example. Debaisieux's (1924) observation is in brief as follows (Fig. 281): Sporoplasms after finding their way into the gall bladder of host fish develop into

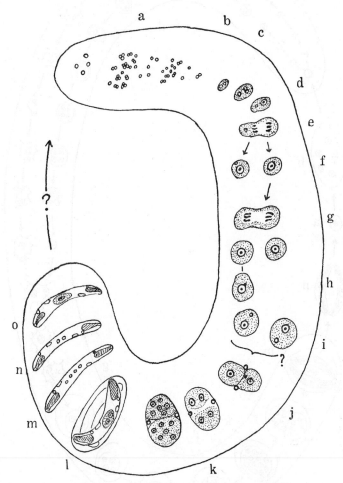

Fig. 281. The development of *Sphaeromyxa sabrazesi* (Debaisieux). a, vegetative nuclei; b, association of two vegetative nuclei; c, the same within a cell; d, primary propagative cell; e, its division; f, secondary propagative cells; g, their division; h, formation of sporocyte; i, two sporocytes; j, formation of pansporoblast; k, pansporoblast at later stages; l, pansporoblast with two spores, the sporoplasm of which contains two nuclei; m, four nuclei in sporoplasm; n, two nuclei remain functional, the other two degenerate, o, fusion of the two nuclei.

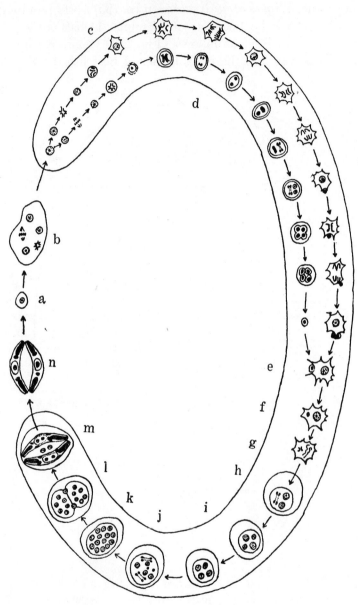

Fig. 282. The development of *Sphaeromyxa sabrazesi* (Naville). a, uninucleate amoebula enters the gall bladder; b, young multinucleate trophozoite; c, development of macrogametes; d, development of microgametes; e, f, plasmogamy; g–m, development of pansporoblast; n, fusion of the two nuclei in the sporoplasm.

large trophozoites containing many nuclei (*a, b*) two vegetative nuclei become surrounded by a cytoplasmic mass (*c*) and this develops into a primary propagative cell (*d*) which divides (3 chromosomes are noted) (*e*) and forms secondary propagative cells (*f*). A binucleate sporocyte is formed from the latter by unequal nuclear division (*g-i*) and two sporocytes unite to form a tetranucleate pansporoblast (*j*) which develops into two spores (*k, l*). Sporoplasm shows first two nuclei (*l*), but later four (*m*), of which

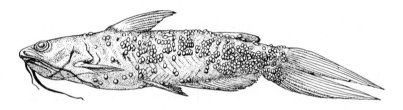

Fig. 283. A channel cat, heavily infected with *Henneguya exilis*, ×½ (Kudo).

two degenerate (*n*) and the other two fuse into one nucleus (*o*). On the other hand, according to Naville (1930) a uninucleate amoebula (Fig. 282, *a*) enters the gall bladder and develops into multinucleate trophozoite in which nuclear division reveals four chromosomes (*b*); within the trophozoite, macrogametes and microgametes are independently formed, during which process, chromosome number is reduced into half (2) (*c, d*); plasogamy between a macrogamete and a microgamete results in production of a binucleate pansporoblast (*e, f*), from which repeated nuclear division (*g-l*) forms two spores (*m*); each of the two nuclei of the sporoplasm is haploid and the diploid number is restored when the two nuclei fuse into one (*n*).

The site of infection by Myxosporida varies among different species. They have been found in almost all kinds of tissues and organs of host fish, although each species has its special site of infection in one to several species of fish. The gills and gall bladder are most frequently parasitized by Myxosporida in freshwater fishes, while the gall bladder and urinary bladder of marine fishes harbor one or more species of Myxosporida. When the infection is concentrated in the fins or integument, the resulting changes are quite conspicuous (Fig. 283). The infection in the gills is usually manifest by whitish pustules which can be frequently detected with the unaided eye. When the wall of the alimentary canal, mesentery, liver, and other organs are attacked, one sees considerable changes in them. Heavy myxosporidan infection of the gall bladder or urinary bladder of the host fish may cause abnormal appearance and coloration or unusual enlargement of the organ, but under ordinary circumstances the infection is detected only by a microscopical examination of its contents. Certain histological changes in the host fish have been mentioned elsewhere (p. 33).

Severe epidemic diseases of fishes are frequently found to be due to myxosporidan infections. According to Davis (1924), the "wormy" halibut of the Pacific coast of North America is due to *Unicapsula muscularis* (Fig. 285), which invades the muscular tissue of the host fish. The "boil disease" of the barbel, *Barbus barbus* and others, of European waters, is caused by *Myxobolus pfeifferi* (Keysselitz, 1908). *Myxosoma cerebralis*, which attacks the supporting tissues of salmonid fish, is known to be responsible for the so-called "twist disease" (Plehn, 1904), which is often fatal especially to young fishes and occurs in an epidemic form. *Henneguya salminicola* invades the body muscles of various species of Pacific salmon and produces opaque white cysts, 3-6 mm. in diameter; it is thus responsible for the so-called "tapioca disease" of salmon (Fish, 1939). *Kudoa thyrsites* (p. 791) attacks the body muscle fibers of the barracouta in which the infected muscles become liquefied. This condition is known as "milky barracouta" or "pap snoek" and may affect as much as 5 per cent of the commercial catches (Willis, 1949). (Taxonomy, Gurley, 1894, Thélohan, 1895, Auerbach, 1910, Kudo, 1920, 1933, Tripathi, 1948, Meglitsch, 1960; development, Kudo, 1920, Naville, 1927, 1930, Noble, 1944; species from North America, Gurley, 1894, Mavor, 1915, 1916, Davis, 1917, Kudo, 1920-1944, Jameson, 1929, 1931, Meglitsch, 1937-1947a, Fantham *et al.*, 1939, 1940, Noble, 1939, 1941, Rice and Jahn, 1943; from South America, da Cunha and Fonseca, 1917, 1918, Nemeczek, 1926, Pinto, 1928, Guimarães, 1931; from Europe, Thélohan, 1895, Cépède, 1906, Auerbach, 1910, 1912, Parisi, 1912, Jameson, 1913, Georgévitch, 1916-1936,

Dunkerly, 1921, Petruschewsky, 1932, Jaczo, 1940; from Asia, Fujita, 1923, 1927, Chakravarty, 1939, 1943, Chakravarty and Basu, 1948, Tripathi, 1953; from Siberia, Bauer, 1948, Achmerov, 1954, Bogdanov, 1961.)

Myxosporida are divided into two suborders.

Suborder 1. **Unipolarina** Tripathi

One, two, three, four or six polar capsules are located at or near the anterior end of the spore; in some genera, the polar capsules may be widely separated from each other or located in the central part of the spore, but the polar filaments are attached to or near the anterior tip. Coelozoic or histozoic. Nine families.

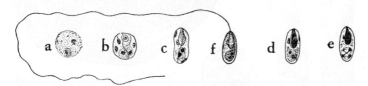

FIG. 284. *Coccomyxa morovi.* a, a young binucleate trophozoite; b–e, development of sporoblast; f, a spore with the extruded polar filament, ×665 (Léger and Hesse).

Family 1. **Coccomyxidae** Léger and Hesse

Spore with one polar capsule; sporoplasm without idoinophilous vacuole; coelozoic or histozoic.

Genus **Coccomyxa** L. and H. Spore ellipsoidal; circular in cross-section; polar capsule pyriform; coelozoic. Probably a borderline form between Myxosporida and Microsporida (Léger and Hesse, 1907).

C. morovi L. and H. (Fig. 284, *a–f*). In the gall bladder of *Clupea pilchardus;* spores 14μ by 5-6μ (Georgévitch, 1926).

Genus **Unicapsula** Davis. Spherical spore with 1 polar capsule; shell-valves asymmetrical; sutural line sinuous; histozoic in marine fish.

U. muscularis D. (Fig. 285). Spores about 6μ in diameter; 2 uninucleate sporoplasms; in the muscle of halibut; Pacific coast of North America; the cause of "wormy" halibut (Davis, 1924).

Family 2. **Ceratomyxidae** Doflein

Spores are laterally prolonged; therefore the largest diameter of the spore is at right angles to sutural plane; two polar capsules, one in each shell-valve; no iodinophile vacuole in sporoplasm; coelozoic.

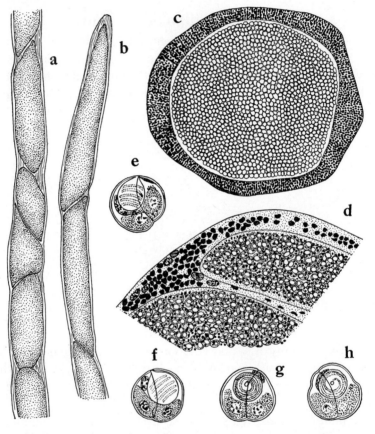

FIG. 285. *Unicapsula muscularis* (Davis). a, b, infected muscle fibers, ×20; c, cross-section of an infected muscle, ×190; d, part of a section of an infected muscle, ×575; e–h, spores, ×2500.

Genus **Ceratomyxa** Thélohan. Spore arched; shell-valves conical and hollow; breadth more than twice the sutural diameter; coelozoic in marine fish, except *C. shasta* Noble (1950) which was found "widely distributed in viscera" of fingerling rainbow trout, *Salmo gairdneri*. (Numerous species.)

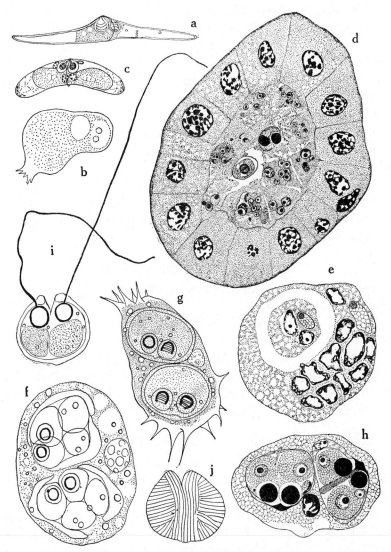

Fig. 286. a, *Ceratomyxa mesospora*, ×1000 (Davis); b, c, *C. hopkinsi*, ×1000 (Jameson); d–j, *Leptotheca ohlmacheri* (d, section of a uriniferous tubule of *Rana pipiens*, with trophozoites and spores, ×800; e, a trophozoite with a bud; f–h, disporous trophozoites; i, a spore with extruded polar filaments; j, surface view of spore, ×1500) (Kudo).

784 Protozoology

C. *mesospora* Davis (Fig. 286, *a*). In the gall bladder of *Cestracion zygaena;* spores 8μ in sutural diameter and 50-65μ wide.

C. *hopkinsi* Jameson (Fig. 286, *b, c*). In the gall bladder of *Parophrys vetulus, Microstomus pacificus* and *Citharichthys xanthostigmus;* trophozoites disporous; spores 5.7-7.5μ in sutural diameter and 28.8-39μ broad.

Genus **Leptotheca** Thélohan. Shell-valves hemispherical; breadth less than twice the sutural diameter; coelozoic in gall bladder or urinary bladder of marine fish and in renal tubules of amphibians. (Numerous species.)

L. *ohlmacheri* (Gurley) (Fig. 286, *d-j*). In the uriniferous tubules of kidney of frogs and toads; spores 9.5-12μ in sutural diameter and 13-14.5μ wide; with two uninucleate sporoplasms (Kudo, 1922).

Family 3. **Trilosporidae** Tripathi

Spore with three polar capsules; sporoplasm without an iodinophile vacuole; coelozoic.

Genus **Trilospora** Noble (1939). Spore triangular with concave sides in anterior end view; ovoidal in front view. (One species.)

T. *californica* N. Spores 7.2μ in sutural diameter by 16μ broad; polar capsules 3μ by 1.5μ; often four polar capsules instead of three; in the gall bladder of *Typhlogobius californiensis* and *Gibbonsia elegans elegans* (Noble, 1939).

Family 4. **Wardiidae** Kudo

Spore triangular, pyramidal or cordiform; two polar capsules, one in each shell-valve; no iodinophile vacuole in sporoplasm; histozoic or coelozoic.

Genus **Wardia** Kudo. Spore isosceles triangular with two convex sides in front view; oval in side view; shell-valves with reticulate ridges and fringe-like posterior processes; two large spherical polar capsules in the central part; histozoic.

W. *ovinocua* K. (Fig. 287, *a*). In the ovary of *Lepomis humilis;* spores 9-11μ in sutural diameter and 10-12μ wide.

Genus **Mitraspora** Fujita. Spores pyramidal with two convex sides; oval in end and side views; one pyriform polar capsule in each of the two shell-valves; shell longitudinally striated; with or without posterior filaments; coelozoic in host's kidney.

M. elongata Kudo. In the kidney of *Apomotis cyanellus;* spores 15-17μ by 5-6μ.

Genus **Myxoproteus** Doflein. Spore cordiform; with or without processes on shell; in urinary bladder of marine fish.

M. cordiformis Davis (Fig. 287, *b*). In the urinary bladder of *Chaetodipterus faber;* spores 12μ by 10-11μ.

Family 5. **Sphaerosporidae** Davis

Spores spherical or subspherical; two polar capsules, one in each shell-valve; no iodinophile vacuole in sporoplasm; coelozoic or histozoic.

Genus **Sphaerospora** Thélohan. Spores spherical or subspherical; sutural line straight; coelozoic or histozoic in marine or fresh-water fishes.

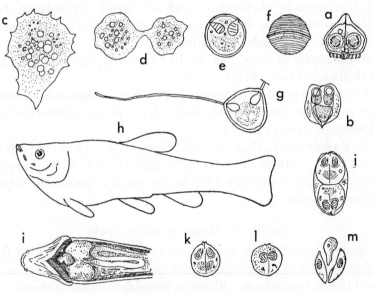

FIG. 287. a, *Wardia ovinocua,* ×1330 (Kudo); b, *Myxoproteus cordiformis,* ×1000 (Davis); c–g, *Sphaerospora polymorpha* (Kudo) (c, a trophozoite in life, ×1530; d, stage in plasmotomy, ×700; e, f, two views of spore in life; g, a spore with the extruded filaments); h–m, *S. tincae* (Léger) (h, external appearance of a heavily infected young tench; i, internal appearance, ×⅔; j, mature pansporoblast; k, l, spores; m, germination of spore, ×1000).

S. polymorpha Davis (Figs. 287, *c-g*). In the urinary bladder of toadfish, *Opsanus tau* and *O. beta*. Trophozoites amoeboid with conical pseudopodia; up to 100μ long, the majority being 20-50μ long; plasmotomy; disporoblastic; disporous or polysporous. Spores spheroidal; shell-valves finely striated; polar capsules divergent; fresh spores measure 7.5-9.5μ by 7-8μ. The trophozoites suffer frequently infection by *Nosema notabilis* (p. 813). (Development and hyperparasitism, Kudo, 1944.)

S. tincae Plehn (*S. pernicialis* Léger) (Fig. 287, *h-m*). In the kidney and other viscera of *Tinca tinca* in France and Germany; cause of epidemic disease among young tench; disease is manifest by great distension of anterior portion of abdomen and up-turned mouth: infection fatal through rupture of abdominal wall; spores 7-8.75μ in diameter (Léger, 1929).

Genus **Sinuolinea** Davis. Spore spherical or subspherical; sutural line sinuous; shell-valves without lateral appendages; two polar capsules some distance away from the anterior end; coelozoic in urinary bladder of marine fish.

S. dimorpha D. (Fig. 289, *g*). In *Cynoscion regalis;* spores 15μ in diameter (Davis, 1917).

Genus **Davisia** Laird. Spore spherical or subspherical; sutural line curved or sinuous; each shell-valve with an attached lateral appendage; coelozoic.

D. diplocrepis L. (Fig. 289, *a*). Spore ovoid with curved lateral appendages; sutural diameter 9-12μ and breadth 12-14μ; appendages 10-14μ long by 2.8-3.8μ; in the urinary bladder of *Diplocrepis puniceus*.

Family 6. **Myxosomatidae** Poche

Spore circular, ovoid or pyriform in front view; somewhat flattened parallel to sutural plane; two pyriform polar capsules in sutural plane; sporoplasm without iodinophile vacuole; histozoic.

Genus **Myxosoma** Thélohan (*Lentospora* Plehn). Spore circular, oval or ellipsoid in front view; lenticular in profile; two pyriform polar capsules under sutural line; histozoic in freshwater or marine fish. This genus can be distinguished from genus Myxobolus by the absence of an iodinophile vacuole in sporoplasm. (Several species.)

M. catostomi Kudo (Figs. 61; 280). In the muscle and connective tissue of *Catostomus commersonii;* spores 13-15μ by 10-11.5μ (Kudo, 1926).

M. cerebralis (Hofer) (Fig. 288, *a*). In the cartilage and perichondrium of salmonid fish; young fish are especially affected by infection, the disease being known as the "twist-disease" (Drehkrankheit); spores 6-10μ in diameter. This organism has been found to occur in young trout in eastern United States (Hoffman *et al.*, 1962).

M. funduli Kudo. In the gills of Fundulus; spherical cysts up to 360μ by 264μ; spores pyriform, 14μ by 8μ by 6μ; polar capsules 8μ by 2μ (Kudo, 1918). (Other species, Bond, 1938-1939.)

Family 7. **Myxobolidae** Thélohan

Spore subspherical, ovoid or ellipsoid; somewhat flattened parallel to sutural plane; one or two pyriform polar capsules under sutural line; sporoplasm with a conspicuous iodinophile vacuole; histozoic or coelozoic.

Genus **Myxobolus** Bütschli (*Disparospora,* Achemirov). Spore subspherical, ovoid, ellipsoid or pyriform in front view; without posterior process; two polar capsules pyriform; iodinophile vacuole in sporoplasm; histozoic in freshwater fish. (Numerous species.)

M. pfeifferi Thélohan (Fig. 288, *e, f*). In the muscle and connective tissue of body and various organs of *Barbus barbus, B. fluviatilis,* and *B. plebejus;* tumor up to a diameter of 7 cm; most of infected fish die from the effect (Keysselitz); spores 12-12.5μ by 10-10.5μ.

M. orbiculatus Kudo (Fig. 288, *g-i*). In muscle of *Notropis gilberti;* spores 9-10μ in diameter by 6.5-7μ thick.

M. conspicuus K. (Fig. 288, *j, k*). In corium of head of *Moxostoma breviceps;* tumors 1/2-4 mm.; spores 9-11.5μ by 6.5-8μ (Kudo, 1929).

M. intestinalis K. (Fig. 1, *a*). In the intestinal wall of *Pomoxis sparoides;* (fixed unstained) spores, 12-13μ by 10-12.5μ; the histological changes brought about by this protozoan have been mentioned elsewhere (p. 33) (Kudo, 1929).

M. squamosus K. (Fig. 288, *l-o*). In connective tissue below scales of *Hybopsis kentuckiensis;* spore circular in front view, 8-9μ in diameter, 4.5-5μ thick.

Genus **Thelohanellus** Kudo. Pyriform spores, each with one polar capsule; sporoplasm with an iodinophilous vacuole; histozoic in freshwater fish. (Many species.)

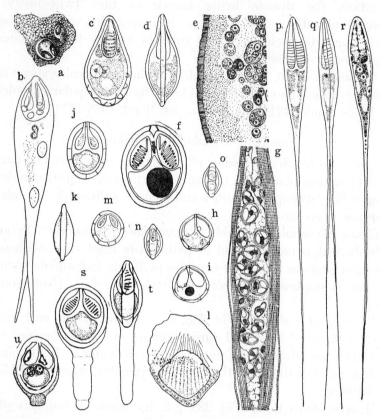

Fig. 288. a, *Myxosoma cerebralis*, showing two views of spore, ×800 (Plehn); b, a spore of *Agarella gracilis*, ×1660 (Dunkerly); c, d, front and side views of fresh spores of *Thelohanellus notatus*, ×1530 (Kudo); e, f, *Myxobolus pfeifferi* (Keysselitz) (e, Part of section of a cyst; f, a spore treated with iodine solution, ×1780); g–i, *M. orbiculatus* (Kudo) (g, infected host's muscle, ×600; h, a fresh spore; i, Lugol-treated spore, ×1000); j, k, views of fresh spores of *M. conspicuus*, ×1530 (Kudo); l–o, *M. squamosus* (l, a cyst under a scale, ×6.5; m–o, views of fresh spores, ×1530); p–r, spores of *Henneguya exilis*, ×1530; s–u, spores of *Unicauda clavicauda*, ×1530 (s, t, fresh spores; u, a stained spore without the process (Kudo)).

T. notatus (Mavor) (Figs. 1, *b;* 288, *c, d*). In subdermal connective tissue of *Pimephales notatus, Cliola vigilax, Notropis cornutus, N. blennius,* and *Leuciscus rutilus;* tumor up to 7 mm. in diameter; spores 17-18μ by 7.5-10μ; host tissue surrounding the organism becomes so greatly changed that it appears as an epithelium (p. 33) (Debaisieux, 1925; Kudo, 1929, 1934).

Genus **Hoferellus** Berg (*Hoferia Doflein*). Spore pyramidal; two posterior spinous processes; histozoic; sporoplasm with an iodinophile vacuole.

H. cyprini (Doflein). Spore 10-12μ by 8μ; 9-10 longitudinal striae on each shell-valve; in lumen and epithelial cells of renal tubules of kidney of *Cyprinus carpio.*

Genus **Henneguya** Thélohan (*Myxobilatus* Davis). Spore circular or ovoidal in front view; flattened; two polar capsules at anterior end; each shell-valve prolonged posteriorly into a long process; sporoplasm with an iodinophilous vacuole; mostly histozoic in freshwater fish. Davis (1944) established Myxobilatus based upon asymmetrical shell-valves and the presence of one polar capsule in each valve, but in practice, the detection of the sutural line is exceedingly difficult. Therefore, it is considered a synonym of Henneguya. (Many species.)

H. exilis Kudo (Figs. 283; 288, *p-r*). In gills and integument of *Ictalurus punctatus;* cysts up to 3 mm. in diameter, conspicuous; spores, total length 60-70μ, spore proper 18-20μ long by 4-5μ wide by 3-3.5μ thick (Kudo, 1929, 1934).

H. mictospora Kudo. In the urinary bladder of *Lepomis* spp. and *Micropterus salmoides;* spores 13.5-15μ long, 8-9μ wide, 6-7.5μ thick; caudal prolongation 30-40μ long.

Genus **Unicauda** Davis. The spore is similar to that of *Henneguya,* but with a single caudal appendage which is not an extension of the shell-valves. (Davis, 1944).

U. clavicauda (Kudo) (Fig. 288, *s-u*). In the subdermal connective tissue of the minnow, *Notropis blennius;* oblong or ellipsoid cysts, 1-1.5 mm. in the longest diameter; spores 10.5-11.5μ by 8.5-9.5μ by 6μ; appendage 20-30μ by 3-6.5μ (Kudo, 1934).

Genus **Trigonosporous** Hoshina. Spore nearly isosceles triangular, with broadly rounded anterior end and flattened posterior end in front-view; spindle-shaped in end-view; two polar capsules; two shell-valves, each drawn out into two long slender processes which are cross-connected by two filaments running parallel to posterior margin of spore; iodinophile vacuole in sporoplasm; one histozoic species.

T. acanthogobii H. (Fig. 289, *b, c*). Spore 7.3-10.5μ in sutural

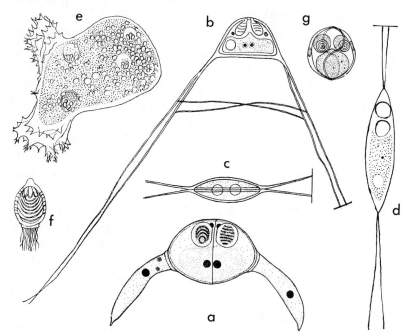

Fig. 289. a, spore of *Davisia diplocrepis* in neutral red, ×1680 (Laird); b, c, *Trigonosporous acanthogobii,* ×1060 (b, front and c, posterior end views) (Hoshina); d, spore of *Neohenneguya tetraradiata* (Tripathi); e, f, *Chloromyxum leydigi* (e, a trophozoite, ×500; f, spore, ×1000 (Thélohan); g, spore of *Sinuolinea dimorpha,* ×620 (Davis).

diameter, 11-18μ broad and 4.7-6μ thick; posterior filamentous processes 50-99μ long; connecting filaments 13-19μ long; the organism forms cysts, 0.6-0.8 mm. by 0.4-0.7 mm., in the gills of a blackish water fish, *Acanthogobius flavimanus* (Hoshina, 1952).

Genus **Neohenneguya** Tripathi (1953). Spore fusiform with two equally long prolongations at both ends; spherical polar capsules, tendem in position, but near one end; with an iodinophile vacuole; histozoic. (One species.)

N. tetraradiata T. (Fig. 289, *d*). In the gills of *Odontamblyopus rubicundus;* cysts measured 700μ in diameter; young trophozoites are irregular in shape and contained one developing spore; spore 16.2-21.6μ in length (excluding the prolongations) and 5.4μ in diameter; prolongations 63-72μ long; polar capsules 2-2.7μ in diameter; iodinophile vacuole 1.5-2.4μ in diameter.

Family 8. **Chloromyxidae** Thélohan

Spore with four polar capsules; sporoplasm without an iodinophile vacuole; coelozoic or histozoic.

Genus **Chloromyxum** Mingazzini (*Caudomyxum* Bauer). Spore spherical or oval; shell-valves, often with ridges; with or without posterior processes; coelozoic or histozoic in freshwater or marine fish and also in amphibians. (Numerous species.)

C. leydigi M. (Fig. 289, *e, f*). In the gall-bladder of various species of Raja, Torpedo and Cestracion; spores 6-9μ by 5-6μ; widely distributed. (Structure and development, Erdmann, 1917; Naville, 1927.)

C. trijugum Kudo (Fig. 290, *a*). In the gall-bladder of *Xenotis megalotis* and *Pomoxis sparoides;* spores 8-10μ by 5-7μ.

C. histolytica Prudhomme and Pantaléon (1959). This organism is said to be the cause of "milky salmon" in *Salmo salar.*

Genus **Kudoa** Meglitsch (*Neochloromyxum*, Matsumoto). Spore quadrate or stallate in anterior end view; shell delicate; sutural lines of shell-valves usually indistinct; histozoic. Several species.

K. clupeidae (Hahn) (Fig. 290, *b-e*). In the body muscles of *Clupea harengus, Brevoortia tyrannus,* etc., spores 5.1μ by 6.4μ; polar capsules 1.5μ by 1μ (Meglitsch, 1947, 1947a). Nigrelli (1946) found this species in the ocean pout (*Macrozoares americanus*). Sindermann (1959) found it common in one- and two-year-old herring in southern Gulf of Maine and Gulf of St. Lawrence.

K. thyrsites (Gilchrist) (Fig. 290, *f-i*). In the body muscles of the barracouta, *Thyrsites atum,* in Australia and Africa; pyramidal spores 6-7μ high and 12-17μ wide; two uninucleate sporoplasms; polar capsules homogeneous in appearance (Willis, 1949). (Effect on host, p. 780.)

Genus **Agarella** Dunkerly. Spore ovoidal, flattened; with two long posterior processes; histozoic. (One species.)

A. gracilis D. (Fig. 288, *b*). In the testis of South American lungfish, *Lepidosiren paradoxa* (Dunkerly, 1915, 1925).

Family 9. Hexacapsulidae

Spore with six polar capsules; histozoic.

Genus **Hexacapsula** Arai and Matsumoto. Spore hexagonal and stellate in anterior end view; shell composed of six valves; sutural lines indistinct; other stages unknown; histozoic. (One species.)

H. neothunni A. and M. (Fig. 290, *j, k*). In the trunk muscle of the yellowfin tuna, *Neothunnus inacropterus;* apparently the cause of "jellied tuna"; trophozoites and other stages unknown; spore 5.3-7.3μ in sutural diameter and 9-13μ broad.

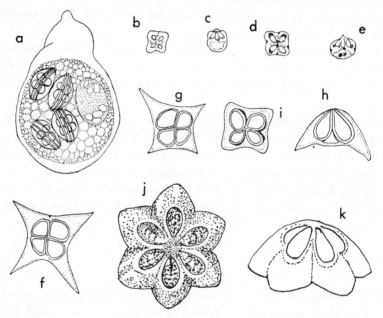

FIG. 290. a, a sporulating trophozoite of *Chloromyxum trijugum,* ×1130 (Kudo); b–e, *Kudoa clupeidae* (Meglitsch) (b, c, two views of fresh spores, ×1240; d, e, stained spores, ×1430); f–i, *K. thyrsites* (Willis) (f–h, preserved spores; i, a spore from section); j, k, different views of fresh spores of *Hexacapsula neothunni,* ×3000 (Arai and Matsumoto).

Suborder 2. **Bipolarina** Tripathi

Spore elongate along the sutural plane; two polar capsules are widely separated from each other, one capsule being located near each end and opening at its tip; coelozoic or histozoic. One family.

Family 1. **Myxidiidae** Thélohan

Spore fusiform or ellipsoid; one polar capsule at or near each end; sporoplasm without idoinophilous vacuole; histozoic or coelozoic.

Genus **Myxidium** Bütschli (*Cystodiscus* Lutz). Spore fusiform with pointed or rounded ends; sutural line is typically straight

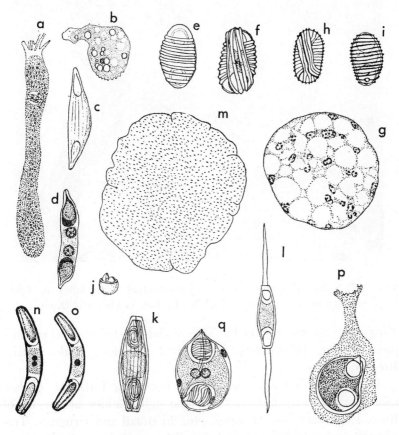

FIG. 291. a–d, *Myxidium lieberkühni* (a trophozoite, ×220; b, a small trophozoite, ×1000; c, d, spores, ×1400) (Kudo); e, f, *M. immersum*, ×1400 (Kudo); g–i, *M. serotinum* (Kudo) (g, a stained young trophozoite, ×1530; h, i, two views of spore, ×915); j–l, *Sphaeromyxa balbianii* (j, ×⅔; k, a spore, ×1400 (Davis); l, a spore with extruded filaments, ×840 (Thélohan)); m–o, *S. sabrazesi* (m, trophozoite, ×10; n, o, spores, ×1000 (Schröder); p, q, *Zschokkella hildae* (p, ×600; q, ×1060) (Auerbach).

and coincides with, or is at an acute angle to, the axis of the spore; polar filament filamentous and not ribbon-like; coelozoic or histozoic. (Numerous species.)

M. *lieberkühni* Bütschli (Fig. 291, *a-d*). In urinary bladder of

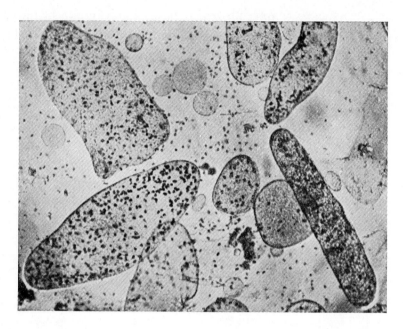

FIG. 292. Scattered spores, young and sporulating trophozoites of *Myxid-ium serotinum,* as seen in the bile of a frog in life, ×64 (Kudo).

Esox spp.; spores 18-20μ by 5-6μ; widely distributed. (Development, Cohn, 1896, Debaisieux, 1916; division, Kudo, 1921a, Bremer, 1922.)

M. *immersum* Lutz (*Cystodiscus immersus* Lutz; M. *lindoy-ense* Carini). (Fig. 291, *e, f*). In the gall bladder of species of Bufo, Leptodactylus, Atelopus, etc.; in Brazil and Uruguay. Trophozoites circular to oval, and very thin; up to 4 mm. in diameter; disporoblastic; polysporous. Spores 11.8-13.3μ by 7.5-8.6μ; shell-valves marked with two longitudinal and seven to nine transverse ridges (Cordero, 1919; Kudo and Sprague, 1940).

M. *serotinum* Kudo and Sprague (Figs. 291, *g-i;* 292). In the gall bladder of *Bufo terrestris, Rana pipiens, R. clamitans* and R.

sphenocephala; in the United States. Trophozoites up to 6.5 by 1.8 mm., extremely thin; cytoplasm highly alveolated; endogenous budding; disporoblastic; polysporous. Spores 16-18μ by 9μ; shell-valve with two to four longitudinal and ten to thirteen transverse ridges (Kudo, 1943).

M. kudoi Meglitsch. In gall-bladder of *Ictalurus furcatus;* trophozoites large disc-like up to 1 mm. in diameter; spores 8.5-12μ long by 4-6μ (Meglitsch, 1937).

Genus **Sphaeromyxa** Thélohan. Spore fusiform with truncate ends; polar filaments short, ribbon-like; trophozoites large and discoid; coelozoic in marine fish. (Several species.)

S. balbianii T. (Fig. 291, *j-l*). In gall bladder of Motella and other marine fish in Europe and of Siphostoma in the United States; spores 15-20μ by 5-6μ (Naville, 1930).

S. sabrazesi Laveran and Mesnil (Figs. 281; 282; 291, *m-o*). In the gall bladder of Hippocampus, Motella, etc.; spores 22-28μ by 3-4μ (Debaisieux, 1925; Naville, 1930).

Genus **Zschokkella** Auerbach. Spore semi-circular in front view; fusiform in profile; circular in cross-section; ends pointed obliquely; polar capsules large, spherical; sutural line usually in S-form, coelozoic in fish or amphibians. A few species.

Z. hildae A. (Fig. 291, *p, q*). In urinary bladder of *Gadus* spp.; spores 16-29μ by 13-18μ (Auerbach, 1910).

Order 2. **Actinomyxida** Stolc

The spores of Actinomyxida have unique structure and appearance. The spore is covered by two membranes. The conspicuous outer membrane (epispore) is composed of three valves which may be drawn out into various processes. The inner membrane is very delicate, and may not be easily seen in many spores. There are three polar capsules, each with a short, coiled polar filament. One to many sporoplasms occur in each spore. In the grown stage, the pansporoblast develops into eight sporoblasts which in turn develop into eight spores. (Development, Janiszewska, 1957.)

The Actinomyxida inhabit the body cavity and gut epithelium of Oligochaeta (Tubificidae and Lumbricidae) and Sipunculoidea. They have been less frequently studied and, therefore, are not so well known as Myxosporida. (Four families, Janiszewska, 1957.)

Family 1. **Tetractinomyxidae** Poche

Spore with a tetrahedric epispore without processes and contains a single binucleate sporoplasm; parasitic in the coelom of Petalostoma.

Genus **Tetractinomyxon** Ikeda. In the coelom of the sipunculid *Petalostoma minutum;* spores tetrahedron, without processes; trophozoite a rounded body, when mature; pansporoblast develops eight spores; each spore when mature with a binucleate sporo-

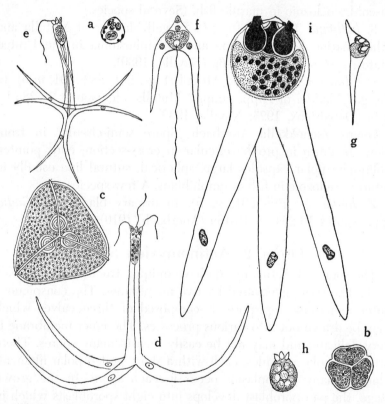

FIG. 293. a, *Tetractinomyxon intermedium,* ×800 (Ikeda); b, *Sphaeractinomyxon stolci,* ×600 (Caullery and Mesnil); c, *S. gigas,* ×665 (Granata); d, *Triactinomyxon ignotum,* ×165 (Léger); e, *Hexactinomyxon psammoryctis,* ×300 (Stolc); f, g, *Synactinomyxon tubificis,* ×600 (Stolc); h, *Neoactinomyxum globosum,* ×860 (Granata); i, *Guyenotia sphaerulosa,* ×2095 (Naville).

plasm; seemingly a borderline forms between Myxosporida and Actinomyxida. Two species.

T. intermedium I. (Fig. 293, *a*). Spherical pansporoblasts 20-25μ in diameter; spores 7-8μ in diameter; in coelom of the sipunculid, *Petalostoma minutum* (Ikeda, 1912).

Family 2. **Sphaeractinomyxidae** Janiszewska

Spore spheroidal; epispore without any distinct processes; with numerous sporoplasms. Parasites of Oligochaeta.

Genus **Sphaeractinomyxon** Caullery and Mesnil. In the coelom and gut epithelium of oligochaetes; spores rounded, without any processes; in an early stage of development, there two unincleate bodies surrounded by a binucleate envelope; two inner cells multiply into sixteen cells which unite in pairs; nucleus of zygote of sporont divides first into two; one of the nuclei divides into six which form three shell-valves and three polar capsules, while the other nucleus together with a portion of cytoplasm remains outside the envelope, and undergoes multiplication; the multinucleate mass migrates into the spore; and later divides into a large number of uninucleate sporoplasms which, when the spores gain entrance into a new host, begin development. (Four species.)

S. stolci C. and M. (Fig. 293, *b*). Spore spheroidal; 17-25μ in diameter; in the coelom of *Clitellis arenarius* and *Hemitubifex benedii*.

S. gigas Granata (Fig. 293, *c*). Spore slightly rounded triangular in polar view; 32-45μ in diameter; in the coelom of *Limnodrilus hoffmeisteri* (Granata, 1925).

Genus **Neoactinomyxon** Granata. Spore spheroidal; epispore without any processes, distended to hemisphere; many sporoplasms. (One species.)

N. globosum G. (Fig. 293, *h*). In the gut epithelium of *Limnodrilus udekemianus;* spore with 16 sporoplasms, measures 15-20μ by 12-16μ (Granata, 1925; Jírovec, 1940).

Family 3. **Triactinomyxidae**

Epispore is drawn out into three conspicuous processes which may or may not unite with those of other seven spores; parasitic in Oligochaeta.

Genus **Triactinomyxon** Stolc. Each of three valves drawn out into a long process, the whole epispore anchor-like; sporoplasm with 8-100 nuclei, in gut epithelium of Oligochaeta.

T. ignotum S. (Fig. 293, *d*). Spore 110-165μ by 11-15μ; sporoplasm with eight nuclei; in *Tubifex tubifex*.

T. naidanum Naidu. Sporoplasm with 12 nuclei; (Dimensions not given); in the gut of *Nais communis punjabensis* (Naidu, 1956).

T. magnum Granata. Spore about 500μ in the largest dimension; sporoplasm with sixteen nuclei; in *Limnodrilus udekemianus*.

T. legeri Mackinnon and Adams. Spore 90-140μ by 11-16μ; sporoplasm with twenty-four nuclei; in *Tubifex tubifex*.

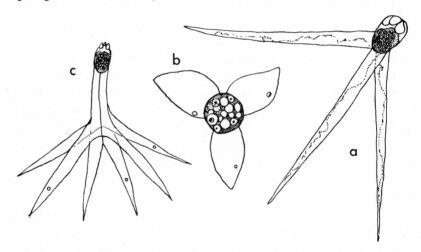

Fig. 294. a, a spore of *Echinactinomyxon radiatum*, ×550; b, a spore of *Aurantiactinomyxon raabei innioris*, ×780; c, a spore of *Hexactinomyxon hedvigi*, ×210 (Janiszewska).

Genus **Guyenotia** Naville. Epispore with equally long finger-like processes that are longer than the diameter of spore; sporoplasm with thirty-two nuclei; in oligochaetes.

G. sphaerulosa N. (Fig. 293, *i*). Spore spheroid, 15μ in diameter; processes of mature spores 40μ long; sporoplasm with thirty-two nuclei; in the gut epithelium of *Tubifex tubifex*.

Genus **Raabeia** Janiszewska. Epispore is drawn out into three long, attenuated processes, without forming a style; sporoplasm barrel or flask in shape, and contains many nuclei; in the coelom of Oligochaeta (Janiszewska, 1955).

R. *gorlicensis* J. Epispore with three long processes, about 170μ long, tapering to a sharp point and curved upwards; three protruding polar capsules divergent; sporoplasm with up to thirty-two nuclei; in the coelom of *Tubifex tubifex* (Janiszewska, 1955, 1957).

Genus **Echinactinomyxon** J. Epispore styleless; three equally long processes tapering to a sharp point and straight or curved downward; in the gut epithelium of oligochaetes (Janiszewska, 1957).

E. *radiatum* J. (Fig. 294, *a*). Three epispore processes 125μ long; sporoplasm 27-30μ long and contains up to thirty-two (?) nuclei; in the gut epithelium of *Tubifex tubifex*.

Genus **Aurantiactinomyxon** J. Epispore styleless, with three processes, broad and leaf-like; sporoplasm with about sixteen nuclei; uninucleate sporozoite fusiform; in the gut epithelium of oligochaetes.

A. *raabei iunioris* J. (Fig. 294, *b*). Episporal processes 25-30μ long; episporal cavity spherical, 17μ in diameter; sporoplasm gives rise to about 16 fusiform sporozoites measuring 3μ by 1.5μ; in the gut epithelium of *Limnodrilus hoffmeisteri*.

Genus **Synactinomyxon** Štolc. Epispore with two long and a short processes with attenuate end; eight spores bound together by the short processes within the pansporoblast; sporoplasm with many nuclei; in the gut epithelium of oligochaetes.

S. *tubificis* S. (Fig. 293, *f*, *g*). In the gut epithelium of *Tubifex tubifex*.

Genus **Siedleckiella** Janiszewska. Epispore tripodal with obtuse or finger-like processes; all three processes of a spore joined with those of other spores so that eight spores make a hexahedrid reticulum; sporoplasm with more than 100 nuclei; in the gut epithelium of oligochaetes.

S. *silesica* J. Spore 185-205μ in diameter; sporoplasm 30-40μ in the largest diameter, contains more than 100 nuclei; in the gut epithelium of *Tubifex* sp. (Janoszewska, 1952).

Genus **Antonactinomyxon** J. Epispore with three clavate processes; the end of an expanded process of one spore joined with three processes of three other spores so that eight spores form a tridimensional reticulum; sporoplasm with many nuclei; in the coelom of oligochaetes.

A. antonii (J.). Epispore processes 130μ long; in the coelom of *Limnodrilus claparedeanus* (Janiszewska, 1955, 1955a, 1957).

Family 4. **Hexactinomyxidae**

Epispore with more than three processes.

Genus **Hexactinomyxon** Štolc. Spore anchor-shaped with three

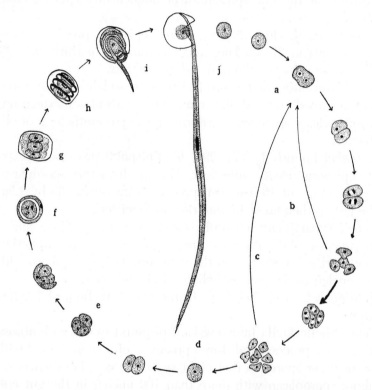

Fig. 295. Diagram illustrating the probable development of Helicosporida, × about 1600 (Keilin). a–c, schizont and schizogony; d, sporont(?); e, three stages in formation of four-celled stage; f, hypothetical stage; g, young spore before the spiral filament is formed; h, mature spore; i, j, opening of spore and liberation of sporoplasms. a–h, in living host larva; i, j, in dead host body.

double (six in all) processes which taper to a sharp point; sporoplasm with many nuclei; in the gut epithelium of oligochaetes.

H. psammoryctis S. (Fig. 293, *e*). In the gut epithelium of *Psammoryctes balbatus*.

H. hedvigi Janiszewska (Fig. 294, *c*). Epispore style shorter and processes broader than those of the species just mentioned; style together with the sporoplasm measures 136-140µ, the processes arising from the style 152-180µ long; sporoplasm barrel-shaped; sporoplasm with many nuclei; in *Tubifex tubifex* (Janiszewska, 1955).

Order 3. **Helicosporida** Kudo

This order has been created to include the interesting organism, Helicosporidium, observed by Keilin. Although quite peculiar in the structure of its spore, the organism seems to be best placed in the Cnidosporidia.

The minute spore (Fig. 295) is composed of a thin membrane of one piece and of three uninucleate sporoplasms, around which is coiled a long thick filament. Young trophozoites are found in the host tissues or body cavity. They undergo schizogony, at the end of which uninucleate sporonts become differentiated. A sporont divides apparently twice and thus forms four small cells which develop into a spore. The complete life-history is still unknown.

Genus **Helicosporidium** Keilin. Parasitic in arthropods; schizogony and sporogony; spore with central sporoplasms and a single thick coiled filament. (One species, Keilin, 1921.)

H. parasiticum K. (Fig. 295). In body cavity, fat body, and nervous tissue of larvae of *Dasyhelea obscura* and *Mycetobia pallipes* (Diptera), and *Hericia hericia* (Acarina), all of which inhabit wounds of elm and horse-chestnut trees; schizonts minute; spores 5-6µ in diameter; extruded filament 60-65µ by 1µ thick.

References

ACHMEROV, A. H., 1954. The conjugate species of a new genus of Myxosporidia. C. R. Acad. Sci., URSS, 97:1101.

ARAI, Y., and MATSUMOTO, K., 1953. On a new sporozoan, *Hexacapsula neothunni* gen. et sp. nov. from the muscle of yellowfin tuna, *Neothunnus macropterus*. Bull. Japan. Soc. Sci. Fisheries, 18:293.

AUERBACH, M., 1910. Die Cnidosporidien. Leipzig.

—— 1912. Studien über die Myxosporidien der norwegischen Seefische und ihre Verbreitung. Zool. Jahrb. Syst., 34:1.

AWERINZEW, S., 1913. Ergebnisse der Untersuchungen über parasitische Protozoen der tropischen Region Afrikas. III. Zool. Anz., 42:151.

BAUER, O. N., 1948. Parasites of fishes of River Lena (Russian). Trans. All-Union Sci. Res. Inst. Lake & River Fisheries, Leningrad, 27:157.

BOGDANOV, E. A., 1961. An endemic disease of salmonids in Sakhalin Island (Russian). Boklady Akad. Nauk., SSSR, 134:785.

BOND, F. F., 1938. Cnidosporidia from *Fundulcs heteroclitus.* Tr. Am. Micr. Soc., 57:107.

—— 1938a. The doubtful relationship of Sporozoa to the ulcers of *Fundulus heteroclitus.* J. Parasit., 24:207.

—— 1939. The seasonal incidence of myxosporidian parasites infecting *Fundulus heteroclitus.* Tr. Am. Micr. Soc., 58:156.

BREMER, H., 1922. Studien über Kernbau und Kernteilung von *Myxidium lieberkühni* Bütschli. Arch. Protist., 45:273.

BÜTSCHLI, O., 1881. Beiträge zur Kenntnis der Fischpsorospermien. Ztschr. wiss. Zool., 35:620.

—— 1882. Myxosporidia. Bronn's Klassen und Ordnungen der Protozoa. 1:590.

CARINI, A., 1932. *Myxidium lindoyense* n. sp., etc. Rev. Biol. Hyg., 3:83.

CAULLERY, M., and MESNIL, F., 1905. Recherches sur les Actinomyxidies. I. Arch. Protist., 6:272.

CÉPÈDE, C., 1906. Myxosporidies des poissons des Alpes Françaises. Ann. l'Uni. Grenoble, 18:57.

CHAKRAVARTY, M., 1939. Studies on Myxosporidia from the fishes of Bengal, etc. Arch. Protist., 92:169.

—— 1943. Studies on Myxosporidia from the common food fishes of Bengal. Proc. Indian Acad. Sci., 18:21.

—— and BASU, S. P., 1948. Observations on some myxosporidians parasitic in fishes, etc. Proc. Zool. Soc. Bengal, 1:23.

COHN, L., 1896. Ueber die Myxosporidien von *Esox lucius* und *Perca fluviatilis.* Zool. Jahrb. Anat., 9:227.

CORDERO, E. H., 1919. *Cystodiscus immersus* Lutz: Mixosporidio de los batracios del Uruguay. Physis, 4:403.

DA CUNHA, A. M., and DA FONSECA, O., 1917-1918. Sobre os myxosporidios dos peixes brazileiros. I-IV. Brazil Medico, 31:321-32, 414.

DAVIS, H. S., 1917. The Myxosporidia of the Beaufort region. Bull. U. S. Bureau Fish., 35:199.

—— 1924. A new myxosporidian parasite, the cause of "wormy" halibut. Rep. U. S. Comm. Fisheries for 1923, App. 8.

—— 1944. A revision of the genus Henneguya with descriptions of two new species. Tr. Am. Micr. Soc., 63:311.

DEBAISIEUX, P., 1918. Notes sur le *Myxidium lieberkühni.* La Cellule, 30:281.

———— 1924. *Sphaeromyxa sabrazesi* Laveran et Mesnil. *Ibid.*, 35:269.

———— 1925. Études sur les Myxosporidies. III. Arch. zool. exper. gén., 64:353.

DUNKERLY, J. S., 1915. *Agarella gracilis*, etc. Proc. Roy. Phys. Soc., 19:213.

———— 1921. Fish Myxosporidia from Plymouth. Parasitology, 12:328.

———— 1925. The development and relationship of the Myxosporidia. Quart. J. Micr. Sc., 69:185.

ERDMANN, R., 1917. *Chloromyxum leydigi* und seine Beziehungen zu anderen Myxosporidien. Arch. Protist., 36:276.

FANTHAM, H. B., PORTER, A., and RICHARDSON, L. R., 1939. Some Myxosporidia found in certain freshwater fishes in Quebec Province, Canada. Parasitology, 31:1.

———— ———— ———— 1940. Some more Myxosporidia observed in Canadian fishes. *Ibid.*, 32:333.

FISH, F. F., 1939. Observations on *Henneguya salminicola*, etc. J. Parasit., 25:169.

FUJITA, T., 1912. Notes on new sporozoan parasites of fishes. Zool. Anz., 30:259.

———— 1923. Studies on Myxosporidia of Japan. J. Coll. Agr. Sapporo, 10:191.

———— 1924. Studies on myxosporidian infection of the crucian carp. Japan J. Zool., 1:45.

———— 1927. Studies on Myxosporidia of Japan. J. Col. Agr. Sapporo, 16:229.

GEORGÉVITCH, J., 1916. Note sur les myxosporidies recueillie as Roscoff. Bull. soc. zool. France, 41:86.

———— 1917. Recherches sur le developpement de *Ceratomyxa herouardi*. Arch. zool. exper. gén., 56:375.

———— 1926. Sur la Coccomyxa de la sardine. *Ibid.*, 65(N.-R):57.

———— 1936. Nouvelles études sur les myxosporidies. Bull. l'Acad. Sc. Math. Natur. Belgrade, B, 3:87.

GRANATA, L., 1925. Gli Attinomissidi. Arch. Protist., 50:139.

GUIMARÃES, J. R. A., 1931. Myxosporideos da ichtiofauna brasileira. Fac. Med. São Paulo Thesis. 50 pp.

GURLEY, R. R., 1894. The Myxosporidia or psorosperms of fishes, etc. Rep. U. S. Fish. Comm., 26: 65.

HOFFMAN, G. L., et al., 1962. Whirling disease of trouts caused by *Myxosoma cerebralis* in the United States. Spec. Sci. Rep. Fisheries, No. 427.

HOSHINA, T., 1952. Notes on some myxosporidian parasites of fishes of Japan. J. Tokyo Univ. Fisheries, 39:69.

IKEDA, I., 1912. Studies on some sporozoan parasites of Sipunculoids. I. Arch. Protist., 25:240.

JACZO, I., 1940. Beiträge zur Kenntnis der Myxosporidien der Balaton-Fische. I. Arb. ungarisch. Biol. Forschungsinst., 12:277.

JAMESON, A. P., 1913. A note on some Myxosporidia collected at Monaco. Bull. l'Inst. Océan., 273:1.

—— 1929. Myxosporidia from Californian fishes. J. Parasitol., 16:59.

—— 1931. Notes on California Myxosporidia. *Ibid.*, 18:59.

JANISZEWSKA, J., 1953. *Siedleckiella silesica* n.g., n.sp. Actinomyxidia. Zool. Polon., 6:49.

—— 1955. *Siedleckiella antonii* sp.n. Remarks on the sporogenesis in the genus Siedleckiella and in other Actinomyxidia. *Ibid.*, 6:88.

—— 1955a. Actinomyxidia—Morphology, ecology, history of investigations, systematics, development. Acta Paras. Pol., 2:405.

—— 1957. Actinomyxidia II. Zool. Pol., 8:3.

JÍROVEC, O., 1940. Zur Kenntnis einiger in Oligochäten parasitierenden Protisten. II. Arch. Protist., 94:212.

JOHNSTON, T. H., and BANCROFT, M. J., 1919. Some new sporozoan parasites of Queensland freshwater fish. Proc. Roy. Soc. N. S. Wales, 52:520.

KEILIN, D., 1921. On the life history of *Helicosporidium parasiticum* n.g., n. sp., etc. Parasitology, 13:97.

KEYSSELITZ, G., 1908. Ueber durch Sporozoen (Myxospordien) hervorgerufene pathologische Veränderungen. Verh. Ges. deutsch. Natur. Aertze, 79:452.

KUDO, R. R., 1918. Contributions to the study of parasitic Protozoa. IV. J. Parasit., 4:141.

—— 1920. Studies on Myxosporidia. Illinois Biol. Monogr., 5:245.

—— 1921. On the nature of structures characteristic of cnidosporidian spores. Tr. Am. Micr. Soc., 40:59.

—— 1921a. On some Protozoa parasitic in freshwater fishes of New York. J. Parasit., 7:166.

—— 1922. On the morphology and life history of a myxosporidian, *Leptotheca ohlmacheri*, etc. Parasitology, 14:221.

—— 1926. On *Myxosoma catostomi* Kudo, 1923, etc. Arch. Protist., 56:90.

—— 1929. Histozoic Myxosporidia found in freshwater fishes of Illinois. *Ibid.*, 65:364.

—— 1933. A taxonomic consideration of Myxosporidia. Tr. Am. Micr. Soc., 52:195.

—— 1934. Studies on some protozoan parasites of fishes of Illinois. Illinois Biol. Monogr., 13:1.

—— 1943. Further observations on the protozoan, *Myxidium serotinum*, etc. J. Morphol., 72:263.

—— 1944. The morphology and development of *Nosema notabilis* Kudo, and of its host, *Sphaerospora polymorpha* Davis, etc. Illinois Biol. Monogr., 20:1.

—— and SPRAGUE, V., 1940. On *Myxidium immersum* and *M. serotinum* n. sp., etc. Rev. Med. Trop. Parasit. Bact. Clin. Lab., Havana, 6:65.

LAIRD, M., 1953. The protozoa of New Zealand intertidal zone fishes. Tr. Roy. Soc., New Zealand, 81:79.

LÉGER, L., 1929. Une nouvelle maladie parasitaire funeste aux élevages de tanche, "la sphérosporose." Trav. Lab. Hydr. Pisc. Uni. Grenoble, 21:7.

—— and HESSE, E., 1907. Sur une nouvelle myxosporidie parasite de la sardine. C. R. Acad. Sci., 145:85.

LOM, J., and VÁVRA, J., 1962. A proposal to the classification of the subphylum Cnidospora. System. Zool., 11:172.

LUTZ, A., 1889. Ueber ein Myxosporidium aus der Gallenblase brasilianischer Batrachier. Centralbl. Bakt., 5:84.

MATSUMOTO, K., 1954. On the two new Myxosporidia *Chloromyxum musculoliquefaciens* sp. nov. and *Neochloromyxum cruciformum* gen. et sp. nov., from the jellied muscle of swordfish *Xiphias gladius* Linné and common Japanese sea-bass *Lateolabrax japonicus*. Bull. Japan. Soc. Sci. Fisheries, 20:469.

MAVOR, J. W., 1915. Studies on the Sporozoa of the fishes of the St. Andrew's region. Ann. Rep. Dep. Marine Fish., 47 (Suppl.):25.

—— 1916. Studies on the protozoan parasites of the fishes of the Georgian Bay. Tr. Roy. Soc. Canada, Ser. 3, 10:63.

MEGLITSCH, P. A., 1937. On some new and known Myxosporidia of the fishes of Illinois. J. Parasit., 23:467.

—— 1947. Studies on Myxosporidia from the Beaufort region. I. *Ibid.*, 33:265.

—— 1947a. II. *Ibid.*, 33:271.

—— 1960. Some coelozoic Myxosporidia from New Zealand fishes. I. Tr. Roy. Soc. New Zealand, 88(2):265.

NAIDU, K. V., 1956. A new species of actinomyxid sporozoan parasitic in a fresh-water oligochaete. J. Protozool., 3:209.

NAVILLE, A., 1927. Le cycle chromosomique, la fecondation et la reduction chromatique de *Chloromyxum leydigi*. Ann. Inst. Océan., 4:177.

—— 1930. Recherches sur le sexualité chez les myxosporidies. Arch. Protist., 69:327.

—— 1930a. Le cycle chromosomique d'une nouvelle actinomyxidie: *Guyenotia sphaerulosa* n. g., n. sp. Quart. J. Micr. Sci., 73:547.

NEMECZEK, A., 1926. Beiträge zur Kenntnis der Myxosporidienfauna Brasiliens. Arch. Protist., 54:137.

NIGRELLI, R. F., 1946. Parasites and diseases of the ocean pout, *Macrozoarces americanus*. Bull. Bingham Ocean. Collect., 9:187.

NOBLE, E. R., 1939. Myxosporidia from tide pool fishes of California. J. Parasit., 25:359.

—— 1941. On distribution relationships between California tide pool fishes and their myxosporidian parasites. *Ibid.*, 27:409.

—— 1941a. Nuclear cycles in the life history of the protozoan genus Ceratomyxa. J. Morphol., 69:455.

—— 1944. Life cycles in the Myxosporidia. Quart. Rev. Biol., 19:213.

PARISI, B., 1912. Primo contributo alla distribuzione geografica dei missosporidi in Italia. Atti. Soc. Ital. Sc. Nat., 50:283.

PETRUSCHEWSKY, G. K., 1932. Zur Systematik und Cytologie der Myxosporidia aus einigen Fischen des Weissen Meeres. Arch. Protist., 78: 543.

PINTO, C., 1928. Myxosporideos e outros protozoarios intestinaes de peixes observadoes na America do Sul. Arch. Inst. Biol. São Paulo, 1:1.

PREHN, M., 1904. Ueber die Drehkrankheit der Salmoniden. Arch. Protist., 5:145.

—— 1925. Eine neue Schleienkrankheit. Fisch.-Zeit., 28:299.

PRUDHOMME, M., and PANTALÉON, J., 1959. Sur un cas de myxosporidiose du saumon. Bull. Acad. Vét. France, 32:137.

RICE, V. J., and JAHN, T. L., 1943. Myxosporidian parasites from the gills of some fishes of the Okoboji region. Proc. Iowa Acad. Sci., 50:313.

—— —— 1943a. Internal myxosporidian infections of some fishes of Okoboji region. *Ibid.*, 50:323.

SCHRÖDER, O., 1907. Beiträge zur Entwicklungsgeschichte der Myxosporidien, *Sphaeromyxa sabrazesi*. Arch. Protist., 9:359.

SINDERMANN, C. J., 1959. Zoogeography of sea herring parasites. J. Parasit., 45, Supple.:34.

SOUTHWELL, T., and PRASHAD, B., 1918. Parasites of Indian fishes etc. Rec. Indian Mus., 15:341.

ŠTOLC, A., 1899. Actinomyxidies, nouveau groupe de Mésozoaires parent des Myxosporidies. Bull. Intern. l'Acad. Sci. de Boheme, 22:1.

THÉLOHAN, P., 1892. Observation sur les myxosporidies et essai de classification de ces organisms. Bull. Soc. Philom., 4:165.

—— 1895. Recherches sur les myxosporidies. Bull. Sc. Fr. Belg., 26:100.

TRIPATHI, Y. R., 1948. Some new Myxosporidia from Plymouth with a proposed new classification of the order. Parasitology, 39:110.

—— 1953. Studies on parasites of Indian fishes. I. Record Indian Mus., 50:63.

WARD, H. B., 1919. Notes on North American Myxosporidia. J. Parasit., 6:49.

WILLIS, A. G., 1949. On the vegatative forms and life history of *Chloromyxum thysites* Gilchrist and its doubtful systematic position. Australian J. Sci. Res., Series B. Biol. Sci., 2:379.

Chapter 30
Order 4. **Microsporida** Balbiani

THE organisms placed in this order are far more widely distributed as cytozoic parasites among various animal phyla than either Myxosporida or Actinomyxida, having been found parasitic in protozoa themselves up to lower vertebrates. However, they are typically parasites of arthropods and fishes. These organisms invade and undergo asexual reproduction and sporogony within the host cells which show usually an enormous hypertrophy of both cytoplasm and nucleus (Fig. 297), a characteristic feature of the microsporidan infection.

The microsporidan spores are relatively small and simple in appearance compared with those of Myxosporida or Actinomyxida. The spores of a given species vary somewhat in dimensions (Kudo, 1924), especially if the species belong to the genera in which varied number of spores are formed in a sporont. Some workers such as Walters (1958), Kramer, (1960a), etc., hold that the spore dimensions are affected by various conditions, but until a more precise characterization of microsporidan species becomes available, the form and dimensions of the spore must be taken into consideration for species identification (Thomson, 1960a).

Microsporidan spores vary from spherical to cylindrical in shape and are highly refractile. Viewed in life, a clear area may often be found near one end and sometimes at both ends. No other structures are recognized. But each spore contains a sporoplasm and a long fine polar filament. The latter structure can be made to extrude from the spore and its presence identifies the organism as a true microsporidan. Morganthaler (1920) demonstrated the tubular nature of the filament of *Nosema apis* and the emergence of a fluid mass and a sporoplasm (?) from open end. Oshima (1937) suggested after studying the filament extrusion of *N. bombycis* that the sporoplasm is directly inoculated through the filament into the host gut-epithelium. Gibbs (1953, 1956),

807

Bailey (1955a), Kramer (1960) and West (1960) agree with this view.

On the other hand, Korke (1916) noted an apparent sporoplasm attached to the end of a filament in *N. ctenocephali* (*N. pulicis*) and considered that the filament may conduct the amoebula to a

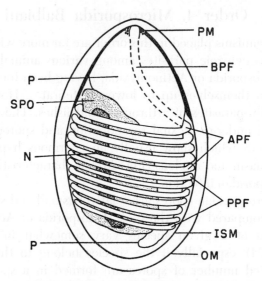

Fig. 296. A diagram of a microsporidan spore viewed under electron microscope (Kudo and Daniels). APF, anterior polar filament; BPF, basal portion of the filament; ISM, inner surface membrane; N, nucleus; OM, outer membrane; p, polaroplast; PM, polar mass and polaroplast membrane; PPF, posterior polar filament; SPO, sporoplasm.

distant part of the tissue and thus ensure an advance of the parasite into fresh area. Dissanaike (1955, 1957a), Dissanaike and Canning (1957) and Canning (1957) believe that the end of the filament is attached to the sporoplasm and the force of the extrusion of the filament "drags out" the sporoplasm. The recent electron microscope studies have not settled this point, as Weiser, (1959) and Huger (1960) find the filament solid, while Lom and Vavra, (1961) and Kudo and Daniels (1963) hold the filament tubular. It appears certain that the polar filament is coiled spirally along the inner surface of the spore membrane, the sporoplasm is located within this coil and there is a swelling substance surrounding the basal portion of the filament (Fig. 296).

When such spores are taken into the digestive tract of a specific host, the polar filaments are extruded and the sporoplasms emerge as amoebulae and enter the gut epithelium and then into the blood stream or body cavity and reach the specific site of in-

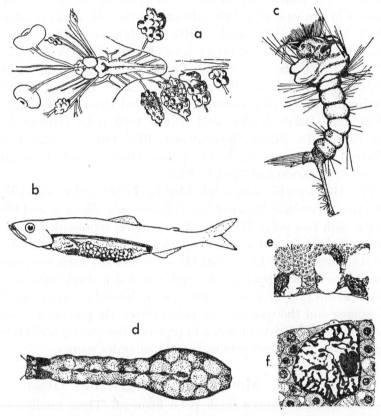

FIG. 297. Effects of microsporidan infections upon hosts, a, the central nervous system of *Lophius piscatoris* infected by *Nosema lophii* (Doflein); b, a smelt infected by *Glugea hertwigi*, ×2/3 (Schrader); c, a Culex larva infected by *Thelohania opacita*, ×14 (Kudo); d, a simulium larva infected by *T. multispora*, ×10 (Strickland); e,f, normal and hypertrophied nuclei of the adipose tissue cells of larval *Culex pipiens*, the latter due to a heavy infection by *Stempellia magna*, ×1330 (Kudo).

fection. There, they grow and undergo repeated binary or multiple division, filling the host cells with continuously increasing schizonts, bringing about often extremely hypertrophied nuclei of host cells (Fig. 297). In the trophozoites develop a variable number of sporonts, each of which produces a number of spores characteristic of each genus.

When infected heavily, the host animals die as a direct result of infection which may occur in an epidemic form as is well known in the case of pebrine disease of the silk worm (Pasteur, 1870; Stempell, 1909; Kudo, 1916; Hutchinson, 1920; Jameson, 1922), nosema-disease of honey bees (Zander, 1911; White, 1919; Farrar, 1947), etc. (Taxonomy, Léger and Hesse, 1922, Kudo, 1924b, Jírovec, 1936, Weiser, 1947, 1961, Thomson, 1960a, Codreanu, 1961; spore structure, Léger and Hesse, 1916a, Kudo, 1920, 1921, 1924b, Kohler, 1921, Vavra, 1960; polar filament, Kudo, 1913, 1918, 1924b, Morgenthaler, 1922, Oshima, 1927, 1937; electron microscope study of spores, Huger, 1960, Lom and Vavra, 1961, Kudo and Daniels, 1963.)

The Microsporida was subdivided by Léger and Hesse (1922) into two suborders: Monocnidea with one polar filament and Dicnidea with two polar filaments. The latter is represented by one genus *Telomyxa* and a single species, *glugeiformis* which had been discovered by Léger and Hesse. It has not been seen until recently when Codreanu (1961) rediscovered it. Codreanu found that Telomyxa spores are "conjugate or bilocular" spores or diplospores and the sporonts are polysporous. He proposed Monocytosporea and Polycytosporea to replace Monocnidea and Dicnidea. The present work retains the old suborder names.

Suborder 1. **Monocnidina** Léger and Hesse
Each spore contains a single polar filament. (Three families.)

Family 1. **Nosematidae** Labbé
Spores are oval, pyriform, ellipsoid or subcylindrical (length less than four times the width). The majority of Microsporida belong to this family.

Genus **Nosema** Nägeli. Each sporont develops into a single spore. Numerous species.

N. bombycis N. (Fig. 298, *a, b*). In all tissue cells of the embryo, larva, pupa and adult of the silkworm, *Bombyx mori*. Oval spores 3-4μ by 1.5-2μ; polar filament up to 98μ or more in length. Advanced infection in host silkworm is characterized by the appearance of numerous minute brownish-black spots scattered over the body surface, which gave rise to such names as "pébrine", "Fleckenkrankheit," etc. Heavily infected larvae cannot spin cocoons and perish without pupation. The organisms invade, and develop in, ova so that newly hatched larvae are already infected with this microsporidan. Viable spores introduced *per os* bring about infections in *Arctia caja* (Stempell, 1909), *Margarnia pyloalis, Chilo simplex* (Oshima, 1935), and *Hyphantria cunea* (Kudo and DeCoursey, 1940). (Morphology and development, Stempell, 1909; Kudo, 1924b.)

N. bryozoides (Korotneff) (Fig. 298, *c, d*). In *Plumatella (Alcyonella) gungosa* and *P. repens* (Bryozoa). The infection occurs in the testicular cells which are attached to the funiculus; infected host cells become later detached from it and float in the body cavity and finally disintegrate and set free the spores. Spores ellipsoid, 7-10μ by 5-6μ (Braem, 1911; Schröder, 1914).

N. apis Zander (Fig. 298, *e-g*). In the midgut wall and Malpighian tubules of adult honey bees, *Apis mellifica*. The parasites are encountered most frequently in workers, although drones and queens are also susceptible to infection. Spores elongate oval, 4-6μ by 2-4μ; polar filaments long, and often shows two sections of different undulations (Fig. 289, *g*) (Kudo, 1921a). The ovary of experimentally infected queen bees undergoes various degrees of degeneration depending on the extent of the gut infection (Fyg, 1945; Farrar, 1947; Hassanein, 1951), though the eggs are free from the parasites. The disease is known as nosema-disease and appears to be quite destructive especially in package bees. The organism is capable of infecting thirteen widely different insects (Fantham and Porter, 1913). (Morphology and development, Zander, 1909, Fantham and Porter, 1912; epidemiology and control of infection, Bailey, 1955; fumagillin for Nosema control, Farrar, 1954; general review, Bailey, 1959.)

N. cyclopis Kudo (Fig. 298, *h, i*). In *Cyclops fuscus;* spores 4.5μ by 3μ (Kudo, 1921b).

N. anophelis K. (Fig. 298, *j*, *k*). In the gut epithelium and fat
bodies of the larvae and adult of *Anopheles quadrimaculatus;*
spores elongate ovoid, 4.7-5.8μ by 2.3-3.2μ (Kudo, 1924b); also
in *A. maculipennis* (Missiroli, 1928).

N. aedis K. (Fig. 298, *l*, *m*). In the adipose tissue of a larval
Aedes aegypti; spores broadly pyriform, 7.5-9μ by 4-5μ.

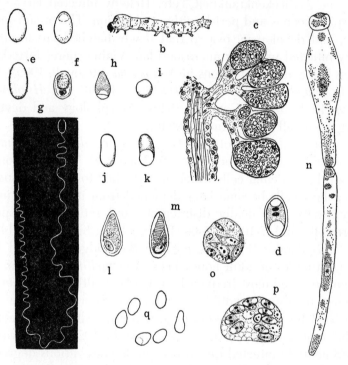

Fig. 298. a, b, *Nosema bombycis* (Kudo) (a, fresh spores, ×1500; b, a
heavily infected silkworm larva showing characteristic dots on integument,
×⅔); c, d, *N. bryozoides* (c, infected funiculus, ×270 (Braem); d, a stained
spore, ×1200 (Schröder)); e–g, *N. apis* (Kudo) (e, a fresh spore; f, a stained
spore, ×1560; g, a spore with the extruded polar filament as seen in dark
field, ×800); h, i, views of fresh spores of *N. cyclopis*, ×1560 (Kudo); j, k,
fresh spores of *N. anophelis*, ×1600 (Kudo); 1, m, preserved and stained
spores of *N. aedis*, ×1530 (Kudo); n, *Carcinoecetes conformis,* a gregarine,
infected by schizonts and spores of *Nosema frenzelinae* (Léger and Duboscq);
o–q, *Nosema notabilis*, ×1400 (Kudo) (o, a stained trophozoite of
Sphaerospora polymorpha, a myxosporidan, infected by six trophozoites of
Nosema notabilis; p, another host trophozoite in which nine spores and
two trophozoites of *N. notabilis* occur; q, six fresh spores of *N. notabilis*).

N. frenzelinae Léger and Duboscq (Fig. 298, *n*). In the cytoplasm of the cephaline gregarine, *Carcinoecetes conformis,* parasitic in the gastric caeca and intestine of *Pachygrapsus marmoratus;* spores about 2.8μ long; extruded polar filament up to 25μ long (Léger and Duboscq, 1909).

N. notabilis Kudo (Fig. 298, *o-q*). In the trophozoite of the myxosporidan, *Sphaerospora polymorpha* (p. 786) which inhabits the urinary bladder of *Opsanus tau* and *O. beta.* The host fish remain free from the microsporidan infection. The entire development takes place in the cytoplasm of the host trophozoites. Trophozoites small binucleate, multiply by binary fission. Spores ovoid to ellipsoid; sporoplasm binucleate; fresh spores 2.9-4μ by 1.4-2.5μ; extruded polar filament 45-62μ. When heavily infected, the host myxosporidan trophozoites degenerate and disintegrate. A unique example of hyperparasitism in which two cnidosporidans are involved (Kudo, 1944).

N. helminthorum Moniez. In the helminths, *Moniezia expansa* and *M. benedeni* of sheep and *M.* sp. from a buffalo calf; spores oval, 5.8-6.8μ by 3.3μ; not pathogenic to hosts (Dissanaike, 1957, 1957a).

N. whitei Weiser. In the fat body of larvae and pupae of the flour beetle, *Tribolium castaneum;* spores oval, 4.5-5μ by 2-2.5μ; polar filament longer than 160μ. Infection *per os. T. confusum* and *T. destructor* are refractory (Weiser, 1953, 1961). Weiser considers that *N. buckleyi* Dissanaike (1955) is identical with this species.

N. sp. Cort, Hussey and Ameel. In the larvae of twelve species of strigeoid trematodes; spore 3.9-6μ by 2-4μ; host snails are uninfected. (Experimental infection by feeding the spores to host snails, Cort *et al.,* 1960.)

Genus **Glugea** Thélohan. Each sporont develops into two spores; the infected host cells become extremely hypertrophied, and transform themselves into the so-called **Glugea cysts** (Figs. 297, *b;* 299, *e*). Many species (Kudo, 1924b; Jírovec, 1936; Weiser, 1961).

G. anomala (Moniez) (Fig. 299, *a-f*). In *Gasterosteus aculeatus, G. pungitus* (sticklebacks) and *Gobius minutus;* Cysts conspicuous, up to about 5 mm. in diameter; host cells are extremely hypertrophied; spores 4-6μ by 2-3μ. Morphology and sporogony (Stempell, 1904; Weissenberg, 1913; Debaisieux, 1920).

G. *mülleri* Pfeiffer. In the muscles of *Gammarus pulex* and *G. locusta;* spores 5-6μ by 2-3μ (Debaisieux, 1919).

G. *hertwigi* Weissenberg (Figs. 297, *b;* 299, *g, h*). In the smelt, *Osmerus mordax* and *O. eparlanus.* Schrader (1921) found the intentine the primary site of infection, the cysts varying in size, up to 3 mm. in diameter; as the cysts grow in the mucosa, they come to lie immediately under the peritoneum. Spores measure 4-5.5μ by 2-2.5μ. Fantham, Porter and Richardson (1941) found the cysts in the serous membrane of the hind gut; as the spores were 3.5-4.6μ by 1.5-2μ, they named the organism *Glugea hertwigi* var. *canadensis.* (Morphology and spore-formation, Weissenberg, 1911, 1913; Schrader, 1921.)

Genus **Perezia** Léger and Duboscq. Each sporont produces two spores as in Glugea, but infected host cells are not hypertrophied. A few species.

P. *mesnili* Paillot (Fig. 299, *i*). In cells of silk glands and Malpighian tubules of larvae of *Pieris brassicae;* spores 3.4μ by 1.5-2μ (Paillot, 1918, 1929).

P. *lankesteriae* Léger and Duboscq (Fig. 299, *j*). In the cytoplasm of the gregarine, *Lankesteria ascidiae,* parasitic in the intestine of the tunicate, *Ciona intestinalis.* It attacks only the gregarine which are free in the lumen of the gut; the host nucleus does not undergo hypertrophy; ovoid spores 2.5μ long.

P. *pyraustae* Paillot. In the Malpighian tubules, silk glands and ovary of the larvae and adults of the European corn-borer, *Pyrausta nubilalis;* spores 3.2-4.7 (4.2)μ by 1.8-2.6 (2.1)μ (Kramer, 1959).

P. *fumiferanae* Thomson. In the gut epithelium, muscles, Malpighian tubules, fat bodies, silk glands and gonads of the spruce budworm, *Choristoneura fumiferana;* spores 3-5μ by 2μ; polar filament 65-105μ long. In a natural population, 40 per cent infection was noted. Transmission through mouth and eggs. Experimentally, the organism infects the jack pine budworm, *C. pinus* in which a very similar microsporidan was found to occur in natural populations (Thomson, 1955-59).

P. *disstriae* T. Spores 4-5μ by 2μ: in the silk glands, mid-gut epithelium, Malpighian tubules, fat bodies, etc., of the forest tent catapillar, *Malacosoma disstria* (Thomson, 1959a).

Genus **Gurleya** Doflein. Each sporont develops into four sporo-

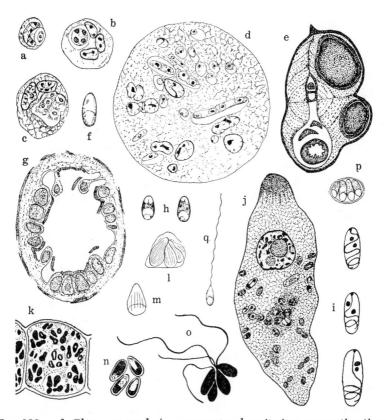

FIG. 299. a–f, *Glugea anomala* (a, a young trophozoite in a connective tissue cell of the intestine of a young host fish, seven days after feeding on spores; b, c, more advanced stages; d, a later stage, the host cell being multinucleated and 41µ in diameter, ×1000 (Weissenberg); e, section of an infected *Gasterosteus aculeatus*, showing two large cysts (Thélohan); f, a fresh spore, ×1500 (Stempell)); g, h, *G. hertwigi* (Schrader) (g, cross-section of the infected intestine of a smelt, ×14; h, 2 spores); i, 3 stained spores of *Perezia mesnili*, ×2265 (Palliot); j, section of *Lankesteria ascidiae*, a gregarine, infected by *P. lankesteriae*, ×900 (Léger and Duboscq); k–o, *Gurleya tetraspora* (k, infected hypodermal cells of Moina, ×660 (Jírovec); l, a mature sporont; m, a fresh spore (Doflein); n, stained spores; o, spores with extruded polar filaments (Jírovec)); p, q, a sporont and a spore with the extruded filament of *Gurleya richardi*, ×1200 (Cépède).

blasts and finally into four spores. A few species. (Species, Kudo, 1924b; Jírovec, 1936; Weiser, 1947, 1961.)

G. *tetraspora* D. (Fig. 299, k-o). In the hypodermal cells of *Daphnia maxima* and *Moina rectirostris;* spores pyriform, 2.8-3.4μ by 1.4-1.6μ (Jírovec, 1942). The infected host appears opaque white.

G. *richardi* Cépède (Fig. 299, *p, q*). In *Diaptomus castor;* spores 4-6μ by 2.8μ.

G. *dispersa* Codreau. In adipo-phagocytes of *Artemia salina;* spores pyriform, 5-5.9μ long (Codreau, 1957). Codreau further described three other species: *Nosema exigua, Glugea artemiae* and *Plistophora myotropha,* from the same host animal.

Genus **Thelohania** Henneguy. Each sporont develops into eight sporoblasts and ultimately into eight spores; sporont membrane may degenerate at different times during spore formation. (Numerous species.)

T. *legeri* Hesse (*T. illinoisensis* Kudo) (Fig. 300, *a-e*). In the fat bodies of the larvae of several species of Anopheles; spores 4-6μ by 3-4μ; heavily infected larvae die without pupation; widely distributed. Spore-formation (Kudo, 1924).

T. *opacita* Kudo (Figs. 297, *c;* 300, *f, g*). In the adipose tissue of the larvae of Culex mosquitoes; spores 5.5-6μ by 3.5-4μ (Kudo, 1922, 1924a).

T. *californica* Kellen and Lipa. Resembles the last-mentioned species; but spores ovate and larger, 7.9μ by 5μ (Kellen and Lipa, 1960); in the adipose tissue of the larvae of *Culex tarsalis;* transmission transovarian. (Structure of spore, Kudo and Daniels, 1963, (Fig. 296.) Kellen and Wills (1962) described eight new species of Thelohania from California mosquitoes.

T. *reniformis* Kudo and Hetherington (Fig. 300, *h*). In the gut cells of the nematode, *Protospirura muris,* in mice; reniform spores 3-4μ by 1.5-1.8μ (Kudo and Hetherington, 1922).

T. *hyphantriae* Weiser. In the fat body of the larvae of *Hyphantria cunea;* spores oval, 4-5μ by 2μ (Weiser and Veber, 1957; Weiser, 1961).

Genus **Stempellia** Léger and Hesse. Each sporont produces one, two, four, or eight sporoblasts and finally one, two, four, or eight spores.

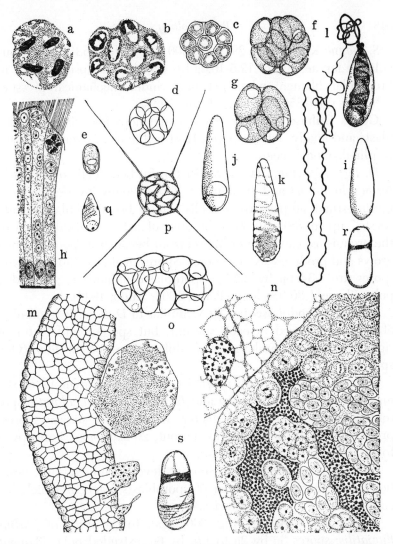

Fig. 300. a–e, *Thelohania legeri*, ×1570 (Kudo) (a, b, stained sporogonic stages; c, d, mature sporonts; e, a fresh spore); f, g, mature octosporous and tetrasporous sporonts of *T. opacita*, ×1570 (Kudo); h, gut epithelial cells of Protospirura infected by *T. reniformis*, ×1040 (Kudo and Hetherington); i-l, *Stempellia magna*, ×1570 (Kudo) (i, j, fresh spores; k, slightly pressed spore in Lugol; l, a spore with the nearly completely extruded polar filament, stained after Fontana); m–o, *Duboscqia legeri* (Kudo) (m, the mid-gut of *Reticulitermes flavipes* with an enlarged and two uninfected fat bodies, ×57; n, portion of an infected and two uninfected fat body cells of the termite in section; o, mature sporont in life, ×1530); p, q, *Trichoduboscqia epeori* (Léger) (p, a mature sporont, ×1330; q, a fresh spore, ×2670); r, s, stained spores of *Plistophora longifilis*, ×1280 (Schuberg).

S. magna Kudo (Figs. 297, *e, f;* 300, *i-l*). In fat-bodies of various culicine larvae; spores 12.5-16.5µ by 4-5µ; polar capsule visible in life; polar filament when extruded under mechanical pressure measures up to 350-400µ long (Kudo, 1925a).

Genus **Duboscqia** Pérez. Sporont develops into sixteen sporoblasts and finally sixteen spores. Host-cell nuclei extremely hypertrophied. One species.

D. legeri P. (Fig. 300, *m-o*). In the fat-body cells of *Reticulitermes lucifugus* and *R. flavipes.* Trophozoites invade the perimidintestinal adipose tissue cells which become enlarged into "cysts," up to 660µ by 300µ, because of active multiplication of the organisms; each binucleate schizont becomes a sporont which grows and produces sixteen spores. Spores ovoid to ellipsoid; fresh spores are 4.3-5.9µ by 2.2-3µ; sporoplasm uninucleate; extruded polar filament 80-95µ long (Pérez, 1908; Kudo, 1942).

Genus **Trichoduboscqia** Léger. Similar to *Duboscqia* in number of spores produced in each sporont; but sporont with 4 (or 3) rigid transparent prolongations, difficult to see in life. (One species.)

T. epeori L. (Fig. 300, *p, q*). In fat-bodies of nymphs of the mayflies, *Epeorus torrentium* and *Rhithrogena semicolorata;* sporonts spherical, 9-10µ in diameter, with usually sixteen spores; prolongations of membrane in sporont, 20-22µ long; spores pyriform, 3.5-4µ long (Léger, 1926).

Genus **Plistophora** Gurley. Sporont develops into variable number (often more than sixteen of sporoblasts, each of which becomes a spore. (Several species.)

P. longifilis Schuberg (Fig. 300, *r, s*). In the testis of *Barbus fluviatilis;* spores 3µ by 2µ to 12µ by 6µ; extruded polar filament up to 510µ long.

P. kudoi Sprague and Ramsey. In the epithelial cells of the mid-gut of *Blatta orientalis;* fresh spores about 3.2µ by 1.75µ; polar filament 25-50µ long.

P. sp. Spangenberg and Claybrook. In *Hydra littoralis;* 2-3µ by 1-2µ; polar filament 28-38µ long; transovarian infection; the host hydra does not suffer from infection.

Genus **Pyrotheca** Hesse. Schizogony and sporogony unknown; spores elongate pyriform, anterior end attenuated, posterior end rounded, slightly curved; sporoplasm in posterior region, with one to two nuclei; polar capsule large. (One species, Hesse, 1935.)

P. incurvata H. (Fig. 301, *a, b*). In fat-bodies and haemocoele of *Megacylcops viridis;* spores 14μ by 3μ; polar filament 130μ long.

Family 2. **Coccosporidae** Kudo

Spores are spherical or subspherical.

Genus **Coccospora** Kudo (*Cocconema,* Léger and Hesse). Spore spherical or subspherical. (Several species, Léger and Hesse, 1921, 1922; Kudo, 1925b.)

C. slavinae (L. and H.) (Fig. 301, *c, d*). In gut-epithelium of *Slavina appendiculata;* spores about 3μ in diameter.

Family 3. **Mrazekiidae** Léger and Hesse

Spores are tubular or cylindrical; straight or curved; the diameter is less than one-fifth the length.

Genus **Mrazekia** L. and H. (*Myxocystis,* Mrazek). Spore, tubular and straight; a long or short process at one extremity (Léger and Hesse, 1916). (Species, Jírovec, 1936a).

M. caudata L. and H. (Fig. 301, *e, f*). In the lymphocytes of *Tubifex tubifex;* spore cylindrical, 16-18μ by 1.3-1.4μ, with a long process. (Development, Lom, 1958.)

Genus **Bacillidium** Janda. Spore cylindrical, but without any process; one end narrowed in a few species (Janda, 1928). (Several species, Jírovec, 1936a.)

B. criodrili J. (Fig. 301, *g*). In the lymphocytes in the posterior portion of the body cavity and nephridia of *Criodrilus lacuum;* infected lymphocytes become hypertrophied from 15μ to 200-400μ in diameter; the infected part of the body appears yellowish; spores 20-22μ by 1μ (Janda); 15.5-17μ by 1.2-1.4μ up to 24-25μ by 1.6μ (commonly 18-20μ by 1.4-1.5) (Jírovec).

B. limnodrili Jírovec (Fig. 301, *h, i*). In lymphocytes within gonads of *Limnodrilus claparedeanus;* spores 22-24μ by 1.5μ (Jírovec, 1936a).

Genus **Cougourdella** Hesse. Spore cylindrical, with an enlarged extremity, resembling the fruit of *Lagenaria cougourda.* Three species (Hesse, 1935).

C. magna H. (Fig. 301, *j, k*). In haemocoele and fat body of *Megacyclops viridis;* spores 18μ by 3μ; polar filament 110μ long; sporoplasm with one to two nuclei or two uninucleate sporoplasms.

Genus **Octosporea** Flu. Spore cylindrical; more or less curved; ends similar. (Six species, Jírovec, 1936a.)

O. muscae-domesticae F. (Fig. 301, *l*). In gut and germ cells of Musca and Drosophila; spores 5-8μ long (Chatton and Krempf, 1911).

Genus **Spiroglugea** (*Spironema*) Léger and Hesse. Spore tubular and spirally curved; polar capsule large. (One species.)

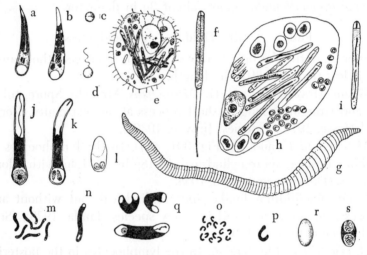

F ig. 301. a, b, stained spores of *Pyrotheca incurvata,* ×1330 (Hesse); c, d, spores of *Coccospora slavinae,* the latter with extruded filament, ×1330 (Léger and Hesse); e, f, *Mrazekia caudata* (e, an infected host cell, ×465 (Mrazek); f, a spore, ×1165 (Léger and Hesse)); g, *Criodrilus lacuum,* infected by *Bacillidium criodrili,* showing the enlarged posterior region, ×⅔ (Janda); h, i, *B. limnodrili* (Jírovec) (h, trophozoites and spores of the microsporidan in a host lymphocyte, ×600; i, a stained spore, ×930); j, k, stained spores of *Cougourdella magna,* ×1330 (Hesse); l, a spore of *Octospora muscae-domesticae,* ×1430 (Chatton and Krempf); m, n, spores of *Spiroglugea octospora* (Léger and Hesse) (m, ×665; n, ×2000); o, p, spores of *Toxoglugea vibrio* (Léger and Hesse) (o, ×665; p, ×2000); q, stained spores of *T. gerridis,* ×2000 (Poisson); r, s, a fresh and a stained spore of *Telomyxa glugeiformis,* ×2000 (Léger and Hesse).

S. octospora L. and H. (Fig. 301, *m, n*). In fat body of larvae of *Ceratopogon* sp.; spores 8-8.5μ by 1μ.

Genus **Toxoglugea** (*Toxonema*) Léger and Hesse. Minute spore curved or arched in semi-circle. (Four species, Poisson, 1941.)

T. vibrio L. and H. (Fig. 301, *o, p*). In the fat body of *Ceratopogon* sp.; spores 3.5μ by less than 0.3μ.

T. gerridis Poisson (Fig. 301, *q*). In the fat body of the bug, *Aquarius najas;* sporont gives rise to eight sporoblasts and then to eight spores; also monosporous; microspores 4.5μ by 0.8μ, the polar filament 40-50μ long; macrospores 7-8μ long.

Suborder 2. **Dicnidina** Léger and Hesse

Each spore contains a single polar filament, but two spores are united (thus "conjugate" or "bilocular," according to Codreanu (1961); sporont produces eight to thirty-two spores.

Family **Telomyxidae** L. and H.

Diplospore; polar filaments are extruded from the two ends.

Genus **Telomyxa** L. and H. Conjugated or bilocular spores; with a coiled polar filament at each end; the equatorial zone bulges out; sporont forms eight to thirty-two spores (Codreanu). Léger and Hesse (1910) defined: spores with two polar capsules, one at each end; sporont develops into eight, sixteen or more spores. (One species.)

T. glugeiformis L. and H. (Fig. 301, *r, s*). In the fat body of the nymphal *Ephemera vulgata;* spores 6.5μ and 4μ (Léger and Hesse, 1910); diplospores ellipsoid, with bulged equatorial zone; 6-7μ long; with a large circular area at each of the rounded ends; polar filament when extruded measures about 60μ; in the nymph of *Ephemera danica* (Codreanu, 1961).

References

BAILEY, L., 1955. The epidemiology and control of Nosema disease of the honey bee. Ann. Appl. Biol., 53:379.

———— 1955a. The infection of the ventriculus of the adult honey bee by *Nosema apis*. Parasitology, 45:86.

———— 1959. Infectious diseases of the honey bee. Rep. Rothamst. Exp. Station for 1959, p. 204.

BORCHERT, A., 1930. Nosemainfektion. Arch. Bienenk., 11:1.

BRAEM, F., 1911. Beiträge zur Kenntnis der Fauna Turkestans. VII, Trav. Soc. Imp. Nat., St. Petersbg., 42:1.

CHATTON, E., and KREMPF, A., 1911. Sur le cycle évolutif et la position systématique des protistes du genre Octosporea, etc. Bull. soc. zool. France, 36:172.

CODREANU, M. R., 1957. Sur quatre espèces nouvelles de microsporidies parasites de l'*Artemia salina* du Roumanie. Ann. Sci. Nat. Zool. (11) 19:561.

———— 1961. Sur la structure bicellulaire des spores de Telomyxa cf. glugeiformis Léger et Hesse, 1910, parasite des nymphes d'Ephemera (France, Roumanie) et les nouveaux sous-ordres des Microsporidies, Monocytosporea nov. et Polycytosporea nov. C. R. Acad. Sci., 253:1613.

CORT, W. W., *et al.*, 1960. Studies on a microsporidian hyperparasites of strigeoid trematodes. I and II. J. Parasit., 46:317 and 327.

DEBAISIEUX, P., 1919. Études sur les microsporidies. II, III. La Cellule, 30:153.

———— 1920. IV. *Ibid.*, 30:215.

———— 1928. Études cytologiques sur quelques microsporidies. *Ibid.*, 38:389.

DISSANAIKE, A. S., 1955. On protozoa hyperparasitic in helminths, etc. J. Helminthol., 31:47.

———— 1957a. The morphology and life-cycle of *Nosema helminthorum* Moniez. Parasitology, 47:335.

———— and CANNING, E. U., 1957. The mode of emergence of the sporoplasm in Microsporidia, etc. *Ibid.*, 47:92.

DOFLEIN, F., 1898. Studien zur Naturgeschichte der Protozoen. III. Zool. Jahrb. Anat., 11:281.

FANTHAM, H. B., and PORTER, A., 1912. The morphology and life history of *Nosema apis,* etc. Ann. Trop. Med. Parasit., 6:163.

———— ———— and RICHARDSON, L. R., 1941. Some Microsporidia found in certain fishes and insects in Eastern Canada. Parasitology, 33:186.

FARRAR, C. L., 1947. Nosema losses in package bees as related to queen supersedure and honey yields. J. Econ. Entom., 40:333.

———— 1954. Fumagillin for Nosema control in package bees. Amer. Bee J., 94:52.

FOÀ, ANNA, 1924. Modificazione al ciclo morfologico e biologico del *Nosema bombycis* Nägeli. Boll. Lab. Zool. Gen. Agr. Portici, 17:147.

FYG, W., 1945. Die Einfluss der Nosema-Infektion auf die Eierstöcke der Bienenkönigin. Schweiz. Bien-Zeit., 68:67.

GIBBS, A. J., 1953. *Gurleya* sp., etc. Parasitology, 43-143.

———— 1956. *Perezia* sp., etc. *Ibid.*, 46:48.

HASSANEIN, M. H., 1951. Studies on the effect of infection with *Nosema apis* on the physiology of the queen honey-bee. Quart. J. Micr. Sci., 92:225.

HESSE, E., 1904. *Thelohania legeri* n. sp., microsporidie nouvelle, parasite des larves *d'Anopheles maculipennis* Meig. C. R. Soc. Biol., 57:570.

—— 1904a. Sur le developpement de *Thelohania legeri. Ibid.*, 57:571.

—— 1935. Sur quelques microsporidies parasites de *Megacyclops viridis.* Arch. zool. exper. gén., 75:651.

HUGER, A., 1960. Electron microscope study on the cytology of a microsporidian spore by means of ultrathin sectioning. J. Insect Pathol., 2:84.

HUTCHINSON, C. M., 1920. Pebrine in India. Mem. Dept. Agr. India, 1:177.

JAMESON, A. P., 1922. Report on the diseases of silkworms in India. Superint. Gov. Print., Calcutta, India. 165 pp.

JANDA, V., 1928. Ueber Microorganismen aus der Leibeshöhle von *Criodrilus lacuum* Hoffm. und eigenartige Neubildungen in der Körperwand dieses Tieres. Arch. Protist., 63:84.

JÍROVEC, O., 1936. Studien über Microsporidien. Mem. Soc. Zool. Tchéc. Prague, 4:1.

—— 1936a. Zur Kenntnis von in Oligochäten parasitierenden Microsporidien aus der Familie Mrazekiidae. Arch. Protist., 87:314.

—— 1942. Zur Kenntnis einiger Cladoceren-Parasiten. II. Zool. Anz., 140:129.

KELLEN, W. R., and LIPA, J. J., 1960. *Thelohania californica* n. sp., etc, J. Insect Pathol., 2:1.

—— and WILLS, W., 1962. New Thelohania from California mosquitoes. *Ibid.*, 4:41.

KOHLER, A., 1921. Ueber die chemische Zusammensetzung der Sporenschale von *Nosema apis.* Zool. Anz., 53:85.

KORKE, V. T., 1916. On a Nosema *(Nosema pulicis* n. sp.), etc. Indian J. Med. Res., 3:725.

KOROTONEFF, A., 1892. Myxosporidium bryozoides. Ztschr. wiss. Zool., 53:591.

KRAMER, J. P., 1959. Studies on the morphology and life history of *Perezia pyraustae* Paillot. Tr. Am. Micr. Soc., 78:336.

—— 1960. Observations on the emergence of the microsporidian sporoplasm. J. Insect Pathol., 2:433.

—— 1960a. Variations among the spores of the microsporidian *Perezia pyraustae.* Am. Midland Nat., 64:485.

KUDO, R. R., 1913. Eine neue Methode die Sporen von *Nosema bombycis* Nägeli mit ihren ausgeschnellten Polfäden dauerhaft zu präparieren, etc. Zool. Anz., 41:368.

—— 1916 Contribution to the study of parasitic Protozoa. II. Bull. Seric. Exp. St., 1:31.

—— 1918. Experiments on the extrusion of polar filaments of cnidosporidian spores. J. Parasit., 4:141.

—— 1920. On the structure of some microsporidian spores. *Ibid.*, 6:178.

—— 1921. Studies on Microsporidia, with special reference to those parasitic in mosquitoes. J. Morphol., 35:123.

—— 1921a. Notes on *Nosema apis.* J. Parasit., 7:85.

—— 1921b. Microsporidia parasitic in copepods. *Ibid.,* 7:137.

—— 1921c. On the nature of structures characteristic of cnidosporidian spores. Tr. Am. Micr. Soc., 40:59.

—— 1922. Studies on Microsporidia parasitic in mosquitoes. II. J. Parasit., 8:70.

—— 1924. III. Arch. Protist., 49:147.

—— 1924a. VI. J. Parasit., 11:84.

—— 1924b. A biologic and taxonomic study of the Microsporidia. Illinois Biol. Monogr., 9:79.

—— 1925. Studies on Microsporidia parasitics in mosquitoes. IV. Centralbl. Bakt. Orig., 96:428.

—— 1925a. V. Biol. Bull., 48:112.

—— 1925b. Microsporidia. Science, 61:366.

—— 1942. On the microsporidian, *Duboscqia legeri,* parasitic in *Reticulitermes flavipes.* J. Morphol., 71:307.

—— 1944. Morphology and development of *Nosema notabilis,* etc. Illinois Biol. Monogr., 20:1.

—— and Daniels, E. W. 1963. An electron microscope study of a microsporidian, Thelohania californica. J. Protozool., 10:112.

—— and DeCoursey, J. D., 1940. Experimental infection of *Hyphantria cunea* with *Nosema bombycis.* J. Parasit., 26:123.

—— and Hetherington, D. C., 1922. Notes on a microsporidian parasite of a nematode. *Ibid.,* 8:129.

Labbé, A., 1899. Sporozoa. Das Tierreich, Lief. 5, 180 pp.

Léger, L., 1926. Sur *Trichoduboscqia epeori* Léger. Trav. Lab. Hydro. Pisc., 18:1.

—— and Duboscq, O. 1909. Microsporidie parasite de Frenzelina. Arch. Protist., 17:117.

—— —— 1909a. *Perezia lankesteriae,* etc. Arch. zool. exper. 1(N.-R):89.

—— and Hesse, E., 1910. Cnidosporidies des larves d'éphémères. C. R. Acad. Sci., 150:411.

—— —— 1916. Mrazekia, genre nouveau de microsporidies à spores tubuleuses. C. R. soc. biol., 79:345.

—— —— 1916a. Sur la structure de la spore des microsporidies. *Ibid.,* 79:1049.

—— —— 1921. Microsporidies à spores sphériques. C. R. Acad. Sci., 173:1419.

—— —— 1922. Microsporidies bactériformes et essai de systèmatique du group. *Ibid.,* 174:327.

Lom, J., 1958. Contribution to the development of *Mrazekia caudata,* etc. Czech. Parasit., 5:147.

—— and Vavra, J., 1961. Ultrastructure of the spores of the fish parasite *Plistophora hyphessobryconis*. Wiad. Parazytol., 7:828.

Missiroli, A., 1928. Alcuni protozoi parassiti dell' "Anopheles maculipennis." Riv. Malariol., 7:1.

Morgenthaler, O., 1922. Der Polfaden von *Nosema apis*. Arch. Bienenk., 4:53.

Nägeli, K. W., 1857. Ueber die neue Krankheit der Seidenraupe und verwandte Organismen. Bot. Zeit., 15:760.

Ohmori, J., 1912. Zur Kenntnis des Pébrine-Erreger, *Nosema bombycis*. Arb. kaiserl. Gesundh., 40:108.

Ohshima, K., 1927. A preliminary note on the structure of the polar filament of *Nosema bombycis*, etc. Ann. Zool. Japan., 11:235.

—— 1935. Infection of *Chilo simplex* by *Nosema bombycis* and function of the haemo-lymphocyte. J. Zool. Soc. Japan, 47:607.

—— 1937. On the function of the polar filament of *Nosema bombycis*. Parasitology, 29:220.

Paillot, A., 1918. Deux microsporidies nouvelles parasites des chenilles de *Pieris brassicae*. C. R. Soc. Biol., 81:66.

—— 1929. Contribution a l'étude des microsporidies parasites de *Pieris brassicae*. Arch. d'Anat. Micros., 25:242.

Pasteur, L., 1870. Étude sur la maladie des vers à soie. Paris.

Pérez, C., 1908. Sur *Duboscqia legeri*, microsporidie nouvelle parasite du *Termes lucifugus*, etc. C. R. Soc. Biol., 65:631.

Poisson, R., 1941. Les microsporidies parasites des insectes hémiptères. IV. Arch. zool. expér. gén., 82(N.-R):30.

Schrader, F., 1921. A microsporidian occurring in the smelt. J. Parasit., 7:151.

Schröder, O., 1914. Beiträge zur Kenntnis einiger Microsporidien. Zool. Anz., 43:320.

Schuberg, A., 1910. Ueber Mikrosporidien aus dem Hoden der Barbe und durch sie verursachte Hypertrophie der Kerne. Arb. kaiserl. Gesundh., 33:401.

Spangenberg, D. B., and Claybrook, D. L., 1961. Infection of Hydra by Microsporidia. J. Protozool., 8:151.

Sprague, V., and Ramsey, J., 1942. Further observations on *Plistophora kudoi*, etc. J. Parasit., 28:399.

Stempell, W., 1904. Ueber *Nosema anomalum*. Arch. Protist., 4:1.

—— 1909. Ueber *Nosema bombycis*. Ibid., 16:281.

Thomson, H. M., 1955. *Perezia fumiferanae* n. sp., etc. J. Parasit., 41:416.

—— 1958. Some aspects of the epidemiology of a microsporidian parasite of the spruce budworm, etc. Canad. J. Zool., 36:309.

—— 1959. A microsporidian infection in the jack-pine budworms, etc. Ibid., 37:117.

—— 1959a. A microsporidian parasite of the forest tent catapillar, etc. Ibid., 37:217.

———— 1960. Variation of some of the characteristics used to distinguish between species of Microsporidia. I. J. Insect Pathol., 2:147.

———— 1960a. A list and brief description of the Microsporidia infecting insects. *Ibid.*, 2:346.

VAVRA, J., 1959. Beitrag zur Cytologie einige Mikrosporidien. Vest. Cesk. Spol. Zool., 23:247.

WALTERS, V. A., 1958. Structure, hatching and size variation of the spores in a species of Nosema, etc. Parasitology, 48:113.

WEISER, J., 1947. Klič k určovani Mikrosporidii. Acta Soc. Sci. Nat. Moravicae, 18:1.

———— 1959. *Nosema laphygmae* n. sp., etc. J. Insect Pathol., 1:52.

———— 1961. Die Mikrosporidien als Parasiten der Insekten. Beih. Zeitschr. angew. Entomol., No. 17.

———— and VEBER, J., 1957. Die Mikrosporidie *Thelohania hyphantriae* Weiser, etc. Zeitschr. angew. Entomol., 40:55.

WEISSENBERG, R., 1911. Ueber einige Mikrosporidien aus Fischen. Sitz.-ber Gesell. naturf. Freunde, Berlin, p. 344.

———— 1913. Beiträge zur Kenntnis des Zeugungskreises der Mikrosporidien, etc. Arch. mikr. Anat., 82:81.

WEST, A. F., JR., 1960. The biology of a species of Nosema parasitic in the flour beetle *Tribolium confusum*. J. Parasit., 46:747.

WHITE, G. F., 1919. Nosema-disease. Bull. U. S. Dept. Agr., No. 780.

ZANDER, E., 1909. Tierische Parasiten als Krankheitserreger bei der Biene. Leipz. Bienenzeit., 24:147.

———— 1911. Krankheit und Schädlinge der erwachsenen Bienen. Handbuch der Bienenkunde. II. 42 pp.

Chapter 31

Subphylum 2. **Ciliophora** Doflein

THE protozoa placed in this subphylum possess cilia, cirri or other compound ciliary structures which serve as organelles of locomotion. Two kinds of nuclei, macronucleus and micronucleus, are present in all without exception. Nutrition is holozoic or saprozoic. Asexual reproduction is by binary fission or budding, and sexual reproduction is by conjugation or autogamy in which micronuclei play an important rôle. The majority are free-living, but some groups are composed of parasitic forms.

The subphylum is divided into two classes: Ciliata and Suctoria. The latter do not possess cilia or cirri in the mature stage, as they are attached to foreign objects and feed by means of tentacles. However, young larvae produced by budding are ciliated and capable of free swimming. Guilcher (1951) showed in several species that the aciliated adult organism possesses infraciliature and a part of the ciliature gives rise to the ciliation of the newly formed bud, which indicates their possible descent from holotrichous ciliates. Fauré-Fremiet (1950) considered Suctoria as an order of Holotricha. The present work recognizes the close affinity between Suctoria and Holotricha, but retains the class Suctoria. Except for certain modifications, the system proposed by Fauré-Fremiet and expanded by Corliss is adopted here.

Class 1. **Ciliata** Perty

The ciliates inhabit all sorts of fresh, brackish and salt waters by free-swimming, creeping or being attached to various objects. They feed on minute plant and animal organisms which ordinarily abound in the water, thus their nutrition is holozoic or saprozoic.

The ciliates vary in size from a few microns to 2 mm. or more in length. The anterior and posterior extremities or poles are permanently differentiated. In mouth-bearing forms, oral and aboral

827

surfaces are distinguishable, while in numerous creeping forms the dorsal and ventral sides can be noted.

The body is covered by the pellicle which is ordinarily uniformly thin and covers the entire body surface very closely. The cytoplasm is differentiated into the ectoplasm and endoplasm. The ectoplasm is the peripheral zone located beneath the pellicle and contains kinetosomes and associated organelles, while the endoplasm contains the nuclei, food vacuoles, contractile vacuoles, mitochondria and various cell inclusions.

Except a small number of astomatous forms, all ciliates possess a cytostome which varies in its location, morphology and associated ciliature. It leads into the cytopharynx or gullet, which ends in the deeper part of the endoplasm. In the cytopharynx, there may be present one or more membranes to facilitate intaking of the food. The wall of the cytopharynx may contain trichites or some other fibrillar structures. When the cytostome is not at the anterior pole, there occurs a peristome or oral groove which begins at or near the anterior end and extends posteriorly. The peristome is ciliated so that food particles are swept down along it and ultimately into the cytostome which is situated at its posterior end. Solid waste materials are extruded from the cytopyge or cell-anus which is usually noticeable only at the time of defecation.

A comparatively small number of the ciliated protozoa which are placed in genus Opalina and three allied genera, live exclusively in the intestine of amphibians, reptiles and fish. They show characteristics that are quite different from the ciliates; namely, they contain two or many nuclei of one type; they reproduce sexually by syngamy or complete fusion of two gametes and not by conjugation; and their asexual reproduction is longitudinal fission or plasmotomy.

Because of these differences, Metcalf (1918, 1923) divided the Ciliata into two groups: Protociliata for Opalina and three other genera, and Euciliata for all other ciliates. Former editions followed Metcalf's scheme. In establishing Protociliata, Metcalf wrote: "It seems not improbable that the Opalinidae and Trichonympha may have arisen from similar ancestors. At least Trichonympha shows us among the Flagellata a highly ciliate form like the Opalinidae and the lower of the Euciliata. It seems still more probable that the Euciliata arose from ancestors which, like Protoopalina to-day, had become disturbed in their relations of mitosis and fission, and that they passed through a similar pseudonucleated condition to a condition of true binucleation and finally reached their present structure with nuclei of two sorts. . . ."

Recent studies have not brought to light any new data which might support Metcalf's view quoted above. Grassé (1952), Jírovec *et al.* (1953) and others transferred the opalinids from the Ciliata into Mastigophora. In the present work, the opalinids are excluded from the Ciliata and dealt with separately at the end of the ciliates (p. 1029).

The class Ciliata is divided into three subclasses: Holotricha, Spirotricha (p. 959) and Peritricha (p. 1016).

Subclass 1. **Holotricha** Stein

These ciliates show uniform ciliation over the body surface, though in Chonotrichida, there is no somatic ciliature in mature forms. Peristomal ciliation is not conspicuous and their is no cytostome in Astomatida. Nutrition is holozoic or saprozoic. Encystment is common. The holotrichous ciliates are widely and abundantly distributed in all sorts of fresh, brackish and salt waters; a few are parasitic. Seven orders.

Order 1. **Gymnostomatida** Bütschli

Cytostome opens directly to the outside; cytopharynx with trichites; body ciliature simple; no oral cilia.

Suborder 1. **Rhabdophorina** Fauré-Fremiet

Cytostome located at the anterior end or laterally; trichites in the cytopharyngeal wall (rhabdophorine); uniform body ciliation; many carnivorous forms. (Eleven families.)

Family 1. **Holophryidae** Perty

Cytostome located at or near the anterior end; body ciliation is uniform and complete.

Genus **Holophrya** Ehrenberg. Oval, globose or ellipsoidal; ciliation uniform; sometimes longer cilia at the anterior or posterior region; cytostome circular, simple, without any ciliary ring around it; cytopharynx with or without trichites or trichocysts; fresh or salt water. (Numerous species.)

H. simplex Schewiakoff (Fig. 302, *a*). Ellipsoidal; eighteen to twenty ciliary rows; cilia uniformly long; cytostome small; cytopharynx without trichocysts or trichites; contractile vacuole and cytopyge posterior; macronucleus large, round; 34μ by 18μ; fresh water.

Genus **Lagynophrya** Kahl. Resembles *Holophrya;* small elongate ovoid to short cylindrical; one side convex, the other more or less flattened; cytopharynx terminates anteriorly in a small cone-like process which may or may not be distinct; stagnant fresh or salt water. (Several species.)

L. mutans K. (Fig. 302, *b*). Body plastic; oval to cylindrical; colorless; narrowly striated; oral cone hemispherical without any trichocysts; body about 90μ long, when contracted about 65μ in diameter; among decaying leaves in fresh water.

Genus **Spasmostoma** Kahl. Somewhat similar to *Holophrya;* cytostome with flaps which beat alternately; ciliation uniform.

S. viride K. (Fig. 302, *c*). Spherical or oval; always with green food vacuoles containing Euglena and allied flagellates; cytostome at anterior end; cytopharynx with trichocysts, which are extensible at the time when food is taken in; cilia on about twenty rows, near cytostome somewhat longer; macronucleus round; body 50-75μ long; sapropelic.

Genus **Urotricha** Claraparède and Lachmann (*Balanitozoon,* Stokes). Body oval to ellipsoidal or conical; with 1 or more longer caudal cilia; ciliation uniform, except in posterior region which may be without cilia; cytostome at or near anterior end, surrounded by ring of heavier cilia; contractile vacuole, posterior; macronucleus spherical; fresh water.

U. agilis (Stokes) (Fig. 302, *d*). Body small; about 15-20μ long; swimming as well as leaping movement; standing fresh water with sphagnum.

U. farcta C. and L. (Fig. 302, *e*). Body 20-30μ long; fresh water. Kahl considers *U. parvula* Penard and *Balanitozoon gyrans* Stokes are identical with this species.

Genus **Plagiocampa** Schewiakoff. Ovoid, spindle-form or cylin-

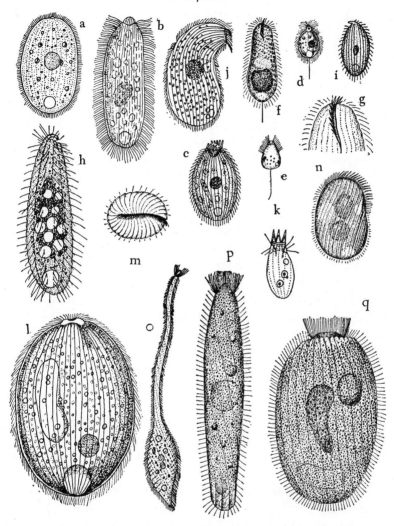

FIG. 302. a, *Holophrya simplex*, ×800 (Roux); b, *Lagynophrya mutans*,
×380 (Kahl); c, *Spasmostoma viride*, ×330 Kahl); d, *Urotricha agilis*,
×530 (Stokes); e, *U. farcta*, ×470 (Lieberkühn); f, g, *Plagiocampa marina* (Noland) (f, ×400; g, anterior end, ×670); h, *Chilophrya utahensis*,
×840 (Pack); i, *C. labiata*, ×500 (Edmondson); j, *Platyophrya lata*,
×280 (Kahl); k, *Stephanopogon colpoda*, non-ciliate side, ×500 (Kahl);
l, *Prorodon discolor*, ×330 (Bütschli); m, n, *Placus socialis*, ×530
(Noland); o, *Lacrymaria olor*, ×170 (Roux); p, *L. lagenula* (contracted),
×400; q, *L. coronata* (contracted), ×530 (both Calkins).

drical; slightly asymmetrical; cytostome at anterior end in a slit; right ridge thickened and lip-like, with about eight long cilia; with or without long caudal cilium; fresh or salt water. (Several species.)

P. marina Kahl (Fig. 302, *f*, *g*). Cylindrical; oval macronucleus central; contractile vacuole terminal; a caudal cilium; 55-90µ long; salt water; Florida (Noland).

Genus **Chilophrya** Kahl. Ovoid or ellipsoid; cytostome at anterior end, surrounded by protrusible rods; on one side there is a lip-like ectoplasmic projection; fresh or salt water.

C. (Prorodon) utahensis (Pack) (Fig. 302, *h*). Body ellipsoid, somewhat asymmetrical; comparatively small number of furrows; ciliation uniform; a finger-like process in front of cytostome; macronucleus small, central; contractile vacuole terminal; endoplasm with zoochlorellae; encystment common; cysts highly sensitive to light; 50µ long; Great Salt Lake, Utah (Pack).

C. (Urotricha) labiata (Edmondson) (Fig. 302, *i*). Body ovoid; a lip-like process in front of cytostome; macronucleus oblong, central; contractile vacuole terminal; 30µ long; fresh water.

Genus **Platyophrya** Kahl. Compressed; flask-like or elongate ovoid; asymmetrical; dorsal surface convex, ventral surface flat or partly concave; spiral striation; position and direction of cytostome variable; macronucleus round; contractile vacuole terminal; fresh water.

P. lata K. (Fig. 302, *j*). Highly compressed; colorless; many striae; on left edge of cytostome five to six cirrus-like projections and on right edge many short bristles; 105µ long; fresh water with sphagnum.

Genus **Stephanopogon** Entz. Somewhat resembles *Platyophrya;* compressed; cytostome at anterior extremity which is drawn out; cytostome surrounded by lobed membranous structures; salt water.

S. colpoda E. (Fig. 302, *k*). Longitudinal striae on 'neck' four to eight in number; two contractile vacuoles; 50-70µ long; creeping movement; salt water among algae.

Genus **Prorodon** Ehrenberg (*Rhagadostoma,* Kahl). Ovoid to cylindrical; ciliation uniform, with sometimes longer caudal cilia; oral basket made up of double trichites which end deep in ectoplasm, oval in cross-section; contractile vacuole terminal; macronucleus massive, spherical or oval; fresh or salt water. Numerous species.

P. discolor E. (Fig. 302, *l*). Ovoidal; 45-55 ciliary rows; macronucleus ellipsoid; micronucleus hemispherical; contractile vacuole terminal; 100-130µ long; fresh water; Kahl (1930) states that it occurs also in brackish water containing 2.5 per cent salt; sapropelic form in salt water is said to possess often long caudal cilia.

P. griseus Claparède and Lachmann. Oblong; 165-200µ long; fresh water.

Genus **Placus** Cohn (*Spathidiopis*, Fabre-Domergue; *Thoracophrya*, Kahl). Body small; ellipsoid or ovoid; somewhat compressed; pellicle with conspicuous spiral furrows; cytostome a narrow slit at anterior extremity; with strong cilia on right margin of slit; cytopyge a long narrow slit with cilia on both sides; macronucleus ellipsoid to sausage-form; contractile vacuole posterior; salt, brackish or fresh water.

P. socialis (Fabre-Domergue) (Fig. 302, *m, n*). 40-50µ by 28-32µ, about 22µ thick; salt water; Florida (Noland, 1937).

Genus **Lacrymaria** Ehrenberg. Polymorphic; cylindrical, spindle- or flask-shaped; with a long contractile proboscis; cytostome round; ciliary rows meridional or spiral to right; near cytostome a ring-like constriction with a circle of longer cilia; cytopharynx usually distinct; contractile vacuole terminal; fresh or salt water. (Numerous species.)

L. olor (Müller) (Fig. 302, *o*). Elongate; highly contractile; two macronuclei; two contractile vacuoles; extended forms 400-500µ up to 1.2 mm. long; when dividing, long neck is formed sidewise so that it appears as oblique division (Penard), however, Bovee (1957) noted the division transverse; fresh and salt water.

L. lagenula Claparède and Lachmann (Fig. 302, *p*). Body flask-shape; neck highly extensible; striation distinct, spiral when contracted; macronucleus short sausage-like or horseshoe-shape; endoplasm granulated; body 70µ long, up to 150µ (Kahl); salt water.

L. coronata C. and L. (Fig. 302, *q*). Large; neck extensible;

body form variable, but usually with bluntly rounded posterior end; endoplasm appears dark; striae spiral; 85-100μ long; salt and brackish water.

Genus **Enchelys** Hill. Flask-shape; anterior end obliquely truncate; cytostome slit-like, rarely round; fresh or salt water. (Several species, Fauré-Fremiet, 1944.)

E. curvilata (Smith) (Fig. 303, *a*). Elongate ovoid; posterior end rounded; longitudinal striation; macronucleus band-form; contractile vacuole terminal; endoplasm yellowish, granulated; about 150μ long; fresh water among algae.

Genus **Crobylura** André. Body when extended spindle-form, with truncate ends; when contracted, thimble-form; cilia short and thick; several long caudal cilia; slit-like cytostome at anterior end; no apparent cytopharynx; macronucleus irregularly rounded, hard to stain; micronucleus not observed; contracile vacuole latero-posterior; fresh water. (One species.)

C. pelagica A. (Fig. 303, *b*). Body 65-95μ long; in freshwater plankton.

Genus **Microregma** Kahl. Small, ovoid; dorsal side convex; vertral side flat; with a small slit-like cytostome near anterior end; with or without caudal bristle; fresh or salt water.

M. (Enchelys) auduboni (Smith) (Fig. 303, *c*). Body plastic; coarsely ciliated; caudal bristle thin; cytostome at anterior end, surrounded by longer cilia; cytopharynx small with trichocysts; round macronucleus central; contractile vacuole near posterior end; 40-55μ; fresh water.

Genus **Chaenea** Quennerstedt. Elongate; anterior end drawn out into a narrow truncated 'head'; but without any ring furrow; 'head' spirally or longitudinally furrowed; often with longer cilia directed anteriorly; cytostome terminal, not lateral; cytopharynx with trichocysts; body striation meridional, or slightly right spiral; macronucleus often distributed; fresh or salt water.

C. limicola Lauterborn (Fig. 303, *d*). Anterior half of body broad; posterior end drawn out into a point; contractile; cytopharynx with trichocysts; many trichocysts in endoplasm; contractile vacuoles in a row; 130-150μ long; stagnant fresh water.

Genus **Pithothorax** Kahl. Slender, barrel-shaped; with firm pellicle; a fairly long caudal bristle, contractile vacuole in posterior half; ciliation coarse and not over entire body surface; resembles *Coleps;* fresh water.

P. ovatus K. (Fig. 303, *e*). Caudal bristle breaks off easily; body 30µ long; fresh water among decaying vegetation.

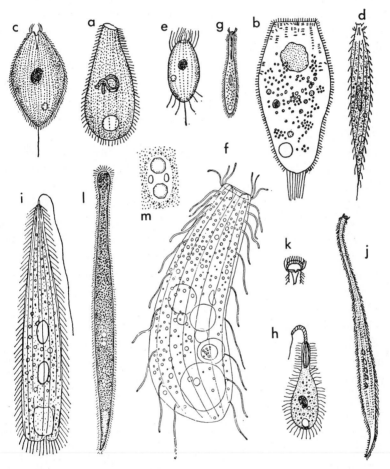

FIG. 303. a, *Enchelys curvilata,* ×200 (Smith); b, *Crobylura pelagica,* ×500 (André); c, *Microregma auduboni,* ×500 (Smith); d, *Chaenea limicola,* ×310 (Penard); e, *Pithothorax ovatus,* ×550 (Kahl); f, *Rhopalophrya salina,* ×1040 (Kirby); g, *Trachelophyllum clavatum,* ×100 (Stokes); h, *Ileonema dispar,* ×190 (Stokes); i, *I. ciliata,* ×800 (Roux); j, k, *Trachelocerca phaenicopterus* (Kahl); l, m, *T. subviridis* (Noland) (l, whole organism, ×155; m, nucleus, ×480).

Genus **Rhopalophrya** Kahl. Cylindrical; furrows widely separated; slightly asymmetrical; curved ventrally; dorsal surface convex; ventral surface flat or slightly concave; anterior end with 'neck'; two spherical macronuclei; fresh or salt water; sapropelic.

R. salina Kirby (Fig. 303, *f*). Cylindrical, tapering gradually to a truncated anterior end, slightly curved ventrally; cilia (6-10µ long) sparsely distributed; two macronuclei, spherical; 29-55µ long; 16-21µ in diameter; in concentrated brine (salts "34.8 per cent; pH 9.48") from Searles Lake; California (Kirby, 1934).

Genus **Enchelyodon** Claparède and Lachmann, Elongated; cylindrical, ovoid or flask-shaped; some with head-like prolongation; cytopharynx with trichites; cilia long at anterior end; fresh or salt water. (Several species.)

E. californicus Kahl. 120-130µ long; elongate ovoid to nearly cylindrical; not distinctly flattened; macronucleus horseshoe-like, with a large micronucleus; in mosses; California.

Genus **Trachelophyllum** Claparède and Lachmann. Elongate; flattened; flexible, ribbon-like; anterior end neck-like and tip truncate; cytopharynx narrow, round in cross-section, with trichocysts; ciliary rows widely apart; two macronuclei, each with a micronucleus; contractile vacuole terminal; fresh or salt water. (Several species.)

T. clavatum Stokes (Fig. 303, *g*). About 200µ long; fresh water.

Genus **Ileonema** Stokes (*Monomastix* Roux). Body flattened; flask-shaped; somewhat similar to *Trachelophyllum,* but there is a remarkable flagellum-like process extending from anterior end; cytopharynx with trichocysts; fresh water.

I. dispar S. (Fig. 303, *h*). Highly contractile; an anterior flagellum half body length, whose basal portion spirally furrowed; cytostome at base of the flagellum; cytopharynx spindle-form with trichites; two contractile vacuoles and cytopyge posterior; ovoid macronucleus; movement slow creeping; about 120µ long; fresh water among algae.

I. ciliata Roux (Fig. 303,*i*). 75µ by 14µ; fresh water.

Genus **Trachelocerca** Ehrenberg. Elongate, vermiform or flask-shaped; more or less extensible, with drawn-out anterior end; without any ring-furrow which marks the 'head' of *Lacrymaria,* and when contracted pellicular striae not spiral and no neck as is the case with *Chaenea;* salt water. Many species. (Nuclear form change, Raikov, 1958.)

T. phoenicopterus Cohn (Fig. 303, *j, k*). Elongate; extensible and contractile; neck and tail distinct when contracted; cytostome at anterior end, surrounded by a ridge containing indistinctly visible short trichocysts, cytopharynx with trichocysts; macronuclei made up of four radially arranged endosomes suspended in the nucleoplasm (Gruber, Kahl); six compact micronuclei, located in the central part of the macronucleus (Raikov, 1958); contractile vacuoles in a row, rarely seen; Salt water (Calkins).

T. subviridis Sauerbrey (Fig. 303, *l, m*). Highly extensible and contractile; nucleus contains peculiar crystal-like bodies; size variable; when extended 320-480µ long; salt water. Noland (1937) observed the organism in a salt spring in Florida.

Family 2. **Colepidae** Ehrenberg

Cytostome at the anterior end; body barrel-shaped; with pellicular plates; uniform ciliation.

Genus **Coleps** Nitzsch. Body-form constant, barrel-shaped; with regularly arranged ectoplasmic plates; cytostome at anterior end, surrounded by slightly longer cilia; often spinous projections at or near posterior end; one or more long caudal cilia, often overlooked; fresh or salt water. Many species (Noland, 1925, 1937; Kahl, 1930).

C. hirtus (Müller) (Fig. 304, *a*). 40-65µ long; fifteen to twenty rows of platelets; three posterior processes; fresh water.

C. elongatus Ehrenberg (Fig. 304, *b*). 40-55µ long; slender; about thirteen rows (Noland, 1925) or 14-17 rows (Kahl) of platelets; three posterior processes; fresh water.

C. bicuspis Noland (Fig. 304, *c*). About 55µ long; sixteen rows of platelets; two posterior processes; fresh water.

C. octospinus N. (Fig. 304, *d*). 80-110µ long; eight posterior spines; about twenty-four rows of platelets; Geiman (1931) found this organism in an acid march pond and noted variation in number and location of accessory spines; fresh water.

C. spiralis N. (Fig. 304, *e*). About twenty-three longitudinal rows of platelets slightly spirally twisted; posterior spines drawn together; a long caudal cilium; about 50μ long; salt water; Florida (Noland, 1937).

C. heteracanthus N. (Fig. 304, *f*). Anterior processes only on one side; posterior spines; caudal cilium; about 90μ by 35μ; salt water; Florida.

Genus **Tiarina** Bergh. Somewhat similar to *Coleps,* but posterior end tapering to a point; salt water.

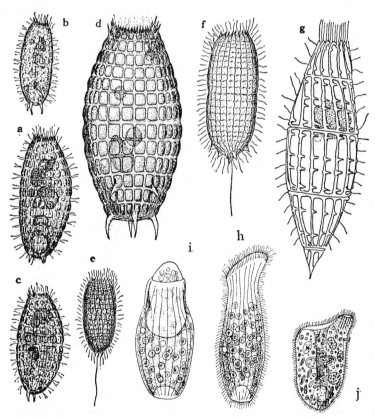

FIG. 304. a, *Coleps hirtus,* ×530 (Noland); b, *C. elongatus,* ×530 (Noland); c, *C. bicuspis,* ×530 (Noland); d, *C. octospinus,* ×530 (Noland); e, *C. spiralis,* ×400 (Noland); f, *C. heteracanthus,* ×400 (Noland); g, *Tiarina fusus,* ×530 (Fauré-Fremiet); h, i, *Spathidium spathula,* ×200 (Woodruff and Spencer); j, *Spathidioides sulcata,* ×260 (Brodsky).

T. fusus (Claparède and Lachmann) (Fig. 304, *g*). 85-135µ long.

Family 3. **Spathidiidae** Kahl

Slit-like cytostome at the anterior end, on non-ciliated ridge; uniform body ciliation.

Genus **Spathidium** Dujardin. Flask- or sack-shaped; compressed; anterior region slightly narrowed into a neck, and truncate; ciliation uniform; cytostome occupies whole anterior end; contractile vacuole posterior; macronucleus elongate; several micronuclei; trichocysts around cytostome and scattered throughout; fresh or salt water. Numerous species. (Trichites, Dragesco, 1952; gigantism and species formation, Wenzel, 1955.)

S. spathula Müller (Figs. 22, *c*; 304, *h*, *i*). Up to 250µ long; fresh water. (Morphology and food-capture, Woodruff and Spencer, 1922; conjugation, Woodruff and Spencer, 1924.)

Genus **Spathidioides** Brodsky. Somewhat similar to *Spathidium;* but oral ridge highly flattened on ventral side and conspicuously developed into a wart-like swelling on dorsal side; this knob contains trichocysts; sapropelic.

S. sulcata B. (Fig. 304, *j*). 65-85µ long; posterior end pointed, highly flattened; anterior end elevated at one side where cytostome and cytopharynx with ten rods are located.

Genus **Enchelydium** Kahl. Somewhat similar to *Spathidium;* but oral ridge forms a swollen ring with trichocysts; the ridge circular or elongated in cross-section; when swimming, the organisms appear as if cytostome is opened; with dorsal bristle; fresh water.

E. fusidens K. (Fig. 305, *a*). Cylindrical, contractile; cilia dense and rather long; macronucleus reniform, often appears as composed of two spherical parts; contractile vacuole terminal; oral ring with spindle-like trichocysts; food vacuoles not seen; extended body 110µ long; contracted 75µ; sapropelic.

Genus **Homalozoon** Stokes. Vermiform, flattened; one side with short cilia in twelve longitudinal rows; the opposite side without cilia, but marked by a conspicuous longitudinal ridge; macronucleus moniliform; several micronuclei; many contracile vacuoles in a row; fresh water.

H. vermiculare S. (Fig. 305, *b*). Body 150-1500µ long; standing fresh water (Weinreb, 1955).

Genus **Cranotheridium** Schewiakoff. Spathidium-like organisms; anterior end obliquely truncate, near the extended side of which is located the cytostome; cytopharynx surrounded by a group of trichites; fresh water.

C. taeniatum S. (Fig. 305, c). Anterior end flattened; with a group of trichites; macronucleus long band-form; with many micronuclei; contractile vacuole terminal; ciliation and striation close; colorless; movement slow; about 170µ long; fresh water.

Genus **Penardiella** Kahl. Ellipsoid, somewhat compressed; oral

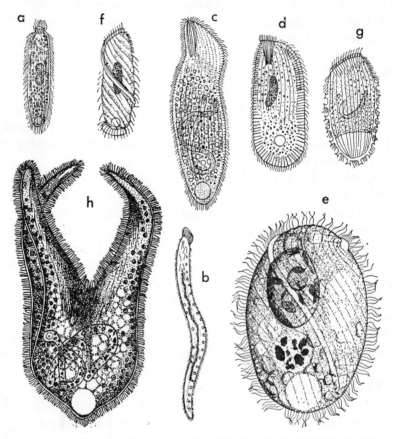

FIG. 305. a, *Enchelydium fusidens*, ×240 (Kahl); b, *Homalozoon vermiculare*, ×80 (Stokes); c, *Cranotheridium taeniatum*, ×300 (Schewiakoff); d, *Penardiella crassa*, ×210 (Kahl); e, *Perispira ovum*, ×665 (Dewey and Kidder); f, *P. strephosoma*, ×280 (Kahl); g, *Legendrea bellerophon*, ×190 (Penard); h, *Teuthophrys trisula*, ×330 (Wenrich).

ridge slightly oblique; a girdle with trichocysts encircling the body; fresh water.

P. crassa (Penard) (Fig. 305, *d*). Elongate ellipsoid, flattened; trichocysts in posterior portion of girdle are longer and those in the dorsal region are fewer in number and shorter; macronucleus sausage-form; contractile vacuole posterior, in front of the girdle; body 160µ by 50µ; sapropelic.

Genus **Perispira** Stein. Ovoid or cylindrical; oral ridge turns right-spirally down to posterior end.

P. ovum S. (Fig. 305, *e*). Oval; starved individuals 30-60µ by 20-45µ, well-fed forms 65-120µ by 50-110µ; spiral ridge complete turn; cytostome in the anterior end of the ridge, with a number of delicate trichites; ovoid to elongate macronucleus; a micronucleus; a terminal contractile vacuole; in fresh water (Dewey and Kidder, 1940). The ciliate was cultured bacteria-free by feeding on sterile *Euglena gracilis*.

P. strephosoma Stokes (Fig. 305, *f*). Oval to cylindrical; about 85µ long; standing water with sphagnum.

Genus **Legendrea** Fauré-Fremiet. Ellipsoid or ovoid; a peripheral zone with small tentacular processes bearing trichocysts.

L. bellerophon Penard (Fig. 305, *g*). 100-180µ; fresh water.

Genus **Teuthophrys** Chatton and Beauchamp. Body rounded posteriorly, anterior end with three radially equidistant, spirally curved arms (counter-clockwise when viewed from poserior end); the depressions between arms form furrows; cytostome apical, at the inner bases of arms; contractile vacuole terminal; ciliation uniform, except the inner surfaces of arms where longer cilia as well as trichocysts are present; with zoochlorellae; macronucleus rope-shaped and wound; micronucleus unobserved. (One species.)

T. trisula C. and B. (Fig. 305, *h*). 150-300µ long; length: width 3:1-2:1; ponds in Pennsylvania and California (Wenrich, 1929).

Family 4. **Metacystidae** Kahl

Cytostome terminal; ciliation uniform; pellicle enclosing a peculiar alveolar zone; usually encased in a pseudochitinous lorica.

Genus **Metacystis** Cohn. Oblong; ciliation general, except posterior end; ciliary circle around cytostome; usually one caudal cilium; with a large posterior vesicle containing turbid fluid.

M. truncata C. (Fig. 306, *a*). Elongate, not much difference in

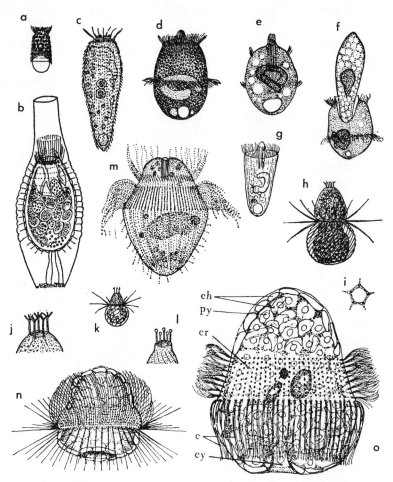

FIG. 306. a, *Metacystis truncata*, ×270 (Cohn); b, *Vasicola ciliata*, ×250 (Kahl); c, *Pelatractus grandis*, ×170 (Penard); d-f, *Didinium nasutum*, ×170 (Kudo); g, *D. balbianii*, ×290 (Bütschli); h-j, *Mesodinium pulex* (h, ×670; i, oral view; j, oral tentacles, ×1330) (Noland); k, l, *M. acarus* (k, ×670; 1, oral tentacles, ×1330) (Noland); m, *Askenasia faurei*, ×530 (Fauré-Fremiet); n, o, *Cyclotrichium meunieri* (Bary and Stuckey). n, diagram of organism in life; o, a composite figure from stained specimens (c, cirri; ch, chromatophores; cr, ciliary row; cy, "cytostome"; py, pyrenoid).

body width at different levels; with about twelve furrow rings; body length up to 30μ; salt water.

Genus **Vasicola** Tatem (*Pelamphora* Lauterborn). Ovoid with caudal cilia; lorica flask-shape, highly ringed; cytostome at anterior end, its lip with four rows of long cilia; body surface with shorter cilia; macronucleus round, central, with a micronucleus; contractile vacuole near macronucleus; fresh or salt water.

V. ciliata T. (*Pelamphora bütschlii* L.) (Fig. 306, *b*). Body about 100μ long; sapropelic in fresh water.

Genus **Pelatractus** Kahl. Somewhat similar to *Vasicola;* but without lorica or caudal cilia; with a terminal vacuole; without lip of *Vasicola;* sapropelic.

P. (Vasicola) grandis (Penard) (Fig. 306, *c*). Free-swimming; elongated fusiform; numerous contractile vacuoles on one side; body 125-220μ long; sapropelic in fresh water.

Family 5. **Didiniidae** Poche

Cytostome apical; one or more ciliary zones encircling the body; no oral cilia, but in some, a ring of tentacles occur around the cytostome; cytopharynx with trichites.

Genus **Didinium** Stein (*Monodinium* Fabre-Domergue). Barrel-shaped; one to several girdles of cilia (pectinellae); expansible cytostome at the tip of a proboscis, supported by a dense layer of long trichites; macronucleus horseshoe-shaped; two to three and occasionally four micronuclei, close to macronucleus; contractile vacuole terminal; fresh or salt water. (Several species.)

D. nasutum (Müller) (Figs. 22, *e, f;* 42; 78; 93; 306, *d-f*). 80-200μ long; endoplasm highly granulated; with two girdles of pectinelles; feeds on Paramecium; spherical cysts (Fig. 75) with three walls, 60-80μ in diameter; fresh water. (Morphology, Thon, 1905; Calkins, 1915, Beers, 1935; encystment, food requirement and conjugation (Beers, 1927, 1930, 1933, 1935; Burbanck, and Eisen, 1960; longevity of cysts, Beers, 1937; excystment, Beers, 1945, 1946 (Fig. 78); fibrillar structures, ten Kate, 1927; meiosis in conjugation (p. 245) Prandtl, 1906.)

D. balbianii (Fabre-Domergue) (Fig. 306, *g*). 60-100μ long; a single girdle of pectinelles near anterior end; fresh water.

Genus **Mesodinium** Stein. Ovoid; an equatorial furrow marks conical anterior and spherical posterior parts; in the furrow are inserted two (or one) rings of strong cilia; one directed anteriorly and the other posteriorly; with tentacle-like retractile processes around the cytostome; fresh and salt water.

M. *pulex* (Claparède and Lachmann) (Fig. 306, *h-j*). Oral tentacles with trifurcate tips; body 20-31µ long; salt water; Florida. Noland states that the freshwater forms are 21-38µ long.

M. *acarus* Stein (Fig. 306, *k, l*). Oral tentacles with capitate tip; 10-16µ long; salt water, Florida (Noland, 1937).

Genus **Askenasia** Blochmann. Resembles *Didinium;* ovoid; with two closely arranged rings of long cilia; anterior ring made up of some sixty pectinelles which are directed anteriorly; posterior ring composed of about the same number of long cilia directed posteriorly and arranged parallel to body surface; fresh or salt water.

A. *faurei* Kahl (Fig. 306, *m*). Body oval, anterior end broadly rounded; posterior region conical; pectinelles about 13µ long; the second band (10µ) of long cilia; an ellipsoid macronucleus; a micronucleus; body about 58-60µ long; fresh water.

Genus **Cyclotrichium** Meunier. Body spheroid to ellipsoid with a large non-ciliated oral field which is surrounded by a pectinelle-ring, one end dome-like, and the other truncate; macronucleus sausage-shaped; in salt water.

C. *meunieri* Powers (Fig. 306, *n, o*). Anterior end broadly rounded; posterior region conical; cytostome obscure; oral funnel at anterior end in a depression; broad ciliated band at about middle; ectoplasm with concave chromatophore (covered with haematochrome) plates on surface, below which numerous pyrenoids occur in vacuoles; endoplasm with numerous granules; 25-42µ by 18-34µ; Powers (1932) found that the 'red water' in Frenchman Bay in Maine was caused by the swarming of this organism. The same author held later that this ciliate may be the same as *Mesodinium rubrum* as observed by Leegaard (1920).

Bary and Stuckey (1950) found this organism in an extensive area of brownish-maroon water in Wellington harbour in April and August, 1948. Their description follows: body 22-47µ by 19-41µ; anterior half dome-like, posterior half expanded; posterior end truncate; "cytostome"; greenish-maroon chromatophores close to body surface; no ingested food material.

Family 6. **Actinobolinidae** Kahl

Cytostome at the anterior end; retractile tentacles in addition to uniform body ciliature.

Genus **Actinobolina** Strand (*Actinobolus* Stein). Ovate or spherical; ciliation uniform; extensible tentacles among cilia; contractile vacuole terminal; macronucleus curved band; fresh water.

A. vorax (Wenrich) (Fig. 307, *a*). Body 100-200μ long; elongate oval to spheroid; yellowish brown in color; cytostome at anterior end; contractile vacuole terminal; macronucleus rope-like; thirty to sixty ciliary rows; about thirty tentacles in each ciliary row; tentacles may be extended to twice the diameter of the body or be completely withdrawn; feeds chiefly on rotifers which stop all movements as though completely paralyzed upon coming in contact with the tentacles (Wenrich, 1929a).

Genus **Dactylochlamys** Lauterborn Body spindle-form, though variable; posterior end drawn out into tail; pellicle with eight to twelve undulating spiral ridges on which tentacle-like processes and long cilia are alternately situated; these processes are retractile (Kahl) and similar in structure to those of Suctoria; cytostome has not been detected; possibly allied to Suctoria; fresh water. (One species.)

D. pisciformis L. (Fig. 307, *b*). Body 80-120μ long.

Genus **Enchelyomorpha** Kahl. Conical, compressed; posterior end broadly rounded; anterior portion narrow; cilia on ring-furrows; anterior half with unretractile short tentacles; cytostome not noted; macronucleus with a central endosome surrounded by spherules; contractile vacuole terminal large.

E. vermicularis (Smith) (Fig. 307, *c*). Body 30-45μ long; fresh and brackish water.

Family 7. **Bütschliidae** Poche

Cytostome at or near the anterior end; ciliation uniform or in a few zones; with refractile concrement vacuole (Fig. 33, *d*) in anterior part of the body; in the alimentary canal of herbivorous animals.

Genus **Bütschlia** Schuberg. Ovoid, anterior end truncate, posterior end rounded; cytostome at anterior end, surrounded by long cilia; thick ectoplasm at anterior end; macronucleus spherical; micronucleus (?); concretion vacuole; ciliation uniform; in stomach of cattle.

B. *parva* S. (Fig. 307, *d*). 30-50μ by 20-30μ Conjugation (Dogiel, 1928).

Genus **Blepharoprosthium** Bundle. Pyriform, anterior half contractile, ciliated; caudal cilia; macronucleus reniform; in the caecum and colon of horse.

B. *pireum* B. (Fig. 307, *e*). 54-86μ by 34-52μ (Hsiung, 1930a).

Genus **Didesmis** Fiorentini. Anterior end neck-like, with large cytostome; anterior and posterior ends ciliated; macronucleus ellipsoid; in the caecum and colon of horse. (Species, Hsiung, 1930a.)

D. *quadrata* F. (Fig. 307, *f*). 50-90μ by 33-68μ; with a deep dorsal groove.

Genus **Blepharosphaera** Bundle. Spherical or ellipsoidal; ciliation uniform except in posterior region; caudal cilia; in the caecum and colon of horse.

B. *intestinalis* B. (Fig. 307, *g*). 38-74μ in diameter (Hsiung, 1930a).

Genus **Blepharoconus** Gassovsky. Oval; small cytostome; cilia on anterior one-third to one-half; caudal cilia; macronucleus ovoid; three contractile vacuoles; cytopharynx with rods; in the colon of horse.

B. *cervicalis* Hsiung (Fig. 307, *h*). 56-83μ by 48-70μ; Iowa (Hsiung, 1930, 1930a).

Genus **Bundleia** da Cunha and Muniz. Ellipsoid; cytostome small; cilia at anterior and posterior ends, posterior cilia much less numerous; in the caecum and colon of horse.

B. *postciliata* (Bundle) (Fig. 307, *i*). 30-56μ by 17-32μ (Hsiung, 1930a).

Genus **Polymorphella** Corliss (*Polymorpha*, Dogiel). Flask-shaped; ciliation on anterior region, a few caudal cilia; macronucleus disc-shaped; contractile vacuole terminal; in the caecum and colon of horse.

P. *ampulla* D. (Fig. 308, *a*). 22-36μ by 13-21μ (Hsiung, 1930a).

Genus **Holophryoides** Gassovsky. Oval, with comparatively large cytostome at anterior end; ciliation uniform; macronucleus small, ellipsoid; contractile vacuole subterminal; in the colon and caecum of horse.

H. ovalis (Fiorentini) (Fig. 308, *b*). 95-140μ by 65-90μ.

Genus **Blepharozoum** Gassovsky. Ellipsoid, with attenuated

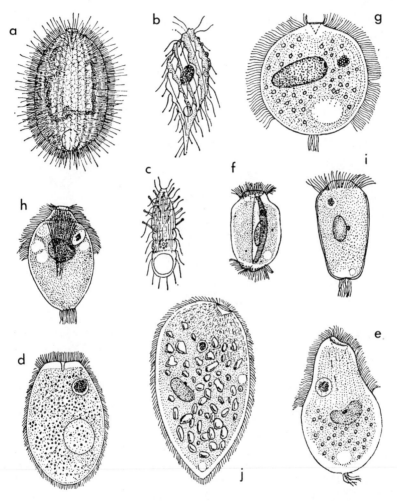

Fig. 307. a, *Actinobolina vorax*, ×300 (Wenrich); b, *Dactylochlamys pisciformis*, ×330 (Kahl); c, *Enchelyomorpha vermicularis*, ×670 (Kahl); d, *Bütschlia parva*, ×670 (Schuberg); e, *Blepharoprosthium pireum*, ×470; f, *Didesmis quadrata*, ×270; g, *Blepharosphaera intestinalis*, ×600; h, *Blepharoconus cervicalis*, ×360; i, *Bundleia postciliata*, ×530 (Hsiung); j, *Blepharozoum zonatum*, ×200 (Gassovsky).

posterior end; ciliation uniform; cytostome near anterior tip; two
contractile vacuoles; macronucleus small, reniform; in caecum of
horse.

B. zonatum G. (Fig. 307, j). 230-245μ by 115-122μ (Hsiung,
1930a).

Genus **Prorodonopsis** Gassovsky. Pyriform; ciliation uniform;
three contractile vacuoles; macronucleus sausage-shaped; in the
colon of horse.

P. coli G. (Fig. 308, c). 55-67μ by 38-45μ (Hsiung, 1930a).

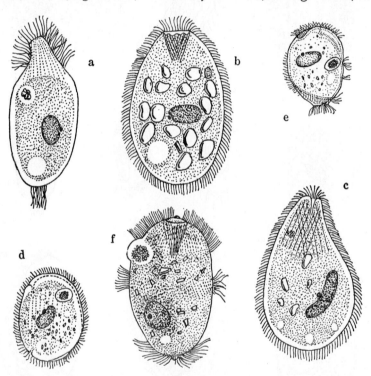

FIG. 308. a, *Polymorphella ampulla*, ×1170 (Hsiung); b, *Holophryoides
ovalis*, ×410 (Gassovsky); c, *Prorodonopsis coli*, ×700 (Gassovsky); d,
Paraisotrichopsis composita, ×450 (Hsiung); e, *Sulcoarcus pellucidulus*,
×410 (Hsiung); f, *Alloiozona trizona*, ×450 Hsiung).

Genus **Paraisotrichopsis** Gassovsky. Body uniformly ciliated;
spiral groove from anterior to posterior end; in the caecum of
horse.

P. composita G. (Fig. 308, d). 43-56μ by 31-40μ (Hsiung,
1930a).

Genus **Sulcoarcus** Hsiung. Ovoid, compressed; a short spiral groove at anterior end; cytostome at ventral end of the groove; cytopyge terminal; concretion vacuole mid-ventral, contractile vacuole posterior to it; cilia on groove, posterior end and mid-ventral region (Hsiung, 1935).

S. *pellucidulus* H. (Fig. 308, *e*). 33-56µ by 30-40µ; in faeces of mule.

Genus **Alloiozona** Hsiung. Cilia in three (anterior, equatorial and posterior) zones; in the caecum and colon of horse (Hsiung, 1930, 1930a).

A. *trizona* H. (Fig. 308, *f*). 50-90µ by 30-60µ.

Genus **Ampullacula** Hsiung. Flask-shaped; posterior half bearing fine, short cilia; neck with longer cilia; in the caecum of horse.

A. *ampulla* (Fiorentini). About 110µ by 40µ (Hsiung, 1930a).

References

ANIGSTEIN, L., 1911. Ueber zwei neue marine Ciliaten. Arch. Protist., 24: 127.

BARY, B. M., and STUCKEY, R. G., 1950. An occurrence in Wellington Harbour of *Cyclotrichium meuvieri* Powers, a ciliate causing red water, etc. Tr. Roy. Soc. New Zealand, 78:86.

BEERS, C. D., 1927. Factors involved in encystment in the ciliate *Didinium nasutum*. J. Morphol. Physiol., 43:499.

———— 1930. On the possibility of indefinite reproduction in the ciliate Didinium, etc. Am. Nat., 63:125.

———— 1933. Diet in relation to depression and recovery in the ciliate *Didinium nasutum*. Arch. Protist., 79:101.

———— 1935. Structural changes during encystment and excystment. etc. *Ibid.*, 84:133.

———— 1945.The encystment process in the ciliate *Didinium nasutum*. J. Elisha Mitchell Sci. Soc., 61:264.

———— 1946. Excystment in *Didinium nasutum,* with special reference to the rôle of bacteria. J. Exper. Zool., 103:201.

BOVEE, E. C., 1957. The binary fission of *Lacrymaria olor*. J. Protozool., 4:248.

BURBANCK, W. D., and EISEN, J. D., 1960. The inadequacy of monobacterially fed *Paramecium aurelia* as food for *Didinium nasutum*. J. Protozool., 7:201.

BÜTSCHLI, O., 1887-1889. Protozoa. In: Bronn's Klassen und Ordnungen des Thier-reichs. I.

CALKINS, G. N., 1915. *Didinium nasutum*. I. J. Exper. Zool., 19:225.

CHATTON, E., and BEAUCHAMP, P. D., 1923. *Teuthophrys trisulca*, etc. Arch. zool. exper. gén., 61(N. et R.):123.

CORLISS, J. O., 1955. The opalinid infusorians: flagellates or ciliates? J. Protozool., 2:107.

——— 1961. The ciliated protozoa. Pergamon Press.

DEWEY, V., and KIDDER, G. W., 1940. Growth studies on ciliates. VI. Biol. Bull., 79:255.

DOFLEIN, F., and REICHENOW, E., 1949-1953. Lehrbuch der Protozoenkunde. 6th ed. Jena.

DOGIEL, V., 1928. Ueber die Conjugation von *Bütschlia parva*. Arch. Protist., 62:80.

DRAGESCO, J., 1952. Sur la structure des trichocystes toxique des infusoires holotriches gymnostomes. Bull. Micros. Appliquée, 2:92.

FAURÉ-FREMIET, E., 1944. Polymorphisme de l'*Enchelys mutans*. Bull. soc. zool. France, 69:212.

——— 1950. Morphologie comparée et systématique des Ciliés. *Ibid.*, 75:109.

——— 1950a. Ecologie des ciliés psammophiles littoraux. Bull. biol. France Belgique, 84:35.

———, STOLKOWSKI, J., and DUCORNET, J., 1948. Étude expérimentale de la calcification tégumentaire chez un infusoire cilié *Coleps hirtus*. Biochem. Biophys. Acta, 2:668.

GEIMAN, Q. M., 1931. Morphological variations in *Coleps octospinus*. Tr. Am. Micr. Soc., 50:136.

GUILCHER, Y., 1950. Morphogenèse et morphologie comparée chez les ciliés gemmipares: chonotriches et tentaculifères. Ann. Biol., Paris, 26:465.

——— 1951. Contribution à l'étude des ciliés gemmipares, chonotriches et tentaculifères. Ann. Sci. nat., zool. (sér. 11), 13:33.

HSIUNG, T. S., 1930. Some new ciliates from the large intestine of the horse. Tr. Am. Micr. Soc., 49:34.

——— 1930a. A monograph on the Protozoa of the large intestine of the horse. Iowa State Coll. J. Sci., 4:356.

——— 1935. On some new ciliates from the mule, etc. Bull. Fan Mem. Inst. Biol., 6:81.

KAHL, A., 1926. Neue und wenige bekannte Formen der holotrichen und heterotrichen Ciliaten. Arch. Protist., 55:197.

——— 1927. Neue und ergänzende Beobachtungen holotricher Ciliaten. I. *Ibid.*, 60:34.

——— 1930. Urtiere oder Protozoa. I. Dahl's Die Tierwelt Deutschlands, etc. Part. 18:1.

——— 1930a. Neue und ergänzende Beobachtungen holotricher Infusorien. II. Arch. Protist., 70:313.

KENT, W. S., 1880-1882. A manual of Infusoria. London.

KIRBY, H. Jr., 1934. Some ciliates from salt marshes in California. Arch. Protistenk., 82:114.

LEEGAARD, C., 1920. Microplankton from the Finnish waters during the month of May, 1912. Acta Soc. Sci. Fenn. Helsingfors, 48:5:1.

NOLAND, L. E., 1925. A review of the genus Coleps with descriptions of two new species. Tr. Am. Micr. Soc., 44:3.

—— 1937. Observations on marine ciliates of the Gulf coast of Florida. *Ibid.*, 56:160.

PENARD, E., 1922. Études sur les infusoires d'eau douce. Geneva.

POWERS, P. B. A., 1932. *Cyclotrichium meunieri*, etc. Biol. Bull., 63:74.

PRANDTL, H., 1906. Die Konjugation von *Didinium nasutum*. Arch. Protist., 7:251.

RAIKOV, I. B., 1958. Der Formwechsel des Kernapparatus einiger niederer Ciliaten. I. Die Gattung Trachelocerca. Arch. Protist., 103:129.

ROUILLER, C., *et al.*, 1957. The pharyngeal protein fibers of the ciliates. In: Sjöstrand and Rhodin's Electron Microscopy. (Proc. Stockholm conference, September, 1956), p. 216.

ROUX, J., 1901. Faune infusorienne des eaux stagnantes de environs de Genevè. Mém. cour. fac. sc. l'Uni. Geneva, 148 pp.

STEIN, F., 1859. Der Organismus der Infusionsthiere, etc. I. Leipzig.

—— 1867. Der Organismus der Infusionsthiere. II. Leipzig.

STOKES, A. C., 1888. A preliminary contribution toward a history of the freshwater Infusoria of the United States. J. Trenton Nat. Hist. Soc., 1:71.

—— 1890. Notices of new fresh-water Infusoria. Proc. Am. Phil. Soc., 28:74.

TEN KATE, C. G. B., 1927. Ueber das Fibrillensystem der Ciliaten. Arch. Protist., 57:362.

THON, K., 1905. Ueber den feineren Bau von Didinium, etc. *Ibid.*, 5:281.

WEINREB, S., 1955. *Homalozoon vermiculare*. I, II. J. Protozool., 2:59, 67.

WENRICH, D. H., 1929. Observations on some freshwater ciliates. I. Tr. Am. Micr. Soc., 48:221.

—— 1929a. The structure and behavior of *Actinobolus vorax*. Biol. Bull., 56:390.

WENZEL, F., 1955. Ueber eine Artenstehung innerhalb der Gattung Spathidium. Arch. Protist., 100:515.

WOODRUFF, L. L., and SPENCER, H., 1922. Studies on *Spathidium spathula*. I. J. Exper. Zool., 35:189.

—— —— 1924. II. *Ibid.*, 39:133.

Chapter 32

Order 1. Gymnostomatida Bütschli (*continued*)
Suborder 1. Rhabdophorina Fauré-Fremiet (Continued)
Family 8. Amphileptidae Bütschli

CYTOSTOME slit-like, located anteriorly, but not at the anterior tip; body often flask-shaped, laterally compressed; ciliation uniform.

Genus **Amphileptus** Ehrenberg. Flask-shaped; somewhat compressed; ciliation uniform and complete; slit-like cytostome not reaching the middle of body, without trichocyst-borders; many contractile vacuoles; two or more macronuclei; fresh or salt water.

A. claparedei Stein (*A. meleagris* Claparède and Lachmann) (Fig. 309, *a*). Slightly flattened; broadly flask-shaped; with bluntly pointed posterior and neck-like anterior end; cytostome about two-fifths from ventral margin; trichocysts indistinct; dorsal ciliary rows also not distinct; contractile vacuoles irregularly distributed; 120-150µ long; fresh and salt water, on stalks of Zoothamnium, Carchesium, Epistylis, etc.

A. branchiarum Wenrich (Fig. 309, *b*). On the integument and gills of frog tadpoles; swimming individuals killed with iodine, 100-135µ by 40-60µ (Wenrich, 1924).

Genus **Litonotus** Wrzesniowski (*Lionotus*, Bütschli; *Hemiophrys*, W.). Flask-shape; elongate, flattened; anterior region neck-like; cilia only on right side; without trichocyst-borders; cytostome with trichocysts; one (terminal) or many (in 1-2 rows) contractile vacuoles; two macronuclei; one micronucleus; fresh or salt water.

L. fasciola Ehrenberg (Fig. 309, *c*). Elongate flask in form; hyaline; with flattened neck and tail, both of which are moderately contractile; posterior end bluntly rounded; without trichocysts; neck stout, bent toward the dorsal side; cytostome a long slit; contractile vacuole posterior; two spherical macronuclei between

852

which a micronucleus is located; 100µ long; fresh water and probably also in salt water.

Genus **Loxophyllum** Dujardin (*Opisthodon* Stein). Generally similar to *Lionotus* in appearance; but ventral side with a hyaline

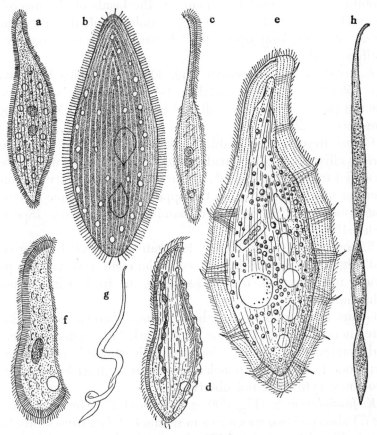

Fig. 309. a, *Amphileptus claparedei*, ×370 (Roux); b, *A. branchiarum*, flattened, ×490 (Wenrich); c, *Litonotus fasciola*, ×540 (Kahl); d, *Loxophyllum meleagris*, ×120 (Penard); e, *L. setigerum*, ×570 (Sauerbrey); f, *Bryophyllum vorax*, ×360 (Stokes); g, h, *Kentrophoros fasciolatum* (g, ×50; h, ×110) (Noland).

border, reaching posterior end and bearing trichocysts; dorsal side with either similar trichocyst-border or with trichocyst-warts; macronucleus a single mass or moniliform; contractile vacuole, one to many; fresh or salt water. (Species, Vuksanovic, 1959.)

L. meleagris D. (Fig. 309, *d*). Form and size highly variable; flask-shape to broad leaf-like; broad ventral seam with trichocysts and often undulating; dorsal seam narrow and near its edge, groups of trichocysts in wart-like protuberances; macronucleus moniliform; micronuclei, as many as the beads of the macronucleus (Penard, 1922); contractile vacuole terminal, with a long canal; 300-400μ long, up to 700μ (Penard, 1922); feeds mainly on rotifers; fresh water.

L. setigerum Quennerstedt (Fig. 309, *e*). 100-350μ long; average 150μ by 60μ; form variable; 1-4 macronuclei; several contractile vacuoles in a row; salt and brackish water. (Morphology, Sauerbrey, 1928.)

Genus **Bryophyllum** Kahl. Similar to *Loxophyllum;* but uniformly ciliated on both broad surfaces; ventral ridge with closely arranged trichocysts, extends to the posterior extremity and ends there or may continue on to the opposite side for some distance; macronucleus ovoid to coiled bandform; in mosses. (Species, Gelei, 1933.)

B. vorax Stokes (Fig. 309, *f*). Elongate; trichocyst-bearing ventral ridge turns up a little on dorsal side; contractile vacuole posterior; macronucleus oval; 130μ long; in fresh water among sphagnum.

Genus **Kentrophoros** Sauerbrey (*Centrophorella*, Kahl). Elongate, nematode-like; anterior end greatly attenuated; posterior end pointed; body surface longitudinally striated; ciliation uniform; one to three macronuclei; numerous contractile vacuoles in two rows; cytostome not observed.

K. fasciolatum S. (Fig. 309, *g*, *h*). About 270μ by 38μ. Noland (1937) observed two specimens in sediment taken from sandy bottom in Florida; contracted 650μ long; extended 1 mm. long.

K. lanceolata Fauré-Fremict. Ribbon-like; 460-520μ by 40μ; ventral side ciliated; dorsal side covered with dark sulphur bacteria (Caulobacteria), except the extremities; five to six spherical micronuclei, about 4μ in diameter; on sandy flat of Cape Cod (Fauré-Fremiet, 1951).

Family 9. **Tracheliidae** Ehrenberg

Circular cytostome is located some distance from the anterior end, often at the base of proboscis; body ciliation uniform.

Genus **Trachelius** Schrank. Oval to spherical; anterior end drawn out into a relatively short finger-like process or a snout; posterior end rounded; round cytostome at base of neck; cytopharynx with trichites; contractile vacuoles many; macronucleus simple or bandform; fresh water.

T. ovum Ehrenberg (Fig. 310, *a*). Spheroidal to ellipsoid; right side flattened and with a longitudinal groove; left side convex; proboscis about one-fourth to one-half the body length; cilia short and closely set; numerous contractile vacuoles; macronucleus short sausage-form, often divided into spherules; endoplasm penetrated by branching cytoplasmic skeins or bands and often with numerous small brown excretion granules; 200-400μ long; fresh water.

Genus **Dileptus** Dujardin. Elongate; snout or neck-like prolongation conspicuous; somewhat bent dorsally; along convex ventral side of neck many rows of trichocysts; a row of strong cilia; dorsal surface with three rows of short bristles; cytostome surrounded by a ring; cytopharynx with long trichocysts; posterior end drawn out into a tail; contractile vacuoles, two or more; body ciliation uniform; macronucleus bandform, moniliform or divided into numerous independent bodies; fresh or salt water. (Many species.)

D. americanus Kahl (Fig. 310, *b*). Proboscis bent dorsally sickle-like; macronucleus made up of two sausage-shaped or often horseshoe-shaped parts; two contractile vacuoles on dorsal side; 200μ long; in mosses.

D. anser Müller (Fig. 310, *c*). Proboscis slightly flattened; macronucleus divided into 100 or more discoid bodies; sixteen or more vesicular micronuclei (Jones, 1951); contractile vacuoles in a row on the aboral surface, with two to three in proboscis; 250-400μ, sometimes up to 600μ long; in fresh water. (Culture, encystment and excystment, Jones, 1951); electron microscope study, Dumont, 1961.)

D. monilatus (Stokes). Resembles *D. anser*, but the macronucleus moniliform (Jones and Beers, 1953).

Genus **Paradileptus** Wenrich (*Tentaculifera* Sokoloff). Body broader at the level of cytostome; with a wide peristomal field that bears the cytostome and is surrounded for two-thirds to three-fourths its circumference by a raised rim which is continuous anteriorly with the spirally wound proboscis; trichocyst-zone traversing the rim and anterior edge of proboscis; contractile vacuoles small, numerous, distributed; macronucleus segmented; fresh water (Wenrich, 1929a).

P. conicus W. (Fig. 310, *d*). 100-200µ by 50-100µ.

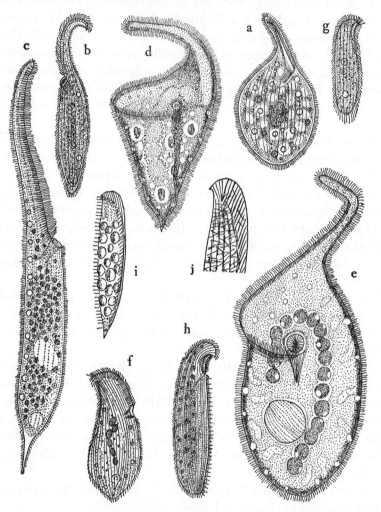

FIG. 310. a, *Trachelius ovum*, ×130 (Roux); b, *Dileptus americanus*, ×250 (Kahl); c, *D. anser*, ×310 (Hayes); d, *Paradileptus conicus*, ×340 (Wenrich); e, *P. robustus*, ×340 (Wenrich); f, *Branchioecetes gammari* ×200 (Penard); g, *Loxodes vorax*, ×190 (Stokes); h, *L. magnus*, ×80 (Kahl); i, j, *Remanella rugosa* (i, dorsal side, ×130; j, anterior part showing the endoskeleton) (Kahl).

P. robustus W. (Fig. 310, *e*). 180-450μ long.

P. estensis Canella. 600-800μ long; feeds on rotifers (Canella, 1951).

Genus **Branchioecetes** Kahl. Preoral part somewhat like that of *Amphileptus,* and bent dorsally; ventral side of neck with two rows of trichocysts; cytostome at posterior end of neck; cytopharynx with trichocysts; ectocommensals on Asellus or Gammarus.

B. gammari Penard (Fig. 310, *f*). 130-200μ long; on Gammarus.

Family 10. **Loxodidae** Bütschli

Slit-like cytostome on the slightly concave side of the body; body laterally compressed; ciliation only on the right side.

Genus **Loxodes** Ehrenberg. Lancet-like; strongly compressed; anterior end curved ventrally, and usually pointed; right side slightly convex; uniform ciliation on about twelve longitudinal rows; ectoplasm appears brownish, because of closely arranged brownish protrichocysts; endoplasm reticulated; two or more vesicular macronuclei; one or more micronuclei; two to twenty-five Müller's vesicles (p. 99; Fig. 33, *a, b*). in dorsal region; fresh water.

L. vorax Stokes (Fig. 310, *g*). 125-140μ long; yellowish brown, a row of slightly longer cilia; sapropelic in standing fresh water.

L magnus S. (Fig. 310, *h*). Extended about 700μ long; dark brown; twelve to twenty or more Müller's vesicles in a row along dorsal border; standing pond water.

Genus **Remanella** Kahl. Similar to *Loxodes* in general appearance; but with endoskeleton consisting of 12-20μ long spindleform needles lying below broad ciliated surface in 3-5 longitudinal strings connected with fibrils; Müller's vesicles (Fig. 33, *c*) in some, said to be different from those of Loxodes (Kahl); sandy shore of sea.

R. rugosa K. (Fig. 310, *i, j*). 200-300μ long.

Family 11. **Pycnothricidae** Poche

Body ciliation uniform; the ectoplasm is thick and conspicuous, cytostome vary in position; if it is located posteriorly, a furrow or groove connects it with the anterior end of the body; parasitic in the alimentary canal of mammals.

Genus **Pycnothrix** Schubotz. Large, elongate; with broadly rounded anterior and narrowed posterior end; somewhat flattened; short thick cilia throughout; ectoplasm thick; macronucleus spherical, in anterior one-sixth; micronucleus (?); two longitudinal grooves, one beginning on each side near anterior end, united at the notched posterior end; a series of apertures in grooves considered as cytostomes; at posterior one-third, an aperture gives rise to branching canals running through endoplasm, and is considered as excretory in function. (One species.)

P. monocystoides S. (Fig. 311, *a*). 300μ-2 mm. long; in the colon of *Procavia capensis* and *P. brucei.*

Genus **Nicollella** Chatton and Pérard. Elongate; a narrow

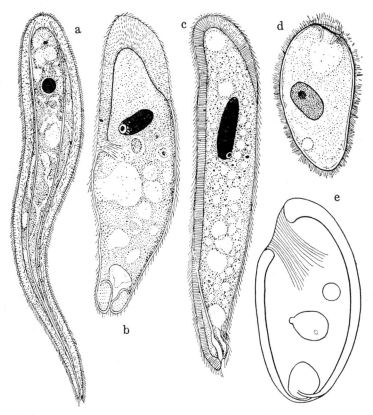

Fig. 311. a, *Pycnothrix monocystoides,* ×50; b, *Nicollella ctenodactyli,* ×170; c, *Collinella gundi,* ×170 (Chatton and Pérard); d, *Buxtonella sulcata,* ×395 (Jameson); e, *Taliaferria clarki,* ×500 (Hegner and Rees).

groove extends from the anterior end to cytostome, located at the middle of body; bilobed posteriorly; contractile vacuole terminal; macronucleus ellipsoid, anterior; a micronucleus; ectoplasm thick anteriorly; ciliation uniform (Chatton and Pérard, 1921). (One species.)

N. ctenodactyli C. and P. (Fig. 311, *b*). 70-550μ by 40-150μ; in the colon of *Ctenodactylus gundi.*

Genus **Collinella** Chatton and Pérard. More elongate than *Nicollella;* uniform ciliation; a groove extends from end to end; cytostome at posterior end of the groove; contractile vacuole terminal; macronucleus much elongated, central or posterior (Chatton and Pérard, 1921). One species.)

C. gundi C. and P. (Fig. 311, *c*). 550-600μ by 100μ; in the colon of *Ctenodactylus gundi.*

Genus **Buxtonella** Jameson. Ovoid; a prominent curved groove bordered by two ridges from end to end; cytostome near anterior end; uniform ciliation; in the caecum of cattle (Jameson, 1926). (One species.)

B. sulcata J. (Fig. 311, *d*). 55-124μ by 40-72μ.

Genus **Taliaferria** Hegner and Rees. Body ovate; circular in cross-section; ectoplasm is two-layered and thick; ciliation uniform; cytostome anterior, subterminal; macronucleus and a closely attached micronucleus near center; two contractile vacuoles; cytopyge (Hegner and Rees, 1933). (One species.)

T. clarki H. and R. (Fig. 311, *e*). 83-146μ by 42-83μ; in the caecum and colon of the red spider monkey (*Ateles geoffroyi*).

Suborder 2. **Cyrtophorina** Fauré-Fremiet

Cytostome is located mid-ventrally in the anterior half of the body; cytopharynx with fused armature of trichites (cyrtophorine); body often dorso-ventrally flattened; body ciliation is confined to the ventral surface.

Family 1. **Dysteriidae** Claparède and Lachmann

A characteristic stylus on ventral side near the posterior end which is used for attachment to substrate; ciliation only on the ventral surface.

Genus **Dysteria** Huxley (*Ervilia* Dujardin; *Iduna, Aegyria*

Claparède and Lachmann; *Cypridium* Kent). Ovate, dorsal surface convex, ventral surface flat or concave; left ventral side with nonciliated ventral plate; postoral ciliation is continuation of preoral to right of cytostome and parallel to right margin; cytostome in a furrow near right side; posterior stylus or spine conspicuous; macronucleus spheroid or ovoid, central; with a micronucleus; usually two contractile vacuoles; fresh or salt water. (Numerous species.)

D. calkinsi Kahl (*D. lanceolata* Calkins) (Fig. 312, *a*). About 45μ by 27μ; salt water; Woods Hole.

Genus **Trochilia** Dujardin. Similar to *Dysteria;* but ciliation on the ventral side in an arched zone; fresh or salt water. (Several species.)

T. palustris Stein (Fig. 312, *b*). 25μ long; fresh water.

Genus **Trichopus** Claparède and Lachmann. Ovoid, narrowed posteriorly; cytostome anterior one-fourth, bordered by a collar; about forty ciliary rows; antapical pit with a fringe of membranelles occurs at the posterior end; fixation by vesicles and a secretory ampoule; the vibrating fringe aids in spinning of the glutinous secretion; macronucleus ovoid; contractile vacuole dorso-lateral; marine.

T. lachmanni Fauré-Fremiet. 70-90μ long (Fauré-Fremiet, 1957).

Genus **Trochilioides** Kahl. Rounded at anterior end, narrowed posteriorly; right side more convex than left; cytostome anterior with cytopharynx and preoral membrane; conspicuous longitudinal bands on right half with longitudinal striae, becoming shorter toward left; fresh or salt water.

T. recta K. (Fig. 312, *c*). 40-50μ long; sapropelic in fresh and brackish water.

Genus **Hartmannula** Poche (*Onychodactylus* Entz). Ventral surface uniformly ciliated; cytopharynx with short rods; in salt water.

H. entzi Kahl (Fig. 312, *d*). 80-140μ long; salt water.

Family 2. **Chlamydodontidae** Claus

Body rigid, flattened dorso-ventrally; body ciliation confined to ventral surface; no stylus.

Genus **Chlamydodon** Ehrenberg. Ellipsoid, reniform, elongate triangular, etc.; cilia only on ventral surface, anterior cilia longer; cytostome elongate oval and covered with a membrane bearing a slit; oral basket made up of closely arranged rods with apical processes; along lateral margin, there is a characteristic striped band which is a canalicule of unknown function; fresh or salt water.

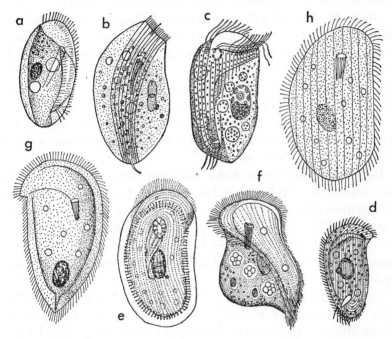

Fig. 312. a, *Dysteria calkinsi,* ×540 (Calkins); b, *Trochilia palustris,* ×1070 (Roux); c, *Trochilioides recta,* ×740 (Kahl); d, *Hartmannula entzi,* ×220 (Entz); e, *Chlamydodon mnemosyne,* ×520 (MacDougall); f, *Phascolodon vorticella,* ×340 (Stein); g, *Chilodonella caudata,* ×1000 (Stokes); h, *C. fluviatilis,* ×800 (Stokes).

C. mnemosyne E. (Fig. 312, *e*). Ellipsoid or reniform; right side convex, left side concave; ventral side flat, dorsal side greatly convex; a band of trichites, 'railroad track,' parallel to body outline; oral basket with eight to ten rods; macronucleus oval; four to five contractile vacuoles distributed; 60-90µ long; salt water. MacDougall (1928) observed it in the brackish water at Woods Hole and studied its neuromotor system. (Ciliature, Fauré-Fremiet, 1950.)

C. pedarius Kaneda. Reniform, flattened; 110-138µ by 50-70µ;

5-12 contractile vacuoles in two longitudinal rows; cytopyge right-posterior; about forty ciliary rows; eleven to thirteen trichites in the oral basket; the striped band, encircling the ventral surface, is composed of numerous semicircular trichites (capable of movement), arranged in a cross-tie fashion and make a canalicular structure; in brackish water (Kaneda, 1953-1961).

Genus **Phascolodon** Stein. Ovoid; with broad anterior end and bluntly pointed posterior end; ventral side concave or flat, dorsal side convex; ciliated field on ventral surface narrowed laterally behind cytostome, forming V-shaped ciliated area (about twelve rows); cytostome ellipsoid with oral basket; macronucleus oval with a micronucleus; two contractile vacuoles; fresh water.

P. vorticella S. (Fig. 312, *f*). 80-110μ long; cytostome covered by a slit-bearing membrane; with two preoral membranes; macronucleus ovoid; fresh water.

Genus **Cryptopharynx** Kahl. Ellipsoid, anterior third bent to left; ventral surface flat, dorsal surface with hump; spiral interciliary furrows ridged; oval cytostome at anterior end; no cytopharynx; dorsal hump yellowish, granulated with gelatinous cover; two macronuclei; one micronucleus; two contractile vacuoles, one posterior and the other toward left side at the bend of body. (One species.)

C. setigerus K. (Fig. 313, *a, b*). Elongate ellipsoid; anterior region bent to left; ventral surface flat, dorsal surface with a hump; about fifteen ventral ciliary rows; two vesicular macronuclei and one micronucleus dorso-central; 33-96μ by 21-45μ (Kirby). Kirby (1934) found the organism in salt marsh pools (salinity 1.2-9.7 per cent) with purple bacteria; California.

Genus **Chilodonella** Strand (*Chilodon* Ehrenberg). Ovoid; dorso-ventrally flattened; dorsal surface convex, ventral surface flat; ventral surface with ciliary rows; anteriorly flattened dorsal surface with a cross-row of bristles; cytostome round; oral basket conspicuous, protrusible; macronucleus rounded; contractile vacuoles variable in number; fresh or salt water or ectocommensal on fish and amphipods. (Many species.)

C. cucullulus (Müller) (*Chilodon steini*, Blochmann) (Figs. 56; 313, *c-e*). Nineteen to twenty ventral ciliary rows; oral basket with about twelve rods and with three preoral membranes; macronucleus oval, a characteristic concentric structure; micronucleus small; body 100-300µ long, most often 130-150µ long; fresh and brackish water. (Conjugation, Ivanić, 1933; ciliature, Fauré-Fremiet, 1950.)

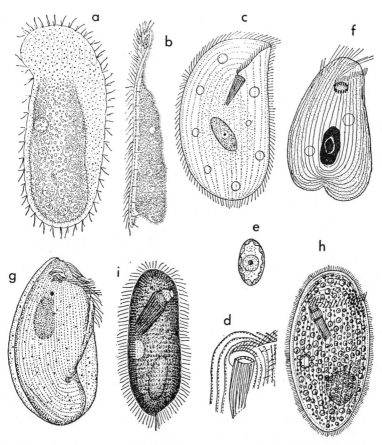

Fig. 313. a, b, *Cryptopharynx setigerus*, ×650 (Kirby); c-e, *Chilodonella cucullulus* (c, ×270 (Stein); d, oral region; e, nucleus (Penard)); f, *C. cyprini*, ×670 (Moroff); g, *Allosphaerium palustris*, ×1000 (Kidder and Summers); h, *Nassula aurea*, ×190 (Schewiakoff); i, *Paranassula microstoma*, ×400 (Noland).

C. caudata (Stokes) (Fig. 312, g). About 42µ long; standing water.

C. fluviatilis (S.) (Fig. 312, h). About 50µ long; fresh water.

C. uncinata (Ehrenberg) (Fig. 98). 50-90μ long; about eleven ventral ciliary rows; some seven dorsal bristles; widely distributed in various freshwater bodies; several varieties. (Conjugation, MacDougall, 1935.)

C. cyprini (Moroff) (Fig. 313, *f*). 50-70μ by 30-40μ; in the integument and gills of cyprinoid fishes; the organism, if freed from the host body, dies in twelve to twenty-four hours. (Ciliation and silver-line system, Krascheninnikow, 1934, 1953.)

C. longipharynx Kidder and Summers. 17-21μ (average 19μ) long; cytopharynx long, reaches posterior end; ectocommensal on amphipods, *Talorchestia longicornis* and *Orchestia palustris;* Woods Hole (Kidder and Summers, 1935).

C. hyalina K. and S. 40μ (36-47μ) long; ectocommensal on *Orchestia agilis;* Woods Hole.

C. rotunda K. and S. 29 μ(27-34μ) long; ectocommensal on *Orchestia agilis;* Woods Hole.

Genus **Allosphaerium** Kidder and Summers. Oval; right side concave, left side more or less flat; body highly flattened; arched dorsal surface devoid of cilia; ventral surface slightly concave with twelve to twenty-seven ciliary rows; right and left margin of ventral surface with a pellicular fold; cytostome anterior-ventral, oval or irregular, surrounded by ridge on posterior border, extending to left margin; three groups of ciliary membranes extending out of cytostome; macronucleus oval, central or anterior; a micronucleus; two (or one) contractile vacuoles; a refractile spherule regularly present in posterior portion of endoplasm; ectocommensal on the carapace and gills of amphipods.

A. palustris K. and S. (Fig. 313, *g*). 46-59μ long; 27 ventral ciliary rows; on *Orchestia palustris* and *Talorchestia longicornis;* Woods Hole.

A. sulcatum K. and S. 24-32μ long; twelve ciliary rows; on the carapace of *Orchestia agilis* and *O. palustris;* Woods Hole.

A. granulosum K. and S. 32-42μ long; rotund; seventeen ciliary rows; cytoplasm granulated; on carapace of *Orchestia agilis* and *O. palustris;* Woods Hole.

A. caudatum K. and S. Resembles *A. palustris;* 35-45μ long; fourteen ciliary rows; one contractile vacuole; ectoplasm at posterior end, drawn out into a shelf; on *Orchestia agilis;* Woods Hole.

A. convexa K. and S. 24-36µ long; seventeen ciliary rows; on the carapace and gill lamellae of *Talorchestia longicornis;* Woods Hole.

Family 3. Nassulidae Bütschli

Body ciliation complete, but dorsal ciliation usually less dense than that of the ventral surface.

Genus **Nassula** Ehrenberg. Oval to elongate; ventral surface flat, dorsal surface convex; usually brightly colored, due to food material; cystostome one-third to one-fourth from anterior end; body often bent to left near cytostome; opening of oral basket deep, in a vestibule with a membrane; macronucleus spherical or ovoid, central; a single micronucleus; contractile vacuole large, with accessory vacuoles and opens ventrally through a tubule-pore; fresh or salt water. (Many species.)

N. aurea E. (Fig. 313, *h*). 200-250µ long; fresh and brackish water (Kahl).

Genus **Paranassula** Kahl. Similar in general appearance to *Nassula;* but with preoral and dorsal suture line; longer caudal cilia on dorsal suture; pharyngeal basket not funnel-like, with sixteen to eighteen trichites; about seventy-five ciliary rows; trichocysts especially in anterior region.

P. microstoma (Claparède and Lachmann) (Fig. 313, *i*). Pellicle roughened by a criss-cross of longitudinal and circular furrows; macronucleus elongate oval, posterior; contractile vacuole near middle and right-dorsal; about 80-95µ long; salt water; Florida (Noland).

Genus **Cyclogramma** Perty. Somewhat resembling *Nassula;* but conspicuous oral basket in pyriform depression and opens toward left on ventral surface; depression with a short row of small membranes at its anterior edge; trichocysts usually better developed than in Nassula; fresh water.

C. trichocystis (Stokes) (Fig. 314, *a*). Body colorless or slightly rose-colored; trichocysts thick and obliquely arranged; one contractile vacuole; usually full of blue-green food vacuoles; actively motile; about 60µ long; in fresh water among algae.

Genus **Chilodontopsis** Blochmann. Elongate ellipsoid; colorless; ventral surface flattened, dorsal surface slightly convex; both sides ciliated; oral basket without vestibule; cytostome with a

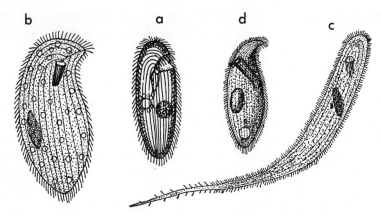

FIG. 314. a, *Cyclogramma trichocystis*, ×510 (Stokes); b, *Chilodontopsis vorax*, ×200 (Stokes); c, *Eucamptocerca longa*, ×320 (da Cunha); d, *Orthodonella hamatus*, ×160 (Entz).

membranous ring; usually with a postoral ciliary furrow; fresh water.

C. vorax (Stokes) (Fig. 314, *b*). Elongate ellipsoid; anterior region slightly curved to left; snout fairly distinct; oral basket with about sixteen rods; several contractile vacuoles distributed, a large one terminal; macronucleus large, lenticular, granulated; with a closely attached micronucleus; 50-160μ long; fresh water.

Genus **Eucamptocerca** da Cunha. Elongate; posterior part drawn out into a caudal prolongation; dorso-ventrally flattened; ciliation on both sides; round cytostome with oral basket in anterior ventral surface. (One species.)

E. longa da C. (Fig. 314, *c*). 300μ by 25μ; macronucleus ovoid, with a micronucleus; contractile vacuole (?); in brackish water (salt content 3 per cent); Brazil.

Genus **Orthodonella** Bhatia (*Orthodon*, Gruber). Oval; contractile; colorless; much flattened; anterior region curved toward left; striation on both dorsal and ventral sides; cytostome toward right border; oral basket long; macronucleus oval; contractile vacuole terminal; fresh or salt water.

O. hamatus G. (Fig. 314, *d*). Extended 200-260μ long, contracted 90-150μ long; flask-shaped; oral basket with sixteen trichites; salt water.

Family 4. **Clathrostomatidae** Kahl

Body ciliation uniform; body not flattened; cytostome ventral, in an oral vestibule somewhat resembling that of trichostomes; provisionally placed here.

Genus **Clathrostoma** Penard. Ellipsoid; with an oval pit in anterior half of the flattened ventral surface, in which occur three to five concentric rows of shorter cilia; cytostome a long slit located at the bottom of this pit; with a basket composed of long fibrils on the outer edge of the pit; in fresh water.

C. viminale P. (Fig. 317, *k*). Resembles a small *Frontonia leucas;* macronucleus short sausage-form; four micronuclei in a compact group; endoplasm with excretion crystals; five preoral ciliary rows; 130-180μ long; in fresh water.

References

CHATTON, E., and PÉRARD, C., 1921. Les Nicollelidae, infusoires intestinaux des gondis et des damans, etc. Bull. biol. France et Belgique, 55:87.

CANELLA, M. F., 1951. Contributi alla conoscenza dei Ciliati. II. Ann. Univ. Ferrara, Sez. 3, Biol. Anim., 1:81.

DA CUNHA, A. M., 1914. Beitrag zur Kenntnis der Protozoenfauna Brasiliens. Mem. Inst. Oswaldo Cruz, 6:169.

DUMONT, J. N., 1961. Observations on the fine structure of the ciliate *Dileptus anser.* J. Protozool., 8:392.

FAURÉ-FREMIET, E., 1950. Méchanisme de la morphogenèse chez quelques ciliés gymnostomes hypostomiens. Arch. Anat. Micro. Morph. Expér., 39:1.

———— 1951. The marine sand-dwelling ciliates of Cape Cod. Biol. Bull., 100:59.

———— 1954. Réorganisation du type endomixique ches les Loxodidae et chez les Centrophorella. J. Protozool., 1:20.

———— 1957. *Trichopus lachmanni* n. sp., structure et morphogenèse. *Ibid.,* 4:145.

GELEI, J. v., 1933. Beiträge zur Ciliatenfauna der Umgebung von Szeged. II. Arch. Protist., 81:201.

HEGNER, R. W., and REES, C. W., 1933. *Taliaferria clarki,* etc. Tr. Am. Micr. Soc., 52:317.

IVANIĆ, M., 1933. Die Conjugation von *Chilodon cucullulus.* Arch. Protist., 79:313.

JAMESON, A. P., 1926. A ciliate, *Buxtonella sulcata,* etc. Parasitology, 18:182.

JONES, E. E. JR., 1951. Encystment, excystment, and the nuclear cycle in the ciliate *Dileptus anser*. J. El. Mitch. Sci. Soc., 67:205.

—— and BEERS, C. D., 1953. Some observations on structure and behavior in the ciliate *Dileptus monilatus*. I. Ibid., 69:42.

KAHL, A., 1930-1935. Utiere oder Protozoa. I. In Dahl's Die Tierwelt Deutschlands. Parts: 18, 21, 25, 30. Jena.

KANEDA, M., 1953. *Chlamydodon pedarius* n.sp. J. Sci. Hiroshima University, B. 1, 14:51.

—— 1960. The structure and reorganization of the macronucleus during the binary fission of *Chlamydodon pedarius*. Japan. J. Zool., 12:477.

—— 1960a. The morphology and morphogenesis of the cortical structures during binary fission of *Chlamydodon pedarius*. J. Sci. Hiroshima University, B. 1, 18:265.

—— 1960b. Phase contrast microscopy of cytoplasmic organelles in the gymnostome ciliate *Chlamydodon pedarius*. J. Protozool., 7:306.

—— 1961. Fine structure of the macronucleus of the gymnostome ciliate *Chlamydodon pedarius*. Japan. J. Genetics, 36:223.

KENT, W. S., 1880-1882. A manual of Infusoria. London.

KIDDER, G. W., and SUMMERS, F. M., 1935. Taxonomic and cytological studies on the ciliates associated with the amphipod family, etc. I. Biol. Bull., 68:51.

KIRBY, H., JR., 1934. Some ciliates from salt marshes in California. Arch. Protist., 82:114.

KRASCHENINNIKOW, S., 1934. Ueber die Cilienanordnung bei *Chilodonella cyprini*, etc. Ann. Protist., 4:135.

—— 1953. The silver-line system of *Chilodonella cyprini*. J. Morphol., 92:79.

MacDOUGALL, M. S., 1935. Cytological studies of the genus Chilodonella, etc. I. Arch. Protist., 84:198.

NOLAND, L. E., 1937. Observations on marine ciliates of the Gulf coast of Florida. Tr. Am. Micr. Soc., 56:160.

PENARD, E., 1922. Études sur les infusoires d'eau douce. Geneva.

SAUERBREY, E., 1928. Beobachtungen über einige neue oder wenig bekannte marine Ciliaten. Arch. Protist., 62:355.

STEIN, F., 1867. Der Organismus der Infusionstiere. Vol. 2.

STOKES, A. C., 1888. A preliminary contribution toward a history of the freshwater Infusoria of the United States. J. Trenton Nat. Hist. Soc., 1:71.

VUKSANOVIC, A., 1959. Contribution to the study of the genus Loxophyllum (Rumanian). Rev. Biol. (Ruman.), 4:165.

WENRICH, D. H., 1924. A new protozoan parasite. *Amphileptus branchiarum*, etc. Tr. Am. Micr. Soc., 43:191.

—— 1929. Observations on some freshwater ciliates. II. *Ibid.*, 48:352.

Chapter 33

Order 2. **Trichostomatida** Bütschli

B ODY ciliation uniform, but in some it may be asymmetrical; ciliature in vestibule, but no buccal ciliature. (Fifteen families.)

Family 1. **Colpodidae** Poche

Cytostome occurs ventrally in the anterior half of the body, at the base of ciliated vestibule; body often reniform, with ciliary rows making asymmetrical patterns; division within reproductive cysts. (Phylogeny of the family, Stout, 1960.)

Genus **Colpoda** Müller. Reniform; compressed; right border semi-circular; posterior half of the left border often convex; oral funnel in the middle of flattened ventral side; cytostome is displaced to the right of the median plane, which leads into peristome cavity and gives rise dorsally to a diagonal groove; right edge of cytostome bears a ciliated area, but no protruding membrane; macronucleus spherical or oval, central; a compact micronucleus; a contractile vacuole terminal; in fresh water. Many species. Burt (1940) made a comparative study of five species, which are mentioned here.

C. cucullus M. (Fig. 315, *a*). 40-110μ long; anterior keel with eight to ten indentations; twenty-nine to thirty-four ciliary grooves; cilia mostly paired; macronucleus with a stellate endosome; trichocysts rod-form; usually with abundant food vacuoles; in fresh water with decaying plants. (Respiration, Pigon, 1959; cysts, Pigon and Edstrom, 1961.)

C. inflata (Stokes) (Fig. 315, *b*). 35-90μ long; anterior keel with six to eight indentations; number of ciliary grooves (or meridians) twenty-one to twenty-four; cilia mostly in pairs; in fresh water among vegetation.

C. maupasi Enriques (Fig. 315, *c*). 35-90μ long; cytostome about one-fourth from the anterior end; anterior keel with five indentations; sixteen to eighteen meridians; in fresh water. Bensonhurst strain has been reported to produce several types of

cysts, due probably to nutrition, age and size of the trophic forms
(Padnos, Jakowska and Nigrelli, 1954). (Serological study and
temperature on cyst transformation, Padnos, 1962, 1962a.)

C. aspera Kahl (Fig. 315, *d*). 12-42μ long; cytostome about one-
third from the anterior end; fourteen to sixteen meridians; anterior keel with five indentations; in fresh water.

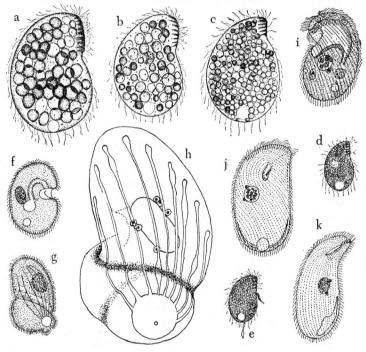

FIG. 315. a, *Colpoda cucullus;* b, *C. inflata;* c, *C. maupasi;* d, *C. aspera;*
e, *C. steini,* all ×330 (Burt); f, g, *Tillina magna,* ×100 (Bresslau); h, *T.*
canalifera, ×330 (Turner); i, *Bresslaua vorax,* ×100 (Kahl); j, *Bryophrya*
bavariensis, ×280 (Kahl); k, *Woodruffia rostrata,* ×190 (Kahl).

C. steini Maupas (Fig. 315, *e*). 15-42μ; cytostome about two-
fifths from the anterior end, and with a bundle of long membranellae; five to six preoral ridges; paired and single cilia; one
pair of long caudal cilia; twelve meridians; in fresh water. The
organism can live in various organs of the land slug, *Agriolimax*
agrestis (Reynolds, 1936).

C. duodenaria Taylor and Furgason. 20-40μ (9-60μ) long; twelve longitudinal ciliary rows; three postoral rows; two long cilia at the posterior end; long cilia project out from the cytostome along its posterior margin, forming a "beard"; a contractile vacuole terminal; macronucleus ovoid, with crescentic micronucleus; division into two to eight individuals in division cyst; but no divison in trophozoite stage; bacteria-feeder; fresh water. (Encystment, Taylor and Strickland, 1939; identity, Burt, 1940.)

Genus **Tillina** Gruber. Similar to *Colpoda* in general appearance and structure; but cytopharynx a long curved, ciliated tube; in fresh water.

T. magna G. (Fig. 315, *f, g*). 180-200μ long (Gruber), up to 400μ long (Bresslau); macronucleus oval to rod-shape; micronuclei vesicular, highly variable in number (2-16) (Beers); a contractile vacuole terminal, with six long collecting canals; division cyst produces four individuals; in stagnant water and also coprozoic. (Morphology, Gregory, 1909; Beers, 1944, 1945; encystment and excystment, Beers, 1945, 1946, 1946a, Bridgeman, 1957.)

T. canalifera Turner (Figs. 27; 315, *h*). 150-200μ by 100-150μ; resembles *magna;* but macronucleus ellipsoid, about one-third the body length; four to fourteen micronuclei, clustered around the macronucleus; a terminal contractile vacuole with seven to nine long permanent collecting canals; cytoplasm with 3-7μ long refractile rods; in fresh water (Turner, 1937). (Cytoplasmic inclusions, Turner, 1940.)

Genus **Bresslaua** Kahl. General body form resembles *Colpoda;* but cytopharynx large and occupies the entire anterior half.

B. vorax K. (Fig. 315, *i*). Dimorphic; microstome form, pyriform to reniform, plastic, with a latero-ventral invagination, 35-100μ by 16-50μ; macrostome form, oval in dorsal view; rigid with convex dorsal and flattened ventral surface, up to 250μ long; the prominent cleft in the left anterior quarter forms the anterior margin of vestibular opening; one macronucleus and one micronucleus; encystment (Stout, 1960).

Genus **Bryophrya** Kahl. Ovoid to ellipsoid; anterior end more or less bent toward left side; cytostome median, about one-third from anterior end, its right edge continues in horseshoe form around the posterior end and half of the left edge; anterior portion of left edge of the cytostome with posteriorly directed membrane; macronucleus oval or spherical; micronuclei; in fresh water.

B. *bavariensis* K. (Fig. 315, *j*). 50-120μ long.

Genus **Woodruffia** Kahl. Form similar to *Chilodonella* (p. 862); highly flattened snout bent toward left; cytostome, a narrow diagonal slit, its left edge with a membranous structure and its right edge with densely standing short cilia; macronucleus spherical; several (?) micronuclei; contractile vacuole flattened, terminal; in salt water.

W. *rostrata* K. (Fig. 315, *k*). 120-180μ long; salt water culture with Oscillatoria.

W. *metabolica* Johnson and Larson (1938). Pyriform; 85-400μ long; division cysts 85-155μ in diameter; resting cysts 40-62μ in diameter; in freshwater ponds. Johnson and Evans (1939, 1940) find two types of protective cysts in this ciliate: "stable" and "unstable" cysts, formation of both of which depends upon the absence of food. These cysts have three membranes: a thin innermost endocyst, a rigid mesocyst and a gelatinous outer ectocyst. The protoplasmic mass of the stable cyst is smaller, and free from vacuoles, and its ectocyst is thick, while that of the unstable cyst is larger, contains at least one fluid vacuole and its ectocyst is very thin. Crowding, feeding on starved Paramecium, increasing the temperature, and increasing the salt concentration of the medium, are said to influence the formation of unstable cysts. The two authors (1941) further reported that when free-swimming individuals were subjected, in the absence of food, to extremes of temperature, high concentrations of hydrogen-ion, and low oxygen tension, unstable cysts were formed; when the oxygen tension decreased, the tendency to encyst increased, even when ample food was present. The unstable cysts are said to remain viable for six months. Excystment is induced by changing the balanced salt solution, by replacing it with distilled water and by lowering temperature from 30° to 20°C.

Family 2. **Microthoracidae** Wrzesniowski

Cytostome and vestibule on ventral surface; body ciliation is much reduced; body with ribbed rigid pellicle, compressed laterally.

Genus **Microthorax** Engelmann (*Kreyella,* Kahl). Small, flattened; with delicate keeled armor which is more or less pointed anteriorly and rounded posteriorly; ventral armor with three ciliary rows; oral depression posterior-ventral, with a stiff ectoplasmic lip on right side, below which there is a small membrane, and with a small tooth on left margin; no cytopharynx; macronucleus spherical; two contractile vacuoles; in fresh water. (Many species.)

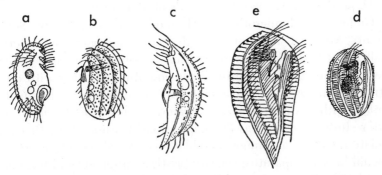

Fig. 316. a, *Microthorax simulans,* ×620 (Kahl); b, *Leptopharynx sphagnetorum,* ×570 (Kahl); c, *Drepanomonas dentata,* ×540 (Penard); d, e, *Pseudomicrothorax agilis* (d, ×340; e, oral area, ×670) (Kahl).

M. simulans Kahl (Fig. 316, *a*). 30-35μ long; decaying plant infusion, also in moss.

Genus **Leptopharynx** Mermod (*Trichopelma,* Levander). Compressed; surface with longitudinal furrows, seen as lines in end-view; course ciliation throughout; cytostome toward left edge about 1/3 from the anterior end; cytopharynx tubular; macronucleus spheroid, central; two contractile vacuoles; fresh water.

L. sphagnetorum (Levander) (Fig. 316, *b*). 25-40μ long; in fresh water.

L. agilis (Savoie). 30-40μ (Savoie, 1957).

Genus **Drepanomonas** Fresenius (*Drepanoceras,* Stein). Highly flattened; aboral surface convex; oral surface flat or concave; with a few deep longitudinal furrows; ciliation sparse; cytostome and a small cytopharynx simple, near the middle of body; fresh water. (Several species.)

D. dentata F. (Fig. 316, *c*). With a small process near cytostome; two rows of ciliary furrows on both oral and aboral sur-

874 *Protozoology*

faces; cilia on both ends of oral surface; 40-65μ long; in fresh water.

Genus **Discotricha** Tuffrau. Ovate; papillae with trichocysts on dorsal and ventral surfaces, except the oral region; ventral surface ciliated, bearing nine rows of cirri in two fields; mouth toward left under a semi-rigid plate; contractile vacuole large, posterior; vacuolar pit large; macronucleus irregularly hemispherical; sand-dwelling. (Closely related to Chlamydodon, Tiffrau, 1954.)

D. papillifera T. 40μ long; feeds on diatomes.

Genus **Pseudomicrothorax** Mermod (*Craspedothorax*, Sondheim). More or less compressed; cytostome opens in anterior half toward left side, in a depression surrounded by ciliary rows; body surface marked with a broad longitudinal ridge with cross striation; furrows canal-like; cilia on ventral side; cytopharynx tubular, with elastic rods; fresh water. Corliss (1958) points out, after studying *P. dubius* (Thompson and Corliss, 1958) that this ciliate possesses a gymnostome-like pharyngeal basket; a tetrahymenal buccal apparatus; sensory bristles; flattened rigid form and restricted ciliature, resembling those of certain spirotrichs and suggested its transfer to Hymenostomatida.

P. agilis M. (Fig. 316, *d, e*). Ellipsoid; 48-58μ long; in fresh water.

Family 3. **Isotrichidae** Bütschli

Cytostome located at or near the anterior end; body ciliation complete and uniform; in stomach of ruminants.

Genus **Isotricha** Stein. Ovoid; flattened; dense longitudinal ciliary rows; cytostome at or near anterior end; several contractile vacuoles; reniform macronucleus and a micronucleus connected with, and suspended by, fibrils, karyophore; locomotion with posterior end directed forward; in the stomach of cattle and sheep.

I. prostoma S. (Fig. 317, *a*). 80-195μ by 53-85μ; the organism selects and ingests certain rod-form bacteria only among many types of rumen bacteria (Gutierrez, 1958). (Cytology, Campbell, 1929.)

I. intestinalis S. (Fig. 317, *b*). 97-130μ by 68-88μ.

Genus **Dasytricha** Schuberg. Oval, flattened; cilia in longitudinal spiral rows; no karyophore; in the stomach of cattle.

D. ruminantium S. (Fig. 317, *c*). 50-75μ by 30-40μ.

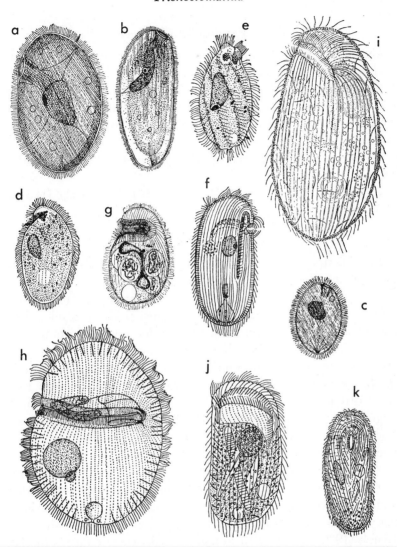

FIG. 317. a, *Isotricha prostoma*, ×500 (Becker and Talbott); b, *I. intest-inalis*, ×500; c, *Dasytricha ruminantium*, ×330 (Becker and Talbott); d, *Paraisotricha colpoidea*, ×270; e, *P. beckeri*, ×360 (Hsiung); f, *Plagiopyla nasuta*, ×340 (Kahl); g, *P. minuta*, ×400 (Powers); h, *Lechriopyla mystax*, ×340 (Lynch); i, *Sonderia pharyngea*, ×500 (Kirby); j, *S. vorax*, ×310 (Kahl); k, *Clathrostoma viminale*, ×220 (Penard).

Family 4. **Paraisotrichidae** da Cunha

Cytostome near the anterior end; with a concrement vacuole; body ciliation uniform and complete.

Genus **Paraisotricha** Fiorentini. Uniformly ciliated in more or less spiral longitudinal rows; longer cilia at anterior end; cytostome near anterior tip; contractile vacuole posterior; in the caecum and colon of horse.

P. colpoidea F. (Fig. 317, *d*). 70-100μ by 42-60μ. (Conjugation, Dogiel, 1930.)

P. beckeri Hsiung (Fig. 317, *e*). 52-98μ by 30-52μ (Hsiung, 1930, 1930a).

Family 5. **Plagiopylidae** Schewiakoff

Body flattened; ciliation uniform; cytostome central at the end of peristome, located in the anterior half of body.

Genus **Plagiopyla** Stein. Peristome a broad ventrally opened groove from which body ciliation begins; peristomal cilia short, except a zone of longer cilia at anterior end; cytostome near median line at the end of the peristome; cytopharynx long; a peculiar 'stripe band' located on dorsal surface has usually its origin in the peristomal groove, after taking an anterior course for a short distance, curves back and runs down posteriorly near right edge and terminates about one-third the body length from posterior end; macronucleus rounded; a micronucleus; contractile vacuole terminal; free-living or endozoic.

P. nasuta S. (Fig. 317, *f*). Ovoid; tapering anteriorly; peristome at right angels or slightly oblique to the edge; trichocysts at right angles to body surface; macronucleus round to irregular in shape; body about 100μ (80-180μ) long; sapropelic in brackish water. Lynch (1930) observed this ciliate in salt water cultures in California and found it to be 70-114μ by 31-56μ by 22-37μ.

P. minuta Powers (Fig. 317, *g*). 50-75μ by 36-46μ; in the intestine of *Strongylocentrotus droebachiensis;* the Bay of Fundy (Powers, 1933). Morphology (Beers, 1954).

Genus **Lechriopyla** Lynch Similar to *Plagiopyla;* but with a large internal organella, *furcula,* embracing the vestibule from right, and a large crescentic motorium at left end of peristome; in the intestine of sea-urchins.

L. mystax L. (Fig. 317, *h*). 113-174μ long; in the gut of *Strongylocentrotus purpuratus* and *S. franciscanus;* California.

Genus **Sonderia** Kahl. Similar to *Plagiopyla* in general appearance; ellipsoid; flattened; peristome small and varied; body covered by 2-4μ thick gelatinous envelope which regulates osmosis, since no contractile vacuole occurs (Kahl); with or without a striped band; trichocysts slanting posteriorly; in salt or brackish water. Kirby (1934) showed that several species of the genus are common in the pools and ditches in salt marshes of California, salinities of which range 3.5-10 per cent or even up to 15-20 per cent.

S. pharyngea Kirby (Fig. 317, *i*). Ovoid to ellipsoid; flattened; 84-110μ by 48-65μ; gelatinous layer about 2μ thick, with bacteria; about 60 longitudinal ciliary rows, each with two borders; peristome about 35μ long, at anterior end, oblique; with closely set cilia from the opposite inner surfaces; cytopharynx conspicuous; spherical macronulceus anterior, with a micronucleus; trichocysts (7-9μ long) distributed sparsely and unevenly, oblique to body surface; a group of bristle-like cilia at posterior end; often brightly colored because of food material; in salt marsh, California.

S. vorax Kahl (Fig. 317, *j*). Broadly ellipsoid; size variable, 70-180μ long; ventral surface flattened; posterior border of peristomal cavity extending anteriorly; in salt marsh; California (Kirby, 1934).

Family 6. **Marynidae** Poche

Vestibule at the anterior end; with gelatinous lorica; swims backward.

Genus **Maryna** Gruber. Peristome makes a complete circle, thus the cone is entirely separated from anterior edge of body; cytostome left ventral, elongate slit; ridge also with a slit; gelatinous lorica dichotomous.

M. socialis G. (Fig. 318, *a, b*). About 150μ long; in infusion made from long-dried mud.

Genus **Mycterothrix** Lauterborn (*Trichorhynchus,* Balbiani). Anterior cone continuous on dorsal side with body ridge; hence free edge of body only on ventral side; no ventral slit.

M. erlangeri L. (Fig. 318, *c*). Nearly spherical with zoochlorellae; 50-55μ by 40-50μ; fresh water.

Family 7. **Cyathodiniidae** da Cunha

Cytostome at the base of a triangular peristome-like non-ciliated field; body ciliation confined to the anterior half of body.

Genus **Cyathodinium** da Cunha. Conical or pyriform; broad cytostome occupies the entire anterior end and extends posteriorly one-fourth to three-fourths the body length; deep with prominent ridges; oral cilia in a single row on left ridge; body cilia comparatively long, confined to anterior half; macronucleus round or ellipsoid; a micronucleus; one to several contractile vacuoles; in the caecum and colon of guinea pigs.

C. conicum da C. Inverted cone; 50-80μ by 20-30μ; in the caecum of *Cavia aperea* and *C. porcella*.

C. piriforme da C. (Fig. 318, *d*). Typical form inverted pyri-

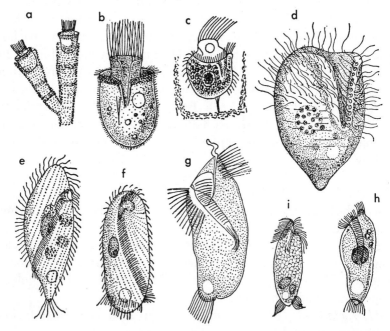

FIG. 318. a, b, *Maryna socialis* (a, ×40; b, ×160) (Gruber); c, *Mycterothrix erlangeri*, ×310 (Kahl); d, *Cyathodinium piriforme*, ×1290 (Lucas); e, *Spirozona caudata*, ×370 (Kahl); f, *Trichospira inversa*, ×360 (Kahl); g, *Blepharocorys uncinata*, ×540 (Reichenow); h, *B. bovis*, ×850 (Dogiel); i, *Charonina equi*, ×570 (Hsiung).

form; second form conical with tapering anterior end; contractile vacuole posterior; 30-40μ by 20-30μ; in the caecum of *Cavia aperea* and *C. porcella.* (Occurrence and cytology, Lucas, 1932, 1932a; Nie, 1950.)

Family 8. **Spirozonidae** Kahl

In addition to uniform ciliation, a special zone of cilia from cytostome to posterior end; a tuft of long cilia at the posterior end.

Genus **Spirozona** Kahl. Short spindle-form; anterior end truncate, posterior region drawn out to a rounded end, with a group of longer cilia; spiral ciliation; beginning near right posterior third the central ciliary row runs over ridge to left and then reaches the cytostome; other rows are parallel to it; cytostome in anterior one-fourth, with cytopharynx; ellipsoid macronucleus nearly central; contractile vacuole terminal; fresh water, sapropelic.

S. caudata K. (Fig. 318, *e*). 80-100μ long.

Family 9. **Trichospiridae** Kahl

Body ciliation complete; with a band of closely set cilia from the vestibular region to posterior end; without any tuft of cilia.

Genus **Trichospira** Roux. Body cylindrical; posterior end rounded, anterior end conical in profile, where the cytostome surrounded by two spiral rows of cilia, is located; a special ciliary band beginning in the cytostomal region runs down on ventral side, turns spirally to left and circles partially posterior region of body; ciliary rows parallel to it; macronucleus oval, with a micronucleus; contractile vacuole posterior; fresh water, sapropelic.

T. inversa (Claparède and Lachmann) (Fig. 318, *f*). 70-100μ long.

Family 10. **Blepharocoridae** Hsiung

Cytostome near anterior end; body ciliation much reduced to tufts occurring near anterior and posterior ends.

Genus **Blepharocorys** Bundle. Oral groove deep, near anterior end; three (oral, dorsal and ventral) ciliary zones at anterior end; a caudal ciliary zone; in the caecum and colon of horse or stomach of cattle. (Many species.)

B. uncinata Fiorentini (*B. equi* Schumacher) (Fig. 318, *g*). With a screw-like anterior process; 55-74µ by 22-30µ; in the caecum and colon of horse (Hsiung, 1930a).

B. bovis Dogiel (Fig. 318, *h*). 23-37µ by 10-17µ; in the stomach of cattle (Dogiel, 1926).

Genus **Charonina** Strand (*Charon,* Jameson). Two caudal ciliary zones; in the colon of horse or in stomach of ruminants.

C. equi Hsiung (Fig. 318, *i*). 30-48µ by 10-14µ; in the colon of horse (Hsiung, 1930, 1930a).

Family 11. **Balantidiidae** Doflein and Reichenow

Vestibule near the anterior end, with cytostome at its base; body ciliation uniform and complete; in the alimentary canal of invertebrates and vertebrates.

Genus **Balantidium** Claparède and Lachmann (*Balantidiopsis* Bütschli; *Balantioides* Alexeieff). Oval, ellipsoid to subcylindrical; peristome begins at or near anterior end; cytopharynx not well developed; longitudinal ciliation uniform; macronucleus elongated; a micronucleus; contractile vacuole and cytopyge terminal; in the gut of vertebrates and invertebrates. (Numerous species. Hegner, 1934; Kudo and Meglitsch, 1938.) This genus has hitherto been placed in order Heterotrichida, but is transferred to Trichostomatida, following the views that had been expressed by Fauré-Fremiet (1955) and Krascheninnikow and Wenrich (1958).

B. coli Malmsten (Fig. 319). Ovoid; 40-80µ by 30-60µ, but length varies 30-150µ; body covered by many slightly oblique longitudinal rows of cilia; peristome small near anterior tip, lined with coarser cilia; inconspicuous cytostome and cytopharynx are located at the end of peristome; two contractile vacuoles, one terminal, the other near the middle of body; macronucleus sausage-shape and a vesicular micronucleus; cytopyge near the posterior tip; food particles are of various kinds, including erythrocytes and other host cell fragments, starch grains, etc. The trophozoites multiply by binary fission. Conjugation has been described by Brumpt (1909), Jameson (1927), Scott (1927), Nelson (1934), etc. Svensson (1955) also observed frequent conjugation "in the first weeks after isolation, during some weeks after mixing two strains and also, but less markedly, after transfer into different bacterial mixtures. Thus conjugation seems to be a means for causing homogeneity in a culture. When stability is reached with regular growth of the balantidia in good balance with the concomitant

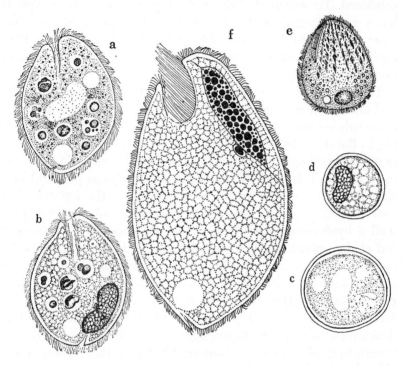

FIG. 319. a–d, *Balantidium coli*, ×530 (a, a living trophozoite; b, a stained one; c, a fresh and d, stained cyst) (Kudo); e, *B. duodeni*, ×170 (Stein); f, *B. praenucleatum*, ×950 (Kudo and Meglitsch).

bacteria, conjugation becomes rare or disappears entirely." Abnormal infraciliature (Krascheninnikow, 1959).

The cysts are circular to ovoid in outline; slightly yellowish or greenish and refractile; 40-60µ in diameter; cyst wall made up of two membranes; cytoplasm hyaline; macronucleus and a contractile vacuole are usually seen.

This ciliate lives in the colon and caecum of man and causes balantidiosis or balantidial dysentery. Strong (1904) made the first histological study of the infection. The organisms invade the tissues and blood vessels of the mucosa and submucosa. At the beginning there is hyperaemia with punctiform haemorrhages, and later vascular dilatation, round cell infiltration, eosinophilia, etc., develop in the infected area. Finally deep-seated ulcers are produced. The balantidial dysentery is usually of chronic type. It has a wide geographical distribution. In the United States a few cases of infections have been observed in recent years. In the Philippine Islands, more cases have been noticed than anywhere else.

This ciliate is a very common parasite in the intestine of pigs, and also of chimpanzee and orang-outang. In pigs, the organism ordinarily confines itself to the lumen of the intestine, but according to Ratcliffe (1934), when the host animals become infected by organisms belonging to Salmonella, it invades and ulcerates the intestinal wall. In experimental infection in guinea pigs and rats, the ciliates are usually found in the lumen of the intestine, but Uribe (1958) reported that the organisms invade the intestinal wall if hyaluronidase preparation and *Bacillus coli* were inoculated at the same time. The cysts developing in pigs appear to become the chief source of infection, since balantidial dysentery is more commonly found among those who come in contact with pigs or pig intestine. The cysts remain viable for weeks in pig faeces in moist and dark places, though they are easily killed by desiccation or exposure to sun light. The cysts may reach human mouth in food or in water contaminated with them, through unclean hands of persons who come in contact with faeces or intestine of pigs, and in some cases perhaps through uncooked sausage.

B. suis McDonald. Ellipsoid; 35-120µ by 20-60µ; macronucleus more elongate than that of *B. coli;* in the intestine of pigs (McDonald, 1922). Levine (1940) through a series of culture studies, has come to consider that *B. coli* and *B. suis* are only morphological variations due to the nutritional condition and that *B. suis* is synonymous with *B. coli.* Lamy and Roux (1950) observed both forms in cultures started with single individuals, and considered the elongate *suis* as conjugants and the oval *coli* as vegetative forms.

B. caviae Neiva, da Cunha and Travassos. In the caecum of guinea-pig. (Morphology, Scott, 1927; Nie, 1950.)

Other domestic and wild animals harbor various species of Balantidium.

B. duodeni Stein (Fig. 319, *e*). 70-80μ by 55-60μ; in the intestine of the frog.

B. praenucleatum Kudo and Meglitsch (Fig. 319, *f*). 42-127μ long, 32-102μ thick, 25-80μ wide; macronucleus close to anterior end; in the colon of *Blatta orientalis* (Kudo and Meglitsch, 1938).

Family 12. **Entorhipidiidae** Madsen

A frontal "lobe" over cytostome; ciliation uniform; body compressed laterally; in the gut of sea urchin.

Genus **Entorhipidium** Lynch. Triangular in general outline; colorless; large, 155-350, long; flattened; posterior end drawn out, with a bristle; anterior end bent to left; cytostome in depression close to left anterior border, with long cilia; with or without a cross-groove from preoral region; cytopharynx inconspicuous; trichocysts; macronucleus oval to sausage-form; one to several micronuclei; several (excretory) vacuoles left-ventral; in intestine of the sea urchin *Strongylocentrotus purpuratus*. (Four species.)

E. echini L. (Fig. 320, *a*). About 250μ by 125μ; California.

Genus **Entodiscus** Madsen. Broadly or narrowly lancet-like, without narrowed posterior portion; cytostome small on left narrow side, about two-fifths the body length from anterior end; without trichocysts; macronucleus central, with a micronucleus; contractile vacuole subterminal; swimming movement rapid without interruption. Two species. (Morphology, Powers, 1933, 1933a.) Some authors consider that Entodiscus and Biggaria belong to Tetrahymenina (Corliss, 1961).

E. indomitus M. (Fig. 320, *b*). 80-117μ by 20-23μ; in the intestine of *Strongylocentrotus droebachiensis*.

E. borealis (Hentschel) (Fig. 320, *c*). Oval; cytostome nearer anterior end; 105-170μ by 60-115μ; in the gut of *Strongylocentrotus droebachiensis* and *Echinus esculentus;* Powers (1933) studied this species in the first-named host from Maine, and found a supporting rod which is imbedded in the margin along the right wall of the oral cavity and which he named *stomatostyle*.

Genus **Biggaria** Kahl. Scoop-like form; anterior two-thirds thin, posterior region thickened, terminating in a rudder-like style;

cilia in longitudinal rows; longer cilia on caudal prolongation; cytostome in the posterior half, opening into a vestibule, into which long cilia project from the roof; aperture to cytopharynx with two membranes; contractile vacuole subterminal; in the intestine of sea-urchins.

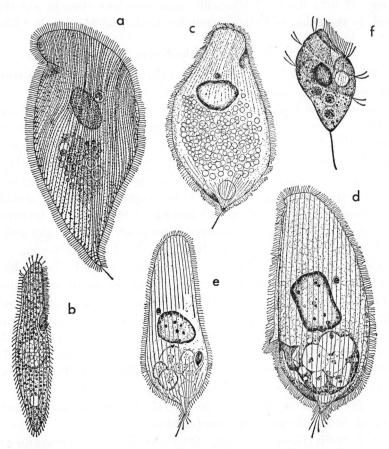

FIG. 320. a, *Entorhipidium echini,* ×270 (Lynch); b, *Entodiscus indomitus,* ×380 (Madsen); c, *E. borealis,* ×380 (Powers); d, *Biggaria bermudense,* ×380 (Powers); e, *B. echinometris,* ×380 (Powers); f, *Trimyema compressum,* ×410 (Lackey).

B. bermudense (Biggar) (Fig. 320, *d*). 90-185μ by 48-82μ; in *Lytechinus variegatus;* Bermuda (Biggar), North Carolina (Powers). Powers (1935) found the organism at Tortugas in *Lytechinus var-*

iegatus, Centrechinus antillarum, Echinometra lucunter, Trip-neustes esculentus and *Astrophyga magnifica.*

B. *echinometris* (B.) (Fig. 320, *e*). 80-195μ by 33-70μ; in *Echinometris subangularis* (Bermuda) and *Lytechinus variegatus* (North Carolina).

Family 13. **Trimyemidae** Kahl

Cytostome near the anterior end; with a long caudal cilium; cilia in three to four spiral rows on the anterior body surface.

Genus **Trimyema** Lackey. Ovoid, more or less flattened; anterior end bluntly pointed, posterior end similar or rounded; with a long caudal cilium; cilia on three to four spiral rows which are usually located in the anterior half of body; round cytostome near anterior end with a small cytopharynx; spherical macronucleus central with a small micronucleus; one contractile vacuole; active swimmer; fresh or salt water.

T. *compressum* L. (Fig. 320, *f*). About 65μ by 35μ; Lackey found it in Imhoff tank; fresh and salt water (Kahl). Klein (1930) studied its silverline system.

Family 14. **Conidophryidae** Mohr and LeVeque

Mature individuals without cilia, attached to hairs of crustaceans; immature (ciliated) swimming forms produced by budding are impaled upon the hairs through cytostome.

Genus **Conidophrys** Chatton and Lwoff (Fig. 321). Trophont or the form attached to host's appendages (*a*), cylindrical, with a thick pellicle; contents divide into two or three (and up to several) smaller bodies which develop into tomites or free-swimming individuals (*b*); when the latter come in contact with the ends of the secretory hairs of the host, they become attached through their cytopharynx (*c*) and lost their cilia; during the development into the cucurbitoid mature stage (*f, g*), the organism passes through lacrymoid (*d*) and spheroid (*e*) stages; on freshwater amphipods and isopods (Chatton and Lwoff, 1934, 1936).

C. *pilisuctor* C. and L. (Fig. 322). Lacrymoid trophont 12-15μ by 6-7μ; cucurbitoid forms 50-60μ long; free-swimming tomites 12-14μ in diameter by 6-7μ high, ciliated and possess a comparatively long cytopharynx; nourishment of trophont through host's hairs; in amphipods and isopods, especially on *Corophium acherusicum,* France. Mohr and LeVeque (1948) found it on the wood-boring isopods, *Limnoria lignorum* and *Corophium acherusicum* in California.

Family 15. **Coelosomididae** Corliss

Vestibule at the anterior end; a cytostome at its base; body ciliation uniform and complete.

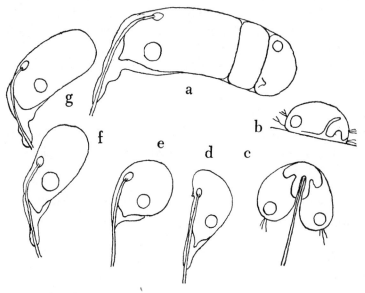

Fig. 321. The developmental cycle of *Conidiophrys pilisuctor* (Chatton and Lwoff). a, trophont with two tomites; b, freed tomite; c, tomite becoming attached to host's hair; d, lacrymoid trophont; e, spheroid stage; f, g, cucurbitoid stage.

Genus *Coelosomides* Strand (*Coelosoma,* Anigstein). Body cylindrical; ciliation uniform; a conspicuous ciliated vestibule occurs at anterior end; endoplasm vacuolated; marine.

C. marina (Anigstein) (Fig. 323, *a, b*). About 200μ long; central endoplasm highly vacuolated; periphery finely reticulated; macronucleus elongate; micronucleus compact (Anigstein, 1911; Fauré-Fremiet, 1950).

Genus **Pseudoprorodon** Blochmann. Body ovoid; usually flattened; one side convex, the other concave; ectoplasm conspicuously alveolated; trichocysts grouped; one or more contractile vacuoles posterior-lateral or distributed, with many pores; macronucleus elongate; cytopharynx with trichites; fresh or salt water.

P. farctus (Claparède and Lachmann) (Figs. 323, *c*). Ellipsoid; cytostome surrounded by long trichites; contractile vacuole posterior, with secondary vacuoles; macronucleus elongate; body 150-200μ long; fresh water.

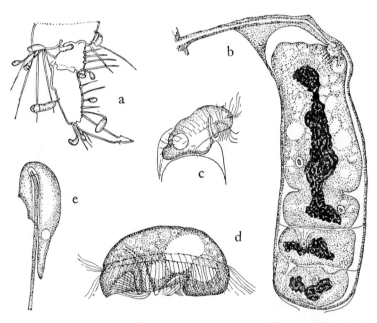

FIG. 322. *Conidiophrys pilisuctor* (Chatton and Lwoff). a, trophonts of all ages on an appendage of *Corophium acherusicum;* b, a stained mature trophont with two formed and one developing tomites, ×1330; c, a tomite emerging from trophont, ×1330; d, a living tomite, ×2230; e, newly attached lacrymoid trophont, ×1330.

Genus **Bursella** Schmidt. Oval; anterior end broadly and obliquely truncate where a large ciliated groove-like pit occurs; ridges of pit contractile; cilia short; macronucleus, spherical to ellipsoidal; several micronuclei; endoplasm reticulated; with symbiotic algae; ectoplasm with trichocysts; fresh water.

B. spumosa S. 240-560μ long; freshwater pond.

Genus **Paraspathidium** Noland. Body cyclindrical; cytostome

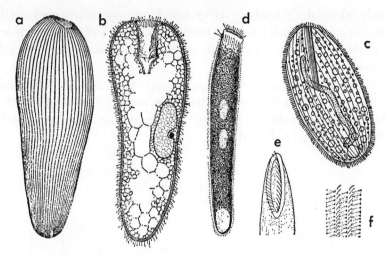

FIG. 323. a, b, *Coelosomides marina*, ×245 (Fauré-Fremiet) (a, silver-impregnated surface view; b, optical section); c, *Pseudoprorodon farctus*, ×270 (Roux); d–f, *Paraspathidium trichostomum* (Noland) (d, ×130; e, oral region, ×400; f, portion of pellicle, ×1000).

an elongate slit, bordered on one side by strong cilia and on the other by weaker cilia and a shelf-like, nonundulatory membrane; two longer cilia on dorsal edge near anterior tip; anterior one-third compressed; posterior two-thirds nearly cylindrical; two oval macronuclei, each with a micronucleus; cytoplasm filled with numerous refractile granules; about seventy rows of cilia; contractile vacuole terminal; salt water. (One species.)

P. trichostomum N. (Fig. 323, *d-f*). About 220μ long; macronuclei 44μ long each; salt water; Florida (Noland, 1937).

References

BEERS, C. D., 1944. The maintenance of vitality in pure lines of the ciliate *Tillina magna*. Am. Nat., 78:68.

——— 1945. Some factors affecting excystment in the ciliate *Tillina magna*. Physiol. Zool., 18:80.

——— 1946. History of the nuclei of *Tillina magna* during division and encystment. J. Morphol., 78:181.

——— 1946a. *Tillina magna*: etc. Biol. Bull., 91:256.

——— 1954. *Plagiopyla minuta* and *Euplotes balteatus*, etc. J. Protozool., 1:86.

——— 1961. The obligate commensal ciliates of *Strongylocentrotus droebachiensis*, etc. Biol. Bull., 121:69.

BRIDGMAN, A. J., 1957. Studies on dried cysts of *Tillina magna*. J. Protozool., 3:17.

BRUMPT, E., 1909. Demonstration du role pathogene du *Balantidium coli* et conjugaison de cet infusoire. C. R. Soc. Biol., 67:103.

BURT, R. L., 1940. Specific analysis of the genus Colpoda with special reference to the standardization of experimental material. Tr. Am. Micr. Soc., 59:414.

———, KIDDER, G. W., and CLAFF, C. L., 1941. Nuclear reorganization in the family Colpodidae. J. Morphol., 69:537.

CAMPBELL, A. S., 1929. The structure of *Isotricha prostoma*. Arch. Protist., 66:331.

CHATTON, E., and LWOFF, A., 1934. Sur un infusoire parasite des poils sécreteurs des crustacés Edriophtalmes et la famille nouvelle des Pilisuctoridae. C. R. Acad. Sci., 199:696.

——— ——— 1936. Les Pilisuctoridae. Bull. biol. France et Belg., 70:86.

CLAFF, C. L., DEWEY, VIRGINIA C., and KIDDER, G. W., 1941. Feeding mechanisms and nutrition in three species of Bresslaua. Biol. Bull., 81:221.

CORLISS, J. O., 1961. The ciliated protozoa. Pergamon Press.

DOGIEL, V., 1926. Une nouvelle espéce du genre Blepharocorys, *B. bovis*, etc. Ann Parasitol., 4:61.

——— 1930. Die prospektive Potenz der Syncaryonderivate an der Conjugation von Paraisotricha erläutert. Arch. Protist., 70:497.

FAURÉ-FREMIET, E., 1955. La position systematique de genre Balantidium. J. Protozool., 2:54.

GREGORY, L. H., 1909. Observations on the life history of *Tillina magna*. J. Exper. Zool., 6:383.

GUTIERREZ, J., 1958. Observations on bacterial feeding by the rumen ciliate *Isotricha prostoma*. J. Protozool., 5:122.

HEGNER, R. W., 1934. Specificity in the genus Balantidium based on size and shape, etc. Amer. J. Hyg., 19:38.

HSIUNG, T. S., 1930. Some new ciliates from the large intestine of the horse. Tr. Am. Micr. Soc., 49:34.

——— 1930a. A monograph on the Protozoa of the large intestine of the horse. Iowa State Coll. J. Sci., 4:356.

JAMESON, A. P., 1927. The behavior of *Balantidium coli* in cultures. Parasitology, 19:411.

JOHNSON, W. H., and EVANS, F. R., 1939. A study of encystment in the ciliate, *Woodruffia metabolica*. Arch. Protist., 92:91.

——— ——— 1940. Environmental factors affecting cystment in *Woodruffia metabolica*. Phyiol. Zool., 13:102.

——— ——— 1941. A further study of environmental factors affecting cystment in *Woodruffia metabolica*. *Ibid.*, 14:227.

——— and LARSON, E., 1938. Studies on the morphology and life history of *Woodruffia metabolica*, n. sp. Arch. Protist., 90:383.

KIDDER, G. W., and CLAFF, C. L., 1938. Cytological investigations of *Colpoda cucullus*. Biol. Bull., 74:178.

KIRBY, H. JR., 1934. Some ciliates from salt marshes in California. Arch. Protist., 82:114.

KRASCHENINNIKOW, S., 1959. Abnormal infraciliature of *Balantidium coli* and *B. caviae*(?) and some morphological observations on these species. J. Protozool., 6:61.

―――― and WENRICH, D. H., 1958. Some observations on the morphology and division of *Balantidium coli* and *B. caviae*(?). *Ibid.*, 5:196.

KUDO, R. R., and MEGLITSCH, P. A., 1938. On *Balantidium praenucleatum*, etc. Arch. Protist., 91:111.

LACKEY, J. B., 1925. The fauna of Imhoff tanks. New Jersey Agr. Exper. Station, no. 417.

LAMY, L., and ROUX, H., 1950. Remarques morphologiques, biologiques et spécifiques sur les Balantidium de culture. Bull. Soc. Path. Exot., 43:422.

LEVINE, N. D., 1940. The effect of food intake upon the dimensions of Balantidium from swine in culture. Amer. J. Hyg., 32:81.

LUCAS, M. S., 1932. A study of *Cyathodinium piriforme*. Arch. Protist., 77:64.

―――― 1932a. The cytoplasmic phases of rejuvenescence and fission in *Cyathodinium piriforme*. II. *Ibid.*, 77:406.

LYNCH, J., 1929. Studies on the ciliates from the intestine of Strongylocentrotus. I. Univ. Cal. Publ. Zool., 33:27.

―――― 1930. II. *Ibid.*, 33:307.

McDONALD, J. D., 1922. On *Balantidium coli* and *B. suis* (sp. nov.). Univ. Cal. Publ. Zool., 20:243.

MOHR, J. L., and LeVEQUE, J. A., 1948. Occurrence of Conidiophrys, etc. J. Parasit., 34:253.

NELSON, E. C., 1934. Observations and experiments on conjugation of the Balantidium from the chimpanzee. Amer. J. Hyg., 20:106.

PADNOS, M., 1962. Serological studies on protozoa. I. J. Protozool., 9:7.

―――― 1962a. Cytology of cold induced transformation of octogenic reproductive cysts to resting cysts in *Colpoda maupasi*. *Ibid.*, 9:13.

――――, JAKOWSKA, S., and NIGRELLI, R. F., 1954. Morphology and life history of *Colpoda maupasi*, etc. *Ibid.*, 1:13.

PIGON, A., 1959. Respiration of *Colpoda cucullus* during active life and encystment. J. Protozool., 6:303.

―――― and EDSTROM, J. E., 1961. Excystment ability, respiratory metabolism, and RNA content in two types of resting cysts of *Colpoda cucullus*. *Ibid.*, 8:257.

POWERS, P. B. A., 1933. Studies on the ciliates from sea urchins. I. Biol. Bull., 65:106.

―――― 1933a. II. *Ibid.*, 65:122.

——— 1935. Studies on the ciliates of sea urchins. Papers Tortugas Lab., 29:293.

REYNOLDS, B. D., 1936. *Colpoda steini,* a facultative parasite of the land slug, *Agriolimax agrestis.* J. Parasit., 22:48.

SAVOIE, A., 1957. Le cilié *Trichopelma agilis* n. sp. J. Protozool., 4:276.

SCOTT, M. J., 1927. Studies on the Balantidium from the guinea-pig. J. Morphol. & Physiol., 44:417.

STOUT, J. D., 1960. Morphogenesis in the ciliate *Bresslaua vorax* Kahl and the phylogeny of the Colpodidae. J. Protozool., 7:26.

STRONG, R. P., 1904. The clinical and pathological significance of *Balantidium coli.* Bur. Gov. Lab. Manila. Biol. Lab. Bull., no. 26.

TAYLOR, C. V,, and FURGASON, W. H., 1938. Structural analysis of *Colpoda duodenaria* sp. nov. Arch. Protist., 90:320.

——— and STRICKLAND, A. G., 1939. Reactions of *Colpoda duodenaria* to environmental factors. II. Physiol. Zool., 12:219.

TURNER, J. P., 1937. Studies on the ciliate *Tillina canalifera* n. sp. Tr. Am. Micr. Soc., 56:447.

——— 1940. Cytoplasmic inclusions in the ciliate *Tillina canalifera.* Arch. Protist., 93:255.

TUFFRAU, M., 1954. *Discotricha papillifera* n.g., n. sp. J. Protozool., 1:183.

Chapter 34

Order 3. Hymenostomatida Delage and Hérouard

V ENTRALLY located buccal cavity with an undulating membrane on right and three membranelles on left; body ciliation uniform. This order is subdivided into three suborders: Tetrahymenina, Peniculina (p. 902) and Pleuronematina (p. 913).

Suborder. 1. Tetrahymenina Fauré-Fremiet

Tetrahymenal oral ciliature inconspicuous; vestibule seldom present; body usually small. (Four families.)

Family 1. Ophryoglenidae Kent

A refractile "watch-glass" body is located in the cytoplasm near the buccal cavity; uniform body ciliation.

Genus **Ophryoglena** Ehrenberg. Ellipsoidal to cylindrical; ends rounded or attenuated; preoral depression in form of '6' due to an ectoplasmic membrane extending from the left edge, cilia on the right edge; tetrahymenal buccal ciliation; one (or two) contractile vacuole with long radiating canals, opens on right ventral side; macronucleus of various shape with several endosomes; a micronucleus; fresh or salt water or parasitic. (Species, Canella and Trincas, 1961.)

O. collini Lichtenstein (Fig. 324, *a*). Pyriform; macronucleus horseshoe-shape; 200-300µ by 120-230µ; in the caecum of Baetis larvae.

O. parasitica André. Ovoid; dark; micronucleus (?); 170-350µ by 180-200µ; in the gastrovascular cavity of *Dendrocoelum lacteum.*

O. pyriformis Rossolimo (Fig. 324, *b*). Flask-shape; 240-300µ long; in the gastrovascular cavity of various Turbellaria.

O. intestinalis R. (Fig. 324, *c*). Up to 1.5 mm. by 450-500µ; smallest 60µ long; in the gastrovascular cavity of *Dicotylus* sp.

O. atra Lieberkühn. Oval, posterior end broadly rounded; 300-500µ long; grayish; filled with globules; cytostome near anterior end; macronucleus elongated; a contractile vacuole; trichocysts; stagnant fresh water.

Genus **Ichthyophthirius** Fouquet. Body ovoid; ciliation uniform; vestibule inconspicuous; cytostome near the anterior end;

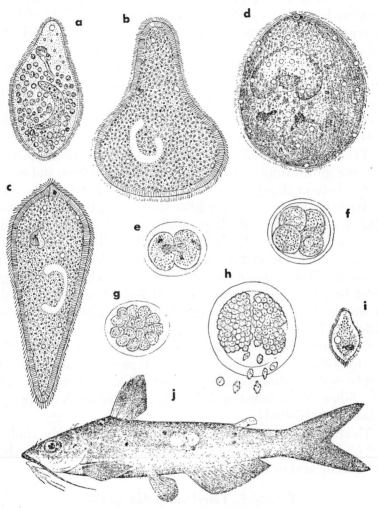

Fig. 324. a, *Ophryoglena collini*, ×150 (Lichtenstein); b, *O. pyriformis*, ×180 (Rossolimo); c, *O. intestinalis*, ×55 (Rossolimo); d–j, *Ichthyophthirius multifiliis* (d, a large trophozoite, ×75 (Bütschli); e–h, development within cysts; i, a young individual, ×400 (Fouquet); j, a catfish, *Ameiurus albidus*, heavily infected by the ciliate (Stiles, 1894)).

tetrahymnal apparatus (Mugard); cytoplasm with various inclusions; macronucleus horseshoe-shaped; there is no reproduction within the host body; multiplication within cyst which occurs after dropping off the fish skin and in which numerous (up to 1000) ciliated bodies (30-45μ in diameter) are produced; conjugation has been reported; parasitic in the integument of freshwater fishes; in aquarium, host fish may suffer death; widely distributed.

I. multifiliis F. (Fig. 324, *d-j*). 100-1000μ long; ovoid; produces pustules in the epidermis or gills; cytostome is large, 30-40μ in diameter. Pearson (1932) and Kudo (1934) reported extensive infections in large open ponds in Indiana and Illinois and Butcher (1941, 1943) noted infections in many yearling trout in hatcheries in 1939 and 1940. MacLennan (1935, 1935a, 1937, 1942) observed that the grown trophozoites leave the host epithelium and encyst on the bottom of aquarium; the cytostome is absorbed; the body protoplasm divides into 100-1000 small spherical ciliated cells, 18-22μ in diameter, which presently metamorphose into elongated forms, measuring about 40μ by 10μ. These young cilates break through the cyst wall and seek new host fish by active swimming. The young ciliates are able to attack the fish integument for at least ninety-six hours, though their infectivity decreases markedly after forty-eight hours. (Fibrillar structures, ten Kate, 1927.)

I. marinus Sikama. Similar to the above, but somewhat smaller; up to 450μ by 360μ; macronucleus typically constricted into beads; cyst membrane thicker; on the gill and skin of many marine fishes; heavy infection is said to result in death of host fish (Sikama, 1938, 1961).

These and other ectoparasitic protozoa of fishes can be eradicated by bathing the infected fish in weak solutions of sodium chloride, acetic acid or formaldehyde for five to ten minutes.

Family 2. **Tetrahymenidae** Corliss

With tetrahymenal buccal ciliature; body ciliation uniform; one or two post-oral ciliary rows are stomatogenous.

Genus **Tetrahymena** Furgason (1941). Pyriform; small forms; uniform ciliation; ciliary rows or meridians seventeen to forty-two; two postoral meridians; preoral suture straight; cytostome small, close to anterior end, pyriform; its axis parallel to body axis; inconspicuous ectoplasmic ridge or flange on the left margin of mouth; an undulating membrane on right side and three membranellae on left in the cytostome; a single contractile vacuole; macronucleus ovoid; micronucleus absent in some; a long caudal cilium in *setifera;* in fresh water or parasitic. Corliss (1961) lists the following genera as probable synonyms of Tetrahymena: Lambornella, Leptoglena, Leucophrydium, Leucophrys, Paraglaucoma, Protobalantidium, Ptyxidium, Trichoda, etc. (Comparative study of strains and allied forms, Corliss, 1952, 1952a; as facultative parasites, Thompson, 1958; literature, Corliss, 1954; biology, Elliott, 1959a.)

T. pyriformis (Ehrenberg) (*T. geleii* Furgason) (Fig. 325, *a-d*). fifty-nine strains (Corliss, 1952a;) 40-60µ long; seventeen to twenty-three ciliary meridians; pyriform cytostome about one-tenth the body length; with or without micronucleus; bacteria-feeder; in fresh water. Kozloff (1956) brought about experimental infections in the garden slug, *Deroceras reticulatum* with micronucleate and amicronucleate *pyriformis.* The sites of infection were found to be digestive glands, intestine, and perivisceral fluid. There were no morphological changes, except somewhat smaller in size. (Taxonomy, Corliss, 1952, 1952a; axenic culture, Kidder, 1941; electron microscope study, Roth and Minick, 1961, Metz and Westfall, 1954; effects of antibiotics, Loefer, 1951, 1952, Blumberg and Loefer, 1952, Loefer and Matney, 1952, Gross, 1955; distribution, Elliott and Hayes, 1955, Gruchy, 1955; meiosis, Ray, 1956; incidence of haploidy, Elliott and Clark, 1956; temperature and division rate, Prescott, 1957; serological types, Loefer, Owen and Christensen, 1958; enzyme systems, Seaman, 1954; genetics of pyridoxine and serine mutants, Elliott and Clark, 1958; senility, Nanney, 1959; mating, Outoka, 1961.)

T. vorax (Kidder, Lilly and Claff) (*Glaucoma vorax* K. L. and C.) (Fig. 41). Form and size vary; bacteria-feeders elongate pyriform, 50-75µ long; saprozoic forms fusiform, 30-70µ long, decreasing in size with the age of culture; sterile particle-feeders, 60-80µ long; carnivores and cannibals broadly pyriform, 100-250µ long; nineteen to twenty-one ciliary meridians; macronucleus ovoid, central; in carnivores, outline irregular; apparently without micronucleus; pond water. Polymorphism (Williams, 1960).

T. limacis (Warren). In liver and other visceral organs of the slug *Deroceras reticulatum* (*D. agreste*); 33-68 (55)µ by 18-35 (27)µ;

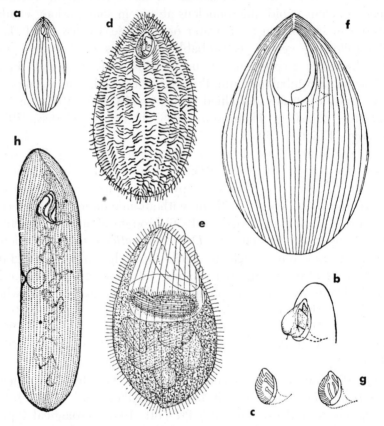

Fig. 325. a–d, *Tetrahymena pyriformis* (a, ×535 (Kidder); b, c, cytostomal structure; d, oral view, ×950 (Furgason)); e–g, *T. patula* (e, a well-fed animal, ×280 (Maupas); f, ×535 (Kidder); g cytostome (Furgason)); h, *Deltopylum rhabdoides*, ×665 (Fauré-Fremiet and Mugard).

those from cultures measure 28-68 (44)µ by 17-42 (27µ); the parasitic phase is cucumber-shaped with apiculate anterior end; the free-living organisms are pyriform, somewhat pointed anteriorly; cytostome at about one-fourth from the anterior end, with an undulating membrane and three membranelles; 32-40 ciliary rows (Kozloff, 1946). As was mentioned above, experimentally *pyriformis* can infect slugs without changing morphological characteristics except the size. The ciliate was further found in terrestrial gastropods, *Monadenia fidelis* and *Prophysaon andersoni* (Kozloff, 1956). Corliss (1952) held that Kozloff's form is different from *limacis* and called it *T. faurei*. (Number of ciliary rows, Kozloff, 1956a.)

T. rostrata (Kahl). 60-80µ long; in fresh water (often in dead rotifers); Kozloff (1957) found this species in the renal organ of the garden slug, *Deroceras reticulatum* and established axenic cultures. (Ecology and life cycle, Stout, 1954.)

T. (Leucophrys) patula (Ehrenberg) (Fig. 325, *e-g*). Broadly pyriform; 80-160µ long; occasionally small forms occur; cytostome pyriform, about one-third the body length; forty to forty-five ciliary meridians; macronucleus irregularly ovoid; a micronucleus attached to macronucleus; carnivorous, but may be cultured in sterile media (Kidder); fresh water. (Morphogenesis, Fauré-Fremiet, 1948.)

Genus **Deltopylum** Fauré-Fremiet and Mugard. Cylindrical; uniform ciliation on about seventy ciliary rows; a triangular cytostome in the anterior fourth, with a paroral undulating membrane on right and three adoral membranes; a contractile vacuole on mid-right side, a pore being located in a depression of pellicle above it; macronucleus irregularly ribbon-like; five or six micronuclei; in fresh water (Fauré-Fremiet and Mugard, 1946).

D. rhabdoïdes F. and M. (Fig. 325, *h*). Cylindrical; 150-180µ by 40-45µ; anterior end slightly attenuated and curved, posterior end rounded; the organism grows well on the gut of Chironomus larvae in laboratory.

Genus **Colpidium** Stein. Elongate reniform; ciliary meridians variable in number, but typically one postoral meridian; small triangular cytostome one-fourth from anterior end toward right side; a small ectoplasmic flange along right border of cytostome which shows an undulating membrane on right and three membranellae on left; rounded macronucleus; a micronucleus; a contractile vacuole; fresh or salt water or parasitic.

C. colpoda (Ehrenberg) (*Tillina helia* Stokes) (Figs. 10, *c*; 326, *a*, *b*). Elongate reniform; 90-150μ long; cytostome about one-tenth the body length; 55-60 ciliary meridians; preoral suture curved to left; macronucleus oval, central; a micronucleus; fresh water. Bacteria-free culture (Kidder, 1941); division (Kidder and Diller, 1934); effect of food bacteria on division (Burbank, 1942).

C. campylum (Stokes) (Fig. 326, *c*). Elongate reniform; twenty-seven to thirty ciliary meridians; preoral suture curved to right; 50-70μ long; in fresh and brackish water. Division (Kidder and Diller, 1934).

C. striatum S. Similar to the last species; contractile vacuole further posterior; 50μ long; in standing water.

C. echini (Russo) (Fig. 326, *d*). In the intestinal caeca of *Strongylocentrotus lividus;* 37-64(55)μ by 21-28(25)μ; 24 longitudinal ciliary rows; cytostome at anterior third (Powers, 1933).

Genus **Glaucoma** Ehrenberg (*Dallasia* Stokes). Ovoid or ellipsoid; cytostome about one-fourth the body length, near anterior end, ellipsoid; cytostome with an inconspicuous undulating membrane on right and three membranellae on left; ectoplasmic ridge on right border of mouth; ciliation uniform; thirty to forty ciliary meridians; seven postoral meridians; macronucleus rounded; a micronucleus; a contractile vacuole; with or without one or more caudal bristles; fresh water. (Oral structure, Corliss, 1954; morphogenesis in *chattoni*, Frankel, 1960, 1961; nutrition, Holtz *et al.*, 1961.)

G. scintillans E. (Fig. 326, *e*, *f*). Ovate with rounded ends; 45-75μ long; U-shaped cytostome, about one-fourth the body length, oblique; ectoplasmic flange and three membranellae conspicuous; a contractile vacuole in posterior one-third; macronucleus oval, central; a micronucleus; bacteria-feeder; in fresh water. (Bacteria-free culture, Kidder, 1941; division, Kidder and Diller, 1934.)

Genus **Monochilum** Schewiakoff. Ovoid to ellipsoid; medium large; uniform and dense ciliation in rows; oblong cytostome left of median line, in about one-fourth the body length from anterior end; short cytopharynx conical, with an undulating membrane; contractile vacuole near middle; in fresh water.

M. frontatum S. (Fig. 326, *g*). Anterior end broader; ventrally flattened, dorsally somewhat convex; macronucleus ellipsoid; a micronucleus; feeds on algae; 80μ by 30μ.

Genus **Dichilum** Schewiakoff. Similar to *Monochilum;* but membrane on both edges of the cytostome; in fresh or salt water. *D. cuneiforme* S. (Fig. 326, *h*). Ellipsoid; cytostome about one-fifth the body length from anterior end; right membrane larger

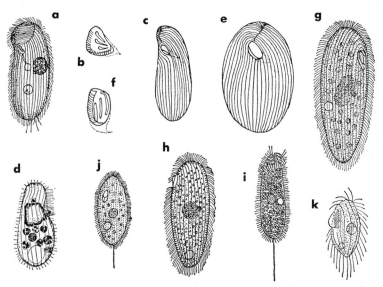

FIG. 326. a, b, *Colpidium colpoda* (a, ×180 (Kahl); b, cytostome (Furgason)); c, *C. campylum,* ×535 (Kidder); d, *C. echini,* ×385 (Powers); e, f, *Glaucoma scintillans* (e, a diagram (Kidder); f, cytostome (Furgason)); g, *Monochilum frontatum,* ×440 (Schewiakoff); h, *Dichilum cuneiforme,* ×700 (Schewiakoff); i, *Loxocephalus plagius,* ×380 (Stokes); j, *Saprophilus agitatus,* ×450 (Stokes); k, *S. muscorum,* ×440 (Kahl).

than left; small cytopharynx; macronucleus ellipsoid; about 40μ by 24μ; in fresh water.

Genus **Loxocephalus** Eberhard. Ovoid to cylindrical; sometimes compressed; crescentic cytostome on slightly flattened area near anterior end, with two membranes; often a zone of cilia around body; usually one (or more) long caudal cilium; endoplasm granulated, yellowish to dark brown; macronucleus ovoid; a single contractile vacuole; in fresh or brackish water. (Many species.)

L. plagius (Stokes) (Fig. 326, *i*). 50-65μ long; nearly cylindrical;

fifteen to sixteen ciliary rows; endoplasm usually darkly colored; in fresh water among decaying vegetation.

Genus **Saprophilus** Stokes. Ovoid to pyriform; compressed; tetrahymenal cytostome in the anterior one-fourth to one-third of the body; macronucleus spherical; contractile vacuole posterior; in fresh water.

S. agitatus S. (Fig. 326, *j*). Ellipsoidal; ends bluntly pointed; plastic; close striation; about 40μ long; in decomposing matter in fresh water.

S. muscorum Kahl (Fig. 326, *k*). Cytostome large; trichocysts; contractile vacuole with a distinct canal; one post-oral and fifteen primary meridians, eight of which meet ventrally at the preoral suture; a long caudal cilium; body 25-35μ long; in fresh water and soil (Stout, 1956).

Family 3. **Cohnilembidae** Kahl

Buccal cavity elongate; three adoral membranelles are shifted to right, against the undulating membrane; body ciliation uniform and complete.

Genus **Cohnilembus** Kahl (*Lembus*, Cohn). Slender spindleform; flexible; peristome from anterior end to the middle of body or longer, curved to right, with two membranes on right edge; a caudal cilium or a few longer cilia at posterior end; macronucleus oval, central; in salt or fresh water, some parasitic.

C. fusiformis (C.) (Fig. 327, *a*). Striation spiral; peristome about one-sixth the body length; a few cilia at posterior end; oval macronucleus central; contractile vacuole posterior; about 60μ long; in fresh water.

C. caeci Powers (Fig. 327, *b*). About 32-92μ long; in the intestine of *Tripneustes esculentus* and other echinoids; Tortugas.

Genus **Anophrys** Cohn. Cigar-shaped; flexible; longitudinal ciliary rows; peristome begins near the anterior end, parallel to body axis and about one-third the body length; a row of free cilia on right edge of peristome; cytostome inconspicuous; spherical macronucleus central; contractile vacuole terminal; in the intestine of sea-urchins.

A. elongata Biggar (Fig. 327, *c*). About 96μ long (Powers); 166μ long (Biggar); in the gut of *Lytechinus variegatus* and *Echinometris subangularis;* Bermuda (Biggar); Powers (1935) found this species also in the hosts mentioned for *Biggaria bermudense*.

A. *aglycus* Powers (Fig. 327, *d*). 56-120μ by 16-35μ; in the gut of *Centrechinus antillarum* and *Echinometra lucunter;* Tortugas (Powers, 1935).

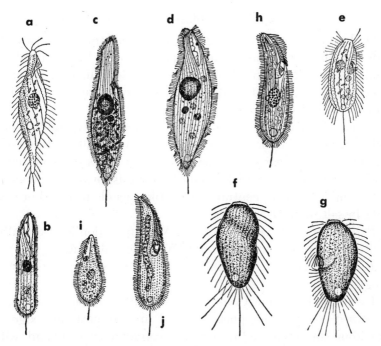

Fig. 327. a, *Cohnilembus fusiformis,* ×560 (Kahl); b, *C. caeci,* ×390 (Powers); c, *Anophrys elongata,* ×390 (Powers); d, *A. aglycus,* ×390 (Powers); e, *Uronema marinum,* ×490 (Kahl); f, g, *U. pluricaudatum,* ×940 (Noland); h, *Philaster digitiformis,* ×220 (Kahl); i, *P. armata,* ×240 (Kahl); j, *Helicostoma buddenbrocki,* ×190 (Kahl).

Genus **Uronema** Dujardin. Oval to elongate ovoid; slightly flattened; anterior region not ciliated; inconspicuous peristome with ciliated right edge; cytostome on the ventral side close to left border in the anterior half, with a small tongue-like membrane; cytopharynx indistinct; macronucleus spherical, central; contractile vacuole terminal; in salt or fresh water. (Comparison with Cyclidium, Párducz, 1940.)

U. *marinum* D. (Fig. 327, *e*). 30-50μ long; in salt water among algae. Structure (Párducz, 1939).

U. pluricaudatum Norland (Fig. 327, *f, g).* Body appears to be twisted in dorsal view, due to a spiral depression that runs obliquely down toward cytostome; with about eight caudal cilia; in salt water; Florida (Noland, 1937).

Family 4. **Philasteridae** Kahl

Oral groove elongate triangular; adoral ciliature is tetrahymenal; but the three membranelles are replaced by three ciliary fields: deltoid, trapezoid and falciform fields (Mugard); stomatogenesis involves multiplication of kinetosomes at the posterior end of paroral membrane.

Genus **Philaster** Fabre-Domergue (*Philasterides,* Kahl). Body cylindrical; peristome about one-third to two-fifths the body length, broader near cytostome and with a series of longer cilia; uniform ciliation; a caudal cilium; trichocysts; oval macronucleus with a micronucleus, central; contractile vacuole terminal or central; in salt or fresh water.

P. digitiformis F-D. (Fig. 327, *h*). Anterior region bent dorsally; contractile vacuole terminal; 100-150μ long; salt water.

P. armata (Kahl). (Fig. 327, *i*). Anterior end straight; 70-80μ long; fresh water.

Genus **Helicostoma** Cohn. Similar to *Philaster* in general appearance; preoral side-pouch curved around posterior edge of persitome and separated from it by a refractile curved band; with or without a pigment spot near cytostome; macronucleus oval or band-form; contractile vacuole terminal; in salt water.

H. buddenbrocki Kahl (Fig. 327, *j*). 130-200μ long; in salt and brackish water.

Suborder 2. **Peniculina** Fauré-Fremiet

Oral ciliature characterized by peniculi located deep in the buccal cavity; often with ciliated vestibule; body uniformly ciliated. (Two families.)

Family 1. **Parameciidae** Dujardin

Oral groove leads into the buccal cavity; buccal ciliature composed of endoral membrane, two peniculi and quadri-partite ciliary organelle; body ciliation uniform; trichocysts.

Genus **Paramecium** Hill (*Paramaecium* Müller). Cigar- or foot-shaped; circular or ellipsoid in cross section; with a single macronucleus and one to several vesicular or compact micronuclei; peristome long, broad, and slightly oblique; in fresh or brackish water. Several species. (Comparative morphology, Wenrich, 1928a; Wichterman, 1953; ciliary arrangement, Lieberman, 1928; pellicular structure, Gelei, 1939; excretory system, Gelei, 1939a; spiral movement, Bullington, 1930; cultivation, Wichterman, 1949; buccal structure, Yusa, 1957.)

P. caudatum Ehrenberg (Figs. 22, *a, b;* 45, *a-e;* 55; 85; 328, *a*). 180-300µ long; with a compact micronucleus, a massive macronucleus; two contractile vacuoles on aboral surface; posterior end bluntly pointed; in fresh water. The most widely distributed species. (Cytology and physiology, Müller, 1932; contractile vacuoles, Dimitrowa, 1928; cytopharynx, Gelei, 1934; calcium and iron, Kruszynski, 1939; nuclear variation, Diller, 1940; re-conjugation, Diller, 1942; food vacuoles, Bozler, 1924; electron microscope study, Watanabe and Tsuda, 1957; oxygen poisoning, Wittner, 1957; effect of ethanol, Pace and Hoagland, 1954; effect of urea on fission, Miyake, 1955.)

P. aurelia E. (Figs. 2, *g, h;* 60; 91; 102; 103; 104; 328, *b*). 120-180µ long; two small vesicular micronuclei, a massive macronucleus; two contractile vacuoles on aboral surface; posterior end more rounded than *P. caudatum;* in fresh water. (Nutrition, Phelps, 1934; Soldo and van Wagtendonk, 1961; genetics, Beale, 1954; radiation effects, Powers, 1955; growth factor, Conner *et al.,* 1953; amacronucleation, Margolin, 1954; buccal organelles in division and conjugation, Porter, 1960; autogamy and hemixis, Diller, 1936.)

P. multimicronucleatum Powers and Mitchell (Figs. 19; 20; 30; 31; 328, *c*). The largest species, 200-330µ long; three to seven contractile vacuoles; four or more vesicular micronuclei; a single macronucleus; in fresh water. (Cytology and physiology Müller, 1932; axenic culture, Johnson and Baker, 1942; nutrition, Johnson and Miller, 1956, Miller and Johnson, 1957, 1960; relation to Oikomonas and bacteria in culture, Hardin, 1944; mating types, Giese, 1957.)

P. bursaria (Ehrenberg) (Figs. 86, 90, 328, *d*). Foot-shaped, somewhat compressed; about 100-150µ by 50-60µ; green with zoochlorellae as symbionts; a compact micronucleus; a macronucleus; two contractile vacuoles; in fresh water. (Relation between Chlorella and host, Parker, 1926, Pringsheim, 1928; nuclear change and gullet organelles, Ehret and Powers, 1955, 1957; micronuclear variation, Woodruff, 1931; axenic culture, Loefer, 1936; removal of symbionts, Jennings, 1938, Wichterman, 1948, Schulze, 1951.)

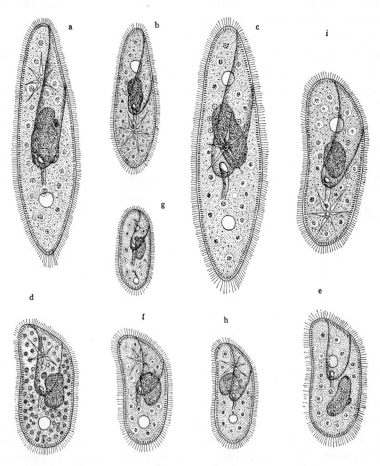

Fɪɢ. 328. Semi-diagrammatic drawings of nine species of Paramecium in oral surface view, showing distinguishing characteristics taken from fresh and stained specimens, ×230 (several authors), a, *P. caudatum;* b, *P. aurelia;* c, *P. multimicronucleatum;* d, *P. bursaria;* e, *P. putrinum;* f, *P. calkinsi;* g, *P. trichium;* h, *P. polycaryum;* i, *P. woodruffi.*

P. putrinum Claparède and Lachmann (Fig. 328, *e*). Similar to *P. bursaria,* but a single contractile vacuole and an elongated macronucleus; no zoochlorellae; 80-150μ long; in fresh water.

P. calkinsi Woodruff (Fig. 328, *f*). Foot-shaped; posterior end broadly rounded; 100-150μ by 50μ; two vesicular micronuclei; two contractile vacuoles; in fresh, brackish and salt water. Ecology, Morphology, mating types (Wichterman, 1951).

P. trichium Stokes (Fig. 328, *g*). Oblong; somewhat compressed; 50-105 (80-90)μ long; a compact micronucleus; two contractile vacuoles deeply situated, each with a convoluted outlet; in fresh water. (Structure and division, Wenrich, 1926; conjugation (p. 226), Diller, 1948, 1949, Yankovskii, 1961.)

P. polycaryum Woodruff and Spencer (Fig. 328, *h*). Form similar to *P. bursaria;* 70-110μ long; two contractile vacuoles; three to eight vesicular micronuclei; in fresh water. (Autogamy and conjugation, Diller, 1954, 1958.)

P. woodruffi Wenrich (Fig. 328, *i*). Similar to *P. polycaryum;* 150-210μ long; two contractile vacuoles; three to four vesicular micronuclei; brackish water (Wenrich, 1928).

P. jenningsi Diller and Earl. Resembles *aurelia* in general morphology; 115-218μ long; two microsnuclei larger than those of *aurelia;* macronuclear anlagen with long persisting chromatinic centers (Diller and Earl, 1958).

Although Paramecium occurs widely in various freshwater bodies throughout the world and has been studied extensively by numerous investigators by mass or pedigree culture method, there are only a few observations concerning the process of encystment. Bütschli considered that Paramecium was one of the protozoa in which encystment did not occur. Stages in encystment were however observed in *P. bursaria* (by Prowazek) and in *P. putrinum* (by Lindner). In recent years, four observers reported their findings on the encystment of Paramecium. Curtis and Guthrie (1927) give figures in their textbook of zoology, showing the process (in *P. caudatum?*) (Fig. 329, *a-c*), while Cleveland (1927) injected Paramecium culture into the rectum of frogs and observed that the ciliate encysted with a thin membrane. Michelson (1928) found that if *P. caudatum* is kept in Knopagar medium, the organism becomes ellipsoidal under certain condi-

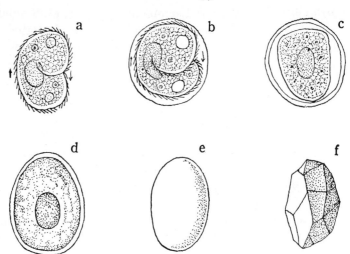

FIG. 329. a–c, encystment in a species of Paramecium (Curtis and Guthrie); d–f, encystment of *P. caudatum,* ×380 (Michelson).

tions, later spherical to oval, losing all organellae except the nuclei, and develops a thick membrane; the fully formed cyst is elongated and angular, and resembles a sand particle (Fig. 329, *f*). Michelson considers its resemblance to a sand grain as the chief cause of the cyst having been overlooked by workers. In all these cases, it may however be added that excystment has not been established.

Genus **Physalophrya** Kahl. Without peristome; but cytostome located near the anterior half of body, resembles much that of *Paramecium;* although there is no membrane, a ciliary row occurs in the left dorsal wall of cytopharynx; in fresh water. Taxonomic status is not clear; but because of its general resemblance to Paramecium, the genus with only one species is mentioned here.

P. spumosa (Penard) (Fig. 330, *a*). Oval to cylindrical; highly plastic; cytoplasm reticulated; numerous contractile vacuoles; 150-320μ long; in fresh water.

Family 2. **Frontoniidae** Kahl

Vestibule shallow or absent; compound ciliary organelles in the buccal cavity; cytostome expansive, but ill-defined in the herbivorous genera; body generally large; ciliation complete; a single contractile vacuole. Corliss (1961) assigned four genera (Frontonia, Disematostoma, Espejoia and Stokesia) to it and placed Urocentrum and Cinetochilum in two separate families. Unfortunately, information on the adoral ciliary organelles of the many of the genera which Kahl (1931) grouped originally under this familial name still remains to be incomplete. Therefore, they are provisionally retained in this family.

Genus **Frontonia** Ehrenberg. Ovoid to ellipsoid; anterior end more broadly rounded than posterior end; flattened; oral groove lies in anterior third or more or less flattened ventral surface, to right of median line; lancet-like with pointed anterior and truncate posterior end; left edge is more curved than right edge, and posteriorly becomes a prominent ectoplasmic lip; cytostome with a complex organization (on left edge a large undulating membrane composed of three layers, each being made up of four rows of cilia; on right, semi-membranous groups of cilia; three outer rows of cilia from the postoral suture; along this suture ectoplasm is discontinuous so that large food matter is taken in; with a small triangular ciliated field posterior to cytostome and left of suture); a long narrow postoral groove which is ordinarily nearly closed; cytopharynx with numerous strong fibrils; ciliary rows close and uniform; ectoplasm with numerous fusiform trichocysts; macronucleus oval; one to several micronuclei; one to two contractile vacuoles, with collecting canals and an external pore; in fresh or salt water. (Species identification and movement, Bullington, 1939; trichocysts, Krüger, 1931.)

F. leucas E. (Figs. 2, *i, j;* 330, *b, c*). 150-600μ long; feeds on filamentous algae, but may take in Arcella and even large amoebae (Beers, 1933); among algae in fresh water. (Nuclear apparatus and autogamy, Devi, 1960, 1961.)

F. branchiostomae Codreanu (Fig. 330, *d*). 75-100μ by 55-95μ; commensal in the branchial cavity of Amphioxus.

F. depressa (Stokes) Penard. Oval, flattened; 60-80μ by 30-40μ; oval mouth, with three vestibular meridians on right, an undulating membrane on left and a membranella in the middle; forty to fifty meridians, with short cilia; body surface smooth, though with trichocysts; cysts show little dedifferentiation; fresh water, moss and soil (Stout, 1956a).

F. microstoma Kahl. 200-300μ long; in brackish water (Roque, 1961).

Genus **Disematostoma** Lauterborn. Somewhat similar to *Frontonia;* pyriform; with broadly rounded, truncate or concave ante-

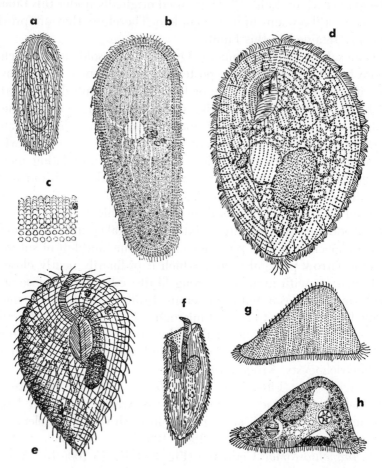

FIG. 330. a, *Physalophrya spumosa,* ×160 (Penard); b, c, *Frontonia leucas* (Bullington) (b, aboral view showing the contractile vacuole, collecting canals, macronucleus, 4 micronuclei and trichocysts, ×220; c, part of pellicle with wart-like projections over trichocysts); d, *F. branchiostomae,* ×490 (Codreanu); e, *Disematostoma butschlii,* ×340 (Kahl), f, *Espejoia musicola,* ×300 (Penard); g, h, *Stokesia vernalis,* ×340 (Wenrich).

rior end and bluntly pointed narrow posterior end; preoral canal wide; a dorsal ridge in posterior region of body; macronucleus sausage-form; a micronucleus; contractile vacuole in middle of body, with long collecting canals; in fresh water.

D. bütschlii L. (Fig. 330, *e*). 135-155μ long; with or without zoochlorellae; in fresh water.

D. colpidioides von Gelei. Reniform, twisted; 100-160μ long (Tuffrau and Sovoie, 1961).

Genus **Espejoia** Bürger. Ellipsoid; anterior end obliquely truncate; large cytostome at anterior end; postoral groove on ventral side, one-fourth to one-third the body length; a conspicuous membrane on the left edge of groove; in gelatinous envelope of eggs of insects and molluscs.

E. musicola (Penard) (Fig. 330, *f*). Elongate; right side flat, left side convex; 80-100μ long (Penard); 70-80μ long and dimorphic (Fauré-Fremiet and Mugard, 1949).

Genus **Stokesia** Wenrich. Oblique cone with rounded angles; flat anterior surface uniformly ciliated; with peristome bearing zones of longer cilia, at the bottom of which is located the cytostome; a girdle of longer cilia around the organism in the region of its greatest diameter; pellicle finely striated; with zoochlorellae; trichocysts; free-swimming; in freshwater pond. One species (Wenrich, 1929).

S. vernalis W (Fig. 330, *g, h*). 100-160μ in diameter; macronucleus; two to four micronuclei; fresh water.

Genus **Lembadion** Perty. Oval; dorsal side convex, ventral side concave; cytostome three-fourths to four-fifths the body length; on its left with a large membrane composed of many ciliary rows and on its right, numerous narrow rows of short free cilia; an undulating membrane and ciliary rows near posterior end; contractile vacuole in mid-dorsal region with a long tubule opening at posterior-right side; close ciliation uniform; macronucleus ellipsoid, subterminal; a micronucleus; long caudal cilia; in fresh water.

L. bullinum P. (Fig. 331, *a*). 120-200μ long; posterior cilia 40-50μ long.

Genus **Malacophrys** Kahl. Ellipsoid or cylindrical; plastic; cilia uniformly close-set in longitudinal rows; slit-like cytostome at anterior extremity; in fresh water.

M. rotans K. (Fig. 331, *b*). Oval; close and dense ciliation; spherical macronucleus central; a micronucleus; a single contractile vacuole; body 40-50µ long; fresh water.

Genus **Cryptochilum** Maupas. Ellipsoid; with rounded anterior end, posterior end pointed in profile; highly compressed; uniform and close ciliation; cytostome near middle; one or more longer cilia at posterior end; contractile vacuole posterior; macronucleus round; a micronucleus; commensal. (Several species, Powers, 1933, 1935.)

C. echini M. (Fig. 331, *c*). 70-140µ long; in the gut of *Echinus lividus*.

Genus **Eurychilum** André. Elongate ellipsoid; anterior end somewhat narrowed; cilia short; dense ciliation not in rows; contractile vacuole terminal; macronucleus band-form; cytostome about two-fifths from anterior end and toward right, with a strong undulating membrane on left; no cytopharynx; actively swimming. (One species.)

E. actiniae A. (Fig. 331, *d*). About 155µ long; in gastrovascular cavity of *Sagartia parasitica*.

Genus **Balanonema** Kahl. Body cylindrical; but with plug-like ends; cytostome difficult to see; a caudal cilium; macronucleus oval; contractile vacuole; ciliation uniform or broken in the middle zone; fresh water.

B. biceps (Penard) (Fig. 331, *e*). Ellipsoid; no cilia in the middle region; contractile vacuole central; macronucleus posterior to it; 42-50µ long.

Genus **Platynematum** Kahl. Ovoid or ellipsoid; highly flattened; with a long caudal cilium; contractile vacuole posterior-right; small cytostome more or less toward right side, with two outer membranes; ciliary furrows horseshoe-shaped; in fresh or salt water.

P. sociale (Penard) (Fig. 331, *f*). Anterior half more flattened; ventral side concave; cytostome in the anterior third; yellowish and granulated; 30-50µ long; sapropelic in fresh and brackish water.

Genus **Dexiotrichides** Kahl. Reniform; compressed; cytostome near middle, with two membranes; long cilia sparse; a special oblique row of cilia; a single caudal cilium; contractile vacuole terminal; spheroidal macronucleus anterior; a micronucleus; in fresh water. (One species.)

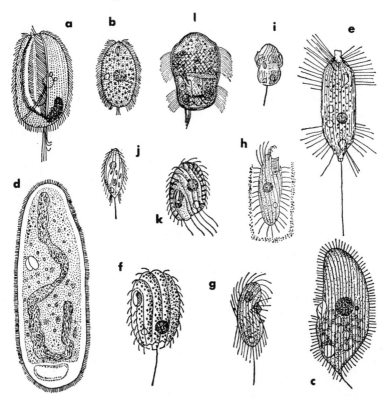

Fig. 331. a, *Lembadion bullinum*, ×170; b, *Malacophrys rotans*, ×500 (Kahl); c, *Crytochilidium echini*, ×380 (Powers); d, *Eurychilum actiniae*, ×300 (André); e, *Balanonema biceps*, ×600 (Penard); f, *Platynematum sociale*, ×500; g, *Dexiotrichides centralis*, ×500; h, *Cyrtolophosis mucicola*, ×670; i, *Urozona bütschllii*, ×440; j, *Homalogastra setosa*, ×450; k, *Cinetochilum margaritaceum*, ×440 (Kahl); l, *Urocentrum turbo*, ×200 (Bütschli).

D. centralis (Stokes) (Fig. 331, g). About 30–45µ long; in decaying vegetable matter.

Genus **Cyrtolophosis** Stokes. Ovoid or ellipsoid; with a mucilaginous envelope in which it lives, but from which it emerges freely; cytostome near anterior end with a pocket-forming membrane; on right side a short row of special stiff cilia, bent ventrally; sparse ciliation spiral to posterior-left; spherical macronucleus central; a contractile vacuole; in fresh water.

C. mucicola S. (Fig. 331, *h*). 25-28μ long; in infusion of leaves.

Genus **Urozona** Schewiakoff. Ovoid, both ends broadly rounded; a distinct constriction in the ciliated middle region; ciliary band composed of five to six rows of cilia, directed anteriorly and arranged longitudinally; cytostome with a membrane; rounded macronucleus and a micronucleus posterior; contractile vacuole subterminal; in fresh water.

U. bütschlii S. (Fig. 331, *i*). 20-25μ long (Kahl); 30-40μ (Schewiakoff); in stagnant water.

Genus **Homalogastra** Kahl. Broad fusiform; furrows spiral to left; a long caudal cilium; a group of cilia on right and left side of it; macronucleus spherical, anterior; contractile vacuole posterior; in fresh water.

H. setosa K. (Fig. 331, *j*). About 30μ long; fresh water.

Genus **Cinetochilum** Perty. Oval to ellipsoid; highly flattened; cilia on flat ventral surface only; cytostome right of median line in posterior half, with a membrane on both edges which form a pocket; oblique non-ciliated postoral field leads to left posterior end; with 3-4 caudal cilia; macronucleus spherical, central; contractile vacuole terminal; in fresh or salt water. (Neuroneme system, Gelei, 1940.)

C. margaritaceum P. (Fig. 331, *k*). 15-45μ long; in fresh and brackish water.

Genus **Urocentrum** Nitzsch. Short cocoon-shaped, constricted in the middle; ventral surface flat; two broad girdles of cilia; fused cilia at posterior end; with a zone of short cilia in the constricted area; cytopharynx with a stiff ectoplasmic membrane which separates two undulating membranes; macronucleus horseshoe-shaped, posterior; a micronucleus; contractile vacuole terminal with eight long collecting canals which reach the middle of body; in fresh water.

U. turbo (Müller) (Fig. 331, *l*). 50-80μ long; unique movement. Fission (Kidder and Diller, 1934).

Suborder 3. **Pleuronematina** Fauré-Fremiet

With a conspicuous undulating membrane; no vestibule; cytostome in the middle section or posterior half of body; body ciliation sparse.

Family **Pleuronematidae** Kent

With the characters of the suborder.

Genus **Pleuronema** Dujardin. Ovoid to ellipsoid; peristome begins at anterior end and extends for two-thirds the body length; a conspicuous membrane at both edges; semicircular swelling to left near oral area; no cytopharynx; close striation longitudinal; one to many posterior stiff sensory cilia; macronucleus round or oval; a micronucleus; a contractile vacuole; trichocysts in some species; fresh or salt water, also commensal in freshwater mussels.

P. crassum D. (Fig. 332, *a*). 70-120μ long; somewhat compressed; Woods Hole (Calkins).

P. anodontae Kahl (Fig. 332, *b*). About 55μ long; posterior cilium about one-half the body length; in Sphaerium, Anodonta.

P. setigerum Calkins (Fig. 332, *c*). Ellipsoid; flattened; ventral surface slightly concave; about twenty-five ciliary rows; 38-50μ long (Noland); in salt water; Massachusettes, Florida.

P. coronatum Kent (Fig. 332, *d*). Elongate ovoid; both ends equally rounded; caudal cilia long; about forty ciliary rows; 47-75μ long (Noland, 1937); in fresh and salt water; Florida.

P. marinum D. (Fig. 332, *e*). Elongate ovoid; trichocysts distinct; caudal cilia medium long; about fifty ciliary rows; 51-126μ long (Noland); in salt water; Florida.

Genus **Cyclidium** Müller. Small, 15-60μ long; ovoid; usually with refractile pellicle; with a caudal cilium; peristome near right side; on its right edge occurs a membrane which forms a pocket around cytostomal groove and on its left edge either free cilia or a membrane which unites with that on right; no semicircular swelling on left of oral region; round macronucleus with a micronucleus; contractile vacuole posterior; fresh or salt water. Numerous species. (Three species in sea urchin, Powers, 1935; comparison with Uronema, Párducz, 1940.)

C. litomesum Stokes (Fig. 332, *f*). About 40μ long; dorsal surface

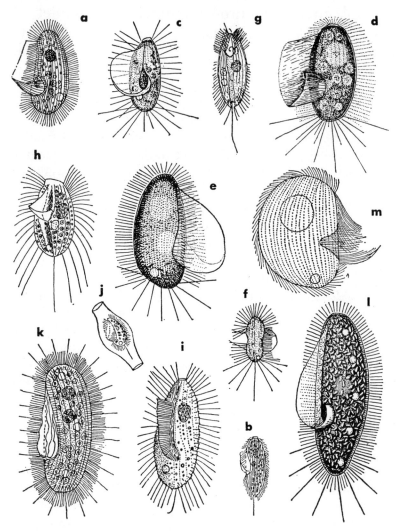

FIG. 332. a, *Pleuronema crassum*, ×240; b, *P. anodontae*, ×290 (Kahl); c, *P. setigerum*, ×540; d, *P. coronatum*, ×540; e, *P. marinum*, ×400 (Noland); f, *Cyclidium litomesum*, ×300 (Stokes); g, *Cristigera phoenix*, ×500 (Penard); h, *C. media*, ×400 (Kahl); i, *Ctedoctema acanthocrypta*, ×840 (Kahl); j, *Calyptotricha pleuronemoides*, ×180 (Kahl); k, *Histiobalantium natans*, ×420 (Kahl); l, *H. semisetatum*, ×270 (Noland); m, *Pleurocoptes hydractiniae*, ×470 (Wallengren).

slightly convex with a depression in middle; ventral surface more or less concave; cilia long; in fresh water.

Genus **Cristigera** Roux. Similar to *Cyclidium;* much compressed; with a postoral depression; peristome closer to mid-ventral line; fresh or salt water. (Several species.)

C. phoenix Penard (Fig. 332, *g*). 35-50μ long; fresh water.

C. media Kahl (Fig. 332, *h*). 45-50μ long; in salt water.

Genus **Ctedoctema** Stokes. Similar to *Cyclidium* in body form; peristome nearer median line, diagonally right to left; right peristomal ridge with a sail-like membrane which surrounds the cytostome at its posterior end; trichocysts throughout; fresh water.

C. acanthocrypta S. (Fig. 332, *i*). Ovoid; anterior end truncate; macronucleus round, anterior; about 35μ long; in fresh water among vegetation.

Genus **Calyptotricha** Phillips. Somewhat resembles *Pleuronema* or *Cyclidium;* but dwelling in a lorica which is opened at both ends; with zoochlorellae; fresh water.

C. pleuronemoides P. (Fig. 332, *j*). Lorica about 85μ high; body about 50μ long; Kellicott's (1885) form is more elongated; in fresh water.

Genus **Histiobalantium** Stokes. Ovoid; ventral side flattened; ciliation uniform; long stiff cilia distributed over the body surface; peristome deep; both anterior and posterior regions with a well-developed membrane, connected with the undulating membrane; macronucleus in two parts; one to two micronuclei; several contractile vacuoles distributed; fresh water.

H. natans (Claparède and Lachmann) (Fig. 332, *k*). 70-110μ long.

H. semisetatum Noland (Fig. 332, *l*). Elongate ellipsoid; posterior end bluntly rounded; macronucleus spherical; longer cilia on posterior half only; contractile vacuoles on dorsal side; 126-205μ long; salt water; Florida (Noland, 1937).

Genus **Pleurocoptes** Wallengren. Ovoid, dorsal side hemispherical, ventral side flattened; peristome large, reaching the posterior one-third; cytopharynx indistinct; longer cilia along peristome; macronucleus spherical; several micronuclei; contractile vacuole terminal; ectocommensal.

P. hydractiniae W. (Fig. 332, *m*). 60-70μ long; on *Hydractinia echinata.*

References

BEALE, G. H., 1954. The genetics of *Paramecium aurelia*. Cambridge Monograph in Exper. Biol. No. 2. Cambridge University.

BEERS, C. D., 1933. The ingestion of large amoebae by the ciliate *Frontonia leucas*. J. El. Mitchell Sci. Soc., 48:223.

BOZLER, E., 1924. Ueber die Morphologie der Ernaehrungsorganelle und die Physiologie der Nahrungsaufnahme bei *Paramecium caudatum*. Arch. Protist., 49:163.

BULLINGTON, W. E., 1930. A study of spiraling in the ciliate Frontonia with a review of the genus, etc. Arch. Protist., 92:10.

BURBANK, W. D., 1942. *Physiology of the ciliate Colpidium colpoda*. I. Physiol. Zool., 15:342.

BUTCHER, A. D., 1941. Outbreaks of white spots or ichthyophthiriasis (*Ichthyophthirius multifiliis*) at the hatcheries, etc. Proc. Roy. Soc. Victoria, 53:126.

———— 1943. Observations on some phases of the life cycle of *Ichthyophthirius multifiliis*, etc. Australian Zool., 10:125.

CLEVELAND, L. R., 1927. The encystment of Paramecium in the recta of frogs. Science, 66:221.

CONNER, R. L., *et al.*, 1953. The isolation from lemon juice of a growth factor of steroid nature required for the growth of a strain of *Paramecium aurelia*. J. Gen. Microbiol., 9:435.

CORLISS, J. O., 1952. Comparative studies on holotrichous ciliates in the Colpidium-Glaucoma-Leucophrys-Tetrahymena group. I. Tr. Am. Micr. Soc., 71:159.

———— 1952a. Review of the genus Tetrahymena. Proc. Soc. Protoz., 3:3.

———— 1958. The systematic position of *Pseudomicrothorax dubius*, etc. J. Protozool., 5:184.

CURTIS, W. C., and GUTHRIE, M. J., 1927. Textbook of general zoology. New York.

DILLER, W. F., 1936. Nuclear reorganization processes in *Paramecium aurelia*, etc. J. Morphol., 59:11.

———— 1940. Nuclear variation in *Paramecium caudatum. Ibid.*, 66:605.

———— 1942. Re-conjugation in *Paramecium caudatum. Ibid.*, 70:229.

———— 1948. Nuclear behavior of *Paramecium trichium* during conjugation. *Ibid.*, 82:1.

———— 1949. An abbreviated conjugation process in *Paramecium trichium*. Biol. Bull., 97:331.

———— 1954. Autogamy in *Paramecium polycaryum*. J. Protozool., 1:60.

———— 1958. Studies on conjugation in *Paramecium polycaryum. Ibid.*, 5:282.

———— and EARL, P. R., 1958. *Paramecium jenningsi* n. sp. *Ibid.*, 5:155.

DIMITROWA, A., 1928. Untersuchungen über die überzahligen pulsierenden Vakuolen bei *Paramecium caudatum*. Arch. Protist., 64:462.

Egelhaaf, A., 1955. Cytologisch-entwicklungsphysiologische Untersuchungen zur Konjugation von *Paramecium bursaria*. Arch. Protist., 100: 447.

Ehret, C. F., and Powers, E. L., 1955. Macronuclear and nucleolar development in *Paramecium bursaria*. Exper. Cell. Res., 9:241.

———— ———— 1957. The organization of gullet organelle in *Paramecium bursaria*. J. Protozool., 4:55.

Fauré-Fremiet, E., 1948. Doublets homopolaires et régulation morphogénétique chez le cilié *Leucophrys patula*. Arch. d'Anat. Micr. Morph. Exp., 37:183.

———— and Mugard, H., 1946. Sur un infusoire holotriche histiophage, *Deltopylum rhabdoïdes* n. g., n. sp. Bull. soc. zool. France., 71:161.

———— ———— 1949. Le dimorphisme de *Espejoia mucicola*. Hydrobiologia, 1:379.

Furgason, W. H., 1940. The significant cytostomal pattern of the "Glaucoma-Colipidium group," and a proposed new genus and species, *Tetrahymena geleii*. Arch. Protist., 94:224.

Gelei, J. v., 1934. Der feinere Bau des Cytopharynx von Paramecium und seine systematische Bedeutung. Arch. Protistenk., 82:331.

———— 1939. Das äussere Stützgerüstsystem des Paramecium-körpers. *Ibid.*, 92:245.

———— 1939a. Neue Beiträge zum Bau und zu der Funktion des Exkretionssystems von Paramecium. *Ibid.*, 92:385.

———— 1940. Cinetochilum und sein Neuronemensystem. *Ibid.*, 94:57.

Haas, G., 1934. Beiträge zur Kenntnis der Cytologie von *Ichthyophthirius multifiliis*. Arch. Protist., 82:88.

Hardin, G., 1944. Symbiosis of Paramecium and Oikomonas. Ecology, 25: 304.

Jennings, H. S., 1938. Sex reaction types and their interrelations in *Paramecium bursaria*. I. Proc. Nat. Acad. Sci., 24:112.

Kahl, A., 1931. Urtiere oder Protozoa. Dahl's Die Tierwelt Deutschlands, etc. Part 21.

Kidder, G. W., 1941. Growth studies on ciliates. VII. Biol. Bull., 85:50.

———— and Diller, W. F., 1934. Observations on the binary fission of four species of common free-living ciliates, etc. *Ibid.*, 67:201.

————, Lilly, D. M., and Claff, C. L., 1940. Growth studies on ciliates. IV. *Ibid.*, 78:9.

Köster, W., 1933. Untersuchungen ueber Teilung und Conjugation bei *Paramecium multimicronucleatum*. Arch. Protist., 80:410.

Kozloff, E. N., 1946. The morphology and systematic position of a holotrichous ciliate parasitizing *Deroceras agreste*. J. Morphol., 79:445.

Krüger, F., 1931. Dunkelfelduntersuchungen über den Bau der Trichocysten von *Frontonia leucas*. Arch. Protist., 74:207.

Kruszynski, J., 1939. Mikrochemische Untersuchungen des veraschten *Paramecium caudatum*. *Ibid.*, 92:1.

Kudo, R. R., 1934. Studies on some protozoan parasites of fishes of Illinois. Illinois Biol. Monogr., 13:1.

Lieberman, P. R., 1929. Ciliary arrangement in different species of Paramecium. Tr. Am. Micr. Soc., 48:1.

Loefer, J. B., 1936. Bacteria-free culture of *Paramecium bursaria* and concentration of the medium as a factor in growth. J. Exper. Zool., 72: 387.

MacLennan, R. F., 1935. Observations on the life cycle of Ichthyophthirius, etc. Northwest Sci., 9, 3 pp.

———— 1935a. Dedifferentiation and redifferentiation in Ichthyophthirius. I. Arch. Protist., 86:191.

———— 1937. Growth in the ciliate Ichthyophthirius. I. J. Exper. Zool., 76: 423.

———— 1942. II. *Ibid.*, 91:1.

Margolin, P., 1954. A method for obtaining amacronucleated animals in *Paramecium aurelia*. J. Protozool., 1:174.

Michelson, E., 1928. Existenzbedingungen und Cystenbildung bei *Paramecium caudatum*. Arch. Protist., 61:167.

Miller, C. A., and Johnson, W. H., 1957. A purine and pyrimidine requirement for *Paramecium multimicronucleatum*. J. Protozool., 4:200.

———— ———— 1960. Nutrition of Paramecium: a fatty acid requirement. *Ibid.*, 7:297.

Miyake, A., 1955. The effect of urea on binary fission in *Paramecium caudatum*. J. Inst. Polytech., Osaka City University, D, 6:43.

Mohr, J. L., and LeVeque, J. A., 1948. Occurrence of Conidophrys, etc. J. Parasit., 34:253.

Müller, W., 1932. Cytologische und vergleichend-physiologische Untersuchungen über *Paramecium multimicronucleatum* und *P. caudatum*, etc. Arch. Protist., 78:361.

Nie, D., 1950. Morphology and taxonomy of the intestinal protozoa of the guinea-pigs, *Cavia porcella*. J. Morphol., 86:381.

Noland, L. E., 1937. Observations on marine ciliates of the Gulf coast of Florida. Tr. Am. Micr. Soc., 56:160.

Pace, D. M., and Hoagland, R. A., 1954. The effects of ethanol on conjugation and division in *Paramecium caudatum*. J. Protozool., 1:83.

Párducz, B., 1939. Körperbau und einige Lebenserscheinungen von *Uronema marinum*. Arch. Protist., 92:283.

———— 1940. Verwandtschaftliche Beziehungen zwischen den Gattungen Uronema und Cyclidium. *Ibid.*, 93:185.

Parker, R. C., 1926. Symbiosis in *Paramecium bursaria*. J. Exper. Zool., 46:1.

Pearson, N. E., 1932. Ichthyophthiriasis among the fishes of a pond in Indianapolis. Proc. Indiana Acad. Sci., 41:455.

Porter, E. D., 1960. The buccal organelles in *Paramecium aurelia*, etc. J. Protozool., 7:211.

POWERS, E. L., 1955. Radiation effects in Paramecium. Ann. N. Y. Acad. Sci., 59:619.

PRINGSHEIM, E. G., 1928. Physiologische Untersuchungen an *Paramecium bursaria*. Arch. Protist., 64:289.

ROSSOLIMO, L. L., 1926. Parasitische Infusorien aus dem Baikal-See. Arch. Protist., 54:468.

SCHULZE, K. L., 1951. Experimentelle Untersuchungen ueber die Chlorellen-Symbiose bei Ciliaten. Biol. Generalis, 19:281.

SIKAMA, Y., 1938. Ueber die Weisspünktchenkrankheit bei Seefischen. J. Shanghai Sci. Inst., 4:113.

———— 1961. On a new species of Ichthyophthirius found in marine fishes. Sci. Rep. Yokosuka City Mus., 6:66.

SOLDO, A. T., and VAN WAGTENDONK, W. J., 1961. Nitrogen metabolism in *Paramecium aurelia*. J. Protozool., 8:41.

STRANGHÖNER, E., 1932. Teilungsrate und Kernreorganisationsprozess bei *Paramecium multimicronucleatum*. Arch. Protist., 78:302.

TARANTOLA, V. A., and VAN WAGTENDONK, W. J., 1959. Further nutritional requirements of *Paramecium aurelia*. J. Protozool., 6:189.

WATANABE, K., and TSUDA, S., 1957. The fine structure of pellicle and pharynx of Paramecium observed with electron microscope. Zool. Mag. (Tokyo), 66:183.

WENRICH, D. H., 1926. The structure and division of *Paramecium trichium*. J. Morphol. Physiol., 43:81.

———— 1928. *Paramecium woodruffi* n. sp. Tr. Am. Micr. Soc., 47:256.

———— 1928a. Eight well-defined species of Paramecium. *Ibid.*, 47:275.

WICHTERMAN, R., 1948. The biological effects of x-rays on mating types and conjugation of *Paramecium bursaria*. Biol. Bull., 94:113.

———— 1949. The collection, cultivation, and sterilization of Paramecium. Proc. Penn. Acad. Sci., 23:151.

———— 1951. The ecology, cultivation, structural characteristics and mating types of *Paramecium calkinsi*. *Ibid.*, 25:51.

———— 1953. The biology of Paramecium. New York.

———— 1954. The common occurrence of micronuclear variation during binary fission in an unusual race of *Paramecium caudatum*. J. Protozool., 1:54.

WITTNER, M., 1957. Inhibition and reversal of oxygen poisoning in Paramecium. J. Protozool., 4:24.

WOODRUFF, L. L., 1921. The structure, life history and intrageneric relationships of *Paramecium calkinsi*, sp. nov. Biol. Bull., 41:171.

———— 1931. Micronuclear variation in *Paramecium bursaria*. Quart. J. Micr. Sci., 74:537.

YANKOVSKII, A. V., 1961. Conjugation in *Paramecium trichium*. I. Tsitologiia (Russian), 2:581.

YUSA, A., 1957. The morphology and morphogenesis of the buccal organelles in Paramecium, etc. J. Protozool., 4:128.

Chapter 35

Order 4. **Thigmotrichida** Chatton and Lwoff

THESE ciliates possess thigmotactic cilia in the anterior half of body; buccal ciliature and cytostome in the posterior half or at the posterior end; some without cytostome, but with a sucker or suctorial tentacle; body ciliation uniform or much reduced. Parasites or commensals on or in bivalve mollusks. Though appearing heterogeneous, Chatton and Lwoff (1949) maintain that there is a phylogenetic unity among them, which condition has been brought about by degenerative influence because of similar conditions of habitat. (Taxonomy, Jarocki and Raabe, 1932; Chatton and Lwoff, 1949; Raabe, 1959.) Following Chatton and Lwoff (1939), the order is divided into two suborders: Rhynchodina and Arhynchodina (p. 929).

Suborder 1. **Rhynchodina** Chatton and Lwoff

Cytosome lacking, but with anterior suctorial tentacle; body ciliature mostly reduced, absent in some. (Three families.)

Family 1. **Ancistrocomidae** Chatton and Lwoff

Ovate to pyriform; suctorial tentacle; body and thigmotactic ciliation confined to anterior part of body.

Genus **Ancistrocoma** C. and L. (*Parachaenia* Kofoid and Bush). Elongate pyriform with attentuated anterior end; somewhat flattened dorso-ventrally; a contractile suctorial tentacle at the anterior tip, which is used for attachment to the epithelium of host, and which continues internally as a long curved canal; longitudinal ciliation on dorso-lateral and ventral sides, beginning at the anterior end; parasitic in the gills and palps of mollusks. (Taxonomy, Kozloff, 1946b; Chatton and Lwoff, 1950.)

A. pelseneeri C. and L. (*Parachaenia myae* Kofoid and Bush) (Fig. 333, *a*). Body 50-83 (62)µ by 14-20(16)µ by 11-16(12.5)µ; fourteen ciliary rows on dorso-lateral and ventral surfaces; five rows on the ventral side extend only two-thirds from the anterior end; tentacle continues internally for about two-thirds of body, curved; macronucleus sausage-shaped; a single micronucleus; on the gills and palps of mussels: *Mya arenaria, M. irus, M. inconspicua, M. nasuta, M. secta, Cryptomya californica* (Kozloff, 1946b).

Genus **Hypocomagalma** Jarocki and Raabe. Ovoid or pyriform with attenuated anterior end; asymmetrical; twenty-two to twenty-four ciliary rows which do not reach the posterior end; a suctorial tentacle at the anterior end; on mollusks.

H. pholadidis Kozloff (Fig. 333, *b*). 63-89μ by 18-25μ by 16-21μ; anterior end bent ventrally; twenty-four or twenty-five ciliary rows; one or more contractile vacuoles; macronucleus sausage-shaped; a single micronucleus; parasitic in the epithelium of the gills and palps of *Pholadidea penita* (Kozloff, 1946b).

Genus **Syringopharynx** Collin. Elongate ovoid, narrowed anteriorly; a suctorial tentacle at anterior end; fourteen ciliary rows (six dorsal, six ventral and two lateral); on mollusks (Collin, 1914).

S. pterotrachae C. Body 55μ by 25μ; macronucleus elongate band; on the gills of *Pterotracha coronata* (Chatton and Lwoff, 1950).

Genus **Goniocoma** Chatton and Lwoff. Ovoid with attenuated anterior end; end of suctorial tentacle extremely slender; twenty-seven to twenty-nine ciliary rows; of the fourteen dorsal rows, the median row is very short and the rows on either side of it are progressively longer; ventral rows pass over the posterior end and terminate on dorsal surface; on the gills of mollusks.

G. macomae (C. and L.). Body 33-39μ by 13-18μ; a comparatively voluminous micronucleus; on the gills and palps of *Macoma balthica* (Chatton and Lwoff, 1950).

Genus **Holocoma** Chatton and Lwoff. Cylindrical; ventral surface convex; tentacle at anterior end; nineteen to twenty-three ciliary rows; six to ten median dorsal rows relatively short, seven left and six right rows long; on the gills of mollusks.

H. primigenius C. and L. (Fig. 333, *c*). Elongated body 41-59μ by 15μ; ventral surface convex; elongate macronucleus; on the gills of *Macoma balthica* (Chatton and Lwoff, 1950).

Genus **Insignicoma** Kozloff. Elongate pyriform; a contractile tentacle with internal canalicule; median ciliary rows on anterior half of ventral surface; two right ciliary rows; left rows short and closely set; an inverted V-shaped row of long cilia on left-lateral surface at about the middle of body; on mollusks.

I. venusta K. (Fig. 333, *d*). 42-52μ by 18-21μ by 15-18μ; fifteen median, two right and sixteen to seventeen left ciliary rows; macronucleus ovoid; micronucleus spherical; on the gills and palps of *Botula californiensis* (Kozloff, 1946a).

Genus **Raabella** Chatton and Lwoff. Three groups of ciliary rows; eight to eleven short median rows; six to eleven longer rows

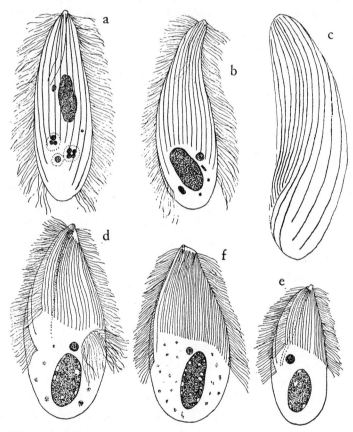

Fig. 333. a, ventral view of a stained *Ancistrocoma pelseneeri*, ×1120 (Kozloff); b, *Hypocomagalma pholadidis*, ×840 (Kozloff); c, ciliature as viewed from right side of *Holocoma primigenius*, ×1130 (Chatton and Lwoff); d, ventral view of *Insignicoma venusta*, ×1245; e, *Raabella botulae*, ×1245; f, *Crebricoma kozloffi*, ×755 (Kozloff).

on left-lateral side; two longer rows on the right side; on mollusks.

R. *botulae* (Kozloff (Fig. 333, *e*). 31-39μ by 14-17μ by 12-14μ; eleven median rows; eleven closely set left rows; two longer right rows; macronucleus ovoid to sausage-shaped; spherical micronucleus; on the gills and palps of *Botula californiensis* (Kozloff, 1946a).

Genus **Crebricoma** Kozloff. Pyriform; anterior suctorial tentacle; many ciliary rows, the majority of which are closely set; two long rows on the right side; anterior terminals of the rows make a V-shaped suture; on the gills of mollusks.

C. *kozloffi* Chatton and Lwoff (*C. carinata* K.) (Fig. 333, *f*). Boyd 58-71μ by 27-39μ by 22-31μ; two ciliary rows on right side long, about two-thirds the body length; more than thirty rows of closely set cilia (one-half to two-thirds the body length and longer toward left); macronucleus ellipsoid; on the gills and palps of *Mytilus edulis* (Kozloff, 1946; Chatton and Lwoff, 1950).

Genus **Hypocomides** Chatton and Lwoff. Elongate; some twenty-three ciliary rows; about twenty median rows, short; two longer rows on right; a short curved row near the posterior end; on mollusks.

H. *mediolariae* C. and L. (Fig. 334, *a*). 27-50μ by 15-27μ; on the gills of *Mediolaria marmorata* (Chatton and Lwoff, 1922).

Genus **Anisocomides** Chatton and Lwoff. Body ovoid, slightly flattened; twelve ciliary rows; two short median rows and five additional rows which are progressively longer toward left; a short oblique row, posterior to the outermost left row; four right rows much longer; on the gills of mollusks.

A *zyrpheae* (C. and L.) (Fig. 334, *b*). 19-38μ by 10-15μ by 7-10μ; on the gills of *Pholas (Zyrphea) crispata* (Chatton and Lwoff, 1926).

Genus **Hypocomatidium** Jarocki and Raabe. Similar to *Anisocomides*, but without the short posterior ciliary row; on the gills of mollusks.

H. *sphaerii* J. and R. Ovoid; 30-45μ by 14-18μ by 9-12μ; nine ciliary rows; five rows on left-ventral and four on right; on the gills of *Sphaerium corneum and S. rivicola* (Jarocki and Raabe, 1932).

Genus **Isocomides** Chatton and Lwoff. Elongated; fourteen to eighteen ciliary rows on anterior two-thirds of the ventral surface; six to seven on right and eight to eleven on left; in addition, there is a short transverse row with a dozen long cilia, posterior to other rows; on mussels.

I. mytili (C. and L.) (Fig. 334, *c*). 57-64μ by 20-22μ; on the gills of *Mytilus edulis* (Chatton and Lwoff, 1922).

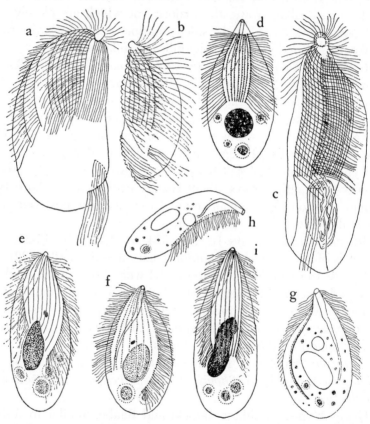

FIG. 334. a, *Hypocomides mediolariae*, ×1000; b, left side view of *Anisocomides zyrpheae* in life, ×1065; c, *Isocomides mytili* in life, ×1000 (Chatton and Lwoff); d, *Hypocomina tegularum*, ×1245; e, *Heterocinetopsis goniobasidis*, ×1145; f–h, *Hypocomella phoronopsidis*, ×1300 (f, ventral view of a stained specimen; g, h, dorsal and right side views in life)); i, *Enerthecoma kozloffi*, ×1145 (Kozloff).

Genus **Hypocomina** Chatton and Lwoff. Ovoid to pyriform; an anterior tenatcle; eight to ten ciliary rows about half the body-length and starting a little distance away from the anterior tip; on mollusks.

H. tegularum Kozloff (Fig. 334, *d*). 26-36μ by 12-17μ by 9-12μ; anterior end bent ventrally; nine ciliary rows, five rows on right being slightly longer than the other four; spherical macronucleus; parasitic on the ctenidium of *Tegula brunnae* (Kozloff, 1946).

Genus **Heterocinetopsis** Jarocki. Body elongate, flattened dorso-ventrally; a contractile tentacle, its canalicule extending one-third to two-thirds the body length; ten to twelve ciliary rows; the median rows about one-half the body length, the rows toward left being progressively longer; on mollusks (Jarocki, 1935).

H. goniobasidis (Kozloff) Fig. 334, *e*). 36-48μ by 15-20μ by 11-14μ; ten ciliary rows; macronucleus pyriform; ovoid micronucleus inconspicuous; parasitic on the epithelium of the gills and mantle of *Goniobasis plicifera silicula* (Kozloff, 1946c).

Genus **Hypocomella** Chatton and Lwoff (*Hypocomidium* Raabe). Pyriform, asymmetrical, flattened; a long retractile tentacle; seven to thirteen ciliary rows on the ventral surface, three rows on left being progressively longer; on mollusks (Chatton and Lwoff, 1922, 1950).

H. phoronopsidis (Kozloff) (Fig. 334, *f-h*). 26-37μ by 11-16μ by 6.5-11μ; eight ventral ciliary rows; ovoid macronucleus and micronucleus; on the tentacles of *Phoronopsis viridis* (Kozloff, 1945a).

Genus **Enerthecoma** Jarocki. Pyriform, symmetrical; eight ciliary rows on the ventral side; three on left are somewhat separated from five others and closely set; on the gills of mollusks.

E. kozloffi Chatton and Lwoff (Fig. 334, *i*). 32-56μ by 13-21μ by 10-13μ; eight ciliary rows about two-thirds the body length; macronucleus elongate; micronucleus fusiform; on the gills of *Viviparus fasciatus* and *V. malleatus* (Kozloff, 1946c; Chatton and Lwoff, 1950).

Genus **Ignotocoma** Kozloff. Pyriform with pointed anterior end; ciliated surface concave; cilia on two fields; seven rows in right and eight rows in left field; an ellipsoid macronucleus and a micronucleus; sucker and intracytoplasmic canal as in other genera; a contractile vacuole close to concave surface; on polychaetes (Kozloff, 1961).

I. sabellarum K. 23-33μ by 12-16μ by 8.5-10μ; on the peristomal cirri of the polychaetes: *Schizobranchia insignis* and *Eudistylia polymorpha* (Kozloff, 1961).

Genus **Cepedella** Poyarkoff. Pyriform with a pointed anterior end; macronucleus globular; without contractile vacuole.

C. hepatica P. Body 16-26μ long; in the liver of *Sphaerium corneum*. It occurs also in *Pisidium (Eupisidium) obtusale* (Dobrzanska, 1959).

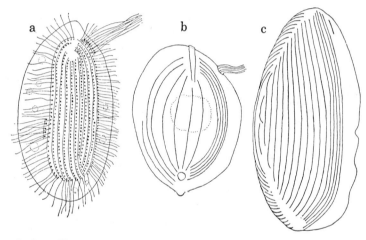

FIG. 335. a, *Hypocoma parasitica*, ×1350; b, *Heterocoma hyperparasitica*, ×1200; c, ciliature of *Parahypocoma collini*, as seen from left-ventral side in life (Chatton and Lwoff).

Family 2. **Hypocomidae** Bütschli

Body somewhat flattened; suctoral tentacle is located near anterior end or in antero-dorsal region; ciliation on dorsal side.

Genus **Hypocoma** Grüber. Dorsal side convex; ventral side flattened; a suctorial tentacle at the anterior end; about 13 ciliary rows; an adoral zone, a short row (eight granules) at anterior-left side; on colonial Protozoa.

H. parasitica G. (Fig. 335, *a*). 30-38μ by 18-20μ by 18μ; thirteen ciliary rows on the flattened surface; adoral zone, a short row; eleven general ciliary rows; macronucleus horseshoe-shape; a large central food vacuole; on solitary or colonial peritrichs such as Vorticella, Zoothamnium, etc. (Chatton and Lwoff, 1950).

Genus **Heterocoma** Chatton and Lwoff. Body ovoid; ventral side flattened; suctorial tentacle antero-ventral; thirteen ciliary rows make an ellipsoidal field; an adoral zone, five closely-set rows on left and seven widely separated rows on right; in the brachial cavity of Salpa (Chatton and Lwoff, 1939).

H. hyperparasitica C. and L. (Fig. 335, *b*). Body ovoid, with bluntly pointed posterior end; about 44μ long; a large food vacuole in cytoplasm; in the brachial cavity of *Salpa mucronata-democratica* (Chatton and Lwoff, 1950).

Genus **Parahypocoma** Chatton and Lwoff. Ellipsoid; highly flattened; anterior end tapers slightly; twenty-nine to thirty-four ciliary rows; the adoral zone as in the other two genera; a comparatively short suctorial tentacle at anterior end; macronucleus horseshoe-shaped; a large central food vacuole; parasitic in ascidians.

P. collini C. and L. (Fig. 335, *c*). In *Ascidia mentula* and *Ciona intestinalis* (Chatton and Lwoff, 1950).

Family 3. **Sphenophryidae** Chatton and Lwoff

Mature individuals without cilia; larva formed by budding, with several rows of cilia; suctorial tentacle.

Genus **Sphenophrya** Chatton and Lwoff. Body elongated, "quarter orange-" or banana-shaped; attached to the gills of host mollusks by a suctorial tentacle; adult stage without cilia; ciliature is reduced to infraciliature of two groups; multiplication by budding; larvae are ciliated; on the gills of mollusks (Chatton and Lwoff, 1921).

S. dosiniae C. and L. (Fig. 336, *a-c*). Body 120μ by 15-20μ; young embryo ciliated; on the gills of *Dosinia exoleta, Venus ovata, Corbula gibba,* etc. (France); *Mactra solidissima* (Woods Hole) (Chatton and Lwoff, 1950).

Genus **Pelecyophrya** Chatton and Lwoff. Body hatchet-shaped, laterally compressed; posterior end rounded; a large "sucker" at the anterior end; infraciliature in two groups, five on right and

four on left; multiplication by budding; on the gills of mollusks
(Chatton and Lwoff, 1922).

P. tapetis C. and L. (Fig. 336, *d*). Body 23-25µ by about 10µ;
macronucleus spherical; ovoid micronucleus; cytoplasm contains
framents of host cells including nuclei; conjugation; on the gills of
Tapes aureus (Chatton and Lwoff, 1950).

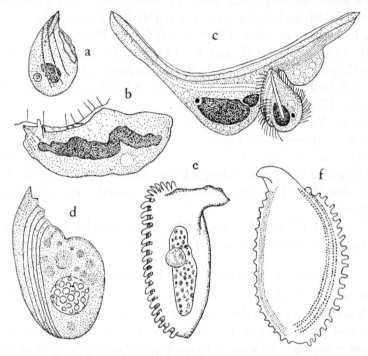

Fig. 336, a–c, *Sphaenophrya dosiniae* (a, a young embryo; b, a growing
individual attached to an epithelial cell of the host by a suctorial tentacle;
c, an individual from which a bud is ready to separate); d, a side view of
Pelecyophrya tapetis in life; e, f, *Gargarius gargarius*, ×1200 (e, in life,
showing a macronucleus and a micronucleus; f, diagram showing the
ciliature) (Chatton and Lwoff).

Genus **Gargarius** Chatton and Lwoff. Dorso-ventrally flattened;
with a "horn" near the anterior end; sucker occupies the entire
ventral surface, its margin showing papillous extensions; on the
ventral surface there are two groups of infraciliature; four rows
on each side; on Mytilus (Chatton and Lwoff, 1934).

G. gargarius C. and L. (Fig. 336, *e*, f). Body about 35µ long;

ciliated embryos formed by budding or unequal division; macronucleus elongate; micronucleus spherical; on *Mytilus edulis* (Chatton and Lwoff, 1950).

Suborder 1. **Arhynchodina** Corliss

With cytostome and adoral ciliature; body ciliation uniform. (Four families.)

Family 1. **Conchophthiridae** Kahl

Functional cytostome in the posterior half of body; body compressed laterally; with uniform ciliation.

Genus **Conchophthirus** Stein. Oval to ellipsoid; flattened; right margin concave at cytostomal region, left margin convex; ventral surface somewhat flattened, dorsal surface convex; cytostome on right side near middle in a depression with an undulating membrane; anterior third with muciferous granules (Beers, 1962); macronucleus; micronucleus; contractile vacuole opens through a canal to right side; in the mantle cavity and gills of various mussels. (Species, Kidder, 1934, 1934a, Uyemura, 1934, 1935; morphology, Raabe, 1932, 1934, Kidder, 1934; cytochemistry, Chakravarty *et al.*, 1959.)

C. anodontae (Ehrenberg) (Fig. 337, *a*). Ovoid; cytostome in anterior third, with an overhanging projection in front; cytopharynx, surrounded by circular fibrils, continues down as a fine, distensible tubule, to near the macronucleus; with peristomal basket; ciliary grooves orginate in a wide ventral suture near anterior end; anterior region filled with refractile granules; macronucleus posterior; contractile vacuole between nuclei and peristome, with a slit-like aperture (Fig. 29); 65-125μ by 47-86μ; in the mantle cavity, gills and on non-ciliated surface of palps of *Elliptio complanatus;* Woods Hole.

C. magna Kidder (Fig. 337, *b*). Much larger; 123-204μ by 63-116μ; closer ciliation; anterior one-third filled with smaller granules; irregularly outlined macronucleus, 25-30μ in diameter, central; two (or one) micronuclei; aperture for contractile vacuole large; mantle cavity of *Elliptio complanatus;* Massachusetts.

C. mytili de Morgan (Fig. 59). Reniform; 130-220μ by 76-161μ; peristomal groove on the right side; trichocysts conspicuous along

frontal margin; macronucleus oval; two micronuclei; on the foot
of the common mussel, *Mytilus edulis*. The organism lives for
forty-eight to seventy-two hours in pure sea water if kept at

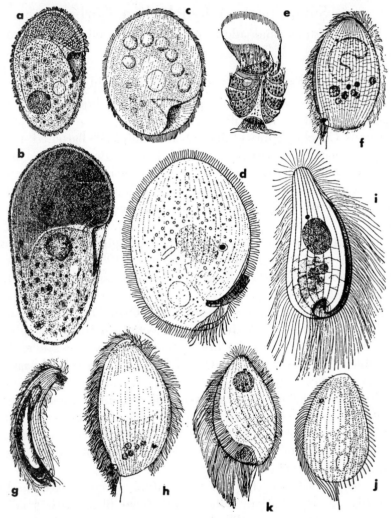

Fig. 337. a, *Concophthirus anodontae;* b, *C. magna,* ×300 (Kidder); c,
Myxophyllum steenstrupi, ×280 (Raabe); d, *Cochliophilus depressus,*
×600 (Kozloff); e, *Hemispeira asteriasi,* ×705 (Wallengren); f, g, two
views of *Ancistrum mytili,* ×500; h, *A. isseli,* ×500 (Kidder); i, *A.
japonica,* ×100 (Uyemura); j, *Eupoterion pernix,* ×500 (MacLennan
and Connell); k, *Ancistrina ovata,* ×630 (Cheissin).

about 14°C. (Beers, 1959). (Division and conjugation, Kidder, 1933 *b, c*.)

Genus **Myxophyllum** Raabe. Oval or spheroid; pellicle elastic and flexible; peristome on posterior right, without undulating membrane; 7 macronuclei; a micronucleus; ciliation uniform; in the slime covering land pulmonates.

M. steenstrupi (Stein) (Fig. 337, *c*). 120μ by 100-120μ; on *Succinea putris*, etc.

Genus **Cochliophilus** Kozloff. Ovoid and compressed; peristome in right-posterior fourth of the body; membrane-like fine cilia overlie a series of thick cilia from the anterior end of the peristome to cytostome; longitudinal rows of cilia; a vesicular micronucleus; an ovoid macronucleus; a contractile vacuole; in mollusks.

C. depressus K. (Fig. 337, *d*). About 93μ by 63μ by 15μ; fifty-two to fifty-six ciliary rows; peristomal membraneous cilia motile; macronucleus oblong; in the mantle cavity of the pulmonate snail, *Phytia setifer* in San Francisco Bay (Kozloff, 1945).

Family 2. **Thigmophryidae** Chatton and Lwoff

Cytostome posterior; body ciliation uniform; thigmotactic field is made up of closely set cilia derived from several somatic rows.

Genus **Thigmophrya** Chatton and Lwoff. Body elongate; round or oblong in optical cross section; cytostome in posterior third; contractile vacuole opens in cytopharynx; on the gills or palps of lamellibranchs.

T. macomae C. and L. Elongate ovoid; flattened; ventral surface slightly concave; oral funnel opened; contractile vacuole opens at the bottom of cytopharynx; numerous ciliary rows; about 110μ by 40μ; on the gills of *Macoma (Tellina) balthica* (Chatton and Lwoff, 1923).

Family 3. **Hemispeiridae** König

Cytostome posterior; adoral ciliature; cilia in the anterior portions of somatic rows are thigmotactic; body ciliation typically thick.

Genus **Hemispeira** Fabre-Domergue (*Hemispeiropsis,* König). Nearly spherical; flattened; longitudinal non-ciliated furrow on ventral surface, which encircles thigmotactic posterior cilia; four to five cross-furrows of cilia: a huge adoral membrane at anterior end; macronucleus, micronucleus large; contractile vacuole, anterior-right; commensal.

H. asteriasi F.-D. (Fig. 337, *e*). 20-30µ long; ectocommensal on *Asterias glacialis* (Wallengren, 1895).

Genus **Protophrya**, Kofoid (*Isselina* Cépède). Ellipsoid to pyriform; spherical macronucleus; cytostome close to the posterior end. (Taxonomy, Raabe, 1949; ciliation, Chatton and Lwoff, 1949.)

P. ovicola K. About 60µ long; in the uterus and brood-sac of the mollusks, *Littorina rudis* and *L. obtusata* (Kofoid, 1903).

Genus **Ancistrum** Maupas (*Ancistruma* Strand). Ovoid, pyriform or somewhat irregular; flattened; right side with more numerous large cilia than the left; peristome on right side; cytostome near posterior extremity; macronucleus round or sausage-shape, central; a micronucleus; contractile vacuole posterior; commensal in the mantle cavity of various marine mussels. Many species. (Morphology, reproduction, Kidder, 1933, 1933a; species, Raabe, 1959).

A. mytili (Quennerstedt) (Figs. 18; 337, *f*, *g*). Oval; dorsal surface convex, ventral surface concave; dorsal edge of peristome curves around the cytostome; peristomal floor folded and protruding; longitudinal ciliary rows on both surfaces; three rows of long cilia on peristomal edges; macronucleus sausage-form; a compact micronucleus anterior; 52-74µ by 20-38µ. Kidder (1933) found it in abundance in the mantle cavity of *Mytilus edulis at* Woods Hole and New York.

A. isseli Kahl (Fig. 337, *h*). Bluntly pointed at both ends; 70-88µ by 31-54µ. Kidder (1933) observed it abundantly in the mantle cavity of the solitary mussel, *Modiolus modiolus*, Massachusetts and New York, and studied its conjugation and nuclear reorganization.

A. japonica Uyemura (Fig. 337, *i*). Body oval or elongate pyriform; 55-76(67)µ by 14-29(20)µ; subspherical macronucleus conspicuous; a compact micronucleus; usually a single contractile vacuole, posterior; in the mantle cavity of marine mussels; *Meritrix meritrix, Paphia philippinarum, Cyclina sinensis, Mactra veneriformis, M. sulcataria,* and *Dosinia bilnulata* (Uyemura, 1937).

Genus **Eupoterion** MacLennan and Connell. Small ovoid; slightly compressed; cilia short, in longitudinal rows; rows of long cilia in peristome on mid-ventral surface and extend posteriorly, making a half turn to left around cytostome; small conical cytostome lies in postero-ventral margin of body; contractile vacuole terminal; large round macronucleus anterior; a micronucleus; commensal.

E. pernix M. and C. (Fig. 337, *j*). Forty-six to forty-eight ciliary rows; six rows of heavy peristomal cilia; 38-56µ long; in the intestinal contents of the mask limpet, *Acmaea persona;* California.

Genus **Ancistrumina** Raabe (*Ancistrina* Cheissin). Ovoid; anterior end attenuated; peristomal field along narrow right side; fifteen to eighteen ciliary rows parallel to peristomal ridges; cytostome right-posterior, marked with oral ring, with a membrane and a zone of membranellae; right ridge of peristome marked by two adoral ciliary rows; macronucleus anterior, spheroidal; a micronucleus; commensal. (Taxonomy, Raabe, 1959.)

A. ovata C. (Fig. 337, *k*). 38-48µ by 15-20µ; in the mantle cavity of molluscs: *Benedictia biacalensis, B. limneoides* and *Choanomphalus* sp.

Genus **Ancistrella** Cheissin. Elongate; ends rounded; ventral surface less convex than dorsal surface; sixteen to seventeen longitudinal ciliary rows; ciliation uniform, except anterior-dorsal region, bearing bristle-like longer cilia; two adoral ciliary rows on right of peristome, curved dorsally behind cytostome; contractile vacuole posterior; macronucleus single or divided into as many as seven parts; micronucleus; commensal.

A. choanomphali C. (Fig. 338, *a*). 55-90µ by 18-20µ; in the mantle cavity of *Choanomphalus* sp.

Genus **Ancistrospira** Chatton and Lwoff. Ciliation meridional to spiral; peristome right spiral; commensal.

A. veneris C. and L. 50-60µ by 22-28µ; ovoid, anterior end pointed; ciliary rows meridional; thigmotactic field on the left side, sharply marked from body ciliation; on the gills of *Venus fasciata.*

Genus **Boveria** Stevens (*Tiarella* Cheissin). Conical; cytostome at posterior end; peristome spiral posteriorly; macronucleus oval, in anterior half; a micronucleus; contractile vacuole posterior; ectocommensal on gills of various marine animals such as Teredo, Bankia, Tellina, Capsa and Holothuria. Several species.

B. teredinidi Pickard (Fig. 338, *b*). 27-173μ by 12-31μ; on gills of *Teredo navalis;* California (Pickard, 1927).

Genus **Plagiospira** Issel. Conical; anterior end attentuated; peristome runs spirally from middle of body to cytostome, with long cilia; macronucleus oval, anterior; a micronucleus; contractile vacuole near middle of body; somewhat spirally arranged striae widely apart on right side; commensal.

P. crinita I. (Fig. 338, *c*). 32-58μ by 18-34μ; in *Cardita calyculata* and *Loripes lacteus.*

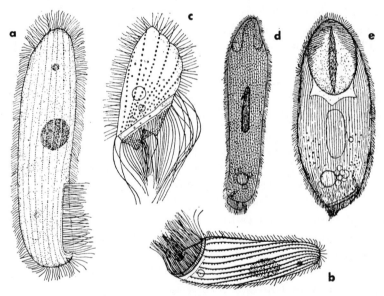

Fig. 338. a, *Ancistrella choanomphali,* ×840 (Cheissin); b, *Boveria teredinidi,* ×550 (Pickard); c, *Plagiospira crinita,* ×740 (Issel); d, *Histerocineta eiseniae,* ×250 (Beers); e, *Ptychostomum bacteriophilum,* ×500 (Miyashita).

Family 4. **Hysterocinetidae** Diesing

Cytostome at the posterior end; a sucker at the anterior end; body flattened; ciliation uniform and dense. (Morphology, Raabe, 1949; Kozloff, 1960.) Inclusion of this family in the present suborder is provisional, since its affinity to other forms is not yet clear.

Beers (1938) who placed it in Hymenostomata, in agreement with Cheissin (1931), states that the nutrition is in part saprozoic, and that the organisms are in the process of acquiring the saprozoic and astomatous condition.

Genus **Hysterocineta** Diesing (*Ladopsis* Cheissen). Elongate; flattened; flexible, an inverted V- or U-shaped sucker conspicuously present in antero-ventral margin; ciliation uniform; cytostome and cytopharynx at the posterior end; an undulating membrane along peristome which borders the posterior margin of body; macronucleus elongate; a micronucleus; contractile vacuole posterior; in the intestine of gastropods and oligochaetes. Four species. (Taconomy, Jarocki, 1934; Beers, 1938; Raabe, 1949.)

H. eiseniae Beers (Fig. 338, *d*). 190-210μ by 35-40μ (Beers); 195μ by 41μ to 230μ by 45μ (Kozloff); cytostome not functional; endoplasm with small granules; macronucleus 45-50μ long; sucker inverted V, about 25-30μ long; in the intestine of *Eisenia lönnbergi* (Beers, 1938; Kozloff, 1960).

Genus **Ptychostomum** Stein (*Lada* Vejdovsky). Sucker circular or ovoid; macronucleus ovoid or reniform, not elongate; in oligochaetes and mollusks. Several species. (Taxonomy, Studitsky, 1932, Raabe, 1949; morphology, Kozloff, 1960.)

P. bacteriophilum Miyashita (Fig. 338, *e*). Elongate oval; 70-130μ by 30-45μ; sucker oval and large, about 50μ in diameter; macronucleus ellipsoid; endoplasm with numerous rods (symbiotic bacteria?); in the freshwater oligochaete, *Criodrilus* sp.

Genus **Protoptychostomum** Raabe. Oval; compressed laterally; anterior left side slightly thickened with parallel myonemes; macronucleus ovoid; two contractile vacuoles (Raabe, 1949a).

P. simplex (André) (*Anoplophrya simplex*). In the intestine of *Eiseniella tetraedra;* 87-93μ by 50-62μ.

References

BEERS, C. D., 1938. *Hysterocineta eiseniae*, etc. Arch. Protist., 91:516.
———— 1959. Some observations on the autecology of the ciliate *Conchophthirus mytili.* J. El. Mitchell Sci. Soc., 75:3.
———— 1962. The chemical nature and function of the endoplasmic granules of the ciliate *Conchophthirus curtus.* J. Protozool., 9:364.
CHAKRAVARTY, M. M., *et al.*, 1959. Morphological and cytochemical studies in *Conchophthirus lamellidens*, etc. Proc. Zool. Soc., 12:41.

CHATTON, E., and LWOFF, A., 1922. Sur l'évolution des infusoires des lamellibranches, etc. C. R. Acad. Sci., 175:787.

—— —— 1923. Sur l'évolution des infusoires des lamellibranches. *Ibid.*, 177:81.

—— —— 1926. Diagnoses de ciliés thigmotriches nouveaux. Bull. Soc. Zool. Fr., 51:345.

—— —— 1939. Sur le suçoir des infusoires thigmotriches rhyncoidés, etc. C. R. Acad. Sci., 209:333.

—— —— 1949. Recherches sur les ciliés thigmotrichs. I. Arch. zool. exper. gén., 86:169.

—— —— 1950. II. *Ibid.*, 86:393.

CHEISSIN, E., 1931. Infusorien Ancistridae und Boveriidae aus dem Baikalsee. Arch. Protist., 73:280.

DOBRZANSKA, J., 1959. *Cepedella hepatica* Poyarkoff, etc., Bull. Acad. Polonaise Sci., Sci. Biol., 7:189.

JAROCKI, J., 1934. Two new hypocomid ciliates. *Heterocineta janickii*, etc. Mem. Acad. Pol. Sci. Lett. Cl. Math. Nat. Series B, Sc. Nat., p. 167.

—— and RAABE, Z., 1932. Ueber drei neue Infusorien-Genera der Familie Hypocomidae, etc. Bull. Acad. Pol. Sc. Lett. Ser. B. Sci. Nat. (II), p. 29.

KIDDER, G. W., 1933. On the genus Ancistruma. I. Biol. Bull., 64:1.

—— 1933a. II. Arch. Protist., 81:1.

—— 1933b. Studies on *Conchophthirius mytili*. I. *Ibid.*, 79:1.

—— 1933c. II. *Ibid.*, 79:25.

—— 1933d. *Conchophthirius caryoclada* sp. nov. Biol. Bull., 65:175.

—— 1934. Studies on the ciliates from freshwater mussels. I. *Ibid.*, 66:69.

—— 1934a. II. *Ibid.*, 66:286.

KÖNIG, A., 1894. *Hemispeiropsis comatulae*, etc. Sitzb. kais. Akad. Wiss., Wien. M.-N. Cl., 103:55.

KOFOID, C. A., 1903. On the structure of *Protophrya ovicola*, etc. Mark Anniv. Vol., Harvard Uni., p. 111.

—— and BUSH, M., 1936. The life cycle of *Parachaenia myae*, etc. Bull. Mus. Roy. Hist. Nat., 12:1.

KOZLOFF, E. N., 1945. *Cochliophilus depressus*, etc. Biol. Bull., 89:95.

—— 1945a. *Heterocineta phoronopsidis*, etc. *Ibid.*, 89:180.

—— 1946. Studies on ciliates of the family Ancistrocomidae, etc. I. *Ibid.*, 90:1.

—— 1946a. II. *Ibid.*, 90:200.

—— 1946b. III. *Ibid.*, 91:189.

—— 1946c. IV. *Ibid.*, 91:200.

—— 1960. Morphological studies on holotrichous ciliates of the family Hysterocinetidae. I. J. Protozool., 7:41.

—— 1961. A new genus and two new species of Ancistrocomid ciliates, etc. *Ibid.*, 8:60.

MacLennan, R. F., and Connell, F. H., 1931. The morphology of *Eupoterion pernix*. Univ. Cal. Publ. Zool., 36:141.

Miyashita, Y., 1933. Drei neue parasitische Infusorien aus dem Darme einer japanischen Süsswasseroligochaete. Ann. Zool. Japon., 14:127.

Mjassnikowa, M., 1930. *Sphenophrya sphaerii*, etc. Arch. Protist., 71:255.

Pickard, E. A., 1927. The neuromotor apparatus of *Boveria teredinidi*, etc. Univ. Cal. Publ. Zool., 29:405.

Raabe, Z., 1934. Weitere Untersuchungen an einigen Arten des Genus Conchophthirus. Mém. Acad. Pol. Sci. Lett. Ser. B, 10:221.

———— 1949. Recherches sur les ciliés thigmotriches. III. Ann. Univ. Marie Curie-Skl. Sec. C, 4:119.

———— 1949a. Studies on the family Hysterocinetidae Diesing. Ann. Mus. Zool. Polonici, 14:21.

———— 1959. Recherches sur les ciliés thigmotriches. VI. Acta Parasit. Polonica, 7:215.

Stevens, N. M., 1903. Further studies on the ciliate Infusoria, Licnophora and Boveria. Arch. Protist., 3:1.

Studitsky, A. N., 1932. Ueber die Morphologie, Cytologie und Systematik von *Ptychostomum chattoni*. *Ibid.*, 76:188.

Uyemura, M., 1934. Ueber einige neue Ciliaten aus dem Darmkanal von japanischen Echinoidien. I. Sci. Rep. Tokyo Bunrika Daigaku, 1:181.

———— 1935. Ueber drei in der Süsswassermuschel lebende Ciliaten (Conchophthirius). *Ibid.*, 2:89.

———— 1937. Studies on ciliates from marine mussels in Japan. I. *Ibid.*, 3:115.

Wallengren, H., 1895. Studier öfver ciliata infusorier. II. 77 pp. Lund.

Chapter 36

Order 5. **Chonotrichida** Wallengren

T HE body is vase-shaped; anterior peristome and collar or funnel with ciliature derived from the ventral cilia of swimming larva; mature individuals without body ciliation, though infraciliature persists; asexual reproduction by budding; conjugation is known in a few forms; attached to aquatic crustaceans directly or by a non-contractile stalk. Studies by several French workers indicate a close affinity between chonotrichs and gymnotrichs. The present work follows Fauré-Fremiet and Corliss in placing the order here. (Taxonomy, Kahl, 1935; Guilcher, 1951; morphology and development, Guilcher, 1951.) (Three families.)

Family 1. **Spirochonidae** Stein

Funnel at anterior end spiralled.

Genus **Spirochona** Stein. Peristome funnel spirally wound; ciliary zone on floor of the spiral furrow; attached to Gammarus in fresh water. (Many species, Swarczewsky, 1928.)

S. *gemmipara* S. (Fig. 339, *a*). 80-120µ long; attached to the gillplates of *Gammarus pulex* and other species. (Morphology, Guilcher 1951; conjugation, Tuffrau, 1953.)

Family 2. **Stylochonidae** Mohr

Funnel not spiral in form; with or without stalk.

Genus **Stylochona** Kent. Peristomal funnel with an inner funnel. One species.

S. *coronata* K. (Fig. 339, *b*). About 60µ long; on marine Gammarus.

Genus **Kentrochona** Rompel (*Kentrochonopsis,* Doflein). Peristomal funnel wide, simple, membranous; with or without a few (2) spines.

K. *nebaliae* R. (Fig. 339, *c*). About 40µ long; much flattened, with its broad side attached by means of gelatinous substance to epi- and exo-podite of *Nebalia geoffroyi;* salt water.

938

Genus **Trichochona** Mohr. Elongate; with a long stalk; pellicle thick; a single and simple funnel; two ciliary patches, one parallel to funnel rim and the other diagonal in the deep part of funnel; one macronucleus; one to four micronuclei; budding; marine. (One species, Mohr, 1948.)

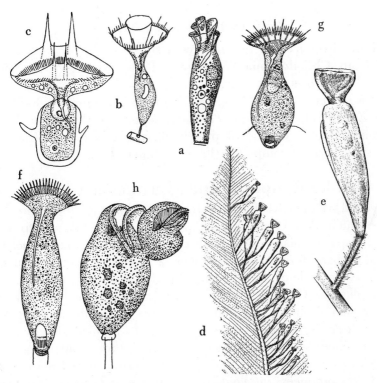

Fig. 339. a, *Spirochona gemmipara*, ×300 (Hertwig); b, *Stylochona coronata*, ×400 (Kent); c, *Kentrochona nebaliae*, ×970 (Rompel); d, e, *Trichochona lecythoides* (Mohr) (d, a portion of a host's appendage with 16 attached organisms, ×110; e, an individual, ×405); f, *Heliochona scheuteni*, ×550 (Wallengren); g, *H. sessilis*, ×510 (Wallengren); h, *Chilodochona quennerstedti*, ×400 (Wallengren).

T. lecythoides M. (Fig. 339, *d, e*). Body 35-86μ by 3-28μ; funnel 8-21.5μ high; stalk 16-51μ long; peristomal funnel with horizontal ciliary lines, up to thirty-two; diagonal lines about twenty; on the appendages of the marine crustacean, *Amphithoë* sp.

Genus **Heliochona** Plate. Peristomal funnel with numerous needle-like spines. (Taxonomy, Wallengren, 1895; Guilcher, 1950.)

H. scheutini Stein (Fig. 339, *f*). About 80-90µ long; on appendages of *Gammarus locusta;* salt water.

H. sessilis Plate (Fig. 339, *g*). About 60µ long; on *Gammarus locusta;* salt water. ·

Family 3. **Chilodochonidae** Wallengren

Funnel reduced to two lips; with stalk.

Genus **Chilodochona** Wallengren. Peristome drawn out into two lips; with a long stalk.

C. quennerstedti W. (Fig. 339, *h*). 60-115µ long; stalk, 40-160µ; on *Ebalia turnefacta* and *Portunus depurator;* the peduncle is composed of a bundle of protein fibers secreted by intracytoplasmic granular ampullae (Fauré-Fremiet *et al.*, 1956).

Order 6. **Apostomatida** Chatton and Lwoff

ASYMMETRICAL forms with a rosette-like cytostome through which liquid or small solid particles are taken into the body; sparse ciliary rows spiral; adoral rows short; macronucleus oval to band-form; a micronucleus; a single contractile vacuole.

The life-cycle of the ciliates grouped here appears to be highly complex and Chatton and Lwoff (1935) distinguished several developmental phases (Fig. 340), as follows: (1) **Trophont** or vegetative phase: right-spiral ciliary rows; nucleus pushed aside by food bodies; body grows, but does not divide. (2) **Protomont:** transitory stage between 1 and 3 in which the organism does not nourish itself, but produces "vitelloid" reserve plates; nucleus central, condensed; ciliary rows become straight. (3) **Tomont:** the body undergoes division usually in encysted condition into more or less a large number of small ciliated individuals. (4) **Protomite:** a stage in which a renewed torsion begins, and which leads to tomite stage. (5) **Tomite:** small free-swimming and non-feeding stage, but serves for distribution. (6) **Phoront:** a stage which is produced by a tomite when it becomes attached to a crustacean and encysts; within the cyst a complete transformation to trophont takes place.

Family **Foettingeriidae** Chatton

Genus **Foettingeria** Caullery and Mesnil. Trophonts large, up to 1 mm. long; sublenticular, anterior end attenuated; dorsal sur-

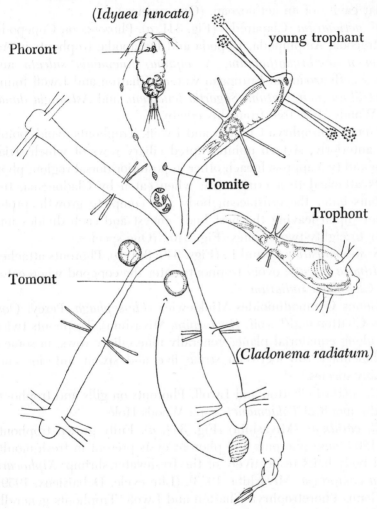

(Idyaea funcata)

Phoront

young trophant

Tomite

Trophont

Tomont

(Cladonema radiatum)

FIG. 340. Diagram illustrating the life-cycle of *Spirophrya subparasitica* (Chatton and Lwoff).

face convex, ventral surface concave; right side less convex than left side; nine spiral ciliary rows nearly evenly arranged; in gastrovascular cavity of various actinozoans; tomont on outer surface of host body, gives rise to numerous tomites with meridional ciliary rows; each tomite becomes a phoront by encysting on a crustacean, and develops into a trophont when taken into gastrovascular cavity of an actinozoan. (One species.)

F. actiniarum (Claparède) (Fig. 341, *a*). Phoronts on Copepoda, Ostracoda, Amphipoda, Isopoda and Decapoda; trophonts in *Actinia mesembryanthemum*, *A. equina*, *Anemonia sulcata* and other actinozoans in European waters; Chatton and Lwoff found *Metridium marginatum*, *Sagartia leucolena* and *Astrangia danae* of Woods Hole free from this ciliate.

Genus **Spirophrya** Chatton and Lwoff. Trophonts ovoid, pointed anteriorly; sixteen uninterrupted ciliary rows of which striae one and two approach each other in posterior-dorsal region; phoronts attached to a crustacean; when eaten by Cladonema, trophonts enter the crustacean body and complete growth; protomonts upon leaving the host body encyst and each divides into four to eighty-two tomites (Fig. 340). (One species.)

S. subparasitica C. and L. (Figs. 340; 341, *b*). Phoronts attached to *Idyaea furcata;* ovoid trophonts enter the copepod when eaten by *Cladnema radiatum*.

Genus **Gymnodinioides** Minkiewicz (*Physophaga*, Percy; *Oospira* Chatton and Lwoff; *Hyalospira*, Miyashita). Trophonts twisted along equatorial plane; generally nine ciliary rows, in some a rudimentary row between striae five and six at anterior end. (Many species.)

G. calkinsi Chatton and Lwoff. Phoronts on gills and trophonts in the moult of *Palaemonetes* sp.; Woods Hole.

G. caridinae (Miyashita) (Fig. 344, *a*). Fully grown trophonts 80-120μ long; phoronts and phoront cysts present in fresh moults and body hairs respectively of the freshwater shrimp, *Xiphocaridina compressa* (Miyashita, 1933). (Life cycle, Debaisieux, 1959.)

Genus **Phoretrophrya** Chatton and Lwoff. Trophonts generally with nine ciliary rows; striae one, two, and three, close to one another. (One species.)

P. nebaliae C. and L. (Fig. 341, *c*). Phoronts and tomonts on appendages, and trophonts in the moult, of *Nebalia geoffroyi*.

Genus **Synophrya** Chatton and Lwoff Trophonts and tomonts

similar to those of *Gymnodinioides;* but development highly complicated. (One species.)

S. *hypertrophica* C. and L. (Fig. 341, *d*). Phoronts in branchial lamellae, and trophonts in the moult, of *Portunus depurator*, etc.

Genus **Ophiurespira** Chatton and Lwoff. Trophonts ovoid; ten ciliary rows; striae nine and ten interrupted. (One species.)

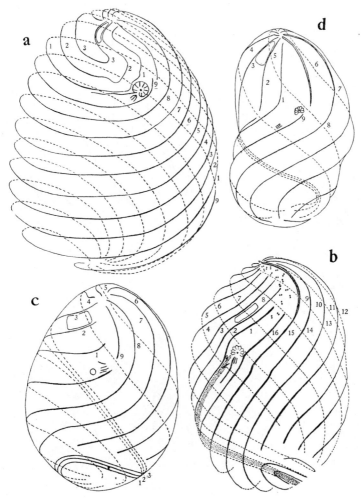

FIG. 341. a, *Foettingeria actiniarum*, a trophont; b, *Spirophrya subparasitica*, a trophont, ×1000; c, *Phoretrophrya nebaliae*, ×1180; d, *Synophrya hypertrophica* (Chatton and Lwoff).

O. weilli C. and L. (Fig. 342, *a*). Trophonts in the intestine of *Ophiothrix fragilis* and *Amphiura squamata* (Ophiuroidea).

Genus **Photorophrya** Chatton and Lwoff. Trophonts small; ciliation approximately that of *Ophiurespira;* massive macronucleus; with peculiar trichocysts comparable with the nematocysts of Polykrikos (p. 386); ecto- or endo-parasitic in encysted stages of other apostomeans. (Several species.)

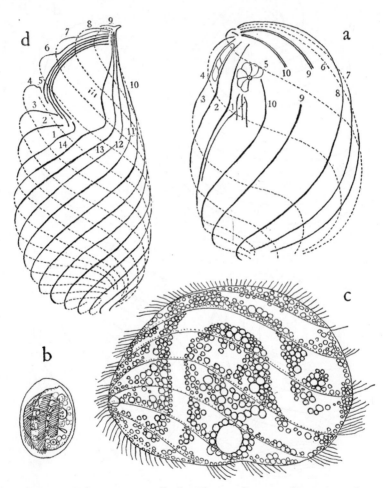

Fig. 342. a, *Ophiurespira weilli;* b, *Photorophrya insidiosa,* a trophont in a phoront of Gymnodinioides, ×800; c, *Vampyrophrya pelagica,* a trophont, ×740; d, *Pericaryon cesticola,* a trophont (Chatton and Lwoff).

P. insidiosa C. and L. (Fig. 342, *b*). Phoronts, trophonts and tomites in phoronts of *Gymnodinioides*.

Genus **Polyspira** Minkiewicz. Trophonts reniform; nine rows and several extra rows; striae one to four and five to nine with two others in two bands.

P. delagei M. (Fig. 343, *a*). Phoronts on gills and trophonts in the moult of *Eupagurus berhardus*.

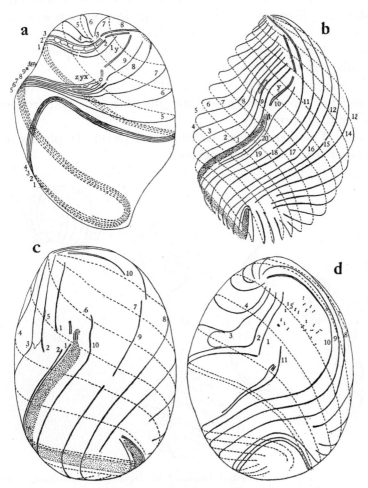

FIG. 343. a, *Polyspira delagei;* b, *Calospira minkiewiczi,* a trophont, ×1300; c, *Vampyrophrya pelagica,* d, *Traumatiophtora punctata,* ×1300 (Chatton and Lwoff).

Genus **Pericaryon** Chatton. Trophonts ellipsoid; fourteen ciliary rows.

P. cesticola C. (Fig. 342, *d*). Trophonts in the gastrovascular cavity of the ctenophore, *Cestus veneris;* other stages unknown.

Genus **Calospira** Chatton and Lwoff. Trophonts resemble those of *Spirophrya;* twenty ciliary rows; macronucleus twisted bandform; a micronucleus.

C. minkiewiczi C. and L. (Fig. 343, *b*). Phoronts attached to integument of *Harpacticus gracilis* (copepod); trophonts in its fresh carcass; tomonts and tomites in water.

Genus **Vampyrophrya** Chatton and Lwoff. Trophonts ovoid; ten ciliary rows. (One species.)

V. pelagica C. and L. (Fig. 342, *c;* 343, *c*). Phoronts on *Paracalanus parvus, Clausocalanus furcatus, etc.,* and trophonts in their fresh carcasses.

Genus **Traumatiophtora** Chatton and Lwoff. Trophonts oval; eleven ciliary rows. (One species.)

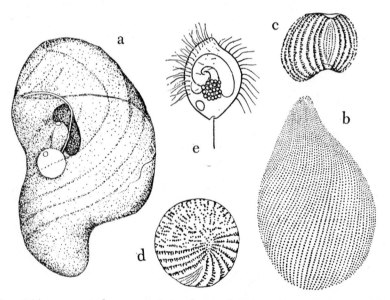

FIG. 344. a, a newly excysted trophont of *Hyalospira caridinae,* ×1000 (Miyashita); b–e, *Cryptocaryum halosydnae* (Fauré-Fremiet and Mugard) (b, the infraciliature of trophont, ×450; c, tomont of third or fourth generation; d, anterior end view; e, tomite in life, ×800).

T. punctata C. and L. (Fig. 343, *d*). Trophonts in fresh carcass of *Acartia clausi*.

Genus **Cyrtocaryum** Fauré-Fremiet and Mugard. Trophont, astomous; external appearance resembles *Anoplophrya* (p. 948); macronucleus reticulate as in *Foettingeria;* liberated in sea water; no encystment, but multiplication in free state; differentiation of an oral ciliary field.

C. halosydnae F. and M. (Fig. 344, *b-e*). Trophont in the lateral caeca of the digestive tube of *Halosydna gelatinosa;* pyriform, 90-120μ by 65-80μ; with about sixty slightly spiral ciliary rows; cilia in the anterior region strongly thygmotactic. When freed in the sea water, no encystment occurs, but division into eight to sixteen subspherical individuals in chain, takes place. Tomont 45μ long; tomites 20μ by 16μ, asymmetrical, with a long caudal bristle.

References

CHATTON, E., and LWOFF, A., 1935. Les ciliés apostomes. Arch. zool. exper. gén., 77:1.

CORLISS, J. O., 1961. The ciliated protozoa. Pergamon Press.

DEBAISIEUX, P., 1959. Ciliates apostomes parasites de Palaemon. La Cellule, 60:331.

FAURÉ-FREMIET, E., and MUGARD, H., 1949. Un infusoire apostome parasite d'un polychète: etc. C. R. Acad. Sci., 228:1753.

———— ROUILLER, C., *et al.*, 1956. Structure et origine du pédoncule chez Chilodochona. J. Protozool., 3:188.

GUILCHER, Y., 1951. Contribution à l'étude des ciliés gemmipares, chonotriches et tentaculiferes. Ann. Sci. Nat., Zool., Ser. 11., 13:33.

KAHL, A., 1935. Urtiere oder Protozoa. Dahl's Die Tierwelt Deutschlands, etc. Part 30.

MIYASHITA, Y., 1933. Studies on freshwater foettingeriid ciliate, *Hyalospira caridinae.* Japan J. Zool., 4:439.

MOHR, J. L., 1948. *Trichochona lecythroides,* a new genus and species, etc. Allan Hancock Found. Publ., Occasional Papers, no. 5.

SWARCZEWSKY, B., 1928. Zur Kenntnis der Baikalprotistenfauna. Arch. Protist., 64:44.

TUFFRAU, M., 1953. Les processus cytologiques de la conjugaison chez *Spirochona gemmipara* Stein. Bull. Biol. France et Belg., 87:314.

WALLENGREN, H., 1895. Studier öfver ciliata infusorier. II. 77 pp. Univ. Lund.

Chapter 37
Order 7. **Astomatida** Schewiakoff

THE ciliates placed in this order possess no cytostome; endo-skeleton or holdfast organelles may occur in some forms. The body ciliation is usually uniform. Asexual reproduction by binary fission or budding which often results in chain formation. Sexual reproduction is conjugation; encystment takes place in some.

These organisms are parasitic in the invertebrates living in fresh or salt water. (Taxonomy, Cépède, 1910, 1923, Cheissin, 1930, Heidenreich, 1935, Delphy, 1936, Puytorac, 1954; cytology, Rossolimo and Perzewa, 1929, Puytorac, 1954, 1961.) (Four families.)

Family 1. **Anoplophryidae** Cépède

Body ciliation uniform and complete; no holdfast organelles; no skeletal structures.

Genus **Anoplophrya** Stein. Oval, elongate, ellipsoid or cylindrical; macronucleus ovoid to cylindrical; micronucleus small; one to several contractile vacuoles; ciliation dense and uniform; in coelom and gut of Annelida and Crustacea. (Numerous species, Rossolimo, 1926.)

A. *marylandensis* Conklin (Fig. 345, *a*). 36-72μ by 16-42μ; in the intestine of *Lumbricus terrestris* and *Helodrilus caliginosus;* Baltimore, Maryland (Conklin, 1930).

A. *orchestii* Summers and Kidder (Fig. 345, *b*). Polymorphic according to size; pyriform to broadly ovoid; seven to forty-five ciliary rows meridional, unequally spaced, and more on one surface; macronucleus voluminous, a compact micronucleus; body 6-68μ long; in the sandflea, *Orchestia agilis;* Woods Hole, Massachusetts (Summers and Kidder, 1936).

Genus **Rhizocaryum** Caullery and Mesnil. With hollowed ventral surface which serves for attachment; macronucleus drawn out like a tree-root. (One species.)

R. concavum C. and M. (Fig. 345, *c*). In the gut of *Polydora caeca* and *P. flava* (polychaetes).

Genus **Metaphrya** Ikeda. Pyriform, anterior end bent slightly to one side; twelve longitudinal ciliary furrows; below ectoplasm, a

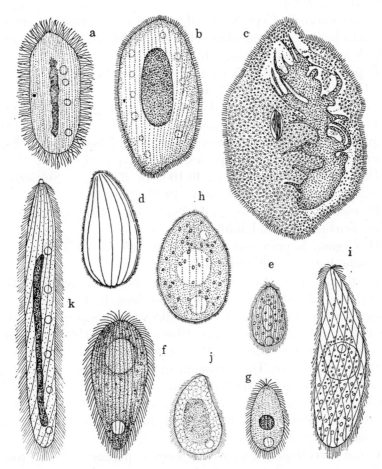

Fig. 345. a, *Anoplophrya marylandensis*, ×500 (Conklin); b, *A. orchestii*, ×500 (Summers and Kidder); c, *Rhizocaryum concavum*, ×670 (Cépède); d, *Metaphrya sagittae*, ×120 (Ikeda); e, *Perezella pelagica*, ×340 (Cépède); f, *Dogielella sphaerii*, ×470 (Poljansky); g, *D. minuta*, ×670 (Poljansky); h, *D. virginia*, ×670 (Kepner and Carroll); i, *Orchitophrya stellarum*, ×870; j, *Kofoidella eleutheriae*, ×270; k, *Bütschliella opheliae*, ×350 (Cépède).

layer of refringent materials; endoplasm sparse; macronucleus basket-like, large, with a spacious hollow; a micronucleus; no contractile vacuoles. (One species.)

M. *sagittae* I. (Fig. 345, d). About 250μ by 130μ; in the body cavity of *Sagitta* sp.

Genus **Perezella** Cépède. Ovoid; ventral surface concave, serves for attachment; macronucleus ellipsoid; contractile vacuole terminal; longitudinally, uniformly, ciliated. (A few species.)

P. *pelagica* C. (Fig. 345, e). In the coelom of copepods (Ascartia, Clausia, Paracalanus); about 48μ long.

Genus **Dogielella** Poljansky. Pyriform; longitudinal ciliary rows; contractile vacuole terminal; macronucleus spherical, with a spherical or elliptical micronucleus; in the parenchyma of flatworms or mollusks. (Four species, Poljansky, 1925.)

D. *sphaerii* P. (Fig. 345, f). 40-100μ by 25-54μ: in *Sphaerium corneum*. Conjugation (Poljansky, 1926).

D. *minuta* P. (Fig. 345, g). 12-28μ by up to 20μ; in *Stenostomum leucops* (Platyhelminthes).

D. *virginia* Kepner and Carroll (Fig. 345, h). 40-50μ long; in the same host animal; Virginia.

D. *renalis* Kay. Elongate pyriform, but extremely plastic; 61-184μ by 27-82μ; spherical macronucleus in the middle of body; one micronucleus; a contractile vacuole anterior; in the renal organ of *Physella* sp. (Kay, 1946). Kozloff (1954) apparently observed the same species in the excretory tubule of the limpet, *Ferrissia peninsulae*. The arrangement of its cilia and contractile vacuole pores is closely related to that of Tetrahymena and considered that it evolved from a ciliate of this type by loss of the mouth. He established *Curimostoma* and proposed to place it provisionally in Tetrahymenidae.

Genus **Orchitophrya** Cépède. Elongate pyriform; ciliary rows oblique; macronucleus spherical, central. (One species.)

O. *stellarum* C. (Fig. 345, i). In gonads of the echinoderm, *Asteracanthion* (*Asterias*) *rubens;* 35-65μ long.

Genus **Kofoidella** Cépède. Pyriform; macronucleus broadly oval; contractile vacuole, subterminal. (One species.)

K. *eleutheriae* C. (Fig. 345, j). In gastrovascular cavity of the medusa, *Eleutheria dichotoma;* 30-80μ long.

Genus **Herpetophrya** Siedlecki. Ovoid; with a pointed, mobile, tactile, non-ciliated cone; macronucleus globular; without contractile vacuole. (One species.)

H. astomata S. In coelom of Polymnia (annelid).

Genus **Bütschliella** Awerinzew. Elongate with pointed anterior end, with non-ciliated retractile anterior cap; cilia in about ten slightly spiral rows; macronucleus band-form; several contractile vacuoles in a longitudinal row. (Several species.)

B. opheliae A. (Fig. 345, *k*). In *Ophelia limacina;* 280-360μ by 35-50μ.

B. chaetogastri Penard. Elongate lanceolate, slightly flattened; longitudinal rows of long cilia; cytoplasm colorless; macronucleus elongate; micronucleus voluminous, vesicular; without contractile vacuole; 60-120μ long; in the oesophagus of *Chaetogaster* sp.

Genus **Spirobutschliella** Hovasse (1950). Elongate fusiform with rounded extremities; ciliation uniform and in spiral rows; anterior tip not ciliated; pellicle thick; macronucleus, a long spindle reaching the both ends of the body; a median micronucleus; in the intestine of Annelida.

S. clignyi (Cépède). In the mid-gut of *Potamoceros triqueter;* 180-550μ by about 50μ; micronucleus fusiform, 6-10μ long; often infected by a microsporidan, *Gurleya nova* Hovasse (Fjeld, 1956).

Genus **Protanoplophrya** Miyashita. Similar to *Anoplophrya;* but with rudimentary oral apparatus, a long slit, an undulating membrane and cytopharynx in anterior region of body; macronucleus elongate band; numerous contractile vacuoles. (One species.)

P. stomata Miyashita (Fig. 346, *a*). Cylindrical; up to 1.5 mm. by about 70μ; in hind-gut of *Viviparus japonicus* and *V. malleatus.*

Family 2. **Haptophyridae** Cépède

With or without attaching organelles; usually with a sucking organelle; contractile vacuole a long canal.

Genus **Haptophrya** Stein. Elongate; uniformly ciliated; anterior end with a neck-like constriction; a circular sucker surrounded by one to two rows of cilia. (A few species.)

H. michiganensis Woodhead (Fig. 346, *b*). 1.1-1.6 mm. long; in the gut of the four-toed salamander, *Hemidactylium scutatum;*

Michigan. (Cytology, Bush, 1933; contractile canal, MacLennan, 1944.)

H. virginiensis Meyer. 354µ by 95µ; macronucleus about one-third of the body length; in the intestine of *Rana palustris.*

Genus **Steinella** Cépède. Anterior end broad; sucker-like depression without encircling cilia, but with two chitinous hooks. (One species.)

S. uncinata (Schultze). Up to 200µ long; in gastrovascular cavity of *Planaria ulvae, Gunda segmentata* and *Proceros* sp.

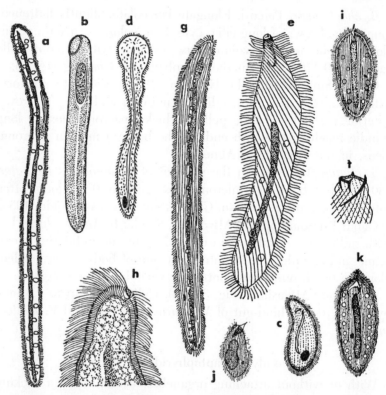

Fig. 346. *Protanoplophrya stomata,* ×100 (Miyashita); b, *Haptophrya michiganensis,* ×35 (Woodhead); c, *Lachmannella recurva,* ×100 (Cépède); d, *Sieboldiellina planariarum,* ×100 (Cépède); e, f, *Intoshellina poljanskyi* (e, ×300; f, ×870) (Cheissin); g, h, *Monodontophrya kijenskiji* (g, ×100; h, anterior end, ×870) (Cheissin); i, *Maupasella nova,* ×280 (Cépède); j, *Schultzellina mucronata,* ×670; k, *Metaradiophrya lumbrici,* ×140 (Cépède.)

Genus **Lachmannella** Cépède. With a chitinous hook at anterior end; elongate pyriform, anterior end curved; ciliation longitudinal and dense. (One species.)

L. recurva (Claparède and Lachmann) (Fig. 346, *c*). In the gastrovascular cavity of *Planaria limacina;* about 200μ long.

Genus **Sieboldiellina** Collin. Vermiform, with neck-like constriction; simple sucker at anterior end. (One species.)

S. planariarum (Siebold) (Fig. 346, *d*). Up to 700μ long; in gastrovascular cavity of various fresh- and salt-water turbellarians, most frequently *Planaria torva.*

Family 3. **Intoshellinidae** Cépède

Body ellipsoid to elongate, with uniform ciliation; contractile vacuoles, not canal-like; with holdfast or attaching organelles.

Genus **Intoshellina** Cépède. Elongate; ciliary rows slightly spiral; macronucleus voluminous, highly elongate; five to seven contractile vacuoles scattered in posterior region; a complicated attaching organella at anterior end (Fig. 346, *f*); vestigial cytopharynx.

I. poljanskyi Cheissin (Fig. 346, *e*, *f*). 170-280μ long; in the intestine of *Limnodrilus arenarius.*

Genus **Monodontophrya** Vejdowsky. Elongate; anterior end with thick ectoplasm; attaching organella at anterior end, with fibrils; macronucleus elongate; numerous contractile vacuoles in a longitudinal row.

M. kijenskiji Cheissin (Fig. 346, *g*, *h*). 400-800μ long; in anterior portion of intestine of *Tubifex inflatus.*

Genus **Maupasella** Cépède. Ellipsoid; close longitudinal ciliary rows; with a spinous attaching organella at anterior end, with fibrils; contractile vacuoles in two irregular rows; macronucleus elongate. One species.

M. nova C. (Fig. 346, *i*). 70-130μ long; in the intestine of *Allolobophora caliginosa* (annelid). (Supplementary chromatic body, Keilin. 1920.)

Genus **Schultzellina** Cépède. Similar to *Maupasella;* but with attaching organella set obliquely; macronucleus voluminous, reniform.

S. mucronata C. (Fig. 346, *j*). In the intestine of *Allurus tetraedurus* (annelid).

Genus **Hoplitophrya** Stein. Slender, elongate; elongated macronucleus; a micronucleus; a single longitudinal row of many contractile vacuoles on the dorsal side; a single median spicule with a small pointed tooth at its anterior end; in the intestine of oligochaetes. (Several species.)

H. secans S. Elongated; 160-500μ by 20-35μ; fifteen to thirty contractile vacuoles in a row; spicule 10-15μ long; in the intestine of *Lumbricus variegatus*.

H. criodrili Miyashita (Fig. 347, *a*). Ellipsoid, slightly flattened; 90-130μ by 45-60μ; periphery of endoplasm highly granulated; attaching organelle about 25μ long; macronucleus bandform; two rows of contractile vacuoles; in the anterior half of the gut of an oligochaete, *Criodrilus* sp.

Genus **Radiophrya** Rossolimo. Elongate, often with satellites; attaching organelle about 25μ long; macronucleus bandform; plasmic fibrils; macronucleus a narrow long band; a single row of many small contractile vacuoles, close to the nucleus. (Many species.)

R. hoplites R. (Fig. 347, *b*, *c*). 100-1000μ long; in the intestine of Lamprodrilus, Teleuscolex, Styloscolex, and other oligochaetes.

Genus **Metaradiophrya** Heidenreich. Ovoid to ellipsoid; with two lateral rows of contractile vacuoles; with a hook attached to a long shaft; ectoplasmic fibers supporting the hook; in the intestine of oligochaetes. Several species. (EM study, Puytorac, 1961.)

M. lumbrici (Dujardin) (Fig. 346, *k*). 120-140μ by 60-70μ; in the intestine of *Lumbricus terrestris*, *L. rubellus* and *Eisenia foetida*. (Morphology, Williams, 1942; argyrome, Puytorac, 1951.)

M. asymmetrica Beers. 115-150μ by 55-70μ; hook 10μ long; shaft 25-30μ by 2μ in antero-lateral margin in ectoplasm; twenty-five to thirty supporting fibrils; two rows of four vacuoles each, which do not contract regularly *in vitro;* in the intestine (middle third) of *Eisenia lönnbergi* (Beers, 1938).

Genus **Protoradiophrya** Rossolimo. Elongate; near anterior end a shallow depression along which is found a spicule which may be split posteriorly. (A few species.)

P. fissispiculata Cheissin (Fig. 347, *d*). 180-350μ long; in the anterior portion of intestine of *Styloscolex* sp.

Genus **Mrazekiella** Kijenskij. Elongate; anterior portion broad

with sucker-like depression, posterior region cylindrical; anterior
end with attaching organella composed of arrowhead and skeletal
ribs; macronucleus an elongate band; contractile vacuoles distributed. (A few species.)

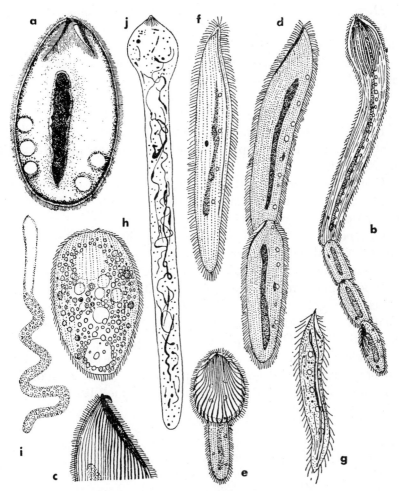

Fig. 347. a, *Hoplitophrya criodrili*, ×500 (Miyashita); b, c, *Radiophrya
hoplites* (Cheissin) (b, ×130; c, ×300); d, *Protoradiophrya fissispiculata*,
×330 (Cheissin); e, *Mrazekiella intermedia*, ×210; f, *Mesnilella rostrata*,
×470 (Cheissin); g, *M. clavata*, ×290 (Penard); h, *Opalinopsis sepiolae*,
×670 (Gonder); i, j, *Chromidina elegans* (i, ×330 (Chatton and Lwoff);
j, ×220 (Wermel)).

M. intermedia Cheissin (Fig. 347, *e*). 180-260μ long; in the anterior portion of intestine of *Branchiura coccinea*.

Genus **Mesnilella** Cépède. Elongate; with one or more long spicules imbedded in endoplasm; contractile vacuoles in one to two rows. (Numerous species.)

M. rostrata Rossolimo (Fig. 347, *f*). 100-1200μ long; in the intestine of various oligochaetes (Styloscolex, Teleuscolex, Lamprodrilus, Agriodrilus, etc.).

M. clavata Leidy (Fig. 347, *g*). 100-200μ long; in the intestine of *Lumbricus variegatus*.

Family 4. **Opalinopsidae** Hartog

Body elongate; macronucleus fragmented or of network; aberrant forms. Provisionally placed here; some authors place them in Apostomatida.

Genus **Opalinopsis** Foettinger. Oval or ellipsoid; macronucleus fragmented; ciliation uniform and close; parasite in the liver of cephalopods. (A few species.)

O. sepiolae F. (Fig. 347, *h*). 40-80μ long; in the liver of *Sepiola rondeletii* and *Octopus tetracirrhus*.

Genus **Chromidina** Gonder (*Benedenia* Foettinger). Elongate; anterior region broader, end pointed; uniform ciliation; macronucleus in irregular network distributed throughout body; micronucleus osbcure; budding and encystment; Cheissin holds that this is identical with Opalinopsis. (One species.)

C. elegans Foettinger (Fig. 347, *i*, *j*). 500-1500μ by about 30-60μ in kidney and gonad of cephalopods: Sepia, Loligo, Illex and Spirula (Jepps, 1931). (Morphology, Wermel, 1928.)

References

BEERS, C. D., 1938. Structure and division in the astomatous ciliate *Metaradiophrya asymmetrica* n. sp. J. El. Mitchell Sci. Soc., 54:111.

BUSH, M., 1933. The morphology of the ciliate *Haptophrya michiganensis*, etc. Tr. Am. Micr. Soc., 52:223.

CÉPÈDE, C., 1910. Recherches sur les infusoires astomes: etc. Arch. zool. exper, gén., Sér. 5, 3:341.

———— 1923. V, VI. Bull. Soc. Zool. France, 48:105.

CHEISSIN, E., 1930. Morphologische und systematische Studien ueber Astomata aus dem Baikalsee. Arch. Protist., 70:531.

CONKLIN, C., 1930. *Anoplophrya marylandensis*, etc. Biol. Bull., 58:176.

CORLISS, J. O., 1961. The ciliated protozoa. Pergamon Press.

DELPHY, J., 1936. Sur les Anoplophryimorphes. III. Bull. Mus. Nat. d'hist. nat., 8:516

FAURÉ-FREMIET, E., 1950. Morphologie comparée et systematique des ciliés. Bull. soc. zool. France, 75:109.

―――― 1961. Documents et observationes écologiques et pratiques sur la culture des infusoires ciliés. Hydrobiol., 18:300.

FJELD, P., 1956. *Spirobutschliella clignyi* from Oslofjord. J. Protozool., 3:63.

HEIDENREICH, E., 1935. Untersuchungen an parasitischen Ciliaten aus Anneliden. I, II. Arch. Protist., 84:315.

HOVASSE, R., 1950. *Spirobutschliella chattoni,* etc. Bull. Inst. Océanogr., no. 962.

JEPPS, M. W., 1931. On parasitic ciliate from Spirula. Danish "Dana"-Exp. 1920-1922. Oceanogr. Rep., 8:35.

KAY, M. W., 1946. Observations on *Dogielella renalis, etc.* J. Parasit., 32:197.

KEILIN, D., 1920. On the occurrence of a supplementary chromatic body in *Maupasella nova,* etc. Parasitology, 12:92.

KOZLOFF, E. N., 1954. Studies on an astomatous ciliate from a freshwater limpet, *Ferressia peninsulae.* J. Protozool., 1:200.

MACLENNAN, R. F., 1944. The pulsatory cycle of the contractile canal in the ciliate Haptophrya. Tr. Am. Micr. Soc., 63:187

MEYER, S. L., 1939. Description of *Haptophrya virginiensis,* etc. J. Parasit., 25:141.

MIYASHITA, Y., 1933. Drei neue parasitische Infusorien aus dem Darme einer japanischen Süsswasseroligochaete. Ann. Zool. Japon., 14:127.

POLJANSKIJ, J. I., 1925. Drei neue parasitsche Infusorien aus dem Parenchym eninger Mollusken und Turbellarien. Arch. Protist., 52:381.

―――― 1926. Die conjugation von *Dogielella sphaerii. Ibid.,* 53:407.

PUYTORAC, P. DE, 1951. Sur le présence d'un argyrome chez quelques ciliés astomes. Arch. zool. exper. gén., 88(N.-R):49.

―――― 1954. Contribution à l'étude cytologique et taxonomique des infusoires astomes. Ann. Sci. Nat., Zool., 18:85.

―――― 1959. Observations sur quelques ciliés astomes des oligochètes du Lac d'Ohrid. I. J. Protozool., 6:157.

―――― 1960. II. *Ibid.,* 7:278.

―――― 1961. Complément à l'étude de l'ultrastructure des ciliés de genre Metaradiophrya. Arch. Anat. Micr. Morphol., Exper., 50:35.

RAABE, Z., 1949. Recherches sur les ciliés thigmotriches. IV. Ann. Univ. Maria Curie-Sklodowska, Sec. C., 4:195.

ROSSOLIMO, L. L., 1926. Parasitische Infusorien aus dem Baikalsee. Arch. Protist., 54:468.

―――― and PERZEWA, T. A., 1929. Zur Kenntnis einiger astomen Infusorien: etc. *Ibid.,* 67:237.

Summers, F. M., and Kidder, G. W., 1936. Taxonomic and cytological studies on the ciliates associated with the amphipod family Orchestiidae from the Woods Hole district. *Ibid.*, 86:379.

Wermel, E. W., 1928. Untersuchungen ueber *Chromidina elegans. Ibid.*, 64:419.

Williams, G. W., 1942. Observations on several species of Metaradiophrya. J. Morphol., 70:545.

Woodhead, A. E., 1928. *Haptophrya michiganensis* sp.nov. J. Parasit., 14: 177.

Chapter 38

Subclass 2. Spirotricha Bütschli

WITH well-developed adoral zone of membranelles which winds clockwise to cytostome; body ciliation is sparse except Heterotrichida; cirri in Hypotrichida.

The subclass is divided into six orders: Heterotrichida, Oligotrichida (p. 979), Tintinnida (p. 981), Entodiniomorphida (p. 982), Hypotrichida (p. 996), and Odontostomatida (p. 1011).

Order 1. Heterotrichida Stein

Body ciliation uniform; in Peritromidae the dorsal surface is without, or with a few, cilia; in Licnophoridae cilia are on the edge of attaching disk. (Ten families.)

Family 1. Bursariidae Perty

Body ciliation complete and uniform; peristome a large funnel-like cavity.

Genus **Bursaria** Müller. Ovoid; anterior end truncate, potserior end broadly rounded; dorsal surface convex, ventral surface flattened; deep peristome begins at anterior end and reaches about central part of body, where it gives rise to cytostome and cytopharynx, which is bent to left; lengthwise fold divides peristome into two chambers; striation longitudinal; ciliation complete and uniform; macronucleus band-form; many micronuclei; many contractile vacuoles distributed along lateral and posterior borders; cysts with a double envelope; fresh water. (Cytology and conjugation, Poljansky, 1934); division, Schmähl, 1926; fibrils, Peschkowsky, 1927.)

B. truncatella M. (Fig. 348, *a*). 500-1000μ long; macronucleus a long rod; ten to thirty-four vesicular micronuclei; fission mostly during night; feeds on various protozoa; cysts 120-200μ in diameter; macronucleus becomes coiled and intertwined; fresh water. (Peristome formation, Schmähl, 1926; encystment, Beers, 1948, Miyake, 1957.)

959

Genus **Thylacidium** Schewiakoff. Similar to *Bursaria* in general appearance; but smaller in size; peristome simple in structure without longitudinal fold; with zoochlorellae; fresh water. (One species.)

T. truncatum S. (Fig. 348, *b*). 60-100μ long.

Genus **Bursaridium** Lauterborn. Similar to *Bursaria;* peristome funnel turns to right; fresh water.

B. difficile Kahl (Fig. 348, *c*). Anterior end truncate, cytopharynx slanting toward right; about 130μ long.

Family 2. **Metopidae** Kahl

Body ciliation uniform; peristome usually curves to posterior-right; adoral zone of membranelles straight to spiral; anterior part of body is twisted.

Genus **Metopus** Claparède and Lachmann. Body form changeable; when extended oblong or fusiform; peristome conspicuous, slightly spirally diagonal, beginning at the anterior end and reaching the middle of body; when contracted, peristome much spirally coiled; cytopharynx short; body ciliation uniform, longitudinal or in some, spiral; longer cilia at ends; conspicuous contractile vacuole terminal; macronucleus ovoid to elongate; fresh or salt water; sapropelic, some parasitic. (Numerous species.)

M. es Müller (*M. sigmoides*, C. and L.) (Figs. 89; 348, *d*). 120-200μ long; sapropelic. Noland's (1927) study on its conjugation has been described (p. 239). (Morphology, physiology, Schulze, 1959.)

M. striatus McMurrich (Fig. 348, e). 80-120μ long; fresh water.

M. fuscus Kahl (Fig. 348, *f*). 180-300μ long by 60μ wide and 40μ thick; fresh water.

M. circumlabens Biggar (Fig. 348, *g*). 70-165μ by 50-75μ; in the digestive tract of sea urchins, *Diadema setosum* and *Echinometris subangularis* in Bermuda (Biggar, 1932; Lucas, 1934); in *Centrechinus antillarum*, etc., in Tortugas (Powers, 1935); in *Diadema setosum* and *Echinometra oblonga* in Japan (Uyemura, 1933).

Genus **Spirorhynchus** da Cunha. Fusiform; somewhat flattened; anterior end drawn out and curved toward left; posterior end also drawn out; spiral peristome; cytopharynx small with an undulating membrane; cilia uniformly long; contractile vacuole posterior; longitudinally striated; body surface with closely adhering bacteria (Kirby); three spherical macronuclei; micronucleus (?); in brackish water (da Cunha, 1915).

S. *verrucosus* da C. (Fig. 348, *h*). 122-140μ by 20-22μ. Kirby (1934) observed it in salt marsh with 3 per cent salinity; California.

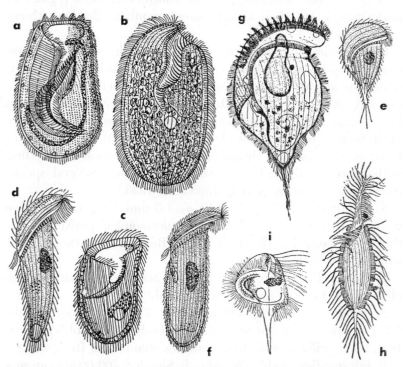

FIG. 348. a, *Bursaria truncatella*, ×60 (Kahl); b, *Thylacidium truncatum*, ×440 (Schewiakoff); c, *Bursaridium difficile*, ×210 (Kahl); d, *Metopus es*, ×260 e, *M striatus*, ×220; f, *M. fuscus*, ×150 (Kahl); g, *M. circumlabens*, ×370 (Powers); h, *Spirorhynchus verrucosus*, ×360 (Kirby); i, *Caenomorpha medusula*, ×200 (Blochmann).

Genus **Caenomorpha** Perty (*Gyrocoris* Stein). Bell-shaped; carapaced ectoplasm in some species bears protrichocysts; strong marginal zone of about eight rows of cilia; one to two dorsal rows of longer cilia and a dense spiral field around caudal prolongation; peristome long; cytostome posterior; cytopharynx directed anteriorly; a single elongate or two spherical macronuclei; a micronucleus; fresh or salt water (sapropelic). (Several species.)

C. medusula P. (Fig. 348, *i*). 150µ by 130µ; fresh and brackish water. (Several varieties.)

Family 3. **Spirostomatidae** Kent

Body elongate and large, highly contractile; some pigmented; uniform body ciliation; adoral zone of membranelles long, though inconspicuous.

Genus **Spirostomum** Ehrenberg. Elongated; cylindrical; somewhat compressed; ectoplasm with highly developed myonemes which are arranged lengthwise independent of ciliary rows, hence highly contractile; yellowish to brown; excretory vacuole terminal large, with a long dorsal canal; macronucleus either ovoid or chain form; cilia short; rows longitudinal; caudal cilia are thigmotactic, secrete mucous threads (Jennings); peristome closely lined with short membranellae; fresh or salt water. Several species. (Cytology, Seshachar and Padmavathi, 1956.)

S. ambiguum E. (Figs. 40, 349, *a*). 1-3 mm. long; macronucleus composed of many beads; many micronuclei; peristome two-thirds the body length; fresh water. Regeneration (Seyd, 1936); irritability (Blättner, 1926).

S. minus Roux (Fig. 349, *b*). 500-800µ long; macronucleus chain-form; in fresh and salt water (Kahl).

S. loxodes Stokes (Fig. 349, *c*). About 300µ long (length: width, 6-7:1); peristome about one-third the body length; oblique striation; longer cilia at ends; macronucleus chain-form; fresh water.

S. intermedium Kahl (Fig. 349, *d*). Slender; 400-600µ long; macronucleus chain-form; fresh water.

S. teres Claparède and Lachmann (Fig. 349, *e*). 150-400µ long; macronucleus oval; in fresh water and also reported from salt water.

S. filum (E.) (Fig. 349, *f*). Peristome one-fourth the body length; posterior end drawn out; 200-300µ up to 700µ long; fresh water.

Genus **Gruberia** Kahl. Similar to *Spirostomum* in general appearance; but posterior end drawn out; slightly contractile; contractile vacuole posterior; macronucleus compact or beaded; salt water.

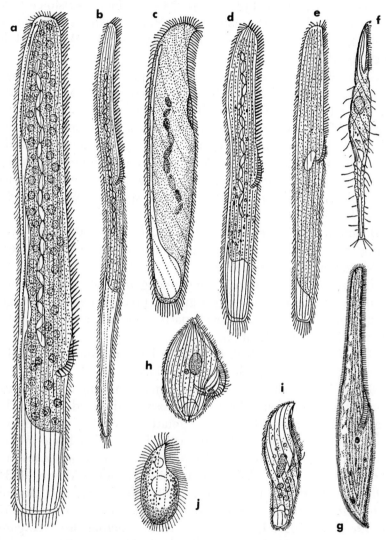

Fig. 349. a, *Spirostomum ambiguum*, ×65; b, *S. minus*, ×140; c, *S. loxodes*, ×240 (Stokes); d, *S. intermedium*, ×140; e, *S. teres*, ×200 (Kahl); f, *S. filum*, ×190 (Penard); g, *Gruberia calkinsi*, ×140 (Beltran); h, *Blepharisma lateritium*, ×160; i, *B. steini*, ×340 (Penard); j, *Protocruzia pigerrima*, ×390 (Faria, da Cunha and Pinto).

G. calkinsi Beltrán (Fig. 349, *g*). 200-800μ long; peristome two-thirds the body length; many (contractile?) vacuoles distributed; moniliform macronucleus; many micronuclei; Woods Hole (Beltrán, 1933).

Genus **Blepharisma** Perty. Pyriform, spindle-form or ellipsoid; somewhat narrowed anteriorly; compressed; peristome on the left border, which is twisted to right at posterior end and connected with oral funnel with membrane; in front of cytostome a two-layered undulating membrane on right edge; ciliary rows longitudinal; ciliation dense; contractile vacuole and cytopyge terminal; macronucleus one or divided into several parts; several species rose-colored; fresh or salt water. Many species. (Nuclear variation, Hirshfield *et al.*, 1963).

B. lateritium (Ehrenberg) (Fig. 349, *h*). 130-200μ long; pyriform; macronucleus oval; a micronucleus; rose-colored; fresh water among decaying leaves.

B. persicinum P. 80-120μ long; elongate oval; posterior end pointed; left peristomal edge sigmoid; preoral membrane large; macronucleus in three to seven parts; rose-colored; fresh water among decaying vegetation.

B. steini Kahl (Fig. 349, *i*). 80-200μ long; macronucleus ovoid; reddish to colorless; fresh water in sphagnum.

B. undulans Stein. 150-300μ long; macronucleus in two parts; undulating membrane long; cytopharynx directed posteriorly; fresh water among decaying vegetation. Suzuki (1954) distinguishes three subspecies, based on the form and behavior of the macronucleus during division: *B.u.undulans*, *B.u.americanus* and *B.u.japonicus* (Fig. 350). (Contractile vacuole, Moore, 1934; influence of light on color, Giese, 1938; morphology and physiology, Stolte, 1924; macronucleus in division, Suzuki, 1957, McLoughlin, 1957, Helson *et al.*, 1959: macronuclear reorganization, Young, 1939; multiconjugation, Weisz, 1950a; zoopurpurpin, Weisz, 1950; conjugation, Hirshfield and Pecora, 1956, Suzuki, 1957, Bhandary, 1960; races, Seshachar and Bhandary, 1962.)

Genus **Protocruzia** Faria, da Cunha and Pinto. Peristome does not turn right, leads directly into cytostome; convex left side not ciliated, but bears bristles; flat right side with three to five faintly marked ciliary rows; peristome begins at pointed anterior end and extends one-fourth to one-third the body length; cytopharynx (?); macronucleus simple; contractile vacuole subterminal; salt water.

P. pigerrima Cohn (Fig. 349, *j*). About 20µ (da Cunha); 50-60µ long (Kahl); peristome one-fourth to one-third the body length; salt water.

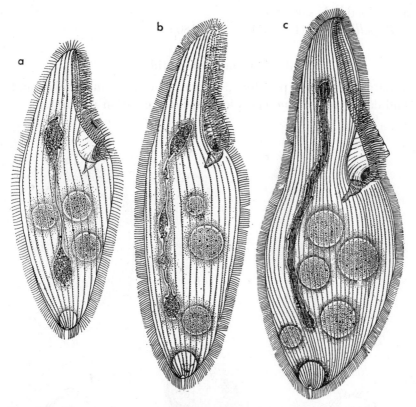

Fig. 350. Three subspecies of *Blepharisma undulans* (Suzuki). a, *B. u. undulans,* ×250; b, *B. u. americanus,* ×320; c, *B. u. japonicus,* ×250.

Genus **Phacodinium** Prowazek. Oval; marked grooves on body surface; cilia in cirrus-like fused groups; peristome long on left margin; cytostome posterior; contractile vacuole terminal; macronucleus horseshoe-shape; five to nine micronuclei; fresh water. (One species.)

P. metschnicoffi (Certes) (Fig. 351, *a*). About 100µ long.

Genus **Pseudoblepharisma** Kahl. Body form intermediate between *Spirostomum* and *Blepharisma;* right peristomal edge with two rows of cilia; fresh water.

P. tenuis K. (Fig. 351, *b*). 100-200μ long.

Genus **Parablepharisma** Kahl. Similar to *Blepharisma;* but peristome-bearing anterior half narrowed neck-like and pointed; ectoplasm covered with gelatinous layer in which symbiotic bacteria are imbedded; salt water.

P. pellitum K. (Fig. 351, *c*). 120-180μ long.

Family 4. **Condylostomatidae** Kahl

Large triangular peristome; adoral zone of membranelles and undulating membrane conspicuous; body ciliation uniform.

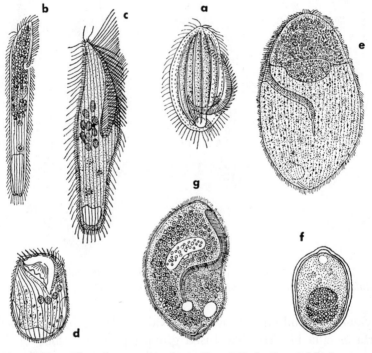

FIG. 351. a, *Phacodinium metschnicoffi*, ×270; b, *Pseudoblepharisma tenuis*, ×310; c, *Parablepharisma pellitum*, ×340 (Kahl); d, *Condylostoma vorticella*, ×120 (Penard); e, f, *Nyctotherus ovalis*, ×340 (Kudo); g, *N. cordiformis*, ×170 (Stein).

Genus **Condylostoma** Bory. Ellipsoid; anterior end truncate, posterior end rounded or bluntly pointed; slightly flattened; peristome wide at anterior end and V-shaped, peristomal field not ciliated; a large membrane on right edge and adoral zone on left; macronucleus moniliform; one to several contractile vacuoles often with canal; cytopyge posterior; fresh or salt water. (Many species, Spiegel, 1926.)

C. vorticella (Ehrenberg) (Fig. 351, *d*). 100-200μ long; fresh water.

C. patens (Müller). 250-550μ long; salt water; Woods Hole (Calkins).

Family 5. **Plagiotomidae** Bütschli

Oval to renifrom; body ciliation uniform and dense; adoral membranelles conspicuous; inhabitants in the alimentary canal of invertebrates and vertebrates.

Genus **Plagiotoma** Dujardin. Elongated oval; peristome begins at anterior end; greatly flattened; macronucleus bandform; in the gut of oligochaetes.

P. lumbrici D. In earthworms.

Genus **Nyctotherus** Leidy. Oval or reniform; compressed; peristome begins at anterior end, turns slightly to right and ends in cytostome located midway between the ends; cytopharynx runs dorsally and posteriorly, a long tube with undulating membrane; ciliary rows longitudinal and close-set; massive macronucleus in anterior half with a micronucleus; in some, nuclei are suspended by a karyophore; endoplasm with discoid glycogen bodies, especially in anterior region, hence yellowish to brown; contractile vacuole and cytopyge terminal; in the colon of Amphibia and various invertebrates. (Numerous species, Geiman and Wichterman, 1937, Wichterman, 1938, Carini, 1938-1945.)

N. ovalis L. (Figs. 3; 351, *e, f*). Ovoid; anterior half compressed; macronucleus elongate, at right angles to dorso-ventral axis at anterior one-third; micronucleus in front of macronucleus; distinct karyophore; glycogen bodies; 90-185μ by 62-95μ; giant forms up to 360μ by 240μ; cysts 72-106μ by 58-80μ; in the colon of cockroaches. The chromatin spherules of the macronucleus are often very large (p. 48). (Fibrillar structure, ten Kate, 1927, King *et al.*, 1961; nuclear structure, Kudo, 1936.)

N. cordiformis (Ehrenberg) (Figs. 87; 351, *g*). 60-200μ by 40-140μ; ovoid; micronucleus behind macronucleus; no karyophore; in the colon of frogs and toads. Higgins (1929) notes that there are certain differences between American and European forms and that the organisms exhibit a great variety of form and size in the tadpoles of various frogs, although those of adult frogs are relatively constant in form. (Life cycle, Wichterman, 1936 (p. 237); tactile cilia, Fernandez-Galiano, 1948; fibrillar structure, ten Kate, 1927.)

Family 6. **Stentoridae** Carus

Body trumpet-shaped; a conspicuous adoral zone of membranelles extends to the anterior end of body; body ciliation uniform; highly contractile.

Genus **Stentor** Oken. When extended, trumpet-shaped or cylindrical; highly contractile; some with mucilaginous lorica; usually oval to pyriform while swimming; conspicuous peristomal field frontal; adoral zone encircles peristome in a spiral form, leaving a narrow gap on ventral side; the zone and field sink toward cytostome and the former continues into cytopharynx; macronucleus round, oval or elongated, in a single mass or moniliform; contractile vacuole anterior-left; free-swimming or attached; fresh water. (Ectoplasmic structure, Fauré-Fremiet and Rouiller, 1955.)

S. coeruleus Ehrenberg (Figs. 14; 352, *a*). Fully extended body 1-2 mm. long; anterior end greatly expanded; the beautiful blue color is due to a pigment, stentorin, lodged in interstriation granules (p. 51); macronucleus moniliform; fresh water. (Biology of Stentor, Tartar, 1961; body and nuclear size, Burnside, 1929; effect of environment, Stolte, 1922; cytology, Dierks, 1926; Weisz, 1949; regeneration, Schwartz, 1935; Weisz, 1948-1951, Tartar, 1953-1956; vertical distribution, Sprugel, 1951; equivalence of nuclear nodes, Tartar, 1957.)

S. polymorphus (Müller) (Fig. 352, *b*). Colorless; with symbiotic Chlorella 1-2 mm. long when extended; macronucleus beaded; anterior end expanded.

S. mülleri (Bory) (Fig. 352, *c*). Colorless; with zoochlorellae; 2-3 mm. long; anterior end expanded; posterior portion drawn out into stalk, often housed in a gelatinous tube; on body surface 3-4 longer and stiff cilia grouped among cilia; macronucleus moniliform.

S. *roeseli* Ehrenberg (Fig. 352, *d*). 0.5-1 mm. long; anterior end expanded; body surface with groups of longer cilia; posterior portion drawn out and often housed in a gelatinous tube; macronucleus long band-form.

S. *igneus* E. (Fig. 352, *e*). Rose-colored or colorless; 200-400μ long; macronucleus oval; ciliation uniform.

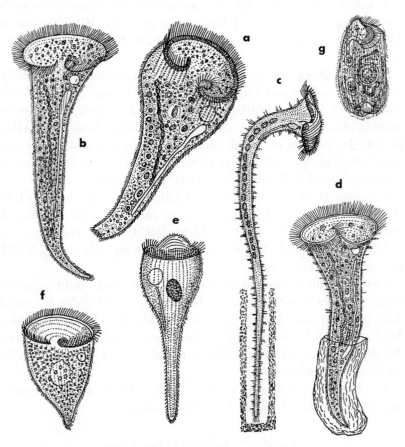

Fig. 352. a, *Stentor coeruleus*, somewhat contracted, ×70 (Roux); b, S. *polymorphus*, ×60 (Roux); c, S. *mülleri*, ×50 (Kahl); d, S. *roeseli*, ×75 (Roux); e, S. *igneus*, ×160 (Kahl); f, S. *amethystinus*, ×100 (Kahl); g, *Climacostomum vireus*, ×100 (Stein).

S. niger (Müller) Yellowish or brown; macronucleus oval; 200-300μ long. Pigment (Barbier, 1956).

S. multiformis (M.) Dark blue to bluish green; anterior end not expanded; 150-200μ long; macronucleus oval.

S. amethystinus Leidy (Fig. 352, *f*). Habitually pyriform (contracted); amethyst-blue; with zoochlorellae; 300-600μ long; macronucleus oval.

S. pyriformis Johnson. When extended 500μ long; anterior end 200μ in diameter.

S. introversus Tartar. Adoral zone retractable; blue-green; with 70-90 ectoplasmic stripes of blue-green granules; 103-306μ (up to 450μ) by 135-288μ; macronucleus moniliform; fresh water (Tartar, 1958).

Genus **Fabrea** Henneguy. Pyriform; posterior end broadly rounded, anterior end bluntly pointed; peristome extends down from anterior end two-fifths or more the body length, its posterior portion closely wound; peculiar black spot beneath membranellae in anterior portion of spiral adoral zone, composed of numerous pigment granules; without contractile vacuole; macronucleus, a sausage-shaped body or in four parts; in salt water.

F. salina H. (Fig. 353, *a*, *b*). 120-220μ by 67-125μ (Kirby); 130-450μ by 70-200μ (Henneguy); cysts ovoidal, with gelatinous envelope; 89-111μ by 72-105μ. Kirby (1934) found the organism in ditches and pools in salt marshes, showing salinities 7.5-20.1 per cent in California.

Genus **Climacostomum** Stein. Oval; flattened; right edge of peristome without membrane, left edge, semicircular or spiral with a strong adoral zone; peristomal field ciliated; cytopharynx a long curved tube with a longitudinal row of cilia; macronucleus bandform; contractile vacuole terminal, with two long canals; fresh or brackish water.

C. virens (Ehrenberg) (Fig. 352, *g*). 100-300μ long; with or without zoochlorellae; fresh and brackish water.

Family 7. **Folliculinidae** Dons

With a pseudochitinous lorica which is attached to marine plants or animals; anteriorly the body is extended into two wings bearing the adoral zone of membranelles; body ciliation uniform; in binary fission within lorica, the anterior half swims away as a larva which later becomes sessile by secretion of a lorica.

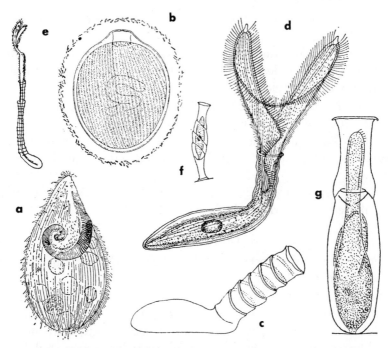

FIG. 353. a, b, *Fabrea salina* (Kirby) (a, trophozoite, ×170; b, cyst, ×330); c, side-view of the lorica of a folliculinan, ×150 (Andrews); d, *Folliculina moebiusi*, ×170 (Stein); e, *F. producta*, ×110 (Wright); f, *Pseudofolliculina arctica*, ×50 (Dons); g, *Parafolliculina violacea*, ×230 (Andrews).

Genus **Folliculina** Lamarck. Horny or chitinous lorica (Fig. 353, c) attached on broad surface; neck of the lorica oblique to perpendicular; sometimes with a collar or spiral ridge; neck uniform in diameter; in salt or fresh water. (Species, Andrews, 1914, 1921, 1923, Sahrhage, 1916; test secretion, Dewey, 1939.)

F. moebiusi Kahl (Fig. 353, d). Lorica about 500μ long.

F. producta (Wright) (Fig. 353, e). Lorica yellowish brown; 250μ long; neck often long; Atlantic coast.

F. boltoni Kent. Lorica about 200μ; lorica and body blue

green; aperture only slightly enlarged; short neck oblique or upright; in fresh water (Hamilton, 1950, 1952).

Genus **Microfolliculina** Dons. Posterior end or sides of lorica with sack-like protuberances.

M. limnoriae (Giard). Lorica dark blue; pellicle faintly striated; salt water.

Genus **Pseudofolliculina** Dons. Lorica attached with its posterior end; more or less vertical; without ring-furrow in middle; with or without style; salt water.

P. arctica D. (Fig. 353, *f*). Lorica about 430μ high, with spiral ridge; off Norwegian coast 15-28 m. deep.

Genus **Parafolliculina** Dons. Neck of lorica with a basal swelling; attached either with posterior end or on a lateral surface; salt water.

P. violacea (Giard) (Fig. 353, *g*). Total length 225-288μ; widely distributed in salt water (Andrews, 1921, 1942).

Family 8. **Clevelandellidae** Kidder

Peristome at the posterior end; conical in shape with uniform ciliation; in wood-feeding roaches.

Genus **Clevelandella** Kidder (*Clevelandia* K.). Elongate pyriform or spear-shaped; posterior region drawn out, at the end of which peristome and cytostome are located; body more or less flexible; completely ciliated; one macronucleus supported by a karyophore; a micronucleus; a contractile vacuole at posterior left, near cytopyge; endocommensals in the colon of wood-feeding roaches, *Panesthia javanica* and *P. spadica*. (Several species.)

C. panesthiae K. (Fig. 354, *a, b*). Broadly fusiform with bluntly pointed anterior end and truncate posterior end; 87-156 (123)μ by 53-78 (62)μ; peristomal projection about one-fifth the body length; peristome is nearly enclosed; macronucleus massive; a vesicular micronucleus on its anterior border; karyophore separates the endoplasm into two parts: anterior part with glycogenous platelets, posterior part with numerous food particles; often parasitized by Sphaerita (p. 1074); in the colon of *Panesthia javanica* and *P. spadica* (Kidder, 1937, 1938).

Genus **Paraclevelandia** Kidder. Elongate pyriform; body rigid; posterior end truncated obliquely to left; no peristomal projection; one macronucleus and one micronucleus; at anterior end, there is a sac connected with the karyophore, which is said to be a "macronuclear reservoir"; endocommensals.

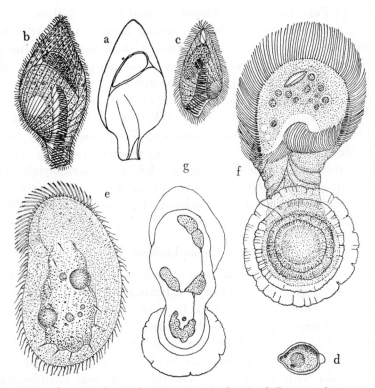

FIG. 354. a, b, ventral and dorsal views of *Clevelandella panesthiae*, ×300; c, d, *Paraclevelandia brevis* (c, ventral view, ×760; d, a cyst, ×740) (Kidder); e, *Peritromus californicus*, ×360 (Kirby); f, *Lichnophora macfarlandi*, ×420; g, *L. conklini*, ×340 (Stevens).

P. brevis K. (Fig. 354, c, d). Conical in shape; 16-38 (38)μ by 9-21 (19)μ; macronucleus spherical to elongate ellipsoid; micronucleus comparatively large, retains nuclear stains longer than macronucleus; anterior sac may sometimes be absent; cysts, 14-19μ long; ovoid; with a spherical macronucleus and a micronucleus; in the colon of *Panesthia javanica* and *P. spadica* (Kidder, 1938).

Family 9. **Peritromidae** Stein

Dorso-ventrally flattened; cilia confined to ventral surface; adoral zone of membranelles conspicuous at the anterior end; cytostome on left near the middle of body.

Genus **Peritromus** Stein. Ovoid; ventral surface flattened, dorsal surface with hump of irregular outline bearing a few stiff cilia; ciliary rows only on ventral surface; a small undulating membrane at posterior end of peristome; short marginal spines; two macro- and two micro-nuclei; salt water.

P. emmae S. 90-100μ long; creeping on bottom; Woods Hole.

P. californicus Kirby (Fig. 354, *e*). Peristome short; left margin slightly concave; dorsal hump with wart-like protuberances, bearing spines (about 12μ long); sixteen to nineteen or more ventral ciliary rows; two spherical macronuclei, one anterior right and the other posterior left of hump; micronuclei 4 (2-5); 89-165μ by 60-96μ; salt marsh pools with salinity "1.2-6 per cent" in California (Kirby, 1934).

Family 10. **Licnophoridae** Stevens

No body ciliation; adoral zone of membranelles well developed on the anterior disk; posterior disk bears membranous cilia, serving for attachment; commensals in marine animals.

Genus **Licnophora** Claparède. Discoid; body roughly divisible into basal disc, neck and oral disc; basal disc for attachment, with several concentric ciliary coronas; neck flattened, contractile narrowed part with or without a ventral furrow and fibril-bundles (both running from oral groove to basal disc); oral disc highly flattened, round or ovoid; edge with membranelle zone which extends to pharyngeal funnel; macronucleus long chain-form; without contractile vacuole; commensal in salt water animals.

L. macfarlandi Stevens (Fig. 354, *f*). Average 90-110μ by 45-60μ; diameter of basal disc 40-45μ; basal disc circular; macronuclei in twenty-five to thirty-five parts in four groups; commensal in the respiratory tree of *Stichopus californicus* (Stevens, 1901). (Morphology, fission and regeneration, Balmuth, 1941, 1942.)

L. conklini S. (Fig. 354, *g*). 100-135μ long; commensal in *Crepidula plana* of Atlantic coast.

References

ANDREWS, E. A., 1914. The bottle-animalcule, Folliculina: ecological notes. Biol. Bull., 26:262.

———— 1921. American Follinulinas: taxonomic notes. Am. Nat., 55:347.

———— 1923. Folliculina: case making, anatomy and transforma ion. J. Morphol., 38:207.

———— 1942. *Parafolliculina violacea* at Woods Hole. Biol. Bull., 83:91.

BALAMUTH, W., 1941. Studies on the organization of ciliate Protozoa. I. J. Morphol., 68:241.

———— 1942. II. J. Exper. Zool., 91:15.

BARBIER, M., *et al.*, 1956. Sur les pigments du cilié *Stentor niger*. C. R. Acad. Sci., 242:2182.

BEERS, C. D., 1948. Encystment in the ciliate *Bursaria truncatella*. Biol. Bull., 94:86.

BELTRÁN, E., 1933. *Gruberia calkinsi*, etc. *Ibid.*, 64:21.

BHANDARY, A. V., 1959. Cytology of an Indian race of *Blepharisma undulans* (Stein). J. Protozool., 6:333.

———— 1960. Conjugation in *Blepharisma u. americanum*. *Ibid.*, 7:250.

BIGGAR, R. B., 1922. Studies on ciliates from Bermuda sea urchins. J. Parasit., 18:252.

BLÄTTNER, H., 1926. Beiträge zur Reizphysiologie von *Spirostomum ambiguum*. Arch. Protist., 53:253.

CARINI, A., 1938. Sobre um Nyctotherus da intestino de um grillotalpideo. Arch. Biol., 22, 1 p.

———— 1938a. Sobre um Nyctotherus da intestino da "Testudo tabulata." *Ibid.*, 22, 2 pp.

———— 1939. Sobre um Nyctotherus da cloaca de uma Amphisbaena. *Ibid.*, 23, 1 p.

———— 1940. Contribuição ao estudo dos nictoteros dos batraquios do Brasil. *Ibid.*, 24, 15 pp.

———— 1945. Sobre um Nyctotherus do *Crossodatylus gaudichaudi*. *Ibid.*, 29, 2 pp.

DA CUNHA, A. M., 1915. *Spirorhynchus verrucosus*, etc. Brazil Medico, 19:3.

DEWEY, V. C., 1939. Test secretion in two species of Folliculina. Biol. Bull., 77:448.

DIERKS, K., 1926. Untersuchungen über die Morphologie und Physiologie des *Stentor coeruleus*, etc. Arch. Protist., 54:1.

FAURÉ-FREMIET, E., and ROUILLER, C., 1955. Microscopie électronique des structures ectoplasmiques ches les ciliés du genre Stentor. C. R. Acad. Sci., 241:678.

FERNANDEZ-GALIANO, D., 1948. Los cilios tactiles de *Nyctotherus cordiformis*. Bol. Real Soc. Espan. Hist. Nat., 46:219.

GEIMAN, Q. M., and WICHTERMAN, R., 1937. Intestinal Protozoa from Galapagos tortoises. J. Morphol., 23:331.

HAMILTON, J. M., 1950. A folliculinid from northwestern Iowa. Science, 111:288.

———— 1952. Studies on loricate Ciliophora. Proc. Iowa Acad. Sci., 58:469.

HEGNER, R. W., 1940. *Nyctotherus beltrani*, etc. J. Parasitol., 26:315.

HELSON, L., *et al.*, 1959. Macronuclear changes in a strain of *Blepharisma undulans* during the division cycle. J. Protozool., 6:131.

HIGGINS, H. T., 1929. Variation in the Nyctotherus found in frog and toad tadpoles and adults. Tr. Am. Micr. Soc., 48:141.

HIRSHFIELD, H. I., *et al.*, 1963. Macronuclear variability of Blepharisma associated with growth. Internat. Soc. Cell Biol., 2:27.

———— and PECORA, P., 1956. Studies of isolated Blepharisma and Blepharisma fragments. J. Protozool., 3:14.

KAHL, A., 1932. Urtiere oder Protozoa. Dahl's Die Tierwelt Deutschlands, etc. Part 15.

KIDDER, G. W., 1937. The intestinal Protozoa of the wood-feeding roach Panesthia. Parasitology, 29:163.

———— 1938. Nuclear reorganization without cell division in *Paraclevelandia simplex*, etc. Arch. Protist., 91:69.

KING, R. L., *et al.*, 1961. The ciliature and infraciliature of *Nyctotherus ovalis* Leidy. J. Protozool., 8:98.

KIRBY, H., JR., 1934. Some ciliates from salt marshes in California. Arch. Protist., 82:114.

KUDO, R. R., 1936. Studies on *Nyctotherus ovalis*, with special reference to its nuclear structure. *Ibid.*, 87:10.

LUCAS, M. S., 1934. Ciliates from Bermuda sea urchins. I. J. Roy. Micr. Soc., 54:79.

McLOUGHLIN, D. K., 1957. Macronuclear morphogenesis during division of *Blepharisma undulans*. J. Protozool., 4:150.

MIYAKE, A., 1957. Artificial induction of excystment in *Bursaria truncatella*. Physiol. & Ecol., 7:123.

MOORE, I., 1934. Morphology of the contractile vacuole and cloacal region of *Blepharisma undulans*. J. Exper. Zool., 69:59.

NEIVA, A., *et al.*, 1914. Parasitologische Beiträge. Mem. Inst. Oswaldo Cruz, 6:180.

NIE, D., 1950. Morphology and taxonomy of the intestinal Protozoa of the guinea-pig, *Cavia porcella*. J. Morphol., 86:381.

NOLAND, L. E., 1927. Conjugation in the ciliate, *Metopus sigmoides*. J. Morphol. Physiol., 44:341.

PESCHKOWSKY, L., 1927. Skelettgebilde bei Infusorien. Arch. Protist., 56:31.

POLJANSKY, G.: 1934. Geschlechtsprozesse bei *Bursaria truncatella*. *Ibid.*, 81:420.

POWERS, P. B. A., 1936. Studies on the ciliates of sea urchins, etc. Papers Tortugas Lab., 29:293.

SAHRHAGE, H., 1916. Ueber die Organisation und die Teilungsvorgang des Flaschentierchens (*Folliculina ampulla*). Arch. Protist., 37:139.

SCHMÄHL, O., 1926. Die Neubildung des Peristoms bei Teilung von *Bursaria truncatella. Ibid.,* 54:359.

SCHULTZE, E., 1959. Morphologische, cytologische und oekologischephysiologische Untersuchungen an Faulschlammciliaten, etc. Arch. Protist., 103:371.

SCHWARTZ, V., 1935. Versuche über Regeneration und Kerndimorphismus bei *Stentor coeruleus. Ibid.,* 85:100.

SESHACHAR, B. R., and BHANDARY, A. V., 1962. Observations on the life-cycle of a new race of *Blepharisma undulans* from India. J. Protozool., 9:265.

———— and PADMAVATHI, P. B., 1956. The cytology of a new species of Spirostomum. *Ibid.,* 3:145.

SEYD, E. L., 1936. Studies on the regulation of *Spirostomum ambiguum.* Arch. Protist., 86:454.

SPIEGEL, A., 1926. Einige neue marine Ciliaten. Arch. Protist., 55:184.

SPRUGEL, G., JR., 1951. Vertical distribution of *Stentor coeruleus,* etc. Ecology, 32:147.

STEINBURG, P., 1959. The cause of giantism and cannibalism in *Blepharisma undulans.* Biol. Rev. City Coll. N.Y., 21:4.

STEVENS, N. M., 1901. Studies on ciliate Infusoria. Proc. Cal. Acad. Sci., Ser. 3, 3:1.

———— 1903. Further studies on the ciliate Infusoria, Licnophora and Boveria. Arch. Protist., 3:1.

STOLTE, H.-A., 1922. Der Einfluss der Umwelt auf Macronucleus und Plasma von *Stentor coeruleus. Ibid.,* 45:344.

———— 1924. Morphologische und physiologische Untersuchungen an *Blepharisma undulans. Ibid.,* 48:245.

SUZUKI, S., 1954. Taxonomic studies on *Blepharisma undulans* with special reference to the macronuclear variation. J. Sci. Hiroshima University, Series B. Div. 1, 15: No. 7.

———— 1957. Conjugation in *Blepharisma undulans japonicus* with special reference to the nuclear phenomena. Bull. Yamagata University, Nat. Sci., 4:43.

———— 1957a. Parthenogenetic conjugation in *Blepharisma undulans japonicus. Ibid.,* 4:69.

———— 1957b. Morphogenesis in the regeneration of Blepharisma, etc. *Ibid.,* 4:85.

TARTAR, V., 1953. Chimeras and nuclear transplantations in ciliates, *Stentor coeruleus* × *S. polymorphus.* J. Exper. Zool., 124:63.

———— 1954. Reaction of *Stentor coeruleus* to homoplastic grafting. *Ibid.,* 127:511.

———— 1956. Grafting experiments concerning primordium formation in *Stentor coeruleus. Ibid.,* 131:75.

———— 1957. Equivalence of macronuclear nodes. *Ibid.,* 135:387.

———— 1958. *Stentor introversus,* n. sp. J. Protozool., 5:93.

———— 1961. The biology of Stentor. Pergamon Press.

TEN KATE, C. G. B., 1927. Ueber das Fibrillensystem der Ciliaten. Arch. Protist., 57:362.

UYEMURA, M., 1933. On two ciliates: *Entorhipidium tenue* and *Metopus circumlabens,* etc. J. Nat. Hist. Soc. Tokio, 31:5 pp.

WEISZ, P. B., 1948. Time, polarity, size and nuclear content in the regeneration of Stentor fragments. J. Exper. Zool., 107:269.

———— 1948a. Regeneration in Stentor and the gradient theory. *Ibid.,* 109: 439.

———— 1949. A cytochemical and cytological study of differentiation in normal and reorganizational stages of *Stentor coeruleus.* J. Morphol., 84:335.

———— 1950. On the mitochondria nature of the pigmented granules in Stentor and Blepharisma. *Ibid.,* 86:177.

———— 1950a. Multiconjugation in Blepharisma. Biol. Bull., 98:242.

———— 1951. An experimental analysis of morphogenesis in *Stentor coeruleus.* J. Exper. Zool., 116:231.

WICHTERMAN, R., 1936. Division and conjugation in *Nyctotherus cordiformis,* etc. J. Morphol., 60:563.

———— 1938. The present state of knowledge concerning the existence of species of Nyctotherus living in man. Am. J. Trop. Med., 18:67.

WOODRUFF, L. L., 1935. Physiological significance of conjugation in *Blepharisma undulans.* J. Exper. Zool., 70:287.

YOUNG, D., 1939. Macronuclear reorganization in *Blepharisma undulans.* J. Morphol., 64:297.

Chapter 39

Order 2. **Oligotrichida** Bütschli

BODY ciliation is greatly reduced or absent; adoral membranelles are well developed and extend beyond the anterior end of body; body small. (Two families.)

Family 1. **Halteriidae** Claparède and Lachmann

Adoral membranelles well developed; bristles or tufts of cilia on body; in all types of water.

Genus **Halteria** Dujardin. Spherical or broadly fusiform; anterior border bears conspicuous adoral zone; oral part of peristome with a small membrane on right edge and cirri on left; with an equatorial zone of small oblique grooves, each bearing three long cirri or bristles, macronucleus oval; a micronucleus; contractile vacuole left of cytostome; fresh water. (Several species, Szabó, 1935.)

H. grandinella (Müller) (Fig. 355, *a*). About seven bristle-bearing grooves; fifteen frontal and seven adoral membranellae; 20-40μ long. Kahl (1932) distinguishes two varieties; var. *cirrifera* (Fig. 355, *b*), 25-50μ long, with huge cirri instead of fine body cirri; and var. *chlorelligera* (Fig. 355, *c*), 40-50μ long, with bristles and large zoochlorellae; fresh water. (Division, Fauré-Fremiet, 1953.)

Genus **Strombidium** Claparède and Lachmann. Ovoid to spherical; adoral zone very conspicuous two to four conspicuous sickleform frontal membranellae and adoral membranellae extend down cytopharynx, the first section surrounding an apical process; no body bristles or cirri; trichocysts; macronucleus oval or band-form; a micronucleus; a contractile vacuole; salt or fresh water. Numerous species. (Division, Fauré-Fremiet, 1953; Kormos and Kormos, 1958.)

S. calkinsi Fauré-Fremiet (Fig. 355, *d*). 35-60μ long; brackish and salt water; Calkins (1902) first observed it at Woods Hole.

Genus **Tontonia** Fauré-Fremiet. With well-developed apical collar; a long cytoplasmic (contractile) caudal process; salt water.

T. gracillima F.-F. (Fig. 355, *e*). 48-52μ long; caudal process 250-300μ long; macronucleus moniliform; with zoochlorellae.

Family 2. **Strobilidiidae** Kahl

Adoral membranelles form a spiral crown at the anterior end; the majority are marine; some in brackish or fresh water.

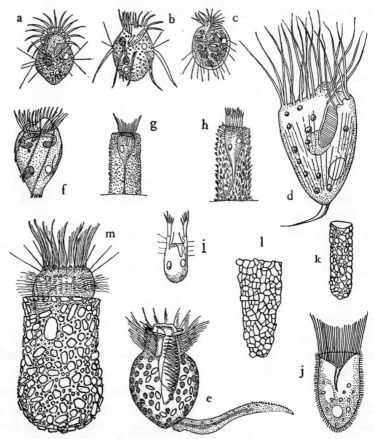

Fig. 355. a, *Halteria grandinella*, ×490 (Kahl); b, *H. g.* var. *cirrifera*, ×370 (Kahl); c, *H. g.* var. *chlorelligera*, ×260 (Kahl); d, *Strombidium calkinsi*, ×900 (Calkins); e, *Tontonia gracillima*, ×540 (Fauré-Fremiet); f, *Strobilidium gyrans*, ×340 (Kahl); g, *Tintinnidium fluviatile*, ×140 (Kent); h, i, *T. semiciliatum*, ×140 (Sterki); j, *Strombidinopsis gyrans*, ×270 (Kent); k, *Tintinnopsis cylindrata*, ×440 (Daday); l, *T. illinoisensis*, ×420 (Hempel); m, Codonella cratera, ×540 (Fauré-Fremiet).

Genus **Strobilidium** Schewiakoff. Pyriform or turnip-shaped; cytostome at anterior end; without cytopharynx; horseshoe-shaped macronucleus anterior; a micronucleus; a contractile vacuole; fresh or salt water. (Several species, Busch, 1921.)

S. gyrans (Stokes) (Fig. 355, *f*). Lateral border with rounded elevation near anterior end, posterior end truncate; 40-70μ long; in standing fresh water.

Order 3. **Tintinnida** Kofoid and Campbell

The loricate ciliates placed in this order are, as a rule, conical or trumpet-shaped and are attached by the adhesive aboral end of the body to the base of the lorica. The adoral zone of membranelles is well developed; the body ciliation is sparse or lacking; usually two macronuclei. The majority are pelagic marine organisms, though some are found in brackish water and a few in fresh water.

Kofoid and Campbell (1929, 1939) described many genera and species which were mainly based upon the morphological characteristics of the lorica. A few species are mentioned here. (Toxonomy, Hofker, 1932; species, Campbell, 1942, Balech, 1942-1951, Rampi, 1950, Silva, 1950; evolution, Kofoid, 1930; lorica formation, Busch, 1925; in Mediterranean Sea, Balech, 1959.)

Genus **Tintinnidium** Stein. Elongated lorica, highly irregular in form; soft; aboral end closed or with a minute opening; wall viscous and freely agglomerates foreign bodies; salt or fresh water.

T. fluviatile Stein (Fig. 355, *g*). Lorica 100-200μ by 45μ; on vegetation in fresh water.

T. semiciliatum Sterki (Fig. 355, *h*, *i*). 40-60μ long; on plants in fresh water.

Genus **Strombidinopsis** Kent. Lorica often absent; ovate or pyriform; frontal border with numerous long cirrus-like cilia; body covered by fine cilia; contractile vacuole posterior; fresh water.

S. gyrans K. (Fig. 355, *j*). 30-80μ long; fresh water pond.

Genus **Tintinnopsis** Stein. Lorica bowl-shaped; always with a broad aperture; aboral end closed; wall thin and covered with foreign bodies; salt or fresh water. (Species, Balech, 1945.)

T. cylindrata Kofoid and Campbell (Fig. 355, *k*). Lorica 40-50μ long; in lakes.

T. illinoisensis Hempel (Fig. 355, *l*). Lorica 59μ long; in rivers.

Genus **Codonella** Haeckel. Lorica urn- to pot-shaped; sharply divided externally and internally into a collar and bowl; collar without spiral structure; in fresh water.

C. cratera (Leidy) (Fig. 355, *m*). Lorica 60-70μ by 40μ; a number of varieties are often mentioned.

Order 4. **Entodiniomorphida** Reichenow

Adoral zone of membranelles conspicuous; in addition a dorsal zone of membranelles; in some, other membranellar tufts; endocommensals in the rumen of ruminants and in the intestine of certain other herbivores. (Two families.)

Family 1. **Ophryoscolecidae** Stein

With the adoral zone and a dorsal zone of membranelles; in the rumen of ruminants; Sharp (1914) employed "forma" to distinguish forms in Entodinium with common characteristics, differing in certain others, which scheme was extended to the whole family by Dogiel (1927). It is most probable that many species are varieties of a single species as judged by the work of Poljansky and Strelkow (1934); but since information is still incomplete, the present work ranks various formae with species, in agreement with Kofoid and MacLennan (1930).

The views regarding the relationship between these ciliates and host animals are still not in agreement. Some workers are inclined to think it commensalism, while others consider it true symbiosis (Becker, Schulz and Emmerson, 1930; Mowry and Becker, 1930; Sugden, 1953; Hungate, 1955). (Morphology, Bretschneider, 1934, 1935; contractile vacuoles, MacLennan, 1933; conjugation, Dogiel, 1925; numbers in cattle stomach, Dogiel and Fedorowa, 1929; fauna in African antelopes, Dogiel, 1932; in yaks, Dogiel, 1934; in Indian goat Das-Gupta, 1935; in Indian ox, Kofoid and MacLennan, 1930, 1932, 1933; in gaur, Kofoid and Christenson, 1934; in sheep, wild sheep and goat, Ferber and Fedorowa, 1929, Bush and Kofoid, 1948; in white-tailed deer, Zielyk, 1961.)

Genus **Ophryoscolex** Stein. Ovoid; with adoral and dorsal zones of membranellae; dorsal zone some distance behind anterior end, encircling three-fourths the body circumference at middle, broken on right ventral side; three skeletal plates extend over the body length on right-ventral side; nine to fifteen contractile vacuoles in two (anterior and posterior) circles; macronucleus simple, elongate; in the stomach of cattle, sheep, goat and wild sheep *(Ovis orientalis cycloceros)*. (Several species, Kofoid and MacLennan, 1933; neuromotor system, Fernandez, 1949.)

Dogiel (1927) designated the following species as three formae of *O. caudatus* Eberlein.

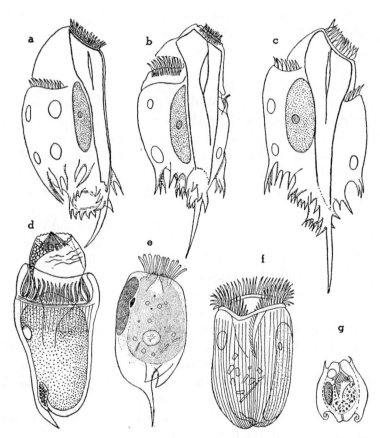

Fig. 356. a, *Ophryoscolex bicoronatus*, ×340 (Dogiel); b, *O. caudatus*, ×310 (Dogiel); c, *O. quadricoronatus*, ×340 (Dogiel); d, *Caloscolex cuspidatus*, ×310 (Dogiel); e, *Entodinium caudatum*, ×500 (Becker and Talbott); f, *E. bursa*, ×390 (Schuberg); g, *Amphacanthus ovum-rajae*, ×350 (Dogiel).

O. bicoronatus Dogiel (Fig. 356, *a*). 120-170μ by 81-90μ; primary spine 38-58μ long; in sheep.

O. caudatus Eberlein (Fig. 356, *b*). 137-162μ by 80-98μ; preanal spines 47-60μ long; in sheep, goat, and cattle.

O. quadricoronatus Dogiel (Fig. 356, *c*). 128-180μ by 86-100μ; preanal spines 48-63μ long; in sheep and *Ovis orientalis cycloceros.*

Genus **Caloscolex** Dogiel. Ovoid; anterior end truncate, posterior end rounded with or without processes; two zones of membranellae; dorsal zone encircles the body completely; three skeletal plates variously modified; seven contractile vacuoles in a single circle; nucleus elongate; in the stomach of *Camelus dromedarius.* (Several species.)

C. cuspidatus D. (Fig. 356, *d*). 130-160μ by 73-90μ.

Genus **Entodinium** Stein. Without dorsal zone; adoral zone at truncate anterior end; without skeleton; contractile vacuole anterior; macronucleus, cylindrical or sausage-form, dorsal; micronucleus anterior to middle and on left-ventral side of macronucleus; in cattle, sheep, deer and reindeer. (Numerous species, Kofoid and MacLennan, 1930; MacLennan, 1935; Lubinsky, 1958; Zielyk, 1961.)

E. caudatum S. (Fig. 356, *e*). 50-80μ long; in cattle and sheep.

E. bursa S. (Fig. 356, *f*). 55-114μ by 37-78μ (Schuberg); 80μ by 60μ (Becker and Talbott); in the stomach of cattle.

Genus **Amphacanthus** Dogiel. Similar to *Entodinium;* but spinous processes at both anterior and posterior ends; in stomach of *Camelus dromedarius.* (One species.)

A. ovum-rajae D. (Fig. 356, *g*). 46-55μ by 32-48μ.

Genus **Eodinium** Kofoid and MacLennan. Dorsal zone on the same level as adoral zone; without skeleton; macronucleus a straight, rod-like body beneath dorsal surface; two contractile vacuoles; in cattle and sheep. Several species.

E. lobatum K. and M. (Fig. 357, *a*). 44-60μ by 29-37μ; in *Bos indicus* (Kofoid and MacLennan, 1932).

Genus **Diplodinium** Schuberg. Adoral and dorsal zones at the same level; without skeletal plates; macronucleus beneath right side, its anterior third bent ventrally at an angle of 30°-90°; two contractile vacuoles; in cattle, antelope, *Camelus dromedarius,* reindeer, goat. (Numerous species, Kofoid and MacLennan, **1932.**)

D. dentatum (Stein) (Fig. 357, *b*). 65-82µ by 40-50µ; in cattle (including *Bos indicus*).

Genus **Eremoplastron** Kofoid and MacLennan. Adoral and dor-

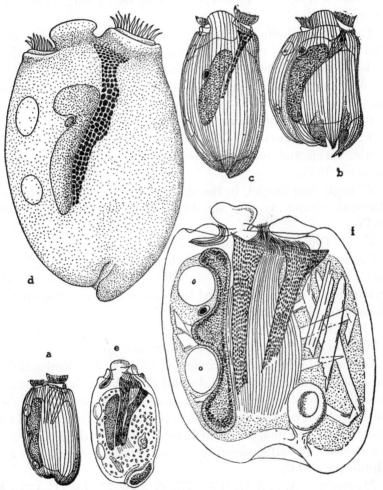

Fig. 357. a, *Eodinium lobatum*, ×540 (Kofoid and MacLennan); b, *Diplodinium dentatum*, ×250 (Kofoid and MacLennan); c, *Eremoplastron bovis*, ×550 (Kofoid and MacLennan); d, *Eudiplodinium maggii*, ×500 (Dogiel); e, *Diploplastron affine*, ×320 (Dogiel); f, *Metadinium medium*, ×320 (Dogiel).

sal zones at anterior end; a single narrow skeletal plate beneath right surface; triangular or rod-like macronucleus, anterior end of which is often bent ventrally; two contractile vacuoles; in cattle, antelope, sheep, reindeer. (Numerous species, Kofoid and Mac-Lennan, 1932.)

E. bovis (Dogiel) (Fig. 357, *c*). 52-100µ by 34-50µ; in cattle and sheep.

Genus **Eudiplodinium** Dogiel. Adoral and dorsal zones at anterior end; a single, narrow, skeletal plate beneath right surface; rod-like macronucleus with its anterior end enlarged to form a hook opening dorsally; pellicle and ectoplasm thick; two contractile vacuoles with heavy membranes and prominent pores; in cattle. (Species, Kofoid and MacLennan, 1932.)

E. maggii (Fiorentini) (Fig. 357, *d*). 104-255µ by 63-170µ; in cattle, sheep and reindeer. Neuromotor apparatus (Fernández-Galiano, 1955).

E. neglectum Dogiel. In the stomach of *Rangifer tarandus* and *Alces americana*. Average size 81.3µ by 66.2µ (Krascheninnikow, 1955).

Genus **Diploplastron** Kofoid and MacLennan. Adoral and dorsal zones at anterior end; two skeletal plates beneath right surface; macronucleus narrow; rod-like; two contractile vacuoles below dorsal surface, separated from macronucleus. (One species, Kofoid and MacLennan, 1932.)

D. affine (Dogiel and Fedorowa) (Fig. 357, *e*.) 88-120µ by 47-65µ; in the stomach of cattle, sheep, and goat.

Genus **Metadinium** Awerinzew and Mutafowa. Adoral and dorsal zones at anterior end; two skeletal plates beneath right surface sometimes fused posteriorly; macronucleus with two to three dorsal lobes; two contractile vacuoles; pellicle and ectoplasm thick; conspicuous oesophageal fibrils beneath dorsal and right sides; in the stomach of cattle, sheep, goat, and reindeer (Awerinzew and Mutafowa, 1914).

M. medium A. and M. (Fig. 357, *f*). 180-272µ by 111-175µ; in cattle.

Genus **Polyplastron** Dogiel. Adoral and dorsal zones at anterior end; two skeletal plates beneath right surface, separate or fused; three longitudinal plates beneath left surface, with anterior ends connected by cross bars; contractile vacuoles beneath dorsal surface in a longitudinal row, also with additional vacuoles; in the stomach of cattle and sheep. Species (Kofoid and MacLennan, 1932).

P. multivesiculatum (D. and Fedorowa) (Fig. 358, *a*). 120-190μ by 78-140μ; in cattle and sheep. MacLennan (1934) found that the skeletal plates are made up of small, roughly prismatic blocks of glycogen, each with a central granule. Infraciliature (Fernandez-Galiano, 1958).

Genus **Elytroplastron** Kofoid and MacLennan. Two zones at anterior end, two skeletal plates beneath right surface, a small plate

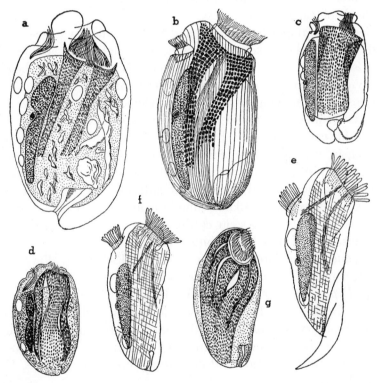

Fig. 358. a, *Polyplastron multivesiculatum,* ×360 (Dogiel); b, *Elytroplastron hegneri,* ×340 (Dogiel); c, *Ostracodinium dentatum,* ×440 (Dogiel); d, *Enoploplastron triloricatum,* ×370 (Dogiel); e, *Epidinium caudatum,* ×340 (Becker and Talbott); f, *E. ecaudatum,* ×340 (Becker and Talbott); g, *Epiplastron africanum,* ×300 (Dogiel).

beneath ventral surface, and a long plate below left side; pellicle and ectoplasm thick; conspicuous fibrils beneath dorsal-right side. (One species.)

E. hegneri (Becker and Talbott) (Fig. 358, *b*). 100-160μ by 67-97μ; in cattle, sheep, *Buffelus bubalus* and *Bos indicus* (Becker, 1933).

Genus **Ostracodinium** Dogiel. Two zones at anterior end; broad skeletal plate beneath right side; two to six contractile vacuoles in a dorsal row; cytopharyngeal fibrils thick, extend to posterior end; in cattle, sheep, antelope, steenbok, and reindeer. (Numerous species, Kofoid and MacLennan, 1932.)

O. dentatum (Fiorentini) (Fig. 358, *c*). 52-110μ by 31-68μ; in the stomach of cattle.

Genus **Enoploplastron** Kofoid and MacLennan. Two zones near anterior end; three skeletal plates beneath right-ventral side, either separate or partly fused; two contractile vacuoles; heavy pharyngeal fibrils; in cattle, reindeer and antelope.

E. triloricatum (Dogiel) (Fig. 358, *d*). Dogiel (1927) mentions size differences among those occurring in different host species, as follows: in cattle, 85-112μ by 51-70μ; in reindeer, 75-103μ by 40-58μ; in antelope (*Rhaphiceros* sp.), 60-110μ by 37-56μ.

Genus **Epidinium** Crawley. Elongate; twisted around the main axis; two zones; dorsal zone not at anterior end; three skeletal plates, with secondary plates; simple macronucleus club-shaped; two contractile vacuoles; in cattle, sheep, reindeer, camels, etc. (Species, Kofoid and MacLennan, 1932.)

E. caudatum (Fiorentini) (Fig. 358, *e*). 113-151μ by 45-61μ; in cattle, camels, *Cervus canadensis* and reindeer.

E. (Diplodinium) ecaudatum (F). Figs. 16; 358, *f*). 112-140μ by 40-60μ (Becker and Talbott); in cattle, sheep, and reindeer. The classical observation of Sharp (1914) on its neuromotor system has been described elsewhere (p. 71). (A more recent study, Timothée, 1952; isolation and *in vitro* culture for eight days, Oxford, 1958.)

Genus **Epiplastron** Kofoid and MacLennan. Elongate; two zones; dorsal zone not at anterior end; five skeletal plates, with secondary plates; macronucleus simple, elongate; two contractile vacuoles; in antelopes.

E. africanum (Dogiel) (Fig. 358, *g*). 90-140µ by 30-55µ; in *Rhaphiceros* sp.

Genus **Ophisthotrichum** Buisson. Two zones; dorsal zone at middle or near posterior end of body; one-piece skeletal plate well developed; two contractile vacuoles posterior; conjugation (Dogiel); in many African antelopes. (One species.)

O. janus (Dogiel) (*O. thomasi* B.) (Fig. 359, *a*). 90-150µ by 42-60µ. Conjugation (Dogiel, 1925).

Genus **Cunhaia** Hasselmann. Cytostome near anterior end, with adoral zone; dorsal zone on one-third of anterior-dorsal surface; two contractile vacuoles; skelton (?); in the caecum of guinea pig, *Cavia aperea*. (One species.)

C. curvata H. (Fig. 359, *b*). 60-80µ by 30-40µ; in Brazil.

Family 2. **Cycloposthiidae** Poche

In addition to the adoral and dorsal zones of membranelles, other groups of membranelles in the posterior and caudal region; in horses; also in anthropoid apes and rhinoceros.

Genus **Cycloposthium** Bundle. Large, elongate barrel-shaped; cytostome in center of a retractile conical elevation at anterior end; adoral zone conspicuous; an open ring-zone of membranellae near posterior end on both dorsal and ventral sides; pellicle ridged; skeleton club-shaped; several contractile vacuoles in a row along bandform macronucleus; in the caecum and colon of horse. (Many species, Hsiung, 1930; cytology, Strelkow, 1929, 1932.)

C. bipalmatum (Fiorentini) (Fig. 359, *c*). 80-127µ by 35-57µ. (Conjugation, Dogiel, 1925.)

C. dentiferum Gassovsky (Fig. 359, *d*). 140-222µ by 80-110µ.

Genus **Spirodinium** Fiorentini. Elongate, more or less fusiform; adoral zone at anterior end; anterior ciliary zone encircles the body at least once; a posterior ciliary arch, only one-half spiral; a dorsal cavity of unknown function (Davis, 1941), lined with stiff rods; in the colon and caecum of the horse. (Species, Hsiung, 1930, 1935.)

S. equi F. (Fig. 359, *e*). 82-196µ by 46-108µ; widely distributed. (Morphology, Hsiung, 1935a, Davis, 1941; division, Davis, 1941.)

Genus **Triadinium** Fiorentini. More or less helmet-shaped; compressed; adoral zone at anterior end; two posterior (ventral and dorsal) zones; with or without a caudal projection; in the caecum and colon of horse. (Species, Hsiung, 1935.)

T. caudatum F. (Fig. 359, *f*). 59-86μ by 50-68μ.

T. galea Gassovsky. 59-78μ by 50-60μ.

T. minimum G. (Fig. 359, *g*). 35-58μ by 30-40μ.

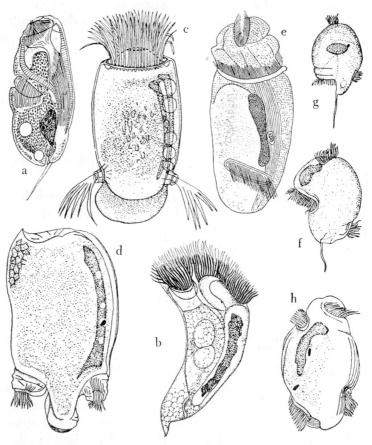

FIG. 359. a, *Ophisthotrichum janus*, ×370 (Dogiel); b, *Cunhaia curvata*, ×670 (Hasselmann); c, *Cycloposthium bipalmatum*, ×300 (Bundle); d, *C. dentiferum*, ×270 (Hsiung), e, *Spirodinium equi*, ×350 (Davis); f, *Triadinium caudatum*, ×300 (Hsiung); g, *T. minimum*, ×440 (Hsiung); h, *Tetratoxum unifasciculatum*, ×280 (Hsiung).

Genus **Tetratoxum** Gassovsky. Slightly compressed; two anterior and two posterior zones of membranellae; in the colon of horse. (Species, Hsiung, 1930.)

T. unifasciculatum (Fiorentini) (Fig. 359, *h*). 88-186μ by 60-108μ; widely distributed. (Morphology and micronuclear division, Davis, 1941a.)

T. escavatum Hsiung. 95-135μ by 55-90μ.

T. parvum H. 67-98μ by 39-52μ.

Genus **Tripalmaria** Gassovsky (*Tricaudalia,* Buisson). Adoral zone at anterior end; two dorsal and one ventral-posterior zones in tuft-form; macronucleus inverted U-shape; in the colon of horse. (Cytology, Strelkow, 1932.)

T. dogieli G. (Fig. 360, *a*). 77-123μ by 47-62μ (Hsiung, 1930).

Genus **Triplumaria** Hoare. Adoral zone; two dorsal and one ventral cirrose tufts (caudals); skeleton, composed of polygonal plates arranged in a single layer, surrounds the body except the dorsal surface; dorsal groove supported by rod-like skeleton; macronucleus elongate sausage-form, with a micronucleus attached to its dorsal surface near middle; about six contractile vacuoles arranged in line along dorsal surface of body; in the intestine of Indian rhinoceros (Hoare, 1937).

T. hamertoni H. 129-207μ long, 65-82μ thick, 4-39μ broad; endo-commensal in the intestine of *Rhinoceros unicornis* in Zoological Garden in London.

Genus **Cochliatoxum** Gassovsky. Adoral zone near anterior end; three additional zones, one antero-dorsal, one postero-dorsal and one postero-ventral; macronucleus with curved anterior end; in the colon of horse. (One species.)

C. periachtum G. (Fig. 360, *b*). 210-370μ by 130-210μ (Hsuing, 1930).

Genus **Ditoxum** Gassovsky. Large adoral zone near anterior end; two dorsal (anterior and posterior) zones; macronucleus curved club-shaped; in the colon of horse (Hsiung, 1935).

D. funinucleum G. (Fig. 360, *c*). 135-203μ by 70-101μ.

Genus **Troglodytella** Brumpt and Joyeux. Ellipsoid; flattened; adoral zone; three additional zones (anterior zone continuous or not continuous on ventral surface; posterior zone continuous on dorsal surface; between them a small zone on each side); skeletal

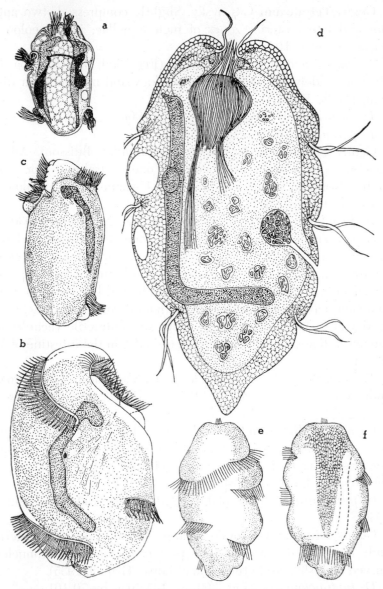

Fig. 360. a, *Tripalmaria dogieli,* ×180 (Gassovsky); b, *Cochliatoxum peri-achtum,* ×270 (Hsiung); c, *Ditoxum funinucleum,* ×270 (Hsiung); d–f, *Troglodytella abrassarti* (d, ×670 (Swezey); e, ventral and f, dorsal view, ×210 (Brumpt and Joyeux)).

plates in anterior region; macronucleus L-form; contractile vacuoles in two circles; in the colon of anthropoid apes.

T. abrassarti B. and J. (Fig. 360, *d-f*). About 145-220µ by 120-160µ; in the colon of chimpanzees (Brumpt and Joyeux, 1912). Reichenow (1920) distinguished var. *acuminata* on the basis of the drawn-out posterior end, which was found by Swezey (1932) to be a variant of *T. abrassarti.* (Cytology, Swezey, 1934; cultivation, Nelson, 1932; Swezey, 1935.)

T. gorillae Reichenow. 200-280µ by 120-160µ; in the colon of gorilla; with anterior zone not reaching the right side.

References

AWERINZEW, S., and MUTAFOWA, R., 1914. Material zur Kenntnis der Infusorien aus dem Magen der Wiederkäuer. Arch. Protist., 33:109.

BALECH, E., 1942. Tintinnoineos del Estrecho le Maire. Physis, 19:245.

——— 1945. Tintinnoinea de Atlantida. Comm. Mus. Argent. Cien. Nat. Ser. Cien. Zool., no. 7.

——— 1951. Nuevos datos aobre Tintinnoinea de Argentina y Uruguay. Physis, 20:291.

——— 1959. Tintinnoinea del Mediterraneo. Trab. Inst. Español Oceanogr., no. 28.

BECKER, E. R., 1933. Concerning *Elytroplastron hegneri*. Tr. Am. Micr. Soc., 52:217.

———, SCHULZ, J. A., and EMMERSON, M. A., 1930. Experiments on the physiological relationships between the stomach Infusoria of ruminants and their hosts. Iowa State Coll. J. Sci., 4:215.

——— and TALBOTT, M., 1927. The protozoan fauna of the rumen and reticulum of American cattle. *Ibid.*, 1:345.

BRETSCHNEIDER, L. H. 1934. Beiträge zur Strukturlehre der Ophryoscoleciden. II. Arch. Protist., 82:298.

BRUMPT, E., and JOYEUX, C., 1912. Sur un infusoire nouveau parasite du Chimpanzé, etc. Bull. Soc. Path. Exot., 5:499.

BUSCH, W., 1921. Studien über Ciliaten des nordatlantischen Ozeans und Schwarzen Meers. I. Arch. Protist., 42:364.

——— 1925. Beitrag zur Kenntnis der Gehäusbildung bei den Tintinnidae, etc. *Ibid.*, 58:183.

BUSH, MILDRED, and KOFOID, C. A., 1948. Ciliates from Sierra Nevada bighorn, etc. Univ. Cal. Publ. Zool., 53:237.

CRAWLEY, H., 1923. Evolution in the ciliate family Ophryoscolecidae. Proc. Acad. Nat. Sci., Philadelphia, 75:393.

DA CUNHA, A. M., 1914. Ueber die Ziliaten, welche in Brasilien im Magen von Rindern und Schafen vorkommen. Mem. Inst. Oswaldo Cruz., 6:58.

DAS-GUPTA, M., 1935. Preliminary observations on the protozoan fauna of the rumen of the Indian goat, etc. Arch. Protist., 85:153.

DAVIS, T. G., 1941. Morphology and division in *Spirodinium equi*, J. Morphol., 69:225.

——— 1941a. Morphology and division in *Tetratoxum unifasciculatum*. Tr. Am. Micr. Soc., 60:441.

DOGIEL, V., 1925. Die Geschlechtsprozesse bei Infusorien etc., Arch. Protist. 50:283.

——— 1927. Morphologie der Familie Ophryoscolecidae. *Ibid.*, 59:1.

——— 1932. Beschreibung einiger neuer Vertreter der Familie Ophryoscolecidae, etc. *Ibid.*, 77:92.

——— 1934. Angaben über die Ophryoscolecide, etc. *Ibid.*, 82:290.

——— and FEDOROWA, T. 1929. Ueber die Zahl der Infusorien im Wiederkäuermagen. Zentralbl. Bakt. I. Orig., 112:135.

FAURÉ-FREMIET, E., 1953. Le bipartition énantiotrope chez les ciliés oligotriches. Arch. Anat. Micros. Morph. Expér., 42:209.

FERBER, K. E., and FEDOROWA, T., 1929. Zählung und Teilungsquote der Infusorien im Pansen der Wiederkäuer. Biol. Zentralbl., 49:321.

FERNANDEZ-GALIANO, D., 1949. Sobre el aparato neuromotor y otras estructuras protoplasmaticas de *"Ophryoscolex purkinjei."* Trab. Inst. Cien. Nat. J.d. Acosta, 2:257.

——— 1955. El aparato neuromotor de *"Eudiplodinium maggii."* Bol. Real. Soc. Española N.H., 53:53.

——— 1958. La infraciliacion en *Polyplastron multivesiculatum*, etc. *Ibid.*, 56:89.

HOARE, C. A., 1937. A new cycloposthiid ciliate, etc. Parasitology, 29:559.

HOFKER, J., 1932. Studien über Tintinnoidea. Arch. Protist., 75:315.

HSIUNG, T. S., 1930. A monograph on the Protozoa of the large intestine of the horse. Iowa State Coll. J. Sci., 4:356.

——— 1935. Notes on the known species of Triadinium, etc. Bull. Fan Mem. Inst. Biol., 6:21.

——— 1935a. On some new ciliates from the mule, etc. *Ibid.*, 6:81.

HUNGATE, R. W., 1955. Mutualistic intestinal protozoa. In Hutner and Lwoff: Biochemistry and physiology of protozoa, 2:159.

KAHL, A., 1932. Urtiere oder Protozoa. I. Dahl's Die Tierwelt Deutschlands, etc. Part 25.

KOFOID, C. A., 1930. Factors in the evolution of the pelagic ciliata, the Tintinnoinea. Contr. Marine Biol., Stanford Univ., 39 pp.

——— 1935. On two remarkable ciliate protozoa from the caecum of the Indian elephant. Proc. Nat. Acad. Sci., 21:501.

——— and CAMPBELL, A. S., 1929. A conspectus of the marine and freshwater Ciliata, belonging to the suborder Tintinnoinea, etc. Univ. Cal. Publ. Zool., 34:1.

——— ——— 1939. The Tintinnoinea. Bull. Mus. Comp. Zool. Harvard, 84:1.

—— and CHRISTENSON, J. F., 1934. Ciliates from *Bos Gaurus.* Uni. California Publ. Zool., 39:341.

—— and MacLENNAN, R. F., 1930. Ciliates from *Bos indicus.* I. *Ibid.,* 33:471.

—— —— 1932. II. *Ibid.,* 37:53.

—— —— 1933. III. *Ibid.,* 39:1.

KORMOS, J., and KORMOS, K., 1958. Die Zellteilungstypen der Protozoen. Acta Biol. Acad. Sci. Hungaricae, 8:127.

KRASCHENINNIKOW, S., 1955. Observations on the morphology and division of *Eudiplodinium neglectum,* etc. J. Protozool., 2:124.

LATTEUR, B., 1958. Les ciliates Polydiniinae. La Cellule, 59:271.

LUBINSKY, G., 1958. Ophryoscolecidae of reindeer from the Canadian arctic. I. Entodiniinae. Canad. J. Zool., 36:819.

MacLENNAN, R. F., 1933. The pulsatory cycles of the contractile vacuoles in the Ophryoscolecidae, etc. Uni. California Publ. Zool. 39:205.

—— 1935. Ciliates from the stomach of musk-deer. Tr. Am. Micr. Soc., 54:181.

MOWRY, H. A., and BECKER, E. R., 1930. Experiments on the biology of Infusoria inhabiting the rumen of goats. Iowa State Coll. J. Sci., 5:35.

NELSON, E. C., 1932. The cultivation of a species of Troglodytella, etc. Science, 75:317.

OXFORD, A. E., 1955. The rumen ciliate protozoa, etc. Exper. Parasitol., 4:569.

—— 1958. Bloat in cattle. IX. New Zealand J. Agr. Res., 1:809.

RAMPI, L., 1950. I Tintinnoidi della acque di Monaco, etc. Bull. l'Inst. Océanogr., no. 965.

REICHENOW, E., 1920. Den Wiederkäuer-Infusorien verwandte Formen aus Gorilla und Schimpanse. Arch. Protist., 41:1.

SILVA, ESTELA DE S. E., 1950. Les Tintinnides de la baie de Cascais (Portugal). Bull. l'Inst. Oceanogr., no. 974.

STRELKOW, A., 1929. Morphologische Studien über oligotriche Infusorien aus dem Darme des Pferdes. I. Arch. Protist., 68:503.

—— 1932. II, III. *Ibid.,* 75:191.

SUGDEN, B., 1953. The cultivation and metabolism of oligotrich protozoa from the sheep's rumen. J. Gen. Microbiol., 9:44.

SWEZEY, W. W., 1932. The transition of *Troglodytella abrassarti* and *T. a. acuminata,* intestinal ciliates of the chimpanzee. J. Parasit., 19:12.

—— 1934. Cytology of *Troglodytella abrassarti, etc.* J. Morphol., 56:621.

—— 1935. Cultivation of *Troglodytella abrassarti,* etc. J. Parasitol., 21:10.

SZABÓ, M., 1935. Neuere Beiträge zur Kenntnis der Gattung Halteria. Arch. Protist., 86:307.

TIMOTHÉE, C., 1952. Ye système fibrillaire d'*Epidinium ecaudatum* Crawley. Ann. Sci. Nat. Zool., 11 Sér., 14:375.

ZIELYK, M. W., 1961. Ophryoscolecid fauna from the stomach of the white-tailed deer, etc. J. Protozool., 8:33.

Chapter 40

Order 5. Hypotrichida Stein

THE members of this suborder are, as a rule, flattened and strong cilia or cirri are restricted to the ventral surface. Except the family Aspidiscidae, the dorsal surface possesses rows of short slightly moveable tactile bristles. The peristome is very large with a well-developed adoral zone. The cirri on the ventral surface are called, according to their location, frontals, ventrals, marginals, anals (transversals), and caudals, as was mentioned before (Fig. 11, *b*). Asexual reproduction is by binary fission and sexual reproduction by conjugation. Enycystment is common. Mostly free-living in fresh, brackish or salt water; a few parasitic. (Four families.)

Family 1. Oxytrichidae Ehrenberg

Adoral zone well developed; cirri on ventral surface; in some ventrals reduced; in others cirri in rows; with right and left marginals.

Genus **Oxytricha** Bory (*Histrio*, Sterki; *Opisthotricha*, Kent; *Steinia*, Diesing). Ellipsoid; flexible; ventral surface flattened, dorsal surface convex; eight frontals; five ventrals; five anals; short caudals; marginals may or may not be continuous along posterior border; macronucleus in two parts, rarely single or in four parts; fresh or salt water. (Numerous species, Horváth, 1933; neuromotor system, Lund, 1935.)

O. fallax Stein (Fig. 361, *a*). Posterior region broadly rounded; about 150μ long; fresh water. Amicronucleate race (Reynolds, 1932).

O. bifaria Stokes (Fig. 361, *b*). Right side convex; left side flattened; posterior end pointed; about 250μ long; fresh water infusion.

O. ludibunda S. (Fig. 361, *c*). Ellipsoid; flexible; 100μ long; fresh water among sphagnum.

O. setigera S. (Fig. 361, *d*). Elongate ellipsoid; 5 frontals; ventrals shifted anteriorly; 50µ long; fresh water.

Genus **Tachysoma** Stokes (*Actinotricha* Cohn). Flexible; frontals eight to ten, of which anterior three are usually the largest; five ventrals scattered; five anals; marginals at some distance from lateral borders, interrupted posteriorly; fresh or salt water.

T. parvistyla S. (Fig. 361, *e*). Ten frontals scattered; about 63µ long; in shallow freshwater pools.

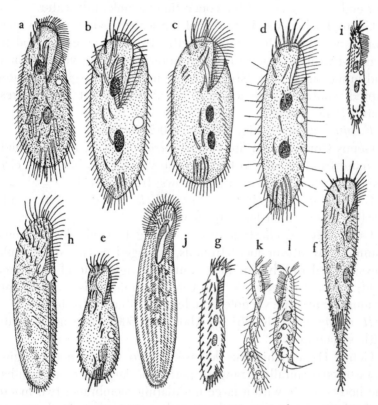

FIG. 361. a, *Oxytricha fallax,* ×230 (Stein); b, *O. bifaria,* ×180 (Stokes); c, *O. ludibunda,* ×400 (Stokes); d, *O. setigera,* ×870 (Stokes); e, *Tachysoma parvistyla,* ×490 (Stokes); *Urosoma caudata,* ×250 (Stokes); g, *Amphisiella thiophaga,* ×380 (Kahl); h, *Eschaneustyla brachytona,* ×240 (Stokes); i, *Gonostomum strenuum,* ×160 (Engelmann); j, *Hemicycliostyla sphagni,* ×100 (Stokes); k, l, *Cladotricha koltzowii* (k, ×170; l, ×300) (Kahl).

Genus **Urosoma** Kowalewski. Similar to *Oxytricha;* but posterior portion drawn out and much narrowed; fresh water.

U. caudata (Stokes) (Fig. 361, *f*). 200-250μ long; pond water.

Genus **Amphisiella** Gourret and Roeser. With a single row of ventrals and two marginal rows; salt or fresh water. Several species.

A. thiophaga (Kahl) (Fig. 361, *g*). 70-100μ long; salt water.

A. lithophora Fauré-Fremiet (1954). 120-135μ by 26-31μ; anterior end constricted; with a concretion vacuole; salt water.

Genus **Eschaneustyla** Stokes. Elliptical or ovate; narrow peristome one-third the body length; frontals numerous, about twenty-two in addition to two at anterior margin; ventrals small and numerous in three oblique rows; no anals; marginals uninterrupted; contractile vacuole a long canal near left border; fresh water. (One species.)

E. brachytona S. (Fig. 361, *h*). 170-220μ long.

Genus **Gonostomum** Sterki (*Plagiotricha*, Kent). Flexible; eight or more frontals; one to two oblique ventral rows of short cirri; four or five anals; two marginal rows; fresh water.

G. strenuum (Engelmann) (Fig. 361, *i*). Elongate; with caudal bristles; about 150μ long; fresh water.

Genus **Hemicycliostyla** Stokes. Elongate oval; flexible; ends rounded; twenty or more frontals, arranged in two semicircular rows; adoral row begins near center on right side of peristomal field; ventral surface entirely covered with fine cilia; no anals; one or more contractile vacuoles; nucleus distributed; fresh water.

H. sphagni S. (Fig. 361, *j*). About 400-500μ long; marsh water with sphagnum.

Genus **Hypotrichidium** Ilowaisky. Two ventral and marginal rows of cirri spirally arranged; peristome large, extends one-half the body length, with a large undulating membrane; two macro- and micro-nuclei; contractile vacuole anterior-left; fresh water.

H. conicum I. (Fig. 362, *a*). 90-150μ long.

Genus **Cladotricha** Gajevskaja. Elongate band-form; anterior end rounded, posterior end rounded or attenuated; frontals only two featherly cirri; macronucleus spheroidal; micronucleus; without contractile vacuole; salt water, with 5 to 20 per cent salt content (One species.)

C. koltzowii G. (Fig. 361, *k*, *l*). Band-form up to about 200μ long; posteriorly attenuated forms up to about 100μ long.

Genus **Psilotricha** Stein. Oval to ellipsoid; frontals and anals undifferentiated; ventrals and marginals long cirri, few; ventrals in two rows and a rudimentary row toward left; with or without zoochlorellae; fresh water. (A few species.)

P. acuminata S. (Fig. 362, *b*). 80-100μ long.

Genus **Kahlia** Horvath. Frontal margin with three to four strong cirri; five to eight ventral longitudinal rows; marginals; sapropelic in fresh water.

K. acrobates H. (Fig. 362, *c*). 100-200μ long; soil infusion.

Genus **Uroleptus** Ehrenberg. Elongate body drawn out into a tail-like portion; three frontals; two to four rows of ventral cirri; marginals; no anals; sometimes rose- or violet-colored; fresh or salt water. (Many species.)

U. limnetis Stokes (Fig. 362, *d*). About 200μ long; fresh water among vegetation.

U. longicaudatus S. (Fig. 362, *e*). About 200μ long; marsh water with sphagnum.

U. halseyi Calkins (Fig. 362, *f*). About 160μ by 20μ; peristome one-sixth to one-seventh the body length; three ventrals; macronucleus divided into many (up to twenty-six) parts; two (1-3) micronuclei; fresh water (Calkins, 1930).

Genus **Uroleptopsis** Kahl. Ventrals in two uninterrupted rows; salt water. (A few species.)

U. citrina K. (Fig. 362, *g*). Elongate; flexible; ectoplasm with pale-yellow ringed bodies which give the organism yellowish color; marginals discontinuous posteriorly; two contractile vacuoles near left border; 150-250μ long; salt water.

Genus **Strongylidium** Sterki. Two to five ventral rows of cirri; marginals spirally arranged; three to six frontals; two or more macronuclei; fresh or salt water. (Many species.)

S. californicum Kahl (Fig. 362, *h*). Four to five frontals; macronuclei about thirty in number; four micronuclei; contractile vacuole with short canals; about 250μ long; fresh water among vegetation.

Genus **Stichotricha** Perty. Slender ovoid or fusiform; peristome-bearing part narrowed; not flexible; usually four spiral rows of cirri; sometimes tube-dwelling, and then in groups; fresh or salt water. (Many species.)

S. *secunda* P. (Fig. 362, *i*). 130-200μ long; in fresh water.

S. *intermedia* Froud (Fig. 362, *j*). Solitary; non-loricate; 40-170μ long, two-fifths of which is a bent proboscis; two rows of

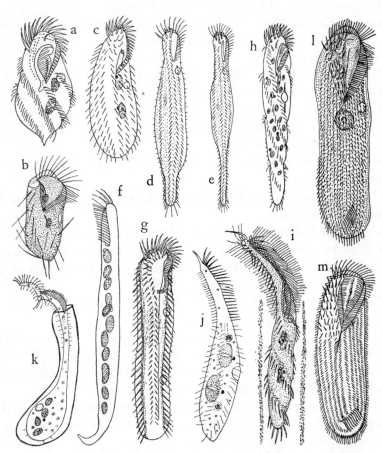

Fig. 362. a, *Hypotrichidium conicum*, ×200 (Kahl); b, *Psilotricha acuminata*, ×230 (Stein); c, *Kahlia acrobates*, ×240 (Kahl); d, *Uroleptus limnetis*, ×240 (Stokes); e, *U. longicaudatus*, ×240 (Stokes); f, *U. halseyi*, ×470 (Calkins); g, *Uroleptopsis citrina*, ×260 (Kahl); h, *Strongylidium californicum*, ×200 (Kahl); i, *Stichotricha secunda*, ×340 (Kahl); j, *S. intermedia* (Froud); k, *Chaetospira mülleri* (Froud); l, *Urostyla grandis*, ×140 (Stein); m, *U. trichogaster*, ×150 (Kahl).

body cilia; two rows of dorsal cilia, 5μ long; among Lemna in fresh water (Froud, 1949).

Genus **Chaetospira** Lachmann. Similar to *Stichotricha;* but peristome-bearing part flexible; fresh or salt water.

C. mülleri L. (Fig. 362, *k*). Flask-shaped, 60-200μ long, in a lorica; cytostome at the base of proboscis; a single (two or more) micronucleus; macronucleus in two to eight parts; ingested diatoms lose color in ten minutes; Bodo is immobilized in less than one minute; binary fission; the anterior individual remains in the lorica, while the posterior individual (averaging 46μ long) swims away and sooner or later becomes attached to substrate; cysts pyriform, 35-55μ by 15-20μ; among Lemna in fresh water (Froud, 1949).

Genus **Urostyla** Ehrenberg. Ellipsoid; flexible; ends rounded; flattened ventral surface with four to ten rows of small cirri and two marginal rows; three or more frontals; five to twelve anals; macronucleus a single body or in many parts; fresh or salt water. (Numerous species.)

U. grandis E. (Figs. 52; 362, *l*). 300-400μ long; macronucleus in 100 or more parts; six to eight micronuclei; fresh water. (Nuclei, Raabe, 1946, 1947.)

U. trichogaster Stokes (Fig. 362, *m*). 250-330μ long; fresh water.

U. caudata S. (Fig. 363, *a*). Elongate ellipsoid; flexible; narrowed anterior part bent to left; peristome one-third the body length; macronucleus in many parts; contractile vacuoles near left margin; about 600μ long; fresh water with sphagnum.

U. polymicronucleata Merriman. Elliptical with broadly rounded ends; flexible; 225μ by 65μ; opaque, green or brown because of the ingested diatoms; three large and ten small frontals; four ventral rows of cirri; marginals; macronucleus in two parts; three to eleven micronuclei (Merriman, 1937).

U. coei Turner. Elliptical, with a more pointed posterior end; 200μ by 50μ; four rows of ventral cirri, the right row being the longest; five frontals; macronucleus in two masses; four micronuclei (Turner, 1939).

Genus **Kerona** Ehrenberg. Reniform; no caudals; six oblique rows of ventral cirri; commensal. (One species.)

K. polyporum E. (Fig. 363, *b*). 120-200μ long; commensal on Hydra.

Genus **Keronopsis** Penard. Two ventral rows of cirri reaching frontal field; caudals variable; macronucleus usually in several (rarely two) parts; fresh or salt water. (Numerous species.)

K. rubra (Ehrenberg) (Fig. 363, *c*). Reddish; 200-300μ long; salt water.

Genus **Epiclintes** Stein. Elongate; spoon-shaped; flattened ventral surface with more than two rows of cirri; two marginal rows; frontals undifferentiated; anals; no caudals; salt or fresh water. (A few species.)

E. pluvialis Smith (Fig. 363, *d*). About 375μ long; fresh water.

Genus **Holosticha** Wrzesniowski. Three frontals along anterior margin; two ventral and two marginal rows of cirri; anals; fresh or salt water. (Numerous species).

H. vernalis Stokes (Fig. 363, *e*). Seven anals; about 180μ long; shallow pools with algae.

H. hymenophora S. (Fig. 363, *f*). Five anals; two contractile vacuoles; 160-200μ long; shallow pools.

Genus **Paraholosticha** Kahl. Elongate-oval; flexible; ventral cirri in two parallel oblique rows; with a row of stiff cirri along frontal margin, posterior to it two short rows of cirri; marginals continuous or interrupted at posterior border; fresh water.

P. hebicola K. (Fig. 363, *g*). 150-190μ long; fresh water among algae.

Genus **Trichotaxis** Stokes. Similar to *Holosticha;* but with three rows of ventral cirri; fresh or salt water.

T. stagnatilis S. (Fig. 363, *h*). About 160μ long; ellipsoid; in fresh water among decaying vegetation.

Genus **Balladyna** Kowalewski. Ellipsoid; frontals not well developed or lacking; one ventral and two marginal rows of cirri; long dorsal and lateral stiff cirri; fresh water.

B. elongata Roux (Fig. 363, *i*). 32-35μ by 11-12μ; fresh water among plants and detritus.

Genus **Pleurotricha** Stein. Oblong to ellipsoid; marginals continuous; eight frontals; three to four ventrals; seven anals of which two are more posterior; two rows of ventral cirri; between ventrals and marginals one to three rows of few coarse cilia; fresh water.

P. lanceolata (Ehrenberg) (Fig. 363, *j*). 100-165µ long; 2 macro- and 2 micro-nuclei. Manwell (1928) studied its conjugation, division, encystment and nuclear variation. (Encystment, Penn, 1935; excystment, Jeffries, 1956.)

Genus **Gastrostyla** Engelmann. Frontals distributed except three along the frontal margin; ventrals irregular; five anals; macronucleus divided into two to eight parts; fresh or salt water. (Morphology and physiology, Weyer, 1930.)

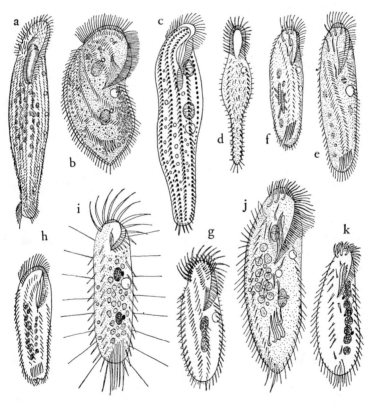

FIG. 363. a, *Urostyla caudata*, ×90 (Stokes); b, *Kerona polyporum*, ×200 (Stein); c, *Keronopsis rubra*, ×270 (Entz); d, *Epiclintes pluvialis*, ×100 (Smith); e, *Holosticha vernalis*, ×220 (Stokes); f, *H. hymenophora*, ×180 (Stokes); g, *Paraholosticha herbicola*, ×200 (Kahl); h, *Trichotaxis stagnatilis*, ×190 (Stokes); i, *Balladyna elongata*, ×800 (Roux); j, *Pleurotricha lanceolata*, ×250 (Stein); k, *Gastrostyla muscorum*. ×200 (Kahl).

G. muscorum Kahl (Fig. 363, *k*). 130-200μ long; macronucleus in eight parts; fresh water in vegetation.

Genus **Stylonychia** Ehrenberg. Ovoid to reniform; not flexible; ventral surface flat, dorsal surface convex; eight frontals; five ventrals; five anals; marginals; three caudals; with short dorsal bristles; fresh or salt water. (Many species).

S. mytilus (Müller) (Fig. 364, *a*). 100-300μ long; fresh, brackish and salt water. (Encystment, von Brand, 1923; vitamins and hormones on division, Lichtenberg, 1955.)

S. pustulata E. (Figs. 95; 364, *b*). About 150μ long; fresh water. (Cytology, Hall, 1931; division and reorganization, Summers, 1935; enzymes, Hunter, 1959.)

S. putrina Stokes (Fig. 364, *c*). 125-150μ long; fresh water. (Mating types, Downs, 1959.)

S. notophora S. (Fig. 364, *d*). About 125μ long; standing water.

Genus **Onychodromus** Stein. Not flexible; somewhat rectangular; anterior end truncate, posterior end rounded; ventral surface flat; dorsal surface convex; peristome broadly triangular in ventral view; three frontals; three rows of cirri parallel to the right edge of peristome; five to six anals; marginals uninterrupted; four to eight macronuclei; contractile vacuole; fresh water. (One species.)

O. grandis S. (Fig. 364, *e*). 100-300μ long.

Genus **Onychodromopsis** Stokes. Similar to *Onychodromus;* but flexible; six frontals of which the anterior three are the largest; fresh water. (One species.)

O. flexilis S. (Fig. 364, *f*). 90-125μ long; standing pond water.

Family 2. **Euplotidae** Ehrenberg

Peristome large with well-developed adoral zone; ventrals in group and reduced; anals of five cirri conspicuous.

Genus **Euplotes** Ehrenberg. Inflexible body ovoid; ventral surface flattened, dorsal surface convex; longitudinally ridged; peristome broadly triangular; frontal part of adoral zone lies in flat furrow; nine or more frontal-ventrals; five anals; four scattered caudals; macronucleus band-like; a micronucleus; contractile vacuole posterior; fresh or salt water. (Comparative morphology,

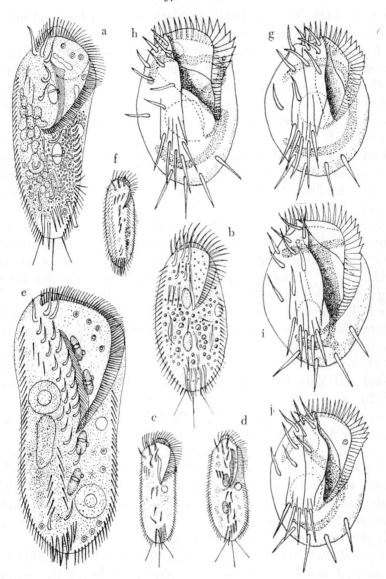

FIG. 364. a, *Stylonychia mytilus,* ×200 (Stein); b, *S. pustulata,* ×400 (Roux); c, *S. putrina,* ×200 (Stokes); d, *S. notophora,* ×200 (Stokes); e, *Onychodromus grandis,* ×230 (Stein); f, *Onychodromopsis flexilis,* ×240 (Stokes); g, *Euplotes patella,* ×420 (Pierson); h, *E. eurystomus,* ×330 (Pierson); i, *E. woodruffi,* ×310 (Pierson); j, *E. aediculatus,* ×290 (Pierson).

Pierson, 1943, Tuffrau, 1960; symbiotic bacteria, Fauré-Fremiet, 1952; marine species, Borror, 1962.)

E. *patella* (Müller) (Fig. 364, g). Subcircular to elliptical; average dimensions 91µ by 52µ; nine frontral-ventrals; aboral surface with six prominent ridges with rows of bristles embedded in rosettes of granules; peritome narrow; peristomal plate small triangle; macronucleus simple C-form band; micronucleus near anterior-left end; membranellae straight; posterior end of cytopharynx anterior to, and to left of, the fifth anal cirrus; post-pharyngeal sac; fresh and brackish water. (Doubles and amicronucleates, Kimball, 1941; mating type (p. 232); nutrition, Lily and Henry, 1956.)

E. *eurystomus* Wrzesniowski (Fig. 364, h). Elongated ellipsoid; length 100-195µ; average dimensions 138µ by 78µ; nine frontal-ventrals; no aboral ridges, but seven rows of bristles; peristome wide, deep; peristomal depression sigmoid; membranellae forming sigmoid curve; end of cytopharynx far to left and anterior to the fifth anal cirrus; post-pharyngeal sac; macronucleus 3-shaped; micronucleus near flattened anterior corner of macronucleus; fresh and brackish water. (Division and conjugation, Turner, 1930; neuromotor system, Turner, 1933, Hammond, 1937, Hammond and Kofoid, 1937; DNA synthesis, Kimball and Prescott, 1962.)

E. *woodruffi* Gaw (Fig. 364, i). Oval; length 120-165µ; average dimensions 140µ by 90µ; nine frontal-ventrals; aboral surface often with eight low ridges; peristome wide, with a small peristomal plate; end of cytopharynx almost below the median ridge; fourth ridge between anal cirri often extends to anterior end of body; post-pharyngeal sac; macronucleus consistently T-shaped; micronucleus anterior-right; brackish (with salinity 2.30 parts of salt per 1000) and fresh water (Gaw, 1939).

E. *aediculatus* Pierson (Fig. 364, j). Elliptical; length 110-165µ; average dimensions 132µ by 84µ; nine frontal-ventrals; aboral surface usually without ridges, but with about six rows of bristles; peristome narrow; peristomal plate long triangular, drawn out posteriorly; a niche midway on the right border of peristome; anal cirri often form a straight transverse line; fourth ridge between anals may reach anterior end of body; macronucleus C-shape with a flattened part in the left-anterior region; micronucleus some distance from macronucleus at anterior-left region; post-pharyngeal sac; fresh and brackish (salinity 2.30 parts of salt per 1000) water.

E. plumipes Stokes. Similar to *E. eurystomus*. About 125μ long; fresh water.

E. carinatus S. (Fig. 365, *a*). About 70μ by 50μ; fresh water.

E. charon (Müller) (Fig. 365, *b*). 70-90μ long; salt water.

E. leticiensis Bovee (1957). Oval, with a triangular wing on left; 87-203(155)μ by 67-194(130)μ; two right caudals furcated (Bovee).

Genus **Euplotidium** Noland. Cylindrical; nine frontal-ventrals in two rows toward right; five anals; a groove extends backward from oral region to ventral side, in which the left-most anal cirrus lies; peristome opened widely at anterior end, but covered posteriorly by a transparent, curved, flap-like membrane; adoral zone made up of about eighty membranellae; longitudinal ridges (carinae), three dorsal and two lateral; a row of protrichocysts under each carina; a broad zone of protrichocysts in antero-dorsal region; cytoplasm densely granulated; salt water. (One species, Noland, 1937.)

E. agitatum N. (Fig. 365, *c, d*). 65-95μ long; erratic movement rapid; observed in half-dead sponges in Florida.

Genus **Certesia** Fabre-Domergue. Ellipsoid; flattened; dorsal surface slightly convex, ventral surface flat or concave; five frontals at anterior border; seven ventrals; five anals; no caudals; marginals small in number; four macronuclei; salt water. (One species.)

C. quadrinucleata F.-D. (Fig. 365, *e*). 70-100μ by about 45μ. (Morphology, Sauerbrey, 1928.)

Genus **Diophrys** Dujardin. Peristome relatively large, often reaching anals; seven to nine frontal-ventrals; five anals; three strong cirri right-dorsal near posterior margin; salt water.

D. appendiculata (Ehrenberg) (Fig. 365, *f*). 60-100μ long; salt water; Woods Hole (Calkins). (Division and reorganization, Summers, 1935.)

Genus **Uronychia** Stein. Without frontals and ventrals; five anals; three right-dorsal cirri (as in *Diophrys*); two left-ventral cirri near posterior margin; peristome, oval with a large undu-

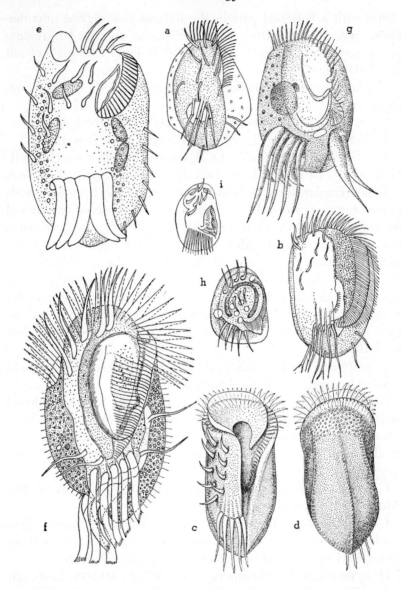

FIG. 365. a, *Euplotes carinatus*, ×430 (Stokes); b, *E. charon*, ×440 (Kahl); c, d, *Euplotidium agitatum*, ×540 (Noland); e, *Certesia quadrinucleata*, ×670 (Sauerbrey); f, *Diophrys appendiculata*, ×570 (Wallengren); g, *Uronychia setigera*, ×870 (Calkins); h, *Aspidisca lynceus*, ×300 (Stein); i, *A. polystyla*, ×290 (Kahl).

lating membrane on right edge; salt water. (Several species.)

U. setigera Calkins (Fig. 365, *g*). 40μ by 25μ; salt water; Woods Hole.

Genus **Gastrocirrhus** Lepsi. Anterior end truncate with a ring of cilia; posterior end bluntly pointed; slightly flattened; a wide peristome leading to cytostome, with undulating membrane on left; sixteen cirri on ventral surface arranged on right and posterior margins (Lepsi) or six frontrals, five ventrals, five caudals (Bullington); marine. Apparently intermediate between Heterotrichida and Hypotrichida (Lepsi).

G. stentoreus Bullington (Fig. 366, *a*). About 104μ by 71-81μ; dark granulated cytoplasm; active jumping as well as swimming movement; in Tortugas (Bullington, 1940).

G. adhaerens Fauré-Fremiet. Height and peristome diameter about 100μ; marine.

Family 3. **Paraeuplotidae** Wichterman

Well-developed adoral zone; cirri reduced to caudals; with a few tufts of cirri.

Genus **Paraeuplotes** Wichterman. Ovoid; ventral surface slightly concave, dorsal surface highly convex and bare, but with one ridge; frontal and adoral zones well developed; ventral surface with a semicircular ciliary ring on the right half, posterior half of which is marked by a plate and with two ciliary tufts, near the middle of anterior half; five to six caudal cirri; macronucleus curved band-form; a terminal contractile vacuole; zooxanthellae, but no food vacuole in cytoplasm; marine, on the coral.

P. tortugensis W. (Fig. 366, *b*, *c*). Subcircular to ovoid; average individuals 85μ by 75μ; ciliary plate 37μ long, with longer cilia; adoral zone reaches nearly the posterior end; "micronucleus not clearly differentiated" (Wichterman); five to six caudal cirri about 13μ long; zooxanthellae yellowish brown, about 12μ in diameter, fill the body; found on *Eunicea crassa* (coral); Tortugas, Florida.

Genus **Euplotaspis** Chatton and Seguela. Ellipsoid; ventral surface flat or slightly concave; dorsal surface convex; membranellae and cirri with fringed tips; peristome very long; ten frontal-ventrals; five anals; three or four caudals difficult to see in life; dorsal surface without striae or ciliary processes; macronucleus arched band; a single micronucleus. (One species, Chatton and Seguela, 1936.)

E. cionaecola C. and S. (Fig. 366, *d*). 60-70μ by 45-55μ; in the branchial cavity of the ascidian, *Ciona intestinalis.*

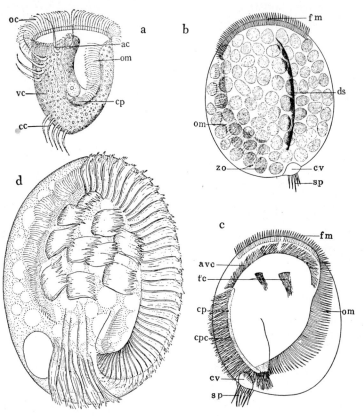

Fɪɢ. 366. a, ventral view of *Gastrocirrhus stentoreus,* ×330 (Bullington) (ac, anterior cirri; cc, caudal cirri; cp, cytopharynx; oc, oral cilia; om, oral membrane; vc, ventral cirri); b, c, dorsal and ventral views of *Paraeuplotes tortugensis,* ×490 (Wichterman) (avc, anterior ventral cilia; cp, ciliary plate; cpc, ciliary plate cilia; cv, contractile vacuole; ds, dorsal swelling; fm, frontal membranellae; om, adoral membranellae; sp. caudel cirri; tc, tufts of cilia; zo, zoothanthellae); d, ventral view of *Euplotaspis cionaecola,* ×1285 (Chatton and Seguela).

Family 4. **Aspidiscidae** Stein

Adoral zone poorly developed; cirri reduced in number and are limited to frontals and anals.

Genus **Aspidisca** Ehrenberg. Small; ovoid; inflexible; right and dorsal side convex, ventral side flattened; dorsal surface conspicuously ridged; adoral zone reduced or rudimentary; seven frontalventrals; five to twelve anals; macronucleus horseshoe-shaped or occasionally in two rounded parts; contractile vacuole posterior; fresh or salt water. (Numerous species.)

A. lynceus E. (Figs. 58; 365, *h*). 30-50μ long; fresh water. (Division and reorganization, Summers, 1935).

A. polystyla Stein (Fig. 365, *i*). About 50μ long; marine; Woods Hole (Calkins).

Order 6. **Odontostomatida** Sawaya
(*Ctenostomatida* Lauterborn)

The small carapaced ciliates placed in this order are laterally compressed, and wedge-shaped, with a sparse ciliation. The adoral membranelles are reduced to eight. All are polysaprobic; majority are fresh water inhabitants. This order had been known as Ctenostomata but as pointed out by Corliss, Sawaya had found that the name was preoccupied and proposed Odontostomatida for it. (Three families.)

Family 1. **Epalxellidae** Corliss

Anterior row of cilia on the left side; four posterior rows on left and at least two posterior rows on right side.

Genus **Epalxella** C. (*Epalxis*, Roux). Rounded triangular; anterior end pointed toward ventral surface, posterior end irregularly truncate; dorsal surface more convex; right carapace with one dorsal and one ventral ciliary row in posterior region; usually four (2-3) median teeth; all anal teeth without spine; with comb-like structures posterior to oral aperture; one to two oval macronuclei dorsal; contractile vacuole posterior-ventral; sapropelic in fresh or salt water. (Many species.)

E. mirabilis (Roux) (Fig. 367, *a*). 38-45μ by 27-30μ; fresh water.

Genus **Saprodinium** Lauterborn. Similar to *Epalxella;* but some (left and right) of anal teeth with spines; sapropelic in fresh or salt water. (Several species.)

S. dentatum L. (Fig. 367, *b*). 60-80μ long; fresh water (Lackey, 1925).

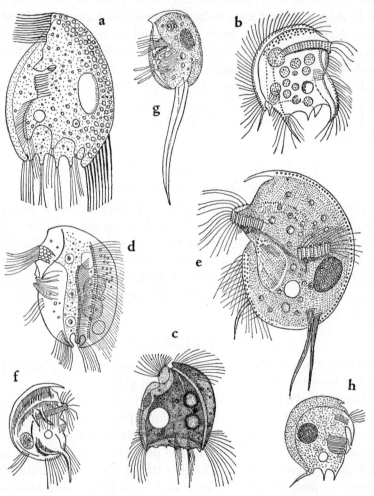

Fig. 367. a, *Epalxis mirabilis*, ×1200 (Roux); b., *Saprodinium dentatum*, ×430 (Kahl); c, *S. putrinium*, ×470 (Lackey); d, *Pelodinium reniforme*, ×600 (Lauterborn); e, f, *Discomorpha pectinata*, (e, ×500; f, ×220) (Kahl); g, *Mylestoma bipartitum*, ×470 (Kahl); h, *Atopodinium fibulatum*, ×520 (Kahl).

S. putrinium Lackey (Fig. 367, *c*). 50μ long, 40μ wide, about 15μ thick; in Imhoff tanks.

Genus **Pelodinium** Lauterborn. Right carapace with two median rows of cilia, its median anal teeth fused into one so that there are only three teeth. (One species.)

P. reniforme L. (Fig. 367, *d*). 40-50μ long; sapropelic.

Family 2. **Discomorphellidae** Corliss

The dorsal keel ends anteriorly in a spine and posteriorly continues down to the ventral surface; body ciliation consists of two ventral rows and two tufts of cilia on posterior left side.

Genus **Discomorphella** Corliss (*Discomorpha* Levander). Oval; ventrally directed anterior spine long; posterior end without teeth or ridges; ciliated bands on both lateral surfaces; two spines on right side; two cirrus-like groups on posterior-left; sapropelic. (A few species.)

D. pectinata (L.) (Fig. 367, *e, f*). 70-90μ long; sapropelic.

Family 3. **Mylestomidae** Kahl

Body ciliation is much reduced; conspicuous posterior tufts of cilia.

Genus **Mylestoma** Kahl. Posterior margin without any indentation, though sometimes a small one on right side, but none on left; three often long ribbon-like cirri on peristome; fresh or salt water. (Several species.)

M. bipartitum (Gourret and Roesner) (Fig. 367, *g*). 35-50μ long; two caudal processes; salt water.

Genus **Atopodinium** Kahl. Posterior left side with one large, and right side with two indentations; macronucleus spherical; sapropelic.

A. fibulatum K. (Fig. 367, *h*). 40-50μ long.

References

Borror, A. C., 1962. *Euplotes minuta* Yocum. J. Protozool., 9:271.

Bovee, E. C., 1957. *Euplotes leticiensis* n. sp., etc. *Ibid.*, 4:124.

Brand, T. v., 1923. Die Encystierung bei *Vorticella microstoma* und hypotrichen Infusorien Arch. Protist., 47:59.

Bullington, W. E., 1940. Some ciliates from Tortugas. Carnegie Inst. Wash. Publ. No. 517.

CALKINS, G. N., 1902. Marine Protozoa from Woods Hole. Bull. U. S. Fish Comm., 21:415.

——— 1930. *Uroleptus halseyi.* II. Arch. Protist., 69:151.

CHATTON, E., and SEGUELA, J., 1936. Un hypotriche de la branchie de *Ciona intestinalis,* etc. Bull. soc. zool. France, 61:232.

DOWNS, L. E., 1959. Mating types and their determination in *Stylonychia putrina.* J. Protozool., 6:285.

FAURÉ-FREMIET, E., 1952. Symbiontes bactériens des ciliés du genre Euplotes. C. R. Acad. Sci., 235:402.

——— 1954. *Gastrocirrhus adhaerens* n. sp. Ann. Acad. Brasil. Ciencias, 26:163.

——— 1954a. *Amphisiella lithophora* n. sp., etc. Bull. Soc. Zool. France, 79:473.

FROUD, J., 1949. Observations on hypotrichous ciliates: the genera Stichotricha and Chaetospira. Quart. J. Micr. Sci., 90:141.

GAW, H. Z., 1939. *Euplotes woodruffi* sp. n. Arch. Protist., 93:1.

HALL, R. P., 1931. Vacuome and Golgi apparatus in the ciliate, Stylonychia. Ztschr. Zellforsch. mikr. Anat., 13:770.

HAMMOND, D. M., 1937. The neuromotor system of *Euplotes patella* during binary fission and conjugation. Quart. J. Micr. Sci., 79:507.

——— and KOFOID, C. A., 1937. The continuity of structure and function in the neuromotor system of *Euplotes patella,* etc. Proc. Am. Philos. Soc., 77:207.

HASHIMOTO, K., 1961. Stomatogenesis and formation of cirri in fragments of *Oxytricha fallax.* J. Protozool., 8:433.

HORVÁTH, J. v., 1933. Beiträge zur hypotrichen Fauna der Umgebung von Szeged. I. Arch. Protist., 80:281.

HUNTER, N. W., 1959. Enzyme systems of *Stylonychia pustulata.* II. J. Protozool., 6:100.

KAHL, A., 1932. Ctenostomata (Lauterborn) n. Subord. Arch. Protist., 77:231.

——— 1932a. Urtiere oder Protozoa. Dahl's Die Tierwelt Deutschlands, etc. Part 25.

KENT, S., 1881-1882. A manual of Infusoria. London.

KIMBALL, R. F., and PRESCOTT, D. M., 1962. DNA synthesis and distribution during growth and amitosis of the macronucleus of Euplotes. J. Protozool., 9:88.

LEPSI, J., 1928. Un nouveau protozoaire marine: *Gastrocirrhus intermedius.* Ann. Protist., 1:195.

LICHTENBERG, E., 1955. Die Beeinflussung der Lebensfaehigkeit und Teilungrate von *Stylonychia mytilus* durch Aussenfactoren und Wirkstoffe. Arch. Protist., 100:395.

LILLY, D. M., and HENRY, S. M., 1956. Supplementary factors in the nutrition of Euplotes. J. Protozool., 3:200.

LUND, E. E., 1935. The neuromotor system of Oxytricha. J. Morphol., 58: 257.

MANWELL, R. D., 1928. Conjugation, division and encystment in *Pleurotricha lanceolata*. Biol. Bull., 54:417.

MERRIMAN, D., 1937. Description of *Urostyla polymicronucleata*. Arch. Protist., 88:427.

NOLAND, L. E., 1937. Observations on marine ciliates of the Gulf coast of Florida. Tr. Am. Micr. Soc., 56:160.

PENN, A. B. K., 1935. Factors which control encystment in *Pleurotricha lanceolata*. Arch. Protist., 84:101.

PIERSON, B. F., 1943. A comparative morphological study of several species of Euplotes, etc. J. Morphol., 72:125.

RAABE, H., 1946. L'appareil nucléaire d'*Urostyla grandis*. I. Ann. Univ. Marie Curie-Sklodowska, 1:1.

────── 1947. II. *Ibid.*, 1:133.

REYNOLDS, M. E. C., 1932. Regeneration in an amicronucleate infusorian. J. Exper. Zool., 62:327.

ROUX, J., 1901. Faune infusorienne des eaux stagnantes des environs de Genevè. Mém. cour. l'Uni. Genevè, Geneva.

SAUERBREY, E., 1928. Beobachtungen über einige neue oder wenig bekannte marine Ciliaten. Arch. Protist., 62:353.

STEIN, F., 1867. Der Organismus der Infusionstiere. Vol. 2.

STOKES, A. C., 1888. A preliminary contribution toward a history of the freshwater Infusoria of the United States. J. Trenton Nat. Hist. Soc., 1:71.

SUMMERS, F. M., 1935. The division and reorganization of the macronuclei of *Aspidisca lynceus*, etc. Arch. Protist., 85:173.

TUFFRAU, M., 1960. Revision de genre Euplotes, fondee sur la comparison des structures superficielles. Hydrobiologia, 15:1.

TURNER, J. P., 1930. Division and conjugation in *Euplotes patella*, etc. Univ. Cal. Publ. Zool., 33:193.

────── 1939. A new species of hypotrichous ciliate, *Urostyla coei*. Tr. Am. Micr. Soc., 58:395.

WALLENGREN, H., 1900. Studier öfver Ciliata infusorier. IV. Kongl. Fysio. Säll. Handl., 11:2:1.

WEYER, G., 1930. Untersuchungen über die Morphologie und Physiologie des Formwechsel der *Gastrostyla steini*. Arch. Protist., 71:139.

WICHTERMAN, R., 1942. A new ciliate from a coral of Tortugas, etc. Carnegie Inst. Wash. Publ., 524:105.

Chapter 41

Subclass 3. **Peritricha** Stein

THE peritrichs show a much enlarged disc-like anterior region with prominent adoral ciliature which winds counterclockwise to the cytostome. Mature individuals are without body ciliation and usually attached to substrate by a stalk formed by a scopula. Telotrochs (Fig. 368) or free-swimming larvae are provided with a ciliary girdle. Asexual reproduction is by binary fission and sexual reproduction by conjugation (p. 238) or autogamy. Large colonial forms occur. (Taxonomy, Kahl, 1935, Stiller, 1939, 1940, Nenninger, 1948, Mattes, 1950, Kralik, 1961; structure of stalk, Precht, 1935, Rouiller *et al.*, 1956.)

Order **Peritrichida** Stein

The order is divided into two suborders: Sessilina and Mobilina.

Suborder 1. **Sessilina** Kahl

Mature individuals attached to the substrate with contractile or non-contractile stalks; no body ciliature; telotroch formation; often colonial. (Seven families.)

Family 1. **Vorticellidae** Ehrenberg

Bell-shaped body; with contractile stalks; sessile when mature; often colonial.

Genus **Vorticella** Linnaeus. Inverted bell-form; colorless, yellowish, or greenish; peristome more or less outwardly extended; pellicle sometimes annulated; with a contractile stalk, macronucleus band-form; micronucleus; one to two contractile vacuoles; solitary; in fresh or salt water, attached to submerged objects and aquatic plants or animals. Numerous species. (Taxonomy, Noland and Finley, 1931, Kahl, 1935, Nenninger, 1948; movements of food vacuoles, Hall and Dunihue, 1931.)

V. campanula Ehrenberg (Fig. 368, *a-c*). Usually in groups; endoplasm filled with refractile reserve granules; vestibule very

1016

large with an outer pharyngeal membrane; 50-157μ by 35-99μ; peristome 60-125μ wide; stalk 50-415μ by 5.6-12μ; fresh water.

V. convallaria (L.) (Fig. 368, *d, e*). Resembles the last-named species; but anterior end somewhat narrow; usually without re-fractile granules in endoplasm; 50-95μ by 35-53μ; peristome 55-75μ wide; stalk 25-460μ by 4-6.5μ; fresh water. Axenic culture (Levine, 1959).

V. microstoma Ehrenberg (Figs. 88; 368, *f-p*). 35-83μ by 22-50μ; peristome 12-25μ wide; stalk 20-385μ by 1.5-4μ; com-

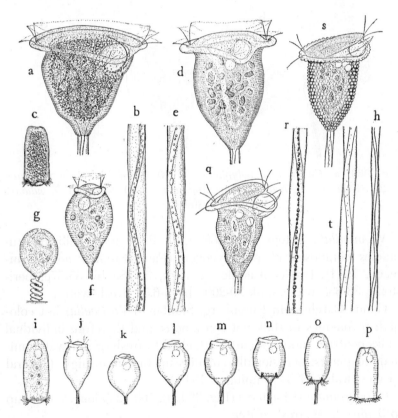

FIG. 368. a–c, *Vorticella campanula* (a, ×400; b, part of stalk, ×800; c, telotroch, ×200); d, e, *V. convallaria* (d, ×400; e, ×800); f-p, *V. micro-stoma* (f, g, ×400; h, ×840; i, telotroch, ×400; j–p, telotroch-formation *in vitro*, ×270); q, r, *V. picta* (q, ×400; r, ×800); s, t, *V. monilata* (s, ×400; t, ×800) (Noland and Finley).

mon in fresh-water infusion. (Conjugation, Finley, 1943; encystment, Brand, 1923; chromatographic analysis, Finley and Williams, 1955; cultures, Finley *et al.*, 1959.)

V. *picta* (E.) (Fig. 368, *q, r*). 41-63μ by 20-37μ; peristome 35-50μ; stalk 205-500μ by 4-7μ; two contractile vacuoles; with refractile granules in stalk; fresh water.

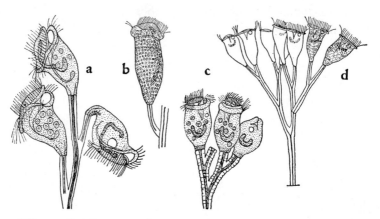

Fɪɢ. 369. a, *Carchesium polypinum*, ×200 (Stein); b, *C. granulatum*, ×220 (Kellicott); c, *Zoothamnium arbuscula*, ×200 (Stein); d, *Z. adamsi*, ×150 (Stokes).

V. *monilata* Tatem (Fig. 368, *s, t*). Body with pellicular tubercles composed of paraglycogen (Fauré-Fremiet and Thaureaux, 1944); two contractile vacuoles; 50-78μ by 35-57μ; peristome 36-63μ wide; stalk 50-200μ by 5-6.5μ; fresh water.

Genus **Carchesium** Ehrenberg. Similar to *Vorticella;* but colonial; myonemes in stalk not continuous, and therefore individual stalks contract independently; attached to fresh or salt water animals or plants; occasionally colonies up to 4 mm. high. (Several species, Kahl, 1935; Nenninger, 1948.)

C. *polypinum* (Linnaeus) (Fig. 369, *a*). 100-125μ long; colony up to 3 mm. long; fresh water.

C. *granulatum* Kellicott (Fig. 369, *b*). About 100μ long; two contractile vacuoles anterior; on Cambarus and aquatic plants.

Genus **Zoothamnium** Bory. Similar to *Carchesium;* but myonemes (Fig. 15) of all stalks of a colony are continuous with one another, so that the entire colony contracts or expands simultaneously; fresh or salt water; colonies sometimes several millimeters high. (Numerous species, Kahl, 1935; Nenninger, 1948; development, Summers, 1938, 1938a.)

Z. arbuscula Ehrenberg. (Fig. 369, *c*). 40-60μ long; colony up to more than 6 mm. high; fresh water. (Morphology and life cycle, Furssenko, 1929.)

Z. adamsi Stokes (Fig. 369, *d*). About 60μ long; colony about 250μ high; attached to Cladophora.

Family 2. **Epistylidae** Kent

Stalk not contractile; solitary or colonical.

Genus **Epistylis** Ehrenberg. Inverted bell-form; individuals usually on dichotomous non-contractile stalk, forming large colonies; attached to fresh or salt water animals. (Numerous species, Nenninger 1948; conversion of DNA to RNA, Seshachar and Dass, 1953.)

E. plicatilis E. (Fig. 370, *a*). 110-162μ long (Nenninger); colony often up to 3 mm. high; in fresh water.

E. fugitans Kellicott. 50-60μ long; attached to Sida in early spring.

E. cambari K. (Fig. 370, *b*, *c*). About 50μ long; attached to the gills of Cambarus.

E. niagarae K. (Fig. 370, *d*). Expanded body about 160μ long; peristomal ring prominent; flat cap makes a slight angle with the ring; bandform macronucleus transverse to long axis, in the anterior third; gullet with ciliated wall; forty to fifty in a colony; attached to the antennae and body surface of crayfish (Kellicott, 1883) or to painted and snapping turtles (Bishop and Jahn, 1941).

Genus **Rhabdostyla** Kent. Similar to *Epistylis;* but solitary with a non-contractile stalk; attached to aquatic animals in fresh or salt water. (Numerous species, Nenninger, 1948.)

R. vernalis Stokes (Fig. 370, *e*). About 50μ long; attached to Cyclops and Cypris in pools in early spring. (Sexual differentiation, Finley, 1952.)

Genus **Opisthostyla** Stokes. Similar to *Rhabdostyla;* but stalk long, is bent at its point of attachment to submerged object, and acts like a spring; fresh or salt water (Nenninger, 1948).

O. annulata S. (Fig. 370, *f*). Body about 23µ long; fresh water.

Genus **Campanella** Goldfuss. Similar to *Epistylis;* but adoral double zone turns four to six times; fresh water.

C. umbellaria (Linnaeus) (Fig. 370, *g*). Colony may reach several millimeters in height; individuals 130-250µ long (Kent).

Genus **Pyxidium** Kent. Stalk simple, not branching; peristome even when fully opened, not constricted from the body proper;

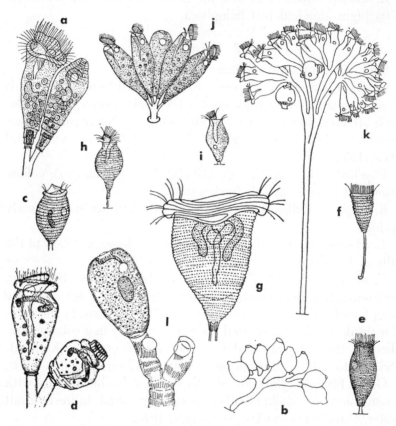

FIG. 370. a, *Epistylis plicatilis*, ×200 (Stein); b, c, *E. cambari* (Kellicott) (b, ×140; c, ×340); d, *E. niagarae*, ×150 (Bishop and Jahn); e, *Rhabdostyla vernalis*, ×320; f, *Opisthostyla annulata*, ×440 (Stokes); g, *Campanella umbellaria*, ×180 (Schröder); h, *Pyxidium vernale*, ×240; i, *P. urceolatum*, ×140 (Stokes); j, *Opercularia stenostoma*, ×140 (Udekem), k, *O. plicatilis*, ×40 (Stokes); 1, *Operculariella parasitica*, ×245 (Stammer).

frontal disk small, oblique, supported by style-like slender process arising from peristome; attached to freshwater animals and in vegetation. (Taxonomy, Nenninger.)

P. vernale Stokes (Fig. 370, *h*). Solitary or few together; 70-85µ long; fresh water among algae.

P. urceolatum S. (Fig. 370, *i*). About 90µ long; fresh water on plants.

Genus **Opercularia** Stein. Individuals similar to *Pyxidium;* but short stalk dichotomous; peristome border like a band.

O. stenostoma S. (Fig. 370, *j*). When extended, up to 125µ long; attached to *Asellus aquaticus* and others.

O. plicatilis Stokes (Fig. 370, *k*). About 254µ long; colony 1.25-2.5 mm. high; pond water.

Genus **Operculariella** Stammer. Fixed stalk, branched, short and rigid; peristome small, without border, smooth; without disk or frontal cilia; vestibule large (Stammer, 1948).

O. parasitica S. (Fig. 370, *l*). 100-110µ long; barrel-shaped; peristome opening only one-fourth the body breadth; macronucleus about 30µ long; parasitic in the oesophagus of *Dytiscus marginalis, Acilius sulcatus, Hydaticus transversalis, Graphoderes zonatus* and *G. bilineatus.*

Family 3. **Scyphidiidae** Kahl

Without stalk, but attached to substrate by scopula.

Genus **Scyphidia** Dujardin. Cylindrical; posterior end attached to submerged objects or aquatic animals; body usually cross-striated; fresh or salt water. (Species, Nenninger, 1948; Hirshfield, 1949.)

S. amphibiarum Nenninger (Fig. 371, *a*). On tadpoles; about 76µ long.

S. ubiquita Hirshfield. In mantle cavities of various species of limpets and turbans; about 80µ by 30µ when expanded (Hirshfield, 1949).

Genus **Paravorticella** Kahl. Similar to *Scyphidia;* but posterior portion is much elongated and contractile; salt water, attached or parasitic.

P. clymenellae (Shumway) (Fig. 371, *b*). 100µ long; in the colon of the annelid, *Clymenella torquata;* Woods Hole.

Genus **Glossatella** Bütschli. With a large adoral membrane; often attached to fish and amphibian larvae.

G. tintinnabulum (Kent) (Fig. 371, *c*). 30-43µ long; attached to the epidermis and gills of young Triton.

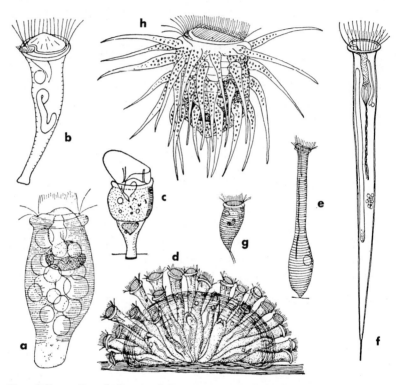

FIG. 371. a, *Scyphidia amphibiarum,* ×570 (Nenninger); b, *Paravorticella clymenellae,* ×65 (Shumway); c, *Glossatella tintinnabulum,* ×610 (Penard); d, *Ophridium sessile,* ×65 (Kent); e, *O. vernalis,* ×160 (Stokes); f, *O. ectatum,* ×160 (Mast); g, *Astylozoon fallax,* ×170 (Engelmann); h, *Hastatella aesculacantha,* ×580 (Jarocki).

Genus **Ambiphrya** Raabe. Cylindrical; macronucleus a long ribbon; on gills of fishes.

A. miri R. 45µ by 25µ; on the gills of *Nerophis ophidion* (Raabe, 1952).

Genus **Ellobiophrya** Chatton and Lwoff. Posterior end drawn out into two arm-like processes by means of which the organism holds fast to the gill bars of the mussel, *Donax vittatus.* (One species.)

E. donacis C. and L. (Fig. 372, *a*). 50μ by 40μ excluding the processes.

Family 4. **Ophrydiidae** Kent

Elongate body with a long "neck"; solitary or colonial within a gelatinous mass; often with zoochlorellae.

Genus **Ophrydium** Ehrenberg (*Gerda* Claparède and Lachmann). Cylindrical with a contractile neck; posterior end pointed or rounded; variable number of individuals in a common mucilaginous mass; pellicle usually cross-striated; fresh water.

O. sessile Kent. (Fig. 371, *d*). Fully extended body up to 300μ long; colorless or slightly brownish; ovoid colony up to 5 mm. by three mm.; attached to freshwater plants.

O. vernalis (Stokes) (Fig. 371*e*). About 250μ long; highly contractile; in shallow freshwater ponds in early spring (Stokes).

O. ectatum Mast (Fig. 371, *f*). 225-400μ long; with many zoochlorellae; colony up to 3 mm. in diameter; in fresh water (Mast, 1944).

Family 5. **Astylozoonidae** Kahl

Without stalk and not sessile; swimming with peristome-bearing end forward.

Genus **Astylozoon** Engelmann (*Geleiella* Stiller). Free-swimming; pyriform or conical; aboral end attenuated, with one to two thigmotactic stiff cilia; pellicle smooth or furrowed; with or without gelatinous envelop; in fresh water. (A few species.)

A. fallax E. (Fig. 371, *g*). 70-100μ; fresh water.

Genus **Hastatella** Erlanger. Free-swimming; body surface with 2-4 rings of long conical ectoplasmic processes; fresh water.

H. aesculacantha Jarocki and Jacubowska (Fig. 371, *h*). 30-52μ by 24-40μ; in stagnant water.

Genus **Telotrochidium** Kent (*Opisthonecta* Fauré-Fremiet). Roughly conical with rounded ends; a ring of long cilia close to the aboral end; two parallel rows of adoral zone; a papilla with about twelve long cilia, just above the opening into the vestibule; macronucleus sausage-form; three contractile vacuoles connected with the cytopharynx; fresh water. No stalked stages have been found; but some workers place this in Epistylidae.

T. henneguyi (F.-F.) (Fig. 372, *b, c*). 148-170μ long; cysts about 57μ in diameter; sometimes infected by a parasitic suctorian, *En-*

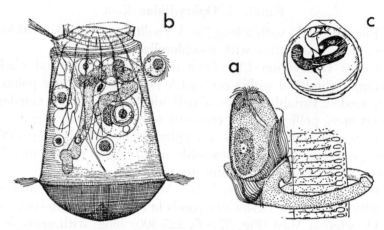

FIG. 372. a, *Ellobiophrya donacis,* ×900 (Chatton and Lwoff); b, c, *Telotrochidium henneguyi* (b, ×335 (Lynch and Noble); c, a cyst in life, ×340 (Rosenberg)).

dosphaera engelmanni (Lynch and Noble, 1931) (p. 1041). (Conjugation, Rosenberg, 1940; neuromotor system, Kofoid and Rosenberg, 1940; encystment, Rosenberg, 1938; nuclear changes in cysts, Dodd, 1962.)

Family 6. **Vaginicolidae** Kent

With a lorica; body attached by a short stalk to the base; body extends beyond lorica.

Genus **Vaginicola** Lamarck. Lorica without stalk, attached to substratum directly with its posterior end; body elongate and cylindrical; fresh or salt water. (Numerous species, Swarczewsky, 1930.)

V. leptosoma Stokes (Fig. 373, *a*). Lorica about 160μ high; when extended, about one-third of body protruding; on algae in pond water.

V. annulata S. (Fig. 373, *b*). Lorica about 120μ high; below middle, a ring-like elevation; anterior one-third of body protruding, when extended; pond water.

Genus **Cothurnia** Ehrenberg. Similar to *Vaginicola;* but lorica stands on a short stalk; fresh or salt water. (Numerous species, Swarczewsky, 1930.)

C. *canthocampti* Stokes (Fig. 373, *c*). Lorica about 80μ high; on *Canthocamptus minutus.*

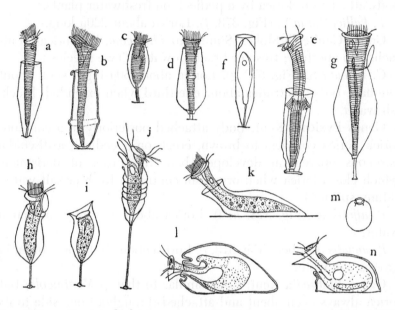

FIG. 373. a, *Vaginicola leptosoma,* ×130 (Stokes); b, *V. annulata,* ×170 (Stokes); c, *Cothurnia canthocampti,* ×150 (Stokes); d, *C. annulata,* ×340 (Stokes); e, *Thuricola folliculata,* ×110 (Kahl); f, *Thuricolopsis kellicottiana,* ×110 (Stokes); g, *Caulicola valvata,* ×760 (Stokes); h, i, *Pyxicola affinis,* ×170 (Kent); j, *P. socialis,* ×170 (Kent); k, *Platycola longicollis,* ×200 (De Fromentel); l, *Lagenophrys vaginicola,* ×380 (Penard); m, *L. patina,* ×150 (Stokes); n, *L. labiata,* ×340 (Penard).

C. *annulata* S. (Fig. 373, *d*). Lorica about 55μ high; fresh water.

C. *variabilis* Kellicott. On the gills of *Cambarus* spp. (Hamilton, 1952).

Genus **Thuricola** Kent. Body and lorica as in *Vaginicola;* but lorica with a simple or complex valve-like apparatus which closes obliquely after the manner of a door when protoplasmic body contracts; salt or fresh water.

T. folliculata (Müller) (Fig. 373, *e*). Lorica 127-170μ high (Kent); 160-200μ high (Kahl); salt and fresh water.

Genus **Thuricolopsis** Stokes. Loricia with an internal, narrow, flexible valve-rest, adherent to lorica wall and projecting across cavity to receive and support the descended valve; protoplasmic body attached to lorica by a pedicel; on freshwater plants.

T. kellicottiana S. (Fig. 373, *f*). Lorica about 220μ long.

Genus **Caulicola** Stokes. Similar to *Thuricola;* but lorica-lid attached to aperture; fresh or brackish water. (Two species.)

C. valvata S. (Fig. 373 *g*). Lorica about 50μ high; stalk about one-half; body protrudes about one-third when extended; brackish water.

Genus **Pyxicola** Kent. Body attached posteriorly to a corneous lorica; lorica colorless to brown, erect, on a pedicel; a discoidal corneous operculum developed beneath border of peristome, which closes lorica when organism contracts; fresh or salt water. (Many species.)

P. affinis K. (Fig. 373, *h, i*). Lorica about 85μ long; in marsh water.

P. socialis (Gruber) (Fig. 373, *j*). Lorica about 100μ long; often in groups; salt water.

Genus **Platycola** Kent. Body similar to that of *Vaginicola;* but lorica always decumbent and attached throughout one side to its fulcrum of support; fresh or salt water. (Many species.)

P. longicollis K. (Fig. 373, *k*). Lorica yellow to brown when older; about 126μ long; fresh water.

Family 7. **Lagenophryidae** Bütschli

With a lorica; with or without a short stalk; when extended the peristomal portion protrudes beyond lorica.

Genus **Lagenophrys** Stein. Lorica with flattened adhering surface, short neck and convex surface; "striped body" connects body with lorica near aperture; attached to fresh or salt water animals. (Many species, Swarczewsky, 1930; biology, Awerinzew, 1936.)

L. vaginicola S. (Fig. 373, *l*). Lorica 70μ by 48μ; attached to caudal bristles and appendages of *Cyclops minutus* and *Cantho-camptus* sp.

L. patina Stokes (Fig. 373, *m*). Lorica 55μ by 50μ; on Gam-marus.

L. labiata S. (Fig. 373, *n*). Lorica 60μ by 55μ; on Gammarus.

L. lunatus Imamura. Lorica 50-60μ in diameter; on *Leander paucidens* and *Palaemon varians* (Imamura, 1940). (Morphology and reproduction, Debaisieux, 1958.)

Suborder 2. **Mobilina** Kahl

Free-swimming; without stalk; body axis shortened; with a highly developed attaching organelle on the aboral end. (One family.)

Family **Urceolariidae** Stein

Genus **Urceolaria** Lamarck. Peristome more or less oblique; horny corona of attaching disc with obliquely placed simple den-

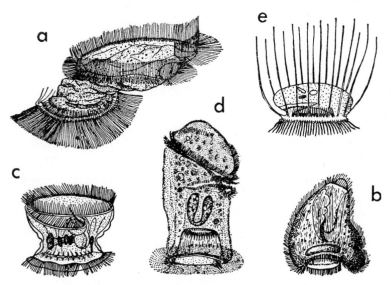

FIG. 374. a, *Urceolaria mitra*, ×270 (Wallengren); b, *U. paradoxa*, ×215 (Claparède and Lachmann); c, *Trichodina pediculus*, ×425 (James-Clark); d, *T. urinicola*, ×470 (Fulton); e, *Cyclochaeta spongil-lae*, ×460 (Jackson).

ticles without radial processes; commensals. (Morphology, Wallengren, 1897; E M study, Fauré-Fremiet *et al.*, 1956.)

U. mitra (Siebold) (Fig. 374, *a*). 80-140μ long; on planarians.

U. paradoxa Claparède and Lachmann (Fig. 374, *b*). 70-80μ in diameter; colonial forms; in the respiratory cavity of *Cyclostoma elegans*.

U. karyolobia Hirshfield. 45-50μ in diameter, 20-30μ high; macronucleus lobate and conspicuous; in the mantle cavity of limpets, *Lottia gigantea* and *Acmaea* spp. (Hirshfield, 1949).

Genus **Trichodina** Ehrenberg. Body barrel-shaped; with well-developed adhesive basal disk, and a skeletal ring with radially arranged denticles composed of distally projecting blades and medially extending spines; adoral ciliary row spirals one to three times; without cirri; commensals on, or parasitic in, aquatic animals. (Many species, Mueller, 1932, 1937, Davis, 1947, Lom, 1958, Uzmann and Stickney, 1954, Raabe and Raabe, 1959; biometry, Fauré-Fremiet, 1943.)

T. pediculus (O. F. Müller) (Fig. 374, *c*). Body discoid, often with a median constriction; diameter 35-60μ, height 25-55μ; adhesive disk 27.5-47.5μ in diameter; adoral spiral complete, describing an arc of more than 360° (35-45μ in diameter); sausage-shaped macronucleus; a single micronucleus; adhesive disk with a denticulate ring consisting of denticles (22-25) with curved hooks and long rays extending to the middle of the disk; on Hydra and on the gills of Necturus and Triturus larvae, although Raabe considers hydra the sole host. (Morphology, Raabe, 1958, 1959.)

T. domerguei (Wallengren). Discoid, slightly constricted; diameter 40-70μ; height 35-60μ; adoral spiral complete, 35-50μ in diameter; macronucleus sausage-shaped with a single micronucleus; denticulate ring with denticles with long curved hooks (20-34) and medium long rays which do not reach the center of the adhesive disc (variations occur). Many varieties. (On fishes, Triton, Unio, etc.)

T. urinicola Fulton (Fig. 374, *d*). Body small in diameter; 50-90μ high; twenty-eight to thirty-six hooks; diameter of adhesive disk 30-55μ; diameter of denticulate ring, 18-30μ; denticles twenty-six to thirty-one; rays fuse with central argentophile area; in the urinary bladder of various amphibians.

T. myicola Uzmann and Stickney. Bell-shaped to discoid; average diameter 81μ; height 31-86μ; two parallel adoral rows of cilia, forming "semi-membrane"; diameter of denticulate ring 29-46μ; number of denticles 26-36; diameter of the basal disk 42-79μ; on the oral region of the marine bivalve, *Mya arenaria* (Uzmann and Stickney, 1954).

Genus **Cylochaeta** Jackson. Saucer-form; peristomal surface parallel to the basal disc; upper surface with numerous flat wrinkles; basal disc composed of cuticular rings, velum, a ring of standing cirri, and membranellae; commensal on, or parasitic in, fresh or salt water animals. (Several species.)

C. spongillae J. (Fig. 374, *e*). About 60μ in diameter; in interstices of *Spongilla fluviatilis*.

Order **Opalinida** Poche

In this group are placed those ciliated protozoa which were formerly designated as Protociliata. However, as was already mentioned (p. 828), they differ wholly from the ciliates considered in foregoing chapters, except that they are ciliated.

The opalinids are all endocommensals in the large intestine of Salientia, only a few having been found in urodeles, reptiles and fish (Metcalf, 1923, 1940). The body is uniformly covered by cilia which are arranged in oblique longitudinal rows. There is no cytostome and nutrition is saprozoic. they contain two or many nuclei of one kind. Their sexual reproduction is sexual fusion and not by conjugation. They multiply by longitudinal fission or plasmotomy. Therefore this group is now removed from the class Ciliata, and retained here provisionally. (Taxonomy, Metcalf, 1920, 1923, 1940; geographical distribution, Metcalf, 1920, 1929, 1940; cytology and development, van Overbeek de Meyer, 1929; species, Bhatia and Gulati, 1927, Carini, 1938-1942, Beltrán, 1941, 1941a; culture, Yang, 1960.)

Family **Opalinidae** Claus

Genus **Opalina** Purkinje and Valentin. Highly flattened; multinucleate; in amphibians. (Numerous species, Metcalf, 1923, 1940; growth and nuclear division, Hegner and Wu, 1921; cytology, ten Kate, 1927; Dutta, 1958.)

O. hylaxena Metcalf (Fig. 375, *a*). In *Hyla versicolor;* large individuals about 420μ long, 125μ wide, 28μ thick. (Several subspecies, Metcalf.)

O. obtrigonoidea M. (Fig. 375, *b-d*). 400-840μ long, 175-180μ wide, 20-25μ thick; in various species of frogs and toads (Rana, Hyla, Bufo, Gastrophryne, etc.), North America. (Numerous subspecies, Metcalf.)

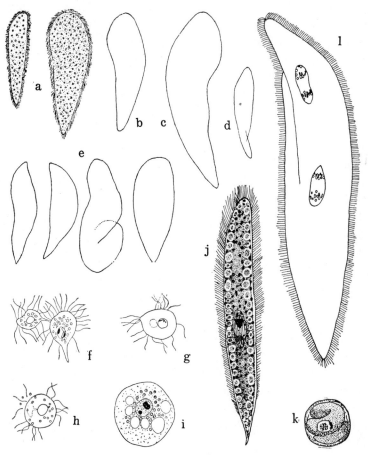

FIG. 375. a–i, l, Metcalf; j, k, Léger and Duboscq. a, two individuals of *Opalina hylaxena*, ×78; b–d, three individuals of *O. obtrigonoidea*, ×78 (b, from *Bufo fowleri;* c, from *Rana pipiens;* d, from *R. palustris*); e, four individuals of *Cepedea cantabrigensis*, ×78; f–i, stages in sexual reproduction in *Protoopalina intestinalis;* j, k, *P. saturnalis*, ×500; l, *P. mitotica*, ×240.

O. carolinensis M. 90-400μ by 32-170μ; in *Rana pipiens sphenocephala*.

O. pickeringii M. 200-333μ by 68-100μ; in *Hyla pickeringii*.

O. oregonensis M. 526μ by 123μ; in *Hyla regilla*.

O. spiralis M. 300-355μ long, 130-140μ wide, 25-42μ thick; in *Bufo compactilis*.

O. chorophili M. About 470μ by 100μ; in *Chorophilus triseriatus*.

O. kennicotii M. About 240μ by 85μ; in *Rana areolata*.

Genus **Cepedea** Metcalf. Cylindrical or pyriform; circular in cross-section; multinucleate; all in Amphibia. Numerous species. (Cytology, Fernandez, 1947.)

C. cantabrigensis M. (Fig. 375, *e*). About 350μ by 84μ; in *Rana cantabrigensis*.

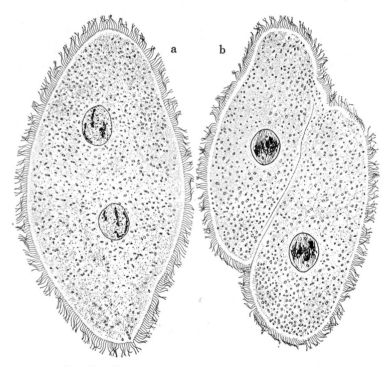

Fig. 376. *Zelleriella elliptica*, stained specimens, ×440 (Chen). a, a typical vegetative individual; b, an individual which is nearly completely divided, the nuclei being at early metaphase.

C. hawaiensis M. 170-200μ by 43-60μ; in *Rana catesbeiana;* Hawaii.

C. obovoidea M. About 315μ by 98μ; in *Bufo lentiginosus.*

C. floridensis M. About 230μ by 89μ; in *Scaphiopus albus.*

Genus **Protoopalina** Metcalf. Cylindrical or spindle-shaped, circular in cross-section; two nuclei; in the colon of various species of Amphibia with one exception. (Numerous species.)

P. intestinalis (Stein) (Fig. 375, *f-i*). About 330μ by 68μ; in *Bombina bombina,* and *B. pachypa;* Europe. (Cytochemistry, Sakhanova, 1962.)

P. saturnalis Léger and Duboscq (Fig. 375, *j, k*). In the marine fish, *Box boops;* 100-152μ by 22-60μ.

P. mitotica M. (Fig. 375, *l*). 300μ by 37μ; in *Ambystoma tigrinum.*

Genus **Zelleriella** Metcalf. Greatly flattened; two similar nuclei; all in Amphibia. Numerous species. (Cytology, Chen, 1948.)

Z. scaphiopodos M. In *Scaphiopus solitarius;* about 150μ long, 90μ broad, 13μ thick.

Z. antilliensis M. About 180μ long, 113μ wide, 32μ thich; in *Bufo marinus.*

Z. hirsuta M. About 113μ long, 60μ wide, 22μ thick; in *Bufo cognatus.*

Z. elliptica Chen. (Fig. 376). In *Bufo valliceps;* average dimensions 184μ by 91μ. Chen (1948) distinguishes four other species from the same host, all of which possess twenty-four chromosomes.

References

AWERINZEW, G. W., 1936. Zur Biologie des Infusors Lagenophrys. Arch. Protist., 87:131.

BELTRÁN, E., 1941. Opalinidos parasitos en anfibios mexicanos. Rev. Soc. Mexicana Hist. Nat., 2:127.

———— 1941a. *Zelleriella leptodeirae* sp. nov., etc. *Ibid.,* 2:267.

BHATIA, B. L., and GULATI, A. N., 1927. On some parasitic ciliates from Indian frogs, toads, etc. Arch. Protist., 57:85.

BISHOP, E. L. JR., and JAHN, T. L., 1941. Observations on colonial peritrichs of the Okoboji region. Proc. Iowa Acad. Sci., 48:417.

BRAND, T. v., 1923. Die Encystierung ben *Vorticella microstoma* und hypotrichen Infusorien. Arch. Protist., 47:59.

CARINI, A., 1938. Contribuição ao conhecimento das "Opalinidae" dos batráquios do Brasil. II. Bol. Biol., N.S., 3:147.

―― 1938a. *Zelleriella corniola*, etc. Arch. Biol., 22:1.

―― 1940. Contribuição ao conhecimento das "Opalinidae" dos batráquios do Brasil. *Ibid.*, 24, 5 pp.

―― 1942. Sobre uma *Zelleriella* do cecum do *Siphonops annulatus*. *Ibid.*, 26, 2 pp.

CHEN, T. T., 1948. Chromosomes in Opalinidae with special reference to their behavior, morphology, etc. J. Morphol., 88:281.

DAVIS, H. S., 1947. Studies of the protozoan parasites of freshwater fishes. Fish. Bull., Fish Wildlife Service, 51, no. 41.

DEBAISIEUX, P., 1959. *Lagenophrys lunatus*. La Cellule, 59:361.

DODD, E. E., 1962. A cytochemical investigation of the unclear changes in the cyst cycle of *Telotrichidium henneguyi*. J. Protozool., 9:93.

DUTTA, G. P., 1958. Histochemical studies of protozoa. I. Res. Bull. Panjab Univ., no. 143:97.

FAURÉ-FREMIET, E., 1943. Étude biométrique de quelques trichodines. Bull. Soc. Zool. France, 68:158.

―― and MUGARD, H., 1946. Une trichodine parasite endovésicale chez *Rana esculenta*. *Ibid.*, 71:36.

―― and THAUREAUX, J., 1944. Les globules de "paraglycogène" chez *Balantidium elongatum* et *Vorticella monilata*. *Ibid.*, 69:3.

―― *et al.*, 1956. L'appareil squelettique et nyoide des Urceolaires: étude au microscope électronique. Bull. Soc. Zool. France, 81:77.

FINLEY, H. E., 1939. Sexual differentiation in peritrichous ciliates. J. Morphol., 91:569.

―― 1943. The conjugation of *Vorticella microstoma*. Tr. Am. Micr. Soc., 62:97.

―― *et al.*, 1955. Chromatographic analysis of the asexual and sexual stages of a ciliate. J. Protozool., 2:13.

―― *et al.*, 1959. Non-axenic and axenic growth of *Vorticella microstoma*. *Ibid.*, 6:201.

FULTON, J. F. JR., 1923. *Trichodina pediculus* and a new closely related species. Proc. Boston Soc. Nat. Hist., 37:1.

FURSSENKO, A., 1929. Lebenscyclus und Morphologie von *Zoothamnium arbuscula*. Arch. Protist., 67:376.

GRASSÉ, P.-P., 1952. Traité de Zoologie. I. Fasc. 1. Paris.

HAMILTON, J. M., 1952. Studies on loricate Ciliophora. I. Tr. Am. Micr. Soc., 71:382.

HEGNER, R. W., 1932. Observations and experiments on the opalinid ciliates of the green frog. J. Parasit., 18:274.

―― and WU, H. F., 1921. An analysis of the relation between growth and nuclear division in a parasitic infusorian, *Oplaina* sp. Am. Nat., 55:335.

HIRSHFIELD, H., 1949. The morphology of *Urceolaria karyolobia* sp. nov., etc. J. Morphol., 85:1.

IMAMURA, T., 1940. Two species of Lagenophrys from Sapporo. Anno. Zool. Japon., 19: 267.

JAROCKI, J., and JAKUBOWSKA,W., 1927. Eine neue, solitär freischwimmende Peritriche, *Hastatella aesculacantha* n. sp. Zool. Anz., 73:270.

KAHL, A., 1935. Peritricha und Chonotricha. In Dahl's Die Tierwelt Deutschlands, etc. Part 30:651.

KENT, S., 1881-1882. A manual of Infusoria.

KOFOID, C. A., and ROSENBERG, L. E., 1940. The neuromotor system of *Opisthonecta henneguyi*. Proc. Am. Philos. Soc., 82:421.

KRALIK, U., 1961. Beiträge zur Systematik von peritrichen Ciliaten. Intern. Rev. Ges. Hydrobiol., 46:65.

LEVINE, L., 1959. Axenizing *Vorticella convallaria*. J. Protozool., 6:169.

LOM, J., 1958. A contribution to the systematics and morphology of endo-parasitic trichodinids from amphibians, with a proposal of uniform specific characteristics. *Ibid.*, 5:251.

———— 1961. Ectoparasitic trichodinids from freshwater fish in Czechoslo-vakia. Vestnik Cesk Spol. Zool., 25:215.

MAST, S. O., 1944. A new peritrich belonging to the genus Ophrydium. Tr. Am. Micr. Soc., 63:181.

MATTHES, D., 1950. Beitrag zur Peritrichenfauna der Umgebung Erlangens. Zool. Jahrbuch., Syst., 79:437.

METCALF, M. M., 1909. Opalina. Arch. Protist., 13:195.

———— 1920. Upon an important method of studying problems of relation-ship and of geographical distribution. Proc. Nat. Acad. Sci., 6: 432.

———— 1920a. The classification of the Opalinidae. Science, 52:135.

———— 1923. The opalinid ciliate infusorians. Smithsonian Inst. U. S. Nat. Mus., Bull., 120:1.

———— 1928. The bell-toads and their opalinid parasites. Am. Nat., 62:5.

———— 1929. Parasites and the aid they given in problems of taxonomy, geographical distribution and paleogeography. Smithsonian Misc. Coll., 81: no. 8.

———— 1940. Further studies on the opalinid ciliate infusorians and their hosts. Proc. U. S. Nat. Mus., 87:465.

MUELLER, J. F., 1932. *Trichodina renicola*, a cilate parasite of the urinary tract of *Esox niger*. Roosevelt Wild Life Ann., 3:139.

———— 1937. Some species of Trichodina, etc. Tr. Am. Micr. Soc., 61:177.

NENNINGER, U., 1948. Die Peritrichen der Umgebung von Erlangen, etc. Zool. Jahrb. Syst., 77:169.

NOLAND, L. E., and FINLEY, H. E., 1931. Studies on the taxonomy of the genus Vorticella. Tr. Am. Micr. Soc., 50:81.

PENARD, E., 1922. Étude sur les infusoires d'eau douce. Geneva.

PRECHT, H., 1935. Die Struktur des Stieles bei den Sessilia. Arch. Protist., 85:234.

RAABE, J., and RAABE, Z., 1959. Urceolariidae of molluscs of the Baltic Sea. Acta Parasitol. Polonica, 7:453.

RAABE, Z., 1952. *Ambiphrya miri* g. n., sp.n. Ann. Univ. Marie Curie Skl., 6:339.

—— 1958. On some species of Trichodina of gills of Adriatic fishes. Acta Parasitol. Polonica, 6:355.

—— 1959. *Trichodina pediculus* and *T. domerguei. Ibid.*, 7:189.

—— 1959a. Urceolariidae of gills of Gobiidae and Cottidae from Baltic Sea. *Ibid.*, 7:441.

Rosenberg, L. E., 1938. Cyst stages of *Opisthonecta henneguyi*. Tr. Am. Micr. Soc., 57:147.

—— 1940. Conjugation in *Ophisthonecta henneguyi*, etc. Proc. Am. Philos. Soc., 82:437.

Rouiller, C., *et al.*, 1956. Origin ciliaire des fibrilles scléro protéiques pédonuculaires chez les ciliés péritriches. Exper. Cell. Res., 11:527.

Sakhanova, K. M., 1962. Morphological and cytochemical research on *Protoopalina intestinalis* (Russ.). Tsitologiya, 3:577.

Seshachar, B. R., and Dass, C. M. S., 1953. Evidence for the conversion of DNA to RNA in *Epistylis articulata*. Exper. Cell. Res., 5:248.

Stammer, H.-J., 1948. Eine neue eigenartige endoparasitische Peritriche, *Operculariella parasitica* n. g., n. sp. Zool. Jahrb. Syst., 77:163.

Stiller, J., 1939. Die Peritrichenfauna der Nordsee bei Helgoland. Arch. Protist., 92: 415.

—— 1940. Beitrag zur Peritrichenfauna des grossen Plöner Sees in Holstein. Arch. Hydrobiol., 36:263.

Stokes, A. C., 1888. A preliminary contribution toward a history of the freshwater Infusoria of the United States. J. Trenton Nat. Hist. Soc., 1:71.

Summers, F. M., 1938. Some aspects of normal development in the colonial ciliate *Zoothamnium alterans*. Biol. Bull., 74:117.

—— 1938a. Form regulation in *Zoothamnium alterans. Ibid.*, 74:130.

Swarczewsky, B., 1930. Zur Kenntnis der Baikalprotistenfauna. IV. Arch. Protist., 69:455.

ten Kate, C. G. B., 1927. Ueber das Fibrillensystem der Ciliaten. Arch. Protist., 57:362.

Thompson, S., *et al.*, 1947. *Syphidia ameiuri*, n. sp., etc. Tr. Am. Micr. Soc., 66:315.

Uzmann, J. R., and Stickney, A. P., 1954. *Trichodina myicola* n. sp., etc. J. Protozool., 1:149.

van Overbeek de Meyer, G. A. W., 1929. Beiträge zur Wachstums- und Plasmadifferenzierungs-Erscheinungen an *Opalina ranarum*. Arch. Protist., 66:207.

Wallengren, H., 1897. Studier öfver ciliata Infusorier. III. Särtryck Fysiogr. Sällsk. Handl., 8:1.

—— 1897a. Zur Kenntnis der Gattung Trichodina. Biol. Centralbl., 17:55.

Wetzel, A., 1925. Vergleichend cytologische Untersuchungen an Ciliaten, Arch. Protistenk, 51:209.

Yang, W. C. T., 1960. On the continuous culture of opalinids. J. Parasit., 46:32.

Chapter 42

Class 2. Suctoria Claparède and Lachmann

S UCTORIA do not possess any cilia in the mature stage, but young larvae which are produced by budding are ciliated. When the free-swimming larva becomes attached to substrate, a non-contractile stalk is produced by the scopula, cilia are lost and tentacles develop. The body is of varied shape and is covered by a pellicle and occasionally possesses a lorica. There is no cytostome, and the food-capturing is carried on exclusively by the tentacles. Tentacles are of two kinds; one is suctorial in function and bears a rounded knob, while the other is for piercing through the body of a prey and is pointed. The tentacles may be confined to limited areas or may be distributed over the entire body surface. The suctorians feed on chiefly ciliates and nutrition is thus holozoic.

Asexual reproduction is by budding. The buds are formed by either exogenous or endogenous gemmation. Sexual reproduction by conjugation has been described in several genera. While the ciliate affinities of Suctoria had been recognized for a long time, Guilcher's (1951) studies revealed the persistence of infraciliature of a holotrichous type among the suctorians, which led Fauré-Fremiet and Corliss to consider Suctorida as an order of Holotricha.

Order Suctorida Claparède and Lachmann

The order is subdivided into seven families.

Family 1. Acinetidae Bütschli

Suctorial tentacles in fascicles; with or without lorica; with or without stalk; budding endogenous.

Genus **Acineta** Ehrenberg. Lorica more or less flattened; usually with stalk; tentacles in two (one or three) fascicles; body completely or partly filling lorica; larva with ciliated band or

1036

completely ciliated; fresh or salt water. (Numerous species, Swarczewsky, 1928a.)

A. tuberosa E. (Fig. 377, *a*). Lorica 50-100μ high; with stalk; salt and brackish water. (Budding and metamorphosis, Kormos, 1957.)

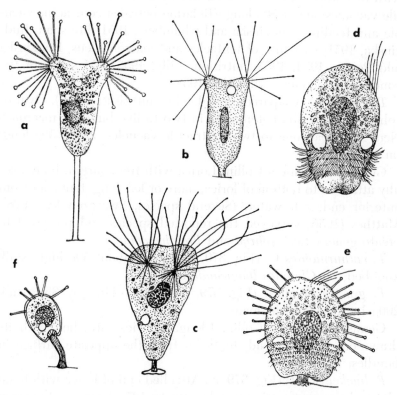

FIG. 377. a, *Acineta tuberosa*, ×670 (Calkins); b, *A. lacustris*, ×200 (Stokes); c-e, *Tokophrya infusionum* (c, ×400; d, a free-swimming bud; e, a young attached form, ×800) (Collin); f, *T. cyclopum*, a young individual, ×500 (Collin).

A. cuspidata Stokes. Lorica cup-shaped; front end wih two opposing sharp points; lorica 32-42μ high; on Oedogonium in fresh water.

A. lacustris S. (Fig. 377, *b*). Lorica elongate ovoid; flattened; 75-185μ high; on Anacharis in pond.

Genus **Tokophrya** Bütschli. Pyriform or pyramidal; without lorica; tentacles in one to four fascicles on anterior surface; stalk not rigid; simple endogenous budding; fresh water. (Several species.)

T. infusionum (Stein) (Fig. 377, *c-e*). Inverted pyramid; stalk with or without attaching disk; macronucleus oval; two contractile vacuoles; about 60μ long. (Relation between contractile vacuole and feeding, Rudzinska and Chambers, 1951; life span, Rudzinska, 1951; structure of tentacles and macronucleus, Rudzinska and Porter, 1954, 1955; nutrition, Lilly, 1953; EM study of the contractile vacuole, Rudzinska, 1958.)

T. cyclopum (Claparéde and Lachmann) (Fig. 377, *f*). Oval or spherical; stalk short; tentacles in two to five bundles; macronucleus spherical; one to two contractile vacuoles; about 50μ long; on Cyclops, etc.

Genus **Thecacineta** Collin. Lorica with free margin; body usually attached to bottom of lorica, more or less long; tentacles from anterior end; salt water. (Several species, Swarczewsky, 1928.) Matthes (1956) surveyed the reported species and proposed to create genera *Loricaphrya* and *Praethecacineta*. .

T. cothurnioides C. (Fig. 379, *a*). Lorica about 50μ high; stalk knobbed; on *Cletodes longicaudatus*.

T. gracilis (Wailes) (Fig. 379, *b*). Lorica 110μ by 35μ; stalk 200μ by 4μ; on hydrozoans.

Genus **Periacineta** Collin. Elongate lorica; attached with its drawn-out posterior end; tentacles from the opposite surface in bundles; fresh water.

P. buckei (Kent) (Fig. 379, *c*). Attached end of lorica with basal plate; three contractile vacuoles; up to 125μ long; on *Lymnaea stagnalis* and *Ranatra linearis*. Matthes (1954a) considers this a Discophrya.

Genus **Hallezia** Sand. Without lorica; with or without a short stalk; tentacles in bundles; fresh water.

H. brachypoda (Stokes) (Fig. 379, *d*). 34-42μ in diameter; in standing water among leaves.

Genus **Solenophrya** Claparède and Lachmann. Lorica attached directly with its under side; body usually not filling lorica; tentacles in fascicles; fresh water.

S. micraster Penard (Fig. 378). Lorica pentagonal viewed from above; five fascicles of tentacles; body diameter, 15-50μ; lorica 17.5-53μ high; fresh water. (Morphology, development, feeding processes, Hull, 1954, 1954a.)

S. *inclusa* Stokes. Lorica subspherical; about 44μ in diameter; standing fresh water.

S. *pera* S. Lorica satchel-form; about 40-45μ high; body about 35μ long; standing fresh water.

Genus **Acinetopsis** Robin. Lorica in close contact with body on sides; stalked; one to six large retractile tentacles and numerous small tentacles from apical end; mainly salt water.

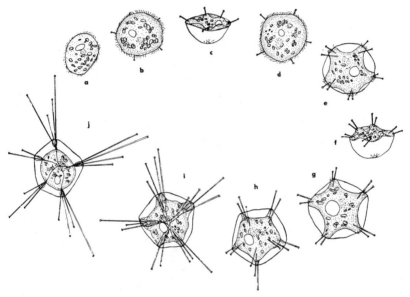

FIG. 378. Developmental stages of *Solenophrya micraster,* ×330 (Hull). a, free-swimming stage; b, attachment to substrate.

A. *tentaculata* Root (Fig. 379, *e, f*). Lorica 187μ high; stalk 287μ long; large tentacles up to 500μ long; body about 138μ by 100μ; on *Obelia commissuralis* and *O. geniculata;* Woods Hole (Root, 1922).

Genus **Tachyblaston** Martin. Lorica with short stalk; tentacles distributed on anterior surface; nucleus oval; salt water. (One species.)

T. *ephelotensis* M. (Fig. 379, *g, h*). Lorica 30-93μ high; stalk 20-30μ long; attached to *Ephelota gemmipara.*

Genus **Dactylophrya** Collin. Cup-like lorica, filled with the protoplasmic body; with a short stalk; twelve to fifteen arm-like tentacles from anterior surface; salt water. (One species.)

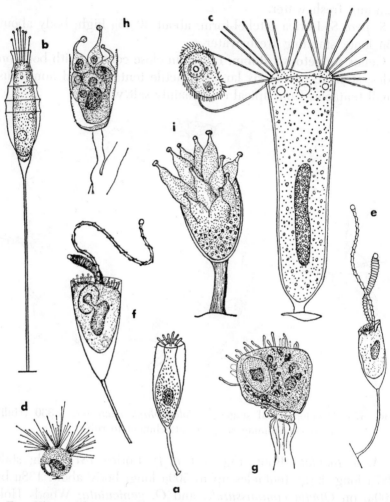

Fig. 379. a, *Thecacineta cothurnioides*, ×400 (Collin); b, *T. gracilis*, ×270 (Wailes); c, *Periacineta buckei*, feeding on Chilodonella, ×530 (Collin); d, *Hallezia brachypoda*, ×200 (Stokes); e, f, *Acinetopsis tentaculata* (e, ×130; f. ×230) (Root); g, h, *Tachyblaston epheloten-sis* (g, a young individual in Ephelota, ×260; h, mature form, ×500, Martin); i, *Dactylophrya roscovita*, ×830 (Collin).

D. roscovita C. (Fig. 379, *i*). About 40µ long excluding stalk; on the hydrozoan, *Diphasia attenuata*.

Genus **Pseudogemma** Collin. Attached with a short stalk to larger suctorians; without tentacles; endogenous budding; swarmer with four ciliary bands; salt water.

P. pachystyla C. (Fig. 380, *a*). About 30µ long; stalk 3-4µ wide; swarmer 15µ by 9µ; on *Acineta tuberosa*.

Genus **Endosphaera** Engelmann. Spherical without lorica; without tentacles; budding endogenous; swarmer with three equatorial ciliary bands; parasitic in Peritricha; fresh and salt water.

E. engelmanni Entz (Fig. 380, *b*). 15-41µ in diameter; imbedded in the host's cytoplasm; swarmer 13-19µ in diameter; in *Telotrochidium henneguyi* (p. 1023), and other peritrichs.

Genus **Allantosoma** Gassovsky. With neither lorica nor stalk; elongate; one or more tentacles at ends; macronucleus oval or spherical; compact micronucleus; a single contractile vacuole; cytoplasm often filled with small spheroidal bodies; development unknown; in mammalian intestine. (Species, Hsiung, 1930.)

A. intestinalis G. (Fig. 380, *c*). 33-60µ by 18-37µ; attached to various ciliates living in the caecum and colon of horse.

A. dicorniger Hsiung (Fig. 380, *d*). 20-33µ by 10-20µ; unattached; in the colon of horse (Hsiung, 1928).

A. brevicorniger H. (Fig. 380, *e*). 23-36µ by 7-11µ; attached to various ciliates in the caecum and colon of horse.

Genus **Anarma** Goodrich and Jahn. Radially or somewhat bilaterally symmetrical; without stalk or lorica; attached directly or by a short protoplasmic process to substratum; one to two fascicles of capitate tentacles; multiplication by external budding near base or by a single internal ciliated bud; conjugation; ectocommensal on *Chrysemys picta bellii* (Goodrich and Jahn, 1943).

A. multiruga G. and J. (Fig. 380, *f*, *g*). Body cylindrical, 70-150µ by 35-70µ; body surface with seven or eight longitudinal folds; pellicle thin; cytoplasm granulated; nucleus ribbon-form; two to six contractile vacuoles, each with a permanent canal and a pore; attached directly or indirectly to the carapace and plastron of the turtle.

Genus **Squalorophrya** Goodrich and Jahn. Elongate; radially symmetrical; lorica, rigid, close-fitting, covered with debris; with a stalk; capitate tentacles at distal end; ectocommensal on *Chrysemys picta bellii*.

S. *macrostyla* G. and J. (Fig. 380, *h, i*). Cylindrical, with 4 longitudinal grooves; body about 90μ by 40μ; striated stalk, short and thick, about 30μ long; lorica highly viscous with debris; nucleus ovoid to elongate, sometimes Y-shaped; two contractile

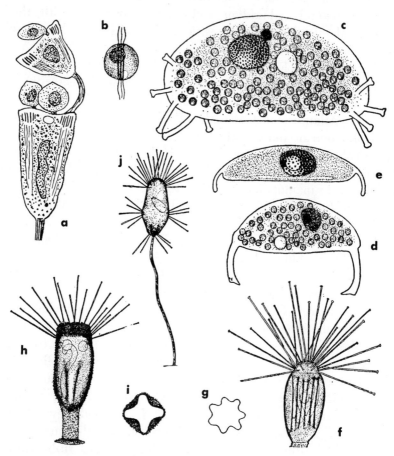

Fig. 380. a, *Pseudogemma pachystyla*, ×400 (Collin); b, *Endosphaera engelmanni*, ×500 (Lynch and Noble); c, *Allantosoma intestinalis*, ×1050 (Hsiung); d, *A. dicorniger*, ×1300 (Hsiung); e, *A. brevicorniger*, ×1400 (Hsiung); f, g, *Anarma multiruga*, ×230; h, i, *Squalophrya macrostyla*, ×670; j, *Multifasciculatum elegans*, ×660 (Goodrich and Jahn).

vacuoles, each with a permanent canal and a pore; on *Chrysemys picta bellii.*

Genus **Multifasciculatum** Goodrich and Jahn. Radially or bilaterally symmetrical; stalked; without lorica; pellicle thin; several fascicles of tentacles on distal, lateral and proximal regions of body; ectocommensal on *Chrysemys picta bellii.*

M. elegans G. and J. (Fig. 380, *j*). Body ovoid; 50-90μ by 20-50μ; stalk striated, about 150-270μ long; tentacles in four groups; nucleus ovoid; one to three contractile vacuoles; attached to the plastron of the turtle.

Family 2. **Dendrosomatidae** Fraipont

Body form irregular or branching; suctorial tentacles in clusters; without stalk; budding endogenous.

Genus **Dendrosoma** Ehrenberg. Dendritic; often large; nucleus band-form, branched; numerous contractile vacuoles; fresh water. (Taxonomy and morphology, Gönnert, 1935.)

D. radians E. (Fig. 381, *a*). Brownish; 1.2-2.5 mm. high; on vegetation. (Morphology, Gönnert.)

Genus **Trichophrya** Claparède and Lachmann *(Platophrya* Gönnert). Body small; rounded or elongate, but variable; without stalk; tentacles in fascicles, not branching; simple or multiple endogenous budding; fresh or salt water.

T. epistylidis C. and L. (*T. sinuosa* Stokes) (Fig. 381, *b*). Form irregular; with many fascicles of tentacles; nucleus band-form, curved; numerous vacuoles; up to 240μ long; on Epistylis, etc., in fresh water. (Morphology, Gönnert.)

T. salparum Entz (Fig. 381, *c*). On various tunicates such as *Molgula manhattensis;* 40-60μ long; tentacles in two groups; salt water; Woods Hole (Calkins).

T. columbiae Wailes (Fig. 381, *d*). 60-75μ by 40-48μ in diameter; cylindrical; tentacles at ends; nucleus spherical; in marine plankton; Vancouver (Wailes).

T. micropteri Davis (Fig. 381, *e*). Body elongate, irregular or rounded; up to 30-40μ long by 10-12μ; fully extended tentacles 10-12μ long; cytoplasm often filled with yellow to orange spherules; a single micronucleus; a single contractile vacuole; attached to the gill of small mouth black bass, *Micropterus dolomieu.*

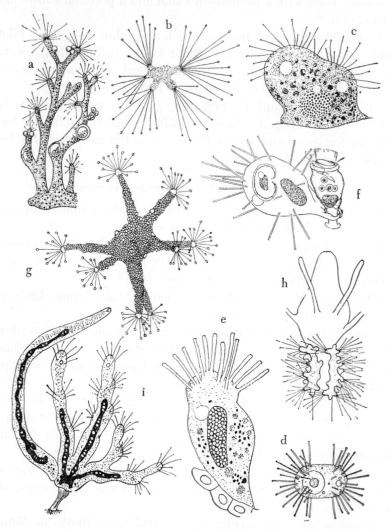

FIG. 381. a, *Dendrosoma radians*, ×35 (Kent); b, *Trichophrya epistylidis* ×250 (Stokes); c, *T. salparum*, ×170 (Collin); d, *T. columbiae*, ×200 (Wailes); e, *T. micropteri*, ×650 (Davis); f, *Erastophrya chattoni* (Fauré-Fremiet); g, *Astrophrya arenaria*, ×65 (Awerinzew); h, *Lernaeophrya capitata*, ×35 (Pérez); i, *Dendrosomides paguri*, ×200 (Collin).

Davis (1942) states that when abundantly present, the suctorian may cause serious injury to the host.

T. intermedia Prost (1952). On the gills of young *Salmo salar*.

Genus **Erastophrya** Fauré-Fremiet. Pyriform; distributed tentacles; posterior end drawn out into two "arms" by means of which the organism grasps the stalk of a peritrich; fresh water (Fauré-Fremiet, 1943). (One species.)

E. chattoni F.-F. (Fig. 381, *f*). Body up to 130μ long; macronucleus spherical to sausage form; a single micronucleus; a contractile vacuole; endogenous budding, gemma about 40μ long; a commensal on *Glossatella piscicola*.

Genus **Astrophrya** Awerinzew. Stellate; central portion drawn out into eight elongate processes, each with a fascicle of tentacles; body covered by sand grains and other objects. (One species.)

A. arenaria A. (Fig. 381, *g*). 145-188μ in diameter; processes 80-190μ long; in Volga river plankton.

Genus **Lernaeophrya** Pérez. Body large; with numerous short prolongations, bearing very long multifasciculate tentacles; nucleus branched; brackish water. (One species.)

L. capitata P. (Fig. 381, *h*). Attached to the hydrozoan, *Cordylophora lacustris* in brackish water; 400-500μ long; tentacles 400μ long. (Morphology, Gönnert.)

Genus **Staurophrya** Zacharias. Rounded body drawn out into six processes.

S. elegans Z. (Fig. 382, *a*). Tentacles not capitate; macronucleus round; one to two contractile vacuoles; about 50μ in diameter; in fresh water.

Swarczewsky (1928) established the following genera for the forms he had found in Lake Baikal: *Baikalophrya, Stylophrya, Baikalodendron* and *Gorgonosoma*.

Family 3. **Ophryodendridae** Stein

Suctorial tentacles on ends of "arms"; with or without short stalk; exogenous budding.

Genus **Ophryodendron** Claparède and Lachmann. With one long or three to six shorter retractile processes, bearing suctorial tentacles; on Crustacea, Annelida, etc.; salt water. (Several species.)

O. porcellanum Kent (Fig. 382, *b*). 60-100μ long; on *Porcellana platycheles*, etc.

O. belgicum Fraipont (Fig. 382, *c*). 38-114μ long; vermicular form 100μ; on Bryozoa and hydrozoans; Vancouver (Wailes).

Genus **Dendrosomides** Collin. Branched body similar to Dendrosoma, but with a peduncle; reproduction by budding of vermicular form; salt water. (One species.)

D. paguri C. (Fig. 381, *i*). 200-300μ long; vermicular forms 350μ long; on the crabs, *Eupagurus excavatus* and *E. cuanensis.*

Genus **Rhabdophrya** Chatton and Collin. Elongate, rod-form; with short peduncle, not branched; tentacles distributed over entire surface; macronucleus ellipsoid; micronucleus small; two to three contractile vacuoles; salt or brackish water. (Several species.)

R. trimorpha C. and C. (Fig. 382, *d*). Up to 150μ long; on the copepod, *Cletodes longicaudatus.*

Family 4. **Dendrocometidae** Haeckel

Suctorial tentacles localized on branched arms; without stalk; budding endogenous.

Genus **Dendrocometes** Stein. Body rounded; with variable number of branched arms; fresh water. (Taxonomy, Swarczewsky, 1928a.)

D. paradoxus S. (Fig. 382, *e*). Up to 100μ long; on *Gammarus pulex, G. puteanus*, etc. (Morphology and biology, Pestel, 1932.)

Genus **Stylocometes** Stein. Arms not branched; tentacles fingerlike, not branched; macronucleus elongate; two to three micronuclei.

S. digitatus (Claparède and Lachmann). Up to 110μ long; on the gills of *Asellus aquaticus* and on *Aphrydium versatile.* (Morphology and conjugation, Skreb-Guilcher, 1955.)

Genus **Dendrocometides** Swarczewsky. Body more or less arched; suctorial tentacles slender, pointed and simple or branched; attached to crustaceans on its broad and circular surface (Swarczewsky, 1928a).

D. priscus S. (Fig. 382, *f*). Diameter 60-65μ, height 18-20μ; on *Acanthogammarus albus;* Lake Baikal.

Genus **Discosoma** S. Discoid; circular in front view; short and pointed tentacles radially arranged, four or six in each row; gemmation, endogenous and simple.

D. tenella S. (Fig. 382, *g*). Diameter 75μ, height 10μ; on *Acanthogammarus victorii*, etc.; Lake Baikal.

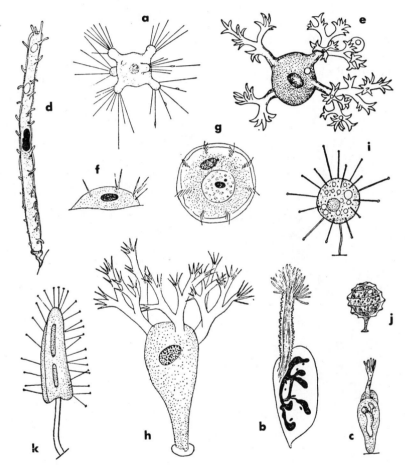

Fig. 382. a, *Staurophrya elegans*, ×200 (Zacharias); b, *Ophryodendron porcellanum*, ×220 (Collin); c, *O. belgicum*, ×270 (Fraipont); d, *Rhabdophrya trimorpha*, ×430 (Collin); e, *Dendrocometes paradoxus*, ×270 (Wrzesnowski); f, *Dendrocometides priscus*, ×220; g, *Discosoma tenella*, ×220; h, *Cometodendron clavatum*, ×220 (Swarczewsky); i, j, *Podophrya fixa* (i, ×400 (Wales); j, ×220 (Collin)); k, *P. elongata*, ×240 (Wailes).

Genus **Cometodendron** S. Body elongate; attached to substrate by a "foot," well-developed arms; short and pointed tentacles at the ends of arms; simple endogenous gemmation.

C. clavatum S. (Fig. 382, *h*). 150μ by 40-50μ; the foot 20-22μ; on *Acanthogammarus victorii*, etc.; Lake Baikal.

Family 5. **Podophryidae** Haeckel

Suctorial tentacles not in clusters; with or without lorica; with or without stalk; budding exogenous.

Genus **Podophrya** Ehrenberg. Subspherical; normally with a

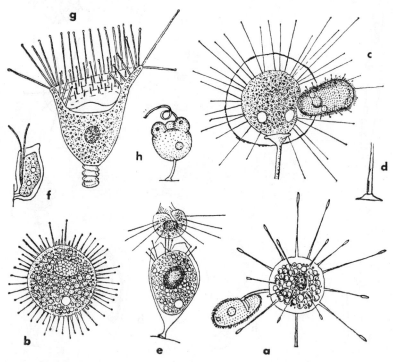

Fig. 383. a, *Parapodophrya typha*, ×270 (Kahl); b, *Sphaerophrya soliformis*, ×200 (Lauterborn); c, d, *Paracineta limbata* (c, a bud is ready to leave; d, basal part of stalk), ×460 (Collin); e, *Metacineta mystacina*, capturing Halteria, ×400 (Collin); f, *Urnula epistylidis*, ×140 (Claparède and Lachmann); g, *Lecanophrya drosera*, ×390 (Kahl); h, *Ophryocephalus capitatum*, ×200 (Wailes).

rigid stalk; suctorial tentacles in fascicles or distributed on entire body surface; encystment common; fresh or salt water. (Many species.)

P. fixa Müller (Fig. 382, *i, j*). Spherical; tentacles of various lengths; stalked; nucleus spheroid; one contractile vacuole; 10-28μ long; fresh water.

P. collini Root. Ovoid; stalked; thirty to sixty capitate tentacles, distributed; nucleus spherical; one contractile vacuole; 40-50μ in diameter; in swamp (Root, 1914). (Reproduction, Hull, 1956, Kormos, 1956; ingestion mechanism, Hull, 1961.)

P. elongata Wailes (Fig. 382, *k*). Elongate; flattened; with a pedicel; tentacles distributed; nucleus cylindrical; 95-105μ long; stalk 65-85μ by 7-9μ; on the marine copepod, *Euchaeta japonica;* Vancouver.

P. parasitica Fauré-Fremiet. Parasitic on the posterior region of a ciliate *Nassula ornata;* spherical, 25-50μ in diameter (Fauré-Fremiet, 1945).

Genus **Parapodophrya** Kahl. Spherical; tentacles radiating, a few long, more or less conical at proximal portion; stalk thin; salt water.

P. typha K. (Fig. 383, *a*). 50-60μ in diameter; salt water (Kahl, 1931).

Genus **Sphaerophrya** Claparède and Lachmann. Spherical, without stalk; with or without distributed tentacles; multiplication by binary fission or exogenous budding; fresh water, free-living or parasitic.

S. soliformis Lauterborn (Fig. 383, *b*). Spherical; numerous tentacles about one-fourth to one-third the body diameter; a contractile vacuole; nucleus oval; diameter about 100μ; sapropelic.

S. magna Maupas. Spherical; about 50μ in diameter; numerous tentacles of different length; nucleus spheroid; standing fresh water with decaying vegetation.

S. stentoris M. Parasitic in Stentor; larvae ciliated on posterior end; the other end with capitate tentacles; nucleus spheroid; 2 contractile vacuoles; about 50μ long.

Genus **Paracineta** Collin. Spherical to ellipsoidal; tentacles distributed; mostly in salt water, a few in fresh water.

P. limbata (Maupas) (Fig. 383, *c, d*). With or without gelatinous envelope; 20-50μ in diameter; larva with many ciliated bands, contractile; on plants and animals in salt water.

Genus **Metacineta** Bütschli. Lorica funnel-shaped, lower end drawn out for attachment; tentacles grouped at anterior end; nucleus spherical; one contractile vacuole. (One species.)

M. mystacina (Ehrenberg) (Fig. 383, *e*). Lorica up to 700μ long; in fresh and salt water.

Genus **Urnula** Claparède and Lachmann. Lorica colorless; lower end pointed, attached; aperture narrowed, round or triangular; body more or less filling lorica; one to two (up to five) long active tentacles; nucleus central, oval; one or more contractile vacuoles; fresh water.

U. epistylidis C. and L. (Fig. 383, *f*). Up to 80μ long; on Epistylis, Dendrosoma, etc.

Genus **Lecanophrya** Kahl. Body rounded rectangular in cross section; anterior region bowl-shaped; somewhat rigid tentacles located on the inner surface of bowl; salt water.

L. drosera K. (Fig. 383, *g*). 40-70μ high; hollow stalk; tentacles in 3-5 indistinct rows; attached to the antennae of the copepod, *Nitocra typica*.

Genus **Ophryocephalus** Wailes. Spheroidal, stalked; a single long mobile, capitate tentacle; multiplication by multiple exogenous budding from apical region; on *Ephelota gemmipara* and *E. coronata* (p. 1053); salt water. (One species.)

O. capitatum W. (Fig. 383, *h*). About 55μ long; tentacle up to 100μ by 1.5-5μ; Vancouver.

Family 6. **Discophryidae** Collin

Small number of tentacles; without lorica; with a short, stout stalk; thick pellicle; budding endogenous.

Genus **Discophrya** Lachmann. Elongate; a short stout pedicel with a plate; tentacles evenly distributed on anterior surface or in fascicles; contractile vacuoles, each with a canalicule leading to body surface; mainly fresh water. (Several species, Swarczewsky, 1928b; generic consideration, Matthes, 1954; species on aquatic insects, Matthes, 1953, 1954.)

D. elongata (Claparède and L.) (Fig. 384, *a*). Cylindrical; tentacles on anterior end and in two posterior fascicles; stalk striated; about 80μ long; on the shell of *Paldina vivipara* in fresh water.

Genus **Thaumatophrya** Collin. Spherical; long stalk; tentacles distributed, tapering toward distal end; salt water. (One species.)

T. trold (Claparède and Lachmann) (Fig. 384, *b*). About 75μ in diameter.

Genus **Rhynchophrya** Collin. Oblong; bilaterally symmetrical; a short striated stalk; one main long and a few shorter tentacles;

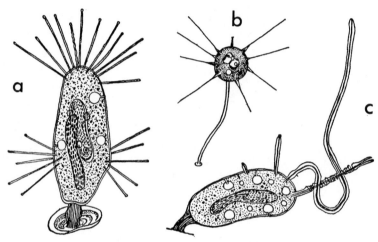

FIG. 384. a, *Discophrya elongata*, ×440 (Collin); b, *Thaumatophrya trold*, ×1150 (Claparède and Lachmann); c, *Rhynchophrya palpans*, ×440 (Collin).

six to ten contractile vacuoles, each with a canalicule leading to outside; fresh water. (One species.)

R. palpans C. (Fig. 384, *c*). 85μ by 50μ; tentacles retractile, 10-200μ long; stalk 20μ by 10μ; on *Hydrophilus piceus*.

Genus **Choanophrya** Hartog. Spheroidal to oval; stalked; ten to twelve tentacles; tubular, expansible at distal end to engulf voluminous food particles; macronucleus oval to spherical, a micronucleus; fresh water. (One species.)

C. infundibulifera H. (Fig. 386, *a*). 65μ by 60μ; fully extended tentacles 200μ long; on *Cyclops ornatus*. Tentacles and feeding (Farkas, 1924).

Genus **Rhyncheta** Zenker. Protoplasmic body attached directly

to an aquatic animal; with a long mobile tentacle bearing a
sucker at its end.

R. cyclopum Z. (Fig. 386, *b, c*). About 170μ long; on Cyclops.
Genus **Heliophrya** Saedeleer and Tellier. Without lorica or

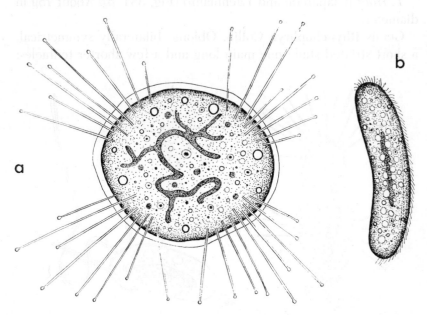

Fig. 385. *Heliophrya erharsi*, ×430. a, adult; b, a swimming individual
(Dragesco *et al.*).

stalk, attached to substrate with its entire basal surface which is
characteristically spread out; larva is lenticular with equatorial
ciliation (Matthes, 1954).

H. erhardi (Rieder) (Fig. 385). Body rounded or irregularly quadrangular; pellicular expansion up to 12μ wide; 8-20 long tentacles in each of the 4 fascicles; macronucleus dendritic; many
micronuclei (Dragesco *et al.*, 1955).

Family 7. **Ephelotidae** Kent

With suctorial and prehensile tentacles; with or without lorica;
with stalk; budding exogenous.

Genus **Ephelota** Wright. Without lorica; stalk stout, often
striated; suctorial and prehensile tentacles distributed; macronu-

cleus usually elongate, curved; on hydroids, bryozoans, algae, etc.; salt water. (Numerous species.)

E. *gemmipara* Hertwig (Fig. 386, *d*). About 250μ by 220μ; stalk up to 1.5 mm. long; on hydroids, bryozoans, etc.

E. *coronata* Kent (Fig. 386, *e*). Flattened; 90-200μ long; stalk longitudinally striated (Kent); on hydroids, bryozoans, algae, etc.

E. *plana* Wailes (Fig. 386, *f*). 150-320μ by 100-150μ; stalk

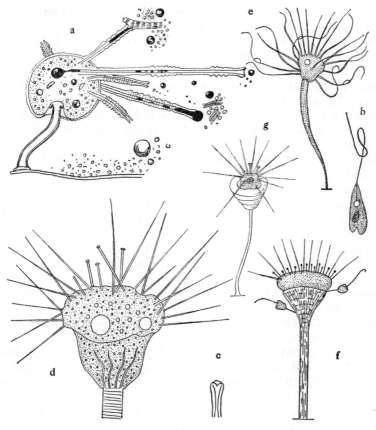

Fɪɢ. 386. a, *Choanophrya infundibulifera*, feeding on disintegrating part of a Cyclops, ×400 (Collin); b, c, *Rhyncheta cyclopum* (b, ×100; c, end of tentacle, ×400) (Zenker); d, *Ephelota gemmipara*, ×200 (Hertwig); e, E. *coronata*, ×140 (Kent); f, E. *plana*, front view, with two attached Ophryocephalus, ×35 (Wailes); g, *Podocyathus diadema*, ×200 (Kent).

100μ-1 mm. long; on bryozoans and crustaceans. (Conjugation, Grell, 1953; EM structure of the tentacles, Rouiller *et al.*, 1956.)

Genus **Podocyathus** Kent. It differs from *Ephelota* in having a conspicuous lorica; salt water. (One species.)

P. diadema K. (Fig. 386, *g*). Lorica about 42μ long; on bryozoans, hydrozoans, etc.

References

CANELLA, M. F., 1957. Studi e ricerche sui tentaculiferi nel quadro della biologia generale. Ann. Univ. Ferrara, N.S., 1:259.

COLLIN, B., 1911. Études monographique sur les Acinétiéns. I. Arch. zool. exper. gén., Sér. 5, 8:421.

——— 1912. II. *Ibid.*, 51:1.

DAVIS, H. S., 1942. A suctorian parasite of the small mouth black bass, etc. Tr. Am. Micr. Soc., 61:309.

DRAGESCO, J., *et al.*, 1955. Morphologie et biologie d'un Tentaculifère peu connu: *Heliophrya erhardi* (Rieder) Matthes. Bull. Microsc. Appl., 5:105.

FARKAS, B., (1924). Beiträge zur Kenntnis der Suctorien. Arch. Protist., 48:125.

FAURÉ-FREMIET, E., 1943. Commensalisme et adaption chez un acinétien: *Erastophrya chattoni*, etc. Bull. soc. zool. Fr., 68:145.

——— 1945. *Podophrya parasitica* nov. sp. Bull. Biol. France et Belg., 79:85.

GÖNNERT, R., 1935. Ueber Systematik, Morphologie, Entwicklungsgeschichte und Parasiten einiger Dendrosomidae, etc. Arch. Protist., 86:113.

GOODRICH, J. P., and JAHN, T. L., 1943. Epizoic Suctoria from turtles. Tr. Am. Micr. Soc., 62:245.

GRELL, K. G., 1953. Die Konjugation von *Ephelota gemmipara*. Arch. Protist., 98:287.

GUILCHER, Y., 1951. Contribution à l'étude des ciliés gemmipares, chonotriches et tentaculifères. Ann. des Sc. Nat., Zool., Sér. 11, 13:33.

HSIUNG, T. S., 1928. Suctoria of the large intestine of the horse. Iowa State Coll. J. Sci., 3:101.

——— 1930. A monograph on the Protozoa of the large intestine of the horse. *Ibid.*, 4:356.

HULL, R. W., 1954. The morphology and life cycle of *Solenophrya micraster* Penard. J. Protozool., 1:93.

——— 1954a. Feeding processes in *Solenophrya micraster*. *Ibid.*, 1:178.

——— 1956. Reproductive potential of isolated *Podophrya collini*. *Ibid.*, 3 (suppl.):11.

———— 1961. Studies on suctorian protozoa: the mechanism of ingestion of prey cytoplasm. *Ibid.*, 8:351.

KAHL, A., 1931. Ueber die verwandtschaftlichen Beziehungen der Suctorien zu den Prostomen Infusorien. Arch. Protist., 73:424.

———— 1934. Suctoria. Grimpe's Die Tierwelt der Nord- und Ostsee. Part 26. Leipzig.

KENT, S., 1881-1882. A manual of the Infusoria.

KORMOS, J., 1938. Bau und Funktion der Saugroehrchen der Suctorien. All. Közl., 35:130.

KORMOS, J., and K., 1956. Neue Untersuchungen ueber den Geschlechtsdimorphismus der Prodiscophrya. Acta Biol. Sci. Hung., 7:109.

———— ———— 1956a. Determination in der Entwicklung der Suctoria. *Ibid.*, 7:365.

———— ———— 1958. Aeussere und innere Konjugation. *Ibid.*, 8:104.

———— ———— 1961. Phylogenetische Wertung konvergenter und divergenter Eigenschaften bei den Suktorien. *Ibid.*, 11:335.

LILLY, D. M., 1953. The nutrition of carnivorous protozoa. Ann. N.Y. Acad. Sci., 56:910.

MATTHES, D., 1953. Suktorienstudien. V. Zeitschr. Morph. Oekol. Tier., 42:307.

———— 1954. Beitrag zur Kenntnis der Gattung Discophrya Lachmann. Arch. Protist., 99:187.

———— 1954a. *Discophrya buckei* (Kent), eine formenreiche Art, etc. Zool. Anz., 153:242.

———— 1954b. Suktorienstudien VI. Arch. Protistenk., 100:143.

———— 1956. VIII. *Ibid.*, 101:477.

PALINCSAR, E. E., 1959. Nutritional effects of nucleic acids and nucleic acid components on cultures of the suctorian *Podophrya collini*. J. Protozool., 6:195.

PESTEL, B., 1932. Beiträge zur Morphologie und Biologie des *Dendrocometes paradoxus.* Arch. Protist., 75:403.

PROST, M., 1952. *Trichophrya intermedia* sp. n. on the gills of salmon-fry. Ann. Univ. Maria Curie, 6:379.

ROOT, F. M., 1914. Reproduction and reactions to food in the suctorian, *Podophrya collini* n. sp. Arch. Protist., 35:164.

———— 1922. A new suctorian from Woods Hole. Tr. Am. Micr. Soc., 41:77.

ROUILLER, C., et al., 1956. Les tentacules d'Ephelota; étude au microscope electronique. J. Protozool., 3:194.

RUDZINSKA, M. A., 1951. The effect of overfeeding and starvation on the life span and reproduction of *Tokophrya infusionum*, etc. J. Gerontol., 6, Suppl. 3:144.

———— 1958. An electron microscope study of the contractile vacuole in *Tokophrya infusionum*. J. Biophys. Biochem. Cytol., 4:195.

———— and CHAMBERS, R., 1951. The activity of the contractile vacuole in a suctorian (*Tokophrya infusionum*). Biol. Bull., 100:49.

———— and PORTER, K. R., 1954. Electron microscope study of intact tentacles and disc in *Tokophrya infusionum*. Experientia, 10:460.

———— ———— 1955. Observations on the fine structure of the macronucleus of *Tokophrya infusionum*. J. Biophys. Biochem. Cytol., 1:421.

SKREB-GUILCHER, Y., 1955. Quelques remarques sur un protozoaire cilié le Tentaculifère, *Stylocometes digitatus* Stein. Bull. Micros. Appliquée, 5:118.

SWARCZEWSKY, B., 1928. Zur Kenntnis der Baikalprotistenfauna. I. Arch. Protist., 61:349.

———— 1928a. II. *Ibid.*, 62:41.

———— 1928b. III. *Ibid.*, 63:1.

———— 1928c. IV. *Ibid.*, 63:362.

WAILES, G. H., 1928. Dinoflagellates and Protozoa from British Columbia. Vancouver Museum Notes, 3:25.

Chapter 43

Collection, Cultivation, and Observation of Protozoa

Collection

IN THE foregoing chapters it has been shown that various species of protozoa have characteristic habitats and that many of free-living forms are widely distributed in bodies of water: fresh, brackish, and salt; while the parasitic forms are confined to specific host animals. Of free-living protozoa many species may occur in large numbers within a small area under favorable conditions, but the majority are present in comparatively small numbers. If one who has become acquainted with the representative forms intends to make collection, it is well to carry a dissecting microscope in order to avoid bringing back numerous jars containing much water, but few organisms. Submerged plants, decaying leaves, surface scum, ooze, etc., should be examined under the microscope. When desired forms are found, they should be collected together with a quantity of water in which they occur.

When the material is brought into the laboratory, it is often necessary to concentrate the organisms in a relatively small volume of water. For this purpose, the water may partly be filtered rapidly through a fine milling cloth and the residue quickly poured back into a suitable container before filtration is completed. The container should be placed in a cool moderately lighted room to allow the organisms to become established in the new environment. Stigma-bearing Phytomastigia will then be collected in a few hours on the side of the container, facing the strongest light, and the members of Sarcodina will be found among the debris on the bottom. Many forms will not only live long, but also multiply in such a container.

For obtaining large freshwater amoebae, fill several finger bowls with the collected material and place one or two rice grains to each. After a few days, examine the bottom surface of the bowls under a binocular dissecting microscope. If amoebae were included in the collection, they will be found particularly around the rice grains. Pipette them off and begin separate cultures (p. 1061).

In order to collect parasitic protozoa, one must, of course, find the host organisms that harbor them. Various species of tadpoles, frogs, cockroaches, termites, etc., which are of common occurrence or easily obtained and which are hosts to numerous species of protozoa, are useful material for class work.

Intestinal protozoa of man are usually studied in the faeces of an infected person. Natural movement should be collected. Do not use oily purgatives in obtaining faecal specimens, as they make the microscopical examination difficult by the presence of numerous oil droplets. The receptacle must be thoroughly cleaned and dry, and provided with a cover. Urine or water must be excluded completely. The faeces must be examined as soon as possible, since the active trophozoites degenerate quickly once leaving the human intestine. If dysenteric or diarrhoeic stools are to be examined, they must not be older than one hour or two. If this is not possible, wrap the container with woolen cloth while transporting, the organisms may live for several hours. Care must, however, be exercised during the microscopical examination, as there will be present unavoidably a large number of degenerating forms. If the stool is formed and normal, it would contain usually encysted forms and no trophozoites, if the host is infected by a protozoan, unless mucus, pus, or blood is present in it. Examination of such faeces can be delayed, as the cysts are quite resistant (p. 536).

Certain protozoa are now kept in cultures by many workers. Consult the catalogue prepared by the Committee on cultures, Society of Protozoologists (1958) and culture collection of algae (Starr, 1956, 1958).

Cultivation

For extensive study or for class work, a large number of certain species of protozoa are frequently needed. Detection and diagnosis of human protozoa are often more satisfactorily made by culture method than by microscopical examination of the collected material. Success in culturing protozoa depends upon several factors. First, an abundant supply of proper food material must be made available. For example, several species of Paramecium live almost exclusively on bacterial organisms, while Didinium and allied ciliates depend upon Paramecium and other ciliates as sources of food supply. For cultivating chromatophore-bearing forms successfully, good light and proper kinds and amount of inorganic substances are necessary. In the second place, the temperature and chemical constituents of the culture medium must be adjusted to suit individual species. As a rule, lower temperatures seem to be much more favorable for culture than higher temperatures, although this is naturally not the case with those parasitic in homoiothermal animals. Furthermore, proper hydrogen ion concentration of the culture must be maintained. In the third place, both protozoa and metazoa which prey upon the forms under cultivation must be excluded from the culture. For instance, it is necessary to remove *Didinium nasutum* in order to obtain a rich culture of Paramecium. For successful culture of *Amoeba proteus,* Aeolosoma, Daphnia, Cyclops, etc., must be excluded from the culture.

Mixed cultures of many free-living protozoa are easily maintained by adding from time to time a small amount of ripe hay-infusion or dried lettuce powder to the collected water mentioned before. Chilomonas, Peranema, Bodo, Arcella, Amoeba, Paramecium, Colpoda, Stylonychia, Euplotes, etc., often multiply in such cultures. To obtain a large number of a single species, individuals are taken out under a binocular dissecting microscope by means of a finely drawn-out pipette and transferred to a suitable culture medium. Such a culture is called a *mass* or *stock culture.* If a culture is started with a single individual, the resulting population makes up a *clone* or a *pure line.*

Aside from the cultures of blood-inhabiting protozoa and of some 100 free-living forms, the protozoan cultures are by no means "pure" cultures in the bacteriological sense, even if only one species of protozoa is present, since bacteria and other microorganisms are invariably abundantly present in them.

A. Free-living Protozoa

To deal with all the culture media employed by numerous workers for various free-living Protozoa is beyond the scope of the present work. Here only a few examples will be given. For further information, the reader is referred to Bělař (1928), Needham *et al.* (1937), etc.

Chromatophore-bearing flagellates. There are a number of culture fluids. Two examples:

(a)
Peptone or tryptone	2.0	gm.
KH₂PO₄	0.25	gm.
MgSO₄	0.25	gm.
KCl	0.25	gm.
FeCl₃	trace	
Sodium acetate	2.0	gm.
Pyrex distilled water	1000	cc.

(b)
Peptone or tryptone	2.5	gm.
KNO₃	0.5	gm.
KH₂PO₄	0.5	gm.
MgSO₄	0.1	gm.
NaCl	0.1	gm.
Sodium acetate	2.5	gm.
Dextrose	2.0	gm.
Pyrex distilled water	1000	cc.

Chilomonas and other colorless flagellates. A number of culture fluids have been advocated. A simple yet satisfactory one is as follows: Fill a finger bowl with about 150 cc. of glass distilled water and place four rice grains on the bottom. Let the dish stand for a few days, and then introduce with a pipette a number of desired flagellates from a mass culture into it. Cover the bowl and keep it at about 20°C. (Soil-water culture, Pringsheim, 1946a.)

Mast (1939) used the following media for *Chilomonas paramecium.*

(a) Glucose-peptone solution:

Peptone	8 gm.
Glucose	2 gm.
Water	1000 cc.

(b) Acetate-ammonium solution:

Sodium acetate	1.5 gm.
Ammonium chloride	0.46 gm.
Ammonium sulphate	0.1 gm.
Dipotassium hydrogen phosphate	0.2 gm.
Magnesium chloride	0.01 gm.
Calcium chloride	0.012 gm.
Water	1000 cc.

Amoeba proteus and other freshwater amoebae. Fill a finger bowl with 200 cc. of glass distilled water, and place four rice grains in it. After a few days, seed with amoebae (p. 1058), add about 5 cc. of Chilomonas culture, and cover the bowl with a glass cover. In about two weeks, a ring of amoebae will be found around each rice grain, and if Chilomonas do not overmultiply, the amoebae will be found abundantly in another two weeks. If properly maintained, subcultures may be made every four to six weeks. Chalkley (1930) advocates substitution of the plain water with a salt solution which is composed of

NaCl	0.1 gm.
KCl	0.004 gm.
$CaCl_2$	0.006 gm.
Glass distilled water	1000 cc.

If the culture water becomes turbid, make subcultures or pour off the water and fill with fresh distilled water or the solution. Culture should be kept at 18-22°C.

Hahnert (1932) used the following culture solution:

KCl	0.004 gm.
$CaCl_2$	0.004 gm.
$CaH_4(PO_4)_2$	0.002 gm.
$Mg_3(PO_4)_2$	0.002 gm.
$Ca_3(PO_4)_2$	0.002 gm.
Pyrex distilled water	1000 cc.

Pelomyxa carolinensis. These amoebae grow well in a finger bowl with 150 cc. of redistilled water with 1-3 rice grains to which large numbers of Paramecium are added daily. Pace and

Belda (1944) advocate the following solution instead of distilled water:

K_2HPO_4	0.08 gm.
KH_2PO_4	0.08 gm.
$CaCl_2$	0.104 gm.
$Mg_3(PO_4)_2 \cdot 4H_2O$	0.002 gm.
Pyrex distilled water	1000 cc.

Small mono- or di-phasic amoebae. Musgrave and Clegg's medium, modified by Walker, is as follows:

Agar	2.5 gm.
NaCl	0.05 gm.
Liebig's beef-extract	0.05 gm.
Normal NaOH	2 cc.
Distilled water	100 cc.

Arcella and other Testacea. The testaceans commonly multiply in a mixed culture for several weeks after the collection was made. Hegner's method for Arcella: Pond water with weeds is shaken up violently and filtered through eight thicknesses of cheese cloth, which prevents the passage of coarse particles. The filtrate is distributed among Petri dishes, and when suspended particles have settled down to the bottom, specimens of Arcella are introduced. This will serve also for Difflugia and other testaceans. Hay or rice infusion is also a good culture medium for these organisms.

Foraminiferida. Lee *et al.*, (1961) give useful information on laboratory culture.

Actinophrys and Actinosphaerium Bělař cultivated these heliozoans successfully in Knop's solution:

Magnesium sulphate	0.25 gm.
Calcium nitrate	1 gm.
Potassium phosphate	0.25 gm.
Potassium chloride	0.12 gm.
Iron chloride	trace
Distilled water	1000 cc.

Freshwater ciliates. They are easily cultivated in a weak infusion of hay, bread, cracker, lettuce leaf, etc. The battery jars containing the infusions should be left standing uncovered for a few days to allow a rich bacterial growth in them. Seed them with material such as submerged leaves or surface scum containing the ciliates. If desired, culture may be started with a single individual in a watch glass. (Collection, cultivation and sterilization of Paramecium, Wichterman, 1949; cultivation of soil protozoa, Singh, 1955.)

Axenic Culture. An increasing number of flagellates and ciliates have been successfully cultured axenically. In order to eliminate associated organisms, repeated washing, dilution and migration methods had been used (Glaser and Coria, 1930; Claff, 1940; Kidder, 1941; Slater, 1955), and antibiotics have, in recent years, been used for elimination of bacteria and other organisms for initiation and continuance of axenic cultures. For example, Balamuth (1962) used a mixture of penicillin, streptomycin, terramycin, magnamycin, polymyxin B for sterilization of Entamoeba cultures. Elliott and Clark (1956) maintained axenic cultures of *Tetrahymena pyriformis* by addition of penicillin and streptomycin. Effects of antibiotics on *Entamoeba histolytica* (Phillips, 1951).

Free-living Phytomastigia. Many media are known. (See Pringsheim, 1926, 1937, 1946; Hall, 1937, 1941; Hutner and Provasoli, 1951, etc.)

Storm and Hutner (1953) reported an excellent growth of *Peranema trichophorum* in pure culture by using a medium, composed of liquid whole milk 1.0 cc., soil extract 10 cc. and distilled water 89 cc.

Tetrahymena and allied forms. Kidder, Dewey and Parks use a basal medium as quoted below:

	γ per ml		γ per ml
DL-alanine	110	DL-threonine	326
L-arginine	206	L-tryptophane	72
L-aspartic acid	122	DL-valine	162
Glycine	10		
L-glutamic acid	233	Ca pantothenate	0.10
L-histidine	87	Nicotinamide	0.10
DL-isoleucine	276	Pyridoxine HCl	1.00
L-leucine	344	Pyridoxal HCl	0.10
L-lysine	272	Pyridoxamine HCl	0.10
DL-methionine	248	Riboflavin	0.10
L-phenylalanine	160	Pteroylglutamic acid	0.01
L-proline	250	Thiamine HCl	1.00
DL-serine	394	Biotin (free acid)	0.0005
		Choline Cl	1.00

	γ per ml		γ per ml
MgSo$_4$·7H$_2$O	100	Guanylic acid	30
Fe(NH$_4$)$_2$(SO$_4$)$_2$·6H$_2$O	25	Adenylic acid	20
MnCl$_2$·4H$_2$O	0.5	Cytidylic acid	25
ZnCl$_2$	0.05	Uracil	10
CaCl$_2$·2H$_2$O	50		
CuCl$_2$·2H$_2$O	5	Dextrose	2,500
FeCl$_3$·6H$_2$O	1,25	Na acetate	1,000
K$_2$HPO$_4$	1,000	Tween 85	700
KH$_2$OP$_4$	1,000	Protogen	1 unit

Paramecium aurelia. See van Wagtendonk *et al.,* (1953).

P. multimicronucleatum. See Johnson and Baker (1942); Johnson and Miller (1956).

B. Parasitic Protozoa

Intestinal flagellates of man. There are numerous media which have been used successfully by several investigators.

(a) Ovo-mucoid medium (Hogue, 1921). White of two eggs are broken in a sterile flask with beads. Add 200 cc. of 0.7% NaCl solution and cook the whole for thirty minutes over a boiling water bath, shaking the mixture constantly. Filter through a coarse cheese cloth and through cotton-wool with the aid of a suction pump. Put 6 cc. of the filtrate in each test tube. Autoclave the tubes for twenty minutes under fifteen pounds pressure. After cooling, a small amount of fresh faecal material containing the flagellates is introduced into the tubes. Incubate at 37°C.

(b) Sodium chloride sheep serum water (Hogue, 1922). Composed of 100 cc. sterile 0.95% NaCl and 10-15 cc. of sterile sheep serum water (dilution 1:3). 15 cc. to each tube. *Trichomonas hominis, T. tenax,* and *Retortamonas intestinalis* grow well.

Trichomonas vaginalis. Johnson and Trussell (1943) reported the following mixture the most suitable medium:

Bacto-peptone	32 gm.
Bacto-agar	1.6 gm.
Cysteine HCl	2.4 gm.
Maltose	1.6 gm.
Difco liver infusion	320 cc.
Ringer's solution	960 cc.
NaOH(N/1)	11-13 cc.

Heat the mixture in a water bath to melt the agar; filter through a coarse paper; add 0.7 cc. of 0.5 per cent aqueous methylene blue; adjust pH to 5.8-6.0 with N/1 HCL or NaOH; tube 8 cc.; autoclave. After cooling, add aseptically 2 cc. of sterile (filtered) human serum. Incubate at least four days; store at room temperature for two to three weeks or as long as an amber "anaerobic" zone is apparent.

Termite flagellates. Trager's (1934) media are as follows:

	Solution A gm. per liter water	Solution U gm. per liter water
NaCl	1.169	2.164
NaHCO$_3$	0.840	0.773
Na$_3$C$_6$H$_5$O$_7 \cdot$2H$_2$O (citrate)	2.943	1.509
NaH$_2$PO$_4 \cdot$H$_2$O	0.690	0
KCl	0.745	0
KH$_2$PO$_4$	0	1.784
CaCl$_2$	0.111	0.083
MgSO$_4$	0	0.048

In solution A, *Trichomonas* sp. and *Tricercomitus termopsidis* were cultivated. For *Trichomonas termopsidis,* a small amount of Loeffler's blood serum and cellulose were added. All three flagellates were cultured for over three years. In solution U to which 0.01 per cent blood serum, cellulose and charcoal, were added, *Trichonympha sphaerica* (from *Termopsis angusticollis*) grew well and multiplied up to two weeks, although *T. campanula* and *T. collaris* failed to do so. The culture in a test tube was inoculated with the entire hindgut of a termite and kept at room temperature.

Lophomonas blattarum and L. striata. A mixture of one sterile egg-white and 100 cc. of sterile Ringer's solution, to which a small amount of yeast cake is added, is an excellent culture medium. Incubation at room temperature; subcultures every four to six days.

Trypanosoma and Leishmania. Novy, MacNeal and Nicolle (NNN) medium: 14 gm. of agar and 6 gm. of NaCl are dissolved by heating in 900 cc. of distilled water. When the mixture cools to about 50°C., 50-100 cc. of sterile defibrinated rabbit blood is gently added and carefully mixed so as to prevent the formation of bubbles. The blood agar is now distributed among sterile test tubes to the height of about 3 cm., and the tubes are left slanted until the medium becomes solid. The tubes are then incubated at 37°C. for twenty-four hours to determine sterility and further to hasten the formation of condensation water (pH 7.6). Sterile blood or splenic puncture containing *Trypanosoma cruzi* or Leishmania is introduced by a sterile pipette to the condensation water in which organisms multiply. Incubation at 37°C. for trypanosomes and at 20-24°C. for Leishmania.

For cultivating *T. gambiense* and *T. rhodesiense,* Tobie, von Brand and Mehlman (1950) used the following medium:

(a) Base. 1.5 gm. Bacto-beef, 2.5 gm. Bacto-peptone, 4 gm. sodium chloride and 7.5 gm. Bacto-agar, are dissolved in 500 cc. distilled water. After adjusting pH to 7.2-7.4 with NaOH, autoclave at fifteen pounds pressure for twenty minutes. Cool this to about 45°C., then add whole rabbit blood which had been inactivated at 56°C. for thirty minutes, in the proportion of 25 cc. blood to 75 cc. base, using 0.5 per cent sterile sodium citrate to prevent the coagulation. This base is placed in test tubes (5 cc. each and slanted) or in flasks (25 cc.), and allowed to solidify.

(b) Liquid phase. Sterile Locke's solution. This is added in amounts of 2 cc. (to test tubes) or 10-15 cc. (to flasks), and cotton plugs are applied. The trypanosomes are said to grow well and to reach the peak population in ten to fourteen days.

Entamoebae. Various species have been cultured *in vitro* in association with living bacteria. Griffin and his coworkers (1949-1950) found that for *Entamoeba histolytica* and *E. terrapinae,* in addition to living bacteria, rice starch grains of particular size and certain factors from serum are necessary. Phillips (1950-1953) obtained cultures of *E. histolytica* in association with *Trypanosoma cruzi.* Recently, Diamond (1961) reported fifty successful transfers over a period of six months of axenic cultures of *E. histolytica* in a diphasic medium. (Nutrition of parasitic amoebae, M. Lwoff, 1951.)

Entamoeba barreti. Barret and Smith (1924) used a mixture of nine parts of 0.5% NaCl and one part of human blood serum. Incubation at 10-15°C.

E. invadens. Ratcliffe and Geiman (1938) used a mixture of gastric mucin 0.3 gm., "ground alum" salt 0.5 gm., and distilled water 100 cc. About 2 mg. of sterile rice starch is added to each culture tube at the time of inoculation. Culture at 20-30°C. and sub-culture every seven days.

E. histolytica and other amoebae of man. The first successful culture was made by Boeck and Drbohlav (1925) who used the following media.

(a) Locke-egg-serum (LES) medium. The contents of four eggs (washed and dipped in alcohol) are mixed with, and broken in, 50 cc. of Locke's solution in a sterile flask with beads. The solution is made up as follows:

NaCl	9 gm.
CaCl$_2$	0.2 gm.
KCl	0.4 gm.
NaHCO$_3$	0.2 gm.
Glucose	2.5 gm.
Distilled water	1000 cc.

The emulsion is now tubed so that when coagulated by heat, there is 1-1.5 inches of slant. These tubes are now slanted and heated at 70°C. until the medium becomes solidified. They are then autoclaved for twenty minutes at fifteen pounds pressure (temperature must be raised and lowered slowly). After cooling the slant is covered with a mixture of eight parts of sterile Locke's solution and one part of sterile inactivated human blood serum. The tubes are next incubated to determine sterility. The culture tubes are inoculated with a small amount of faecal matter containing active trophozoites. Incubation at 37°C. Yorke and Adams (1926) obtained rich cultures by inoculating this medium with washed and concentrated cysts of *E. histolytica* in twenty-four hours.

(b) Locke-egg-albumin (LEA) medium. The serum in LES medium is replaced by 1 per cent solution of crystallized egg albumin in Locke's solution which has been sterilized by passage through a Berkefeld filter.

Dobell and Laidlaw (1926) used Ringer's solution instead of Locke's.

(c) Ringer-egg-serum (RES) or Ringer-egg-albumin (REA) medium. Solid medium is the same as that of (a) or (b), but made up in Ringer's solution which is composed of

NaCl	9 gm.
KCl	0.2 gm.
CaCl₂	0.2 gm.
Distilled water	1000 cc.

The covering liquid is serum-Ringer or egg-albumin. The latter is prepared by breaking one egg white in 250 cc. of Ringer's solution which is passed through a Seitz filter. Before inoculating with amoebae, a small amount of sterile solid rice-starch (dry-heated at 180°C. for one hour) is added to the culture tube.

(d) Horse-serum-serum (HSS) or Horse-serum-egg-albumin (HSA) medium. Whole horse-serum, sterilized by filtration, is tubed and slanted at 80°C. for about sixty to seventy minutes (do not heat longer). When the slants have cooled, they are covered with diluted serum or egg-albumin given for (c). The tubes are incubated for sterility and sterile rice-starch is added immediately before inoculation. Frye and Meleny (1939) substituted the liquid portion of this medium by 0.5% solution of Lily liver extract No. 343 in 0.85% NaCl.

(e) Liver-agar-serum (LAS) medium. Cleveland and Sanders (1930) used the following medium:

Liver infusion agar	
(Difco dehydrated)	30 gm.
Glass distilled water	1000 cc.

The medium is tubed, autoclaved, and slanted. The slants are covered with a 1:6 dilution of sterile fresh horse serum in 0.85% NaCl solution. A 5 mm. loop of sterile rice flour or powdered unpolished rice is added to each tube. In making subculture, remove two or three drops of the rice flour debris from the bottom with a sterile pipette.

(f) Egg-yolk-saline medium (Balamuth and Sandza, 1944). Two eggs are hard-boiled. Upon cooling, the egg white is discarded and the yolks are crumbled in a beaker containing 125 ml of 0.8 per cent sodium chloride solution. The mixture is boiled for ten minutes, and after replacement of evaporated water the infusion is filtered by suction pump and restored to 125 ml. The filtrate is autoclaved twenty minutes at fifteen pounds pressure. Upon cooling, a slight precipitation of yolk settles, and is removed by simple filtration, after which 125 ml. of N/15 phosphate buffer (pH 7.5) is added, making the total salt concentration N/30 phosphate solution in 0.4 per cent sodium chloride. This final mixture is tubed in 5 ml. amounts, autoclaved as before, and then is stored under refrigeration until use. Before introducing amoebae a loop of sterile rice starch is added to each tube.

To inhibit bacterial growth in cultures of Entamoeba, various antibiotics have been tried. For example, Spingarn and Edelman (1947) found that when streptomycin was added in the amount of 1000-3000 units per cc. to culture of *E. histolytica,* the survival of the amoebae in culture was prolonged from an average of eight days to 33.7 days, which effect was apparently due to the inhibition of bacteria.

Encystment of *Entamoeba histolytica* is usually brought about by first cultivating the organisms in starch-free media and then by transferring them into media with starch. Balamuth (1951) recommends a diphasic medium of the following composition: 2 gm. of Wilson liver concentrate powder is brought to boiling in 80 ml. distilled water and filtered. Then 6.4 ml. of 0.25 molar Na_3PO_4. 12 H_2O and 7.6 ml. of 1.0 molar potassium phosphate buffer (in the ratio of 4.7 parts K_2HPO_4 to 0.3 part KH_2PO_4) are added. By adding distilled water in a volumetric flask bring the mixture to 100 ml. Transfer it to a beaker and add 3 gm. Bacto-agar. Heat gently until agar dissolves; then autoclave for twenty minutes at fifteen pounds pressure. The pH should be about 7.2. The overlay is prepared by mixing double-strength eggyolk and normal horse serum (10:1) and rise starch is added last. (Technique, Brooke, 1958.)

Plasmodium. Bass and John's (1912) culture is as follows: 10 cc. of defibrinated human blood containing Plasmodium and 0.1 cc. of 50 per cent sterile dextrose solution are mixed in test tubes and incubated at 37-39°C. In the culture, the organisms develop in the upper layer of erythrocytes. Since that time a number of investigators have undertaken cultivation of different species of Plasmodium. (For information the reader is referred to Geiman, Anfinsen *et al.*, 1946; Trager, 1950, and Manwell and Brody, 1950.)

Balantidium coli. Barret and Yarbrough (1921) first cultivated this ciliate in a medium consisting of sixteen parts of 0.5% NaCl and one part of inactivated human blood serum. The medium is tubed. Inoculation of a small amount of the faecal matter containing the trophozoites is made into the bottom of the tubes. Incubation at 37°C. Maximum development is reached in forty-eight to seventy-two hours. Subcultures are made every second day. Rees used a mixture of sixteen parts of Ringer's solution and one part of Loeffler's dehydrated blood serum.

Atchley (1935) employed a medium composed of four parts of Ringer's solution and one part of faeces, which is filtered after twenty-four hours, centrifuged and sterilized by passage through a Seitz filter. Nelson (1940) also used one part of caecal contents of pig in nine parts of Ringer's solution, which mixture is passed through a sieve and then filtered through a thick absorbent cotton. Balantidium which shows positive geotropism, is freed of faecal debris by passage downward through cotton in V-tube. The ciliates are introduced into the culture tubes. Incubation at 37°C. Subcultures are made every seven to twenty-two days. Nelson found that autoclaved medium is unsuitable until a living bacterial population has been established. Balantidium can also be cultivated in the media given for the intestinal amoebae.

Microscopical Examination

Protozoa should be studied as far as possible in life. Permanent preparations while indispensable in revealing many intracellular structures, cannot replace fresh preparations. The microscopic slides of standard size, 3″ by 1″, should be of white glass and preferably thin. The so-called No. 1 slides measure about 0.75 mm. in thickness. For darkfield illumination thin slides are essential. No. 1 coverglasses should be used for both fresh and permanent preparations. They are about 130-170μ thick. The most convenient size of the coverglass is about seven-eighths square inch which many prefer to circular ones.

The slides and coverglasses must be thoroughly cleaned before being used. Immerse them in concentrated mineral acids (nitric acid is best fitted) for ten minutes. Pour off the acid, wash the slides and coverglasses for about ten minutes in running water, rinse in distilled water, and keep them in 95 per cent alcohol. When needed they are dried one by one with clean cheese cloth. Handle slides and covers with a pair of forceps. If thumb and fingers are used, hold them edgewise.

A. Fresh Preparations

In making fresh preparations with large protozoa care must be exercised to avoid pressure of the coverglass on the organisms as this will cause deformities. If small bits of detritus or debris are included in the preparation, the coverglass will be supported by them and the organisms will not be subjected to any pressure. Although ordinary slides are used most frequently, it is sometimes advisable to use a depression slide especially for prolonged observation. To make a preparation with this slide, a small drop of water containing specimens is placed in the center of a coverglass, and is covered by a small circular coverglass (about 1 cm. in diameter), which in turn is covered by a depression slide with a thin coat of vaseline along the edge of the depression, so as to make an air-tight compartment. In turning over the whole, care must be taken to prevent the smaller circular cover from touching any part of the slide, as this would cause the water to run down into the depression. Nemeczek (1926) seems to have been the first one who used the second coverglass for this preparation. If the protozoa to be examined are large and observation can be made under a low power objective, the small coverglass should be omitted.

As far as possible examine fresh preparations with low power objectives. The lower the magnification, the brighter and the larger the field. The microscopical objects can quickly and easily be measured, if an ocular micrometer division has been calculated in combination with different objectives.

The free-living ciliates swim about so actively as to make their observation difficult. However, an actively swimming ciliate will sooner or later come to stop upon coming in contact with various debris, air bubbles or margin of the coverglass to allow a study of its structure. Various reagents recommended for retardation of swimming movements of ciliates, bring about deformities in the organisms and therefore, must not be used; but a drop of saturated or 10 per cent solution of methyl cellulose may be added to a ciliate preparation to retard the active movement of the organism without causing any visible abnormality (Marsland, 1943). A weak solution of nickel sulphate retards the movement of cilia and flagella (Tartar, 1950) and in Paramecium movement is said to be completely arrested in two hours and recovery is complete on returning to normal culture fluid.

The phase microscope is highly useful for observation of cilia, flagella and other filamentous organelles as well as various internal structures in living protozoa.

When treated with highly diluted solutions of certain dyes, living protozoa exhibit some of their organellae or inclusions stained without apparent injury to the organisms. These **vital stains** are usually prepared in absolute alcohol solutions. A small amount is uniformly applied to the slide and allowed to dry, before water containing Protozoa is placed on it. Congo red (1:1,000) is used as an indicator, as its red color of the salt changes blue in weak acids. Janus Green B (1:10,000-50,000) stains mitochondria. Methylene blue (1:10,000 or more) stains cytoplasmic granules, nucleus, cytoplasmic processes, etc., Neutral red (1:3,000-30,000) is an indicator: yellowish red (alkaline), cherry red (weak acid), and blue (strong acid). It also stains nucleus slightly. Golgi bodies are studied in it, though its specificity for this structure is not clear.

Parasitic protozoa should be studied in the tissue or body fluids in which they occur. When they are too small in amount to make a suitable preparation, one of the following solutions may be used.

Physiological salt solution. Widely used concentrations of NaCl solutions are 0.5-0.7 per cent for cold-blooded animals and 0.8-0.9 per cent for warm-blooded animals.

Ringer's solution. The one Dobell advocated has been given already (p. 1068). Another frequently used solution consists of

NaCl	0.8 gm.
KCl	0.02 gm.
CaCl$_2$	0.02 gm.
(NaHCO$_3$	0.02 gm.)
Glass distilled water	100 cc.

For demonstrating organellae, the following reagents which kill the protozoa upon application, may be used on living protozoa.

Lugol's solution. This is made up of potassium iodide 1.5 gm., water 25 cc., and iodine 1 gm. The solution deteriorates easily. Flagella and cilia stain clearly. Glycogen bodies stain ordinarily reddish brown. Cysts of intestinal protozoa are more easily studied in Lugol's solution.

Sudan III and IV. 2 per cent absolute alcohol solution diluted before use with the same amount of 45 per cent alcohol. Neutral fats are stained red.

Methyl green. 1 per cent solution in 1 per cent acetic acid solution makes an excellent nuclear stain.

Nigrosin. 10 per cent solution if used in smears and air-dried makes the pellicular patterns of flagellates and ciliates stand out clearly.

In the case of **faecal examination** if the stool is dysenteric, a small portion is placed by a tooth-pick or platinum loop on a slide and covered with a cover glass. Before placing the cover, all large particles must be removed quickly so that the smear will be uniformly thin. Smears of diarrhoeic stools can be made in a similar way. But if the faecal material is formed or semiformed, a small drop of warm (37°C.) 0.85% NaCl solution is first placed on the slide, and a small portion of the faeces, particularly mucus, pus or blood, is emulsified in it. The whole is covered by a coverglass. The faecal smear should not be too thick or too thin for a satisfactory observation. If the smear is too thick, it will be impossible to distinguish objects clearly, and on the other hand, if it is too thin, there will be much time lost in observing widely scattered protozoa. The optimum thickness of the smear is one through which the print of this page can be read.

The success in faecal examination for intestinal protozoa depends almost entirely on continued practice, since the faecal matter contains myriads of objects which may resemble protozoa (Fig. 387, *c-h*). Aside from certain coprozoic protozoa (p. 25) which appear in old faeces, *Blastocystis hominis* (Fig. 387, *c-f*) occur in almost all faeces. This organism which is considered to be a fungus and harmless to its host, is usually spherical and measures about 5-25μ in diameter. Within a very thin membrane, there is a narrow peripheral cytoplasmic layer in which one or two nuclei and several refractile granules are present. The cytoplasmic ring encloses a large homogeneous body which is somewhat eosinophile, but not iodinophile. In some the cytoplasm may be more abundant and the inclusion body smaller. Dividing forms appear peanut-shaped. (Blastocytis, Grassé 1926; Reyer, 1939.)

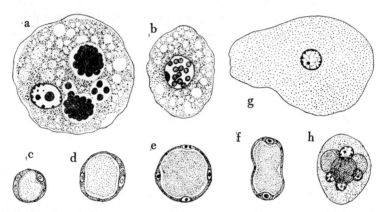

Fig. 387. a, Sphaerita in a stained trophozoite of *Entamoeba coli;* b, Nucleophaga in a stained trophozoite of *Iodamoeba bütschlii;* c, d, *Blastocystis hominis* (in an unstained smear); e, f, stained *Blastocystis hominis;* g, an epithelial cell from a faecal smear; h, a polymorphonuclear leucocyte with three ingested erythrocytes. All ×1150 (Kudo).

In a number of parasitic protozoa, there occur foreign organisms which may be mistaken for food inclusions or chromatin. They are vegetable organisms which were named by Dangeard as *Sphaerita* and *Nucleophaga* (Fig. 387, *a, b*). The former occurs in the cytoplasm and the latter in the nucleus of the host protozoan. These parasites are spherical and about 0.5-1µ in diameter; they are found most frequently in spherical masses composed of varying numbers of individuals. Nucleophaga appears to destroy the host nucleus. Degenerating epithelial cells or leucocytes (Fig. 387, *g, h*) may simulate parasitic amoebae. Fishes and birds are often infected by Coccidida and when they are consumed as food, the oocysts pass the alimentary canal unchanged and appear in the stools. (Sphaerita, Chatton and Brodsky, 1909; Mattes, 1924;

Becker, 1926; Sassuchin, 1928; Sassuchin *et al.*, 1930; Jahn, 1933; Kirby, 1941.)

The cysts of intestinal protozoa are, as a rule, distributed throughout the formed faeces and difficult to detect in small portions of the voided specimens. Flecks of mucus in the fluid stool obtained by use of a saline purge may contain more numerous cysts than naturally passed one. Several methods for concentrating cysts for microscopical examination are known. The simplest one is to emulsify thoroughly a small mass of faeces about the size of a lump sugar in a dish by adding a small amount of once-boiled tap water. Add to it about 500 cc. of water and pour the whole emulsion into a glass cylinder, and let it stand for about fifteen minutes. Remove the scum floating on the surface and draw off the turbid fluid into another cylinder, leaving the sediment and a little fluid just above it untouched. The majority of cysts are suspended in the drawn-off portion of the emulsion. Centrifuge the fluid, pour off the supernatant fluid and add water. Centrifuge again. Repeat this three times until the supernatant fluid becomes clear. The sediment will be found to contain more numerous cysts than small sample specimens. Bijlmer (1948) finds the following method the most satisfactory. Suspend a fleck of faeces about the size of a pea in a dish with some 33 per cent $ZnSO_4$. If much debris appear on the surface, filter through a layer of cheese-cloth. The fluid is decanted into a centrifuge tube, and some more $ZnSO_4$ solution is added to half a centimeter from the top. After centrifuging for two minutes, lift a loopful of material from the surface and place on a slide. (Technique in laboratory diagnosis, Brooke, 1958.)

B. Permanent Preparations

Permanent preparations are employed, as was stated before, to supplement, and not to supplant, fresh preparations. Smear preparations are more frequently studied, while section preparations are indispensable in extensive studies of protozoa. Various fixatives and stains produce different results, care must be exercised in making and evaluating permanent preparations. Diversity of stained objects (Wenrich, 1941).

a. Smear Preparations

Smears are made either on coverglasses or slides. However, coverglass-smears are more properly fixed and require a smaller amount of reagents than slide-smears. Greater care must be exercised in handling coverglasses, as they are easily broken. Large free-living protozoa do not frequently adhere to the glass, since there is not enough albuminous substance in the culture fluid. If a small drop of fresh egg-white emulsified in sterile distilled water is smeared on the coverglass very thinly with the tip of a clean finger, before mounting material for smear, more specimens will adhere to and remain on the coverglass upon the completion of the preparation. Let the smear lie horizontally for five to ten minutes or longer.

Parasitic protozoa live in media rich in albuminous substances, and therefore, easily adhere to the coverglass in smear. Make uniformly thin smears on coverglasses. If the smears are made from dysenteric or fluid stools, they should be fixed almost immediately. Smears made from diarrhoeic or formed stools by emulsifying in warm salt solution, should be left for a few minutes. In any case, do not let the smear become dry except a narrow marginal zone.

The smears are fixed next. The most commonly used fixative for protozoa is **Schaudinn's** fluid. This is made up as follows:

Cold saturated mercuric bichloride (6-7%)	66 cc.
Absolute or 95% alcohol	33 cc.
Glacial acetic acid	1 cc.

The first two can be kept mixed without deterioration, but the acid must be added just before fixation. Fix at room temperature or warmed to 50°C. The fixative is placed in a square Petri dish and the smear is gently dropped on it with the smeared surface facing downward. With a little experience, air bubbles can be avoided and make the smear float on the surface of the fixative. After about one minute, turn it around and let it stay on the bottom of the dish for five to ten more minutes. In case the smear is too thick, a thin coat of vaseline on the upper side of the coverglass will make it float. About six coverglass-smears may be fixed in the dish simultaneously.

The coverglass-smears are now transferred to a Columbia staining jar for coverglasses, containing 70 per cent alcohol for ten minutes, followed by two changes for similar length of time. Transfer the smears next to 50 per cent alcohol for five minutes, and then to a jar with water, which is now placed under gently running tap water for fifteen minutes. Rinse them in distilled water and stain.

Other fixatives frequently used for protozoa are as follows:

Bouin's fluid

Picric acid (saturated)	75 cc.
Formaldehyde	25 cc.
Glacial acetic acid	5 cc.

Fixation for five to thirty minutes; wash with 70 per cent alcohol until picric acid is completely washed away from the smears.

Sublimate-acetic

Saturated sublimate solution	100 cc.
Glacial acetic acid	2 cc.

This is the original fixative for Feulgen's nucleal reaction (p. 1079). Fixation and after-treatment similar to Schaudinn's fluid.

Carnoy's fluid

Absolute alcohol	30 cc.
Glacial acetic acid	10 cc.

Fixation for five to thirty minutes; wash in 95 per cent alcohol.

Osmium tetroxide

The vapor from or the solution itself of 1 per cent Osmium tetroxide may be used. Fixation in two to five minutes; wash in running water.

Flemming's fluid

1% chromic acid	30 cc.
2% osmium tetroxide	8 cc.
Glacial acetic acid	2 cc.

Fixation for ten to fifty minutes; wash for one hour or longer in running water.

Champy's fluid

1% chromic acid	7 parts
3% potassium dichromate	7 parts
2% osmium tetroxide	4 parts

Da Fano's fluid

Formaldehyde (40%)	15 cc.
Cobalt nitrate	1 gm.
distilled water	100 cc.

The most commonly used stain is **Heidenhain's** iron haematoxylin, as it is dependable and gives a clear nuclear picture, although it is unsatisfactory for voluminous organisms or smears of uneven thickness. It requires a mordant, ammonio-ferric sulphate (iron alum) and a dye, haematoxylin. Crystals of iron alum become yellow and opaque very easily. Select clear violet crystals and prepare 2 per cent aqueous solution. Haematoxylin solution must be well "ripe." The most convenient way of preparing it is to make 10 per cent absolute alcohol solution as it does not require ripening. By diluting this stock solution with distilled water, prepare 0.5 or 1 per cent slightly alcoholic solution which will be ready for immediate and repeated use. Smears are left in the mordant in a jar for one to three hours or longer. Wash them with running water for five minutes and rinse in distilled water. Place the smears now in haematoxylin for one to three hours or longer. After brief washing in water, the smears are decolorized in Petri dish in a diluted iron alum, 0.5% HCl in water or 50 per cent alcohol, or saturated aqueous solution of picric acid under the microscope. Upon completion, the smears are washed thoroughly in running water for about thirty minutes. Rinse them in distilled water. Transfer them through ascending series of alcohol (50 to 95 per cent). If counterstaining with eosin in desired, dip the smears which were taken out from 70 per cent alcohol, in 1 per cent eosin in 95 per cent alcohol for a few seconds, and then in 95 per cent plain alcohol. After two passages through absolute alcohol and through xylol, the smears are mounted one by one on a slide in a small drop of mixture of Canada balsam and xylol. The finished preparations are placed in a drying oven at about 60°C. for a few days.

Other stains that are often used are as follows:

Delafield's haematoxylin. If the stock solution is diluted to 1:5-10, a slow, but progressive staining which requires no decolorization may be made; but if stock solution is used, stain for one to sixteen hours, and decolorize in 0.5% HCl water or alcohol. If mounted in a neutral mounting medium, the staining remains true for a long time.

Mayer's paracarmine. In slightly acidified 70 per cent alcohol solution, it is excellent for staining large protozoa. If over-stained, decolorize with 0.5% HCl alcohol.

Giemsa's stain. Shake the stock solution bottle well. By means of a stopper-pipette dilute the stock with neutral distilled water (5-10 drops to 10 cc.). Smears fixed in Schaudinn's fluid and washed in neutral distilled water are stained in this solution for ten minutes to six hours to overnight. Rinse them thoroughly in neutral distilled water and transfer them through the following jars in order (about five minutes in each): (a) acetone alone; (b) acetone: xylol, 8:2; (c) acetone: xylol, 5:5; (d) acetone: xylol, 2:8; (e) two changes of xylol. The smears are now mounted in cedar wood oil (which is used for immersion objectives) and the preparations should be allowed to dry for a longer time than the balsam-mounted preparations.

Feulgen's nucleal reaction. The following solutions are needed.

(a) N HCl solution. This is prepared by mixing 82.5 cc. of HCl (specific gravity 1.19) and 1000 cc. of distilled water.

(b) Fuchsin-sodium bisulphite. Dissolve 1 gm. of powdered fuchsin (basic fuchsin, diamant fuchsin or parafuchsin) in 200 cc. of distilled water which has been brought to boiling point. After frequent shaking for about five minutes, filter the solution when cooled down to 50°C. into a bottle and add 20 cc. HCl solution (a). Cool the solution further down to about 25°C. and add 1 gm. of anhydrous sodium bisulphite. Apply stopper tightly. Decolorization of the solution will be completed in a few hours, but keep the bottle in a dark place for at least twenty-four hours before using it.

(c) Sulphurous water:

Distilled or tap water	200 cc.
10% anhydrous sodium	
bisulphite	10 cc.
HCl solution (a)	10 cc.

Feulgen's reaction is used to detect thymonucleic acid, a constituent of chromatin. By a partial hydrolysis, certain purinbodies in the acid are split into aldehydes which show a sharp Schiff's reaction upon coming in contact with fuchsin-sodium bisulphite. Thus this is a reaction, and not a staining method. Smears fixed in sublimate-acetic or Schaudinn's fluid are brought down to running water, after being placed for about twenty-four hours in 95 per cent alcohol. Immerse them in cold HCl (a) for one minute, then place them in HCl kept at 60°C. (over a microburner or in an incubator) for five minutes, quickly immerse in cold HCl. After rapidly rinsing in distilled water, place the smears in solution (b) for thirty minutes to three hours. There is no overstaining. The smears are then washed in three changes (at least two minutes in each) of solution (c). Wash them in running water for thirty minutes. If counterstaining is desired, dip in 0.1 per cent light green solution and rinse again in water. The smears are now dehydrated through a series of alcohol in the usual manner and mounted in Canada balsam (Feulgen and Rossenbeck, 1924; Feulgen-Brauns, 1924; Feulgen, 1926; Coleman, 1938; Stowell, 1945; Ely and Ross, 1949).

Silver-impregnation methods. Since Klein (1926) applied silver nitrate in demonstrating the silver-line system of ciliates, various modifications have been proposed.

Dry silver method (Klein, 1926, 1958). Air-dried cover glass smears are placed for six to eight minutes in a 2 per cent solution of silver nitrate and thoroughly washed. The smears are exposed to bright daylight on a white background for two to eight hours with occasional checking under the microscope as reduction of silver nitrate results in darkening. The smears are then washed thoroughly and air-dried in a vertical position. They are finally mounted in Canada balsam.

Wet silver method (modified after Gelei and Horváth, 1931). The ciliates are fixed in a centrifuge tube for five to ten minutes in sublimate-formaldehyde solution, composed of saturated corrosive sublimate 95 cc. and formaldehyde 5 cc. The specimens are now washed twice in nonchlorinated water and once in distilled water; they are then treated in 1.5-2 per cent solution of silver nitrate for five to twenty minutes. Without washing, the specimens in the tube are exposed to direct sunlight for ten to sixty minutes in distilled water, after which the specimens are washed four to six times in distilled water, one minute each. Passing through a gradually ascending alcohol series and xylol, the specimens are mounted in Canada balsam.

Chatton-Lwoff method (Chatton and Lwoff, 1936; Corliss, 1952). Fix the ciliates with Champy's fluid for one to three minutes, then in Da Fano's solution for several hours. Place the specimens on a slide, eliminate excess fluid and embed them in warm (35-45°C.) gelatin containing 0.05 per cent sodium chloride. Refrigerate in a moist chamber until the gelatin has set, and then immerse 10-20 minutes in 3 per cent aqueous silver nitrate at 5-10°C. Wash with cold distilled water, submerge the preparation in cold (below 10°C.) water to a depth of 3-4 cm. over white paper and expose to a strong light for ten to thirty minutes. Remove the preparation to 70 per cent alcohol, followed by usual dehydration. After clearing, the preparation is mounted in Canada balsam. Silver is deposited on various structures which appear black in the mounted specimens.

Fontana's method. For staining filamentous structures such as the extruded polar filament of microsporidian spores, this method is the most satisfactory one. After air-drying the smears are fixed for five minutes in a mixture of formaldehyde, 20 cc.; glacial acetic acid, 1 cc.; and distilled water, 100 cc. After washing in running water, the smears are placed in the following mordant composed of equal parts of 5 per cent tannic acid and 1 per cent carbolic acid, for about two minutes at about 60°C. Wash the smears in water and place them for three to five minutes in 0.25 per cent solution of silver nitrate warmed to 60°C., to which ammonia has been added drop by drop until a grayish brown cloud appeared. Wash thoroughly and air-dry. After passing through 95 per cent and absolute alcohol, and xylol, the smears are mounted in Canada balsam.

b. Blood Film Preparations

Thin film. The finger tip or ear lobe is cleaned with 70 per cent alcohol. Prick it with an aseptic blood lancet or a sterilized needle. Wipe off the first drop with gauze and receive the second drop on a clean slide about half an inch from one end (Fig. 388, *1*). Use care not to let the slide touch the finger or ear-lobe itself. Quickly bring a second slide, one corner of which had been cut away, to the inner margin of the blood drop (*1*), and let the blood spread

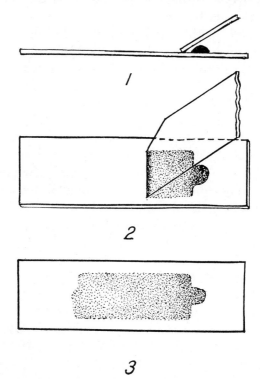

Fig. 388. Diagrams showing how a thin blood film is made on a slide.

along the edge of the second slide. Next push the second slide over the surface of the first slide at an angle of about 45° toward the other end (*2*). Thus a thin film of blood is spread over the slide (*3*). Let the slide lie horizontally and dry, under a cover to prevent dust particles falling on it and to keep away flies or other insects. If properly made, the film is made up of a single layer of blood cells.

Thick films. Often parasites are so few that to find them in a thin film involves a great deal of time. In such cases, a thick film is advocated. For this, two to four drops of blood are placed in the central half-inch square area, and spread them into an even layer with a needle or with a corner of a slide. Let the film dry. With a little practice, a satisfactory thick smear can be made. It will take two hours or more to dry. Do not dry by heat, but placing it in an incubator at 37°C. will hasten the drying. When thoroughly dry, immerse it in water and dehaemoglobinize it. Air dry again.

Thin and thick film. Often it is time-saving if thin and thick films are made on a single slide. Place a single drop of blood near the center and make a thin film of it toward one end of the slide. Make a small thick smear in the center of the other half of the slide. Dry. When thoroughly dry, immerse the thick film part in distilled water and dehaemoglobinize it. Let the slide dry.

Blood smears must be stained as soon as possible to insure a proper staining, as lapse of time or summer heat will often cause poor staining especially of thick films. Of several blood stains, Giemsa's and Wright's stains are used here. For staining with **Giemsa's** stain, the thin film is fixed in absolute methyl alcohol for five minutes. Rinse the slide well in neutral distilled water. After shaking the stock bottle (obtained from reliable makers) well, dilute it with neutral distilled water in a ratio of one drop of stain to 1-2 cc. of water. Mix the solution and the blood film is placed in it for 0.5-2 hours or longer if needed. Rinse the slide thoroughly in neutral distilled water and wipe on water with a tissue paper from the underside and edges of the slide. Let the slide stand on end to dry. When thoroughly dry, place a drop of xylol and a drop of cedar wood oil (used for immersion objectives) and cover with a coverglass. The mounting medium should be absolutely neutral. Do not use Canada balsam for mounting, as acid in it promptly spoils the staining.

For **Wright's** stain, fixation is not necessary. With a medicine dropper, cover the dried blood film with drops of undiluted Wright's stain, and let the film stand horizontally for three to five minutes; then the same number of drops of neutral distilled water is added to the stain and the whole is left for ten to thirty minutes. The stain is then poured off and the film is rinsed in neutral distilled water. Dry. Mount in xylol and cedar wood oil. Use of coverglass on a stained blood film is advocated, since a cedar wood oil mounted slide allows the use of dry objectives which in the hand of an experienced worker would give enough magnification for species determination of Plasmodium, and which will very clearly reveal any trypanosomes present in the film. Furthermore, the film is protected against scratches, and contamination by many objects which may bring about confusion in detecting looked-for organisms.

Films made from splenic punctures for Leishmania or Trypanosoma are similarly treated and prepared.

c. Section Preparations

Paraffin sections should be made according to usual histological technique. Fixatives and stains are the same as those mentioned for smear preparations.

References

BALAMUTH, W., 1951. Biological studies on *Entamoeba histolytica*. III. J. Infect. Dis., 88:230.

――― 1962. Effects of some environmental factors upon growth and encystation of *Entamoeba invadens*. J. Parasit., 48:101.

――― and SANDZA, J. G., 1944. Simple, standardized culture medium for physiological studies on *Entamoeba histolytica*. Proc. Soc. Exper. Biol. Med., 57:161.

BASS, C. C., and JOHNS, F. M., 1912. The cultivation of malarial plasmodia (*Plasmodium vivax* and *P. falciparum*) *in vitro*. J. Exper. Med. 16: 567.

BECKER, E. R., 1926. *Endamoeba citelli*, etc. Biol. Bull., 50:444.

BĚLAŘ, K., 1928. Untersuchungen der Protozoen. Methodik. wiss. Biol., 1.

BIJLMER, J., 1948. On the recovery of Protozoa and eggs of some species of helminths in human faeces. J. Parasit., 34:101.

BOECK, W. C., and DRBOHLAV, J., 1925. The cultivation of *Endamoeba histolytica*. Amer. J. Hyg., 5:371.

BROOKE, M. M., 1958. Amebiasis. Methods in laboratory diagnosis. U. S. Dept. Health Education and Welfare, 67 pp.

CHATTON, E., and BRODSKY, A., 1909. Les parasitisme d'une Chytridinée du genre Sphaerita, etc. Arch. Protist., 17:1.

――― and LWOFF, A., 1936. Techniques pour l'étude de protozoaires, spécialement de leurs structures superficielles (cinétome et argyrome). Bull. Soc. Fr. Micr., 5:25.

CLAFF, C. L., 1940. A migration-dilution apparatus for the sterilization of Protozoa. Physiol. Zool., 13:334.

CLEVELAND, L. R., and SANDERS, E. P., 1930. Encystation, multiple fission without encystment, etc. Arch. Protist., 70:223.

COLEMAN, L. C., 1938. Preparation of leuco basic fuchsin for use in the Feulgen reaction. Stain Tech., 13:123.

Committee on cultures, the Society of Protozoologists. 1958. A catalogue of laboratory strains of free-living and parasitic protozoa. J. Protozool., 5:1.

CORLISS, J. O., 1952. Silver impregnation of ciliated protozoa by the Chatton-Lwoff technic. Stain Technol., 28:97.

DIAMOND, L. S., 1961. Axenic cultivation of *Entamoeba histolytica*. Science, 134:336.

DOBELL, C., and LAIDLAW, P. P., 1926. On the cultivation of *Entamoeba histolytica* and some other entozoic amoebae. Parasitology, 18:283.

ELY, J. O., and Ross, M. H., 1949. Nucleic acids and the Feulgen reaction. Anat. Rec., 104:103.

FEULGEN, R., 1926. Die Nuclealfärbung. Abderharden's Handb. biol. Arbeitsmeth., Abt. 5, 2:1055.

——— and ROSSENBECK, H., 1924. Mikroskopisch-chemischer Nachweis einer Nucleinsäure von Typus der Thymonucleinsäure und die darauf beruhende elektive Färbung von Zellkernen in mikroskopischen Präparaten. Ztschr. physiol. Chem., 135:203.

FEULGEN-BRAUNS, F., 1924. Untersuchungen über Nuclealfärbung. Pflüger's Arch. gesamt. Physiol., 203:415.

GEIMAN, Q. M., ANFINSEN, C. B., et al., 1946. Studies on malarial parasites. VII. J. Exper. Med., 84:583.

GELEI, J. v., and HORVÁTH, P., 1931. Eine nasse Silber-bzw. Gold-methode für die Herstellung der reizleitenden Elemente bei den Ciliaten. Ztschr. wiss. Mikr., 48:9.

GLASER, R. W., and CORIA, N. A., 1930. Methods for the pure culture of certain Protozoa. J. Exper. Med., 51:787.

GRASSÉ, P. P., 1926. Contributions à l'étude des flagellés parasites. Arch. zool. exper. gén., 65:345.

GRIFFIN, A. M., and McCARTEN, W. G., 1949. Sterols and fatty acids in the nutrition of entozoic amoebae in cultures. Proc. Soc. Exper. Biol., 72:645.

——— and MICHINI, L. J., 1950. Some sources of variability in cultures of entozoic amoebae. II. J. Parasit., 36:247.

GURR, E., 1956. A practical manual of medical and biological staining techniques. 2 ed. N.Y.

HAHNERT, W. F., 1932. Studies on the chemical needs of *Amoeba proteus*: a cultural method. Biol. Bull., 62:205.

HALL, R. P., 1937. Growth of free-living Protozoa in pure cultures. In Needham *et al.*: Culture methods for invertebrate animals.

———— 1941. Food requirements and other factors influencing growth of Protozoa in pure cultures. In Calkins and Summers: Protozoa in biological research.

HOGUE, M. J., 1921. The cultivation of *Trichomonas hominis*. Am. J. Trop. Med., 1:211.

———— 1922. A study of *Trichomonas hominis*, its cultivation, etc. Bull. Johns Hopkins Hosp., 33:437.

HUTNER, S. H., and PROVASOLI, L., 1951. The phytoflagellates. In: Lwoffs Biochemistry and physiology of protozoa. 1.

JAHN, T. L., 1933. On certain parasites of Phacus and Euglena: *Sphaerita phaci*, sp. nov. Arch. Protist., 79:349.

JOHNSON, J. G., and TRUSSELL, R. E., 1943. Experimental basis for the chemotherapy of *Trichomonas vaginalis* infestations. Proc. Soc. Exper. Biol. Med., 54:245.

JOHNSON, W. H., and BAKER, E. G. S., 1942. The sterile culture of *Paramecium multimicronucleatum*. Science, 95:333.

———— and MILLER, C. A., 1956. A further analysis of the nutrition of Paramecium. J. Protozool., 3:221.

KIDDER, G. W., 1941. The technique and significance of control in protozoan culture. In Calkins and Summers: Protozoa in biological research.

———— DEWEY, V. C., and PARKS, R. E. JR., 1951. Studies on inorganic requirements of Tetrahymena. Physiol. Zool., 24:69.

KIRBY, H. JR., 1941. Organisms living on and in Protozoa. In Calkins and Summers: Protozoa in biological research.

———— 1950. Materials and methods in the study of protozoa. Berkeley, California.

KLEIN, B. M., 1926. Ergebnisse mit einer Silbermethode bei Ciliaten. Arch. Protist., 56:243.

———— 1958. The dry silver method and its proper use. J. Protozool., 5:99.

LEE, J. J., et al., 1961. Growth and physiology of Foraminifera in the laboratory. Micropaleontology, 7:461.

LWOFF, M., 1951. Nutrition of parasitic amoebae. In A. Lwoff: Biochemistry and physiology of protozoa. Vol. 1. p. 235.

MANWELL, R. D., and BRODY, G., 1950. Survival and growth of four species of avian plasmodia on the Harvard culture medium. J. Nat. Mal. Soc., 9:132.

MARSLAND, D. A., 1943. Quieting Paramecium for the elementary student. Science, 98:414.

MATTES, O., 1924. Ueber Chytridineen im Plasma und Kern von *Amoeba sphaeronucleus* und *A. terricola*. Arch. Protist., 47:413.

NEEDHAM, J. G., GALTSOFF, P. S., LUTZ, F. E., and WELCH, P. S., 1937. Culture methods for invertebrate animals. Ithaca, N.Y.

PHILLIPS, B. P., 1950. Cultivation of *Entamoeba histolytica* with *Trypanosoma cruzi*. Science, 111:8.

—— 1951. Measurements of direct amebicidal potential, etc. Am. J. Trop. Med., 31:561.

—— 1953. Studies on the cultivation of *Endamoeba histolytica* with *Trypanosoma cruzi*. Ann. N. Y. Acad. Sci., 56:1028.

PRINGSHEIM, E. G., 1926. Kulturversuche mit chlorophyllführenden Mikroorganismen. V. Beitr. Biol. Pfl., 14:283.

—— 1937. Beiträge zur Physiologie saprotropher Algen und Flagellaten. III. Planta, 27:61.

—— 1946. Pure cultures of algae. Cambridge.

—— 1946a. The biphasic or soil-water culture method for growing algae and Flagellata. J. Ecol., 33:193.

RATCLIFFE, H. L., and GEIMAN, Q. M., 1938. Spontaneous and experimental amoebic infection in reptiles. Arch. Path., 25:160.

REYER, W., 1939. Ueber die Vermehrung von Blastocystis in der Kultur. Arch. Protist., 92:226.

SASSUCHIN, D. N., 1928. Zur Frage über die Parasiten der Protozoen. *Ibid.*, 64:61.

——, POPOFF, P. P., KUDRJEWZEW, W. A., and BOGENKO, W. P., 1930. Ueber parasitische Infektion bei Darmprotozoen. *Ibid.*, 71:229.

SINGH, B. N., 1955. Culturing soil protozoa and estimating their numbers in soil. Soil Zool. Proc. Univ. Nottingham II Easter Sch. Agr. Sci., p. 403.

SLATER, J. V., 1955. Some observations on the cultivation and sterilization of protozoa. Tr. Am. Micr. Soc., 74:80.

SPINGARN, C. L., and EDELMAN, M. H., 1947. The prolongation of the viability of cultures of *E. histolytica* by the addition of streptomycin. J. Parasit., 33:416.

STARR, R. C., 1956. Culture collection of algae at Indiana University. Lloydia, 19:129.

—— 1958. Recent additions to the culture collection of algae at Indiana University. J. Protozool., 5:232.

STORM, J., and HUTNER, S. H., 1953. Nutrition of Peranema. Ann. N. Y. Acad. Sci., 56:901.

STOWELL, R E., 1945. Feulgen reaction for thymonucleic acid. Stain Tech., 20:45.

TARTAR, V., 1950. Methods for the study and cultivation of protozoa. In Studies honoring T. Kincaid, Univ. Wash. Press.

TAYLOR, C. V., and VAN WAGTENDONK, W. J., 1941. Growth studies of *Colpoda duodenaria*. I. Physiol. Zool., 14:431.

THOMAS, R., 1953. L'action anaesthesique du sulfate de nickel sur *Paramecium caudatum*. Bull. Micr. App., 3:73.

TOBIE, ELEANOR, J., BRAND, T., and MEHLMAN, B., 1950. Cultural and physiological observations on *Trypanosoma rhodesiense* and *T. gambiense*. J. Parasit., 36:48.

TRAGER, W., 1934. The cultivation of a cellulose-digesting flagellate, *Tricho-*

monas termopsidis, and of certain other termite Protozoa. Biol. Bull., 66:182.

———— 1950. Studies on the extracellular cultivation of an intracellular parasite (avian malaria). I. J. Exper. Med., 92:349.

VAN WAGTENDONK, W. J., *et al.,* 1953. Growth requirement of *Paramecium aurelia* var. 4, etc. Ann. N. Y. Acad. Sci., 56:929.

WENRICH, D. H., 1941. The morphology of some Protozoan parasites in relation to microtechnique J. Parasit., 27:1.

WENYON, C. M., 1926. Protozoology. London and Baltimore.

WICHTERMAN, R., 1949. The collection, cultivation, and sterilization of Paramecium. Proc. Penn. Acad. Sci., 23:151.

Author Index

A

Aaronson, S., 131, 165, 314, 322
Aberle, S. D., 733, 749
Achmerov, A. H., 781, 801
Actor, P., 423, 431,
Adams, A. R. D., 537, 538, 539, 564, 1067
Adams, J. A., 653, 673
Adamson, A. M., 395
Adler, S., 422, 431, 633, 673
Ahlstrom, E. H., 316, 322
Alden, R. H., 121, 168
Alexander, G., 28, 189
Alexeieff, A., 91, 105, 454, 472, 768, 771
Allee, W. C., 134, 165
Allegre, C. F. 355, 366, 649, 673
Allen, Ena A., 701, 708
Allen, M. B., 102, 105
Allen W. E., 390, 392
Allison, A. C., 727, 749
Allman, G. J., 16, 85
Altenberg, E., 288, 290
Alvey, C. H., 123, 169
Amaral, A. D. F., 548, 555
Amrein, Y. U., 698, 708
Andai, G., 440, 454
Anderson, A. P., 423, 436
Anderson, E., 58, 61, 62, 80, 105, 417, 431, 471, 474, 607, 615
Anderson H. H., 164, 165
Andre, J., 225, 228, 266
Andresen, N., 93, 127, 129, 141, 142, 146, 165, 525, 526, 555
Andrews, Bess J., 480, 493
Andrews, E. A., 971, 972, 975
Andrews, J., 452, 456, 470, 474, 689, 693, 708
Andrews, J. M., 749
Andrews, Mary N., 469, 474
Anfinsen, C. B., 1070, 1085
Angerer, C. A., 522, 555
Anigstein, L. 849, 886
Aragão, H. B., 338, 346, 706, 708, 742
Arai, Y., 792, 801

Arantes, J. B., 749
Arcichovskij, V., 51, 105
Aris, F. W., 725, 750
Arndt, A., 190
Asami, K., 468, 474
Ashcroft, M. T., 413, 431
Atchley, F. O., 749, 1070
Auerbach, M., 774, 780, 795, 801
Awerinzew, G. W., 1026, 1032
Awerinzew, S., 802, 986, 993

B

Babudieri, B., 767, 771
Bach, M. K., 353, 366
Backhouse, T. C., 749
Baernstein, H. D., 414, 431
Bailey, L. 808, 811, 821
Bairati, A., 52, 106, 152, 165
Baker, E. G. S., 1064, 1086
Baker, H., 10, 16
Baker, H., 131, 165, 314, 322
Baker, J. R., 419, 431, 742, 749, 753
Balamuth, W., 212, 253, 539, 555, 974, 975, 1063, 1068, 1069, 1084
Balbiani, G., 12, 14, 16
Balech, E., 357, 366, 372, 388, 392, 393, 981, 993
Ball, G. H., 544, 557, 645, 646, 666, 673, 677, 701, 706, 708, 709, 736, 749, 767, 771
Ballantine, D., 372, 393
Bancroft, M. J., 804
Baner, J. R., 708
Baraban, L., 767, 771
Barbagallo, P., 531, 556
Barbier, M., 51, 106, 970, 975
Barker, H. A., 211, 215, 253, 254, 266
Barksdale, W. L. 697, 708
Barnes, W. B., 469, 474
Barnett, S.F., 749
Barrera, A., 740, 757
Barret, H. P., 15, 16, 546, 555, 1066, 1070
Barrow, J. H., Jr., 419, 431, 547, 555, 764, 771

1089

Subject Index

1113

1172 *Protozoology*

Ulivina—*continued*
  rynchoboli, 645
Ultraviolet rays, 161
Undulating membrane, 61, 63, 66, 409
Unicapsula, 781
  muscularis, 781
Unicauda, 789
  clavicauda, 789
Unipolarina, 781–792
Unstable cyst, 212,
Uradiophora, 651
  cuenoti, 651
Urceolaria, 1027–1028
  karyolobia, 1028
  mitra, 1028
  paradoxa, 1028
Urceolariidae, 1027–1029
Urecolus, 360
  cyclostomus, 360
  sabulosus, 360
Urea, 145, 188
Urease, 127
Urechis caupo, 645
Urinympha, 489
  talea, 489
Urnula, 1050
  epistylidis, 1050
Urocentrum, 912
  turbo, 912
Uroglena, 315–316
  volvox, 316
Uroglenopsis, 316
  americana, 316
  europaea, 316
Uroleptopsis, 999
  citrina, 999
Uroleptus, 999
  halseyi, 999
  limnetis, 999
  longicaudatus, 999
  mobilis, 274
Uronema, 901
  marinum, 901
  pluricaudatum, 902
Uronychia, 1007, 1009
  setigera, 1009
Urophagus, 454
  rostratus, 454
Urosoma, 998

Urosoma—*continued*
  caudata, 998
Urospora, 640
  chiridotae, 640
  hardyi, 640
Urosporidae, 640–642
Urosporidium, 764
  fuliginosum, 764
Urostyla, 1001
  caudata, 1001
  coei, 1001
  grandis, 1001
  polymicronucleata, 1001
  trichogaster, 1001
Urotricha, 830
  agilis, 830
  farcta, 830
  labiata, 832
  parvula, 830
Urozona, 912
  bütschlii, 912
Uta stansburiana hesperis, 701

V

Vacuolaria, 365
  virescens, 365
Vacuome, 91
Vaginicola, 1024
  annulata, 1024–1025
  leptostoma, 1024
Vaginicolidae, 1024–1026
Vahlkampfia, 37, 527
  limax, 527
  patuxent, 527
Valvulina, 596
Valvulinidae, 596
Vampyrella, 500
  lateritia, 500
Vampyrellidae, 499–505
Vampyrophrya, 946
  pelagica, 946
Varanus salvator, 547
  varius, 547
Variation in Protozoa, 268–290
Vasicola, 843
  ciliata, 843
  grandis, 843
Vaucheria, 497
Vectors, 410, 411, 413, 414, 421, 422,